Free Radicals in Biology and Medicine

THIRD EDITION

Barry Halliwell

Professor of Medical Biochemistry, King's College, London
Co-Director, International Antioxidant Research Centre
Co-Director, Neurodegenerative Disease Research Centre
Research Professor, University of California, Davis, Medical Center
Visiting Professor of Biochemistry, National University of Singapore

and

John M.C. Gutteridge

Director, Oxygen Chemistry Laboratory, Directorate of Anaesthesia and Critical Care,
Royal Brompton and Harefield NHS Trust, London
Visiting Professor, Pharmacology group, King's College, London
Visiting Professor, Department of Biochemistry, Osaka University Medical School, Osaka

This book has been approved by the Committee of the International
Society for Free Radical Research as an authoritative statement
of the current state of research in this field

OXFORD

UNIVERSITY PRESS

OXFORD
UNIVERSITY PRESS

Great Clarendon Street, Oxford OX2 6DP

Oxford University Press is a department of the University of Oxford.
It furthers the University's objective of excellence in research, scholarship,
and education by publishing worldwide in

Oxford New York

Athens Auckland Bangkok Bogotá Buenos Aires Calcutta
Cape Town Chennai Dar es Salaam Delhi Florence Hong Kong Istanbul
Karachi Kuala Lumpur Madrid Melbourne Mexico City Mumbai
Nairobi Paris São Paulo Singapore Taipei Tokyo Toronto Warsaw

with associated companies in Berlin Ibadan

Oxford is a registered trade mark of Oxford University Press
in the UK and in certain other countries

Published in the United States
by Oxford University Press, Inc., New York

First edition published 1985
Second edition published 1989
This edition published 1999
Reprinted 2000

A catalogue record for this book is available from the British Library

Library of Congress Cataloging in Publication Data
(Data applied for)

ISBN 0 19 850045 9 (Hbk)
ISBN 0 19 850044 0 (Pbk)

Printed by Thomson Press (India) Ltd

Preface to the third edition

When we first sat down (in January 1996) to write a third edition, we optimistically assumed that it could be finished by June. We underestimated the pace of advance in this subject: almost every section has needed extensive updating. Sometimes our job was easier: new data have often clarified the previously obscure, especially data obtained using the techniques of modern molecular, cellular, and structural biology. More often, however, new data have transformed the way in which scientists think about a particular topic, requiring complete rewriting. We have tried to maintain the essential simplicity of our approach and readability of the text, and hope that we have succeeded.

Several reviewers of the second edition requested specific referencing in the text. This is more difficult than it seems because many of the statements made are distilled from several published papers and interpreted through the scientific experiences (or prejudices) of the authors. Citing every relevant paper would generate a book three times the length, out-of-date within six months and too expensive for most people to buy. As a compromise each major statement in the text is now provided with at least one reference, either cited as a superscript in the text or placed in the legend of an adjacent figure or table. References quoted are not always the seminal ones, as we have tried to choose references (often reviews or recent papers) that should make it easy for the reader to access further literature. We hope that experts will forgive us if they find their pet paper uncited.

During the rewriting of the second edition, we felt like painters of the Forth Bridge. Having completed the third edition, we feel like early retirement.

London   B.H.
September 1998   J.M.C.G.

Preface to the second edition

The explosive growth of interest in free radicals, and the enormous amount of research work undertaken since 1984, has necessitated the writing of a second edition of this book after only three years. During the extensive rewriting necessary, we have sometimes felt like painters of the Forth Bridge. Production of the revised edition has been helped by information and critical comments provided by the following scientists, to whom we are very grateful. Any remaining errors are the entire responsibility of the authors, however.

B.N. Ames	T. Connors	D.J. Hockley	Philips Analytical
B. Anderton	F. Corongiu	R.L. Hoult	E.A. Porta
B.M. Babior	J.T. Curnutte	V. Kagan	W.A. Pryor
D.R. Blake	C. Dahlgren	D. Leake	C. Rice-Evans
J.M. Braughler	A.T. Diplock	T. Lindahl	G. Rotilio
L. Breimer	E.A. Dratz	M. Matsuo	L.L. Smith
G. Burton	H. Esterbauer	M.J. Mitchinson	T.F. Slater
R. Cammack	I. Fridovich	D.P.R. Muller	Y. Sugiura
C.J. Chesterton	E. Getzoff	H.J. Okamoto	S.P. Wolff
			R.L. Willson

London
October 1988

B.H.
J.M.C.G.

Preface to the first edition

The importance of radical reactions in radiation damage, food preservation, combustion, and in the rubber and paint industry, has been known for many years to people in the respective fields, but it has rarely been appreciated by biologists and clinicians. The interest in radicals shown by the latter groups has been raised recently by the discovery of the importance of radical reactions in normal body chemistry and in the mode of action of many toxins. The discoveries of hypoxic cell sensitizers that potentiate radiation-induced radical damage to cancerous tumours, of the enzyme superoxide dismutase, and of the mechanism of action of such toxins as paraquat and carbon tetrachloride provide major examples of this importance.

Any expanding field attracts the charlatans, such as those who make money out of proposing that consuming radical scavengers will make you live for ever or that taking tablets containing superoxide dismutase will enhance your health and sex life. In evaluating these and other less obviously silly claims, it is useful to understand the basic chemistry of radical reactions.

This book is aimed mainly at biologists and clinicians. It assumes virtually no knowledge of chemistry and attempts to lead the reader as painlessly as possible into an understanding of what free radicals are, how they are generated, and how they can react. Having established this basis, the role of radical reactions in several biological systems is critically evaluated in the hope that the careful techniques needed to *prove* their importance will become more widely used. We believe that free-radical chemists should also find these latter chapters useful.

London   B.H.
                                                                                 J.M.C.G.

Acknowledgements

In writing the third edition, we have again received invaluable help from experts who commented on and criticized sections of the draft text and/or provided data, diagrams, and photographs. Special thanks are due to the following (in alphabetical order), but the responsibility for any errors or omissions in the text is entirely that of the authors.

David Adams (Birmingham, UK)
Adriano Aguzzi (Zurich)
Randy Allen (Lubbock)
Bruce Ames (Berkeley)
Brian Anderton (London)
Daniel Aneshansley (New York)
Kumi Arakane (Tokyo)
Ohara Augusto (São Paulo)
Bernie Babior (La Jolla)
Jim Barber (London)
Roberto Bolli (Louisville)
Tony Breen (Cambridge)
George Britton (Liverpool)
Robert H. Brown (Boston)
Gary Buettner (Iowa City)
Roy Burdon (Strathclyde)
John Butler (Manchester)
Tony Campbell (Cardiff)
Luis Candeias (Northwood)
Keri Carpenter (Cambridge)
Ho Zoon Chae (Bethesda)
Britton Chance (Philadelphia)
Augustine Choi (Baltimore)
Andrew Collins (Aberdeen)
Mario Comporti (Siena)
Francesco Corongiu (Bari)
Carroll Cross (Davis)
Merit Cudkowicz (Boston)
C. Dahlgren (Umea)
Vincent Daniels (London)
Arthur Dannenberg (Baltimore)
Ann Dewar (London)
Tony Diplock (London)
Miral Dizdaroglu (Bethesda)

Jason Eiserich (Birmingham, Alabama)
Charles Epstein (San Francisco)
Hermann Esterbauer†
Pat Evans (London)
Philip Evans (Cambridge)
Timothy Evans (London)
Leopold Flohe (Braunschweig)
Bob Floyd (Oklahoma)
Marc Fontecave (Grenoble)
Chris Foote (Los Angeles)
Joseph Formica (Richmond, Virginia)
Irwin Fridovich (Durham, N. Carolina)
Elizabeth Getzoff (La Jolla)
Fred Gey (Switzerland)
Irving Goldberg (Boston)
Daniel Hinshaw (Ann Arbor)
David Hockley (London)
Paul Hopkins (Seattle)
Laurence Hurley (Austin)
Paul Hyslop (Indianapolis)
L. Jackson Roberts II (Nashville)
Balyanaraman Kalyanaraman (Milwaukee)
Harparkash Kaur (London)
Herbert Kayden (New York)
Norman Krinsky (Boston)
Massaki Kurata (Gifu)
Frans Kuypers (Oakland)
Debra Laskin (Piscataway)
Howard Leese (York)
Dan Liebler (Tucson)
Sam Louie (Davis)
Jennifer Martin (Queensland)

Ron Mason (Research Triangle Park)

Joe McCord (Denver)

Alton Meister[†]

David Metzler (Iowa City)

Malcolm Mitchinson (Cambridge)

Robin Mockett (Dallas)

Jason Morrow (Nashville)

Rex Munday (Hamilton)

John Murphy (Strathclyde)

Etsuo Niki (Tokyo)

Alberto Noronha-Dutra (London)

James Olson (Iowa City)

Sten Orrenius (Stockholm)

Bill Pryor (Baton Rouge)

Rafael Radi (Montevideo)

Sue Goo Rhee (Bethesda)

Margaret Rice (New York)

Catherine Rice-Evans (London)

Joe Rotilio (Rome)

Aziz Sancar (Chapel Hill)

Tony Segal (London)

Helmut Sies (Dusseldorf)

William Smith (East Lansing)

Raj Sohal (Dallas)

Jeremy Spencer (London)

Earl Stadtman (Bethesda)

Daniel Steenkamp (Cape Town)

Roland Stocker (Sydney)

Masatoshi Suzuki (Tokyo)

Martyn Symons (London)

Csaba Szabó (Cincinnati)

John Tainer (La Jolla)

Naoyuki Taniguchi (Osaka)

Alison Telfer (London)

Maret Traber (Berkeley)

M.-C. Tsai (Taipei)

Robert Turesky (Lausanne)

Fulvio Ursini (Udine)

Taurus Wah (Hong Kong)

Peter Wardman (Northwood)

Martin Warren (London)

Sigmund Weitzman (Chicago)

Albrecht Wendel (Konstanz)

Matthew Whiteman (London)

[†]Regretfully deceased: great losses to this research field.

J.M.C.G. thanks his family (Pushpa, Samantha, and Mark) for their continuing support and encouragement. B.H. is particularly indebted to Yvonne D'Souza-Rauto for her excellent typing skills.

Contents

Plate section falls between pages 128 and 129

Abbreviations

AAPH	azobis(2-amidino-propane)hydro-chloride	CoQ	coenzyme Q
		COX-1	cyclooxygenase 1
		COX-2	cyclooxygenase 2
ABTS	2, 2′-Azinobis (3-ethylbenzothia zoline 6-sulphonate)	CSF	cerebrospinal fluid
		DABCO	1,4-diazabicyclooctane
		DAG	diacylglycerol
AGE	advanced glycation end-product	DBD	diaminobenzidine
		DCFH-DA	dichlorofluorescein diacetate
ALS	amyotrophic lateral sclerosis	DETAPAC	diethylenetriamine-pentaacetic acid
AMPA	α-amino-3-hydroxy-5-methyl-4-isoxa-zole-4-propionate	DHA	docosahexaenoic acid or dehydroascorbate
AMVN	2,2′-azobis(2,4-dimethylvaleronitrile)	DHB	dihydroxybenzoate
		DHETE	dihydroxyeicosa-tetraenoic acid
apoE	apolipoprotein E	DHF	dihydroxyfumarate
AP site	apurinic (or apyrimidinic) site	DHLA	dihydrolipoate
		DHR	dihydrorhodamine
ARDS	adult respiratory distress syndrome	DMPO	5,5-dimethylpyrro-line-N-oxide
ARE	antioxidant response element	DMSO	dimethylsulphoxide
		DOPA	dihydroxy-phenylalanine
AZT	azidodeoxythymidine		
BDI	bleomycin-detectable iron	DPPH	1,1-diphenyl-2-picryl-hydrazyl
BHA	butylated hydroxyanisole	DTNB	5,5′-dithiobis(2-nitro-benzoic acid)
BHT	butylated hydroxytoluene	ECD	electrochemical detection
BSE	bovine spongiform encephalopathy	EDRF	endothelium-derived relaxing factor
CF	cystic fibrosis	EGF	epidermal growth factor
CFTR	cystic fibrosis trans-membrane conduc-tance regulator	EPR	electron paramagnetic resonance
CGD	chronic granulomatous disease	ESR	electron spin resonance
CHO cell	Chinese hamster ovary cell	ETYA	eicosatetraynoic acid
CJD	Creutzfeldt–Jakob disease	FA	Friedreich's ataxia
COPD	chronic obstructive pulmonary disease	FAD	flavin adenine dinucleotide

FAPy	formamidopyrimidine	IRE	iron responsive elements
FGF	fibroblast growth factor	IRP	iron regulatory protein
FMN	flavin mononucleotide	KA	kainic acid
GABA	γ-aminobutyric acid	LCAT	lecithin–cholesterol acyltransferase
GC–MS	gas chromatography–mass spectrometry	LDL	low-density lipoprotein
GFP	green fluorescent protein	LFA-1	lymphocyte function antigen 1
GPX	glutathione peroxidase	LTA_4,	
GSH	glutathione	LTB_4...	leukotriene A_4, B_4, etc.
GST	glutathione S-transferase	MAO	monoamine oxidase
HAP	huntingtin-associated protein	MAP	mitogen-activated protein
HDL	high-density lipoprotein	MCP-1	monocyte chemo-attractant protein-1
HETE	hydroxyeicosa-tetraenoic acid	M-CSF	macrophage colony-stimulating factor
HHE	*trans*-4-hydroxy-2-hexenal	MDA	malondialdehyde
HHT	12-hydroxy-5,8,10-heptadecatrienoic acid	MELAS	mitochondrial encephalo-myopathy, lactic acidosis and stroke-like episodes
HIV	human immuno-deficiency virus	MEOS	microsomal ethanol oxidizing system
HLA	human leukocyte antigen	MHC	major histocompatibility complex
HNE	4-hydroxy-2-nonenal	MODY	maturity onset diabetes of the young
HNF	hepatic nuclear factor		
HPD	haematoporphyrin derivative	MPG	mercaptopropionyl-glycine
HPETE	hydroperoxyeicosa-tetraenoic acid	MPO	myeloperoxidase
		MPP^+	1-methyl-4-phenyl-pyridinium ion
HPLC	high performance liquid chromatography	MPTP	1-methyl-4-phenyl-1,2,3,6-tetra-hydropyridine
HRP	horseradish peroxidase		
HSF	heat-shock transcription factor	MRP	multi-drug resistance associated protein
IBD	inflammatory bowel disease	MS	multiple sclerosis
ICAM	intercellular adhesion molecule	MSA	multiple system atrophy
		NAAQS	national ambient air quality standard
ICE	interleukin-1β converting enzyme	NAD^+	nicotinamide adenine dinucleotide
IGF	insulin-like growth factor		
IP_3	inositol 1,2,5-triphosphate	$NADP^+$	nicotinamide adenine dinucleotide phosphate

NBT	nitroblue tetrazolium
NCL	neuronal ceroid lipofuscinosis
NGF	nerve growth factor
NMDA	*N*-methyl-D-aspartate
NOS	nitric oxide synthase (eNOS, endothelial NOS; iNOS, inducible NOS; nNOS, neuronal NOS)
NSAID	non-steroidal anti-inflammatory drug
NTA	nitrilotriacetate
8-OHdG	8-hydroxydeoxy guanosine
8-OHG	8-hydroxyguanine
PAF	platelet-activating factor
PARP	poly(ADP–ribose) polymerase
PBN	phenyl-*tert*-butyl-nitrone
PCP	pentachlorophenol
PD	Parkinson's disease
PDGF	platelet-derived growth factor
PDI	protein-disulphide isomerase
PECAM-1	platelet endothelial cell adhesion molecule 1
PEG	polyethylene glycol
PFL	pyruvate–formate lyase
PGE_2, F_2, G_2	prostaglandin E_2 etc. ...
PGI_2	prostacyclin
PHGPX	phospholipid hydro-peroxide gluta-thione peroxidase
PKC	protein kinase C
PMA	phorbol myristate acetate
PNDA	*p*-nitrosodimethyl-aniline
PO_2	Partial pressure of oxygen
POBN	α-(4-pyridyl 1-oxide)-*N-tert*-butylnitrone
p.p.b.	parts per billion
p.p.m.	parts per million
PSNP	progressive supra-nuclear palsy
PUFA	polyunsaturated fatty acid
PUVA	psoralen–ultraviolet A
RA	rheumatoid arthritis
RAGE	receptor for AGE
RDA	recommended dietary allowance
RFI	relative fluorescence intensity
RNS	reactive nitrogen species
RPE	retinal pigment epithelium
ROP	retinopathy of prematurity
ROS	reactive oxygen species
SAM	*S*-adenosylmethionine
SAR	systemic acquired resistance
SDA	semidehydroascorbate
SIM	selected ion monitoring
SNAP	*S*-nitroso-*N*-acetyl-penicillamine
SOD	superoxide dismutase
SRS-A	slow-reacting substance A
TBA	thiobarbituric acid
TCDD	2,3,7,8-tetrachloro-dibenzo-*p*-dioxin
TCHQ	tetrachlorohydro-quinone
TGF	transforming growth factor
TIMP	tissue inhibitor of metalloproteinases
TSA	thiol-specific antioxidant
TXA_2, TXB_2	thromboxanes A_2 and B_2
UV	ultraviolet
VCAM	vascular cell adhesion molecule
VLDL	very-low-density lipoprotein

1

Oxygen is a toxic gas—an introduction to oxygen toxicity and reactive oxygen species

1.1 The history of oxygen: a major air pollutant

The element oxygen (chemical symbol O) exists in air as a diatomic molecule, O_2, which strictly should be called dioxygen. Over 99% of the O_2 in the atmosphere is the isotope[a] oxygen-16 but there are traces of oxygen-17 (about 0.04%) and oxygen-18 (about 0.2%). Except for certain anaerobic and aero-tolerant unicelluar organisms, all animals, plants and bacteria require O_2 for efficient production of energy by the use of O_2-dependent electron-transport chains, such as those in the mitochondria of eukaryotic cells. This need for O_2 obscures the fact that O_2 is a toxic mutagenic gas as well as a serious fire risk; aerobes survive because they have antioxidant defences to protect against it.

O_2 appeared in significant amounts in the Earth's atmosphere over 2.5×10^9 years ago (Table 1.1), and geological evidence suggests that this was due to the evolution of photosynthesis by blue-green algae (cyanobacteria). As they split water to obtain the hydrogen needed to drive metabolic reductions, these bacteria released tonnes of O_2 into the atmosphere. The inexorable rise in atmospheric O_2 concentrations was advantageous in one way, in that it led to the formation of the ozone (O_3) layer in the stratosphere. The ability of O_3 and O_2 to filter much of the intense solar ultraviolet (UV-C) radiation helped living organisms leave the sea and colonize the land, but O_2 itself must have placed a severe stress on the organisms present.

When living organisms first appeared on the Earth, they did so under an atmosphere containing very little O_2, i.e. they were essentially anaerobes. Anaerobic microorganisms still survive to this day, but their growth is inhibited and they can often be killed by exposure to 21% O_2, the current atmospheric level. As the O_2 content of the atmosphere rose, many primitive organisms must have died out. Present-day anaerobes are presumably the descendants of those primitive organisms that followed the evolutionary path of 'adapting' to rising atmospheric O_2 levels by restricting themselves to environments into which the O_2 did not penetrate (Fig. 1.1). However, other organisms began the evolutionary process of evolving antioxidant defence systems to protect against O_2 toxicity. In retrospect, this was a fruitful path to follow. Organisms that tolerated the presence of O_2 could also evolve to use it for metabolic transformations (e.g. oxidase, oxygenase and hydroxylase enzymes, such as

---

[a] See notes at the end of each chapter.

Table 1.1. Some of the main events preceding the appearance of humans on earth

Approximate time (million years ago)	Event
3500	Intense solar radiation bombards the surface of the Earth. Free-radical chemistry contributes to the formation of the first complex organic molecules. Anaerobic life begins, forming by-products such as sulphide, nitrite and alcohols.
>2500	Blue-green algae (cyanobacteria) acquire the ability to split water and release O_2: $2H_2O \rightarrow 4H + O_2\uparrow$
1300	Oxygen levels in the atmosphere reach 1%. Primitive anaerobic organisms disappear or retreat to oxygen-free areas. More complex cells with nuclei (**eukaryotes**) begin to evolve. Eukaryotes and cyanobacteria develop into green leaf plants. Eukaryotes and prokaryotes able to reduce O_2 to H_2O eventually develop into animals. Emergence of multicellular organisms.
500	Oxygen levels in the atmosphere reach 10%. Ozone layer screens out much UV light and facilitates emergence of life forms from the sea.
65	Primates appear.
5	Humans appear. Atmospheric oxygen levels reach 21% (of dry air).

Adapted from Harman, D. (1986) *Free Radicals, Ageing and Degenerative Diseases.* Alan R. Liss, Inc.

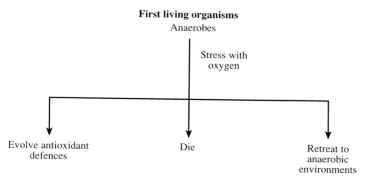

Fig. 1.1. Evolutionary adaptations to the appearance of O_2. Development of antioxidant defences allowed evolution of O_2-using enzymes and electron-transport chains enabling oxidation of food material more efficiently. Aerobic respiration produces much more energy per unit mass of food, allowing the development of complex multicellular organisms. The bigger organisms then have to develop mechanisms for delivering O_2 at the right level to their cells, so that many cells are shielded from the full brunt of 21% O_2.

tyrosine hydroxylase or cytochromes P450) and for efficient energy production by using electron-transport chains with O_2 as the terminal electron acceptor, such as those present in mitochondria. Mitochondria make over 80% of the ATP needed by mammalian cells, and the lethal effects of inhibiting this, e.g. by cyanide, show how important the mitochondria are.

The evolution of efficient energy production allowed the development of complex multicellular organisms, which also needed systems to ensure that the O_2 could be distributed throughout the organism. One advantage of evolving such systems is that delivery of O_2 to cells can be controlled: for example, most cells in the human body are never exposed to the full force of atmospheric O_2 (Fig. 1.2). There must then be mechanisms for monitoring O_2 levels in the body and altering respiration rate and blood flow to control such levels (see Section 6.8.3).

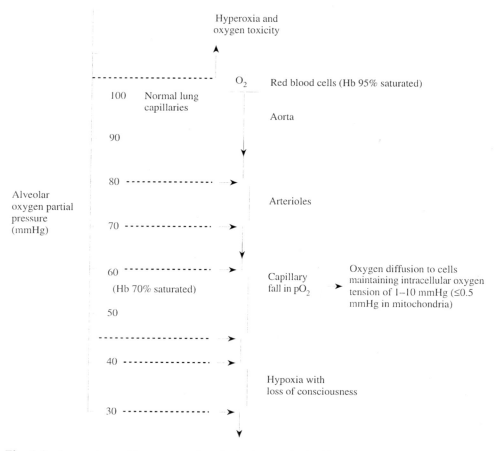

Fig. 1.2. Approximate O_2 concentrations in the human body. Note that most cells are exposed only to fairly low O_2 concentrations: this may be regarded as an antioxidant defence mechanism, although it also renders aerobic cells vulnerable to interruptions of the transport mechanism. HB, haemoglobin.

1.2   Oxygen today

Oxygen is now the most prevalent element in the Earth's crust (atomic abundance 53.8%) and the percentage of O_2 in the atmosphere has reached 21%. The barometric pressure of dry air at sea level is 760 mm mercury,[b] giving an O_2 partial pressure of about 159 mmHg.

Oxygen levels may have been even higher at periods in the Earth's history. It has been proposed that in the mid–to–late Devonian period, O_2 increased from about 18% to 20%, but then rose sharply to 35% by the late Carboniferous as plant life flourished, CO_2 levels fell drastically and great deposits of coal and oil were formed (Fig. 1.3). This increased O_2 concentration may have permitted insects (whose O_2 distribution system depends largely on diffusion) to become larger. For example, the giant Carboniferous dragonfly *Meganeura monyi* had a thoracic diameter of about 2.8 cm (compared with about 1 cm maximum for present–day dragonflies). Most of the various insects that attained exceptionally large body sizes during the Carboniferous did not persist after the Permian, when O_2 concentrations fell again. The plants and animals existing in Carboniferous times must presumably have had enhanced antioxidant defences, which would be fascinating to study if these species could ever be resurrected.

1.2.1   *Oxygen in water and organic solvents*

Oxygen is also found dissolved in seas, lakes, rivers and other bodies of water. The solubility of O_2 in sea water exposed to air at 10 °C corresponds to a concentration of 0.284 mmol/dm³,[c] and decreases at higher temperatures (e.g. 0.212 mmol/dm³ at 25 °C). Oxygen is more soluble in fresh water, e.g. for distilled water: 0.258 mmol/dm³ at 25 °C, 0.355 mmol/dm³ at 10 °C. Of course, the O_2 concentration experienced by living cells within a multicellular organism will depend on how far the O_2 has to move in order to get to them as well as on how quickly they consume it. Mitochondrial respiration can function well at low O_2 concentrations, so one way of diminishing O_2 toxicity has been to decrease the concentration to which cells within the body are exposed (Fig. 1.2). For example, the O_2 tension in human venous blood is only around 40 mmHg (about 53 µmol/dm³ O_2). Within most or all eukaryotic cells, there is an O_2 gradient, decreasing in concentration from the cell membrane to the oxygen–consuming mitochondria (Fig. 1.2). Under physiological conditions, O_2 is five to eight times more soluble in organic solvents than in water, a point worth bearing in mind when considering oxidative damage to the hydrophobic interior of biological membranes (Chapter 4).

1.3   Oxygen and anaerobes

As the O_2 content of the atmosphere rose, it exposed living organisms to O_2 toxicity: oxidations in the cell harmful to the organism and in some cases lethal. There was considerable pressure upon organisms to evolve protective mechanisms against O_2 toxicity, or to retreat to environments that the O_2 did not

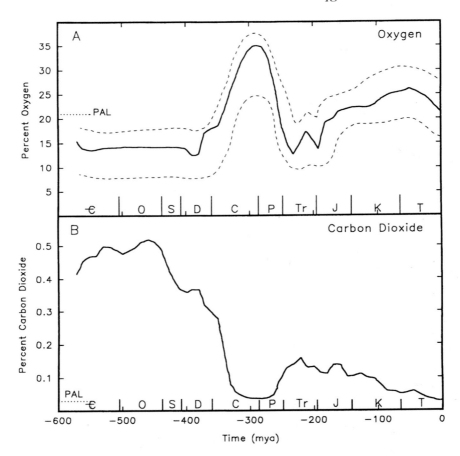

Fig. 1.3. Changes in O_2 and CO_2 during the Earth's history. (a) Calculated palaeoatmospheric O_2 concentration showing the estimate (solid line) and its range (dashed lines). Between the mid-to-late Devonian (380–360 million years ago (mya)), O_2 increased from ~18 to 20%, and then rose sharply to ~35% by late Carboniferous (286 mya). The present atmospheric level (PAL) of 21% is indicated. Oxygen steadily declined throughout the Permian (286–250 mya) and dropped to about 15% by the end of the Palaeozoic (250 mya). The model is based on the exchange rate of fixed carbon between the atmosphere, ocean and sediments and on various assumptions. (b) Palaeozoic atmospheric CO_2 model. This gas was present in relatively large amounts in the Ordovician–Silurian, fell precipitously during the Devonian–Carboniferous, and increased in the late Permian. The minimum value shown for the late Carboniferous and early Permian approximates present CO_2 levels (about 0.036%; dotted line). Data from *Nature* **375**, 118 (1995) by courtesy of Dr J.B. Graham and his colleagues and Macmillan.

penetrate (Fig. 1.1). Studies of O_2 toxicity in present-day anaerobes may show us what happened to the numerous primitive species that failed to adapt and were lost during evolution.

The term 'anaerobic organism' covers a wide range of biological variation.[25,26,28] There are 'strict' anaerobes such as the bacterium *Treponema denticola* and several *Clostridium* spp. that will grow in the laboratory only if O_2 is virtually absent. Indeed, the treatment of gas gangrene due to *Clostridium* infections by exposure of the patient to pure O_2 at pressures higher than atmospheric (**hyperbaric oxygen therapy**) is based on the known sensitivity of these anaerobes to O_2. As discussed below, however, hyperbaric treatment is not without problems.

'Moderate' anaerobes can grow under atmospheres containing up to about 10% O_2 (e.g. *Bacteroides fragilis* or *Clostridium novyi* Type A), whereas **micro-aerophiles**, such as *Campylobacter jejuni* (a major cause of diarrhoea in humans) and *Treponema pallidum* (the causative agent of syphilis), require a low concentration of O_2 for growth but cannot tolerate 21%. Even 'strict anaerobes' display a wide spectrum of O_2 tolerance. Some are killed by even a brief exposure to O_2 whereas for others O_2 inhibits growth but does not kill the cells, e.g. *Methanobacterium* AZ ceases growth at 0.01 p.p.m. O_2 but survives exposure for several days to 7 p.p.m. dissolved O_2, equivalent to an atmospheric concentration of 20%. Sometimes the induction of antioxidant defence systems is involved in such tolerance, as discussed in Chapter 3.

Many terrestrial and aquatic environments develop a low enough O_2 concentration to harbour anaerobes. For example, in the human mouth, strict anaerobes can be cultured from pockets in the gums, from decaying teeth, and from the deeper layers of dental plaque (e.g. *T. denticola*); whereas less strict anaerobes and microaerophiles can be found in the more superficial layers of plaque on the teeth.[26] The human colon (over 90% of faecal bacteria are anaerobes), rotting material, polluted waters and gangrenous wounds all provide places for anaerobic bacteria to thrive.

1.3.1 *Why does oxygen injure anaerobes?*

The damaging effects of O_2 on strict anaerobes seem to be due to the oxidation of essential cellular components.[28] Anaerobes thrive in reducing environments and, by oxidizing such essential metabolic intermediates as thiols, iron–sulphur proteins, and reduced pteridines, the O_2 can 'drain away' the reducing equivalents that are needed for biosynthetic reactions within the cell. These oxidations often simultaneously reduce O_2 to oxygen free radicals and other toxic oxygen-derived species (see below). Some enzymes in anaerobes are inhibited directly by O_2, e.g. the nitrogen-fixing enzyme **nitrogenase** of *Clostridium pasteurianum* is inactivated by oxidation of essential components at its active site. Nitrogenase, which catalyses reduction of atmospheric nitrogen (N_2) to ammonia (NH_3), is essential for growth of the organism in environments poor in nitrogen compounds.

Indeed, all known nitrogenase enzymes are inactivated by O_2 to some extent.[14] Surprisingly, perhaps, not all nitrogen-fixing species are strict anaerobes. Indeed, a study of nitrogen-fixing organisms has shown a variety of ways around the problem of the O_2-sensitivity of nitrogenases. *C. pasteurianum* adopts a simple solution and keeps away from O_2. Several aerotolerant, nitrogen-fixing bacteria can surround themselves with a thick capsule to restrict the entry of O_2; this strategy of 'antioxidant defence' is also used[9] by certain streptococci that do not fix N_2. Some photosynthetic N_2-fixing blue-green algae locate their nitrogenase in specialized, thick-walled, O_2-resistant cells known as **heterocysts**.[14] In the root nodules of leguminous plants that engage in a symbiotic relationship with nitrogen-fixing bacteria, an O_2-binding protein, **leghaemoglobin** is present, apparently to control the free O_2 concentration and so to prevent the nitrogen-fixing system of the bacteroids in the nodule from being damaged. The photosynthetic organism *Gloeocapsa* contains both nitrogenase and an O_2-evolving photosynthetic apparatus within the same cell, but its life cycle is such that nitrogenase is only highly active when the rate of photosynthesis is low.[14]

Anaerobes can teach us a great deal about the evolution of protective mechanisms against O_2 toxicity and we will consider them again in Chapter 3, when reviewing the various protective mechanisms thought to exist.

1.4   Oxygen and aerobes

1.4.1   *Oxygen transport in mammals*

Complex multicellular organisms such as mammals have evolved mechanisms to ensure that O_2 is delivered to all the cells that need it. Some O_2 travels dissolved in blood plasma, but the solubility of O_2 in water at body temperature is limited (Section 1.2.1). Most O_2 carried in the blood is transported by **haemoglobin**. The haemoglobin molecule has four protein subunits, two α-chains and two β-chains. Each chain carries a **haem** group (Fig. 1.4) with iron in the Fe^{2+} state. The O_2 reversibly attaches to haem, binding at high O_2 concentrations as blood flows through the lungs and dissociating again at the lower levels of O_2 in the tissues (Fig. 1.2).

$$Fe^{2+} + O_2 \rightleftharpoons Fe^{2+} - O_2$$

Only the ferrous form (Fe^{2+}) of haemoglobin can bind O_2. Haemoglobin also helps to control pH in body fluids and it may act as a carrier of nitric oxide.[23] The β-subunits of haemoglobin contain cysteine residues at position 93 and traces of adducts of nitric oxide with the thiol (–SH) group of this cysteine can be detected in oxyhaemoglobin, whereas traces of nitric oxide are bound to the haem in deoxyhaemoglobin *in vivo*. The physiological significance of this **S-nitrosohaemoglobin** is at present uncertain. Nitric oxide can bind to haem in many other haem-containing proteins (see Section 2.4.6).

Heart and many other muscles (sometimes called 'red muscles') contain the pigment **myoglobin**, which contains only a single polypeptide chain (bearing one haem), whose structure resembles that of the haemoglobin subunits. Myoglobin appears to act as a store of O_2 in tissues: it binds O_2 more tightly than haemoglobin and only releases it if O_2 tension becomes very low. Leghaemoglobin in the root nodule may perform a similar function (Section 1.3.1).

Haemoglobin within erythrocytes delivers O_2 to cells in aerobic tissues, permitting efficient energy production and allowing the operation of enzymes

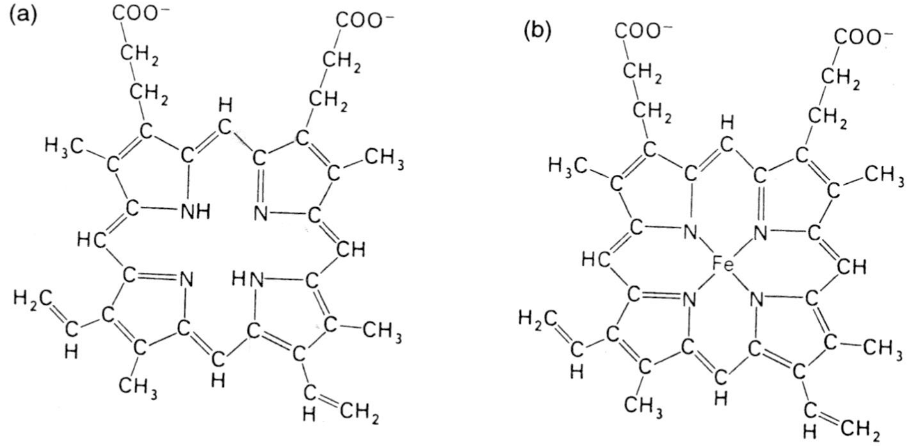

Fig. 1.4. The structure of protoporphyrin IX (a) and of iron–protoporphyrin IX or haem (b). The porphyrin ring contains four pyrrole units linked by methene (–CH=) bridges. The haem ring gives haemoglobin and myoglobin their distinctive reddish-brown colour. In haemoglobin and myoglobin the Fe^{2+} is coordinated to the four nitrogens in the protoporphyrin IX ring. It also ligands to a histidine residue from the protein, and O_2 forms the sixth ligand in the oxy-proteins.

that catalyse O_2-dependent metabolic transformations. Haem can be synthesized in mammalian cells and by many bacteria, including some anaerobes. Two steps in this pathway, catalysed by the enzymes **coproporphyrinogen oxidase** and **protoporphyrinogen oxidase**, use O_2 in eukaryotes, but other electron acceptors, such as $NADP^+$, in some anaerobic bacteria.[6]

1.4.2   Oxygen sensing

The **carotid body** is an organ found in mammals, birds and some fish that senses O_2 levels in the blood and alters the respiration rate accordingly. In humans, it is located at the bifurcation of the carotid artery.[2] Similarly, low O_2 levels increase production of the protein hormone **erythropoietin** in the kidney and to a lesser extent in the liver. Erythropoietin acts to increase the net output of red blood cells in the bone marrow. In both cases, the 'O₂ sensor' is thought to be a haem protein. The mechanism of action of the carotid body is considered further in Section 6.8.3.

1.4.3   Mitochondrial electron transport

About 85–90% of the O_2 taken up by animals is utilized by the mitochondria; these organelles are the major source of ATP in animals, in non-photosynthetic plant tissues and in leaves in the dark.[18,31] The essence of metabolic energy production is that food materials are oxidized: they lose electrons, which are accepted by electron carriers (Fig. 1.5), such as nicotinamide adenine dinucleotide (NAD^+) and flavins (flavin mononucleotide (FMN) and flavin adenine dinucleotide (FAD)). The resulting reduced nicotinamide adenine dinucleotide (NADH) and reduced flavins ($FMNH_2$ and $FADH_2$) are re-oxidized by O_2 in mitochondria, producing large amounts of ATP. Oxidation is catalysed in a stepwise fashion so that the energy is released gradually. This is achieved by the **electron transport chain** present in the inner mitochondrial membrane (Fig. 1.6). Electrons pass from NADH to non-haem iron proteins—these accept the electrons by converting their bound iron from Fe(III) to Fe^{2+}. They pass on the electrons by re-oxidizing to Fe(III). Later in the chain, the cytochrome proteins work in the same way, except that their iron ions are bound to haem rings (Fig. 1.4). Cytochromes accept electrons by forming Fe^{2+}–haem, and release them again by oxidizing to Fe(III)–haem. In this respect, they are very different from haemoglobin which only works in the Fe^{2+} form.

The part of the electron-transport chain that actually uses O_2 is the terminal oxidase enzyme, **cytochrome oxidase**. It removes one electron from each of four reduced (Fe^{2+}–haem) cytochrome c molecules, oxidizing them to ferric cytochrome c. It adds the four electrons on to O_2; the overall reaction is

$$O_2 + 4H^+ + 4e^- \rightarrow 2H_2O$$

However, it is chemically impossible to add four electrons to O_2 at once—it must be done in stages. Hence cytochrome oxidase is a very complex system, because it catalyses several reduction steps. For example, the bovine enzyme has 13 different protein subunits and a relative molecular mass of 200 000. Since partially reduced oxygen species are damaging (see below) the enzyme must also keep them safely bound to its active sites until they are fully converted to water.

Cytochrome oxidase has both haem iron, and copper ions, bound to it. These metals play key roles in O_2 reduction and safe binding of partially reduced oxygen intermediates. Mammalian cytochrome oxidase has a very high affinity for O_2: it still works well at O_2 concentrations of less than 1 mmHg.[31,34] Hence energy production by mitochondria can continue at low O_2 concentrations. This is important—fully oxygenated blood leaving the lungs has an O_2 concentration of some 100 mmHg, but this falls rapidly in the tissues as oxyhaemoglobin unloads O_2. Cells deep within a tissue may experience extracellular O_2 concentrations of only 5–15 mmHg, and O_2 still has to cross the cell to reach mitochondria. Hence the O_2 concentration within actively respiring mitochondria may be very low (Fig. 1.2). This sequestration of partially reduced O_2 species by cytochrome oxidase and the operation of cells at fairly low intracellular pO_2 may be regarded as attempts to minimize damage by O_2, i.e. as antioxidant defences.

1.4.4 *Bacterial electron transport chains*

Many bacteria, such as *Salmonella typhimurium* and *Escherichia coli*, can grow under both aerobic and anaerobic conditions. *E. coli* can adapt its respiratory

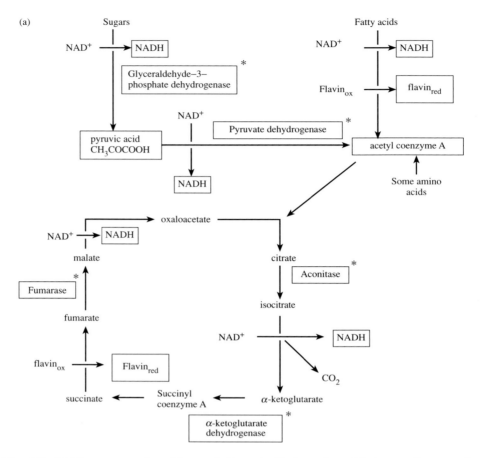

Fig. 1.5. (a) The essence of aerobic metabolism: the Krebs cycle and its inputs. Enzymes that are potential targets of damage by oxygen radicals are highlighted by a*. Amino acids can often generate various intermediates of the Krebs cycle instead of, or in addition to, acetyl coenzyme A. Glyceraldehyde-3-phosphate dehydrogenase is an essential enzyme in **glycolysis**, the pathway converting glucose to pyruvic acid. (b) Structures of the electron carriers. NAD^+ accepts two electrons

$$NAD^+ + H^+ + 2e^- \rightleftharpoons NADH$$

and cells contain 'pools' of NAD^+ and NADH; the ratio of their concentrations is one factor regulating the activity of NAD^+-using enzymes. The active part of NAD^+ is derived from the B-vitamin nicotinamide. $NADP^+$, which differs from NAD^+ by one phosphate group, plays different metabolic roles (see Section 3.8) but is oxidized and reduced by the same mechanism. **Flavoproteins** contain derivatives of the vitamin riboflavin, either FMN or FAD. They also accept two electrons.

$$FP + 2e^- + 2H^+ (flavin_{ox}) \rightleftharpoons FPH_2 (flavin_{red})$$

FMN or FAD are usually attached to the active sites of the enzymes that use them, not free as a pool. Most enzymes in the Krebs cycle use NAD^+ as ultimate electron acceptor but one enzyme in fatty acid metabolism and the Krebs cycle enzyme **succinate dehydrogenase**

$$succinate + FAD \rightleftharpoons fumarate + FADH_2$$

are flavoproteins.

(b)

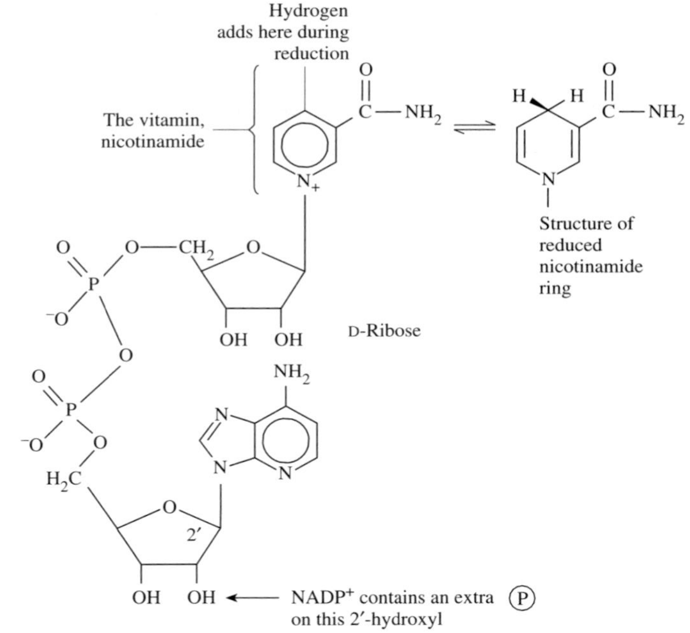

Hydrogen adds here during reduction

The vitamin, nicotinamide

Structure of reduced nicotinamide ring

D-Ribose

NADP⁺ contains an extra (P) on this 2′-hydroxyl

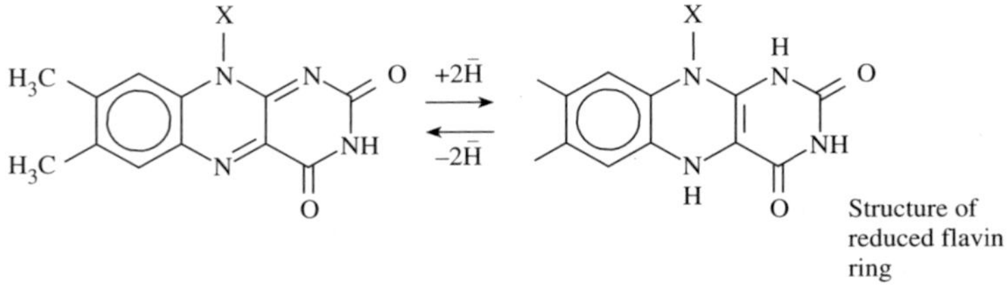

+2H̄

−2H̄

Structure of reduced flavin ring

Riboflavin: X is — CH₂(CHOH)₃CH₂OH

FMN: X is — CH₂(CHOH)₃CH₂OP=O

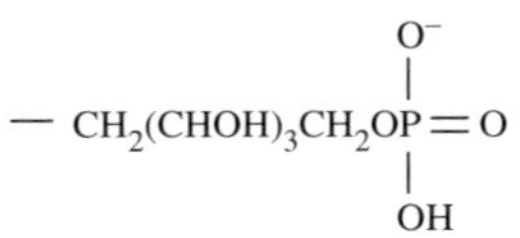

FAD: X is — CH₂OP—O—P—O—CH₂   Adenine ring

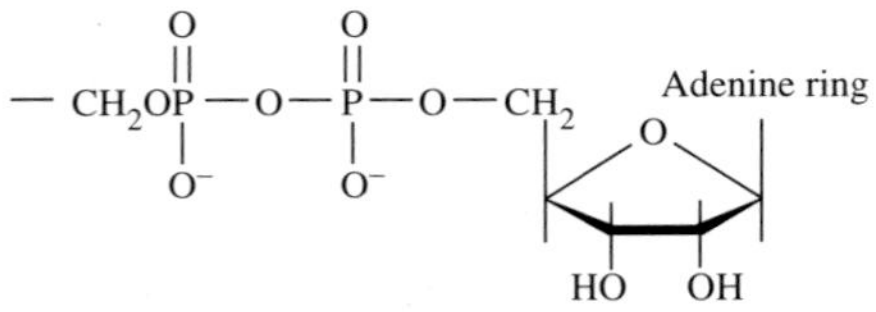

Fig. 1.5 (b)

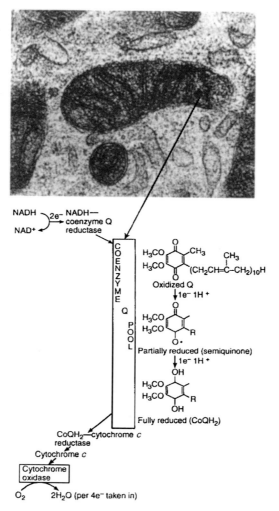

Fig. 1.6. The electron-transport chain of animal mitochondria (photograph shows human lung mitochondria, courtesy of Ann Dewar). Mitochondria have an outer membrane and an infolded inner membrane which contains the electron-transport chain. The central space of mitochondria (the **matrix**) contains many enzymes, including those of the Krebs cycle (Fig. 1.5), and those that break down fatty acids to acetyl coenzyme A. During the operation of the electron-transport chain, NADH is oxidized to NAD^+ by a multienzyme complex known as **NADH coenzyme Q reductase** (sometimes called **NADH dehydrogenase** or **complex I**) and the two electrons released are eventually passed on to coenzyme Q. The number of 'repeats' in the coenzyme Q side chain is variable; it is most often 10 in the human (CoQ_{10}, shown in figure) but nine in rats. The NADH coenzyme Q reductase complex contains flavoproteins (FMN at active site) and **non–haem–iron proteins** (iron ions are present but not in haem). **Coenzyme Q (ubiquinone)** accepts one electron to form a semiquinone (free radical) or two electrons to form a fully reduced (**ubiquinol**) form. The ubisemiquinone radical can be detected in respiring mitochondria. Coenzyme Q also accepts electrons from various reduced flavoproteins generated by the Krebs cycle (succinate dehydrogenase) and β-oxidation

chain to work with different terminal electron acceptors.[6] In the presence of O_2, substrates such as glucose are oxidized by metabolic pathways including the Krebs cycle (Fig. 1.4) and electrons are fed into an electron-transport chain containing quinones, *b*-type cytochromes and a cytochrome oxidase, either cytochrome bo oxidase or cytochrome bd oxidase. The former (unlike mammalian cytochrome oxidase) has a low affinity for O_2 but a high maximum velocity; it could serve to consume O_2 rapidly when O_2 is abundant (another antioxidant defence perhaps?). At lower O_2 levels cytochrome bd oxidase, with a hundred-fold greater affinity for O_2, is used. Indeed, O_2 concentration regulates both the function and amount of these two cytochrome oxidases. When O_2 is absent, for example, levels of cytochrome bo fall whereas those of bd rise. Anaerobic *E. coli* cells can produce at least five different terminal electron acceptor systems, depending on what is available to accept electrons. **Fumarate reductase** donates electrons to fumarate to give succinate, two **nitrate reductases** reduce nitrate (NO_3^-) to nitrite (NO_2^-) and even **sulphoxide reductase** (which works on a range of sulphoxides, including dimethylsulphoxide) and **amine-N-oxide reductase** have been identified. If no electron acceptor is available, *E. coli* switches to fermentation as a (much less efficient) source of energy.

1.5   Oxidases and oxygenases in aerobes

Most of the 10–15% of O_2 taken up by aerobic eukaryotes that is not consumed by mitochondria is used by various oxidase and oxygenase enzymes, and also by direct chemical ('non-enzymic') oxidation reactions. For example, the enzyme **D-amino acid oxidase** uses O_2 to oxidize unwanted D-amino acids. **Xanthine oxidase** uses O_2 to oxidize xanthine and hypoxanthine into

---

of fatty acids. The enzyme **dihydro-orotic acid dehydrogenase**, which catalyses a step in pyrimidine synthesis, also feeds electrons into the electron-transport chain at several points in the region of CoQ. The complex of proteins transferring electrons from succinate to CoQ is called **complex II**. From Q the electrons pass through another multienzyme complex (**coenzyme Q-cytochrome *c* reductase**, or **complex III**, which contains an iron–sulphur protein plus cytochromes *b* and c_1), on to cytochrome *c*. **Cytochromes** are haem proteins which accept electrons by allowing Fe(III) at the centre of the haem ring to be reduced to Fe^{2+}, i.e. they can accept one electron at a time per molecule.

$$\text{cytochrome–Fe(III)} + e^- \rightleftharpoons \text{cytochrome–Fe}^{2+}$$

Different cytochromes, designated by small letters, contain different proteins and different haem groups. Finally, reduced cytochrome *c* is re-oxidized by a multienzyme complex, **cytochrome *c* oxidase** (**complex IV**) which contains cytochrome *a*, cytochrome a_3 and copper ions. For every four electrons taken in by this complex, one oxygen molecule is fully reduced to two molecules of water. As electrons travel through the electron-transport chain, protons cross the inner mitochondrial membrane into the inter-membrane space, creating a pH difference plus a charge difference (an **electrochemical gradient**). The energy of this gradient is used by an **ATP synthase** enzyme (**complex V**) to generate ATP. ATP is transferred across the inner mitochondrial membrane in exchange for ADP by an **adenine nucleotide translocase** and thus made available to the rest of the cell. The outer mitochondrial membrane is much more permeable to metabolites than the inner membrane.

uric acid. In animal tissues, cells are embedded in a matrix containing, among many other compounds, the protein **collagen** which gives strength and flexibility. When collagen is being synthesized, **proline** and **lysine hydroxy-lase** enzymes use O_2 to help put essential hydroxyl (−OH) groups on to the amino acids proline and lysine in the protein. Synthesis of the essential hormones **epinephrine (adrenalin)** and **norepinephrine (noradrenalin)** begins with O_2-dependent addition of an −OH group to the amino acid tyrosine by a **tyrosine hydroxylase** enzyme. In general, these O_2-using enzymes bind O_2 much less efficiently than does cytochrome oxidase.[11] Hence oxidases and hydroxylases are often O_2-limited in their action at normal cellular O_2 concentrations.

1.5.1   *Cytochromes P450*

The endoplasmic reticulum (and often other organelles) of many animal and plant tissues contain cytochromes known collectively as cytochrome P450.[17] Over 150 genes encoding the **P450 superfamily** have been described (Table 1.2 gives some examples). The name P450 was originally given because the reduced forms of the cytochromes bind carbon monoxide to produce a complex that absorbs light strongly at 450 nm. Cytochromes P450 are involved in the oxidation of a wide range of substrates at the expense of O_2. One atom of the oxygen enters the substrate and the other forms water, such a reaction being known as a **mono–oxygenase** or **mixed–function oxidase** reaction. The functioning of cytochromes P450 requires a reducing agent (RH_2), and the overall reaction catalysed can be represented by the following equation, in which AH is the substrate:

$$AH + O_2 + RH_2 \rightarrow AOH + R + H_2O$$

P450s are haem proteins containing a single polypeptide chain: four ligands to the iron are provided by the haem and a fifth as a thiolate anion (S^-) from a cysteine residue. In the 'resting' enzymes, the sixth ligand is water.

Liver endoplasmic reticulum is especially rich in P450s, which metabolize a large number of chemicals. Some of these compounds can increase synthesis of one or more P450 forms when fed to animals. One such inducer is the barbiturate **phenobarbital**, hydroxylation of which increases its solubility and aids its excretion from the body. Excessive intake of ethanol by mammals increases the ability of liver microsomal fractions to oxidize this substance, because of increased synthesis of a specific form of cytochrome P450, **CYP2E1** or the **ethanol-inducible cytochrome P450**. CYP2E1 can oxidize ethanol, other alcohols, acetone and some other organic solvents, including carbon tetrachloride (Table 1.2).

Substrates for cytochromes P450 include insecticides such as heptachlor and aldrin, hydrocarbons such as benzpyrene and toluene and drugs such as phenacetin, amphetamine, methadone and paracetamol (Table 1.2).[17] Usually the product of reaction with P450 is less toxic than the substrate, but this is not always the case; it is the P450-derived metabolic products of

Table 1.2. A few of the many forms of cytochrome P450 found in humans

Family	Examples of subfamilies	Examples of isoenzymes	Examples of inducers	Representative substrates
CYP1 (polycyclic aromatic hydrocarbon-inducible)	CYP1A	CYP1A1 (aryl hydrocarbon hydroxylase)	3-methylcholanthrene, 2,3,7,8-tetrachloro-p-dibenzodioxin (TCDD)	7-ethoxyresorufin, benzpyrene, benzanthracene
CYP2 (phenobarbital family)	CYP2B	CYP2B1	phenobarbital, aroclor	olefins, acetylenes, dimethylbenzanthrene, benzphetamine
		CYP2B2	phenobarbital	dimethylbenzanthracene
	CYP2E	CYP2E1	ethanol, ether, acetone, dimethylsulphoxide	ethanol, other alcohols, toluene, xylenes, chlorzoxazone (a muscle relaxant)
CYP3 (steroid-inducible)	CYP3A	CYP3A4	rifampicin	methadone, oestradiol, testosterone
CYP4	CYP4A		clofibrate and some other peroxisome proliferators (Section 9.13.5)	prostaglandins, fatty acids

Other families include CYP11 (mitochondrial proteins; steroid metabolism), CYP17 (steroid 17α-hydroxylase), CYP21 (steroid 21-hydroxylase) and CYP19 (aromatase). The types of P450 present vary between different organisms, tissues and cells and can be affected by induction or inhibition of expression. A major form in human liver and intestine is CYP3A4, which can be up to 50% of total P450. Each P450 gene family (denoted by a number) is divided into subfamilies (denoted by a capital letter) and the members of each subfamily are identified by numbers.

paracetamol (see Section 8.9) and of carcinogenic hydrocarbons such as benzpyrene (Section 9.13.2) that cause much of the cellular damage produced by these compounds.

In the liver the electrons required by the P450 system are donated by NADPH via a flavoprotein enzyme **NADPH-cytochrome P450 reductase**, each molecule of which contains one FMN and one FAD. Figure 1.7 shows

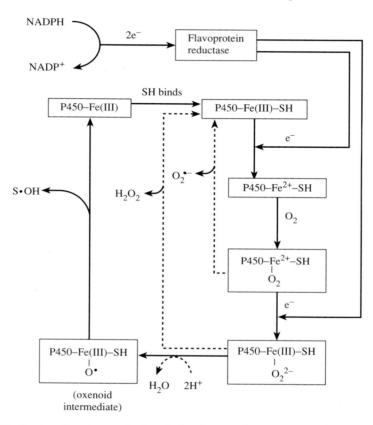

Fig. 1.7. Mechanism for substrate hydroxylation by cytochrome P450 in liver. SH represents the substrate molecule. The mechanism of P450 hydroxylation appears similar in other tissues or bacteria, but the source of reducing power used is different. The one shown is for liver endoplasmic reticulum. Capture of the second electron generates an $Fe(III)$–P450–substrate–O_2^{2-} complex, which appears to protonate and undergo O–O bond cleavage releasing H_2O and leaving a single oxygen atom bound to the iron. This oxo-iron species has been variously written as $[FeO]^{3+}$, $Fe(V)=O$, $Fe(IV)=O^{\bullet}$ or $[Fe(IV)=O]^{\bullet+}$ and may be an oxo-iron porphyrin radical cation similar to compound I in horseradish peroxidase, $[Fe(IV)=O]^{\bullet+}$ (see Section 3.16.4). Subsequent insertion of an atom of oxygen into the substrate (the $[FeO]^{3+}$ intermediate abstracts a hydrogen atom from the substrate, yielding a carbon radical and iron-bound hydroxyl radical that immediately recombines with the carbon radical) regenerates $Fe(III)$–P450. The P450 reaction cycle can also generate $O_2^{\bullet-}$ and H_2O_2, shown by the dotted lines.

a possible mechanism for substrate hydroxylation by P450, but the nature of the actual hydroxylating species at the active site of these proteins is still a subject of debate (see figure legend). Adrenal cortex mitochondria contain a cytochrome P450 which is involved in the hydroxylation of cholesterol to give the adrenal steroid hormones (e.g. aldosterone, hydrocortisone and corticosterone) but the electrons it requires are donated by a non-haem-iron protein known as **adrenodoxin**. A flavoprotein enzyme transfers electrons from NADPH to adrenodoxin.

Cytochrome P450 is also found in some bacteria, e.g. in *Pseudomonas putida* a P450 serves to hydroxylate camphor. Here electrons are supplied by the non-haem-iron protein **putidaredoxin**, which is kept reduced at the expense of NADH by a flavoprotein enzyme. The hydroxylated camphor can then be metabolized by the cells to provide energy.

1.6 Oxygen toxicity in aerobes

1.6.1 *Oxygen toxicity in bacteria and plants*

Despite its many advantages, even 21% O_2 causes damage to aerobes.[2,5,32] We are not usually aware of this until we attempt to measure oxidative damage in aerobes under ambient O_2, and find that it can be detected (Chapters 4 and 5). Oxygen toxicity has more usually been studied as the effects of exposing organisms to elevated O_2 concentrations.[2] Oxygen supplied at concentrations greater than those in normal air has been known for decades to be toxic to plants and animals and to bacteria such as *E. coli*. Indeed, studies of bacterial chemotaxis to O_2 (**aerotaxis**) show that several strains swim away from regions of high O_2 concentration and tend to settle in regions of optimal 'redox state' for their growth. *E. coli* also moves away from solutions containing low levels of hydrogen peroxide or hypochlorous acid, which might enable it to avoid engulfment by phagocytes (Section 6.7).[3]

Plots of the logarithm of survival time against the logarithm of the O_2 pressure have shown inverse, approximately linear relationships, for protozoa, mice, rats, rabbits, fish and insects.[2] Plant tissues are damaged at O_2 concentrations above normal;[18] there is an inhibition of chloroplast development, decrease in seed viability and root growth, membrane damage, and shrivelling and shedding of leaves. Green plants produce O_2 during photosynthesis and can expose themselves and their surroundings to it, e.g. O_2 bubbles from some aquatic plants have been reported to interfere with the breeding of mosquitoes. Hence photosynthetic plant tissues have a particular problem with O_2 toxicity, which is considered in Section 7.5.

Figure 1.8 shows an example of O_2 toxicity: exposure of *E. coli* to high-pressure oxygen causes immediate growth inhibition. This can be temporarily reversed by adding certain amino acids (such as valine) to the culture medium, suggesting that one deleterious effect of excess O_2 is to block their biosynthesis. However, even when these amino acids are added, growth soon ceases again due to interference with other metabolic processes, such as the synthesis of

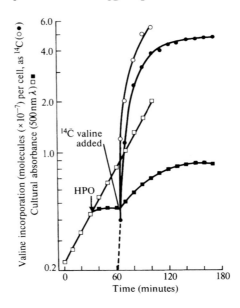

Fig. 1.8. Inhibition of the growth of *E. coli* cells by exposure to high-pressure oxygen. The growth medium was mineral salts, glucose and amino acids (no valine) at 37 °C. At the point marked HPO, the atmosphere was changed from air to one of 80% O_2 at 5 atm total pressure. At the point indicated, valine was added and growth was restored. Closed symbols: HPO experiment; open symbols: normal air control. Growth soon ceases again, however, because of damage to other enzymes, e.g. those involved in the biosynthesis of NAD^+. From Brown and Yein (1978) *Biochem. Biophys. Res. Commun.* **85**, 1219, by courtesy of the authors and Academic Press.

essential enzyme cofactors. For example, the enzyme **quinolinate synthetase**,[15] which catalyses a key step in the biosynthesis of NAD^+ and $NADP^+$, is inactivated at high O_2 concentrations.

Another well-known effect of O_2 is to enhance the damaging effects of ionizing radiation[d] upon bacterial, plant and animal cells; Fig. 1.9 illustrates this effect using cultured Chinese hamster ovary cells exposed to X-rays.

1.6.2 *Oxygen toxicity in humans and other animals*

The toxicity of O_2 to humans has been of interest in relation to diving, underwater swimming, design of the gas supply in spacecraft and submarines, and in the use of hyperbaric O_2 in the treatment of cancer, infections, multiple sclerosis and lung diseases.[2,8,12,30] Rises in the O_2 partial pressure to which an organism is subjected can be due not only to an increase in the percentage of O_2 in the air but also, as in diving, to an increase in the total pressure.

High-pressure O_2 frequently causes acute central nervous system toxicity in animals, producing convulsions. Oxygen at 1 atm does not usually produce such convulsions in animals, but O_2 concentrations of 50% or above,

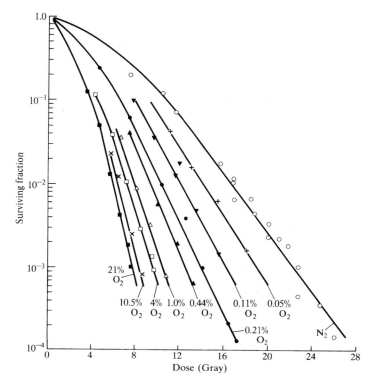

Fig. 1.9. The 'oxygen effect' in exposing cultured Chinese hamster ovary cells to ionizing radiation. Figure by courtesy of Dr H.B. Michaels.

corresponding to an inspired partial pressure of 360 mmHg, gradually damage the lungs. For example, adult rats usually show severe lung damage within 72 h of exposure to pure O_2.[2] Exposure of humans to pure O_2 at 1 atm for as little as 6 h can lead to chest soreness, cough and a sore throat in some people; 24 h exposure leads in all cases to damage to the alveoli of the lungs. This is manifested at first as oedema, mainly due to damage to the capillary/interstitial barrier, thus allowing proteins to leak from the blood plasma into the interstitial space. The endothelial cells lining the capillaries are highly sensitive to elevated O_2.

Further O_2 exposure causes the death of alveolar epithelial cells, and the penetration of protein-rich fluid into the alveoli themselves, seriously interfering with gas exchange. The lung is presumably a primary target of O_2 toxicity because it is exposed to higher O_2 levels than other body tissues (Fig. 1.2). The epithelium lining the alveoli of the lung is a monolayer containing the very thin so-called **type I cells** and the more cuboidal **type II cells**. The latter secrete **surfactant**, an agent that lowers surface tension and permits the alveoli to function. Type I cells are specialized to aid diffusion of gas from alveoli to

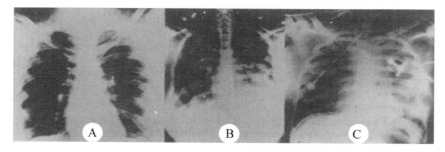

Fig. 1.10. Oxygen toxicity in humans. (A) Initial chest X-ray of a patient with mild respiratory discomfort after administration of pure O_2 for a non-pulmonary condition. There are no visible significant abnormalities. (B) Further exposure causing X-ray-visible damage with diffuse, irregular pulmonary densities of various sizes in both lungs. (C) Late radiological manifestations of pulmonary O_2 toxicity with extension and joining up of the lesions. Radiological manifestations are due to fluid accumulation (oedema), atelectasis (incomplete expansion or collapse of alveoli) and accumulation of cellular debris in alveolar spaces and in the terminal airways. There is laboured, gasping breathing often accompanied by frothy, bloody sputum. The damaged lungs cannot absorb sufficient O_2 for the body, resulting in cyanosis. From Huber, GL and Drath, DB Chapter 14 in Ref. 16, by courtesy of the authors and the publisher.

the blood capillaries surrounding the alveoli: their surface area is very great and so type I cells cover the majority of the alveolar surface. Type I cells seem more easily damaged and killed by excess O_2 (and many other toxins) than type II cells; after reversible injury, type II cells usually proliferate and differentiate into type I cells. However, type II cells can also be injured by excess O_2, e.g. surfactant production can be inhibited.

If too much destruction of lung cells occurs, the laying down of inelastic fibrous material in the lung (**fibrosis**) will eventually occur, permanently impairing gas exchange.[2,8,12] Figure 1.10 shows the gradual development of pulmonary O_2 toxicity as seen on chest X-rays. Some clinical and experimental observations suggest that O_2 may worsen lung damage caused by other means even at concentrations thought to be 'safe', i.e. if the lungs are already injured, elevated O_2 can make the injury worse.

1.6.3   *Retinopathy of prematurity*

Other tissues do not escape damage when animals are exposed to high O_2 concentrations, however. High O_2 levels can cause a general 'stress reaction' in animals, which stimulates the action of some endocrine glands.[2,29] Removal of, for example, the thyroid gland decreases the toxic effects of O_2 in some animals whereas administration of thyroxine, cortisone or adrenalin can worsen O_2 toxicity. Exposure of pregnant animals to elevated O_2 concentrations has been reported to increase the incidence of foetal abnormalities.

The form of blindness originally known as **retrolental fibroplasia** (from the Latin for 'formation of fibrous tissue behind the lens') but now more usually called **retinopathy of prematurity** (ROP) arose abruptly in the early 1940s among human infants born prematurely, and quickly became widespread. Not until 1954 was it realized that this disease is associated with the use of high O_2 concentrations in incubators for premature babies, and more careful control of O_2 use (continuous monitoring, with supplemental O_2 given only when needed to maintain blood O_2 levels) together with supplementation of babies with the antioxidant α-tocopherol (Section 3.22.7) has greatly decreased the severity of ROP. However, the problem has not gone away, since the most premature babies (those with birth-weights of 1000 g or less) need high O_2 levels in order to survive at all.[29a]

A phenomenon similar to ROP can be reproduced in newborn rats, rabbits and kittens exposed to elevated O_2. The excess O_2 appears to inhibit the growth of retinal blood vessels. On return to a normal atmosphere the resulting hypoxia induces retinal cells to secrete **angiogenic factors** that cause excessive regrowth of blood capillaries, which sometimes occurs to an extent that causes detachment of the retina and subsequent blindness. The new vessels lack structural integrity and often bleed. Agents that prevent the action of these angiogenic factors (e.g. by blocking the receptors on vascular endothelial cells to which they bind) are under investigation as protective agents.[19]

1.6.4 *Factors affecting oxygen toxicity*

The damaging effects of O_2 on aerobic organisms vary considerably with the type of organism used, its age, physiological state and diet.[2,20] Different tissues (and cells within tissues, e.g. type I versus type II cells in the lung) are affected in different ways. For example, buoyancy in many fish is regulated by the **swim bladder**, which is filled mainly with O_2 plus smaller amounts of N_2 and CO_2. With increasing pressure at higher depths, many fish keep the swim bladder volume constant by increasing its O_2 content, often up to 90% of the total gas. For example, the effective O_2 concentration in the swim-bladder of the **rat–tail fish** at a depth of 3000 m is 2500 times greater than ambient, yet the bladder remains undamaged. The fish as a whole cannot tolerate anywhere approaching this O_2 concentration, and so its swim bladder (and the **gas gland** responsible for filling it with O_2) must be specially protected (this is discussed further in Chapter 3). Cold-blooded animals, such as turtles and crocodiles, are fairly resistant to O_2 toxicity at low environmental temperatures, but become more sensitive at higher temperatures.[2] Newborn rats appear more resistant to O_2 than adult rats, but begin to lose this advantage when about 30 days old.

Oxygen toxicity is also influenced by the presence in the diet of varying amounts of vitamins A, E and C, metals such as zinc, iron and copper, synthetic antioxidants (now added to many human foods), and polyunsaturated fatty acids. For example, adult female Sprague–Dawley rats fed on a fat-free test diet supplemented with cod-liver oil could tolerate pure O_2 much better than if the supplement consisted of coconut oil.[24]

1.7   What causes the toxic effects of oxygen?

Perhaps the earliest suggestion made to explain O_2 toxicity was that O_2 inhibits cellular enzymes.[2,20] Indeed, direct inhibition by O_2 is thought to account for the loss of nitrogenase activity in O_2-exposed *C. pasteurianum* (Section 1.3.1 above). Another good example of the direct effect of O_2 anaerobes comes from green plants. During photosynthesis, illuminated green plants fix CO_2 into sugars by a complex metabolic pathway known as the **Calvin cycle**. The first enzyme in this pathway, **ribulose bisphosphate carboxylase**, catalyses reaction of CO_2 with a five-carbon sugar (ribulose 1,5-bisphosphate) to produce two molecules of phosphoglyceric acid (Section 7.5). Oxygen is an alternative substrate for this enzyme, competitive with CO_2, and so at elevated O_2 concentrations there is less CO_2 fixation and hence less plant growth.[18]

In general, however, the rates of direct inactivation of enzymes by O_2 in aerobic cells are too slow and too limited in extent to account for the rate at which toxic effects develop; most enzymes are totally unaffected by O_2. In 1954, Rebecca Gershman and Daniel L. Gilbert drew a parallel between the effects of O_2 and those of ionizing radiation and proposed that most of the damaging effects of O_2 could be attributed to the formation of free **oxygen radicals**.[16]

Figure 1.8 shows that the inhibition of growth observed on exposing *E. coli* to high-pressure O_2 can be relieved by adding valine to the culture medium. Valine synthesis is impaired because of a rapid inhibition of the enzyme **dihydroxyacid dehydratase**, which catalyses a reaction in the metabolic pathway leading to valine. This is probably not a direct inhibition by O_2, but rather by oxygen radicals such as superoxide[13] (Chapter 3). Other enzymes inactivated by superoxide radical ($O_2^{\bullet -}$) in *E. coli* exposed to high pressure O_2 include the Krebs cycle enzymes **aconitase** and **fumarase** (Fig. 1.5). *E. coli* contains three fumarases: fumarases A and B are inactivated by $O_2^{\bullet -}$ whereas fumarase C is not. Levels of fumarase C increase when *E. coli* is exposed to oxidizing conditions, perhaps as a 'replacement' for the superoxide-sensitive fumarases A and B. Aconitase may also be an important target of damage by $O_2^{\bullet -}$ in mammalian tissues exposed to excess O_2.[28a] The onset of O_2-induced convulsions in animals is correlated with a decrease in the cerebral content of the neurotransmitter GABA (γ-aminobutyric acid), perhaps because of an inhibition of the enzyme **glutamate decarboxylase** (glutamate $\rightarrow$ GABA + CO_2) by O_2; it has not, however, been shown that the enzyme inhibition *in vivo* is due to O_2 itself rather than to the effects of an increased production of oxygen radicals.[20]

1.8   What is a free radical?

In order to understand the discussion in this section, it is essential to appreciate what is meant by such chemical terms as 'covalent bond', 'Pauli principle', 'atomic orbital', 'antibonding molecular orbital', 'spin quantum number' and

'Hund's rule'. Readers requiring explanation of such terms are advised to consult Appendix I before reading further in this chapter.

There are several definitions of the term 'free radical', as well as debates about whether the term 'free' is superfluous. We adopt a simple definition: **a free radical is any species capable of independent existence** (hence the term 'free') **that contains one or more unpaired electrons**. An unpaired electron is one that occupies an atomic or molecular orbital by itself. A superscript dot after the formula is usually used to donate free radical species.

The presence of one or more unpaired electrons usually causes free radicals to be attracted slightly to a magnetic field (i.e. to be **paramagnetic**), and sometimes makes them highly reactive, although the chemical reactivity of radicals varies over a wide spectrum. Consideration of our broad definition shows that there are many free radicals in chemistry and biology. The simplest is atomic hydrogen; ^1_1H has only one electron, which must therefore be unpaired. Radicals can be formed by the loss of a single electron from a non-radical,

$$X \rightarrow e^- + X^{\bullet +}$$

or by the gain of a single electron by a non-radical.

$$Y + e^- \rightarrow Y^{\bullet -}$$

Radicals can be formed when a covalent bond is broken if one electron from each of the pair shared remains with each atom, a process known as **homolytic fission**.[35] The energy required to dissociate the covalent bond can be provided by heat, electromagnetic radiation or other means, as will be discussed further in subsequent chapters. Many covalent bonds only dissociate at high temperatures, e.g. 450–600 °C is often required to rupture C–C, C–H or C–O bonds. Many studies of free-radical reactions have been carried out in the gas phase at high temperatures; combustion is well known to chemists as a free-radical process.

If A and B are two atoms covalently bonded (: representing the electron pair), homolytic fission can be written as:

$$A:B \rightarrow A^\bullet + B^\bullet$$

For example, homolytic fission of one of the O–H covalent bonds in the H_2O molecule will yield a hydrogen radical ($H^\bullet$) and a hydroxyl radical (usually written as $OH^\bullet$; although some authors write it as $^\bullet OH$, presumably to emphasize the location of the unpaired electron on the oxygen). The opposite of homolytic fission is **heterolytic fission**, in which one atom receives both electrons when a covalent bond breaks, i.e.

$$A:B \rightarrow A^- + B^+$$

A receives both electrons. This gives A a negative charge and B is left with a positive charge. Heterolytic fission of water gives the hydrogen ion H^+ and the hydroxide ion OH^-. In fact, pure water is very slightly ionized in this way

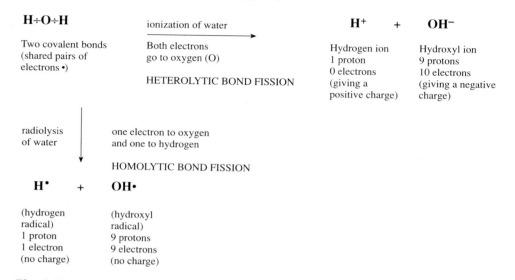

Fig. 1.11. Homolytic or heterolytic fission of water. Water can be split into ions (with paired electrons, or no electrons at all in the case of H^+) or free radicals (with unpaired electrons).

and contains $10^{-7}\,mol/dm^3$ each of H^+ and OH^- ions at 25 °C. **Hydroxyl radical** and **hydroxyl ion** are repeatedly confused in the biomedical literature: Fig. 1.11 emphasizes the structural difference between them.

Let us now examine O_2 and species derived from it to see how far they fit into our definition of free radicals.

1.9   Oxygen and its derivatives

Inspection of Fig. 1.12 shows that the diatomic oxygen molecule qualifies under our definition as a free radical: it has two unpaired electrons, each located in a different $\pi^\star$ antibonding orbital. These two electrons have the same spin quantum number (or, as is often written, they have **parallel spins**). This is the most stable state, or **ground state**, of oxygen. Oxygen can act as an oxidizing agent (for basic definitions see Table 1.3). However, if O_2 attempts to oxidize another atom or molecule by accepting a pair of electrons from it, both of these electrons must be of antiparallel spin so as to fit in to the vacant spaces in the $\pi^\star$ orbitals (Fig. 1.12). A pair of electrons in an atomic or molecular orbital would not meet this criterion, however, since they would have opposite spins in accordance with Pauli's principle. This imposes a restriction on electron transfer which tends to make O_2 accept its electrons one at a time, and contributes to explaining why O_2 reacts sluggishly with many non-radicals.[16] Theoretically, the complex organic compounds of the human body should immediately combust in the O_2 of the air but the spin restriction and other factors slow this down, fortunately!

Table 1.3. Basic definitions

Term	Definition	Example
Oxidation	gain in oxygen	$C + O_2 \rightarrow CO_2$ (carbon is oxidized to carbon dioxide)
	loss of electrons	$Na \rightarrow Na^+ + e^-$ (a sodium atom is oxidized to a sodium ion)
		$O_2^{\bullet -} \rightarrow O_2 + e^-$ (a superoxide radical is oxidized to oxygen)
Reduction		
	loss of oxygen	$CO_2 + C \rightarrow 2CO$ (CO_2 is reduced to carbon monoxide; C is oxidized to CO)
	gain of hydrogen	$C + 2H_2 \rightarrow CH_4$ (carbon is reduced to methane)
	gain of electrons	$Cl + e^- \rightarrow Cl^-$ (a chlorine atom is reduced to a chloride ion)
		$O_2 + e^- \rightarrow O_2^{\bullet -}$ (oxygen is reduced to superoxide radical)
Oxidizing agent	oxidizes another chemical by taking electrons from it, or by taking hydrogen, or by adding oxygen	
Reducing agent	reduces another chemical by supplying electrons to it, by supplying hydrogen or by removing oxygen	

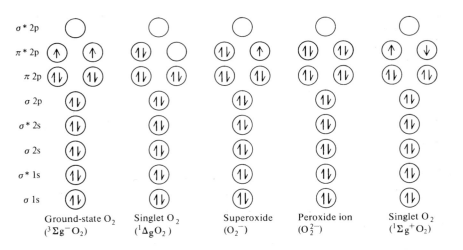

Fig. 1.12. A simplified version of bonding in the diatomic oxygen molecule and its derivatives. Atomic oxygen (atomic number 8) has a total of eight electrons. In O_2 16 electrons are present.

1.9.1 *Singlet oxygen*

More reactive forms of oxygen, known as **singlet oxygens**, can be generated by an input of energy (see Chapter 2 for details of how this is done). The $^1\Delta g O_2$ state (Fig. 1.12) has an energy 93.6 kJ (22.4 kcal)[e] above the ground state. The $^1\Sigma g^+ O_2$ state is even more reactive, 157 kJ (37.5 kcal) above the ground state. By our definition, $^1\Delta gO_2$ is not a radical; there are no unpaired electrons. In both forms of singlet oxygen the spin restriction is removed and so the oxidizing ability of the oxygen is greatly increased.

1.9.2   *Superoxide radical*

If a single electron is added to the ground-state O_2 molecule, it must enter one of the $\pi^\star$ antibonding orbitals (Fig. 1.12). The product is **superoxide radical**, $O_2^{\bullet-}$. With only one unpaired electron, superoxide is less of a radical than is O_2 itself, despite its 'super' name. Indeed, some authors write it as O_2^{-} rather than $O_2^{\bullet-}$. The chemistry of $O_2^{\bullet-}$ is considered in detail in Chapter 2.

Addition of another electron to $O_2^{\bullet-}$ will give O_2^{2-}, the **peroxide ion** which, as may be seen from Fig. 1.12, is not a radical. Since the extra electrons in $O_2^{\bullet-}$ and O_2^{2-} are entering antibonding orbitals, the strength of the oxygen–oxygen bond is decreasing (see Appendix I for an explanation if needed). In ground-state O_2 the atoms are effectively bonded by two covalent bonds, but in $O_2^{\bullet-}$ only by one-and-a-half bonds (there is an extra electron in an antibonding orbital), and in O_2^{2-} by one bond only. Hence the oxygen–oxygen bond in O_2^{2-} is much weaker. Addition of another two electrons to O_2^{2-} would eliminate the bond entirely since they would go into the $\sigma^\star 2p$ orbitals, so giving $2O^{2-}$ species. Usually in biological systems the two-electron reduction product of O_2 is hydrogen peroxide (H_2O_2), and the four-electron product, water. To summarize:

$$O_2 \xrightarrow[\text{reduction}]{\text{one-electron}} O_2^{\bullet-}$$

$$O_2 \xrightarrow[\substack{\text{reduction}\\ \text{(plus } 2H^+)}]{\text{two-electron}} H_2O_2 \text{ (protonated form of } O_2^{2-})$$

$$O_2 \xrightarrow[\substack{\text{reduction}\\ \text{(plus } 4H^+)}]{\text{four-electron}} 2H_2O \text{ (protonated form of } O^{2-})$$

1.9.3   *Ozone*

Ozone (O_3), a triatomic pale-blue gas, is an important shield for solar radiation in the higher reaches of the atmosphere.[36] Unlike O_2, it is not a free radical and is **diamagnetic**, i.e. it is weakly repelled by a magnetic field. Ozone is produced by the photodissociation of molecular O_2 into oxygen atoms, which then react with O_2 molecules

$$O_2 \xrightarrow{\text{solar energy}} 2O$$

$$O_2 + O \rightarrow O_3$$

The two oxygen–oxygen bonds in the ozone molecule are of equal length and intermediate in nature between those of an oxygen–oxygen single bond and a double bond. Ozone has an irritating pungent smell and severely damages the lungs. It is a much more powerful oxidizing agent than ground-state oxygen. Significant amounts of ozone can form in the lower atmosphere in urban air as a result of a series of complex photochemical events resulting from air pollution.

Much O_3 toxicity appears due to direct oxidations by O_3, although free radicals can also be involved (Section 8.10.2). Recently there has been concern over the use of fluorinated hydrocarbons in, for example, aerosol sprays because they may help to deplete the ozone layer in the upper atmosphere.[36] Homolytic photodissociation of chlorofluorocarbon gases such as CF_2Cl_2 and $CFCl_3$ produces **chlorine radicals** (chlorine atoms) in the atmosphere, which react with ozone by such reactions as

$$Cl^\bullet + O_3 \rightarrow O_2 + ClO$$

1.10   Questions of terminology: oxygen-derived species, reactive oxygen species and oxidants

Reactive oxygen species (ROS) is a collective term often used by scientists to include not only the oxygen radicals ($O_2^{\bullet-}$ and $OH^\bullet$) but also some non-radical derivatives of O_2 (Table 1.4). These include H_2O_2, hypochlorous acid (HOCl, an oxidizing and chlorinating agent produced by activated phagocytes) and ozone (O_3). 'Reactive' is, of course, a relative term; neither $O_2^{\bullet-}$ nor H_2O_2 is particularly reactive in aqueous solution. Hence, some authors use the term 'oxygen-derived species' instead. Another popular collective term is 'oxidants'. However, since $O_2^{\bullet-}$ and H_2O_2 can act as both oxidizing and reducing agents in different systems in aqueous solution (Chapter 2), we prefer to avoid that term.

1.11   Sources of superoxide in aerobes[13a]

One of the most popular theories to explain O_2 toxicity has been the **superoxide theory of O_2 toxicity**, which proposes that O_2 toxicity is due to

Table 1.4. Reactive oxygen species

Radicals	Non-radicals
Superoxide, $O_2^{\bullet-}$	Hydrogen peroxide, H_2O_2
Hydroxyl, $OH^\bullet$	Hypochlorous acid, $HOCl^a$
Peroxyl, $RO_2^\bullet$	Ozone, O_3
Alkoxyl, $RO^\bullet$	Singlet oxygen $^1\Delta g$
Hydroperoxyl, $HO_2^\bullet$	Peroxynitrite, $ONOO^{-b}$

[a] Could equally well be called a 'reactive chlorinating species'. Discussed in Section 2.5.2.
[b] Discussed in Section 2.5.4. Could equally well be called a 'reactive nitrogen species'.

over-production of $O_2^{\bullet -}$. We will examine this theory in detail in Chapter 3 after discussing the chemistry of $O_2^{\bullet -}$ in Chapter 2, but it is convenient here to review the mechanisms by which $O_2^{\bullet -}$ can be produced.

1.11.1   Enzymes

Some $O_2^{\bullet -}$ is produced 'deliberately' *in vivo*, e.g. by activated phagocytic cells (Section 6.7). Several enzymes have been identified that are capable of reducing O_2 to $O_2^{\bullet -}$ (Table 1.5).[11] One of the most studied is **xanthine oxidase** (Fig. 1.13). However, most of the xanthine/hypoxanthine oxidation *in vivo* is catalysed by a xanthine dehydrogenase enzyme which transfers electrons from the substrates on to NAD^+ rather than on to O_2. Xanthine dehydrogenase becomes converted to xanthine oxidase during its purification, either because of attack by proteolytic enzymes or by oxidation of thiol ($-SH$) groups. Thus the metabolism of xanthine or hypoxanthine by this enzyme, which is largely located in the cytosol of animal and plant tissues, would not normally produce $O_2^{\bullet -}$ *in vivo*. However, the dehydrogenase–oxidase conversion can occur when tissues are injured (Section 9.5). Xanthine dehydrogenase from human breast milk has also been reported[30a] to catalyse O_2-dependent oxidation of NADH, generating $O_2^{\bullet -}$.

1.11.2   Auto-oxidation reactions

Several biologically-important molecules oxidize in the presence of O_2 to yield $O_2^{\bullet -}$; these include glyceraldehyde, $FMNH_2$, $FADH_2$ (the reduced forms of FMN and FAD), the hormones adrenalin and noradrenalin, L-DOPA (dihydroxyphenylalanine), the neurotransmitter dopamine, tetrahydropteridines and thiol compounds such as cysteine (e.g. Ref. 10). In general, the compounds slowly reduce O_2 to generate $O_2^{\bullet -}$, which then further oxidizes the compound, setting up a complex chain reaction. Tetrahydropteridines act as cofactors for several oxygenase enzymes, including nitric oxide synthase, phenylalanine hydroxylase and tyrosine hydroxylase. Oxidation of adrenalin and of photochemically reduced flavins has been employed as a source of $O_2^{\bullet -}$ in the laboratory (Section 3.2.4).

Since O_2 is poorly reactive, how can these 'auto-oxidations' begin? Most of them are greatly accelerated by the addition of ions of such transition metals as iron and copper. Since it is impossible to free laboratory solutions from contaminating metal ions completely, it is possible that all 'auto-oxidations' are in fact metal ion-catalysed.

Anaerobic bacteria are often grown in the laboratory in culture media containing auto-oxidizable molecules, such as thiols. This can confuse studies of O_2 toxicity: on exposure of the cultures to O_2, media constituents can be oxidized to generate $O_2^{\bullet -}$ and other free radicals that could lead to damage to the bacteria.[7]

An interesting example of auto-oxidation reactions is the **Russell effect** (Section 2.5.1).

Table 1.5. Some enzymes that generate the superoxide radical

Enzyme	Location	Comments/representative references
Peroxidases (non-specific)	Plants and bacteria, some animal tissues (e.g. phagocyte myeloperoxidase, thyroid peroxidase)	$O_2^{\bullet-}$ produced during the oxidase reaction (Sections 3.16 and 6.6.1)
Cellobiose oxidase	White-rot fungus (Sporotrichum pulverulentum)	Contains FAD and a b-type cytochrome. Oxidizes a range of disaccharides, $O_2^{\bullet-}$ is primary reduced oxygen product (Section 6.6.4). See Fig. 1.13, also Section 9.5
Xanthine oxidase Nitropropane dioxygenase	Intestine, ischaemic tissues Hansenula mrakii (a yeast)	Catalyses oxidation of 2-nitropropane into acetone. $O_2^{\bullet-}$ is produced and involved in the catalytic mechanism. Superoxide dismutase (SOD) inhibits the oxidation (J. Biol. Chem. 253, 226 (1978)).
Nitric oxide synthase (NOS)	Most mammalian cells	Generates nitric oxide from the amino acid L-arginine. Reported to release $O_2^{\bullet-}$ when arginine levels are low (J. Biol. Chem. 269, 12589 (1994))
Indoleamine 2,3-dioxygenase	Most animal tissues, especially small intestine, not liver. Activity of enzyme in lung increases during viral infection or after injection of bacterial endotoxin (30- to 100-fold) but is not increased by exposure to elevated O_2 concentrations.	Cleaves the indole ring of tryptophan and several related compounds such as serotonin. $O_2^{\bullet-}$ produced and involved in the catalytic mechanism, hence SOD inhibits the oxidation. Decreasing the SOD activity of isolated rabbit intestinal cells by treatment with the SOD inhibitor diethyldithiocarbamate increased tryptophan degradation, as did addition of xanthine, consistent with a role for $O_2^{\bullet-}$ in intact cells (J. Biol. Chem. 264, 1616 (1989); 252, 2774 (1977))
Tryptophan dioxygenase	Liver	Same reaction as above but specific for tryptophan
Galactose oxidase	Fungi (e.g. Dactylium dendroides)	Copper-containing enzyme. $O_2^{\bullet-}$ thought to be produced and involved in the catalytic mechanism. Oxidizes a –CH$_2$OH group of the sugar galactose to –CHO (J. Am. Chem. Soc. 100, 1899 (1978); Annu. Rev. Biochem. 58, 257 (1989)
Aldehyde oxidase	Liver	Contains molybdenum, iron. Broad substrate specificity; oxidizes a wide range of aldehydes and other compounds, produces $O_2^{\bullet-}$ (Q. Rev. Biophys. 21, 299 (1988))

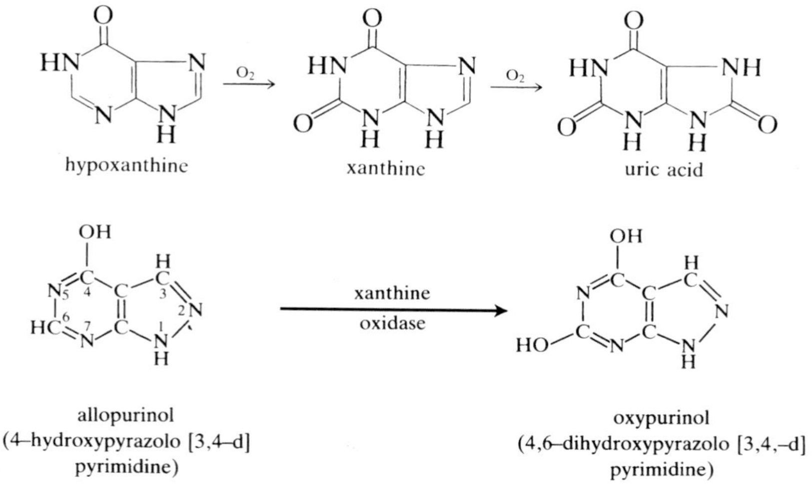

Fig. 1.13. Xanthine oxidase catalyses oxidation of both hypoxanthine and xanthine to uric acid whilst reducing O_2 to both $O_2^{\bullet-}$ and H_2O_2. It will also act on many other substrates, including acetaldehyde (ethanal, CH_3CHO). A powerful inhibitor of xanthine oxidase is the structurally related compound **allopurinol**. Allopurinol is oxidized by the enzyme to give **oxypurinol**, which binds tightly to the active site and causes the inhibition. Hence allopurinol has been called a 'suicide substrate' of xanthine oxidase. Allopurinol is widely used in clinical medicine to inhibit uric acid accumulation in conditions such as gout, and oxypurinol is a major metabolite of allopurinol in the human body. Commercial preparations of the water-soluble vitamin **folic acid** often inhibit xanthine oxidase. However, this inhibition is mostly caused by a contaminant of the folic acid, **pterinaldehyde** (2-amino-4-hydroxypteridine-6-aldehyde). Commercial xanthine oxidase is often used as a laboratory source of $O_2^{\bullet-}$, e.g. in assays of superoxide dismutase (Chapter 3). The commercial enzyme is often obtained from cream, and the purification process employed by some manufacturers involves the use of proteolytic enzymes to free the oxidase from the milk fat globule membranes. Sometimes these proteases are still present in the final preparation and this must be carefully checked for (e.g. one report of the damaging effects of $O_2^{\bullet-}$ from a xanthine–xanthine oxidase system upon chloroplast membranes turned out to be an effect of the trypsin contamination of the enzyme preparation). Phospholipases may also contaminate commercial xanthine oxidase preparations, and chelating agents such as EDTA are often present.

1.11.3 *Haem proteins*

The iron in the haem rings of haemoglobin and myoglobin is in the Fe^{2+} state, and essentially remains so when O_2 binds. However, some delocalization of the electron takes place (for an explanation of this see Appendix I) and an intermediate structure results.

$$\text{haem–}Fe^{2+}\text{–}O_2 \leftrightarrow \text{haem–}Fe(\text{III})\text{–}O_2^{\bullet-}$$

The bonding is intermediate between Fe^{2+} bonded to O_2, and $Fe(\text{III})$ bonded to the superoxide radical. Every so often a molecule of oxyhaemoglobin undergoes decomposition and releases $O_2^{\bullet-}$ (this is an oversimplification of the actual mechanism by which $O_2^{\bullet-}$ is released but it is sufficient for our purpose):[1,4]

$$haem-Fe^{2+}-O_2 \rightarrow O_2^{\bullet-} + haem-Fe(III)$$

The product with Fe(III) present in the haem ring is unable to bind O_2 and is thus biologically inactive; it is known as **methaemoglobin**. (The equivalent for myoglobin is **metmyoglobin**.) It has been estimated that about 3% of the haemoglobin present in human erythrocytes undergoes such oxidation every day, and so these cells are exposed to a constant flux of $O_2^{\bullet-}$. Chapter 8 explores how they cope with this.

Haemoglobin and myoglobin oxidation is speeded up by the presence of nitrite (NO_2^-) ion or of certain transition metal ions, especially copper.[33] The presence of large amounts of nitrate (NO_3^-) in the water supply of some rural areas, due to excessive use of inorganic fertilizers, can cause problems in young bottle-fed babies: the NO_3^- in the water used to make up feeds is reduced by gut bacteria to NO_2^-, which is then absorbed and causes sufficient methaemoglobin formation to interfere with oxygenation of the body tissues. Several abnormal mutant haemoglobins oxidize much more readily than normal, as do the isolated α- or β-chains which accumulate in the diseases known as **thalassaemias**; see Section 3.18.4).[37] Oxymyoglobin, and the oxygenated form of the leghaemoglobin present in the root nodules of leguminous plants, also release $O_2^{\bullet-}$ ions. Several bacteria and yeasts contain haemoglobin- or myoglobin-like proteins, often dimeric haem-containing proteins or monomeric haem flavoproteins. For example, the **flavohaemoglobin** found in *E. coli*, which contains a haem and FAD, readily reduces O_2 to $O_2^{\bullet-}$ *in vitro*.[27]

1.11.4   *Mitochondrial electron transport*

Probably the most important source of $O_2^{\bullet-}$ *in vivo* in many (or perhaps all) aerobic cells is the electron transport chains. They are present in many bacterial membranes and within mitochondria and endoplasmic reticulum in nucleated (eukaryotic) cells. (The chloroplast electron transport chain is discussed in Chapter 7.)

Whereas cytochrome oxidase releases no detectable oxygen radicals into free solution, some earlier components of the mitochondrial electron transport chain do leak a few electrons directly on to O_2, whilst passing the great bulk of them on to the next component in the chain.[34] This leakage generates $O_2^{\bullet-}$. Whereas the activity of mammalian cytochrome oxidase is O_2-saturated at very low O_2 tensions, the rate of electron leakage (and hence $O_2^{\bullet-}$ production) by mitochondria is increased at elevated O_2 concentrations. For example, in slices of rat lung exposed to air, about 9% of total O_2 uptake led to $O_2^{\bullet-}$ formation, the rest being due to cytochrome oxidase activity. In an atmosphere containing 85% O_2, however, $O_2^{\bullet-}$ formation accounted for 18% of the total O_2 uptake.

At physiological O_2 levels, it has been suggested that about 1–3% of the O_2 reduced in mitochondria may form $O_2^{\bullet-}$. This low rate of leakage is probably due to low intra-mitochondrial O_2 concentrations, and to the arrangement of electron carriers into complexes (Fig. 1.6) that facilitate electron movement to the next component of the chain rather than electron escape to O_2. It follows that damage to mitochondrial organization can favour leakage and increase $O_2^{\bullet-}$ production. There is considerable debate over where exactly leakage takes place: *b*-type cytochromes, constituents of complex I and the radical form of coenzyme Q (Fig. 1.6) have all been implicated.[34]

Mitochondrial DNA

The production of oxygen radicals by respiring mitochondria may contribute to damage to mitochondrial proteins, lipids and DNA, which seems to increase with age.[21,32] Each mitochondrion contains up to 10 double-stranded circular DNA molecules, containing only 16 549 base pairs per molecule.[31] Mitochondrial DNA encodes only 13 of the mitochondrial proteins (e.g. of the 26 subunits in complex I, only seven are encoded by mitochondrial DNA, the rest by nuclear DNA), but they are still essential for mitochondrial function. It also encodes one protein in complex III, three in complex IV, two in complex V (the ATPase), 22 transfer RNAs and two ribosomal RNAs. Unlike nuclear DNA, mitochondrial DNA is not coated with histone proteins and so may be susceptible to damage. Indeed, mutations in mitochondrial DNA (especially deletions of base sequences and substitutions) accumulate in old tissues. Mitochondrial mutations have been associated with a wide range of human diseases, including **Leber's hereditary optic neuropathy** (degeneration of the optic nerve) and **MELAS** (mitochondrial encephalomyopathy, lactic acidosis and stroke-like episodes). The former is caused by mutations in the genes encoding constituents of the electron transport chain and the latter by mutations in the genes encoding a transfer RNA carrying the amino acid leucine.[31]

1.11.5   *Bacterial superoxide production*

Studies on *E. coli* showed that some cytosolic enzymes can generate $O_2^{\bullet-}$, but the major proportion of $O_2^{\bullet-}$ comes from the electron transport chain, which has been estimated to generate $3O_2^{\bullet-}$ per 10^5 electrons transferred under physiological conditions, which corresponds to $5\,\mu M$ $O_2^{\bullet-}$ production per second. Obviously, these numbers will be affected by the precise growth conditions. Again, the terminal oxidases appear not to be important sources of $O_2^{\bullet-}$, most of which seems to arise from fumarate reductase and NADH dehydrogenase activities. Fumarate reductase activities in aerobic *E. coli* are low (Section 1.4.4) but still seem to generate a significant percentage of the total $O_2^{\bullet-}$ produced.[13,22]

1.11.6   *Endoplasmic reticulum*

Isolated subcellular fractions containing endoplasmic reticulum (**microsomal fractions**) from various tissues have been shown to produce $O_2^{\bullet-}$ (and H_2O_2)

rapidly when incubated with NADPH. The rate of production is increased at elevated O_2 concentrations. These ROS largely arise from the cytochrome P450 system, in two ways.[17] First, the intermediates in the catalytic cycle can be short-circuited in such a way that O_2 is reduced to $O_2^{\bullet-}/H_2O_2$ instead of being added to the substrate. This side-reaction is often referred to as **uncoupling** or the 'oxidase activity' of P450 (Fig. 1.7). Phenobarbital-inducible (CYP2B) and ethanol-inducible (CYP2E1) cytochromes P450 exhibit especially high rates of O_2 reduction. Some substrates also facilitate these reactions, e.g. 1,1,1-trichloroethane binds to P450 and starts the reaction cycle but is a poor substrate for oxygenation, leading to ROS release. Hexobarbital has a similar, but less marked, effect. For the cytochrome P450 of *P. putida*, very little $O_2^{\bullet-}$ or H_2O_2 is released with camphor as substrate. However, the smaller substrate norcamphor fits more loosely in the active site, and only about 12% of the electrons fed into the P450 are used in substrate hydroxylation, the rest forming $O_2^{\bullet-}$ and H_2O_2.

Second, electrons may escape to O_2 from the flavins in the NADPH–P450 reductase enzyme. In addition, liver endoplasmic reticulum contains an enzyme system, **desaturase**, that introduces carbon–carbon double bonds into fatty acids. The system requires O_2, NADH or NADPH, and a special cytochrome known as cytochrome b_5. Electrons from NAD(P)H are transferred to cytochrome b_5 by a flavoprotein enzyme, and reduced cytochrome b_5 then donates electrons to the desaturase enzyme. Both cytochrome b_5 and the flavoprotein can leak electrons on to O_2 to make $O_2^{\bullet-}$, and this might be an additional source of $O_2^{\bullet-}$ *in vivo*.

1.11.7 *The nucleus*

The membrane surrounding the cell nucleus also contains an electron transport chain, of unknown function, that can 'leak' electrons to give $O_2^{\bullet-}$, at a rate increasing with O_2 concentration, in the presence of NADH or NADPH. It resembles the microsomal electron transport system and may be of especial importance *in vivo* because the oxygen radicals that it generates are close to the nuclear DNA.

1.11.8 *Quantification*

Production of $O_2^{\bullet-}$ seems to occur within all aerobic cells, to an extent dependent on O_2 concentration. About 0.5% of electrons from respiratory substrates in *E. coli* have been estimated to form $O_2^{\bullet-}$. In mitochondria, 1–3% of electrons are said to form $O_2^{\bullet-}$. These numbers do not seem large, but it must be remembered that aerobic organisms consume a lot of O_2 during respiration. That ROS are quantitatively significant products of aerobic metabolism is illustrated by the calculation below.

In the next chapter, we will discuss in detail the chemistry of free radicals and other 'reactive species' important in biological systems, including ROS.

Box 1.1

How much superoxide is made in the human body?

An adult at rest utilizes $3.5\,ml\ O_2/kg/min$ or $352.8\,l/day$ (assuming $70\,kg$ body mass) or $14.7\,mol/day$. If 1% makes $O_2^{\bullet-}$ this is $0.147\,mol/day$ or $53.66\,mol/year$ or about $1.7\,kg/year$ (of $O_2^{\bullet-}$). During bodily exertion, this could increase (with O_2 uptake) up to 10-fold, assuming that the 1% figure still applied. This calculation is taken from *Nutr. Rev.* **52** 255 (1994).

References

1. Balagopalakrishna, C *et al.* (1996) Production of $O_2^{\bullet-}$ from hemoglobin-bound O_2 under hypoxic conditions. *Biochemistry* **35**, 6393.
2. Balentine, JD (1982) *Pathology of Oxygen Toxicity.* Academic Press, New York.
3. Benov, L and Fridovich, I (1996) *E. coli* exhibits negative chemotaxis in gradients of H_2O_2, OCl^- and N-chlorotaurine: products of the respiratory burst of phagocytic cells. *Proc. Natl. Acad. Sci. USA* **93**, 4999.
4. Brantley, RE Jr *et al.* (1993) The mechanism of autoxidation of myoglobin. *J. Biol. Chem.* **268**, 6995.
5. Bruyninckx, WJ *et al.* (1978) Are physiological oxygen concentrations mutagenic? *Nature* **274**, 606.
6. Bunn, HF and Poyton, RO (1996) O_2 sensing and molecular adaptation to hypoxia. *Physiol. Rev.* **76**, 839.
7. Carlsson, J *et al.* (1978) H_2O_2 and $O_2^{\bullet-}$ formation in anaerobic broth media exposed to atmospheric O_2. *Appl. Env. Microbiol.* **36**, 223.
8. Clark, JM (1988) Pulmonary limits of oxygen tolerance in man. *Exp. Lung. Res.* **14**, 897.
9. Cleary, PP and Larkin, A (1979) Hyaluronic acid capsule: strategy for oxygen resistance in group A streptococci. *J. Bacteriol.* **140**, 1090.
10. Davis, MD *et al.* (1988) The auto-oxidation of tetrahydrobiopterin. *Eur. J. Biochem.* **173**, 345.
11. De Groot, H and Littauer, A (1989) Hypoxia, reactive O_2 and cell injury. *Free Rad. Biol. Med.* **173**, 541.
12. Deneke, SM and Fanburg, BL (1980) Normobaric oxygen toxicity of the lung. *New Engl. J. Med.* **303**, 76.
13. Flint, DH *et al.* (1993) The inactivation of Fe–S cluster containing hydro-lyases by $O_2^{\bullet-}$. *J. Biol. Chem.* **268**, 22369.
13a. Fridovich, I (1995) Superoxide radical and SOD. *Annu. Rev. Biochem.* **64**, 97.
14. Gallon, JR (1981) The oxygen-sensitivity of nitrogenase: a problem for biochemists and micro-organisms. *Trends Biochem. Sci.* January 1981, 19.
15. Gardner, PR and Fridovich, I (1991) Quinolinate synthetase: the O_2-sensitive site of *de novo* $NAD(P)^+$ biosynthesis. *Arch. Biochem. Biophys.* **284**, 106.
16. Gilbert, DL (ed.) (1981) *Oxygen and Living Processes: an Inter-disciplinary Approach.* Springer, New York.
17. Goeptar, AR *et al.* (1995) Oxygen and xenobiotic reductase activities of cytochrome P450. *Crit. Rev. Toxicol.* **25**, 25.
18. Halliwell, B (1984) *Chloroplast Metabolism.* Oxford University Press, Oxford.
19. Hammes, HP *et al.* (1996) Subcutaneous injection of a cyclic peptide antagonistofvitronectin receptor-type integrins inhibits retinal neovascularization. *Nature Med.* **2**, 529.

20. Haugaard, N (1968) Cellular mechanisms of oxygen toxicity. *Physiol. Rev.* **48**, 229.

21. Hayakawa, M *et al.* (1992) Age-associated oxygen damage and mutations in mitochondrial DNA in human hearts. *Biochem. Biophys. Res. Commun.* **189**, 979.

22. Imlay, JA (1995) A metabolic enzyme that rapidly produces $O_2^{\bullet-}$, fumarate reductase of *E. coli. J. Biol. Chem.* **270**, 19767.

23. Jia, L *et al.* (1996) S-Nitrosohaemoglobin: a dynamic activity of blood involved in vascular control. *Nature* **380**, 221.

24. Kehrer, JP and Autor AP (1978) The effect of dietary fatty acids on the composition of adult rat lung lipids: relationship to oxygen toxicity. *Toxicol. Appl. Pharmacol.* **44**, 423.

25. Krieg, NR and Hoffman, PS (1986) Microaerophily and oxygen toxicity. *Annu. Rev. Microbiol.* **40**, 107.

26. Marquis, RE (1995) Oxygen metabolism, oxidative stress and acid–base physiology of dental plaque biofilms. *J. Indust. Microbiol.* **15**, 198.

27. Membrillo-Hernandez, J *et al.* (1996) The flavohaemoglobin (HMP) of *E. coli* generates $O_2^{\bullet-}$ *in vitro* and causes oxidative stress *in vivo*. *FEBS Lett.* **382**, 141.

28. Morris JG (1976) Oxygen and the obligate anaerobe. *J. Appl. Bacteriol.* **40**, 229.

28a. Morton RL *et al.* (1998) Loss of lung mitochondrial aconitase activity due to hyperoxia in bronchopulmonary dysplasia in primates. *Am. J. Physiol.* **274**, L127.

29. Neriishi, K and Frank, L (1994) Castration prolongs tolerance of young male rats to pulmonary O_2 toxicity. *Am. J. Physiol.* **247**, R475.

29a. Raju, TNK *et al.* (1997) Vitamin E prophylaxis to reduce ROP: A reappraisal of published trials. *J. Pediat.* 131, 844.

30. Roth, RN and Weiss, LD (1994) Hyperbaric O_2 and wound healing. *Clin. Dermatol.* **12**, 141.

30a. Sanders, SA *et al.* (1997) NADH oxidase activity of human xanthine oxidoreductase. *Eur. J. Biochem.* **245**, 541.

31. Schapira, AHV (1996) Entering the powerhouse. *Odyssey* **2**, 8.

32. Shigenaga, MK *et al.* (1994) Oxidative damage and mitochondrial decay in ageing. *Proc. Natl Acad. Sci. USA* **91**, 10771.

33. Tomoda, A. *et al.* (1981) Involvement of $O_2^{\bullet-}$ in the reaction mechanism of haemoglobin oxidation by nitrite. *Biochem. J.* **193**, 169.

34. Turrens, JF (1997) Superoxide production by the mitochondrial respiratory chain. *Biosci. Rep.* **17**, 3.

35. von Sonntag, C (1987) *The Chemical Basis of Radiation Biology*. Taylor & Francis, London.

36. Wellburn, AR *et al.* (1994) The relative implications of O_3 formation both in the stratosphere and the trophosphere. *Proc. R. Soc. Edin.* **102B**, 33.

37. Winterbourn, CC *et al.* (1976) Reactions involving $O_2^{\bullet-}$ and normal and unstable haemoglobins. *Biochem. J.* **155**, 493.

Notes

[a]This term is explained in Appendix I, Section A1.1.

[b]1 atm $= 760$ mmHg $= 0.1013$ megapascals (approved SI unit).

[c]This term is explained in Appendix I, Section A.3.

[d]Ionization radiation is discussed in detail in Section 8.16.

[e]Energies should be expressed in kilojoules (kJ) but the use of kilocalories is still widespread. 1 kcal is 4.18 kJ.

2

The chemistry of free radicals and related 'reactive species'

2.1 Introduction

As we saw in Chapter 1, a free radical can be defined as any species containing one or more unpaired electrons. Although Chapter 1 focused upon oxygen radicals and other oxygen-derived species, a wide range of free radicals can be made in living systems. Table 2.1 gives some examples. For example, thiol (R–SH) compounds oxidize in the presence of transition metal ions to form, among other products, **thiyl radicals**, RS$^\bullet$.

$$RSH + Cu^{2+} \rightarrow RS^\bullet + Cu^+ + H^+$$

Thiyl radicals are **sulphur-centred** radicals. Homolytic fission of chlorine (Cl_2) gas gives **chlorine radicals** (identical with chlorine atoms); for example, UV light can fragment the covalent bond

$$Cl_2 \xrightarrow[\text{light}]{\text{UV}} 2Cl^\bullet$$

Carbon-centred radicals (Table 2.1) are important intermediates in lipid peroxidation. **Transition-metal ions** qualify as free radicals under the broad definition given above. Some **oxides of nitrogen** ($NO^\bullet$, $NO_2^\bullet$) are also free radicals: indeed, just as the term 'reactive oxygen species' has been introduced into biology (Chapter 1), so has the term **reactive nitrogen species** (Table 2.2).

2.2 How do radicals react?

Reactivity depends upon (a) which radical is studied and (b) with what that radical is presented. If two free radicals meet, they can join their unpaired electrons to form a covalent bond. Thus, atomic hydrogen forms diatomic hydrogen:

$$H^\bullet + H^\bullet \rightarrow H_2$$

A more biologically relevant example is the very fast reaction of nitric oxide radical ($NO^\bullet$) and $O_2^{\bullet-}$ to form a non-radical product, **peroxynitrite**.[4]

$$NO^\bullet + O_2^{\bullet-} \rightarrow ONOO^- \text{ (peroxynitrite)}$$

However, when a free radical reacts with a non-radical, a new radical results, and **chain reactions** can be set up:

1. A radical ($X^\bullet$) may add on to another molecule. The adduct must still have an unpaired electron.

$$X^\bullet + Y \rightarrow [X\text{–}Y]^\bullet$$

Table 2.1. Examples of free radicals

Name	Formula	Comments/examples
Hydrogen atom	$H^\bullet$	The simplest free radical
Trichloromethyl	$CCl_3^\bullet$	A carbon-centred radical (i.e. the unpaired electron resides on carbon). $CCl_3^\bullet$ is formed during metabolism of CCl_4 in the liver and contributes to the toxic effects of this solvent (Section 8.2). Carbon radicals usually react rapidly with O_2 to make peroxyl radicals, e.g. $$CCl_3^\bullet + O_2 \rightarrow CCl_3O_2^\bullet$$
Superoxide	$O_2^{\bullet -}$	An oxygen-centred radical
Hydroxyl	$OH^\bullet$	A highly reactive oxygen-centred radical; attacks all biomolecules
Thiyl/perthiyl	$RS^\bullet/RSS^\bullet$	A group of radicals that have unpaired electrons residing on sulphur
Peroxyl, alkoxyl	$RO_2^\bullet$, $RO^\bullet$	Oxygen-centred radicals formed (among other routes) during the breakdown of organic peroxides and reaction of carbon radicals with $O_2(RO_2^\bullet)$
Oxides of nitrogen	$NO^\bullet$, $NO_2^\bullet$	Nitric oxide is formed *in vivo* from the amino acid L-arginine; nitrogen dioxide is made when $NO^\bullet$ reacts with O_2; both are found in polluted air and smoke from burning organic materials, e.g. cigarette smoke (Section 8.11)
Nitrogen-centred radicals	$C_6H_5N{=}N^\bullet$	Formed during oxidation of phenylhydrazine by erythrocytes (Section 7.3.2) e.g. phenyldiazine radical
Transition-metal ions	Fe, Cu, etc.	Ability to change oxidation numbers by one allows them to accept/donate single electrons; hence they are often powerful catalysts of free-radical reactions

Example: $OH^\bullet$ adds to guanine in DNA; the initial product is an 8-hydroxyguanine radical (Chapter 4).

2. A radical may be a reducing agent, donating a single electron to a non-radical. The recipient then has an unpaired electron

$$X^\bullet + Y \rightarrow X^+ + Y^{\bullet -}$$

Example: $CO_2^{\bullet -}$ reduces Cu^+ to Cu

$$CO_2^{\bullet -} + Cu^+ \rightarrow CO_2 + Cu$$

3. A radical may be an oxidizing agent, accepting a single electron from a non-radical. The non-radical must then have an unpaired electron left

Table 2.2. Reactive nitrogen species

Radicals	Nitric oxide ($NO^\bullet$), nitrogen dioxide ($NO_2^\bullet$)
Non-radicals	Nitrous acid (HNO_2), dinitrogen trioxide (N_2O_3), dinitrogen tetroxide (N_2O_4), nitronium (nitryl) ion (NO_2^+), peroxynitrite ($ONOO^-$), peroxynitrous acid ($ONOOH$), alkyl peroxynitrites ($ROONO$), nitroxyl anion (NO^-), nitrosyl cation (NO^+) and nitryl chloride (NO_2Cl)

Peroxynitrite is often included as both a reactive nitrogen species (RNS) and a reactive oxygen species (ROS) (Table 1.3). 'Reactive' is a relative term: $NO^\bullet$ (like superoxide radical, $O_2^{\bullet-}$) has limited reactivity whereas hydroxyl radical ($OH^\bullet$) reacts with everything. The other RNS/ROS have reactivities intermediate between these extremes. $NO^\bullet$ and $O_2^{\bullet-}$ are similar in other respects: they both have important physiological actions but are toxic in excess, often by generating other more reactive ROS/RNS. Ironically, $O_2^{\bullet-}$ is one of the few molecules with which $NO^\bullet$ reacts quickly (and vice versa).

behind. *Example*: Hydroxyl radical oxidizes the sedative drug promethazine to the radical cation

$$PR + OH^\bullet \rightarrow PR^{\bullet+} + OH^-$$

4. A radical may abstract a hydrogen atom from a C–H bond. As the hydrogen atom has only one electron, an unpaired electron must be left on the carbon. *Example*: Hydroxyl radical abstracts hydrogen from a hydrocarbon side-chain of a fatty acid to initiate lipid peroxidation (Chapter 4).

$$\text{$>$CH} + OH^\bullet \longrightarrow \text{$>$C}^\bullet + H_2O$$

Since most biological molecules are non-radicals, the generation of reactive radicals such as $OH^\bullet$ *in vivo* usually sets off chain reactions. For example, attack of reactive radicals upon fatty acid side-chains in membranes and lipoproteins can abstract hydrogen, leaving a carbon-centred radical

$$\text{$>$CH} + OH^\bullet \longrightarrow \text{$>$C}^\bullet + H_2O$$

Carbon-centred radicals usually react with O_2 to make peroxyl radicals

$$\text{$>$C}^\bullet + O_2 \longrightarrow \text{$>$C}-O-O^\bullet$$

which propagate the chain reaction of **lipid peroxidation** (Chapter 4). In the absence of O_2, carbon radicals might cross-link, two unpaired electrons forming a covalent bond

2.3   Radical chemistry: thermodynamics and kinetics

Classical **thermodynamics** deals with the possibilities of chemical reactions occurring: are they possible or impossible? A reaction is thermodynamically possible if its **free energy** change, ΔG, is negative. For example, thermodynamic calculations have thrown light upon feasible/unfeasible reactions in the chemistry of peroxynitrite (Section 2.5.4). If a reaction is possible, then the rate at which it actually occurs becomes important; this is the area of chemical **kinetics**. A thermodynamically possible reaction may be very slow, or may not even occur at all. Catalysts can only speed up thermodynamically-possible reactions. Living systems are subject to the laws of chemistry, e.g. no enzyme can evolve to catalyse a thermodynamically impossible reaction in isolation.

Application of thermodynamic principles to living organisms is fraught with difficulty, however, since tabulated numerical values of thermodynamic parameters such as ΔG in the chemical literature often refer to reaction conditions inappropriate to living organisms. In addition, biochemical reactions are very closely inter-linked; for example, an 'impossible' reaction can be coupled to a possible one so that the combination is feasible.

2.3.1   *Oxidation and reduction*

A frequently used quantity in understanding free-radical chemistry is **reduction potential**, a thermodynamic parameter which determines the feasibility that compound X can chemically reduce compound Y. For example, a compound is said to be **autoxidizable** if its reduced form can be oxidized by O_2. Reduction potentials can predict which compounds should or should not be autoxidizable. Even when autoxidation is thermodynamically possible, its rate is often slow in the absence of catalysts (most often transition–metal ions).

The chemical reference standard for reduction potentials is the **standard hydrogen electrode**, which contains a platinum electrode dipped in a $1\,M$ ($1\,mol/dm^3$) solution of H^+ ions and exposed to hydrogen (H_2) gas at one atmosphere pressure at a temperature of $25\,°C$. A reversible reaction occurs:

$$\tfrac{1}{2}H_2 \rightleftharpoons H^+ + e^-$$

If this electrode system is connected to a system containing a zinc rod in a $1\,M$ solution of Zn^{2+} ions, electrons flow from the Zn/Zn^{2+} half-cell into the hydrogen electrode, i.e. the overall reaction is

$$Zn \rightarrow Zn^{2+} + 2e^- \text{ (zinc is oxidized)}$$

$$\underline{2H^+ + 2e^- \rightarrow H_2 \text{ (H}^+ \text{ is reduced)}}$$

$$\text{Net: } Zn + 2H^+ \rightarrow Zn^{2+} + H_2$$

The measured voltage is given a negative value. If a copper rod/1 M Cu^{2+} solution is connected to the hydrogen electrode, electrons flow the other way, i.e.

$$Cu^{2+} + 2e^- \rightarrow Cu \text{ (Cu}^{2+} \text{ is reduced)}$$

$$\underline{H_2 \rightarrow 2H^+ + 2e^- \text{ (H}_2 \text{ is oxidized)}}$$

$$\text{Net: } Cu^{2+} + H_2 \rightarrow Cu + 2H^+$$

The measured voltage is given a positive value. These measured voltages are called **standard reduction potentials**, symbolized by $E°$. A system with a negative $E°$ should reduce (i.e. donate electrons to) one that has a positive one. Thus if we connect the Cu/Cu^{2+} half-cell to the Zn/Zn^{2+} half-cell, the latter should reduce the former, i.e. the predicted reaction is

$$Zn \rightarrow Zn^{2+} + 2e^- \text{ (Zn is oxidized)}$$

$$\underline{Cu^{2+} + 2e^- \rightarrow Cu \text{ (Cu}^{2+} \text{ is reduced)}}$$

$$\text{Net: } Zn + Cu^{2+} \rightarrow Zn^{2+} + Cu$$

In living systems, redox reagents are not separated by electric wires and salt bridges into half-cells. Nevertheless, the same principle applies: a system with a negative $E°$ should reduce a system with a less negative, zero, or positive $E°$. To illustrate this, consider Table 2.3, a list of $E°$ values for some biologically relevant species (the values are corrected for pH so as to be the value at pH 7.0, often written as $E°'$).[10] At the top of this list is the hydrated electron (e_{aq}^-), formed by radiolysis of water (Section 2.3.3). This should reduce everything else below it on the list (which has less negative, zero or positive $E°'$ values). Thus hydrated electrons are thermodynamically capable of reducing paraquat to paraquat radical, ferric EDTA to ferrous EDTA and oxygen to $O_2^{\bullet-}$. At the bottom of the list is the highly oxidizing **hydroxyl radical**. This is thermodynamically capable of oxidizing everything else on the list, i.e. everything else should donate electrons to the $OH^\bullet$ system.

Table 2.3 can also be used to predict the directions of reactions away from these extremes. Thus the ascorbate/ascorbyl radical system is capable of reducing the tocopheryl radical/α-tocopherol system, which has a more positive $E°'$. Hence the reaction (ascorbate is ionized at pH 7.4)

$$H^+ + \text{ascorbate}^- + \alpha Toc^\bullet \rightarrow \alpha TocH + \text{ascorbate}^{\bullet-}$$

is thermodynamically possible and ascorbate regenerates vitamin E. Ubiquinol is capable of doing the same. Similarly, $O_2^{\bullet-}$ is capable of reducing ferric

Table 2.3. Some biologically relevant standard reduction potentials

	Half-cell	Standard reduction potential (V)
Highly reducing	H_2O/hydrated electron (e_{aq}^-)	-2.84
	$CO_2/CO_2^{\bullet-}$	-1.80
	O_2, $H^+/HO_2^{\bullet}$	-0.46
	Paraquat/paraquat$^{\bullet-}$	-0.45
	Fe^{3+}–transferrin/Fe^{2+}–transferrin	-0.4 (pH 7.3)
	$O_2/O_2^{\bullet-}$	-0.33
	NAD^+, $H^+/NADH$	-0.32
	FAD, $2H^+/FADH_2$	-0.18
	Fe^{3+}–ferritin/ferritin, Fe^{2+}	-0.19
	Dehydroascorbate/ascorbate$^{\bullet-}$	-0.17
	Ubiquinone, H^+/ubisemiquinone	-0.04
	Fe^{3+}–ADP/Fe^{2+}–ADP	~ 0.1
	Fe^{3+}–citrate/Fe^{2+}–citrate	~ 0.1
	Fe^{3+}–EDTA/Fe^{2+}–EDTA	0.12
	Ubisemiquinone, H^+/ubiquinol	0.2
	Ferricytochrome c/ferrocytochrome c	0.26
	ascorbate$^{\bullet-}$, H^+/ascorbate$^-$	0.28
	$NAD^{\bullet}$, $H^+/NADH$	0.30
	H_2O_2, H^+/H_2O, $OH^{\bullet}$	0.32
	$\alpha T^{\bullet}$, $H^+/\alpha TH$ (α-tocopherol)	0.5
	$HU^{\bullet-}$, H^+/UH_2^- (urate)	0.59
	$RS^{\bullet}/RS^-$ (cysteine)	0.92
	$O_2^{\bullet-}$, $2H^+/H_2O_2$	0.94
	$RO_2^{\bullet}$, $H^+/ROOH$	~ 0.77–1.44^a
	$HO_2^{\bullet}$, H^+/H_2O_2	1.06
Highly oxidizing	$RO^{\bullet}$, H^+/ROH (aliphatic alkoxyl)	1.60 (results variable)
	$OH^{\bullet}$, H^+/H_2O	2.31

Data largely selected from the extensive compilation by Buettner (Ref. 10). Data are corrected to a pH of 7.0 unless otherwise stated, i.e. they are $E^{\circ\prime}$ values rather than E° values. Values quoted in the literature often vary slightly. For each couple the oxidized species is on the left and the reduced species on the right.
^{a}As an example, the value for $CCl_3OO^{\bullet}$, H^+/CCl_3OOH is 1.19 V.

chelates, e.g.

$$O_2^{\bullet-} + Fe(III)\text{–}EDTA \rightarrow O_2 + Fe^{2+}\text{–}EDTA$$

Some reactions are, by contrast, thermodynamically impossible ('unlikely' is perhaps a more realistic term—see the caveats below). Thus superoxide is unlikely[2] to reduce ferric transferrin to ferrous transferrin ($E^{\circ\prime} = -0.4$ V), but it might reduce ferric iron in ferritin to Fe^{2+}.

Caveats

One should not get too excited by the predictive value of E°, $E^{\circ\prime}$ or other thermodynamic parameters, for two reasons. First, reaction conditions make a big difference. Consider the half-cell Zn/Zn^{2+}: the E° refers to a solution with $1\ mol/dm^3$ of Zn^{2+} ions. At lower Zn^{2+} concentrations, the electron donating capacity will rise, as the equilibrium

$$Zn \rightleftharpoons Zn^{2+} + 2e^-$$

shifts to the right. At higher Zn^{2+} concentrations, the electron-donating capacity will fall. If protons are involved in the reaction, pH is obviously an important determining factor. Although the values in Table 2.3 are corrected to pH 7, pH in living organisms can vary widely, from <2 in the gastric juice to >8 in the stroma of illuminated chloroplasts. Thus 'real' reduction potentials in biological systems can differ enormously from standard values. Temperature is often very different from $25\,^\circ C$. To correct E° values for the effects of concentration and temperature (T) the **Nernst equation** is used

$$\text{'effective' reduction potential} = E^\circ + \frac{RT}{nF} \log_{10} \frac{[\text{oxidized}]}{[\text{reduced}]}$$

The second caveat is that E° values predict what is *feasible*, but not what necessarily occurs. An impossible reaction should never occur, but a feasible one might not either. Rates of reaction depend upon temperature and concentration of reactants, and reacting molecules need to have a certain energy when they collide to break the first bonds and get the reaction going. If this required **activation energy** is very high, the reaction can be very slow or may not occur at all.

Thermodynamics of oxygen reduction

In aqueous solution, O_2 is an excellent oxidizing agent; the $E^{\circ\prime}$ for the $4e^-$ reduction to water

$$O_2 + 4H^+ + 4e^- \rightarrow 2H_2O$$

is about $0.8\,V$. Thus thermodynamically, the human body should immediately be oxidized by O_2 in the air, but there is a large activation energy for this process, fortunately! In crematoria, bodies are heated to a sufficiently high temperature to overcome this, and will then burn fiercely. We saw in Chapter 1 that *Escherichia coli* can utilize several electron acceptors to drive its metabolism; and seems to prefer the one with the highest reduction potential. Thus it 'chooses' O_2 ($E^{\circ\prime} = 0.8\,V$) over nitrate ($E^{\circ\prime} = 0.43\,V$) and nitrate over fumarate ($E^{\circ\prime} = 0.03\,V$). Similarly: the mitochondrial electron-transport chain is a gradient of reduction potential from negative to positive; $-0.32\,V$ ($NAD^+/NADH$), $-0.18\,V$ ($FAD/FADH_2$), $-0.04\,V$ (ubiquinol/ubisemiquinone), $0.26\,V$ (oxidized/reduced cytochrome c).

We have already seen that direct $4e^-$ oxidations by O_2 are very slow (Chapter 1). Reactions of O_2 are more likely to proceed in one- or two-electron steps. From Table 2.3, it may be seen that conversion of O_2 to $O_2^{\bullet-}$ seems to need quite powerful reducing systems

$$O_2 + e^- \rightarrow O_2^{\bullet-} \qquad E^{\circ\prime} = -0.33\,V$$

It must be remembered, however, that O_2 concentration is an important variable: this value refers to 1 atm of O_2 (as does the value of $+0.8\,V$ for $O_2/2H_2O$).[44] At lower (physiological) O_2 levels the equilibrium

$$O_2 + e^- \rightleftharpoons O_2^{\bullet-}$$

will move to the left and the reduction potential will rise (i.e. become less negative). For a 1 M concentration of O_2, the value rises to $-0.16\,V$, and it will rise further at lower O_2 levels. Most of the oxidizing power of O_2 does not become available until the third electron reduction, generating $OH^{\bullet}$. The $E^{\circ\prime}$ values are

$$O_2^{\bullet-} + e^- + 2H^+ \rightleftharpoons H_2O_2 \qquad\qquad +0.94\,V$$
$$H_2O_2 + e^- + H^+ \rightleftharpoons H_2O + OH^{\bullet} \qquad +0.32\,V$$
$$OH^{\bullet} + e^- + H^+ \rightleftharpoons H_2O \qquad\qquad +2.31\,V$$

The rates of reaction of most 'autoxidizable molecules' with O_2 are thermo-dynamically very favourable, but usually very slow. They can, however, be catalysed by traces of transition–metal ions, which can promote single electron transfer to O_2.

2.3.2 Reaction rates and rate constants

The rate of a thermodynamically possible reaction depends on temperature, activation energy and concentration of reactants. Rates can be measured either by following the loss of the starting materials (**reactants**), or by following the formation of the products. Reaction rate is then simply defined as the amount of product formed in unit time, or as the amount of reactant used up in unit time. Time is usually quoted in seconds (s) and the amounts in moles (mol).

The rate of a reaction will obviously depend on the concentration of reactants present. To take a simple case, suppose 1 mol of compound A in solution in a volume of 1 dm^3 is reacting to form another substance B:

$$A \rightarrow B$$

Suppose further that after 1 s, 0.01 mol of A have been converted into B. The reaction rate (R) can then be expressed either as 0.01 mol of B formed in 1 dm^3 in 1 s, or as 0.01 mol of A used up in 1 dm^3 in 1 s.

The mathematical relationship between the rate of a reaction and the concentration of the reactants is known as the **rate law**. In this case it is likely that R is proportional to the concentration of A, expressed in molar (mol/dm^3) terms. This is mathematically equivalent to saying that R is equal to the concentration of A multiplied by a constant, the **rate constant** for the reaction, i.e. the rate law is $R = k_1[A]$, where k_1 is the rate constant at the temperature of the experiment, and [A] means the molar concentration of A. Once the reaction has started, A is used up, [A] falls and so R will fall. Hence rate measurements are often made in the first few seconds of a reaction so that the concentration of reactants has not changed significantly from that originally present (so-called **initial rate measurements**). Rate constants, and hence rates of reactions, increase as temperature is raised, so should always be quoted for a specified temperature (Table 2.4).

In the rate law $R = k_1[A]$, the rate of the reaction depends only on the first power of the concentration of A; another way of saying this is that the reaction is **first order with respect to A**. The rate constant k_1 is called a **first-order rate constant** with units of s^{-1}.

Now consider another reaction in which there are two different reactants, e.g. $A + B \rightarrow$ products. This type of equation often represents the reaction of a free radical ($A^{\bullet}$) with some other molecule (B) and it usually follows the rate law $R = k_2[A^{\bullet}][B]$. The reaction is first order with respect to $A^{\bullet}$, first order with respect to B and second order overall; k_2 is a **second-order rate constant**, with units of $dm^3 mol^{-1} s^{-1}$ ($M^{-1} s^{-1}$).

As an example of the information that can be gleaned from published rate constants, let us look at the formation of hydroxyl radicals ($OH^{\bullet}$) from H_2O_2 in the presence of either Fe^{2+} or Cu^+ ions. The published approximate second-order rate constants[a] are given below and show that the reaction of H_2O_2 with Fe^{2+} is much slower than the equivalent reaction with copper ions (it should be noted that these rate constants are affected by the molecule to which the iron is chelated and by temperature; see reference 20 and Table 2.4).

$$H_2O_2 + Fe^{2+} \rightarrow Fe(III) + OH^- + OH^{\bullet} \quad k_2 = 76\,M^{-1}s^{-1}$$
$$\text{(Fenton reaction)}$$

$$H_2O_2 + Cu^+ \rightarrow Cu^{2+} + OH^- + OH^{\bullet} \quad k_2 = 4.7 \times 10^3\,M^{-1}s^{-1}$$

If equal concentrations of H_2O_2 are mixed with equal concentrations of Fe^{2+} or Cu^+, the initial rate of $OH^{\bullet}$ formation in the latter case will be greater by

Table 2.4. Rate constants for reactions of physiological iron chelates with various peroxides

Chelate	Peroxide	k ($M^{-1}s^{-1}$)	
		25 °C	37 °C
Fe^{2+}–ATP	H_2O_2	6.7×10^3	1.6×10^4
	t-butylhydroperoxide	1.3×10^3	2.7×10^3
	cumene hydroperoxide	3.1×10^3	6.5×10^3
Fe^{2+}–citrate	H_2O_2	4.9×10^3	not determined
	t-butylhydroperoxide	1.8×10^3	3.4×10^3
	cumene hydroperoxide	2.2×10^3	4.2×10^3

Data abstracted from Rush and Koppenol (1990) *FEBS Lett.* **275**, 114.

a factor of $4.7 \times 10^3/76$, i.e. 61.8. Values of the rate constant can be applied to see how quickly a reaction might occur under biological conditions. For example, the intracellular concentrations of H_2O_2 and Fe^{2+} are likely to be very low. Let us assume they are in the micromolar (10^{-6} M) range. If 1 μM H_2O_2 comes into contact with 1 μM Fe^{2+}, how much $OH^\bullet$ radical will be formed? The rate law is:

$$R = k_2[H_2O_2][Fe^{2+}]$$

$$= 76(10^{-6})(10^{-6}) = 7.6 \times 10^{-11} \text{ mol dm}^{-3}\text{s}^{-1}$$

This seems a tiny figure, but remember that one mole of a substance contains 6.023×10^{23} molecules (**Avogadro's number**). Hence the *number* of hydroxyl radicals formed per dm^3 per second is 4.58×10^{13}—much more impressive! If the cell volume is 10^{-12}–10^{-11} dm^3 (average volumes for a liver cell) this still means 46–458 hydroxyl radicals formed per cell every second. Of course, as the reaction proceeds both Fe^{2+} and H_2O_2 will be used up and the rate of $OH^\bullet$ production will fall unless they are continuously replenished. Thus, even reactions with low rate-constants (such as the Fenton reaction) can be biologically important if they produce highly reactive products. Rate constants for reaction of biological iron chelates with H_2O_2 may be higher (Table 2.4).

2.3.3 *Measurement of reaction rates for radical reactions*

Many radical reactions proceed extremely quickly and so special techniques are required to measure their rates. Two techniques have commonly been used, **stopped flow** and **pulse radiolysis**.

Pulse radiolysis

Pulse radiolysis[11] of solutions, a technique first introduced in the 1960s, has allowed the direct observation of numerous free radical reactions and can also be used to determine pK_a values[b] and reduction potentials. In pulse radiolysis, the compound to be studied is dissolved (often in water, sometimes in organic solvents, e.g. methanol) and placed in a reaction cell. Radicals are formed directly in the cell by a rapid (10^{-6}–10^{-10} s) 'pulse' of high energy electrons, e.g. from a linear accelerator (Fig. 2.1), and then attack the solute. The resulting solute radicals are monitored over time (typically 10^{-9}–10 s).

By the appropriate choice of experimental conditions, specific radicals can be generated and their reactions followed.[82] Since many radicals absorb light or fluoresce at wavelengths different from their parent compound, the progress of the reaction can be followed by changes in the absorbance/fluorescence spectra (other methods such as electrical conductivity or electron spin resonance can be used if needed, but absorbance studies are still popular). During radiolysis of dilute aqueous solutions, most of the energy is absorbed by the water to produce ionization and excitation within 10^{-16} s:

$$2H_2O \rightarrow H_2O^+ + e^- + H_2O\star$$

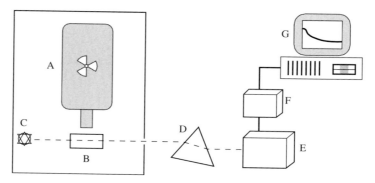

Fig. 2.1. Pulse radiolysis. Ionizing radiation from an accelerator (A) in a shielded room is used to irradiate the sample in a cell (B). Light from a lamp (C) passes through the cell, a monochromator (D), photomultiplier (E), digitizer (F) and computer (G) which displays the spectra. Diagram by courtesy of Dr John Butler. Radiolysis of deaerated water at neutral pH generates $OH^\bullet$ (0.27), $H^\bullet$ (0.06) and e_{aq}^- (0.27). The yields in brackets are expressed as **G-values** in micromolar radicals produced per Gray, where the **Gray** (Gy) is the unit of absorbed radiation dose, equal to $1\,J/kg$. Diatomic hydrogen (0.05) and H_2O_2 (0.07) are also generated.

where e^- represents an electron and $H_2O\star$ an excited-state water molecule. Such excited molecules undergo homolytic fission in 10^{-14}–10^{-13} s to give hydrogen atoms (hydrogen radicals) and hydroxyl radicals:

$$H_2O\star \rightarrow H^\bullet + OH^\bullet$$

Within the same time-scale H_2O^+ also reacts to give $OH^\bullet$

$$H_2O^+ + H_2O \rightarrow H_3O^+ + OH^\bullet$$

The electrons become surrounded by clusters of water molecules within 10^{-12}–10^{-11} s. These **hydrated electrons** are powerful reducing agents (Table 2.3) and are written as e_{aq}^- where 'aq' is an abbreviation for 'aqueous'. Hence three different radicals are produced on 'pulsing' an aqueous solution: $H^\bullet$, $OH^\bullet$ and e_{aq}^-. These radicals are initially formed in clusters called **spurs**, micro-regions of high radical concentration. Reactions within the spur, which give rise to H_2O_2 and H_2, are over in 10^{-8} s and afterwards the distribution of radical species is essentially homogeneous.[82]

Alterations of pH, and addition of various compounds, can 'select' a particular radical for further study. For example, if the aqueous solution is saturated with nitrous oxide (N_2O) gas before pulsing, e_{aq}^- are removed by the reaction:

$$e_{aq}^- + N_2O + H^+ \rightarrow N_2 + OH^\bullet$$

and converted into hydroxyl radicals. By contrast, if the solution is saturated with O_2 and also contains sodium formate (HCOONa), the following reactions occur to produce reducing radicals, $CO_2^{\bullet -}$ and $O_2^{\bullet -}$ (Table 2.3)

$$e_{aq}^- + O_2 \rightarrow O_2^{\bullet -}$$

$$H^{\bullet} + HCOO^- \rightarrow H_2 + CO_2^{\bullet -}$$

$$OH^{\bullet} + HCOO^- \rightarrow H_2O + CO_2^{\bullet -}$$

The carbon dioxide radical is powerfully reducing (Table 2.3) and converts O_2 to $O_2^{\bullet -}$

$$CO_2^{\bullet -} + O_2 \rightarrow O_2^{\bullet -} + CO_2$$

Thus relatively 'clean' sources of $OH^{\bullet}$ or $O_2^{\bullet -}$ can be produced and their reactions with various compounds can be studied.

This technique has been especially useful in investigating the rates and mechanisms of reactions of $OH^{\bullet}$ and $O_2^{\bullet -}$ with biological molecules.[11] Often the reaction is observed directly by following the rise in absorbance of a reaction product, or the loss of absorbance of a reactant. If this is not possible, 'competition methods' may be used to measure reaction rates. For example, $OH^{\bullet}$ (which has scarcely any absorbance above 250 nm) reacts with thio-cyanate ion (SCN^-) to give a product strongly absorbing around 500 nm, $(SCN)_2^{\bullet -}$,

$$OH^{\bullet} + SCN^- \rightarrow OH^- + SCN^{\bullet}$$

$$SCN^{\bullet} + SCN^- \rightleftharpoons (SCN)_2^{\bullet -}$$

and the rate constant for chromogen formation is known. If another compound (X) that reacts with $OH^{\bullet}$ is added, then it will intercept some of the $OH^{\bullet}$, and the absorbance change due to the $(SCN)_2^{\bullet -}$ at 480 nm will be smaller. Knowing the concentrations of SCN^- and X, and the above rate constant, the rate constant for the reaction between X and $OH^{\bullet}$ can be calculated.

Pulse radiolysis can be used to study many types of radical, including those formed by one-electron reduction of quinones, and the ascorbyl radical. The absorbance spectrum and properties of ascorbyl radical can be observed by generating it in a pulse radiolysis apparatus, either by reducing dehydroascorbic acid with e_{aq}^- or by oxidizing ascorbate with the $OH^{\bullet}$ radical:

$$ascorbate^- + OH^{\bullet} \rightarrow OH^- + ascorbate^{\bullet}$$

or with another oxidizing radical such as $Br_2^{\bullet -}$, formed by adding bromide (Br^-) ions.

$$OH^{\bullet} + 2Br^- \rightarrow OH^- + Br_2^{\bullet -}$$

Reduction potentials can easily be determined by pulse radiolysis, e.g. by studying the changes in concentration when a standard system of known potential ($S/S^{\bullet -}$) is mixed with the unknown ($A/A^{\bullet -}$) and the mixture allowed to approach equilibrium

$$A + S^{\bullet -} \rightleftharpoons A^{\bullet -} + S$$

Stopped-flow methods

Stopped-flow methods can be used to study radicals other than those that can be generated by radiolysis, or where the rates of reactions are too slow to be measured conveniently by pulse radiolysis, yet too fast to be measured by standard biochemical techniques (e.g. in the millisecond range). Solutions of the compounds to be reacted are contained in separate syringes, connected to a quartz reaction cell. To initiate the reaction the plungers are pushed so that the syringe contents are forced simultaneously into the reaction cell, where they mix and react. Absorbance (or other) changes can be measured and recorded and so reaction rates can be calculated. For example, a solution of superoxide ion (as its potassium salt, $K^+O_2^{\bullet-}$) in an organic solvent can be placed in one syringe and mixed with a compound in aqueous solution from the other to measure the rate of reaction. Stopped-flow has also been used to study many reactions of ROS/RNS, such as reactions of hypochlorous acid[25] and peroxynitrite. The rate constants for the reactions between potassium ferrate, an Fe(VI) species (K_2FeO_4), and phenolic compounds have also been studied by stopped flow and found to be $\sim 10^7\,M^{-1}\,s^{-1}$.[64,65]

2.4 Chemistry of biologically important radicals

2.4.1 *Transition metals*

Almost all the metals in the first row of the d-block in the periodic table[c] contain unpaired electrons and can thus qualify as free radicals under the broad definition used in this book: the exception is zinc. Indeed, it is particularly appropriate to consider transition-metal ions as radicals, since most of their biological effects, whether beneficial (Table 2.5) or deleterious (e.g. catalysis of unwanted free radical reactions such as 'autoxidations' and Fenton chemistry) involve their ability to accept and donate single electrons.

Iron

For example, iron has two common oxidation numbers in which the electronic configurations are as follows[c]

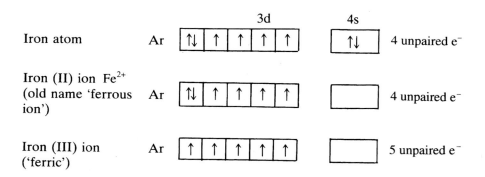

A ferryl (iron(IV)) species also exists; for example, a ferryl haem species is generated when certain haem proteins react with H_2O_2, and ferryl species are involved in catalysis by peroxidases and cytochromes P450 (Sections 1.5.1 and 3.16). In solution in the presence of air, the iron(III) state is the most stable, whereas iron(II) salts are weakly reducing and ferryl compounds are powerful oxidizing agents. If a solution of an iron(II) salt, e.g. ferrous sulphate ($FeSO_4$), is left exposed to the air, it slowly oxidizes to the iron(III) state. The Fe^{2+} undergoes one-electron oxidation, and O_2 dissolved in the solution is reduced to superoxide radical[37]

$$Fe^{2+} + O_2 \rightleftharpoons [Fe^{2+} - O_2 \rightleftharpoons Fe(III) - O_2^{\bullet-}] \rightleftharpoons Fe(III) + O_2^{\bullet-}$$
intermediate complexes

However, this instability of Fe^{2+} compounds is markedly affected by binding of ligands to the iron. In Fe^{2+}, the five 3d orbitals all have the same energy and the electron configuration follows Hund's rule, as shown above (the iron has four unpaired electrons and is said to be in the **high-spin state**). Surrounding the Fe^{2+} with six ligands can split the degeneracy of the 3d orbitals: if this splitting is large the electrons can pair up into the lower-energy orbitals. The Fe^{2+} is then in a **low-spin state**, with no unpaired electrons. Fe^{2+} is then more difficult to oxidize. This state is present in oxyhaemoglobin and oxymyoglobin. By contrast, Fe^{2+} in the deoxy-proteins is in the high-spin state.

Fe(VI) species also exist; for example, potassium ferrate, K_2FeO_4, is a strong oxidizing agent. One-electron reduction of ferrate(VI) (e.g. by $O_2^{\bullet-}$) yields an Fe(V) species, which is also highly reactive, e.g. it oxidizes amino acids:[64,65]

$$Fe(V) + amino\ acid \rightarrow Fe(III) + NH_3 + \alpha\text{-keto acid}$$

Copper
Copper has two common oxidation numbers, copper(I) and copper(II), formerly known as 'cuprous' and 'cupric' respectively:

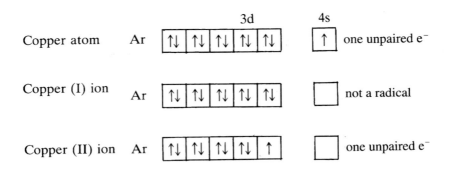

Table 2.5. Biological importance of some d-block elements

Metal	Biochemical significance	Selected references
Copper (Cu)	Essential in human diet. Required for enzymes such as SOD, cytochrome oxidase, lysine oxidase, dopamine–β–hydroxylase and caeruloplasmin. There is about 80 mg Cu in total in adult humans (highest concentrations in liver and brain). Toxic in excess.	See text
Zinc (Zn)	Non-transition element, fixed oxidation number of $2(Zn^{2+})$. 2–3 g present in adult human body (level second only to iron). Often suggested that Zn can act as an antioxidant by displacing iron ions from their binding sites and inhibiting iron–dependent radical reactions. Essential in human diet; found in RNA polymerase, carbonic anhydrase, CuZnSOD, 'zinc fingers'. Plasma zinc ≈ 0.112 mg/100 ml. Toxic in excess.	Science **271**, 1081 (1996)
Vanadium (V)	Essential in animals and suspected to be so in humans. Accumulated in large amounts in some tunicates. Suspected to be involved in regulation of glucose metabolism. Vanadate inhibits strongly the ATPase enzyme which exchanges Na^+ and K^+ ions across cell membranes and may also affect protein kinase and phosphatase enzymes.	Mol. Cell Biochem. **153**, 17 (1995)
Chromium (Cr)	Possibly essential in human diet, suggested to be involved in regulation of glucose metabolism. 'Normal' serum Cr is 1–5 ng/ml. Chromates can damage DNA.	FASEB J. **9**, 1650 (1995)
Manganese (Mn)	Essential in diet (normal blood level 9 μg/ml). Needed for mitochondrial SODs, arginase, also activates a number of hydrolase and carboxylase enzymes. Free and total Mn contents in liver cells of fed rats estimated as about 0.71 and 34 nmol/ml of cell water, respectively. Toxic in excess, in brain can cause a Parkinsonian-type syndrome (Section 8.6.4).	Fed. Proc. **45**, 2817 (1986)

Element	Description	Reference
Iron (Fe)	Essential in human diet: deficiency causes simple anaemia. Most abundant transition metal in human body (about 4–5 g present). Normal serum iron in males is ~ 0.127 mg/100 ml, mostly bound to transferrin. Regulation of iron content of the body is achieved by controlling iron uptake in the gut. Needed for haemoglobin, myoglobin, cyclo-oxygenases, cytochromes, many hydroxylase/oxidase enzymes, ribonucleotide reductase, aconitase, succinate dehydrogenase, catalase and many others.	See text
Cobalt (Co)	Essential as a component of vitamin B_{12} but little else known. Cobalt compounds can cause DNA damage.	*Nutr. Rev.* **43**, 97 (1985)
Nickel (Ni)	Possibly essential in animals, requirement in humans not yet established. Carcinogenic in excess. Found in urease in plant cells and in several bacterial enzymes, such as hydrogenases and carbon monoxide dehydrogenase. Some suggestions that nickel-containing SOD enzymes exist.	See Sections 3.2 and 3.3; also *Biochemistry* **26**, 4901 (1987)
Tungsten (W)	No reports of involvement in animal metabolism. Toxic when salts administered in large amounts. Tungsten-containing enzymes reported in several anaerobic bacteria, e.g. aldehyde–ferredoxin oxidoreductase in *Pyrococus furiosus*.	*J. Biol. Inorg. Chem.* **1**, 292 (1996)
Molybdenum (Mo)	Essential in trace amounts in human diet for some flavin metalloenzymes, e.g. xanthine dehydrogenase, aldehyde oxidase and sulphite oxidase.	*Q. Rev. Biophys.* **21**, 299 (1988)

[a]For an explanation see Appendix II.

Again, the one-electron difference between Cu^+ and Cu^{2+} allows copper to take part in radical reactions. Under appropriate conditions, for example, copper salts can both accept electrons from, and donate electrons to, the superoxide radical $O_2^{\bullet-}$, in very fast reactions[5]

$$Cu^{2+} + O_2^{\bullet-} \rightarrow Cu^+ + O_2 \qquad\qquad k = (5-8) \times 10^9 \, M^{-1} s^{-1}$$

$$H^+ + Cu^+ + HO_2^{\bullet} \rightarrow Cu^{2+} + H_2O_2 \qquad\qquad k \approx 10^9 \, M^{-1} s^{-1}$$

$$\underline{O_2^{\bullet-} + Cu^+ + H_2O \rightarrow Cu^{2+} + OH^- + HO_2^- \quad k \approx 10^{10} \, M^{-1} s^{-1}}$$

Net: $O_2^{\bullet-} + O_2^{\bullet-} + 2H^+ \rightarrow H_2O_2 + O_2$

The copper salt, by changing its oxidation number, is catalysing the net combination of two $O_2^{\bullet-}$ radicals and two H^+ ions to form H_2O_2 and O_2, i.e. it is catalysing the dismutation of $O_2^{\bullet-}$ (Section 2.4.3 below).

Manganese

Manganese has a stable oxidation number in aqueous solution of Mn(II); more-oxidized species such as Mn(III), Mn(IV) and Mn(VII) also exist. Again radical reactions are possible,[5] e.g.

$$Mn^{2+} + O_2^{\bullet-} \rightarrow Mn(O_2)^+$$

$$Mn(O_2)^+ + O_2^{\bullet-} + 2H^+ \rightarrow Mn^{2+} + H_2O_2 + O_2$$

$$Mn(O_2)^+ + 2H^+ \rightarrow H_2O_2 + Mn(III)$$

Overall, to an extent affected by the ligand bound to the manganese, Mn^{2+} is capable of catalysing the dismutation of $O_2^{\bullet-}$ to H_2O_2 (Section 2.4.3 below), at least *in vitro*. By contrast with Fe^{2+} and Cu^+, Mn^{2+} does not react with H_2O_2 to form $OH^{\bullet}$ at a measurable rate.[37]

Zinc

Zinc has a single oxidation number of 2 and does not directly promote radical reactions. The same is true of the non-transition metal aluminium, with a fixed oxidation number of 3, Al(III). It has been suggested that zinc may even inhibit some radical reactions *in vivo* (Table 2.5).

Transition-metal ions as catalysts of free-radical reactions

The variable oxidation number of transition metals helps them to be effective catalysts of reactions involving oxidation and reduction, and they are used for this purpose at the active sites of many enzymes (Table 2.5). The single-electron transfers promoted by metals can overcome the spin restriction on direct reaction of O_2 with non-radical species (Chapter 1). The potential danger is that, unless their availability is carefully controlled, transition metals will catalyse unwanted free-radical reactions. For example, the human body contains many autoxidizable compounds, such as NADH, NADPH, thiols, reduced pteridines, adrenalin, noradrenalin, dihydroxyphenylalanine (L-DOPA) and its metabolites, such as dopamine. All of these are thermodynamically capable of reducing O_2 to $O_2^{\bullet -}$. Fortunately, the rates of these reactions are very low. However, transition metals catalyse such autoxidations: mixtures of thiols, catecholamines, ascorbate, etc., with iron or copper ions will often cause free-radical damage to biomolecules. The reduction potentials of transition-metal ions depend very much on the ligand to the metal, and thus can be altered in different enzymes to allow the same metal to catalyse different reactions. For example, iron–EDTA has an $E^{\circ\prime}$ value close to zero (Table 2.3), whereas transferrin-bound iron is more difficult to reduce (-0.4 V) and cytochrome *c*-bound haem iron much easier ($+0.26$ V).

The Fenton reaction

Fenton chemistry is a prime example of damaging free-radical reactions catalysed by transition metals. A mixture of H_2O_2 with an Fe^{2+} salt oxidizes many different organic molecules. Fenton was the first to report oxidation of an organic compound (tartaric acid) by this system (in 1876). Over 120 years later we are still debating the mechanism of oxidation of organic compounds by Fe^{2+}/H_2O_2.[37,77] It probably involves several oxidizing species, the best characterized being hydroxyl radical:

$$Fe^{2+} + H_2O_2 \rightarrow \text{intermediate complex(es)} \rightarrow Fe(III) + OH^{\bullet} + OH^{-}$$

The reaction is favoured since $E^{\circ\prime}$ for $H_2O_2/OH^{\bullet}$ is ~ 0.32 V, whereas $E^{\circ\prime}$ for aqueous iron ions at pH 7.0 is ~ 0.11 V (Table 2.3). The chemical identity of the oxidizing species additional to $OH^{\bullet}$ formed during Fenton chemistry is uncertain; although Fe(IV) species such as $(Fe=O)^{2+}$ are often suggested, there is no clear evidence for their existence in Fenton systems (by contrast, haem-associated ferryl species are well-established intermediates in catalysis by haem peroxidase enzymes, for example). Ferryl may possibly be the intermediate complex leading to $OH^{\bullet}$ generation.

Traces of Fe(III) might be able to react further with H_2O_2, although this is slower than the reaction of H_2O_2 with Fe^{2+} at physiological pH and very much depends on the ligand to the iron:

$$Fe(III) + H_2O_2 \rightarrow \text{intermediate complex(es)} \rightarrow Fe^{2+} + O_2^{\bullet -} + 2H^{+}$$

Generation of $OH^{\bullet}$ by mixtures of certain Fe(III) chelates (e.g. ferric EDTA) and H_2O_2 appears to involve $O_2^{\bullet -}$, since it is inhibited by superoxide dismutase (SOD).[35] The reaction is often written as above, but no direct evidence for this chemistry has been obtained. By contrast, SOD has no effect on $OH^{\bullet}$ generation by Fe^{2+}/H_2O_2 mixtures. Even more reactions are possible in Fenton systems; e.g.

$$OH^{\bullet} + H_2O_2 \rightarrow H_2O + H^{+} + O_2^{\bullet -}$$
$$O_2^{\bullet -} + Fe(III) \rightarrow Fe^{2+} + O_2$$
$$HO_2^{\bullet} + Fe^{2+} + H^{+} \rightarrow Fe(III) + H_2O_2$$
$$OH^{\bullet} + Fe^{2+} \rightarrow Fe(III) + HO^{-}$$
$$HO_2^{\bullet} + Fe(III) \rightarrow Fe^{2+} + H^{+} + O_2$$

Which of these reactions is predominant depends on the experimental conditions, especially pH and the H_2O_2 : iron concentration ratio as well as what else is added to the reaction mixture. For example, if there is nothing else added for OH· to react with, it can react with H_2O_2 or oxidize Fe^{2+} to Fe(III) ($k_2 \approx 3.5 \times 10^8 \, M^{-1} s^{-1}$), so that a high concentration of Fe^{2+} in the Fenton system can actually depress the yield of OH·.

Thus this simple Fenton mixture of Fe^{2+} and H_2O_2, which almost certainly forms in biological systems under certain circumstances (Chapter 3), can provoke a whole series of radical reactions. The overall sum of these, unless some other reagent is added, is an iron-catalysed decomposition of H_2O_2

$$2H_2O_2 \xrightarrow[\text{catalyst}]{\text{Fe salt}} O_2 + 2H_2O$$

Any added organic molecule will be attacked by OH·, generating radicals that can undergo further reactions, often including oxidations or reductions of the iron ions.

Iron chelates and Fenton chemistry[83a]

The rate constant for reaction of $Fe^{2+}_{(aq)}$ with H_2O_2 is low ($< 100 \, M^{-1} s^{-1}$) but much affected by ligands to the iron (Table 2.4). For example, the rate constant for reaction of Fe^{2+}–ATP with H_2O_2 at pH 7.2 and 25 °C was quoted as $6.6 \times 10^3 \, M^{-1} s^{-1}$, that for Fe^{2+}–ADP as $1.1 \times 10^4 \, M^{-1} s^{-1}$ and that for Fe^{2+}–citrate as $4.9 \times 10^3 \, M^{-1} s^{-1}$. Another important factor is that the extreme reactivity of OH· (Section 2.4.2) means that once formed, it can 'backbite' onto the ligand to the iron, i.e. the ligand, being at a high local concentration, is an easy target of attack by OH·. For example, iron–EDTA chelates are good Fenton reagents, but some of the EDTA is destroyed during the reaction. Obviously, the chance of this happening with any given iron chelate will depend upon its rate of reaction with OH· and the geometry of the iron–ligand complex. This has led to much confusion in the literature about metal-ion chelators as 'inhibitors of Fenton chemistry'. From simple chemical principles, an iron chelator that decreases the observed rate of OH· production from an iron/H_2O_2 system could do so by:

(a) altering the reduction potential of iron to disfavour reaction with H_2O_2;
(b) blocking available sites on the iron to which the H_2O_2 might attach;
(c) promoting oxidation of Fe^{2+} to Fe(III), which is less reactive with H_2O_2; this will probably result in a short 'burst' of superoxide production

$$\text{chelate–}Fe^{2+} + O_2 \rightarrow \text{chelate–Fe(III)} + O_2^{-}$$

(d) intercepting OH· before it can escape from the vicinity of the iron chelate, i.e. the ligand absorbs the OH·;[36]
(e) intercepting OH· precursors (such as ferryl);
(f) if the ligand is in excess over the iron, some of the unbound ligand could scavenge OH· in free solution;
(g) any combination of the above actions.

These principles are important in the design of chelators for therapeutic use (Section 10.6). Chelators of types (a) and (b) might be preferable, since types (d), (e) and (f) will not prevent the reaction indefinitely. They are destroyed as it proceeds and their oxidation will give chelator-derived radicals which might exert unpleasant biological effects, not necessarily revealed by standard toxicological testing of the chelator itself.

2.4.2   Hydroxyl radical

Generation

Hydroxyl radical can be generated in biologically relevant systems by multiple reactions. One is Fenton chemistry (see above). UV-induced homolytic fission of the O–O bond in H_2O_2 makes $OH^{\bullet}$

$$H–O–O–H \xrightarrow{\ UV\ } 2OH^{\bullet}$$

and this could conceivably happen to H_2O_2 generated in sunlight-exposed skin (Section 7.9). In the laboratory, steady-state generation of $OH^{\bullet}$ is frequently carried out using a low-pressure mercury discharge lamp, which emits at 253.7 nm. Hydroxyl radicals can be generated from ozone (Section 8.10). In addition, $OH^{\bullet}$ has been suggested to arise during ethanol metabolism (Section 8.8) and perhaps during peroxynitrous acid decomposition (Section 2.5.4).

Other sources of $OH^{\bullet}$ include:

(1) Ionizing radiation. Since the major constituent of living cells is water, exposure to high-energy radiation such as γ-rays will result in $OH^{\bullet}$ production.[82] Hydroxyl radicals are responsible for a large part of the damage done to cellular DNA, proteins and lipids by ionizing radiation. DNA damage, especially double-strand breaks, is considered to be an important damaging event, especially as double-strand breaks cannot easily be repaired by the cell. Oxygen, normally present in most biological systems, aggravates the damage done by ionizing radiation (Chapter 1). Since living organisms are exposed to 'background' levels of radiation from cosmic rays, natural radioactivity in rocks, release of radioactive radon gas and pollution of the environment with man-made radioactive isotopes, it is likely that some $OH^{\bullet}$ is always formed *in vivo*. Indeed, environmental radiation might account for some of the free-radical damage to biomolecules detected even in healthy organisms (Chapter 4).

(2) From hypochlorous acid[25] reacting with $O_2^{\bullet -}$:

$$HOCl + O_2^{\bullet -} \rightarrow O_2 + Cl^- + OH^{\bullet}$$

The rate constant of this reaction, at $7.5 \times 10^6 \, M^{-1} s^{-1}$, is greater than that for reaction of Fe^{2+} with H_2O_2. HOCl can also react with certain iron chelates, such as ferrocyanide and probably with Fe^{2+} itself, to generate $OH^{\bullet}$.

$$HOCl + Fe(CN)_6^{4-} \rightarrow OH^{\bullet} + Cl^- + Fe(CN)_6^{3-}$$

It is possible that the direct reaction of HOCl with $O_2^{\bullet-}$ as represented above could be catalysed by traces of iron ions, i.e. $O_2^{\bullet-}$ reduces iron ions and Fe^{2+} then converts HOCl to $OH^{\bullet}$ and Cl^-.

(3) Ultrasound, lithotripsy and freeze-drying.[62,79] Ultrasonication of aqueous solutions has been shown to produce both $OH^{\bullet}$ radicals and hydrogen atoms. Ultrasound causes the formation, growth and collapse of gas bubbles, an event called **acoustic cavitation**. Transient cavitations lead, upon collapse, to 'hot spots' in which temperatures of several thousand degrees and pressures of hundreds of atmospheres co-exist. They cause homolytic fission of H_2O to $H^{\bullet}$ and $OH^{\bullet}$. Unlike radiolysis, hydrated electrons are not normally produced. Whether significant levels of radical production occur during the use of ultrasound in medical diagnostic imaging is unknown, although formation of radicals during ultrasonication of human amniotic fluid *in vitro* has been demonstrated.[16] Sonication of lipids in the preparation of liposomes for studies of lipid peroxidation (Section 4.9) is also likely to lead to generation of free radicals and oxidation of lipids.

Kidney stones can be treated by shattering them using the technique of **extracorporeal shock–wave lithotripsy**.[49] The high-energy shock waves delivered by the lithotripter have been shown to produce radicals *in vitro*, but again there is no evidence for any harm produced by such radicals *in vivo*.

The processes of freezing, drying and freeze-drying (**lyophilization**) can generate radicals capable of damaging many biological molecules. This is a particular problem in the manufacture of proteins and similar molecules for biological use,[d] especially when they are to be used as reference standards or for intravenous infusion upon solubilization. Methionine, cysteine, histidine and tryptophan are the most susceptible to oxidative modification (discussed further in Chapter 4).

It is often, but incorrectly, assumed that the freeze drying process is a mild drying treatment of materials that have often been painstakingly separated, purified and characterized. The problem is often twofold in nature: the protein can be directly damaged by radicals, and it can sometimes be modified by secondary products generated by radical attack upon molecules (usually carbohydrates) added during the lyophilization procedure as protective agents. For example, the effect of lyophilization on the conformation of catalase has been studied in some detail. Mammalian catalase normally exists as a tetramer that has both catalase activity, and peroxidase activity towards a limited range of substrates, such as ethanol. Lyophilization of native catalase produces a conformationally altered but not completely denatured catalase monomer, which retains significant catalase but can have increased peroxidase activities (discussed further in Section 3.7).

Heat-treatment of proteins to inactivate viruses can also cause oxidative damage, e.g. loss of $-SH$ groups.

(4) The decomposition of *N*-hydroxy-2-thiopyridone has been suggested as a simple laboratory source of $OH^{\bullet}$, although the reaction also produces a sulphur radical (Fig. 2.2), whose reactivity must be considered.

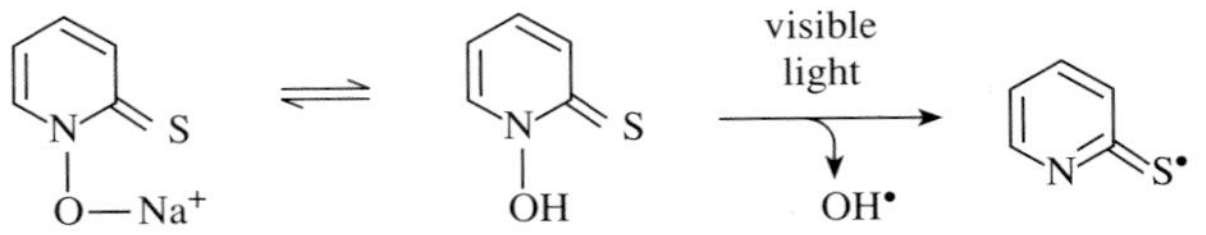

Fig. 2.2. Formation of hydroxyl radical from *N*-hydroxy-2-thiopyridone. Adapted from *Anal. Biochem.* **206**, 309 (1992) by courtesy of Dr T.A. Dix and Academic Press. The compound releases OH• on exposure to visible light. If the –OH group is replaced by –OR, then alkoxyl (RO•) radicals can be generated (*Free Rad. Biol. Med.* **24**, 234, 1998).

Chemistry[5,82]

Hydroxyl radical has a nondescript absorbance spectrum, with a weak absorbance peak at 230 nm. It ionizes at very alkaline pH values:

$$OH^\bullet \rightleftharpoons O^- + H^+ \quad pK_a \approx 12$$

but this is not biologically significant. If OH• radicals meet, they can form dimers, giving hydrogen peroxide:

$$OH^\bullet + OH^\bullet \rightarrow H_2O_2 \quad k = 5 \times 10^9 \, M^{-1} s^{-1}$$

Although this reaction has a high rate constant, it is unlikely to occur *in vivo* because the steady-state concentration of OH• is effectively zero. As soon as OH• is formed, it reacts with molecules in its immediate vicinity, with equally high rate constants (Table 2.6).

Rate constants for OH• reactions have mainly been determined by pulse radiolysis, although other methods exist, such as the deoxyribose method (Chapter 5). Hydroxyl radical reacts very quickly with almost every type of molecule found in living cells: sugars, amino acids, phospholipids, DNA bases and organic acids (for a list of rate constants see Table 2.6). Indeed, OH• is the most reactive oxygen radical known, with a highly positive reduction potential (Table 2.3). For example, it is difficult to demonstrate OH• reactions *in vitro* in solutions containing Tris buffer, since OH• attacks this buffer rapidly and a Tris-derived radical is produced. One end-product of OH• attack on Tris is formaldehyde.[68] The Good buffers (MES, MOPS, etc.) are also powerful OH• scavengers (Table 2.6).

Multiple papers have been published showing that some putative antioxidant is an 'OH• scavenger' when assayed *in vitro*. In fact, almost everything reacts so quickly with OH• that scavenging of this radical is an unlikely mechanism of action for any antioxidant *in vivo*: huge concentrations would be needed to compete with biological molecules for any OH• generated. The sugar alcohol **mannitol** is often used as an 'OH• scavenger' in laboratory experiments, but its rate constant for reaction with OH• is equal to or less than that of many biomolecules (Table 2.6).

Reactions of OH• can be classified into three main types: **hydrogen abstraction**, **addition** and **electron transfer**. These reactions illustrate an important principle of radical chemistry: reaction of a free radical with a

Table 2.6. Second-order rate constants for reactions of the hydroxyl radical

Compound tested	pH	Rate constant ($M^{-1}s^{-1}$)
Carbonate ion, CO_3^{2-}	10.7	2.0×10^8
Bicarbonate ion, HCO_3^-	6.5	1.0×10^7
Fe^{2+}	2.1	3.5×10^8
H_2O_2	7	4.5×10^7
Adenine	7.4	3.0×10^9
Adenosine	7.7	2.5×10^9
AMP	5.4	1.8×10^9
Arginine	7	2.1×10^9
Ascorbic acid	1	7.2×10^9
Benzene	7	3.2×10^9
Benzoic acid	3	4.3×10^9
Butan-l-ol (*n*-butanol)	7	2.2×10^9
Butylated hydroxyanisole		6×10^9
Butylated hydroxytoluene		6×10^9
Catalase	—	2.6×10^{11}
Citric acid	1	3.0×10^7
Cysteine	1	7.9×10^9
Cystine	2	3.2×10^9
Cytidine	2	2.0×10^9
Cytosine	7	2.9×10^9
Deoxyguanylic acid	7	4.1×10^9
Deoxyribose	7.4	3.1×10^9
Desferrioxamine		1.3×10^{10}
Dimethylsulphoxide		3.5×10^9
EDTA		2.8×10^9
Ethanol	7	7.2×10^8
Formate		3.5×10^9
Glucose	7	1.0×10^9
Glutamic acid	2	7.9×10^7
Glutathione	1	8.8×10^9
Glycylgycine	2	7.8×10^7
Glycyltyrosine	2	5.6×10^9
Guanine	—	1.0×10^{10}
Haemoglobin	—	3.6×10^{10}
HEPES		5.1×10^9
Histidine	6.7	3.0×10^9
Hydroxyproline	2	2.1×10^8
Lactate ion	9	4.8×10^9
Lecithin	—	5.0×10^8
Mannitol	7	2.7×10^9
Methanol	7	4.7×10^8
Methionine	7	5.1×10^9
Nicotinic acid	—	6.3×10^8
Phenol	7	4.2×10^9
Phenylalanine	6	3.5×10^9
Propan-l-ol	7	1.5×10^9

Table 2.6. (*Continued*)

Compound tested	pH	Rate constant ($M^{-1}s^{-1}$)
Pyridoxal phosphate	—	1.6×10^9
Ribonuclease	—	1.9×10^{10}
Ribose	7	1.2×10^9
Plasma albumin	—	$>10^{10}$
Thiourea	7	4.7×10^9
Thymine	7	3.1×10^9
Tricine		1.6×10^9
TRIS		1.1×10^9
Tryptophan	6	8.5×10^9
Uracil	7	3.1×10^9
Urea	9	$<7.0 \times 10^5$

Values are mostly taken from Anbar and Neta (1967) *Int. J. Appl. Radiat. Isotopes* **18**, 493; some are from Halliwell *et al.* (1987) *Anal. Biochem.* **165**, 215. The value for mannitol is from *Free Radic. Res. Commun.* **4**, 259 (1988). Note that hydroxyl radical reacts with all biomolecules so rapidly that often the reaction is limited only by the rate at which OH$^\bullet$ contacts the molecule in solution (**a diffusion–controlled rate**). Urea is a rare exception: its rate constant is low for a reaction with OH$^\bullet$ but probably still high in absolute terms when compared with that of many other reactions (e.g. Fenton reactions: Table 2.4).

non-radical species produces a different free radical (Section 2.2). In principle, this new free radical could be more, less or equally reactive than the original radical (Section 2.2). Radicals produced by reactions with OH$^\bullet$ are usually less reactive, however, since OH$^\bullet$ is such an aggressive species (Table 2.6).

An example of hydrogen abstraction is the reaction of OH$^\bullet$ with alcohols. The OH$^\bullet$ abstracts a hydrogen atom (H$^\bullet$) and combines with it to form water, leaving behind an unpaired electron on the carbon atom, e.g. for ethanol:

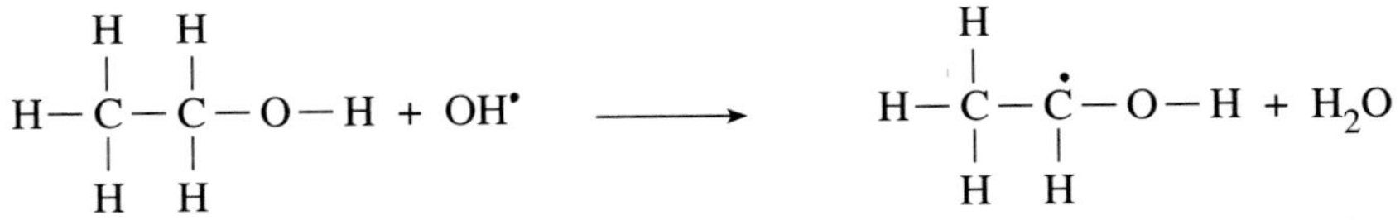

$$\underset{\substack{| \quad | \\ H \quad H}}{\overset{\substack{H \quad H \\ | \quad |}}{H-C-C-O-H}} + OH^\bullet \longrightarrow \underset{\substack{| \quad | \\ H \quad H}}{\overset{\substack{H \\ | \quad \bullet}}{H-C-\overset{}{C}-O-H}} + H_2O$$

hydroxyethyl radical

Further reactions of the carbon radical can then occur, e.g. reaction with oxygen to give peroxyl radicals:

$$^\bullet CH_2OH + O_2 \longrightarrow {}^\bullet O_2CH_2OH$$

peroxyl radical

or (if O_2 levels are low) the joining-up of two radicals to form a non-radical product, the two unpaired electrons between them forming a covalent bond:

$$CH_3\overset{\bullet}{C}HOH + CH_3\overset{\bullet}{C}HOH \longrightarrow \begin{array}{c} CH_3CHOH \\ | \\ CH_3CHOH \end{array}$$

An important, biologically relevant example of H-abstraction by $OH^{\bullet}$ is its ability to initiate lipid peroxidation (Section 4.9).

The reaction of $OH^{\bullet}$ with aromatic compounds often proceeds by addition. For example, $OH^{\bullet}$ adds to the purine base guanine in DNA to form an 8-hydroxyguanine radical (Section 4.6). Similarly, $OH^{\bullet}$ can add on across a double bond in the pyrimidine base thymine. The thymine radical then undergoes a series of further reactions, including reaction with O_2 to give a thymine peroxyl radical. Thus if $OH^{\bullet}$ is generated adjacent to DNA it damages the bases (and deoxyribose sugar) and induces strand breakage (Section 4.6). Hydroxyl radicals can also add on across double bonds.

$$\begin{array}{c} \diagup \\ \diagup \end{array}C{=}C\begin{array}{c} \diagdown \\ \diagdown \end{array} + OH^{\bullet} \longrightarrow \begin{array}{c} OH \\ | \\ \diagup \\ \diagup \end{array}C{-}\overset{\bullet}{C}\begin{array}{c} \diagup \\ \diagdown \end{array}$$

Hydroxyl radicals can take part in electron-transfer reactions e.g. with halide ions

$$Cl^- + OH^{\bullet} \rightarrow Cl^{\bullet} + OH^-$$
$$Cl^{\bullet} + Cl^- \rightarrow Cl_2^{\bullet -}$$

and with nitrite ion

$$NO_2^{\bullet -} + OH^{\bullet} \rightarrow NO_2^{\bullet} + OH^-$$

Reaction of $OH^{\bullet}$ with carbonate ion (CO_3^{2-}) produces carbonate radicals ($CO_3^{\bullet -}$), which are powerful oxidizing agents.

2.4.3   *Superoxide radical*[5,51,67]

Superoxide, by comparison with $OH^{\bullet}$, is far less reactive with non-radical species in aqueous solution. It does react quickly, however, with some other radicals, such as $NO^{\bullet}$ (Section 2.5.4 below), certain iron–sulphur clusters in enzymes (Section 3.6) and certain **phenoxyl radicals**, e.g. that formed by abstracting hydrogen from the $-OH$ group of the amino acid tyrosine[41] ($k = 1.5 \times 10^9 \, M^{-1} \, s^{-1}$).

The reactivity of $O_2^{\bullet -}$ with non-radicals varies considerably depending on whether studies are carried out in organic solvents or in aqueous solution. pH is also an important determinant in the latter case. The pK_a of the reaction

$$HO_2^{\bullet} \rightleftharpoons H^+ + O_2^{\bullet -}$$

is approximately 4.8. Table 2.3 shows that the $HO_2^\bullet$ radical (**hydroperoxyl**) is a more powerful reducing agent than $O_2^{\bullet-}$ ($E^{o\prime}$ values of -0.46 and $-0.33\,V$, respectively). Although at the pH of most body tissues the ratio of $[O_2^{\bullet-}]/[HO_2^\bullet]$ will be large (100/1 at pH 6.8, 1000/1 at pH 7.8), the high reactivity of $HO_2^\bullet$ and its uncharged nature, which might allow it to cross membranes more readily than the charged $O_2^{\bullet-}$, have combined to maintain interest in this species.

Production of superoxide in the laboratory

Various methods are available to generate $O_2^{\bullet-}$ for chemical studies:

(a) O_2 may be reduced electrochemically in an appropriate electrolytic cell in the presence of an organic solvent such as dimethylsulphoxide or acetonitrile.

(b) Tetramethylammonium superoxide, an ionic salt of formula $(CH_3)_4N^+O_2^{\bullet-}$, can be dissolved in a number of organic solvents.

(c) If potassium metal is burned in oxygen, the ionic compound potassium superoxide, $K^+O_2^{\bullet-}$, is obtained (although usually it is far from pure). KO_2 is slightly soluble in organic solvents and its solubility can be increased by the addition of compounds called **crown ethers**. Essentially these are cyclic molecules, a hole in the centre of which can bind K^+. They are very soluble in organic solvents and so 'drag into solution' the central K^+ ion together with its associated $O_2^{\bullet-}$. A popular crown ether in KO_2 experiments is **dicyclohexyl-18-crown-6** (Fig. 2.3). The reaction of $O_2^{\bullet-}$ with another compound added to the organic solvent can be observed, or the KO_2-containing organic solvent can be mixed with an aqueous solution of the compound, e.g. in stopped–flow experiments (Section 2.3.3 above). Superoxide dissolved in the above organic solvents is stable if water is kept away, but in aqueous solution it disappears rapidly.

(d) $O_2^{\bullet-}$ may be generated by pulse radiolysis (Fig. 2.1). Its presence may be detected by observing its UV–absorbance spectrum (maximal absorption around 245 nm; for $HO_2^\bullet$ 225 nm) or its electron spin resonance spectrum (Chapter 4).

(e) Enzymes and photochemical reactions are often used to generate $O_2^{\bullet-}$. Mixtures of xanthine or hypoxanthine with xanthine oxidase are widely

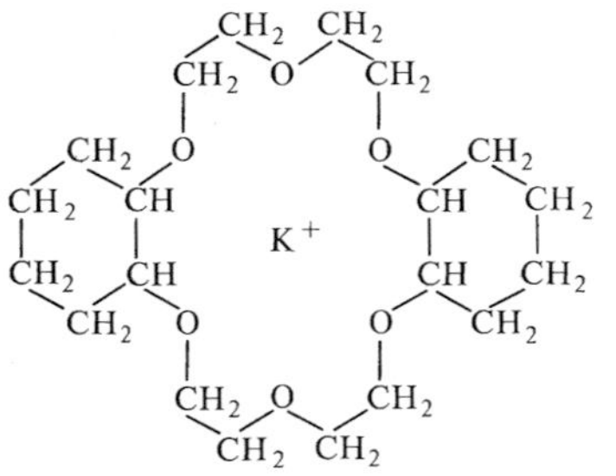

Fig. 2.3. Structure of dicyclohexyl-18-crown-6. The K^+ ion fits into the central 'hole'.

used in biological laboratories, but this enzyme also makes H_2O_2; some points to consider in its use are listed in Fig. 1.13.

(f) Some azo compounds can decompose[40a] to generate $O_2^{\bullet -}$

$$X-N=N-Y \rightarrow X^{\bullet} + N_2 + Y^{\bullet}$$

$$X^{\bullet}(Y^{\bullet}) + O_2 \rightarrow X^+(Y^+) + O_2^{\bullet -}$$

Reactions of superoxide

The rapid disappearance of $O_2^{\bullet -}$ in aqueous solution is due to the **dismutation reaction**. The overall reaction may be represented as follows; it may be seen that one $O_2^{\bullet -}$ is oxidized (to O_2) and another is reduced (to H_2O_2)—the chemical definition of dismutation is a reaction in which the same species is both oxidized and reduced

$$O_2^{\bullet -} + O_2^{\bullet -} + 2H^+ \rightarrow H_2O_2 + O_2$$

However, it is extremely unlikely that four molecules ($2O_2^{\bullet -}$ and $2H^+$) could simultaneously collide in solution, and the rate constant for this reaction as written is at or close to zero ($<0.3\,M^{-1}s^{-1}$). By contrast, the reaction:

$$HO_2^{\bullet} + O_2^{\bullet -} + H^+ \rightarrow H_2O_2 + O_2$$

has $k_2 = 9.7 \times 10^7\,M^{-1}s^{-1}$, and the reaction:

$$HO_2^{\bullet} + HO_2^{\bullet} \rightarrow H_2O_2 + O_2$$

has $k_2 = 8.3 \times 10^5\,M^{-1}s^{-1}$. Hence the dismutation under physiological conditions usually proceeds by protonation of $O_2^{\bullet -}$ followed by reaction of $HO_2^{\bullet}$ with $O_2^{\bullet -}$.

Dismutation is thus most rapid at the acidic pH values needed to protonate $O_2^{\bullet -}$ and will become slower at more alkaline pH values, when the concentration of $HO_2^{\bullet}$ in equilibrium with a given concentration of $O_2^{\bullet -}$ decreases. For example, it may be calculated that in aqueous solution the dismutation reaction will have a 'rate constant' of about $10^2\,M^{-1}s^{-1}$ at pH 11 and about $5 \times 10^5\,M^{-1}s^{-1}$ at pH 7.0. Any molecule that reacts with $O_2^{\bullet -}$ in aqueous solution will be competing for $O_2^{\bullet -}$ with the dismutation reaction. It also follows that any aqueous system generating $O_2^{\bullet -}$ will additionally produce H_2O_2, unless all the $O_2^{\bullet -}$ is intercepted by some other molecule.

$O_2^{\bullet -}$ in aqueous solution can act as a reducing agent, i.e. a donor of electrons. For example, it reduces the haem protein cytochrome *c*.

$$cyt\ c\ (Fe(III)) + O_2^{\bullet -} \rightarrow O_2 + cyt\ c\ (Fe^{2+})$$

and the chloroplast copper-containing protein plastocyanin (Chapter 7):

$$plastocyanin\ (Cu^{2+}) + O_2^{\bullet -} \rightarrow O_2 + plastocyanin\ (Cu^+)$$

Superoxide reduces the yellow dye **nitroblue tetrazolium** (NBT^{2+}) to produce a blue product called **formazan**, although the reaction mechanism is complex (Fig. 2.4). Ability to reduce cytochrome *c* and NBT^{2+} is used in assays of SOD activity (Section 3.2.4).

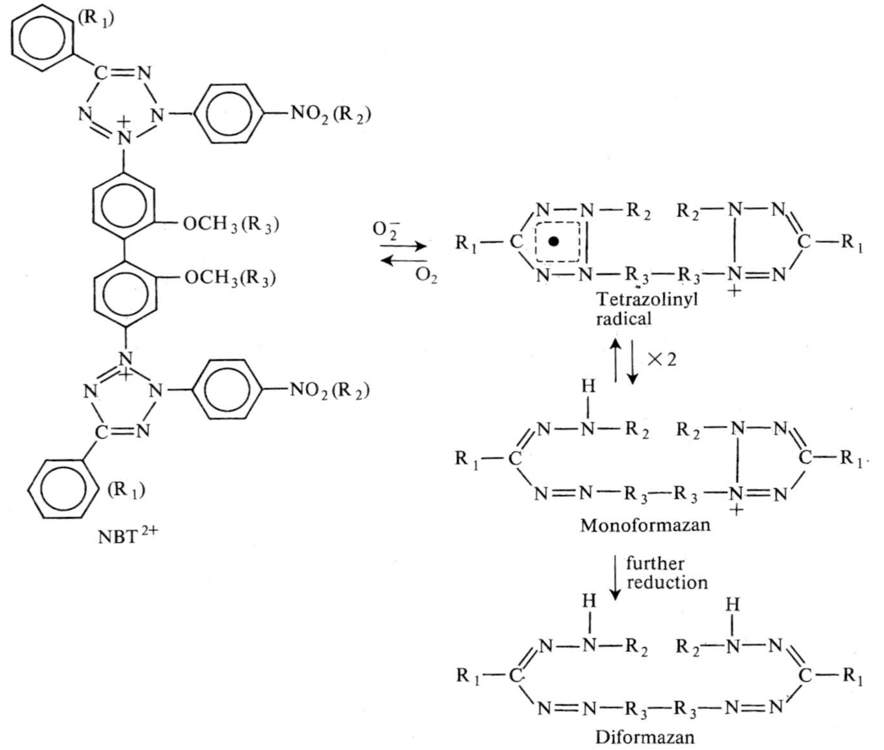

Fig. 2.4. Reduction of NBT^{2+} by superoxide radical. Nitroblue tetrazolium (NBT) is a ditetrazolium salt which can be completely reduced to diformazan by addition of four electrons, with formation of intermediate free radicals. It is probable that the tetrazolinyl radicals can also react with O_2 to form $O_2^{\bullet-}$. Hence high O_2 concentrations depress formazan production from $O_2^{\bullet-}$ plus NBT^{2+}. Reduction of NBT^{2+} should not, therefore, be used as a specific detector of $O_2^{\bullet-}$ in biological systems, since its reduction to a radical by other mechanisms can lead to $O_2^{\bullet-}$ formation (see *J. Phys. Chem.* **84**, 830 (1980); *Arch. Biochem. Biophys.* **318**, 408 (1995)).

Superoxide can also act as an oxidizing agent, e.g. it can oxidize ascorbate:

$$AH_2 + O_2^{\bullet-} \rightarrow A^{\bullet-} + H_2O_2$$

The rate constant has been quoted as $2.7 \times 10^5 \, M^{-1} s^{-1}$ at 25 °C and pH 7.4. Superoxide does not oxidize NADPH or NADH at measurable rates. However, it can interact with NADH bound to the active site of the enzyme lactate dehydrogenase to form an $NAD^{\bullet}$ radical

$$\text{enzyme–NADH} + O_2^{\bullet-} + H^+ \rightarrow \text{enzyme–NAD}^{\bullet} + H_2O_2$$

Unlike $O_2^{\bullet-}$, $HO_2^{\bullet}$ can oxidize NADH directly ($k_2 = 1.8 \times 10^5 \, M^{-1} s^{-1}$). It reduces cytochrome c with a rate constant of 2×10^6 (as compared with $O_2^{\bullet-}$ at $2.6 \times 10^5 \, M^{-1} s^{-1}$). NADH bound at the active site of the enzyme

glyceraldehyde-3-phosphate dehydrogenase[e] is oxidized more rapidly by $HO_2^\bullet$ ($k_2 = 2 \times 10^7 \, M^{-1} s^{-1}$), although this enzyme, unlike lactate dehydrogenase, does not promote reaction of $O_2^{\bullet-}$ with bound NADH.

In general, however, $O_2^{\bullet-}$ in aqueous solution at pH 7.4 is not highly reactive. Its rates of reaction with DNA, lipids, amino acids and most other metabolites are very slow, and may be zero. The direct biological damage that can be caused by $O_2^{\bullet-}$ is highly selective and often involves its reaction with other radicals, e.g. $NO^\bullet$ or iron ions in iron–sulphur proteins.

Superoxide–iron interactions[5,36,37]

There has been much confusion about the interactions of $O_2^{\bullet-}$ with iron, but work in Bielski's laboratory has established the following rate constants for reaction of $O_2^{\bullet-}$ with aqueous iron salts

$$Fe^{2+} + HO_2^\bullet + H^+ \rightarrow Fe(III) + H_2O_2 \quad 1.2 \times 10^6 \, M^{-1} s^{-1}$$

$$Fe^{2+} + O_2^{\bullet-} + H^+ \rightarrow Fe(III) + HO_2^- \quad 1 \times 10^7 \, M^{-1} s^{-1}$$

$$Fe(III) + O_2^{\bullet-} \rightarrow Fe^{2+} + O_2 \quad\quad\quad 1.5 \times 10^8 \, M^{-1} s^{-1}$$

$$Fe(III) + HO_2^\bullet \rightarrow Fe^{2+} + H^+ + O_2 \quad < 10^3 \, M^{-1} s^{-1}$$

Thus $O_2^{\bullet-}$ can reduce Fe(III) but also oxidize Fe^{2+}. The former reaction may proceed through intermediate species, such as **perferryl**

$$Fe(III) + O_2^{\bullet-} \rightleftharpoons [Fe^{2+}-O_2 \leftrightarrow Fe(III)-O_2^{\bullet-}] \rightleftharpoons Fe^{2+} + O_2$$

The values of the rate constants are much affected by binding ligands to the iron. For example, Fe(III) bound to the chelating agent EDTA is still reduced by $O_2^{\bullet-}$:

$$Fe(III)-EDTA + O_2^{\bullet-} \rightleftharpoons (\text{intermediate complexes}) \rightleftharpoons Fe^{2+}-EDTA + O_2$$

$$k_2 = 1.3 \times 10^6 \, M^{-1} s^{-1} \quad \text{at pH 7}$$

whereas Fe(III) attached to the chelators transferrin, lactoferrin or desferrioxamine is reduced much more slowly, if at all, as predicted by the relative reduction potentials (Tables 2.3 and 2.7). Reduction of Fe(III) chelates of citrate, ADP and diethylenetriamine penta-acetic acid (DETAPAC) by $O_2^{\bullet-}$ is also possible (Table 2.7) but the rate constants appear fairly low (e.g. $6 \times 10^3 \, M^{-1} s^{-1}$ for Fe(III)–DETAPAC).

Reduction of Fe(III) by $O_2^{\bullet-}$ can accelerate the Fenton reaction, giving a **superoxide-assisted Fenton reaction**

$$Fe^{2+} + H_2O_2 \rightarrow OH^\bullet + OH^- + Fe(III)$$

$$Fe(III) + O_2^{\bullet-} \rightarrow Fe^{2+} + O_2$$

$$\text{Net:} \quad H_2O_2 + O_2^{\bullet-} \xrightarrow[\text{catalyst}]{Fe} OH^\bullet + OH^- + O_2$$

Table 2.7. Some standard reduction potentials for iron chelators in relation to their interactions with superoxide

Couple	$E^{\circ\prime}$ (V)
$O_2/O_2^{\bullet -}$	-0.33
Likely reductions	
Fe(III)/Fe^{2+} (aq)	$+0.11$
Fe(III)–(1,10 phenanthroline)$_3$/Fe^{2+}–(1,10 phenanthroline)$_3$	$+1.15$
Fe(III)–citrate/Fe^{2+}–citrate	$+0.1$
Fe(III)–ADP/Fe^{2+}–ADP	$+0.1$
Fe(III)–DETAPAC/Fe^{2+}–DETAPAC	$+0.03$
Fe(III)–ferritin/Fe^{2+}–ferritin	-0.19
Fe(III)–cytochrome c/Fe^{2+}–cytochrome c	$+0.26$
Unlikely reductions	
Fe(III)–transferrin/Fe^{2+}–transferrin	-0.4
Fe(III)–desferrioxamine/Fe^{2+}–desferrioxamine	-0.45

Values taken from Ref. 10. Remember that these are thermodynamic data; reactant concentrations and rates of reaction must be considered. Also, remember that the $E^{\circ\prime}$ for $O_2/O_2^{\bullet -}$ is given for one atmosphere O_2 (Section 2.3.1). Thus Fe(III)–EDTA is reduced by $O_2^{\bullet -}$ faster than is Fe(III)–DETAPAC even though both reductions are thermodynamically favourable. DETAPAC is diethylenetriaminepenta-acetic acid.

Other reducing agents, such as ascorbate, can also accelerate OH$^\bullet$ generation: the **Udenfriend system** for hydroxylating aromatic compounds is a mixture of Fe(III)–EDTA, H$_2$O$_2$ and ascorbate. Ascorbate stimulates OH$^\bullet$ generation by recycling Fe(III) to Fe^{2+}, but an excess of ascorbate can decrease the yield of OH$^\bullet$ by scavenging it (rate constant $> 10^9 \, \text{M}^{-1}\text{s}^{-1}$). Several reactions of $O_2^{\bullet -}$ *in vivo*, e.g. its destruction of Fe–S clusters in certain enzymes, can contribute towards providing iron ions needed for Fenton chemistry. Thus, in addition to direct biological damage, $O_2^{\bullet -}$ can also cause indirect damage by facilitating OH$^\bullet$ generation.

Semiquinones and quinones

Superoxide can reduce quinones and oxidize diphenols. Often semiquinones can reduce O_2 to $O_2^{\bullet -}$ (Fig. 2.5). Essentially these are equilibrium reactions. However, dismutation of $O_2^{\bullet -}$ in aqueous solution can favour the reaction of semiquinones with O_2 by removing the $O_2^{\bullet -}$ from the reaction mixture.

Organic solvents[51,67,76]

When superoxide is dissolved in organic solvents, its abilities to act as a base (H$^+$ acceptor) and as a reducing agent are increased. For example, it can reduce dissolved sulphur dioxide (SO$_2$) gas in organic solvents but not in aqueous solution:

$$SO_2 + O_2^{\bullet -} \rightarrow O_2 + SO_2^{\bullet -}$$

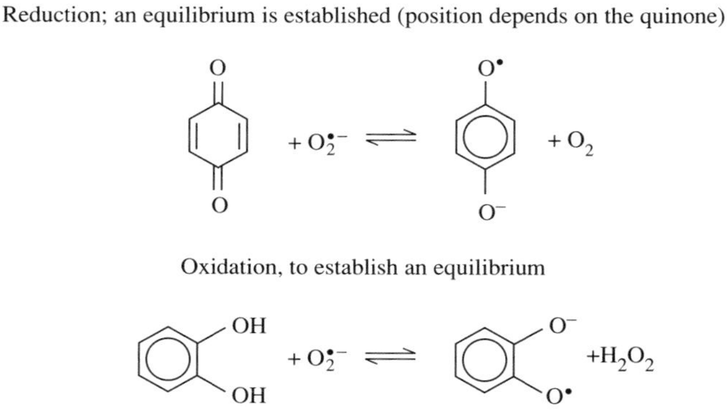

Reduction; an equilibrium is established (position depends on the quinone)

Oxidation, to establish an equilibrium

Fig. 2.5. Reaction of superoxide with quinones and diphenols. Some quinones can be reduced to semiquinones by $O_2^{\bullet-}$ and some diphenols oxidized to semiquinones. Reactions are often reversible, i.e. semiquinones can reduce O_2 to $O_2^{\bullet-}$. Above: reduction of benzoquinone (rate constant $\approx 10^9\,M^{-1}\,s^{-1}$); below: oxidation of catechol (rate constant $\approx 10^9\,M^{-1}\,s^{-1}$).

Also, if protons are not readily available, then dismutation is prevented and the $O_2^{\bullet-}$ persists longer. Further, it acts as a much better **nucleophile**, an agent that attacks centres of positive charge in another molecule. Consider, for example, an ester molecule of general formula:

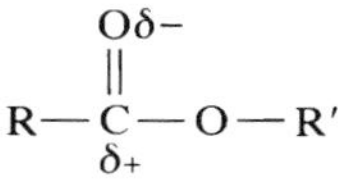

where R and R′ are hydrocarbon groups. Since oxygen is more electronegative than carbon, the carbonyl group is slightly polarized (Appendix I, Section A1.2.3) Superoxide will be attracted to the δ^+ charge and will attack the molecule. One possible reaction mechanism is:

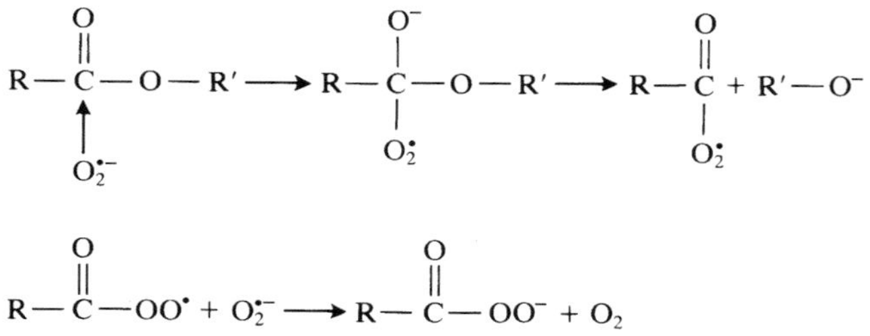

Superoxide in organic media can nucleophilically displace chloride ion from chlorinated hydrocarbons such as chloroform (trichloromethane, $CHCl_3$), tetrachloromethane (carbon tetrachloride, CCl_4), hexachlorobenzene (C_6Cl_6), and from some polychlorobiphenyls, important environmental toxins. For example, in the case of CCl_4 the reaction:

$$CCl_4 + O_2^{\bullet -} \rightarrow CCl_3O_2^{\bullet} + Cl^-$$

is followed by further displacements. By contrast, the nucleophilicity of $O_2^{\bullet -}$ in aqueous solution is low, in part because $O_2^{\bullet -}$ is rapidly removed by dismutation and also because hydration of the $O_2^{\bullet -}$ decreases its charge density by increasing its effective size (see Appendix I, Section A1.2.3 for further explanation if needed).

Superoxide in organic solvents usually only acts as an oxidizing agent towards compounds that can donate H^+ ions, such as ascorbate, catechol and α-tocopherol. Tocopherol (TocH) is slowly oxidized by $O_2^{\bullet -}$ in organic solvents to give tocopheroxyl radical (Toc$^{\bullet}$); suggested reactions include:

$O_2^{\bullet -} + TocH \rightarrow Toc^- + HO_2^{\bullet}$ (deprotonation by $O_2^{\bullet -}$ giving tocopherol ion)

$HO_2^{\bullet} + TocH \rightarrow H_2O_2 + Toc^{\bullet}$ (oxidation of tocopherol by $HO_2^{\bullet}$)

$O_2 + Toc^- \rightarrow O_2^{\bullet -} + Toc^{\bullet}$

$2Toc^{\bullet} \rightarrow$ dimer and other products

In aqueous solution, however, $O_2^{\bullet -}$ probably does not react with α-tocopherol at a significant rate.

2.4.4 *Peroxyl and alkoxyl radicals*[28,56,82]

In general, peroxyl ($RO_2^{\bullet}$) and alkoxyl ($RO^{\bullet}$) radicals are good oxidizing agents, having highly positive $E^{\circ\prime}$ values (Table 2.3), although $RO^{\bullet}$ radicals formed in biological systems often undergo rapid molecular rearrangement to other radical species. Indeed $HO_2^{\bullet}$, protonated $O_2^{\bullet -}$, can be regarded as the simplest peroxyl radical. For example, $RO_2^{\bullet}$ radicals oxidize ascorbate and NADH, the latter leading to $O_2^{\bullet -}$ formation in the presence of O_2

$$RO_2^{\bullet} + NADH \rightarrow RO_2H + NAD^{\bullet}$$

$$NAD^{\bullet} + O_2 \rightarrow NAD^+ + O_2^{\bullet -} \quad k \approx 10^9 \, M^{-1} s^{-1}$$

$RO^{\bullet}$ and $RO_2^{\bullet}$ radicals can abstract $H^{\bullet}$ from other molecules, a reaction important in lipid peroxidation (see Section 4.9). Some $RO_2^{\bullet}$ break down to liberate $O_2^{\bullet -}$, e.g. α-hydroxyalkylperoxyl radicals. For example, when glucose reacts with $OH^{\bullet}$, six different $RO_2^{\bullet}$ radicals are formed, since $H^{\bullet}$ abstraction by $OH^{\bullet}$ can occur at any of the $OH^{\bullet}$ groups. Five of these eliminate $O_2^{\bullet -}$ rapidly, e.g.

$$R-\underset{\underset{OH}{|}}{\overset{\overset{R^1}{|}}{C}}-O_2^{\bullet} \rightarrow R-\overset{\overset{O}{||}}{C}-R + H^+ + O_2^{\bullet -}$$

Peroxyl radicals can react with each other, e.g. by the **Russell mechanism**, to generate some singlet O_2 ($^1O_2$):

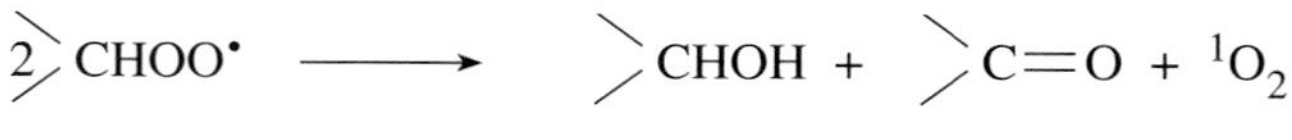

$$2 \overset{>}{} CHOO^\bullet \longrightarrow \overset{>}{} CHOH + \overset{>}{} C{=}O + {}^1O_2$$

Aromatic alkoxyl and peroxyl radicals tend to be less reactive, since electrons can be delocalized into the benzene ring (see Appendix I, Section A1.2.4, if further explanation is needed). For example, abstraction of $H^\bullet$ from the $-OH$ group on tyrosine generates tyrosyl ($TyrO^\bullet$) radical. When $TyrO^\bullet$ is generated in biological systems, it often cross-links to give **bityrosine** (Fig. 2.6).

Generation of $RO_2^\bullet/RO^\bullet$ radicals

Attack of $OH^\bullet$ upon organic compounds often generates carbon-centred radicals. The preferred fate of many of these, under aerobic conditions, is direct reaction with O_2 (rate constants often $>10^9 \, M^{-1} s^{-1}$)

$$R^\bullet + O_2 \rightarrow RO_2^\bullet$$

Decomposition of organic peroxides (ROOH) can generate $RO_2^\bullet$ and $RO^\bullet$ and the latter can also be generated from substituted N-hydroxy-2-thiopyridones (Fig. 2.2). Most peroxides are stable at room temperature, but

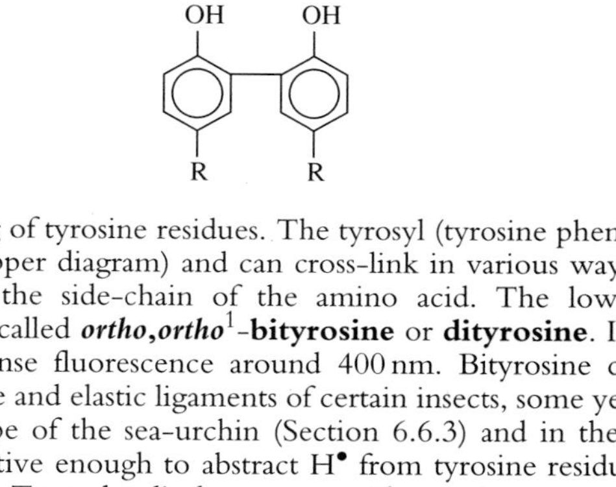

Fig. 2.6. Cross-linking of tyrosine residues. The tyrosyl (tyrosine phenoxyl) radical has various resonance structures (upper diagram) and can cross-link in various ways forming 2,2′, 2,4′ and 4,4′ bityrosines. R is the side-chain of the amino acid. The lower diagram shows **2,2′ bityrosine**, sometimes called *ortho,ortho*[1]-**bityrosine** or **dityrosine**. In alkaline solutions, this compound has an intense fluorescence around 400 nm. Bityrosine cross-links are found in proteins from the cuticle and elastic ligaments of certain insects, some yeast and fungal cell walls, the fertilization envelope of the sea-urchin (Section 6.6.3) and in the thyroid gland (Section 6.6.1). Any radical reactive enough to abstract $H^\bullet$ from tyrosine residues in proteins may lead to bityrosine formation. Tyrosyl radical can react with ascorbate, cysteine, GSH and $O_2^{\bullet-}$ (see text), e.g. $TyrO^\bullet + R{-}SH \rightarrow TyrOH + R{-}S^\bullet$.

they can be decomposed by heating, exposure to UV light (in some cases) or by the addition of transition metal ions e.g. for iron

$$ROOH + Fe(III) \rightarrow RO_2^{\bullet} + Fe^{2+} + H^+$$

$$ROOH + Fe^{2+} \rightarrow RO^{\bullet} + OH^- + Fe(III)$$

These reactions account for much of the stimulation of lipid peroxidation by transition-metal ions in biological systems (Chapter 4). $HO_2^{\bullet}$ can also convert peroxides into $RO_2^{\bullet}$ radicals by $OH^{\bullet}$ abstraction

$$HO_2^{\bullet} + ROOH \rightarrow RO_2^{\bullet} + H_2O_2$$

Azo initiators[53] can be used in the laboratory to generate $RO_2^{\bullet}$ radicals (and $O_2^{\bullet -}$ radicals[40a]) for studies of lipid peroxidation and measurements of anti-oxidant activity (Fig. 2.7). AAPH is water-soluble whereas AMVN is hydro-phobic and can partition into lipids to generate free radicals in the lipid phase. Both decompose at a temperature-controlled rate to give carbon-centred radicals

$$A-N=N-A \rightarrow N_2 + 2A^{\bullet}$$

which react rapidly with O_2 to give peroxyl radicals

$$A^{\bullet} + O_2 \rightarrow AO_2^{\bullet}$$

The carbon-centred radicals are capable of reacting directly with certain biological molecules, perhaps including DNA and albumin −SH groups.[73] When using azo initiators as a source of $RO_2^{\bullet}$ it is therefore essential to ensure that sufficient O_2 is present to ensure complete conversion of the carbon radicals to $RO_2^{\bullet}$.

Peroxyl radicals derived from azo initiators can induce peroxidation of lipids and can damage proteins, e.g. they inactivate the enzyme **lysozyme**.[56] The ability of various antioxidants to protect against azo-initiator-induced lipid

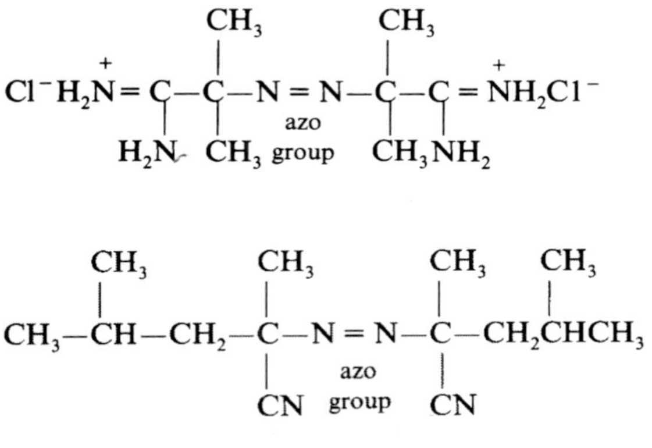

Fig. 2.7. Structure of the 'azo initiators' 2,2′-azobis(2-amidinopropane)hydrochloride (AAPH, above) and 2,2′-azobis(2,4-dimethylvaleronitrile) (AMVN, below).

peroxidation or protein damage is frequently used to assess antioxidant activity, e.g. in the TRAP assay (Section 5.15).

Trichloromethylperoxyl radical[3] can easily be generated in the laboratory by radiolysis of an aqueous mixture of propan-2-ol and tetrachloromethane, CCl_4

$$e_{aq}^- + CCl_4 \rightarrow CCl_3^\bullet + Cl^-$$

$$OH^\bullet + CH_3CHOHCH_3 \rightarrow CH_3\overset{\bullet}{C}OHCH_3 + H_2O$$

$$CH_3\overset{\bullet}{C}OHCH_3 + CCl_4 \rightarrow CH_3COCH_3 + H^+ + CCl_3^\bullet + Cl^-$$

$$CCl_3^\bullet + O_2 \rightarrow CCl_3O_2^\bullet$$

It is a very oxidizing radical ($E^{\circ\prime} = 1.19\,V$) and reacts readily with many antioxidants.

2.4.5 *Sulphur radicals*[28,82,84]

In vivo, thiols (especially reduced glutathione, GSH) are often regarded as antioxidant agents, since they protect protein $-SH$ groups against oxidation and can scavenge oxygen radicals and some other 'reactive species' such as hypochlorous acid and peroxynitrous acid (Chapter 3). However, thiols can themselves generate free radicals.

Formation

Thiyl radicals are formed when thiols react with many carbon-centred radicals

$$RSH + {\Large{\diagdown}}\!\!{\diagup}C^\bullet \longrightarrow {\Large{\diagdown}}\!\!{\diagup}CH + RS^\bullet$$

and with several oxygen radicals, including $OH^\bullet$, $RO^\bullet$, $RO_2^\bullet$ and (at a **very** much lower rate) $O_2^{\bullet-}$.

$$RSH + OH^\bullet \rightarrow RS^\bullet + H_2O$$

$$RSH + RO_2^\bullet \rightarrow RS^\bullet + ROOH$$

The damaging actions of several toxins, including **sporidesmin, gliotoxin** and **diphenyl disulphide**, to animals may involve thiyl radicals (Section 7.3). Thiyl radicals also form when certain thiols react with nitrogen dioxide ($NO_2^\bullet$) or are exposed to peroxynitrite (Section 2.5.4) and during the oxidation of thiols by peroxidases and other haem proteins (Section 3.16.5).

Thiyl radicals are also generated by reaction of thiols with transition metal ions e.g.

$$RSH + Fe(III) \rightarrow RS^\bullet + Fe^{2+} + H^+$$

$$RSH + Cu^{2+} \rightarrow RS^\bullet + Cu^+ + H^+$$

and by the homolytic fission of disulphides, including disulphide bridges in proteins:

$$\text{cysteine-S–S-cysteine} \rightarrow \text{cysteine-S}^{\bullet} + {}^{\bullet}\text{S-cysteine}$$

Human finger-nails are composed largely of α-keratin, a protein rich in disulphide bonds, and sulphur-centred radicals can be produced by repeated cutting of finger-nails. Indeed, grinding of proteins, especially at low temperatures, is a well-established way of generating free radicals, and this can be a problem in the food industry, e.g. in flour milling. Sulphur, carbon and, in the presence of O_2, $RO_2^{\bullet}$ radicals can result.[79]

Reactions

It has often been assumed that $RS^{\bullet}$ radicals are essentially inert and will disappear by dimerization, e.g.

$$\text{GS}^{\bullet} + \text{GS}^{\bullet} \rightarrow \text{GSSG} \quad k = 1.5 \times 10^9 \, \text{M}^{-1}\text{s}^{-1}$$

However, this may be unlikely *in vivo* because steady-state levels of $RS^{\bullet}$ would normally be expected to be low and thus radicals unlikely to meet. GSH levels in cells are in the millimolar range. The pK_a of the GSH thiol group is 9.2

$$\text{GSH} \rightleftharpoons \text{GS}^- + \text{H}^+$$

and so about 1–2% of GSH is ionized at pH 7.4. $GS^{\bullet}$ can react rapidly with GS^-

$$\text{GS}^{\bullet} + \text{GS}^- \rightarrow \text{GSSG}^{\bullet -} \quad k = 8 \times 10^8 \, \text{M}^{-1}\text{s}^{-1}$$

Unlike the oxidizing $GS^{\bullet}$ radical (reduction potential ≈ 0.9 V; Table 2.3), $GSSG^{\bullet -}$ is powerfully reducing ($E^{\circ\prime}$ for the $GSSG/GSSG^{\bullet -}$ couple is about -1.5 V). Thus $GSSG^{\bullet -}$ can reduce metal ions, and also O_2, forming $O_2^{\bullet -}$

$$\text{GSSG}^{\bullet -} + \text{O}_2 \rightarrow \text{GSSG} + \text{O}_2^{\bullet -} \quad k = 5 \times 10^8 \, \text{M}^{-1}\text{s}^{-1}$$

Ascorbate also reacts with $RS^{\bullet}$ radicals

$$\text{AH}^- + \text{RS}^{\bullet} \rightarrow \text{RSH} + \text{A}^{\bullet -} \quad k \approx 5 \times 10^8 \, \text{M}^{-1}\text{s}^{-1} \text{ (for GS}^{\bullet})$$

An alternative reaction of thiyl radicals is formation of peroxyl radicals by reaction with O_2, e.g. for $GS^{\bullet}$

$$\text{GS}^{\bullet} + \text{O}_2 \rightleftharpoons \text{GSOO}^{\bullet} \quad k \approx 3 \times 10^7 \, \text{M}^{-1}\text{s}^{-1}$$

Thiyl peroxyl radicals are unstable and rapidly form other species in the presence of O_2. For example, for thiyl radicals derived from GSH, the amino acid cysteine or the anti-inflammatory drug **penicillamine** (Section 9.7.4) the reaction with O_2 generates $RSOO^{\bullet}$, which can react with more thiol to produce **sulphinyl radical** ($RSO^{\bullet}$) or isomerize (in a light-dependent reaction) to **sulphonyl radical** ($RSO_2^{\bullet}$) which then reacts with O_2 to give

$RSO_2OO^{\bullet}$, **sulphonyl peroxyl**. Thus for cysteine

$$CysS^{\bullet} + O_2 \longrightarrow CysSOO^{\bullet}$$

$$CysSOO^{\bullet} + CysSH \longrightarrow CysSO^{\bullet} + CysSOH$$

$$CysSOO^{\bullet} \xrightarrow{h\gamma} CysSO_2^{\bullet} \ (cysS^+ \overset{O^{\bullet}}{\underset{O^-}{\diagdown\!\!\!\diagup}}\ )$$

$$CysSO_2^{\bullet} + O_2 \longrightarrow CysSO_2OO^{\bullet}$$

$$\begin{matrix} & O \\ & \parallel \\ (CysSOO^{\bullet}) \\ & \parallel \\ & O \end{matrix}$$

Similarly, end products of GSH oxidation by oxygen radicals under aerobic conditions include GSSG, sulphenic acid (GSOH) and sulphonic acid (GSO_3H).

$$GSO^{\bullet} \xrightarrow{\text{reduction}} \underset{\text{sulphenic acid}}{GSOH}$$

$$GSO_2OO^{\bullet} \xrightarrow{\text{reduction}} GSO_2OOH \xrightarrow{\text{reduction}} \underset{\text{sulphonic acid}}{GSO_3H} + H_2O$$

Thiyl radicals can also oxidize NAD(P)H to $NAD(P)^{\bullet}$ radicals, e.g.

$$RS^{\bullet} + NADH \rightarrow RS^- + NAD^{\bullet} + H^+ \quad k = 2.3 \times 10^8 \ \text{(for GSH)}$$

which, if O_2 is present, is likely to be followed by

$$NAD^{\bullet} + O_2 \rightarrow NAD^+ + O_2^{\bullet -} \quad k = 1.9 \times 10^9 \, M^{-1} s^{-1}$$

Additionally, it has been shown[69] that cysteine thiyl radicals ($CysS^{\bullet}$) can abstract hydrogen from linoleic, linolenic and arachidonic acids with rate constants of 10^6–$10^7 \, M^{-1} s^{-1}$, initiating fatty acid peroxidation. Whether this happens *in vivo* is uncertain. It is the reverse of the reaction of carbon-centred radicals with R–SH, discussed at the beginning of this section.

Hence oxidizing thiols generate a whole series of potentially cytotoxic oxygen, sulphur and oxysulphur radicals. Indeed, it was shown many years ago[66] that mixtures of cysteine and copper ions are toxic to mammalian cells, and later work used electron spin resonance techniques (Chapter 4) to spin-trap sulphur radicals in such systems. Thiols may also be involved in the oxidation of low-density lipoproteins during atherosclerosis; indeed, high plasma levels of the thiol **homocysteine** are a risk factor for this process. Whether the damaging effects of plasma homocysteine involve free radicals is unknown.

2.4.6 *Nitric oxide*

Basic chemistry[4,27]

Nitric oxide (officially called **nitrogen monoxide**) is a colourless gas. It is moderately soluble in water (up to $2\,\text{mM}$ at $20\,°\text{C}$) and (like O_2) even more soluble in organic solvents. Hence $NO^{\bullet}$ can diffuse readily between and within cells. Nitric oxide has an unpaired electron in a $\pi^{\star}2p$ antibonding orbital (Appendix I): thus it is a paramagnetic molecule and a free radical. If the unpaired electron is removed by one-electron oxidation, **nitrosonium cation**, NO^{+}, is produced. One-electron reduction would give **nitroxyl anion**, NO^{-}. Table 2.8 gives the reduction potentials. Nitroxyl is a reactive, short-lived species, e.g. it can react with $NO^{\bullet}$ to give **nitrous oxide**, N_2O and possibly hydroxyl radical

$$NO^{-} + NO^{\bullet} \rightarrow ONNO^{\bullet -}$$

$$ONNO^{\bullet -} + NO^{\bullet} \rightarrow N_2O + NO_2^{-}$$

$$ONNO^{\bullet -} + H^{+} \rightarrow N_2O + OH^{\bullet}$$

NO^{-} can also react with O_2 to give peroxynitrite

$$NO^{-} + O_2 \rightarrow ONOO^{-}$$

On exposure to air, nitric oxide reacts with O_2 to form the brown gas nitrogen dioxide ($NO_2^{\bullet}$), which is a far more reactive free radical than $NO^{\bullet}$. The overall reaction is

$$2NO^{\bullet} + O_2 \rightarrow 2NO_2^{\bullet}$$

Table 2.8. Some reduction potentials of the reactive nitrogen species

Couple	$E^{\circ\prime}$ (V)
$OH^{\bullet}/H_2O$	2.31
$ONOOH^{\star}/NO_2^{\bullet}$	2.10
$NO_2^{+}/NO_2^{\bullet}$	1.60
$NO^{+}/NO^{\bullet}$	1.21
$NO_2^{\bullet}/NO_2^{-}$	0.99
$NO^{\bullet}/NO^{-}$ (triplet)	0.39

Selected from the compilation in Koppenol (1996) In Weir *et al.* (eds), *Nitric Oxide and Radicals in the Pulmonary Vasculature*, p. 358. Futura Publishing Co., Armonk, NY. These values apply to pH 7.0 and 1 M gas concentrations. Peroxynitrous acid can act as both a one- or two-electron oxidizing agent ($E^{\circ\prime} = 1.4\,\text{V}$ and $0.99\,\text{V}$ for ONOOH, $H^{+}/NO_2^{\bullet}$, H_2O and ONOOH, H^{+}/NO_2^{-}, H_2O, respectively). The $E^{\circ\prime}$ of activated *trans*-peroxynitrous acid (ONOOH* above) is close to that of the hydroxyl radical.

and it follows a **third-order** rate law, i.e.

$$R = k[NO]^2[O_2]$$

Essentially the same rate law applies to $NO^{\bullet}$ and O_2 in solution. This rate law has interesting biological implications. The rate of $NO^{\bullet}$ oxidation depends upon the square of $NO^{\bullet}$ concentration. The half-life of $1\,\mu M$ $NO^{\bullet}$ is about $12\,min$ in air-saturated solutions and doubles with each 50% decrease in $NO^{\bullet}$ concentration. Physiological levels of $NO^{\bullet}$ may be in the $1-10\,nM$ range and *in vivo* O_2 concentrations are also low (Fig. 1.2). Thus if reaction of $NO^{\bullet}$ with O_2 to form $NO_2^{\bullet}$ were its only fate *in vivo*, the lifetime of $NO^{\bullet}$ would be hours. The oxidation of $NO^{\bullet}$ dissolved in aqueous solutions produces mainly nitrite ion (NO_2^-); the overall equation is

$$4NO^{\bullet} + O_2 + 2H_2O \rightarrow 4H^+ + 4NO_2^-$$

and may be the sum of the equations

$$2NO^{\bullet} + O_2 \rightleftharpoons 2NO_2^{\bullet}$$

$$NO_2^{\bullet} + NO^{\bullet} \rightleftharpoons N_2O_3 \text{ (addition of two radicals)}$$

$$N_2O_3 + 2OH^- \rightleftharpoons 2NO_2^- + H_2O$$

although the role of $NO_2^{\bullet}$ is uncertain.

Physiological roles[48,75]

Interest in $NO^{\bullet}$ arose because of the discovery of its multiple important physiological roles (Table 2.9). Nitric oxide readily binds certain transition metal ions, and many of its physiological effects are exerted as a result of its initial binding to Fe^{2+} haem groups in the enzyme **guanylate cyclase**. For example, $NO^{\bullet}$ synthesized by the **vascular endothelial cells** that line the interior of blood vessels presumably diffuses in all directions, but some of it will reach the underlying smooth muscle, bind to guanylate cyclase and activate it. As a result more **cyclic GMP** is made, which lowers intracellular free Ca^{2+} and relaxes the muscle, dilating the vessel and lowering blood pressure. Much of the $NO^{\bullet}$ generated *in vivo* is eventually lost by interaction with the haem groups of haemoglobin. $NO^{\bullet}$ forms stable complexes with Fe^{2+} and ferrous compounds

$$Fe^{2+} + NO^{\bullet} \rightleftharpoons Fe^{2+}NO$$

including deoxyhaemoglobin (to give $HbFe^{2+}NO$). With oxyhaemoglobin, traces of $NO^{\bullet}$ can bind to $-SH$ groups (Chapter 1). Larger amounts of $NO^{\bullet}$ can cause haemoglobin oxidation to methaemoglobin ($HbFe(III)$) and nitrate. Once $NO^{\bullet}$ has entered an erythrocyte, its half-life will be $<1\,\mu s$. Cells are rarely more than $10\,\mu m$ from the nearest blood vessel, a distance that $NO^{\bullet}$ might diffuse in $<1\,s$.

Basal plasma NO_3^- level in humans on a low NO_3^- diet has been reported as about $30\,\mu M$, although it can rise rapidly after intake of NO_3-rich food.

Table 2.9. Some examples of the physiology and pathology of nitric oxide

Physiological role	Excess implicated in tissue injury in	Effect of 'knock-out'[a] of the relevant gene in mice
Nervous system		
Response to excitatory amino acids (especially glutamate); neurotransmission/neuromodulation; synaptic plasticity (strengthening of synapses that are most often used; plays a role in long-term memory)	Epilepsy, stroke, excitotoxicity (implicated in multiple neurodegenerative diseases)	nNOS: obstruction of the pylorus (the muscle controlling entry of food into in the stomach); less sensitive to brain ischaemia–reperfusion damage; inappropriate, excessive sexual and aggressive behaviour
Vascular system		
Control of blood pressure; inhibition of platelet aggregation; killing of foreign organisms (e.g. *Leishmania*, *Trypanosoma*, *Plasmodium*, *Mycobacteria*, *Listeria*, *Toxoplasma* spp.); see *J. Clin. Invest.* **99**, 2818 (1997)	Septic shock (vasodilation, low blood pressure); chronic inflammation (rheumatoid arthritis, ulcerative colitis); sequelae of chronic infection, including increased cancer risk; transplant rejection	iNOS: increased susceptibility to tuberculosis, *Listeria* and *Leishmania* infection, and lymphoma cell proliferation (e.g. *Proc. Natl Acad. Sci. USA* **94**, 5243 (1997)); resistant to endotoxin-induced hypotension and carrageenan-induced inflammation
Other systems		
Penile erection, bladder control, lung vasodilation, gastrointestinal function (e.g. peristalsis)	Asthma	eNOS: deficient vasodilation in response to acetylcholine; hypertension

NO$^\bullet$ has been implicated in multiple physiological processes, some of which are listed above, yet excess production of NO$^\bullet$ may cause cell injury.[72] Often this excess production results from the action of iNOS enzymes. In rodent models of disease, phagocytes (especially macrophages) are the usual cells over-producing NO$^\bullet$, but the ability of human phagocytes to make NO$^\bullet$ appears limited. Cell injury by NO$^\bullet$ can be direct (e.g. inhibition of ribonucleotide reductase or cytochrome oxidase) or it can involve conversion of NO$^\bullet$ into other reactive nitrogen species, such as peroxynitrite (Section 2.5.4). iNOS is sometimes called NOS2; eNOS is sometimes called NOS3; and nNOS is sometimes called NOS1.
[a] For an explanation of transgenic 'knock-out' technology see Appendix II.

This basal level presumably results from NO$^\bullet$ metabolism.[85] Nitrate is eventually excreted in the urine. Binding of NO$^\bullet$ to the haem in cytochromes P450 can inhibit their activity. NO$^\bullet$ binds to the haem iron in both ferrous and ferric states, causing reversible inhibition.

Synthesis[48,72] of NO$^\bullet$

NO$^\bullet$ is mainly synthesized in living organisms by the action of a group of enzymes called **nitric oxide synthases** (NOSs), which convert the amino acid L-arginine into NO$^\bullet$ and another amino acid, L-citrulline. Oxygen is required and the NOSs contain four cofactors: FAD, FMN, tetrahydrobiopterin (all compounds whose reduced forms can autoxidize to make oxygen radicals) and haem; the haem centre has spectral properties resembling those of cytochromes P450. Like P450s, NOSs can bind NO$^\bullet$ and this decreases their enzyme activity. NOS plus its cofactors in the absence of arginine may lead to $O_2^{\bullet-}$ generation (Table 1.5) but the physiological significance of this is uncertain. Conversion of L-arginine to NO$^\bullet$ is a five-electron oxidation; the electrons are supplied by NADPH.

The activity of NOS is carefully regulated in healthy tissues. There are three types: **neuronal NOS** (nNOS) was originally identified in nervous-system tissues and is present all the time in the cells. **Endothelial NOS** (eNOS) is also expressed constitutively in endothelial cells and synthesizes the NO$^\bullet$ needed for regulation of blood pressure (Table 2.9). Nitric oxide from vascular endothelium was originally identified as an **endothelium–derived relaxing factor** (EDRF) a 'factor' that relaxes blood vessels. The NOS enzymes in healthy tissues require Ca^{2+} and the **calmodulin** proteins for their action, so that the low intracellular 'free' Ca^{2+} within cells can restrict NOS activity (Chapter 4). At sites of chronic inflammation, an 'extra' NOS is often present, called **inducible NOS** (iNOS). iNOS was first identified in macrophages and liver cells after treatment with endotoxin or certain cytokines.[f] iNOSs bind calmodulin extremely tightly and their activity is essentially Ca^{2+}-independent. They catalyse rapid NO$^\bullet$ generation, generating much higher localized concentrations than normal, perhaps as high as the micromolar range.

Having carefully explained NOS nomenclature, we must point out that it is now outdated, because different NOSs occur at multiple locations in the human body. For example, nNOS is found in a variety of neurones in both the central and peripheral nervous system, but eNOS is expressed in some neurones. Although eNOS is constitutive, its levels can be 'induced' in vascular endothelium, e.g. by shear stress.[80] iNOS may occur in normal epithelium in the lung. nNOS can also be induced in neurones by certain treatments. Excess NO$^\bullet$ production, often (but not always) involving iNOS, may occur in certain diseases (Table 2.9), and has been suspected as a major contributor to the disease pathology. In rats and mice, phagocytes (especially macrophages) are prolific producers of NO$^\bullet$ during inflammation, but human phagocytes are more reluctant to generate NO$^\bullet$, at least *in vitro* and may require prolonged exposure to cytokines.[1,86]

Nitrates, nitrites and NO• production

Most dietary nitrates are found in vegetables, to an extent depending on several factors, including soil composition and fertilizer use. Nitrate is reduced to nitrite in the gut, and NO_2^- absorbed. In addition, nitrites and nitrates have been used since ancient times (and still are used) as preserving and curing agents for meats; usually bacon, sausages and tinned meats. Nitrites inhibit the growth of many bacteria in foods, particularly the anaerobe *Clostridium botulinum*, which causes a potentially lethal form of food poisoning. In the absence of nitrites, meat darkens in colour because the Fe^{2+}–haem in myoglobin oxidizes to give dull-red metmyoglobin. NO• binds to myoglobin to give a red complex that slows oxidation.

$$MbFe^{2+}-O_2 + NO^\bullet \rightarrow MbFe^{2+}NO + O_2$$

Despite these advantages, one of the concerns about the use of nitrites is that during cooking they might react with amines to generate **nitrosamines**, potential carcinogens.

$$\underset{\text{secondary amine}}{R_2NH} + HNO_2 \rightarrow \underset{\text{nitrosamine}}{R_2-N-N=O} + H_2O$$

The stomach contents are very acidic and dietary NO_2^- can form **nitrous acid** (HNO_2) that can decompose to oxides of nitrogen. In addition, the salivary glands concentrate NO_3^- from plasma and excrete it into saliva, where it is reduced to NO_2^- by oral bacteria. Average salivary nitrite levels in fasting subjects have been reported as $100\,\mu M$. Perhaps NO• generation is an antibacterial mechanism in the stomach.[19] Nitrous acid is a powerful deaminating agent and its mutagenicity has been debated (Section 4.6). NO_2^- is rapidly oxidized to NO_3^- in animals, probably by interaction with haemoglobin, generating NO_3^- and methaemoglobin.

Free radical chemistry[4,61] of NO•

Nitric oxide is generally unreactive with most non-radicals; for example, the rate of its reaction with thiol (–SH) compounds is low. To generate **thionitrites** (more often called **nitrosothiols**), NO• is usually first converted to a higher oxide of nitrogen (e.g. N_2O_3) or to peroxynitrite (Section 2.5.4). Another source is via thiyl radicals:

$$RS^\bullet + NO^\bullet \rightarrow RSNO \text{ (radical addition)}$$

Nitrosothiols (RSNO) detected in animal blood include *S*-nitrosoalbumin, nitrosohaemoglobin (NO can attach to cysteine 93 of the β-chains) and some of the much less stable **S-nitrosocysteine**. The total concentration of RSNO in plasma is usually around $1\,\mu M$, mostly nitrosoalbumin. Lung lining fluids have much higher concentrations of GSH than does plasma (Section 3.19) and $\sim 0.3\,\mu M$ GSNO has been detected in human lung fluid.

By contrast, NO• reacts fast with many other radicals. For example, its rate constant for reaction with tyrosyl (TyrO•) radical is $> 10^9\,M^{-1}s^{-1}$. Tyrosine radicals are essential at the active sites of some enzymes, in particular **ribonucleotide reductase** (Section 6.2). One of the toxic effects of excess $NO^{\bullet}$ upon cells can be inhibition of this enzyme, apparently because the $NO^{\bullet}$ reacts with the tyrosine radical needed for catalytic activity.[46]

$NO^{\bullet}$ also reacts rapidly with $O_2^{\bullet-}$

$$O_2^{\bullet-} + NO^{\bullet} \rightarrow \underset{\text{peroxynitrite}}{ONOO^-} \quad k > 10^9 \, M^{-1} s^{-1}$$

with peroxyl radicals

$$RO_2^{\bullet} + NO^{\bullet} \rightarrow \underset{\text{alkyl peroxynitrite}}{ROONO} \quad k > 10^9 \, M^{-1} s^{-1}$$

with $OH^{\bullet}$

$$NO^{\bullet} + OH^{\bullet} \rightarrow HNO_2 \quad k > 10^9 \, M^{-1} s^{-1}$$

with radicals from amino acids other than tyrosine (e.g. tryptophan) and with ferryl haem species generated when haem proteins are treated with H_2O_2 (Section 3.18.3). Reactions of $NO^{\bullet}$ with $RO_2^{\bullet}$ and with haem protein/H_2O_2 systems can inhibit damaging effects of these species, and this might be regarded as an antioxidant effect of $NO^{\bullet}$.

Realization of the biological significance of $ONOO^-$ and ROONO has given birth to a novel field, that of the **reactive nitrogen species** (Table 2.2). For example, exposure of cells to excess $NO^{\bullet}$ can cause covalent modification of $-SH$ groups on glyceraldehyde-3-phosphate dehydrogenase and damage to iron–sulphur proteins in mitochondria. However, both these effects are probably due to $NO^{\bullet}$-derived oxidation products (e.g. N_2O_3, $ONOO^-$) rather than $NO^{\bullet}$ itself. Direct toxic effects of excess $NO^{\bullet}$ include inhibition of ribonucleotide reductase (see above) and cytochrome oxidase; $NO^{\bullet}$ competitively inhibits binding of O_2 to this essential mitochondrial multienzyme complex.[9] $NO^{\bullet}$ also reacts[70a] with ferrocytochrome c to give Fe(III) cytochrome c plus NO^-, which can react with O_2 to form $ONOO^-$. Rat liver mitochondria contain NOS,[79a] which suggests that $NO^{\bullet}$ might help regulate mitochondrial function under some circumstances.

Nitric oxide donors

Nitrovasodilators have been used for more than a century as treatment for diseases with abnormal vasoconstriction (e.g. **angina pectoris**) and sometimes (e.g. **amyl nitrite**) as recreational drugs. They appear to act largely or entirely as $NO^{\bullet}$ donors. These drugs include organic nitrates and nitrites (e.g. **nitroglycerine** and amyl nitrite), inorganic nitroso compounds (e.g. **nitroprusside**), **sydnonimines** such as linsidomine (sometimes called **SIN-1**), SNAP and other nitroso thiols (Table 2.10; Fig. 2.9).

Many of these compounds are often used as $NO^{\bullet}$ donors in laboratory experiments. A popular example is **S-nitroso-N-acetylpenicillamine** (SNAP; Fig. 2.8). An important question is how RSNO species generate $NO^{\bullet}$. Simple homolytic fission would yield a (potentially reactive) thiyl radical

$$RSNO \rightarrow RS^{\bullet} + NO^{\bullet}$$

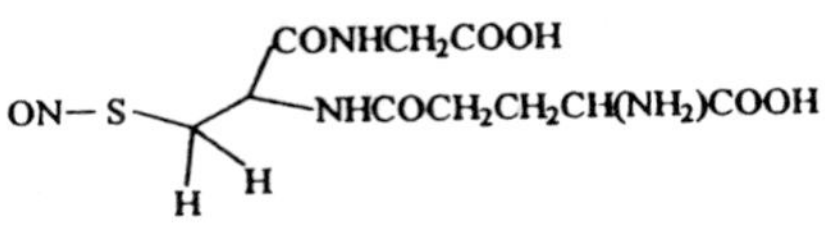

S-Nitrosoglutathione
(GSNO)

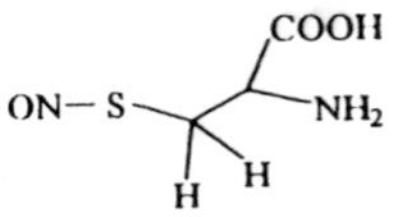

S-Nitroso-L-cysteine
(CySNO)

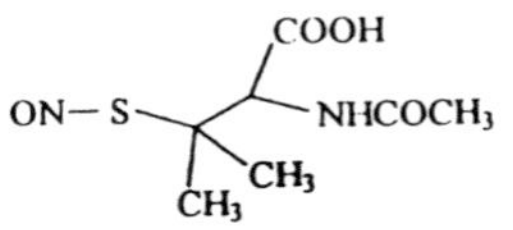

S-Nitroso N-acetyl-DL-penicillamine
(SNAP)

Fig. 2.8. Some nitrosothiols.

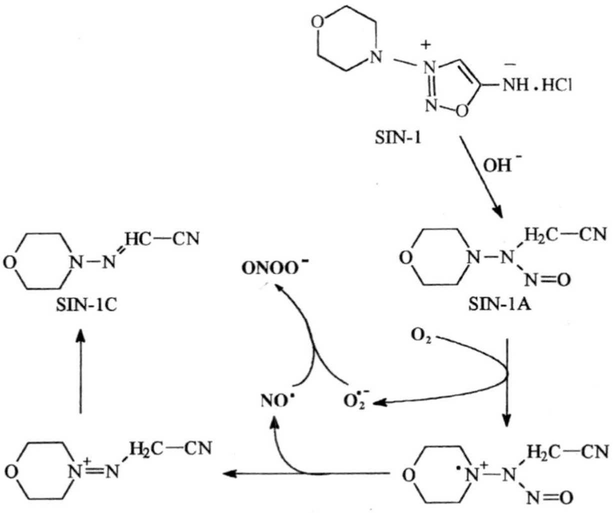

Fig. 2.9. The structure of SIN-1. SIN-1 generates $O_2^{\bullet -}$ and NO$^\bullet$, leading to peroxynitrite generation.

Table 2.10. Some nitric oxide donors

Type of compound	Examples	Comments
Metal nitrosyl complexes	Sodium nitroprusside, (SNP) $Na_2Fe(CN)_5NO$ Ruthenium nitrosylpentachloride, (RNP) $K_2Ru(Cl)_5NO$	$NO^\bullet$ forms one of the ligands to iron. SNP is widely used in pharmacology but it is not always clear how it makes $NO^\bullet$: this can involve photochemical decomposition or reductive breakdown $[Fe(CN)_5NO]^{2-} \xrightarrow{e^-} [Fe(CN)_5]^{3-} + NO^\bullet$ Nitroprusside itself can react with certain biomolecules and its metabolism by microsomes can lead to $O_2^{\bullet-}$ generation (*Biochim. Biophys. Acta* **1289**, 195, 1996). RNP is photolabile and has been used as a source of 'caged $NO^\bullet$'.
S–Nitrosothiols (thionitrites) R–S–N=O	Nitrosocysteine (half-life in seconds) GSNO ($t_{1/2} \approx 160\,h$) S-Nitroso-N-acetyl-DL-penicillamine, SNAP ($t_{1/2} \approx 1\,h$)	Generated by reaction of several RNS; (but not $NO^\bullet$ directly) with thiols. Found *in vivo*, e.g. S-nitrosoalbumin and GSNO; their decomposition is accelerated by light, transition metal ions (especially copper) and reducing agents $2RSNO \xrightarrow[\text{or light}]{2e^-} RSSR + 2NO^\bullet$ $Cu^+ + RSNO \rightarrow [RSNO\cdot Cu]^+ \rightarrow RS^- + NO^\bullet + Cu^{2+}$ $Cu^{2+} + RS^- \rightarrow Cu^+ + RS^\bullet$ $RSNO + RS^\bullet \rightarrow RSSR + NO^\bullet$ Note the production of thiyl radicals Transnitrosation reactions can also occur between thiols, $RSNO + R'S^- \rightleftharpoons RS^- + R'SNO$ including thiols on proteins such as albumin and thioredoxin (Section 3.13)

SIN-1	See Fig. 2.9	Decomposition generates *both* $O_2^{\bullet-}$ and $NO^{\bullet}$, therefore likely to produce $ONOO^-$. Breakdown is accelerated by light. Active metabolite of the anti-angina drug **molsidomine**
Organic nitrates/nitrites	Nitroglycerine (glycerol trinitrate ester) amyl nitrite $[(CH_3)_2CH(CH_2)_2-O-N=O]$	They have been used in medicine and/or as recreational drugs for many years. Reduction (e.g. by cytochromes P450) releases $NO^{\bullet}$ producing vasodilation. Some GSH transferases can catalyse reaction of organic nitrites with RSH to give nitrosothiols (Section 3.10)
NONOates contain the $[N(O)NO]^-$ functional group.	DETA: $R_1=R_2=H_2NCH_2CH_2$, $t_{1/2} \approx 20$ h at 37 °C Spermine: $R_1=H_2N(CH_2)_3\overset{+}{N}H_2(CH_2)_4$, $R_2=H_2N(CH_2)_3$, $t_{1/2}=39$ min PAPA: $R_1=CH_3CH_2CH_2$, $R_2=H_3\overset{+}{N}(CH_2)_3$, $t_{1/2}=15$ min MAHMA: $R_1=CH_3\overset{+}{N}H_2(CH_2)_6$, $R_2=CH_3$, $t_{1/2}=1-2$ min	Generate $NO^{\bullet}$ at variable rates depending on the structure. $t_{1/2}$ values vary according to reaction conditions: the values given are only illustrative

$$(R_1-N-N \overset{\displaystyle R_2 \quad O}{=\!\!=} N-O^- \, Na^+$$

For detailed references consult Packer, L. (ed.) (1996) *Methods in Enzymology*, vol. 268. Many compounds have been used to generate $NO^{\bullet}$ for chemical or pharmacological studies; some are used as $NO^{\bullet}$ donors *in vivo*. Often the mechanism of $NO^{\bullet}$ generation is unclear and is influenced by reaction conditions. One must be aware of what else can be generated (e.g. $RS^{\bullet}$ radicals) by some $NO^{\bullet}$ donors.

Decomposition of many RSNO is catalyzed by transition–metal ions, especially copper, or is light–dependent (e.g. nitroprusside). Hence care must be taken during physiological experiments in attributing effects of nitrosothiols (or other $NO^\bullet$ donors) simply to 'provision of $NO^\bullet$'; their chemistry is complex (Table 2.10).

2.5   Chemistry of biologically important non–radicals

2.5.1   *Hydrogen peroxide*

Hydrogen peroxide (H_2O_2) is a pale–blue covalent viscous liquid, boiling point 150 °C. Its structure is shown in Fig. 2.10. Hydrogen peroxide is toxic to most cells at levels in the $10-100\,\mu M$ range. Indeed, the weak antiseptic activity of honey, which has been used in wound treatment since ancient times, is partly due to the formation of H_2O_2 by enzymes contained within it.[87] In addition, honey contains several other antimicrobial agents such as traces of **propolis**, a resin collected from certain trees.[71] Also, royal jelly produced by bees contains a potent antibacterial protein, **royalisin**.

Hydrogen peroxide mixes readily with water and is very diffusible within and between cells *in vivo*. Physiologists recognized years ago that water crosses the lipid bilayer of cell membranes much more quickly than would be expected by simple diffusion of such a polar molecule.[15] Although some water does appear to diffuse across the lipid bilayer, much of it appears to pass through membrane water channels known as the **aquaporins**. It seems likely that H_2O_2 can also traverse these channels, although more research on this point is needed.

Several enzymes found *in vivo* can generate H_2O_2, including xanthine, urate and D–amino acid oxidases (Chapter 1). In addition, any biological system that generates $O_2^{\bullet-}$ will also produce H_2O_2 by $O_2^{\bullet-}$ dismutation (unless all the $O_2^{\bullet-}$ is intercepted by some other molecule, e.g. a high concentration of cytochrome *c*). H_2O_2 is only a weak oxidizing and reducing agent and is generally poorly reactive. For example, no oxidation occurs when DNA, lipids or most proteins are incubated with H_2O_2, even at millimolar levels. H_2O_2 appears capable of inactivating a few enzymes directly, usually by oxidation of labile essential thiol (−SH) groups at the active site. **Glyceraldehyde–3–phosphate**

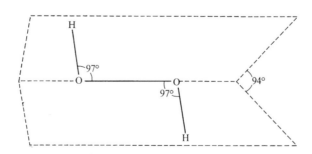

Fig. 2.10. The structure of hydrogen peroxide.

dehydrogenase,[8] an enzyme of the glycolytic pathway, is inactivated, apparently directly, in cells treated with exogenous H_2O_2 (Chapter 4). Chloroplast fructose bisphosphatase is similarly inactivated by H_2O_2 (Chapter 7).

H_2O_2 can also oxidize certain keto-acids such as pyruvate (CH_3COCOO^-) and 2-oxoglutarate. Indeed, high concentrations of keto acids added to culture media can protect cells against H_2O_2 (Section 3.21.2). Thus despite its poor reactivity, H_2O_2 can be cytotoxic and at high concentrations is often used as a disinfectant. Some bacterial strains are very sensitive to H_2O_2, and many animal cells in culture are injured or killed if H_2O_2 is added to the culture medium at concentrations at or above the $10–100\,\mu M$ range. For example, H_2O_2 produced by the pathogen *Mycoplasma pneumoniae* can attack epithelial cells in the trachea. Co-infection of human lymphocytes with the HIV-1 virus and a mycoplasma leads to cell damage involving H_2O_2 generation.[13]

Some cellular damage by H_2O_2 is direct, e.g. attacking glyceraldehyde-3-phosphate dehydrogenase, yet addition of H_2O_2 to cells frequently leads to lipid, DNA and protein oxidation that cannot be mediated by H_2O_2 alone. H_2O_2 can cross cell membranes rapidly and, once inside, can probably react with iron, and possibly copper, ions to form much more damaging species such as $OH^\bullet$. Hydroxyl radicals account for much of the damage done to DNA in H_2O_2-treated cells.[74] In addition, the conversion of H_2O_2 to $OH^\bullet$ can be achieved by ultraviolet light

$$H_2O_2 \rightarrow 2OH^\bullet \text{ (homolytic fission)}$$

H_2O_2 can degrade certain haem proteins (including myoglobin, haemoglobin and cytochrome *c*) to release iron ions.[34] At lower H_2O_2: haem protein ratios, oxidizing species on the protein (ferryl haem plus amino acid radicals) can also cause biological damage (Section 3.18.3).

The Russell effect[17]

In 1896 Becquerel found that uranium compounds could affect a photographic plate wrapped in light-proof paper. This early description of radioactivity excited much curiosity amongst scientists, one of whom was William Russell. In the course of his experiments, Russell made a mask of perforated zinc sheet and placed it between the uranium salts and the photographic plate, intending to produce a shadow of the zinc sheet on the film. To his surprise, he found that the zinc sheet affected the plate more than the uranium salts and maintained this activity even in their absence. Russell later found that many organic materials are also capable of forming a developable image on film.

The **Russell effect**, as it has come to be known, is shown by many freshly abraded metals, notably magnesium, cadmium, nickel, aluminium and lead. Wood, leaves, essential oils, drying oils and paper can all give images, particularly if they have been previously exposed to light or, in the case of wood, if it has been recently charred. Russell found that H_2O_2 was also capable of affecting a photographic plate even if present in minute quantities. Indeed, the Russell effect seems to be due to H_2O_2 or organic peroxides produced in minute amounts from the material tested.

Most modern photographic films are much less sensitive to H_2O_2, but they can be used to show the Russell effect if specially pre-treated. The detection of room temperature degradation of materials of antiquity is of great importance for development of new conservation methods. In cases where autoxidation is the principal degradation mechanism, the Russell effect is one sensitive method of detecting it.

2.5.2   *Hypochlorous acid*[25,70]

Hypochlorous acid is produced by the enzyme **myeloperoxidase** (MPO) in activated neutrophils (Section 6.7.2)

$$H_2O_2 + Cl^- \xrightarrow{\text{MPO}} HOCl + OH^-$$

It is a weak acid, with a pK_a of approximately 7.5. Hence HOCl is about 50% ionized at pH 7.4

$$HOCl \rightleftharpoons H^+ + OCl^-$$

Hypochlorite (OCl^-) absorbs light at 292 nm (molar extinction coefficient, $\varepsilon = 350\,dm^3\,mol^{-1}\,cm^{-1}$) and so concentrations of HOCl can easily be determined by adding alkali and reading the absorbance, or by adding acid and reading the absorbance of HOCl at 235 nm ($\varepsilon = 100\,dm^3\,mol^{-1}\,cm^{-1}$). Studies with HOCl are facilitated by the ease with which it can be made in the laboratory, by simple acidification of commercial solutions of sodium hypochlorite (Na^+OCl^-), constituents of many bleaches. HOCl readily decomposes to liberate chlorine (Cl_2) gas

$$HOCl + H^+ + Cl^- \rightleftharpoons Cl_2 + H_2O$$

Although its importance in bacterial killing by phagocytes is uncertain (Chapter 6), HOCl has attracted much attention because of its high reactivity (Table 2.11) and ability to damage biomolecules, both directly and by decomposing to form chlorine. HOCl is a powerful two-electron oxidizing agent. For example, addition of HOCl to bacteria such as *E. coli* inhibits ATP synthesis by damaging electron transport chain components and the ATP synthase.[38] In animal systems, HOCl could attack many targets. It can inactivate α_1-**antiproteinase**, an important inhibitor of proteolytic enzymes such as **elastase** in human body fluids (Chapter 9), within seconds at pH 7.4. HOCl attacks a methionine residue on the α_1-antiproteinase, which is converted to **methionine sulphoxide. Thrombomodulin,**[31] a glycoprotein present in the membrane of endothelial cells that regulates the blood coagulation pathway by modifying the action of thrombin, can also be inactivated by HOCl, again by attack on methionine. Whether these proteins really are major targets of attack by HOCl *in vivo* is uncertain, since HOCl reacts with many other biomolecules that are present at higher concentrations.

HOCl addition can oxidize thiols, ascorbate, NAD(P)H and lead to chlorination of DNA bases (especially pyrimidines) and tyrosine residues in proteins

Table 2.11. Some rate constants for reactions of hypochlorous acid

Compound	pH	k (M^{-1}s^{-1})
Monochlorodimedon[a]	5.0	7×10^6
NADH	7.0	$> 2 \times 10^5$
Taurine	7.0	4.8×10^5
GSH[b]	7.4	$> 10^7$
Ascorbate	7.4	$\sim 6 \times 10^6$
Ferrocyanide, Fe(CN)$_6^{4-}$	7.0	~ 220
Fe^{2+}	~ 4	1.7×10^4
Fe(II)–citrate	5.0	1.3×10^4
Superoxide	5.0	7.5×10^6

Data from Folkes *et al.* (1995) *Arch. Biochem. Biophys.* **323**, 120. HOCl is a powerful two-electron oxidizing agent; $E^{\circ\prime}$ for the couple HOCl/H$_2$O, Cl$^-$ is about 1.1 V. Note that oxidations/chlorinations caused by HOCl addition are not necessarily achieved by this molecule; it may decompose to give such species as Cl$^+$, Cl$_2$, etc.
[a]This compound, full name 1,1-dimethyl-4-chloro-3,5-cyclohexane-dione, is chlorinated by HOCl; the resulting rise in absorbance at 290 nm is often used to assay HOCl production by myeloperoxidase.
[b]Reaction of GSH with HOCl gives a mixture of products, including GSSG, GSO$_2$SG (glutathione thiosulphonate) and a glutathione sulphonamide (*Biochem. J.* **326**, 87 (1997)).

(generating **3-chlorotyrosine**). Metal ions held in proteins by thiolate (S$^-$) ligands can be released after HOCl treatment, e.g. Zn^{2+} is lost from metal-lothionein.[24] HOCl seems able to cross membranes, causing damage to membrane proteins on its passage (especially to −SH groups and methionine residues) and, if any survives to enter the cytoplasm, to intracellular constituents.

Chlorohydrins, chloramines and hydroxyl radical from HOCl[12,40,58,81]
HOCl can add across double bonds

to give **chlorohydrins**, which are formed when HOCl is added to unsaturated lipids. Similar products (**bromohydrins**) are formed when **hypobromous acid** (HOBr), also produced by certain phagocytes (Section 6.7.2), is added to lipids. Addition of HOCl can also oxidize and chlorinate cholesterol.

Reaction of HOCl with O$_2^{\bullet-}$ generates OH$^\bullet$

$$HOCl + O_2^{\bullet-} \rightarrow O_2 + Cl^- + OH^\bullet$$

as does its reaction with iron ions (Section 2.4.2)

$$HOCl + Fe^{2+} \rightarrow Fe(III) + Cl^- + OH^\bullet$$

Indeed, Fenton himself originally reported that the H_2O_2 in his $Fe^{2+}/H_2O_2/$ tartaric acid system could be replaced by 'chlorine water'. The killing of some bacteria by radiolysis in N_2O-saturated solutions can be greatly increased by the presence of Cl^- ions,[89] apparently partly due to HOCl formation.

Reaction of HOCl with **taurine** (2-aminoethanesulphonic acid), which is present at high levels in many mammalian tissues as an end-product of the metabolism of sulphur-containing amino acids, generates **taurine chloramines**

$$R{-}NH_2 + HOCl \rightarrow RNHCl + H_2O$$

which can also inactivate α_1-antiproteinase. These chloramines were initially identified as 'long-lived oxidants' generated by reaction of HOCl with biological fluids.[84a] Similar products are formed from HOCl plus other amines. Both HOCl and chloramines have been reported to interfere with DNA repair processes in cells. $HOCl/OCl^-$ might also participate in formation of singlet O_2 (Section 2.5.3) and nitryl chloride (Section 2.5.4).

2.5.3   *Singlet oxygen*[26]

We saw in Chapter 1 that two singlet states of O_2 exist. In both, the spin restriction that slows reaction of O_2 with non-radicals is removed, so that singlet oxygens are much more oxidizing than ground-state O_2. The $^1\Sigma g^+$ state of oxygen rapidly decays to the $^1\Delta g$ state, so only the latter is usually considered in biological systems. Hence references to 'singlet oxygen' in this book can be taken to refer to the $^1\Delta g$ state. Although $^1\Delta g O_2$ is not a free radical (Chapter 1) it can be formed in some radical reactions and can lead to others.

Singlet O_2 in solution 'deactivates' by transferring its excitation energy to the solvent, so its lifetime depends on which solvent it is generated in. Thus its lifetime in H_2O is about 3.8 μs, that in deuterium oxide (D_2O) 62 μs, that in hexane 31 μs and that in hexafluorobenzene (C_6F_6) 3900 μs.

Sources of singlet O_2[26]

Light-dependent reactions   Singlet oxygen is most often generated in the laboratory by **photosensitization reactions**. If certain molecules are illuminated with light of the correct wavelength, they absorb it and the energy raises the molecule into an **excited state**. The excitation energy can then be transferred onto an adjacent O_2 molecule, converting it to the singlet state whilst the photosensitizer molecule returns to its ground state. Popular sensitizers of singlet oxygen formation in the laboratory include the dyes **acridine orange**, **methylene blue**, **rose Bengal** and **toluidine blue**; but many biological compounds are also effective *in vitro*, such as the water-soluble vitamin riboflavin and its derivatives FMN and FAD. Also effective are chlorophylls *a* and *b*, the bile pigment **bilirubin, retinal** (a pigment found in the eye) and various porphyrins, both free and bound to proteins (Figs 2.12 and 2.13).

The singlet oxygen produced on illumination of these substances can attack the photosensitizer molecule itself and/or react with any other molecules present. The chemical changes thereby produced are known as **photodynamic effects**. Hence illuminated solutions of flavins lose their orange colour and chlorophylls their green colour as they are attacked (**photobleaching**). Reactions of this type cause the dyes in clothes and curtains and the paint on cars to fade when exposed to sunlight. Indeed, exposure of cells in culture to high-intensity visible light causes damage, especially to the mitochondria, which are rich in haem proteins and flavin-containing proteins. Haem-containing enzymes such as catalase are also inactivated.[23] Some cell culture media contain riboflavin, which aggravates these effects, since the intensity of fluorescent lighting in many laboratories is sufficient to cause flavin-sensitized singlet O_2 formation.[54]

Type I and II reactions[26]

Not all photosensitization damage need arise via singlet O_2, since the excited state of the photosensitizer can often cause damage itself. Such damage is said to occur by a **type I mechanism** as opposed to that caused by singlet O_2, described as a **type II mechanism**. These two mechanisms may operate simultaneously and the relative importance of each depends on the target molecule, the efficiency of energy transfer from the sensitizer to O_2, and on the O_2 concentration. In addition, illumination of several photosensitizers, including porphyrins and acridine dyes, in aqueous solution has been reported to produce $OH^\bullet$ and $O_2^{\bullet-}$ in addition to $^1O_2$. Thus it must never be assumed that $^1O_2$ is the species responsible for any damage observed.[47]

Biological damage by photosensitization

Photosensitization reactions, usually involving $^1O_2$, are important in many biological situations, especially in chloroplasts, which contain chlorophylls and a high O_2 concentration (Section 7.5). The rod cells of the retina in the eye contain retinal (Fig. 2.11), which can sensitize singlet O_2 formation, damaging itself and the lipids around it if the retina is exposed to high light intensities for long periods (Section 7.7). It has also been suggested that the lens contains sensitizers of $^1O_2$ formation. Illumination of milk or milk products can cause development of 'off-flavours' as the riboflavin present photosensitizes the degradation of milk proteins and lipids.[45]

Several diseases can lead to excessive singlet O_2 formation. For example, the **porphyrias** (a term derived from *porphuros*, Greek for purple) are diseases caused by defects in the biosynthesis of haem (Fig. 2.12). In some of these diseases (the **cutaneous porphyrias**), porphyrins accumulate in the skin, exposure of which to light causes damage leading to unpleasant eruptions, scarring and thickening (Fig. 2.13). The severity of the damage depends upon

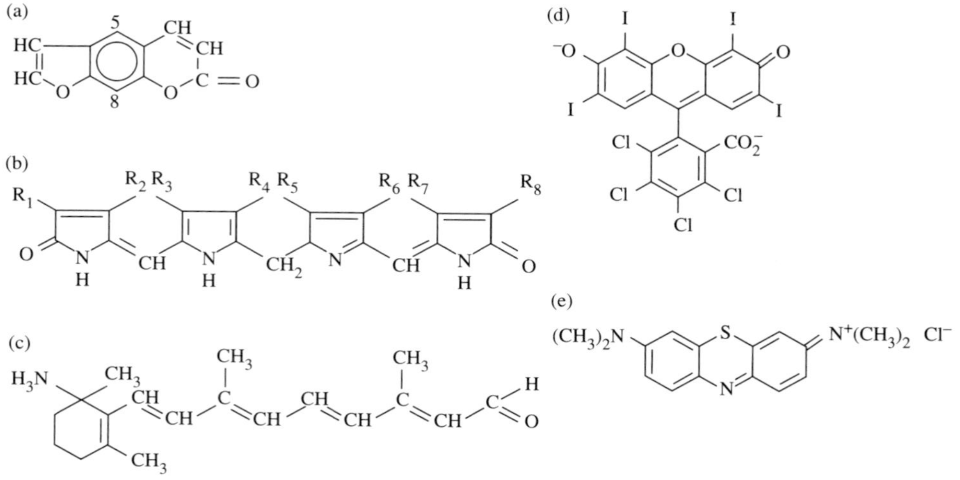

Fig. 2.11. Compounds that can sensitize formation of singlet O_2 when illuminated with light of the correct wavelength. (a) Psoralen (psoralen derivatives are used in treating certain skin diseases), (b) bilirubin, (c) retinal, (d) rose Bengal, (e) methylene blue, chloride salt.

which porphyrin is accumulated and thus differs in different types of porphyria. Oral β-carotene, a singlet O_2 quencher, can offer some protection against this damage (Section 3.23). The British king George III (1738–1820) is believed to have suffered from an acute porphyria syndrome (possibly acute intermittent porphyria; Fig. 2.12) causing periodic severe attacks of abdominal pain, vomiting, psychiatric disturbances and reddish-brown or purple urine, but not skin lesions. The discoloration is due to autoxidation of excreted porphyrin precursors. Even healthy skin can be exposed to small amounts of porphyrins generated by bacteria (Section 7.9).

Several drugs have been shown to sensitize singlet O_2 formation *in vitro*, including some phenothiazines (used as tranquillizers), tetracycline antibiotics[39] and anti-inflammatory drugs, such as **carprofen**.[7] The compound benoxaprofen was introduced as an anti-inflammatory agent but is no longer used, in part because it can cause phototoxicity. Various constituents of sunscreens and perfumes have also been suggested to be photosensitizers.

Photosensitization reactions are also of importance in veterinary medicine.[14] Digestion of chlorophyll in ruminants forms the pigment **phylloerythrin**, which is absorbed from the gut and excreted in the bile. Liver damage or malfunction prevents excretion and can allow phylloerythrin to enter the bloodstream and deposit in the tissues. When the animals are exposed to sunlight the pigment in the skin causes damage that results in reddening and swelling. The fungal product **sporidesmin** achieves a similar effect by inhibiting bile production. The buckwheat plant synthesizes a compound that sensitizes singlet O_2 formation and is toxic to farm animals. The same is true

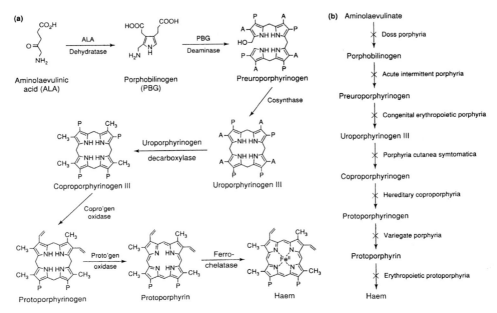

Fig. 2.12. The biosynthesis of haem. Different porphyrias result from inborn errors in different enzymes in the metabolic pathway. For example, hereditary coproporphyria and variegate porphyria are frequently associated with skin lesions (coproporphyrin accumulates in the former, protoporphyrinogen in the latter). In **porphyria cutanea tarda** uroporphyrinogen decarboxylase is decreased and skin lesions are common. This disease is not always inherited: it can result from exposure to certain toxins (Section 8.3). Loss of ferrochelatase activity also leads to skin damage, due to protoporphyrin accumulation. From *Trends Biochem. Sci.* **21**, 231 (1996), by courtesy of Dr Martin Warren and his colleagues and Elsevier Publishers. **5-Aminolaevulinic acid**, which accumulates in acute intermittent porphyria and also in lead poisoning (Section 8.12.4) is an autoxidizable molecule; rats chronically treated showed increased levels of 8-hydroxy-2'-deoxyguanosine in liver DNA (*Carcinogenesis* **15**, 2241 (1994)).

of the St John's wort plant which produces the sensitizer **hypericin**. Japanese scientists have reported a light-induced dermatitis in some people who ingested large quantities of tablets made from the alga *Chlorella*, and attributed the damage to a product derived from chlorophyll. Over-consumption of hypericin tablets, marketed in some health food stores, might have the same effect.

Uses of photosensitization

Sometimes photochemical effects are useful. A well-established case is the treatment of the jaundice often developed by premature infants soon after birth.[22] The yellow colour is due to the accumulation in the skin of the pigment **bilirubin** (Fig. 2.11). Bilirubin is derived from the breakdown of the haemoglobin in unwanted red blood cells, and from the destruction of other haem proteins by the enzyme **haem oxygenase** (Section 3.20). Bilirubin travels in the blood tightly bound to the plasma protein albumin, which can bind two molecules of bilirubin per molecule of protein. The liver takes up

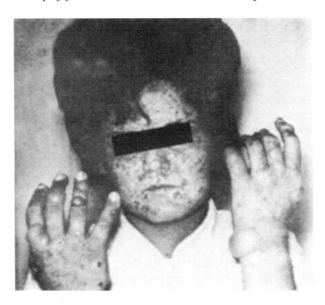

Fig. 2.13. Skin and tissue damage in a porphyria patient. Note the blistering and swelling. From *Angew-Chem. Int. Ed.* **21**, 343 (1982), with permission.

bilirubin and converts it into a water-soluble product by the action of the enzyme **glucuronyl transferase**, which catalyses the reaction:

bilirubin + 2UDP–glucuronic acid → bilirubin diglucuronide + 2UDP

The diglucuronide is disposed of by excretion into the bile. In premature babies insufficient glucuronyl transferase is often present in the liver, so that the lipid-soluble bilirubin accumulates in the blood, and deposits in tissues with a high lipid content, such as the brain, where it can cause irreversible damage. The jaundice can be ameliorated by careful exposure of the babies to blue light from a sunlamp, whereupon the bilirubin deposited in the skin sensitizes its own destruction in a reaction that involves $^1O_2$. In addition, and probably more important in quantitative terms, is a light-induced rearrangement (**photoisomerization**) of the structure of bilirubin to give water-soluble products that can be excreted.

Another example of the application of photosensitization reactions in medicine is the use of **psoralens** (Fig. 2.11), a class of powerful photosensitizers produced by some plants (e.g. celery), in the treatment of skin diseases such as psoriasis.[55] The treatment consists of the combined application of ultraviolet light (wavelength range 320–400 nm) and a psoralen, and is often referred to as PUVA therapy (psoralen ultraviolet A). PUVA therapy is helpful, but there is debate about its safety, e.g. whether it increases the risk of skin cancer.

Certain porphyrins are taken up by cancerous tumours.[57] After injection of a mixture of porphyrin derivatives known as HPD (for **haematoporphyrin**

derivative), fluorescent products are accumulated by tumour tissues and this can be used to detect the presence of the tumour by observing the fluorescence. Irradiation with light of a wavelength absorbed by HPD can damage the tumour. Such reactions are being developed for use in cancer chemotherapy, especially for skin cancer, bladder cancer and some forms of lung cancer, in which the tumour can be illuminated with light from a fibre-optic bronchoscope. Both OH$^\bullet$ and $^1O_2$ appear to be involved in HPD-mediated tumour damage. HPD is a mixture of different porphyrin derivatives and current clinical work uses a less complex mixture (**Photofrin II**) enriched in the tumour-localizing fractions.

Chemical sources of singlet O_2[26,42]

One well-established system often used to generate singlet O_2 in the laboratory is a mixture of H_2O_2 and the hypochlorite ion OCl$^-$, the ionized form of hypochlorous acid (Section 2.5.2)

$$OCl^- + H_2O_2 \rightarrow Cl^- + H_2O + O_2 \text{ (singlet)}$$

The singlet O_2 arises from the H_2O_2. Hypobromite (OBr$^-$) can react similarly. The above reaction might be biologically relevant, since OCl$^-$ is formed by the enzyme myeloperoxidase (Section 2.5.2). However, HOCl/OCl$^-$ is itself highly reactive, which has caused problems in attempts to detect singlet O_2 formation from it. Most 'singlet O_2 detectors' can be directly oxidized by HOCl/OCl$^-$.

The decomposition of the compound **potassium peroxochromate**, K_3CrO_8, in aqueous solution has also been used to produce singlet O_2. Its usefulness is limited, since it also generates $O_2^{\bullet-}$ and OH$^\bullet$ radicals as it decomposes.[59] **Peroxomolybdates** can produce $^1O_2$.[26] Reaction of ozone with several biomolecules generates some $^1O_2$ (Section 8.10.2), as does the reaction of peroxynitrite with H_2O_2[18] (Section 2.5.4) and the self-reaction of peroxyl radicals by a Russell mechanism (Section 2.4.4). For example, with trichloromethylperoxyl radical

$$2CCl_3O_2^\bullet \rightarrow Cl_3COOCCl_3 + {}^1O_2$$

A 'clean' method of producing $^1O_2$ without other reactive oxygen species is the decomposition of aromatic endoperoxides, such as the substituted naphthalene endoperoxides:

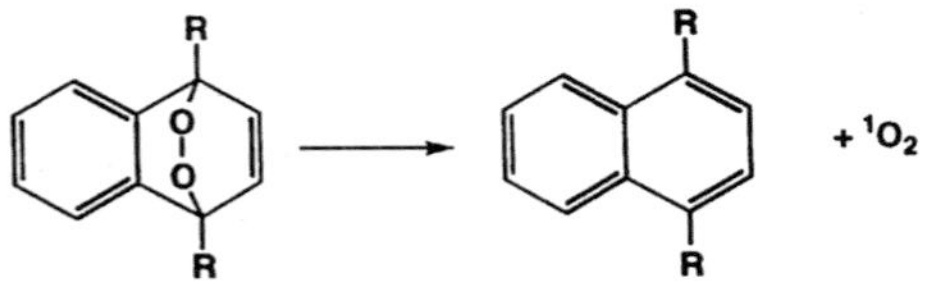

Generation of $^1O_2$ in this way has been used to assess the effects of $^1O_2$ on antioxidants in human blood plasma.[83] Exposure of plasma to a water-soluble

Compound	Structure	Comments
DABCO (1,4-diazabicyclooctane)	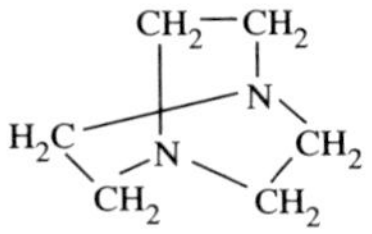	Quencher
Azide ion	N_3^-	Mostly quenching
α-Tocopherol (vitamin E)	See Chapter 3	Mostly quenching but some chemical reaction to give

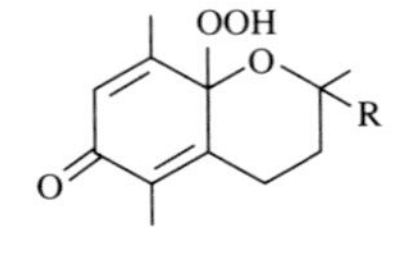

which decomposes to various products including α-tocopherylquinone (R is the side chain)

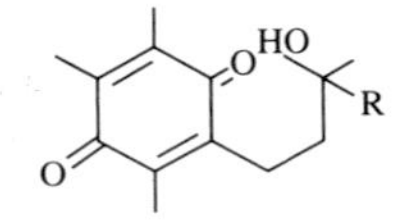

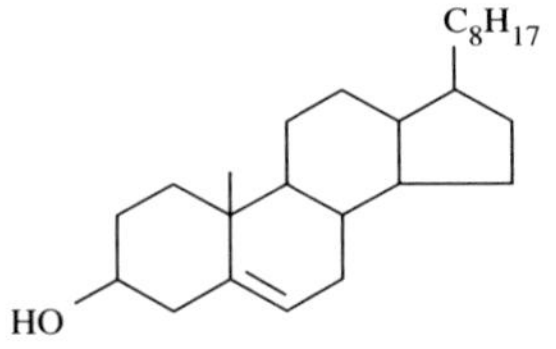

Major product is the 5α-hydroperoxide

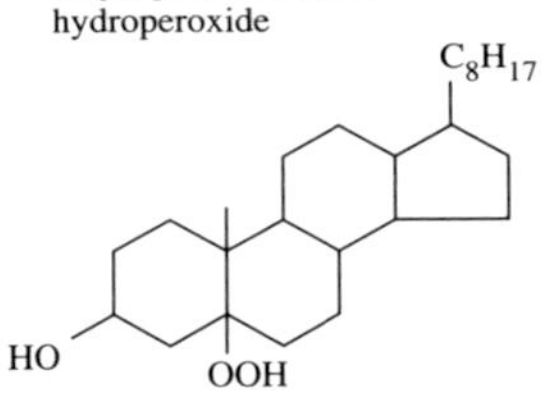

Cholesterol — Mostly quenching, some (very complex!) chemical reactions

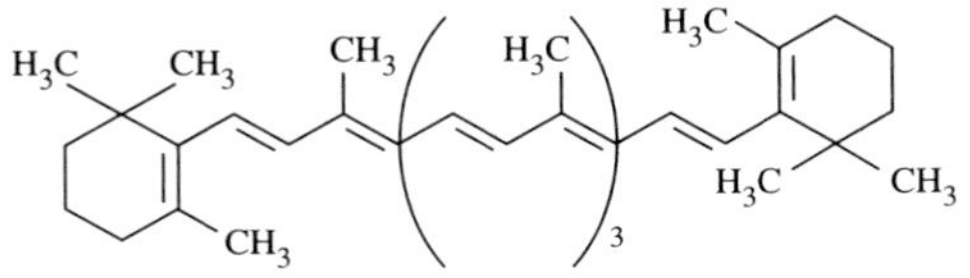

β-Carotene

Diphenylisobenzofuran (other furans react similarly) — Endoperoxide formation, accompanied by loss of the light absorption at 415 nm

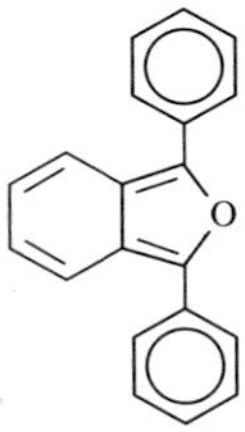

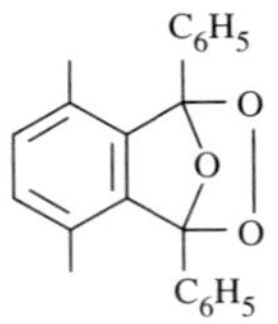

Fig. 2.14. Reactions of singlet O_2 with various biomolecules, antioxidants and 'singlet O_2 quenchers/scavengers'.

Compound	Structure	Comments
DNA		Complex mixture of products from all types of reaction with purine and pyrimidine bases, e.g. 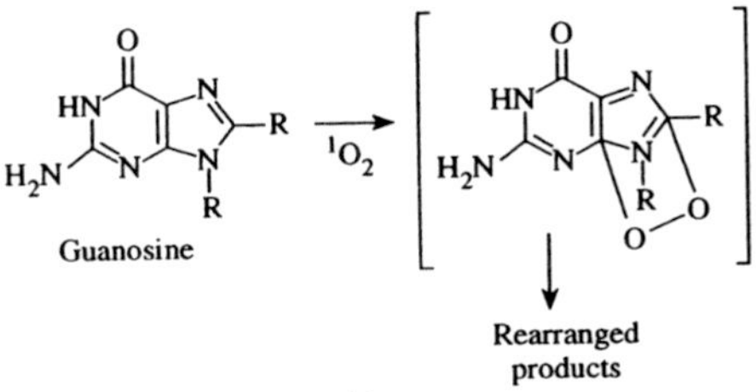

Guanosine

Rearranged products

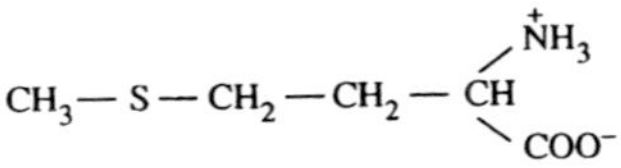

Forms methionine sulphoxide

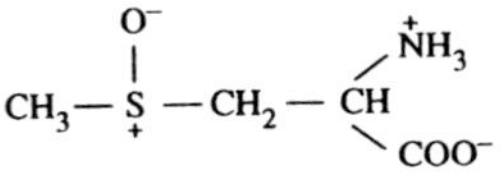

Methionine

$CH_3 - S - CH_2 - CH_2 - CH \overset{\overset{+}{N}H_3}{\underset{COO^-}{}}$

Cysteine

$HS - CH_2 - CH \overset{\overset{+}{N}H_3}{\underset{COO^-}{}}$ (R—SH)

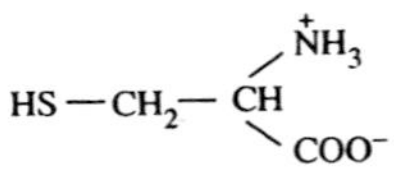

Reaction not well characterized: both disulphides (R—S—S—R) and sulphonic acids (R—SO$_3$H) are produced

Histidine

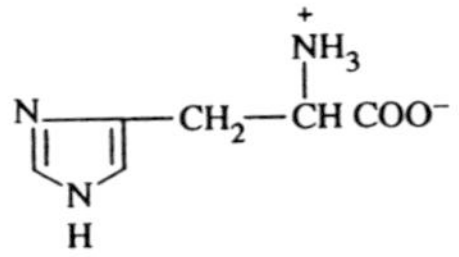

Probably reacts first to give an endoperoxide

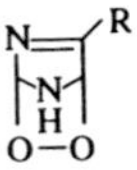

which decomposes to a complex mixture of products

Tryptophan

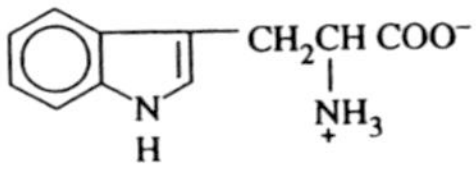

Reacts by several mechanisms, including an ene-reaction to give initially

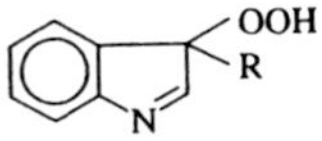

(R is the side-chain) and also formation of a dioxetane

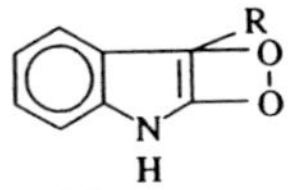

which decomposes to *N*-formylkynurenine

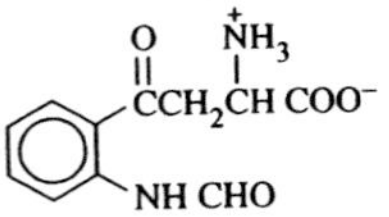

Ascorbate

Produces unstable hydroperoxides

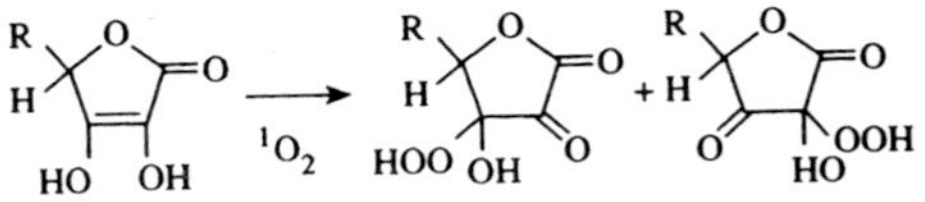

Fig. 2.14. (*Continued*)

$^1O_2$ generator led to depletion of ascorbate, urate, protein −SH groups and bilirubin, but not of α-tocopherol or β-carotene. Lipid-soluble $^1O_2$ generators gave higher rates of loss of bilirubin, protein thiols, ubiquinol and ascorbate, but lower rates for urate.

Reactions of singlet oxygen[26]

Singlet O_2 can interact with other molecules in essentially two ways: it can either react chemically with them, or else it can transfer its excitation energy, returning to the ground state while the other molecule enters an excited state. The latter phenomenon is known as **quenching** of singlet O_2. Several molecules are used in the laboratory as 'singlet O_2 quenchers/scavengers'; they include histidine, DABCO, azide and diphenylisobenzofuran (Chapter 5). Sometimes both events occur with the same molecule (Fig. 2.14).

The best studied chemical reactions of singlet oxygen are those involving compounds that contain carbon−carbon double bonds. Such bonds are present in many biological molecules, including carotenoids, chlorophyll and fatty acids. Compounds containing two double bonds separated by a single bond (known as **conjugated double bonds**) often react to give **endoperoxides**.

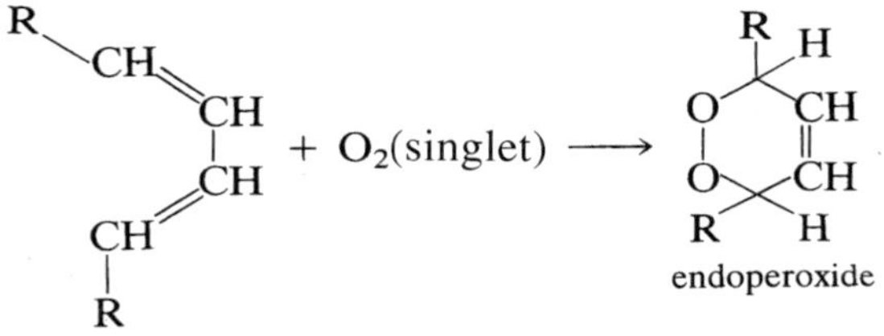

Diphenylisobenzofuran reacts in this way (Fig. 2.14). If one double bond is present, the **ene-reaction** can occur—the singlet oxygen adds on and the double bond changes position:

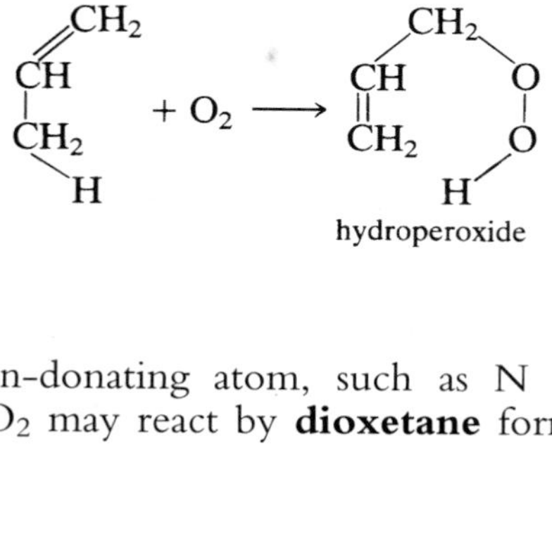

If there is an electron-donating atom, such as N or S, adjacent to the double bond, singlet O_2 may react by **dioxetane** formation. Dioxetanes are unstable and decompose to give compounds containing carbonyl group, $\diagup\!\!\!\!C{=}O$, e.g.

dioxetane   carbonyl compounds

Tryptophan can react in this way (Fig. 2.14).

2.5.4   *Peroxynitrite*[4,61]

$NO^\bullet$, oxidation products of $NO^\bullet$ (e.g. N_2O_3, NO_2^-, NO_3^-) and reactive oxygen species (ROS) are produced *in vivo*, so the question of possible interactions between them arises. For example, it has been shown *in vitro* that mixtures of HOCl and NO_2^- can form **nitryl chloride** (NO_2Cl), an oxidizing, nitrating and chlorinating agent.[21] H_2O_2 and $NO^\bullet$ have been shown to exert synergistic effects in some systems, although the chemistry of their interaction is uncertain. For example, H_2O_2 markedly enhances the inhibitory effect of $NO^\bullet$ on platelet aggregation.[52] H_2O_2 can either increase or decrease the cytotoxicity of $NO^\bullet$ to cultured cells, depending on the cell type and experimental system, the $NO^\bullet$ donor used, concentrations and whether cells were pre-exposed to $NO^\bullet$.[88]

To date, however, most attention has been given to **peroxynitrite** (officially called **oxoperoxonitrate (1−)**), which can form (among other reactions; see Table 2.12) by the combination of $NO^\bullet$ and $O_2^{\bullet -}$

$$NO^\bullet + O_2^{\bullet -} \rightarrow ONOO^-\quad k \approx 7 \times 10^9\,M^{-1}s^{-1}$$

This reaction is much faster than reaction of $NO^\bullet$ with haem compounds ($k < 10^8\,M^{-1}s^{-1}$) and is comparable to the rate at which $O_2^{\bullet -}$ reacts with SOD enzymes. The reaction of $NO^\bullet$ and $O_2^{\bullet -}$ is biologically significant, for at least two reasons. First, $NO^\bullet$ and $O_2^{\bullet -}$ can antagonize each other's biological actions. It has been known for years that $NO^\bullet$-mediated effects are often enhanced by adding SOD to physiological preparations.[48] This was first demonstrated in bioassays of EDRF: adding SOD slowed the rate of loss of the relaxing factor. *In vivo*, excess $O_2^{\bullet -}$ production in or close to vascular endothelium can cause vasoconstriction. Indeed, excess endothelial $O_2^{\bullet -}$ generation has been suggested to be one factor causing **hypertension** (abnormally high blood pressure).[50] Similarly, if an injury system is $O_2^{\bullet -}$-dependent, $NO^\bullet$ can protect by removing $O_2^{\bullet -}$. $NO^\bullet$ can inhibit lipid peroxidation in some systems by removing chain-propagating $RO_2^\bullet$ radicals (Chapter 4).[63]

A second consideration is the consequences of $ONOO^-$ generation. One molecular form of protonated $ONOO^-$ (ONOOH) is a powerfully oxidizing

Table 2.12. Some methods for laboratory preparation of peroxynitrite

Method	Basis of method	Comments
Reaction of H_2O_2 with HNO_2, made from $NaNO_2$ and HCl	$HNO_2 + H_2O_2 \rightarrow HOONO + H_2O$; HOONO has a $t_{1/2}$ of ~ 1 s but can be stabilized by rapidly adding an excess of NaOH	Currently the most popular method. Product contaminated with Cl^-, NO_2^-, NaOH and H_2O_2. 'Decomposed control' often made by adding solution to phosphate (or other) buffer at pH 7.4; $ONOO^-$ protonates to ONOOH and decomposes within seconds whereas other products will remain. Alternatively, H_2O_2 may be removed using manganese dioxide (MnO_2), although with loss of some $ONOO^-$ and contamination by metal ions
Hydroxylamine method	NH_2OH in alkali plus EDTA or DTPA is bubbled with O_2 and oxidized to NO^-, which reacts with O_2 to give $ONOO^-$ $NH_2OH + OH^- + O_2 \rightarrow$ $H_2O_2 + NO^- + H_2O$	H_2O_2 also produced. Some NO_2^-, alkali and NH_2OH remain as contaminants
Irradiation of NO_3^-	Exposure of $NaNO_3$ or KNO_3 to short-wavelength UV light causes some isomerization to peroxynitrite	Poor yield ($\sim 0.3\%$), some NO_2^- also formed. High NO_3^- levels remain. Formation of peroxynitrite from NO_3^- in Martian soil may have caused false-positive results in tests for 'life' carried out by the *Viking* lander probe
Continuous generation Ozone/azide system	Several methods, most popular is decomposition of SIN-1 $N_3^- + 2O_3 \rightarrow ONOO^- + N_2O + O_2$ Solution of sodium azide bubbled with O_3 at 0–4 °C in weakly alkaline solution	Generates both $O_2^{\bullet-}$ and $NO^\bullet$ in equimolar amounts (Fig. 2.8) Solutions are less alkaline than the above methods and contain little, if any H_2O_2, but presence of unreacted azide can be a problem in biological systems
Reaction of H_2O_2 with alkyl nitrites	H_2O_2 in alkali reacts with RONO (e.g. isoamyl nitrite) $H_2O_2 + OH^- +$ $RONO \rightarrow ONOO^- + ROH + H_2O$	An alcohol (ROH) contaminates the reaction mixture. One approach to avoid this is to mix alkaline H_2O_2 with a water-insoluble (e.g. isoamyl) nitrite; $ONOO^-$ stays in the aqueous phase whereas isoamyl alcohol remains in the organic phase and can be separated. Low NO_2^- and H_2O_2 contamination if H_2O_2 and alkyl nitrite used in equimolar amounts
Reaction of $NO^\bullet$ with potassium superoxide	Gaseous $NO^\bullet$ is passed over solid KO_2 $KO_2 + NO^\bullet \rightarrow K^+ONOO^-$	Product can be extracted into alkaline solution, but dismutation of unreacted KO_2 can generate some H_2O_2
Reaction of $O_2^{\bullet-}$ with nitrosothiols	$RSNO + O_2^{\bullet-} \rightarrow RSH + NO^\bullet + O_2$ $O_2^{\bullet-} + NO^\bullet \rightarrow ONOO^-$	Superoxide decomposes RSNO to $NO^\bullet$, which then reacts with another $O_2^{\bullet-}$. See *J. Biol. Chem.* **273**, 7828 (1998).

For references see *Methods Enzymol.* **233**, 229 (1994), *Methods Enzymol.* **269**, 285 (1996) and *Prog. Inorg. Chem.* **41**, 599 (1994).

cytotoxic species ($E^{\circ\prime}$ approaching that of $OH^\bullet$; Table 2.8). The pK_a of $ONOO^-$ is very difficult to determine because of the rapid loss of ONOOH: literature values range from 5.1 to 7.0. Addition of $ONOO^-$ to cells, tissues or body fluids will lead to its rapid protonation, followed by ONOOH-dependent depletion of −SH groups and other antioxidants, oxidation of lipids,[33] DNA strand breakage, nitration and deamination of DNA bases (especially guanine)[90] and nitration of aromatic amino acid residues in proteins. The most studied reaction in proteins has been conversion of tyrosine to **3-nitrotyrosine**,[4] but tryptophan and phenylalanine can also be nitrated. Methionine is oxidized to its sulphoxide; addition of $ONOO^-$ (like that of HOCl) causes inactivation of α_1-antiproteinase. Nitration of tyrosine residues *in vivo* is widely used as a bioassay indicative of $ONOO^-$ generation (Chapter 5). Nitration of tyrosine residues could conceivably lead to enzyme inactivation (e.g. decreased activity of *E. coli* glutamine synthetase and bovine glutathione reductase after exposure to $ONOO^-$ has been reported) and to interference with signal transduction, e.g. nitration of tyrosines might block phosphorylation by tyrosine kinases.[6,29,32] Other protein targets of attack by ONOOH may include glutathione transferases, manganese SOD, structural proteins such as actin and neurofilament L, prostacyclin synthase[91] and the copper transport protein **caeruloplasmin**, which is attacked by ONOOH to cause release of copper ions.[78]

How does peroxynitrite cause damage?[4,61]

The $ONOO^-$ anion is fairly unreactive and solutions of it in alkali are stable for weeks if kept frozen. Peroxynitrite is usually synthesized in the laboratory by reacting nitrite with H_2O_2 at low pH (Table 2.12)

$$NO_2^- + H^+ \rightarrow HONO \text{ (nitrous acid)}$$

$$HONO + H_2O_2 \rightarrow HOONO + H_2O$$

and immediately stabilizing the $ONOO^-$ by adding excess sodium hydroxide. These solutions can be standardized by using the absorbance of $ONOO^-$ at 302 nm in alkali ($\varepsilon = 1670\,M^{-1}\,cm^{-1}$). However, one must be aware that the '$ONOO^-$' obtained is contaminated with NO_2^-, NO_3^- and H_2O_2 as well as excess alkali (causing large pH rises if it is added to poorly buffered solutions). Problems also exist with other laboratory methods of making $ONOO^-$ (Table 2.12). A popular control experiment is to compare the effect of the $ONOO^-$ preparation with 'decomposed $ONOO^-$'; $ONOO^-$ is added to buffer at physiological pH. After a few minutes, all the $ONOO^-$ has decomposed whereas the other products remain. However, the possibility that two agents could be involved in causing damage must not be forgotten (a cytotoxic effect could involve both H_2O_2 and $ONOO^-$, for example). Reaction of $ONOO^-$ with H_2O_2 at pH 7.4 has been reported to generate $^1O_2$. Indeed, CO_2 reacts with $ONOO^-$ ($k = 3 \times 10^4\,M^{-1}\,s^{-1}$) to form an adduct, **nitrosoperoxycarbonate**, $O=N-OOCO_2^-$, which may account for some of the reactions

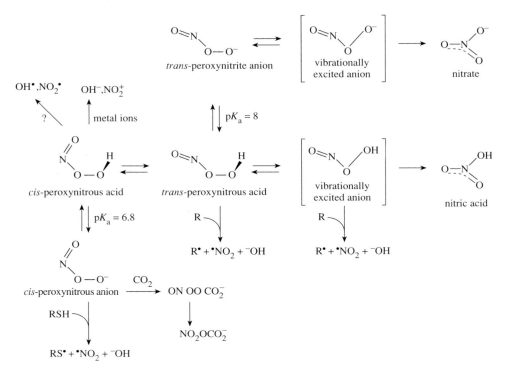

Fig. 2.15. The complexities of peroxynitrite. Homolytic fission of ONOOH to OH$^\bullet$ and NO$_2^\bullet$ seems, at best, only a minor reaction pathway. Certain metal complexes might cause heterolytic fission to NO$_2^+$ and OH$^-$, explaining why they accelerate nitration. At pH 7.4 about 20% of ONOO$^-$ will exist as ONOOH (pK_a=6.8), which rapidly decomposes. Species able to cause damaging oxidations include *cis*-ONOO$^-$ and *cis*-ONOOH (limited reactivity, but might oxidize, for example, −SH compounds), the *trans* isomers and the excited states of those *trans* isomers, especially of ONOOH. Activated *trans*-ONOOH has a reduction potential close to that of OH$^\bullet$ (Table 2.8) and so is potentially highly oxidizing. Some added biomolecules (including thiols) may be oxidized by these various species to radicals, generating NO$_2^\bullet$ also, e.g. *trans*-ONOOH+R → R$^\bullet$+NO$_2^\bullet$+OH$^-$. Diagram adapted from Augusto, O and Radi, R (1996) In *Biothiols in Health and Disease* (ed. Packer, L and Cadenas, E), p. 89, by courtesy of Drs Radi and Augusto and the publishers. Reaction of CO$_2$ with ONOO$^-$ appears to generate O=N−OOCO$_2^-$, which has been suggested to rearrange to form the reactive nitrocarbonate ion (*Arch. Biochem. Biophys.* **327**, 335 (1996)).

attributed to ONOO$^-$ (Fig. 2.15), such as nitration of tyrosine, and of guanine in DNA.[90] Animal body fluids have high concentrations of CO$_2$/HCO$_3^-$ (25 mM HCO$_3^-$ in human blood plasma) suggesting that this may be an important fate of ONOO$^-$ *in vivo*.

At physiological pH, ONOO$^-$ rapidly protonates to peroxynitrous acid (HO−O−N=O) and rearranges, forming nitrate as the major end-product. Some NO$_2^-$ is also generated. Addition of ONOO$^-$ to aromatic compounds leads to both hydroxylation and nitration. Initially, it was suggested that

ONOOH undergoes homolytic fission to generate $OH^{\bullet}$ and nitrogen dioxide radicals

$$ONOOH \rightarrow NO_2^{\bullet} + OH^{\bullet}$$

However, some thermodynamic arguments suggest that this reaction is disfavoured, whereas others suggest it is a preferred reaction pathway.[47a] Experimental data are equally confused. Some groups appear to have detected $OH^{\bullet}$ from decomposing $ONOO^-$ at physiological pH, but assays can be confounded because $ONOO^-$ interferes with the methods, e.g. spin-trapping of $OH^{\bullet}$ with DMPO is affected because of direct reaction of ONOOH with the spin adduct; products of aromatic hydroxylation can be oxidized by $ONOO^-$; Chapter 4). Even when these problems have been overcome, the amount of $OH^{\bullet}$ detected appears to be small[43] or zero, depending on the research group. Some other $OH^{\bullet}$ traps may well be oxidized to '$OH^{\bullet}$-like products' by ONOOH.

Addition of certain metal chelates such as Fe(III)–EDTA or the enzyme copper–zinc superoxide dismutase (CuZnSOD) can increase the yield of aromatic nitrated products, perhaps by promoting heterolytic fission of ONOOH to **nitronium ion**, an excellent nitrating species. Bovine CuZnSOD can be nitrated by $ONOO^-$ addition, for example

$$ONOOH \rightarrow NO_2^+ + OH^-$$

An additional proposal is that an 'activated transition state form' of peroxynitrous acid on the pathway to NO_3^- is the major reactive oxidizing/hydroxylating/nitrating species. It has been suggested that ONOOH exists in *cis* and *trans* forms[g] with different pK_a values (6.8 and 8.0) and that an excited state of the *trans* isomer is the active species. The more stable *cis* form predominates at high pH, but protonation removes repulsion of the negative charges and allows formation of more of the *trans* form, which has the higher pK_a and so can re-ionize. The *trans* forms of ONOOH and $ONOO^-$ can undergo transition to a vibrationally excited state, which involves bending of the N–O–O bond and lengthening of the O–O bond. These activated intermediates can cause damage and/or rearrange to nitrate/nitric acid. The complexities of $ONOO^-$ chemistry are summarized in Fig. 2.15.

Toxicity of nitrotyrosine?

Nitrotyrosine itself could be toxic, e.g. by undergoing redox cycling (Section 8.14.1), by interfering with signal transduction or by becoming incorporated into the microtubule protein *tubulin* and distorting the cytoskeleton, leading eventually to cell death (J. Eiserich, personal communication).

Nitric oxide and superoxide: a balance

By a series of complex mechanisms, $ONOO^-$ can lead to biological damage. Just as $O_2^{\bullet -}$ reacts rapidly with $NO^{\bullet}$, it might also react with $NO_2^{\bullet}$, so yet other species could exist

$$O_2^{\bullet -} + NO_2^{\bullet} \rightarrow O_2NOO^-$$

Nevertheless, the presence of excess $NO^{\bullet}$ can sometimes inhibit damage by $ONOO^-$. For example, whereas equal fluxes of $O_2^{\bullet-}$ and $NO^{\bullet}$ can stimulate lipid peroxidation (via $ONOO^-$), a high ratio of $NO^{\bullet}$ to $O_2^{\bullet-}$ can decrease the peroxidation because of the ability of $NO^{\bullet}$ to scavenge peroxyl and alkoxyl radicals.[63] Such reactions yield nitro, nitroso and nitrated lipid adducts, whose biological significance remains to be elucidated. Thus, for example, in the normal vascular endothelium the $NO^{\bullet}/O_2^{\bullet-}$ ratio experienced by the endothelial cells is probably high. Lowering that ratio might favour pro-oxidant (via $ONOO^-$) rather than antioxidant effects of $NO^{\bullet}$. Peroxynitrite can react with thiols to generate some $NO^{\bullet}$ (Section 3.9). By contrast, exposure of mitochondria to excess $NO^{\bullet}$ inhibits cytochrome oxidase, and the back-up of electrons in the electron transport chain could then result in more $O_2^{\bullet-}$ being produced by leakage of electrons from the electron carriers.[60] A result would be $ONOO^-$ formation, and damage to other sites in mitochondria. $NO^{\bullet}$ has also been reported to suppress phagocyte adherence to vascular endothelium and decrease phagocyte $O_2^{\bullet-}$ production (Chapter 9), to lead to inhibition of some antioxidant defence enzymes (e.g. glutathione peroxidase) and to activation of others (e.g. haem oxygenase), to inhibit P450s (and thus their potential to generate ROS)[30] and to diminish activation of NF-κB (Chapters 3 and 4). Nothing is simple in this area!

References

1. Albina, JE (1995) On the expression of nos by human macrophages. Why no NO? *J. Leuk. Biol.* **58**, 643.
2. Aruoma, OI and Halliwell, B (1987) Superoxide-dependent and ascorbate-dependent formation of hydroxyl radicals from hydrogen peroxide in the presence of iron. Are lactoferrin and transferrin promoters of hydroxyl radical generation? *Biochem. J.* **241**, 273.
3. Aruoma, OI *et al.* (1995) Reaction of plant-derived and synthetic antioxidants with trichloromethylperoxyl radicals. *Free Rad. Res.* **22**, 187.
4. Beckman, JS and Koppenol, WH (1996) $NO^{\bullet}$, $O_2^{\bullet-}$ and $ONOO^-$: the good, the bad, and the ugly. *Am. J. Physiol.* **271**, C1424.
5. Bielski, BHJ and Cabelli, DE (1995) Chapter 3 in Foote, CS *et al.*, eds (1995) *Active Oxygen in Chemistry.* Blackie, London.
6. Berlett, BS *et al.* (1996) Peroxynitrite-mediated nitration of tyrosine residues in *E. coli* glutamine synthetase mimics adenylation: relevance to signal transduction. *Proc. Natl Acad. Sci. USA* **93**, 1776.
7. Bosca, F *et al.* (1997) Photophysical and photochemical characterization of a photosensitizing drug: a combined steady state photolysis and laser flash photolysis study on carprofen. *Chem. Res. Toxicol.* **10**, 820.
8. Brodie, AE and Reed, DJ (1987) Reversible oxidation of glyceraldehyde 3-phosphate dehydrogenase thiols in human lung carcinoma cells by H_2O_2. *Biochem. Biophys. Res. Commun.* **148**, 120.
9. Brown, GC and Cooper, CE (1994) Nanomolar concentrations of NO reversibly inhibit synaptosomal respiration by competing with O_2 at cytochrome oxidase. *FEBS Lett.* **356**, 295.

10. Buettner, GR (1993) The pecking order of free radicals and antioxidants: lipid peroxidation , α-tocopherol and ascorbate. *Arch. Biochem. Biophys.* **300**, 535.

11. Butler, J and Land, E (1996) Pulse radiolysis. In Punchard, NA and Kelly, F (eds) *Free Radicals. A Practical Approach*, p. 47. IRL Press, Oxford.

12. Carr, AC *et al.* (1996) Peroxidase-mediated bromination of unsaturated fatty acids to form bromohydrins. *Arch. Biochem. Biophys.* **327**, 227.

13. Chochola, J *et al.* (1995) Release of H_2O_2 from human T cell lines and normal lymphocytes co-infected with HIV-1 and mycoplasma. *Free Rad. Res.* **23**, 197.

14. Clare, NJ (1955) Photosensitisation in animals. *Adv. Vet. Sci. Comp. Med.* **2**, 182.

15. Connolly, DL *et al.* (1996) Water channels in health and disease. *Lancet* **347**, 210.

16. Crum, LA *et al.* (1987) Free radical production in amniotic fluid and blood plasma by medical ultrasound. *J. Ultrasound Med.* **6**, 643.

17. Daniels, V (1984) The Russell effect—a review of its possible uses in conservation and the scientific examination of materials. *Stud. Conserv.* **29**, 57.

18. Di Mascio, P *et al.* (1994) Singlet molecular O_2 production in the reaction of peroxynitrite with H_2O_2. *FEBS Lett.* **355**, 287.

19. Dykhuizen, RS *et al.* (1996) Antimicrobial effect of acidified NO_2^- on gut pathogens: importance of dietary NO_3^- in host defense. *Antimicrob. Ag. Chemother.* **40**, 1422.

20. Egan, TJ *et al.* (1992) Catalysis of the Haber–Weiss reaction by Fe–DETAPAC. *J. Inorg. Biochem.* **48**, 241.

21. Eiserich, JP *et al.* (1996) Formation of nitrating and chlorinating species by reaction of NO_2^- with HOCl. *J. Biol. Chem.* **271**, 19199.

22. Ennever, JF *et al.* (1987) Rapid clearance of a structural isomer of bilirubin during phototherapy. *J. Clin. Invest.* **79**, 1674.

23. Feierabend, J and Engel, S (1986) Photoinactivation of catalase *in vitro* and in leaves. *Arch. Biochem. Biophys.* **251**, 567.

24. Fliss, H and Menard, M (1991) HOCl-induced mobilization of zinc from metalloproteins. *Arch. Biochem. Biophys.* **287**, 175.

25. Folkes, LK *et al.* (1995) Kinetics and mechanisms of HOCl reactions. *Arch. Biochem. Biophys.* **323**, 120.

26. Foote, CS *et al.* (eds) (1995) *Active Oxygen in Chemistry*. Blackie, London.

27. Ford, PC *et al.* (1993) Autoxidation kinetics of aqueous NO. *FEBS Lett.* **326**, 1.

28. Forni, LG and Willson, RL (1986) Thiyl and phenoxyl free radicals and NADH. *Biochem. J.* **240**, 897.

29. Francescutti, D *et al.* (1996) Peroxynitrite modification of glutathione reductase: modeling studies and kinetic evidence suggest the modification of tyrosines at the GSSG binding site. *Protein Eng.* **9**, 189.

30. Gergel, D *et al.* (1997) Inhibition of rat and human CYP2E1 catalytic activity and reactive O_2 radical formation by NO. *Arch. Biochem. Biophys.* **337**, 239.

31. Glaser, CB *et al.* (1992) Oxidation of a specific methionine in thrombomodulin by activated neutrophil products blocks cofactor activity. *J. Clin. Invest.* **90**, 2565.

32. Gow, AJ *et al.* (1996) Effect of $ONOO^-$ induced protein modifications on tyrosine phosphorylation and degradation. *FEBS Lett.* **385**, 63.

33. Graham, A *et al.* (1993) Peroxynitrite modification of LDL leads to recognition by the macrophage scavenger receptor. *FEBS Lett.* **330**, 181.

34. Gutteridge, JMC (1986) Iron promoters of the Fenton reaction and lipid peroxidation can be released from haemoglobin by peroxides. *FEBS Lett.* **201**, 291.

35. Gutteridge, JMC (1990) $O_2^{\bullet-}$-dependent formation of $OH^{\bullet}$ from Fe(III) complexes and H_2O_2: an evaluation of 14 iron chelators. *Free Rad. Res. Commun.* **9**, 119.

36. Gutteridge, JMC *et al.* (1990) ADP–Fe as a Fenton reactant: radical reactions detected by spin trapping, H abstraction and aromatic hydroxylation. *Arch. Biochem. Biophys.* **277**, 422.

37. Halliwell, B and Gutteridge, JMC (1992) Biologically relevant metal ion-dependent $OH^{\bullet}$ generation. An update. *FEBS Lett.* **307**, 108.

38. Hannum, DM *et al.* (1995) Subunit sites of oxidative inactivation of *E. coli* F_1-ATPase by HOCl. *Biochem. Biophys. Res. Commun.* **212**, 868.

39. Hassan, T and Khan, AU (1986) Phototoxity of the tetracyclines: photo-sensitized emission of singlet delta oxygen. *Proc. Natl Acad. Sci. USA* **83**, 4604.

40. Heinecke, JW *et al.* (1994) Cholesterol chlorohydrin synthesis by the myeloperoxidase–$H_2O_2^-Cl^-$ system: potential markers for lipoproteins oxidatively damaged by phagocytes. *Biochemistry* **33**, 10127.

40a. Ingold, KU *et al.* (1997) Invention of the first azo compound to serve as a $O_2^{\bullet-}$ thermal source under physiological conditions. *J. Am. Chem. Soc.* **119**, 12364.

41. Jin, F *et al.* (1993) The $O_2^{\bullet-}$ radical reacts with tyrosine-derived phenoxyl radicals by addition rather than by electron transfer. *J. Chem. Soc. Perkin Trans.* **2**, 1583.

42. Kanofsky, JR (1989) Bromine derivatives of amino acids as intermediates in the peroxidase-catalyzed formation of singlet O_2. *Arch. Biochem. Biophys.* **274**, 229.

43. Kaur, H *et al.* (1997) Peroxynitrite-dependent aromatic hydroxylation and nitration of salicylate and phenylanine. Is $OH^{\bullet}$ involved? *Free Rad. Res.* **26**, 71.

44. Koppenol, WH and Butler, J (1985) Energetics of interconversion reactions of oxyradicals. *Adv. Free Rad. Biol. Med.* **1**, 91.

45. Korycka-Dahl, M and Richardson, T (1977) Photogeneration of $O_2^{\bullet-}$ in serum of bovine milk and in model systems containing riboflavin and amino acids. *J. Dairy Sci.* **61**, 400.

46. Lepoivre, M *et al.* (1994) Quenching of the tyrosyl free radical of ribonucleotide reductase by NO. *J. Biol. Chem.* **269**, 21891.

47. Martin, JP and Logsdon, N (1987) The role of oxygen radicals in dye-mediated photodynamic effects in *E. coli*. B. *J. Biol. Chem.* **262**, 7213.

47a. Merényi, G and Lind, J (1998) Free radical formation in the peroxynitrous acid (ONOOH)/peroxynitrite ($ONOO^-$) system. *Chem. Res. Tox.* **11**, 243.

48. Moncada, S and Higgs, EA (1995) Molecular mechanisms and therapeutic strategies related to NO. *FASEB J.* **9**, 1319.

49. Morgan, TR *et al.* (1988) Free radical production by high-energy shock waves—comparison with ionizing radiation. *J. Urol.* **139**, 186.

50. Nakazono, K *et al.* (1991) Does $O_2^{\bullet-}$ underlie the pathogenesis of hypertension? *Proc. Natl Acad. Sci. USA* **88**, 10045.

51. Nanni, EJ *et al.* (1980) Does $O_2^{\bullet-}$ oxidize catechol, α-tocopherol and ascorbic acid by direct electron transfer? *J. Am. Chem. Soc.* **102**, 4481.

52. Naseem, KM and Bruckdorfer, KR (1995) H_2O_2 at low concentrations strongly enhances the inhibitory effect of NO on platelets. *Biochem. J.* **310**, 149.

53. Niki, E (1987) Antioxidants in relation to lipid peroxidation. *Chem. Phys. Lipids* **44**, 227.

54. Parshad, R *et al.* (1978) Fluorescent light-induced chromosome damage and its prevention in mouse cells in culture. *Proc. Natl Acad. Sci. USA* **75**, 1830.

55. Pathak, MA and Joshi, PC (1984) Production of active oxygen species ($^1O_2$ and O_2^-) by psoralens and ultraviolet radiation. *Biochim. Biophys. Acta* **798**, 115.

56. Paya, M *et al.* (1992) Peroxyl radical scavenging by a series of coumarins. *Free Rad. Res. Commun.* **17**, 293.

57. Penning, LC and Dubbelman, TMAR (1994) Fundamentals of photodynamic therapy: cellular and biochemical aspects. *Anti-Cancer Drugs* **5**, 139.

58. Pero, RW *et al.* (1996) Hypochlorous acid/N-chloramines are naturally produced DNA repair inhibitors. *Carcinogenesis* **17**, 13.

59. Peters, JW *et al.* (1975) An investigation of potassium perchromate as a source of singlet O_2. *J. Am. Chem. Soc.* **97**, 3299.

60. Poderoso, JJ *et al.* (1996) NO inhibits electron transfer and increases $O_2^{\bullet-}$ production in rat heart mitochondria and submitochondrial particles. *Arch. Biochem. Biophys.* **328**, 85.

61. Pryor, WA and Squadrito, GL (1995) The chemistry of peroxynitrite: a product from the reaction of NO with $O_2^{\bullet-}$. *Am. J. Physiol.* **268**, L699.

62. Riesz, P *et al.* (1990) Sonochemistry of volatile and non–volatile solutes in aqueous solutions: epr and spin trapping studies. *Ultrasonics* **28**, 295.

63. Rubbo, H *et al.* (1995) Inhibition of lipoxygenase-dependent liposome and LDL oxidation: termination of radical chain propagation reactions and formation of nitrogen-containing oxidized lipids. *Arch. Biochem. Biophys.* **324**, 15.

64. Rush, JD *et al.* (1995) The oxidation of phenol by ferrate(VI) and ferrate(V). A pulse radiolysis and stopped-flow study. *Free Rad. Res.* **22**, 349.

65. Rush, JD *et al.* (1996) Reaction of ferrate(VI)/ferrate(V) with H_2O_2 and $O_2^{\bullet-}$ — a stopped-flow and premix pulse radiolysis study. *Free Rad. Res.* **24**, 187.

66. Saez, G *et al.* (1982) The production of free radicals during the autoxidation of cysteine and their effect on isolated rat hepatocytes. *Biochim. Biophys. Acta* **719**, 24.

67. Sawyer, DT and Valentine, JS (1981) How super is superoxide? *Acc. Chem. Res.* **14**, 393.

68. Schäcker, M *et al.* (1991) Oxidation of Tris to 1-carbon compounds in a radical-producing model system, in microsomes, in hepatocytes and in rats. *Free Rad. Res. Commun.* **11**, 339.

69. Schöneich, C *et al.* (1989) Thiyl radical attack on polyunsaturated fatty acids: a possible route to lipid peroxidation. *Biochem. Biophys. Res. Commun.* **101**, 113.

70. Schraufstatter, IU *et al.* (1990) Mechanisms of hypochlorite injury to target cells. *J. Clin. Invest.* **85**, 554.

70a. Sharpe, ME and Cooper, CE (1998) Reactions of NO with mitochondrial cytochrome *c*: a novel mechanism for the formation of NO^- and $ONOO^-$. *Biochem. J.* **332**, 9.

71. Siess, M-H *et al.* (1996) Flavonoids and honey and propolis. *J. Agr. Food Chem.* **44**, 2297.

72. Snyder, SH (1995) No endothelial NO. *Nature* **377**, 196.

73. Soriani, M *et al.* (1994) Antioxidant potential of anaerobic human plasma: role of serum albumin and thiols as scavengers of carbon radicals. *Arch. Biochem. Biophys.* **312**, 180.

74. Spencer, JPE *et al.* (1995) DNA damage in human respiratory tract epithelial cells. *FEBS Lett.* **375**, 179.

75. Stamler, JS (1994) Redox signaling: nitrosylation and related target interactions of NO. *Cell* **78**, 931.

76. Sugimoto, H *et al.* (1987) Oxygenation of polychloro aromatic hydrocarbons by $O_2^{\bullet-}$ in aprotic media. *J. Am. Chem. Soc.* **109**, 8081.
77. Sutton, HC and Winterbourn, CC (1989) On the participation of higher oxidation states of iron and copper in Fenton reactions. *Free Rad. Biol. Med.* **6**, 53.
78. Swain, JA *et al.* (1994) Peroxynitrite releases copper from caeruloplasmin: implications for atherosclerosis. *FEBS Lett.* **342**, 49.
79. Symons, MCR (1996) Radicals generated by bone cutting and fracture. *Free Rad. Biol. Med.* **20**, 831.
79a. Tatoyan, A and Giulivi, C (1998) Purification and characterization of a NOS from rat liver mitochondria. *J. Biol. Chem.* **273**, 11044.
80. Uematsu, M *et al.* (1995) Regulation of endothelial cell nos mRNA expression by shear stress. *Am. J. Physiol.* **269**, C1371.
81. van den Berg, JJM *et al.* (1993) HOCl-mediated modification of cholesterol and phospholipid: analysis of reaction products by gas-chromatography–mass spectrometry. *J. Lipid Res.* **34**, 2005.
82. von Sonntag, C (1987) *The Chemical Basis of Radiation Biology.* Taylor & Francis, London.
83. Wagner, JR *et al.* (1993) The oxidation of blood plasma and LDL components by chemically generated singlet O_2. *J. Biol. Chem.* **268**, 18502.
83a. Walling, C *et al.* (1975) Kinetics of the decomposition of H_2O_2 catalyzed by ferric EDTA complex. *Proc. Natl. Acad. Sci. US* **72**, 140.
84. Wardman, P and von Sonntag, C (1995) Kinetic factors that control the fate of thiyl radicals in cells. *Methods Enzymol.* **251**, 31.
84a. Weiss, SJ (1989) Tissue destruction by neutrophils. *N. Engl. J. Med.* **320**, 365.
85. Wennmalm, A *et al.* (1993) Metabolism and excretion of NO in humans. *Circul. Res.* **73**, 1121.
86. Wheeler, MA *et al.* (1997) Bacterial infection induces NOS in human neutrophils. *J. Clin. Invest.* **99**, 110.
87. White, JW *et al.* (1963) The identification of inhibine, the antibacterial factor in honey. *Biochim. Biophys. Acta* **73**, 57.
88. Wink, DA *et al.* (1996) The effect of various NO-donor agents on H_2O_2-mediated toxicity. *Arch. Biochem. Biophys.* **331**, 241.
89. Wolcott, RG *et al.* (1994) Bactericidal potency of $OH^{\bullet}$ in physiological environments. *J. Biol. Chem.* **269**, 9721.
90. Yermilov, V *et al.* (1996) Effects of CO_2/HCO_3^- on induction of DNA single-strand breaks and formation of 8-nitroguanine, 8-oxoguanine and base-propenal mediated by $ONOO^-$. *FEBS Lett.* **399**, 67.
91. Zou, MH and Ullrich, V (1996) Peroxynitrite formed by simultaneous generation of $NO^{\bullet}$ and $O_2^{\bullet-}$ selectively inhibits bovine aortic prostacyclin synthase. *FEBS Lett.* **382**, 101.

Notes

[a]See Ref. 144 in Chapter 3 and *Methods Enzymol.* **286**, 1 (1990).
[b]This term is explained in Appendix I, Section A1.4.
[c]For further explanation see Appendix I, Section A1.1.
[d]Also see Li *et al.* (1995) in the reference list of Chapter 4.
[e]For a discussion of the metabolic significance of this enzyme see Fig. 1.5.
[f]This term is explained in Section 4.16.
[g]This term is explained in Section 4.9.

3

Antioxidant defences

3.1 Introduction

We saw in Chapter 1 that O_2 is a poisonous molecule: aerobes only survive in its presence because they have evolved **antioxidant defences**. Defence against reactive nitrogen species (RNS) such as peroxynitrite is also needed.

Some mobile organisms avoid O_2 toxicity by swimming away from regions of high O_2 tension. In several bacteria, including *Salmonella typhimurium* and *Escherichia coli*, there is an intracellular redox sensor which measures the redox state of the constituents of the respiratory chain and transmits a signal to the flagellae involved in swimming. Another adaptation may have been to evolve electron-transport chains that minimize $O_2^{\bullet -}$ production by packing redox constituents together in such a way that makes 'escape' of electrons to O_2 less likely. For example, cytochrome oxidase catalyses the stepwise four-electron reduction of O_2 to $2H_2O$ without releasing reactive oxygen species (ROS; Section 1.4.3). Yet another adaptation in multicellular organisms may have been to use the minimum intra-corporeal O_2 levels that permit aerobic respiration to operate. Intracellular (and especially intramitochondrial) O_2 concentrations are much lower than ambient O_2, which may tend to decrease oxidative damage *in vivo* (Chapter 1). Food manufacturers exploit this technique when they seal foods under nitrogen or in vacuum packs.

Antioxidant defences comprise:

(a) agents that catalytically remove free radicals and other 'reactive species'. Examples are the enzymes superoxide dismutase, catalase, peroxidase and 'thiol-specific antioxidants'.
(b) proteins that minimize the availability of pro-oxidants such as iron ions, copper ions and haem. Examples are transferrins, haptoglobins, haemopexin and metallothionein. This category includes proteins that oxidize ferrous ions, such as caeruloplasmin.
(c) proteins that protect biomolecules against damage (including oxidative damage) by other mechanisms, e.g. heat shock proteins.
(d) low-molecular-mass agents that scavenge ROS and RNS. Examples are glutathione, α-tocopherol and (possibly) bilirubin and uric acid. Some low-molecular-mass antioxidants come from the diet, especially ascorbic acid and α-tocopherol. There is an intimate relationship between nutrition and antioxidant defence (Chapter 10).

The composition of antioxidant defences differs from tissue to tissue and cell-type to cell-type (possibly even from cell to cell of the same type) in a given tissue. Extracellular fluids have different protective mechanisms from the intracellular environment (Section 3.19 below). Antioxidant defences can often be induced by exposure of organisms to ROS/RNS and to cellular signal molecules such as cytokines (Chapter 4). In recent years, scientists have realized that antioxidant defences are incomplete (i.e. they do not prevent damage completely), since oxidative damage to DNA, proteins, lipids and small molecules can be demonstrated in living systems under ambient O_2. Hence even 21% O_2 is toxic (Table 4.1). Thus some authors classify the **repair systems** needed to remove damaged molecules as antioxidant defences. Repair is considered in detail in Chapter 4, since an understanding of how it is achieved requires a knowledge of the chemistry of oxidative damage to different biomolecules, also considered in Chapter 4.

3.1.1   *What is an antioxidant? A problem of definition*

Antioxidant is a term widely used but rarely defined: indeed, it is surprisingly difficult to define clearly. Food technologists use antioxidants to inhibit lipid peroxidation and consequent rancidity in food materials, so they often think of antioxidants as inhibitors of lipid peroxidation. Museum curators use 'anti-oxidants' to preserve organic artefacts.[57] Polymer scientists use 'antioxidants' to control polymerization in the manufacture of rubber, plastics and paint and for the protection of clear plastics against ultraviolet light. Combustion is a free-radical process: the oil industry makes extensive use of antioxidants and a knowledge of free-radical mechanisms in the design of better automobile fuels and lubricating oils. All these scientists have their own views on what a good antioxidant should be.

What about living organisms? When ROS and RNS are generated *in vivo*, many antioxidants come into play.[107] Their relative importance depends upon:

- which ROS/RNS is generated
- how it is generated
- where it is generated, and
- what target of damage is measured.

For example, when human body fluids are exposed to the toxic oxidizing gases ozone (O_3) or nitrogen dioxide (NO_2), uric acid appears to be a protective antioxidant (Chapter 8). By contrast, urate provides little protection against damage to blood plasma constituents by hypochlorous acid (HOCl). If the oxidative stress is the same but a different damage target is measured, different answers result. For example, exposure of human blood plasma to gas-phase cigarette smoke causes peroxidation of plasma lipids, which is inhibited by ascorbate (Chapter 8). By contrast, ascorbate does not protect against damage to plasma proteins by cigarette smoke, as measured by the carbonyl assay. ROS/RNS are capable of damaging many targets *in vivo*, including lipids, proteins and DNA (Chapter 4).

To encompass these various complexities, we have proposed a broad defin-ition of an antioxidant as 'any substance that, when present at low concentra-tions compared with those of an oxidizable substrate, significantly delays or prevents oxidation of that substrate'. The term 'oxidizable substrate' includes every type of molecule found *in vivo*. This definition emphasizes the importance of the damage target studied and the source of ROS/RNS used when antioxidant action is examined *in vitro*. There is also no universal 'best' antioxidant, as illustrated by the cigarette-smoke experiments mentioned above: ascorbate protects well against lipid damage but not against protein damage.

Let us now examine the various antioxidant systems in detail, bearing in mind two questions; (i) what can they do? and (ii) what evidence exists that this action is important *in vivo*?

3.2 Antioxidant defence enzymes: superoxide dismutases

The discovery of superoxide dismutase (SOD) enzymes provided much of the basis for our current understanding of antioxidant defence systems, since it led to the postulation of the superoxide theory of oxygen toxicity (Chapter 1).[89] Hence it is appropriate to consider SODs first.

3.2.1 *Copper–zinc SOD*[29a]

In 1938, Mann and Keilin described a blue-green protein containing copper (**haemocuprein**) that they had isolated from bovine blood. In 1953, a similar protein was isolated from horse liver and named **hepatocuprein**. Other proteins of this type were later isolated, such as **cerebrocuprein** from brain. In 1970, it was discovered that the erythrocyte protein contains zinc as well as copper. No enzymic function was detected in any of these proteins, so it was often suggested that they served as metal stores. However, in 1969 McCord and Fridovich reported that the erythrocyte protein is able to remove the superoxide radical catalytically, i.e. it functions as a superoxide dismutase enzyme.[89] Figure 3.1 illustrates this ability using the technique of pulse radiolysis (Section 2.3.3). Despite an intensive search, no other substrate on which SOD enzymes act catalytically has been discovered, i.e. they appear specific for catalytic removal of superoxide.

Copper–zinc-containing superoxide dismutases (CuZnSODs) are unusually stable. In purifying CuZnSOD from erythrocytes, the cells are lysed and haemoglobin removed by treatment with chloroform and ethanol, followed by centrifugation. The enzyme enters the organic phase, from which it can be precipitated by addition of cold acetone and then further purified by ion-exchange chromatography. Not many enzymes will tolerate such procedures. CuZnSODs are also fairly resistant to heating, to attack by proteases, and to denaturation by such reagents as guanidinium chloride, sodium dodecyl sulphate (SDS) and urea.[89]

CuZnSOD in eukaryotes and prokaryotes

Subsequent studies revealed that CuZnSODs are present in virtually all eukaryotic cells. In animal cells, most CuZnSOD is located in the cytosol, but some appears present in lysosomes, nucleus and the space between inner and

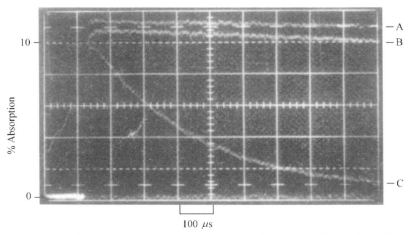

100 μs

Fig. 3.1. The catalytic action of superoxide dismutase as demonstrated by pulse radiolysis. The oscilloscope traces show the decay at pH 8.8 of $O_2^{\bullet -}$ radical (initial concentration 32 μmol/l) as followed by the loss of its absorbance at 250 nm. Trace A, spontaneous dismutation of $O_2^{\bullet -}$; trace C, plus 2 μmol/l of SOD; trace B, as for trace C but SOD boiled for 5 min to destroy enzyme activity. Data from *Biochim. Biophys. Acta* **268**, 605 (1972) by courtesy of Professor G. Rotilio and Elsevier, Amsterdam.

outer mitochondrial membranes. Peroxisomes[a] have also been reported to contain some CuZnSOD.[89,237] CuZnSODs have usually been thought to be much less common in prokaryotic cells such as bacteria or cyanobacteria, although this view may need revision since examples of their presence are accumulating fast. The first to be discovered was a CuZnSOD in the luminescent bacterium *Photobacterium leiognathi*. This organism exists in a symbiotic relationship with the ponyfish, occupying a special gland and imparting a characteristic luminescence to the fish. Comparison of the amino acid composition of the bacterial CuZnSOD with that of higher organisms shows that it is closely related to fish CuZnSOD enzymes. It was therefore initially suggested that the bacterium obtained the gene for its CuZnSOD by gene transfer from its host fish, although there is doubt about this.[89]

CuZnSODs have since been detected in many other bacteria. Thus the free-living (non-symbiotic) bacterium *Caulobacter crescentus* CB15 contains a CuZnSOD, whose sequence suggests that it is related to eukaryotic CuZnSODs with the addition of a 'leader amino acid sequence' typical of exported proteins. This suggests an extracellular location for the CuZnSOD, perhaps the periplasmic space (i.e. the space between the bacterial cell wall and cell membrane). The SOD activity of *E. coli* has been studied for years by many research groups, but only recently has *E. coli* been found to contain low levels of CuZnSOD in its periplasmic space.[23,29] CuZnSOD has also been reported in *Brucella abortus*, *Salmonella typhimurium* and several *Haemophilus* species. It may help to defend the bacteria against external sources of ROS, such as activated phagocytes.[83a] Genes encoding CuZnSOD have been detected using primers[b] in several other bacteria, including *Legionella pneumophila*, the causative agent of **Legionnaires' disease**.

Structure and catalytic ability of CuZnSOD

The CuZnSOD enzymes so far isolated from eukaryotes have relative molecular masses of about 32 000 and contain two protein subunits, each of which bears an active site containing one copper and one zinc ion.[89] A different type of CuZnSOD, **extracellular SOD** (EC-SOD) has been described and is considered in Section 3.19.4 below. Bacterial CuZnSODs, like the eukaryotic enzymes, are usually dimers with one Cu per subunit, although a report that the *E. coli* enzyme is a monomer has appeared.[23]

All CuZnSODs catalyse the same reaction: they greatly accelerate the dismutation of $O_2^{\bullet-}$ (Fig. 3.1)

$$O_2^{\bullet-} + O_2^{\bullet-} + 2H^+ \rightarrow H_2O_2 + O_2 \text{ (ground-state)}$$

Whereas the overall rate constant for the uncatalysed dismutation of $O_2^{\bullet-}$ depends strongly on the pH of the solution (Section 2.4.3) and is about $5 \times 10^5 \, M^{-1} s^{-1}$ at physiological pH, the reaction catalysed by bovine erythrocyte CuZnSOD is almost independent of pH in the range 5.3–9.5, and the rate constant for reaction of $O_2^{\bullet-}$ with the active site is about $1.6 \times 10^9 \, M^{-1} s^{-1}$.

The copper ions in CuZnSODs appear to function in the dismutation reaction by undergoing alternate oxidation and reduction, i.e.

$$\text{Enzyme–Cu}^{2+} + O_2^{\bullet-} \rightarrow \text{Enzyme–Cu}^+ + O_2$$

$$\underline{\text{Enzyme–Cu}^+ + O_2^{\bullet-} + 2H^+ \rightarrow \text{Enzyme–Cu}^{2+} + H_2O_2}$$

$$\text{Net reaction: } O_2^{\bullet-} + O_2^{\bullet-} + 2H^+ \rightarrow H_2O_2 + O_2$$

The Zn^{2+} does not function in the catalytic cycle but helps stabilize the enzyme—this conclusion is drawn from experiments in which the metals are removed from the active sites and replaced either singly or together. Indeed, the regaining of SOD activity on addition of copper ions to the metal-free CuZnSOD enzyme (**apoenzyme**) has been used to measure trace amounts of copper. In general, ions of other transition metals, such as Mn^{2+}, cannot replace the copper to yield a functional enzyme, but cobalt, mercury, or cadmium ions can replace Zn^{2+} in increasing enzyme stability.[89] If the Cu^{2+} is replaced by cobalt ions (Co^{2+}), however, the enzyme has been claimed to still catalyse $O_2^{\bullet-}$ dismutation,[188] although with a rate constant of only $4.8 \times 10^6 \, M^{-1} s^{-1}$.

CuZnSOD structure

The complete amino acid sequences (and, in some cases, three-dimensional-structures) of CuZnSODs from several plants and animals have been determined and in general they are all very similar.[89] Bovine erythrocyte CuZnSOD was the first CuZnSOD to have its detailed structure determined. Each of the

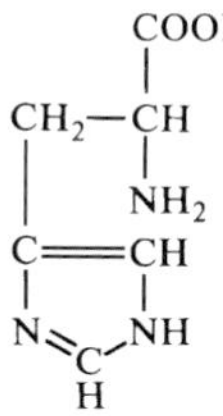

$$
\begin{array}{c}
\text{COOH} \\
| \\
\text{CH}_2\text{—CH} \\
|\qquad| \\
\qquad\text{NH}_2 \\
\text{C}\!\!=\!\!\text{CH} \\
|\qquad| \\
\text{N}\!\!\underset{\displaystyle\underset{\text{H}}{\text{C}}}{\diagdown}\!\!\text{NH}
\end{array}
$$

Fig. 3.2. Structure of the amino acid histidine. The ring structure is known as the **imidazole ring** and contains two nitrogen atoms. Each has five electrons in its outermost shell, three of which are being used in covalent bonding. The remaining two constitute a **lone pair** (Appendix I), and can act as ligands to metal ions.

two subunits is composed primarily of eight antiparallel strands of β-pleated sheet structure that form a flattened cylinder, plus 'loops'. The copper ion is held at the active site by interaction with the nitrogens in the imidazole ring structures (Fig. 3.2) of four histidine residues (numbers 44, 46, 61 and 118 in the amino acid sequence), whereas the zinc ion is bridged to the copper by interaction with the imidazole of His61 and it also interacts with His69 and His78 and the carboxyl ($-COO^-$) group of Asp81. His61, which interacts with both metals, may be involved in supplying the protons needed for the dismutation reaction. Human CuZnSOD shows a similar structure: it is a dimeric enzyme with ellipsoidal dimensions of about $30 \times 40 \times 70\,\text{Å}$. Each subunit contains 153 amino acid residues plus one zinc and one copper ion. Figure 3.3 (see plate section) shows aspects of the subunit structure.

In CuZnSODs, much of the surface of each subunit is negatively charged, repelling $O_2^{\bullet-}$, except for positively charged 'tracks' which lead into the active site (Fig. 3.4, see plate section). A similar arrangement probably exists in the manganese and iron SODs (see below). Hence $O_2^{\bullet-}$ approaching a subunit seems to be 'guided' into the active site. Chemical modification of these positively charged amino acid side-chains decreases enzyme activity, and high ionic strength also diminishes enzyme activity by interfering with this **electrostatic facilitation** of $O_2^{\bullet-}$ dismutation. Variants of SOD have been developed in the laboratory that work faster than the wild-type enzyme by enhancing this mechanism (Fig. 3.4, part a).

Inhibitors of CuZnSOD

Cyanide is a powerful inhibitor of CuZnSODs. These enzymes are also inactivated on incubation with diethyldithiocarbamate

$$
(\text{CH}_3\text{CH}_2)_2\text{N}\text{—}\underset{\displaystyle\underset{\|}{\text{S}}}{\text{C}}\text{—SH}
$$

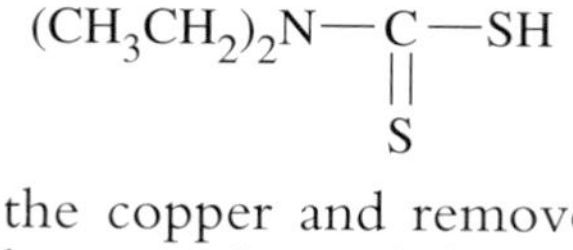

a compound that binds to the copper and removes it from the active sites. Diethyldithiocarbamate has been used to inhibit CuZnSOD activity in many eukaryotic cells. For example, 3 h after 1.5 g of diethyldithiocarbamate per kilogram of body weight was injected into mice, the SOD activity of blood had decreased by 86%, that of liver by 71% and that of brain by 48%.[119]

However, caution should be exercised in the use of diethyldithiocarbamate since it inhibits a number of other copper-containing enzymes as well. Being a thiol, it also exerts several direct antioxidant properties, can generate $RS^\bullet$ radicals and can chelate metals, e.g. copper ions. Hence dithiocarbamates have complex effects *in vivo*, including an ability to block apoptosis (Section 4.17.2) in some cell systems.

Isoenzymes of CuZnSOD

It is possible to visualize SOD enzymes after gel electrophoresis (Fig. 3.5). Inhibition by cyanide can be used to identify CuZnSOD enzymes. Electrophoresis of tissue extracts, or even of purified SOD enzymes, often shows the presence of multiple bands; for example, cow liver extract shows seven bands of CuZnSOD activity. Caution must be exercised in attributing such multiple bands to the presence of SOD isoenzymes since they might arise by attack on the SOD protein by proteolytic enzymes present in the extract. For example, storage of purified *Neurospora crassa* CuZnSOD at low temperatures causes it to show multiple bands on subsequent electrophoresis. If, for example, metal ions were lost from the enzyme this would increase its net negative

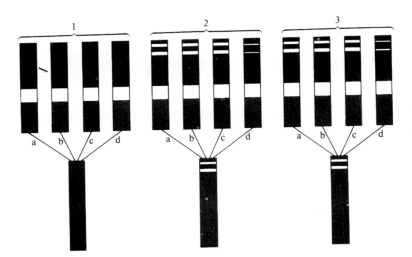

Fig. 3.5. Visualization of SOD on polyacrylamide gels. Protein is applied to the gel and electrophoresis carried out. The gel is then soaked in a solution of nitroblue tetrazolium (NBT) and exposed to an $O_2^{\bullet-}$-generating system. $O_2^{\bullet-}$ reduces NBT to the blue-coloured formazan (Chapter 2) and so the gel turns blue except at the points where SOD activity is located. The enzyme removes the $O_2^{\bullet-}$ and prevents formazan production, and a white 'achromatic zone' is detected (Beauchamp and Fridovich (1971) *Anal. Biochem.* **44**, 276). The figure shows the pattern obtained in polyacrylamide gel electrophoresis of extracts from (a) brain, (b) heart, (c) liver and (d) lung of (1) rat, (2) mouse, (3) chicken. Upper panel, no cyanide added. Lower panel, 2 mM CN^- ions present, which inhibit CuZnSODs. Figure taken from De Rosa *et al.* (1979), *Biochim. Biophys. Acta* **566**, 32, by courtesy of the authors and Elsevier, Amsterdam. The enzyme activity **tetrazolium oxidase**, extensively studied by geneticists, is identical with SOD (*Arch. Biochem. Biophys.* **159**, 738 (1973)). Other proteins that react with $O_2^{\bullet-}$, including some copper and iron proteins (e.g. caeruloplasmin, peroxidase) can also produce achromatic bands.

charge. Some animal CuZnSODs, e.g. the human (see legend to Fig. 3.4) and chicken-liver enzymes, contain cysteine $-SH$ groups and can oxidize and aggregate on storage, as well as forming mixed disulphides with other thiols, such as GSH.[224]

However, purification of enzymes has proved the existence of two slightly different forms of CuZnSOD in wheat seeds, and isoenzymes have also been shown to exist in several other organisms (e.g. *Drosophila*). Indeed a variant of human CuZnSOD known as **SOD-2** has been found to occur in Northern Sweden and Northern Finland.[163] Most of the population is homozygous for 'normal' SOD (**SOD-1**), but there are some heterozygotes with both SOD-1 and SOD-2, and a very few SOD-2 homozygotes. SOD-2 can be separated from SOD-1 by electrophoresis at high pH. SOD-2 is rarely detected in other populations, except in one of the Orkney islands; perhaps the Vikings may have introduced the SOD-2 gene on one of their rampages. SOD-2 has also been found in some Mormons in Utah, USA; between 1850 and 1905, more than 30 000 Mormon converts left Scandinavia for the USA.[60] The SOD-2 enzyme has a slightly lower specific activity than SOD-1, but SOD-2 homozygotes appear to suffer no ill-effects.[163] The term SOD-2 is sometimes used in the literature to describe mitochondrial MnSOD (next section) but we prefer to avoid it because of the risk of confusion.

3.2.2   *Manganese SODs*

The SOD first isolated from *E. coli* was entirely unlike CuZnSOD:[89] It was pink rather than blue-green, not inhibited by cyanide or diethyldithiocarba-mate, had a relative molecular mass of 40 000 rather than 32 000, was destroyed by treatment with chloroform plus ethanol (and hence did not survive the typical purification methods for CuZnSOD) and contained manganese at its active site, this being in the Mn(III) state in the 'resting' enzyme. Despite these differences, MnSODs catalyse essentially the same reaction as CuZnSODs. The (simplified) reaction mechanism can be written as

$$Mn(III) + O_2^{\bullet -} \rightleftharpoons [Mn(III)-O_2^{\bullet -}] \rightarrow Mn^{2+} + O_2$$

$$Mn^{2+} + O_2^{\bullet -} \rightleftharpoons [Mn^{2+}-O_2^{\bullet -}] + 2H^+ \rightarrow Mn(III) + H_2O_2$$

At pH 7.0 the rates of $O_2^{\bullet -}$ dismutation for CuZnSOD and MnSOD are similar, but unlike many CuZnSODs the rates for MnSODs decrease at alkaline pH (e.g. for the *E. coli* enzymes, the rate constant at pH 7.8 is 1.8×10^9, but $0.33 \times 10^9 \, M^{-1} \, s^{-1}$ at pH 10.2). Thus assays of tissues for SOD activity at high pH underestimate the amount of MnSOD present in relation to CuZnSOD. Manganese SODs are also more labile to denaturation by heat or chemicals, such as detergents, than CuZnSODs.

MnSOD location

MnSODs are widespread in bacteria, plants and animals (Table 3.1). In most animal tissues and yeast, MnSOD is largely (if not entirely) located in the mitochondria.[89] For example, the CN^--insensitive SOD bands shown in

Fig. 3.5 are attributable to the Mn-containing enzyme. The relative activities of MnSOD and CuZnSOD depend on the tissue and on the species; one obvious variable is the number of mitochondria present. Mammalian erythrocytes (with no mitochondria) contain no MnSOD, MnSOD is about 10% of total SOD activity in rat liver. In a normal growth medium, the fungus *Dactylium dendroides* contains 80% of its SOD activity as CuZnSOD and 20% as MnSOD. However, if its supply of copper ions is restricted, more MnSOD is synthesized to maintain the total cellular SOD activity approximately constant.[232] This 'extra' MnSOD appears in the cytosol. Similarly, rises in CuZnSOD and falls in MnSOD activity have been observed in the livers of chickens fed on a Mn-restricted diet.[63]

Some crustacea may be an exception to the general rule that MnSOD is mitochondrial.[41a] Thus the blue crab *Callinectes sapidus* contains MnSODs in both mitochondria and cytosol, but no CuZnSOD.

Table 3.1. Some organisms from which manganese SOD has been purified

Organism	No. of subunits	Mol Mn/mol enzyme
Higher organisms		
maize	4	2
bovine adrenal cells	4	2
bovine heart mitochondria	4	2
luminous fungus (*Pleurotus olearius*)	4	2
pea (*Pisum sativum*)	4	1
chicken liver	4	2
rat liver	4	4
human liver	4	4
Saccharomyces cerevisiae	4	4
bullfrog (*Rana catesbeiana*)	4	4
blue crab (*Callinectes sapidus*)	2	2
Bacteria		
Halobacterium halobium	2	1–2
Rhodopseudomonas spheroides	2	1
E. coli	2	1
Bacillus stearothermophilus	2	1
Mycobacterium phlei	4	2
Mycobacterium lepraemurium	2	1
Thermus thermophilus	4	2
Streptococcus faecilis	2	1
Streptococcus mutans	2	1–2
Propionibacterium shermanii	2	3
Bacillus subtilis	2	1
Serratia marcescens	2	1–2
Gluconobacter cerinus	2	1
Acholeplasma laidlawii	2	1
Actinomyces naeslundii	4	2.3
Pseudomonas carboxydohydrogena	–	1

MnSOD structure

MnSODs from higher organisms usually contain four protein subunits and usually have 0.5 or 1.0 ions of Mn per subunit (Table 3.1). Figure 3.6 (see plate section) shows the subunit structure of human MnSOD as determined by X-ray crystallography. By contrast, most, but not all, of the bacterial enzymes have two subunits (Table 3.2). Removal of the manganese from the active site of MnSODs causes loss of catalytic activity; the manganese cannot usually be replaced by any other transition-metal ion, including iron, to yield a functional enzyme (the **cambialistic** enzymes are exceptions: Section 3.2.3 below).

The amino acid sequences of MnSODs, whether from animals, plants or bacteria, are similar to each other and unrelated to the CuZnSOD sequences.

Table 3.2. Some organisms from which iron SOD has been purified

Organism	No. of subunits	Mol Fe/mol enzyme
Bacteria		
Streptococcus mutans	2	1–2
E. coli	2	1–1.8
Desulphovibrio desulphuricans	2	1–2
Thiobacillus denitrificans	2	1
Chromatium vinosum	2	2
Photobacterium leiognathi	2	1[a]
Pseudomonas ovalis	2	1–2
Helicobacter pylori	2	—
Methanobacterium bryantii	4	2–3
Thermoplasma acidophilum[b]	4	2
Azotobacter vinelandii	2	2
Bacillus megaterium	2	1
Mycobacterium tuberculosis	4	4
Propionibacterium shermanii	2	2
Sulpholobus acidocaldarius	2	1
Aquifex pyrophilus	4	~3
Other organisms		
tomato (*Lycopersicon esculentum*)	2	1–2
mustard (*Brassica campestris*)	2	1–2
water-lily (*Nuphar luteum*)	2	1
Porphyridium cruentum (red alga)	2	1
Spirulina platensis (blue-green alga)	2	1
Plectonema boryanum (blue-green alga)	2	1
Anacystis nidulans (blue-green alga)	2	1
Euglena gracilis (green alga)	not reported	1
Crithidia fasciculata (trypanosome)	2	2–3
Ginkgo biloba	2	1

[a]Plus some 'non-specifically bound' iron.
[b]This enzyme has very low activity and the question has been raised as to whether or not it is truly a FeSOD.

This is consistent with the **endosymbiotic theory** for the origin of mitochondria, which suggests that these organelles evolved from a symbiosis between a primitive eukaryote (with CuZnSOD) and a prokaryote (with MnSOD) that eventually became incorporated into the eukaryotic cytoplasm, wrapped in a membrane (which gave rise to the outer mitochondrial membrane).[89]

3.2.3 *Iron and cambialistic SODs*

From the bacterium *E. coli*, four SOD enzymes can be purified, including periplasmic CuZnSOD and intracellular MnSOD. A third SOD was found to be an iron-containing enzyme (**FeSOD**), and similar enzymes were found subsequently in several other bacteria, algae and higher plants (Table 3.2). The fourth *E. coli* SOD is a hybrid enzyme containing subunits of the manganese enzyme and of the iron enzyme in the same dimeric molecule.[89] In *E. coli*, both FeSOD and MnSOD are found in the cell matrix—an early report that FeSOD is located in the periplasmic space was retracted, although CuZnSOD is located there.

Iron-containing SODs usually contain two protein subunits, although some tetrameric enzymes have been described (Table 3.2). The dimeric enzymes usually contain one or two ions of iron per molecule of enzyme. The iron in 'resting' FeSODs is Fe(III) and it probably oscillates between the Fe(III) and Fe^{2+} states during the catalytic cycle, i.e.

$$Fe(III)-enzyme + O_2^{\bullet -} \rightarrow Fe^{2+}-enzyme + O_2$$

$$\frac{Fe^{2+}-enzyme + O_2^{\bullet -} + 2H^+ \rightarrow Fe(III)-enzyme + H_2O_2}{\text{Net reaction: } O_2^{\bullet -} + O_2^{\bullet -} + 2H^+ \rightarrow H_2O_2 + O_2}$$

although this is probably an oversimplification of the mechanisms. Like MnSODs, FeSODs show decreased catalytic activity at high pH values (compared with pH 7) and are not inhibited by CN^-. The rate constants for reaction with $O_2^{\bullet -}$ are slightly lower for FeSODs than for the other types of SOD.

The amino acid sequences of FeSODs are similar to those of MnSODs, and different from CuZnSOD sequences. This explains how a hybrid SOD can occur in *E. coli*. Despite this structural similarity, MnSODs and FeSODs are usually only effective with the correct metal at the active site. However, the SODs from several bacteria, such as *Bacteroides fragilis*, *Streptococcus mutans* and *Propionibacterium shermanii*, appear to be active with either metal present and are often called **cambialistic SODs**.[90,166] The tertiary structure of FeSODs from several organisms, including *Pseudomonas ovalis* and *E. coli* have been determined by X-ray crystallography, as has that of the cambialistic SOD from *P. shermanii*.

Location of FeSODs

Some bacteria (e.g. *E. coli*) contain both FeSOD and MnSOD, whereas others contain only one enzyme.[89] For example, *Bacillus cereus* contains only FeSOD, and *Streptococcus sanguis* only MnSOD. *Photobacterium leiognathi* contains a FeSOD in addition to its CuZnSOD, whereas the non-symbiotic, free-living, strain *Photobacterium sepia* contains a FeSOD but no CuZnSOD. The aerobic bacterium *Nocardia asteroides* has been reported to produce a SOD that contains both manganese and iron ions.

No animal tissues have been found to contain FeSOD, but some higher plant tissues do (Table 3.2). Mitochondria from mustard leaves apparently contain CuZnSOD in the intermembrane space, and MnSOD in the matrix, but the FeSOD appears to be located in the chloroplasts.

There is much interest in the information about the process of evolution that may be obtained by investigating SOD enzymes, e.g. in relation to the origin of mitochondria. For example, the bacterium *Paracoccus denitrificans* shares many structural and biochemical features with mitochondria, and it has been proposed that both it and mitochondria might have evolved from a common ancestral bacterium, i.e. that *P. denitrificans* resembles the symbiotic bacterium that 'fused' with the primitive eukaryote. Consistent with this, *P. denitrificans* contains MnSOD. A Cu-containing protein with some CN^--sensitive SOD activity has also been isolated from *P. denitrificans* but it should not necessarily be assumed that it is related to a CuZnSOD since a number of other copper proteins react with $O_2^{\bullet-}$ radical (Section 3.1.3).

Table 3.3. Representative data for SOD activity in human tissues

Tissue	CuZnSOD[a] (μg/mg protein)	CuZnSOD activity[b] (units/g wet wt)	MnSOD activity[b] (units/g wet wt)
Cerebral grey-matter	3.7	ND	ND
Liver	4.71	106 900	2260
Erythrocytes	0.52[c]	—	—
Renal cortex	1.93 ⎫	24 800	1510
Renal medulla	1.31 ⎭	(total kidney)	(total kidney)
Thyroid	0.38	10 700	276
Testis	2.16	ND	ND
Cardiac muscle	1.82	ND	ND
Gastric mucosa	0.94	Gut 10 000	358
Pituitary	0.99		
Pancreas	0.39	8630	778
Lung	0.47	7500	86
Thoracic aorta	—	7040	86

[a]An immunological method was used which measures the enzyme protein rather than enzyme activity. Data abstracted from Hart *et al.* (1973) *Clin. Chim. Acta* **36**, 125.
[b]Enzyme activity determined by the KO_2 method at high pH, which has low sensitivity for MnSOD. Data from Marklund *et al.* (1984) *Biochem. J.* **222**, 649, and *Arterioscler. Thromb. Vasc. Biol.* **15**, 2032 (1995). Results were obtained from subjects who had died after accidents. Extracellular SOD (EC-SOD) levels in human thoracic aorta were $\sim$6440 units/g; the mean for other tissues is 63–1260 as compared with a mean of 7500–30,200 for CuZnSOD (Section 3.19.4).
[c]Erythrocyte value per milligram of haemoglobin.

3.2.4   *Assays of SOD*

Measuring SOD can involve measuring protein and/or enzyme activity.[19] Immunological methods for detecting CuZnSOD and MnSOD proteins in animal tissues have been developed. Since these two enzymes are very different the antibodies do not cross-react. Table 3.3 shows some results for the amount of CuZnSOD protein in different human tissues. Because of the limited availability of human tissues and the different assays used, these results should be taken only as guidelines, but they do show an especial concentration of SOD in liver, which is broadly consistent with the more extensive data from animal studies (Table 3.4).

Direct determination of SOD activity can be carried out by pulse radiolysis (Fig. 3.1), which has been especially useful in investigations of the mechanism of enzyme action. Similarly, the loss of the ultraviolet absorbance of $O_2^{\bullet-}$ when KO_2 is added to an aqueous solution can be observed in a spectrophotometer

Table 3.4. Examples of the measurement of SOD activities in animal tissues

Animal	Assay	Tissue	Total SOD activity (units/mg protein)	Reference
Mouse	disproportionation of KO_2 in alkaline solution (one unit of SOD causes $O_2^{\bullet-}$ to decay at the rate of $0.1/s^1$ in a 3 ml reaction volume)	pancreatic islets liver kidney erythrocytes heart brain skeletal muscle	331 660 582 52 390 408 282	*Biochem. J.* **199**, 393 (1981)
Rat	riboflavin–light–NBT system (one unit of SOD inhibits NBT reduction by 50%)	liver adrenal kidney erythrocytes spleen heart pancreas (whole) brain lung stomach intestine ovary thymus	22 20 13 4 5 9 1.5 3 3 7 3 2 1	*Biochem. J.* **150**, 31 (1975)
Rat	xanthine–xanthine oxidase–cytochrome *c* method (see text)	adipose tissue	11	*FEBS Lett.* **114**, 42

(Table 3.4) although this method can only be used at alkaline pH values when the rate of non-enzymic $O_2^{\bullet-}$ dismutation is low. Any assay carried out at alkaline pH will underestimate the amounts of FeSOD or MnSOD activity in relation to that of CuZnSOD, and appropriate corrections must be introduced.

However, most laboratories use the so-called **indirect assay methods**[19] for SOD activity. In these, $O_2^{\bullet-}$ is generated and allowed to react with a detector molecule (Fig. 3.7). SOD, by removing the $O_2^{\bullet-}$, inhibits reaction with the detector. In their original work on erythrocyte SOD, McCord and Fridovich used an assay of this type; $O_2^{\bullet-}$ was generated by a mixture of the enzyme xanthine oxidase and its substrate xanthine (Chapter 1), and was detected by its ability to reduce cytochrome c, which causes a rise in absorbance at 550 nm. SOD, by removing $O_2^{\bullet-}$, inhibits the absorbance change. One unit of SOD activity was defined as the amount that inhibits cytochrome c reduction by 50% under their assay conditions. Hence the units of SOD activity quoted in the literature bear no relation to quoted units for other enzymes (1 enzyme unit is normally defined as that amount catalysing transformation of 1 µmol of substrate per minute) and will be different if a different assay is used.

One problem with indirect assays of SOD is that a false SOD activity can be registered by any agent that inhibits $O_2^{\bullet-}$ generation (Fig. 3.7), e.g. by inhibiting the xanthine oxidase enzyme. One advantage of the xanthine oxidase system is that this artifact can be checked for by measuring the production of uric acid in the system as an index of xanthine oxidase activity, and such a control is essential. Many tissue extracts contain cytochrome oxidase, which can oxidize reduced cytochrome c and so interfere with SOD assay. Chemical modification of the cytochrome c by attachment of acetyl (CH_3CO-) groups to some of its amino-acid side-chains prevents it from being a substrate for cytochrome oxidase but still allows reaction with $O_2^{\bullet-}$. Hydrogen peroxide produced directly by xanthine oxidase and also from $O_2^{\bullet-}$ dismutation can slowly re-oxidize reduced cytochrome c. This is not usually a problem in assays of SOD by the xanthine–xanthine oxidase method, but if necessary some catalase can be added to the reaction mixture. It is essential to ensure that any commercial catalase used is not itself contaminated with SOD; commercial cytochrome c also often contains traces of SOD. Peroxynitrite can also oxidize ferrous cytochrome c.

Detector molecules for $O_2^{\bullet-}$ other than cytochrome c can be used (Table 3.5). For example, NBT is reduced by $O_2^{\bullet-}$ to a deep-blue-coloured

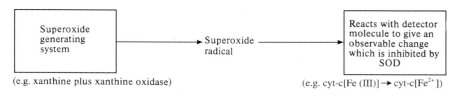

Fig. 3.7. Principles of the indirect assay methods for SOD activity.

formazan (Chapter 2), and adrenalin is oxidized by $O_2^{\bullet-}$ to form a pink product, **adrenochrome**. The ability of $O_2^{\bullet-}$ to oxidize NADH in the presence of lactate dehydrogenase (Chapter 2) has also been used. The compound **luminol** emits light when exposed to $O_2^{\bullet-}$ and other ROS, as does **luciferin** (although the mechanisms are very complex; Section 6.7.2) and they have been used as detector molecules. Inhibition of autoxidation reactions (e.g. of adrenalin or pyrogallol) has also been used (Table 3.5); one problem is that tissue extracts can contain transition metal ions that accelerate such autoxidations.

Each of these methods has its problems.[19,33] In some, it is difficult to check for artifactual effects of a putative $O_2^{\bullet-}$ scavenger on the rate of $O_2^{\bullet-}$ generation. Our laboratory routinely uses inhibition of NBT reduction by a xanthine–xanthine oxidase system to assay SOD in tissue extracts. Although this avoids problems with cytochrome oxidase, it must be remembered that the reaction of NBT with $O_2^{\bullet-}$ is complex (Chapter 2). Although formazan is only sparingly soluble in water, its precipitation can be avoided by keeping absorbance changes fairly low.

Whatever assay is used, it should first be calibrated with pure SOD enzyme, and a known amount of SOD enzyme, added to the crude tissue extract being examined, should be quantitatively detected on subsequent assay. The scientist performing the assay should also think carefully about possible artefacts, not only interference with $O_2^{\bullet-}$ generation but also with $O_2^{\bullet-}$ detection.[33,114] To take one example, the compound **pamoic acid** appeared to inhibit SOD activity in a number of indirect assays, but was found to be interfering with the assays. Reducing agents in tissue extracts can directly interact with certain detector molecules: this is a particular problem with cytochrome *c*, which is easily reduced by ascorbate and thiols. No one assay will be suitable for all systems; for example, assays using NBT reduction cannot be applied to ocular tissues because of the presence of enzyme systems that reduce NBT directly. The units quoted are different for each type of assay, but Tables 3.4 and 3.5 show that the relative amount of SOD activity in different body organs is broadly similar in different mammals.

Distinguishing between different types of SOD

CuZnSODs are inhibited by CN^-, whereas FeSOD and MnSOD are not.[89] Inhibition by CN^- can therefore be used to identify CuZnSOD activity in assays of tissue homogenates or on polyacrylamide gels (e.g. Fig. 3.5).

Both CuZnSOD and FeSOD are inactivated on prolonged exposure to H_2O_2, whereas MnSOD is not. Thus incubating a cell extract with H_2O_2 will inactivate FeSOD but not MnSOD and can be used to distinguish the two. The rate of inactivation of CuZnSOD by H_2O_2 is greater at higher pH values. It may be due to reduction of Cu^{2+} to Cu^+ at the active site by H_2O_2, followed by a reaction of Cu^+ with H_2O_2 to generate $OH^\bullet$ (or an oxidizing oxo-copper complex) which then oxidizes one of the histidine residues essential for the catalytic mechanism (His118 in bovine erythrocyte CuZnSOD), converting it to **2-oxohistidine**.[254]

Table 3.5. Some of the indirect methods that have been used to measure SOD activity

Source of superoxide	Detector of superoxide	Reaction measured[a]
Xanthine–xanthine oxidase	cytochrome c	reduction, ΔA
	nitroblue tetrazolium (NBT)[b]	reduction, ΔA
	luminol, luciferin	light emission
	adrenalin	oxidation, ΔA
	NADH + lactate dehydrogenase	oxidation, ΔA at 340 nm
	hydroxylamine[c]	nitrite (NO_2^-) formation (colorimetric method)
	spin traps	ESR signal (Chapter 5)
Auto-oxidation reactions[d]	adrenalin	oxidation, ΔA
	sulphite	O_2 uptake
	pyrogallol	O_2 uptake, or ΔA
	6-hydroxydopamine	oxidation, ΔA
Directly added $K^+ O_2^-$	—	loss of $O_2^{\bullet -}$, ΔA in UV
	NBT	reduction, ΔA
	cytochrome c	reduction, ΔA
	tetranitromethane	reduction, ΔA
Illuminated flavins[e]	NBT	reduction, ΔA; O_2 uptake (SOD accelerates)
	dianisidine	oxidation, ΔA ('positive' assay)[f]
NADH + phenazine methosulphate[g]	NBT	reduction, ΔA

[a] ΔA: reaction results in an absorbance change that can be measured using a spectrophotometer.

[b] High O_2 concentrations can decrease NBT reduction by $O_2^{\bullet -}$ (Chapter 2).

[c] Hydroxylamine itself reacts slowly, if at all, with $O_2^{\bullet -}$, and nitrite formation may require hydroxyl radicals. Thus the chemistry of this assay is very complex.

[d] A number of compounds have been shown to oxidize in solution with simultaneous production of $O_2^{\bullet -}$; these include 6-hydroxydopamine, pyrogallol, the sulphite ion (SO_3^{2-}), and adrenalin (at alkaline pH). $O_2^{\bullet -}$, once formed, participates in the oxidation of further molecules, so that addition of SOD slows down the observed rates of oxidation. The rates of these oxidations are often greatly accelerated by the presence of transition-metal ions, however, and this can cause problems in the assay of crude extracts containing traces of such ions unless metal-ion chelators are added.

[e] Illumination of a riboflavin solution in the presence of either EDTA or of the amino acid methionine causes a reduction of the flavin. It then re-oxidizes and simultaneously reduces oxygen to $O_2^{\bullet -}$, which is allowed to react with a detector molecule such as NBT. SOD will inhibit the formazan production. Flavin photochemistry is extremely complicated, however, and singlet oxygen is also produced (Chapter 2). In a variation on this assay, an oxygen electrode is used to measure the rate of O_2 consumption during photochemical generation of $O_2^{\bullet -}$ in the presence of NBT. Reduction of the dye by $O_2^{\bullet -}$ is accompanied by stoichiometric oxygen-production, i.e. $NBT + O_2^{\bullet -} \rightarrow NBT$ radical $+ O_2$. On addition of SOD, two $O_2^{\bullet -}$ molecules are required to make one oxygen molecule, and the rate of oxygen uptake increases.

Table 3.5. (*Footnotes continued*)

---

[f]A solution containing riboflavin and the detector molecule *ortho*-dianisidine is illuminated, whereupon the detector is slowly oxidized, accompanied by an absorbance change at 460 nm. Addition of SOD increases the rate of dianisidine oxidation because it removes $O_2^{\bullet-}$, which interacts with an intermediate dianisidine radical and thereby decreases the net rate of oxidation. The assay is called 'positive' because addition of the SOD causes the absorbance change to increase instead of reducing it. At alkaline pH values, SOD accelerates the oxidation of **haematoxylin**, a dye used by histologists as a 'stain'. This reaction has also been proposed as a positive assay for SOD.
[g]Not recommended; see *J. Am. Chem. Soc.* (1982) **104**, 1666.

Another method for distinguishing between different SOD types exploits the fact that FeSODs are generally more sensitive to inhibition by azide, N_3^-. For example, at pH 7.8 10 mM azide inhibits CuZn, Mn and FeSODs by about 10%, 30% and 70%, respectively. There is some variation, however; for example, the *Methanobacterium bryantii* FeSOD is less sensitive to azide than other FeSODs, and CuZnSOD from tomato leaves appears more sensitive to azide than other CuZnSODs.

A third approach has been to remove the metals from SOD proteins in cell extracts, and then to add either Fe^{2+} or Mn^{2+} back to the extract. If a particular band of enzyme activity observed on electrophoresis before metal removal reappears on addition of, say, Fe^{2+}, then it most likely represented a FeSOD. Despite the close structural similarities between FeSODs and MnSODs, most of them will only work with the correct metal at the active site. However, the existence of the cambialistic enzymes means that such results must be interpreted with caution.

3.3 Using SOD enzymes as probes for superoxide

The specificity of SOD for reaction with $O_2^{\bullet-}$ has frequently been used to probe for the involvement of this radical in biological systems. Although $O_2^{\bullet-}$ appears specific for *catalytic* removal of $O_2^{\bullet-}$, the SOD proteins (like any other protein) can react directly with certain ROS/RNS. Thus SODs react with $OH^{\bullet}$, peroxyl/alkoxyl radicals and singlet oxygen because they contain histidine and other side chains that react with these species (Chapter 2). CuZnSOD interacts with peroxynitrite, decomposing it to a nitrating species (Section 2.5.4). For bovine CuZnSOD, one SOD molecule catalyses nitration of a second molecule (or of any other nitratable protein molecules added to the reaction mixture). Nitration of bovine CuZnSOD occurs at Tyr108, close to the copper ion in the active site. MnSOD also catalyses nitration and is slowly inactivated on addition of ONOO[-].[26] Indeed, nitration and inactivation of MnSOD has been observed in 'rejected' human kidney transplants (Chapter 9).

Thus if a large quantity of SOD (or any other protein) is added to a system producing ROS or RNS, an artifactual inhibition of damage due to direct scavenging might result. Controls with heat-denatured SOD, other proteins (e.g. albumin) in equimolar amounts or SOD apoenzymes (the protein with the metals removed from the active site) can usually be performed to address this issue.[c] A similar point applies to the use of other enzymes, such as catalase to implicate H_2O_2 involvement in an observed reaction: any protein at high concentrations can act as a 'general' ROS/RNS scavenger.

Inhibition by SOD must also be interpreted with caution in systems containing quinones and semiquinones. As we saw in Chapter 2, many semiquinones react reversibly with oxygen:

$$semiquinone + O_2 \rightleftharpoons quinone + O_2^{\bullet -}$$

The equilibrium will tend to move to the right because of the non-enzymic dismutation of $O_2^{\bullet -}$. Addition of SOD, by removing $O_2^{\bullet -}$ much faster, can decrease the steady-state concentration of semiquinone. Hence a reaction that is caused by the semiquinone might be mistakenly attributed to $O_2^{\bullet -}$ as a result of the inhibition on addition of SOD. An example is provided by early studies on the inhibition of quinone (vitamin K)-dependent carboxylation reactions by SOD; the inhibition by SOD is not evidence that $O_2^{\bullet -}$ is involved in the carboxylation (Section 6.5).

A similar artefact can arise using $O_2^{\bullet -}$-generating systems. Frequently, an autoxidizable compound, such as **dihydroxyfumarate** (DHF), is added to a cell/tissue and causes damage. SOD inhibits the damage, a result often interpreted to mean that damage is caused by $O_2^{\bullet -}$. However, $O_2^{\bullet -}$ is an essential intermediate in DHF oxidation,[106] and so SOD inhibits that oxidation. Thus *any product of DHF oxidation* could be causing the damage, not necessarily $O_2^{\bullet -}$.

It is also worth noting that several copper-containing proteins other than CuZnSOD can react with $O_2^{\bullet -}$, as can complexes of Cu^{2+} ions with non-metalloproteins such as immunoglobulin G and albumin (Chapter 10). At physiological pH, the rate constant for $O_2^{\bullet -}$ dismutation by each active site of CuZnSOD is about $1.6 \times 10^9 \, M^{-1} s^{-1}$ and the reaction is catalytic (Fig. 3.1). For other copper proteins, the rate constants are much lower; some representative literature values are $2 \times 10^7 \, M^{-1} s^{-1}$ for cytochrome c oxidase, $7 \times 10^5 \, M^{-1} s^{-1}$ for caeruloplasmin and $3 \times 10^6 \, M^{-1} s^{-1}$ for galactose oxidase. More importantly, reactions of these proteins with $O_2^{\bullet -}$ are non-catalytic under physiological conditions; the $O_2^{\bullet -}$ is merely reducing Cu^{2+} at the active site to Cu^+ and the reaction then stops. Several haem proteins interact with $O_2^{\bullet -}$, including haemoglobin (Chapter 7) and peroxidases (Section 3.16 below).

3.3.1   *Are there more SODs to come?*

There have been several recent reports of SOD enzymes containing nickel, e.g. in *Streptomyces griseus* and *S. coelicolor*. More information is awaited with interest.[136]

3.4 Evidence for the physiological importance of superoxide dismutases

The discovery of SOD enzymes led to **the superoxide theory of oxygen toxicity,** which proposes that $O_2^{\bullet-}$ is a major factor in O_2 toxicity and that SODs are an essential defence against it. It is now generally accepted that the biological role of SOD is to scavenge $O_2^{\bullet-}$, which is known to be generated *in vivo* in amounts increasing with O_2 exposure (Chapter 1).[89]

Perhaps the best way of proving that SOD is important would be to eliminate it from living organisms and observe the consequences. Techniques of modern molecular and cell biology (reviewed briefly in Appendix II for the non-initiates) have been used to perform such experiments.

3.4.1 *Gene knockouts in bacteria and yeasts*

E. coli has two main SODs, FeSOD and MnSOD (the periplasmic CuZnSOD contributes only a little to total SOD activity). Touati *et al.*[82] inactivated the genes encoding MnSOD and FeSOD in *E. coli*, generating a bacterium initially thought to have no SOD activity (we now know that periplasmic CuZnSOD was still present). Essentially, non-functional mutated genes carried on a plasmid were exchanged for the 'correct' genes on the *E. coli* chromosome. The properties of this mutant gave direct evidence for an important role of SOD; it would not grow aerobically on a minimal glucose medium. Growth could be restored by removing O_2, by restoring SOD production to the cells as a result of introducing DNA bearing a gene coding for *any* SOD (even for mammalian CuZnSOD), or by enriching the growth medium of the bacteria by providing 20 different amino acids. Indeed, it was previously known (Chapter 1) that certain amino acids could diminish the growth-inhibiting effects of hyperbaric O_2 upon *E. coli*.

Even with this supplementation, the SOD-deficient mutant grew only half as fast as normal *E. coli* and its membranes were leaky to certain ions. It was also more sensitive to damage by increased O_2 concentrations or by H_2O_2. The SOD-deficient *E. coli* mutants also showed increased mutation rates during aerobic (but not anaerobic) growth on rich media. Addition of diethyl-dithiocarbamate to inhibit periplasmic CuZnSOD inhibited growth even on a rich medium.[29]

The steady-state level of $O_2^{\bullet-}$ in aerobic *E. coli* in the log-phase stage of growth has been estimated at 20–40 picomolar ($1 \, pM = 10^{-12} \, M$), rising to about 300 pM in the double mutant.[85] By contrast, intracellular H_2O_2 in wild-type *E. coli* has been estimated as 100–200 nM.[95] It appears that a $O_2^{\bullet-}$ concentration of a few hundred picomolar is sufficient to lead to DNA damage, hypersensitivity to H_2O_2 and impairment of metabolic activity by selective inactivation of enzymes, such as aconitases and dihydroxyacid dehydratase (Chapter 1).

Similar studies have been carried out on other organisms. *Pseudomonas aeruginosa* lacking both SOD types grows only poorly aerobically.[115]

Inactivating the gene for MnSOD in baker's yeast, *Saccharomyces cerevisiae*, causes it to become hypersensitive to damage by O_2.[255] However, MnSOD-deficient yeast in which the electron-transport chain was also absent (**Rho°** **state**) were more resistant, consistent with the view that electron transport is an important source of $O_2^{\bullet-}$ *in vivo*.[100] Expression of *E. coli* FeSOD in the yeast mitochondria (but not in yeast cytosol) offered protection.[18] Similarly, yeast lacking CuZnSOD is hypersensitive to O_2 and will not grow aerobically unless supplemented with such amino acids as lysine and methionine.[52]

3.4.2 *Transgenic animals*

With transgenic animal technology, it is possible to increase the level of (overexpress) foreign proteins or enzymes and to examine the consequences. Appendix II explains the technology (and the considerations to bear in mind when interpreting the results) for those who need it. Transgenic technology is most often applied to mice; for example, mice carrying the human CuZnSOD gene and expressing human CuZnSOD protein in various tissues have been produced. Expression of human CuZnSOD occurs in addition to that of mouse CuZnSOD, thus raising the total CuZnSOD activity. Such mice show increased resistance to O_2 toxicity and certain other toxins, consistent with the superoxide theory of O_2 toxicity. However, they also show certain abnormalities (Box 3.1).[198] Transgenic mice overexpressing human MnSOD in the lung have been generated (without affecting CuZnSOD, catalase or glutathione peroxidase activities) and found to be more resistant to lung damage by 95% O_2 than normal mice.[267] Transgenic strains of the fruit fly, *Drosophila*, overexpressing bovine CuZnSOD were more resistant to hyperoxia and to paraquat, an $O_2^{\bullet-}$ generator (Chapter 8) than control flies.[213] By contrast, *Drosophila* lacking CuZnSOD show decreased lifespan, increased toxicity of O_2 and infertility.

Even more dramatic data are provided using animals lacking SOD.[152] Gene knockout techniques using embryonic stem cells (Appendix II) have been used to generate mice completely lacking MnSOD. In one study, most of these animals died within the first 10 days of life with cardiac abnormalities (Fig. 3.8, see plate section), fat accumulation in liver and skeletal muscle and metabolic acidosis. Severe mitochondrial damage in heart and, to a lesser extent, in other tissues is evidenced by decreases in succinate dehydrogenase and aconitase activities (Fig. 3.8), consistent with an essential role of MnSOD in maintaining normal mitochondrial function. MnSOD-defective strains that survive longer than this quickly succumb to a variety of pathologies, including severe anaemia and neurodegeneration. Indeed, treatment of MnSOD$^-$ mice with low molecular mass scavengers of $O_2^{\bullet-}$ keeps them alive longer, whereupon they suffer severe brain degeneration,[168a] perhaps because the scavengers do not enter the brain.

Mice defective in CuZnSOD have also been obtained.[123b,209] When young, they appear normal (although they are more sensitive to the $O_2^{\bullet-}$ generating

---

Box 3.1
Consequences of expressing human CuZnSOD in mice

1. Transgenic mice with elevated levels of CuZn-SOD
   (a) are more resistant than controls to O_2 toxicity under some experimental conditions;
   (b) are more resistant than controls to certain toxins, e.g. the diabetogenic agent alloxan and the neurotoxins MPTP, methamphetamine, methylene-dioxyamphetamine, methylenedioxymethylamphetamine, 3-nitropropanoic acid and 6-hydroxydopamine;
   (c) suffer more brain infarction after cerebral ischaemia–reperfusion (in young mice), but less in adult mice;
   (d) still show O_2-induced retinopathy (when exposed to 90% O_2 during the first 5 days of life);
   (e) show abnormal neuromuscular junctions in the tongue, altered serotonin metabolism and premature involution of the thymus;
   (f) may show some of the other neurological defects characteristic of Down's syndrome.
2. Pregnant, diabetic (streptozotocin-treated) animals show less fetal damage by elevated glucose.
3. Peritoneal macrophages show decreased microbicidal capacity.
4. Mid-brain neurones from transgenic pups survive better in culture.
5. Isolated hearts are less susceptible to ischaemia–reperfusion injury in some model systems, as is intestine *in vivo*.

For representative references see *Pediat. Res.* **39**, 204 (1996), *J. Neurochem.* **65**, 919 (1995), *ibid.* **67**, 1383 (1996), *J. Physiol.* **84**, 53 (1990), *J. Immunol.* **156**, 1578 (1996), *Am. J. Obstet. Gynecol.* **173**, 1036 (1995), *Br. J. Ophthalmol.* **80**, 429 (1996), *Am. J. Physiol.* **273**, C1130 (1997), *Neuroscience* **85**, 907 (1998).

---

toxin, paraquat), but as they age, neurological damage and cancers have been claimed by some scientists to develop at an accelerated rate, and the mice have reproductive problems. Further studies on these mice are awaited with interest.

3.4.3 *Induction experiments*

The molecular genetic approaches show clearly that SOD enzymes (especially MnSOD) are important in maintaining normal aerobic life, although they raise questions about optimum cellular levels of SOD (Box 3.1) and the targets damaged by excess $O_2^{\bullet-}$.

Much supporting evidence confirms an important role for SODs. Exposure of organisms to elevated O_2 should, according to the superoxide theory of O_2 toxicity, cause them to form more $O_2^{\bullet-}$ *in vivo* and this might lead to synthesis of more SOD if insufficient is present to cope with the increased $O_2^{\bullet-}$ generation. Data support this view. For example, elevated O_2 increases the total SOD activity in *E. coli* cells, due to increased synthesis of MnSOD. *E. coli* cells grown under an atmosphere of pure O_2 were more resistant to the toxic effects of high-pressure O_2 than were cells grown under air. Similarly, it is possible to increase the SOD activity of *E. coli* by incubating it with compounds that increase intracellular $O_2^{\bullet -}$ generation, such as streptonigrin, paraquat, juglone, menadione, pyocyanine, methylene blue and phenazine methosulphate (for more details of these compounds see Chapter 8.) Strains of *E. coli* with elevated SOD activity are resistant to the toxic effects of these $O_2^{\bullet -}$-generating compounds. Similarly, strains with elevated SOD resulting from exposure to these compounds are less sensitive to O_2 toxicity.[89] Of course, elevated O_2 and many $O_2^{\bullet -}$-generating compounds often also induce H_2O_2-degrading enzymes.

Similar induction experiments correlating SOD activity with O_2 exposure have been carried out in many other bacteria, plants and animals. For example, a mutant strain of the alga *Chlorella* with increased SOD activity was more resistant to elevated O_2 and to streptonigrin than was the wild-type. The O_2 produced by the photosynthetic activity of certain symbiotic algae has been observed to increase SOD activity in their animal host, the sea anemone *Anthopleura elegantissima*. Anemones that contain symbiotic algae have greater SOD and catalase activities than anemones which do not. Hence the food provided to the anemone by the photosynthetic 'guest' has to be balanced against the cost of increasing antioxidant defences in the 'host'.[73]

3.4.4 *SOD and oxygen toxicity in animals*

If adult rats are placed in pure O_2 they rapidly develop lung damage and often die after 60–72 h. If, however, rats are exposed to somewhat lower O_2 concentrations (e.g. 85% for 7 days) they can adapt to survive for longer when subsequently exposed to 100% O_2. Several years ago it was shown that this adaptation to hyperoxia is correlated with an increased content of SOD, as measured in lung homogenates.[87] Catalase, glutathione reductase and glutathione peroxidase activities are also increased in the lungs, as is the level of glutathione (GSH). If rats are pretreated with diethyldithiocarbamate to inhibit CuZnSOD, the toxic effects of high O_2 tensions can be enhanced, although care must be exercised in interpreting this observation since diethyldithiocarbamate inhibits several other enzymes. Mice, guinea pigs and hamsters show little induction of SOD after exposure to 85% O_2, and do not become tolerant to pure O_2.

It must be pointed out, however, that lung is a complex tissue containing many different cell types. Damage caused by exposure to O_2 affects some cells more than others (Section 1.6.2) and leads to phagocyte influx into the lung. For example, O_2 exposure can kill type I cells and lead to proliferation of type II cells, which have higher antioxidant defences. Hence changes in enzyme activities as assayed in homogenates of whole lungs may be due to changes in cell populations, or might underestimate very large enzyme changes taking place in only a few cell types.

Newborn rats and rabbits appear more resistant to O_2 toxicity than adult rats.[87,113] This may occur because the SOD activity of their lungs is increased and maintained more effectively under high O_2 than it is in adults. Another possibility is that neonatal cells make less $O_2^{\bullet-}$ than adult cells under O_2 stress. Catalase and glutathione peroxidase activities increase rapidly also. If induction of these three enzymes is prevented (e.g. by injection of protein synthesis inhibitors), newborn rats become highly sensitive to hyperoxia. Newborn rats kept in pathogen-free environments have lower SOD activities in lung than normal, and they are more sensitive to O_2 toxicity. Treatment of adult rats with low levels of bacterial lipopolysaccharide (endotoxin) enhances their resistance to O_2 and induces an increase in SOD, catalase and glutathione peroxidase activities in the lungs. It also lowers levels of certain P450s, a source of $O_2^{\bullet-}$ *in vivo.*

The above data on the correlation between SOD activity and resistance to O_2 toxicity are consistent with the idea that SOD is an important antioxidant, supporting and extending the genetic experiments. Since catalase and per-oxidase activities often increase as well, they also seem important. Indeed, anti-oxidant defence enzymes operate as a balanced, coordinated system. Supporting this view is the observation that injection of liposomes containing SOD into rats has limited protective effects against O_2 toxicity, but including catalase gives much more protection.[253] It must also be realized that hyperoxia can increase expression of many other proteins.

Hibernating animals show greatly decreased O_2 consumption and body temperature, yet activities of SOD and other antioxidant defences in certain tissues may be higher than normal. It has been proposed that this is a 'preparation' for the intense metabolic activity that occurs during awakening.[44]

3.5   The superoxide theory of oxygen toxicity: a critique

3.5.1   *Anaerobes with SOD*

An early question about the essentiality of SOD came from studies of anaerobes, it being suggested that organisms living without O_2 would not make $O_2^{\bullet-}$ and thus would not need a SOD. Indeed, many anaerobes lack SOD activity. However, some do contain SOD (Table 3.6). When SOD is present, it is usually FeSOD in small amounts (compare SOD activities with those of aerobically grown *E. coli* K12 in Table 3.6).

Does the presence of SOD in some anaerobes mean that SOD has functions other than $O_2^{\bullet-}$ removal? Not necessarily, since the word 'anaerobe' covers a wide spectrum of oxygen tolerance (Chapter 1).[164] Several anaerobes can survive brief exposure to O_2, albeit with growth inhibition. It seems reasonable to propose that SOD is present within them in order to aid survival during such exposures. Consistent with this interpretation, growth of *E. coli* under strictly anaerobic conditions for several generations causes loss of MnSOD, but FeSOD remains; it is a constitutive enzyme. On re-exposure to O_2, the MnSOD is promptly resynthesized. The FeSOD seems to aid survival when O_2 is restored until MnSOD can be made.[89]

Some scientists have found a correlation between O_2 tolerance and the SOD content of anaerobes, but others have not. This is perhaps not surprising, since

Table 3.6. Some anaerobic bacteria that contain SOD activity

Anaerobe studied	Type of SOD	Activity (units/mg protein)
Chlorobium thiosulphatophilum	FeSOD	14
Chromatium species	FeSOD	0.6
Desulphovibrio desulphuricans	FeSOD	0.6
Clostridium perfringens	FeSOD	15.6
Bacteroides distasonis	FeSOD	0.1–0.4 (most strains)
	FeSOD	3.2–3.9 (strain ATCC 8503)
Actinomyces naeslundii	MnSOD	—
Propionibacterium shermanii	FeSOD (but produces MnSOD if grown under Fe-restricted conditions)	
Bacteroides fragilis	FeSOD	—
Bacteroides thetaiotaomicron	FeSOD	—
Methanobacterium thermoautotrophicum	FeSOD	—

SOD activity of bacteria can vary enormously, depending on the growth medium and position in the growth curve, so do not take the values too literally. For comparison, aerobically grown *E. coli* contained 44 units SOD/mg protein under the same assay conditions. Most anaerobes examined contain no SOD activity. Assays of crude extracts for SOD activity can also be confounded by the presence of proteins that scavenge $O_2^{\bullet-}$ non-catalytically, e.g. peroxidases (Section 3.16) and the protein **rubrerythrin** in *C. perfringens* (*J. Bacteriol.* **178**, 7152 (1996)).

the bacterial contents of catalase, peroxidase and other enzymes will also be important. Even if SOD is absent, **NADH oxidase** enzymes can reduce O_2 to water and remove it from the immediate environment of the bacteria (Section 3.16 below). Several anaerobes, such as *Bacteroides fragilis*, synthesize more SOD if oxygen is present.

3.5.2 *Aerobes lacking SOD*

The biological role of SOD has also been questioned as a result of the discovery of aerobic organisms that contain no SOD. A few *Leptospira* strains and three virulent strains of the aerobic gonococcus *Neisseria gonorrhoeae* were found to be SOD-negative.[8] However, they are exceptionally rich in catalase and 'non-specific peroxidase' enzymes (Section 3.16 below). *N. gonorrhoeae* colonizes human mucous membranes and provokes a strong neutrophil response, although neutrophil killing of this organism is inefficient. Exposure of *N. gonorrhoeae* to H_2O_2 or neutrophils produces a further rise in catalase activity.[274] *Mycoplasma pneumoniae*[5] contains $O_2^{\bullet-}$-generating systems, but neither SOD, nor catalase, activities.

If metal-ion-catalysed $OH^{\bullet}$ formation is a significant contributor to the toxicity of $O_2^{\bullet-}$-generating systems (Section 3.6 below), it follows that protection against $OH^{\bullet}$ formation could be achieved by the efficient removal

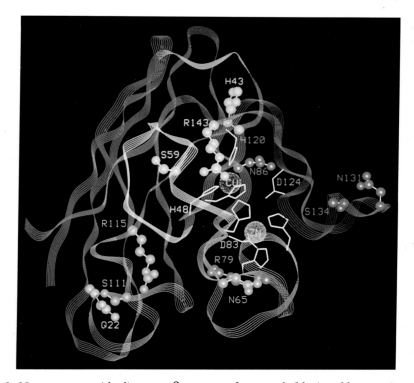

Fig. 3.3. Human superoxide dismutase β-structure framework (blue) and loop regions, along with conformationally important side chains.

The α-carbon backbone is shown as a ribbon coloured to highlight the various loops. Critical side chains that form multiple side-chain–main-chain hydrogen bonds are shown in ball-and-stick representation, labelled by residue number and the one-letter amino acid code and colour-coded to match the loops they stabilize: Gln22 for the purple insertion loop II, His43 for the light-blue Greek-key connection (loop III); Ser59 and Arg143 for the yellow disulphide subloop (IV); Asn65 and Arg79 for the gold Zn-ligand subloop (IV); Ser111 and Arg115 for the green Greek-key loop (VI); and Asn86, Asn131 and Ser134 for the red electrostatic loop (VII). Metal ligands His48, His120 and Asp83 also make side-chain to main-chain hydrogen bonds to the loop regions. The copper ligands (**His46, His48, His63 and His120**) form distorted square-planar geometry, while the zinc ligands (**His63, His71, His80 and Asp83**) are tetrahedral. The copper and zinc are linked directly by the **bridging histidine** (His63) and indirectly by the side-chain carboxylate of buried Asp124, which hydrogen bonds to both a copper- and a zinc-ligating histidine. Photograph by courtesy of Drs John Tainer and Elizabeth Getzoff and the publishers (see *Proc. Natl Acad. USA* **89**, 6109 (1992)).

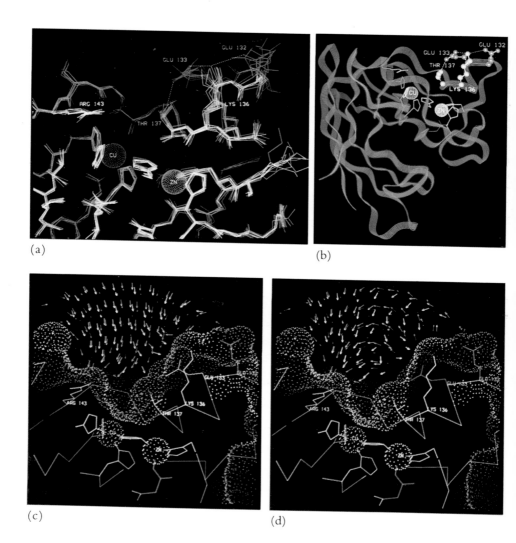

(a)

(b)

(c)

(d)

Fig. 3.4. Network of human SOD residues involved in electrostatic recognition of superoxide: Glu132, Glu133, Lys136 and Thr137. The colour-code is: red, most negative; yellow, negative; green, neutral; light blue, positive; blue, most positive. Copper (gold) and zinc (blue) ions are shown as spheres. The SOD enzyme shown here, and used as a wild-type control for construction of electrostatic mutants, is the double mutant Cys6→Ala, Cys111→Ser. This mutant maintains the activity and rate-versus-pH profile of wild-type human SOD but is more stable. (a) Superimposed copies of the SOD active site. Yellow dotted lines (upper right) show hydrogen bonds linking side chains of electrostatically important residues: Thr137 (green, centre) to Glu133 (red, top centre), to Lys136 (blue, upper right), to Glu132 (red, upper right). All other residues are shown colour-coded by subunit, and appear white where superimposed. (b) Overview of mutated residues in the electrostatic loop and their structural relationship to the active site and overall β-barrel protein fold (blue ribbon following polypeptide chain) in one subunit of the SOD homodimer. Electrostatically important side chains Glu132, Glu133, Lys136 and Thr137 (coloured and oriented as in panel a) form a hydrogen-bonding network (ball-and-stick models with dashed hydrogen bonds) on the solvent-exposed helical turn of the electrostatic loop overhanging the active-site channel (upper right). This helix is orientated so that its C-terminal carbonyl oxygen atoms and helical dipole stabilize and are stabilized by zinc binding. Side chains of copper and zinc ligands and of active-site Arg143 are shown as lines. (c) Cross-section of the active site channel for wild-type human SOD, with colour-coded molecular surface dots and arrows indicating the direction of the electrostatic force on the negative charge of $O_2^{\bullet-}$. (d) Corresponding view for (Glu133→Gln) mutated SOD. Neutralizing negatively charged Glu by converting it to uncharged glutamine increases the reaction rate by guiding $O_2^{\bullet-}$ more directly downward to the catalytic copper ion. Photographs by courtesy of Drs John Tainer and Elizabeth Getzoff, Scripps Institute and Macmillan Journals; see *Nature* **358**, 347 (1992).

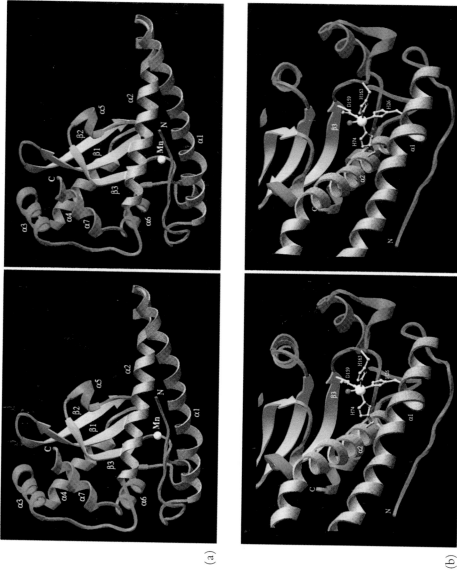

Fig. 3.6. Structure of human mitochondrial MnSOD. Managanese ions are shown as light pink spheres. (A) MnSOD subunit coloured to emphasize secondary structure and organization. The N-terminal domain (bottom) is made up of the blue N-terminal loop and two long purple α-helices (α1 and α2). The C-terminal α/β domain (top) is composed of five blue α-helices (α3–α7) and three yellow β-strands (β1–β3). The manganese lies between the two domains. MnSOD subunits have approximate dimensions $40 \times 47 \times 49$ Å. (B) Active site geometry of MnSOD. The five manganese ligands are drawn in a ball-and-stick representation. Amino acids from both domains contribute to the active site, His26 in α1 and His 74 in α2 from the N-terminal domain and Asp159 in β3 and His163 from the C-terminal domain. The fifth coordination site is occupied by a water molecule (blue sphere). The four active sites of the MnSOD tetramer are grouped in pairs across the dimer interfaces, with residues Glu162 and Tyr166 from one subunit contributing to the active site of the neighbouring subunit. Photographs by courtesy of Drs John Tainer and Elizabeth Getzoff, and Cell Press (see *Cell* **71**, 107 (1992)).

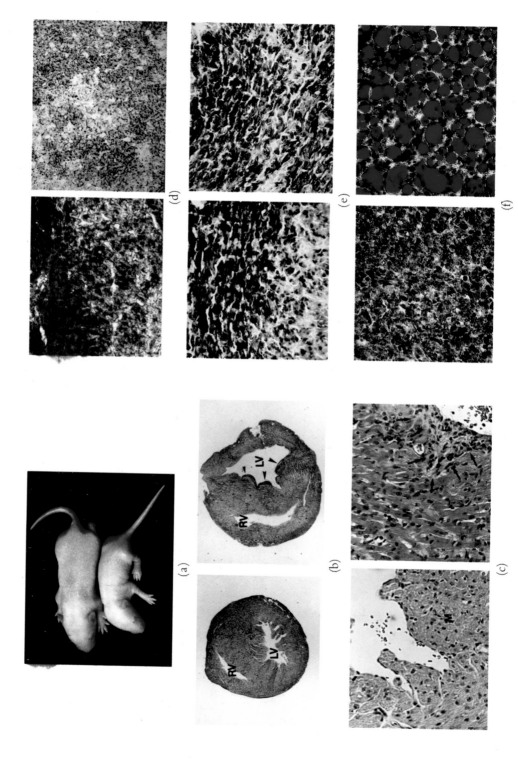

(a)

(b)

(c)

(d)

(e)

(f)

Fig. 3.8. Gross appearance and histopathology of MnSOD-deficient mice. (a) MnSOD −/− (bottom) and +/+ (top) mice, 6 days old. Note the smaller size and slightly yellow tinge of the mutant animal. (b) Transverse section of +/+ (left) and −/− (right) hearts showing left (LV) and right ventricles (RV). The left ventricular cavity of −/− mice is enlarged, and the free wall is thinner than that of +/+ mice. (c) Haematoxylin and eosin-stained, paraffin-embedded, sections of the ventricles of +/+ (left) and −/− (right) mice showing fibrous thickening of the endocardium (E, arrows) in −/− mice. M, myocardium, 100×. (d) Succinate dehydrogenase (SDH) staining in frozen sections of +/+ (left) and −/− (right) hearts showing the absence of functional SDH in the −/− heart. 400×. (e) Cytochrome c oxidase (COX) staining in frozen sections of +/+ (left) and −/− (right) hearts showing comparable levels of functional COX, 400×. (f) Oil Red O stain for the detection of lipid deposits in frozen sections of liver from +/+ (left) and −/− (right) mice, 400×. A dramatic increase in the number of lipid vesicles is present in −/− liver. Original magnifications are as indicated. Reprinted from Li et al. (1995) Nature Genet. **11**, 376, by courtesy of Prof. C.J. Epstein and Macmillan.

```
                              BBBBBB                                                                    HHHHHHHHHHHH
Mouse GPX:            MCAA.RLSAAA.......QSTVYAFSARPLTGGEPVSLGSLRGKVLLIENVASL GTTI DYTEWNDLQKRLGPRGLVVLGFPCNQFGHQENG
Rat GPX:             MSAA.RLSAVA.......QSTVYAFSARPLAGGEPVSLGSLRGKVLLIENVASL GTTI DYTEWNDLQKRLGPRGLVVLGFPCNQFGHQENG
Rabbit GPX:          MCAA.RWAAA........QS.VYSFSAHPLAGGEPVNLGSLRGKVLLIENVASL GTTI DYTQMNELQERLGPRGLVVLGFPCNQFGHQENA
Bovine GPX:          MCAAQR.SAAALAAAAPRTVYAFSARPLAGGEPFNLSSLRGKVLLIENVASL GTTV DYTQMNELQRRLGPRGLVVLGFPCNQFGHQENA
Human GPX:           MCAA.RLAAAA.......QS.VYAFSARPLAGGEPVSLGSLRGKVLLIENVASL GTTV DYTQMNELQRRLGPRGLVVLGFPCNQFGHQENA
Human g1GPX:                    MAFIAKSFYDLSAISLDG.EXVDFNTFGRAVI LENVASL DFTQLNELQCRF.PRLVVLGFPCNQFGHQENCQ
Human pGPX:          MARILQASCLLSILLAGFVSQSRGQEKSKMDCHGGISGTIVEYGALTIDGEEYIPFKQYAGKYVLFVNVASY GLTG.QYIELNALQEEGPFGVIVLGFPCNQFGK QEPGE
Rat pGPX:            MSRILRASCLLSILLAGFVPPGRGQEKSKTDCHGGMSGTIVEYGALTIDGEEYIPFKQYAGKY1LFVNVASY GLTG.QYIELNALQEEGPFGVIVLGFPCNQFGK QEPGE
Bovine pGPX:         MARLFRASCLLSILLAGFIPPSDGQEKSKTDCHAGVGGTIVEYGALTIDGEEYIPFKQYRGKHVFVNVATYCGLTI.QYPELNALQODLKOFGLVILGFPCNQFGG QEPGS
Mouse eGPXH:         NVTELRVFYLVPLLLASYVQTTPREKMKMDCYKDVKGTIVDYEALSLNGKDIPFKQYRGKHVFVNVATYCGLTI.QYPELNALQODLKOFGLVILGFPCNQFGG QEPGS
Rat eGPXH:           MAIQLRVFYLVPLLLASYVQTTPREKKMKDCYKDVKGTIVNYEALSNGKERIPFKQYRGKHVFWNVATYCGLT.QYPELNALQEEAGPYGLVILGFPCNQFGG QEPGS
M.fascicularis eGPXH: NTTQLRVHHLPLLLACFVOTSPKQETMKDCHKDEKGTIVDYEALANKNEYVPFKQYVGKHILFVNVASFCGLTA.QYPELALQEEALRPGFNVSVLGFPCNQFGG QEPGD
Nematode GPXH:       MTQOFWGPCLFSLFMAVLAOETLDPQKSKVDCMKGVAGTIVEYGANTLDGGEYVOFOQYAGKHILFVNVATYCAYTM.QYRDFNPILESNSMGTLNIGFPCNQFYL QEPAE
Pig PHGPX:                  MCASRDDWRCARSMHE FSAKDIOG.HMVNLDKYRGYVCIVTWNVASQ.AOKSLAEYRNKVLLIVWNATYCAYTM.QYRDFNPILESNSMGTLNIGFPCNQFYL QEPAE
Rat PHGPX:                  MCASRDDWRCARSMHE FAAKDIOG.HMVCLDKYRGCVCIVTWNVASQ GKTDYNYTQLVDLHARYAECGLRIAFPCNQFGR QEPGS
Human PHGPX:                MCASRDDWRCARSMHE FSAKDIOG.HMVNLDKYRGFVCIVTWNVASQ GKTEVNYTQLVDLHARYAECGLRIAFPCNQFGR QEPGS
S.mansoni GPX:              MSSSHKSWNS1YEFTVKDING.VDVSLEKYRGHVCLIVVACK GA TPKNYRQLQEMHTRLVGKGLRILAFPCNQFGG QEPWA
N.sylvestris GPXH:          MASQSSKPOSIYDFTVKDAG.NDVDLSIYKGKLIIVNVASOG1TNSNYTELSOLYDKYKNOGLEIIAFPCNQFGG QEPGS
C.sinensis GPXH:            MASQSKTS WHDFTVKDAKG.QDVDLSIYKGKLLIIVNVASQG1TNSNYTELSOLYDKYKNOGLEIIAFPCNQFGG QEPGS
E.coli GPXH:         MSAQLILSHMVLLOLIVAQLGPKIGKQCEITNOTVYDFYOVOMLNG.AOKSLAEYRNKVLLIVWNATYCAYTM.QYEQLENIQXAWVDRGFMVLGFPCNQFLE QEPGD
                              BBBBBBB                                                          BBBBBBBB

                              HHHHHHHHHH                    BBBBBBB                                                      BBBBBB                      HHHHHHHHH
Mouse GPX:           NEEILNSLKYV PGGGFEP....NFTLFEKCEVNGEKAHPLFTFLRNALPT. PSDDPTALMTDPKYIIWSPVCRNDIA NFEKFLVGPDGVPVRRYSRRFRTIDIEPDIETLLSQQSGNS
Rat GPX:             NEEILNSLKYV PGGGFEP....NFTLFEKCEVNGEKAHPLFTFLRNALPA. PSDDPTALMTDPKYIIWSPVCRNDIS SFEKFLVGPDGVPVRRYSRRFRTIDIEPDIEALLSKOPSNP
Rabbit GPX:          NEEILNCLKYV PGGGFEP....NFMLFQKCEVNGAKASPLFA LREALPP. PSDDPTALMTDPKFITWSPVCRNDVS SFEKFLVGPDGVPVRRYSRRFPTIDIEPDIQALLSKGSGGA
Bovine GPX:          NEEILNSLKYV PGGGFEP....NFMLFEKCEVNGAGAHPLFAFLREVLPT. PSDDATALMTDPKFITWSPVCRNDVS NFEKFLVGPDGVPVLRYSRRFFOTIDIEPDIEALLSQGPSCA
Human GPX:           NEEILNSLKYV PGGGYQP....TFTLVQKCEVNGQNEHPVFAYLKDKLPY. PYDDPFSLMTDPKLIIWSPVRRSDVA FEKFLIGPEGEPFRRYSRTFPTINIEPDIKRLLKVA1
Human g1GPX:         NEEILPTLKYV PGGGFVP....NFQLFEKGDIMGEKEQKFYTFLKNSCP. DRLFWEPMKVHDIR GRLFWEPMKIHDIR TTVSNVKMDILSYMRROAALGVKRK
Human pGPX:          NSEILPSLKYV PGGGFVP....NFQLFEKGDIWGEKEQKFYTFLKNSCP. DRLFWEPMKVHDIR TTVSNVKMDILSYMRROAALGARGK
Rat pGPX:            NSEILATLKYV PGGGFTP....NFQLFEKGDIWGEKEQKFYTFLKNSCP. PTAELLGSP. GRLFWEPMKIHDIR TTVSNVKMDILSYMRROAALGARGK
Bovine pGPX:         NLEILPGLKYV PGKGFLP....NFQLFEKGDVNGENEQKIFTFLKRSCP. PTSEILGSP. KHTSWEPIKVHDIR TTVNSVKMDILTYMRRRAVWEAKGK
Mouse eGPXH:         NKEILPGLKYV PGGGFVP....NFQLFEKGDVNGEKEQKITFLKRSCP. HPSETVVMS. KHTSWEPIKVHDIR NFETFLVGPNGVPVMRWFHQAPVSTVKSDIMAYLSHFKTI
Rat eGPXH:           NSEILLGLKYV PGKGFLP....NFQLFEKGDVNGDNEQKVFSFLKSSCP. PTSEILGTF. KSISWDPVKVHDIR NFSKFLVDROGQPVKRYSPTTAPVDIEGDIMELLEKK
M.fascicularis eGPXH: NHELLSGLKYV PGHGMEPHKNHHIFGKLEVNGENDHPLYKFLKERCP. PTVPVIGKR. EHLFWDPMKVHDIR NFSKFLVDKEGNVDRYSPTTPASMEKDIKKLLGVA
Nematode GPXH:       DAEI......KEFAAGYNVK.....FDMFSKICVNGDDAHPLMKWMKVO. PKGRGMLG. HQLIYDPIGTNDVI NFSKFLVDKEGNVVERYAPTTSPLSIEKDIKKLLETA
Pig PHGPX:           NQEI......KEFAAGYNVK.....FDMYSKICVNGDDAHPLWKWMKVO. PKGRGMLG. NAIK TPTINIEPDIKRLLKVA1
Rat PHGPX:           NEEI......KEFAAGYNVK.....FDMFSKICVNGDDAHPLWKWMKIO. PKGRGILG. NAIK
Human PHGPX:         EAEI......KKFVTEKYGV.....OFDMFSKIKVNGSDADDLYKFLKSR. OHGTLT. NMIK
S.mansoni GPX:       IEEI......ONMVCTRFKA.....EYPIFDKVDVNGDNAAPLYKFLKSS...KGGFFG. DSIK
N.sylvestris GPXH:   NEQI......OEFACTRFKA.....EEPIFDKVDVNGDNAAPLYKHLKSS...KGGLFG. DSIK
C.sinensis GPXH:     DEEI.....KTYCTTTWGV.....TFPMFSKIEVNGEGRHPLYOKLIAAAPTAVAPEESGFYARMVSGRAPLYPDDIL SPDMTPEDPIVMESIKLALAK
E.coli GPXH:                                                                                                  HHHHHHHHHH
                          HHHHHHHH                                                                               BBBBBB                      BBBBBB
```

Fig. 3.13. Amino acid sequence alignments of the glutathione peroxidase superfamily. Residues essential at the active site, selenocysteine (position 52 in 'classical' bovine glutathione peroxidase), glutamine and tryptophan (the 'catalytic triad') are marked in red. The four arginine residues (57, 103, 184, 185) and the lysine (91) (numbering is for the bovine enzyme) involved in GSH binding are shown in blue. Green denotes the 'signal peptides' directing export of plasma (cGPXH) and phospholipid hydroperoxide enzymes. HHH, α-helices, βββ, pleated sheets in the bovine enzyme structure. Diagram from *Methods Enzymol.* **252**, 48 (1996) by courtesy of Professors F. Ursini and L. Flohé and of Academic Press. Also see Flohé, L. (1994) 'The Boss's Hobby—Glutathione Peroxidases', in Annual Report of the Gesellschaft für Biotechnologische Forschung mbH, Braunschweig.

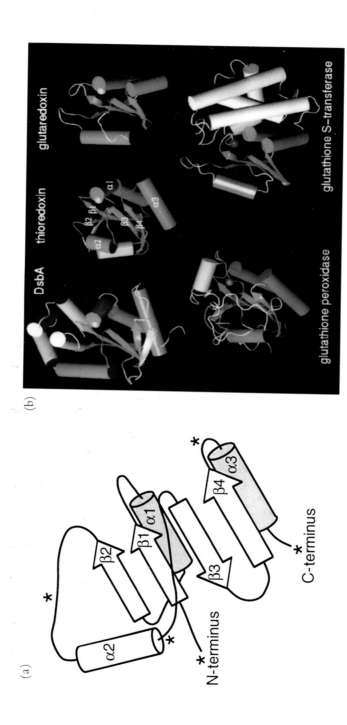

Fig. 3.16. The thioredoxin superfamily. Structural compositions of five proteins with thioredoxin folds. (a) The thioredoxin 'fold' contains around 80 residues and involves β-sheets (β1–4) and α-helices (α1–3). An N-terminal β1α1β2 motif is linked to a C-terminal β3β4α3 motif by a loop of residues also containing a helix (α2). (b) Structural comparison of five proteins to show the presence of thioredoxin folds (in green). **DsbA** is a protein disulphide oxidant from *E.coli*, present in the periplasmic space and involved in the folding of proteins exported from the cytoplasm. The atoms that interact with the cysteine residues of a substrate are shown as coloured balls; yellow is a sulphur atom of the more N-terminal cysteine in thioredoxin, DsbA and glutaredoxin, pink is the selenocysteine residue of bovine erythrocyte glutathione peroxidase and red the tyrosine hydroxyl of rat liver mu class glutathione S-transferase. Adapted from *Structure* **3**, 245 (1995) by courtesy of Dr. J.L. Martin and the publishers.

Fig. 3.30. Carotenoids in nature soothe our eyes and lift our spirits, whether in the petals of flowers, the plumage of birds or in the delicate hues of some fruits and vegetables. Photo and quotation courtesy of Professors J.A. Olson and N.I. Krinsky and the *FASEB Journal* (see *FASEB J.* **9**, 1547 (1995)). Plant phenols such as anthocyanins also contribute blue, red and purple colours to flowers, fruits and leaves.

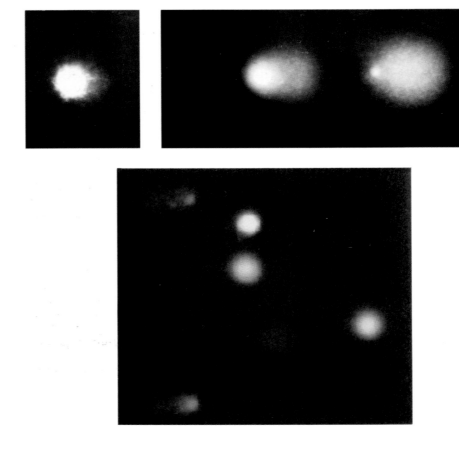

Fig. 4.8. The 'comet' assay. The assay detects strand breaks in DNA. Human lymphocytes (before or after H_2O_2 treatment) were suspended in low melting point agarose gel in buffer at 37°C and pipetted on to a frosted glass microscope slide precoated with 1% high melting point agarose, similarly prepared in buffer. The agarose was allowed to set by incubating at 4°C for 10 min, and the slides were then immersed in lysis solution (1% Triton X-100, 2.5 M NaCl, 0.1 M Na_2EDTA, 10 mM Tris, pH 10.0) at 4°C for 1 h to remove cellular membranes, proteins, etc. Slides were then placed in a horizontal electrophoresis tank containing 0.3 M NaOH and 1 mM Na_2EDTA at 4°C for 40 min before electrophoresis in the same solution at 25 V for 30 min at an ambient temperature of 4°C (temperature of running solution not exceeding 15°C). DNA, being negatively charged, is attracted to the anode. The presence of breaks allows supercoiled DNA to relax and move more quickly to the anode to form a tail (hence the name comet assay), and the fraction of DNA in the tail reflects the frequency of breaks. The slides were washed three times for 5 min each with neutralizing buffer (0.4 M Tris–HCl, pH 7.5) at 4°C before staining. Above: the figure (left to right) shows increasing extents of DNA damage. Below: lymphocyte comets treated with H_2O_2 and stained with acridine orange, which stains double-stranded DNA yellow and single-stranded DNA red. Undamaged comets are yellow, whereas tailed comets are red. The comet assay detects strand breaks in DNA. A modification of the assay permits detection of oxidized bases in addition to breaks. In this case, following lysis, the slides are washed three times for 5 min each in endonuclease III buffer (40 mM Hepes-KOH, 0.1 M KCl, 0.5 mM EDTA, 0.2 mg/ml BSA, pH 8.0), drained, and the agarose covered with 50 μl of either endonuclease III in buffer (1 μg protein/ml) or buffer only, sealed with a coverslip, and incubated for 30 min at 37°C. Endonuclease III introduces breaks in the DNA at sites of oxidized pyrimidines. Other DNA repair enzymes can also be used. Alkaline treatment, electrophoresis and neutralization are carried out as before. Photographs kindly donated by Dr Andrew Collins, Rowett Research Institute. The comet assay has been used to show increased oxidative damage to DNA bases in subjects exposed to 2.5 atmospheres of pure O_2 in hyperbaric chambers *(Mutagenesis* **11**, 605 (1996)).

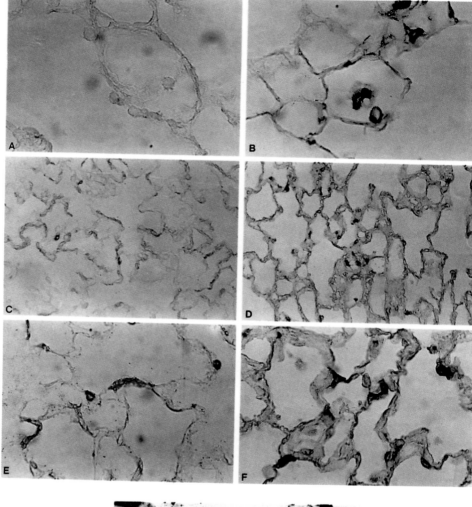

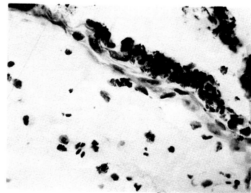

Fig. 5.8. Top panels: Detection of superoxide and of nitrated proteins *in vivo*. Rats were injected with sufficient bacterial lipopolysaccharide to cause septic shock. There was a time-dependent increase in inducible nitric oxide synthase (iNOS) in the lung. Staining of lung sections with an anti-nitrotyrosine antibody showed the presence of this modified amino acid. Perfusion with nitroblue tetrazolium was also carried out; $O_2^{\bullet-}$ reduces it to blue formazan. Lungs from untreated rats (A, C, E) or 48 h after endotoxaemia (B, D, F) were perfused with a phorbol ester to activate the respiratory bust and with nitroblue tetrazolium and fixed. Sections were then stained with an anti-nitrotyrosine antibody (C, D, E, F) or control antiserum (A, B). (A) No staining; (B) formazan production suggestive of $O_2^{\bullet-}$;(C, E) control for anti-nitrotyrosine staining; (D, F) Increased nitrotyrosine and formazan in endotoxaemic animals. Magnification: ×1000 (A,B,E,F) or ×400 (C,D). Data from *J. Leuk. Biol*, **56**, 759 (1994) by courtesy of Dr Debra Laskin and the publishers. Bottom panel: histochemical detection of H_2O_2 production in a healing (3 day) rabbit, inflammatory skin lesion. The leukocytes in the crust are actively producing H_2O_2 (shown by the orange-brown colour). New epithelium is growing beneath the crust. The tissue section was fixed for 4 h at 4 °C and stained with diaminobenzidine for 5 h at 37 °C. Photograph from *J. Leuk. Biol.* **56**, 436 (1994) by courtesy of Dr Arthur M. Dannenberg Jr and the publishers.

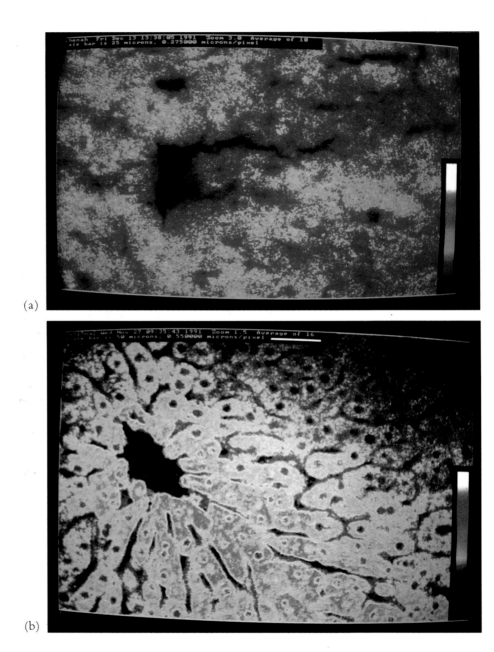

(a)

(b)

Fig. 5.19. Imaging of oxidative stress by confocal laser scanning microscopy. Sections of liver from a control rat (a) or a rat treated with CCl_4 to induce lipid peroxidation (b) were stained with 3-hydroxy-2-naphthoic acid hydrazine reagent, which reacts with carbonyls to give a fluorescent product. Adapted from *Am. J. Pathol.* **142**, 1353 (1993) by courtesy of Professors Mario Comporti and A. Pompella with the permission of the publishers.

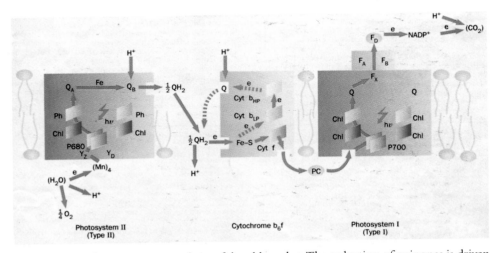

Fig. 7.7. The electron-transport chain of the chloroplast. The reduction of quinones is driven by a type II reaction centre (photosystem II), whereas the reaction centre responsible for oxidizing the reduced quinone and generating reducing power for the Calvin cycle is called photosystem I, a type I reaction centre. Chl, chlorophyll a; Ph, phaeophytin a. A pair of interacting chlorophylls acts as the primary electron donor (P680) for photosystem II, and P700 for photosystem I. Q, Quinone; plastoquinone (a quinone similar to ubiquinone) in photosystem II and phylloquinone (a K vitamin) in photosystem I. Fe, non-haem iron; F_X, F_A and F_B are iron–sulphur centres and F_D is ferredoxin. $(Mn)_4$, the cluster of four manganese ions involved in the H_2O-splitting process; Y_Z and Y_D are tyrosine residues, the former being on the main route of electron transfer. Arrows show electron-transfer paths. Photosystems I and II are connected by the cytochrome b_6-f complex which oxidizes the reduced quinone (QH_2, plastoquinol) and reduces a copper-containing protein, plastocyanin (PC). Because quinones are normally two-electron/two-proton acceptors, the intermediate cytochrome complex accepts two electrons, one reducing an iron sulphur centre (Fe–S) and the other reducing a low potential b-type cytochrome (Cyt b_{LP}). The reduced Fe–S centre donates an electron to PC through a bound cytochrome f (Cyt f). The other electron participates in a quinone-dependent cycle involving high (Cyt b_{HP}) and low (Cyt b_{LP}) potential b-type cytochromes. As electrons flow from Q to plastocyanin, sufficient energy is released to drive the synthesis of ATP from ADP and phosphate. Diagram adapted from *Nature* **370**, 33 (1994), by courtesy of Prof. Jim Barber and the publishers.

Fig. 7.13. Transgenic cotton plants that express chloroplast–localized MnSOD have increased tolerance to chilling-induced oxidative stress. Plants grown to the four-leaf stage in a growth chamber at 28°C day and 15°C night were exposed to chilling temperatures of 15°C day and 4°C night for 5 days. Immature leaves of wild–type cotton plants were visibly damaged (left), whereas those of transgenic cotton plants had no visible damage (right). Photograph from *Plant Physiol.* **107**, 1049 (1995) by courtesy of Dr R.D. Allen and *Plant Physiology.*

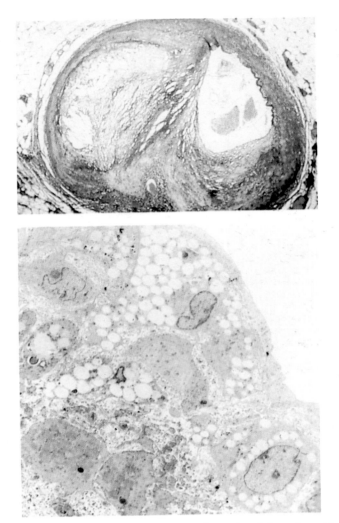

Fig. 9.4. Advanced atherosclerosis. Top: an atherosclerosed and thrombosed human coronary artery obtained during necropsy. Bottom: section through a lesion. Note the lipid–laden foam cells. By courtesy of Dr M.J. Mitchinson and the Upjohn Company.

of either $O_2^{\bullet-}$ or H_2O_2. Both are not necessarily required. Hence one can find organisms with SOD but no catalase or peroxidase activities, such as *Bacillus popilliae*. The gonococcus, by contrast, has exceptionally high catalase activities and can survive without SOD.[8] Consistent with this argument, catalase-negative strains of *Listeria monocytogenes* have increased SOD compared with catalase-positive strains. Given that no protective mechanism is 100% efficient *in vivo*, that FeSOD and CuZnSOD are inactivated by H_2O_2 and that $O_2^{\bullet-}$ can decrease catalase and peroxidase activities (Sections 3.7 and 3.16 below) most organisms have probably evolved to contain both SOD and H_2O_2-removing systems.

3.5.3 *Can manganese replace SOD?*

Several aerotolerant strains of Lactobacillaceae contain SOD, but a few do not, such as *Lactobacillus plantarum*. This organism accumulates manganese ions from its growth medium to an internal concentration of 25 mM or more.[7] If accumulation is prevented by removing manganese from the growth medium, *L. plantarum* will not grow in the presence of O_2. Since chelates of manganese ions with some biomolecules (e.g. polyphosphates) can react with $O_2^{\bullet-}$ (Chapter 2) it has been suggested that these complexes function to catalyse $O_2^{\bullet-}$ removal *in vivo*. *L. plantarum* also possesses a manganese-containing H_2O_2-degrading enzyme (Section 3.7.6 below).

Mutants of *S. cerevisiae* lacking CuZnSOD can be 'rescued' by mutations in other genes, such as the *BSD* (**bypass SOD deficiency**)[149] genes. One of these has been shown to encode a transport protein, mutation of which allows the cells to accumulate manganese. If extracellular manganese is removed, the 'rescue' is inoperative, again suggesting an $O_2^{\bullet-}$-scavenging role for manganese ions.

One feature presumably making it possible to use manganese to remove $O_2^{\bullet-}$ is that manganese ions do not appear to catalyse $OH^\bullet$ formation from H_2O_2 (Chapter 2).

3.6 Why is superoxide cytotoxic?

3.6.1 *Direct damage by superoxide*

The evidence reviewed above shows that SOD is an important antioxidant enzyme. Since SOD appears specific for $O_2^{\bullet-}$ as a substrate, it follows that $O_2^{\bullet-}$ is a species that is deleterious *in vivo*. Is this due to direct damage by $O_2^{\bullet-}$?

When compared with such highly reactive species as $OH^\bullet$, $O_2^{\bullet-}$ seems innocuous in chemical terms. The protonated form of $O_2^{\bullet-}$, $HO_2^\bullet$, is somewhat more reactive than $O_2^{\bullet-}$ itself. For example, $HO_2^\bullet$ can initiate peroxidation of fatty acids (Section 2.4.3), and a small amount of $HO_2^\bullet$ exists in equilibrium with $O_2^{\bullet-}$ even at physiological pH. The pH close to a membrane surface may be more acidic than the pH in bulk solution, so that $HO_2^\bullet$ formation would be favoured. The pH beneath activated macrophages adhering to a surface has been reported to be $\leqslant 5$, and so a considerable amount of any $O_2^{\bullet-}$ that they generate will exist as $HO_2^{\bullet}$. $HO_2^{\bullet}$ should be able to cross membranes as easily as H_2O_2. Much of the $O_2^{\bullet-}$ generated within cells comes from membrane-bound systems (e.g. the electron-transport chains of mitochondria and endoplasmic reticulum), and so $HO_2^{\bullet}$ formed close to the membrane could conceivably produce damage. Any $O_2^{\bullet-}$ that was produced in the hydrophobic membrane interior could be very damaging, since $O_2^{\bullet-}$ is highly reactive in organic solvents (Section 2.4.3). However, it has not been demonstrated that $O_2^{\bullet-}$ or $HO_2^{\bullet}$ mediate direct membrane damage *in vivo*.

Superoxide can decrease the activity of other antioxidant defence enzymes, such as catalase and glutathione peroxidase (Section 3.7 below). In *E. coli* several enzymes are direct targets of damage by $O_2^{\bullet-}$: examples are 6-phosphogluconate dehydratase, aconitase, fumarase and dihydroxyacid dehydratase (Chapter 1).[85] Dihydroxyacid dehydratase, which catalyses the third step in the biosynthesis of branched chain amino acids, contains a $[4Fe-4S]^{2+}$ iron–sulphur cluster at its active site, which is degraded upon exposure to $O_2^{\bullet-}$ or (with a lower rate constant) to O_2 itself. Similarly, *E. coli* aconitase, fumarase A and fumarase B enzymes, which contain similar clusters, are inactivated by O_2 and $O_2^{\bullet-}$; second-order rate constants for reactions of these enzymes with $O_2^{\bullet-}$ are $10^6-10^7 M^{-1}s^{-1}$ whereas with O_2 they are $10^2-10^3 M^{-1}s^{-1}$. Thus energy metabolism in the Krebs cycle is a major target of damage by $O_2^{\bullet-}$. Inactivation is caused by oxidation of the cluster, leading to release of iron ions. Mammalian aconitase[92] has a similar cluster and can be inactivated by $O_2^{\bullet-}$ *in vitro*; it has also been reported that $O_2^{\bullet-}$ can inactivate NADH dehydrogenase in bovine heart submitochondrial particles.[273] Hence energy metabolism may be a target of direct damage by $O_2^{\bullet-}$ in mammalian cells also.

Another important target of damage may be **ribonucleotide reductase**.[93] This enzyme, needed to provide the precursors required for DNA synthesis (Section 6.2), has an essential tyrosine radical at its active site. The reductase is inactivated by exposure to free radicals that can combine with the tyrosine radical: $NO^{\bullet}$ is one example (Chapter 2) and $O_2^{\bullet-}$ may be another. Both *E. coli* and mammalian ribonucleotide reductases have been shown to be inactivated by $O_2^{\bullet-}$-generating systems *in vitro*. Some mammalian creatine kinases are also reported as inactivated by exposure to $O_2^{\bullet-}$. Superoxide may also damage **calcineurin** (Chapter 4), a protein involved in signal transduction.

3.6.2 *Cytotoxicity of superoxide-derived species*

Hydrogen peroxide and peroxynitrite

In addition to direct damage by $O_2^{\bullet-}$, $O_2^{\bullet-}$ could be cytotoxic by generating more-reactive species. Dismutation of $O_2^{\bullet-}$ generates H_2O_2, but H_2O_2 is poorly reactive at physiological levels, although it can attack some enzymes (Section 2.5). A more reactive and cytotoxic species is **peroxynitrite**, produced by the very fast reaction of $O_2^{\bullet-}$ with $NO^{\bullet}$. Peroxynitrite generates a wide range of noxious species under physiological conditions (Section 2.5.4).

For example, $ONOO^-$ inactivates[139a] creatine kinase more efficiently than does $O_2^{\bullet-}$.

Hydroxyl radical

Another species to be considered is **hydroxyl radical**. Many studies *in vitro* that showed the ability of $O_2^{\bullet-}$-generating systems to kill cells and damage biomolecules found that protection was achieved by adding not only SOD, but also catalase and 'OH$^{\bullet}$ scavengers' (such as mannitol, formate, thiourea and dimethylsulphoxide (DMSO)).[108] At first, it was suggested that $O_2^{\bullet-}$ and H_2O_2 react to form $OH^{\bullet}$, according to the overall equation:

$$H_2O_2 + O_2^{\bullet-} \rightarrow O_2 + OH^{\bullet} + OH^-$$

Indeed, formation of $OH^{\bullet}$ in a wide range of $O_2^{\bullet-}$-generating systems has been detected by a variety of methods, including electron spin resonance (ESR) spin trapping and aromatic hydroxylation as well as less specific techniques (Chapter 5). The above reaction was postulated by F. Haber and J. Weiss in 1934 and has become known as the **Haber–Weiss reaction**. However, the rate constant for the reaction in aqueous solution is virtually zero. Nevertheless, $OH^{\bullet}$ formation can be accounted for if the Haber–Weiss reaction is **catalysed** by transition metal ions as first suggested by Weiss in 1935.

$$\text{oxidized metal complex} + O_2^{\bullet-} \rightarrow \text{reduced metal complex} + O_2$$

$$\text{reduced metal complex} + H_2O_2 \rightarrow OH^{\bullet} + OH^- + \text{oxidized metal complex}$$

$$\text{Net: } O_2^{\bullet-} + H_2O_2 \xrightarrow[\text{catalyst}]{\text{metal}} O_2 + OH^{\bullet} + OH^-$$

Transition-metal ions, especially iron ions, contaminate most biochemical reagents. Chelates of such metals as chromium (Cr^{2+}), nickel (Ni^{2+}), cobalt (Co^{2+}), titanium (Ti^{3+}) and vanadium (vanadyl) can participate in $OH^{\bullet}$ formation *in vitro* and $OH^{\bullet}$ may be involved in the toxic effects of such metal compounds *in vivo* (Chapter 8). However, most attention has focused on iron and copper as potential mediators of $OH^{\bullet}$ generation under normal physiological conditions.[108]

$$Fe(III) + O_2^{\bullet-} \rightarrow Fe^{2+} + O_2$$

$$Fe^{2+} + H_2O_2 \rightarrow OH^{\bullet} + OH^- + Fe(III)$$

or

$$Cu^{2+} + O_2^{\bullet-} \rightarrow Cu^+ + O_2$$

$$Cu^+ + H_2O_2 \rightarrow Cu^{2+} + OH^{\bullet} + OH^-$$

$$\text{Net: } O_2^{\bullet-} + H_2O_2 \xrightarrow[\text{catalyst}]{\text{metal}} O_2 + OH^{\bullet} + OH^{\bullet}$$

(the iron-catalyzed or copper-catalyzed Haber–Weiss reactions).

Production of OH$^\bullet$ *in vitro* by $O_2^{\bullet-}$-generating systems containing iron or copper ions (added or present as contaminants in the reagents) is inhibited by catalase, which removes the necessary H_2O_2. It is also inhibited by SOD and by certain metal ion-chelating agents (Chapter 10).

Catalytic metals in vivo?

However, the availability of 'catalytic' iron and copper *in vivo* is very restricted; indeed, this restriction is an important antioxidant defence system (Section 3.18 below). Nevertheless, increased generation of $O_2^{\bullet-}$ and H_2O_2 can *create* conditions favourable for OH$^\bullet$ formation.[d] For example, $O_2^{\bullet-}$ can release iron ions from mammalian ferritin (Section 3.18.4 below) and from iron–sulphur clusters in enzymes, whereas H_2O_2 can degrade haem in haem proteins to release iron (Chapter 2). Peroxynitrite displaces iron from Fe–S proteins and degrades caeruloplasmin to release copper (Section 2.5.4). Increased availability of iron has been measured in cultured cells using a fluorescent probe[38] that detects Fe^{2+}. Exposure to H_2O_2 or organic peroxides caused rises in the iron concentration detected by the probe. Destruction of cells leads to metal ion liberation, which can promote free radical damage in the surrounding tissue.[108,248]

Hydroxyl radicals, once generated, react with the molecules in their immediate vicinity. The much lower reactivity of $O_2^{\bullet-}$ and H_2O_2 means that they can diffuse away from their sites of formation, leading to OH$^\bullet$ generation in different parts of the cell whenever they meet a 'spare' catalytic metal ion. Hence the toxicity of $O_2^{\bullet-}$ and H_2O_2 can be influenced by the availability and distribution of metal ions. Indeed, studies on the bacterium *Staphylococcus aureus* showed many years ago that increasing the iron content of the cells increased their susceptibility to killing by H_2O_2, and cell-permeable OH$^\bullet$ scavengers offered some protection.[212]

Manipulations of the intracellular iron content of *E. coli* have led to similar conclusions.[250] *E. coli* has a high-affinity iron uptake system, involving about 30 genes whose transcription is increased if cell iron content is low. Iron can be taken up as Fe(III) bound to **siderophores** (bacterially produced iron-chelating agents), or directly as Fe^{2+}. The whole system is regulated by the *fur* (**ferric uptake regulator**) gene. The protein encoded by *fur* binds Fe^{2+} and the complex binds to a DNA base sequence (the **iron box**) found in the promoter regions[e] of many genes regulated by iron. The *fur*–iron complex binds to these iron boxes and blocks transcription. *E. coli fur* mutants lacking this DNA-binding ability accumulate excess iron and become hypersensitive to H_2O_2, especially so if MnSOD and FeSOD are also absent. A double mutant, defective in both *fur* and DNA repair, is not viable under aerobic conditions. It can be 'rescued' by addition of an iron chelator (ferrozine), further mutations interfering with iron uptake systems, over-expression of a bacterial ferritin, increasing its SOD activity or adding the OH$^\bullet$ scavengers DMSO or thiourea. *Fur* mutants are also hypersensitive to hypochlorous acid, suggesting that reaction of this species with iron to form OH$^\bullet$ (Section 2.5.2) might be significant *in vivo*.[71]

Superoxide and iron

The only role apparent for $O_2^{\bullet-}$ in the iron/copper-catalysed Haber–Weiss reaction is to regenerate reduced metal ions (Fe^{2+} or Cu^+) which then react with H_2O_2 to form $OH^\bullet$. Hence the iron-catalysed process can equally well be described as a **superoxide-assisted Fenton reaction**. Cells contain many reducing agents at millimolar concentrations (e.g. GSH, NAD(P)H, cysteine, ascorbic acid) that can reduce Fe(III) and Cu^{2+}, which suggests that a role for $O_2^{\bullet-}$ (only at picomolar or nanomolar concentrations) is unlikely. However, the interactions of iron and copper ions with NADH, NADPH and thiol compounds (such as GSH and cysteine) to generate $OH^\bullet$ are often inhibited by SOD *in vitro*, i.e. the $OH^\bullet$ generation is superoxide-dependent.[108,217] By contrast, when ascorbate is the reducing agent, SOD does not prevent $OH^\bullet$ production. Hence the biological importance of $O_2^{\bullet-}$ as a reductant of transition metal ions is uncertain; it may be more important in providing the necessary catalytic metal ions.

Whatever the mechanism, much evidence supports a role of $O_2^{\bullet-}$ in facilitating the toxicity of H_2O_2. Externally added SOD does not usually protect mammalian cells against the toxicity of H_2O_2, but it can do so if allowed to enter the cells.[147] We have already noted that *E. coli* lacking FeSOD and MnSOD is hypersensitive to H_2O_2; killing of *E. coli* K12 by low (2–3 mM) concentrations of H_2O_2 appears to involve DNA damage and it is inhibited by the metal ion chelator *o*-phenanthroline. Since neither $O_2^{\bullet-}$ nor H_2O_2 reacts with DNA, these and other observations suggest that H_2O_2-dependent DNA damage in most cells is mediated by metal-catalysed $OH^\bullet$ formation.[180] This conclusion is supported by examination of the chemical changes in DNA, which show a pattern characteristic of attack by $OH^\bullet$ (Chapters 4 and 5). It seems that H_2O_2 crosses the plasma membrane and cytosol, penetrates the nucleus and interacts with transition metal ions bound to DNA to form $OH^\bullet$. These metals might always be bound to DNA, or metals released from metalloproteins by the action of $O_2^{\bullet-}/H_2O_2$ could then bind to DNA, making it a target of damage by **site-specific $OH^\bullet$ generation**.

Singlet oxygen

Singlet O_2 has also been suggested to mediate the toxicity of $O_2^{\bullet-}$; there are claims that it is produced in the non-enzymic dismutation of $O_2^{\bullet-}$ and the metal-catalysed Haber–Weiss reaction. However, data supporting these proposals are sparse. Indeed, in large amounts, $O_2^{\bullet-}$ can quench $^1O_2$ by electron transfer

$$^1O_2 + O_2^{\bullet-} \rightarrow O_2 + O_2^{\bullet-}$$

Addition of limited amounts of water to suspensions of potassium superoxide ($K^+O_2^{\bullet-}$) in chlorinated hydrocarbon solvents[129] has been shown to produce $^1\Delta gO_2$. However, $O_2^{\bullet-}$ reacts with chlorinated hydrocarbons (Section 2.4.3) and the following reactions (generation of peroxyl radicals followed by their

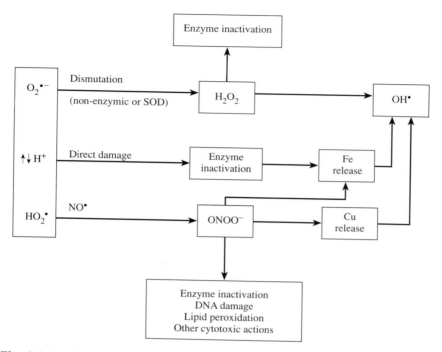

Fig. 3.9. Mechanisms of superoxide-dependent damage to biomolecules: a summary.

self-reactions) might account for singlet oxygen ($^1O_2$) production

$$O_2^{\bullet-} + CCl_4 \rightarrow CCl_3O_2^{\bullet} + Cl^-$$

$$CCl_3O_2^{\bullet} + CCl_3O_2^{\bullet} \rightarrow 2\,^1O_2 + CCl_3CCl_3$$

$$O_2^{\bullet-} + CCl_3O_2^{\bullet} \rightarrow CCl_3O_2^- + \,^1O_2$$

Figure 3.9 summarizes various mechanisms of $O_2^{\bullet-}$ toxicity.

3.7 Antioxidant defence enzymes: catalases

Dismutation of $O_2^{\bullet-}$ generates H_2O_2, a species also generated by several oxidase enzymes *in vivo*, including xanthine, urate and D-amino acid oxidases.[51] Hydrogen peroxide is usually removed in aerobes by two types of enzyme. The **catalases** directly catalyse decomposition of H_2O_2 to ground-state O_2

$$2H_2O_2 \rightarrow 2H_2O + O_2$$

Peroxidase enzymes remove H_2O_2 by using it to oxidize another substrate (written SH_2 below)

$$SH_2 + H_2O_2 \rightarrow S + 2H_2O$$

Most aerobic cells contain catalase activity,[51] although a few do not, such as the bacterium *Bacillus popilliae*, *Mycoplasma pneumoniae*, the green alga *Euglena*, several parasitic helminths (e.g. the liver fluke) and the blue-green alga *Gloeocapsa*. A few anaerobic bacteria, such as *Propionibacterium shermanii*, contain catalase, but most do not. In animals catalase is present in all major body organs, being especially concentrated in liver. Catalase in erythrocytes may help protect them against H_2O_2 generated by dismutation of $O_2^{\bullet-}$ generated by haemoglobin autoxidation (Section 7.2). Since H_2O_2 diffuses readily, erythrocytes can also protect other tissues against oxidative damage by 'absorbing' H_2O_2.[265] The brain, heart and skeletal muscle contain lower levels of catalase than liver (Table 3.7) although the activity varies between cell types: we have already mentioned the difficulty of interpreting measurements on homogenates of whole tissues. Plants have multiple catalases, encoded by several genes, e.g. in maize three isoenzymes encoded by three separate genes have been identified (Chapter 7).

Table 3.7. Catalase and glutathione peroxidase activities in human tissues

Tissue	Individual	Catalase activity $(mg^{-1}$ protein)	Glutathione peroxidase activity $(mg^{-1}$ protein)
Liver	A	1300	190
	B	1500	120
Erythrocytes	A	990	19
	B	1300	19
Kidney cortex	A	430	140
	B	110	87
Adrenal gland	B	300	120
Kidney medulla	A	700	90
	B	220	73
Spleen	A	56	50
Lymph node	A	120	160
Pancreas	A	100	43
	B	120	110
Lung	A	210	53
	B	180	54
Heart	A	54	69
Skeletal muscle	A	36	38
	B	25	22
Brain (grey matter)	A	11	71
	B	3	66
Brain (white matter)	A	20	76
Adipose tissue	A	270	77
	B	560	89

Data from Marklund *et al.* (1982) *Cancer Res.* **42**, 1955. Glutathione peroxidase was assayed with an organic substrate (see text). Results are expressed as enzyme activity/mg protein. Two individuals, denoted A and B, were used as sources of tissue samples.

3.7.1   Catalase structure

Animal catalases consist of four protein subunits, each of which contains a ferric haem group bound to its active site.[210] The haem groups are buried in non-polar pockets, connected to the surface by narrow channels, thus preventing most molecules larger than H_2O_2 from gaining access. Each subunit usually has one molecule of NADPH bound to it. Dissociation of catalase into its subunits, which easily occurs on storage, freeze-drying or exposure of the enzyme to acid or alkali, causes loss of catalase activity.[233] Commercial catalase preparations can be contaminated with these partial denaturation products, as well as with SOD, antioxidant 'stabilizers' such as thymol and even with endotoxin. This should be borne in mind in using commercial catalase as a 'probe' for the involvement of H_2O_2 in a reaction, especially in whole animals.[96]

E. coli has two catalases,[223] one of which (known as **hydroperoxidase II**, HPII) is a hexamer with one haem per subunit (but without bound NADPH). HPII is encoded by the *kat*E gene. By contrast, **hydroperoxidase I** (HPI), encoded by the *kat*G gene, is a tetramer which shows both catalase and a wide range of peroxidase activities, i.e. it is a **bifunctional catalase–peroxidase**. HPI is expressed under both aerobic and anaerobic conditions and levels increase if H_2O_2 is present. Similar enzymes are present in other bacteria.

The three-dimensional structures of several catalases have been determined by X-ray crystallography.[210] For example, each subunit of bovine liver catalase consists of a large antiparallel β-pleated sheet domain with helical insertions, followed by a smaller domain containing four α-helices. The haem group is buried at least $20\,\text{Å}$ below the molecular surface, only accessible by a channel lined with hydrophobic residues. Ligands to the haem iron are provided by Tyr357, His74 and Asp147.

3.7.2   Catalase reaction mechanism

The catalase reaction mechanism may be written as follows. It is, like that of SOD, essentially a dismutation (disproportionation); one H_2O_2 is reduced to H_2O and the other oxidized to O_2

$$\text{catalase–Fe(III)} + H_2O_2 \xrightarrow{k_1} \text{compound I} + H_2O$$

$$\text{compound I} + H_2O_2 \xrightarrow{k_2} \text{catalase–Fe(III)} + H_2O + O_2$$

For rat liver catalase, the rate constants, k_1 and k_2, have values of $1.7 \times 10^7\,\text{M}^{-1}\,\text{s}^{-1}$ and $2.6 \times 10^7\,\text{M}^{-1}\,\text{s}^{-1}$, respectively.[51] Formation of compound I leads to characteristic changes in the absorbance spectrum of catalase (Fig. 3.10), which have been used to assess rates of H_2O_2 production in perfused organs (Chapter 5). The exact structure of compound I is uncertain—the iron is oxidized from Fe(III) to a nominal valency of Fe(V). However, an Fe(IV) oxoporphyrin-cation radical, $(\text{haem}^{\bullet\,+})\text{Fe(IV)O}$, is probably formed, i.e.

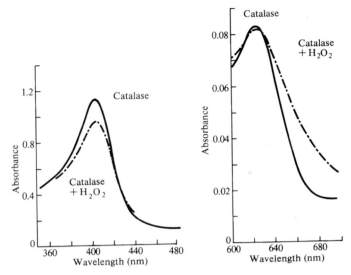

Fig. 3.10. Absorbance spectra of purified rat liver catalase and catalase compound I. The large absorbance of catalase around 400 nm is known as the **Soret band**, a feature of most haem proteins.

the iron is Fe(IV) and the extra oxidizing capacity is 'parked' by one–electron oxidation of the haem. Compound I receives two electrons from H_2O_2 to reform ferric catalase.

Catalase **compound II** can be produced by one electron reduction of compound I[150] and is an inactive Fe(IV) species, catalase–(haem)Fe(IV)O, without the porphyrin radical. Electron transfer can occur within compound I itself, generating an amino acid radical on the protein, or from an external electron donor. One role of the bound NADPH[121] may be to reconvert compound II to ferric catalase, releasing $NADP^+$.

It is very difficult to saturate catalase with H_2O_2—its maximal velocity (V_{max}) for the destruction of H_2O_2 is enormous. However, the above equations show that complete removal of H_2O_2 requires the impact of two H_2O_2 molecules upon a single catalase active site, which becomes less likely as H_2O_2 concentrations fall. The amount of compound I present in a mixture of catalase and H_2O_2 depends on their relative concentrations and on the rate constants k_1 and k_2. Chance *et al.*[51] calculated that at all reasonable concentrations, the rate of removal of H_2O_2 is given by the equation:

$$\text{moles } H_2O_2 \text{ used } (1^{-1}s^{-1}) = 2k_2[H_2O_2][\text{compound I}]$$

$$= 2k_1[H_2O_2][\text{free catalase}]$$

Thus if the concentration of H_2O_2 is fixed, the initial rate of its removal will be proportional to the concentration of catalase present and hence will be higher in liver than in, say, brain or heart (Table 3.7). Similarly, for a given concentration of catalase, the initial rate of H_2O_2 removal will be proportional to the H_2O_2 concentration. As a result, the specific activities (μmol H_2O_2 decomposed/min/mg protein) quoted by manufacturers for their catalase preparations are meaningless unless they describe exactly how the assay was done. Hydrogen peroxide decomposition can be followed by the loss of its absorbance at 240 nm, or by measuring the release of O_2 by using an oxygen electrode. Other assays for H_2O_2 are described in Chapter 5.

3.7.3   Catalase inhibitors

Catalase can be inhibited by azide, cyanide and HOCl, but these are non-specific: they inhibit many other enzymes. A more useful inhibitor[58] is **aminotriazole** (Fig. 3.11) which inhibits catalase when administered to animals or plants. Its inhibitory action is exerted on compound I, so aminotriazole will only inhibit catalase if H_2O_2 is present. Aminotriazole acts by modifying a histidine ligand to the haem (His74 in the bovine enzyme). Peroxide-dependent inhibition of catalase by aminotriazole has been used *in vitro* and *in vivo* to measure rates of H_2O_2 production (Chapter 5). Indeed, the observation that aminotriazole inhibits catalase when it is fed to whole animals or plants shows that H_2O_2 is produced *in vivo*.[236]

3.7.4   Peroxidatic activity of catalase

Mammalian catalases can also catalyse certain peroxidase-type reactions,[51] restricted by accessibility of substrates to the haem. Compound I will oxidize the alcohols methanol (CH_3OH) and ethanol (CH_3CH_2OH) to their corresponding aldehydes HCHO (formaldehyde or methanal) and CH_3CHO (acetaldehyde or ethanal), but propanols or butanols are much poorer substrates. Formic acid (HCOOH) can be oxidized to CO_2 by the peroxidase action of compound I. Catalase can also oxidize nitrite (NO_2^-) into nitrate (NO_3^-) *in vitro* and it has been suggested that it can oxidize elemental mercury (Hg) absorbed into the human body to form Hg^{2+} ions.

The presence of peroxidatic substrates for catalase *in vivo* will decrease the concentration of compound I, causing more free catalase to be formed, and this is another variable that must be considered in assessing how quickly H_2O_2 can

Fig. 3.11. Structure of aminotriazole. The full name of this compound is 3-amino-1,2,4-triazole.

be removed *in vivo*. The separated catalase subunits show little catalase activity, but have peroxidase activity on a much wider range of substrates, including NADH, because the active site haem is now more accessible.

The drug **cyanamide** ($H_2N-C\equiv N$) is converted *in vivo* into a product that inhibits the enzyme aldehyde dehydrogenase, which oxidizes acetaldehyde generated by ethanol metabolism (Chapter 8). Consumption of ethanol after taking cyanamide causes ethanal accumulation and unpleasant symptoms, and so cyanamide is used to deter alcohol consumption. It has been suggested that catalase is responsible for oxidizing cyanamide to the product that inhibits aldehyde dehydrogenase: during this oxidation, the catalase itself loses activity.[62]

3.7.5 Subcellular location of catalase

The catalase activity of animal and plant tissues is largely or completely located in subcellular organelles bounded by a single membrane and known as **peroxisomes**.[51] Peroxisomes contain many of the cellular enzymes that generate H_2O_2, such as glycolate oxidase, urate oxidase (not in primates) and the flavoprotein dehydrogenases involved in the β-oxidation of fatty acids, a metabolic pathway that operates in both mitochondria and peroxisomes in animal tissues. In mitochondria the flavoproteins involved in β-oxidation donate electrons to the electron-transport chain (Chapter 1), but in peroxisomes they react with O_2 to give H_2O_2. It seems logical that these enzymes have been packaged into an organelle with high capacity to destroy H_2O_2. Some CuZnSOD may also be present in peroxisomes.

Mitochondria (at least in liver), chloroplasts and the endoplasmic reticulum contain little, if any, catalase. Hence any H_2O_2 that they generate *in vivo* cannot be disposed of by catalase, unless H_2O_2 diffuses to the peroxisomes. Although much of the catalase activity detected in homogenates of animal and plant tissues is found not to be organelle-associated, this is probably due to the rupture of peroxisomes, which are fragile organelles, during the homogenization. However, some non-peroxisomal catalase may occur in the livers of a few animals, such as guinea-pigs.[43] Rat heart mitochondria contain some catalase in the matrix.[205] *S. cerevisiae* contains two catalases, one (**catalase A**) in peroxisomes and the other (**catalase T**) in the cytosol. Isolated subcellular fractions, especially microsomes, may be heavily contaminated with catalase activity and this can confuse experimental results.

3.7.6 Manganese-containing catalases

Several microorganisms, such as some strains of *Pediococcus*, *Streptococcus*, and *Lactobacillus*, contain **pseudocatalase**, an H_2O_2-degrading enzyme that is insensitive to inhibition by azide or cyanide and does not contain haem. Purification of pseudocatalase from *Lactobacillus plantarum*[32] showed that it had a relative molecular mass of around 172 000 and consisted of five subunits, each of which contains one Mn(III) ion. A similar enzyme was found in *Thermo-leophilum album*, but contained only four subunits.[4] It is thought that the Mn(III) becomes oxidized to a Mn(V) state by H_2O_2, so that the H_2O_2 is decomposed by a reaction mechanism similar to that for haem-containing catalases:

$$catalase-Mn(III) + H_2O_2 + 2H^+ \rightarrow catalase-Mn(V) + 2H_2O$$

$$catalase-Mn(V) + H_2O_2 \rightarrow catalase-Mn(III) + 2H^+ + O_2$$

The Mn-catalase of *L. plantarum* has been shown to be important in protecting this organism against H_2O_2.

3.7.7 *Acatalasaemia*

The gene encoding human catalase is located on chromosome 11; mutations of this gene can result in **acatalasaemia**. In the original description of this condition in Japan, there was a severe deficiency of catalase activity, but the only observable clinical problem was an increased incidence of mouth ulcer-ation. Splicing mutations,[e] or a deletion in exon 4 of the catalase gene, have been identified in different Japanese patients.[123] Another series of mutations has been described in Switzerland and results in an active but unstable enzyme, leading to variable decreases in catalase activity in different tissues: no clinical effects are apparent. *Drosophila* mutants with decreased catalase activity are viable, unless activity is decreased to $< 2\%$ of normal.[160]

Deleting catalase genes in bacteria or yeasts has, in general, much less striking effects on normal growth than deleting SOD genes (Section 3.4.1 above) but the cells usually become more sensitive to added H_2O_2 or H_2O_2-generating toxins. Treatment of *E. coli* with H_2O_2 induces synthesis of HPI but not HPII. It thus seems that catalase is not essential for aerobic life, presumably because in most aerobes other enzymes are present to help deal with H_2O_2. However, catalase is especially important in *N. gonorrhoeae*, as discussed in Section 3.5.2 above.

Aniridia is a disease associated with an increased incidence of both mental retardation and of the type of cancer known as **Wilms' tumour**. Aniridia results from a deletion on chromosome 11 and also leads to decreased catalase activity. The role (if any) of lowered catalase activity in the symptoms of aniridia is unknown.

3.8 Antioxidant defence enzymes: the glutathione peroxidase family[51]

Glutathione peroxidases (GPX) remove H_2O_2 by coupling its reduction to H_2O with oxidation of **reduced glutathione**, GSH.[51]

$$H_2O_2 + 2GSH \rightarrow GSSG + 2H_2O$$

GPX was first discovered (in animal tissues) in 1957. Glutathione peroxidases are not generally present in higher plants or bacteria, although they have been reported in a few algae and fungi. GSH, their substrate, is a low-molecular-mass thiol-containing tripeptide (Fig. 3.12). It is present in animals, plants and many aerobic bacteria (e.g. in *E. coli*), at intracellular concentrations that are often in the millimolar range (Table 3.8), but rarely is it present in anaerobic bacteria. Glutathione peroxidases can be inhibited on incubation with **mercaptosuccinate**.[52a]

H$_2$O$_2$-degrading GPX enzymes are widely distributed in animal tissues (Tables 3.7 and 3.8) and are specific for GSH as a hydrogen donor. However, they can act on peroxides other than H$_2$O$_2$. Thus they can catalyse GSH-dependent reduction of fatty acid hydroperoxides (e.g. linoleic and linolenic acid peroxidation products), cholesterol 7β-hydroperoxide (at a low rate), and various synthetic hydroperoxides such as cumene and *t*-butyl hydroperoxides, which are often used to assay the enzyme *in vitro*. In all cases the peroxide group is reduced to an alcohol,[51] e.g.

$$LOOH + 2GSH \rightarrow GSSG + H_2O + LOH$$

Glutathione peroxidases cannot act upon fatty acid peroxides esterified to lipid molecules in lipoproteins or membranes: they have to be first released by the action of lipase enzymes.

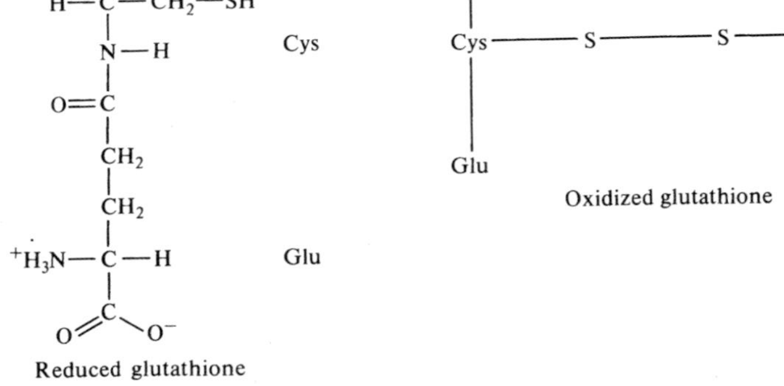

Fig. 3.12. The structures of reduced (GSH) and oxidized (GSSG) glutathione. GSH is a simple tripeptide (glutamic acid–cysteine–glycine). In GSSG, two GSH molecules join together as the –SH groups of cysteine are oxidized to form a disulphide bridge, –S–S–.

Table 3.8. Presence of glutathione and enzymes using it in different organisms. Some examples

System studied	GSH concentration	GSH/GSSG ratio	Glutathione peroxidase	Glutathione reductase
Spinach chloroplasts	3.0 mM	> 10/1	absent	high
Rat tissues				
liver	7–8 mM	> 10/1	high	high
erythrocyte	2 mM	> 10/1	moderate	moderate
heart	2 mM	> 10/1	moderate	moderate
lung	2 mM	> 10/1	moderate	moderate
lens	6–10 mM	> 10/1	moderate	moderate
spleen	4–5 mM	> 10/1	—	—
kidney	4 mM	> 10/1	moderate	moderate
brain	2 mM	> 10/1	moderate	moderate
skeletal muscle	1 mM	> 10.1	low	low
blood plasma	20–30 μM	~ 5/1	low	traces
adipose tissue	$3.2 \mu g/10^6$ cells	> 100/1	low	low
Human tissue				
liver	4 μmol/g wet wt	> 10/1	high	high
kidney	2 μmol/g	> 10/1	high	high
lens	6–10 mM	> 10/1	moderate	moderate
erythrocytes	240 μg/ml blood	> 10/1	moderate	moderate
whole blood	~ 1 mM	> 10/1	high	high
blood plasma[a]	1–3 μM	varies	low	absent
alveolar lining fluid	40–200 μM	varies (usually > 10/1)	traces	traces, or absent
Neurospora crassa	20 μmol/g dry wt	150/1	absent	moderate
E. coli				
aerobically grown	$27 \mu mol/g^b$	> 10/1	absent	moderate
anaerobically grown	7 μmol/g			

[a]Concentrations of cysteine and cystine in plasma are significantly higher (7–8 μM and 80–90 μM, respectively, in humans). Virtually all GSH in whole blood is in the cells, especially erythrocytes.
[b]3.5 mM in log growth: 6.6 mM in stationary phase.
Data compiled from a wide range of publications. The description of enzyme activities is relative rather than absolute. Whenever possible, concentrations are expressed as mmol/l (mM) but these cannot always be calculated from published data. These GSH values are not to be taken too literally, since they (i) may be affected by age, (ii) are different at different times of day in animals and at different points of the growth cycle in bacteria and fungi, (iii) in animal tissues (especially liver), fall on starvation, (iv) vary between the different cell types present in tissues, and (v) may depend on the sex of the animal (e.g. *Free Rad. Biol. Med.* **23**, 648 (1997); *Biochem. J.* **112**, 109 (1969)). GSH : GSSG ratios may vary in different subcellular compartments: ratios of 3 : 1 have been estimated for the interior of the endoplasmic reticulum, where protein folding and disulphide bond formation take place (*Science* **257**, 1496 (1993)).

3.8.1   *Structure of glutathione peroxidases[162a] and reductase*

Glutathione peroxidases consist of four protein subunits, each of which contains one atom of the element **selenium** (Se) at its active site. Selenium is in group VI of the periodic table (see Appendix I), and somewhat resembles sulphur in its chemistry. It is present at the active sites as **selenocysteine**, the amino acid cysteine in which the sulphur atom has been replaced by selenium (R—SeH instead of R—SH). Selenocysteine is encoded in the gene as TGA (UGA in mRNA), also a stop codon (Appendix II). To prevent misreading as 'stop', special conformations of the transcribed mRNA are recognized by a translation factor that ensures insertion of a selenocysteinyl-tRNA by the ribosome. This tRNA is first loaded with serine, which is then converted by an enzyme into selenocysteine.

During the catalytic mechanism of GSX, a **selenol** (protein—Se$^-$) reacts with peroxide to give a **selenenic acid** (protein—SeOH)

$$\text{protein–Se}^- + \text{ROOH} + \text{H}^+ \rightarrow \text{ROH} + \text{protein–SeOH}$$

Glutathione then binds, followed by

$$\text{protein–SeOH–GSH} \rightarrow \text{H}_2\text{O} + \text{protein–Se–SG}$$

The second GSH binds, followed by

$$\text{protein–Se–SG–GSH} \rightarrow \text{protein–SeH–GSSG} \rightarrow \text{protein–Se}^- + \text{H}^+ + \text{GSSG}$$

Traces of selenium are essential in the diet of animals: an important role of dietary selenium is to provide the selenium-containing cofactor for the glutathione peroxidase enzyme family. However, selenium plays other roles. For example, in thyroid hormone biosynthesis it acts as an essential component of a **type I iodothyronine-5′-deiodinase** enzyme, which converts thyroxine (T$_4$) to the more biologically-active thyroid hormone 3,5,3′-triiodothyronine (T$_3$). Several other selenoproteins are present *in vivo*, e.g. plasma from humans and other animals contains **selenoprotein P**, a glycoselenoprotein of unknown function (10 selenocysteines/molecule of protein) that accounts for most of the selenium present in plasma and has been suggested to protect against certain toxins (Chapter 8).

The ratios of reduced to oxidized glutathione (GSH/GSSG) in normal cells are high (Table 3.8) so there must be a mechanism for reducing GSSG back to GSH. This is achieved by **glutathione reductase** enzymes, which catalyse the reaction:

$$\text{GSSG} + \text{NADPH} + \text{H}^+ \rightarrow 2\text{GSH} + \text{NADP}^+$$

The NADPH required is provided in animal and plant tissues by several enzyme systems, but the best known is the **oxidative pentose phosphate**

pathway.[51] The first enzyme in this pathway is **glucose-6-phosphate dehydrogenase**,

$$\text{glucose 6-phosphate} + NADP^+ \rightarrow \text{6-phosphogluconate} + NADPH + H^+$$

followed by 6-phosphogluconate dehydrogenase,

$$\text{6-phosphogluconate} + NADP^+ \rightarrow CO_2 + NADPH + H^+$$
$$+ \text{ribulose 5-phosphate}$$

The rate at which the pentose phosphate pathway operates is controlled by the supply of $NADP^+$ to glucose-6-phosphate dehydrogenase. As glutathione reductase operates and lowers the $NADPH/NADP^+$ ratio, the pentose phosphate pathway speeds up to replace the NADPH.

Glutathione reductases contain two protein subunits,[245] each with the flavin FAD at its active site. Apparently the NADPH reduces the FAD, which then passes its electrons onto a disulphide bridge (−S−S−) between two cysteine residues in the active site. The two −SH groups so formed then interact with GSSG and reduce it to 2GSH, re-forming the protein disulphide.

3.8.2   *A family of enzymes*[162a]

Selenium-containing peroxidases comprise a family of enzymes (the **glutathione peroxidase superfamily**), of which at least four types exist (Fig. 3.13, see plates section). One is the 'classical' glutathione peroxidase (GPX). By contrast, mammalian plasma contains low levels of a different glutathione peroxidase (also found in certain other extracellular fluids such as milk and lung lining fluid), a glycoprotein tetramer. Whether or not it functions as a glutathione peroxidase in plasma is uncertain because of the very low levels of GSH: plasma GSH levels are in the micromolar range (Table 3.8) but the K_m of the enzyme for GSH is in the millimolar range.

Another member of the family is **phospholipid hydroperoxide glutathione peroxidase** (PHGPX), a monomeric protein of relative molecular mass 19 000 that can reduce esterified fatty acid and cholesterol hydroperoxides. Thus it can act upon peroxidized fatty acid residues within membranes and lipoproteins (unlike classical glutathione peroxidase), reducing them to alcohols. This enzyme is also very active in reducing thymine hydroperoxide,[20] formed as a consequence of free radical attack on the DNA base thymine. Selenium-deprivation of rats lowers levels of glutathione peroxidase much more quickly than levels of PHGPX, which has led to the suggestion that maintaining PHGPX activity is more important than for GPX.[127] PHGPX can also act on H_2O_2 and simple organic peroxides.

Another type of tetrameric glutathione peroxidase has been found in the cells lining the gastrointestinal tract. Intestinal glutathione peroxidase (GPX-GI) may serve to metabolize peroxides in ingested food lipids as well as any generated during lipid peroxidation in the intestine itself. GPX-GI is also present in liver, but apparently not in other tissues.

3.8.3 *Cooperation of glutathione peroxidase[f] and catalase in the removal of hydrogen peroxide* in vivo

Skeletal muscle, spermatozoa and certain regions of the brain contain low levels of catalase but more GPX, so the question as to which enzyme is likely to be more important in removing H_2O_2 *in vivo* is easily answered. Most animal tissues contain substantial activities of both GPX and catalase, however,[51] so how do they co-operate with each other?

Mammalian red blood cells contain no subcellular organelles, and both catalase and GPX enzymes float in the cytosol, although it is possible that some catalase might be attached to the inside of the erythrocyte membrane. The normal low rate of production of H_2O_2 in these cells (via haemoglobin and SOD; Chapter 7) seems to be mainly dealt with by GPX, which would explain the limited consequences of acatalasaemia (Section 3.7.7 above). However, catalase does make some contribution, and if the concentration of H_2O_2 is raised, e.g. by supplying erythrocytes with H_2O_2 (or a toxin that increases intracellular H_2O_2 generation), then catalase becomes more important in removing H_2O_2.[91] This makes sense in view of the high K_m of catalase for H_2O_2 and its very high rate of H_2O_2 destruction when high H_2O_2 levels are present.

Liver contains high concentrations of both catalase and GPX. Whereas catalase is largely or entirely in the peroxisomes, GPX is found mainly in the cytosol but also in the matrix of mitochondria (about 10% of the total). The distribution of GSH is similar. Mitochondria lack catalase (except in heart) and apparently lack the enzymes of GSH synthesis. They must import GSH from the cytosol, but do contain glutathione reductase as well as GPX.[168]

This subcellular compartmentalization will obviously affect H_2O_2 removal mechanisms *in vivo*. Thus H_2O_2 produced by say, glycollate oxidase and urate oxidase (peroxisomal enzymes) is largely disposed of by catalase, whereas H_2O_2 arising from mitochondria, the endoplasmic reticulum or soluble (cytosolic) enzymes such as CuZnSOD is acted upon by GPX.[51] The capacity of the glutathione system to cope with H_2O_2 in liver and other tissues depends on the activity of GPX and glutathione reductase, the rate of NADPH supply (e.g. by the pentose phosphate pathway) and the GSH content, which varies at different times of day (Table 3.8). In liver, the glutathione system has a high capacity to deal with H_2O_2 but in lung, eye and muscle the capacity of the system appears more restricted. For example, inhibition of catalase in the eyes of rabbits by aminotriazole administration caused the concentration of H_2O_2 in the aqueous humour of the eye to rise from about 0.06 to 0.15 mM even though glutathione reductase and GPX activities were unaffected. The glutathione system apparently could not cope with the extra load caused by the loss of catalase activity. Feeding young rabbits with aminotriazole can cause cataracts to develop, perhaps due to oxidation of lens proteins (Chapter 7).

The cell nucleus also contains GSH, plus GPX, glutathione reductase and glutathione transferase activities.[239]

3.8.4   Assessing the operation of the glutathione peroxidase system

The rate of operation of the glutathione peroxidase system *in vivo* can be assessed in a number of ways.[51] A common approach is to measure the pentose phosphate pathway activity by supplying 1-^{14}C-labelled glucose to the tissue and measuring the release of radioactive [14]CO_2 in the 6–phosphogluconate dehydrogenase reaction (Section 3.8.1 above). An increased pathway activity has been observed upon exposing several tissues, including isolated perfused rat lung, ox retina and human erythrocytes to elevated O_2, presumably as more NADPH is consumed by glutathione reductase as it deals with increased GSSG production from GPX. NADPH is also needed for the continued functioning of catalase (Section 3.7 above) but this probably consumes only small amounts.

An alternative approach has been to measure GSSG release: if cells are treated with reagents that oxidize internal GSH to GSSG (such as **diamide**, $(CH_3)_2NCON=NCON(CH_3)_2$)[143] they rapidly eject GSSG into the surrounding medium. In intact liver, GSSG is released into the bile.[131] The rate of release of GSSG in perfused organs can be taken as a measure of GPX activity if glucose is omitted from the perfusing medium, so that NADPH cannot be produced by the pentose phosphate pathway for glutathione reductase activity. Exposure of isolated perfused organs (e.g. liver and lung) to elevated O_2 causes increased GSSG release. Care must be taken in attributing this entirely to more H_2O_2 production, since GPX can also act on organic hydroperoxides.[234]

Figure 3.14 shows GSSG release when H_2O_2 at increasing concentrations is infused into a perfused rat liver: the 'saturation' of GSSG release is probably related to the increased action of catalase in removing H_2O_2 at higher concentrations. Inclusion of glycollate in the perfusing medium at physiological concentrations causes only a small increase in GSSG release, suggesting that H_2O_2 generated by the peroxisomal enzyme glycollate oxidase is largely disposed of by catalase in the same organelles. If aminotriazole is used to inhibit catalase, glycollate infusion does cause a marked increase in GSSG release. The undestroyed H_2O_2 presumably diffuses across the peroxisomal membrane and has to be dealt with by GPX.

3.9   Glutathione in metabolism

Apart from its role as a cofactor for the glutathione peroxidase family, GSH is involved in many other metabolic processes, including ascorbic acid metabolism (Section 3.22 below), maintaining communication between cells,[21] and in generally preventing protein −SH groups from oxidizing and cross-linking. It also seems involved in intracellular copper transport;[197] GSH can chelate copper ions and diminish their ability to generate free radicals, or at least to release radicals into solution.[109] GSH is a radioprotective agent (Chapter 8) and

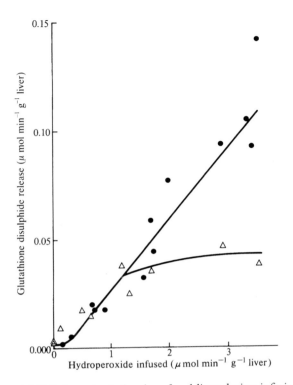

Fig. 3.14. Release of GSSG from the isolated perfused liver during infusion of H₂O₂. H₂O₂ was included in the perfusion medium at the concentrations stated (open triangles). The closed circles show the results when an organic hydroperoxide was infused instead. Diagram by courtesy of Professor Helmut Sies.

a cofactor for several enzymes in different metabolic pathways, including glyoxylases[247], and enzymes involved in leukotriene synthesis (Chapter 6). **Methylglyoxal** (CH_3COCHO) reacts non-enzymically with GSH and the complex is oxidized by the **glyoxylase I** enzyme into S-lactoylglutathione ($CH_3CHOHCOSG$), which is hydrolysed by **glyoxylase II** into GSH and lactate. Methylglyoxal arises 'accidentally' from glyceraldehyde 3-phosphate and dihydroxyacetone phosphate during glycolysis, both non-enzymically and by 'escape' from the active site of the enzyme triose phosphate isomerase. Methylglyoxal is also formed from acetone by cytochrome P4502E1. It must be removed because of its high reactivity with DNA and proteins, e.g. covalent binding to cysteine, arginine and lysine residues.[247]

Glutathione also plays a role in protein folding and the degradation of proteins with disulphide bonds, such as insulin (the first step in insulin removal is cleavage of disulphide bridges linking the two peptide chains). Yet another physiological role of GSH is seen in the freshwater coelenterate *Hydra*, which recognizes wounded prey by the release of GSH from damaged cells; the GSH binds to specific receptors which induce tentacle contractions and mouth opening in the hungry predator.[27]

3.9.1   *Scavenging of reactive species by GSH*

In vitro, GSH can react with $OH^\bullet$, HOCl, peroxynitrite, $RO^\bullet$, $RO_2^\bullet$, carbon-centred radicals and $^1O_2$. Its reaction with free radicals will generate thiyl ($GS^\bullet$) radicals (Chapter 2). $GS^\bullet$ radicals can generate $O_2^{\bullet -}$ by the reaction (Chapter 2)

$$GS^\bullet \xrightarrow{GS^-} GSSG^{\bullet -} \xrightarrow{O_2} GSSG + O_2^{\bullet -}$$

Hence SOD might cooperate with GSH in helping to remove free radicals *in vivo*, as might ascorbate

$$AScH^- + RS^\bullet \rightarrow RSH + ASc^{\bullet -}$$

Since GSH is present at millimolar intracellular concentrations (Table 3.8) scavenging of the above species is feasible *in vivo*. Reaction of GSH with $ONOO^-$ appears to lead to formation of some nitrosothiol (GSNO), which can decompose to regenerate nitric oxide,[257] i.e. GSH can, to some extent, 'recycle' $ONOO^-$ to $NO^\bullet$.

3.9.2   *Glutathione biosynthesis and degradation*

GSH is synthesized in two steps.[168] First, the enzyme γ-**glutamylcysteine synthetase** catalyses dipeptide formation

L-glutamate + L-cysteine + ATP → L-γ-glutamyl-L-cysteine + ADP + P_i

and the product is converted to GSH by **glutathione synthetase**

L-γ-glutamyl-L-cysteine + glycine + ATP → GSH + ADP + P_i

Cells can make the necessary cysteine from methionine, or they can take it up from the surrounding fluids (Table 3.8, footnote). Often they take up the disulphide form (**cystine**) and reduce it to cysteine inside the cell. γ-Glutamylcysteine synthetase is feedback inhibited by GSH (competitively with glutamate) and does not appear saturated at normal cellular levels of cysteine, so that increased cysteine can promote GSH synthesis under certain circumstances. It has been suggested that variations in cysteine can account for the effects of starvation and refeeding on GSH levels (Table 3.8).

Glutamylcysteine synthetase can be inhibited[168] by **buthionine sulphoximine** (BSO), widely used in experiments to deplete cellular GSH levels. BSO has the structure

$$\underset{\underset{NH}{\overset{||}{}}}{CH_3CH_2CH_2CH_2SCH_3CH_2\overset{\displaystyle O}{\overset{\displaystyle ||}{C}}}\underset{COOH}{\overset{NH_2}{CH}}$$

GSH is in a constant state of metabolic turnover, e.g. its half-life has been estimated to be only 4 days in human erythrocytes and 3 h in rat liver. Turnover in red blood cells can involve export of GSSG from the cells into plasma by an ATP-dependent transport system similar to that found in the liver (which excretes GSSG into bile). Liver also secretes GSH constantly into plasma.[131] In cells other than erythrocytes, γ-**glutamyltranspeptidase** (which also degrades GSH conjugates produced by glutathione transferases) can break down GSH. The enzyme is located on the plasma membrane with its active site facing outwards. It acts on extracellular GSH and transfers the glutamate residue on to other amino acids such as cysteine, methionine and glutamine

$$\text{GSH} + \text{amino acid} \rightarrow \gamma\text{-glutamylamino acid} + \text{cysteinyl-glycine}$$

The Cys–Gly dipeptide can then be taken up and hydrolysed, and the γ-glutamylamino acids converted to 5-oxoproline and then to glutamate (Fig. 3.15) or, in the case of Glu–Cys, recycled to GSH. Kidney has a high

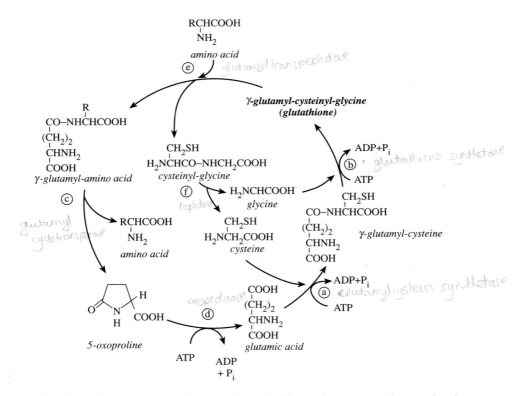

Fig. 3.15. Glutathione synthesis and metabolism. (a) Glutamylcysteine synthetase; (b) glutathione synthetase; (c) γ-glutamyl cyclotransferase; (d) oxoprolinase; (e) γ-glutamyl-transpeptidase; (f) peptidase. Diagram by courtesy of the late Professor Alton Meister and *J. Biol. Chem.*

level of γ-glutamyltranspeptidase, acting upon glutathione–xenobiotic conjugates and upon plasma GSH and GSSG. K_m values of the transpeptidase for GSH and GSSG are in the micromolar range, similar to actual levels of these substrates in plasma (Table 3.8).

The main source of plasma GSH is the liver: export appears to occur by at least two transport mechanisms and distinct carriers export GSH into plasma or into bile.[131] In summary, plasma GSH and GSSG turn over extremely rapidly (half time is a few minutes): they originate largely from the liver and are metabolized by transpeptidase, mostly in kidney. Lung and intestine also have high levels of transpeptidase activity.

3.10   The glutathione S-transferase superfamily

Glutathione is also involved in the metabolism of herbicides, pesticides and **xenobiotics** ('foreign compounds') generally in both animal and plant tissues.[117] For example, maize (corn) leaves contain an enzyme which detoxifies the herbicide **atrazine** by combining it with GSH. Some of the **herbicide safeners**, a class of compounds that protect crops from injury by certain herbicides, act in part by raising GSH levels in the plant.

Many xenobiotics supplied to living organisms are metabolized by conjugation with GSH, catalysed by **glutathione S-transferase** (GST) enzymes

$$RX + GSH \rightarrow RSG + HX$$

Liver is especially rich in these enzymes and the resulting glutathione conjugates are often excreted into bile using ATP-dependent glutathione S-conjugate 'efflux pumps'; the same pumps are involved in the export of GSSG when liver is subjected to oxidative stress.[131] Alternatively, the adducts can be degraded and acetylated to form *N*-acetylcysteine conjugates (**mercapturic acids**) which can be excreted in the urine. Compounds metabolized by glutathione transferases in animals include chloroform, organic nitrates (Section 2.4.6), bromobenzene, aflatoxin, DDT, naphthalene and paracetamol. The presence of large amounts of such xenobiotics can decrease hepatic GSH concentrations, thereby impairing the antioxidant defence capacity of the liver. Some glutathione transferases can metabolize cytotoxic aldehydes produced during lipid peroxidation, such as 4–hydroxynonenal (Section 4.9.10). The most widely used substrate to assay GST activity is **1-chloro-2,4-dinitrobenzene**, which is conjugated with GSH to give *S*-(2,4-dinitrophenyl) glutathione, whose formation is easily assayed spectrophotometrically.

Some GSTs show a glutathione-peroxidase-like activity, with organic hydroperoxides, which was formerly called **non-selenium glutathione peroxidase** (Table 3.9). They catalyse reaction of organic peroxides (but *not* of H_2O_2) with GSH to form GSSG and alcohols. Perfused livers from rats fed a selenium-deficient diet release no GSSG when H_2O_2 is infused, but do so when *tert*-butylhydroperoxide is infused, indicating that these non-selenium enzymes can function in whole organs.[234] The significance of glutathione transferases as peroxide-metabolizing systems *in vivo* is uncertain. They have,

Table 3.9. Selenium and non-selenium glutathione peroxidase activities in different animals

Animal	Organ studied	Non-selenium glutathione peroxidase activity as percentage of total
Rat	adrenal	38
	spleen	0
	liver	35
	lung	0
	heart	0
	kidney	31
	testis	91
Hamster	liver	43
Sheep	liver	81
Pig	liver	67
Chicken	liver	70
Human	liver	84
Rabbit	liver	~50
Mouse	liver	30

Total activity was measured using cumene hydroperoxide as substrate. Results are mostly abstracted from H. Sies *et al.* (1982) *Proceedings of the Third International Symposium on Oxidases and Related Redox Systems* (King, T.E. *et al.*, eds). Pergamon Press, Oxford, p. 169. It should be noted that results can vary widely depending on the strain and sex of animal selected, and on the dietary content of selenium.

however, caused confusion in many laboratories where synthetic organic hydroperoxides have been used to assay 'glutathione peroxidase' levels in biological material, e.g. cumene hydroperoxide will detect both the 'real' selenium GPX and some of the transferases (Table 3.9). It is better to use H_2O_2 as a substrate to assay the former enzymes. Some GSTs can also show phospholipid hydroperoxide glutathione peroxidase activity.[125a]

All eukaryotes have multiple cytosolic and membrane-bound GST iso-enzymes, each with distinct substrate specificities and other properties.[117] As well as their catalytic functions, many GSTs appear to serve as intracellular carrier proteins for haem, bilirubin, bile pigments and steroids, which bind non-enzymically to the proteins. Microsomes contain a GST involved in drug metabolism, and **leukotriene C4 synthase** (involved in leukotriene biosynthesis; Chapter 6) is a GST, although it does not appear to play a role in xenobiotic metabolism. The cytosolic enzymes are dimers.

At least six gene families encode multiple GST subunits for cytosolic enzymes, designated alpha, mu, pi, sigma, kappa and theta GST. A GST can contain two identical subunits, e.g. a dimer of type 1 mu subunits is written as GST M1-1. Subunits can be different, but always come from the same family, e.g. a heterodimer of type 1 and 2 alpha subunits is A1-2. Similarly, P denotes pi, K kappa and T theta. For example, human liver contains GSTs A1-1, A1-2 and A2-2, among others.

The alpha, mu and pi class enzymes are most abundant in mammals, and levels are often increased by exposure to foreign compounds, many of which are substrates. Human 'inborn errors of metabolism' which decrease the activity of certain GSTs have been claimed to be associated with increased risk of some forms of cancer. In alpha, mu, pi and sigma GSTs, a tyrosine residue is important in GSH binding and addition of peroxynitrite can inactivate the proteins by tyrosine nitration.

Usually conjugation with GSH is the first step in detoxifying a xenobiotic. Sometimes, however, the resulting products are *more* damaging, e.g. GSH conjugates of several halogenated hydrocarbons, including dibromo- and dichloroethane, cause kidney damage.

3.11   Mixed disulphides

Most intracellular free glutathione *in vivo* is present as GSH rather than GSSG (Table 3.8), but some may be present as 'mixed' disulphides with other compounds that contain −SH groups, such as cysteine, coenzyme A and the −SH of the cysteine residues of several proteins (including human CuZnSOD and carbonic anhydrase[g] III). A coenzyme A−glutathione disulphide adduct has been reported to be a powerful vasoconstrictor.[225] If R−SH is used to represent these other thiols, then the mixed disulphides have the general formula:

$$Glu - Cys - Gly$$
$$\qquad\quad |$$
$$\qquad\; S-S-R$$

Thiol groups on proteins are often essential for protein stability and/or function (e.g. in enzymes, membrane transport proteins and in the binding of NO$^\bullet$ by haemoglobin (Chapter 1)). Indeed, albumin in blood plasma contributes 0.3–0.5 mM −SH and is the major source of total −SH groups in plasma. The total level of protein−SH in cells may approach or exceed that of GSH. Protein thiol groups (protein−SH) can react with GSSG to form mixed disulphides (a process that has been called **protein-S-thiolation**).

$$GSSG + protein\text{−}SH \rightleftharpoons GSS\text{-}protein + GSH$$

Indeed, incubation of several enzymes with GSSG *in vitro* causes inactivation, probably by forming mixed disulphides.[39,227,246] Examples include adenylate cyclase, chicken liver fatty-acid synthetase and rabbit muscle phosphofructokinase. Mixed disulphides of GSSG with proteins and other thiols accumulate in tissues subjected to oxidative stress, in both mitochondria and cytosol. For example, stimulation of the respiratory burst of human monocytes (Chapter 6) caused S-thiolation of several intracellular proteins,[206] including the glycolytic enzyme glyceraldehyde-3-phosphate dehydrogenase. This enzyme was also S-thiolated when human vascular endothelial cells were exposed to H_2O_2.

GSSG can inhibit protein synthesis in animal cells, and accumulation of GSSG has been suggested to contribute to the inhibition of protein synthesis seen in dehydrated plant tissues, e.g. wheat germ and several mosses.[64]

These actions of GSSG may explain why cells keep intracellular GSSG levels very low (Table 3.8) under normal conditions, and why erythrocytes and organs such as the liver and heart, export GSSG when they are under oxidative stress.

It is also possible that conversion of protein–SH to thiyl radicals by free-radical attack (Chapter 2) can lead to mixed disulphide formation with GSH e.g.

$$\text{protein–SH} + \text{OH}^{\bullet} \rightarrow \text{protein–S}^{\bullet} + \text{H}_2\text{O}$$

$$\text{protein–S}^{\bullet} + \text{GSH} \rightarrow \text{protein–S–S}^{\bullet -}\text{G}$$

$$\text{protein–S–S}^{\bullet -}\text{G} + \text{O}_2 \rightarrow \text{protein–S–S–G} + \text{O}_2^{\bullet -}$$

Selenite (SeO_3^{2-}) can react with GSH to form the selenium equivalent of a mixed disulphide[36]

$$2\text{H}^+ + 4\text{GSH} + \text{SeO}_3^{2-} \rightarrow \text{GSSG} + \underset{\text{selenodiglutathione}}{\text{GSSeSG}} + 3\text{H}_2\text{O}$$

which is a metabolite of selenium in mammalian tissues. Further reduction of GSSeSG leads to **selenide**, Se^{2-}.

3.12 Protein-disulphide isomerase

During the folding of newly synthesized proteins, the correct cysteine–SH residues must join up to form the correct disulphide bridges. **Chaperones** are present within cells to assist generally with protein folding (Section 4.15). However, **protein-disulphide isomerase** (PDI) enzymes, located at high concentrations (possibly millimolar) in the lumen of the endoplasmic reticulum, catalyse the formation of new disulphide bridges and the rearrangement of existing ones by means of exchange with dithiol/disulphide groups at the PDI active site.[262] The activity of PDIs is controlled by the GSH/GSSG ratio, which appears lower within the endoplasmic reticulum than in the rest of the cell, presumably to give optimum conditions for disulphide bridge formation in newly synthesized proteins.[45a]

PDI enzymes from eukaryotes have about 500 amino acid residues, with two regions homologous to the protein **thioredoxin** (Section 3.13 below). The PDI from bacteria, known as **DsbA**, is smaller than animal PDIs and less closely linked to the thioredoxin family. Animal PDI enzymes have been shown to have **dehydroascorbate reductase** activity;[262] as well as acting on proteins they can catalyse the reaction

$$2\text{GSH} + \text{dehydroascorbate} \rightleftharpoons \text{GSSG} + \text{ascorbate}$$

PDI is also involved in the **proline hydroxylase** enzyme, an ascorbate-dependent enzyme which catalyses hydroxylation of proline residues during the synthesis of collagen. Prolyl hydroxylase is a tetramer containing two α- and two β-subunits, the latter being identical to PDI and possibly allowing interaction with ascorbate.

3.13   Thioredoxin

Thioredoxin is a polypeptide, relative molecular mass about 12 000, found in both prokaryotes and eukaryotes. It is widely distributed in mammalian cells, being especially concentrated in the endoplasmic reticulum, but some is also found on the cell surface.

Thioredoxin contains two adjacent $-SH$ groups in its reduced form, that are converted to a disulphide in oxidized thioredoxin. It can undergo redox reactions with multiple proteins.

$$\text{thioredoxin}-(SH)_2 + \text{protein}-S_2 \rightleftharpoons \text{thioredoxin}-S_2 + \text{protein}-(SH)_2$$

Thioredoxin binds to its target protein and, via intermediate formation of a mixed disulphide, reduces the protein disulphide bridge whilst oxidizing its two cysteine$-SH$ groups to a cystine (disulphide). The reduction potential $E^{\circ\prime}$ of *E. coli* thioredoxin is about -0.27 V, indicating its significant reducing potential. Thioredoxin can also catalyse thiol–disulphide exchange between a protein dithiol and another protein disulphide. Features of the thioredoxin structure are seen in several related proteins including glutathione peroxidase, glutathione transferases and PDI (Fig. 3.16, see plate section); all the proteins shown in Fig. 3.16 may have evolved from a common 'thiol-binding protein' ancestor. For example, the 'thioredoxin domains' in PDI constitute its active sites.

Oxidized thioredoxin can be re-reduced *in vivo* by **thioredoxin reductase** enzymes, which contain FAD and show similarities to glutathione reductases, including use of NADPH

$$\text{thioredoxin}-S_2 + NADPH + H^+ \rightarrow NADP^+ + \text{thioredoxin}-(SH)_2$$

Thioredoxin reductase can also reduce PDI and thus achieve an NADPH-dependent reduction of the disulphide bridges of such proteins as insulin. Interestingly, 1-chloro-2-4-dinitrobenzene, used in the assay of GST enzymes, is an irreversible inhibitor of human thioredoxin reductase.[9] Selenite and GSSeSG can also interfere with thioredoxin function by binding to $-SH$ groups on thioredoxin.[36] Thioredoxin reductases isolated from some mammalian cell lines in culture and from human placenta have proved to be selenoproteins, containing selenocysteine as the last-but-one residue at the C-terminus.[94]

PDI enzymes and thioredoxins are members of a group of enzymes often called **thiol–disulphide oxidoreductases**. Another member is **thioltransferase** or **glutaredoxin**, originally discovered as a cofactor of ribonucleotide reductase in *E. coli* (Section 6.2), but later found to be present in many organisms, including animals.[195] Thioltransferases are largely located in the cell cytosol and, unlike thioredoxin, can be reduced directly by GSH. They are presumably involved in catalysing thiol–disulphide interchange with cytosolic proteins, but also have dehydroascorbate reductase activity.

3.13.1   *Thioredoxin and antioxidant defence*

Thioredoxin has appeared in many guises in biology and only now is its central importance to metabolism in animals being appreciated (its key role in plants has been known for decades; Section 7.5). *In vitro*, thioredoxin can stimulate the growth of various cell types in culture. Indeed, several isolated protein 'factors' such as **adult T cell–leukaemia derived factor**,[179] **T–hybridoma (MP-6)-derived B cell stimulatory factor** and **early pregnancy factor** were later shown to be identical with thioredoxin. *In vivo*, thioredoxin acts as a hydrogen donor for ribonucleotide reductase (Chapter 6). It also supplies electrons to **methionine sulphoxide reductase**, an enzyme that repairs oxidative damage to methionine residues in proteins (Section 4.12). Thioredoxin can react directly with H_2O_2, although the metabolic significance of this is unknown.[179] The ability of the dithiol form of thioredoxin to reduce protein disulphides has been known for many years to be important in chloroplasts, where it regulates the enzymes involved in CO_2 fixation (Section 7.5). Thioredoxin can also reactivate oxidized glyceraldehyde-3-phosphate dehydrogenase in mammalian cells subjected to oxidative stress.[84] Another antioxidant role for thioredoxin may be its involvement in the **thiol-specific antioxidant** system.

3.13.2   *Thiol-specific antioxidants*[48–50]

In 1988 it was reported that the yeast *S. cerevisiae* contains a protein of relative molecular mass 25 000 that protects biomolecules against damage by thiyl radicals *in vitro*. This enzyme was assayed by its ability to protect **glutamine synthetase** (an enzyme much used in studies of protein damage by free radicals) against inactivation by a mixture of a synthetic dithiol (**dithiothreitol**) and $FeCl_3$. A similar 'protector protein' was isolated from human red blood cells and shown to protect haemoglobin against oxidation, and DNA against cleavage, by the Fe(III)/thiol pro-oxidant system.

These **thiol-specific antioxidants** (TSAs), as they were first called, probably act by removing thiyl ($RS^\bullet$) radicals, and possibly $RS^\bullet$-derived oxysulphur radicals such as $RSO^\bullet$ and $RSO_2^\bullet$ (Chapter 2). TSAs have also been detected in brain, liver fluke and *Entamoeba histolytica*, a pathogenic protozoan.[199a] Determination of TSA amino acid sequences showed that they have a striking resemblance to one of the subunits of **alkyl hydroperoxide reductase**. This is an enzyme found in bacteria, e.g. *Salmonella typhimurium* and *E. coli*, that uses NADH (or NADPH) to reduce alkyl hydroperoxides (such as thymine, linoleic acid and cumene peroxides) to alcohols. The *S. typhimurium* alkyl hydroperoxide reductase contains two subunits, denoted **AhpF** (with a bound FAD cofactor) and **AhpC**, and it is the latter that resembles the TSAs. Indeed, isolated AhpC shows TSA activity if assayed *in vitro* in the

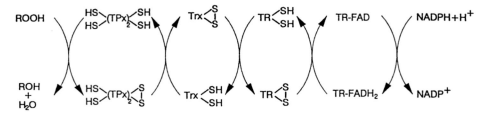

Fig. 3.17. Proposed sequence of events linking thiol-specific antioxidant to thioredoxin. The whole may be regarded as a **thioredoxin-dependent peroxide reductase** or **peroxiredoxin**. Trx, thioredoxin; TR, thioredoxin reductase; TPx, thioredoxin peroxidase. Adapted from *J. Biol. Chem.* **269**, 27670 (1994) by courtesy of Drs Sue Goo Rhee and Ho Zoon Chae and the publisher. RS$^{\bullet}$ radicals may be able to interact directly with the −SH groups on the protein. At least 3 peroxiredoxins may exist in mammalian cells (*J. Biol. Chem.* **243**, 6297, 1998).

glutamine-synthetase inactivation assay. A search of protein sequence databases showed more than 20 additional proteins with sequence similarity, from a wide variety of organisms.

The TSA enzymes may well interact with thioredoxin *in vivo*; both TSA and reduced thioredoxin can react directly with H_2O_2, causing −SH group oxidation. Perhaps *in vivo* TSA is oxidized by H_2O_2, and recycled by reduced thioredoxin (Fig. 3.17). Thus quenching of sulphur radicals may be a 'partial function' of a system evolved to remove H_2O_2 and other peroxides *in vivo*. Where exactly TSAs and thioredoxin-dependent peroxide reductases fit into the overall pattern of cellular antioxidant defence systems remains to be established.

Another possible link is that the glutathione peroxidase found in blood plasma can, at least *in vitro*, use reduced thioredoxin[35] as a preferred substrate to GSH in the reduction of peroxides. The high K_m for GSH has always raised questions about the role of this as a GSH-dependent peroxide removal system (Section 3.8 above). However, how would these enzymes get thioredoxin in blood plasma?

3.14   Evidence for the importance of glutathione and glutathione-metabolizing enzymes *in vivo*

There is considerable evidence to support the view that GSH and enzymes that use it are important *in vivo*. Evidence has come from three lines: the use of inhibitors, the use of organisms defective in GSH metabolism (including transgenic animals and human inborn errors of metabolism) and studies of selenium deficiency in humans and other animals.

3.14.1   *Use of inhibitors*

Buthionine sulphoximine administration can decrease GSH levels. This is not normally lethal to isolated cells or whole organisms but can result in increased sensitivity to ionizing radiation and to toxins normally metabolized by GST enzymes. However, it is much more difficult to deplete mitochondrial GSH than cytosolic GSH. Very severe GSH depletion in animals leads to widespread mitochondrial damage, and is lethal in newborn animals because of multiple organ damage (e.g. to kidney, liver, lung, brain and eye).[168] Tissue injury is ameliorated by replacing GSH, e.g. with a GSH 'delivery agent' such as GSH ethyl or methyl esters (Chapter 10). Mitochondria cannot synthesize GSH and must absorb it from the cytoplasm through an inner membrane transporter.

3.14.2   *Defects in GSH metabolism*[168]

Humans with inborn deficiencies of γ-glutamylcysteine synthetase (very rare), GSH synthetase (less rare), glutathione reductase, γ-glutamyltranspeptidase, 5-oxoprolinase and GPX have been reported. Such individuals often exhibit a tendency to haemolysis and neurological defects. Some patients with glutathione reductase deficiency exhibit early cataract formation; cataracts are also observed in newborn mice treated with buthionine sulphoximine. By contrast, cataract was not observed in transgenic 'knockout' mice lacking GPX.[240] Indeed, these mice appear to develop normally,[123a] although they may be more susceptible to the action of certain toxins such as O_3 and the anthracyclines, and to 'insults' such as myocardial ischaemia–reperfusion. By contrast, transgenic mice lacking γ-glutamyl transpeptidase showed elevated plasma GSH but decreased tissue levels, suffered growth retardation and developed cataracts, and most died between 10 and 18 weeks of age, illustrating the importance of this enzyme in maintaining tissue GSH levels.[156]

In patients with severe GSH synthetase deficiency (**5-oxoprolinuria**) affecting several tissues, the enzyme block leads to decreased feedback inhibition of γ-glutamylcysteine synthetase and thus to increased synthesis of γ-glutamylcysteine. Since the latter is a good substrate of γ-glutamyl cyclotransferase (Fig. 3.15), 5-oxoproline is overproduced leading to life-threatening acidosis. In a mild form of GSH synthetase deficiency apparently affecting only the erythrocyte, the genetic defect leads to synthesis of an unstable GSH synthetase molecule. Replacement of this active but unstable enzyme molecule compensates for the defect in most cells, but not in the erythrocyte (which does not synthesize protein). Several other diseases have been associated with changes in GSH levels, but the contribution made by this to the disease pathology is uncertain (Table 3.10).

Mutant bacteria lacking one of the enzymes involved in GSH synthesis are known; some, but not all, of these mutants are more sensitive to damage by radiation and certain chemicals. Notably, GSH-deficient and glutathione reductase-negative mutants of *E. coli* and yeast are viable. However, GSH-deficient *E. coli* seem to have membranes leaky for K^+ ions and cannot grow in low-potassium media. GSH deficiency may also impair the reassembly of the [Fe–S] cluster-containing enzymes after their inactivation by $O_2^{\bullet-}$ in *E. coli*. Similarly, GSH-deficient yeast are more sensitive to damage by peroxides than controls.

Table 3.10. Some human diseases with reported decreases in glutathione

Disease	Tissue/body fluid	GSH (% of normal)	GSSG (% of normal)	Further discussion in Chapter
AIDS				
(pre-symptomatic)	blood	30	100	9
	alveolar lining fluid	60	100	
ARDS	alveolar lining fluid	<10	478[a]	9
Alcoholic liver	liver	100	ND	8
disease	liver	38–66	100	
	blood	64	ND	
Non-alcoholic	liver	63	100	8
liver disease	liver	100	ND	
	blood	51–66	ND	
Cigarette smokers	red blood cells	126	ND	8
Hereditary	blood	34	ND	—
tyrosinaemia	liver	57	ND	
Idiopathic	lung lining			
pulmonary fibrosis	fluid	23	ND	9
Kwashiorkor	blood	50	ND	4
Marasmus	blood	100	ND	4
Parkinson's disease	substantia nigra	60–72	ND	9
	blood	100	ND	
Wilson's disease	liver	13	356	this chapter

ARDS, adult respiratory distress syndrome; ND, not determined.
Adapted from Uhlig, S and Wendel, A (1992) *Life Sci.* **51**, 1083, by courtesy of Prof. A. Wendel and the publishers.
[a]Data from Bunnell, E and Pacht, ER (1993) *Am. Rev. Resp. Dis.* **148**, 1174.

3.14.3 *Selenium deficiency in animals*

Selenium deficiency in animals produces a variety of diseases that are strikingly similar to those induced by a deficiency of the lipid-soluble chain-breaking antioxidant vitamin E (Section 3.22.7 below),[65a] consistent with the importance of selenium to antioxidant defence. Diseases induced by selenium deficiency include **white muscle disease**, a muscle wasting syndrome (mink, horses, pigs, cattle, sheep), **mulberry heart disease** (pigs) and infertility (many species). Indeed, to a considerable extent, the effects of selenium deficiency can be overcome by giving excess vitamin E and vice versa. Exceptions to this generalization include the observation that selenium cannot protect female rats against foetal reabsorption caused by vitamin E lack, nor can normal dietary vitamin E levels protect rats against the damage to the pancreas that occurs on selenium-deficient diets.

   A combined deficiency of selenium and vitamin E in the diet of animals is usually eventually fatal. Rats fed on such a diet showed elevated plasma levels of F_2-isoprostanes,[12] a marker of lipid peroxidation (Chapter 4). Injection of iron or copper salts into rats fed on diets deficient in both selenium and vitamin E causes a marked increase in the ethane content of the expired breath, a putative index of lipid peroxidation *in vivo* (Chapter 5). The increase is smaller if either vitamin E or selenium is resupplied. Administration of vitamin E to a child with an inborn deficiency of glutathione synthetase in white blood cells was reported to improve the functioning of these cells. Selenium-deficient rats are more susceptible to the toxic effects of elevated O_2 concentrations than normal.

These data suggest that the family of selenium-dependent glutathione peroxidases is important *in vivo*. However, several selenoproteins exist and it remains possible that consequences of selenium deficiency relate to lowered levels of selenoproteins other than glutathione peroxidases, such as seleno-protein thioredoxin reductases (Section 3.13) and selenoprotein P.

3.14.4   *Human selenium deficiency*

The realization that selenium is essential in the human diet came largely from the discovery in the People's Republic of China of a selenium-responsive degenerative heart disease known as **Keshan disease**. There is wasting and enlargement of the heart, heart rhythm disruptions and eventual cardiac arrest in severe cases. The name comes from an episode in 1935 in which 57 out of 286 inhabitants of a village in Keshan county of Heilongjiang province died of the disease. Epidemiological studies showed that the incidence of the disease was correlated with that of various degenerative diseases in animals that are known to be related to selenium deficiency. Both Keshan disease and the animal diseases could be prevented by administration of small doses of sodium selenite (Na_2SeO_3). In all affected areas, selenium concentrations in foodstuffs or in animals eating them were found to be extremely low, as was plasma selenium in the subjects. Whole-blood selenium levels in many countries fall into the $1-3\,\mu M$ range but levels in low-selenium areas of China with Keshan disease were much lower (Table 3.11).

Thus an abnormally low selenium intake appears to be the major factor leading to Keshan disease, but it cannot be the sole cause. For example, Keshan disease is not endemic in all low-selenium areas of China and seasonal variations in the disease incidence do not seem to be caused by changes in selenium intake (or vitamin E intake). One major contributing factor pre-disposing to cardiomyopathy may be infection with **Coxsackie viruses**.[25] Feeding mice a selenium-deficient diet favoured selection and replication of an especially damaging viral subtype in the heart after inoculation of the animals with a normally benign strain.

Since glutathione peroxidases are antioxidant selenoproteins, it seems logical to attribute Keshan disease to a lack of active enzymes. This is an assumption: selenium plays other biochemical roles.[94] Measurements of total selenium and glutathione peroxidase activity in whole blood of New Zealanders showed a good correlation between these two variables at total blood selenium

Table 3.11. Blood selenium concentrations in human populations worldwide

	Concentration in whole blood (μM) (ng/ml in parentheses)
Azerbeijan	1.39 (110)
Canada	2.31 (182)
China, Peoples' Republic of	
high-selenium area with toxicity	40.52 (3200)
high-selenium area without toxicity	5.57 (440)
moderate selenium area (Beijing)	1.20 (95)
low-selenium area without disease	0.34 (27)
low-selenium area with Keshan disease and/or Kashin–Beck disease	0.15 (12)
Egypt	0.86 (68)
Finland	0.71 (56)
Guatemala	0.29 (23)
New Zealand	0.86 (68)
Senegal, Velingara	1.11 (88)
Sweden	1.52 (120)
UK	
1974	4.05 (320)
1986	1.52 (120)
1991	0.96 (76)
Ukraine	5.59 (442)
USA	
South Dakota	3.24 (256)
Ohio	1.98 (157)
Venezuela	
high-selenium area	10.30 (813)
moderate-selenium area	4.50 (355)
Zaire	
Karawa schoolchildren	0.39 (31)
Karawa cretins	0.29 (23)
Businga adults	0.60 (47)
Kikwit adults	2.49 (197)

Data selected from Diplock, AT (1993) *J. Clin. Nutr. Suppl.* **57**, 256S, by courtesy of Professor AT Diplock. Selenium intakes and blood levels have fallen in the UK due to dietary changes (e.g. less use of flour from selenium-rich wheat imported from the USA).

concentrations of $< 1.3\,\mu$M, but not at higher levels. It seems from this and other data that about $1\,\mu$M is the 'threshold figure' at which glutathione peroxidase (at least in blood) reaches maximum activity.

Selenium deficiency has also been implicated in **Kashin–Beck disease**, a disabling joint disease seen in children of age range 5–13 years in such areas as Northern China, North Korea and East Siberia. The name comes from the Russian scientists who first described the disease. Administration of sodium selenite and vitamin E during the early stages of the disease has beneficial effects. The geographical distribution of Kashin–Beck disease resembles (but is not identical with) that of Keshan disease.

A study of blood selenium concentrations led Chinese scientists to conclude that a minimum adequate dietary intake for humans is about 30 µg. Normal daily intakes in advanced countries are in the range 60–200 µg/day and in 1989 the US recommended daily intake was set at 70 µg/day for adult men and 55 µg/day for adult women. Larger amounts (>350–600 µg/day) can produce toxic effects: an early sign of excess selenium intake is deformation and loss of fingernails, toenails and sometimes hair, and a 'garlic breath' smell due to dimethylselenide, $(CH_3)_2Se$. Selenium-accumulating plants such as *Astragalus* (**milk vetch**) species can poison cattle (leading to **blind staggers** and **alkali disease**) and are a particular nuisance to farmers in certain parts of the world.[42] Indeed, in 1295 the explorer Marco Polo described a disease of horses that sounds exactly like selenium toxicity. Human selenium poisoning was reported in 13 persons in the USA who consumed a 'health food supplement' containing over 180 times more selenium than stated on the label.

Selenium deficiency has sometimes been claimed to accompany protein-calorie malnutrition or prolonged intravenous feeding, but often the reports are inconclusive. Residents of low-selenium areas in Finland or New Zealand appear to suffer no obvious ill-effects, although they may well eat, of course, food that was grown in areas of higher selenium. Nevertheless, suggestions remain that low blood selenium might predispose to cardiovascular disease, cancer or complications of pregnancy. The Finnish government began supplementing fertilizers with selenium in 1984, but no dramatic changes in cancer or cardiovascular mortality have been observed to date.

3.14.5   *Conclusion*

The data presented above suggest that GSH and enzymes of its metabolism play important roles *in vivo*: a general feature of the mutants studied is that they are more sensitive to toxins and radiation, and cataract is common in mutant mammals. Nevertheless, studies of bacterial and yeast mutants lacking GSH, or animals lacking 'classical' glutathione peroxidase, show that these agents are not essential for survival (unlike, say, MnSOD). Organisms may well adapt to their loss by upregulating other mechanisms of peroxide removal, for example.

3.15   **Other sulphur-containing compounds possibly involved in antioxidant defence**

GSH plays a key part in eukaryotic metabolism. However, many organisms, especially prokaryotes (some of which do not produce any GSH) and plants, contain other sulphur-containing compounds that could fulfil similar roles. For example, γ-glutamylcysteine is the major thiol in halobacteria,[242] and **homoglutathione** accumulates in some higher plants. Some *Streptomyces* and

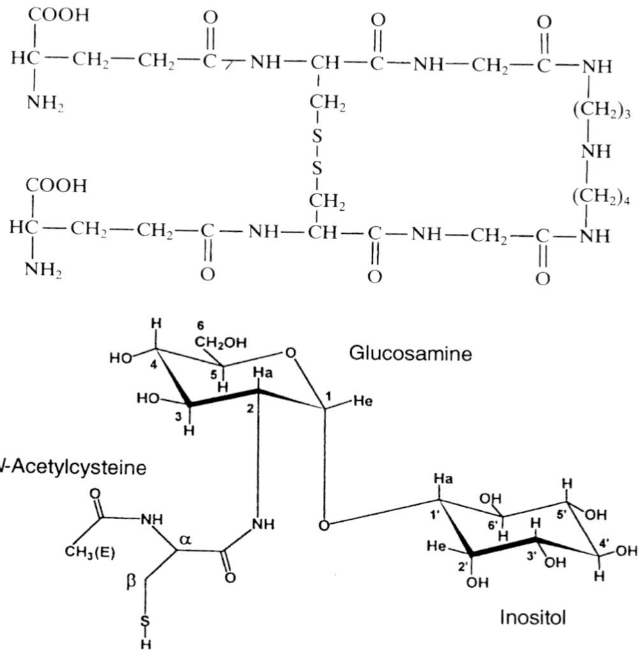

Fig. 3.18. Structures of trypanothione (top) and mycothiol (bottom). Trypanothione consists of glutathione covalently bonded to the polyamine spermidine. The disulphide form is shown and it can be converted by trypanothione reductase plus NADPH to a dithiol and $NADP^+$. Bottom diagram from *Biochem. J.* **325**, 623 (1997) by courtesy of Dr D.J. Steenkamp and the Biochemical Society.

Actinomycetales produce a cysteine derivative, 2-(N-acetylcysteinyl)amido-2-deoxy-α-D-glucopyranosyl-*myo*-inositol (**mycothiol**),[184] apparently as an alternative to GSH (Fig. 3.18). Sea-urchin eggs contain **ovothiol C**, apparently to scavenge H_2O_2 (Section 6.6).

3.15.1   *Trypanothione*[80]

Some parasitic protozoa, such as *Crithidia fasciculata* or *Trypanosoma brucei*, lack GSH but contain **trypanothione**, a molecule which consists of glutathione covalently linked to the polyamine spermidine (Fig. 3.18). The dithiol form of trypanothione scavenges peroxides and reduces disulphides, and is then regenerated by **trypanothione reductase**, which resembles glutathione reductase (e.g. it contains FAD). Most of these organisms lack catalase and glutathione peroxidase, so trypanothione is presumably their major mechanism for removing peroxides. Inhibitors of trypanothione synthesis or reduction might thus be therapeutically useful in treating tropical diseases caused by parasitic trypanosomes, e.g. African sleeping sickness (*T. brucei*), Chagas' disease

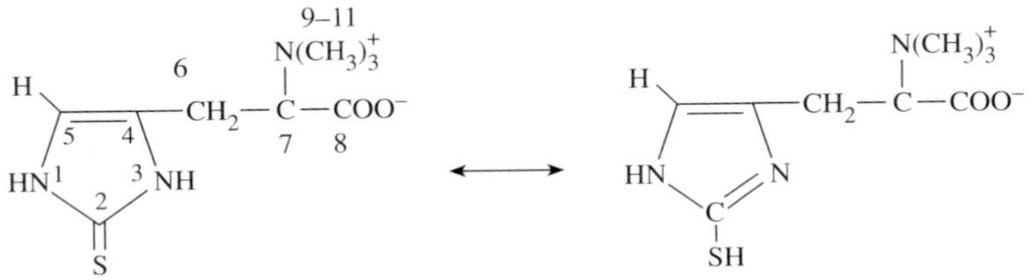

Fig. 3.19. Structure of the thiol (right) and thione (left) forms of ergothioneine. The thione form predominates at pH 7.4.

(*T. cruzi*) and leishmaniasis. Trivalent **arsenicals** (R–As=O) and trivalent **antimonials** (Sb(III)) inhibit trypanothione reductase, so preventing conversion of trypanothione disulphide to the dithiol (Fig. 3.18). Arsenicals have been used in the treatment of trypanosome infections since the pioneering work of Paul Ehrlich. Several quinones and nitro-compounds also interact with the reductase to exert cytotoxic effects (Section 8.14).

3.15.2   *Ergothioneine*[3]

Ergothioneine, which exists as an equilibrium between **thiol** and **thione** forms (Fig. 3.19), was originally discovered in *Claviceps purpurea* (the ergot fungus). Early work on ergothioneine reported its presence in rat erythrocytes and liver and suggested that it arose from dietary plant material, although the distribution of ergothioneine in the plant kingdom has not been fully investigated. The presence of ergothioneine in human, rat and other animal tissues has been confirmed by modern analytical techniques, including HPLC and nuclear magnetic resonance.

The function of ergothioneine in animal and plant tissues is unknown: early suggestions that it is a neurotransmitter have not been convincing. *In vitro*, ergothioneine forms stable complexes with iron and copper ions, and reacts rapidly with HOCl, peroxynitrite, OH$^\bullet$, RO$^\bullet$ and RO$_2^\bullet$ species. Hence it is possible that it could function as an antioxidant.[3] The one-electron oxidation product of ergothioneine can be re-reduced by ascorbate ($k = 6.3 \times 10^8 \, M^{-1} s^{-1}$).[11] More work is required to establish the exact levels, and origin, of ergothioneine in animal tissues and to investigate its putative antioxidant role.

3.16   Antioxidant defence enzymes: other peroxidases

A peroxidase is any enzyme that uses H$_2$O$_2$ to oxidize another substrate. Peroxidases can be specific for a particular substrate (such as GSH for glutathione peroxidase), but most have a broader substrate specificity.

3.16.1   Cytochrome c peroxidase: another specific peroxidase[51]

The haem-containing enzyme **cytochrome c peroxidase** is found between the inner and outer membranes of yeast mitochondria, which contain no catalase (present only in yeast cytosol and peroxisomes) or glutathione peroxidase. Cytochrome c peroxidase is also found in some bacteria.

It reacts rapidly with H_2O_2 to form a stable enzyme–substrate complex that has an absorbance maximum at 419 nm, whereas that of the free enzyme is at 407 nm. H_2O_2 performs a two-electron oxidation of the enzyme. The haem ferric iron is oxidized to an Fe(IV), ferryl, state and the extra electron is accommodated by oxidizing a tryptophan residue (Trp191) in the active site to a radical. Reduced cytochrome c (Fe^{2+}) is then oxidized to the Fe(III) form and the enzyme returns to its resting state:

$$\text{enzyme} + H_2O_2 \rightarrow \text{complex } (\lambda_{max} = 419 \, \text{nm})$$

$$\text{complex} + 2\text{cyt-}c(\text{Fe}^{2+}) \rightarrow \text{enzyme} + 2\text{cyt-}c(\text{Fe}^{3+}) + 2\text{OH}^-$$

An electron from the first cytochrome c neutralizes the tryptophan radical, leaving compound II (still a ferryl haem species) and the second electron re-forms the Fe(III) enzyme. Spectrophotometric measurement of the intermediate complex has been used as a method for measuring rates of H_2O_2 formation (Chapter 5).

3.16.2   NADH peroxidase and oxidase[185,271]

Several bacteria, such as *Lactobacillus casei* and *Enterococcus faecalis*, contain flavoprotein peroxidases, which use H_2O_2 to oxidize NADH into NAD$^+$.

$$\text{NADH} + H_2O_2 \rightarrow \text{NAD}^+ + 2H_2O$$

An aerotolerant mutant of the anaerobe *Clostridium perfringens* was reported to have gained such an NADH peroxidase. This enzyme should not be confused with the **NADH oxidase** enzymes (also flavoproteins) found in some bacteria, in which NADH is oxidized to NAD$^+$, and oxygen reduced to H_2O_2 or to water. Several aerotolerant anaerobic bacteria synthesize NADH oxidase on exposure to O_2; by its action it reduces O_2 to water and so removes O_2 from the environment of the bacteria. One such bacterium is *Treponema denticola*, found in human dental plaque.[164] The aerotolerant plaque organism *Streptococcus mutans*, implicated in the development of dental caries, produces both NADH oxidase and NADH peroxidase, as well as SOD. NADH oxidases are presumably useful because they allow the cell to survive exposure to limited amounts of O_2. However, the NADH required must be provided by metabolism, as is also true for NADH peroxidase. Drainage of cellular reducing equivalents into the NADH peroxidase or oxidase reactions could contribute to the growth-inhibitory effects of O_2.

An H_2O_2-producing NADH oxidase[185] isolated from the facultative compost-dwelling anaerobe *Amphibacillus xylanus* (which lacks catalase) was found to have an amino acid sequence resembling that of the AhpF component of **alkyl hydroperoxide reductase** from *S. typhimurium* (Section 3.13 above). Indeed, AhpF can catalyse reduction of O_2 to H_2O_2 in the presence of NADH. Addition of the AhpC component to AhpF or to the *A. xylanus* NADH oxidase caused them to catalyse O_2 reduction completely to water; both combinations could then catalyse NADH-dependent reduction of H_2O_2 and organic hydroperoxides. A protein resembling AhpC has been identified in the periplasmic space of *E. coli* and suggested to function as a thioredoxin-dependent peroxide-removing system (**thiol peroxidase**).[49] *E. coli* mutants lacking this enzyme are viable, but grow more slowly and are more sensitive to growth inhibition by peroxides or by paraquat.

These studies illustrate the important principle that enzymes purified as NADH oxidases/peroxidases may not actually function as such in the cell; it is easy to detect 'partial' functions of enzymes when assaying crude extracts. A mammalian example of this may be the thiol-specific antioxidants (Section 3.1.13).

3.16.3   *'Non-specific' peroxidases*

Plants and bacteria often harbour haem-containing peroxidases capable of using H_2O_2 to oxidize a wide range of substrates. For example, *E. coli* HP-I is a bifunctional catalase-peroxidase. 'Non-specific' peroxidases are usually assayed in cell extracts by using artificial substrates, which they oxidize in the presence of H_2O_2 to give coloured products. Often the true substrates of these peroxidases *in vivo* have not been identified. Such artificial substrates include **guaiacol** (a phenol produced by certain millipedes, possibly as a defence system),[70] **benzidine** and *o*-**dianisidine**.

'Non-specific' peroxidases have also been found in some animal systems, although they are not widespread. For example, **lactoperoxidase** is found in milk and saliva.[46] It may function to deter the growth of some strains of bacteria since, among many other substrates, lactoperoxidase can oxidize thiocyanate (SCN^-) ions, which are found in both milk and saliva, into **hypothiocyanite** ($OSCN^-$), which is very toxic to some bacterial strains, including *E. coli*, streptococci and *Salmonella typhimurium*. Lactoperoxidase may be one of the factors in milk that protects babies against infections of the gastrointestinal tract. The H_2O_2 that it needs to oxidize SCN^- seems to be derived from some of the other strains of bacteria present (e.g. *Streptococcus sanguis*), which excrete H_2O_2 into their surroundings.[46,218]

Myeloperoxidase, another 'non-specific' peroxidase, is found in phagocytic cells (Chapter 6). **Thyroid peroxidase** serves to oxidize iodide ion (I^-) into iodine atoms and attach them to the thyroid hormones. The H_2O_2 it needs is provided by a system requiring NADPH and Ca^{2+} (Chapter 6). A **uterine peroxidase**[133] has been described in rodents. Its function is unknown and it is possible that the enzyme is produced by eosinophils (Chapter 6) present within uterine tissue.

3.16.4   *Horseradish peroxidase*

Perhaps the most-studied non-specific peroxidase[91a,269] is **horseradish per-oxidase** (HRP), obtained from the roots of the horseradish plant (*Armoracia lapathifolia*). Several different forms of HRP exist, each containing bound carbohydrate and calcium ions, but they all have broad substrate specificity. For example, HRPs will oxidize guaiacol, pyrogallol, CN^- ion, NADH, thiol compounds, phenols and the plant hormone, **indoleacetic acid** (**auxin**). Oxidations by HRP, and probably by most other non-specific peroxidases, usually occur by the following series of reactions,[269] in which SH_2 is the substrate:

$$peroxidase + H_2O_2 \rightarrow compound\ I$$
$$compound\ I + SH_2 \rightarrow SH^\bullet + compound\ II$$
$$compound\ II + SH_2 \rightarrow SH^\bullet + peroxidase$$

The iron in the haem ring of 'resting' peroxidase is in the Fe(III) state. Hydrogen peroxide removes two electrons to give compound I, which contains iron in the Fe(IV) oxidation state (perhaps as Fe(IV)=O), the extra oxidizing capacity being accommodated by one-electron oxidation of the haem to give a radical cation (haem$^{\bullet+}$). The two electrons are replaced in two one-electron steps, in each of which a substrate molecule forms a radical, SH$^\bullet$ (e.g. phenols (ROH) give **phenoxyl**, R–O$^\bullet$, radicals). Hence peroxidase/ H_2O_2 mixtures have been used to generate free radicals in the laboratory[173] from almost every compound under the sun. Compound II is the intermediate state of the enzyme; the haem iron is still in the ferryl state but the haem$^{\bullet+}$ radical has been neutralized by one-electron addition.

The substrate-derived radicals (SH$^\bullet$) can then undergo a disproportionation reaction, one reducing the other to SH_2 and simultaneously being itself oxidized to S.

$$SH^\bullet + SH^\bullet \rightarrow S + SH_2$$

Phenoxyl radicals can alternatively link together to give biphenols, e.g. tyrosine phenoxyl radicals can produce bityrosine (Fig. 2.6).

3.16.5   *Peroxidases as oxidases*[269]

Radicals produced by peroxidase-dependent oxidation of substrates can sometimes reduce O_2 to superoxide

$$SH^\bullet + O_2 \rightarrow S + O_2^{\bullet-} + H^+$$

Superoxide can give H_2O_2 by the dismutation reaction

$$O_2^{\bullet-} + O_2^{\bullet-} + 2H^+ \rightarrow H_2O_2 + O_2$$

Thus, when HRP is oxidizing a substrate whose SH$^\bullet$ radical reduces O_2 to $O_2^{\bullet-}$, only catalytic quantities of H_2O_2 need be added to cause oxidation of the substrate. For example, oxidation of NADH by HRP occurs without addition of H_2O_2, since traces of H_2O_2 are always present in NADH solutions due to autoxidation. HRP-mediated NADH oxidation generates $NAD^\bullet$ radical, which can reduce O_2 to $O_2^{\bullet -}$

$$peroxidase + H_2O_2 \rightarrow compound\ I$$

$$compound\ I + NADH \rightarrow compound\ II + NAD^\bullet + H_2O$$

$$compound\ II + NADH \rightarrow peroxidase + NAD^\bullet + H_2O$$

$$2NAD^\bullet + 2O_2 \rightarrow 2NAD^+ + 2O_2^{\bullet -}$$

$$2O_2^{\bullet -} + 2H^+ \rightarrow H_2O_2 + O_2\ (dismutation)$$

$$\text{overall reaction:}\ 2NADH + O_2 + 2H^+ \rightarrow 2NAD^+ + 2H_2O$$

$NAD^\bullet$ radicals can also join together to give an NAD dimer:

$$NAD^\bullet + NAD^\bullet \rightarrow (NAD)_2$$

but this reaction is slower (second-order rate constant $3 \times 10^7\,M^{-1}\,s^{-1}$) than the reaction of $NAD^\bullet$ with O_2 ($k_2 = 1.0 \times 10^9\,M^{-1}\,s^{-1}$). NADH oxidation is one example of the so-called **oxidase reactions of peroxidase**, as compared with its 'normal' reactions in which equal amounts of SH_2 and H_2O_2 must be provided and no O_2 is consumed.

Superoxide radicals formed by oxidase reactions can combine[169] with HRP itself to generate a form of the enzyme known as **oxyperoxidase**, or **compound III**:

$$enzyme-Fe(III) + O_2^{\bullet -} \rightarrow enzyme\ (Fe^{2+}-O_2)$$

Oxyperoxidase oxidizes NADH only slowly, and so its accumulation during the reaction slows down the overall rate of NADH oxidation. Compound III decays spontaneously, but slowly, to release $O_2^{\bullet -}$.

HRP and other non-specific peroxidases can oxidize thiols into thiyl radicals[110] in the presence of H_2O_2, e.g.

$$H_2O_2 + 2GSH \xrightarrow{HRP} 2H_2O + 2GS^\bullet$$

These radicals can then participate in several reactions that result in O_2 uptake, e.g. for $GS^\bullet$

$$GS^\bullet + \underset{\substack{(ionized \\ form\ of\ GSH)}}{GS^-} \rightarrow GSSG^{\bullet -}$$

$$GSSG^{\bullet -} + O_2 \rightarrow GSSG + O_2^{\bullet -}$$

$$GS^\bullet + O_2 \rightarrow GSO_2^\bullet$$

Another example of an oxidase reaction of peroxidase is its oxidation of **2-nitropropane** ($CH_3CH(NO_2)CH_3$),[200] a compound that has been widely used as a solvent (e.g. in inks, paints and varnishes). It is also found in cigarette smoke; a 'standard' US filterless cigarette generates smoke containing about 1 µg of 2-nitropropane. Again, peroxidases use traces of H_2O_2 to initiate nitropropane oxidation, generating intermediate radicals that reduce O_2 to $O_2^{\bullet-}$. Superoxide then participates in the continued oxidation of nitropropane.

3.16.6   *Why do plants have so much peroxidase?*

The deposition of the rigid polymer **lignin** in plant cell walls involves the polymerization of phenols derived from the aromatic amino acid phenylalanine. Peroxidase bound to the cell walls oxidizes these phenols into **phenoxyl radicals** which polymerize to form the lignin.[161] One source of the required H_2O_2 appears to be the simultaneous oxidation by peroxidase of NADH generated by a malate dehydrogenase enzyme, also bound to the cell walls.

Plant peroxidases are additionally involved in the degradation of the hormone, indoleacetic acid (**auxin**),[238] and thus in the regulation of plant growth. This reaction is more complex than the usual peroxidase reactions (Fig. 3.20).

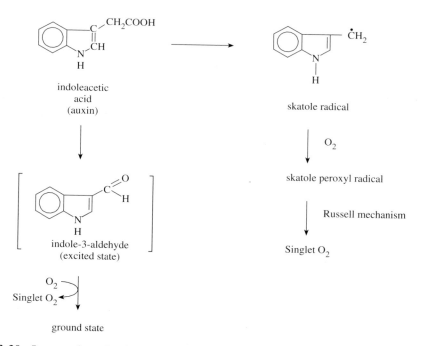

Fig. 3.20. Suggested mechanisms of auxin oxidation by horseradish peroxidase. The carbon-centred skatole radical reacts with O_2 to make a peroxyl radical, and the self-reaction of $RO_2^{\bullet}$ radicals can generate singlet O_2 (Chapter 2). Breakdown of oxidation products of auxin generates indole-3-acetaldehyde, which may be formed in the excited state and produce $^1O_2$ as it decays to the ground state. See *J. Biol. Chem.* **261**, 16860 (1986) and *Biochem. J.* **333**, 223 (1998).

A low-level luminescence (Chapter 5) has been observed from root and stem tissues of a wide variety of plants, and has been suggested to arise in part from reactions carried out by peroxidase. For example, oxidation of auxin can lead to singlet O_2 formation (Fig. 3.20). Manganese-containing and other peroxidases are involved in the degradation of lignin by fungi (Chapter 6).

In general, however, the identity of the *in vivo* substrates of plant and bacterial 'non-specific' peroxidases is unknown, making it difficult to assess their contribution to H_2O_2-removal *in vivo*.

3.16.7 *Chloroperoxidase and bromoperoxidase*

Chloroperoxidase is a non-specific peroxidase,[105] first isolated from the fungus *Caldariomyces fumago*. It contains haem and catalyses the usual 'non-specific' peroxidase reactions. In addition, it can catalyse addition of halogen atoms on to a wide range of substrates in the presence of H_2O_2 and the halide ions, chloride (Cl^-), bromide (Br^-) or iodide (I^-). If SH is the substrate and X^- the halide, these may be written:

$$\text{substrate–H} + X^- + H_2O_2 + H^+ \rightarrow \text{substrate–X} + 2H_2O$$

For example, the amino acid tyrosine is converted to chloro-, bromo- or iodo-tyrosine depending on the halide present.

Many marine organisms are rich in halogenated compounds and contain similar enzymes, e.g. the **purple bleeder sponge** and the **tropical marine sponge**. In some cases only one halide can act as a substrate, e.g. a bromoperoxidase has been isolated from several marine organisms. Bromoperoxidases from a few fungi, several marine algae (e.g. the brown alga *Ascophyllum nodosum*)[59] and the lichen *Xanthoria parietina* have been reported to contain vanadium (apparently as vanadate) at the active site instead of haem.[199]

3.16.8 *Ascorbate peroxidase*

Chloroplasts (Chapter 7), several blue-green algae and the green alga *Euglena* contain no catalase, glutathione peroxidase or 'non-specific' peroxidase activities, but they do contain high activities of an **ascorbate peroxidase** enzyme, which catalyses the overall reaction

$$\text{ascorbate} + H_2O_2 \rightarrow 2H_2O + \text{dehydroascorbate}$$

These enzymes are haem proteins inhibited by cyanide and azide and operate by 'classical' peroxidase mechanisms, i.e. they form ascorbyl radical, which then disproportionates into ascorbate and dehydroascorbate (DHA). Compound I has the absorbance spectrum expected from an Fe(IV)–haem cation radical. An ascorbate peroxidase activity has been reported in the parasite *Trypanosoma cruzi*, which lacks catalase activity. The bacterium *Synechococcus* contains both ascorbate peroxidase and a catalase/peroxidase enzyme similar to *E. coli* HPI.

When incubated in the absence of ascorbate, spinach ascorbate peroxidase is rapidly inactivated, apparently by a reaction of compound I with H_2O_2 that results in destruction of the haem ring.

An ascorbate peroxidase has been discovered in bovine eye, the first demonstration of an enzyme of this type in mammals.[258a]

3.16.9   *Peroxidase 'mimics'*

It has long been known that myoglobin, haemoglobin and a complex of haemoglobin with the haemoglobin-binding proteins in plasma (**haptoglobins**) display some peroxidase activities *in vitro* using H_2O_2 and a suitable electron donor. The peroxidase properties of haemoglobin are widely used as the basis of a diagnostic test for gastrointestinal bleeding (**faecal occult blood test**).

3.17   Antioxidant defence enzymes: co-operation

3.17.1   *The need for co-operation*

The system of antioxidant defence enzymes and their cofactors used by aerobes is complex, but it seems generally true that organisms have only sufficient defences to cope with their normal exposure to O_2. Hyperoxia, or increased free-radical generation by other mechanisms, often induce antioxidant enzymes. Indeed, this is part of the evidence for their importance *in vivo* (Section 3.4.3 above). Mechanisms of induction are considered in Chapter 4.

The antioxidant defences operate as a balanced and coordinated system and each relies on the action of the others. For example, the survival time of rats exposed to pure O_2 was increased about 70% when liposomes containing both catalase and SOD were injected intravenously before and during O_2 exposure; liposomes containing SOD or catalase alone were much less protective.[253] Exposure of rats to pure O_2 at high pressures causes convulsions (Chapter 1); pre-injection of liposomes containing both enzymes delayed the time to onset of convulsions approximately three-fold. Again, liposomes with SOD alone or catalase alone were much less protective.

What is the mechanism of this co-operation? There are several possibilities. If $O_2^{\bullet-}$ is not adequately removed it might inhibit catalase.[150] Catalase is *partially* inhibited by $O_2^{\bullet}$. The reaction,

$$\text{enzyme–Fe(III)} + O_2^{\bullet-} \rightleftharpoons \text{enzyme–Fe}^{2+}\text{–}O_2$$

generates a **ferroxycatalase** that does not rapidly decompose H_2O_2. Formation of ferroxycatalase is similar to generation of peroxidase compound III, which is also poorly reactive (Section 3.16). In addition, $O_2^{\bullet-}$ can reduce catalase compound I to compound II, also poorly active in degrading H_2O_2. Compound II contains ferryl haem, but the haem radical cation is neutralized. It has been suggested[5] that *Mycoplasma pneumoniae*, a pathogen that infects the human respiratory tract, decreases the catalase activity of its target cells by producing $O_2^{\bullet-}$, so that the cells are further damaged by H_2O_2 resulting from dismutation of the $O_2^{\bullet-}$.

Superoxide-generating systems can also inactivate glutathione peroxidase *in vitro* in the absence of GSH. Since most cells are rich in GSH, this reaction might seem unlikely to be important *in vivo*. However, treatment of rats with diethyldithiocarbamate caused a decrease in lung CuZnSOD activity which was followed by a loss of glutathione peroxidase activity in the lung, consistent with $O_2^{\bullet -}$ inactivation of GPX *in vivo*.

Addition of peroxynitrite can also inactivate catalase and glutathione peroxidase, at least *in vitro*. Similarly, H_2O_2 can inactivate CuZnSOD and FeSOD. For example, when wild-type strains of certain blue–green algae are brightly illuminated at high O_2 levels (as sometimes happens in nature)[1] they can be killed. Strains of the alga *Plectonema boryanum* sensitive to this **photo-oxidative death** contain FeSOD as the major cellular SOD, whereas a resistant strain contained mainly MnSOD (H_2O_2-resistant). Under photo-oxidative conditions FeSOD activity rapidly declines, whereas in the resistant strain MnSOD activity is maintained. Inactivation of CuZnSOD by H_2O_2 causes oxidation of histidine[254] at the active site and loss of copper ions, a pro-oxidant, from the enzyme. For example, incubation of human erythrocytes with millimolar levels of H_2O_2 led to inactivation of CuZnSOD and degradation of the modified protein by the proteasome system (Chapter 4).[220]

Further evidence[61] for cooperation of antioxidant defence enzymes is provided by cell transfection studies.[e] Transfection of a gene encoding CuZnSOD into mouse epidermal cells raised CuZnSOD levels in the cells and *sensitized* them to damage upon incubation with a mixture of xanthine and xanthine oxidase, whereas subsequent transfection with genes encoding catalase or glutathione peroxidase rendered the cells equally or more resistant to damage. Hence SOD activity must be in balance with that of H_2O_2-removing enzymes. This has also been demonstrated in other organisms, including *E. coli*.[229] For example, over-expression of *both* SOD and catalase was needed to increase lifespan in transgenic *Drosophila* (Chapter 10).

3.17.2 *Down's syndrome*

One medically important area in which an imbalance of antioxidant defence systems may be important is **trisomy 21**, often called Down's syndrome after the physician John Langdon Down, who described it in 1866. The syndrome is often characterized by mental retardation, morphological abnormalities and an increased rate of development of Alzheimer's disease. The disease is caused by the presence of an extra copy of chromosome 21, resulting from errors in cell division. Trisomy 21 results in increased production of many of the proteins encoded by genes on this chromosome. One of these is CuZnSOD; exactly 1.5 times as much CuZnSOD activity is found in cells from trisomy 21 patients than from normal controls. Indeed, cells from such patients are less readily damaged by *M. pneumoniae* (Section 3.17.1 above).[5]

Is Down's syndrome related to the rise in CuZnSOD, or to one or more of the many other changes that must occur? This question has been examined using transgenic mice (Box 3.1). Animals have been obtained with

1.6- to 6-fold more SOD activity, especially in the brain. Rises in brain catalase activity (but not in glutathione peroxidase) were also observed. The mice appear morphologically normal and are more resistant to several toxins (Box 3.1). However, abnormal neuromuscular junctions were reported in the tongue, apparently similar to those found in tongue muscles of trisomy 21 patients. The mice also showed abnormalities in serotonin metabolism comparable to those reported in patients. However, they did not show the changes in dorsal root ganglia, that are found in patients, i.e. only some of the neurological changes in trisomy 21 occur in the mice. At the moment the contribution, if any, of elevated CuZnSOD to Down's syndrome is an open question.

3.18 Antioxidant defence: sequestration of metal ions

Iron and copper are essential in the human body for the synthesis of a huge range of enzymes and other proteins involved in respiration, O_2 transport, $NO^\bullet$ formation and other redox reactions. Yet these metals are potentially dangerous: their ability to undergo one-electron transfers

$$Fe(III) + e^- \rightleftharpoons Fe^{2+}$$

$$Cu^{2+} + e^- \rightleftharpoons Cu^+$$

enables them to be powerful catalysts of autoxidation reactions (e.g. oxidation of adrenalin, dopamine and ascorbate), conversion of H_2O_2 to $OH^\bullet$ and decomposition of lipid peroxides to reactive peroxyl and alkoxyl radicals (Chapter 4). It is not only 'free' metal ions that are catalytic: haem and certain haem proteins can decompose lipid peroxides and interact with H_2O_2 to cause damage (Section 3.18.3 below). Hence organisms must be careful in how they handle these metals. Like O_2 and $NO^\bullet$, iron and copper are essential but dangerous.[108]

3.18.1 *Iron metabolism*[56]

The average adult human male contains about 4.5 g of iron, about two-thirds of which is present in haemoglobin, and he absorbs about 1 mg of iron per day from the diet and excretes approximately the same amount when in iron balance. Women often have less body iron, partly because of menstruation.

Since the total plasma iron turnover is some 35 mg/day, efficient mechanisms to regulate iron content must exist in the body. Slight disturbances of iron metabolism will lead to either deficiency or overload: both with deleterious consequences (e.g., iron deficiency in growing children can impair mental development). It has been estimated that more than 500 million people in the world are iron-deficient and several million are iron-overloaded. Iron is lost from the body in sweat, faeces (e.g. by loss of unabsorbed iron, and of iron in cells shed from the lining of the gut), urine and blood (especially menstrual bleeding in women). Body iron stores have a marked negative regulatory effect on the efficiency of iron absorption; normally only 10–15% of non-haem iron is absorbed but this increases in iron-deficiency anaemia. Foods such as wheat flour and breakfast cereals are fortified with iron in many countries; sometimes infant formulas and some weaning foods are also. It has been estimated that fortification contributes about 10% of adult iron intake in the UK and more than double this for pre-school children.

Most inorganic iron in food is in the Fe(III) state, the most stable oxidation state for iron. The main site of iron absorption is the duodenum,where it is absorbed as Fe^{2+} by a carrier protein. Hence agents that solubilize and reduce Fe(III) in the diet, such as gastric hydrochloric acid and ascorbate, are believed to facilitate iron absorption. However, many other dietary components influence iron uptake. For example, **phytates**, present in cereals, nuts and legumes, chelate iron and slow its absorption.

Haem iron is absorbed by a different pathway: haem is taken up and the iron removed from it in the mucosal cells, probably by the action of the enzyme, **haem oxygenase**[162] (Section 3.20 below). A much greater percentage of dietary haem iron appears to be absorbed than non-haem iron, and its uptake is not affected by dietary iron-binding agents such as phytates.

Transferrin

Some of the iron absorbed is retained in the gut mucosal cells within ferritin and other proteins and will be lost in the faeces when these cells are shed. The rest enters the circulation bound to **transferrin**. Transferrin also accepts iron released by the destruction of aged red blood cells, e.g. in the spleen (estimated as 20–25 mg/day). Circulating transferrin only accounts for about 3 mg of total body iron, but this iron turns over about 10 times daily. Transferrin is a glycoprotein (relative molecular mass 79 000), mostly synthesized in the liver. Transferrin has N-terminal and C-terminal domains, each of which tightly binds one atom of Fe(III) at pH 7.4. Two monoferric forms of transferrin can therefore exist, with iron bound to the N-terminal or to the C-terminal domains respectively. Tight binding of iron to either domain requires the presence of carbonate (CO_3^{2-}) or bicarbonate (HCO_3^-) anion.[22] In healthy people, transferrin is usually no more than 20–30% loaded with iron. Hence human and other animal blood plasmas have considerable iron-binding capacity and their content of 'free' iron ions is virtually zero. The affinity of transferrin for iron at pH 7.4 is high (stability constant $\approx 10^{22}$) and is slightly higher for the C-terminal domain than the N-terminal domain. The strength of binding decreases at lower pH values.

Transferrin can bind several metal ions other than Fe(III), although with lower affinity, including aluminium (Al(III), stability constant $\approx 10^{12}$) and bismuth ((Bi(III)), stability constant $\approx 10^{19}$).[155]

Other iron-binding proteins

A protein similar to transferrin, known as **lactoferrin**, is found in saliva, vaginal mucus, seminal fluid, tears, bile, nasal secretions, milk (more in human milk than cow's milk) and other secretory fluids and is released by activated neutrophils at sites of inflammation (Chapter 6). Lactoferrin also binds two atoms of Fe(III) per molecule, but binds its iron more tightly at low pH values than does transferrin.[107] Lactoferrin has been suggested[14] (but not proven) to have many roles *in vivo*: it binds to DNA and may influence gene transcription, has various effects on cells of the immune system and may be an antibacterial factor in human milk.

Many similar iron-binding proteins are known. Thus egg white contains **ovotransferrin** (sometimes called **conalbumin**). Melanoma cells can express a **melanotransferrin**, perhaps to aid acquisition of iron to facilitate rapid growth. **Uteroferrin** has been isolated as a purple-coloured, iron-binding protein, which is synthesized in large amounts by the pig uterus after treatment with the hormone progesterone. The purple colour arises from an iron centre at which iron is co-ordinated with one or more tyrosine residues. Utero-ferrin can also act as a phosphatase enzyme, belonging to the type 5 (tartrate-resistant) acid phosphatase class. In humans, this type of enzyme is found in osteoclasts (cells that destroy the matrix of bone and cartilage: Chapter 9) and serum levels increase in diseases where bone resorption is enhanced.[118,159]

Iron within cells[56]

Cells that require iron express **transferrin receptors** on their surface: the more iron they want, the more receptors they make. Receptors bind iron-carrying plasma transferrin, which is internalized by receptor-mediated endocytosis and enters the cytoplasm in a vacuole. (The liver and many cells in culture can also absorb 'low molecular mass' iron and copper chelates, such as iron–citrate.)[203] The contents of the vacuole are acidified by the action of an H^+-pumping ATPase, which weakens iron binding to transferrin. The iron is then removed by iron-binding agents, whose identity is uncertain. They may include citrate, ATP, GTP, inositol phosphates and other phosphate esters. The iron-free transferrin (**apotransferrin**) is then ejected from the cell for re-use. At pH 7.4, the transferrin receptor has a much higher affinity for iron-loaded transferrin than for the apoprotein, so it can collect the iron-bearing protein from the plasma. This order of affinities is reversed at acidic pH, so keeping the apotransferrin on the receptor for ejection from the cell.

The **low molecular mass iron pool** supplies iron for the biosynthesis of essential proteins. For example, mitochondria have been reported to take up iron salts rapidly and may have small 'pools' of non-protein-bound iron ions in the matrix. ATP-dependent uptake systems for iron–citrate and iron–ATP chelates have been reported in rat liver nuclei.[102] However, the size and chemical nature of this 'non-protein-bound iron pool' in cells and organelles is unknown: in 1987, Crichton and Charloteaux-Waters[56] described it as 'appearing somewhat like the Loch Ness Monster, only to disappear from view before its presence or nature can be confirmed.' Studies[38] using fluorescent iron chelators have given some evidence consistent with its presence. Using the fluorescent iron-binding agent, **calcein**, they estimated an average intracellular iron pool (calcein-chelatable) of $0.35\,\mu M$ in human erythroleukaemia cells.

Addition of Fe^{2+} to the cells raised this value, but it returned to normal within 30–60 min. Addition of H_2O_2 or *tert*-butyl hydroperoxide also raised the level, and it took longer (60–90 min) to return to normal.

Ferritin[112]

Most intracellular iron is stored in **ferritin**. Mammalian ferritins consist of a hollow protein shell, 12–13 nm outside diameter (7–8 nm inside diameter), composed of 24 subunits. Each subunit has a relative molecular mass of about 20 000. The shell surrounds an iron core that can hold up to 4500 ions of iron per molecule, but usually has fewer. Traces of other metals can be present in ferritin, including copper.[37]

Iron enters ferritin as Fe^{2+} which is oxidized to Fe(III) and deposited in the core as an insoluble hydrated ferric oxide. The chemistry of the oxidation is uncertain, but phosphate associated with ferritin plays an important role. How cells release iron from ferritin is similarly uncertain; *in vitro*, iron can be released (as Fe^{2+}) by several reducing agents including ascorbate, thiols, urate and reduced flavins. Iron enters and leaves through channels, which are of two types: non-polar and polar (probably much more important).

Mammalian ferritins are made of two subunit types (**H-** and **L-chains**). H- and L-chains are of similar size with some 50% amino acid sequence identity. Ferritin proteins vary widely in H/L ratios, from $H_{24}L_0$ to H_0L_{24}; ratios often change in various diseases. In general, ferritins in liver and spleen are richer in L-subunits than those in heart and brain. By contrast, **bacterioferritins** and plant ferritins are homopolymers; bacterioferritins contain bound haem.

The H-chains have a metal-binding site and can oxidize Fe^{2+} to Fe(III). At low fluxes of iron into ferritin H_2O_2 is produced on the H-chains whereas at high fluxes Fe^{2+} oxidation can also occur on the surface of the growing core. Tyrosyl radicals and $OH^•$ have been detected from ferritin during iron loading but are (fortunately) not stoichiometrically produced as Fe^{2+} is oxidized.

Ferritin can be converted in lysosomes into an insoluble product called **haemosiderin**, probably by proteolytic attack.[112]

Regulation of cellular iron balance[56,112,203]

There is a synchronized regulation of the synthesis of transferrin receptors and ferritin subunits in mammalian cells. This involves cytoplasmic **iron-regulatory proteins** (IRPs) which bind to **iron-responsive elements** (IREs),[24] special base sequences in the mRNAs of ferritin and of transferrin receptor proteins. When iron levels are too low, IRPs bind to IREs. This binding stabilizes transferrin receptor mRNA so that more protein is made. By contrast, it prevents translation of ferritin mRNA. When there is enough cellular iron, mRNA for the transferrin receptor is rapidly degraded. IRP-1 is identical with a cytosolic form of the enzyme **aconitase**. (Aconitase also occurs in mitochondria, in the Krebs cycle.) IRP-1 has aconitase activity when its [4Fe–4S] cluster is present, but is an IRP when iron is low; switching between these forms depends on cellular iron status. Mutations inactivating the IRE in mRNA encoding L-ferritin are associated with elevated plasma ferritin levels and cataract in humans.[24] Another IRP, **IRP-2**, without aconitase activity, has been identified in mammalian cells: the presence of iron seems to facilitate its degradation by the proteasome system (Chapter 4).

IRPs also affect the synthesis of several other iron-containing proteins, including mitochondrial aconitase and δ-aminolaevulinate synthetase, which catalyses the first step in haem synthesis in erythrocyte precursor cells (Fig. 2.12). IRP binding blocks translation of the mRNA encoding this enzyme and thus decreases haem biosynthesis.

3.18.2   *Copper metabolism*[103,111]

Copper ions are powerful catalysts of free radical damage: they convert H_2O_2 to $OH^\bullet$,[144] decompose lipid peroxides, catalyse autoxidation reactions (especially of ascorbate) and are highly effective in causing oxidative DNA damage and stimulating peroxidation of low density lipoproteins (LDL; Chapters 4 and 9). Like iron, copper must be handled carefully *in vivo*.

An average adult human contains about 80 mg copper in total. Copper is absorbed from the diet, probably in the duodenum as chelates with amino acids (such as histidine) or small peptides. Not all the copper entering gut mucosal cells enters the blood: some is stored in these cells and lost from the body when they are shed. The copper that is absorbed apparently enters the blood bound to albumin; each molecule of this plasma protein has one high affinity binding site for copper plus a number of weaker copper ion-binding sites.

Copper-albumin is taken up by the liver, which incorporates copper into the protein **caeruloplasmin**, relative molecular mass about 132 000. Caeruloplasmin is secreted into plasma and unwanted copper is excreted into bile. Caeruloplasmin contains six tightly-bound copper ions and often a seventh, which is less tightly bound. The copper-binding sites cluster together near one face of the molecule. Although the bulk of caeruloplasmin is made in the liver, some is also made in certain cell types in lung and brain.[270] Caeruloplasmin donates copper to copper-requiring cells after binding to cell-surface receptors; endocytosis has been observed in some (but not all) cell types in culture. The intracellular nature of copper released from caeruloplasmin is uncertain, but one suggested carrier is GSH (Section 3.9 above).

Human blood plasma normally contains 200–400 mg/l of caeruloplasmin, accounting for at least 90% of total plasma copper. The remaining plasma copper is often claimed to consist of copper ions bound to albumin, histidine or small peptides. *In vitro*, copper can be removed from such chelates by the chelating agent ***o*-phenanthroline**.[104] However, if mammalian plasma is incubated with phenanthroline, no chelatable copper is detected, which implies that amounts of these non-caeruloplasmin copper chelates in plasma are very small.[78] Chromatography, dialysis or low-temperature storage of plasma or serum can lead to release of copper from caeruloplasmin, which may account for reports of 'non-caeruloplasmin copper' in earlier studies.[104]

Copper release in 'aged' plasma appears to occur by the action of a plasma metalloproteinase on caeruloplasmin.[75] Caeruloplasmin has structural similarities to two proteins involved in blood coagulation, **factors V** and **VIII**. These factors are activated by proteolytic cleavage, and the similarity of caeruloplasmin to them may, in part, explain its sensitivity to proteolysis. Peroxynitrite addition (Chapter 2) to caeruloplasmin also releases copper.

Caeruloplasmin as an oxidase[103]

In vitro, caeruloplasmin can catalyse oxidation of a wide range of polyamine and polyphenol substrates, including catecholamines, but these activities have acidic pH optima and their biological significance is uncertain. Caeruloplasmin also has a **ferroxidase** activity; it oxidizes Fe^{2+} to $Fe(III)$ and can facilitate iron loading on to transferrin (it does not appear to be involved in iron loading into ferritin). The physiological importance of the ferroxidase activity of caeruloplasmin is illustrated by studies on patients with **acaeruloplasminaemia**, due to mutations in the gene encoding this protein. They show low serum iron, elevated serum ferritin and increased iron deposition in the brain and liver. Pathological consequences include diabetes, retinal degeneration and neurological abnormalities. Injection of caeruloplasmin raises serum iron. Similarly, copper deficiency in animals produces an anaemia that does not respond to iron administration but can be corrected by injection of caeruloplasmin.[111]

Ferrous salts spontaneously oxidize at physiological pH, but this produces oxygen radicals:

$$Fe^{2+} + O_2 \rightarrow Fe(III) + \boxed{O_2^{\bullet -}}$$

$$2O_2^{\bullet -} + 2H^+ \rightarrow H_2O_2 + O_2$$

$$Fe^{2+} + H_2O_2 \rightarrow Fe(III) + \boxed{OH^{\bullet}} + OH^-$$

By contrast, caeruloplasmin-catalysed Fe^{2+} oxidation does not: four Fe^{2+} are oxidized and one O_2 reduced to $2H_2O$.

The ferroxidase activity of caeruloplasmin is inhibited by azide and is often called **ferroxidase I**, to distinguish it from other ferroxidases that have been described. In humans caeruloplasmin accounts for virtually all the measurable ferroxidase activity, whereas in certain other animals, e.g. rabbits, plasma additionally contains an azide-resistant ferroxidase called **ferroxidase II**. Decreases in plasma ferroxidase activity in the face of unchanged or raised caeruloplasmin protein levels have been observed in several human diseases, including adult respiratory distress syndrome and rheumatoid arthritis, but the physiological significance of this is uncertain.

3.18.3   *Haem proteins: potential pro-oxidants*

Haemoglobin and myoglobin are normally intracellular proteins, e.g. haemoglobin is packaged within red cells, which are rich in antioxidant defence enzymes (Chapter 7). Both these proteins undergo slow oxidation to form $O_2^{\bullet -}$

and Fe(III) protein. Both can also cause damage when mixed with H_2O_2, which can readily penetrate to the haem centres. Such damage can occur *in vivo* and also during meat processing, causing lipid oxidation and 'off-flavours'.

When haemoglobin or myoglobin are exposed to an excess ($\geqslant 10:1$ molar ratio) of H_2O_2 *in vitro* they are degraded, releasing both haem and iron ions (from haem ring breakdown).[202] Haem and iron so released can stimulate lipid peroxidation,[172] and iron ions can cause $OH^\bullet$ formation from H_2O_2.[108] At lower H_2O_2:protein ratios (e.g. 1:1), haemoglobin and myoglobin are converted to iron(IV) (**haem ferryl**) species. Amino acid radicals are also generated on the protein; tyrosine and tryptophan peroxyl radicals (Trp/TyrOO$^\bullet$) have been identified.[101] Cross-links between haem and tyrosine can also occur.

$$\text{haem Fe(III) protein} + H_2O_2 \rightarrow \text{haem}[Fe(IV)=O]\text{protein}^\bullet + H^+ + H_2O$$

Both the ferryl species and the amino acid radicals can stimulate peroxidation of lipids[124] and oxidize other molecules,[216] including thiols, urate and certain proteins, to form secondary free radical products, e.g. urate radical and RS$^\bullet$. When added to lipids (including lipoproteins), haem proteins in the *absence* of added H_2O_2 can stimulate peroxidation by a mechanism that probably involves the decomposition of pre-existing traces of lipid peroxides in the lipids to alkoxyl and peroxyl radicals.[124,214]

$$\text{haem Fe(III)} + \text{LipOOH} \rightarrow \text{haem Fe}^{2+} + \text{LipOO}^\bullet + H^+$$

$$\text{haem Fe}^{2+} + \text{LipOOH} \rightarrow \text{haem Fe(III)} + \text{LipO}^\bullet + OH^-$$

Large amounts of lipid peroxides can, like H_2O_2, cause release of free iron ions from haem proteins. Hence haem,[172] haemoglobin and myoglobin outside their normal location are potentially damaging molecules. Myoglobin released as a result of severe muscle injury (e.g. crushing of muscles) can, for example, lead to kidney failure, one suggested mechanism being free radical damage involving 'free' iron release. Cytochrome *c* can also be degraded by excess H_2O_2, as can cytochromes P450.[16]

3.18.4   *Evidence that metal ion sequestration is important*

Antibacterial effects[140]

Why do organisms take such care in the handling of iron, copper and other transition metal ions? Bacteria need these metals for growth. Restricting metal availability in body fluids such as plasma, tears, milk, uterine and respiratory tract lining fluids is thus an antibacterial mechanism. Indeed, some bacteria have evolved mechanisms to circumvent this. They may synthesize powerful iron chelators (**siderophores**), able to take iron from transferrin.[170] The lungs of patients with cystic fibrosis often become infected with *Pseudomonas aeruginosa*, which produces an elastase that can cleave transferrin and lactoferrin to release iron.[40] Unfortunately, the cleaved proteins, unlike transferrin itself, can cause $OH^\bullet$ formation from H_2O_2 (Section 8.7).

Diminishing free-radical reactions[107]

A major advantage of metal-ion sequestration is that it diminishes their ability to cause damaging free-radical reactions (Table 3.12). Manipulations of the iron content of *S. aureus* and *E. coli* that illustrate this point have already been discussed (Section 3.6 above). Similarly, several studies have shown that the toxicity of H_2O_2 or organic peroxides to animal and plant cells in culture can be increased by raising their iron content, and decreased by the presence of chelating agents such as phenanthroline or the Fe(III) chelator **desferrioxamine**. Desferrioxamine enters cells inefficiently, but prolonged incubations can cause intracellular iron depletion.

Desferrioxamine has been shown to protect against tissue damage in several animal models of human disease (Chapter 10). By contrast, injection of ferric salts bound to the chelating agent nitrilotriacetate (NTA) into animals causes severe tissue injury,[251] increased free radical damage to lipids, proteins and DNA and leads to renal cancer. Unlike ferrioxamine, ferric NTA is effective at generating $OH^\bullet$ from H_2O_2. NTA has been widely used in detergents (as a chelator to prevent mineral build up) and is also employed in the laboratory[22] to prepare iron-loaded transferrin.

Sequestration of metals into proteins can only be an effective antioxidant defence if the protein-bound metal ions are less effective as free-radical catalysts than 'unchelated' metals or low molecular mass chelates of iron with such molecules as citrate or ADP. Iron correctly bound to the two specific iron-binding sites of lactoferrin or transferrin will not stimulate autoxidations, react with $O_2^{\bullet-}$ or H_2O_2 or decompose lipid peroxides at pH 7.4. However, iron can be released from transferrin to iron chelators at low pH. By contrast, iron in uteroferrin and other purple acid phosphatases may be more accessible to catalyse free-radical reactions.[118]

Iron in the ferritin cores is insoluble and as such is probably not pro-oxidant within the cell. However, it can be mobilized by reducing agents, including $O_2^{\bullet-}$. Nevertheless, the amount of $O_2^{\bullet-}$-mobilizable iron in ferritin is a tiny percentage of the total iron content;[37] it seems that $O_2^{\bullet-}$ cannot mobilize iron from the core itself but can only release iron still within the channels or loosely bound to the protein surface. Much confusion has resulted from studies of commercial ferritin preparations, which are often partially degraded and thus much easier to mobilize iron from. In general, haemosiderin iron is even less available than that in ferritin, essentially because haemosiderin is insoluble.[187] A similar strategy seems to be used by the malarial parasite, which packages unwanted haem (a potential pro-oxidant) as insoluble **haemozoin** (Chapter 7). Superoxide can also release iron from the iron–sulphur clusters present at the active sites of certain enzymes such as aconitase (Chapter 2).

Similarly, the 'tightly bound' caeruloplasmin copper ions are not available to catalyse free-radical damage. There has been some suggestion that the seventh copper, which is more loosely associated, might be able to catalyse free-radical reactions such as the oxidation of LDL, but data are conflicting.[177] The ease of degradation of caeruloplasmin during isolation and the fact

Table 3.12. Role of transition-metal ions in converting less reactive into more reactive species

Starting agent	More reactive species produced on addition of metal ions	Metal involved	Comment
H_2O_2 ($\pm O_2^{\bullet-}$)	$OH^{\bullet}$ (and possibly reactive oxo-metal species)	Fe/Cu	iron-dependent and copper-dependent conversion of H_2O_2 to $OH^{\bullet}$ and other oxidizing species (e.g. ferryl)
HOCl	$OH^{\bullet}$	Fe	Fe^{2+} reacts with HOCl to give Cl^- and $OH^{\bullet}$ (Chapter 2)
Lipid peroxides	peroxyl radicals, alkoxyl radicals, cytotoxic aldehydes	Fe/Cu	see Chapter 4
Thiols (R–SH)	$O_2^{\bullet-}$, H_2O_2, $RS^{\bullet}$, $OH^{\bullet}$	Fe/Cu	oxidation of thiols such as GSH produces thiyl radicals and oxygen radicals (Chapter 2)
NAD(P)H	$NAD(P)^{\bullet}$, $O_2^{\bullet-}$, $OH^{\bullet}$	Fe/Cu	$NAD(P)^{\bullet}$ radicals reduce O_2 to give $O_2^{\bullet-}$; copper is especially good at promoting NAD(P)H oxidation
Ascorbic acid	$OH^{\bullet}$, possibly $O_2^{\bullet-}$, semidehydroascorbate radical	Fe, especially Cu	oxidation of ascorbate produces cytotoxic species
Alloxan, adrenalin, DOPA, dopamine, dihydroxyfumarate, tetrahydrofolates, 6-hydroxydopamine, other 'autoxidizable' compounds	$OH^{\bullet}$, $O_2^{\bullet-}$, carbon-centred or other radicals derived from the toxin	Fe, Cu, Mn, often other metals	most 'autoxidations' are dependent on the presence of traces of transition-metal ions
Peroxynitrite (ONOO$^-$)	NO_2^+ ??	Fe/Cu	transition-metal ions and some metalloproteins (including CuZnSOD) accelerate nitration of aromatic compounds on addition of ONOO$^-$, possibly by facilitating its decomposition to nitrating species such as NO_2^+, nitronium ion

that many commercial preparations are extensively degraded has not helped.

By contrast, copper ions attached to albumin or histidine can still react with H_2O_2 to cause free-radical damage. However, the $OH^\bullet$ or other reactive species produced is largely absorbed by the ligand. The ability of albumin to bind copper ions may be a protective mechanism; among its many other functions, albumin may bind copper and prevent it from associating with lipoproteins and membranes. If $O_2^{\bullet-}$ or H_2O_2 were produced in plasma (e.g. from activated phagocytic cells) they might still react with copper, but this would damage the albumin carrier. Since plasma albumin levels are high and its turnover is rapid, this is unlikely to be of much consequence. In this context, albumin has been called a **sacrificial antioxidant**.[107]

The importance of metal ion sequestration is amply illustrated by the pathology of **iron overload** and **copper overload** diseases. For example, a mouse strain that synthesizes transferrin at less than 1% of the normal rate has been described. Gastrointestinal iron absorption is greater than normal: low molecular mass iron chelates are present in plasma and iron accumulates in many tissues. The mice die soon after birth unless transferrin is injected.[235a]

Iron overload

The amount of iron in the mammalian body is determined by the amount absorbed from the gastrointestinal tract. Once iron has entered, there appears to be no dedicated physiological mechanism for disposing of it (unless one counts shedding of gut lining cells and menstrual bleeding). The consequences of inability to dispose of excess iron are seen in **iron overload**. Acute iron overload can be caused by ingestion of large quantities of iron salts, usually by children who eat 'iron tablets' (containing ferrous salts) prescribed for their parents. There is vomiting, gastrointestinal bleeding and severe shock. Since the introduction of therapy with the chelating agent desferrioxamine, the mortality from acute iron poisoning has fallen in the UK. Large quantities of iron salts, particularly in the Fe^{2+} state, can degrade the protective layer of gastrointestinal mucus and attack the cells underneath.

A more slowly developing iron overload of dietary origin has been described in rural Zimbabwe and other parts of Africa and is related to drinking acidic beer out of iron pots, although a genetic predisposition to iron overload may also be involved.[175] Transient iron overload can occur during cancer chemotherapy (Chapter 9). Iron overload is well documented in patients suffering from the inherited disease, **idiopathic haemochromatosis**.[55] In this disease, the normal tight regulation of gut iron absorption fails, so that more dietary iron than usual is taken up (absorption of cobalt, lead, manganese and zinc is also increased, but the clinical significance of this is uncertain).

The metabolic abnormality causing the increased iron uptake is unknown but the defective gene is located on chromosome 6, in the human leukocyte antigen (HLA) region; hence it has been dubbed **HLA-H**. The protein it encodes is widely distributed in the gastrointestinal tract and other tissues (except brain). The time taken for clinically significant iron overload to develop in haemochromatosis is often 40 or more years and depends to some extent on the diet of the patient. When overload occurs, the iron-binding capacity of transferrin in the plasma is often exceeded, so that **non-transferrin-bound iron**[99,120] is present. At least some, and perhaps all, of this iron is bound to plasma citrate, although other low-molecular-mass chelating agents and albumin may also be involved. The liver attempts to remove non-transferrin-bound iron by rapidly taking it up, and so the liver becomes iron-overloaded.

The pathology resulting from iron overload in haemochromatosis is devastating. It includes liver fibrosis and cirrhosis, an elevated risk of hepatoma (liver cancer), weakness and malaise, weight loss, skin pigmentation, diabetes (since pancreatic β-cells are damaged), cardiac malfunctions and a chronic joint inflammation resembling rheumatoid arthritis. Haemochromatosis is not a rare disease: the prevalence in Australian, American and several European populations has been estimated at about 0.3–0.5%. The heterozygote frequency is about 10%, making this the commonest genetic disorder among Caucasians.

Thalassaemias[140]

Iron overload can also result from medical treatment of other diseases. For example, the **thalassaemias** (named from a Greek word meaning 'the sea') are inborn conditions in which the rate of synthesis of one of the haemoglobin chains is diminished, the prefix α- or β-thalassaemia being used to identify the chain that is synthesized abnormally slowly. Thalassaemias often arise from mutations within gene promoters or in introns or exons, leading to abnormal splicing of mRNA so that a defective protein product is obtained on translation.[e] **Thalassaemia major** (sometimes called **Cooley's anaemia**) and **thalassaemia minor** refer to the homozygous and heterozygous states, respectively.

Untreated patients with thalassaemia major die in infancy, but can be kept alive by regular blood transfusions. Since each unit of blood contains about 200 mg of iron, the patients eventually become iron-overloaded, leading to saturation of transferrin and often the appearance of non-transferrin-bound iron in the blood. Iron accumulates, especially in the liver and spleen. Less iron accumulates in the heart, but this organ is very sensitive to iron. Hence many thalassaemic patients treated by blood transfusion suffer cardiac malfunctions. Similar problems arise in the treatment of other chronic anaemias by transfusion. Treatment of iron overload resulting from idiopathic haemochromatosis is usually by blood-letting (**phlebotomy** or **venesection**), whereas chelating agents are administered to transfused thalassaemic patients in an effort to slow or prevent the accumulation of iron in the body.

Children with β-thalassaemia major in Britain were first given the chelating agent desferrioxamine in 1962, and it has been successful in prolonging their lifespan. Desferrioxamine, a powerful chelator of Fe(III), is isolated from *Streptomyces pilosus*. Desferrioxamine and its Fe(III) complex (**ferrioxamine**) are rapidly excreted, in both urine and bile, so removing iron from the body. Large doses of desferrioxamine are required, it cannot be given by mouth (requiring subcutaneous or intravenous infusion) and it penetrates only slowly into several cell types. Hence there is considerable interest in the development of new chelating agents without the disadvantages of desferrioxamine; candidates include rhodotorulic acid and 1,2-dimethyl-3-hydroxypyrid-4-one. These are discussed further in Chapter 10.

Non-transferrin-bound iron

The non-transferrin-bound iron found in the blood plasma of iron-overloaded haemochromatosis or thalassaemia patients will stimulate lipid peroxidation and the formation of $OH^\bullet$ radicals from H_2O_2 in experiments *in vitro*, which suggests that the pathology of iron overload is related to increased free-radical damage. The sensitivity of the heart to even a small loading with iron might be explained by its relatively poor protection against oxygen radicals: cardiac catalase activity is low and activities of SOD and glutathione peroxidase are only moderate. Other examples of the sensitivity of the heart to radical damage include the cardiomyopathy induced by the anti-tumour drug doxorubicin, which is known to stimulate oxygen radical production (Chapter 9) and the heart lesions seen in Keshan disease, in which lack of dietary selenium causes a drop in tissue glutathione peroxidase activities (Section 3.14.3 above). It has even been suggested that the lower incidence of heart disease in women as compared with men is related to their lower body iron stores (Chapter 10). The pancreatic β-cells, often damaged in haemochromatosis, are also sensitive to free-radical attack (Chapter 8).

Some evidence consistent with this 'radical hypothesis' of damage induced by iron overload has been obtained e.g. iron overloaded patients show elevated levels of certain lipid peroxidation end-products, protein carbonyls,[h] and subnormal levels of vitamins E and C.[272] ESR spectroscopy/spin trapping (Chapter 5) has been used to detect $OH^\bullet$ adducts in the bile of iron-overloaded rats.[128] An additional mechanism of damage is the labilization of lysosomal membranes caused by excessive formation of haemosiderin within them: this lysosomal damage might be a consequence of iron-stimulated lipid peroxidation. However, present evidence, although suggestive, does not rigorously prove that free-radical reactions are a major cause of the pathology of iron overload.

Copper overload

The toxicity of excess copper is illustrated by **Wilson's disease**,[189] an inherited recessive disorder often characterized by low concentrations of caeruloplasmin in the blood. It occurs worldwide with a prevalence of up to 1 in 30000. Copper is deposited in the liver, kidney, cornea and brain, causing damage that leads to lack of co-ordination, tremors and progressive mental retardation.

It seems likely (but is not rigorously proven) that copper-stimulated free-radical reactions are involved in the pathology of Wilson's disease. Plasma levels of ascorbate, vitamin E and urate are sub-normal and increased levels of products of uric acid oxidation (Section 3.21.7 below) and lipid peroxidation have been detected.[189,258] Treatment involves a copper-restricted diet and use of chelating agents, such as the thiol **penicillamine**, that promote copper excretion. Oral administration of zinc salts may also help to prevent copper accumulation, by interfering with the intestinal absorption of copper. Zinc ions (Zn^{2+}) might also compete with copper ions for binding to target sites that could be damaged by free radicals.

The defective gene in Wilson's disease is on chromosome 13 and encodes a **Cu^{2+}-transporting ATPase** enzyme; the gene is strongly expressed in liver and kidney. The low plasma caeruloplasmin levels often seen in Wilson's disease are probably secondary to this defect, which impairs copper transport between cells and extracellular fluids. A related disorder, encoded by a gene on the X chromosome, **Menkes disease**, involves a defect in a similar protein. In patients with this disease, copper export from the liver is normal but that from many other tissues is defective; the Menkes disease gene is expressed in most body tissues except liver. For example, copper enters the intestinal cells but is not transported further, resulting in severe copper deficiency in other body tissues. Patients show a wide range of problems including skeletal defects, aneurysms and degeneration of the nervous system. Impaired biosynthesis of essential copper-containing enzymes (e.g. lysyl oxidase and dopamine-β-hydroxylase) may be involved.

Insufficient dietary intake of copper can also lead to free-radical damage, e.g. by decreasing CuZnSOD activities and caeruloplasmin levels.

3.19 Metal–ion sequestration in different environments

3.19.1 *Intracellular requirements*

Within cells, iron appears to be in constant transit between the 'low molecular mass pool', ferritin and its ultimate destination in functional proteins. The same may be true of copper. Oxidative stress can lead to more iron release (e.g. from ferritin and iron–sulphur clusters by the action of $O_2^{\bullet-}$; from haem proteins by the action of peroxides) and possibly more copper release. This may explain why cells rely on enzymes that remove $O_2^{\bullet-}$ (SODs), H_2O_2 (catalases and peroxidases) and lipid peroxides for their antioxidant defence; it is important to remove these species before they come into contact with transition metal ions and generate more-damaging agents (Table 3.12).[107]

3.19.2 *Metallothioneins*

Metallothioneins are proteins with a low molecular mass (about 6500) and are encoded by genes on chromosome 16 (in humans). They are found in the cytosol of eukaryotic cells, especially in liver, kidney, and gut,[72] and may also be present in the nucleus. Two isoforms, MT-I and MT-II, are found in all animal tissues, and a third isoform, MT-III, in brain. Metallothioneins are rich in sulphur (23–33% of the amino acids are cysteine). They therefore represent a significant portion of total cell protein thiol. Each metallothionein molecule can bind 5–7 ions of such metals as zinc (Zn^{2+}), silver (Ag^+), copper (Cu^+), cadmium (Cd^{2+}) and mercury (Hg^{2+}). Binding of metals to metallothioneins is achieved by association of cysteine –SH groups with the metal ion; for example, Cd^{2+} and Zn^{2+} are linked to four cysteine thiolate ligands (Cys-S$^-$) in a tetrahedral arrangement. Binding of Cu^+ is tighter than that of Cu^{2+} and appears to involve three thiolate ligands. The metallothionein content of liver, kidney, and gut can be increased by injection or oral administration of Cd, Cu or Zn salts, and synthesis of these proteins is also increased by several hormones, including glucocorticoids, glucagon and adrenalin, and by interleukin I produced during inflammation (Chapter 4).

Proposed functions of metallothioneins include storage of heavy metals in a non-toxic form, regulation of cellular copper and zinc metabolism, and control of the absorption of these metals from the gut. For example, cultured mammalian cells that have lost the ability to make metallothioneins are exceptionally sensitive to injury by Cd^{2+} ions, whereas cells overproducing these proteins are more resistant to Cd^{2+}. It has repeatedly been suggested that metallothioneins might exert antioxidant properties. Sequestration of copper ions will diminish radical generation promoted by this metal, Zn^{2+} released from zinc–metallothionein might inhibit lipid peroxidation and the high content of –SH groups in metallothioneins makes them excellent scavengers of singlet O_2 and $OH^\bullet$ radicals, although the resulting sulphur-centred radicals must not be ignored.[81,83] For example, reaction of $OH^\bullet$ with metallothionein can generate a thiyl ($RS^\bullet$) radical which can react with an adjacent thiolate (RS^-) on the protein

$$RS^\bullet + RS^- \rightleftharpoons RSSR^{\bullet -}$$

or combine with O_2 to give the thiyl peroxyl radical

$$RS^\bullet + O_2 \rightarrow RSOO^\bullet$$

$RSSR^{\bullet -}$ can react with O_2,

$$RSSR^{\bullet -} + O_2 \rightarrow O_2^{\bullet -} + RSSR$$

Nevertheless, metallothionein-enriched Chinese hamster ovary (CHO) cells are more resistant to damage by H_2O_2 than are the parent cells, consistent with an antioxidant role for metallothioneins.[54] Similar experiments[228] showed that cells over-expressing metallothionein were more resistant to excess $NO^\bullet$. Transgenic mice defective in metallothionein appeared normal, but cells from them were more sensitive to damage by Cd^{2+}, the artificial organic peroxide *tert*-butylhydroperoxide, or the $O_2^{\bullet -}$-generating toxin paraquat[151] (Chapter 8). *Saccharomyces cerevisiae* strains lacking CuZnSOD are hypersensitive to O_2 and paraquat and will not grow on lactate as a carbon source, but expression of yeast or monkey metallothionein proteins in the presence of copper suppresses the lactate growth defect, as well as some other phenotypic effects of the CuZnSOD absence.[243] Pulse radiolysis experiments[83] showed that the yeast

Cu^+ protein reacts rapidly with $O_2^{\bullet-}$. However, the reaction led to some displacement of Cu^{2+} from the protein and so the overall biological effects of this reaction are uncertain.

3.19.3   *Phytochelatins*

Several plants contain small, cysteine-rich proteins involved in the accumulation and detoxification of cadmium, zinc, copper and other metals. They consist of repetitive γ-glutamylcysteine units and are probably synthesized from GSH. These **phytochelatins** may function in an analogous way to metallothioneins.[97]

3.19.4   *The extracellular environment*[107]

The extracellular fluids of multicellular organisms must protect the following against damage by ROS and RNS:

- the surfaces of the cells with which they are in contact. For example, seminal plasma is important in protecting spermatozoa against ROS generated by phagocytes in semen (Section 7.8)
- essential constituents of the aqueous phase of the fluids (e.g. surfactant in lung lining fluids)
- essential constituents of the lipid phase of the fluids, e.g. plasma lipoproteins.

In animals, one striking contrast to the intracellular environment is that extracellular fluids such as blood plasma, seminal fluid, respiratory tract lining fluids, cerebrospinal fluid (CSF) and synovial fluid, contain little or no catalase activity. The traces of catalase sometimes detected may leak from cells, e.g. slight haemolysis during centrifugation of blood for preparation of plasma. Activities of SOD and GPX (Section 3.8) are low, as are levels of GSH (Table 3.8), except that the fluid lining the lower part of the respiratory tract has much higher GSH levels (up to 400 μM; Table 3.8).

Extracellular superoxide dismutase

Of the limited SOD activity present in extracellular fluids, some is CuZnSOD and MnSOD (possibly leakage from damaged cells), but much of it is a different type of CuZnSOD called **extracellular SOD** (EC-SOD).[2] Plasma EC-SOD activity shows a large interspecies variation, e.g. rabbits, rats, mice and guinea pigs have one or two orders of magnitude more than pigs, dogs and humans.

Unlike intracellular CuZnSOD, EC-SOD has a high relative molecular mass (about 135 000) and is a tetrameric glycoprotein, each subunit containing one Cu and one Zn. There are several forms of EC-SOD (A, B and C); B and C bind to heparin. It seems likely that most EC-SOD is bound to cell surfaces by association with cell surface carbohydrates *in vivo*, especially in the lung and in blood vessel walls. For example, human thoracic aorta has fairly low

CuZnSOD activity but its level of EC-SOD is comparable to that of CuZnSOD. This is unlike most tissues, where CuZnSOD is predominant (Table 3.3). Injection of heparin into animals produces a rise in plasma EC-SOD, presumably by displacing the enzymes from cell surfaces.

EC-SOD may serve to minimize interaction of $O_2^{\bullet-}$ and $NO^{\bullet}$ to form peroxynitrite, e.g. on vascular endothelial cells.[192] Transgenic mouse EC-SOD 'knockouts' appeared normal and had no compensatory changes in any of the other antioxidant defence enzymes, but they developed more lung damage and died faster when exposed to pure O_2, and showed an exaggerated lung inflammation when exposed to ozone at 1.5 p.p.m.[47]

Several parasites, e.g. *Taenia*, *Trichinella*, *Schistoma* and *Ascaris* species,[89] secrete 'normal' CuZnSOD (and sometimes other antioxidant proteins[199a]) on to their surfaces, perhaps to help protect them against damage by host phagocytes.

Metal-ion sequestration

It thus seems that in extracellular fluids, antioxidant defence enzymes are present only at low levels, if at all. Blood lipoproteins contain vitamin E and other lipid-soluble antioxidants. Vitamin C and urate contribute to antioxidant defence of plasma (Sections 3.21 and 3.22 below). ROS/RNS can be generated in the aqueous phase of plasma, e.g. $NO^{\bullet}$ from vascular endothelium, $O_2^{\bullet-}$ and H_2O_2 from enzymes (e.g. xanthine oxidase released from cells), autoxidizing compounds, and activated phagocytes. In whole blood, some $O_2^{\bullet-}$ and H_2O_2 might (like $NO^{\bullet}$) diffuse into erythrocytes for metabolism (indeed H_2O_2 can diffuse into and out of any cell). This argument is supported by the presence of an anion channel permitting $O_2^{\bullet-}$ movement through the erythrocyte membrane. However, for this to happen, it is essential that $O_2^{\bullet-}$ and H_2O_2 can be prevented from reacting to form $OH^{\bullet}$.

The presence of transferrin at high levels in plasma suggests that a major antioxidant defence is to bind transition metal ions in forms that will not stimulate free-radical reactions. The transferrin present in normal human plasma is only 20–30% loaded with iron, so that the content of non-protein bound iron in plasma is effectively nil. Hence if animal plasma or purified transferrin is added to lipids undergoing iron-dependent peroxidation, the peroxidation is inhibited (provided that the iron present does not overwhelm the iron-binding capacity of the transferrin). The protein lactoferrin can act as an antioxidant in the same way and it may be secreted by neutrophils at sites of inflammation to aid in iron sequestration (Chapter 6). Any Fe^{2+} ions released into plasma will be assisted in binding to transferrin by the ferroxidase I activity of caeruloplasmin.

Metal-binding antioxidant defence may be weaker in some other extra-cellular fluids.[107] Thus human CSF contains little transferrin, albumin or caeruloplasmin and its transferrin is at or close to iron saturation (Chapter 9), but CSF has high concentrations of ascorbic acid (about ten times those in plasma). Synovial fluid has lower concentrations of albumin, transferrin and caeruloplasmin than plasma, whereas the fluid that lines the alveoli of the lung also has a low protein content but a higher content of ascorbic acid and much more GSH than plasma. None of these fluids contains much SOD, catalase or glutathione peroxidase but some urate is always present (in primates), although levels are usually lower than in plasma.

Haem binding

Haem and haem proteins can act as pro-oxidants (Section 3.18.3 above), but haem proteins must often be released from damaged cells, e.g. lysed erythrocytes. Plasma contains the haemoglobin-binding proteins known as **haptoglobins**, as well as a haem-binding protein (**haemopexin**).[107] Binding of haemoglobin to haptoglobin, or of haem to haemopexin, decreases the effectiveness of these compounds in stimulating lipid peroxidation (Fig. 3.21). The haemoglobin–haptoglobin or haem–haemopexin complexes are cleared from the circulation by the liver. The haem–haemopexin complex is then dissociated, the haem being taken up by the liver, and the haemopexin returned to the circulation. By contrast, at least part of the haptoglobin–haemoglobin complex taken up by the liver is excreted intact into the bile, which may account for the increases in biliary iron excretion seen in patients with haemolytic anaemias. The rest is degraded.

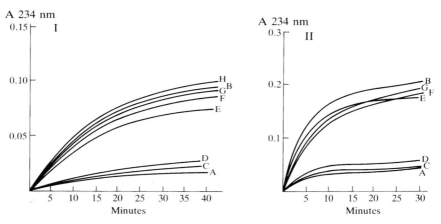

Fig. 3.21. Antioxidant effects of haemopexin and haptoglobin. (I) Haemin-stimulated peroxidation of fatty-acid micelles measured as diene conjugation at 234 nm. Fatty-acid micelles were incubated at pH 7.4 and 25 °C alone (curve A), with haemin (curve B), or plus haemin together with (C) the lipid-soluble chain-breaking antioxidant butylated hydroxytoluene (BHT), (D) apohaemopexin, a powerful inhibitor, (E) albumin, a weak inhibitor, (F) haptoglobin (little effect). (G) desferrioxamine (little effect), or (H) apotransferrin (no effect). (II) Haemoglobin-stimulated lipid peroxidation measured as conjugated dienes. Fatty-acid micelles were incubated at pH 6.4 and 25 °C alone (curve A), with desferrioxamine and methaemoglobin (curve B), or with desferrioxamine and methaemoglobin plus (C) haptoglobin, a powerful inhibitor, (D) BHT, (E) apohaemopexin (no effect), (F) apotransferrin (no effect) or (G) albumin (no effect). Data abstracted from *Biochem. J.* **256**, 861 (1988). Haemopexin can also protect LDL against oxidation by haem or haemoglobin—presumably released haem is involved in the latter case (*Biochemistry* **35**, 13112 (1996)).

Albumin

Albumin may be an important extracellular antioxidant.[107] It binds haem (Fig. 3.21) and copper ions. Albumin also contains an exposed cysteine –SH group (position 34 in the human protein). Since albumin is present in plasma at concentrations approaching 0.5 mM, it provides the bulk of 'total plasma thiols', some of which occur as mixed disulphides. Albumin –SH can react quickly with peroxynitrous acid and HOCl and slowly with H_2O_2. The putative antioxidant **bilirubin** (Section 3.21 below) is transported in an albumin bound form.

Albumin also maintains osmotic pressure and transports multiple drugs, hormones and free fatty acids in the blood. Rats unable to synthesize albumin show increased levels of plasma cholesterol ester hydroperoxides and decreased levels of ascorbate and ubiquinol.[268] A few cases of human **analbuminaemia** have been reported;[261] as in the rat, it is not life-threatening but parameters of oxidative damage have not been studied. Commercial albumin preparations are often partially oxidized at Cys34 and contaminated with metals such as iron, copper and vanadium.[204]

Why are there different antioxidants in the extracellular fluids?

Why should the mechanisms of antioxidant protection in extracellular fluids differ from intracellular mechanisms? During the 'respiratory burst' of phagocytic cells, some $O_2^{\bullet-}$ and H_2O_2 is released into the surrounding fluids and many other cell types also release these species (Chapter 6). Provided that $O_2^{\bullet-}/H_2O_2$ can be stopped from forming destructive $OH^{\bullet}$, they may act as *useful* signals between cells: this concept is discussed further in Chapter 4. Thus $O_2^{\bullet-}$ may be involved in phagocyte recruitment to sites of inflammation, whereas low levels of H_2O_2 may facilitate cell proliferation, the aggregation of platelets and adherence of phagocytes to vascular endothelium. The presence of too much SOD or H_2O_2-removing enzymes might interfere with these putatively 'useful' processes.

At sites of extensive tissue injury, however, cell destruction can lead to release of pro-oxidants such as iron ions, haem and haem proteins, potentiating damage by $O_2^{\bullet-}$ and H_2O_2. Of course, cell damage will also release extracellular antioxidant enzymes to help counter these effects. For example, when lung is injured the increased exudation of plasma proteins into the lining fluid (often measured as an index of damage), plus SOD, catalase, GSH and GPX released from injured cells, may augment local antioxidant defences. Haptoglobin and caeruloplasmin are **acute-phase proteins**; their plasma levels rise as a result of tissue injury (Chapter 4).

3.20   Haem oxygenase

Haem oxygenase[162] is an enzyme, found in the endoplasmic reticulum, that catalyses the breakdown of haem to biliverdin, with the release of iron ions and carbon monoxide (CO), as shown in Fig. 3.22. NADPH–cytochrome

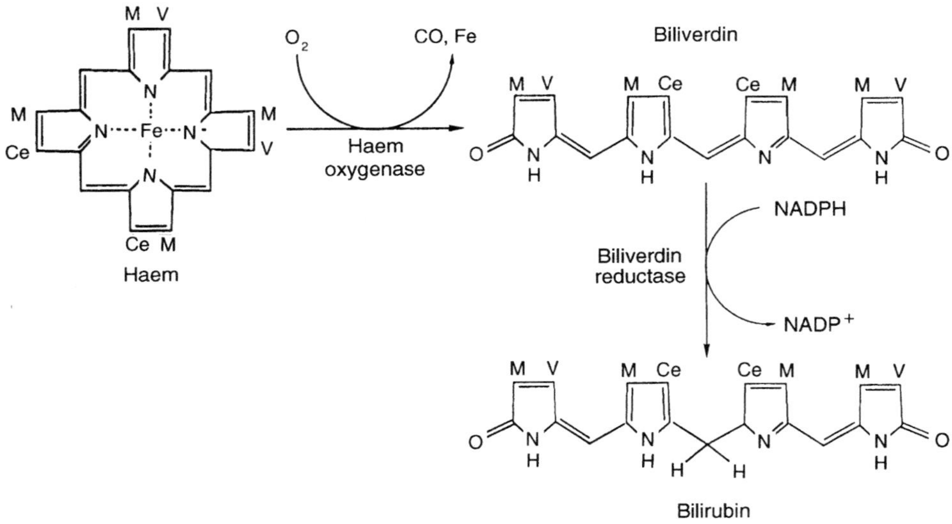

Fig. 3.22. Haem degradation by haem oxygenase. Adapted from *Am. J. Resp. Cell Mol. Biol.* **15**, 9 (1996) by courtesy of Dr Augustine Choi and the publishers.

P450 reductase acts as an electron donor to haem oxygenase. The biliverdin produced is converted to **bilirubin** by the enzyme **biliverdin reductase** in the cytosol. Two isoforms of haem oxygenase have been characterized; a constitutive isoform (**HO-2**) predominant under normal physiological conditions, and a stress-induced[135] isoform (**HO-1**, which is identical with one of the heat shock proteins, **hsp32** and is further discussed in Chapters 4 and 7). HO-2 is present at high levels in the spleen, presumably to destroy haem from the processing of worn-out red blood cells.

The overall effect of haem oxygenase is to remove a pro-oxidant (haem) whilst generating a putative antioxidant (bilirubin) and another pro-oxidant (iron), so we are not quite sure whether to classify it as an antioxidant enzyme. Carbon monoxide has been proposed to serve as a cell signalling molecule, especially in the brain, but the evidence is equivocal as yet. About 80% of the bilirubin excreted in human bile comes from haemoglobin breakdown and much of the rest from cytochrome P450 destruction. Studies on endothelial[17] and skin cells show that the induction of HO-1 by exposing them to haem is also accompanied by a rise in ferritin levels; if this is prevented, exposure of the cells to haem *sensitizes* them to H_2O_2, i.e. haem oxygenase must cooperate with iron sequestration systems to achieve maximum cell protection.

Haem oxygenase can be inhibited by **tin protoporphyrin**, which has been used to study the biological role of this enzyme. Pre-treatment of rats with haemoglobin to induce HO-1 decreased the lethality of subsequent endotoxin treatment, apparently due to the HO-1, since the protection was obviated by tin protoporphyrin.[191]

3.21 Antioxidant protection by low-molecular-mass agents: compounds synthesized *in vivo*

As well as protein antioxidants, several low-molecular-mass compounds are thought to be important in antioxidant defence. These can be divided into compounds made *in vivo*, and compounds obtained from the diet. Let us consider some of the former first, starting with bilirubin. Direct antioxidant effects of GSH have already been considered (Section 3.9.1 above).

3.21.1 *Bilirubin*

Bilirubin is an end-product of haem degradation in mammals (Fig. 3.22). Its precursor biliverdin is blue-green in colour, whereas bilirubin is bright yellow. Biliverdin reductase is not present in birds, reptiles and amphibia, so that they excrete biliverdin rather than bilirubin.

Most bilirubin produced in mammals results from the catabolism of haemoglobin (released from senescent red blood cells), by reticuloendothelial cells in the spleen, liver and bone marrow.[162] Some 20% arises from other haem proteins such as myoglobin, catalase, cytochromes P450 and other cytochromes. Around 500 mmol (275 mg) of bilirubin is produced each day in the human body.

Bilirubin binds tightly to albumin in a 1:1 stoichiometry; other molecules, particularly certain drugs, can compete for albumin binding sites and displace bilirubin. Bilirubin is insoluble in water at physiological pH and binding to albumin prevents uptake by extrahepatic tissues, particularly lipid-rich areas such as the brain, and directs bilirubin to the liver. In the liver the albumin–bilirubin complex dissociates, and bilirubin enters the hepatocyte by a carrier mediated process. It then binds to cytosolic proteins, mainly glutathione S-transferases. Bilirubin undergoes conjugation in the endoplasmic reticulum with glucuronic acid, catalysed by **uridine diphosphoglucuronyl transferase**, forming mono- and di-glucuronides which are water soluble and available for excretion in the bile. Impaired conjugation can lead to jaundice[76] in premature babies (Chapter 2).

In vitro, bilirubin is a powerful scavenger of peroxyl radicals and singlet O_2; perhaps the jaundice of premature babies has some physiological role provided it does not occur to excess.[183] Bilirubin bound to albumin can protect both the protein and albumin-bound fatty acids against free-radical damage. However, there is little direct evidence that bilirubin is an important antioxidant *in vivo*. Measurement of free-radical degradation products of bilirubin is a possible approach. It must not be forgotten that bilirubin can sensitize formation of singlet O_2 in the presence of light and high levels in premature babies are toxic to the brain.

3.21.2   α-Keto acids

Several keto acids, including pyruvate[181] and α-ketoglutarate (often called 2-oxoglutarate) react non-enzymically with H_2O_2 and act as 'H_2O_2 scavengers' when present in cell culture media. It is also likely that they can scavenge HOCl and ONOOH. Decarboxylation of 2-oxoglutarate has been used as an assay for H_2O_2 (Chapter 5). Whether such reactions occur physiologically in animal tissues is uncertain, but the non-enzymic conversion of glyoxylate (OHC·COOH) to formate by H_2O_2 has been suggested to occur in the peroxisomes of illuminated green leaves. Pyruvate is an important food source for the early human embryo and may exert antioxidant effects (Chapter 7).

3.21.3   Sex hormones

The female sex hormones oestradiol, oestrone and oestriol can inhibit lipid peroxidation (including LDL peroxidation) *in vitro* at micromolar concentrations, presumably because they possess phenolic −OH groups and so can act as chain-breaking antioxidants in a way similar to that of vitamin E (Section 3.22.8 below).[132] There is increasing evidence for a beneficial effect of oestrogen administration (e.g. in hormone replacement therapy) in the prevention of cardiovascular disease in post-menopausal women. Are the two effects related?

In general, the concentrations of hormones needed to inhibit peroxidation *in vitro* seem much larger than physiological levels (picomolar range). The phenoxyl radicals produced during antioxidant activity (Fig. 3.27) are reactive and capable of damaging proteins and DNA *in vitro*.[158] Indeed, the synthetic oestrogen **diethylstilboestrol** is a powerful inhibitor of lipid peroxidation at micromolar concentrations *in vitro*, although it is a known carcinogen in animals and humans (an illustration of the care that must be taken when designing antioxidant inhibitors of lipid peroxidation for therapeutic use).[266]

Nevertheless, it has been shown that LDL isolated from post-menopausal women after treatment with skin patches containing 17β-oestradiol, were more resistant to peroxidation (induced by copper ions) *in vitro*.[219] Whether this was a direct antioxidant effect or some other action on lipid metabolism remains to be established.

3.21.4   Melatonin

Similar cautions apply when evaluating claims of an antioxidant role for the hormone **melatonin**. It is produced mainly by the pineal gland at the base of the brain to help regulate circadian rhythms, and has been popularized as a cure for 'jet lag'. Smaller amounts are made in several other tissues.[211] Melatonin is produced by the methylation and acetylation of **serotonin** (Fig. 3.23); blood levels of melatonin are low during the day and increase at night as pineal synthesis accelerates. Melatonin shows antioxidant activity *in vitro* (probably by

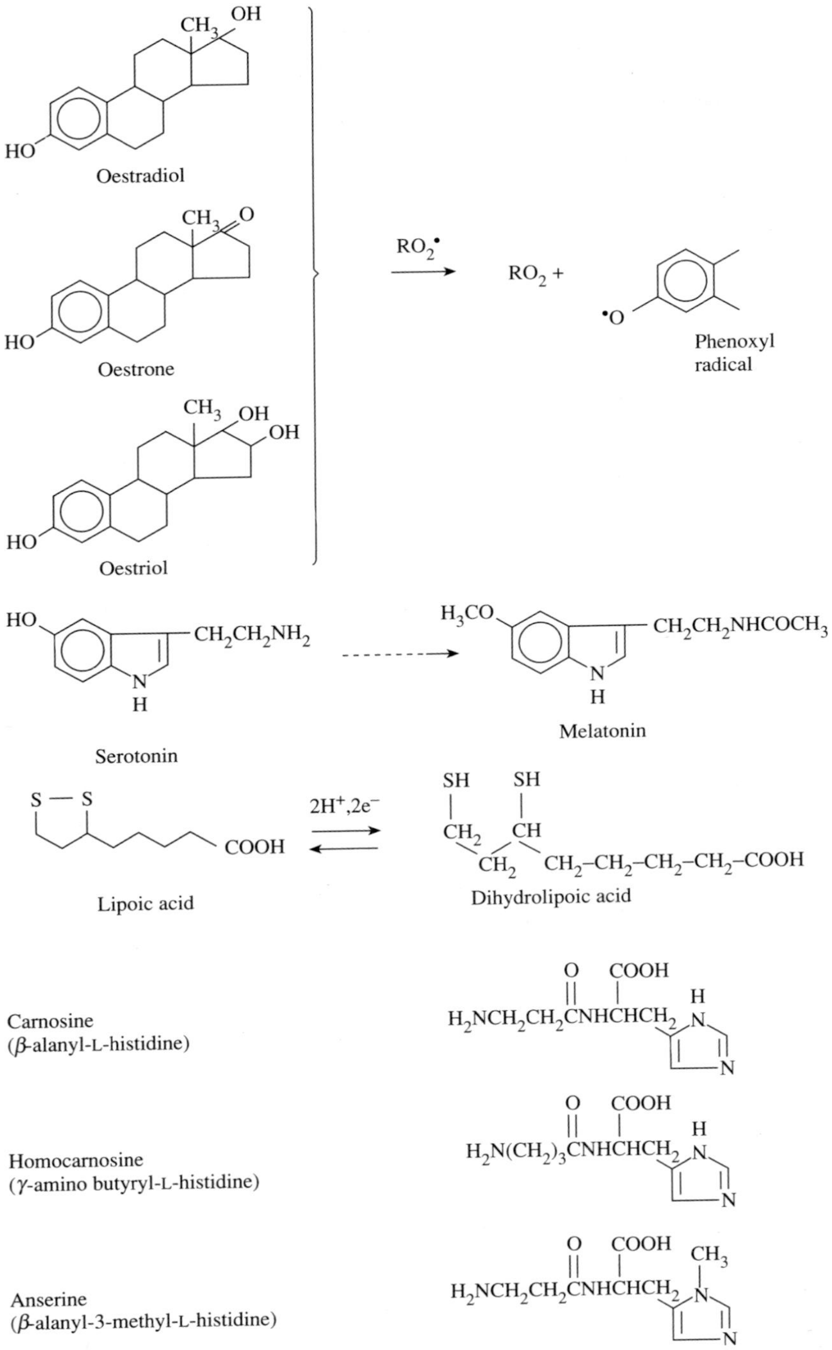

Oestradiol
Oestrone
Oestriol
RO2•
RO2 +
•O
Phenoxyl radical
Serotonin
Melatonin
Lipoic acid
2H+,2e−
Dihydrolipoic acid
Carnosine
(β-alanyl-L-histidine)
Homocarnosine
(γ-amino butyryl-L-histidine)
Anserine
(β-alanyl-3-methyl-L-histidine)

Fig. 3.23. Structures of some putative antioxidants. In lipoamide the –COOH group becomes –CONH2.

donation of hydrogen by the NH group), but at concentrations several orders of magnitude higher than those present *in vivo*. Its precursor, serotonin, is a much better lipid peroxidation inhibitor *in vitro*, as would be expected since it contains a phenolic −OH group which in melatonin is blocked by methylation (Fig. 3.23).[165] Nevertheless, high doses of melatonin do seem able to induce synthesis of antioxidant defence enzymes, such as glutathione peroxidase, in animals and so any *in vivo* antioxidant effects it exerts may be indirect (affecting endogenous antioxidant defences) rather than direct.[211]

3.21.5   *Lipoic acid*[193]

Lipoic acid, **1,2-dithiolane-3-pentanoic acid** (sometimes called **thioctic acid**), is an essential cofactor (as its amide form, **lipoamide**) in the multi-enzyme complexes that catalyse the decarboxylation of α-keto acids, such as pyruvate (to acetyl coenzyme A) and α-ketoglutarate (to succinyl coenzyme A) in the Krebs cycle (Chapter 1). Both the oxidized (disulphide) and reduced (dithiol; Fig. 3.23) forms of lipoic acid show antioxidant properties *in vitro*; they scavenge $RO_2^\bullet$, HOCl, $OH^\bullet$ and ONOOH. Lipoic acid can bind iron and copper ions in forms poorly catalytic for free-radical reactions.

With a reduction potential ($E^{\circ\prime}$) of $-0.32\,V$ for the dehydrolipoic acid (DHLA)/lipoic acid couple, DHLA is a powerful reducing agent. Thus it can reduce GSSG to GSH ($E^{\circ\prime}$ for GSH/GSSG is about $-0.24\,V$), dehydroascorbate to ascorbate and it can regenerate α-tocopherol from the α-tocopheryl radical, either directly or via ascorbate. Indeed, administration of lipoic acid reversed some of the effects produced by α-tocopherol deficiency in mice, and decreased the incidence of cataract in mice treated with buthionine sulphoximine to inhibit GSH synthesis. *In vivo*, a strong reducing agent must also be needed to convert lipoic acid to DHLA; the **lipoamide dehydrogenase** enzymes use NADH or NADPH and show some activity on lipoic acid. Glutathione and thioredoxin reductases may also act on lipoic acid.

Levels of 'free' lipoic acid/DHLA in tissues and body fluids are very low, so it is unlikely to exert antioxidant effects *in vivo*. However, its wide range of antioxidant properties and ability to regenerate other naturally-occurring antioxidants have provoked attempts to use it as a therapeutic antioxidant, e.g. in the treatment of diabetes. The *R*-stereoisomer of lipoic acid seems to promote glucose uptake by muscle. Of course, the use of thiols as antioxidants should be undertaken with consideration of the potential reactivity of the thiyl and oxysulphur radicals that can be produced during their radical-scavenging activities (Chapter 2).

3.21.6   *Coenzyme Q*

Coenzyme Q plays an essential role in the mitochondrial electron-transport chain (Chapter 1), undergoing simultaneous oxidation and reduction via a free-radical intermediate, **ubisemiquinone** ($CoQH^\bullet$). It is also found in other cell membranes and lipoproteins.[77] *In vitro*, ubiquinol ($CoQH_2$) can scavenge $RO_2^{\bullet}$ radicals and thereby inhibit lipid peroxidation

$$RO_2^{\bullet} + CoQH_2 \rightarrow RO_2H + CoQH^{\bullet}$$

The ubiquinol content of LDL is an important factor in their resistance to peroxidation (Chapter 9). Ubiquinol can regenerate α-tocopherol from its radical in lipoproteins and membranes ($E^{\circ\prime} = 0.24\,V$ for $QH/Q^{\bullet}$ and $0.5\,V$ for α-TocOH/α-TocO$^{\bullet}$)

$$\alpha\text{-Toc}^{\bullet} + CoQH_2 \rightarrow CoQH^{\bullet} + \alpha\text{-TocH}$$

The rate constant for scavenging of $RO_2^{\bullet}$ by $CoQH_2$ is about one-tenth that of α-tocopherol, so the recycling mechanism may be more important *in vivo*, since the α-tocopherol and coenzyme Q levels in most cell membranes are broadly comparable. By contrast, LDL contain much less ubiquinol than α-tocopherol (Chapter 9).

The overall contribution of ubiquinol to antioxidant defence *in vivo* is uncertain. It may be especially important in mitochondria, where the electron-transport chain can easily re-oxidize/re-reduce $CoQH^{\bullet}$. However, it has been suggested that mitochondrial ubisemiquinone can be a *source* of $O_2^{\bullet-}$ radicals (Chapter 1) although this may not happen without exposure of it to a source of protons, e.g. by membrane disruption. The enzyme **DT-diaphorase** (Section 8.6) can reduce CoQ to $CoQH_2$ and this has been suggested to be part of its antioxidant function *in vivo*.[31]

A CoQ-deficient strain of *S. cerevisiae* was abnormally sensitive to damage by oxidation products of PUFAs, suggesting further that CoQ can exert anti-oxidant effects in at least some organisms.[67] Some scientists have suggested that the ratio of ubiquinol to ubiquinone in human plasma may be an index of oxidative stress.[268a]

3.21.7   Uric acid

Uric acid is produced by the oxidation of hypoxanthine and xanthine by xanthine oxidase and dehydrogenase enzymes. In most species, the peroxisomal enzyme **urate oxidase** converts urate into allantoin, which is further converted to allantoate and then gloxylate plus urea, all products much more soluble in water than is urate (Fig. 3.24).

However, humans and other primates lack urate oxidase: the gene can be detected in the human genome, but a stop codon is present in one of the exons[e] and the defective gene appears not to be transcribed. Hence urate accumulates in human blood plasma to concentrations normally in the range of $0.2–0.4\,mM$ and is excreted in the urine.[6] It is also present intracellularly and in all other body fluids, usually at somewhat lower levels (e.g. $100–200\,\mu M$ in saliva). At physiological pH almost all uric acid is ionized to urate,[235] bearing a single negative charge, since the pK_a of uric acid is around 5.4. Urate has limited solubility in water: excess production *in vivo* can lead to its crystallization out

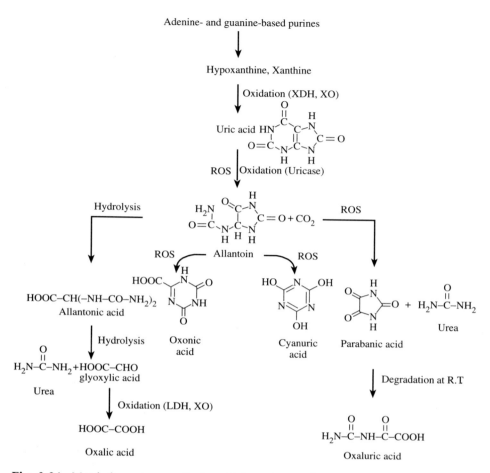

Fig. 3.24. Metabolic pathways of uric acid formation, and degradation by both enzymic and non-enzymic pathways. Diagram by courtesy of Dr H. Kaur. XDH, xanthine dehydrogenase; XO, xanthine oxidase; ROS, reactive oxygen species; RT, room temperature.

of solution. This occurs in **gout**, a disease often treated with the xanthine dehydrogenase/oxidase inhibitor **allopurinol**. The excruciatingly painful character of attacks of gout, caused by joint inflammation triggered by urate crystal deposition, have been recognized since ancient times and afflicted (to name but a few) the Roman emperor Claudius, Henry VIII and Benjamin Franklin. Indeed, transgenic urate oxidase 'knockout' mice show elevated plasma urate and develop kidney stones.[267a]

Strong oxidants, such as $OH^\bullet$, oxidize urate into a free radical, which also bears a single negative charge at pH 7.4, since its pK_a is 3.1. The unpaired electron is delocalized over the purine ring, giving a resonance-stabilized radical which does not appear to react with O_2 to form a peroxyl radical (rate constant $< 10^{-2} M^{-1} s^{-1}$).[235] The reduction potential of the urate/urate radical system at pH 7 is 0.59 V, which is considerably higher than that of ascorbate ($E^{\circ'} = -0.28 V$). Hence ascorbate would be expected to reduce the urate radical:

$$UrH^{\bullet -} + AscH^- \rightarrow UrH^{2-} + Asc^{\bullet -}$$

and this has been demonstrated experimentally (second–order rate constant $\approx 10^6 M^{-1} s^{-1}$). Urate also reacts with organic peroxyl radicals

$$RO_2^{\bullet} + UrH_2^- \rightarrow ROOH + UrH^{\bullet -} + H^+$$

For example, its rate constant for reaction with trichloromethylperoxyl radical ($CCl_3O_2^{\bullet}$) is $3 \times 10^8 M^{-1} s^{-1}$.

In 1981, Ames *et al.*[6] pointed to the fact that urate is a powerful scavenger of $RO_2^{\bullet}$, $OH^{\bullet}$ and singlet O_2 *in vitro*, proposed that its biological function is as an antioxidant and further suggested that the loss of urate oxidase (permitting urate accumulation) was advantageous to primates. Since then, it has been discovered that urate is a powerful scavenger of O_3 and $NO_2^{\bullet}$ and may help to protect against these oxidizing air pollutants in the respiratory tract (Chapter 8). Urate is a substrate for oxidation by haem protein/H_2O_2 systems and might be able to protect against oxidative damage by being preferentially oxidized. Urate also protects proteins against nitration on addition of $ONOO^-$. It can chelate metal ions: it binds iron and copper ions in forms apparently poorly reactive in catalysing free-radical reactions. Consistent with an antioxidant role for urate *in vivo* are the high level present, enough for scavenging of ROS/RNS to be possible, and observations that products of urate degradation by ROS, especially **allantoin** (Fig. 3.24)[98] increase in concentration in patients subjected to oxidative stress, e.g. in Wilson's disease, haemochromatosis and rheumatoid arthritis.

Nevertheless, reaction of urate with haem proteins plus H_2O_2, $RO_2^{\bullet}$, $OH^{\bullet}$ or $ONOO^-$ generates urate radical.[235] This radical may not be biologically innocuous, since it has been shown *in vitro* to lead to inactivation of at least two proteins: yeast alcohol dehydrogenase and human α_1–antiproteinase.[137,263] Inactivation of a_1–antiproteinase can be prevented by the simultaneous presence of ascorbate, presumably as a result of the reaction of $UrH^{\bullet -}$ with ascorbate. Further studies of the possible biological effects of urate radicals, and their interaction with biological antioxidants other than ascorbate (such as thiols), are warranted.

Urate and related purines do not seem to be important antioxidants in plant cells, but some of the **alkaloids** present have antioxidant activity *in vitro*. An example is **boldine**, found in the bark and leaves of the South American tree *Peumus boldo*. The physiological relevance (if any) of these *in vitro* observations is uncertain.

3.21.8   *Histidine-containing dipeptides*

Many mammalian tissues, especially muscle, but also brain, contain millimolar levels of dipeptides composed of histidine plus another amino acid (Fig. 3.24). They include **carnosine**, **homocarnosine** and **anserine**. *In vitro*, these dipeptides can exert antioxidant effects;[13] for example, the imidazole ring chelates copper ions and prevents copper-dependent oxidative damage.[138] They are weak inhibitors of lipid peroxidation; claims of greater inhibitory effects are often due to the ability[10] of these compounds to interfere with measurement of peroxidation end-products by the TBA test (Chapter 5).

By contrast, histidine alone *promotes*[264] iron ion-dependent lipid peroxidation, e.g. in isolated membrane fractions and in muscle homogenates. Histidine also enhances the toxicity of H_2O_2 to mammalian cells in culture.[45,230] The dipeptides do not exert such 'pro-oxidant' effects. Hence it has been suggested[10] that they are 'safer' ways of accumulating histidine at high levels for use as an intracellular buffer (the histidine imidazole ring has a pK_a of about 6) and, perhaps, copper ion chelator. Another suggestion is that these dipeptides help to protect proteins against glycation,[122] since carnosine is rapidly glycated upon incubation with high levels of glucose (or other sugars) *in vitro*.

Enzymes (**carnosinases**) that hydrolyse these dipeptides[260] have been described in various muscles and in brain but it is uncertain if they are specific in this role or are just 'general dipeptidases'. **Carcinine** (β-alanyl-histamine), presumably a decarboxylation product of carnosine, has been detected in several mammalian tissues and shows some antioxidant properties *in vitro*.[13] However, both histidine and the dipeptides were reported[15] to aggravate $OH^\bullet$ production from H_2O_2 by ions (Ni^{2+}) of the carcinogenic metal nickel (Chapter 8). Nevertheless, plants of the genus *Alyssum* can tolerate and accumulate vast amounts of nickel by binding it to histidine.[145]

3.21.9   *Melanins*

Melanins are pigments, found throughout the animal kingdom and in some fungi, that are formed by oxidation and polymerization of tyrosine. This occurs by the action of **tyrosine hydroxylase** and **tyrosinase** enzymes. Tyrosine hydroxylases are iron-containing enzymes that hydroxylate tyrosine to **L-DOPA** (dihydroxyphenylalanine). Tyrosinases are copper-containing proteins which convert tyrosine to L-DOPA and then oxidize L-DOPA to semiquinones and quinones, which polymerize (Fig. 3.25). The end-product of polymerization contains high concentrations of *o*-quinone (oxidizing) and *o*-hydroquinone (reducing) groups as well as semiquinones. The brown or black melanins (**eumelanins**) found in skin help protect against ultraviolet light.

The melanin polymer contains many unpaired electrons left over from the polymerization process. Hence one can regard melanins as large free radicals. The movement of unpaired electrons between different energy levels helps to

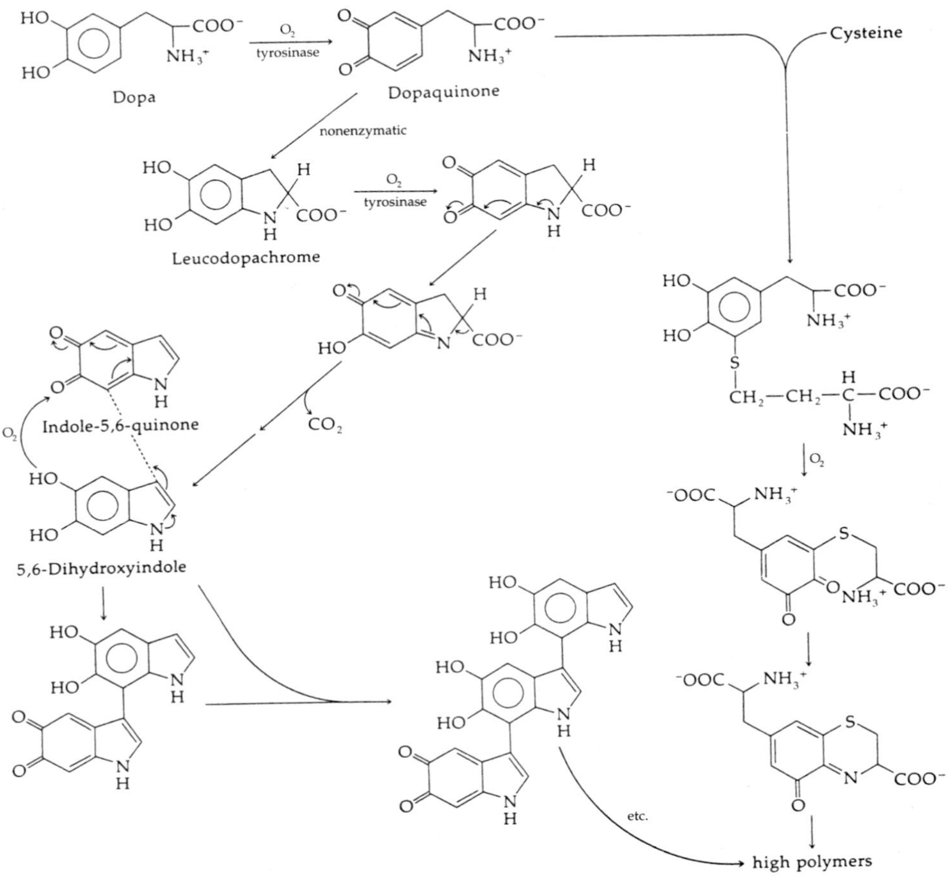

Fig. 3.25. Biosynthesis of melanin. The reaction pathways are complex and involve both enzyme-catalysed and non-enzymic reactions. Cysteine is involved (via cysteinyl-DOPAs) in the synthesis of eumelanin, found in reddish hair. Diagram from *Biochemistry, The Chemical Reactions of Living Cells* by courtesy of Dr David E. Metzler and Academic Press.

absorb ultraviolet radiation. Illumination of eumelanins generates $O_2^{\bullet -}$ within the molecule, but this is very quickly scavenged; $O_2^{\bullet -}$ can reduce melanin quinones to semiquinones, and oxidize hydroquinones, also to semiquinones (Chapter 2).[142,221] Hence, overall, eumelanins are 'radical sinks' for $O_2^{\bullet -}$ and $RO_2^{\bullet}$. The red-brown or yellow pigment found in the skin and hair of fair-skinned, red-headed humans is **pheomelanin**, which is less good as a radical scavenger. On exposure of pheomelanin to strong light it is degraded, and a net formation of $O_2^{\bullet -}$ can be measured.[53] Pheomelanins contain cysteine, which cross-links with dopaquinone during polymerization to give **cysteinyl DOPAs** (Fig. 3.25).

Melanins also bind transition and other metal ions and the binding of iron to neuromelanin is of interest in Parkinson's disease (Chapter 9). Sequestration of metal ions in this way could conceivably contribute to antioxidant defence, but the metal ions are probably still redox-active and could damage the melanin itself, e.g. in the presence of H_2O_2.

It has been suggested that the presence of high levels of melanin contributes to the resistance of some pigmented fungi[126,259] and of the cancer **melanoma** to ionizing radiation. For example, melanized cells of the pathogenic fungus *Cryptococcus neoformans*, which can cause life-threatening infections in AIDS patients, are more resistant to killing by ROS and ROS than non-melanized cells. *C. neoformans* also produces large amounts of mannitol, perhaps as an $OH^\bullet$ scavenger. Melanin might play an antioxidant role in the substantia nigra (Chapter 9) and in the retina (Chapter 7), but the overall physiological importance of melanins as antioxidants is uncertain.

3.22   Antioxidant protection by low-molecular-mass agents: compounds derived from the diet

A huge range of dietary constituents has been suggested to exert antioxidant effects *in vivo*. The health aspects of dietary antioxidants are discussed in Chapter 10; here we will look at their basic chemistry and biochemistry.

3.22.1   *Ascorbic acid (vitamin C)*[186]

Pure ascorbic acid is a white crystalline solid, very soluble in water. It was first isolated from adrenal glands, cabbages and oranges as an 'acidic carbohydrate' by Szent-Györgyi in 1928.

Ascorbic acid has two ionizable −OH groups (Fig. 3.26). Since pK_{a1} is 4.25 and pK_{a2} is 11.8, a mono-anion is the favoured form at physiological pH. Hence we use the name **ascorbate** from now on. Plants and most animals can synthesize ascorbate from glucose, but humans, other primates, guinea-pigs and fruit-bats lost the enzyme required for the terminal step (**gulonolactone oxidase**) and so require ascorbate to be present in the diet, as the water-soluble vitamin, vitamin C. Gulonolactone oxidase catalyses the reaction:

$$\text{L-gulono-}\gamma\text{-lactone} + O_2 \rightarrow \text{L-ascorbate} + H_2O_2$$

Note that H_2O_2 is produced, i.e. high rates of ascorbate synthesis in animals such as the rat could, ironically, impose an oxidative stress. A DNA sequence resembling the L-gulonolactone oxidase gene is present in both the human and guinea-pig genomes but in an extensively mutated, inactive form. In a sense, inability to make ascorbate (like the lack of urate oxidase) is a universal inborn error of metabolism in humans.

Ascorbate is required *in vivo* as a cofactor for at least eight enzymes, of which the best known are **proline hydroxylase** and **lysine hydroxylase**, involved in the biosynthesis of collagen. Both these enzymes contain iron at their active

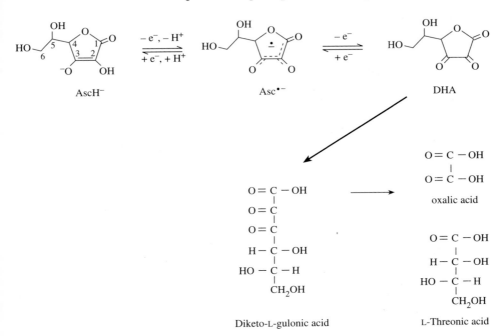

Fig. 3.26. Structure of ascorbic acid and its oxidation and degradation products. At physiological pH the acid form is largely ionized (ascorbate) since the pK_{a1} of ascorbic acid is 4.25.

sites. Collagen synthesized in the absence of ascorbate is insufficiently hydroxylated and does not form fibres properly, giving rise to poor wound-healing and fragility of blood vessels. Ascorbate is also required by the copper-containing enzyme **dopamine-β-hydroxylase**, which converts dopamine into noradrenalin.

Deficiency of ascorbate in the human diet causes **scurvy**. In 1536, the French explorer Jacques Cartier vividly described the nature of this disease, which afflicted all but ten of the 110 men aboard his ships wintering in the frozen St Lawrence river: "The victims' weakened limbs became swollen and discoloured, whilst their putrid gums bled profusely". Nearly 30 years later, the Dutch physician Ronsseus advised that sailors consume oranges to prevent scurvy. In 1639, one of England's leading physicians, John Woodall, recommended lemon juice as an antiscorbutic. James Lind, a Scottish naval surgeon, was the first man in medical history to conduct a controlled clinical trial, whereby he proved that scurvy could be cured by drinking lemon juice. In May 1747, he tested a variety of reputed remedies on 12 scorbutic sailors. Two of them were restricted to a control diet, but each of the others was additionally given one of the substances under trial. The two seamen who were provided with two oranges and a lemon each day made a speedy recovery. The only other sailors to show any signs of recovery were those who had been given cider. Lind observed no improvement in the condition of those who had been given either oil of vitriol (dilute sulphuric acid), vinegar, sea–water (popular folk-remedies for scurvy at the time) or just the control diet.

Lind's conclusions were acted upon by Captain James Cook on his second voyage round the world. Although Cook was at sea for three years, not a single member of his crew died from scurvy, thanks to adequate provision of lemon juice, as well as fresh fruit and vegetables. Surprisingly, it was not until 1795 that the British Admiralty finally agreed to Lind's demands for a regular issue of lemon juice on British ships. The effect of this action was dramatic; in 1780, there had been 1457 cases of scurvy admitted to Haslar naval hospital, but only two admissions took place between 1806 and 1810. Unfortunately, lemon juice was soon replaced by cheaper lime juice in a money-saving exercise rather typical of British governments throughout the ages; hence English sailors were often nicknamed 'limeys'. Scurvy returned and it took over a century to realize that lime juice had only about a quarter of the antiscorbutic activity of the lemon juice which it had displaced.

3.22.2 *Ascorbate as an antioxidant* in vitro[28]

Mammalian cells accumulate ascorbate from tissue fluids against a concentration gradient coupled to uptake of Na^+. Gut absorption of ascorbate is also Na^+-dependent. Several cell types (especially neutrophils) also rapidly take up oxidized ascorbate (DHA) if it is present and may use some of the glucose transport systems to do so.[217a]

The most striking chemical property of ascorbate is its ability to act as a reducing agent. We have already seen (Section 3.18 above) that its ability to reduce Fe(III) to Fe^{2+} may be important in promoting iron uptake in the gut. The observation that dietary ascorbate inhibits the carcinogenic action of several nitroso-compounds fed to animals (Chapter 10) can in part be attributed to its ability to reduce them to inactive forms. Indeed, ascorbate is probably required by certain hydroxylase enzymes in order to keep the iron or copper at the active site in the reduced form necessary for hydroxylation to occur. Perusal of Table 2.3 shows that the reduction potential of ascorbate places it close to the bottom of the 'pecking order' for oxidizing species, i.e. it will tend to reduce more-reactive species such as $OH^\bullet$, $O_2^{\bullet-}$ and urate radical.

Donation of one electron by ascorbate gives the **semidehydroascorbate** (SDA) or **ascorbyl** radical (Fig. 3.26), which can be further oxidized to give **dehydroascorbate** (DHA). Ascorbyl radical is relatively unreactive, being neither strongly oxidizing nor strongly reducing, nor does it seem to reduce O_2 to $O_2^{\bullet-}$ at a high rate (if at all). The poor reactivity of ascorbyl is the essence of many of ascorbate's antioxidant effects: a reactive radical interacts with ascorbate and a much less reactive (ascorbyl) radical is formed. Left to itself, ascorbyl undergoes a disproportionation reaction, regenerating some ascorbate

$$2SDA \rightleftharpoons ascorbate + DHA$$

Dehydroascorbate is unstable and breaks down rapidly in a complex way, eventually producing oxalic and L-threonic acids (Fig. 3.26). Aqueous solutions of ascorbate are stable at pH 7.4 unless transition metal ions are present, which catalyse rapid ascorbate oxidation. Copper salts are excellent catalysts—if you want plenty of vitamin C from vegetables, do not cook them in copper pans! Copper- and iron-induced oxidation of ascorbate produce H_2O_2 and $OH^{\bullet}$; the multiple literature reports of the ability of ascorbate to degrade DNA and damage various animal cells in culture, including cancer cells, can probably be attributed to the formation of ROS in the presence of traces of transition metal ions in the reaction solutions or cell culture media.[108] Ascorbate/Cu^{2+} mixtures inactivate many proteins, probably by formation of $OH^{\bullet}$ and/or oxo-copper ion species (Chapter 2). Iron salt/ascorbate mixtures have been used for decades *in vitro* to induce lipid peroxidation and other free radical damage (Chapter 4).

3.22.3 *Is ascorbate an antioxidant* in vivo?

In vitro, ascorbate has been shown to have a multiplicity of antioxidant properties, protecting various biomolecules against damage by both ROS and RNS (Box 3.2).

The levels of ascorbate found *in vivo* (30–100 µM in human plasma; higher in CSF, aqueous humour of the eye, gastric juice and lung lining fluid: millimolar levels intracellularly in many cell types) are sufficient to exert such antioxidant effects, i.e. an antioxidant action of ascorbate is feasible *in vivo*. However, direct evidence that ascorbate does act as an antioxidant *in vivo* is limited. We know that vitamin C is essential in the human diet; there is an established deficiency disease (scurvy) and the role of ascorbate as a cofactor for several enzymes is well established.

The effects of dietary ascorbate depletion can be studied in experimental animals such as the guinea-pig, or a mutant rat strain (**ODS rats**)[186] that is unable to synthesize ascorbate. The ODS rat was discovered in the laboratories of the Japanese pharmaceutical company Shionogi & Co, as suffering from osteogenic disorders when fed the usual rat chow; hence the name 'osteogenic disorder Shionogi' (ODS) rat. Surprisingly little work has been reported in which 'state-of-the-art' parameters of oxidative damage (Chapter 5) have been measured in animals in relation to ascorbate intake, although studies of ascorbate–vitamin E interactions using deuterated vitamin E have been carried out in guinea-pigs; these studies failed to provide evidence for recycling of vitamin E radicals by ascorbate in these animals.[43a]

Nevertheless, feeding guinea-pigs or ODS rats on a diet restricted in vitamin C has been reported to decrease the vitamin E content of tissues, consistent with an interaction between the two vitamins *in vivo*.[243a] An early study on guinea-pigs showed that a vitamin C-deficient diet led to increased exhalation of pentane and ethane, suggestive of increased lipid peroxidation *in vivo*. Unfortunately, the validity of such hydrocarbon measurements as an index

---

Box 3.2

Ascorbate as an antioxidant *in vitro*

- scavenges $O_2^{\bullet-}$ and $HO_2^{\bullet}$ (overall rate constant $> 10^5\,M^{-1}\,s^{-1}$ at pH 7.4)
- scavenges $OH^{\bullet}$ (rate constant $> 10^9\,M^{-1}\,s^{-1}$, although rate constants for reaction of $OH^{\bullet}$ with most other molecules *in vivo* are comparable)
- scavenges water-soluble peroxyl ($RO_2^{\bullet}$) radicals. Lipophilic ascorbate esters have been developed for use in foods and can scavenge lipid-soluble $RO_2^{\bullet}$ radicals
- scavenges thiyl, and oxysulphur radicals (Chapter 2)
- scavenges ergothioneine-derived radicals (Section 3.15.2)
- is a substrate for ascorbate peroxidase, an enzyme essential for H_2O_2 removal in chloroplasts (Chapter 7)
- prevents damage by radicals arising by attack of $OH^{\bullet}$ or $RO_2^{\bullet}$ upon urate, probably by reacting with urate radicals (Section 3.21.7)
- is a powerful scavenger of hypochlorous acid, peroxynitrous acid and nitrosating agents (Chapter 2).
- inhibits lipid peroxidation induced by haemoglobin- or myoglobin-H_2O_2 mixtures; ascorbate reduces the haem Fe(IV) species back to the Fe^{2+} state, preventing peroxide-dependent oxidations and haem breakdown
- is a powerful scavenger and quencher of singlet O_2 (Chapter 2)
- co-operates with vitamin E; regenerates α-tocopherol from α-tocopheryl radicals in membranes and lipoproteins (Section 3.22.2)
- scavenges nitroxide radicals (Chapter 5)
- infusion or oral administration of gram doses has been shown to improve vascular endothelium-dependent vasodilation in patients with vascular dysfunction, possibly by scavenging oxygen radicals and thus preserving $NO^{\bullet}$ (*Circul. Res.* **93**, 1107, 1993)
- protects plasma lipids against peroxidation induced by activated neutrophils and AAPH-derived peroxyl radicals
- protects membranes and lipoproteins against lipid peroxidation induced by species present in cigarette smoke, but does not inhibit protein carbonyl formation by cigarette smoke in human plasma (Chapter 8)
- is a powerful scavenger of O_3 and $NO_2^{\bullet}$ in human body fluids; it probably protects lung lining fluids against inhaled oxidizing air pollutants (Chapter 8)
- inhibits oxidative damage by scavenging radicals generated from certain drugs (e.g. phenylbutazone)
- protects against phagocyte adhesion to endothelium induced by oxidized LDL in a hamster dorsal skin-fold chamber model

For a list of references to the above observations see *Free Rad. Res.* **25**, 439 (1996).

---

of lipid peroxidation has repeatedly been questioned (Chapter 5). Ascorbate intake (in the range 150–900 mg/kg diet) did not appear to affect lipid peroxidation in ODS rats, as measured by a specific assay. Levels of 8-hydroxydeoxyguanosine, an index of oxidative DNA damage (Chapter 4), were elevated in sperm from adult men seriously deficient in vitamin C and normalized by adding ascorbate to the diet (Chapter 10). Ascorbate also decreased oxidative DNA damage in lymphocytes from smokers (Chapter 8).

Further evidence for an antioxidant role of ascorbate *in vivo* is provided by studies of its depletion under conditions of oxidative stress.[226] Thus ascorbate becomes oxidized to DHA in synovial fluid in the knee-joints of patients with active rheumatoid arthritis and in the lungs of patients with adult respiratory distress syndrome (Chapter 9). Presumably ascorbate is acting to scavenge ROS/RNS (Box 3.2) derived from the many activated phagocytes present. Measurement of ascorbyl radical by ESR has been used as an index of oxidative stress in several systems (Chapter 5). For example, infection with *Helicobacter pylori* predisposes to gastric ulceration and cancer. Gastric mucosa from patients with inflammation associated with this infection contained higher levels of ascorbyl than normal mucosa.[69]

Ascorbate/DHA ratios seem to be kept very high in body fluids and tissues during health, i.e. almost no DHA is present. Indeed, injection of DHA into animals has been reported to induce diabetes. The ascorbate/DHA ratio has been reported to fall in some diseases, including diabetes and rheumatoid arthritis (Chapter 9).

3.22.4 *'Recycling' of ascorbate*[28]

Oxidation of ascorbate by reaction with ROS/RNS in body fluids seems to lead to its depletion, probably by the reactions

$$\text{ascorbate} \xrightarrow{\text{radical attack}} \text{semidehydroascorbate (SDA) radical}$$

$$2\text{SDA} \xrightarrow{\text{disproportionation}} \text{ascorbate} + \text{DHA}$$

$$\text{DHA} \xrightarrow[\text{breakdown}]{\text{rapid non-enzymic}} \text{oxalate, threonate, other oxidation products}$$

In addition, erythrocytes, neutrophils and probably some other cell types take up DHA rapidly and convert it back to (intracellular) ascorbate. Indeed, many tissues possess enzymes that can convert ascorbate radical or DHA back to ascorbate at the expense of GSH or of NADH. One example is **NADH–semidehydroascorbate reductase enzymes** (Table 3.13). Glutathione-dependent **dehydroascorbate reductase** enzymes have been identified in plants and in several mammalian tissues. However, their identity as unique enzymes is uncertain since some of the proteins involved in thiol–disulphide interchange within the cell (such as protein-disulphide isomerase and glutaredoxin; Sections 3.12 and 3.13 above) show dehydroascorbate reductase activity.[262] For example, the 'DHA reductase' purified from human neutrophils was found to be identical with glutaredoxin.

Evidence that GSH and ascorbate interact *in vivo* is provided by studies on animals treated with inhibitors of GSH synthesis.[168] Severe glutathione

206   *Antioxidant defences*

Table 3.13. Semidehydroascorbate reductase activity in rat tissues

Tissue	Enzyme activity (mean $\pm$ SEM)
Adrenal cortex	49.6 ± 2.4
Brain	9.1 ± 0.6
Heart	0
Ileum	3.3 ± 0.3
Kidney	49.3 ± 4.9
Liver	30.9 ± 1.0
Lung	8.9 ± 1.8
Pancreas	16.3 ± 1.1
Skeletal muscle	0
Spleen	6.3 ± 0.3
Testis	11.4 ± 0.3
Thyroid gland	5.8 ± 0.3

Semidehydroascorbate reductase activity was assayed in homogenates of several rat tissues. The enzyme reduces semidehydroascorbate (SDA) to ascorbate at the expense of NADH. Data from Diliberto *et al. J. Neurochem.* (1982) **39**, 563. Enzyme activity is quoted as nmol NADH oxidized/min/mg protein. The high activity present in adrenal gland cortex may be needed because SDA is formed from ascorbate during the dopamine-β-hydroxylase reaction, involved in hormone biosynthesis. The enzyme is membrane bound, including endoplasmic reticulum and outer mitochondrial membrane. At least some of the 'SDA reductase' assayed in crude extracts may be a partial function of other enzyme systems.

depletion in newborn rats and guinea-pigs is lethal, but death can be prevented by high doses of ascorbate (but not DHA). The onset of scurvy in guinea-pigs fed a diet low in ascorbate is delayed by administering GSH precursors. In isolated mouse hepatocytes, lowering GSH levels has been reported to increase the synthesis of ascorbate.

3.22.5 *Pro-oxidant effects of ascorbate*

In vitro, vitamin C can also exert **pro-oxidant** properties. The classic Udenfriend system for making OH$^\bullet$ in the laboratory consists of ferric-EDTA, H_2O_2 and ascorbate. The ascorbate acts as reductant to the iron, easily permitted by the relative reduction potentials (Table 2.3).

$$Fe(III) + ascorbate \rightarrow Fe^{2+} + ascorbate^\bullet$$

$$Fe^{2+} + H_2O_2 \rightarrow Fe(III) + OH^\bullet + OH^-$$

Thus copper– or iron–ascorbate mixtures stimulate free-radical damage to DNA, lipids and proteins *in vitro*.[108] Instillation of high levels of ascorbate with iron or copper ions into the stomach of animals was reported to lead to OH$^\bullet$ generation[128] and the mixture of metal ions and ascorbate in some vitamin pills has been claimed to generate OH$^\bullet$ once the pills dissolve. A mixture of ascorbate and copper ions (which will generate OH$^\bullet$) rapidly inactivates several enzymes, including catalase. Pro-oxidant effects of ascorbate are also well-known to food scientists.[201]

Are these pro-oxidant effects physiologically relevant? The question is very important in relation to attempts to assess the optimal ascorbate content of the human diet and the desirability or otherwise of supplementation (Chapter 10). There is no convincing evidence for toxicity of high-dose ascorbate in healthy humans, but nor are multi-gram doses to be recommended. Ascorbate is ultra-filtered and reabsorbed in the kidney, but the active uptake system has limited capacity so that a dose of 200 mg daily is sufficient to saturate cells and body fluids with ascorbate:[154] any excess is simply excreted. However, heavy smokers may need more vitamin C than this.

It should perhaps first be noted that *in vitro* pro-oxidant effects are not unique to ascorbate; they can be demonstrated with many reducing agents in the presence of transition metal ions, including vitamin E (see below), GSH, NAD(P)H and several plant phenolics.[231] Thus if ascorbate's pro-oxidant effects are relevant *in vivo*, the pro-oxidant effects of these other reductants might also be expected to occur. The key question is the availability of 'catalytic' transition metal ions. This relates to another important nutritional question: what is the optimal dietary intake of iron and copper? Iron is essential for human health, especially in children and pregnant women, but could too much iron intake cause harm, either in the body or in the colon (where unabsorbed dietary iron, copper, vitamin E and plant phenolics will end up)?

In vivo, iron and copper ions are largely sequestered in forms unable to catalyse free-radical reactions (Section 3.18 above). Extracellular fluids have essentially no transition metal ions catalytic for free-radical reactions. Thus the pro-oxidant properties of ascorbate (and the other biological reducing agents) would be very limited. In contrast, cells have low-molecular-mass intracellular 'pools' of iron (and possibly copper). If these came into contact with ascorbate, pro-oxidant effects could occur. Perhaps such effects do occur but are simply masked by the dominant antioxidant effects of ascorbate. If so, how would this balance be affected by raising ascorbate and/or iron levels? Our knowledge of the chemical nature of the 'low molecular mass intracellular iron pool' is limited. We do know that free radical damage to lipids, proteins and DNA occurs *in vivo* even in healthy subjects (Table 4.1).

However, ascorbate may help to facilitate normal iron sequestration:[249] it has been reported that ascorbate enhances ferritin mRNA translation in response to iron in cultured cells by interacting with the IRE-binding proteins (Section 3.18 above), and ascorbate may also decrease the breakdown of ferritin within lysosomes.

3.22.6   *Ascorbate and iron overload disease*

We might perhaps learn something by looking at pathological situations, such as thalassaemia or haemochromatosis (Section 3.18 above). Patients with iron overload can have non-transferrin-bound iron in their plasma that appears capable of catalysing free-radical reactions (*in vitro* at least). It is commonly observed that plasma ascorbate levels in such patients are sub-normal. There are a few published 'case reports' that giving vitamin C to iron-overloaded subjects without administration of an iron chelating agent (such as desferrioxamine) can produce deleterious clinical effects.

Injury to human tissues causes increased availability of transition metal ions that can promote free-radical reactions (Chapter 4) and there are repeated (but controversial) *suggestions* that high body iron and/or copper stores are associated with increased risk of cancer and cardiovascular disease (Chapter 10). Could this be because the more iron or copper is in a tissue, the more is potentially mobilizable to catalyse free-radical reactions after an injury? If this is so, then the pro-oxidant effects of ascorbate might conceivably be aggravated in disease or traumatic injury. Indeed, we have hypothesized[107] that the decline in ascorbate at sites of tissue injury might be beneficial because (i) ascorbate is helping to scavenge ROS/RNS and recycle α-tocopherol and (ii) ascorbate removal minimizes its potential pro-oxidant interactions with metal ions released by tissue damage. Thus one can speculate that giving lots of ascorbate to sick people may not be a good thing. It remains possible that any pro-oxidant properties of ascorbate under these circumstances are still out-weighed by its antioxidant effects.

Another question is whether ascorbate could favour excessive uptake of iron into the human body, since the reduction of ferric ions to Fe^{2+} by ascorbate is believed to facilitate iron uptake in the gut. There is no evidence to support this view in healthy subjects; iron uptake appears to be carefully regulated whatever the ascorbate intake. However, the issue needs to be addressed in relation to haemochromatosis.

3.22.7   *Vitamin E*[65a,252]

Vitamin E, as a scavenger of peroxyl radicals, is probably the most important (but not the only), inhibitor of the free-radical chain reaction of lipid peroxidation (Chapter 4) in animals. However, it must be remembered that initiation of lipid peroxidation (Chapter 4) can be prevented, as an earlier line of defence, by enzymes that scavenge ROS/RNS and proteins that sequester transition metal ions. Sequestration of metal ions also prevents them from decomposing peroxides into chain-propagating peroxyl and alkoxyl radicals. For example, Table 3.14 relates vitamin E to other antioxidants in extracellular fluids.

The name 'vitamin E' does not refer to a particular chemical structure; it is a nutritional term. It was first used to refer to a fat-soluble 'factor' discovered in 1922 to be essential in the diet of rats to permit normal reproduction. Later work showed that vitamin E is essential in the diets of all other animals:[65] lack of it causes a wide variety of symptoms including sterility in male rats, dogs, cocks, rabbits and monkeys, haemolysis in rats and chicks, muscular degeneration in rabbits, guinea-pigs, crocodiles, snakes, lizards, elephants, monkeys, ducks, mice and minks, 'white-muscle disease' in lambs, flamingoes and calves, and degeneration of the cerebellum in chicks. Indeed, the dietary content of vitamin E is one factor that affects the sensitivity of laboratory animals to certain toxins (Chapter 8) or to tissue insults such as ischaemia–reperfusion (Chapter 9).

Short-term absence of vitamin E from the human diet does not cause any specific deficiency disease,[178] although low vitamin E levels in premature babies (Table 3.15) can predispose to haemolytic anaemia, probably due to increased fragility of the erythrocyte membrane. There is no evidence that human muscular dystrophy or multiple sclerosis respond to vitamin E administration. Sources of vitamin E in the human diet include wheat-germ, vegetable oils, margarines, nuts, grains and green leafy vegetables.

3.22.8   Chemistry of vitamin E

Eight naturally-occurring substances have been found to have vitamin E activity in animal tests: *d-α-*, *d-β-*, *d-γ-* and *d-δ-***tocopherols**, and *d-α-*, *d-β-*, *d-γ-* and *d-δ-***tocotrienols**. The name 'tocopherol' comes from the Greek words *tokos* (childbirth) and *phero* (to bring forth). The tocopherols have three asymmetric carbon atoms, giving eight optical isomers. The most effective form biologically in animals is *RRR-α*-tocopherol, formerly called *d-α*-tocopherol (Fig. 3.27). Other tocopherols might fulfil important additional roles in plants (Chapter 7) and a metabolite of *γ*-tocopherol has been proposed to help regulate Na^+ metabolism (as a component of 'natriuretic hormone') in animals.[178a] The *β-*, *γ-* and *δ*-tocopherols appear less important than *α*-tocopherol as antioxidants in humans. Although all these forms are capable of acting as antioxidants, *β-*, *γ-* and *δ*-tocopherols are not retained as well in the body tissues.

The terms '*α*-tocopherol' and 'vitamin E' are now used in the literature almost interchangeably. This is, strictly speaking, incorrect: vitamin E is a nutritional term and the other tocopherols do have vitamin E activity. Synthetic 'vitamin E' (*dl-α-***tocopherol**, sometimes called **all-*rac*-α-tocopherol**) contains about 12.5% of *d-α*-tocopherol, together with seven other tocopherol isomers that are less biologically active (21–90% in animal bioassays).

Tocopherols and tocotrienols inhibit lipid peroxidation largely because they scavenge lipid peroxyl ($LO_2^{\bullet}$) radicals much faster than these radicals can react with adjacent fatty acid side-chains or with membrane proteins. Rate constants for the reaction

$$\alpha\text{-TocH} + LO_2^{\bullet} \rightarrow \alpha\text{-Toc}^{\bullet} + LO_2H$$

Table 3.14. A summary of extracellular antioxidant defences in blood plasma

Defence	Mode of action	Comments
Transferrin, lactoferrin	bind iron and stop its pro-oxidant activity	these proteins are not easily damaged by H_2O_2, HOCl, ONOO$^-$ or lipid peroxides; only release iron ions at acidic pH (especially lactoferrin)
Caeruloplasmin	catalytically oxidizes Fe^{2+} to Fe(III) *without release* of oxygen radical intermediates. Reacts stoichiometrically with $O_2^{\bullet -}$ (reports of greater $O_2^{\bullet -}$-scavenging activity may be due to contamination with EC-SOD; see *Free Rad. Biol. Med.* **2**, 255, 1988).	acute–phase protein
Erythrocytes	can take up $O_2^{\bullet -}$ and H_2O_2 for metabolism by intracellular enzymes	H_2O_2 could diffuse also into platelets, phagocytes, endothelial cells, etc. for metabolism
Albumins	highly soluble small proteins (human 65 000 relative molecular mass, 585 amino acids, 17 disulphide bridges, one free –SH at Cys34). Bind copper tightly and iron weakly. Present at high concentrations (40–60 mg/ml). Possible sacrificial antioxidant. Rapidly scavenges HOCl and peroxynitrous acid. Provides high –SH level in plasma. Binds haem, can help protect lipoproteins against haem–dependent oxidation.	liver synthesis and plasma concentration drop during liver injury, sometimes called 'a negative acute–phase response'

Haptoglobin/ haemopexin	bind free haemoglobin/haem and decrease their pro-oxidant ability	acute-phase proteins
Urate	inhibits lipid peroxidation and scavenges ROS/RNS	can also bind iron and copper ions
Vitamin E	lipid-soluble antioxidants: chain-breaking by trapping peroxyl radicals	major, lipid-soluble, chain-breaking antioxidant in human plasma; important in protecting lipoproteins against oxidation. Interacts with ubiquinol
Glucose	scavenger of $OH^\bullet$ radical; rate constant comparable to that of mannitol. Can exert pro-oxidant effects by glycation of proteins	normal plasma concentration around 4.5 mM, greater soon after carbohydrate-containing meals.
Bilirubin	postulated antioxidant	also a sensitizer of $^1O_2$ production

Extracellular SOD, glutathione peroxidase and catalase activities are not included here; see Section 3.19.4 for a discussion.

Table 3.15. Representative blood tocopherol levels in humans

Subjects	Total tocopherol (μM)
Adults	
young	22
elderly	20
Children (2–12 years)	28
Term infants	18
Premature infants	9
Infants and children with protein-calorie malnutrition	6
Kwashiorkor	7
Gastrointestinal diseases	
cholestatic liver disease	<2
abetalipoproteinaemia	<2
coeliac disease	7
nontropical sprue	6
tropical sprue	9
chronic pancreatitis	6
ulcerative colitis	10
Haemolytic anaemias	
β-thalassaemia major	10
thalassaemia intermedia	5
sickle-cell anaemia	14
glucose-6-phosphate dehydrogenase deficiency	12
hereditary spherocytosis	12
Miscellaneous	
total parenteral nutrition	13
Gaucher's disease	
severe	2
chronic	8

It should be noted that 'normal' values vary among different populations. Because tocopherol is carried by lipoproteins, the plasma lipid content influences the plasma tocopherol level. Hence plasma vitamin E levels are best expressed as a ratio with total plasma lipids. For example, it has been suggested that some or all of the apparent α-tocopherol depletion in adult respiratory distress syndrome is due to lower plasma lipids. For practical purposes total cholesterol or the sum of plasma cholesterol and triglycerides is often used as a basis for comparison. The requirement for vitamin E increases when the intake of PUFAs increases. Attempts have been made to specify a fixed ratio of dietary RRR-α-tocopherol to PUFAs, but this has not been completely satisfactory. When the primary PUFA in the diet is linoleic acid, as in most US diets, a ratio of approximately 0.4 mg RRR-α-tocopherol to 1 g of PUFA has been suggested as adequate for adult humans. As intakes of the common US vegetable oils increase, vitamin E intake increases as well, provided the oils have not been over-used, deteriorated or become rancid. Data by courtesy of the Vitamin E Research and Information Service. Also see *Nutrition* **13**, 450, 1997.

are about $10^6 \, M^{-1} s^{-1}$, some four orders of magnitude faster than those for reaction of $LO_2^{\bullet}$ radicals with lipids ($\sim 10^2 \, M^{-1} s^{-1}$).

In addition, tocopherols both quench and react with singlet O_2 (Chapter 2) and might protect membranes against this species. α-Tocopherol reacts slowly

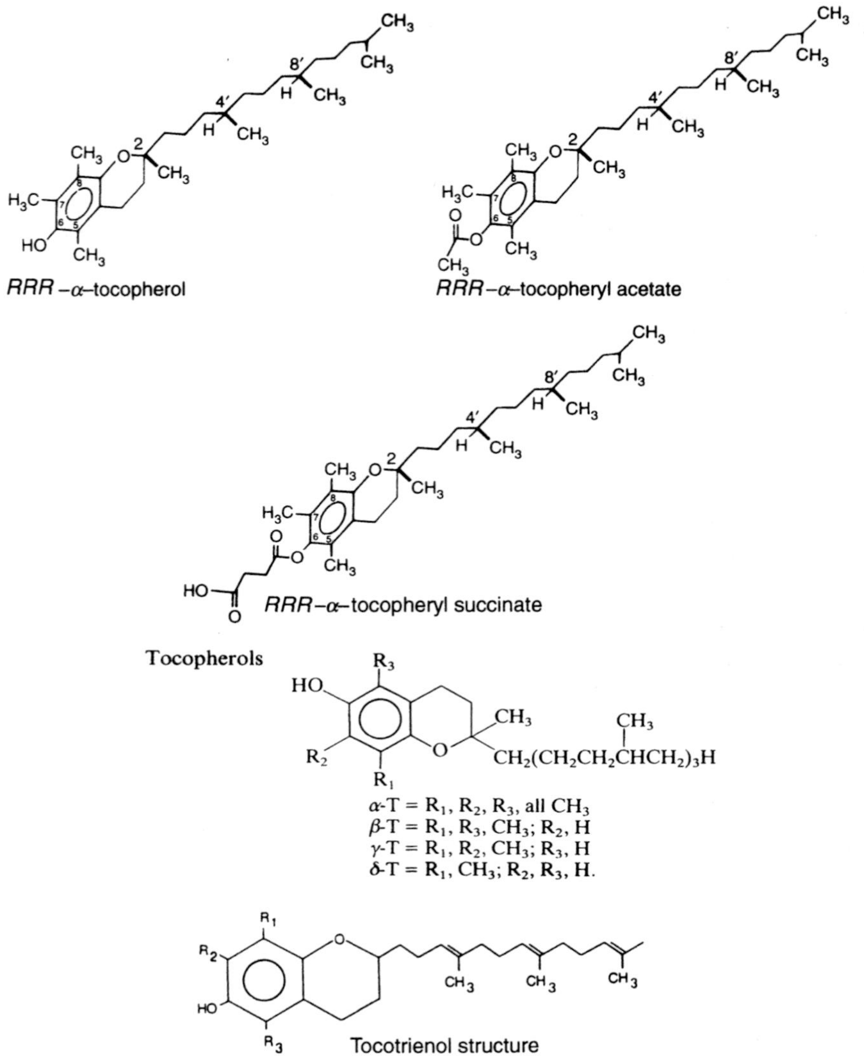

Fig. 3.27. Compounds with vitamin E activity. (Top) Structural formulae of 'natural vitamin E' (*RRR*-α-tocopherol or *d*-α-tocopherol) and its esters with acetic acid (*RRR*-α-tocopheryl acetate) and succinic acid (*RRR*-α-tocopheryl succinate). The esters are often used in commercial vitamin E preparations because they are more stable than vitamin E itself. They are rapidly hydrolysed by esterases in the gut. (From *Tolerance and Safety of Vitamin E*, by H. Kappus and A.T. Diplock, by courtesy of these authors and the Vitamin E Research and Information Service.) Tocopheryl acetate is not usually thought to occur in nature, but may be secreted by the squash beetle, *Epilachna variestis*, during its 'defence response', as a carrier for an irritant chemical (*Experientia* **52**, 616 (1996)). (Middle) Diagrammatic representation of the structures of α-, β-, γ- and δ-tocopherols. All the tocopherols have a **chromanol** ring structure and a **phytol** side-chain which anchors them in the membrane. (Bottom) Basic structure of the tocotrienols, which have three double bonds in the hydrophobic side chain. Side-chain nomenclature as for the tocopherols.

with $O_2^{\bullet -}$ (probably as $HO_2^{\bullet}$; Chapter 2) and, like most other biological molecules, at an almost diffusion-controlled rate with $OH^{\bullet}$. The $\alpha Toc^{\bullet}$ radical might also react with a further peroxyl radical to give non-radical products

$$LO_2^{\bullet} + \alpha Toc^{\bullet} \rightarrow \alpha TocOOL$$

i.e. one molecule of α-tocopherol is, in principle, capable of terminating two peroxidation chains. Products of the above reaction include eight α-substituted tocopherones and epoxy (hydroperoxy) tocopherones: the former

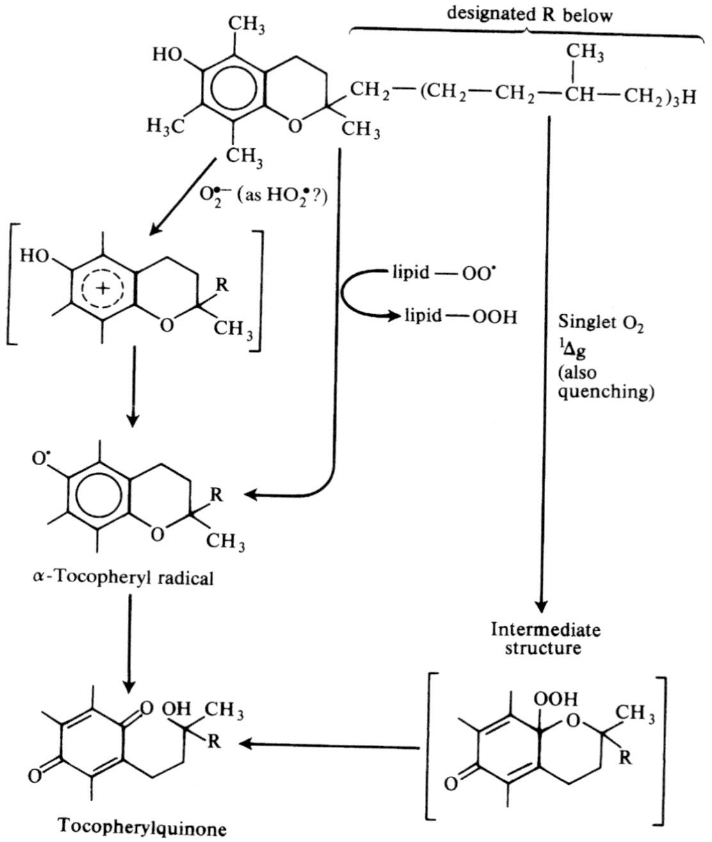

Fig. 3.28. Structure of tocopherol and some of its oxidation products. Tocopheryl radical may be recycled to tocopherol or can undergo further oxidation by a series of mechanisms. A frequent product of further oxidation is tocopherylquinone. Traces of α-tocopherylquinone are found in animal (including human) tissues: it is metabolized by reduction to the hydroquinone (which itself can exert antioxidant properties: *Proc. Natl Acad. Sci. USA* **94**, 7885, 1997) and can be conjugated with glucuronic acid and excreted in the bile or degraded in the kidneys to α-**tocopheronic acid**, followed by conjugation and excretion in urine.

readily hydrolyse to **tocopherylquinone** (Fig. 3.28) and the latter to epoxyquinones.

α-Tocopherol at high levels can exert 'membrane stabilizing' and other effects independent of its antioxidant ability. For example, high intakes of vitamin E supplements may diminish platelet aggregation, perhaps by affecting prostacyclin synthesis. Several studies using isolated hepatocytes have shown that incubation with α-tocopherol succinate ester (Fig. 3.27)[207] can protect against hyperoxia and the effects of toxins better than unesterified α-tocopherol, even though comparable amounts of α-tocopherol appeared inside the cells. Additionally, there is evidence from cell culture work that α-tocopherol can affect the types of fatty acids that become incorporated into membrane lipids, or even cell growth itself. For example, α-tocopherol can decrease proliferation of cultured smooth muscle cells taken from blood vessel walls.[244] However, peroxyl radical scavenging is probably the major antioxidant activity of α-tocopherol at normal tissue levels.

α-Tocopherol is a fat-soluble molecule and therefore tends to concentrate in the interior of membranes and in lipoproteins (Table 3.15 summarizes blood tocopherol levels). For example, mitochondrial membranes contain about one molecule of α-tocopherol per 2100 molecules of phospholipid, but there is much more than this in the chloroplast thylakoid membrane and in the outer segment membranes of the retinal rods. Both those membranes must be especially protected against peroxidation *in vivo* (Chapter 7). The hydrophobic tail of α-tocopherol anchors the molecule in the membrane, positioning the **chromanol ring** containing the phenolic −OH group at the hydrocarbon interface. It is this group which is responsible for $RO_2^{\bullet}$ scavenging activity (Fig. 3.28) and which interacts with the aqueous phase.

3.22.9   *Recycling of α-tocopheryl radicals*[252]

During its action as a chain-breaking antioxidant, α-tocopherol is consumed and converted to the radical form. Physiological mechanisms may exist for reducing the radical back to α-tocopherol. Synergy between vitamin E and vitamin C was first suggested by Al Tappel in 1968. Pulse radiolysis studies confirmed that ascorbate can reduce α-tocopheryl radical back to α-tocopherol with a fairly high rate constant ($\sim 1.5 \times 10^6\,M^{-1}s^{-1}$) and this reaction has since been shown to occur in isolated membranes, cultured cells and lipoproteins, although it has been hard to establish conclusively that it occurs *in vivo*.[43a] Ubiquinol can also recycle the vitamin E radical (Section 3.21.6 above) and there is limited evidence that GSH can do so in some membrane systems. However, it is generally felt that the ascorbate-dependent recycling system is likely to be the most important *in vivo*. Ascorbate is simultaneously converted to ascorbyl.

Alternative fates of the vitamin E radical include conversion to α-tocopheryl-quinone (Fig. 3.28), and on to various metabolites, some of which are excreted in the urine.

3.22.10   *Pro-oxidant effects of α-tocopherol*[201]

Like ascorbate, tocopherols can reduce Fe(III) to Fe^{2+} and Cu^{2+} to Cu^+ and thus they can exert pro-oxidant effects in some *in vitro* systems. Indeed, this iron-reducing ability was the basis of some of the earliest colorimetric methods used to measure vitamin E. In addition, the α-Toc• radical is not completely unreactive with lipids: it can abstract hydrogen from polyunsaturated fatty acids (PUFAs)

$$\alpha\text{-Toc}^\bullet + LH \rightarrow L^\bullet + \alpha\text{-TocH}$$

However, the rate constants are about $5 \times 10^{-2} M^{-1} s^{-1}$, four or five orders of magnitude lower than the rate constants for reaction of peroxyl radicals with PUFAs.[176] Thus if αToc• is generated in a lipid system in the absence of $RO_2^\bullet$ (e.g. if copper ions are added to a tocopherol-containing lipid), αToc• can act as a weak promoter of lipid oxidation. This phenomenon has been observed in the food industry and in studies of the oxidation of LDL *in vitro* (Chapter 9). The recycling of the αToc• radical by ascorbate and other reducing agents presumably largely prevents such occurrences *in vivo*.

Even very high intakes of vitamin E in adults show virtually no toxicity, although they can affect blood coagulation activity by interfering with the action of **vitamin K**. However, at least some of this activity is due to the vitamin E metabolite **α-tocopherylquinone** (Fig. 3.28),[68] which is a more powerful anti-coagulant.

3.22.11   *Processing of dietary vitamin E*[252]

If large oral doses of α-tocopherol are taken, much is not absorbed and is excreted in the faeces. However, a high percentage of the α-tocopherol in foods is absorbed. Being fat-soluble, it enters the body in **chylomicrons** (Fig. 3.29 explains the intricacies of fat metabolism). There appears to be little discrimination between the different forms of vitamin E by the intestine. However, levels of *RRR*-α-tocopherol are higher in plasma because the human liver incorporates it selectively into the **very low density lipoproteins** (VLDLs) that it secretes into the blood. The VLDLs contain lipids that originated from the diet (by uptake of chylomicron remnant particles by the liver) as well as lipids made in the liver itself (Fig. 3.29).

Liver secretion of VLDL contributes to the daily 'turnover' of plasma *RRR*-α-tocopherol. A hepatic **α-tocopherol transfer protein** appears to selectively incorporate *RRR*-α-tocopherol into VLDL. Patients with the autosomal recessive neurodegenerative disease called **ataxia with isolated vitamin E deficiency** (AVED) have an impaired ability to incorporate α-tocopherol into VLDL secreted by the liver, because of mutations in the gene encoding the tocopherol transfer protein.

LDL arise from VLDL after triglyceride has been taken up by tissues. During the conversion of VLDL to LDL in the circulation, a portion of the *RRR*-α-tocopherol remains in the LDL, but some is transferred to HDL.

Equilibration of RRR-α-tocopherol between LDL and HDL occurs because these two lipoproteins readily exchange tocopherol without the assistance of any transfer proteins (Fig. 3.29).

3.22.12   *Evidence for an antioxidant effect of α-tocopherol in vivo*[65a]

In animals, many of the signs of vitamin E deficiency can be partly or completely alleviated by feeding synthetic chain-breaking antioxidants (e.g. ethoxyquin or promethazine) or by raising the selenium content of the diet (Section 3.14 above). It is well-known to veterinary practitioners, zoo-keepers and farmers that feeding unsaturated fats to animals increases their requirement for vitamin E. Thus for every 1% of corn oil fed to young pigs above 4% of the diet, 100 mg extra vitamin E is required. Feeding more vitamin E to pigs, chickens and cows has been reported to increase the stability of their meat against rancidity on storage.[222] Chicks fed on lard, a mainly saturated fat, can remain healthy without vitamin E for weeks.

Tissue samples taken from vitamin E-deficient animals show evidence of peroxidation (e.g. as elevated levels of isoprostanes or, less convincingly, as TBA-reactive material),[i] and tissue homogenates or subcellular fractions from such animals peroxidize more rapidly than normal when incubated *in vitro*. Vitamin E-deficient animals are more sensitive to the toxic effects of pure O_2. Vitamin E-deficient rats exhale more hydrocarbon gases and accumulate fluorescent pigments in certain tissues more rapidly than normal, especially if they are fed a diet rich in PUFAs. Lack of vitamin E in the diets of rodents increases the rate of accumulation of **senescent cell antigen**, a cell-surface protein indicative of 'old' cells. It is a breakdown product of the membrane ion transport protein **band 3** (Chapter 7).

Although a short-term lack of vitamin E in the diet of adult humans does not produce obvious acute signs of disease, it does increase susceptibility to peroxidation of membranes, often revealed by an increased rate of haemolysis when erythrocytes are treated with H_2O_2 *in vitro* (for further discussion of this **peroxide stress haemolysis test** see Chapter 7). Severe depletion of body α-tocopherol stores occurs in adult humans only as a result of abnormalities of fat absorption by the gut after prolonged intravenous feeding, or as a result of some inborn error in vitamin E metabolism (Table 3.15).

For example, in patients suffering the rare inborn error of lipid metabolism known as **abetalipoproteinaemia**, dietary fat is ingested and absorbed, but not transported out of the intestinal mucosal cells, because of an inherited inability to synthesize apoprotein B, an essential component of chylomicrons (Fig. 3.29). Patients with abetalipoproteinaemia have negligible plasma α-tocopherol concentrations and eventually develop neuronal damage, retinal degeneration and abnormally shaped erythrocytes (**acanthocytes**).[178] The neuropathy and retinopathy can be prevented by administering very large oral doses of α-tocopherol (sufficient to ensure some oral absorption). Patients can also be given RRR-α-tocopheryl polyethylene glycol (PEG) succinate, a more hydrophilic molecule in which water soluble PEG is attached to the 'spare'

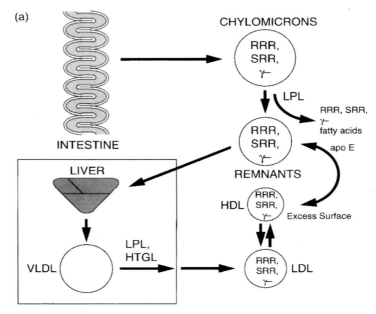

Vitamin E transport during chylomicron catabolism

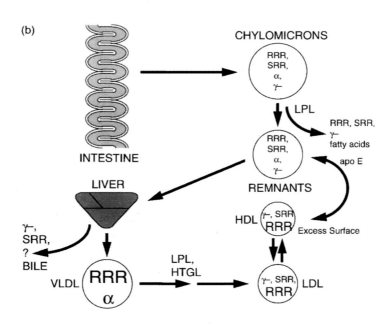

α-Tocopherol transport during VLDL catabolism

Fig. 3.29.

COOH on α-tocopherol succinate. Neurological and retinal disorders have also been observed in some patients with cystic fibrosis or with congenital defects that impair bile production: both conditions again result in impaired fat absorption.

Newborn babies have low concentrations of α-tocopherol in plasma, especially if they are premature (Table 3.15). Their erythrocytes are more susceptible to lipid peroxidation *in vitro*, although this does not normally cause a clinical problem. Sometimes haemolysis is seen *in vivo* in premature babies, and this **haemolytic syndrome of prematurity** responds to α-tocopherol therapy. α-Tocopherol also helps to protect premature babies against retrolental fibroplasia (Chapter 1). The occasional haemolysis occurring in patients with thalassaemia (Section 3.18.4 above) or glucose-6-phosphate dehydrogenase deficiency (Chapter 7) can be decreased by administration of extra oral α-tocopherol and a similar protective effect has been suggested to occur in sickle cell anaemia. In these diseases there is extra 'oxidative stress' or a decrease in other protective mechanisms so that the effects of α-tocopherol are more readily seen.

Fig. 3.29. Human lipoprotein metabolism in relation to that of vitamin E. Lipoproteins are plasma lipid transport vehicles consisting of a hydrophobic core (with triglyceride and cholesterol ester) and a surface permitting interaction with aqueous environments and usually containing cholesterol, phospholipid and proteins. The four major classes are chylomicrons, very low density lipoproteins (VLDL), low density lipoproteins (LDL) and high density lipoproteins (HDLs). (a) The transfers of tocopherols during chylomicron catabolism. The intestine absorbs and processes dietary lipids and secretes chylomicrons, carrying in them various forms of vitamin E (*RRR*- and *SRR*-α-tocopherols, *β*-, *γ*- and *δ*-tocopherols, tocotrienols, etc.) into the lymph, from where they eventually enter the bloodstream. Fat digestion requires bile; defects in bile production impair absorption of fats and fat-soluble vitamins (including vitamin E). Chylomicrons are hydrolysed in the circulation by a **lipoprotein lipase** (LPL) enzyme on the surface of capillaries, resulting in transfer of fatty acids and tocopherols to tissues. The chylomicron remnants can transfer tocopherols to HDL and can acquire apolipoprotein E (apoE), a protein which directs the chylomicron remnants to the liver. HDL tocopherols can transfer to other circulating lipoproteins, such as LDL and VLDL (transfer to circulating VLDL is not shown). (b) The events following hepatic uptake of chylomicron remnants. The liver secretes lipids in nascent VLDL. The hepatic α-**tocopherol transfer protein** preferentially transfers *RRR*-α-tocopherol to VLDL. The large typeface indicates that the plasma lipoproteins are enriched in *RRR*-α-tocopherol by this mechanism. Other forms of tocopherol (such as *γ*-tocopherol or *SRR*-α-tocopherol) are secreted in bile and so their lifetime *in vivo* is shorter than that of *RRR*-α-tocopherol. Vitamin E is not stored in the liver. Once VLDL is secreted in the circulation, both LPL and hepatic triglyceride lipase (HTGL) participate in its conversion to LDL. Only about half of the VLDL is converted to LDL; the remainder is taken up by the liver (not shown). During triglyceride hydrolysis by LPL and HTGL, tocopherol can be transferred to HDL, in a manner analogous to transfer during chylomicron catabolism (transfer not shown). The secretion of *RRR*-α-tocopherol in nascent VLDL by the liver is the mechanism that maintains plasma tocopherol concentrations; the exchange of tocopherols between lipoproteins determines individual lipoprotein concentrations. Diagram and text by courtesy of Drs Herbert Kayden and Maret G. Traber and the *Journal of Lipid Research*.

3.23   Carotenoids: important biological antioxidants?[146]

There is considerable epidemiological evidence that diets rich in fruits, grains and vegetables are protective against several human diseases, especially cardiovascular disease and some types of cancer (Chapter 10). The antioxidant vitamins C and E contribute some of this protective effect, but there are many other constituents that may exert additional antioxidant effects, or protect by completely different mechanisms (Chapter 10). In terms of antioxidants, particular attention has been paid to the carotenoids and plant phenolics, although evidence that they are important antioxidants *in vivo* is limited as yet.

Carotenoids (of which the first to be isolated was from carrots, in 1831) are a group of coloured pigments (usually yellow, red or orange) that are widespread in plant tissues (Fig. 3.30, see plate section). They are also found in some animals (e.g. lobsters) and certain bacteria. Over 600 carotenoids have been described. Carotenoids from the diet can be found in the tissues of humans and some other mammals, but many other animals (e.g. rodents, sheep, hares or elephants) do not normally absorb them. In humans, the largest *amounts* of carotenoids are found in adipose tissue (80–85% of total) and liver (8–12%) but the *concentration* is highest in the corpus luteum of the ovary and in adrenal gland; testis appears enriched in lutein. Lutein and zeaxanthin are present in the macula of the eye (Chapter 7). Plasma carotenoid levels and tissue content vary widely with diet. Human plasma levels are usually in the low micromolar range (e.g. lycopene, 0.5–1.0 μM; β-carotene, 0.3–0.6 μM; α-carotene, 0.05–0.1 μM; lutein, $\sim 0.3\,\mu$M). Human plasma also contains cryptoxanthin and some zeaxanthin (Figs 3.31 and 3.32).

Absorption of dietary carotenoids is incomplete and depends on what food mixtures have been eaten and on how the food has been processed. For example, tomatoes are rich in lycopene, but little is absorbed from raw tomatoes. More is taken up from cooked tomatoes or tomato paste, e.g. on pizzas, although some *trans–cis* isomerization of lycopene occurs during processing.[241] There is also considerable person-to-person variation in absorption.

3.23.1   *Carotenoid chemistry*[41]

The most striking feature of carotenoid structures is the long system of alternating double and single bonds that forms the central part of the molecule (Fig. 3.32). This allows extensive electron delocalization over the entire chain, causing carotenoids to absorb in the visible range (Fig. 3.30). The basic skeleton of carotenoids has 40 carbon atoms and can be modified by cyclization at one or both ends, by reducing certain double bonds and by addition of oxygen-containing functional groups (Fig. 3.31). Carotenoids that contain one or more oxygen functions are known as **xanthophylls**, the parent hydrocarbons as **carotenes**.

Carotenoids are usually known by their trivial names, although a semi-systematic scheme has been devised based on the stem name carotene preceded by two Greek-letter prefixes that indicate the type of end-group present. Seven

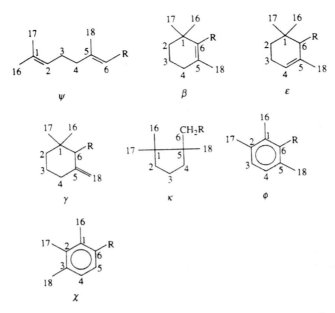

Fig. 3.31. The basic structure of carotenoids.(I) The seven different types of end-group found in natural carotenoids. Thus β-carotene has two β-groups, α-carotene has one β- and one ε-group. Diagram from *FASEB J.* **9**, 1551 (1995) by courtesy of Dr George Britton and the publishers.

types of end-group are known (Fig. 3.31). Thus β-carotene should be called β,β-carotene, and α-carotene should be called β,ε-carotene. Zeaxanthin is β,β-carotene-3,3′-diol. Some carotenoids have a structure with fewer than 40 carbon atoms; these compounds are called **apocarotenoids** when carbon atoms have been lost from the ends of the molecule or **norcarotenoids** when carbon atoms have been lost from within the chain.

In principle, each double bond in the polyene chain could exist as *cis* or *trans* geometric isomers.[j] *Trans* forms are more common in nature, presumably because the most stable form of long polyunsaturated chains is usually a linear, *trans*, extended conformation. *Cis* bonds create kinking of the chain and greatly modify the overall molecular shape. Some *cis* isomers of β-carotene (and other carotenoids) do occur, however, especially 9-*cis*-β-carotene. However, all-*trans*-β-carotene is preferentially absorbed by the human gut. Heating of tomato juice isomerizes some of the lycopene present from all-*trans* to 9-*cis* and 13-*cis* isomers; both *cis* and *trans* forms can be absorbed.

As would be expected from their chemical structures (Figs 3.31 and 3.32), carotenoids are very hydrophobic structures, completely insoluble in water. Carotenoids in mammalian blood are located in the circulating lipoproteins. In tissues they occur within fat stores, in the hydrophobic interior of membranes and bound to hydrophobic domains of certain proteins.

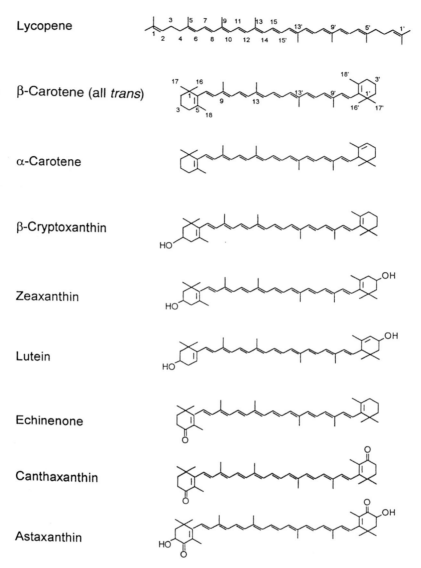

Fig. 3.32. The basic structure of carotenoids. II. Structures of several carotenoids found in plants and animals. Structures by courtesy of Professors Catherine Rice-Evans and Norman Krinsky.

3.23.2   *Metabolic roles of carotenoids*[146]

Carotenoids have been suggested to have beneficial effects on the immune system, but their best-established role is to serve as a precursor of the fat-soluble vitamin known as **vitamin A**, or **retinol**. Vitamin A is essential for cell growth and differentiation, and in vision; vitamin A deficiency is the leading cause of childhood blindness in the world.

Over 50 carotenoids can generate vitamin A, although β-carotene is probably the most important in humans. Indeed, carotenoids are often the major dietary source of vitamin A in humans. Some ingested carotenoids are acted upon by dioxygenase enzymes, largely in the gut but also to some extent in the liver and other organs, to generate the aldehyde **retinal**, which can be reduced to retinol (or oxidized to retinoic acid) in many tissues. As well as this 'central cleavage' of carotenoids by dioxygenases, other carotenoid-cleaving enzymes (causing cleavage at other sites: **excentric cleavage**) have been described.

Epidemiological studies show that high blood levels of carotenoids are associated with a decreased incidence of certain forms of cancer, e.g. lung cancer (Chapter 10). Carotenoids may have direct anti-cancer effects; for example, *in vitro* they can facilitate cell–cell communication via 'gap junctions'[k] (by stimulating synthesis of the **connexin proteins**).[30] Increased communication decreases growth of transformed cells in culture. Carotenoids can decrease tumour production by administered carcinogens in animals (especially skin tumours in animals exposed to ultraviolet light). However, the relevance of these effects *in vivo* is uncertain. One major problem in cell studies is how to deliver the highly insoluble carotenoids to the cell, growing in an aqueous medium. Often such organic solvents as tetrahydrofuran are used. Another possibility to be considered is that high body levels of carotenoids merely reflect diets rich in plants, which contain many anti-cancer agents (Chapter 10).

3.23.3 *Carotenoids as antioxidants*

In plants, carotenoids play a key antioxidant role, helping to prevent the formation of, and quench, ROS (especially singlet O_2) formed during photosynthesis (Chapter 7). Indeed, β-carotene administration is protective against light-induced skin damage in patients with porphyria (Chapter 2).[167] Sunlight depletes β-carotene in skin, consistent with a protective role in normal subjects.[34]

In vitro studies have shown that β-carotene inhibits peroxidation of simple lipid systems at low O_2 concentration, but not at high O_2 concentration. However, studies with LDL show that β-carotene does not protect them against peroxidation whatever the O_2 concentration.[116] Although carotenoids are powerful quenchers/scavengers of singlet O_2 (Chapter 2), how important this would be to healthy animals is uncertain, although $^1O_2$ can form during lipid peroxidation. Lycopene appears to be the best singlet O_2 quencher *in vitro*, although all the carotenoids are very good. Exposure to sunlight can decrease carotenoid levels in plasma and skin, and scavenging by carotenoids could be important in the eye (Chapter 7).

In vitro studies also show the potential of carotenoids to act as free-radical scavengers.[79,157] There have been suggestions that vitamin A can scavenge some free radicals *in vitro* (as would be expected from its double-bond structure). Oxidizing radicals can react with carotenoids by electron transfer, e.g. for nitrogen dioxide reacting with β-carotene (written Car below) a radical cation is produced

$$NO_2^\bullet + Car \rightarrow Car^{\bullet+} + NO_2^- \qquad k \approx 1 \times 10^8\,M^{-1}\,s^{-1}$$

Possible fates of $Car^{\bullet+}$ include dismutation

$$2Car^{\bullet+} \rightleftharpoons Car + Car^{2+}$$

and reaction with ascorbate (if $Car^{\bullet+}$ was present at a membrane surface to interact with the hydrophilic ascorbate)

$$Car^{\bullet+} + Asc \rightarrow Car + Asc^\bullet + H^+$$

Peroxyl radicals can react with carotenoids, e.g.

$$CCl_3O_2^\bullet + Car \rightarrow Car^{\bullet+} + CCl_3O_2^-$$

or by addition reactions

$$Car + RO_2^\bullet \rightarrow [Car-OOR]^\bullet$$

These addition products could intercept another radical, or react with O_2 to give a peroxyl radical

$$[Car-OOR]^\bullet + R^1O_2^\bullet \longrightarrow R^1 - OOCar - OOR$$
$$\downarrow$$
$$\text{decomposition products}$$

$$[Car-OOR]^\bullet + O_2 \rightarrow [OO-Car-OOR]^\bullet$$

Thiyl ($RS^\bullet$) radicals undergo addition reactions with carotenoids (k for $GS^\bullet \approx 2 \times 10^8\,M^{-1}\,s^{-1}$).

$$RS^\bullet + Car \rightarrow [Car-SR]^\bullet$$

possibly followed by O_2 addition to give a peroxyl radical

$$[Car-SR]^\bullet + O_2 \rightleftharpoons [RS-Car-OO]^\bullet$$

Carotenoids might also react with $OH^\bullet$, and perhaps some other species, by hydrogen atom donation

$$OH^\bullet + CarH \rightarrow Car^\bullet + H_2O$$

The carbon-centred ($Car^\bullet$) radicals are fairly stable because of extensive electron delocalization. Hence reaction with O_2 to give peroxyl radicals

$$Car^\bullet + O_2 \rightleftharpoons CarO_2^\bullet$$

is slow. $Car^\bullet$ could then add another radical, e.g.

$$Car^\bullet + RO_2^\bullet \rightarrow Car-OOR \text{ (non-radical product)}$$

Radicals such as $[OO-Car-OOR]^\bullet$, and possibly $CarO_2^\bullet$, could propagate free-radical chain reactions such as lipid peroxidation by abstracting hydrogen. Hence it is possible to explain how O_2 concentration can affect the antioxidant/pro-oxidant properties of carotenoids *in vitro*. Pure carotenoids, even as solids, are susceptible to oxidation and can break down to a complex mixture of products, as is evidenced by loss of the characteristic colour (**bleaching**).

The rate constants for reaction of carotenoids with these various radicals are high, suggesting that carotenoids in membranes could be capable of reacting with them. Whether or not such reactions would protect the surrounding PUFAs and proteins largely depends on the reactivities of the various carotenoid-derived radicals that can be formed (which is affected by the O_2 concentration). Membranes also contain higher levels of vitamin E and ubiquinol as free-radical scavengers, although interactions of vitamin E with carotenoids might occur, e.g. if carotenoids react with tocopheryl radical

$$H^+ + \alpha\text{-Toc}^\bullet + Car \rightarrow \alpha\text{-TocH} + Car^{\bullet +}$$

although, depending on the carotenoid, the reverse reaction may be more favourable.[174a]

It is difficult as yet to be certain whether carotenoids do play any net antioxidant role *in vivo*: their biological effects may be exerted through other mechanisms.

3.24 Plant phenols[215,231]

A phenol contains an $-OH$ group attached to a benzene ring. Plants contain a huge range of phenols (Table 3.16), including, of course, tocopherols and tocotrienols. Many phenols other than vitamin E exert powerful antioxidant effects *in vitro*, inhibiting lipid peroxidation by acting as chain-breaking peroxyl-radical scavengers. Phenols with two adjacent $-OH$ groups, or other chelating structures, can also bind transition metal ions (especially iron and copper) in forms poorly active in promoting free-radical reactions; this chelating ability can interfere with metal absorption from the diet. Phenols can also directly scavenge ROS, such as $OH^\bullet$, $ONOOH$ and $HOCl$. Thus, unlike β-carotene, many plant phenolics are good inhibitors of lipid peroxidation *in vitro*. Sometimes, however, like vitamin C, phenols can reduce transition metal ions and exert pro-oxidant effects *in vitro* although the significance of this *in vivo* is unknown. **Thymol** (Fig. 3.33) is often used as an antiseptic and antioxidant in commercial enzyme suspensions (e.g. of catalase).

3.24.1 *Phenols in the diet*

Within plants, some phenols serve as important biosynthetic precursors (e.g. **caffeic acid** is a precursor of lignin; Chapter 6) whereas others may be produced to absorb ultraviolet radiation. Indeed, mutants of the plant *Arabidopsis* unable to synthesize phenols showed greater oxidative damage by ultraviolet

Table 3.16. Some dietary sources of plant phenolics

Compound	Some sources
Flavanols	
epicatechin	green and black teas
catechin	red wine
epigallocatechin	
epicatechin gallate	
epigallocatechin gallate	
Flavanones	
naringin	peel of citrus fruits
taxifolin	citrus fruits
Flavonols	
kaempferol	endive, leek, broccoli, radish, grapefruit, black tea
quercetin	onion, lettuce, broccoli, cranberry, apple skin, berries, olive, tea, red wine
myricetin	cranberry, grapes, red wine
Flavones	
chrysin	fruit skin
apigenin	celery, parsley
Anthocyanidins	
malvidin	red grapes, red wine
cyanidin	cherry, raspberry, strawberry, grapes
apigenidin	coloured fruit and peels
Phenylpropanoids	
caffeic acid	white grapes, white wine, olives, olive oil, spinach, cabbage, asparagus, coffee
p-coumaric acid	white grapes, white wine, tomatoes, spinach, cabbage, asparagus
chlorogenic acid	apples, pears, cherries, plums, peaches, apricots, blueberries, tomatoes, anise

Adapted from *Free Rad. Biol. Med.* **20**, 933 (1996) by courtesy of Professor C. Rice-Evans and Elsevier (Amsterdam).

light.[148] Other phenols, especially the red/blue **anthocyanins**, and the yellow **aurones** and **chalcones**, may attract pollinating insects. Soybean is rich in tocopherols, isoflavone glycosides (e.g. **genistein** and **daidzein**; Fig. 3.33) and caffeic acid. **Taxifolin** (Fig. 3.34) has been identified as an antioxidant in peanut extracts. An antioxidant activity measured in rice extracts was identified as a *C*-glycosyl flavonoid, **isovitexin**. Sesame seed oil contains **sesamol**, esters of caffeic acid (Fig. 3.33) are present in large amounts in canary seed, and both green and black teas are rich in **catechins** (Fig. 3.35) which contribute much of the *in vitro* antioxidant activity of teas. Catechin, epicatechin and gallic acid are also significant contributors to the total antioxidant activity of red wine.[215] Cottonseed oil contains a yellow polyphenol, **gossypol** (Fig. 3.33), which has spermicidal effects and has undergone clinical trials as a male contraceptive.

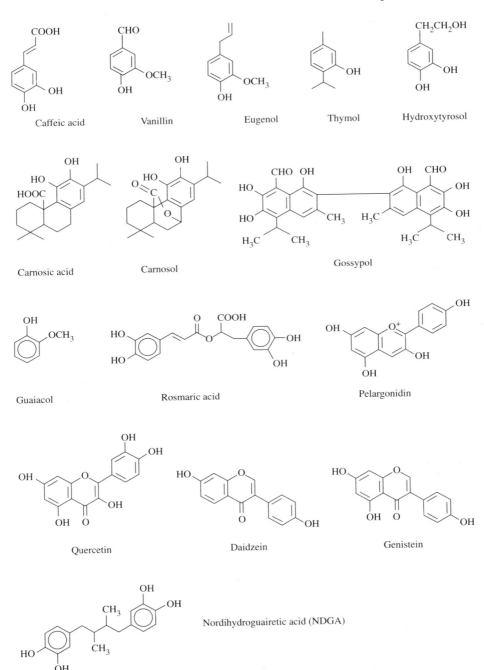

Fig. 3.33. Structure of some plant phenolics. The isoflavones, such as genistein and daidzein, have a weak anti-oestrogen-like activity, and contribute to the ability of certain plants (eaten in excessive amounts) to interfere with reproduction in farm animals.

Generic structure	Flavonoid	Hydroxylation pattern								
		2'	3'	4'	5'	3	5	6	7	8
Flavane	Catechins Meciadonol		•	•		Methoxy	•		•	•
Flavanone	Taxifolin Naringenin Naringin		•	• •		•	• •		• Rhamno-glucoside	
Flavone	Luteolin Apigenin		•	•			•		•	
Flavonol	Quercetin Myricetin Gossypetin Fisetin Cirsiliol Morin Kaempferol Galangin Baicalein Rutin Quercetrin Gossypin	•	• • • • • • •	• • • • • • •	•	• • • • H • • H Rutinose Rhamnose •	• • • • • • • • • • • •	•	• • • • Methoxy • • • • • •	• Glycoside

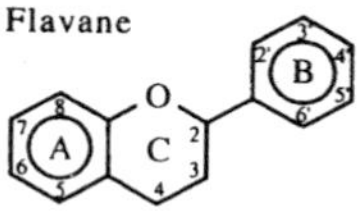

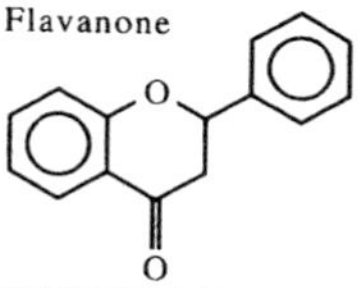

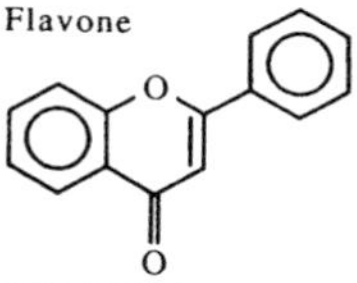

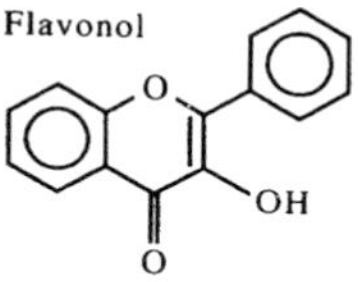

Fig. 3.34. Structure and hydroxylation pattern of some members of the flavonoid family. Taken from *Food Chem. Toxicol.* **33**, 1061 (1995) by courtesy of Dr Joe Formica and Elsevier, Amsterdam. The family members include flavones, isoflavones, 3-hydroxyflavones (flavonols), flavanones, anthocyanidins and chalcones. The flavonoids contain a three-ring structure (A, C, B): variations in the six-membered ring C and type of substituents produce the different members of the flavonoid family.

Herbs and spices have been used for many years to preserve foods;[201] in the book of Exodus in the Old Testament it is stated that spices were added to oil to keep it fresh. Herbs and spices are rich sources of antioxidants: extracts of sage, rosemary, peppers, tarragon, ginger, thyme and oregano all inhibit lipid peroxidation *in vitro*, largely due to the phenols present. Examples are **carnosic acid** and **rosmaric acid** in rosemary, **dehydrozingerone** in ginger and five phenolic acid amides in black pepper. Onion, wine and tea are rich in the flavonoid **quercetin** (Fig. 3.34). Oil of cloves contains **eugenol** (Fig. 3.33), sometimes added to dental materials as an analgesic and antiseptic. Wood smoke contains various phenols, such as **guaiacol** (Fig. 3.33), that can inhibit lipid peroxidation in smoked foods. Olives contain **hydroxytyrosol** (Fig. 3.33). Further examples of the widespread distribution of phenolic compounds in plants are given in Table 3.16.

Plant phenols ingested in the diet are generally thought to be poorly absorbed from the gut by animals and are largely excreted faecally, although there is growing evidence that some are taken up.[125] For example, studies on human volunteers with ileostomies found that about 25% of quercetin was absorbed; but about 50% of the quercetin present in a meal of fried onions was taken up. Often phenols are present in plants attached to sugars (as **glycosides**) in the plant (e.g. most quercetin in onions): these can be hydrolysed in the gut by glycosidase enzymes. However, in the ileostomy study about 17% of quercetin rutinoside was absorbed. Of course, measurement of absorption as the amount eaten minus the amount recovered in ileostomy bags does not necessarily mean that the missing quercetin was delivered to tissues or body fluids at sufficient levels to exert biological effects. Nevertheless, HPLC analysis[153,194] has revealed the presence of several phenols (and their glycosides) in human blood plasma, at total levels that can approach 1 μM.

3.24.2   *Are plant phenols antioxidants* in vivo?

Two recent observations drew attention to the potential *in vivo* antioxidant effects of plant phenols. First, phenols in red wine were found to inhibit LDL oxidation *in vitro*, and it was suggested that they might exert an important cardioprotective effect by limiting LDL oxidation *in vivo*. This was further suggested as an explanation of the lower incidence of heart attacks in certain areas of France (the '**French paradox**') despite the high prevalence of factors promoting cardiovascular disease, such as smoking and high fat intake.[88] Alcohol alone, however, has some cardioprotective effect (at moderate intakes). Teetotallers may be pleased to learn that phenols in cocoa and chocolate can also inhibit LDL oxidation *in vitro*.[139]

Second, an epidemiological study in the Netherlands (the **Zutphen study**)[134] suggested an inverse relation of the incidence of coronary heart disease and stroke in elderly men with the dietary intake of flavonoids (especially quercetin), which originated mainly from tea, fruits (e.g. apples) and vegetables (e.g. onions) in the population examined. Similar studies have since been published from Finland.

Several plant phenols can inhibit LDL oxidation *in vitro*, e.g. caffeic acid and hydroxytyrosol (Fig. 3.33). However, more attention has been paid to the **flavonoids** (such as quercetin), whose general structures and classification are shown in Fig. 3.34. Like other phenols, flavonoids frequently occur as glycosides. For example, quercetin can be linked to the sugar rhamnose (giving **quercetrin**) or rutinose (to give **rutin**). However, glucose is more usually attached.

Flavonoids are widely distributed in plants and other plant products and average total daily consumption in the Netherlands[134] was estimated as at least 23 mg, of which 16 mg/day is quercetin, more than the average daily intake of vitamin E (7–10 mg). *In vitro*, flavonoids are often powerful inhibitors of lipid peroxidation, ROS/RNS scavengers, inhibitors of damage by haem protein/peroxide mixtures, metal ion binding agents and inhibitors of lipoxygenase and cyclooxygenase enzymes. The degree of hydroxylation and relative positions of –OH groups are of primary importance in determining antioxidant ability. In isolated cells, some flavonoids have been reported to exert anti-cancer effects, prevent expression of adhesion molecules and inhibit replication of HIV. In whole animals, administration of flavonoids has been reported to exert various anti-inflammatory and anti-cancer effects.[208]

However, much further research is needed to assess the physiological importance of plant phenolics as antioxidants in the human body. Possible biological effects of any flavonoid-derived radicals generated during *in vivo* antioxidant activity must also be considered, as is illustrated by **gossypol** and **eugenol**. Both inhibit lipid peroxidation *in vitro*, but the resulting radicals can exert cytotoxic effects to other molecular targets such as DNA (again *in vitro*). Quercetin (and possibly other flavonoid-derived) phenoxyl radicals can be reduced back to quercetin by ascorbate *in vitro*.

Not all the biological effects of phenols are necessarily related to antioxidant activity; **genistein** and **daidzein** and perhaps some other flavonoids, oppose oestrogen action and inhibit protein kinases,[86,171] decreasing cell proliferation (see legend to Fig. 3.33). Nor are the effects of phenolics necessarily all beneficial: several phenolics inhibit[66] thyroid peroxidase activity *in vitro*, which could give them a potential anti-thyroid effect if sufficient were absorbed. Many phenolics inhibit certain cytochromes P450. Several flavonoids inhibit protein kinases and genistein may block the growth of new blood vessels (an **anti-angiogenic effect**). Of course, some of these effects could be beneficial in the context of preventing cancer development.

3.24.3   *Herbal medicines*[190,256]

There is also considerable interest in the development of flavonoids and their derivatives for therapeutic use (Chapter 10), e.g. as anti-inflammatory, anti-ischaemic and anti-thrombotic agents. An extract of the ornamental tree *Ginkgo biloba* has been used in Chinese herbal medicine for thousands of years: the extract has antioxidant properties, apparently largely due to the flavonoids present, which include rutin, kaempferol, quercetin and myricetin. Green and black tea decrease tumorigenesis in carcinogen-treated rodents, and so there is much interest in the potential anti-cancer effects of the catechins (Fig. 3.35). Nevertheless, the oxidation of phenolic compounds in tea and coffee[182] on standing generates H_2O_2, which reaches levels of 20–160 µM in coffee and about 50 µM in black tea.

Kampo medicines are traditional medicines in Japan; they are extracts of multiple herbs and contain a complex mixture of phenols and other compounds, including **glycyrrhizin** from roots of the licorice plant, *Glycyrrhiza glabra*. Extracts of **propolis**, a resinous substance collected by bees (Chapter 1) have often been used in herbal medicine, and contain many phenolic and other plant-derived compounds.

It should never be assumed that natural products are safe to administer in large quantities just because they are natural. Examples of noxious agents

gallic acid   catechin   epigallocatechin

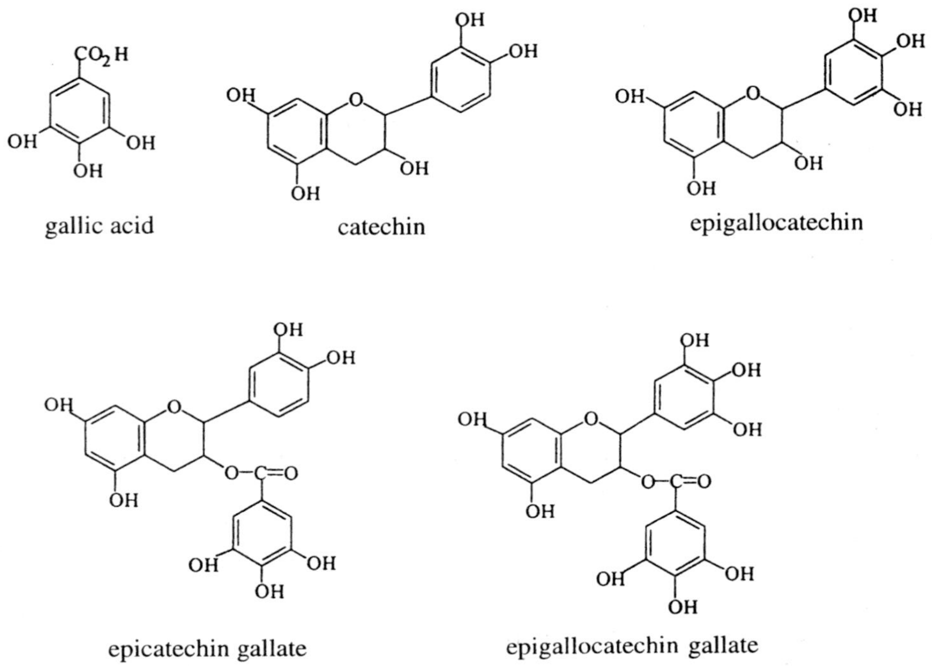

epicatechin gallate   epigallocatechin gallate

Fig. 3.35. Structure of gallic acid and the catechins. Adapted from *Free Rad. Biol. Med.* **20**, 933 (1996) by courtesy of Professor C. Rice-Evans and Elsevier, Amsterdam.

produced by plants include cyanide and aflatoxin (Chapter 9). In addition, nordihydroguairetic acid (Fig. 3.33), a powerful inhibitor of lipoxygenase-catalysed and non-enzymic lipid oxidation isolated from the creosote bush *Larrea divaricata*, is no longer used as a food preservative because of adverse toxicological reports.

References

1. Abeliovich, A and Shilo, M (1972) Photooxidative death in blue-green algae. *J. Bacteriol.* **111**, 682.
2. Abrahamsson, T *et al.* (1992) Vascular bound recombinant ECSOD type C protects against the detrimental effects of $O_2^{\bullet -}$ on endothelium-dependent arterial relaxation. *Circul. Res.* **70**, 264.
3. Akanmu, D *et al.* (1991) The antioxidant action of ergothioneine. *Arch. Biochem. Biophys.* **288**, 10.
4. Allgood, GS and Perry, JJ (1986) Characterization of a Mn-containing catalase from the obligate thermophile. *Thermoleophilum album. J. Bacteriol.* **168**, 563.
5. Almagor, M *et al.* (1984) Role of $O_2^{\bullet -}$ in host cell injury induced by *Mycoplasma pneumoniae* infection. A study in normal and trisomy 21 cells. *J. Clin. Invest.* **73**, 842.

6. Ames, BN *et al.* (1981) Uric acid provides an antioxidant defense in humans against oxidant- and radical-caused aging and cancer. A hypothesis. *Proc. Natl Acad. Sci. USA* **78**, 6858.

7. Archibald, FS (1985) Manganese: its acquisition by and function in the lactic acid bacteria. *CRC Crit. Rev. Microbiol.* **13**, 63.

8. Archibald, FS and Duong, MN (1986) SOD and oxygen toxicity defenses in the genus *Neisseria. Infect. Immun.* **51**, 631.

9. Arner, ESJ *et al.* (1995) 1-Chloro-2,4-dinitrobenzene is an irreversible inhibitor of human thioredoxin reductase. *J. Biol. Chem.* **270**, 3479.

10. Aruoma, OI *et al.* (1989) Carnosine, homocarnosine and anserine: could they act as antioxidants *in vivo? Biochem. J.* **264**, 863.

11. Asmus, KD *et al.* (1996) One-electron oxidation of ergothioneine and analogues investigated by pulse radiolysis: redox reaction involving ergothioneine and vitamin C. *Biochem. J.* **315**, 625.

12. Awad, JA *et al.* (1994) Detection and localization of lipid peroxidation in Se- and vitamin E-deficient rats using F_2-isoprostanes. *J. Nutr.* **124**, 810.

13. Babizhayev, MA *et al.* (1994) L-Carnosine (β-alanyl-L-histidine) and carcinine (β-alanylhistamine) act as natural antioxidants with $OH^\bullet$ scavenging and lipid-peroxidase activities. *Biochem. J.* **304**, 509.

14. Baeuerle, PA (1995) Enter a polypeptide messenger. *Nature* **373**, 661.

15. Bal, W *et al.* (1996) Interactions of nickel(II) with histones: enhancement of 2′-deoxyguanosine oxidation by Ni(II) complexes with CH_3CO-Cys-Ala-Ile-His-NH_2, a putative metal binding sequence of histone H_3. *Chem. Res. Toxicol.* **9**, 535.

16. Baliga, R *et al.* (1996) Evidence for cytochrome P-450 as a source of catalytic iron in myoglobinuric acute renal failure. *Kidney Int.* **49**, 362–369.

17. Balla, J *et al.* (1993) Endothelial-cell heme uptake from heme proteins: induction of sensitization and desensitization to oxidant damage. *Proc. Natl Acad. Sci. USA* **90**, 9285.

18. Balzan, R *et al.* (1995) *E. coli* FeSOD targeted to the mitochondria of yeast cells protects the cells against oxidative stress. *Proc. Natl Acad. Sci USA* **92**, 4219.

19. Bannister, JV and Calabrese, L (1987) Assays for SOD. *Methods Biochem. Anal.* **32**, 279.

20. Bao, Y *et al.* (1997) Reduction of thymine hydroperoxide by phospholipid hydroperoxide glutathione peroxidase and glutathione transferases. *FEBS Lett.* **410**, 210.

21. Barhoumi, R *et al.* (1993) Concurrent analysis of intracellular GSH content and gap junctional intercellular communication. *Cytometry* **14**, 747.

22. Bates, GW and Schlabach, MR (1973) The reaction of ferric salts with transferrin. *J. Biol. Chem.* **248**, 3228.

23. Battistoni, A *et al.* (1996) The CuZnSOD from *E. coli* retains monomeric structure at high protein concentration. *Biochem. J.* **320**, 713.

24. Beaumont, C *et al.* (1995) Mutation in the IRE of the L-ferritin mRNA in a family with dominant hyperferritinaemia and cataract. *Nature Genet.* **11**, 444.

25. Beck, MA *et al.* (1995) Rapid genomic evolution of a non-virulent Coxsackievirus B3 in selenium-deficient mice results in selection of identical virulent isolates. *Nature Med.* **1**, 433.

26. Beckman, JS (1996) Oxidative damage and tyrosine nitration from $ONOO^-$. *Chem. Res. Toxicol.* **9**, 836.

27. Bellis, SL *et al.* (1994) Affinity purification of *Hydra* glutathione binding proteins. *FEBS Lett.* **354**, 320.

28. Bendich, A *et al.* (1986) The antioxidant role of vitamin C. *Adv. Free Rad. Biol. Med.* **2**, 419.

29. Benov, L and Fridovich, I (1996) Functional significance of the CuZnSOD in *E. coli*. *Arch. Biochem. Biophys.* **327**, 249.

29a. Bertini, I *et al.* (1998) Structure and properties of CuZnSODs. *Adv. Inorg. Chem.* **45**, 127.

30. Bertram, JS (1993) Inhibition of chemically-induced neoplastic transformation by carotenoids. *Ann. N.Y. Acad. Sci.* **686**, 161.

31. Beyer, RE *et al.* (1994) The relative essentiality of the antioxidative function of coenzyme Q—the interactive role of DT-diaphorase. *Mol. Asp. Med.* **15**, S117.

32. Beyer, WF, Jr and Fridovich, I (1985) Pseudocatalase from *Lactobacillus plantarum*: evidence for a homopentameric structure containing two atoms of manganese per subunit. *Biochemistry* **24**, 6460.

33. Beyer, WF, Jr and Fridovich, I (1987) Assaying for SOD: some large consequences of minor changes in conditions. *Anal. Biochem.* **161**, 559.

34. Biesalski, HK *et al.* (1996) Effects of controlled exposure of sunlight on plasma and skin levels of β-carotene. *Free Rad. Res.* **24**, 215.

35. Björnstedt, M *et al.* (1994) The thioredoxin and glutaredoxin systems are efficient electron donors to human plasma glutathione peroxidase. *J. Biol. Chem.* **269**, 29382.

36. Björnstedt, M *et al.* (1995) Selenite and selenodiglutathione: reactions with thioredoxin systems. *Methods Enzymol.* **252**, 209.

37. Bolann, BJ and Ulvik, RJ (1993) Stimulated decay of $O_2^{\bullet-}$ caused by ferritin-bound copper. *FEBS Lett.* **328**, 263.

38. Breuer, W *et al.* (1996) Dynamics of the cytosolic chelatable pool of K562 cells. *FEBS Lett.* **382**, 304.

39. Brigelius, R *et al.* (1983) Identification and quantitation of glutathione in hepatic protein mixed disulphides and its relationship to GSSG. *Biochem. Pharmacol.* **32**, 2529.

40. Britigan, BE *et al.* (1993) Transferrin and lactoferrin undergo proteolytic cleavage in the *Pseudomonas aeruginosa*-infected lungs of patients with cystic fibrosis. *Infect. Immun.* **61**, 5049.

41. Britton, G (1995) Structure and properties of carotenoids in relation to function. *FASEB J.* **9**, 1551.

41a. Brouwer, M *et al* (1997) The paradigm that all O_2-respiring eukaryotes have cytosolic CuZnSOD and that MnSOD is localized to the mitochondria does not apply to a large group of marine arthropods. *Biochem.* **36**, 13381.

42. Brown, TA and Shrift, A (1982) Selenium: toxicity and tolerance in higher plants. *Biol. Rev.* **57**, 59.

43. Bulitta, C *et al.* (1996) Cytoplasmic and peroxisomal catalases of the guinea pig liver: evidence for two distinct proteins. *Biochim. Biophys. Acta* **1293**, 55.

43a. Burton, GW *et al* (1990) Biokinetics of dietary RRR-α-tocopherol in the male guinea-pig at three dietary levels of vitamin C and two levels of vitamin E. *Lipids* **25**, 199.

44. Buzadzic, B *et al.* (1990) Antioxidant defenses in the ground squirrel *Citellus citellus*. 2. The effect of hibernation. *Free Rad. Biol. Med.* **9**, 407.

45. Cantoni, O *et al.* (1994) The L-histidine mediated enhancement of H_2O_2-induced cytotoxicity is a general response in cultured mammalian cell lines and is always associated with the formation of DNA double strand breaks. *FEBS Lett.* **353**, 75.

45a. Carelli, S *et al.* (1997) Cysteine and GSH secretion in response to protein disulfide bond formation in the ER. *Science* **277**, 1681.

46. Carlsson, J (1987) Salivary peroxidase: an important part of our defense against O_2 toxicity. *J. Oral Pathol.* **16**, 412.

47. Carlsson, LM *et al.* (1995) Mice lacking ECSOD are more sensitive to hyperoxia. *Proc. Natl Acad. Sci. USA* **92**, 6264.

48. Cha, M-K and Kim, IH (1995) Thioredoxin-linked peroxidase from human red blood cells. *Biochem. Biophys. Res. Commun.* **217**, 900.

49. Cha, M-K *et al.* (1996) Mutation and mutagenesis of thiol peroxidase of *E. coli* and a new type of thiol peroxidase family. *J. Bacteriol.* **178**, 5610.

50. Chae, HZ *et al.* (1994) Cloning and sequencing of thiol-specific antioxidant from mammalian brain: alkyl hydroperoxide reductase and thiol-specific antioxidant define a large family of antioxidant enzymes. *Proc. Natl Acad. Sci. USA* **91**, 7017.

51. Chance, B *et al.* (1979) Hydroperoxide metabolism in mammalian organs. *Physiol. Rev.* **59**, 527.

52. Chang, EC and Kosman, DJ (1990) O_2-dependent methionine auxotrophy in CuZnSOD-deficient mutants of *Saccharomyces cerevisiae*. *J. Bacteriol* **172**, 1840.

52a. Chaudiere, J *et al.* (1984) Mechanism of Se-GPX and its inhibition by mercapto-carboxylic acids and other mercaptans. *J. Biol Chem.* **259**, 1043.

53. Chedekel, MR *et al.* (1978) Photodestruction of pheomelanin: role of O_2. *Proc. Natl Acad. Sci. USA* **75**, 5395.

54. Chubatsu, LS and Meneghini, R (1993) Metallothionein protects DNA from oxidative damage. *Biochem. J.* **291**, 193.

55. Crawford, DHG *et al.* (1996) Factors influencing disease expression in hemo-chromatosis. *Annu. Rev. Nutr.* **16**, 139.

56. Crichton, RR and Charloteaux-Waters, M (1987) Iron transport and storage. *Eur. J. Biochem.* **164**, 485.

57. Daniels, V (1989) Oxidative damage and the preservation of organic artefacts. *Free Rad. Res. Commun.* **5**, 213.

58. Darr, D and Fridovich, I (1986) Irreversible inactivation of catalase by 3-amino-1,2,4-triazole. *Biochem. Pharmacol.* **35**, 3642.

59. De Boer, E *et al.* (1986) Bromoperoxidase from *Ascophyllum nodosum:* a novel class of enzymes containing vanadium as a prosthetic group? *Biochim. Biophys. Acta* **869**, 48.

60. De Croo, S *et al.* (1988) Isoelectric focusing of SOD: report of the unique SOD A*2 allele in a US white population. *Human Hered.* **38**, 1.

61. de Haan, JB *et al.* (1996) Elevation in the ratio of Cu/Zn-SOD to glutathione peroxidase activity induces features of cellular senescence and this effect is mediated by H_2O_2. *Human Mol. Genet.* **5**, 283.

62. De Master, EG *et al.* (1984) The metabolic activation of cyanamide to an inhibitor of aldehyde dehydrogenase is catalyzed by catalase. *Biochem. Biophys. Res. Commun.* **122**, 358.

63. De Rosa, G *et al.* (1980) Regulation of SOD activity by dietary manganese. *J. Nutr.* **110**, 795.

64. Dhindsa, RS (1987) Glutathione status and protein synthesis during drought and subsequent rehydration in *Tortula ruralis*. *Plant Physiol.* **83**, 816.

65. Dierenfeld, ES (1989) Vitamin E deficiency in zoo reptiles, birds and ungulates. *J. Zoo Wildlife Med.* **20**, 3.

65a. Diplock, AT (1985) *Fat-soluble Vitamins* Heineman, London.

66. Divi, RL and Doerge, DR (1996) Inhibition of thyroid peroxidase by dietary flavonoids. *Chem. Res. Toxicol.* **9**, 16.

67. Do, TQ *et al.* (1996) Enhanced sensitivity of ubiquinone-deficient mutants of *S. cerevisiae* to products of autoxidized PUFAs. *Proc. Natl Acad. Sci. USA* **93**, 7534.

68. Dowd, P and Zheng, ZB (1995) On the mechanism of the anticlotting action of vitamin E quinone. *Proc. Natl Acad. Sci. USA* **92**, 8171.

69. Drake, IM *et al.* (1996) Ascorbic acid may protect against human gastric cancer by scavenging mucosal oxygen radicals. *Carcinogenesis* **17**, 559.

70. Duffey, SS and Blum, MS (1977) Phenol and guaiacol: biosynthesis, detoxication and function in a polydesmid millipede, *Oxidus gracilis*. *Insect. Biochem.* **7**, 57.

71. Dukan, S and Touati, D (1996) HOCl stress in *E. coli*: resistance, DNA damage, and comparison with H_2O_2 stress. *J. Bacteriol.* **178**, 6145.

72. Dunn, MA *et al.* (1987) Metallothionein. *Proc. Soc. Exp. Biol. Med.* **185**, 107.

73. Dykens, JA and Shick, JM (1984) Photobiology of the symbiotic sea anemone, *Anthopleura elegantissima*: defences against photodynamic effects, and seasonal photoacclimatization. *Biol. Bull.* **167**, 683.

74. Eddy, L *et al.* (1990) Reduction of ferrylmyoglobin in rat diaphragm. *Am. J. Physiol.* **259**, C995.

75. Ehrenwald, E and Fox, PL (1994) Isolation of nonlabile human ceruloplasmin by chromatographic removal of a plasma metalloproteinase. *Arch. Biochem. Biophys.* **309**, 392.

76. Ennever, JF *et al.* (1987) Rapid clearance of a structural isomer of bilirubin during phototherapy. *J. Clin. Invest.* **79**, 1674.

77. Ernster, L and Dallner, G (1995) Biochemical, physiological and medical aspects of ubiquinone function. *Biochim. Biophys. Acta* **1271**, 195.

78. Evans, PJ *et al.* (1989) Non-caeruloplasmin copper and ferroxidase activity in mammalian serum. *Free Rad. Res. Commun.* **7**, 55.

79. Everett, SA *et al.* (1996) Scavenging of NO_2, thiyl and sulfonyl free radicals by the nutritional antioxidant β-carotene. *J. Biol. Chem.* **271**, 3988.

80. Fairlamb, A (1966) Pathways to drug discovery. *Biochemist* Feb/March issue, pp. 11–16.

81. Fang, X *et al.* (1995) Generation and reactions of the disulphide radical anion derived from metallothionein: a pulse radiolytic study. *Int. J. Rad. Biol.* **68**, 459.

82. Farr, SB *et al.* (1986) Oxygen-dependent mutagenesis in *Escherichia coli* lacking SOD. *Proc. Natl Acad. Sci. USA* **83**, 8268.

83. Felix, K *et al.* (1993) A pulse radiolytic study on the reaction of $OH^{\bullet}$ and $O_2^{\bullet-}$ radicals with yeast Cu(I) thionein. *Biochim. Biophys. Acta* **1203**, 104.

83a. Farrant, JL *et al.* (1997) Bacterial Cu- and Zn-cofactored SOD contributes to the pathogenesis of systemic salmonellosis *Molec. Microbiol.* **25**, 785.

84. Fernando, MR *et al.* (1992) Thioredoxin regenerates proteins inactivated by oxidative stress in endothelial cells. *Eur. J. Biochem.* **209**, 917.

85. Flint, DH *et al.* (1993) The inactivation of Fe–S cluster containing hydro-lyases by $O_2^{\bullet-}$. *J. Biol. Chem.* **268**, 22369.

86. Fotsis, T *et al.* (1993) Genistein, a dietary-derived inhibitor of *in vitro* angiogenesis. *Proc. Natl Acad. Sci. USA* **90**, 2690.

87. Frank, L (1985) Effects of O_2 on the newborn. *Fed. Proc.* **44**, 2328.

88. Frankel, EN *et al.* (1995) Principal phenolic phytochemicals in selected California wines and their antioxidant activity in inhibiting oxidation of human LDL. *J. Agric. Food Chem.* **43**, 890.

89. Fridovich, I (1995) Superoxide radical and superoxide dismutases. *Annu. Rev. Biochem.* **64**, 97.

90. Gabbianelli, R *et al.* (1995) Metal uptake of recombinant cambialistic SOD from *Propionibacterium shermanii* is affected by growth conditions of host *E. coli* cells. *Biochem. Biophys. Res. Commun.* **216**, 841.

91. Gaetani, GF *et al.* (1995) Importance of catalase in the disposal of H_2O_2 within human erythrocytes. *Blood* **84**, 325.

91a. Gajhede, M *et al.* (1997) Crystal structure of horseradish peroxidase C at 2.15 Å resolution. *Nature Struct. Biol.* **4**, 1032.

92. Gardner, PR *et al.* (1994) Aconitase is a sensitive and critical target of O_2 poisoning in cultured mammalian cells and in rat lungs. *Proc. Natl Acad. Sci. USA* **91**, 12248.

93. Gaudu, P *et al.* (1996) The irreversible inactivation of ribonucleotide reductase from *E. coli* by $O_2^{\bullet-}$. *FEBS Lett.* **387**, 137.

94. Gladyshev, VN *et al.* (1996) Selenocysteine, identified as the penultimate C-terminal residue in human T-cell thioredoxin reductase, corresponds to TGA in the human placental gene. *Proc. Natl Acad. Sci. USA* **93**, 6146.

95. González-Flecha, B and Demple, B (1995) Metabolic sources of H_2O_2 in aerobically growing *E. coli*. *J. Biol. Chem.* **270**, 13681.

96. Gordon, T (1986) Purity of catalase preparations: contamination by endotoxin and its role in the inhibition of airway inflammation. *Free Rad. Biol. Med.* **2**, 373.

97. Grill, E *et al.* (1987) Phytochelatins: a class of heavy-metal-binding peptides from plants, are functionally analogous to metallothioneins. *Proc. Natl Acad. Sci. USA* **84**, 439.

98. Grootveld, M and Halliwell B (1987) Measurement of allantoin and uric acid in human body fluids. *Biochem. J.* **243**, 803.

99. Grootveld, M *et al.* (1989) Non-transferrin-bound iron in plasma or serum from patients with idiopathic hemochromatosis. *J. Biol. Chem.* **264**, 4417.

100. Guidot, DM *et al.* (1993) Absence of electron transport (Rh° state) restores growth of a MnSOD-deficient *S. cerevisiae* in hyperoxia. *J. Biol. Chem.* **268**, 26699.

101. Gunther, MR *et al.* (1995) Self-peroxidation of metmyoglobin results in formation of an O_2-reactive tryptophan-centered radical. *J. Biol. Chem.* **270**, 16075.

102. Gurgueira, SA and Meneghini, R (1996) An ATP-dependent iron transport system in isolated rat liver nuclei. *J. Biol. Chem.* **271**, 13616.

103. Gutteridge, JMC and Stocks, J (1981) Caeruloplasmin: physiological and pathological perspectives. *CRC Crit. Rev. Clin. Lab. Sci.* **14**, 257.

104. Gutteridge, JMC *et al.* (1985) The behaviour of caeruloplasmin in stored human extracellular fluids in relation to ferroxidase II activity, lipid peroxidation and phenanthroline-detectable copper. *Biochem. J.* **230**, 517.

105. Hager, LP *et al.* (1966) Chloroperoxidase. II Utilization of halogen anions. *J. Biol. Chem.* **241**, 1769.

106. Halliwell, B (1977) Generation of H_2O_2, $O_2^{\bullet-}$ and OH$^{\bullet}$ during the oxidation of dihydroxyfumaric acid by peroxidase. *Biochem. J.* **163**, 441.

107. Halliwell, B and Gutteridge, JMC (1990) The antioxidants of human extracellular fluids. *Arch. Biochem. Biophys.* **280**, 1.

108. Halliwell, B and Gutteridge, JMC (1990) Role of free radicals and catalytic metal ions in human disease. *Methods Enzymol.* **186**, 1.

109. Hanna, PM and Mason, RP (1992) Direct evidence for inhibition of free radical formation from Cu(I) and H_2O_2 by GSH and other potential ligands using the EPR spin-trapping technique. *Arch. Biochem. Biophys.* **295**, 205.

110. Harman, LS *et al.* (1986) One- and two-electron oxidation of GSH by peroxidases. *J. Biol. Chem.* **261**, 1642.

111. Harris, ED (1995) The iron–copper connection: the link to ceruloplasmin grows stronger. *Nutr. Rev.* **53**, 170.

112. Harrison, PM and Arosio, P (1996) The ferritins: molecular properties, iron storage function and cellular regulation. *Biochim. Biophys. Acta* **1275**, 161.

113. Hass, MA and Massaro, D (1987) Differences in CuZnSOD induction in lungs of neonatal and adult rats. *Am. J. Physiol.* **253**, C66.

114. Hassan, HM *et al.* (1980) Inhibitors of SODs: a cautionary tale. *Arch. Biochem. Biophys.* **199**, 349.

115. Hassett, DJ *et al.* (1995) *Pseudomonas aeruginosa, sod*A *and sod*B mutants defective in manganese- and iron-cofactored SOD activity demonstrate the importance of the iron-cofactored form in aerobic metabolism. *J. Bacteriol.* **177**, 6330.

116. Hatta, A and Frei, B (1995) Oxidative modification and antioxidant protection of human LDL at high and low O_2 partial pressures. *J. Lipid Res.* **36**, 2383.

117. Hayes, JD and Pulford, DJ (1995) The glutathione-S-transferase supergene family: regulation of GST* and the contribution of the isoenzymes to cancer chemoprotection and drug resistance. *Crit. Rev. Biochem. Mol. Biol.* **30**, 445.

118. Hayman, AR and Cox, TM (1994) Purple acid phosphatase of the human macrophage and osteoclast. *J. Biol. Chem.* **269**, 1294.

119. Heikkila, RE *et al.* (1976) *In vivo* inhibition of SOD in mice by diethyl-dithiocarbamate. *J. Biol. Chem.* **251**, 2182.

120. Hershko, C and Peto, TEA (1987) Non-transferrin plasma iron. *Br. J. Haematol.* **66**, 149.

121. Hillar, A *et al.* (1994) NADPH binding and control of catalase compound II formation: comparison of bovine, yeast and *E. coli* enzymes. *Biochem. J.* **300**, 531.

122. Hipkiss, AR *et al.* (1995) Non-enzymatic glycosylation of the dipeptide L-carnosine, a potential anti-protein-cross-linking agent. *FEBS Lett.* **371**, 81.

123. Hirono, A *et al.* (1995) A novel human catalase mutation (358T → del) causing Japanese-type acatalasemia. *Blood Cells Mol. Dis.* **21**, 232.

123a. Ho, YS *et al.* (1997) Mice deficient in cellular glutathione peroxidase develop normally and show no increased sensitivity to hyperoxia. *J. Biol. Chem.* **272**, 16644.

123b. Ho, YS *et al.* (1998) Reduced fertility in female mice lacking CuZnSOD. *J. Biol. Chem.* **273**, 7765.

124. Hogg, N *et al.* (1994) The role of lipid hydroperoxides in the myoglobin-dependent oxidation of LDL. *Arch. Biochem. Biophys.* **314**, 39.

125. Hollman, PCH *et al.* (1995) Absorption of dietary quercetin glycosides and quercetin in healthy ileostomy volunteers. *Am. J. Clin. Nutr.* **62**, 1276.

125a. Hurst R *et al.* (1998) Phospholipid hydroperoxide glutathione peroxidase activity of human glutathione transferases. *Biochem. J.* **332**, 97.

126. Jacobson, ES and Tinnell, SB (1993) Antioxidant function of fungal melanin. *J. Bacteriol.* **175**, 7102.

127. Jotti, A *et al.* (1994) Protective effect of dietary selenium supplementation on delayed cardiotoxicity of adriamycin in rat: is PHGPX but not GPX involved? *Free Rad. Biol. Med.* **16**, 283.

128. Kadiiska, MB *et al.* (1995) Iron supplementation generates OH$^\bullet$ *in vivo. J. Clin. Invest.* **96**, 1653.

129. Kanofsky, JR (1986) Singlet O_2 production in superoxide ion–halocarbon systems. *J. Am. Chem. Soc.* **108**, 2977.

130. Kanofsky, JR (1988) Singlet O_2 production from the peroxidase-catalyzed oxidation of indole-3-acetic acid. *J. Biol. Chem.* **263**, 14171.

131. Kaplowitz, N *et al.* (1996) GSH transporters: molecular characterization and role in GSH homeostasis. *Biol. Chem. Hoppe Seyler* **377**, 267.

132. Keaney, JF, Jr *et al.* (1994) 17β-Estradiol preserves endothelial vasodilator function and limits low-density lipoprotein oxidation in hypercholesterolemic swine. *Circulation* **89**, 2251.

133. Keeping, HS and Lyttle, CR (1984) Monoclonal antibody to rat uterine peroxidase and its use in identification of the peroxidase as being of eosinophil origin. *Biochim. Biophys. Acta* **802**, 399.

134. Keli, SO *et al.* (1996) Dietary flavonoids, antioxidant vitamins and incidence of stroke. The Zutphen study. *Arch. Int. Med.* **154**, 637.

135. Keyse, SM and Tyrrell, RM (1989) Heme oxygenase is the major 32-kDa stress protein induced in human skin fibroblasts by UVA radiation, H_2O_2 and sodium arsenite. *Proc. Natl Acad. Sci. USA* **86**, 99.

136. Kim, E-J *et al.* (1998) Transcriptional and post-transcriptional regulation by nickel of *sod* N gene encoding Ni-containing SOD from *Streptomyces coelicolor* Müller. *Molec. Microbiol.* **27**, 187.

137. Kittridge, KJ and Willson RL (1984) Uric acid substantially enhances the free radical-induced inactivation of alcohol dehydrogenase. *FEBS Lett.* **170**, 162.

138. Kohen, R *et al.* (1988) Antioxidant activity of carnosine, homocarnosine and anserine present in muscle and brain. *Proc. Natl Acad. Sci. USA* **85**, 3175.

139. Kondo, K *et al.* (1996) Inhibition of LDL oxidation by cocoa. *Lancet* **348**, 1514.

139a. Konorev, EA *et al.* (1998) Rapid and irreversible inhibition of creatine kinase by ONOO$^-$ *FEBS Lett.* **427**, 171.

140. Kontoghiorghes, GJ and Weinberg, ED (1995) Iron: mammalian defense systems, mechanisms of disease, and chelation therapy approaches. *Blood Rev.* **9**, 33.

141. Korycka-Dahl, M and Richardson, T (1977) Photogeneration of $O_2^{\bullet-}$ in serum of bovine milk and in model systems containing riboflavin and amino acids. *J. Dairy Sci.* **61**, 400.

142. Korytowski, W *et al.* (1986) Reaction of $O_2^{\bullet-}$ with melanins: electron spin resonance and spin trapping studies. *Biochim. Biophys. Acta* **882**, 145.

143. Kosower, NW and Kosower, EM (1995) Diamide, an oxidant probe for thiols. *Methods Enzymol.* **251**, 123.

144. Kozlov, YN and Berdnikov, VM (1973) Photodecomposition of H_2O_2 in the presence of copper ions. IV. Determination of rate constants of elementary reactions. *Russ. J. Phys. Chem.* **47**, 338.

145. Kramer, U *et al.* (1996) Free histidine as a metal chelator in plants that accumulate nickel. *Nature* **379**, 635.

146. Krinsky, NI (1993) Actions of carotenoids in biological systems. *Annu. Rev. Nutr.* **13**, 561.

147. Kyle, ME *et al.* (1988) Endocytosis of SOD is required in order for the enzyme to protect hepatocytes from the cytotoxicity of H_2O_2. *J. Biol. Chem.* **263**, 3784.

148. Landry, LG *et al.* (1995) *Arabidopsis* mutants lacking phenolic sunscreens exhibit enhanced UV-B injury and oxidative damage. *Plant Physiol.* **109**, 1159.

149. Lapinskas, PJ *et al.* (1995) Mutations in PMR1 suppress oxidative damage in yeast cells lacking SOD. *Mol. Cell. Biol.* **15**, 1382.

150. Lardinois, OM (1995) Reactions of bovine liver catalase with $O_2^{\bullet-}$ and H_2O_2. *Free Rad. Res.* **22**, 251.

151. Lazo, JS *et al.* (1995) Enhanced sensitivity to oxidative stress in cultured embryonic cells from transgenic mice deficient in metallothionein I and II genes. *J. Biol. Chem.* **270**, 5506.

152. Lebovitz, RM *et al.* (1996) Neurodegeneration, myocardial injury and perinatal death in mitochondrial SOD-deficient mice. *Proc. Natl Acad. Sci. USA* **93**· 9782.

153. Lee, MJ *et al.* (1995) Analysis of plasma and urinary tea polyphenols in human subjects. *Cancer Epidemiol. Biomark. Prevent.* **4**, 393.

154. Levine, M *et al.* (1996) Vitamin C pharmacokinetics in healthy volunteers: evidence for a recommended dietary allowance. *Proc. Natl Acad. Sci. USA* **93**, 3704.

155. Li, H *et al.* (1996) Unexpectedly strong binding of a large metal ion (Bi^{3+}) to human serum transferrin. *J. Biol. Chem.* **271**, 9483.

156. Lieberman, MW *et al.* (1996) Growth retardation and cysteine deficiency in γ-glutamyltranspeptidase-deficient mice. *Proc. Natl Acad. Sci. USA* **93**, 7923.

157. Liebler, DC and McClure, TD (1996) Antioxidant reactions of β-carotene: identification of carotenoid-radical adducts. *Chem. Res. Toxicol.* **9**, 8.

158. Liehr, JG (1996) Antioxidant and pro-oxidant properties of estrogens. *J. Lab. Clin. Med.* **128**, 344.

159. Ling, P and Roberts, RM (1993) Uteroferrin and intracellular tartrate-resistant acid phosphatases are the products of the same gene. *J. Biol. Chem.* **268**, 6896.

160. Mackay, WJ and Bewley, GC (1989) The genetics of catalase in *Drosophila melanogaster:* isolation and characterization of acatalasemic mutants. *Genetics* **122**, 643.

161. Mader, M and Füssl, R (1982) Role of peroxidase in lignification of tobacco cells. *J. Biol. Chem.* **70**, 1132.

162. Maines, MD (1988) Heme oxygenase: function, multiplicity, regulatory mechanisms and clinical applications. *FASEB J.* **2**, 2557.

162a. Maiorino, M *et al.* (1995) Probing the presumed catalytic triad of Se-containing peroxidases by mutational analysis of PHGP$_X$ *Biol Chem.* *Hoppe-Seyler* **376**, 651.

163. Marklund, SL *et al.* (1976) A comparison between the common type and rare genetic variant of human CuZnSOD. *Eur. J. Biochem.* **65**, 415.

164. Marquis, RE (1995) Oxygen metabolism, oxidative stress and acid-base physiology of dental plaque biofilms. *J. Indust. Microbiol.* **15**, 198.

165. Marshall, KA *et al.* (1996) Evaluation of the antioxidant activity of melatonin *in vitro. Free Rad. Biol. Med.* **21**, 307.

166. Martin, ME *et al.* (1986) A *Streptococcus mutans* SOD that is active with either manganese or iron as a cofactor. *J. Biol. Chem.* **261**, 9361.

167. Mathews-Roth, MM (1987) Photoprotection by carotenoids. *Fed. Proc.* **46**, 1890.

168.   Meister, A (1995) Mitochondrial changes associated with glutathione deficiency. *Biochim. Biophys. Acta* **1271**, 35.

168a. Melov, S *et al.* (1998) A novel neurological phenotype in mice lacking mitochondrial MnSOD. *Nature Genet.* **18**, 159.

169.   Metodiewa, D and Dunford, HB (1989) The reactions of horseradish peroxidase, lactoperoxidase and myeloperoxidase with enzymatically generated $O_2^{\bullet-}$. *Arch. Biochem. Biophys.* **272**, 245.

170.   Michelsen, PA *et al.* (1982) Ability of *Neisseria gonorrhoeae, N. meningitidis* and commensal *Neisseria* species to obtain iron from lactoferrin. *Infect. Immun.* **35**, 915.

171.   Miksicek, RJ (1993) Commonly occurring plant flavonoids have estrogenic activity. *Mol. Pharmacol.* **44**, 37.

172.   Miller, YI *et al.* (1995) The involvement of LDL in hemin transport potentiates peroxidative damage. *Biochim. Biophys. Acta* **1272**, 119.

173.   Moreno, SNJ *et al.* (1988) Oxidation of cyanide to the cyanyl radical by peroxidase/H_2O_2 systems as determined by spin trapping. *Arch. Biochem. Biophys.* **265**, 267.

174.   Morris, SM and Albright JT (1984) Catalase, glutathione peroxidase, and SOD in the rete mirabile and gas gland epithelium of six species of marine fishes. *J. Exp. Zool.* **232**, 29.

174a. Mortensen, A and Stibsted, LH (1997) Relative stability of carotenoid radical cations and homologue tocopheroxyl radicals. *FEBS Lett.* **417**, 261.

175.   Moyo, VM *et al.* (1997) Traditional beer consumption and the iron status of spouse pairs from a rural community in Zimbabwe. *Blood* **89**, 2159.

176.   Mukai, K *et al.* (1993) Kinetic study of reactions between tocopheroxyl radicals and fatty acids. *Lipids* **28**, 753.

177.   Mukhopadhyay, CK *et al.* (1996) Ceruloplasmin enhances smooth muscle cell- and endothelial cell-mediated LDL oxidation by a $O_2^{\bullet-}$-dependent mechanism. *J. Biol. Chem.* **271**, 14773.

178.   Muller, DPR and Goss-Sampson, MA (1990) Neurochemical, neurophysiological, and neuropathological studies in vitamin E deficiency. *Crit. Rev. Neurobiol.* **5**, 239.

178a. Murray, ED, Jr *et al.* (1997) Endogenous natriuretic factors. *J. Pharmacol. Exp. Ther.* **282**, 657.

179.   Nakamura, H *et al.* (1994) Adult T cell leukaemia-derived factor/human thioredoxin protects endothelial F-2 cell injury caused by activated neutrophils or H_2O_2. *Immunol. Lett.* **42**, 75.

180.   Nassi-Calo, L *et al.* (1989) *o*-Phenanthroline protects mammalian cells from H_2O_2-induced gene mutation and morphological transformation. *Carcinogenesis* **10**, 1055.

181.   Nath, KA *et al.* (1995) α-Ketoacids scavenge H_2O_2 *in vitro* and *in vivo* and reduce menadione-induced DNA injury and cytotoxicity. *Am. J. Physiol.* **268**, C227.

182.   Nehlig, A and Debry, G (1994) Potential genotoxic, mutagenic and antimutagenic effects of coffee: A review. *Mutat. Res.* **317**, 145.

183.   Neuzil, J and Stocker, R (1993) Bilirubin attenuates radical-mediated damage to serum albumin. *FEBS Lett.* **331**, 281.

184.   Newton, GL *et al.* (1995) The structure of U17 isolated from *Streptomyces clavuligerus* and its properties as an antioxidant thiol. *Eur. J. Biochem.* **230**, 821.

185. Niimura, Y *et al.* (1995) *Amphibacillus xylanus* NADH oxidase and *Salmonella typhimurium* alkyl hydroperoxide reductase flavoprotein components show extremely high scavenging activity for both alkyl hydroperoxide and H_2O_2 in the presence of *S. typhimurium* alkyl hydroperoxide reductase 22-kDa protein component. *J. Biol. Chem.* **270**, 25645.

186. Nishikimi, M and Yagi, K (1996) Biochemistry and molecular biology of ascorbic acid biosynthesis. In *Subcellular Biochemistry*, vol. 25 (ed. Harris, RJ). Plenum Press, New York, p. 17.

187. O'Connell, MJ *et al.* (1986) Haemosiderin-like properties of free-radical-modified ferritin. *Biochem. J.* **240**, 297.

188. O'Neill, P *et al.* (1982) Evidence for catalytic dismutation of $O_2^{\bullet-}$ by cobalt(II) derivatives of bovine SOD in aqueous solution as studied by pulse radiolysis. *Biochem. J.* **205**, 181.

189. Ogihara, H *et al.* (1995) Plasma copper and antioxidant status in Wilson's disease. *Pediatr. Res.* **37**, 219.

190. Okuda, T *et al.* (1993) Antioxidant phenolics in oriental medicine. In *Active Oxygens, Lipid Peroxides and Antioxidants* (ed. Yagi, K). Japan Sci. Soc. Press, Tokyo, p. 333.

191. Otterbein, L *et al.* (1995) Hemoglobin provides protection against lethal endotoxemia in rats: the role of heme oxygenase-1. *Am. J. Resp. Cell Mol. Biol.* **13**, 595.

192. Oury, TD *et al.* (1996) Extracellular SOD: a regulator of NO bioavailability. *Lab. Invest.* **75**, 617.

193. Packer, L *et al.* (1995) α-Lipoic acid as a biological antioxidant. *Free Rad. Biol. Med.* **19**, 227.

194. Paganga, G *et al.* (1997) The identification of flavonoids as glycosides in human plasma. *FEBS Lett.* **401**, 78.

195. Park, JB and Levine, M (1996) Purification, cloning and expression of dehydroascorbic acid–reducing activity from human neutrophils: identification as glutaredoxin. *Biochem. J.* **315**, 931.

196. Pathak, MA and Joshi, PC (1984) Production of active oxygen species ($^1O_2$ and O_2^-) by psoralens and ultraviolet radiation. *Biochim. Biophys. Acta* **798**, 115.

197. Pedersen, JZ *et al.* (1996) Cu–glutathione complexes under physiological conditions: structures in solution different from the solid state coordination. *Biometals* **9**, 3.

198. Peled-Kamar, M *et al.* (1997) Oxidative stress mediates impairment of muscle function in transgenic mice with elevated level of wild-type CuZnSOD. *Proc. Natl Acad. Sci. USA* **94**, 3883.

199. Plat, H *et al.* (1987) The bromoperoxidase from the lichen *Xanthoria parietina* as a novel vanadium enzyme. *Biochem. J.* **248**, 277.

199a. Poole, LB *et al.* (1997) Peroxidase activity of a TSA-like antioxidant protein from a pathogenic amoeba. *Free Rad. Biol. Med.* **23**, 955.

200. Porter, DJT and Bright HJ (1983) The mechanism of oxidation of nitroalkanes by horseradish peroxidase. *J. Biol. Chem.* **258**, 9913.

201. Porter, WL (1995) Paradoxical behaviour of antioxidants in food and biological systems. *Tox. Ind. Health* **9**, 93.

202. Prasad, MR *et al.* (1989) Effects of oxyradicals on oxymyoglobin. *Biochem. J.* **263**, 731.

203.   Qian, ZM and Tang, PL (1995) Mechanisms of iron uptake by mammalian cells. *Biochim. Biophys. Acta* **1269**, 205.
204.   Quinlan, GJ *et al.* (1992) Vandadium and copper in clinical solutions of albumin and their potential to damage protein structure. *J. Pharmacol. Sci.* **81**, 611.
205.   Radi, R *et al.* (1991) Detection of catalase in rat heart mitochondria. *J. Biol. Chem.* **266**, 22028.
206.   Ravichandran, V *et al.* (1994) *S*-Thiolation of glyceraldehyde-3-phosphate dehydrogenase induced by the phagocytosis-associated respiratory burst in blood monocytes. *J. Biol. Chem.* **269**, 25010.
207.   Ray, SD and Fariss, MW (1994) Role of cellular energy status in tocopheryl hemisuccinate cytoprotection against ethyl methanesulfonate-induced toxicity. *Arch. Biochem. Biophys.* **311**, 180.
208.   Read, MA (1995) Flavonoids: naturally occurring anti-inflammatory agents. *Am. J. Pathol.* **147**, 235.
209.   Reaume, AG *et al.* (1996) Motor neurones in CuZnSOD-deficient mice develop normally but exhibit enhanced cell death after axonal injury. *Nature Genet.* **13**, 43.
210.   Reid, TJ III *et al.* (1981) Structure and heme environment of beef liver catalase at 2.5 Å resolution. *Proc. Natl Acad. Sci. USA* **78**, 4767.
211.   Reiter, RJ (1997) Antioxidant actions of melatonin. *Adv. Pharmacol.* **38**, 103.
212.   Repine, JE *et al.* (1981) H_2O_2 kills *S. aureus* by reacting with staphylococcal iron to form OH$^\bullet$. *J. Biol. Chem.* **256**, 7094.
213.   Reveillaud, I *et al.* (1992) Stress resistance of *Drosophila* transgenic for bovine CuZnSOD. *Free Rad. Res. Commun.* **17**, 73.
214.   Rice-Evans, C *et al.* (1993) Oxidized LDL induce iron release from activated myoglobin. *FEBS Lett.* **326**, 177.
215.   Rice-Evans, C *et al.* (1996) Structure–antioxidant activity relationships of flavonoids and phenolic acids. *Free Rad. Biol. Med.* **20**, 933.
216.   Romero, FJ *et al.* (1992) The reactivity of thiols and disulfides with different redox states of myoglobin. *J. Biol. Chem.* **267**, 1680.
217.   Rowley, DA and Halliwell, B (1982) Superoxide-dependent formation of OH$^\bullet$ from NADH and NADPH in the presence of iron salts. *FEBS Lett.* **142**, 39.
217a. Rumsey, SC *et al.* (1997) Glucose transporter isoforms GLUTI and GLUT3 transport dehydroascorbic acid. *J. Biol. Chem.* **272**, 18982.
218.   Ryan, CS and Kleinberg, I (1995) Bacteria in human mouths involved in the production and utilization of H_2O_2. *Arch. Oral Biol.* **40**, 753.
219.   Sack, MN *et al.* (1994) Oestrogen and inhibition of oxidation of LDL in postmenopausal women. *Lancet* **343**, 269.
220.   Salo, DC *et al.* (1990) SOD undergoes proteolysis and fragmentation following oxidative modification and inactivation. *J. Biol. Chem.* **265**, 11919.
221.   Sarna, T *et al.* (1986) Interaction of radicals from water radiolysis with melanin. *Biochim. Biophys. Acta* **883**, 162.
222.   Schaefer, DM *et al.* (1995) Supranutritional administration of vitamins E and C improves oxidative stability of beef. *J. Nutr.* **125**, 1792S.
223.   Schellhorn, HE (1994) Regulation of hydroperoxidase (catalase) expression in *E. coli. FEMS Microbiol. Lett.* **131**, 113.
224.   Schinina, ME *et al.* (1996) Amino acid sequence of chicken CuZnSOD and identification of GSH adducts at exposed cysteine residues. *Eur. J. Biochem.* **237**, 433.

225. Schleiter, H *et al.* (1995) Coenzyme A GSSG. A potent vasoconstrictor derived from the adrenal gland. *Circul. Res.* **76**, 675.

226. Schorah, CJ *et al.* (1996) Total vitamin C, ascorbic acid and DHA concentrations in plasma of critically ill patients. *Am. J. Clin. Nutr.* **63**, 760.

227. Schuppe-Koistinen, I *et al.* (1994) Studies on the reversibility of protein S-thiolation in human endothelial cells. *Arch. Biochem. Biophys.* **315**, 226.

228. Schwarz, MA *et al.* (1995) Metallothionein protects against the cytotoxic and DNA-damaging effects of NO$^\bullet$. *Proc. Natl Acad. Sci. USA* **92**, 4452.

229. Scott, MD *et al.* (1987) SOD-rich bacteria. Paradoxical increase in oxidant toxicity. *J. Biol. Chem.* **262**, 3640.

230. Shacter, E *et al.* (1990) DNA damage induced by phorbol ester-stimulated neutrophils is augmented by extracellular cofactors. Role of histidine and metals. *J. Biol. Chem.* **265**, 6693.

231. Shahidi, F and Wanasundara PKJPD (1992) Phenolic antioxidants. *Crit. Rev. Food Sci. Nutr.* **32**, 67.

232. Shatzman, AR and Kosman, DJ (1979) Biosynthesis and cellular distribution of the two SODs of *Dactylium dendroides*. *J. Bacteriol.* **137**, 313.

233. Sichak, SP and Dounce, AL (1987) A study of the catalase monomer produced by lyophilization. *Biochim. Biophys. Acta* **925**, 282.

234. Sies, H *et al.* (1978) Glutathione efflux from perfused rat liver after phenobarbital treatment, during drug oxidations, and in selenium deficiency. *Eur. J. Biochem.* **89**, 113.

235. Simic, MG and Jovanovic, SV (1989) Antioxidation mechanisms of uric acid. *J. Am. Chem. Soc.* **111**, 5778.

235a. Simpson, RJ *et al.* (1992) Non-transferrin bound iron species in the serum of hypotransferrinemic mice *Biochim. Biophys. Acta.* **1156**, 19.

236. Sinet, PM *et al.* (1980) H$_2$O$_2$ production by rat brain *in vivo*. *J. Neurochem.* **34**, 1421.

237. Slot, JW *et al.* (1986) Intracellular localization of the copper–zinc and manganese SODs in rat liver parenchymal cells. *Lab. Invest.* **35**, 363.

238. Smith, AM *et al.* (1982) Oxidation of indole-3-acetic acid by peroxidase: involvement of reduced peroxidase and compound III with O$_2^{\bullet -}$ as a product. *Biochemistry* **21**, 4414.

239. Soboll, S *et al.* (1995) The content of GSH and glutathione-S-transferases and the glutathione peroxidase activity in rat liver nuclei determined by a non-aqueous technique of cell fractionation. *Biochem. J.* **311**, 889.

240. Spector, A *et al.* (1996) Variation in cellular glutathione peroxidase activity in lens epithelial cells, transgenics and knockouts does not significantly change the response to oxidative stress. *Exp. Eye Res.* **62**, 521.

241. Stahl, W and Sies, H (1996) Lycopene: a biologically-important carotenoid for humans? *Arch. Biochem. Biophys.* **336**, 1.

242. Sundquist, AR and Fahey RC (1989) The function of γ-glutamylcysteine and bis-γ-glutamylcystine reductase in *H. halobium*. *J. Biol. Chem.* **264**, 719.

243. Tamai, KT *et al.* (1993) Yeast and mammalian metallothioneins functionally substitute for yeast CuZnSOD. *Proc. Natl Acad. Sci. USA* **90**, 8013.

243a. Tanaka, K *et al.* (1997) Interactions between vitamin C and E are observed in tissues of inherently scorbutic rats. *J. Nutr.* **127**, 2060.

244. Tasinato, A *et al.* (1995) *d*-α-Tocopherol inhibition of vascular smooth muscle cell proliferation occurs at physiological concentrations, correlates with

protein kinase C inhibition, and is independent of its antioxidant properties. *Proc. Natl Acad. Sci. USA* **92**, 12190.

245.   Thienne, R *et al.* (1981) Three-dimensional structure of glutathione reductase at 2 Å resolution. *J. Mol. Biol.* **152**, 763.

246.   Thomas, C *et al.* (1995) Protein sulfhydryls and their role in the antioxidant function of protein-S-thiolation. *Arch. Biochem. Biophys.* **319**, 1.

247.   Thornalley, PJ (1996) Pharmacology of methylglyoxal. *Gen. Pharmacol.* **27**, 565.

248.   Till, GO *et al.* (1985) Lipid peroxidation and acute lung injury after thermal trauma to skin. Evidence of a role for OH$^\bullet$. *Am. J. Pathol.* **119**, 376.

249.   Toth, I and Bridges KR (1995) Ascorbic acid enhances ferritin mRNA translation by an IRP/aconitase swith. *J. Biol. Chem.* **270**, 19540.

250.   Touati, D *et al.* (1995) Lethal oxidative damage and mutagenesis are generated by iron in Δfur mutants of *E. coli*: protective role of SOD. *J. Bacteriol.* **177**, 2305.

251.   Toyokuni, S *et al.* (1995) Treatment of Wistar rats with a renal carcinogen, Fe(III)NTA, causes DNA protein cross-linking between thymine and tyrosine in their renal chromatin. *Int. J. Cancer* **62**, 309.

252.   Traber, MG (1994) Determinants of plasma vitamin E concentrations. *Free Rad. Biol. Med.* **16**, 229.

253.   Turrens, JF *et al.* (1984) Protection against O_2 toxicity by intravenous injection of liposome-entrapped catalase and SOD. *J. Clin. Invest.* **73**, 87.

254.   Uchida, K and Kawakishi, S (1994) Identification of oxidized histidine generated at the active site of CuZnSOD exposed to H_2O_2. *J. Biol. Chem.* **269**, 2405.

255.   Van Loon, APGM *et al.* (1986) A yeast mutant lacking mitochondrial MnSOD is hypersensitive to O_2. *Proc. Natl Acad. Sci. USA* **83**, 3820.

256.   Vaya, J *et al.* (1997) Antioxidant constituents from licorice roots: isolation, structure elucidation and antioxidative capacity towards LDL oxidation. *Free Rad. Biol. Med.* **23**, 302.

257.   Villa, LM *et al.* (1994) Peroxynitrite induces both vasodilation and impaired vascular relaxation in the isolated perfused rat heart. *Proc. Natl Acad. Sci. USA* **91**, 12383.

258.   Von Herbay, A *et al.* (1994) Low vitamin E content in plasma of patients with alcoholic liver disease, haemochromatosis and Wilson's disease. *J. Hepatol.* **20**, 41.

258a. Wada, N *et al.* (1998) Purification and molecular properties of ascorbate peroxidase from bovine eye. *Biochem. Biophys. Res. Common.* **242**, 256.

259.   Wang, Y and Casadevall, A (1994) Susceptibility of melanized and non-melanized *Cryptococcus neoformans* to nitrogen- and oxygen-derived oxidants. *Infect. Immun.* **62**, 3004.

260.   Wassif, WS *et al.* (1994) Serum carnosinase activities in central nervous system disorders. *Clin. Chim. Acta* **225**, 57.

261.   Watkins, S *et al.* (1994) Analbuminemia: three cases resulting from different point mutations in the albumin gene. *Proc. Natl Acad. Sci. USA* **91**, 9417.

262.   Wells, WW *et al.* (1990) Mammalian thioltransferase (glutaredoxin) and protein disulfide isomerase have dehydroascorbate reductase activity. *J. Biol. Chem.* **265**, 15361.

263.   Whiteman, M and Halliwell, B (1996) Protection against ONOO$^-$ -dependent tyrosine nitration and α_1-antiproteinase inactivation by ascorbic acid. A comparison with other biological antioxidants. *Free Rad. Res.* **25**, 275.

264. Winkler, P *et al.* (1984) Selective promotion of Fe^{2+} ion-dependent lipid peroxidation in Ehrlich ascites tumor cells by histidine as compared with other amino acids. *Biochim. Biophys. Acta* **796**, 226.

265. Winterbourn, CC and Stern, A (1987) Human red cells scavenge extracellular H$_2$O$_2$ and inhibit formation of HOCl and OH$^\bullet$. *J. Clin. Invest.* **80**, 1486.

266. Wiseman, H and Halliwell, B (1993) Carcinogenic antioxidants: diethylstilboestrol, hexoestrol and 17α-ethynyloestradiol. *FEBS Lett.* **322**, 159.

267. Wispe, JR *et al.* (1992) Human Mn-SOD in pulmonary epithelial cells of transgenic mice confers protection from O$_2$ injury. *J. Biol. Chem.* **267**, 23937.

267a. Wu, X *et al.* (1994) Hyperuricaemia and urate nephropathy in urate oxidase-deficient mice. *Proc. Natl. Acad. Sci. US* **91**, 742.

268. Yamamoto, Y *et al.* (1992) Comparison of plasma levels of lipid hydroperoxides and antioxidants in hyperlipidemic Nagase analbuminemic rats, Sprague–Dawley rats and humans. *Biochem. Biophys. Res. Commun.* **189**, 518.

268a. Yamashita, S and Yamamoto, Y (1997) Simultaneous detection of ubiquinol and ubiquinone in human plasma as a marker of oxidative stress. *Anal. Biochem.* **250**, 66.

269. Yamazaki, I. and Yokata, K (1973) Oxidation states of peroxidase. *Mol. Cell. Biochem.* **2**, 39.

270. Yang, F *et al.* (1996) Cellular expression of ceruloplasmin in baboon and mouse lung during development and inflammation. *Am. J. Resp. Cell Mol. Biol.* **14**, 161.

271. Yeh, JI *et al.* (1996) Structure of the native cysteine–sulfenic acid redox center of enterococcal NADH peroxidase refined at 2.8 Å resolution. *Biochemistry* **35**, 9951.

272. Young, IS *et al.* (1994) Antioxidant status and lipid peroxidation in hereditary haemochromatosis. *Free Rad. Biol. Med.* **16**, 393.

273. Zhang, Y *et al.* (1990) The oxidative inactivation of mitochondrial electron transport chain components and ATPase. *J. Biol. Chem.* **265**, 16630.

274. Zheng, H *et al.* (1992) Regulation of catalase in *Neisseria gonorrhoeae*. *J. Clin. Invest.* **90**, 1000.

Notes

[a]These organelles are discussed further in Section 3.7.5.
[b]This technology is explained in Appendix II.
[c]The apoenzyme would not be a satisfactory control if ONOO$^-$ is involved, since the Cu is involved in catalysing nitration.
[d]Fenton-type chemistry may produce a variety of reactive species (Section 2.4). For simplicity we refer here only to OH$^\bullet$ (the best-established species).
[e]Appendix II gives further explanation if needed.
[f]'Glutathione peroxidase' refers here to the 'classical' tetrameric enzyme, GPX.
[g]This enzyme is further discussed in Section 4.10.
[h]For an explanation please see Chapter 5.
[i]For a discussion of these methods see Chapters 4 and 5.
[j]If an explanation is needed, please consult Section 4.9.
[k]Gap junctions are specialized structures that form between the plasma membranes of adjacent cells in most animal tissues, connecting the cell cytoplasms together. Small molecules can cross, but not macromolecules. Gap junctions are constructed from protein subunits called **connexins**.

4

Oxidative stress: adaptation, damage, repair and death

4.1 Introduction

In healthy aerobic organisms, production of reactive oxygen species (ROS) and reactive nitrogen species (RNS) is approximately balanced by antioxidant defence systems. The balance is not perfect, however, so that some ROS/RNS-mediated damage occurs continuously and damaged molecules have to be repaired (e.g. DNA) or replaced (e.g. most oxidized proteins). For example, some iron–sulphur proteins essential to the metabolism of *Escherichia coli* may undergo random low-level damage by stray $O_2^{\bullet-}$ radicals and the H_2O_2-removing systems of *E. coli* seem to keep cellular levels at $0.1-0.2\,\mu M$ rather than removing H_2O_2 completely (Chapter 1).[39] Table 4.1 summarizes the evidence for ongoing oxidative damage *in vivo* in animals, including humans.

The term **oxidative stress** is widely used in the free-radical literature but it is rarely defined. In essence, it refers to the situation of a serious imbalance between production of ROS/RNS and antioxidant defence. Sies, who introduced the term from the title of the book he edited in 1985, *Oxidative Stress*,[109] defined it in 1991 in the Introduction to the second edition as 'a disturbance in the prooxidant–antioxidant balance in favour of the former, leading to potential damage'. Such damage is often, again loosely, called **oxidative damage**.

In principle, oxidative stress can result from:

1. Diminished antioxidants, e.g. mutations affecting antioxidant defence enzymes such as copper–zinc superoxide dismutase (CuZnSOD), MnSOD or glutathione peroxidase (reviewed in Chapter 3). Depletions of dietary antioxidants and other essential dietary constituents can also lead to oxidative stress (Table 4.2). Research in Jamaica[38] has shown that children with the protein deficiency disease **kwashiorkor** suffer additional problems related to oxidative stress, including low glutathione (GSH) levels and iron overload (Tables 4.2 and 4.5).
2. Increased production of ROS/RNS, e.g. by exposure to elevated O_2 (Chapter 1), the presence of toxins that are metabolized to produce ROS/RNS (Chapter 8), or excessive activation of 'natural' ROS/RNS systems (e.g. inappropriate activation of phagocytic cells in chronic inflammatory diseases, such as rheumatoid arthritis and ulcerative colitis (Chapter 9)).

4.2 Consequences of oxidative stress: adaptation, damage or stimulation?

Oxidative stress can result in adaptation or cell injury.

Table 4.1. Evidence that damage by RNS and ROS occurs *in vivo*

Target of damage	Evidence
DNA	Low levels of oxidative base damage products are present in DNA isolated from all aerobic cells; levels often increase in animals with chronic inflammatory diseases or subjected to oxidative stress, e.g. smoking. Some base damage products are excreted in urine, presumably resulting from DNA repair processes. Smokers and rheumatoid arthritis patients excrete more 8–hydroxydeoxyguanosine (8–OHdG). Elevated 8–OHdG concentrations are frequently observed in animals treated with toxins; also seen in O_3-exposed plants.
Protein	Attack of ROS upon proteins produces carbonyls and other amino acid modifications (e.g. methionine sulphoxide, 2-oxohistidine, protein peroxides, hydroxylation of tyrosine to DOPA, formylkynurenine). Low levels of carbonyls and certain other products (e.g. *ortho*-tyrosine, valine oxidation products) have been detected in healthy animal tissues and body fluids. Nitrotyrosines, products of attack on tyrosine by RNS, have been detected in atherosclerotic lesions, human plasma and urine; concentrations are higher in body fluids/tissues from patients with chronic inflammatory diseases. In rat skeletal muscle, nitrotyrosine is apparently present in the Ca^{2+}-ATPase system of the sarcoplasmic reticulum and levels increase with age (*FEBS Lett.* **379**, 286 (1996)). Bityrosine has been detected in urine and atherosclerotic lesions.
Lipid	Accumulation of 'age pigments' in tissues. Lipid peroxidation in atherosclerotic lesions. Presence of specific end–products of peroxidation (e.g. isoprostanes) in body fluids (including urine); levels increase in plasma during oxidative stress, e.g. in smokers, CCl_4 treatment of animals, and in premature babies.
Uric acid	Attacked by several ROS to generate allantoin, cyanuric acid, parabanic acid, oxonic acid and other products, which are present in human body fluids. Levels increase in chronic inflammatory/metal overload diseases (Chapter 3). Only applicable to primates, which lack urate oxidase.

Unless referenced above, further details may be found in this chapter, in Chapters 5 and 9 and in *Free Rad. Res.* **25**, 57 (1996).

4.2.1 *Adaptation*

Cells can usually tolerate mild oxidative stress, which often results in up–regulation of the synthesis of antioxidant defence systems in an attempt to restore the oxidant/antioxidant balance. For example, if adult rats are gradually acclimatized to elevated concentrations of oxygen, they can tolerate pure oxygen for much longer than control rats, apparently due to increased synthesis of lung antioxidants (Chapter 3). Another example of adaptation may be some cases of **ischaemic preconditioning**. For example, a brief period of ischaemia

Table 4.2. Some links between dietary constituents and oxidative stress

Type of link	Constituent	Needed for
Direct	Antioxidant nutrients (e.g. vitamins E, C, possibly carotenoids, possibly plant-derived phenols such as flavonoids)	
Indirect	Fe	Catalase, FeSOD, prevention of hypoxia (O_2 transport by haemoglobin, O_2 storage by myoglobin), nitric oxide synthase ($NO^{\bullet}$ often has antioxidant properties)
	Mn	MnSOD; Mn acts as $O_2^{\bullet -}$ scavenger in some organisms (Chapter 2).
	Cu	CuZnSOD, caeruloplasmin, normal iron metabolism
	Zn	CuZnSOD, metallothionein, membrane stabilization, displacement of iron
	Mg	Magnesium is a cofactor for multiple enzymes (e.g. in pentose phosphate pathway). No specific link to antioxidant defence but severe magnesium deficiency in animals increases oxidative damage, especially to the heart (e.g. *Cardiovasc. Res.* **31**, 677 (1996))
	Proteins	Provide amino acids for synthesis of antioxidant defence enzymes, metal binding proteins, albumin (as 'sacrificial antioxidant'), GSH
	Riboflavin	Glutathione reductase (has FAD cofactor)
	Thiamine	Required for **transketolase**, an enzyme in the pentose phosphate pathway
	Selenium	Glutathione peroxidases, selenoprotein P (Chapter 3), other selenoproteins
	Nicotinamide	NAD^+, $NADP^+$, NADPH, NADH: energy metabolism, repair of DNA, PARP, glutathione reductase, maintaining catalase function
	Folic acid	Minimizes level of plasma homocysteine—high levels are a risk factor for cardiovascular disease, perhaps by oxidative damage to endothelium (Chapter 10)
	Inducers of antioxidant defences	Agents that upregulate expression of genes coding antioxidant defence systems (further discussed in Chapter 10)

in pig hearts led to depression of contractile function, and administration of anti-oxidants offered protection.[112] However, repeated brief periods of ischaemia led to quicker return of contractile function, but this adaptive response was blocked in the antioxidant-treated animals. Hence ROS produced by ischaemia–reperfusion caused damage leading to depressed contractility, but also led to a response protective against subsequent insult. In some cases, mild oxidative stress can up-regulate defences so as to protect the cell against much more severe oxidative stress applied subsequently. Thus exposure of *E. coli* to low levels of H_2O_2 can render it resistant to much higher levels.

Mechanisms of adaptation often involve changes in gene expression that result in elevated antioxidant defences. Indeed, the field of 'redox regulation' of gene expression is gaining increasing prominence. Oxidative stress can also *decrease* transcription of certain genes. For example, transcription of the mRNAs encoding certain P450 enzymes is decreased[6] when hepatocytes are treated with H_2O_2, or rats exposed to elevated O_2.

However, adaptation to oxidative stress need not always involve increased antioxidant defences. For example, by culturing HeLa cells (a malignant cell line) at gradually increasing O_2 concentrations over a 21 month period, it was possible to obtain cells capable of growing under 80% O_2, a level lethal to 'normal' HeLa cells.[36] These cells did not show elevated levels of SOD or H_2O_2-removing enzymes, and it was proposed that their increased O_2 tolerance was due to a change in sensitivity of targets normally vulnerable to oxidative damage. A similar phenomenon may occur with the enzyme fumarase in *E. coli* (Chapter 1).

4.2.2 Cell injury

'Injury' is a widely used, but vague, term. It can be defined as 'the result of a chemical or physical stimulus, either in excess or in deficiency, that transiently or permanently alters the homeostasis of the cell'.[85] The response to injury may be reversible: the cell enters a temporary or prolonged altered steady-state which does not lead to cell death. Reversible responses may be transient, or can be early events in response to an insult that eventually leads to irreversible injury. Sometimes reversible responses are sustained; they are then often called the **cellular adaptations** referred to above, which include such phenomena as induction of antioxidant defences, heat shock proteins (Section 4.15 below), or of cytochromes P450 (in response to xenobiotics).

Oxidative stress can cause damage to all types of biomolecule, including DNA, proteins and lipids (lipid peroxidation). In many situations it is unclear which is the most important target, since injury mechanisms overlap widely. The primary cellular target of oxidative stress can vary depending on the cell, the type of stress imposed and how severe the stress is. For example, carbon tetrachloride injures cells primarily by lipid peroxidation (Chapter 8). By contrast, DNA is an important early target of damage when H_2O_2 is added to many mammalian cells;[105] increased DNA strand breakage occurs before detectable lipid peroxidation or oxidative protein damage. We emphasize the word 'detectable' because such conclusions are obviously dependent on the assays used to measure such damage. For example, measurement of protein carbonyls would not detect important early oxidative protein damage by oxidation of essential −SH groups on membrane ion transporters. Methods are considered in detail in Chapter 5.

Excessive DNA strand breakage is associated with depletion of cellular ATP and NAD^+ levels. The latter often occurs because a chromatin-bound enzyme,

poly(ADP–ribose)polymerase (PARP) splits the NAD^+ molecule and transfers the ADP–ribose portion on to nuclear proteins, including itself (**automodification**) (Fig. 4.1). ADP–ribosylation of proteins is thought to facilitate DNA repair. However, excessive activation of PARP (i.e. if there are too many DNA strand breaks) can deplete the NAD^+ pool, interfering with ATP synthesis and perhaps even leading to cell death.[113] Treatment of cells with **nicotinamide**, which maintains NAD^+ levels, can often delay or prevent this cell death. Indeed, cells from transgenic mice with a 'knockout' of the PARP gene were less sensitive to killing upon exposure to excess $NO^{\bullet}$ or to hypoxanthine/xanthine oxidase.[54]

Oxidative stress also has striking effects on cell calcium metabolism, resulting in rises in 'free' intracellular Ca^{2+} levels (Section 4.4. below).

4.2.3   Changes in cell behaviour

The consequences of oxidative stress upon cell behaviour are also variable. Usually, dividing cells halt division until repair (at least of DNA damage) is largely complete. However, low-level oxidative stress (usually achieved by adding xanthine/xanthine oxidase, or H_2O_2) has been shown to stimulate proliferation of many cell types in culture (Fig. 4.2), including vascular smooth muscle cells. By contrast, higher levels of oxidative stress usually decrease cell proliferation in culture (Fig. 4.2) and may have toxic effects.

The mechanism, and relevance *in vivo*, of these pro-proliferative effects of 'low-level oxidative stress' is unclear, although stimulation of fibroblast proliferation by ROS has been suggested to contribute to fibrosis in several human diseases.[95] Oxidative stress can also affect intercellular communication

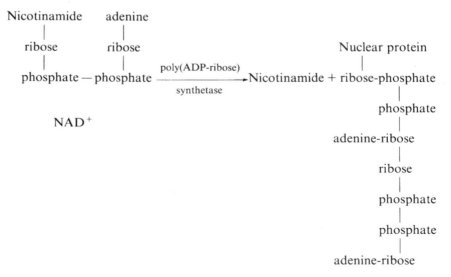

Fig. 4.1. The reaction catalysed by poly(ADP–ribose) polymerase (PARP).

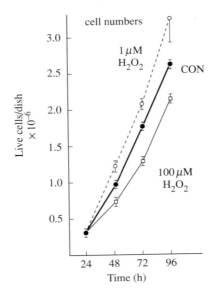

Fig. 4.2. The effect of exogenous H_2O_2 on the growth of baby-hamster kidney fibroblast cells. At 24 h, monolayer cultures were treated with 1 μM H_2O_2, which stimulated growth, or 100 μM H_2O_2, which inhibited it. Data from *Free Rad. Res.* **29**, 121 (1994), by courtesy of Professor Roy Burdon and Harwood Academic Press.

through gap junctions[76] as well as signal transduction pathways. For example, $ONOO^-$ and NO_2Cl (Chapter 2) could conceivably inhibit protein kinase-dependent phosphorylation of tyrosine residues by nitrating them, whereas low levels of H_2O_2 can sometimes activate signal transduction pathways by oxidizing −SH groups or raising Ca^{2+} levels.

4.3   Consequences of oxidative stress: cell death

A cell exposed to severe oxidative stress may die. Cell death can result from multiple mechanisms, such as bleb rupture (Section 4.4 below). Excessive activation of PARP can deplete intracellular $NAD^+/NADP^+$ levels so much that the cell cannot make ATP and dies. This effect has sometimes been called a **suicide response**; since DNA repair is not completely efficient a cell with extensively damaged DNA may 'commit suicide' in the interests of the organism to avoid the risk of becoming a cancerous cell. PARP can be inhibited by several reagents, including **theophylline, theobromine** and **3-aminobenzamide**. They decrease the falls in NAD^+ seen in cells with extensive DNA damage, e.g. after irradiation or toxin treatment, and often maintain cell viability.

Cell death can occur by essentially two mechanisms, **necrosis** and **apoptosis**, although death by mechanisms with features overlapping these pathways is sometimes seen. Necrosis and apoptosis can both result from oxidative stress.

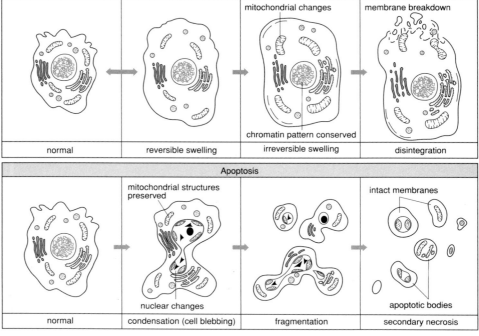

Fig. 4.3. Apoptosis and necrosis. Taken from 'Guide to Cell Proliferation and Apoptosis Methods', by kind permission of Boehringer Mannheim.

For example, in mammalian cells, adding millimolar levels of H_2O_2 often causes death by necrosis, whereas lower levels can trigger apoptosis (in some cell types).[49] During necrotic cell death, the cell swells and ruptures, releasing its contents into the surrounding area and affecting adjacent cells. Contents can include antioxidants such as catalase or GSH, and pro-oxidants such as iron and copper ions. In apoptosis, the cell's own intrinsic 'suicide mechanism' is activated; apoptosing cells do not release their contents and so apoptosis does not, in general, affect surrounding cells (Fig. 4.3).

Let us now examine some of these phenomena in detail.

4.4 Oxidative stress and calcium

4.4.1 *Cell calcium metabolism*[85]

Uncontrolled rises in intracellular 'free' Ca^{2+} can lead to cell injury and death by apoptosis or necrosis. Normally, intracellular 'free' Ca^{2+} levels are low, in the $0.1\,\mu M$ range. Total cell Ca^{2+} is much greater, but it is safely sequestered, e.g. within mitochondria and the endoplasmic reticulum and bound to cytoplasmic proteins. Transient increases in free Ca^{2+} are used in the regulation of physiological processes including cell proliferation in response to several agents. For example changes in Ca^{2+} entry are used by electrically excitable cells to transmit signals (e.g. neurotransmitter release). Many extracellular signalling molecules bind to cell surface receptors and lead to release of Ca^{2+} from the endoplasmic reticulum by stimulating synthesis of an intracellular messenger molecule, **inositol triphosphate** (IP_3). IP_3 opens Ca^{2+} channels in the endoplasmic reticulum membrane and lets Ca^{2+} escape into the cytosol. The IP_3-stimulated Ca^{2+}-release channels resemble those found in the sarcoplasmic reticulum of muscle, which release Ca^{2+} to trigger muscle contraction.

IP_3 is derived from phosphorylated forms of phosphatidylinositol, a membrane phospholipid (Section 4.9 below). Phospho-phosphatidylinositols are cleaved by a **phospholipase C** whose activity is stimulated by a **G-protein**[a] activated when the signal binds to the receptor. The other product of hydrolysis, **diacylglycerol** (DAG) can activate **protein kinase C** (PKC), a Ca^{2+}-dependent enzyme which phosphorylates proteins on serine and threonine −OH groups. The rise in free Ca^{2+} induced by IP_3 appears to lead to movement of PKC from the cytosol to the inner face of the plasma membrane to permit this activation. Normally, IP_3 and DAG are rapidly removed, the signal ceases and the phosphorylated proteins are rapidly dephosphorylated by a family of **phosphatase** enzymes.

The normal, low intracellular Ca^{2+} levels are maintained by the concerted operation of several systems:[85]

(a) plasma membrane ATP-dependent Ca^{2+}-extrusion systems (in most cells). In cells which make extensive use of Ca^{2+} signalling, Na^+−Ca^{2+} exchange transporters are additionally present (e.g. in muscle and nerve). Extracellular Ca^{2+} levels are in the millimolar range in animals, so these Ca^{2+}-export systems need energy to drive them. It comes from ATP hydrolysis, or by linking Ca^{2+} export to entry of Na^+ down its concentration gradient (extracellular Na^+ is far higher than intracellular Na^+). The Na^+ is then pumped out by a plasma membrane ATP-dependent Na^+–K^+ exchange system, often called the **sodium pump**;

(b) uptake of Ca^{2+} into the lumen of the endoplasmic reticulum of cells, by an ATP-dependent pump;

(c) uptake of Ca^{2+} by mitochondria, linked directly to the use of energy from the proton gradient generated by electron transport (Chapter 1). This system seems to be a 'back-up' to the others; it only operates at fairly high intracellular Ca^{2+} levels;

(d) binding of Ca^{2+} to proteins, e.g. calmodulin. Calmodulins are Ca^{2+}-binding proteins found in all eukaryotic cells. They function in the regulation of many enzymes, including nitric oxide synthase (Chapter 2), and cellular processes such as muscle contraction, neurotransmission and cytoskeletal assembly. Calmodulins are rich in methionine residues, and are a potential target of oxidative damage,[120] since methionine is easily oxidized by ROS/RNS. The binding of Ca^{2+} to high-affinity binding sites on calmodulin causes a conformational change which enables the calmodulin to bind to various target proteins in the cell. These include the Ca^{2+}-ATPase in the plasma membrane (which is activated by calmodulin binding), and a family of Ca^{2+}/calmodulin-dependent **protein kinases** which phosphorylate serines and threonines in their substrate proteins, activating such processes as smooth muscle contraction, catecholamine synthesis in certain neurones and the breakdown of glycogen in muscle.

4.4.2 *Dysregulation by oxidative stress*[85]

Oxidative stress dysregulates Ca^{2+} metabolism. Mitochondrial damage by ROS/RNS (e.g. $ONOO^-$) can cause Ca^{2+} release. Peroxides can damage the endoplasmic reticulum Ca^{2+}-uptake system and interfere with Ca^{2+} efflux through the plasma membrane, by leading to oxidation of essential –SH groups on the transmembrane channels. A common consequence of excessive ($>5\,\mu M$) rises in extracellular Ca^{2+} is **membrane blebbing** (Fig. 4.4), caused by disruption of the cytoskeleton. A rise in Ca^{2+} can promote dissociation of actin microfilaments from **α-actinin**, a protein that serves as an intermediate in the association of microfilaments with actin-binding proteins in the plasma membrane. Ca^{2+} can also activate certain proteases (**calpains**) that cleave actin-binding proteins, eliminating the plasma membrane 'anchorage' to the cytoskeleton and allowing it to 'bleb out'. Calpains are present in all mammalian cells and perform several essential functions, but their activity is normally tightly-regulated.[75] Over-activation of calpains has been implicated in many diseases, including muscular dystrophy and Alzheimer's disease. Oxidation of –SH groups on cytoskeletal proteins by ROS/RNS, and falls in ATP (needed for maintenance of cytoskeletal integrity) can also facilitate blebbing. Calpains themselves are cysteine proteases[75] and so might also be subject to oxidation.

(a)

(b)

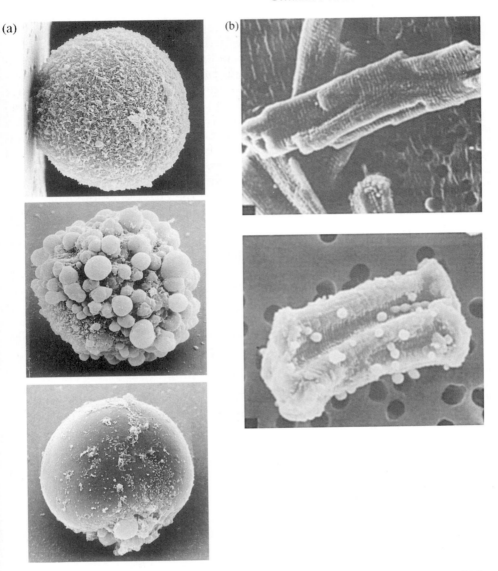

Fig. 4.4. Bleb formation. (a) Hepatocytes were exposed to the quinone menadione, which generates oxidative stress. Top: untreated cell, showing microvillus surface structure. Centre: after exposure to 200 µM menadione for 30 min, showing multiple blebs. Bottom: a hepatocyte at a later stage, dominated by a single bleb which may soon rupture; the remainder of the original cell, with smaller blebs, can be seen at the bottom of the photograph. From *Trends Pharmacol. Sci.* **10**, 282 (1989) by courtesy of Professor Sten Orrenius and Elsevier Science Publishers. (b) Blebbing in rat cardiac myocytes treated with cumene hydroperoxide for 30 min. Top: normal myocytes. Bottom: myocytes after peroxide exposure. Photograph by courtesy of Dr A.A. Noronha-Dutra.

One physiological role of Ca^{2+} is the regulation of nitric oxide synthase (eNOS, nNOS) and **phospholipase A$_2$** activities. The latter enzyme cleaves membrane phospholipids to release arachidonic acid for synthesis of prostaglandins and leukotrienes (Chapter 6). Both enzymes are Ca^{2+}- and calmodulin-dependent and therefore may become abnormally active if there is an excessive rise in intracellular free Ca^{2+}. Thus excess production of free fatty acids and of $NO^\bullet$ can occur. $NO^\bullet$ can cause direct damage to cells (Chapter 2). It can also lead to formation of $ONOO^-$, with multiple potential cytotoxic effects (Chapter 2). Over-production of fatty acids and the hydrolysis products of membrane lipids can lead to enhanced prostaglandin production and exert detergent–like effects on membranes. The rise in Ca^{2+} can also lead to a rise in $O_2^{\bullet-}$ production; in certain cells containing the enzyme xanthine dehydrogenase, activation of calpains can convert this to xanthine oxidase, generating more ROS (Chapters 1 and 9). Rises in Ca^{2+} can also activate **Ca^{2+}-dependent endonucleases** in the cell nucleus to cause DNA fragmentation, an event which is important in apoptosis.

If blebbing proceeds to such an extent that bleb rupture occurs (Fig. 4.4) without immediate resealing, the cell loses its ion gradients and is effectively dead. The cell contents are released into the surrounding area, i.e. necrotic cell death occurs. Death involving excessive rises in Ca^{2+} may be a common feature of multiple noxious stimuli, including oxidative stress and, ironically, O_2 deprivation (Table 4.3).

4.4.3 *The mitochondrial permeability transition*

The inner mitochondrial membrane is impermeable to most molecules; indeed, this is essential to maintain the proton gradient needed for ATP synthesis

Table 4.3. Some agents causing necrotic cell death in which excessive rises in intracellular Ca^{2+} have been implicated

Agent	Comments
O_2 deprivation. Inhibition of ATP-producing pathways (e.g. cyanide, rotenone, iodoacetate, mitochondrial uncouplers such as dinitrophenol)	Leads to severe fall in ATP; plasma membrane and ER Ca^{2+} pumps inoperative, no mitochondrial Ca^{2+} uptake (probably Ca^{2+} release); metabolic inhibition potentiates cell damage by low levels of H_2O_2
A23187 (calcimycin)	Ca^{2+} ionophore, carries Ca^{2+} into the cell, raises Ca^{2+} to extracellular levels
Quinones/semiquinones	Can act both by producing ROS (Chapter 8) and by direct chemical modification of $-SH$ groups, depleting GSH and inhibiting Ca^{2+}-sequestration systems
Neurotoxic metals, e.g. lead, mercury, tin	See Chapter 8

(Chapter 1). Under certain conditions this permeability is lost because 'pores' open in the membrane that allow movement of molecules smaller than about 1500 relative molecular mass.[46] ATP synthesis thus stops.

Pore opening requires elevated Ca^{2+} levels plus another 'inducing agent'; effective agents include organic peroxides, agents oxidizing $-SH$ groups, and peroxynitrite.[65,89] As solutes (including mitochondrial GSH and Ca^{2+}) escape from the matrix through the pore, osmotic imbalance occurs and the mitochondria swell. The pore opening can be blocked by **cyclosporin**, a drug used to prevent transplant rejection.

The physiological role of this 'mitochondrial permeability transition' is unclear, but it provides a mechanism by which large rises in free Ca^{2+} in cells might effectively destroy mitochondrial function.

4.5   Oxidative stress and transition metals

4.5.1   *Iron*

Just as oxidative stress dysregulates Ca^{2+} metabolism, it also interferes with iron (and probably copper) metabolism in cells and tissues. Cells contain a 'low molecular mass' iron pool, potentially catalytic for free-radical reactions (Chapter 3). The overall effect of oxidative stress is to increase the levels of potentially catalytic metals. Mechanisms include the following:

(a) Necrotic cell death causes iron ion release into the surrounding environment (Fig. 4.3). A simple illustration is that homogenates of tissues, especially brain, undergo lipid peroxidation faster than intact tissues, and such peroxidation is largely inhibited by iron ion chelating agents.[43]
(b) Superoxide can release iron from ferritin and iron–sulphur proteins (Chapters 2 and 3).
(c) Peroxides can release iron by degrading haem proteins, e.g. myoglobin, haemoglobin, cytochrome *c* and cytochrome P450[91] (Chapter 3).
(d) Peroxynitrite can attack iron–sulphur proteins, releasing iron. The ability of excess $NO^{\bullet}$ to damage mitochondrial Fe–S centres reported in some papers is probably mediated via $ONOO^-$ formation (Chapter 2).

Thus pre-incubating cells with iron ion-chelating agents can decrease oxidative damage, and cells deficient in SOD are hypersensitive to injury by H_2O_2 (Chapter 3). It is ironic that chelators designed for other metals can sometimes bind iron and alter rates of oxidative damage. For example, **Quin-2** was one of the first fluorescent 'probes' used to monitor intracellular Ca^{2+}. Unfortunately, it is also a powerful chelator of iron ions and the ferric–Quin 2 complex can react[103] with H_2O_2 to form $OH^{\bullet}$. Another 'probe' for Ca^{2+} within cells is **ruthenium red**, which contains ruthenium (Ru) ions. Ru(III) ions can react with H_2O_2[74] to form $OH^{\bullet}$.

4.5.2   *Evidence for dysregulation of iron*

Evidence consistent with changes in intracellular iron in cultured cells has been obtained using the fluorescent iron-sensitive probe **calcein** (Chapter 3). When

K562 cells were treated with $250 \mu M$ H_2O_2 or $100 \mu M$ *t*-butyl hydroperoxide, the signal from the probe indicated a significant rise in intracellular 'free'[b] iron (i.e. iron available for binding to calcein). Iron levels returned to normal in about 1 h.

The **bleomycin assay** was developed as a first attempt to measure the availability in animal body fluids of iron that might promote free-radical reactions.[44] The antibiotic bleomycin requires iron salts in order to degrade DNA (Chapter 9). If other reagents are present in excess, the extent of the DNA degradation is proportional to the amount of iron that can be bound by bleomycin (Fig. 4.5). Hence the assay measures bleomycin-chelatable iron or bleomycin-detectable iron (BDI).

Bleomycin has a fairly low affinity for iron, and cannot remove iron ions at pH 7.4 from pure iron proteins (e.g. lactoferrin, transferrin, ferritin, myoglobin, haemoglobin and catalase). Hence no bleomycin-detectable iron is present in freshly prepared serum or plasma from healthy animals (Table 4.5), because most iron is bound to transferrin, and some to ferritin. BDI is, however, present in lung lining fluid, some sweat samples and in the plasma from patients with iron-overload diseases and certain premature babies (Table 4.5).

BDI presumably represents iron ions bound to low-molecular-mass chelating agents such as citrate, or loosely bound to certain proteins, such as albumin. Several studies upon human body fluids show that the iron measurable in the bleomycin assay often (but not always) correlates with the ability of these fluids

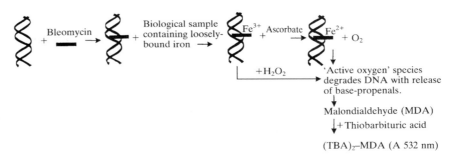

Fig. 4.5. The bleomycin assay. The products of DNA degradation by bleomycin include base propenals which react, on heating with thiobarbituric acid (TBA) at low pH, to form a pink TBA–malondialdehyde adduct. The assay is highly sensitive and readily detects the traces of iron contamination present in most biological reagents. It can also be used to detect traces of iron contamination in fluids for biomedical use (Table 4.5). Some contamination of laboratory reagents can be removed by treating reagents with Chelex resin, which binds both iron and copper salts. However, there is little point in treating water with Chelex and then dissolving other reagents in it; commercial reagents (especially phosphates and phosphate esters) usually contribute much more iron than does the water (Table 4.4). In addition, Chelex treatment usually leaves the solution with an alkaline pH and this limits the treatment of most biological buffers. A more effective procedure (*FEBS Lett.* **214**, 362 (1987)) for the removal of iron ions from biological buffers is to use columns or dialysis sacs containing iron-binding proteins, e.g. transferrin or conalbumin (much cheaper). Careful control of pH in the bleomycin assay is essential (*Methods Enzymol.* **233**, 82 (1994)).

Table 4.4. Iron contamination of laboratory reagents and biomedical fluids

Solution analysed	Iron concentration
Laboratory reagents [a]	
hydrochloric acid, 5.8 M	1–2 μmol/l[b]
saline buffer (67.5 mM Na$_2$HPO$_4$ + 4 mM KCl adjusted to pH 7.4 with HCl)	
fresh	10 μmol/l[b]
old (stored in laboratory for several weeks in a flask covered with Parafilm[c])	18 μmol/l[b]
EDTA, 50 mM	8 μmol/l[b]
sodium formate, 0.5 M	9 μmol/l[b]
urea, 0.5 M	6 μmol/l[b]
thiourea, 0.5 M	3 μmol/l[b]
ascorbic acid, 20 mM	4 μmol/l[b]
Biomedical fluids	
clinical albumin solutions	15–32 μmol/mmol albumin[d]
University of Wisconsin organ preservation fluid (used for preservation of heart and liver before transplantation)	1.9 ± 0.4 μM[e]
Intralipid infusion for premature babies	0–0.6 μM[e]

[a]Solutions were made up in double-distilled water that itself contained no detectable iron. These traces of iron provide a catalyst for free-radical reactions such as lipid peroxidation and OH$^\bullet$ generation (*J. Inorg. Biochem.* **14**, 127 (1981)).
[b]Measured by atomic absorption.
[c]Parafilm can also contain the antioxidant butylated hydroxytoluene and release it into reaction mixtures (*Plant Cell Rep.* **16**, 192 (1996)).
[d]Samples from six different manufacturers; also present were 1–7.9 μmol/mmol copper and 2.8–12.1 μmol/mmol vanadium (*J. Pharm. Sci.* **81**, 611 (1992)).
[e]Measured by the bleomycin assay (*Transplantation* **62**, 1046 (1996)).

to stimulate free-radical reactions. For example, plasma samples from iron-overloaded patients could stimulate peroxidation of lipids *in vitro*, and BDI levels were associated with increased oxidative protein damage in patients undergoing cancer chemotherapy. Application of the bleomycin assay to tissue homogenates also measures iron: if the tissue is first subjected to an insult (e.g. ischaemia–reperfusion) before assay, the levels of BDI measured in homogenates tend to increase.

Other assays to measure 'catalytic' iron have been developed, based on the same principle as the bleomycin assay.[28,53] For example, iron chelatable by desferrioxamine or by nitrilotriacetic acid (NTA) has been measured, usually by HPLC analysis of their iron chelates. For example, ischaemia–reperfusion of rabbit kidneys prior to homogenization increased the amount of desferrioxamine-chelatable iron in the homogenates.[53] NTA and desferrioxamine are much stronger iron-chelating agents than bleomycin and NTA in particular may be capable of removing iron from certain iron proteins. Desferrioxamine is often used to inhibit oxidative stress *in vitro* and *in vivo* (Chapter 10), and

Table 4.5. Concentrations of bleomycin-detectable iron (BDI) in body fluids

Sample	BDI concentration	Comments
Healthy animals (plasma)	0	Never present: has considerable iron-*binding* activity due to unsaturated transferrin
Rheumatoid arthritis		
plasma	0	Never present unless arthritis related to an iron-overload disease
synovial fluid	$3.1 \pm 1.9\,\mu M$	About 40% of fluids show BDI if assayed at pH 5.3 (*Biochem. J.* **245**, 415 (1987))
Fulminant liver failure (plasma)	$2 \pm 2\,\mu M$	Possibly a consequence of iron release from failing liver and/or its inability to make transferrin (*Free Rad. Res.* **20**, 139 (1994))
Acute lymphoblastic leukaemia, after chemotherapy (plasma)	up to $1\,\mu M$	Transferrin saturation in patients pre-chemotherapy higher than normal, BDI sometimes but not usually present. Chemotherapy causes rise in total plasma iron; when transferrin saturation reaches 100% BDI appears (*Cancer Lett.* **94**, 219 (1995))
ARDS patients with multi-organ failure (plasma)	$0.5 \pm 0.2\,\mu M$	Represents 33% of ARDS patients in this study (*Thorax* **49**, 702 (1994))
Lung lining fluid (bronchoalveolar lavage)		
normal controls	$0.3 \pm 0.02\,\mu M$	BDI appears present normally in BAL fluid, which has little transferrin. In severe ARDS, excess entry of plasma proteins such as transferrin due to disruption of alveolar permeability barrier may bind the BDI (*Biochem. Biophys. Res. Commun.* **220**, 1024 (1996))
ARDS patients		
survivors	$0.3 \pm 0.09\,\mu M$	
non-survivors	0	
Cardiopulmonary bypass; extracorporeal blood circulation	$0.3 \pm 0.9\,\mu M$	Perhaps released due to damage to blood cells or anoxic/reperfused tissues (*FEBS Lett.* **328**, 103 (1993); *Free Rad. Res.* **21**, 53 (1994); *Ann. Thoracic Surg.* **60**, 1735 (1995))
blood cardioplegia (plasma)	$0.2 \pm 0.3\,\mu M$	

Mice severely deficient in ability to make transferrin (plasma)	16.0 ± 5.0 µM	Mice die unless transferrin periodically injected (*Biochim. Biophys. Acta* **1156**, 19 (1992))
Patients with iron overload due to idiopathic haemochromatosis (Chapter 3) (plasma)	4.3 ± 6.7 µM	At least some of the iron is bound to citrate (*Clin. Sci.* **68**, 463 (1985); *J. Biol. Chem.* **264**, 4417 (1989))
Umbilical cord blood plasma newborn full-term babies	0.3 ± 0.6 µM[a]	Transferrin saturation is high at birth and may go over 100% in a few, apparently normal babies (*FEBS Lett.* **303**, 210 (1992))
premature babies	~2–5 µM[b]	*Free Rad. Res.* **16**, 285 (1995)
Kwashiorkor (plasma)	1.0–19.5 µM	BDI present in 58% of Kwashiorkor patients (*Eur. J. Clin. Nutr.* **49**, 208 (1995))
Human sweat	0.7 ± 2.5 µM, 4.6 ± 3.0 µM	BDI much more often present in trunk sweat than arm sweat (*Clin. Chim. Acta* **145**, 267 (1985))
Urine, kidney reperfusion injury before ischaemia–reperfusion	138 ± 54 pmol	*Kidney Int.* **34**, 474 (1988)
after ischaemia–reperfusion	1249 ± 506 pmol	
Homogenized human tissues cerebellum	26 ± 5 nmol/mg tissue	*FEBS Lett.* **353**, 246 (1994);
substantia nigra	51 ± 27 nmol/mg tissue	
arterial wall	3 ± 2.4 µM	*Free Rad. Res.* **23**, 465 (1995)
Bacteria (five different strains)	0.003–1.046 µmol/ mg protein	*Biochem. Int.* **8**, 89 (1984)

[a] Only 6/25 positive.
[b] High % of samples positive.

HPLC analysis of its iron(III) chelate, ferrioxamine, provides a useful confirmatory test that it has in fact acted as an iron chelator.[53]

It should be noted that cell culture media, biochemical and clinical reagents are all likely to carry traces of iron contamination (Table 4.4), which must be allowed for in performing the above assays.

4.5.3 *Copper*

As well as Ca^{2+} and iron, oxidative stress may cause increased availability of copper ions. For example, caeruloplasmin[43] is readily degraded by proteases or on exposure to $ONOO^-$ to release copper, although it is more resistant to damage by peroxides (Chapter 3). Copper ions are released when tissues are homogenized, e.g. brain and arterial wall (Table 4.6).

In human blood plasma, essentially all the copper is bound to caeruloplasmin (Chapter 3). However, an assay for non-caeruloplasmin copper, the **phenanthroline assay**, has been developed, somewhat analogous to the bleomycin assay. It is based on the ability of the chelating agent, 1,10-phenanthroline (*o*-phenanthroline) to degrade DNA in the presence of copper ions, O_2 and a reducing agent.[94] Degradation of DNA by the copper–phenanthroline complex involves reduction of Cu^{2+} to Cu^+ by the reducing agent, followed by its reaction with H_2O_2 to form a species (probably $OH^\bullet$) that immediately attacks the adjacent DNA. Degradation results in the release of a product from DNA that reacts, upon heating with thiobarbituric acid at acidic pH, to form a pink chromogen.

If all other reagents are in excess, the technique measures 'available' copper in biological fluids, i.e. copper that can be chelated by phenanthroline. Reagents are treated with Chelex (Fig. 4.5, legend) to remove contaminating copper ions, and azide is added to inactivate catalase (H_2O_2 is needed for the DNA degradation). Any 'available copper' in the test system is chelated by phenanthroline and reduced by added mercaptoethanol; the resulting damage to DNA is proportional to the amount of **phenanthroline-detectable copper**. This assay will detect copper bound to the copper-binding sites of albumin and to amino acids such as histidine, but not copper on caeruloplasmin.

Phenanthroline-detectable copper has been found in numerous biological samples (Table 4.6). It is not present in plasma from healthy animals (which suggests that levels of copper–albumin and copper–amino acid chelates are low, or zero). However, if plasma or serum are allowed to stand at $4\,^\circ C$ for a few days (or frozen for a few months), phenanthroline-detectable copper appears, as caeruloplasmin undergoes degradation (Section 3.18).

4.6 Mechanisms of damage to cellular targets by oxidative stress: DNA

There is increasing evidence that ROS and RNS are involved in the development of cancer, not only by direct effects on DNA but also by affecting signal transduction, cell proliferation, cell death and intercellular

Table 4.6. Phenanthroline-detectable copper in some biological materials

Material	Concentration	Comments
Healthy animals (plasma/serum)	0	Never detected in fresh samples. Amount of non-caeruloplasmin copper must be <0.1 μM. Detected in *stored* samples, probably due to breakdown of caeruloplasmin on storage *Pediat. Res.* **37**, 219 (1995). Variably present in patients treated with copper-chelating agents
Wilson's disease,[a] untreated (plasma)	2.7 μM	
Liver failure (plasma)	1–2 μM	May be due to metal release from failing liver and/or cessation of liver caeruloplasmin synthesis (*Free Rad. Res.* **20**, 139 (1994))
Sweat, human		Sweat also contains iron (Table 4.5) and may be an excretory route for both metals. Their presence facilitates bacterial growth in sweat (*Clin. Chim. Acta* **145**, 267 (1985))
arm	18.1 ± 3.9 μM	
trunk	27.0 ± 3 μM	
Homogenized human brain tissue		Levels of copper released on tissue homogenization broadly comparable to those of BDI (*FEBS Lett.* **353**, 246 (1994); *Free Rad. Res.* **23**, 465 (1995))
cerebellum	21 ± 10 nmol/mg tissue	
substantia nigra	53 ± 18 nmol/mg tissue	
arterial wall	4.9 ± 2.8 μM	
Atherosclerotic material 'gruel' from advanced lesions	~4.0 μM	Copper ions are powerful catalysts of low-density lipoprotein (LDL) oxidation, which may contribute to atherosclerosis (Chapter 9) (*Biochem. J.* **286**, 901 (1992); *FEBS Lett.* **368**, 513 (1995)). Atherosclerotic lesion extracts will stimulate *in vitro* LDL oxidation.
endarterectomy samples	0.3 ± 0.1 nmol/mg protein	
aneurysms	0.3 ± 0.2 nmol/mg protein	

[a] A 'copper-overload' disease (Chapter 3).

communication (Chapter 9). ROS/RNS can lead to DNA damage by direct chemical attack on DNA, and also by indirect mechanisms. Examples of the latter include the activation of Ca^{2+}-dependent endonucleases as a consequence of rises in intracellular free Ca^{2+}, and interference with enzymes that replicate or repair DNA. This section will focus on direct damage to DNA. Such damage can occur to the purine/pyrimidine bases and/or to the deoxyribose sugar (Fig. 4.6).

4.6.1   *DNA and chromatin structure*

DNA is a long, thread-like molecule made up of a large number of **deoxy-ribonucleotides**, each composed of a base, the sugar deoxyribose, and a phosphate group (Fig. 4.6). The two purine bases in DNA are **adenine** and **guanine** and the pyrimidines are **thymine** and **cytosine**. The backbone of DNA consists of deoxyriboses linked by esterifying the −OH group on position 5 of one to that on position 3 of the next by a phosphate group (a **phosphodiester bond**). The presence of multiple phosphate groups gives DNA a substantial negative charge since phosphates are ionized at physiological pH. Hence DNA can readily bind metal cations such as Na^+, K^+, iron and copper ions. In the cell nucleus, two DNA chains are associated into a **double helix** with the hydrophobic purines and pyrimidines inside and the hydrophilic phosphates and deoxyribose sugar units on the outside. The chains run in opposite directions, one $5' \rightarrow 3'$ and the other $3' \rightarrow 5'$. The helix is held together by hydrophobic interactions of the bases and by hydrogen bonds (G≡C, A=T; Fig. 4.6). When DNA is heated to high temperatures these bonds rupture and the strands separate (**DNA denaturation** or **melting**). On cooling, the strands return together (**renaturation** or **annealing**). These properties of DNA are made use of in molecular biology experiments (Appendix II). The double helix contains two kinds of grooves, a **major groove** and a **minor groove**. The sequence of bases on DNA carries the genetic information (Appendix II). *E. coli* DNA contains 4.7×10^6 base pairs and forms a continuous circle whereas human nuclear DNA is linear and has about 3×10^9 base pairs.

In the nuclei of eukaryotic cells, DNA is packaged into **chromosomes** (46 in humans), each of which contains a single length of double-helical DNA. The chromosomes are paired: one in each pair is inherited from the male parent and the other from the mother. There are 22 pairs of **autosomes** (genes on them are autosomally inherited) and two **sex chromosomes** (carrying sex-linked genes). For example, in the sex chromosomes in males a Y chromosome is inherited from the father and an X chromosome from the mother. Each human chromosome contains $50-250 \times 10^6$ base pairs. Proteins (**histones**) containing many basic amino acid residues (which are positively charged at physiological pH) help in the packaging of DNA in chromosomes; DNA is wound around histones to form **nucleosomes**, arranged like beads on a string. The fundamental 'packing unit' of DNA in chromatin is the nucleosome; these are repeating units, each containing about 200 base pairs of DNA and two molecules each of histones H2A, H2B, H3 and H4 (Fig. 4.6). The nucleosomes are separated by flexible linker DNA. The entire complex of

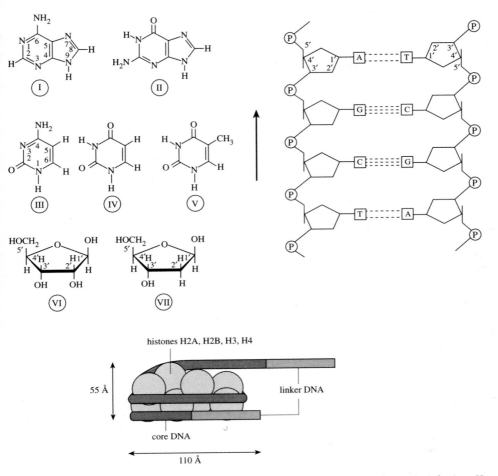

Fig. 4.6. Nucleic acid structure. Top left: structures of DNA and RNA bases. I, Adenine; II, guanine; III, cytosine (all found in both RNA and DNA); IV, uracil (found only in RNA); V, thymine (found only in DNA); VI, D–ribose (the sugar found in RNA); VII, 2–deoxy-D–ribose ('deoxyribose', found in DNA). Top right: hydrogen bonding in DNA. Adenine pairs with thymine by two hydrogen bonds and guanine with cytosine by three hydrogen bonds. A **nucleoside** is a base attached to a sugar (e.g. for a ribose sugar the nucleosides are adenosine, cytosine, guanosine, thymidine; in a deoxyribose sugar they are deoxyadenosine (dA), deoxycytosine (dC), deoxyguanosine (dG) and deoxythymidine (dT)). When a phosphate is further attached, a **nucleotide** results (e.g. adenosine monophosphate (AMP), GMP, CMP, TMP or dAMP, dGMP, dCMP, dTMP). Bottom: schematic diagram of a nucleosome core particle. There are five different histone types, H1, H2A, H2B, H3 and H4. The DNA molecule is wound $1\frac{3}{4}$ turns around a histone octamer (two molecules each of histones H2A, H2B, H3 and H4). Histone H1 (not shown) is bound to the linker DNA and helps to pack together adjacent nucleosomes into further-folded structures. Adapted from *Trends Pharmacol. Sci.* **9**, 402 (1988) by courtesy of Dr L. Hurley and Elsevier, Cambridge.

DNA with histones and other proteins is called **chromatin**. In principle, the default state of DNA in chromosomes is to be fully assembled into nucleosomes. The non-nucleosomal regions are often sites at which gene transcription is initiated and are maintained as non-nucleosomal sites by proteins that regulate gene transcription (**transcription factors**).

4.6.2   *DNA cleavage and replication*

When chromatin is attacked by nuclease enzymes, it is first cleaved at the internucleosomal sites, since the histones protect the rest of the DNA. If the digested DNA is subjected to gel electrophoresis, a set of discrete bands called a **ladder** is observed: the DNA contents of these bands represent DNA fragments as multiples of the basic 200 base-pair unit (Fig. 4.7).

DNA is replicated by enzymes called **DNA polymerases** which use deoxyribonucleoside triphosphates (dATP, dGTP, dCTP and dTTP) to add deoxyribonucleotides, e.g.

$$DNA + dATP \rightarrow (DNA{-}A) + pyrophosphate$$

$$(DNA{-}A) + dGTP \rightarrow (DNA{-}A{-}G) + pyrophosphate$$

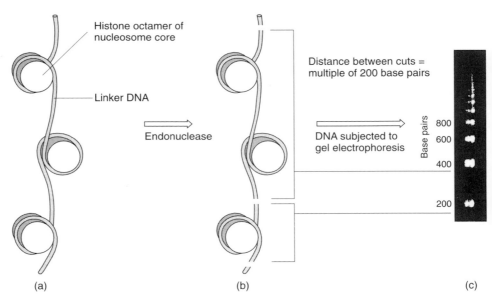

Fig. 4.7. Internucleosomal cleavage of DNA. High molecular weight DNA, with histones arranged like beads on a string (a), is cleaved with an endonuclease which cuts between nucleosomes to give fragments of DNA that are multiples of 180–200 base pairs. When these fragments are separated by gel electrophoresis, a 'ladder' is seen (c). From 'Guide to Cell Proliferation and Apoptosis Methods', by courtesy of Boehringer Mannheim.

DNA polymerases add deoxyribonucleotides to the 3'-OH terminus of a pre-existing DNA chain, the **primer**. They will not copy a single strand of DNA unless a primer is present to give a 'starter point' for the second strand. Polymerases can copy from a complete double helix only if the sugar–phosphate backbone is broken and the helix partly unravelled at the site where copying starts. DNA polymerase catalyses the formation of a phosphodiester bond only if the base on the incoming nucleotide hydrogen bonds with (is **complementary to**) the base on the strand being copied. Once a base has been inserted, most DNA polymerases 'double-check' it and excise it if it is wrong (the **proof-reading** or **error-checking function**). Hence DNA replication is highly accurate (proceeds with a high degree of **fidelity**). Most mutations seem to be caused by chemical attack on, or spontaneous degradation of, the DNA itself rather than errors of the polymerases. By contrast, the RNA polymerases that copy DNA into RNA, and the reverse transcriptases found in some viruses (Appendix II) do not have an error-checking function and make more mistakes.

Eukaryotic cells contain at least five DNA polymerases. In human cells, it is thought that DNA polymerase α starts DNA replication and that polymerases δ (and possibly ε) perform the bulk of it. DNA polymerase β is involved largely in repair. DNA polymerase γ is present in mitochondria and replicates mitochondrial DNA. None of these polymerases can start DNA chains *de novo*: they need a primer. Indeed, in DNA replication a short RNA primer is used, made by an RNA polymerase (**primase**). This primer RNA is removed at the end of replication by the nuclease activity of DNA polymerase.

4.6.3   Telomeres

The need for a primer could cause a problem for eukaryotic chromosomes; the daughter DNA molecule would have an incomplete 5' end and should get shorter with every round of replication. This problem is avoided because the ends of chromosomes contain hundreds of repeats in tandem of a six-nucleotide sequence (AGGGTT in humans). These ends are called **telomeres** (after the Greek *telos*, 'an end'). An RNA–protein complex enzyme called **telomerase** resynthesizes the telomeric sequences using its internal RNA as a template. This is further discussed in Chapter 9 in the context of cancer and ageing.

4.6.4   Damage to DNA by ROS and RNS

Measurement of strand breakage

DNA itself is a very stable molecule but can undergo 'spontaneous' chemical decomposition over a lifetime.[71] Loss of purines (leaving **apurinic sites**) may occur 10^4 times per day in the human genome. Cytosine can slowly lose its amino group (**deaminate**) to generate uracil (Fig. 4.6), whereas 5-methyl-cytosine in DNA (Appendix II) can deaminate to thymine. However, oxidative stress greatly accelerates DNA damage. Thus DNA strand breakage is frequently observed in cells subjected to oxidative stress, which can be done by any of the following methods:

(a) exposure to ionizing radiation, elevated O_2 concentrations, activated phagocytic cells, 'redox cycling' chemicals (Chapter 8), cigarette smoke or ozone;

(b) addition of H_2O_2, peroxynitrite, nitrous acid, organic hydroperoxides;

(c) exposure to 'autoxidizing' chemicals (e.g. dihydroxyfumarate, dopamine, L-DOPA, noradrenalin, adrenalin);

(d) exposure to xanthine oxidase plus its substrates (xanthine or hypoxanthine) (Care must be taken in the use of commercial xanthine oxidase, which is often heavily contaminated with proteases, nucleases and other material directly injurious to cells. A control without substrates is always needed to check for effects of the enzyme preparation itself.);

(e) addition of tumour necrosis factor alpha (TNFα).[92a]

Strand breakage can be measured by several methods (Table 4.7). Most strand breaks detected by these techniques exist as such in the DNA, whereas others are *created* by some of the measurement methods, in which sugar or base damage is converted into strand breakage on exposure of DNA to high pH during the assay (**alkali–labile sites**). As examples of the various methods available, Fig. 4.7 shows the ladder pattern produced by internucleosomal DNA cleavage and Fig. 4.8 (see plate section) shows the use of the 'comet' assay to measure strand breakage in cells treated with H_2O_2. Single-strand breaks are usually efficiently repaired by the cell (see below) but double-strand breaks are more of a problem.

Effects of $NO^{\bullet}$, $O_2^{\bullet-}$, H_2O_2 and $OH^{\bullet}$ on DNA bases

What ROS/RNS damage DNA? Superoxide, nitric oxide or H_2O_2, at physiologically relevant levels, do not appear to react with any of the DNA or RNA bases or with the ribose or deoxyribose sugars at significant rates. By contrast, as might be expected from the high reactivity of hydroxyl radical, exposure of DNA to $OH^{\bullet}$ generates a multitude of products, since it attacks sugars, purines and pyrimidines.[10,116] For example, $OH^{\bullet}$ can add on to guanine at positions 4, 5 or 8 in the purine ring. Addition to C-8 produces a C-8 OH–adduct radical that can be reduced to 8-hydroxy-7,8-dihydroguanine, oxidized to 8-hydroxyguanine or undergo opening of the imidazole ring, followed by one–electron reduction and protonation, to give 2,6–diamino-4-hydroxy-5-formamidopyrimidine, usually abbreviated as FAPyG (Fig. 4.9).

Similarly, $OH^{\bullet}$ can add on to C-4, C-5, or C-8 of adenine. Pyrimidines are also attacked by $OH^{\bullet}$ to give multiple products. Thus, thymine can suffer hydrogen atom abstraction from the ring or from the methyl group. The resulting radicals are converted into various thymine peroxides, which can break down to *cis*- and *trans*-thymine glycols (5,6-dihydroxy-6-hydrothymines), and 5-hydroxy-5-methylhydantoin, 5-hydroxy-6-hydrothymine, 6-hydroxy-5-hydrothymine and 5-(hydroxymethyl) uracil. Cytosine can form several products, including cytosine glycol and 5,6-dihydroxycytosine

Table 4.7. Some of the methods available to measure strand breakage in DNA

Method	Comment
[³H]thymidine incorporation	Cells are incubated with radioactive (tritiated) thymidine, which is incorporated into nuclear DNA. Production of radioactive fragments measured. Only useful for proliferating cells which can incorporate DNA precursors.
Incorporation of BrdU[a]	Non-radioactive alternative to [³H]thymidine incorporation. Incorporated into DNA. Fragments detected using antibodies against DNA containing BrdU.
Enzymatic labelling techniques	Free OH groups created by strand breakage are labelled with modified (e.g. fluorescent) nucleotides in the presence of enzymes, e.g. terminal deoxynucleotidyl transferase, TdT. The incorporated modified bases are detected fluorimetrically or using antibodies. A popular method is **TUNEL** (TdT-mediated X-dUTP nick end-labelling). Requires permeabilization of the cell to allow enzymes in.
Gel electrophoresis	See Fig. 4.7
'Comet' assay (single-cell gel electrophoresis)	See Fig. 4.8
Alkaline unwinding techniques	Differential elution allows detection of strand breaks in DNA, DNA interstrand cross-links and DNA–protein cross-links. The principle is that the DNA double helix unwinds at high pH. Unwinding begins from 'free ends', so is faster if strand breaks are present. Thus retention of unwound DNA on filters can be measured (e.g. if DNA is prelabelled). An adaptation of this method is fluorimetric analysis of DNA unwinding (**FADU**): the fluorescence of a fluorochrome (e.g. bis-benzamide) bound to DNA is used to measure residual duplex DNA after a period of alkaline unwinding.
Use of supercoiled plasmid DNA	Strand breakage causes 'relaxation' to open circle and linear forms, measured as different migration on agarose gel electrophoresis or as increased fluorescence due to more intercalation of **ethidium bromide** in the relaxed DNA. Ethidium is a planar aromatic dye which readily intercalates between base pairs, although this is restricted in intact DNA by its resistance to distortion.
Cell DNA fragmentation ELISA	An antibody-based technique which detects histone-covered DNA fragments in cell cytoplasm; often used to study apoptosis, which generates DNA fragments inside cells.

[a]5-Bromo-2'-deoxyuridine.

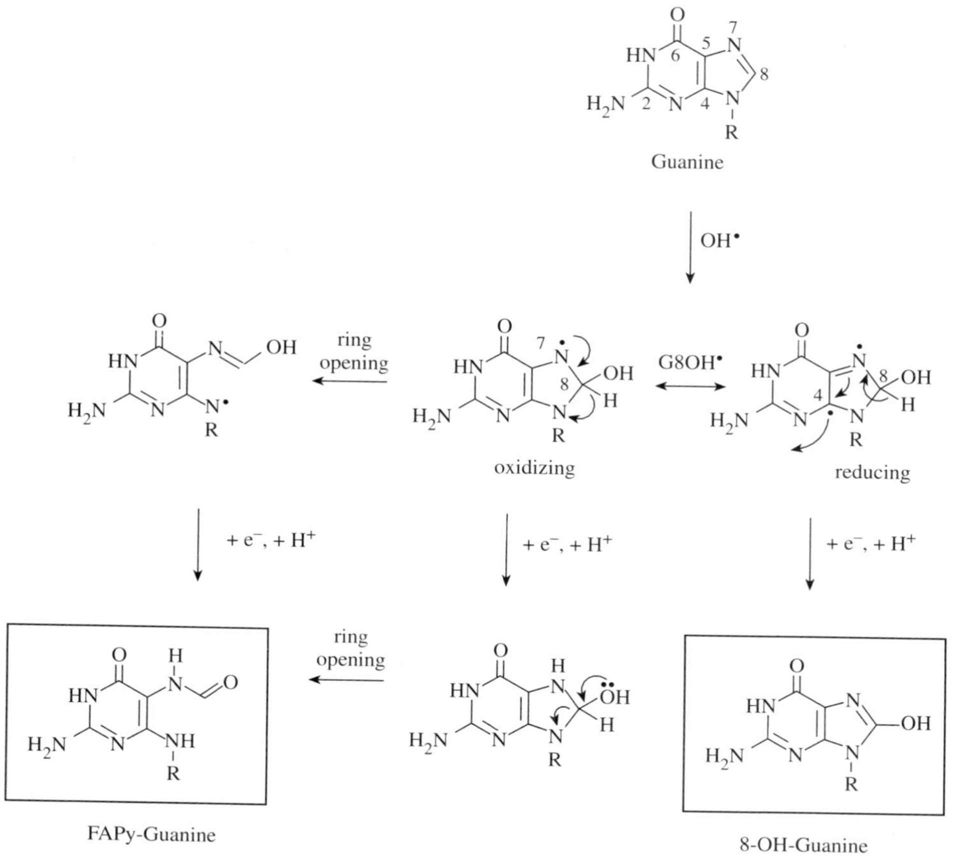

Fig. 4.9. Guanine modification by hydroxyl radicals. Addition of OH• radicals to C-8 of guanine in DNA generates an 8-hydroxyguanine radical (when a radical reacts with a non-radical, a new radical is formed). This 8-hydroxyguanine radical can be oxidized (losing one electron) to 8-hydroxyguanine or reduced to give a ring-opened product. Thus to assess radical attack on C-8 of guanine, both 2,6-diamino-4-hydroxy-5-formamidopyrimidine (FAPyG) and 8-hydroxyguanine should be measured, since reaction conditions can affect the ratio between them, i.e. the same amount of OH• attack on guanine can lead to different levels of 8-hydroxyguanine. Attack of OH• at other positions upon guanine is possible. Adapted from *Free Rad. Biol. Med.* **18**, 1033 (1995) by courtesy of Professor John Murphy, Dr Tony Breen and Elsevier Publishers.

(Fig. 4.10 shows the structures of some of these products). The presence of transition–metal ions affects the rate and mechanism of the peroxide decomposition pathways and hence the ratios of the various end-products.[114]

Sugar damage and DNA–protein cross-linking

The deoxyribose sugar is also fragmented by OH•, yielding a multiplicity of products.[116] All positions in the sugar moiety are susceptible, with the formation of carbon-centred radicals due to hydrogen abstraction by $OH^\bullet$. In the presence of O_2 these convert rapidly to sugar peroxyl radicals, which undergo a series of reactions, including disproportionation, rearrangement, elimination of water and C–C bond fragmentation, to yield a variety of carbonyl products. Some sugar lesions cause DNA strand breakage only on subsequent alkali treatment (Section 4.6.2 above). Irradiated solutions of 2-deoxy-D-ribose produce carbonyls and dicarbonyls which are mutagenic to *Salmonella typhimurium* (in the Ames test). Malondialdehyde is also produced.

Like many carbohydrates, 2-deoxy-D-ribose has a weak iron binding property which makes it particularly vulnerable to attack by $OH^\bullet$ radicals produced during Fenton reactions. The release of MDA from 2-deoxy-D-ribose has been made the basis of an assay for $OH^\bullet$ radical (the deoxyribose assay; Chapter 5). By contrast, copper ions bind to GC-rich sequences in DNA; subsequent reaction with H_2O_2 favours formation of guanine damage products.[101]

Our detailed chemical knowledge of products of $OH^\bullet$ attack upon DNA has largely come from studies by radiation chemists. Not only $OH^\bullet$ but also $H^\bullet$ and e_{aq}^- species can attack DNA. Ionizing radiation is well known to be both mutagenic and carcinogenic and much of the cell damage caused by ionizing radiation involves the formation of $OH^\bullet$ radicals by homolysis of water (Chapter 2). As well as base and sugar modifications and single- and double-strand breaks (double-strand breaks being especially important events in cell damage by radiation), nuclear proteins can be attacked by radicals. The resulting protein-derived radicals can cross-link to base-derived radicals if the two meet in chromatin, giving **DNA–protein cross-links** (e.g. thymine to tyrosine) that interfere with chromatin unfolding, DNA repair, replication and transcription.[3] Double-strand breaks may be caused by a large amount of radiation energy being deposited in one place, causing multiple attacks by $OH^\bullet$ on the same short stretch of DNA.

Effects of scavengers

Hydroxyl-radical scavengers do not always protect DNA against damage by $OH^\bullet$. Often (as with $OH^\bullet$ generated from H_2O_2 reacting with metal ions bound to the DNA), the scavengers cannot intercept the $OH^\bullet$ formed directly upon the DNA. Another reason is that some of the scavenger-derived radicals formed can themselves cause DNA damage.[77] Radicals derived from formate, propan-2-ol, glycerol and (under anoxic conditions) dimethylsulphoxide are able to cause single-strand breaks in DNA. The latter effect appears to be due to methyl ($CH_3^\bullet$) radicals, which can also attack DNA bases. In the presence of O_2, $CH_3^\bullet$ radicals react rapidly to give $CH_3O_2^\bullet$, apparently less damaging.

Singlet O_2 and $ONOO^-$

Whereas $OH^\bullet$ is almost indiscriminate in its attack upon DNA (Fig. 4.10), **singlet O_2** is much more selective. It is inefficient at producing strand breakage, and generates mainly guanine-derived products, including 8-hydroxyguanine and FAPyG. Radicals formed during lipid peroxidation ($RO_2^\bullet$, $RO^\bullet$)

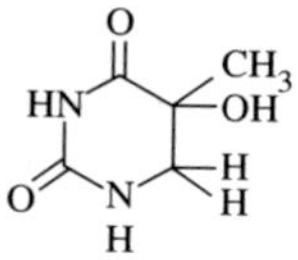

5-Hydroxy-6-hydrothymine

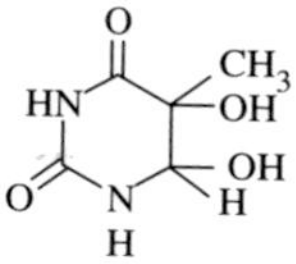

Thymine glycol (*cis-* and *trans-*)

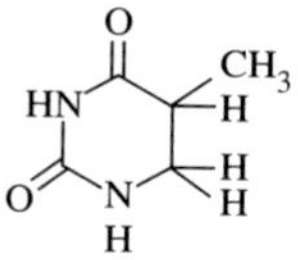

5,6-Dihydrothymine

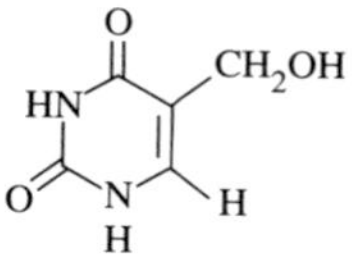

5-Hydroxymethyluracil

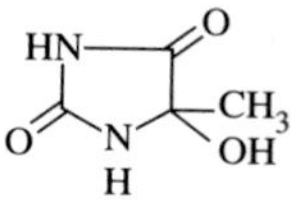

5-Hydroxy-5-methyl-hydantoin

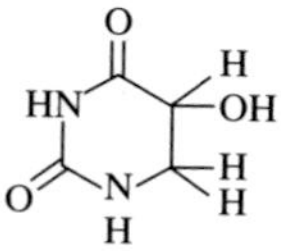

5-Hydroxy-6-hydrouracil

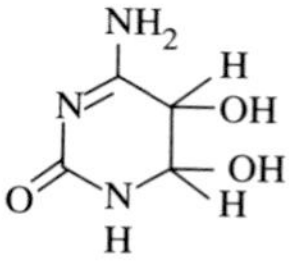

Cytosine glycol

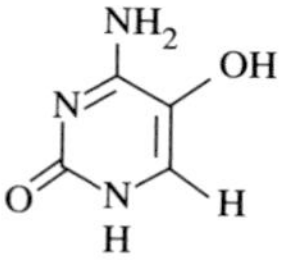

5-Hydroxycytosine

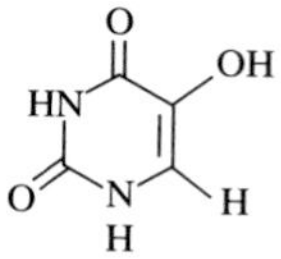

5-Hydroxyuracil

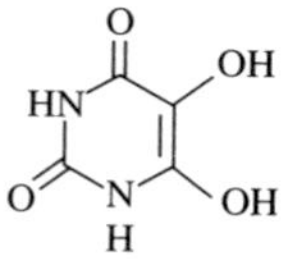

5-6-Dihydroxy-uracil

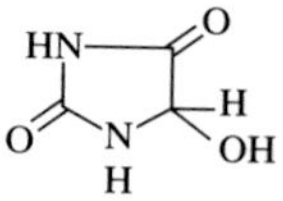

5-Hydroxyhydantoin

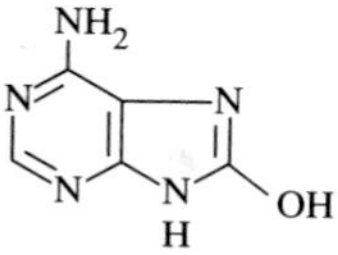

8-Hydroxyadenine

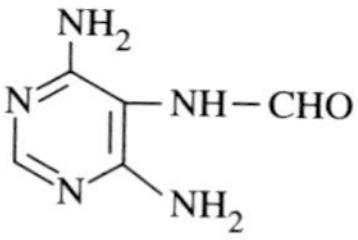

4,6-Diamino-5-formamidopyrimidine

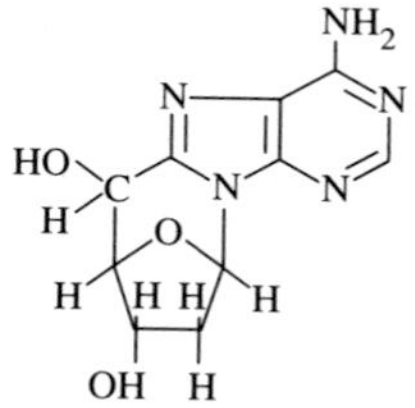

8,5'-Cyclo-2'-deoxyadenosine-(5' R- and 5' S-)

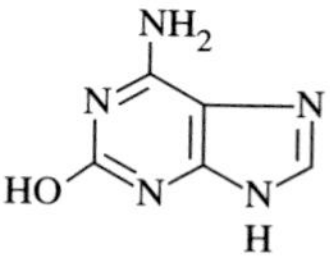

2-Hydroxyadenine

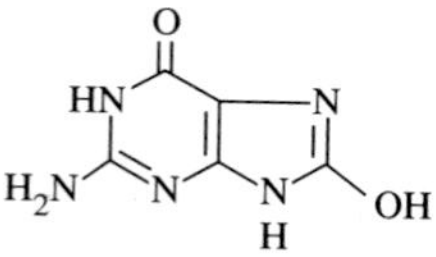

8-Hydroxyguanine

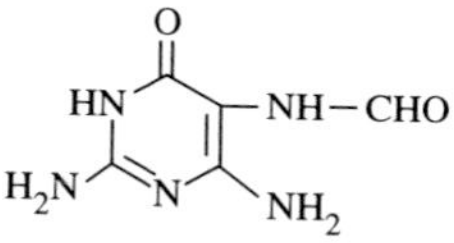

2,6-Diamino-4-hydroxy-5-formamidopyrimidine

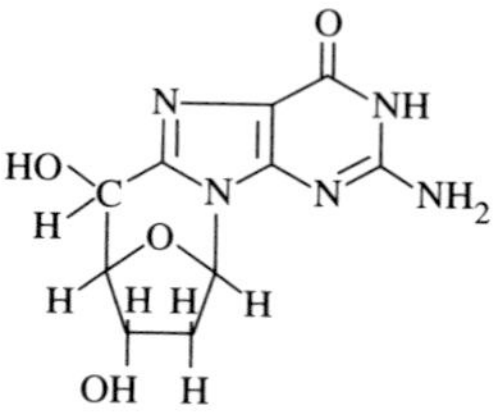

8,5'-Cyclo-2'-deoxyguanosine (5' R- and 5' S-)

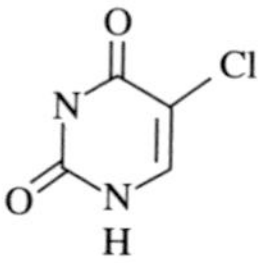

5-Chlorouracil

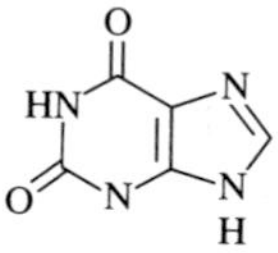

Xanthine

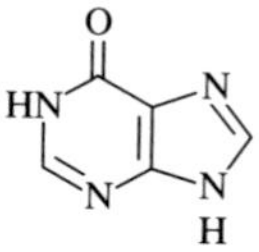

Hypoxanthine

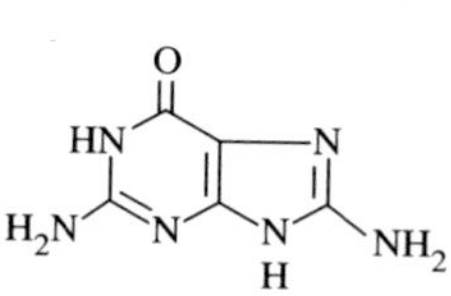

8-Nitroguanine

Fig. 4.10.

can also damage DNA, but the spectrum of products is ill-defined: in general, they do not seem to cause as much damage as $OH^\bullet$ and guanine is again a preferred target. However, AAPH-derived peroxyl radicals (Chapter 2) can oxidize thymine[73] to several products, including hydroxymethyluracil.

RNS ($NO_2^\bullet$, ONOOH, N_2O_3 and HNO_2) can produce nitration, nitrosation and deamination of DNA bases (Fig. 4.10), e.g. **8-nitroguanine** and the deamination products **xanthine** and **hypoxanthine**,[110] can result from addition of $ONOO^-$ to cells (Chapter 2). Conversion of cytosine to uracil is also accelerated by RNS. 8-Nitroguanine is unstable in DNA and spontaneously depurinates, leaving an abasic site.

Exposure of cells to ultraviolet (UV) light can cause DNA damage. First, UV can convert H_2O_2 to $OH^\bullet$,

$$H_2O_2 \xrightarrow{\text{homolysis}} 2OH^\bullet$$

Second, UV itself can cause covalent cross-linking of adjacent pyrimidines on DNA to give **pyrimidine dimers** (Chapter 7).

4.6.5 *Damage to mitochondrial and chloroplast DNA*

In both plants and animals the genetic information encoding the vast majority of proteins is encoded in nuclear DNA. However, chloroplasts and, to a lesser extent, mitochondria contain some DNA. For example, human mitochondrial DNA is a double-helical circular DNA encoding only 13 proteins (Chapter 1). The rest of the proteins needed in mitochondria are made in the cytosol using mRNA from nuclear DNA, and have to be imported.

ROS/RNS can damage mitochondrial and chloroplast DNA; mitochondrial DNA damage has been suggested to be important in several human diseases and in the ageing process (Chapters 1 and 10). Indeed, oxidative DNA base damage (measured as levels of 8-hydroxydeoxyguanosine in DNA) has been detected in rat liver and human brain mitochondrial DNA at steady-state levels several-fold higher than in nuclear DNA.[99] Which ROS or RNS are responsible has not yet been elucidated. This apparent increased oxidative damage to mitochondrial DNA compared with nuclear DNA could be because of the proximity of mitochondrial DNA to ROS generated during electron

---

Fig. 4.10. Some of the products resulting from attack of ROS/RNS upon DNA. The modified bases in the upper diagram (except for 5,6-dihydrothymine) are formed in DNA by attack of hydroxyl radical upon the DNA bases. 5,6-Dihydrothymine is a product of attack of hydrogen atoms or hydrated electrons upon thymine. The bottom shows the structures of 8-nitroguanine, formed by treatment of DNA with peroxynitrite, and of xanthine and hypoxanthine, which can result from deamination of guanine and adenine respectively (although they can be formed by other mechanisms also). 5-Chlorouracil is a chlorinated base produced when DNA is treated with HOCl and subsequently hydrolysed with acid; it probably results from breakdown of a chlorinated cytosine under acidic conditions (*Chem. Res. Toxicol.*, **10**, 1240 (1997)). Most structures by courtesy of Dr Miral Dizdaroglu.

transport, and the fact that mitochondrial DNA is not protected against attack by histones. It is also possible that DNA repair in mitochondria is less rapid than in the nucleus, so that more base damage accumulates.[119]

Intermediate radicals formed during lipid peroxidation, as well as end-products of peroxidation, can also attack DNA and have been suggested to damage mitochondrial DNA, which is in close proximity to the mitochondrial inner membrane. Oxidative damage could contribute to the deletions and other mutations in mitochondrial DNA that accumulate with age at a higher rate than in nuclear DNA (Chapter 1). Damage to mitochondrial DNA may play a role in several neurodegenerative diseases (Chapter 9) and increased mitochondrial DNA damage in atherosclerotic hearts has been reported.

The high O_2 concentrations generated during photosynthesis should also favour damage to chloroplast DNA, but little information is available on this topic. Plant nuclei contain DNA repair enzymes, but the rate of repair of UV-induced lesions (and probably of other lesions) appears slower in the chloroplast than in the nucleus.[18a]

4.6.6   *Why does hydrogen peroxide lead to DNA damage?*

DNA isolated from aerobic bacteria, plants and animals shows low levels of multiple base damage products, a pattern that is consistent with attack by $OH^\bullet$ *in vivo* (Table 4.8). Indeed, Ames[53a] has calculated that there could be up to about 10^3 oxidative damaging events upon the DNA of each cell in the human body every day. This figure, combined with the significant rates of 'spontaneous' loss of purines from DNA and deamination of cytosine to uracil, as well as any errors made during DNA replication, emphasizes the importance of DNA repair mechanisms.

Addition of H_2O_2 to many mammalian cell types produces increased strand breakage within a few minutes (Fig. 4.8) and, in the cases that have been examined, an increase in DNA base modification products (Table 4.8 gives an example). The chemical pattern of this damage (rises in the levels of multiple products from all four bases) is consistent with attack by $OH^\bullet$. Yet H_2O_2 itself, like $O_2^{\bullet-}$, does not damage DNA. How then could $OH^\bullet$ be formed to attack DNA in the nucleus and mitochondria? If $OH^\bullet$ is attacking DNA, it must be produced very close to the DNA since this radical is so reactive that it cannot diffuse from its site of formation (Chapter 2). Metabolic sources of $OH^\bullet$ or '$OH^\bullet$-like' species include the reaction of $O_2^{\bullet-}$ with HOCl and the breakdown of peroxynitrous acid (Chapter 2). However, when DNA is exposed to $ONOO^-$ at pH 7.4 the major products are nitroguanine, hypoxanthine and xanthine (Fig. 4.10) rather than those listed in Table 4.8. Similarly, HOCl rapidly chlorinates DNA bases, forming chloramines and ring chlorination products rather than purine oxidation products (Fig. 4.10).

By far the greatest interest has been in Fenton chemistry as a source of $OH^\bullet$. Pre-treatment of cells with certain metal ion chelating agents (such as *o*-phenanthroline) can prevent damage to DNA by peroxides added subsequently. If Fenton chemistry generates $OH^\bullet$ in the nucleus, then the transition–metal ions

Table 4.8. Baseline levels of DNA base modification in human respiratory tract epithelial cells and the increase in amount after exposure of the cells to 1 mM H_2O_2.

Modified base	Amount in DNA (nmol/mg DNA)	
	Baseline	Increase due to H_2O_2 treatment
5-Hydroxyuracil	0.067 ± 0.020	0.687 ± 0.021
5-Hydroxymethyluracil	0.003 ± 0.000	0.017 ± 0.002
8-Hydroxyguanine	0.700 ± 0.015	4.431 ± 0.561
FAPy-guanine	0.580 ± 0.009	9.012 ± 0.611
8-Hydroxyadenine	0.064 ± 0.014	0.550 ± 0.041
FAPy-adenine	0.258 ± 0.018	24.405 ± 2.452

Results are means of three separate experiments $\pm SD$. H_2O_2 also caused a rapid rise in DNA strand breaks.
Data abstracted from *FEBS Lett.* **374**, 233 (1995).

that convert H_2O_2 into the $OH^\bullet$ must be in very close proximity to DNA *in vivo*. DNA, with its mass of negatively charged phosphate groups, is effectively a large anion capable of binding many cations. Thus iron and copper bind tightly to DNA: the DNA binding affinity of Fe(III) has been calculated to be about 2.1×10^{14} at pH 7.4 and that for Cu^{2+} as 2×10^4. Copper appears to bind preferentially to GC-rich residues in DNA (Section 4.6.4).

Are metals always bound to DNA *in vivo*? Iron and copper are present in the nucleus, but this does not mean that they are in a molecular form that will cause $OH^\bullet$ formation. Another possibility is that oxidative stress causes the release of intracellular iron and/or copper ions (Section 4.5 above) that then bind to DNA, making it a target of H_2O_2 attack. In cultured cells, raising the level of Fe^{2+} in the culture medium leads to increases in steady-state levels of oxidative DNA damage.[121] Either way, the binding of metal ions to DNA favours 'site-specific' $OH^\bullet$ generation, leading to DNA damage that is very difficult for '$OH^\bullet$ scavengers' to protect against. Copper/H_2O_2 reactions may also generate DNA-damaging oxo-copper complexes in addition to $OH^\bullet$ (Chapter 2).

DNA-associated copper ions in cells might also react with certain phenols to produce ROS and oxidized phenolic intermediates capable of binding to DNA.[69] This interaction could cause a range of DNA lesions, including base modification, strand breaks and phenol adducts to the DNA bases, all of which might contribute to the carcinogenicity of certain phenolic compounds. Phenolic compounds that cause DNA damage in the presence of copper ions *in vitro* include 2-hydroxyoestradiol, 2-methoxyoestradiol, diethylstilboestrol and L-DOPA.

It should be noted that DNA damage caused by oxidative stress need not necessarily involve direct attack on DNA by ROS/RNS. Large rises in intracellular free Ca^{2+} caused by oxidative stress can activate nuclear Ca^{2+}-dependent endonucleases, which will fragment DNA. This damage will not be accompanied by chemical changes in the DNA bases, however, and so can easily be distinguished experimentally.

4.6.7   Use of iron and hydrogen peroxide for DNA 'footprinting'

The binding of various proteins to DNA regulates gene expression. Thus, it is often important to identify the DNA sequence to which a given regulatory protein binds (Appendix II). The problem can be approached by the technique of 'footprinting'. Essentially, the DNA–protein complex is treated with a DNA-degrading reagent such as a deoxyribonuclease enzyme. The bound protein protects the part of the DNA to which it is attached, while the rest of the DNA is hydrolysed. The protected part can then be isolated (e.g. by gel electrophoresis) and studied.

The high reactivity of $OH^\bullet$ towards DNA has led to the development of DNA footprinting methods employing this radical.[52] In one such method, reaction of Fe^{2+}-EDTA with H_2O_2 is used to produce $OH^\bullet$. Fe^{2+}-EDTA has an overall negative charge at physiological pH, and so will not bind to DNA (also negatively charged). Any $OH^\bullet$ radicals produced by reaction of Fe^{2+}-EDTA with H_2O_2 that escape scavenging by EDTA itself enter 'free solution' and randomly damage sites in DNA. A bound protein blocks the cleavage of DNA by $OH^\bullet$, allowing identification of the base sequence to which the protein binds. Since $OH^\bullet$ has no marked base or sequence preference for reaction with DNA, it does not have the specificity problems encountered in the use of nuclease enzymes.

Another agent that has been used in DNA footprinting is **methidium propyl EDTA-Fe^{2+}**.[52] Methidium intercalates between the DNA bases with no particular site specificity; attachment of Fe^{2+}-EDTA to it allows DNA strand breakage at the binding sites by Fenton chemistry. By contrast, attachment of Fe^{2+}-EDTA to the intercalating compound **distamycin** results in DNA strand breakage at the sites to which distamycin binds. This approach allows cleavage of DNA at *specific* sites by attaching an agent generating $OH^\bullet$ [Fe^{2+}-EDTA] to a compound that intercalates at specific sites. The copper–phenanthroline complex,[94] which degrades DNA via $OH^\bullet$ (Section 4.5), has also been used in footprinting studies, as has peroxynitrite.[61] Yet another approach is to attach iron chelators to histones, allowing site-specific generation of $OH^\bullet$ within chromatin.

4.6.8   Histidine as an extracellular pro-oxidant for DNA damage

Superoxide-generating systems such as xanthine oxidase and its substrates hypoxanthine and xanthine, as well as activated neutrophils, cause extensive damage to DNA within cells, largely by generation of H_2O_2, which enters cells readily. Certain amino acids, especially histidine, enhance DNA damage when present in the cell culture media. Since EDTA decreases this enhancement of DNA damage, the effect of histidine may involve metal binding. Presumably the histidine binds metal ions and carries them into the cell. L-Histidine also enhances the cytotoxicity of H_2O_2 added directly to cultured mammalian cells, an effect which seems to be associated with induction of DNA double-strand breaks.[107] This latter effect is not inhibited by EDTA but is inhibited by the cell-permeable chelator *o*-phenanthroline. Whether these effects are important *in vivo* remains to be ascertained. Histidine can also promote metal ion-dependent lipid peroxidation (Section 3.21.8).

4.7 Consequences of damage to DNA by ROS/RNS: mutation

DNA strand breakage can lead to activation of PARP, perhaps to an extent that can cause cell death (Sections 4.2 and 4.3 above). Modification of bases in DNA can lead to mutations, either directly or during attempts by the cell to replicate or repair damaged DNA.[101]

For example, RNS deaminate adenine to hypoxanthine, cytosine to uracil and guanine to xanthine. Whereas adenine pairs with thymine, hypoxanthine can pair with cytosine instead. Uracil pairs with adenine rather than guanine, although xanthine still pairs with cytosine. Thus RNS cause **AT ↔ GC transition mutations**, e.g.

$$G - C \quad \xrightarrow[\text{deamination}]{\text{RNS} \atop \text{C}} \quad G ---U \quad \xrightarrow[\text{replication}]{\text{DNA}} \quad G - C$$

$$+$$

$$A ---U \quad \xrightarrow[\text{replication}]{\text{DNA}} \quad A - T$$

$$A - T \quad \xrightarrow[\text{deamination}]{\text{RNS} \atop \text{A}} \quad HX ---T \quad \longrightarrow \quad HX ---C \quad \longrightarrow \quad G - C$$

$$+$$

$$A - T$$

The presence of thymine glycol is a strong block to DNA replication, but it can lead to T → C transition mutations (although the great majority of thymine glycol still hydrogen bonds correctly with adenine). The thymine lesion 5-hydroxymethyluracil also mispairs (again with a low probability) to guanine. 8-Hydroxyguanine leads to **GC → TA transversion** mutations since it can hydrogen bond (with low probability) to adenine. Again, most 8-hydroxyguanine pairs correctly with cytosine.

$$G - C \quad \xrightarrow{\text{ROS}} \quad 80HG - C \quad \longrightarrow \quad 80HG ---A \quad \longrightarrow \quad T - A$$

$$+$$

$$G - C$$

8-Hydroxyadenine has a low probability of mispairing with guanine, but mostly it pairs correctly with thymine. The ring-opened purines (FAPyG, FAPyA) seem to block DNA replication. Some end-products of lipid peroxidation can also bind to DNA and cause mutations (Section 4.9 below).

4.7.1   *Mutagenicity of oxidative base damage*

One approach to examining the overall mutagenic effect of the multiple base changes in DNA caused by ROS is to treat cells with H_2O_2 and look for mutations in particular genes, either in the nuclear or mitochondrial genomes or in transfected DNA.[79] Thus in one study a plasmid was transfected into monkey cells. The cells were treated with H_2O_2 and then incubated to allow repair and replication of the plasmid, and the plasmid was then isolated for examination. The H_2O_2-induced mutations detected included deletions and base substitutions. The distribution of ROS/RNS-induced mutations in DNA is not random; they cluster at so-called **hotspots**. This can be accounted for by several mechanisms, including accessibility of DNA to ROS (for example, inter-nucleosomal DNA is more exposed than is DNA within nucleosomes) and patterns of transition metal ion binding, involved in converting H_2O_2 to DNA-damaging species such as $OH^\bullet$.

Among the mutations identified in mapping studies like the above are CC → TT transitions, so-called **tandem mutations** which have been suggested to be unique markers of DNA damage by UV light or ROS. However, the most common changes observed are C → T transitions, and G → C and G → T transversions. None of these changes is unique to oxidative damage: they can occur as a result of polymerase errors, spontaneous deamination of cytosine to uracil (or 5-methylcytosine to thymine), and the misreplication of certain DNA-carcinogen adducts. However, ROS-induced G → T changes usually involve 8-hydroxydeoxyguanine. In cells exposed to singlet O_2, G → T was the most common mutation, consistent with a role for 8-hydroxydeoxyguanine (since singlet O_2 appears selective for damage at guanine).

Overall, the pattern of mutation generated by oxidative damage depends upon the conformation and base sequence of the gene studied (the mutagenicity of most modified bases is affected by the sequences around them), the efficiencies of repair, the type of DNA polymerase which replicates the gene and the conformation of the surrounding DNA, which can affect the accuracy of copying by polymerases.

4.8   Consequences of damage to DNA by ROS/RNS: DNA repair[102]

The correct functioning of DNA repair enzymes is essential for allowing aerobes to survive without excessive mutation rates. Dividing cells whose DNA becomes damaged usually stop cell division until repair is complete (Chapter 9).

DNA polymerase enzymes usually show high fidelity because of their error-correcting function, but some mistakes are made and so 'mismatch repair' is needed (Fig. 4.11). Spontaneous depurinations/deaminations must be repaired. All aerobic cells generate ROS, and some produce RNS; both can cause potentially-mutagenic changes in DNA, as can UV light. The precursors of DNA (dATP, dGTP, dTTP and dCTP) can also suffer oxidative damage. Most attention has been paid to dGTP, which can be attacked by certain ROS and converted to 8-hydroxy-dGTP, which can then be mistakenly incorporated into DNA by DNA polymerases.[51] The same is true of 8-hydroxyd-ATP.

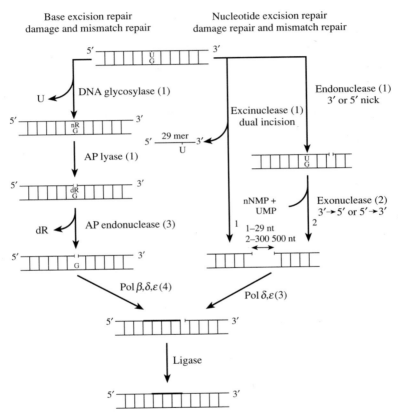

Fig. 4.11. DNA repair. Both damaged bases and mismatches (incorporation of a normal base in the wrong place) are rectified by base or by nucleotide excision pathways. For simplicity, a deamination of cytosine (to uracil) is taken as an example of both mismatch and damage. The repair of mismatch and lesions by base excision follows the same pathway. Removal of damage by nucleotide excision occurs by dual incision, whereas removal of mismatches by nucleotide excision occurs by endonuclease/exonuclease action. The size of repair patches are 1–4 nucleotides for base excision, 27–29 nucleotides for damage repair by nucleotide excision and 300–500 nucleotides for mismatch repair by the nucleotide (general) mismatch repair system. 'Pol', DNA polymerases β, δ, ε. From *J. Biol. Chem.* **270**, 15915 (1995) by courtesy of Dr Aziz Sancar and the publishers.

Adducts of carcinogens bound to DNA (Chapter 9) must also be repaired if possible. One advantage of the DNA double helix is that, if one strand is damaged, the information in the other strand can be used to replace it accurately. Thus double-strand breaks are more damaging because this information is not available. In addition, in a double helix the DNA bases are less exposed to attack and undergo spontaneous deamination/depurination at lower rates.

Some examples of *direct* removal of unwanted changes from DNA are known. One is the DNA **photolyase enzyme** (Section 4.8.2).[88] Methylation of guanine in DNA to O^6-methylguanine by certain carcinogens results in a mutagenic lesion. **O^6-Methylguanine–DNA-methyltransferase**, found in all species, is a 'suicide' enzyme that repairs DNA directly by transferring the O^6-methyl group to a cysteine residue at its active site, simultaneously inactivating itself. Overexpression of this enzyme helps to protect transgenic mice from cancers induced by DNA-alkylating agents. Thymine hydroperoxide is a substrate for glutathione peroxidase and transferase enzymes (Chapter 3), although the biological significance of this reaction (and of the alcohol product) are uncertain. Glutathione is frequently suggested to 'repair' base radicals generated by attack of certain ROS on DNA (Chapter 8).

4.8.1   *Sanitization of the nucleotide pool*[51]

Some years ago, a gene (*mut*T) was identified in *E. coli* which, when mutated, increased the spontaneous occurrence of AT $\leftrightarrow$ CG transversions. Purification of the MutT protein product showed it to be an enzyme that destroys dGTP containing 8-hydroxyguanine, i.e. 8-hydroxy-dGTP. Presumably ROS/RNS can attack dGTP in the DNA precursor pool. This enzyme serves to prevent incorporation of 8-hydroxy-dGTP into DNA, by hydrolysing it to the monophosphate 8-hydroxy-dGMP. A similar enzyme has been detected in mammalian (including human) cells. For mice, exceptionally high enzyme levels are found in embryonic stem cell lines: levels in mouse liver, thymus and large intestine are higher than in other tissues. The enzyme is found in human skin (Chapter 7). In a human T-cell leukaemia line (**Jurkat cells**), about 4% of the total 8-hydroxy-dGTP hydrolase activity was present in mitochondria.[60] 8-Hydroxy-dGMP cannot be rephosphorylated and seems to be further hydrolysed to 8-hydroxy-dG, a urinary excretion product in humans (Chapter 5).

4.8.2   *Repair of pyrimidine dimers*

In UV-exposed DNA, adjacent pyrimidine residues on a DNA strand can be covalently cross-linked (Chapter 7). This distorts the double helix and blocks replication and transcription. In *E. coli*, three enzymes are involved in repair of such lesions. First, a complex of enzymes encoded by the **UVrABC genes** senses the distortion and cuts the damaged DNA strand in two places, eight bases away from the dimer on the 5' side and four bases away on the 3' side. The removal of these 12 bases leaves a gap. DNA polymerase I enters and resynthesizes DNA using information in the intact strand. Finally, the 3' end of the newly synthesized stretch of DNA is sealed to the original portion of the DNA chain by a **DNA ligase** enzyme.[102]

E. coli also contains a **DNA photolyase**.[88] This enzyme binds to distorted DNA, absorbs light in the near–UV and blue spectral regions, and uses the light energy to split the dimer back to its original bases. Photolyases contain two chromogens. One is FAD and the other differs according to the source of the enzyme (in *E. coli* it is a folate derivative). Why such an enzyme is present in an organism whose normal habitat is the dark recesses of the animal colon is an enigma.

Similar enzymes have been found in many eukaryotes, including *Drosophila*, plants, the goldfish *Carassius auratus* and some mammals (the opossum and kangaroo rat; both marsupials) but not humans. In the absence of light, the photolyase can still bind to pyrimidine dimers and it has been suggested that its presence may facilitate repair by other mechanisms. For example, *E. coli* DNA photolyase binds to DNA cross-linked by *cis*-platin (an anti–cancer drug; Chapter 9) and stimulates removal of the cross-link by excision repair *in vitro*.

4.8.3   *Excision repair*[102]

In general, mispaired, oxidized and deaminated bases are removed from DNA by the same types of mechanism. In the first type (the only type for bulky lesions such as pyrimidine dimers and certain carcinogen adducts) the DNA is cut on both sides of the lesion by a multisubunit ATP-dependent **excinuclease** complex. The unwanted base is thus removed as part of an oligonucleotide, a DNA polymerase fills the gap (using information from the undamaged strand of DNA) and a ligase seals the DNA. This process is called **nucleotide excision repair**. In *E. coli* excision repair involves three proteins, **UvrA**, **UvrB** and **UvrC** (Section 4.8.2), whereas in humans many more proteins are involved. Yeast excinuclease is broadly similar to the human complex.

In the second type of repair, the damaged base is removed directly (**base excision repair**). **DNA glycosylase** enzymes hydrolyse the bond linking the abnormal base to the sugar–phosphate backbone. Their action leaves behind an **apurinic** or **apyrimidinic** site (AP site) in DNA. AP sites are mutagenic and act as blocks to transcription and DNA replication. AP sites also arise by spontaneous base loss in DNA (usually depurination), and 8-nitroguanine rapidly depurinates. Enzymes (sometimes part of the glycosylase itself) recognize AP sites and nick the DNA on both sides of the missing base (an **AP lyase** cleaves 3' and an **AP endonuclease** 5' to the AP site), so releasing the AP-deoxyribose. DNA polymerase fills in the one-nucleotide gap with the correct base as dictated by the undamaged complementary strand. Finally, a ligase seals the DNA.

Examples of base excision repair include removal of hypoxanthine from DNA by **hypoxanthine–DNA glycosylase** and removal of uracil by **uracil–DNA glycosylase**, both found in bacterial and human cells. DNA repair enzymes have usually been purified by assaying their ability to act upon a specific base lesion. However, they often have a broader specificity than anticipated when tested against a range of lesions.[26] For example, human uracil glycosylase also recognizes three uracil derivatives generated from oxidative damage to cytosine (5-hydroxyuracil, alloxan and isodialuric acid), while *E. coli* endonuclease III (a glycosylase despite its name) releases a range of thymine and cytosine oxidation products. This enzyme has a [4Fe–4S] cluster at its active site. A **FAPy glycosylase** that removes formamidopyrimidines and 8-hydroxyguanine from DNA has been found in both bacteria (e.g. in *E. coli* it is called **Fpg**) and in mammalian cells. Human AP endonuclease has been purified; as well as this activity, it also activates the transcription factors Fos and Jun (Section 4.13 below), a process involving a cysteine residue in its N-terminal half. Thus it is sometimes called reducing factor 1 (**Ref 1**).

4.8.4   *Repair of 8-hydroxyguanine*

In summary, let us take an example and ask how cells are protected against the mutagenicity of 8-hydroxyguanine. First, excluding 8-hydroxy-dGTP from the precursor pool stops incorporation into DNA. Second, base excision repair (FAPy and other glycosylases) occurs. Third, nucleotide excision repair occurs. Fourth, if 8-hydroxyguanine does get into DNA and mispairs with adenine, another glycosylase[59] removes the mispaired adenine from the 8-hydroxy-guanine–adenine pair. Overall, 8-hydroxyguanine is repaired faster than several other base lesions, at least in certain human cell types (Fig. 4.12).

4.8.5   *Repair of double-strand breaks and mitochondrial repair*

Methods to cope with DNA double-strand breaks also exist; indeed, such breaks occur physiologically during genetic recombination. The double-strand breaks that can be caused by ionizing radiation and other sources of ROS are harder to deal with, necessitating removal of damaged bases/sugars from the broken end, their replacement (without the benefit of an intact strand as template) and finally resealing.[102] Incorrect repair can generate large abnormal-ities in DNA, visible on the chromosome (**chromosomal aberrations**) or can lead to loss of chromosome portions. Multiple proteins are involved in repairing DNA double-strand breaks: one is a **DNA-dependent serine/threonine kinase** that recognizes free DNA ends. Little else is known: our ignorance about removal of DNA–protein cross-links is also profound.

Mitochondria also have DNA repair capacity, e.g. glycosylases that remove lesions by base excision repair. In general, however, repair activity seems lower than in the nucleus.

4.8.6   *Evidence that DNA repair is important*[102]

The fact that DNA isolated from cells of every aerobic organism (including humans) contains low levels of DNA base damage products suggests that repair enzymes do not achieve complete removal of modified bases.

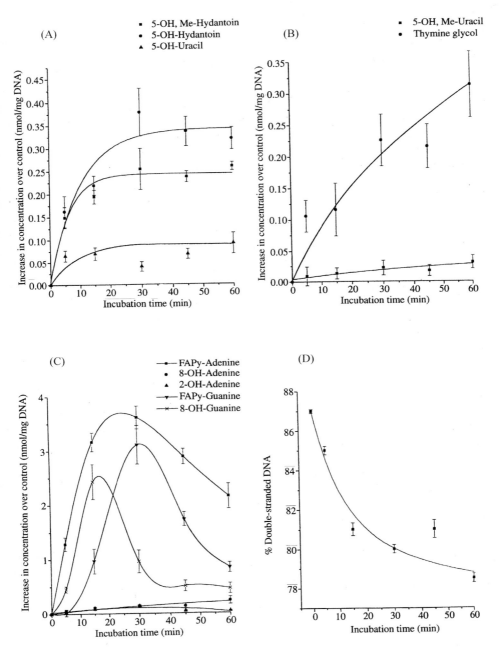

Fig. 4.12. Rates of repair of oxidative DNA base damage. Human cells were treated with a non-lethal dose of H_2O_2 (100 μM). Note the rapid onset of repair of some (C) but not all (A, B) base damage products. While repair proceeds, DNA strand breakage (D) continues to increase, since the repair processes themselves involve formation of strand breaks. Data by courtesy of Dr Jeremy Spencer.

Damage recognition may be the rate–limiting step in repair, since in humans over 10^9 base pairs must be patrolled to locate what are very rare (<1 in 10^6 bases) lesions. Defects in DNA being replicated/transcribed appear to be repaired more rapidly than in DNA in condensed chromatin. In part, this occurs because the DNA is more available to repair enzymes, but some transcription factors have repair activity or interact with repair systems (Section 4.8.3 above). In addition, eukaryotic cells have 'checkpoint' mechanisms that arrest cell cycle progression when DNA damage is detected. Consistent with this limitation on rates of DNA repair, the steady-state levels of one or more base damage products have been observed to increase in many diseases associated with increased ROS/RNS production (Chapter 9). Thus interfering with DNA repair would be expected to cause severe problems.

This is indeed the case. For example, **xeroderma pigmentosum**,[9a] a rare disease in humans in which the skin is severely damaged by sunlight and the risk of skin cancer is high, is usually caused by defects in nucleotide excision repair. At least seven different defects, affecting different proteins important in the repair system, have been identified. Defective DNA mismatch repair is also responsible for non-polyposis colon cancer, one of the most common hereditary cancers. In **severe combined immunodeficient** (scid) mice and in patients suffering **ataxia talangiectasia**, there are defects in the repair of double-strand breaks. The latter is characterized by a high incidence of cancers, and hypersensitivity to ionizing radiation. In **Bloom's syndrome**, marked by photosensitivity, dwarfism and mental retardation, the rate of nick resealing is abnormally low. All these (fortunately rare) diseases illustrate the essentiality of DNA repair mechanisms, consistent with the inability of antioxidant defences to prevent ROS/RNS-related DNA damage completely *in vivo* (Table 4.1).

In bacteria such as *E. coli* (but not generally in mammalian cells), treatment with chemicals that modify DNA increases transcription of genes encoding excision repair enzymes; similar events can result from oxidative stress (Section 4.13 below). This up-regulation of DNA repair systems has often been called the **SOS response**. The products of two genes (*rec*A and *lex*A) are particularly important in controlling this response. *E. coli* cells lacking Fpg or MutY show increased spontaneous mutation rates, and rates are 600- to 1600-fold greater than normal if both genes are defective. *E. coli* lacking uracil DNA glycosylase or MutT protein also suffers more mutations. *Salmonella* strains with defects in DNA repair are more susceptible to damage by low concentrations of H_2O_2 and less virulent to mice than are strains in which both catalase genes (HPI and HPII) have been inactivated.[14] As well as causing direct damage to DNA, ROS/RNS might conceivably interfere with the activity of repair proteins.

4.9   Mechanisms of damage to cellular targets by oxidative stress: lipid peroxidation

4.9.1   *A history of peroxidation: from oils to textiles*

Lipid peroxidation has been defined by A. L. Tappel as 'the oxidative deterioration of polyunsaturated lipids'. Polyunsaturated fatty acids (PUFAs) are those that contain two or more carbon–carbon double bonds, $>C=C<$.

Oxygen-dependent deterioration, leading to **rancidity**, has been recognized since antiquity as a problem in the storage of fats and oils and was often dealt with by using spices (Section 3.24). Rancidity is even more relevant today with the popularity of 'polyunsaturated' margarines and cooking oils, and the importance of paints, plastics, lacquers, waxes and rubber, all of which can undergo oxidative damage.

The first attempts to study this problem began in 1820 when de Saussure, using a simple mercury manometer, observed that a layer of walnut oil on water exposed to air absorbed three times its own volume of air in the course of 8 months. This initial lengthy period was followed by a second phase of rapid air–absorption, the oil taking up 60 times its own volume of air in 10 days. During the following 3 months, the rate of air uptake gradually diminished, so that the oil had eventually taken up 145 times its own volume. Parallel with these changes, the oil became viscous and evil-smelling. Commenting on these experiments a few years later, the famous chemist Berzelius suggested that O_2 uptake might account for not only the autoxidation of oil exposed to air but also a host of similar phenomena. In particular, it was thought that autoxidation might be involved in the spontaneous ignition of wool after its lubrication with linseed oil, a common cause of fires in textile mills at that time. Berzelius also discovered the element selenium, which helps to protect against lipid peroxidation *in vivo* (Chapter 3).

The sequence of reactions which is now recognized as the 'core' of lipid peroxidation was worked out in detail by scientists at the British Rubber Producers' Association research laboratories in the 1940s. The relevance of these reactions to biological systems was not appreciated until much later, however.

4.9.2 Targets of attack: membrane lipids

The membranes that surround cells and cell organelles (such as mitochondria, lysosomes and peroxisomes) contain large amounts of PUFA side-chains. What, therefore, stops us from going rancid ourselves?

The major constituents of biological membranes are lipid and protein, the amount of protein increasing with the number of functions that the membrane performs. In the nerve myelin sheath, which appears to serve largely as an insulator of the nerve axon, only 20% of the dry weight of the membrane is protein, but most membranes have 50% or more protein, and the complex inner mitochondrial membrane (Chapter 1) and chloroplast thylakoid membranes (Chapter 7) have 80% protein. Some proteins are loosely attached to the surface of membranes (**extrinsic proteins**), but most are tightly attached (**intrinsic proteins**), being partially embedded in the membrane, located in the membrane interior or sometimes traversing the membrane. Lipid peroxidation can therefore cause damage to membrane proteins as well as to lipids.

Membrane lipids are generally **amphipathic** molecules, i.e. they contain hydrocarbon regions that tend to cluster together away from water, together with polar parts that like to associate with water. In animal cell membranes

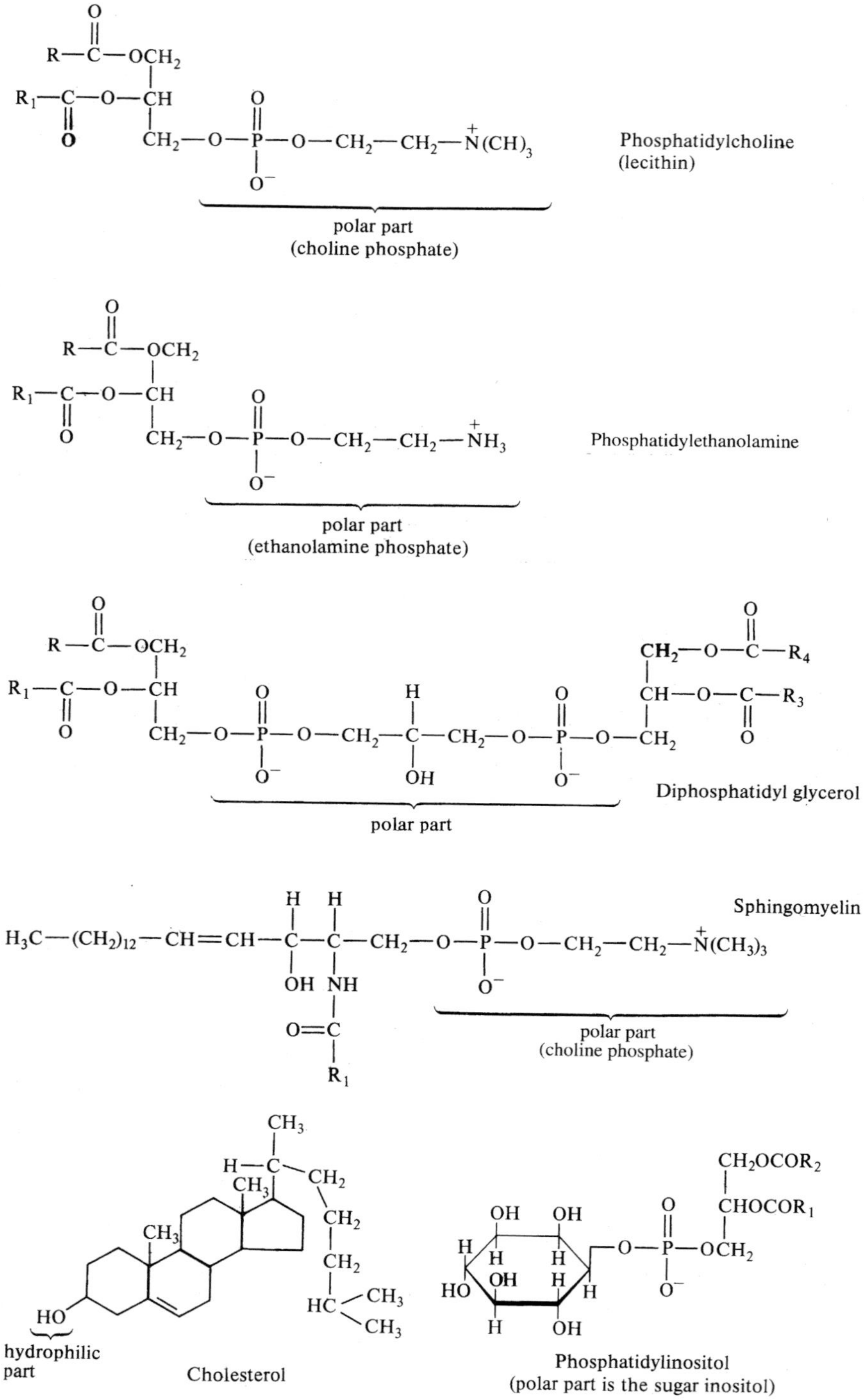
Phosphatidylcholine
(lecithin)
polar part
(choline phosphate)
Phosphatidylethanolamine
polar part
(ethanolamine phosphate)
Diphosphatidyl glycerol
polar part
Sphingomyelin
polar part
(choline phosphate)
hydrophilic
part
Cholesterol
Phosphatidylinositol
(polar part is the sugar inositol)

Fig. 4.13.

the dominant lipids are **phospholipids**, esters based on the alcohol glycerol (Figs 4.13 and 4.14). Some membranes, particularly plasma membranes, contain significant proportions of sphingolipids and of the hydrophobic molecule **cholesterol** (Fig. 4.13). The commonest phospholipid in animal cell membranes is **lecithin** (phosphatidylcholine). By contrast, the membranes of subcellular organelles such as mitochondria or nuclei rarely contain much sphingolipid or cholesterol. Mitochondrial inner membranes contain **cardio-lipin** (Fig. 4.13); it has been suggested that peroxidation of PUFAs on this lipid may contribute to age-related declines in mitochondrial function.[92]

The fatty-acid side-chains of membrane lipids in animal cells have unbranched carbon chains and contain even numbers of carbon atoms, mostly in the range 14–24, and the double bonds are of the *cis* configuration. Their structures and nomenclature are illustrated in Fig. 4.14 and Table 4.9. A double bond in any carbon chain prevents rotation of the groups attached to the carbon atoms forming it, so that they are forced to stay on one side of the double bond or the other, generating *trans* and *cis* configurations (Fig. 4.14). Thus fatty-acid side-chains have 'kinks' in them whenever a *cis* double bond occurs.

Other organisms have different lipid compositions. The composition of bacterial membranes depends very much on the species and even on the culture conditions and stage in the growth cycle. Membrane fractions usually contain 10–30% lipid. In Gram-positive bacteria (i.e. those that take up **Gram's stain**, used by microscopists) phosphatidylglycerol is present, but phosphatidyl-ethanolamine is more common in Gram-negative species (such as *E. coli*). **Mycoplasmas** contain some cholesterol in their membranes.

4.9.3   *Membrane structure*

As the number of double bonds in a fatty-acid molecule increases, its melting point drops. For example, stearic acid (18:0) is solid at room temperature whereas linoleic acid (18:2) is a liquid. Since membrane lipids are amphipathic molecules, on exposure to water they tend to aggregate with their hydrophobic regions clustered together and their hydrophilic regions in contact with H_2O. How this arrangement is achieved depends on the relative amounts of lipid and water. When phospholipids are shaken or sonicated in aqueous solution they form **micelles** (Fig. 4.15), but as more phospholipids are added **liposomes** result, i.e. bags of aqueous solution bounded by a **lipid bilayer**. Liposomes can

---

Fig. 4.13. Lipid molecules found in animals. R_1, R_2, etc. represent long, hydrophobic, fatty-acid side-chains (for structures, see Table 4.9 and Fig. 4.14). In most lipids these are joined by ester bonds to the alcohol glycerol (Fig. 4.14). In sphingomyelins, however, the fatty acids are attached to the $-NH_2$ group of sphingosine (Fig. 4.14). All the lipid molecules shown contain a polar (hydrophilic) part that can interact with water, but in cholesterol this is very small (only an $-OH$ group) so that, overall, cholesterol is a very hydrophobic molecule. Diphosphatidylglycerol (cardiolipin) comprises about 20% of the phospholipids in the inner mitochondrial membrane.

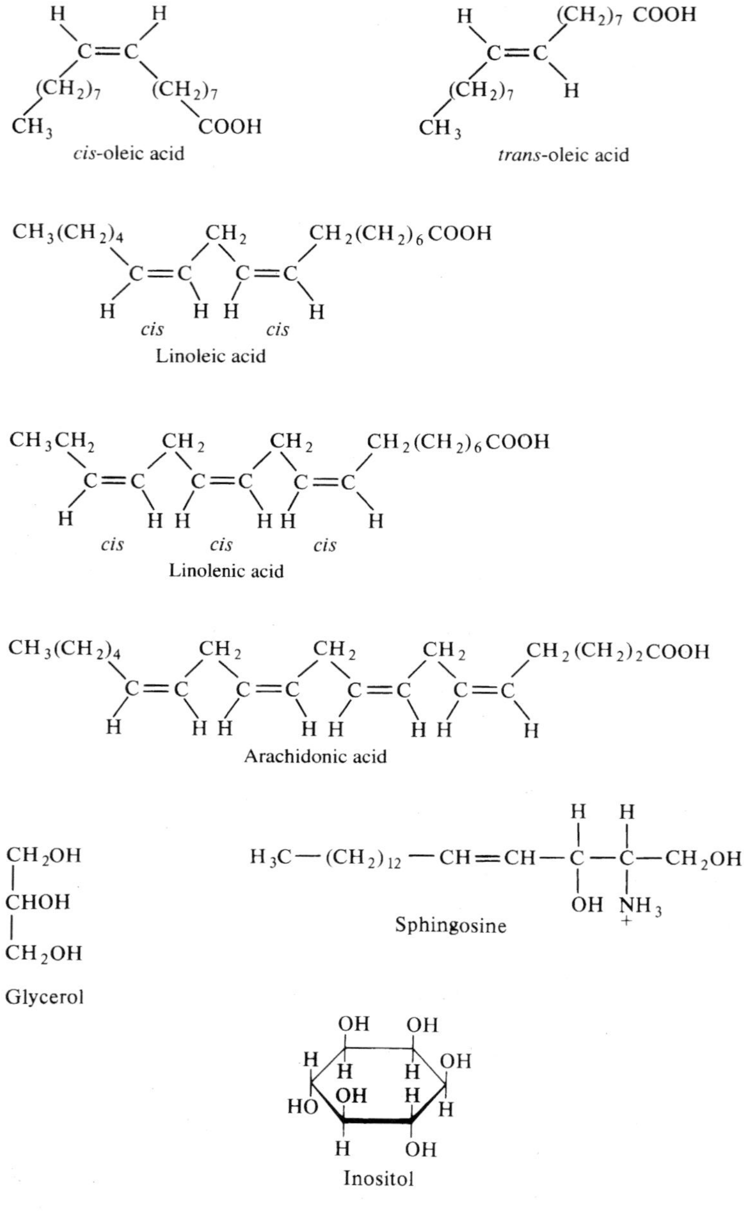

Fig. 4.14. Fatty acids and other 'building blocks' of biological lipids. Phosphorylated inositols are also involved in signal transduction. Top: *cis-* and *trans*-oleic acid ($C_{18:1}$). Note the kink in the chain of the *cis* form. **Linoleic acid** (C_{18}) has two double bonds, at carbons 9 and 11 (where the carbon in the COOH group is counted as carbon number 1). **Linolenic acid** (C_{18}) is polyunsaturated, with double bonds at carbons 9, 12 and 15. **Arachidonic acid** (C_{20}) is polyunsaturated, with double bonds at carbons 5, 8, 11 and 14.

Table 4.9. Some common, naturally occurring fatty acids

Shorthand name[a]	Common name	Examples of occurrence
16:0	Palmitic	Natural fats and oils, especially palm oil
18:0	Stearic	Natural fats and oils, especially beef fat
18:1 (n-9)[b]	Oleic	Natural fats and oils, especially olive oil
18:2 (n-6)	Linoleic	Widespread, many seed oils
18:3 (n-3)	α-linolenic	All plant leaves, some seed oils, e.g. soybean, rapeseed, linseed oils
20:4 (n-6)	Arachidonic	Animal membranes
20:5 (n-3)	Eicosapentaenoic	Fish oils
22:6 (n-3)	Docosahexaenoic[c]	Fish oils, nervous system

[a] Number of carbons in chain : number of double bonds.
[b] The numbering system in parentheses (often used in the nutritional literature) identifies double bonds from the methyl, $-CH_3$, end of the chain. For other nomenclature see Fig. 4.14.
[c] Docosahexaenoic acid is particularly important in the mammalian brain and in the retina of the eye.

be surrounded by a single lipid bilayer (**unilamellar**) or several bilayers (**multilamellar**), as shown in Fig. 4.15. The interior of liposomes contains a portion of the aqueous solution in which they were made, so they are often used as 'parcels' for transporting drugs (including antioxidant enzymes, e.g. see Chapter 1) to target tissues. Liposomes are frequently used in studies of lipid peroxidation.

There is considerable evidence that the lipid bilayer (Figs 4.15 and 4.16) is the basic structure of all cell and organelle membranes, proteins being inserted in different parts of the bilayer. In each half of the lipid bilayer, protein and lipid molecules can diffuse quickly—indeed a lipid molecule in one half of a bilayer can get from one end of 'the average' cell to the other in a few seconds. This **membrane fluidity** is due to the presence of unsaturated and PUFA side-chains, which lower the melting point of the membrane interior so that it effectively gains the chemical nature and viscosity of a light oil. Damage to PUFAs tends to decrease membrane fluidity, which is known to be essential for the proper functioning of biological membranes. By contrast, exchange of lipid molecules between the two halves of the bilayer is rare.

4.9.4   *Targets of attack: fatty acids and lipoproteins*

Dietary fats have to be digested, absorbed and transported around the body (Section 3.22.11). The lipoproteins involved in this process are also a target of oxidative damage. Indeed, peroxidation of **low density lipoproteins** contributes to atherosclerosis (Chapter 9). Fatty acids released by the action of lipases are also potential targets of damage; for example, the albumin protein and albumin-bound bilirubin help protect fatty acids against peroxidation in mammalian blood plasma (Section 3.21.1).

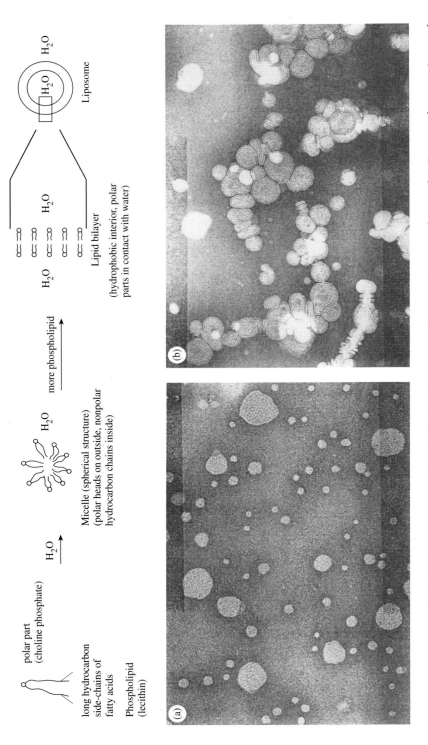

Fig. 4.15. Formation of a lipid bilayer on mixing phospholipids with aqueous solutions. Top: general principles. Bottom: electron micrographs of preparations of (a) unilamellar and (b) multilamellar liposomes.

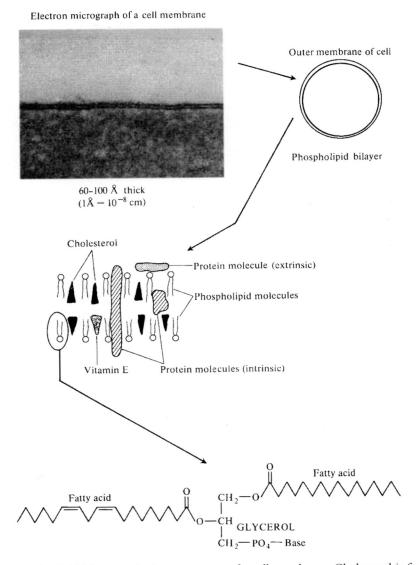

Fig. 4.16. The lipid bilayer as the basic structure of a cell membrane. Cholesterol is found in cell plasma membranes and circulating lipoproteins but not usually in organelle membranes. The membrane shown is that of a sheep red blood cell under the electron microscope.

4.9.5 *How does lipid peroxidation begin?*

Initiation of lipid peroxidation is caused by attack upon a lipid of any species that has sufficient reactivity to abstract a hydrogen atom from a methylene ($-CH_2-$) group. Fatty acids with one or no double bonds are more resistant to such attack than are the PUFAs, yet PUFA side-chains are present in many membranes and lipoproteins.[117]

An adjacent double bond weakens the energy of attachment of the hydrogen atoms present on the next carbon atom (the **allylic** hydrogens), especially if there is a double bond on either side of the $-CH_2-$ (Fig. 4.18), giving **bis–allylic** hydrogens. The reduction potential of a PUFA$^\bullet$/PUFA couple at pH 7 ($E^{\circ\prime}$) has been estimated[62] as about 0.6 V. Hence OH$^\bullet$, HO$_2^\bullet$, RO$^\bullet$ and RO$_2^\bullet$ radicals are (thermodynamically at least) capable of oxidizing PUFAs (see Table 2.3).

Hydroxyl radicals can initiate peroxidation readily[4]

$$-CH_2- + OH^\bullet \rightarrow -\dot{C}H- + H_2O$$

provided that they reach the hydrocarbon side chains without hitting something else first; OH$^\bullet$ generated outside a membrane can also attack extrinsic proteins (e.g. cell surface glycoproteins) and 'head groups' of phospholipids. Hence radiolysis of aqueous solutions, which produces OH$^\bullet$, is well known to stimulate peroxidation of any lipids present; this has been shown not only for biological membranes and fatty acids but also for food lipids (a problem in attempts to sterilize or preserve food by irradiating it). The peroxidation is usually inhibited to some extent by scavengers of OH$^\bullet$, such as mannitol and formate. They will compete with the lipids for any OH$^\bullet$ generated in 'free solution'. However, H$_2$O crosses membranes readily (Chapter 2) and any water undergoing homolysis within the membrane will generate OH$^\bullet$ not accessible to scavengers. The rate constant for reaction of OH$^\bullet$ with artificial lecithin bilayers has been measured[4] as about $5 \times 10^8 \, M^{-1} s^{-1}$.

By contrast, O$_2^{\bullet-}$ is insufficiently reactive to abstract H from lipids; in any case, its charge should preclude it from entering the hydrophobic interior of membranes. Indeed, O$_2^{\bullet-}$ does not readily cross most biological membranes. One exception is the erythrocyte membrane. Here O$_2^{\bullet-}$ can travel via an **anion channel**, present in the membrane to allow the passage of Cl$^-$ and bicarbonate (HCO$_3^-$) ions (Chapter 7). However, O$_2^{\bullet-}$ does not appear to react with any membrane constituents on its passage (Chapter 7).

The protonated form of O$_2^{\bullet-}$, HO$_2^\bullet$, is more reactive (Chapter 2) and can abstract H$^\bullet$ from some isolated fatty acids,[1] such as linoleic, linolenic and arachidonic acids (rate constants 1.2, 1.7 and $3.0 \times 10^3 \, M^{-1} s^{-1}$ respectively)

$$-CH_2- + HO_2^\bullet \rightarrow -\dot{C}H- + H_2O_2$$

HO$_2^\bullet$, being uncharged, should enter membranes more easily than O$_2^{\bullet-}$. Several papers have described HO$_2^\bullet$-dependent peroxidation of liposomes and lipoproteins. In addition to the above reaction, HO$_2^\bullet$ can stimulate peroxidation by reaction with pre-formed lipid hydroperoxides. This reaction, rather than H$^\bullet$ abstraction, may have been the major contributor to the stimulation of peroxidation observed.[1]

$$HO_2^\bullet + ROOH \rightarrow RO_2^\bullet + H_2O_2$$

As well as $RO^\bullet$, $RO_2^\bullet$, $OH^\bullet$ and $HO_2^\bullet$, several iron–oxygen complexes have been suggested[108] to be capable of abstracting H and initiating peroxidation (Section 4.9.7 below).

4.9.6   *Propagation of lipid peroxidation*

Since a hydrogen atom has only one electron, abstraction of $H^\bullet$ from a $-CH_2-$ group leaves behind an unpaired electron on the carbon ($-\overset{\bullet}{C}H-$). $H^\bullet$ abstraction is easiest from bis-allylic methylene groups. For example, the oxidizability of docosahexaenoic acid (22 : 6) *in vitro* is five times greater than that of linoleic acid. One consequence of this is that fish-oil supplements, widely advocated by healthfood stores, undergo oxidation easily because they contain substantial amounts of eicosapentaenoic and docosahexaenoic acids. Controlled clinical trials investigating the alleged benefits of fish-oil supplementation in various diseases have often given conflicting results: one reason may be the extent of oxidation of the preparation that was tested.

The carbon radical is usually stabilized by a molecular rearrangement to form a **conjugated diene** (Fig. 4.17). Carbon radicals can undergo various reactions, e.g. if two of them collided within a membrane they might cross-link the fatty acid side-chains:[32]

$$R-\overset{\bullet}{C}H + R-\overset{\bullet}{C}H \;\rightarrow\; R-CH-CH-R$$

However, the most likely fate of carbon radicals under aerobic conditions is to combine with O_2 (Chapter 2), especially as O_2 is a hydrophobic molecule that concentrates into the interior of membranes (Chapter 1). Reaction with O_2 gives a **peroxyl radical**, $ROO^\bullet$ (or $RO_2^\bullet$), a name sometimes shortened to **peroxy radical**:

$$R^\bullet + O_2 \rightarrow -ROO^\bullet$$

Of course, very low O_2 concentrations might favour self-reaction of carbon-centred radicals,[32] or perhaps their reaction with other membrane components, such as $-SH$ groups on proteins. Hence the O_2 concentration in a biological system can affect the pathway of peroxidation. Formation of peroxyl radicals has been demonstrated during peroxidation of many membrane systems, using spin-trapping methods (Chapter 5).

Peroxyl radicals are capable of abstracting H from another lipid molecule, i.e. an adjacent fatty-acid side-chain:

$$ROO^\bullet + CH \rightarrow -ROOH^c + C^\bullet$$

This is the **propagation stage** of lipid peroxidation. The carbon radical formed can react with O_2 to form another peroxyl radical and so the **chain reaction** of lipid peroxidation can continue (Fig. 4.17). The peroxyl radical combines with the hydrogen atom that it abstracts to give a

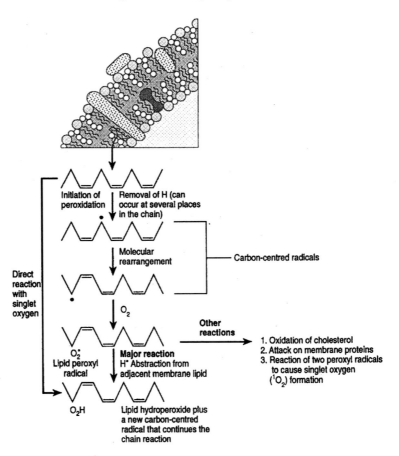

Fig. 4.17. Idealized representation of the initiation and propagation reactions of lipid peroxidation. The peroxidation of a fatty acid with three double bonds is shown.

lipid hydroperoxide (LOOH). This is sometimes shortened to **lipid peroxide**, although the latter term includes cyclic peroxides (e.g. products of singlet O_2 reaction: Chapter 2) as well as LOOH species. Indeed, an alternative fate of peroxyl radicals is to form cyclic peroxides (Figs 4.17 and 4.18).

A single initiation event can lead to formation of multiple molecules of peroxide as a result of the chain reaction. Another complexity is that the initial $H^\bullet$ abstraction from a PUFA can occur at different points on the carbon chain (see legend to Fig. 4.17).[35] Thus peroxidation of linoleic acid gives two hydroperoxides, while that of linolenic acid gives four. Peroxidation of arachidonic acid gives six lipid hydroperoxides as well as cyclic peroxides and other products (Fig. 4.18), including the isoprostanes (Section 4.9.13 below). Eicosapentaenoic acid should give eight hydroperoxides, and docosahexaenoic acid ten.

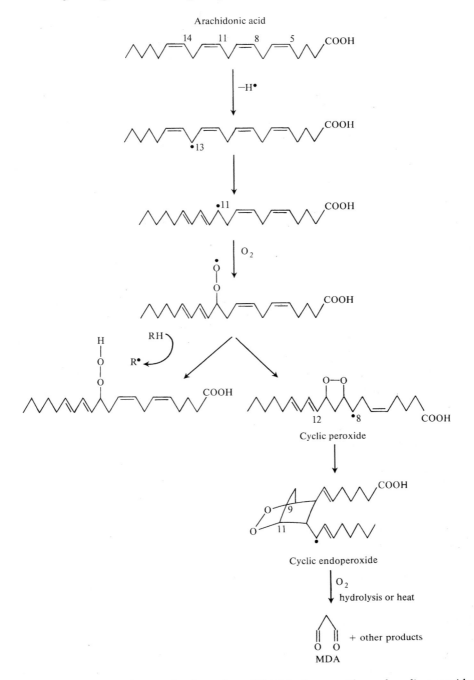

Fig. 4.18. Proposed mechanism for formation of lipid hydroperoxides and cyclic peroxides from arachidonic acid. MDA is malondialdehyde (malonaldehyde). Initial abstraction of an allylic H from C-13 is shown. H can also be abstracted at C-10 or C-7, giving several other peroxide end-products. Peroxyl radicals can attack a double bond in the same chain to generate cyclic peroxide radicals, which can abstract H and also lead to isoprostane formation (Fig. 4.23).

4.9.7   *Iron and lipid peroxidation*

Is OH• involved?

Iron(II) ions can take part in electron-transfer reactions with oxygen (Chapter 1).

$$Fe^{2+} + O_2 \rightleftarrows (Fe^{2+} - O_2 \leftrightarrow Fe(III) - O_2^{\bullet-}) \rightleftarrows Fe(III) + O_2^{\bullet-}$$

Superoxide can dismutate to form H_2O_2, giving all the essential ingredients for formation of OH• radicals:

$$2O_2^{\bullet-} + 2H^+ \rightarrow H_2O_2 + O_2$$

$$Fe^{2+} + H_2O_2 \rightarrow Fe(III) + OH^- + OH^\bullet$$

Thus the addition of an Fe^{2+} salt (or a ferric salt plus a reducing agent, such as ascorbate) to a peroxide-free unsaturated lipid in the presence of O_2 should initiate lipid peroxidation (H• abstraction by OH•). The resulting peroxidation should be inhibitable by H_2O_2-removing enzymes (e.g. catalase), scavengers of OH• and chelating agents that bind iron and prevent its participation in free-radical reactions. Oxidations inhibited by these reagents have been observed upon iron addition to dispersed lipid systems (e.g. fatty acids solubilized by detergents) and ultrapure lipids. However, most scientists, including the authors, find that when catalase or scavengers of OH• are added to membrane fractions (e.g. microsomes), lipoproteins or liposomes undergoing peroxidation stimulated by addition of iron ions, there is no inhibition, despite the fact that OH• radicals can be detected in the reaction mixtures by such techniques as aromatic hydroxylation, spin-trapping and the deoxyribose method (Chapter 5). Formation of OH• *is* inhibited by H_2O_2-scavenging enzymes.[42,48]

Oxo-iron species

It follows that OH• is made from H_2O_2 in these reaction systems, but is not *required* for the peroxidation to take place. The lack of action of OH• scavengers might be interpreted to mean that the required OH• formation is 'site-specific', involving iron ions bound to the membrane, so that any OH• formed reacts immediately with the membrane components and is not amenable to scavenging. However, H_2O_2-removing enzymes should still inhibit. The fact that they do not has led to proposals that initiation of lipid peroxidation by iron salts in the presence of O_2 is achieved by reactive species other than OH•. Ferryl (Chapter 2) is one possibility, although ferryl formation by reaction of Fe^{2+} with H_2O_2 should still require H_2O_2 and inhibition by peroxide-removing enzymes would be expected. Perferryl ($Fe^{2+}-O_2 \leftrightarrow Fe(III)-O_2^{\bullet-}$) could also be involved. However, what little is known of the chemistry of perferryl complexes (Chapter 2) suggests that they are insufficiently reactive to abstract H• or to insert oxygen directly into fatty-acid side-chains.[48] Studies of the kinetics of microsomal or liposomal lipid peroxidation in the presence of Fe^{2+} and/or Fe(III) salts have led to proposals that initiation requires a triple complex of iron(II)/iron(III)/O_2, or at least some specific critical ratio of Fe^{2+} to Fe(III). This proposal explains a number of experimental observations, although attempts to isolate and characterize this complex have been unsuccessful to date. The observation that some other metal ions can replace Fe(III) in stimulating Fe^{2+}-dependent peroxidation in such experiments argues against a requirement for a specific Fe^{2+}–Fe(III) complex.[48]

Peroxide decomposition

However, iron plays another role in lipid peroxidation. Pure lipid peroxides are usually stable at physiological temperatures. They decompose on heating, but also in the presence of transition-metal ions.[48] Iron(II) and certain Fe^{2+} chelates react with lipid peroxides in a similar way to their reaction with H_2O_2, splitting the O–O bond. With H_2O_2, this gives $OH^\bullet$. With a lipid peroxide, ROOH, it produces $RO^\bullet$, an **alkoxyl radical** (sometimes shortened to **alkoxy radical**).

$$\underset{\text{lipid hydroperoxide}}{R\text{--OOH}} + Fe^{2+}\text{--chelate} \rightarrow Fe(III)\text{--chelate} + OH^- + \underset{\text{alkoxyl radical}}{R\text{--O}^\bullet}$$

Alkoxyl radicals can abstract $H^\bullet$ from PUFAs, and from peroxides

$$R\text{--O}^\bullet + L\text{--H} \rightarrow ROH + L^\bullet$$

$$L\text{--OOH} + R\text{--O}^\bullet \rightarrow LOO^\bullet + R\text{--OH}$$

The resulting peroxyl radicals can continue propagation of lipid peroxidation. Fe(III) and certain Fe(III) chelates can decompose peroxides to peroxyl radicals

$$R\text{--OOH} + Fe(III)\text{--chelate} \rightarrow \underset{\text{peroxyl radical}}{RO_2^\bullet} + H^+ + Fe^{2+}\text{--chelate}$$

The reactions of Fe^{2+} ions with lipid hydroperoxides are often faster than their reactions with H_2O_2 (k_2 for $Fe^{2+} + H_2O_2$ is about $76\,M^{-1}s^{-1}$; that for $ROOH + Fe^{2+}$ is about $1.5 \times 10^3\,M^{-1}s^{-1}$); reactions of Fe(III) with hydroperoxides are slower than those of Fe^{2+}. Hence the rate of lipid peroxidation in the presence of Fe(III) can be increased by adding reducing agents, such as ascorbate. Another factor is that Fe^{2+} salts are more soluble at pH 7.4 than Fe(III).

For a given concentration of iron salt, low concentrations of ascorbate stimulate peroxidation, probably by reducing Fe(III) to Fe^{2+}, whereas high concentrations are inhibitory. It has been suggested that high concentrations of ascorbate may reduce some of the $RO_2^\bullet$ radicals directly to hydroperoxides at the membrane surface, and thus interfere with the chain reaction. Alternatively, it may alter the Fe^{2+}/Fe(III) ratio away from the optimum for peroxidation. Ascorbate can also regenerate α-tocopherol in biological membranes (Chapter 3).

Similarly, mixtures of GSH and iron salts have been reported to stimulate lipid peroxidation *in vitro*, but the general view is that GSH is protective *in vivo* because it is a substrate for glutathione peroxidase (Chapter 3), as well as being able to react with various aldehydes produced during peroxidation (such as 4-hydroxy-2-*trans*-nonenal (HNE)) and thus protect the –SH groups of membrane proteins. Consistent with this view, treatment of animals with GSH-depleting agents increases susceptibility to lipid peroxidation.[22]

Although, in general, Fe^{2+} chelates stimulate lipid peroxidation more than Fe(III) chelates, the actual rates of reaction with peroxides depend on the ligand to the metal (see Table 2.4). Indeed, the variable effects of chelating agents on lipid peroxidation can, at least in part, be explained by their ability to influence these different reactions. For example, EDTA can accelerate the reaction of iron ions with H_2O_2 to give $OH^\bullet$ (Chapter 2), but slows their reaction with lipid peroxides.[42] If $O_2^{\bullet-}$ is generated in the reaction system, it can contribute to the peroxidation in two ways. $HO_2^\bullet$ radical can react with ROOH to give $RO_2^\bullet$ (see above). Superoxide can also reduce Fe(III) to Fe^{2+}, although its effects are complex because it is also capable of oxidizing Fe^{2+} back to Fe(III) (Section 2.4.3).

Commercially available lipids are contaminated with lipid peroxides, so that liposomes or micelles made from them already contain traces of lipid peroxides. Sonication, often used to prepare liposomes, introduces peroxides (Chapter 2). When cells are injured, lipid peroxidation is favoured and traces of lipid peroxides are formed enzymically in tissues by cyclo-oxygenase and lipoxygenase enzymes (Chapter 9). Thus membrane fractions isolated from disrupted cells contain lipid peroxides as do lipoprotein fractions isolated from body fluids (Chapter 9). When iron salts or chelates are added to such fractions, the lipid peroxides can be decomposed to generate peroxyl and alkoxyl radicals. Thus the added metal ions are not initiating lipid peroxidation (in the sense of creating peroxide in a completely peroxide-free system) but stimulating peroxidation by decomposing pre-formed peroxides into chain-propagating radicals. It is possible that the putative abilities of ferryl, perferryl and Fe^{2+}/Fe(III)/O_2 complexes to 'initiate' lipid peroxidation are explicable by the abilities of these complexes efficiently to degrade traces of lipid peroxides in the lipid systems that were being studied.

4.9.8   *Which iron chelates stimulate lipid peroxidation?*

We saw in Chapter 2 that, whereas Fe^{2+} ions and certain iron chelates react with H_2O_2 to form $OH^\bullet$, reaction of iron-containing proteins with H_2O_2 and/or $O_2^{\bullet-}$ does not generally result in $OH^\bullet$ formation, unless iron is released from the protein under the assay conditions. The range of iron chelates that can stimulate lipid peroxidation is wider[48] (Table 4.10). It includes not only Fe^{2+} salts and simple chelates (e.g. Fe^{2+}–ADP), but also haem, haem proteins and even the iron-containing enzyme phenylalanine hydroxylase.[56] In studies *in vitro*, ferritin stimulates lipid peroxidation to an extent proportional to its iron content, whereas haemosiderin stimulates much less strongly (on a unit iron basis). The stimulation of lipid peroxidation observed is probably due to the release of iron from the proteins during the assay. For example, stimulation of liposomal peroxidation by ferritin or haemosiderin can be inhibited almost completely by the iron chelator desferrioxamine. Stimulation of peroxidation by myoglobin and haemoglobin can involve peroxide-dependent release of haem and iron from the proteins as well as reactions brought about by ferryl species and amino-acid radicals on the protein itself (Section 3.18.3).

Table 4.10. Biological iron chelates and their possible participation in oxygen radical reactions

Type of iron complex	Will they decompose lipid peroxides to form alkoxyl and/or peroxyl radicals?	Can they form hydroxyl radicals by Fenton chemistry?
Iron bound to		
Phosphate esters (e.g. ADP, ATP)	Yes	Yes
Carbohydrates and organic acids (e.g. citrate, picolinic acid, deoxyribose)	Yes	Yes
DNA	Yes	Yes
Membrane lipids	Yes	Yes
Loosely bound to proteins, e.g. albumin	Yes	Yes
Iron tightly bound to proteins		
(i) Non-haem iron		
Ferritin	Probably no[a]	No[a]
Haemosiderin	No[a]	No[a]
Lactoferrin (iron saturated, 2 mol Fe(III)/mol protein)	Probably no	No (only if iron is released)
Transferrin (iron saturated, 2 mol Fe(III)/mol protein)	Probably no	No (only if iron is released)
(ii) Haem iron		
Haem itself	Yes	Probably no[a]
Haemoglobin	Yes	Yes (when iron is released)
Leghaemoglobin	Yes	Yes (when iron is released)
Myoglobin	Yes	Yes (when iron is released)
Cytochrome *c*	Yes	Yes (when iron is released)
Cytochromes P450	Yes, especially CYP2E1	Yes (when iron is released)
Catalase	Weakly[b]	No[c]

[a]Unless iron is released from protein.
[b]Activity may be due to partial degradation of catalase: catalase subunits show greater peroxidase activity (Chapter 3).
[c]Unless enzyme activity is completely lost and iron is released.

4.9.9   *Copper and other metals as promoters of lipid peroxidation*

Copper ions are powerful promoters of peroxide decomposition;[43] for example, Cu^{2+} is an excellent catalyst of peroxidation of low-density lipoproteins. Like iron, they act mainly by decomposing pre-formed peroxides.

$$ROOH + Cu^{2+} \rightarrow Cu^{+} + RO_2^{\bullet} + H^{+}$$

$$ROOH + Cu^{+} \rightarrow RO^{\bullet} + Cu^{2+} + OH^{-}$$

Certain cobalt(II) chelates can also decompose lipid peroxides, whereas Zn^{2+} and Mn^{2+} ions do not. Many other metal ions cannot themselves stimulate lipid peroxidation but appear to bind to membranes in a way that can facilitate iron-dependent lipid peroxidation under certain reaction conditions; examples are aluminium(III) and lead(II). Their binding to the membrane surface may somehow produce a local 'freezing' of the motion of phospholipid molecules that facilitates propagation reactions. However, results are very much affected by assay conditions and the physiological significance of these effects is unknown.[48]

4.9.10   *Products of peroxide decomposition*

Decomposition of lipid peroxides by heating at high temperatures (as in the heating of oxidized cooking oils) or by exposure to iron or copper ions generates a hugely complex mixture of products, including epoxides, saturated aldehydes (e.g. **hexanal**), unsaturated aldehydes, ketones (e.g. butanones, pentanones, octanones) and hydrocarbons.[29] Thermal homolysis of the O—O bond yields radicals, which can attack other hydroperoxides and PUFAs

$$ROOH \rightarrow RO^{\bullet} + OH^{\bullet}$$

Metal ions allow peroxides to decompose at physiologically relevant temperatures. For example, if Fe^{2+} reacts with a hydroperoxide on the fifth carbon from the methyl end of the fatty acid, **pentane** gas can be produced. This can happen with linoleic acid and arachidonic acid:

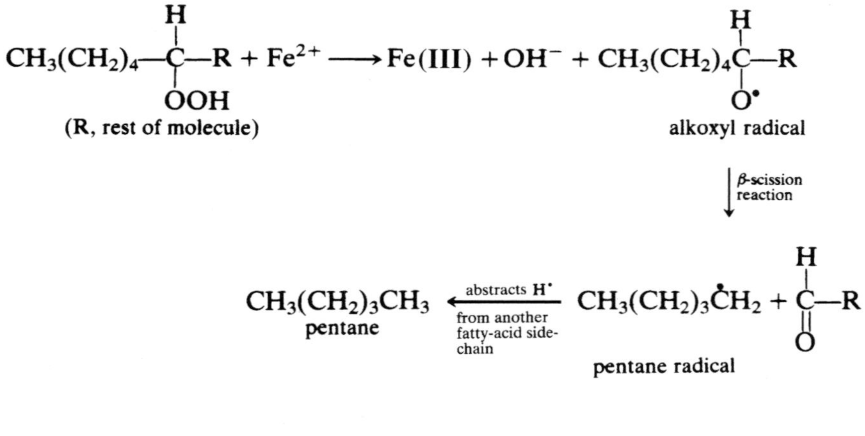

$$CH_3(CH_2)_4 \overset{\overset{\text{H}}{|}}{\underset{\underset{\text{OOH}}{|}}{C}} {-}R + Fe^{2+} \longrightarrow Fe(III) + OH^- + CH_3(CH_2)_4 \overset{\overset{\text{H}}{|}}{\underset{\underset{O^{\bullet}}{|}}{C}} {-}R$$

(R, rest of molecule)   alkoxyl radical

$\downarrow$ β-scission reaction

$$CH_3(CH_2)_3CH_3 \xleftarrow[\substack{\text{from another}\\\text{fatty-acid side-}\\\text{chain}}]{\text{abstracts H}^{\bullet}} CH_3(CH_2)_3\overset{\bullet}{C}H_2 + \overset{\overset{\text{H}}{|}}{\underset{\underset{O}{||}}{C}} {-}R$$

pentane   pentane radical

Ethane (C_2H_6) and **ethylene** (ethene, $H_2C{=}CH_2$) gases are produced in similar reactions from linolenic acid. **β-scission**, shown above, is a well-known reaction of radicals, especially alkoxyl radicals. Cleavage of aldehyde, hydrocarbon or other fragments from peroxidized lipids still leaves an oxidized fragment attached to the parent lipid molecule by an ester bond.

Malondialdehyde

Malondialdehyde, sometimes called **malonaldehyde**, was the focus of attention in lipid peroxidation for many years because it was commonly thought that the thiobarbituric acid (TBA) test, the commonest assay of lipid peroxidation *in vitro*, measures free MDA. In fact, MDA is formed only in small amounts during the peroxidation of most lipids, although larger amounts are produced during the peroxidation of liver microsomes in the presence of iron salts (Chapter 5).

MDA arises largely from peroxidation of PUFAs with more than two double bonds, such as linolenic, arachidonic and docosahexaenoic acids. MDA can also be formed enzymatically during eicosanoid metabolism (Chapter 6). MDA exists in various forms,[29] depending on pH (Fig. 4.19). At physiological pH any 'free' MDA will exist as an enolate anion which has low reactivity toward most amino groups. As the pH falls, however, reactivity is greatly increased. Under physiological conditions, proteins are more readily attacked by MDA than are free amino acids, resulting in modification of several residues, especially lysine, as well as intra- and intermolecular protein cross-links.

$$\text{OHC·CH}_2\text{·CHO} + \text{protein} \begin{matrix} \diagup \text{NH}_2 \\ \diagdown \text{NH}_2 \end{matrix} \longrightarrow \text{protein} \begin{matrix} \diagup \text{NH—CH} \diagdown \\ \diagdown \text{N=CH} \diagup \end{matrix} \text{CH}$$

intramolecular cross-link

$$\text{OHC·CH}_2\text{·CHO} + 2\ \text{protein—NH}_2 \longrightarrow$$

$$\text{protein—NHCH=CH—CH=N—protein}$$

intermolecular cross-link

MDA also reacts with DNA bases (Fig. 4.20) and can introduce mutagenic lesions.[7] Guanine is a preferred target, and the average content of compound C (Fig. 4.20) in human liver DNA was reported as nine adducts per 10^7 bases (Chapter 5). Adducts to adenine have also been identified *in vivo*. The overall contribution of MDA–DNA adducts to mutagenicity *in vivo* is uncertain: indeed pure MDA is poorly mutagenic in bacterial test systems. However, if DNA pre-treated with MDA is expressed in *E. coli*, about a 10-fold increase in mutation frequency is observed: G → T transversions, A → G transitions and C → T transitions are predominant, but some frameshifts and deletions occur.[7]

MDA is rapidly metabolized in mammalian tissues.[29] Aldehyde dehydrogenases oxidize it to a semialdehyde of the dicarboxylic acid **malonic acid**, which decarboxylates to acetaldehyde, a substrate for oxidation by aldehyde dehydrogenases to acetate. Malonic acid is a competitive inhibitor of the succinate dehydrogenase enzyme in mitochondria (Chapter 1).

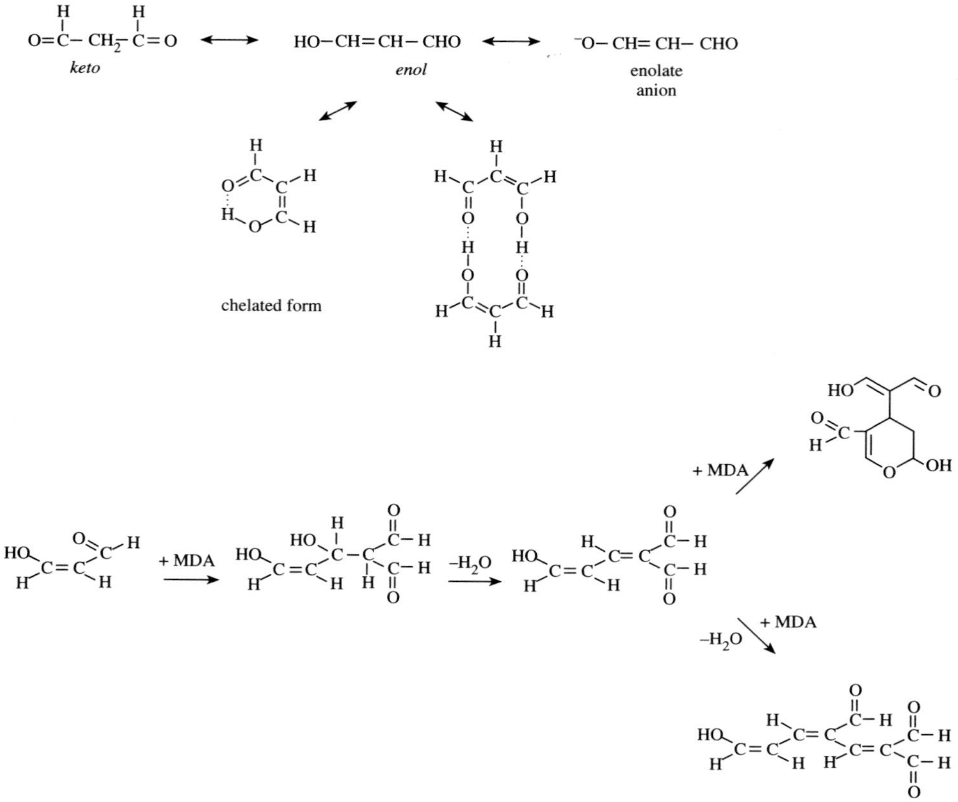

Fig. 4.19. Structures of malondialdehyde (MDA) in aqueous solution. Top: at neutral or alkaline pH, the major (99%) form of MDA is as the enolate anion whereas at acidic values it exists largely as the undissociated *enol* form, β-hydroxyacrolein, in equilibrium with the *keto* form. Intramolecular hydrogen bonds favour formation of a cyclic chelate form, which can also form a dimeric complex. Bottom: like most aldehydes, MDA in aqueous solution is prone to aldol condensations which produce dimers, trimers and larger polymers, many of which are fluorescent (excitation at 365–395 nm, emission at 490 nm). Data abstracted from *Free Rad. Biol. Med.* **11**, 81 (1991) by courtesy of the late Professor Hermann Esterbauer and Elsevier.

4-Hydroxy-2-trans-*nonenal and related unsaturated aldehydes*

4-Hydroxy-2-*trans*-nonenal (HNE) is formed during the peroxidation of *n*-6 PUFAs (Table 4.9), such as linoleic and arachidonic acids, by cleavage of lipid hydroperoxides in the presence of transition-metal ions (Fig. 4.21).

HNE is only one of the many unsaturated aldehydes (**alkenals**) formed during lipid peroxidation; another is **trans-4-hydroxy-2-hexenal** (HHE). Attention was focused upon HHE when it was reported to be the ultimate toxic end-product formed in the liver during the metabolism of pyrrolizidine alkaloids such as **senecionine**,[106] which are contained in many plants and can poison livestock. HHE can also be formed during lipid peroxidation and is

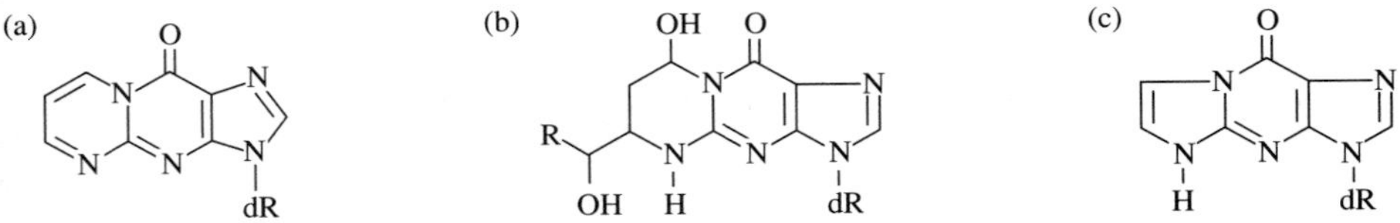

(a)

(b)

(c)

Fig. 4.20. Reaction of cytotoxic aldehydes with DNA bases. The major product of reaction of MDA with guanine in DNA at physiological pH is (a) pyrimido[1,2-α]purine-10(3H)-one (MG), whose structure is shown as a nucleoside (base + deoxyribose) This product has been identified in rat and human DNA at low levels. HNE also reacts with deoxyguanosine, an NH_2 group adding to the double bond of the aldehyde (b). In the presence of peroxides a different sequence occurs whereby $1N^2$-ethenodeoxyguanosine is the stable end-product (c).

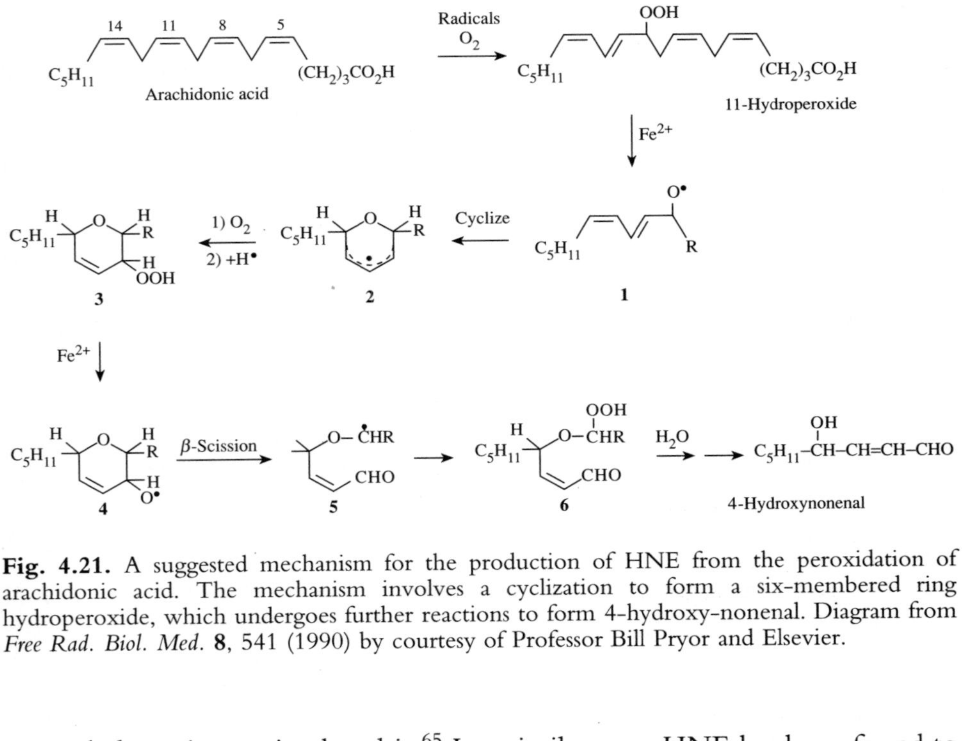

Fig. 4.21. A suggested mechanism for the production of HNE from the peroxidation of arachidonic acid. The mechanism involves a cyclization to form a six-membered ring hydroperoxide, which undergoes further reactions to form 4-hydroxy-nonenal. Diagram from *Free Rad. Biol. Med.* **8**, 541 (1990) by courtesy of Professor Bill Pryor and Elsevier.

particularly toxic to mitochondria.[65] In a similar way, HNE has been found to be one of the toxic agents generated by red algae that can lead to death of reef-dwelling fish such as *Eupomacentrus leucosticus*. It has also been speculated that HNE was one of the toxic agents responsible for the 'Spanish cooking oil syndrome' (Chapter 8).[29]

Since its discovery, numerous research publications have detailed the cyto-toxic properties of HNE, such as cell growth inhibition, genotoxicity, chemo-tactic activity[82] and ability to modify lipoproteins and promote atherosclerosis (Chapter 9).[29] These effects fall into three main categories, depending on the

HNE concentration:

(a) $100 \, \mu M$ or above: acute toxic effects, mitochondrial damage, bleb formation, cell lysis, generally rapid cell death.
(b) $1-20 \, \mu M$: inhibits DNA and protein synthesis, but stimulates phospholipase A_2. Generally inhibits cell proliferation. Can inhibit ADP–ribosyltransferase, the enzyme required for PARP formation, and certain cytochromes P450, e.g. CYP2E1 and CYP1AI. It is possible that HNE can approach or reach such concentrations *in vivo* during oxidative stress.
(c) $0.1 \, \mu M$ or lower: may represent basal levels of HNE in healthy tissues (although probably protein-bound rather than free). *In vitro*, such low HNE levels can stimulate several processes, including phagocyte chemotaxis, and the activities of several enzymes, such as adenylate cyclase, guanylate cyclase and phospholipase C. The physiological role of such stimulations is uncertain as yet.

Hydroxyalkenals owe their chemical reactivity to three main functional groups; the aldehyde group, the double bond and the hydroxyl group (Fig. 4.22).[29] Their ability to react rapidly with thiol (–SH) groups at physiological pH accounts for much of their cytotoxicity. For example, GSH reacts with HNE to form a saturated aldehyde with the GSH residue bound by a **thioether** linkage to carbon atom 3, followed by an intramolecular rearrangement to form a five-membered ring (Fig. 4.22). Thiol groups on proteins as well as amino groups on DNA bases (Fig. 4.20), proteins (e.g. on lysine residues in lipoproteins; Chapter 9) and on the phospholipids phosphatidylethanolamine and phosphatidylserine can also be attacked by HNE. The products of reaction with aminolipids are fluorescent at 430 nm, when excited at 360 nm, properties similar to those of peroxidized lipids and the lipofuscin pigments (Chapter 10). HNE generated *in vivo* rapidly reacts with proteins and peptides and does not stay 'free'; this is one problem in assessing the biological significance of its effects at different 'levels' of addition as indicated above. For example, HNE added to blood plasma (where GSH levels are very low) rapidly binds to –SH groups on albumin.

Cells can also remove HNE. The free aldehyde can be reduced to an alcohol by aldose reductase or oxidized to a carboxylic acid by aldehyde dehydrogenases (Fig. 4.22). HNE can form GSH conjugates, both non-enzymically (see above) and catalysed by certain glutathione transferases (Chapter 3). Conjugates are degraded to mercapturic acids and excreted in urine (Fig. 4.22). Aldehyde–thiol conjugates are in general less toxic by about a factor of 10 than the parent aldehydes, although not completely harmless: they may be somewhat toxic in their own right and/or act as a 'reservoir' that slowly releases HNE.

4.9.11   *Damage to membrane proteins during lipid peroxidation*

Generation within membranes and lipoproteins of peroxyl and alkoxyl radicals, aldehydes (especially HNE) and other products of lipid peroxidation can cause

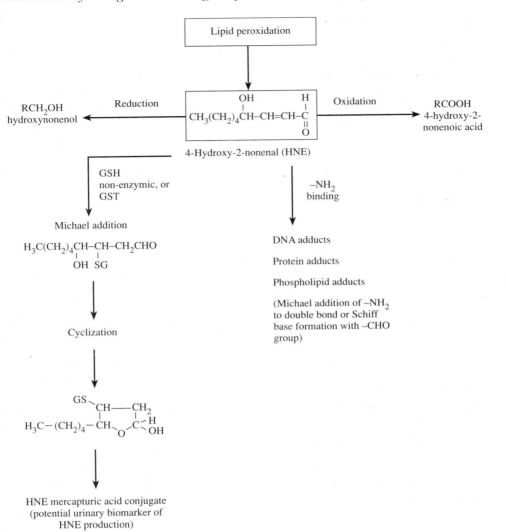

Fig. 4.22. Toxicity and metabolism of 4-hydroxynonenal.

severe damage to the proteins present.[98] Many years ago, it was shown that peroxidation of liver or erythrocyte membranes causes formation of high-molecular–mass protein aggregates within the membrane. The surface receptor molecules that allow cells to respond to hormones and cytokines can be inactivated during lipid peroxidation, as are enzymes such as glucose 6-phosphatase, glycerol–3-phosphate acyl transferase, the Ca^{2+}-ATPase of the endoplasmic reticulum (Section 4.4 above) and the Na^+, K^+-ATPase ('sodium pump') involved in maintenance of correct ion balance within cells. Potassium channels can also be damaged during lipid peroxidation.[27] Since 'voltage regulated' K^+ channels play an essential role in generation of electrical activity in nervous tissue and heart, damage to them can result in irregularities in heartbeat and death of neurones. K^+ channels are also essential transport systems in many cell types. Within mitochondria, both matrix enzymes and constituents of the electron-transport chain can be damaged, and ubiquinone is destroyed.[9]

In general, the overall effects[98] of lipid peroxidation are to decrease membrane fluidity, make it easier for phospholipids to exchange between the two monolayers, increase the 'leakiness' of the membrane bilayer to substances that do not normally cross it other than through specific channels (such as Ca^{2+} ions) and inactivate membrane-bound enzymes. Cross-linking of membrane proteins decreases their lateral and rotational mobility.

Continued oxidation of fatty-acid side-chains and their fragmentation to produce aldehydes and hydrocarbons such as pentane will eventually lead to loss of membrane integrity. For example, rupture of the membranes of lysosomes will spill hydrolytic enzymes into the rest of the cell to cause amplification of damage. Peroxidation of erythrocyte membranes causes them to lose their ability to change shape and squeeze through the smallest capillaries ('deformability') (Chapter 7). It has been suggested that the loss of viability of mammalian spermatozoa on prolonged incubation at $37\,^\circ C$ is due to the accumulation of products of lipid peroxidation (Chapter 7), and the loss of the germinating ability of soybean seeds, stored under warm, damp conditions, has been attributed to the same reason. In some bacteria the DNA is close to or attached to the cell membrane and it can be damaged during peroxidation. Studies on a mutant of *E. coli* that could not synthesize fatty acids, and thus incorporated into its membrane lipids whatever acids it was given in the growth medium, showed that the toxicity of hyperbaric oxygen became greater as the percentage of PUFAs in the membrane was increased. Mitochondrial DNA, which is not coated by histones, can easily be modified by peroxidation products. Damage to 'rough' (ribosome-studded) endoplasmic reticulum by peroxidation will also decrease the ability of the cell to synthesize and export proteins.

4.9.12   *Toxicity of peroxides*

Spin-trapping has been used to detect various free radicals from animals administered lipid hydroperoxides (Chapter 5). Oral administration of large doses of peroxidized fatty acids or lipids to animals leads to deleterious consequences, e.g. heart damage in rats, and 'fatty liver' in both rats and rabbits. In mice, damage to lymphoid tissues has been observed after feeding them with a methyl ester of linoleate hydroperoxide. Overall, the toxicity is low, i.e. animals can tolerate moderate amounts of oxidized lipids in the diet.[87] The gut has mechanisms for removing dietary lipid peroxides (Chapter 3), which presumably help to minimize damaging effects. Some of the aldehydes produced by thermally-induced peroxide decomposition during cooking might be absorbed,[41a] but again are likely to be extensively metabolised, e.g. by conjugation with GSH, by the gut.

4.9.13   *Isoprostanes*[100]

One particular class of toxic products of lipid peroxidation is the **isoprostanes** (Fig. 4.23). Isoprostanes are a series of prostaglandin-like compounds formed during peroxidation of arachidonic acid (similar products are formed from EPA and DHA). Because they are structurally similar to prostaglandin $F_{2\alpha}$ ($PGF_{2\alpha}$: Chapter 7), the compounds shown in Fig. 4.23 are collectively referred to as **F_2-isoprostanes**. Indeed, some commercial antibodies raised against prostaglandins will cross-react with isoprostanes. E_2- and D_2-isoprostanes and iso-thromboxanes have also been identified. A family of free-radical products isomeric to the leukotrienes (**isoleukotrienes**) also exists.

F_2-Isoprostanes are useful 'markers' of lipid peroxidation and can be measured in human blood plasma ($35 \pm 6 \, pg/ml$) and urine ($1600 \pm 600 \, pg/mg$ creatinine) from healthy volunteers, indicative of ongoing lipid peroxidation even in healthy subjects (Table 4.1). The majority of plasma isoprostanes is esterified to phospholipids, but some are 'free'. One of the isoprostanes, **8-epi-$PGF_{2\alpha}$**, is a powerful renal vasoconstrictor, reducing kidney blood flow and glomerular filtration rate by almost half at low nanomolar concentrations. It also exaggerates the response of platelets to agents that promote their aggregation. Isoprostanes are rapidly metabolised and excreted *in vivo*,[6a] so that they must be being continuously produced to maintain a steady-state level in plasma.

Elevated circulating concentrations of F_2-isoprostanes may contribute to the pathology of **hepatorenal syndrome**, an almost uniformly fatal disorder characterized by the development of kidney failure in patients with severe liver disease. Urinary excretion of isoprostanes is elevated in patients with the disease **scleroderma** (Chapter 9). This unpleasant autoimmune disease is characterized by progressive fibrosis of the skin, lungs and gastrointestinal tract.

4.9.14   *Platelet activating factor and lipid peroxidation*

Platelet activating factor (PAF, 1-O-alkyl-2-acetyl-*sn*-glycero-3-phospho-choline), is a phospholipid synthesized by leukocytes, platelets, endothelial cells and some other cell types. It exerts its biological actions at concentrations in the $10^{-10} \, M$ range: these include activation of phagocytes to make ROS, platelet aggregation and marked vascular (e.g. hypotension) and bronchial effects. While useful at physiological levels, over-production of PAF has been suggested to lead to tissue injury in several diseases, including asthma and ischaemia–reperfusion (Chapter 9). Once secreted, PAF is rapidly destroyed by **PAF acetylhydrolase**, an enzyme present within plasma lipoproteins. This enzyme can be inactivated in plasma by ROS. Peroxidation of phosphatidyl-choline can generate fragments that bind to PAF receptors on target cells and exert PAF-like biological activity. Some of these phospholipid fragments are also substrates for PAF acetylhydrolase.[122]

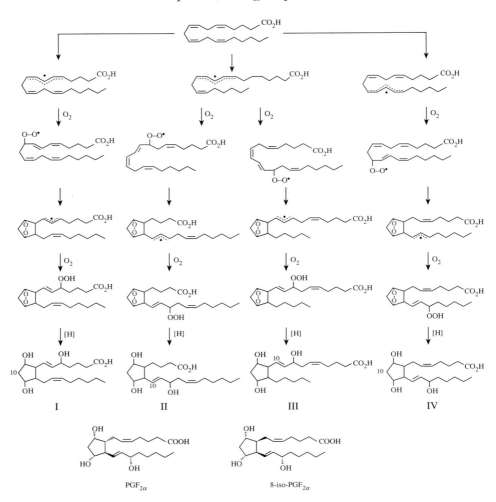

Fig. 4.23. Formation of the F$_2$-isoprostanes from arachidonic acid. Above: attack of free radicals abstracts H$^{\bullet}$ at one of three different positions (Fig. 4.18). Four RO$_2^{\bullet}$ radicals are formed when O$_2$ reacts with the carbon-centred radicals. The RO$_2^{\bullet}$ attack double bonds in the same chain to cyclize the structure and O$_2$ adds again to the new carbon radicals. Reduction gives four isomers (I–IV). For simplicity, stereochemistry is not indicated but each class potentially has 16 stereoisomers, giving 64 isomers in all. Below: these compounds are structurally isomeric with prostaglandin F$_{2\alpha}$. 8-*iso*-PGF$_{2\alpha}$ is sometimes called 8-*epi*-PGF$_{2\alpha}$. Diagram by courtesy of Drs L. Jackson Roberts II and Jason Morrow.

4.9.15   Cholesterol oxidation

Cholesterol in membranes and lipoproteins can become oxidized during lipid peroxidation, generating a mixture of products (Fig. 4.24).[37] There are conflicting reports on the cytotoxicity of cholesterol oxidation products,

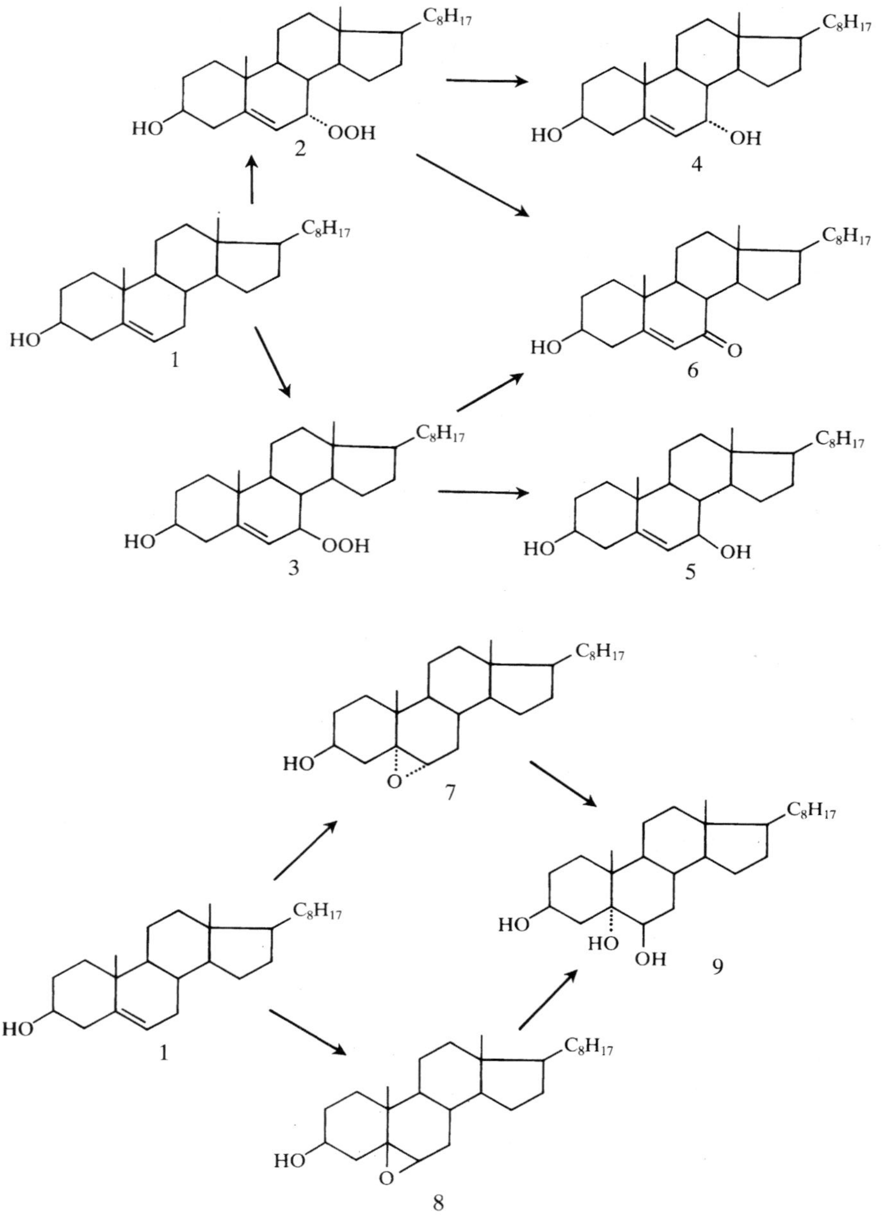

Fig. 4.24. Some products of cholesterol oxidation. 1, Cholesterol; 2 and 3, cholesterol 7–hydroperoxides; 4 and 5, cholest–5–ene–3–β7–diols; 6, 3β–hydroxycholest–5–en–7–one; 7 and 8, 5,6β–epoxy–5β– and 5,6α–epoxy–5α–cholestan–3–β–ols; 9, 5α–cholestane–3β,5,6β–triol. Diagram modified from *Chem. Phys. Lipids* **44**, 87 (1987) by courtesy of Professor L.L. Smith and the publishers. Reaction of singlet O_2 with cholesterol yields primarily the 5α–hydroperoxide (Chapter 2) whereas free–radical oxidation gives 7β and 7α–hydroperoxides and the other products shown.

possibly because most studies have used complex mixtures rather than purified single products. For example, oxidation products of cholesterol have variously been claimed to stimulate or to suppress atherogenesis (Chapter 9).

4.9.16   *Peroxidation of microsomes*

Microsomal fractions prepared from plant or animal tissues undergo lipid peroxidation when incubated with Fe^{2+} salts, or Fe(III) salts plus ascorbate or plus NADPH.[108] Often, iron is added as chelates with ADP, pyrophosphate or EDTA. During microsomal peroxidation, cytochromes b_5 and P450 are attacked, the haem groups being degraded.[104] One product is carbon monoxide, CO. An antibody raised against NADPH–cytochrome P450 reductase (Chapter 1) inhibits NADPH/Fe(III)-dependent peroxidation by more than 90%. The NADPH–cytochrome P450 reductase enzyme, as well as reducing cytochromes P-450, can lead to reduction of some Fe(III) chelates, generating Fe^{2+} and stimulating peroxidation. Hence both cytochrome P450 and its reductase contribute to microsomal lipid peroxidation.

Another way of stimulating peroxidation of microsomal (or other lipids) *in vitro* is to add azo initiators (Chapter 2) or artificial organic hydroperoxides such as **tert-butyl hydroperoxide** or **cumene hydroperoxide**. The decomposition of these peroxides to alkoxyl or peroxyl radicals accelerates the chain reaction of lipid peroxidation.[40] Decomposition is facilitated by metal ions, e.g. by methaemoglobin or (in microsomes) cytochromes P450. CYP2E1 is especially effective. For example, rate constants for the reactions of cumene and *tert*-butylhydroproxides with Fe^{2+}–ATP are 3.1×10^3 and 1.3×10^3 $M^{-1}s^{-1}$ respectively (Chapter 2). The decomposition of cumene and *tert*-butyl hydroperoxides in the presence of a metal ion such as Fe^{2+} can be written:

tert-butyl hydroperoxide   alkoxyl radical

cumene hydroperoxide   alkoxyl radical

The alkoxyl radicals can undergo β-scission, e.g. to give a methyl radical in the case of cumene hydroperoxide

$$PhC(CH_3)_2O^{\bullet} \rightarrow PhC(O)CH_3 + CH_3{}^{\bullet}$$

Cytochromes P450 can carry out both one-electron and two-electron oxidations of organic peroxides.[22a] For cumene hydroperoxide, one-electron reduction of the O—O bond generates an alkoxyl radical that can go on to generate the ketone acetophenone and methane gas (via formation of methyl radical, $CH_3^{\bullet}$). Two-electron reduction generates cumyl alcohol (—OOH reduced to —OH) plus an oxo-iron species of P450 capable of substrate hydroxylation in the absence of O_2 or NADPH. This has been called the **peroxygenase** activity of cytochromes P450. P450 can also degrade peroxidized linoleic acid to a mixture of products, including pentane gas.

Problems to be borne in mind in the study of microsomal lipid peroxidation include the heterogeneity of these fractions (Section 3.7) and the presence of variable amounts of endogenous antioxidants (e.g. vitamin E). Nuclear membranes contain an electron-transport chain (Chapter 1) and undergo peroxidation on incubation with NADPH and Fe(III) salts *in vitro*; and peroxidation of the inner mitochondrial membrane can occur in the presence of iron plus NADH or NADPH.[9] Here iron chelates may be reduced by the NADH–coenzyme Q reductase, complex I.

4.9.17 *Acceleration of lipid peroxidation by species other than oxygen radicals*

Several ROS can accelerate lipid damage. For example, ozone (O_3) oxidizes lipids directly to ozonides, which decompose to aldehydes (Chapter 8). Peroxynitrite stimulates peroxidation (Chapter 2). Nitrogen dioxide ($NO_2^{\bullet}$) can initiate peroxidation both by abstracting $H^{\bullet}$ from PUFAs and by addition reactions (Chapter 8).

Singlet oxygen

Unlike ground-state oxygen, singlet $O_2{}^1\Delta g$ can react directly with carbon–carbon double bonds by an *ene* reaction to give peroxides (Chapter 2). Reaction of singlet O_2 with the double bond between C-12 and C-13 of linoleic acid can produce two hydroperoxides:

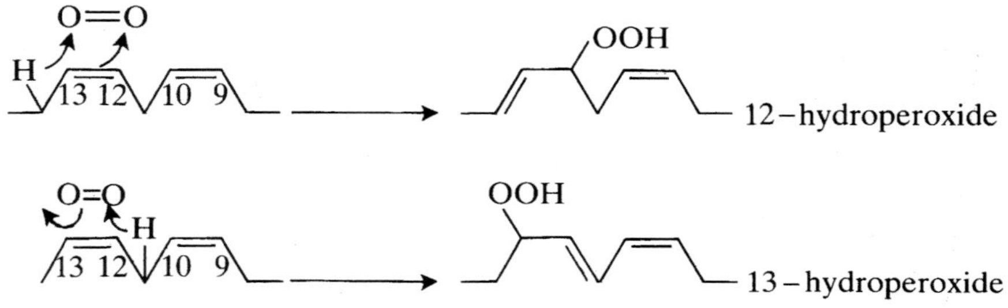

9- and 10-hydroperoxides are also produced by reaction of $^1O_2$ with the double bond between carbons 9 and 10. Thus four products result when linoleic acid is exposed to singlet O_2. By contrast, free-radical peroxidation of linoleate gives mainly, but not exclusively, the 9- and 13-hydroperoxides.

Exposure of lipids to singlet O_2 can cause rapid peroxidation. However, we think it best not to refer to singlet O_2 as *initiating* peroxidation: unless the peroxides are decomposed (e.g. by metal ions) to give peroxyl and alkoxyl radicals, a chain reaction will not begin. The lifetime of singlet oxygen in the hydrophobic interior of membranes will be much greater than in aqueous solution (Chapter 2). Hence illumination of unsaturated fatty acids in the presence of sensitizers of $^1O_2$ formation, such as chlorophyll, rose bengal, methylene blue, bilirubin or porphyrins, induces rapid peroxide formation. Similar effects have been seen with intact membranes.[37]

For example, the rod outer segments in the retina of the eye are rich in PUFAs and form peroxides upon illumination, probably due to $^1O_2$ formation sensitized by retinal (Chapters 2 and 7). Illumination of phospholipid liposomes containing haematin together with cholesterol not only peroxidizes the fatty-acid side-chains but also converts some of the cholesterol into its 5α-hydroperoxide, probably by reaction with $^1O_2$. Illumination of erythrocytes in the presence of porphyrins or bilirubin causes peroxidation, aggregation of membrane proteins, inactivation of enzymes and protein-transport carriers, and eventual haemolysis. Fungi of the genus *Cercospora* produce a toxin known as **cercosporin** that attacks plant cells, but only in the light. It has been suggested that cercosporin sensitizes formation of $^1O_2$, which then generates lipid peroxides in the plant cell, producing disruption of membranes.

Singlet oxygen may be formed during lipid peroxidation and might then cause more lipid peroxide formation. A likely mechanism of $^1O_2$ formation is the reaction of two peroxyl radicals to form a cyclic intermediate which decomposes to give singlet O_2 (Chapter 2):

$$\diagup\!\!\diagup CHO_2^{\bullet} \; + \; \diagup\!\!\diagup CHO_2^{\bullet} \;\longrightarrow\; \diagup\!\!\diagup C{=}O \; + \; \diagup\!\!\diagup C{-}OH \; + \; {}^1O_2$$

carbonyl

This is effectively a **chain termination** reaction of lipid peroxidation, since it removes two peroxyl radicals. Singlet O_2 formation might account for some of the chemiluminescence that accompanies lipid peroxidation (Chapter 5). However, the usefulness of 'singlet O_2 scavengers' in investigating the role played by singlet O_2 in lipid peroxidation has been affected by their lack of specificity: DABCO, diphenylfuran and β-carotene (Chapter 2) can react with $RO_2^{\bullet}$ radicals. Thus the overall contribution of singlet O_2 to lipid peroxidation is uncertain. However, under most physiological/pathological conditions it seems likely to be small; the self-reaction of peroxyl radicals is unlikely to be a favoured reaction until they have accumulated to significant levels within the membrane, i.e. until peroxidation is already extensive.

4.9.18   Peroxidation of other molecules

Membranes, lipoproteins and fatty acids are not the only molecules that contain many double bonds and can be peroxidized under appropriate conditions. Carotenoids can be peroxidized and there is debate about whether or not this can be regarded as an antioxidant property (Chapter 3). The aldehyde retinal can be oxidized as it sensitizes singlet oxygen formation in the eye (Chapter 7). The retinal precursor **retinol** (**vitamin A**) can undergo peroxidation.[31] Exposure of retinol to Fe^{2+} salts *in vitro* causes rapid oxidation to a complex mixture of products. Any excess vitamin A absorbed from the diet is stored in the fat deposits of adipose tissue, but it must be protected against peroxidation (probably by adipose tissue vitamin E).

Several antifungal antibiotics in clinical use (such as **candicidin**, **nystatin** and **amphotericin**) contain conjugated double bonds and are easily oxidized, resulting in loss of the antifungal activity (Chapter 8).

4.9.19   Repair of lipid peroxidation

Peroxides within membranes can be removed (converted to alcohols) by phospholipid hydroperoxide glutathione peroxidase enzymes (Chapter 3). Alternatively, they may be cleaved from membranes by the action of phospholipases, whereupon the released free fatty acid peroxides can be acted upon by 'ordinary' glutathione peroxidase (Fig. 4.25). Rises in Ca^{2+} caused by oxidative stress can activate phospholipase A_2, which can hydrolyse phospholipids (Section 4.4 above). Its degree of 'preference' for oxidized fatty acids is uncertain, however.

When phosphatidylcholine hydroperoxide is added to human plasma, it is reduced to the alcohol, although plasma glutathione peroxidase has almost no GSH available to it (Chapter 3). One reaction contributing to loss of the hydroperoxide is catalysed by **lecithin cholesterol acyltransferase** (**LCAT**),[83] an enzyme found in high-density lipoproteins (HDLs). It reversibly transfers fatty-acid side-chains from phospholipids to cholesterol, generating cholesteryl esters for storage in the HDL core. It also appears to reduce the phosphatidylcholine hydroperoxide to an alcohol, whilst generating a cholesterol ester hydroperoxide.

4.10   Mechanisms of damage to cellular targets by oxidative stress: protein damage

The biological importance of oxidative damage to proteins has only recently been considered in detail, despite the fact that it was shown many years ago that hyperbaric oxygen damage to *E. coli* involves inactivation of specific enzymes (Chapter 1). The delay in focusing attention on proteins perhaps reflects the considerable damage many proteins can sustain without impairment of their observed functions. This is particularly true of enzymes; only when

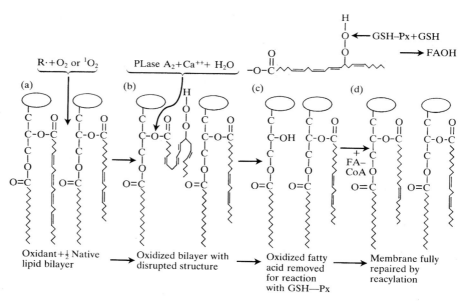

Fig. 4.25. Model for the repair of lipid peroxides. Formation of peroxides in the membrane causes rearrangement as the more-polar hydroperoxides move to the membrane surface. There they are cleaved by phospholipase A$_2$ (PLase A$_2$) in the presence of Ca^{2+} ions. Glutathione peroxidase (GSH-Px) in the surrounding fluid reduces the released fatty acid peroxides to alcohols (FAOH). Repair is completed by reacylation with a fatty-acyl-coenzyme A. Diagram adapted from *Trends Biochem. Sci.* **7**, 31 (1987) by courtesy of Dr. van Kuijk and Elsevier.

essential amino-acid residues at or close to active sites are impaired will 'damage' become evident from activity measurements. For example, **carbonic anhydrase isoenzyme III**[16] is extensively oxidized in the livers of old rats, but its ability to interconvert CO$_2$ and HCO$_3^-$ is only slightly diminished. Carbonic anhydrase III is unique among enzymes of this type because it can also act as a phosphatase, and phosphatase activity does decrease in liver from older rats. Nevertheless, exposure of rat hepatocytes to 100% O$_2$ leads to oxidation of proteins[111] and some loss in glutamine synthetase and glucose-6-phosphate dehydrogenase activities, and rats exposed to hyperbaric oxygen showed more protein damage than lipid peroxidation in the brain.[18] Addition of H$_2$O$_2$ to *E. coli* caused selective damage, i.e. some proteins were oxidized much faster than the bulk of cell protein.[113a]

In the pharmaceutical industry, chemical instability can be a problem in the handling of protein/polypeptide drugs. Methionine and cysteine residues are particularly prone to damage.[68] For example, oxidation of methionine residues in cholecystokinin was a particular nuisance during production of this peptide.

We now know that oxidative damage to proteins is of importance *in vivo* in its own right (damaging receptors, enzymes, signal transduction pathways, transport proteins and the enzymes that maintain low intracellular free Ca^{2+}

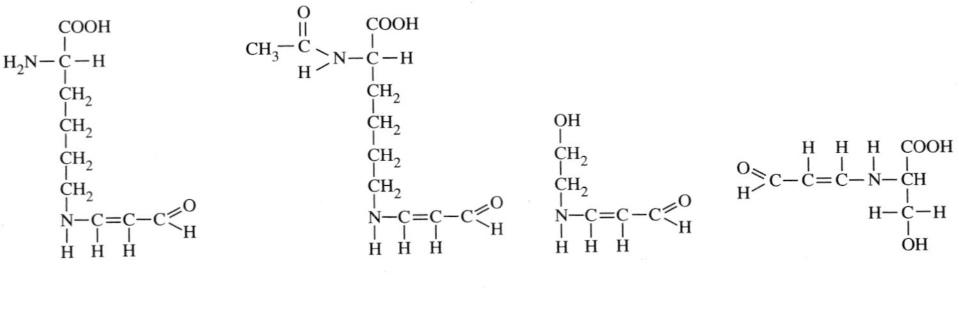

Fig. 4.26. Some products found in urine that arise from MDA-modified proteins. 1, N^{ε}-(2-propenal)lysine; 2, N^{α}-acetyl-N^{ε}-(2-propenal)lysine; 3, N-(2-propenal)etholamine; 4, N-(2-propenal)serine. The lysine adduct (1) and its acetylated form (2) are the major products detected.

levels) but it can also contribute to secondary damage to other biomolecules. Oxidized proteins may be recognized as 'foreign' by the immune system, triggering antibody formation and perhaps autoimmunity,[47] e.g. in rheumatoid arthritis and scleroderma[93] (Chapter 9). Damage to DNA-repair enzymes (endonucleases, ligases, etc.) could raise oxidative DNA damage levels and increase mutation frequency. Damage to DNA polymerases might decrease their fidelity in replicating DNA.

Damage to proteins can occur by direct attack of ROS/RNS upon them, or by 'secondary damage' involving attack by end products of lipid peroxidation, such as MDA and HNE (Section 4.9.10 above). Proteins can also be damaged by glycation reactions (Chapter 9). Various end-products of the degradation of MDA-modified proteins are found in mammalian (including human) urine,[29] such as N^{ε}-(2-propenal)lysine, N^{α}-acetyl-N^{ε}-(2-propenal)lysine, N-(2-propenal)serine and N-(2-propenal)etholamine (Fig. 4.26). These can arise from the metabolism of proteins modified by MDA *in vivo* but a more important source appears to be the ingestion of MDA-modified proteins in the diet: cooked meats are a particularly rich source (Chapter 5). Some HNE-modified proteins can inhibit[33] the action of the multicatalytic protease that degrades abnormal proteins (Section 4.12.3 below).

4.10.1   *Chemistry and significance of protein damage*

Attack of various RNS (e.g. $ONOO^{-}$, $NO_2^{\bullet}$ and NO_2Cl) upon tyrosine (both free and in proteins) leads to production of 3-nitrotyrosine. Nitration of phenylalanine and tryptophan can also occur. Attack of $OH^{\bullet}$ or singlet O_2 (Chapter 2) upon proteins can generate a multiplicity of end-products (Fig. 4.27).

By contrast, H_2O_2 and $O_2^{\bullet-}$ at physiological levels have little or no direct effect on proteins, unless there is an easily-oxidizable accessible −SH group, as in glyceraldehyde-3-phosphate dehydrogenase, and some Calvin cycle enzymes (in chloroplasts), which are targets of inactivation by H_2O_2 *in vivo* (Chapters 1 and 7). Another possible target is **ribonuclease inhibitor**, a cytoplasmic protein rich in cysteine residues. Oxidation of the −SH groups decreases its ability to prevent RNA degradation by certain ribonuclease enzymes.

In the early part of this century, Dakin, Tappel and their colleagues showed that a wide spectrum of carbonyls and peroxides is generated in irradiated proteins. Peroxides can be formed on the peptide backbone, and on the side-chains of several amino-acid residues. Amino-acid peroxides, like lipid peroxides, are fairly stable at physiological temperatures but can decompose to peroxyl and alkoxyl radicals on heating or if transition-metal ions are added.[25] Proteins can bind metal ions, especially iron and copper. Subsequent exposure of them to H_2O_2 generates $OH^•$ which selectively damages the amino-acid residues at the binding site in a site-specific reaction. For example, copper ions can bind to histidine residues on albumin and on apoprotein B on low-density lipoproteins. Subsequent addition of H_2O_2 causes conversion of histidine to 2-oxohistidine (Fig. 4.27).

When amino-acid radicals are generated in a protein, electrons can 'migrate' to other residues, i.e. the final products observed need not necessarily represent the initial sites of free-radical attack upon the protein. For example, methionine radicals can oxidize tryptophan, and tryptophan radicals can oxidize tyrosine: in terms of reduction potentials $E^{○'}Met^•/Met > Trp^•/TrpH > TyrO^•/TyrOH$. Tyrosine radicals can react with several antioxidants and also with $O_2^{•-}$. Thus although $O_2^{•-}$ rarely reacts directly with proteins, its presence can alter[25] the distribution of products generated by exposure to more reactive ROS, such as $OH^•$. For example, if $TyrO^•$ is generated, $O_2^{•-}$ can decrease bityrosine formation; reaction with $O_2^{•-}$ 'diverts' tyrosine radicals and diminishes their cross-linking.

4.10.2   *Damage to specific amino-acid residues*[25]

Cysteine and methionine

Thiol groups are easily oxidized by attack of many ROS/RNS and by direct reaction with transition-metal ions, forming thiyl radicals (Chapter 2).

$$RSH + Cu^{2+} \rightarrow RS^• + H^+ + Cu^+$$

Disulphides may result, or else further oxidation products such as sulphonates. Methionine is readily oxidized by $OH^•$, $^1O_2$, H_2O_2, HOCl, chloramines and $ONOO^-$. The initial product is **methionine sulphoxide**, which can undergo further oxidation to the sulphone (Fig. 4.27). Methionine residues are often essential for the activity of proteins, e.g. for the ability of α_1-antiproteinase to inhibit elastase (Chapter 2).[81] Oxidation of this methionine by HOCl, $ONOO^-$, $RO_2^•$ and other species causes a decrease in the protein's elastase-inhibitory capacity.

Histidine

Loss of histidyl residues in enzymes can often lead to inactivation. For example, *E. coli* glutamine synthetase, which converts glutamate to glutamine, is sensitive to oxidation of histidine.[70] Oxidation of histidine residues seems to

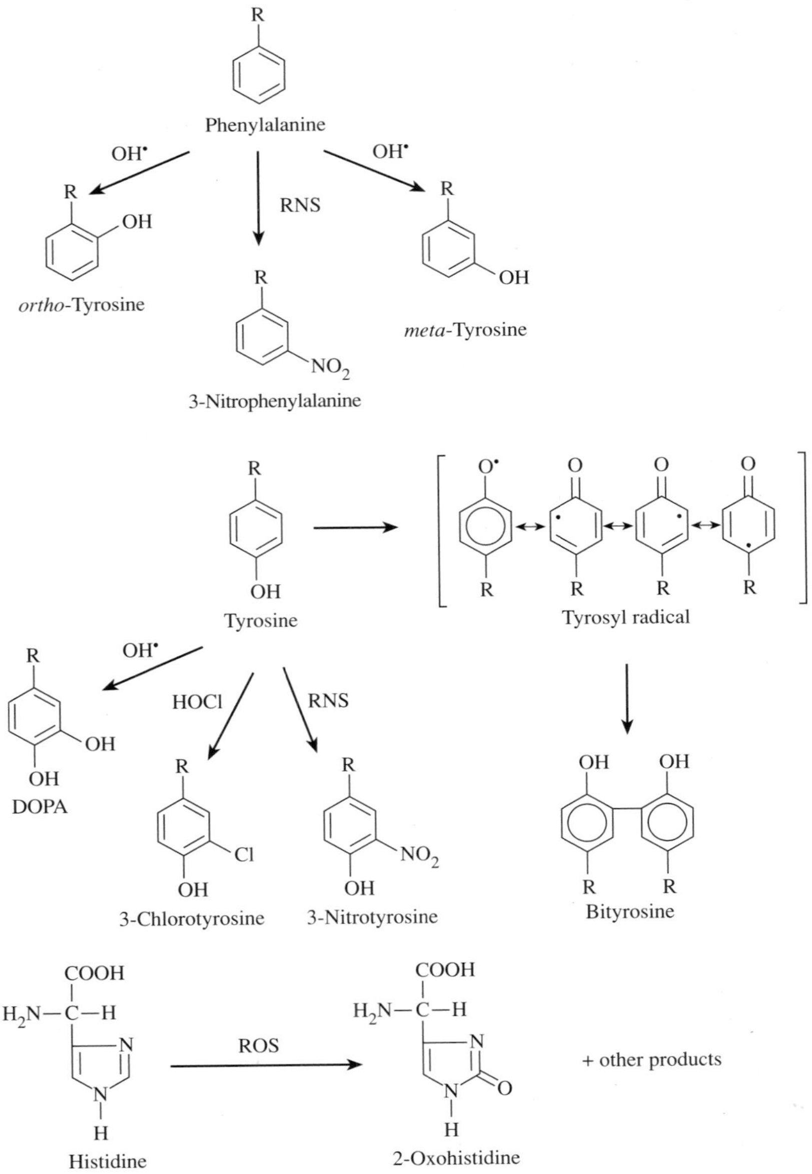

Figure 4.27(a)

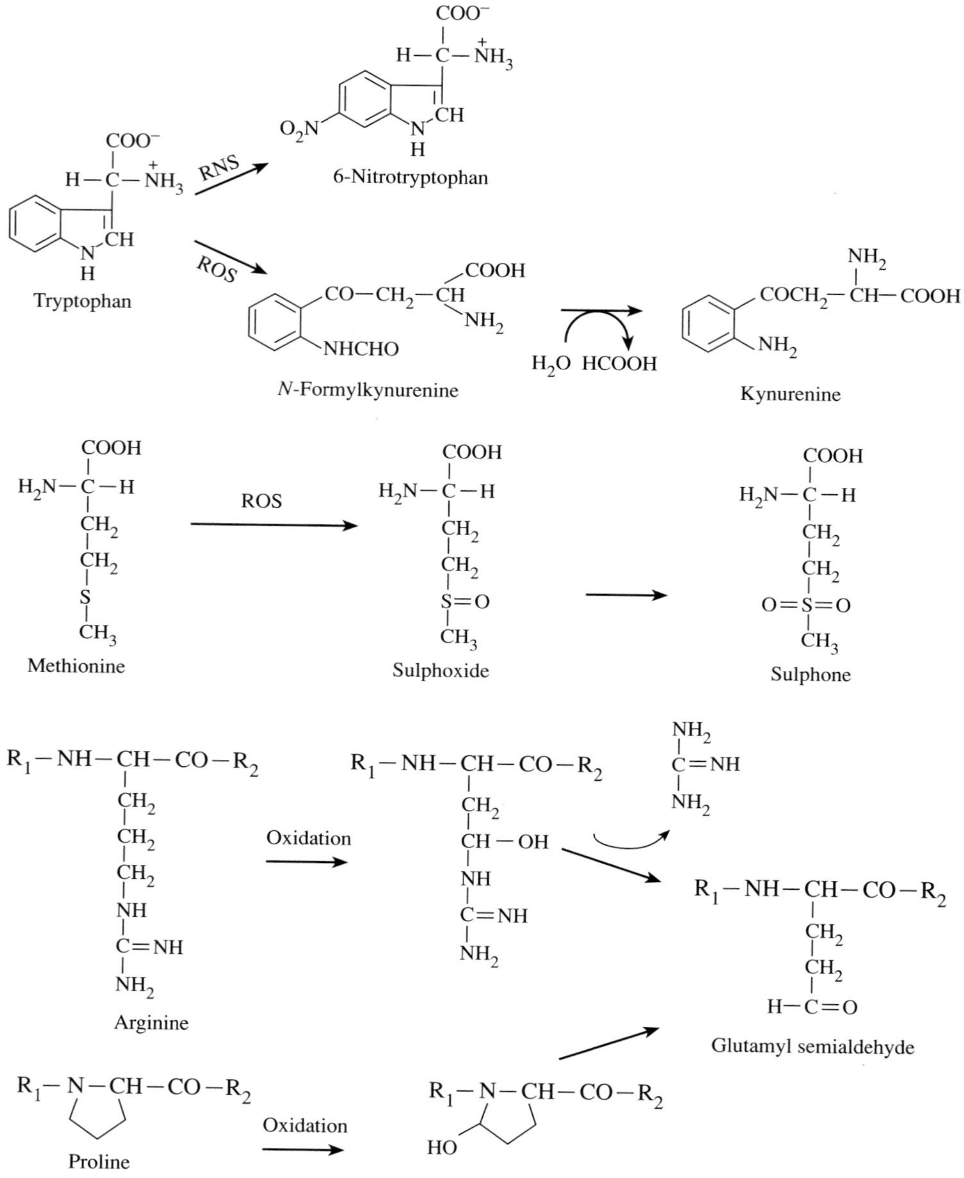

Figure 4.27. Some specific end-products of oxidative damage to amino-acid residues in proteins. Only selected products are shown; many others are known.

be particularly important in protein cross-linking reactions as well as 'marking' proteins for proteolysis (Section 4.12.2 below). Oxidation of histidine can form 2-oxohistidine, among many other products (Fig. 4.27). Histidine has a high reaction rate with singlet O_2, forming several well characterized oxidation products (Chapter 2), which can be detected in proteins exposed to $^1O_2$.

Proline and arginine

Free-radical attack upon proline and arginine residues in proteins can lead to glutamate semialdehyde formation.

Tryptophan

Tryptophan residues in proteins are sensitive to damage by $OH^\bullet$ and some other oxygen radicals, generating peroxyl radicals, peroxides and eventually *N*-formyl-kynurenine and kynurenine as characteristic fluorescent oxidation products (Fig. 4.27). Tryptophan is a powerful singlet O_2 scavenger, and can be fragmented to characteristic products (Chapter 2) in proteins exposed to this species. Tryptophan can also be nitrated by $NO_2^\bullet$ and on addition of peroxynitrite.

Tyrosine

Tyrosine residues in proteins can be attacked by RNS to undergo nitration, and by HOCl, leading to chlorination (Fig. 4.27). Attack of $OH^\bullet$ radicals can hydroxylate tyrosine to **dihydroxyphenylalanine** (DOPA). Tyrosine radicals are important at the active sites of several enzymes, including ribonucleotide reductase (Chapter 6) and appear to be involved in oxidations promoted by mixtures of haem proteins and H_2O_2 (Chapter 3). They can also result from free-radical attack ($OH^\bullet$, $RO_2^\bullet$, $RO^\bullet$) upon proteins. Among the fates of tyrosine radicals is cross-linking to give bityrosine. Nitration, chlorination or cross-linking of tyrosine residues could conceivably inhibit signal transduction by blocking tyrosine phosphorylation. Bityrosine can be detected in human urine, possibly arising by degradation of oxidized proteins. It is also present in atherosclerotic lesions (Chapter 9).

Phenylalanine

Attack of $OH^\bullet$ radicals upon phenylalanine generates several intermediate radicals which can then be converted into *ortho-*, *para-* and *meta*-tyrosines (Fig. 4.27). For example, the presence of *ortho*-tyrosine has been used to detect irradiated food proteins. Phenylalanine can be a target of attack by RNS, generating nitrophenylalanines.

Valine

Valine hydroperoxides are readily formed when proteins are exposed to $OH^\bullet$. Three valine hydroperoxides have been characterised, namely β-hydroperoxyvaline, $(2S,3S)$-γ-hydroperoxyvaline and $(2S,3R)$-γ-hydroperoxyvaline. These can be degraded by transition metals or acted upon by glutathione peroxidase. In each case a degradation product is **valine hydroxide**, which may provide a useful marker for studying $OH^\bullet$ damage to proteins.

4.11 Consequences of oxidative protein damage: interference with cell function

Not only enzymes but also receptors (such as β-adrenoceptors and α_1-adrenergic receptors)[63] and transport proteins can be important early targets of oxidative damage. Damage can occur to proteins involved in maintenance of essential ion gradients between cells and extracellular fluids, such as the Ca^{2+}-ATPase and Ca^{2+}/Na^+ exchange systems that keep intracellular Ca^{2+} levels much lower than extracellular levels (Section 4.4). The Na^+, K^+-ATPase system in the plasma membranes keeps intracellular K^+ high and intracellular Na^+ low when compared with levels in extracellular fluids. It contains catalytically-essential −SH groups that are susceptible to oxidative attack, and is readily inactivated during lipid peroxidation. K^+ channels[27] can also be affected by ROS/RNS and by HNE; they are essential in generation of electrical activity in both neuronal and cardiac systems, but are also important in a variety of other cell types such as T-lymphocytes and epithelial cells.

Alterations of cellular ion balance can produce changes in cell volume,[50] which in turn will affect many aspects of cell function. For example, generation of H_2O_2 (e.g. by adding a substrate for monoamine oxidase) or infusion of *tert*-butyl hydroperoxide in isolated perfused rat liver leads to a net release of K^+ (due to abnormal opening of K^+ channels) and hepatocyte shrinkage. Cell shrinkage in hepatocytes stimulates several metabolic events, including proteolysis, glycogen breakdown, biliary GSSG release and urea synthesis. It decreases others such as protein synthesis, lactate uptake and actin polymerization.

4.12   How organisms deal with oxidative protein damage

4.12.1   *Repair*

Oxidative protein damage, by ROS and RNS, occurs *in vivo* (Table 4.1). Sometimes, 'repair' can take place. Thus GSH and thioredoxin can re-reduce disulphide bridges formed by aberrant oxidative cross-linking of cysteine−SH groups (Chapter 3).

Another important repair system is **peptide methionine sulphoxide reductase**.[81] The action of singlet oxygen, $RO_2^\bullet$, $RO^\bullet$, $ONOO^-$, HOCl or $OH^\bullet$ upon methionine generates methionine sulphoxide (Fig. 4.27). Indeed, lens proteins in cataract contain significant levels of methionine sulphoxide (Chapter 7). *E. coli*, the protozoan *Tetrahymena pyriformis*, yeasts, higher plant and mammalian (including human) tissues have all been found to contain an enzyme that reduces methionine sulphoxide in peptides/proteins back to methionine. It can thus re-activate proteins damaged by previous oxidation of their methionine residues. The source of reducing power used by the enzyme is reduced thioredoxin (Chapter 3), regenerated at the expense of NADPH by thioredoxin reductase. In laboratory assays of methionine-sulphoxide reductase, reduced thioredoxin can be replaced by the synthetic dithiol dithiothreitol.

The importance of the reductase in repairing radical-induced damage to cells cannot yet be evaluated. However, disruption of the gene (*msr*A) encoding it in *E. coli* caused increased sensitivity to H_2O_2 under some (but not all) growth conditions.[81]

4.12.2   *Protein degradation*

When a protein is damaged *in vivo* by ROS or RNS it is often 'marked' for proteolytic degradation and hence removal, thereby increasing protein turnover in cells exposed to oxidative stress. By contrast, heavily oxidized and/or aggregated proteins may resist proteolytic attack and accumulate within cells.[25,41b]
    Although methionine sulphoxide and cysteine → disulphide residues in proteins can be repaired, proteins in which other amino acids have been oxidized or nitrated seem to be effectively irreversibly damaged and are removed. A certain minimum damage seems necessary for recognition of the protein as 'abnormal'. For example, in *E. coli*, oxidation of certain histidine and arginine residues inactivates glutamine synthetase,[70] but oxidation of more residues is needed to cause accelerated degradation by several proteases, such as **protease 50**. An important factor in recognition for destruction seems to be the exposure of hydrophobic regions on the protein surface.
    For example, metal-ion-catalysed oxidation of glucose-6-phosphate dehydrogenase from *Leuconostoc mesenteroides* produced increased surface hydrophobicity, and accelerated its degradation by rat liver. Cross-linking of the same protein with HNE did not increase proteolytic susceptibility or surface hydrophobicity. Indeed, the HNE-linked protein could inhibit degradation of the oxidized protein.[33] Extracts of bacteria, plants and animal cells have been shown to degrade oxidized proteins, and intact cells degrade proteins more quickly when subjected to a limited degree of oxidative stress.[24] The ability of cells to degrade abnormal proteins may decrease with age, since there have been several reports of the accumulation of non-functional proteins during ageing[111] (Chapter 10). Proteolytic systems that recognize oxidatively-modified proteins are present in both the cytosol and in the mitochondria of mammalian cells.[57]

4.12.3   *The proteasome*[57]

Destruction of 'unwanted' (including abnormal) proteins in eukaryotes mostly occurs in large multienzyme complexes called **multicatalytic protease complexes**, or **proteasomes** (although some proteins are degraded in lysosomes). One of the first proteasomes to be isolated was the **20S proteasome**. It has a relative molecular mass about 700,000, and is a cylindrical structure made of a stack of four rings, each containing seven protein subunits. The cylinder has three internal cavities: the central cavity contains several proteolytic sites and access to it is controlled by narrow gates from the two outer cavities. *In vitro*, 20S proteasomes will degrade certain denatured or oxidized proteins; ATP is not required. *In vivo*, however, the 20S proteasome is probably the core of a larger proteasome, the **26S proteasome**; essentially the 20S cylinder plus additional groups of proteins (**19S cap complexes**) attached at each end.
    The 26S proteasome recognizes its targets because they have been marked for degradation by attachment of **ubiquitin**, a 76 amino-acid protein.

Ubiquitination (attachment of multiple ubiquitin molecules) is ATP-dependent and catalysed by a group of enzymes. The 20S proteasome will not attack ubiquitinated proteins; recognition of these is a function of the 19S cap complexes, which have ATPase activity. Some of these cap proteins resemble gene transcription factors. The overall effect is an ATP-dependent degradation of ubiquitinated proteins.

Bacteria do not contain the ubiquitin system. They contain both ATP-dependent and ATP-independent proteases. In bacterial extracts oxidized proteins appear to be degraded by ATP-independent pathways. The same is observed in extracts of mammalian cells, the 20S proteasome or fragments of it apparently being involved (although its precise role has not been pinned down). In erythrocytes, the 26S proteasome is not present but these cells can still degrade oxidized proteins, presumably using the 20S core proteasome.[24]

Physiologically, the 26S proteasome regulates cell division in eukaryotes by acting on **cyclin–dependent protein kinases**. The appearance and disappearance of particular active kinases involved in the cell cycle is regulated not only by their synthesis but also by their degradation by the 26S proteasome. For example, the *Streptomyces* metabolite **lactacystin**[23] inhibits the proteasome and interferes with cell-cycle progression. Proteasomes are responsible for destruction of the inhibitory subunit released during activation of the transcription factor NF-κB (Section 4.13.4 below), and may have many other important biochemical functions.

4.13   Consequences of oxidative stress: adaptation

Exposure of organisms to mild oxidative stress often causes a prompt increase in the synthesis of antioxidant (and other) defence systems. These responses help to protect the cell against the insult, and sometimes against greater oxidative insults applied subsequently. How can this be achieved: how is an increase in ROS/RNS production translated into increased expression of the necessary genes and/or increased mRNA translation needed to raise levels of protective proteins? Our understanding is greatest in the case of certain bacteria, but is rapidly growing for mammalian systems. Some information is also available for yeast and higher plants. Most of the available data focus on regulation of transcription, although it is possible that redox-regulated binding of proteins to mRNA could control translation. For example, newborn rats exposed to elevated O_2 show increased lung catalase activity: this seems to be due not to more transcription of the catalase gene but to stabilization of the catalase mRNA, perhaps promoted by binding of proteins sensitive to redox conditions.[19]

4.13.1   *Bacterial redox regulation:* oxy$R$[25a,66]

Many bacteria respond to mild oxidative stress by becoming resistant to more severe oxidative stress. For example, when *E. coli* or *S. typhimurium* are exposed to moderate levels of H_2O_2, the synthesis of about 30 proteins increases and the cells then become resistant to damage by higher levels of H_2O_2. Some of these proteins overlap with those induced in response to heat shock (Section 4.15 below).

Nine of the proteins induced by H_2O_2 are controlled by one particular gene, called ***oxy*R**. When critical cysteine –SH groups within the protein product of the ***oxy*R** gene become oxidized, the oxidized protein activates the transcription of these nine genes.[66] Both oxidized and reduced *oxy*R can bind to several gene promoters[d] on DNA, but only the oxidized form can activate transcription. Products of activated genes include hydroperoxidase I (product of the *kat*G gene), alkyl hydroperoxide reductase (*ahp*CF gene) and glutathione reductase (*gor* gene). The role of these enzymes in antioxidant defence is discussed in Chapter 3. Also induced by *oxy*R is **Dps**, a protein that binds to DNA and has been hypothesized to protect it against oxidative damage, since deletion of the *dps* gene causes *E. coli* to become hypersensitive to H_2O_2.

The *oxy*R system functions in normal *E. coli* to maintain a steady-state intracellular level of H_2O_2 of about $0.2\,\mu M$ over a wide range of growth conditions. Strains lacking *oxy*R show increased spontaneous mutation rates. Activation of the *oxy*R system also increases the resistance of *E. coli* to HOCl. OxyR activation is reversed by its reduction, probably involving glutaredoxin.[25a]

4.13.2   *Bacterial redox regulation:* soxRS

Excess $O_2^{\bullet-}$ generation within *E. coli* leads to increased H_2O_2 formation and will activate *oxy*R.[25a] It also activates about ten additional genes, including those encoding MnSOD (*sod*A gene), a DNA-repair enzyme (**endonuclease IV**), fumarase C, aconitase A and glucose-6-phosphate dehydrogenase. Fumarase C and aconitase A may be 'back-up' enzymes that are not sensitive to inactivation by $O_2^{\bullet-}$ (Chapter 1). FeSOD is a constitutive enzyme in *E. coli* and is not part of this system (Chapter 3).

The 'oxidative signal' is sensed by the protein encoded by the *sox*R gene.[55] Oxidized SoxR protein activates transcription of the *sox*S gene, whose product binds to the promoters and activates transcription of the genes encoding the proteins listed above. The SoxR protein contains iron–sulphur [2Fe–2S] clusters active in promoting *sox*S transcription only in their oxidized, [2Fe–2S]$^{2+}$, state. Activation of the *sox*RS system also increases resistance to several antibiotics, due to decreased synthesis of a membrane protein, **OmpF**.

The *oxy*R and *sox*RS systems can additionally be activated by exposure of *E. coli* to excess $NO^{\bullet}$. Proteins containing iron–sulphur clusters are also involved in the regulation of iron metabolism in mammalian cells (Section 3.18).

4.13.3   *Bacterial redox regulation: the role of iron*

E. coli has a high-affinity iron-uptake system, comprising proteins encoded by about 30 genes. Their transcription is increased when the cell iron content is low. This system is regulated by the *fur* (**ferric uptake regulator**) gene. If the cells have enough iron, the product of this gene binds Fe^{2+} and the complex binds to a DNA base sequence (the **iron box**) found in the promoter regions of many genes whose expression is regulated by iron. As a result transcription is blocked (Chapter 3). Genes with iron boxes include not only the iron-uptake system but also those encoding the enzymes hydroperoxidase I and II and MnSOD. However, repression of MnSOD gene transcription by *fur* is over-ruled by *sox* activation. Manipulations of *fur* have been used to study the role of iron in promoting oxidative damage in *E. coli* (Chapter 3).

Some metal-chelating agents that can bind iron (such as nalidixic acid, dipyridyl and 1,10-phenanthroline) render *fur* inactive, increase the synthesis of MnSOD by *E. coli* B and even cause the appearance of MnSOD when *E. coli* B is grown under anaerobic conditions.

4.13.4   *Redox regulation in yeast*[122a]

Anaerobically grown *Saccharomyces cerevisiae* responds to the introduction of O_2 by activating the biosynthesis of haem, by a metabolic pathway that has two O_2-dependent enzymes. The binding of haem activates two transcription factors, HAP1 and HAP2/3/4 (where HAP is an abbreviation for **haem activation complex**). HAP binding leads to increased synthesis of multiple proteins, including cytosolic catalase (**catalase T**) and MnSOD (controlled by HAP1). Yeast also has a peroxisomal catalase (**catalase A**), which might also be regulated by this system; neither catalase is transcribed in the absence of O_2, or even under aerobic conditions in mutants deficient in haem biosynthesis.

Both HAP1 and HAP2/3/4 activate expression of a third haem-dependent transcription factor, **ROX1**, which represses transcription of several proteins, including one of the enzymes involved in haem biosynthesis, a form of 'feedback inhibition'. Like bacteria, pre-treatment of yeast with H_2O_2 or $O_2^{\bullet-}$-generating systems causes them to become resistant to higher levels. One gene involved is *yAP1* (which resembles the human transcription factor AP-1; Section 4.13.6 below). *yAP1* seems to control the expression of genes involved in the production and metabolism of glutathione and thioredoxin.

4.13.5   *Redox regulation in mammals: NF-κB*[90]

NF-κB is a complex of proteins that activates the transcription of multiple genes in response to a variety of stimuli. It was the first eukaryotic transcription factor shown to respond directly to oxidative stress, and belongs to the **Rel** family of transcription factors.

Active NF-κB is a heterodimer, usually containing proteins of relative molecular masses 50,000 and 65,000, referred to as **p50** and **p65**. NF-κB is ubiquitously expressed in mammalian cells and usually remains in an inactive, cytoplasmic form. Cytoplasmic retention and lack of activity result from binding to the dimer of an inhibitory subunit, called **IκB**. A wide range of toxicological, pathological or pathogen-related stimuli such as bacterial lipo-polysaccharides, infection with such viruses as human immunodeficiency virus type 1 (HIV-1) and hepatitis B virus (HBV), exposure to UV light, nickel, cobalt, γ-rays or certain inflammatory cytokines (e.g. tumour necrosis factor alpha, TNFα) can lead to activation of NF-κB. Activation results from dissociation of IκB, which is phosphorylated, then ubiquitinated and degraded by the proteasome complex. Activated NF-κB moves into the nucleus, binds to DNA and activates expression of several target genes such as those encoding cytokines, acute phase proteins, iNOS, growth factors, adhesion molecules and cytokine receptors. Prompt resynthesis of IκB by the cell shuts down the signal—presumably IκB enters the nucleus to inactivate the DNA-bound NF-κB.

Most NF-κB-activated genes are involved in what appears to be an early defence network against pathological conditions. For example, NF-κB plays a key role in regulating the response of vascular endothelial cells to stress (Chapter 9). Glucocorticoid hormones decrease NF-κB activation; in some cell types this is achieved by increasing transcription of genes encoding IκB. NF-κB is particularly important in the immune system.

Since many stimuli that activate NF-κB also cause oxidative stress, Pahl and Baeuerle[90] formulated the hypothesis that ROS are a common second-messenger system used by all stimuli to activate NF-κB. Addition of micro-molar amounts of H_2O_2 directly to the culture medium activates NF-κB in several (but not all) cell lines. HOCl can also activate NF-κB in some cells. Many of the data come from studies with the human cell lines Jurkat (a T-lymphoma line) and HeLa (a carcinoma cell line). Incubation with various superoxide- or NO•-generating agents failed to activate the transcription factor. These and other data argue that NF-κB, like the bacterial factor OxyR, responds to peroxides.

Further support for the role of ROS as second messengers that activate NF-κB comes from experiments using antioxidants. A wide range of anti-oxidants (from thiols and catechols to spin traps) can block the activation of NF-κB by any stimulus in certain cell lines. Overexpression of thioredoxin or glutathione peroxidase also prevents NF-κB activation.[64] However, the mech-anism by which ROS cause IκB dissociation and NF-κB activation is uncertain. The initial trigger for IκB degradation is phosphorylation, which leads to binding of ubiquitin and degradation by the proteasome complex. A likely mechanism therefore involves alteration of the activity of tyrosine kinase or phosphatase enzymes. There is growing evidence that some ROS can lead to protein kinase activation[45] (e.g. by raising intracellular Ca^{2+}) and protein phosphatase inactivation.[25b] To take one of many examples, **calcineurin** is a Ca^{2+}/calmodulin-dependent protein phosphatase, particularly important in brain and lymphocytes. Oxidation of iron at its active site decreases its activity.[118]

Activation of NF-κB is one step, but to achieve biological effects the activated transcription factor must then bind to DNA. Binding is also affected by redox conditions, being inhibited by oxidizing agents and potentiated by thiols (including reduced thioredoxin).[20] Hence the overall effects of changes in redox state upon expression of genes under the control of NF-κB can be very variable. Too high a level of ROS or RNS may activate NF-κB but inhibit binding of the active factor to DNA. Other agents that can inhibit

NF-κB activation in certain cells are **caffeic acid phenethyl ester**[84] (a component of propolis from honeybee hives; Chapter 1) and **acetyl-salicylate** (aspirin).[41]

4.13.6   *Redox regulation in mammals: AP-1 and the antioxidant response element*

The transcription factor, AP-1 (**activator protein 1**) also responds to the intracellular 'redox state', and can be affected both by ROS and anti-oxidants.[90,96] AP-1 is a dimer of the products of two genes, *jun* and *fos*. AP-1 exists in two forms: either a homodimer (two c-Jun proteins) or a heterodimer consisting of c-Jun and c-Fos, although many other family members have been described. Jun and Fos were originally described as products of proto-oncogenes. They are thought to function as a link between 'growth' signals received at the cell surface and changes in gene expression (Chapter 9). In 'resting' cells, small amounts of AP-1 exist in the cytoplasm in an inactive form, mainly as phosphorylated c-Jun homodimers. Many different agents inducing oxidative stress, such as H_2O_2, UV light, γ-rays, the cytokine interleukin-1 and phorbol esters, lead to AP-1 activation; there is considerable interest in how this is related to cell proliferation and cancer development (Chapter 9).

Rises in AP-1 activity can be due either to increased synthesis of AP-1 components, or to stimulation of the activity of pre-existing components (by phosphorylation or changes in redox state). H_2O_2 has been shown to increase transcription of *jun* and *fos* in many cell types, but has only a slight effect in activating AP-1. Paradoxically, H_2O_2 has been reported to decrease AP-1 activation by phorbol esters. By contrast, several phenolic antioxidants cause both a significant increase in c-*jun* and c-*fos* mRNA levels, and a strong activation of AP-1 DNA binding.[90,96]

The DNA-binding activity of AP-1 is regulated by a redox mechanism involving cysteine residues in the DNA-binding domain of Fos and Jun: oxidation decreases DNA binding.[20] Oxidation can be reversed by thiols or by a cellular redox/DNA-repair protein called, when first isolated, Ref-1 (**redox factor 1**), which probably interacts with reduced thioredoxin. Ref-1 activity is located on the AP endonuclease protein involved in DNA repair (Section 4.8 above). Hence repair is closely linked to activation of transcription by AP-1.

Antioxidant-induced AP-1 seems to be newly synthesized and the question arises as to which transcription factors are involved. The binding site for a third human redox-regulated transcription factor, **SRF/TCF**, is required for c-*fos* transcription in response to antioxidants. The c-*jun* promoter contains an AP-1-binding site, so that the presence of active AP-1 may increase c-*jun* transcription. Newly synthesized AP-1 binds to DNA target sequences in many promoters, among them being NADP(H)–quinone oxidoreductase (which helps to detoxify cytotoxic quinones; Chapter 8) and a glutathione S-transferase (GST) subunit.[96]

Thus AP-1 responds not only to oxidative stress but also to certain anti-oxidants. It is also up-regulated when some cell types are made hypoxic.

The **antioxidant response element**[96] (ARE), the DNA sequence responsible for activation of the NADP(H)–quinone reductase gene in response to several xenobiotics (including both pro-oxidants and some antioxidants, such as butylated hydroxyanisole) has been found to contain an AP-1 consensus sequence. The ARE of the GST Ya subunit gene in humans forms a less conserved AP-1 site. However, it should be noted that GST enzymes are induced in response to a vast range of xenobiotics and AREs are only one of the mechanisms by which this can be achieved; others include binding of toxins to the **xenobiotic response element**. Reduced thioredoxin may be an important activator of AP-1 *in vivo*, whereas it tends to decrease the activation of NF-κB.[20,90]

4.13.7   *GA-binding protein*[72]

Some constituents of the mitochondrial electron transport chain are encoded in mitochondrial DNA, but the genes for most are in nuclear DNA. Hence it is necessary for nuclear and mitochondrial gene expression to be regulated co-ordinately under certain circumstances, e.g. in response to hypoxia. **GA-binding protein** (GA-BP; sometimes called **nuclear respiratory factor 2**) is a transcription factor that controls expression of the nuclear genes that encode mitochondrial cytochrome oxidase subunits IV and Vb and mitochondrial transcription factor 1 (mtF1). Binding of GA-BP to DNA is inhibited by ROS, apparently due to oxidation of essential −SH groups, suggesting that redox changes can also lead to effects on mitochondrial function.

4.14   Are ROS/RNS important signal molecules *in vivo*?

Table 4.1 shows that antioxidant defences seem to control levels of ROS/RNS rather than eliminate them completely, e.g. OxyR in *E. coli* keeps H_2O_2 levels at $\sim 200\,nM$ (Section 4.13.1). Why is this? Maintaining a large excess of antioxidant defences would have an energy cost—it could be energetically 'cheaper' to repair (DNA) or destroy (proteins, lipids) damaged biomolecules. Also, some ROS probably cannot be protected against. The $OH^\bullet$ that arises from exposure to 'background' radiation will probably inevitably cause damage since it will hit molecules at its site of formation, and the damage has to be repaired. The diseases (e.g. cancer, atherosclerosis; see Chapter 9) that appear to involve continuing biological insult by residual ROS/RNS occur after the reproductive period, so might not be selected against during evolution. The importance of diet-derived antioxidants in maintaining health in later years is increasingly appreciated (Chapter 10).

Another possibility is that ROS/RNS are physiologically useful in controlled amounts, so that living systems have not evolved to eliminate them completely.[5] Phagocyte killing and chemotaxis (Chapter 6), the role of H_2O_2 in thyroid-hormone biosynthesis in animals and lignin synthesis in plants (Chapter 6), and the multiple physiological roles of $NO^\bullet$ (Chapter 2) are clear examples, but there could be many others. Some ROS/RNS-induced mutations may be needed for evolution, allowing response to changing environmental conditions.

B-Lymphocytes have an NADPH oxidase system that releases $O_2^{\bullet-}$, perhaps to allow them to communicate. Fibroblasts and vascular endothelial cells release $O_2^{\bullet-}$ *in vitro* and $O_2^{\bullet-}$ could promote fibroblast proliferation (Chapter 6). Many cells in culture respond to ROS; low levels of ROS have been repeatedly shown to stimulate cell proliferation, whereas higher levels exert inhibitory or cytotoxic actions (e.g. Fig. 4.2). Thus the 'redox balance' seems to be critical in affecting cell behaviour in culture. Redox changes regulate the activity of transcription factors such as AP-1 and NF-κB. Exposure of several cell types to low levels of ROS (especially H_2O_2) causes activation of protein kinases (e.g. PKC) and transcription of 'immediate early' genes such as *fos*. Increased phosphorylation of proteins tends to favour cell proliferation and could involve kinase activation by ROS and/or phosphatase inactivation (Box 4.1). Many

---

Box 4.1
Some stimulatory effects involving reactive oxygen species
(Note that effects have often been described in cell lines in culture)

- activation of NF-κB/AP-1 (see text)
- stimulation of cell proliferation, e.g. fibroblasts (see Fig. 4.2; *Free Rad. Biol. Med.* **22**, 287 (1997))
- *ras* oncogenes stimulate $O_2^{\bullet-}$ production in the NIH-3T3 fibroblast cell line, leading to increased proliferation (*Science* **275**, 1649 (1997))
- H_2O_2 increases affinity of aortic smooth muscle cells for basic fibroblast growth factor (bFGF), stimulating bFGF-dependent proliferation (*FEBS Lett.* **395**, 43 (1996))
- stimulation of sperm function (Chapter 7)
- regulation of apoptosis (Section 4.17)
- response of cells to cytokines, especially TNFα (Section 4.16)
- activation of phospholipase D in endothelial cells; activations of phospholipases C and A$_2$ in many cell types (see text; also *Am. J. Physiol.* **271**, L400 (1996); *Cancer Res.* **49**, 5627 (1989); *J. Lab. Clin. Med.* **125**, 26 (1994))
- up-regulation of adhesion molecules on phagocytes and endothelial cells (Chapters 6 and 9)
- stimulation of adenylate cyclase in mouse smooth muscle cells (via increased phosphorylation) (*Circul. Res.* **77**, 710 (1995))
- mimicry of insulin-like effects on adipocytes (via increased phosphorylation of insulin receptor) (*Proc. Natl Acad. Sci. USA* **84**, 8115 (1987))
- stimulation of rat vascular smooth muscle cells by platelet-derived growth factor (*Science* **270**, 296 (1995))
- stimulation of cyclooxygenase/lipoxygenase activities (Chapter 6)
- regulation of platelet function (Chapter 6)
- regulation of levels of macrophage inflammatory protein-1α (MIP1α) (*J. Biol. Chem.* **271**, 5878 (1996))
- epidermal growth factor-induced tyrosine phosphorylation in epidermoid cancer cells (*J. Biol. Chem.* **272**, 217 (1997))
- H_2O_2 stimulation of the growth of multicellular prostate cancer spheroids (*FEBS Lett.* **419**, 201 (1997)).

protein phosphatases have essential −SH groups, often oxidizable by ROS/ RNS with loss of enzyme activity.

Hence it has repeatedly been suggested that certain ROS might be used as 'signal molecules'. If the binding of a ligand to a receptor first causes more (or less) ROS production, and ROS then exert effects on a signal transduction pathway, then the ROS would be acting, by definition, as second messengers. It seems likely that, if ROS do act in this way, it will be the selectively reactive ones, such as $O_2^{\bullet-}$ and H_2O_2, that are used rather than the indiscriminately reactive $OH^{\bullet}$ (although H_2O_2 might exert selective effects by causing $OH^{\bullet}$ formation at a specific site). Indeed, it has been argued that the low levels of $O_2^{\bullet-}$ and H_2O_2-scavenging enzymes in extracellular fluids (Section 3.19.4) could permit $O_2^{\bullet-}$ and H_2O_2 to act as intercellular signal molecules; sequestration of metal ions in these fluids allows $O_2^{\bullet-}/H_2O_2$ to move around without $OH^{\bullet}$ production by Fenton chemistry. Similarly, $NO^{\bullet}$ and possibly its derivatives (e.g. NO^+) are widely used physiologically, whereas reactive RNS such as $NO_2^{\bullet}$ and $ONOOH$ may be too indiscriminately damaging.

The interactions of RNS and ROS are also important in regulation. H_2O_2 activates NF-κB, but $NO^{\bullet}$ inhibits activation. In vascular endothelium, $O_2^{\bullet-}$ antagonizes the action of $NO^{\bullet}$ and causes vasoconstriction and it has been suggested that $O_2^{\bullet-}/NO^{\bullet}$ balance is one physiological mechanism for regulating vascular tone (although the effect of the $ONOO^-$ produced must be considered). Redox regulation may also be involved in controlling the defence response of injured plants (Section 6.9).

4.14.1   *It can occur, but does it matter?*

In vitro, many effects of ROS have been shown (Box 4.1). But how important are ROS as second messengers *in vivo*? It is, as yet, hard to say. The *oxy*R and *sox*RS systems use ROS as an intracellular signal to induce antioxidant defences. Many transport proteins/receptors/signal transduction systems contain essential −SH groups. Attack upon these by ROS/RNS may oxidize them, often reversibly, and 'send a message' (Fig. 4.28). This does not necessarily mean that this is the usual physiological mechanism by which that message is sent. ROS at low levels generally increase phosphorylation of cellular proteins, but not all increased phosphorylations are due to ROS. For example, human adipocytes respond to H_2O_2 in a similar way to insulin. They also contain a membrane-bound H_2O_2-generating system responsive to insulin (Chapter 6) but there is no evidence as yet that the effects of insulin on fat metabolism *in vivo* are mediated by H_2O_2.

We need to move away from studies of isolated cell systems (often affected by exposure to unphysiological O_2 levels, and complex pro-/anti-oxidant reactions involving such constituents of growth media as iron and copper ions, foetal sera, histidine and thiols) to the situation *in vivo*. Even in the case of NF-κB, where the evidence that ROS (especially H_2O_2) can be common mediators of activation in some cell types is overwhelming, it must be noted that these effects are not demonstrable in other cell lines.[11]

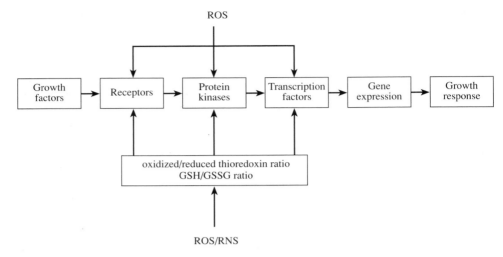

Fig. 4.28. Some possible ways in which ROS/RNS could regulate cell proliferation. Direct effects can occur, or effects mediated by changes in redox state of GSH and/or of thiol proteins such as thioredoxin. Modified from *Free Rad. Biol. Med.* **18**, 788 (1995) by courtesy of Professor Roy Burdon and Elsevier. Effects on protein phosphatases must also be considered (see text).

4.15   Heat-shock and related 'stress-induced' proteins

When cells are subjected to oxidative stress, it is not only the transcription of genes encoding antioxidant defence enzymes that can be increased; levels of many other proteins rise also. Expression of certain other proteins (e.g. cytochromes P450) may simultaneously be decreased.

Cells respond to many other stresses by altering gene transcription. One of the earliest classes of proteins to be studied in this way was the **heat-shock proteins**.[30] After a sudden increase in temperature all cells—from human to microbe—immediately, but transiently, increase activation of a small number of specific genes. Some of these genes encode the heat-shock proteins. This was first observed in the early 1960s in the fruit-fly, *Drosophila*. Chromosomal 'puffing' (due to 'unwrapping' of chromatin, needed to allow gene activation) was observed after exposure of the isolated salivary glands to temperatures slightly above those optimal for normal development of the fly. By 1974 it was clear that this 'puffing' was accompanied by the high-level expression of a unique set of heat-shock proteins.

Investigators soon found that cells produce heat-shock proteins when exposed to diverse 'stresses' such as heavy metals, alcohol and other xenobiotics, and for this reason began to refer to them as **stress proteins**. The concept was thus proposed of a universal response to adverse environmental changes which elicited a cellular defence in times of trouble. A few heat-shock proteins are induced in response to oxidative stress, but much less than for thermal stress.

Heat-shock proteins were identified in animal cells and it became clear that there is considerable resemblance between heat-shock proteins in different species, suggesting a similar protective function in all organisms. Physiologically relevant 'insults' such as fever or ischaemia–reperfusion induce stress proteins and their levels are often increased at sites of inflammation (Table 4.11). Many of the stress proteins are present (at lower levels) in normal or unstressed cells. Inducible protection against such diverse toxic stimuli remained an enigma until it was proposed in 1980 that many of the agents inducing the stress response are protein-denaturing agents, and that accumulation of partially denatured or abnormally folded proteins within a cell would trigger a stress response. Once initiated, the stress proteins would identify and facilitate the refolding (or removal of) damaged proteins. This theory was tested by injecting denatured proteins into cells and inducing the expected stress response.

The ability of oxidative stress to damage proteins could then account for its ability to induce some heat-shock proteins. For example, oxidized low–density lipoproteins have been reported to induce hsp70 in cultured human endothelial cells. Similarly, there are suggestions that thermal induction may sometimes involve ROS. Thus *E. coli* strains lacking both MnSOD and FeSOD are much more sensitive to heat stress than normal strains,[8] as are *S. cerevisiae* mutants lacking SOD and H_2O_2-removing enzymes. Thermal induction of hsp70 in transgenic mice over-expressing glutathione peroxidase was sub-normal, and the mice were less thermotolerant.[78] Overall, however, as stated above, there is only limited overlap between responses to heat and to oxidative stress in most organisms.

4.15.1   *Chaperones*[30]

One of the most prominent heat-shock proteins is the hsp70 family, containing a series of similar proteins with relative molecular masses in the range 66,000–78,000. The essential function of hsp70 is that of a **chaperone**. Cells contain a variety of chaperones, proteins that help guide the correct folding of newly synthesized proteins. Chaperone activity is ATP-dependent. They bind to hydrophobic regions and prevent incorrect folding of, and erroneous interactions between, unfolded polypeptides. *In vivo*, chaperones collaborate with enzymes, such as protein disulphide isomerases (Section 3.12) and peptidyl-prolyl isomerases (which accelerate the *cis/trans* isomerization of peptide bonds next to proline residues) to achieve correct protein conformation.

Hsp70 acts early in the life of a protein, before it leaves the ribosome, and then hands over to **hsp60** to help with later stages of protein folding. A special form of hsp70, **BiP**, helps to fold proteins within the endoplasmic reticulum. Other chaperones occur in different subcellular compartments, including mitochondria. Thus cytoplasmic chaperones eventually 'hand over' proteins destined for organelles to the organellar chaperones; proteins enter organelles in a partially unfolded state and are then helped to fold correctly by the organelle's own hsp70.

Steroid hormones, thyroid hormones and retinoids diffuse directly across cell membranes and bind to intracellular receptor proteins, which then enter the

Table 4.11. Some examples of induction of, and protection by, heat-shock proteins

System	Inducer	Findings	Reference
Human carcinoma cell line	TNFα (apparently acting by ROS generation)	Hsp27 enhances cellular resistance to TNFα; burst of ROS and changes in hsp27 are abolished by over-expression of the glutathione peroxidase gene	*Biochem. J.* **312**, 367 (1995)
Mouse heart	Ischaemia–reperfusion	High-level expression of human hsp70 protects the myocardium	*J. Clin. Invest.* **95**, 1854 (1995)
Murine cells transfected with human hsp70	Glucose deprivation	Cells expressing hsp70 were resistant to injury by glucose deprivation	*J. Clin. Invest.* **91**, 503 (1993)
Human amniotic cells (cultured)	H$_2$O$_2$ treatment	Hsp70 protects against oxidative DNA damage (assessed by 8-hydroxydeoxyguanosine concentration); it is proposed that the hsp70 protein is translocated into the nucleus to protect chromatin DNA or to facilitate DNA repair	*Biochem. Biophys. Res. Commun.* **206**, 548 (1995); also see reference 92a
Human monocytes/ macrophages	Red blood cells	Induce erythrophagocytosis and synthesis of hsp70, 83, 90 and 32 (haem oxygenase) by phagocytic cells; this was prevented by a GSH substitute. ROS from activated phagocytes may interact with haemoglobin to induce heat-shock proteins	*Proc. Natl Acad. Sci. USA* **87**, 1081 (1990)
CHO cells	H$_2$O$_2$ treatment	Overexpression of a phosphorylatable hsp27 protects against F-actin fragmentation	*Cancer Res.* (1996) **56**, 273
Blood vessels	Hypertension	Hsp70 up-regulated; it may be protective against vascular 'stress' caused by high blood pressure. (The immune reaction to hsp could contribute to atherosclerosis.)	*Mol. Med. Today* Sept. 1996, p. 372

Heat-shock proteins can be induced by external stresses, such as heat, heavy metals, organic solvents, ROS (in some cases) and pathophysiological events such as infection and tissue damage. Changes in their expression can also occur during the cell cycle and embryonic growth.

nucleus and act on the promoters of specific genes. The members of this **intracellular receptor superfamily** exist in the inactive cytoplasmic state bound to a heat-shock protein, **hsp90**. Binding of the signal molecules causes hsp90 to dissociate.

In *E. coli*, two molecular chaperone 'machines' have been described, **DnaK–DnaJ–GrpE** and **GroEL–GroES**. DnaK and its two cofactor proteins seem to interact with nascent proteins, which they then hand over to GroEL–GroES (similar to hsp60). It seems that overexpression of all these proteins during stress prevents aggregation of partially denatured proteins and assists their refolding during recovery from stress.

4.15.2   *The role of ubiquitin*[30,57]

The above heat-shock proteins act to preserve, or recover the function of, other proteins during and after stress. Another group of smaller (8500 molecular weight) heat-shock proteins is concerned with the degradation of proteins, presumably helping to dispose of proteins whose denaturation has gone too far. These are the **ubiquitin** proteins, levels of which rise 5- to 7-fold after stress and promote degradation by the proteasome (Section 4.12.3 above).

One can also include metallothioneins (Chapter 3), MnSOD and haem oxygenase[67] in the general family of stress proteins in that they respond to some, but not all, of the classical heat-shock protein inducers.

4.15.3   *Haem oxygenase as a heat-shock protein*

Hsp32, whose level increases in many tissues in response to haem administration, certain metal ions (e.g. cadmium) or to oxidative stress (including UV-irradiation), is identical with one isoform (**HO-1**) of the enzyme **haem oxygenase** (Chapter 3). Several of these inducers appear to act through antioxidant response elements. The promoter of the human HO-1 gene also has binding sites for NF-κB.[67]

4.15.4   *Bacterial stress proteins*

The stress proteins made by certain infectious organisms to help protect them against the host's defence mechanisms (including fever and phagocyte ROS/RNS production) are, ironically, often the major antigens recognized by the host's immune system to mount a defence. For example, the tubercule bacillus *Mycobacterium tuberculosis* responds to menadione by up-regulating heat-shock proteins rather than antioxidant enzymes—unlike *E. coli* and *S. typhimurium*, it does not have a functional *oxyR* system.[34] However, because microbial stress proteins are similar to human stress proteins, the immune system needs highly tuned mechanisms to discern minor differences between the two, and sometimes errors are made. Hence antibodies to heat-shock proteins can be detected in patients with several autoimmune diseases. Several human hsp70 genes are linked to the major histocompatibility antigen complex (MHC) locus. These proteins participate in the early stages of immune responses by presenting foreign antigens to cells of the immune system.

4.15.5   *Heat-shock transcription factor*[30]

Induction of heat-shock proteins involves activation of a **heat-shock transcription factor** (HSF). It binds to DNA sequences (**heat-shock *cis*-elements**) in the promoters of genes encoding heat-shock proteins, to activate transcription. In unstressed cells HSF is present in the cytoplasm and the nucleus (perhaps bound to hsp70) as a monomeric form that has no DNA-binding activity. In response to stress, HSF is released from hsp70 and forms a trimer that binds with high affinity to heat-shock gene promoters. The mechanisms(s) by which stress is sensed leading to HSF activation are not clear: it has been suggested that thermal denaturation of proteins is one signal, i.e. hsp70 could bind to denatured proteins preferentially and so release HSF. HSF can be 'activated' in cells by H_2O_2, but this does not generally result in increased hsp70 transcription or elevated heat-shock protein levels in mammalian cells (in contrast to activation by heat shock). Hence different modes of HSF binding to DNA can occur.

Similarly, in the bacterium *S. typhimurium*,[13] five of the 30 proteins induced by H_2O_2 are also increased by heat shock; induction of one of these by heat depends on *oxy*R. The cytotoxic aldehyde HNE, and several similar aldehydes, up-regulate heat-shock gene expression in cultured cells by activating HSF.[2]

4.16   **Cytokines**[17]

Most of the redox regulation we have discussed so far has been at the cellular level. Can multicellular organisms transmit stress signals *between* cells? For example, if one cell is stressed, can this lead to up-regulation of defences in other cells and tissues? Intercellular information transfer can occur through cell–cell contact (e.g. via gap junctions) and this easily allows information transfer about redox state. Communication between distant cells in complex organisms is partly achieved by hormones such as insulin, thyroid hormones, adrenalin and steroids, which are secreted into the plasma and exert complex effects on many cell types. Many of them alter antioxidant defences (Chapter 1 describes the effects of some hormonal changes on O_2 toxicity). A more localized signalling mechanism is provided by the **cytokines**.

Cytokine is a loosely defined term that encompasses a wide range of soluble polypeptides and glycoproteins. Most cytokines are secreted, but some can be expressed in cell membranes. They tend to act locally by binding to specific high-affinity receptors on their target cells; the receptors for many cytokines have now been identified and their genes cloned. Cytokines exert effects at very low concentrations and are often involved in complex interactions, in which one cytokine affects the secretion of many others. When first discovered, cytokines were usually identified as products of single cell types, e.g. **monokines** from monocytes, **lymphokines** from lymphocytes and **interleukins** from leukocytes. The list soon expanded to include interferons, tumour necrosis factors and cytokines affecting chemotaxis (**chemokines**). It was soon realized that cytokines have multiple cellular sources, a broad (and often overlapping) range of biological activities and some degree of redundancy in their effects, i.e. different cytokines can often affect cells in the same way. One analogy that has been used is that cytokines are 'words' in the language of cellular communication; words mean different things depending on what other words are around them.

Cytokines are involved in local regulation of cell growth and differentiation, signals for the inflammatory/immune response and chemotactic agents. Most cytokines are **pleiotropic**, i.e. they exert a wide range of different effects in various tissues and cells. It is impossible to review all the cytokines here, but we shall briefly mention some that have been particularly implicated in oxidative stress.

4.16.1   *TNFα*[17,92a]

Tumour necrosis factor alpha (TNFα) is a protein, relative molecular mass 17,000, that can be secreted by many cell types (especially monocytes and macrophages but also B- and T-lymphocytes, mast cells, neutrophils and fibroblasts). TNFα acts by binding to two different types of cellular receptor. The name 'tumour necrosis factor' comes from the first identification of TNFα as a factor found in the serum of mice infected with *Bacillus* **Calmette Guerin** (a weakened form of the agent inducing tuberculosis) that could cause haemorrhage and necrosis in transplanted tumours in the same mice. About a century ago, Coley in New York observed that cancer patients repeatedly administered 'bacterial broths' could show haemorrhagic necrosis of their tumours. Later work showed that the active agent was bacterial **endotoxin**, which stimulates production of TNFα. **Endotoxin (lipopolysaccharide)** is a complex glycolipid found in the cell walls of Gram-negative bacteria. Too much endotoxin exposure can cause severe injury, e.g. **endotoxic shock**, and over-production of TNFα is involved. Lower levels of endotoxin exposure can induce defence systems in animals, making them more resistant to further insult.

TNFα was later found to be identical with **cachectin**, a 'body wasting factor' identified in seriously-ill cancer patients. TNFα is strongly pro-inflammatory, e.g. promoting mobilization of fat stores, up-regulation of adhesion molecules resulting in increased phagocyte adherence to vascular endothelium, stimulation of cartilage degradation, bone resorption and phagocyte (especially neutrophil) ROS production. It plays a key part in host defence responses against injury and infection. For example, transgenic 'knock-out' mice lacking TNF receptors are more sensitive to brain damage by certain neurotoxins and after ischaemia–reperfusion.[12] However, excessive, prolonged production of TNFα is thought to contribute to the pathology of several diseases (Chapter 9).

TNFα generally increases oxidative stress both in its target cells and (via promotion of phagocyte ROS production) in surrounding cells. Increased ROS production, in at least some target cells, appears to involve effects of TNFα on mitochondria.[99] This results in up-regulation of antioxidant and other defence proteins, especially MnSOD in mitochondria, but also metallo-thioneins and heat-shock proteins, among others. Thus pre-treatment of animals with low-level TNF (or endotoxin) can protect them against several subsequent insults, including radiation, some cytotoxic drugs and ischaemia–reperfusion. TNFα is a powerful activator of NF-κB in many cell types, in some cases acting via oxidative stress.

4.16.2   *Interleukins*

Interleukin-1 exists in two forms, IL-1α and IL-1β. It is a key mediator of the defence response to injury and infection. It is generally pro-inflammatory and is secreted by monocytes, lymphocytes, macrophages, fibroblasts, keratinocytes and endothelial cells, among others. Often IL-1β is secreted in an inactive form which has to be cleaved by a cysteine protease **interleukin-1β converting enzyme** (ICE) to generate active IL-1β. IL-1 is a key mediator of the acute phase response (Section 4.16.3 below) and is a major contributor to fever production (TNFα also contributes). Again, excess IL-1 has been implicated in tissue damage in several diseases (Chapter 9).

Interleukin-8 is a chemokine (Table 4.12) important in promoting migration, adhesion and activation of neutrophils. It is produced by monocytes and many other cell types, including smooth muscle cells, endothelial cells, lymphocytes, fibroblasts and keratinocytes. Again, over-production has been implicated in several chronic inflammatory diseases. It has been suggested that ROS can stimulate IL-8 production by some cell types.

4.16.3   *The acute-phase response*

A common 'whole body' response to stress in animals is the **acute-phase response**. The body responds to infections and to inflammatory and immuno-logically-mediated diseases with a range of biochemical and clinical alterations. One of the clinical features of the acute-phase response is fever. Biochemical changes include increased synthesis of a group of proteins by the liver (**acute-phase proteins**; Table 4.13), decreased synthesis of albumin by the liver, decreases in plasma iron and zinc content, and negative nitrogen balance. 'Acute' signifies that many of the changes can be measured within hours of the precipitating event. Many features of the acute-phase response can be measured in chronic conditions, such as chronic 'low-grade' infections, rheumatoid arthritis (Chapter 9) and many forms of cancer. The function of some acute-phase proteins (e.g. C-reactive protein (CRP) and serum amyloid A protein (SAA)) is unknown but others are involved in antioxidant defence (Table 4.13).

The acute-phase response is different from the heat-shock responses in that it has evolved only in complex multicellular organisms, and presumably serves to co-ordinate the responses of a variety of cell types to a common external threat. This co-ordinated response depends on intercellular communication, especially involving cytokines.

Table 4.12. Some cytokines

Class	Examples
Interleukins	IL-1α, IL-1β, IL-2 to -7 and IL-9 to -17
Chemokines[a]	α-chemokines, e.g. IL-8 (mainly act on neutrophils); β-chemokines (mainly act on monocytes/macrophages) e.g. monocyte chemoattractant protein 1 (MCP-1); monocyte inflammatory proteins (MIP-1α and -1β); RANTES (regulation-upon-activation, normal T-cell expressed and secreted)
Interferons	Interferons α, β and γ
Tumour necrosis factors	Cachectin (TNFα), lymphotoxins (TNFβ_1, β_2)
Colony-stimulating factors (CSFs)	Granulocyte CSF, macrophage CSF, granulocyte–macrophage CSF
Transforming growth factors	TGFα, TGFβ (several types), activins
Growth factors	Epidermal growth factor (EGF), fibroblast growth factor (FGF), hepatocyte growth factor (HGF), insulin–like growth factor (ILGF), keratinocyte growth factor (KGF), platelet-derived growth factor (PDGF), vascular endothelial growth factor (VEGF) (an angiogenic agent)
Neurotrophic factors	Nerve growth factor (NGF); brain-derived neurotrophic factor (BDNF), ciliary neurotrophic factor (CNTF), neurotrophins

[a]Chemotactic cytokines.
Cytokines play key roles in normal growth and development, inflammatory and other responses to injury and tissue remodelling and repair (e.g. wound healing, angiogenesis). A complex balance between different cytokines regulates the final outcome. Thus some cytokines are often regarded as pro-inflammatory (e.g. IL-1, IL-6, TNFα) whereas others are 'anti-inflammatory' (e.g. IL-1 receptor antagonist, IL-10).

4.17   Consequences of oxidative stress: cell death

At some stage cellular injury may become irreversible: the cell passes a 'point of no return' and dies. For example, membrane blebbing (Fig. 4.4) in its early stages is a reversible event (the cell can recover if the insult is removed), but excess blebbing to the point of rupture is usually irreversible. The principal types of response to irreversible injury are necrosis, apoptosis and terminal differentiation.[85] Measuring irreversible cell injury can be achieved by several methods, but none is straightforward to interpret (Table 4.14).

4.17.1   *Necrosis*

Necrosis is characterized by early cell and organelle swelling, loss of integrity of mitochondrial, plasma membrane, peroxisomal and lysosomal membranes

Table 4.13. Some acute-phase plasma proteins

Protein	Typical increase in concentration	Biological function
Caeruloplasmin	50%	Copper carrier: antioxidant defence (Chapter 3)
C3	50%	Third component of complement
Antiproteases (e.g. α-antiprotease)	2- to 4-fold	Inhibit proteolytic enzymes, such as elastase
Haptoglobin	2- to 4-fold	Binds haemoglobin; antioxidant defence (Chapter 3)
C-reactive protein (CRP)	Several-hundred-fold (very little present normally)	Function unknown[a], but its importance is suggested by the observation that transgenic mice over-expressing CRP are resistant to endotoxin (*Proc. Natl. Acad. Sci. USA* **94**, 2575 (1997))
Serum amyloid A protein (SAA)	Several-hundred-fold	See Chapter 9

[a]The acute-phase response was first recognized in the 1930s. Patients with pneumococcal pneumonia were found to have a protein in their plasma which bound to a polysaccharide from the bacterial cell wall called **C-polysaccharide**. The protein was thereafter called C-reactive protein or CRP. Binding is Ca^{2+}-dependent.

and eventual breakdown of the cell, leading to release of its contents into the surrounding area (Fig. 4.3). Hence the necrotic death of cells affects the surrounding cells, e.g. by the release of lysosomal enzymes, pro-oxidants such as iron and copper ions and antioxidants such as catalase, SOD and GSH.

4.17.2 *Apoptosis*

By contrast, in apoptosis (Fig. 4.3) the earliest changes consist of cell shrinkage, condensation and fragmentation of chromatin. This is usually associated with DNA double-strand breaks in internucleosomal regions, so that the isolated DNA gives a 'ladder pattern' on gel electrophoresis; Fig. 4.7). Other features of apoptosis are collapse of cytoskeletal structure, nuclear fragmentation and eventual **break-up** of the entire cell into **apoptotic bodies**, without rupture of mitochondrial or lysosomal membranes or release of cell contents. Hence apoptosing cells do not affect their neighbours. Apoptotic bodies are phagocytosed by neighbouring cells or by phagocytes.

The DNA fragmentation involves activation of a Ca^{2+}- and Mg^{2+}-dependent endonuclease, perhaps by excessive rises in intracellular Ca^{2+} due to cell injury leading to entry of excess Ca^{2+} into the nucleus. This enzyme is inhibited by Zn^{2+} and **aurintricarboxylic acid** (ATA), agents that block apoptosis (although ATA may also chelate transition-metal ions). Apoptosis of cells is essential during embryonic development, e.g. to eliminate thymocytes that

Table 4.14. Some popular methods used to detect irreversible cell injury[a]

Method	Comments
Vital dyes, e.g. trypan blue, propidium iodide	Cannot penetrate intact plasma membrane but stain the cytoplasm or nucleus of cells with damaged plasma membranes
Tetrazolium salts (e.g. MTT)[b]	Reduced to coloured products (**formazans**) only by metabolically active cells
Release of enzyme, e.g. lactate dehydrogenase (LDH), creatine kinase (CK), glutamate–oxaloacetate transaminase (GOT), glutamate–pyruvate transaminase (GPT)	Release of enzymes from cytosol or other subcellular compartments is indicative of plasma membrane damage. Used in diagnostic medicine to assess damage to liver (GOT, GPT), heart (LDH, CK), etc. Release also possible from reversible injury by the 'pinching off' of blebs, with the rest of the cell remaining viable. Apoptotic cells do not release cytoplasmic contents
^{51}Cr release	Cells must be pre-loaded with ^{51}Cr by incubation with labelled sodium chromate, which binds to intracellular proteins. Radioactive proteins released due to plasma membrane damage.
[^{3}H]thymidine release	Cells pre-loaded with tritiated thymidine, incorporated into DNA. Radioactive DNA release due to severe plasma membrane damage
DNA laddering (Fig. 4.7)	Usually taken as evidence of apoptosis, but is insufficient evidence alone: ladder patterns might also be produced by internucleosomal attack of free radicals. For example, treatment of HepG2 cells with copper–phenanthroline chelates, which bind to DNA and generate OH$^\bullet$, gave a DNA ladder (*Biochem. J.* **317**, 13 (1996)).
Cell replication studies	Measure rate of cell replication (e.g. [^{3}H]thymidine incorporation, antibodies recognizing proteins involved in the cell cycle). Of course, cells no longer able to divide are not necessarily irreversibly injured, i.e. this is an index of cytostasis rather than cytotoxic cell death.

[a]One of the simplest methods, often not used, is simply to look under the microscope and see what the cells are doing!
[b]3–(4,5–dimethylthiazol-2-yl)–2,5–diphenyl-2H–tetrazolium bromide. MTT reduction largely depends on mitochondrial function, but could also be achieved by other cellular reducing systems and perhaps by certain ROS, e.g. $O_2^{\bullet-}$

recognize 'self-antigens' and unwanted cells during tissue remodelling (e.g. loss of the tail in tadpoles). It is often then called **programmed cell death**.

The cell suicide mechanism appears to persist in all cells and can be triggered by certain stresses, or, as many scientists believe, by the simple withdrawal of 'life signals' from the cellular environment.[58] Among insults that trigger apoptosis are interference with energy metabolism (e.g. partial inhibition of mitochondrial electron transport), oxidative stress and the DNA-damaging effects of some drugs used in cancer chemotherapy (Chapter 9). However, these agents can also cause necrosis, depending upon the cell type studied and the level of stress applied (in general, higher stress levels are more likely to favour necrosis). There is considerable speculation that apoptosis plays a role in abnormal cell loss in certain diseases, such as the neurodegenerative diseases and some viral infections. By contrast, failure of cells with damaged DNA to undergo apoptosis can contribute to development of malignancy (Chapter 9).

4.17.3 *Genetics and mechanism of apoptosis*

Apoptosis can be observed in cells from all kinds of higher eukaryotes, from slime moulds and higher plants to vertebrates. For example, apoptosis is required for proper development of the nematode worm *Caenorhabditis elegans*. There are only 1090 cells per worm in this simple model organism, and genetic analysis has shown that only a few genes are required for the death of the 131 of these cells that occurs during normal development. The genes *ced-3* (*cell death abnormal*) and *ced-4* are essential for cell death, whereas the *ced-9* gene product blocks it. In mammalian cells, the *bcl-2* gene (which resembles *ced-9*) has an anti-apoptotic effect:[97] it encodes an integral membrane protein that is present in mitochondria, endoplasmic reticulum and nuclear membrane. The Bcl-2 protein itself has no direct antioxidant activity, but cells containing it are protected to some extent against oxidative stress (and other stresses), probably by several mechanisms, including preventing the release of pro-apoptotic proteins from mitochondria (see below). For example, over-expression of Bcl-2 can protect cells against death by exposure to H_2O_2, *tert*-butyl hydroperoxide, inhibitors of respiration, ionizing radiation or withdrawal of growth factors.

ROS and apoptosis

Many studies on various cell lines have shown an involvement of increased ROS generation or antioxidant depletion in apoptosis,[15] even in response to such stimuli as glucocorticoid hormones, or cytokines such as TNFα and TGFβ. Apoptosis can often be blocked or slowed by antioxidants, and apoptosing cells have a generally more 'oxidized' cytoplasm; they may extrude antioxidants such as GSH, for example.[115] However, oxidative stress is not an obligatory component of apoptosis or for the protective action of Bcl-2: cells cultured even at very low O_2 can be induced to undergo apoptosis, and expression of Bcl-2 still protects.[58] Suggestions as to how Bcl-2 acts include modulating changes in intracellular free Ca^{2+}, blocking release of 'pro-apoptotic' proteins (e.g. cytochrome *c*) from mitochondria and interacting with the protein phosphatase calcineurin. A protein related to Bcl-2, **Bax** (Bcl-2-associated X protein) appears to oppose Bcl-2 action and the Bax:Bcl-2 ratio is one factor that affects whether or not a cell will undergo apoptosis in response to a stress. A family of bax/bcl-2 proteins exists.[97]

The *ced*-3 gene resembles a mammalian gene encoding the ICE enzyme, with cysteine at its active site, which cleaves the IL-1β precursor (Section 4.16 above). ICE is itself synthesized as an inactive proenzyme, requiring cleavage for activation. A whole family (the **ICE family**) of cysteine proteases that cleave adjacent to aspartate residues has been described in mammalian cells and implicated in apoptosis; sometimes they are called **caspases** (cysteine-dependent aspartate-specific proteases).[21] Since apoptosis involves dismantling the cellular architecture (Fig. 4.3), the involvement of a cascade of proteases is not surprising: proteins degraded during apoptosis include histone H1, poly(ADP-ribose) polymerase and β-actin. Some viruses produce proteins (e.g. p35 from baculovirus and CrmA from cowpox virus) that inhibit certain caspases[21] and block apoptosis, presumably permitting continuing viral replication by preventing suicide of the host cells. Zn^{2+} can inhibit caspase-3 as well as the endonuclease. High levels of ROS could conceivably interfere with apoptosis by oxidizing essential –SH groups in active caspases;[49] RNS could also modify –SH groups. Dithiocarbamates, sometimes used as CuZnSOD inhibitors (Section 3.2.1), can also interfere with apoptosis by blocking caspase action.[86]

There is growing evidence that mitochondria often play a key role in apoptosis. Damage to mitochondria can cause release of apoptosis-inducing proteins into the cytosol, where they lead to a cascade of caspase activation.[21] One such protein is cytochrome *c*; direct injection of this protein into cells can promote apoptosis.[67a] Loss of cytochrome *c* from mitochondria will block the electron transport chain, possibly leading to increased $O_2^{\bullet-}$ generation (Chapter 1).

4.18   Summary: what is oxidative stress?

Oxidative stress can produce cell injury by multiple pathways (Fig. 4.29). Often, these overlap and interact in complex ways. For example, damage to DNA by oxidative stress can involve direct oxidative damage, e.g. by $OH^{\bullet}$ production in the nucleus by Fenton chemistry, indirect damage by binding of such end-products of lipid peroxidation as HNE to DNA bases, failure to repair DNA because of oxidative protein damage to polymerases and repair enzymes, and cleavage by nucleases activated by rises in intracellular free Ca^{2+} (Fig. 4.30).

Different antioxidants are needed to protect against these various events, and there is no universal 'marker' of oxidative stress. For example, failure to find lipid peroxidation is not evidence against the occurrence of oxidative

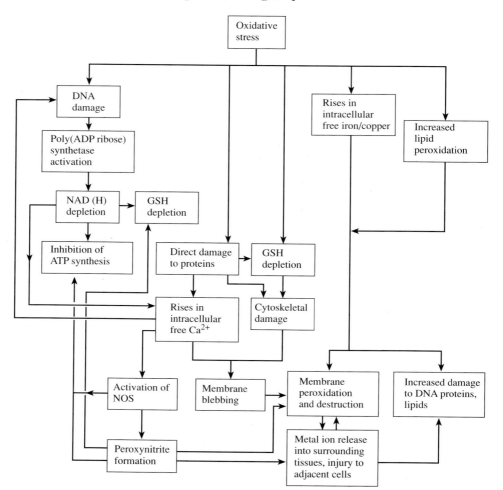

Fig. 4.29. The multiple derangements of cell metabolism that can be caused by oxidative stress. Direct damage to DNA, proteins and/or lipids is possible. Secondary damage can arise when oxidative stress produces rises in 'free' intracellular metal ions, such as Ca^{2+}, Cu^{2+} and Fe^{2+}. Ca^{2+} can stimulate proteases and nucleases, damaging both DNA and the cytoskeleton as well as increasing $NO^{\bullet}$ synthesis; excess $NO^{\bullet}$ can inhibit mitochondrial energy generation and may lead to production of cytotoxic $ONOO^-$. Activation of calpains may lead to increased ROS production, e.g. by conversion of xanthine dehydrogenase to xanthine oxidase.

stress: damage to DNA and proteins is often an important injurious event. Failure of a chain-breaking antioxidant such as vitamin E to protect is also inconclusive, for the same reason. Often, the only evidence that oxidative stress has occurred *in vivo* may be the up-regulation of antioxidant defence systems.

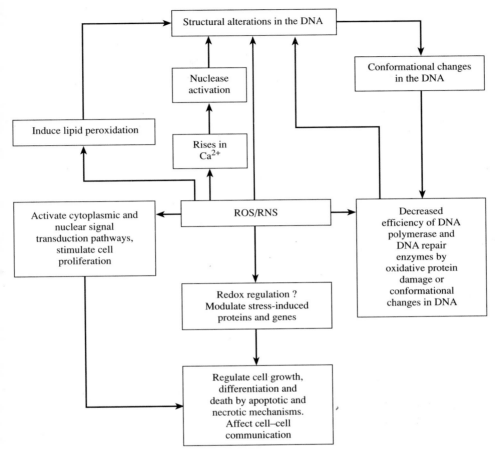

Fig. 4.30. Some of the mechanisms by which oxidative stress can affect DNA.

References

1. Aikens, J and Dix, JA (1991) Perhydroxyl radical (HOO•) initiated lipid peroxidation. *J. Biol. Chem.* **266**, 15091.
2. Allevi, P *et al.* (1995) Structural requirements of aldehydes produced in LPO for the activation of the heat-shock genes in HeLa cells. *Free Rad. Biol. Med.* **18**, 107.
3. Altman, SA *et al.* (1995) Formation of DNA–protein cross-links in cultured mammalian cells upon treatment with iron ions. *Free Rad. Biol. Med.* **19**, 897.
4. Barber, DJW and Thomas, JK (1978) Reactions of radicals with lecithin bilayers. *Radiat. Res.* **74**, 51.
5. Barja, G (1993) Oxygen radicals, a failure or a success of evolution? *Free Rad. Res. Commun.* **18**, 63.
6. Barker, CW *et al.* (1994) Down-regulation of P4501A1 and P4501A2 mRNA expression in isolated hepatocytes by oxidative stress. *J. Biol. Chem.* **269**, 3985.

6a. Basu, S (1998) Metabolism of 8-iso-prostaglandin $F_{2\alpha}$. *FEBS Lett.* **428**, 32.

7. Benamira, M *et al.* (1995) Induction of mutations by replication of MDA-modified M13 DNA in *E. coli*: determination of the extent of DNA modification, genetic requirements for mutagenesis, and types of mutation induced. *Carcinogenesis* **16**, 93.

8. Benov, L and Fridovich, I (1995) SOD protects against aerobic heat shock in *E. coli*. *J. Bacteriol.* **177**, 3344.

9. Bindoli, A (1988) Lipid peroxidation in mitochondria. *Free Rad. Biol. Med.* **5**, 247.

9a. Bonn, D (1998) How DNA-repair pathways may affect cancer risk *Lancet* **351**, 42.

10. Breen, AP and Murphy, JA (1995) Reactions of oxyl radicals with DNA. *Free Rad. Biol. Med.* **18**, 1033.

11. Brennan, P and O'Neill, LAJ (1995) Effects of oxidants and antioxidants on nuclear factor κB activation in 3 different cell lines: evidence against a universal hypothesis involving O_2 radicals. *Biochim. Biophys. Acta* **1260**, 167.

12. Bruce, AJ *et al.* (1996) Altered neuronal and microglial responses to excitotoxic and ischemic brain injury in mice lacking TNF receptors. *Nature Med.* **2**, 788.

13. Buchmeier, NA and Heffron, F (1990) Induction of *Salmonella* stress proteins upon infection of macrophages. *Science* **248**, 730.

14. Buchmeier, NA *et al.* (1995) DNA repair is more important than catalase for *Salmonella* virulence in mice. *J. Clin. Invest.* **95**, 1047.

15. Buttke, TM and Sandstrom, PA (1995) Redox regulation of programmed cell death in lymphocytes. *Free Rad. Res.* **22**, 389.

16. Cabiscol, E and Levine, RL (1995) Carbonic anhydrase III. *J. Biol. Chem.* **270**, 14742.

17. Cassatella, MA (1996) *Cytokines Produced by Polymorphonuclear Neutrophils: Molecular and Biological Aspects.* Springer, Heidelberg.

18. Chavko, M and Harabin, AL (1996) Regional lipid peroxidation and protein oxidation in rat brain after hyperbaric O_2 exposure. *Free Rad. Biol. Med.* **20**, 973.

18a. Chen, JJ *et al.* (1996) Little or no repair of cyclobutyl pyrimidine dimers is observed in the organellar genomes of the young *Arabidopsis* seedling. *Plant Physiol.* **111**, 19.

19. Clerch, LB and Massaro, D (1992) Oxidation–reduction-sensitive binding of lung protein to rat catalase mRMA. *J. Biol. Chem.* **267**, 2853.

20. Clive, DR and Greene, JJ (1996) Cooperation of protein disulphide isomerase and redox environment in the regulation of NF-κB and AP1 binding to DNA. *Cell Biochem. Funct.* **14**, 49.

21. Cohen, GM (1997) Caspases: the executioners of apoptosis. *Biochem. J.* **326**, 1.

22. Comporti, M (1987) Glutathione depleting agents and lipid peroxidation. *Chem. Phys. Lipids* **45**, 143.

22a. Coon, MJ *et al.* (1996) Peroxidative reactions of diversozymes. *FASEB J.* **10**, 428.

23. Craiu, A *et al.* (1997) Lactacystin and *clasto*-lactacystin β-lactone modify multiple proteasome β-subunits and inhibit intracellular protein degradation and MHCI antigen presentation. *J. Biol. Chem.* **272**, 13437.

24. Davies, KJA (1993) Protein modification by oxidants and the role of proteolytic enzymes. *Biochem. Soc. Trans.* **21**, 346.

25. Dean, RT *et al.* (1997) Biochemistry and pathology of radical-mediated protein oxidation. *Biochem. J.* **324**, 1.

25a. Demple, B (1998) A bridge to control. *Science* **279**, 1655.

25b. Denu, JM and Tanner, KG (1998) Specific and reversible inactivation of protein tyrosine phosphatases by H_2O_2. *Biochem.* **37**, 5633.

26. Dizdaroglu, M *et al.* (1996) Novel activities of human uracil DNA *N*-glycosylase for cytosine-derived products of oxidative DNA damage. *Nucleic Acids Res.* **24**, 418.

27. Duprat, F *et al.* (1995) Susceptibility of cloned K^+ channels to ROS. *Proc. Natl Acad. Sci. USA* **92**, 11796.

28. El-Jammal, A and Templeton, DM (1994) Reversed-phase hplc of non-trans-ferrin-bound iron and some hydroxypyridone and hydroxypyrone chelators. *J. Chromatogr.* **B658**, 121.

29. Esterbauer, H *et al.* (1991) Chemistry and biochemistry of 4-hydroxynonenal, MDA and related aldehydes. *Free Rad. Biol. Med.* **11**, 81.

30. Feige, U and Polla, BS (1994) Heat shock proteins: the hsp70 family. *Experientia* **50**, 979.

31. Fisher, D *et al.* (1972) Environmental effects on the autoxidation of retinol. *Biochem. J.* **130**, 259.

32. Frank, H *et al.* (1989) Mass spectrometric detection of cross-linked fatty acids formed during radical-induced lesion of lipid membranes. *Biochem. J.* **260**, 873.

33. Friguet, B *et al.* (1994) Susceptibility of G6PDH modified by HNE and metal-catalyzed oxidation to proteolysis by the multicatalytic protease. *Arch. Biochem. Biophys.* **311**, 168.

34. Garbe, TR *et al.* (1996) Response of *M. tuberculosis* to ROS and RNS. *Mol. Med.* **2**, 1076.

35. Gardner, HW (1989) Oxygen radical chemistry of PUFAs. *Free Rad. Biol. Med.* **7**, 65.

36. Gille, JJP and Joenje, H (1989) Chromosomal instability and progressive loss of chromosomes in HeLa cells during adaptation to hyperoxic growth conditions. *Mutat. Res.* **219**, 225.

37. Girotti, AW (1990) Photodynamic lipid peroxidation in biological systems. *Photochem. Photobiol.* **51**, 497.

38. Golden, MNH (1987) Free radicals in the pathogenesis of Kwashiorkor. *Proc. Nutr. Soc.* **46**, 53.

39. González-Flecha, B and Demple, B (1997) Homeostatic regulation of intra-cellular H_2O_2 concentration in aerobically growing *E. coli*. *J. Bacteriol.* **179**, 382.

40. Greenley, TL and Davies, MJ (1992) Detection of radicals produced by reaction of hydroperoxides with rat liver microsomal fractions. *Biochim. Biophys. Acta* **1116**, 192.

41. Grilli, M *et al.* (1996) Neuroprotection by aspirin and sodium salicylate through blockade of NF-κB activation. *Science* **274**, 1383.

41a. Grootveld, M *et al.* (1998) *In vivo* absorption, metabolism and urinary excretion of α, β-unsaturated aldehydes in experimental animals. *J. Clin. Invest.* **101**, 1210.

41b. Grune, T *et al.* (1998) $ONOO^-$ increases the degradation of aconitase and other cellular proteins by proteasome *J. Biol. Chem.* **273**, 10857.

42. Gutteridge, JMC (1982) The role of $O_2^{\bullet-}$ and $OH^{\bullet}$ in phospholipid peroxidation catalyzed by Fe salts. *FEBS Lett.* **150**, 454.

43. Gutteridge, JMC and Stocks, J (1981) Caeruloplasmin: physiological and pathological perspectives. *CRC Crit. Rev. Clin. Lab. Sci.* **14**, 257.

44. Gutteridge, JMC and Halliwell, B (1987) Radical-promoting loosely-bound iron in biological fluids and the bleomycin assay. *Life Chem. Rep.* **4**, 113.

45. Guyton, KZ *et al.* (1996) Activation of mitogen-activated protein kinase by H_2O_2. *J. Biol. Chem.* **271**, 4138.

46. Halestrap, AP *et al.* (1997) Oxidative stress, thiol reagents and membrane potential modulate the mitochondrial permeability transition by affecting nucleotide binding to the adenine nucleotide translocase. *J. Biol. Chem.* **272**, 3346.

47. Halliwell, B (1982) Production of $O_2^{\bullet -}$, H_2O_2 and $OH^{\bullet}$ by phagocytic cells: a cause of chronic inflammatory disease? *Cell Biol. Int. Rep.* **6**, 529.

48. Halliwell, B and Gutteridge, JMC (1990) Role of free radicals and catalytic metal ions in human disease: an overview. *Methods Enzymol.* **186**, 1.

49. Hampton, MB and Orrenius, S (1997) Dual regulation of caspase activity by H_2O_2: implications for apoptosis. *FEBS Lett.* **414**, 552.

50. Häussinger, D (1996) The role of cellular hydration in the regulation of cell function. *Biochem. J.* **313**, 697.

51. Hayakawa, H *et al.* (1995) Generation and elimination of 8-oxo-7,8-dihydro-2′-deoxyguanosine 5′-triphosphate, a mutagenic substrate for DNA synthesis, in human cells. *Biochemistry* **34**, 89.

52. Hayes, JJ (1995) Chemical probes of DNA structure in chromatin. *Chem. Biol.* **2**, 127.

53. Healing, G *et al.* (1990) Intracellular iron redistribution. An important determinant of reperfusion damage to rabbit kidneys. *Biochem. Pharmacol.* **39**,1239.

53a. Helbock, HJ *et al.* (1998) DNA oxidation matters: the HPLC-electrochemical detection assay of 8-oxo-deoxyguanosine and 8-oxo-guanine. *Proc. Natl. Acad. Sci. US* **95**, 288.

54. Heller, B *et al.* (1995) Inactivation of the PARP gene affects oxygen radical and $NO^{\bullet}$ toxicity in islet cells. *J. Biol. Chem.* **270**, 11176.

55. Hidalgo, E *et al.* (1997) Redox signal transduction via iron–sulfur clusters in the SoxR transcription activator. *Trends Biochem. Sci.* **22**, 207.

56. Hill, MA *et al.* (1988) Reaction of rat liver phenylalanine hydroxylase with fatty acid hydroperoxides. *J. Biol. Chem.* **263**, 5646.

57. Hilt, W and Wolf, DH (1996) Proteasomes: destruction as a programme. *Trends Biochem. Sci.* **21**, 96.

58. Jacobson, MD (1996) ROS and programmed cell death. *Trends Biochem. Sci.* **21**, 83.

59. Kakuma, T *et al.* (1995) Mouse MTH1 protein with 8-oxo-7,8-dihydro-2′-deoxyguanosine 5′-triphosphatase activity that prevents transversion mutation. *J. Biol. Chem.* **270**, 25942.

60. Kang, D *et al.* (1995) Intracellular localization of 8-oxo-dGTPase in human cells, with special reference to the role of the enzyme in mitochondria. *J. Biol. Chem.* **270**, 14659.

61. King, PA *et al.* (1993) 'Footprinting' proteins on DNA with peroxynitrous acid. *Nucleic Acids Res.* **21**, 2473.

62. Koppenol, WH (1990) Oxyradical ractions: from bond-dissociation energies to reduction potentials. *FEBS Lett.* **264**, 165.

63. Kramer, K *et al.* (1986) Influence of lipid peroxidation on β-adrenoceptors. *FEBS Lett.* **198**, 80.

64. Kretz-Remy, C *et al.* (1996) Inhibition of IκB-α phosphorylation and degradation and subsequent NF-κB activation by glutathione peroxidase overexpression. *J. Cell. Biol.* **133**, 1083.

65. Kristal, BS *et al.* (1996) 4-Hydroxyhexenal is a potent inducer of the mitochondrial permeability transition. *J. Biol. Chem.* **271**, 6033.

66. Kullik, I *et al.* (1995) Mutational analysis of the redox-sensitive transcriptional regulator OxyR: regions important for oxidation and transcriptional activation. *J. Bacteriol.* **177**, 1275.

67. Lavrovsky, Y *et al.* (1994) Identification of binding sites for transcription factors NF-κB and AP-2 in the promoter region of the human heme oxygenase I gene. *Proc. Natl Acad. Sci. USA* **91**, 5987.

67a. Li, F *et al.* (1997) Cell-specific induction of apoptosis by microinjection of cytochrome *c*. *J. Biol Chem.* **272**, 30299.

68. Li, S *et al.* (1995) Chemical instability of protein pharmaceuticals: mechanisms of oxidation and strategies for stabilization. *Biotechnol. Bioengng* **48**, 490.

69. Li, Y and Trush, MA (1994) Reactive O_2-dependent DNA damage resulting from the oxidation of phenolic compounds by a copper-redox cycle mechanism. *Cancer Res.* (Suppl.) **54**, 1895S.

70. Liaw, SH *et al.* (1993) A model for oxidative modification of glutamine synthetase, based on crystal structures of mutant H269N and the oxidized enzyme. *Biochemistry* **32**, 7999.

71. Lindahl, T (1996) Endogenous damage to DNA. *Phil. Trans. Roy. Soc. Lond.* **B351**, 1529.

72. Martin, ME *et al.* (1996) Redox regulation of GA-binding protein–αDNA binding activity. *J. Biol. Chem.* **271**, 25617.

73. Martini, M and Termini, J (1997) Peroxy radical oxidation of thymidine. *Chem. Res. Toxicol.* **10**, 234.

74. Meinicke, AR *et al.* (1996) The Ca^{2+} sensor ruthenium red can act as a Fenton-type reagent. *Arch. Biochem. Biophys.* **328**, 239.

75. Ménard, HA and El-Amine, M (1996) The calpain–calpastatin system in rheumatoid arthritis. *Immunol. Today.* December.

76. Mikalsen, SO and Sauner, T (1994) Increased gap junctional intercellular communication in Syrian hamster embryo cells treated with oxidative agents. *Carcinogenesis* **15**, 381.

77. Milligan, JR and Ward, JF (1994) Yield of single-strand breaks due to attack on DNA by scavenger-derived radicals. *Radiat. Res.* **137**, 295.

78. Mirochnitchenko, O *et al.* (1995) Thermosensitive phenotype of transgenic mice overproducing human glutathione peroxidases. *Proc. Natl Acad. Sci. USA* **92**, 8120.

79. Moraes, EC *et al.* (1990) Mutagenesis by H_2O_2 treatment of mammalian cells: a molecular analysis. *Carcinogenesis* **11**, 283.

80. Morgan, RW *et al.* (1986) H_2O_2-inducible proteins in *Salmonella typhimurium* overlap with heat shock and other stress proteins. *Proc. Natl Acad. Sci. USA* **83**, 8059.

81. Moskovitz, J *et al.* (1995) *E. coli* peptide methionine sulfoxide reductase gene: regulation of expression and role in protecting against oxidative damage. *J. Bacteriol.* **177**, 502.

82. Müller, K *et al.* (1996) Cytotoxic and chemotactic potencies of several aldehydic components of oxidized LDL for human monocyte-macrophages. *FEBS Lett.* **388**, 165.

83. Nagata, Y *et al.* (1996) Reaction of phosphatidylcholine hydroperoxide in human plasma: the role of peroxidase and LCAT. *Arch. Biochem. Biophys.* **329**, 24.

84. Natarajan, K *et al.* (1996) Caffeic acid phenethyl ester is a potent and specific inhibitor of activation of nuclear transcription factor NF-κB. *Proc. Natl Acad. Sci. USA* **93**, 9090.

85. Nicotera, P and Orrenius, S (1994) Molecular mechanisms of toxic cell death; an overview. *Methods Toxicol.* **1B**, 23 [and other articles in this volume].

86. Nobel, CSI *et al.* (1997) Mechanism of dithiocarbamate inhibition of apoptosis. *Chem. Res. Toxicol.* **10**, 636.

87. Oarada, M *et al.* (1988) The effect of dietary lipid hydroperoxide on lymphoid tissues in mice. *Biochim. Biophys. Acta* **960**, 229.

88. Özer, Z *et al.* (1995) The other function of DNA photolyase: stimulation of excision repair of chemical damage to DNA. *Biochemistry* **34**, 15886.

89. Packer, MA and Murphy, MP (1995) Peroxynitrite formed by simultaneous $NO^\bullet$ and $O_2^{\bullet-}$ generation causes cyclosporin-A-sensitive mitochondrial Ca^{2+} efflux and depolarisation. *Eur. J. Biochem.* **234**, 231.

90. Pahl, HL and Baeuerle, PA (1994) Oxygen and the control of gene expression. *BioEssays* **16**, 497.

91. Paller, MS and Jacob, HS (1994) Cytochrome P-450 mediates tissue-damaging $OH^\bullet$ formation during reoxygenation of the kidney. *Proc. Natl Acad. Sci. USA* **91**, 7002.

92. Paradies, G *et al.* (1997) Age-dependent decline in the cytochrome *c* oxidase activity in rat heart mitochondria: role of cardiolipin. *FEBS Lett.* **406**, 136.

92a. Park, YM *et al.* (1998) Over expression of HSP25 reduces the level of TNFα-induced oxidative DNA damage biomarker, 80HdG, in L929 cells *J. Cell Physiol.* **174**, 27.

93. Peng, SL *et al.* (1997) Scleroderma: a disease related to damaged proteins? *Nature Med.* **3**, 276.

94. Perrin, DM *et al.* (1996) Oxidative chemical nucleases. *Prog. Nucleic Acid Res. Mol. Biol.* **52**, 123.

94a. Perry, DK *et al* (1997) Zinc is a potent inhibitor of the apoptotic protease, caspase-3 *J. Biol. Chem.* **272**, 18530.

95. Poli, G and Parola, M (1997) Oxidative damage and fibrogenesis. *Free Rad. Biol. Med.* **22**, 287.

96. Prestera, T *et al.* (1995) Parallel induction of heme oxygenase-1 and chemoprotective phase 2 enzymes by electrophiles and antioxidants: regulation by upsteam ARE. *Mol. Med.* **1**, 827.

97. Reed, JC (1997) Double identity for proteins of the Bcl-2 family. *Nature* **387**, 773.

98. Richter, C (1987) Biophysical consequences of lipid peroxidation in membranes. *Chem. Phys. Lipids* **44**, 175.

99. Richter, C *et al.* (1995) Oxidants in mitochondria: from physiology to diseases. *Biochim. Biophys. Acta* **1271**, 67.

100. Roberts, LJ and Morrow, JD (1997) The generation and actions of isoprostanes. *Biochem. Biophys. Acta* **1345**, 121.

101. Rodriguez, M *et al.* (1995) Mapping of Cu/H_2O_2-induced DNA damage at nucleotide resolution in human genomic DNA by ligation-mediated polymerase chain reaction. *J. Biol. Chem.* **270**, 17633.

102. Sancar, A (1996) DNA excision repair. *Annu. Rev. Biochem.* **65**, 43.

103. Sandstrom, BE *et al.* (1994) New roles for quin 2: powerful transition-metal ion chelator that inhibits copper-, but potentiates iron-driven, Fenton-type reactions. *Free Rad. Biol. Med.* **16**, 177.

104. Schacter, BA *et al.* (1972) Hemoprotein catabolism during stimulation of microsomal lipid peroxidation. *Biochim. Biophys. Acta* **279**, 221.

105. Schraufstatter, IU *et al.* (1986) Oxidant injury of cells. *J. Clin. Invest.* **77**, 1312.

106. Segall, HJ *et al.* (1985) *Trans*-4-hydroxy-2-hexenal: a reactive metabolite from the macrocyclic pyrrolizidine alkaloid senecionine. *Science* **229**, 472.

107. Sestili, P *et al.* (1996) AG8 cells, which are highly resistant to H_2O_2, display collateral sensitivity to the combination of H_2O_2 and L-histidine. *Carcinogenesis* **17**, 885.

108. Sevanian, A *et al.* (1990) Microsomal lipid peroxidation: the role of NADPH–cytochrome P450 reductase and cytochrome P450. *Free Rad. Biol. Med.* **8**, 145.

109. Sies, H (1991) *Oxidative Stress II. Oxidants and Antioxidants.* Academic Press, London.

110. Spencer, JPE *et al.* (1996) Base modification and strand breakage in isolated calf thymus DNA and in DNA from human skin epidermal keratinocytes exposed to $ONOO^-$ or 3-morpholinosydnonimine. *Chem. Res. Toxicol.* **9**, 1152.

111. Starke, PE *et al.* (1987) Modification of hepatic proteins in rats exposed to high O_2 concentration. *FASEB J.* **1**, 36.

112. Sun, JZ *et al.* (1996) Evidence for an essential role of ROS in the genesis of late preconditioning against myocardial stunning in conscious pigs. *J. Clin. Invest.* **97**, 562.

113. Szabó, C *et al.* (1996) DNA strand breakage, activation of poly(ADP–ribose) synthetase, and cellular energy depletion are involved in the cytotoxicity in macrophages and smooth muscle cells exposed to peroxynitrite. *Proc. Natl Acad. Sci. USA* **93**, 1753.

113a. Tamarit J *et al.* (1998) Identification of the major oxidatively damaged proteins in *E. coli* exposed to oxidative stress. *J. Biol. Chem.* **273**, 3027.

114. Tofigh, S and Frenkel, K (1989) Effect of metals on nucleoside hydroperoxide, a product of ionizing radiation in DNA. *Free Rad. Biol. Med.* **7**, 131.

115. Van den Dobbelsteen, DJ *et al.* (1996) Rapid and specific efflux of GSH during apoptosis induced by anti-FAS/APO-1 antibody. *J. Biol. Chem.* **271**, 15420.

116. von Sonntag, C (1987) *The Chemical Basis of Radiation Biology.* Taylor and Francis, London.

117. Wagner, BA *et al.* (1994) Free radical-mediated lipid peroxidation in cells: oxidizability is a function of cell lipid *bis*-allylic hydrogen content. *Biochemistry* **33**, 4449.

118. Wang, X *et al.* (1996) SOD protects calcineurin from inactivation. *Nature* **383**, 434.

119. Yakes, FM and Van Houten, B (1997) Mitochondrial DNA damage is more extensive and persists longer than nuclear DNA damage in human cells following oxidative stress. *Proc. Natl Acad. Sci. USA* **94**, 514.

120. Yao, Y *et al.* (1996) Oxidative modification of a carboxyl-terminal vicinal methionine in calmodulin by H_2O_2 inhibits calmodulin–dependent activation of the plasma membrane Ca-ATPase. *Biochemistry* **35**, 2767.

121. Zastawny, TH *et al.* (1995) DNA base modifications and membrane damage in cultured mammalian cells treated with iron ions. *Free Rad. Biol. Med.* **18**, 1013.

122. Zimmerman, GA *et al.* (1995) Oxidatively fragmented phospholipids as inflammatory mediators: the dark side of polyunsaturated lipids. *J. Nutr.* **125**, 1661S.

122a. Zitomer, RS and Lowry, CV (1992) Regulation of gene expression by O_2 in *S. cerevisiae. Microbiol. Rev.* **56**, 1.

Notes

[a]Cells contain multiple regulatory proteins whose activity is regulated by binding of guanosine triphosphate (GTP). Some are monomeric, but the G-proteins are trimeric GTP-binding proteins, i.e. they have three different subunits. G-protein-linked receptors are the largest family of cell-surface receptors, responding to many hormones, neurotransmitters and other signalling molecules. Binding of a signal molecule to the receptor activates the G-protein by causing GDP to fall off and GTP to bind. Eventually the 'switch' is turned off when the G-protein hydrolyses its bound GTP. G-protein-linked receptors most often use either **cyclic AMP** or Ca^{2+} as intracellular 'messengers'.

[b]Iron ions are never free in aqueous solution; they are always liganded to something, if only to H_2O.

[c]Do not confuse this peroxide with a carboxyl group, often written −COOH but with the structure

$$-C\overset{\displaystyle \nearrow O}{\underset{\displaystyle \searrow OH}{}}$$

rather than −C−O−O−H.

[d]If an explanation is needed, please consult Appendix II.

5

Detection of free radicals and other reactive species: trapping and fingerprinting

5.1 Introduction

Free radicals and other 'reactive species' play important roles in living systems and have been implicated in the pathology of many human diseases (Chapter 9) and in the mechanisms of action of several toxins (Chapter 8). One problem in ascertaining the true importance of reactive oxygen species and reactive nitrogen species (ROS/RNS) has been that these evanescent species are difficult to measure *in vivo*.

The only technique that can detect free radicals directly is the spectroscopic technique of **electron spin resonance** (ESR), sometimes called **electron paramagnetic resonance** (EPR). Often, this method is too insensitive to detect directly such radicals as $O_2^{\bullet -}$ and $OH^{\bullet}$ in living systems: direct ESR of biological material can only detect less-reactive radicals, such as ascorbyl radical. One approach to this difficulty has been **trapping**, in which a radical is allowed to react with a trap molecule to give one or more stable products, which are then measured. Mention of trapping usually brings to mind the technique of **spin trapping**, in which the radical reacts with a spin trap to form a more stable radical, which accumulates to a level that does permit detection by ESR. However, there are many other trapping methods: aromatic hydroxylation is one example.

Trapping methods have proved very useful *in vitro* and in studies with whole animals (e.g. in measuring CCl_4-derived radicals; Chapter 8) but their usefulness has been limited in human studies. Another point to consider is that trap molecules perturb the system under investigation. If damage is caused by say, $OH^{\bullet}$, then adding a molecule that traps $OH^{\bullet}$ should decrease the damage, provided that enough $OH^{\bullet}$ is trapped. Indeed, spin traps and other trapping molecules have often been shown to inhibit oxidative damage. In some cases, this has been used as a starting point to design therapeutic antioxidants (Chapter 10).

An alternative to trapping is **fingerprinting** (one could also call it **footprinting**). The principle behind such methods is to measure products of damage by ROS/RNS, i.e. to measure not the species themselves but the damage that they cause. Of course, the end-products must be *specific* markers of oxidative damage.

5.2 ESR and spin trapping[55]

ESR is a technique that is specific for free radicals, since it detects the presence of unpaired electrons. An unpaired electron can have a spin of either $+\frac{1}{2}$ or $-\frac{1}{2}$ (see Appendix I for an explanation if needed) and behaves as a small magnet. If it is exposed to an external magnetic field, it can align itself either parallel or antiparallel (in opposition) to that field, and thus it creates two possible energy levels, which vary with the magnetic field strength. If electromagnetic radiation of the correct energy is applied, it will be absorbed and used to move the electron from the lower energy level to the upper one. Thus an absorption spectrum is obtained, usually in the microwave region of the electromagnetic spectrum. ESR spectrometers are set up to display **first-derivative spectra**, which show not the absorbance but the rate of change of absorbance, so that a point on the derivative curve corresponds to the gradient (slope) at the equivalent point on the absorption plot.

The condition to obtain an absorbance is:

$$\Delta E = g\beta H$$

where ΔE is the energy gap between the two energy levels of the electron, H is the applied magnetic field and β is a constant known as the **Böhr magneton**. The value of g (the **splitting factor**) for a free electron is 2.00232 and most biologically-important radicals have values close to this. If the above equation is obeyed, an absorption spectrum results. For a single electron this can be represented in a stylized way as

but, if presented as its first derivative (as ESR machines do) it will appear as:

A number of atomic nuclei, such as those of hydrogen and nitrogen, also behave like small magnets and will align either parallel or antiparallel to the applied magnetic field. Thus in a hydrogen atom the single unpaired electron will be exposed to two different magnetic fields: the one applied plus that from the nucleus, or the one applied minus that from the nucleus. Thus there will be two energy absorptions and the single line becomes a doublet, i.e.

or, as more usually presented,

If the unpaired electron is close to two hydrogens, each can be
same way with the applied field, in opposite ways, or one in the sam
one opposite, i.e.

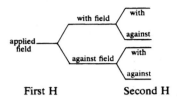

giving a three-line spectrum in which the intensities are in the ratio $1:2:1$, i.e.

In the carbon-centred **methyl radical** ($CH_3^\bullet$) the unpaired electron on the carbon is close to three hydrogens and the ESR spectrum thus contains four lines. The field of each hydrogen nucleus can align with or against the applied magnetic field, and a 'tree' diagram can be used to predict the spectrum, i.e.

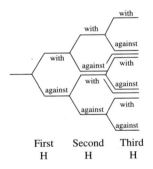

which therefore consists of four lines with intensity ratios $1:3:3:1$, or:

…ESR spectrum of a radical is called the **hyperfine**
…lex in radicals containing many nuclei. A radical
…R spectrum by looking at the g value, hyperfine
… an example, Fig. 5.1 shows the different ESR
…*t*-tocopheroxyl radicals.

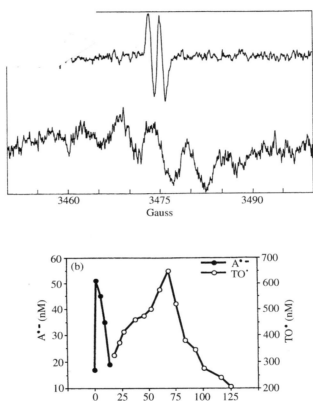

Fig. 5.1. ESR spectra of ascorbate and tocopheroxyl free radicals. Plasma was incubated with a mixture of hypoxanthine and xanthine oxidase (HX/XO), which generates $O_2^{\bullet-}$ and H_2O_2. Prior to the introduction of the free radical generating system, ascorbyl radical levels were extremely low (<20 nM). Upon adding the HX/XO system, ascorbyl radical levels rose rapidly but then fell as ascorbate was depleted (b). Not until ascorbate had almost disappeared did the tocopheroxyl radical appear. (a) Top spectrum: ascorbyl free radical spectrum measured at 1.5 min after adding HX/XO. Modulation amplitude 0.96 gauss, scan rate 50 gauss/168 s, time constant 0.66 s, receiver gain 5×10^6. Bottom spectrum: tocopheroxyl free radical spectrum measured at 70 min. Modulation amplitude 0.96 gauss, scan rate 50 gauss/335 s, time constant 1.3 s, receiver gain 5×10^6. Under these experimental conditions approximately ten times more tocopheroxyl radical is needed to produce an ESR signal above the noise level, largely because tocopheroxyl radical has a broad seven-line spectrum whereas the ascorbyl radical is a narrow doublet. Data from *Free Rad. Biol. Med.* **14**, 649 (1993) by courtesy of Dr Gary Buettner and Elsevier.

The sensitivity of ESR can be as high as 10^{-10} M, sufficient to detect species such as the ascorbate and tocopheroxyl radicals (Fig. 5.1) in biological systems, but not enough to detect species such as $O_2^{\bullet-}$ and $OH^{\bullet}$ unless extremely low temperatures are used to minimize their reactions. To detect these and other transient radicals, several techniques are available. In chemical studies, flow systems can be used, whereby the radicals are continuously generated to maintain a steady-state concentration (Section 2.3.3). Another approach is to generate the radical in a frozen transparent solid matrix which prevents it from colliding with other species and undergoing reaction. In 1969, a 'rapid freezing' ESR technique enabled the first identification of $O_2^{\bullet-}$ production by an enzyme-catalysed reaction (xanthine oxidase catalysing the oxidation of xanthine).[34] By allowing the matrix to warm up, reactions of the radical can then be observed. Immobilized radical spectra are often hard to identify in complex biological material, and so freeze-quenching techniques are not often used. Free radicals are readily detected from mechanical disruption of fingernails, bone, cartilage and tooth enamel (Chapter 2).

5.2.1   Spin trapping[28,43]

In this technique, a reactive radical is allowed to react with a trap to produce a long-lived radical. Reaction of nitroso (R—NO) compounds with radicals often produces **nitroxide** (sometimes called **aminoxyl**) radicals that have a long lifetime

$$R-N=O \qquad + \; R''^{\bullet} \qquad \longrightarrow \qquad \begin{matrix} R \\ | \\ N-O^{\bullet} \\ | \\ R' \end{matrix}$$

R symbolizes   reactive   nitroxide radical
'rest of molecule'   radical   (fairly stable)

Their stability is associated with electron delocalization between the nitrogen and oxygen atoms

$$N-O^{\bullet} \leftrightarrow \overset{+}{N}-O^{-}$$

Nitrone traps also produce nitroxide radicals

$$\begin{matrix} H & O^{-} \\ | & | \\ R-C{=}\overset{+}{N}-R' \end{matrix} \; + \; R'''^{\bullet} \rightarrow \begin{matrix} H & O^{\bullet} \\ | & | \\ R-C-N-R' \\ | \\ R'' \end{matrix}$$

The ESR spectra of nitroxides have a main triplet (1 : 1 : 1) splitting due to interaction of the unpaired electron with the nitrogen nucleus of the nitroxide group. Secondary splittings arise from magnetic nuclei in the trapped radical and sometimes from other magnetic nuclei in the spin trap. In nitroso spin traps

Table 5.1. A selection of the 'spin-traps' that have been used in biological systems

Name	Abbreviation	Structure
tert-Nitrosobutane (nitroso-*tert*-butane)	tNB (NtB)	
α-Phenyl-*tert*-butylnitrone	PBN	
5,5-Dimethylpyrroline-*N*-oxide	DMPO	
tert-Butylnitrosobenzene	BNB	
α-(4-Pyridyl-1-oxide)-*N*-*tert*-butylnitrone	4-POBN	
3,5-Dibromo-4-nitroso-benzenesulphonic acid	DBNBS	

Spin traps also vary in their hydrophobicity; for example, DMPO (octanol: water partition coefficient 0.08) or POBN (0.09) will be much less useful in trapping radicals within membranes or lipoproteins than such species as PBN (partition coefficient 10.4).

the radical detected adds directly to the nitrogen, whereas with nitrones it adds to the carbon adjacent to the nitrogen. With nitroso traps the trapped radical can thus more easily influence the ESR spectrum and usually generates hyperfine splittings due to the trapped radical, whereas with nitrone traps the spectra tend to be broadly similar whatever the radical trapped. However, nitroso compounds often give less stable adducts than nitrones, especially when oxygen radicals are trapped. Table 5.2 shows some of the trapping molecules that have been used; 5,5-dimethylpyrroline-*N*-oxide (DMPO) and α-phenyl-*tert*-butylnitrone (PBN) are especially popular. The 'ideal' trap should react rapidly and specifically with the radical one wishes to study, and produce a product that is (i) chemically stable, (ii) not metabolized by living systems and (iii) has a unique ESR spectrum. Another factor to be considered is where the probe will localize in a biological system; for example, PBN is much more hydrophobic than DMPO (Table 5.1).

5.2.2 DMPO and PBN

Spin-trapping methods have often been used to detect the presence of $O_2^{\bullet-}$ and $OH^\bullet$ in biological systems, and also the formation of organic radicals during lipid peroxidation (Chapter 4). None of the spin traps at present in use is ideal,[51] although better ones are being developed.

For example, DMPO reacts with both $OH^\bullet$ and $O_2^{\bullet-}$ radicals to form products with different ESR spectra (Fig. 5.2). The rate constants for the reactions are very different, however, at approximately $10\,M^{-1}\,s^{-1}$ for $O_2^{\bullet-}$

Table 5.2. Formation of hydroxyl radicals during the oxidation of xanthine by xanthine oxidase in the presence of an iron chelate

Reagent added	Rate constant for reaction with $OH^\bullet(M^{-1}s^{-1})$	Amount (nmol) of hydroxylated products formed	Percentage inhibition of hydroxylation
None	—	102	—
Mannitol (5 mM)	2.7×10^9	64	37
Sodium formate (5 mM)	3.7×10^9	49	52
Thiourea (5 mM)	4.7×10^9	24	76
Urea (5 mM)	$<7.0 \times 10^5$	102	0

Experiments were carried out at pH 7.4 using $OH^\bullet$ generated by reaction of $O_2^{\bullet-}$ and H_2O_2 (from oxidation of xanthine by xanthine oxidase) with ferric EDTA. (*FEBS Lett.* **92**, 321 (1978)). Any $OH^\bullet$ formed was detected by measuring the formation of hydroxylated products from salicylate (colorimetrically). Hydroxyl radicals generated by iron–EDTA that escape direct reaction with iron–EDTA and other reagents are equally 'available' to the salicylate and to other added scavengers of $OH^\bullet$. Hence the amount of hydroxylation can be decreased by adding '$OH^\bullet$ scavengers'. The more quickly these 'scavengers' react with $OH^\bullet$, the more inhibition of the formation of hydroxylated products they should produce. However, published rate constants are variable and $OH^\bullet$ scavengers often cannot protect against damage by site-specific reactions (see text).

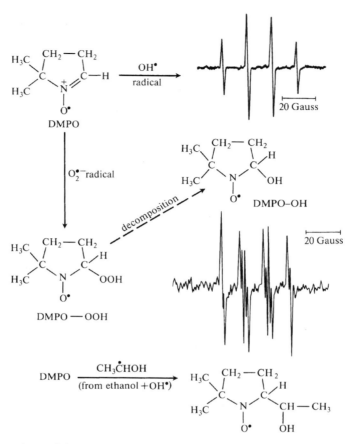

Fig. 5.2. Reactions of the spin trap DMPO with superoxide, hydroxyl and ethanol radicals. The ESR spectra of the DMPO–OH and DMPO–OOH adducts are shown diagrammatically to illustrate the key differences between them.

and $3.4 \times 10^9 \, M^{-1} s^{-1}$ for OH$^\bullet$, i.e. OH$^\bullet$ is trapped far more quickly than $O_2^{\bullet -}$. Hence efficient trapping of $O_2^{\bullet -}$ requires higher DMPO concentrations.[51] Unfortunately, the product of reaction of DMPO with $O_2^{\bullet -}$ (DMPO–OOH) is unstable and decomposes to give several products, including DMPO–OH, the same product generated by direct reaction with OH$^\bullet$ (Fig. 5.2). Glutathione peroxidase has been reported to convert DMPO–OOH to DMPO–OH. Even worse, decomposition of the DMPO–OOH adduct has been claimed to release some OH$^\bullet$, i.e. addition of DMPO to a system producing only $O_2^{\bullet -}$ might *cause* small amounts of OH$^\bullet$ to be formed.[51]

There are solutions to these problems. If the DMPO–OH signal observed is due to trapping of OH$^\bullet$ by DMPO, then addition of excess ethanol should abolish the signal. Ethanol scavenges OH$^\bullet$ and will compete with DMPO; at high enough levels, it prevents any OH$^\bullet$ from reacting with DMPO. In addition, reaction of ethanol with OH$^\bullet$ produces a **hydroxyethyl radical** (Fig. 5.2) which reacts with DMPO to give a new radical with a different ESR spectrum. This spectrum should appear as the OH$^\bullet$ signal is lost. If, however, the DMPO–OH signal arose from trapping of $O_2^{\bullet-}$ and decomposition of the DMPO–OOH adduct, then ethanol would not abolish it, since it does not scavenge $O_2^{\bullet-}$.

Dimethylsulphoxide (DMSO) may be used instead of ethanol: its reaction with OH$^\bullet$ produces a methyl ($CH_3^\bullet$) radical that forms a characteristic free-radical adduct with DMPO:

$$(CH_3)_2S{=}O + OH^\bullet \longrightarrow \underset{\underset{OH}{|}}{CH_3S}{=}O + CH_3^\bullet$$

Thus if OH$^\bullet$ was really being trapped, addition of DMSO should result in the DMPO–OH signal decreasing in parallel with a rise in the DMPO–CH_3 signal. However, the $CH_3^\bullet$ radical also reacts with O_2, much faster than it reacts with DMPO (rate constants $\approx 3.7 \times 10^9$ and $10^6 \, M^{-1} s^{-1}$ respectively) and so elevated O_2 will compete with DMPO for the $CH_3^\bullet$ produced. The ability of spin traps to identify organic-solvent derived radicals should be borne in mind when tissues or body fluids are treated with such solvents to extract lipophilic radicals for ESR studies.[10] For example, chloroform/methanol mixtures are widely used to extract lipids, but methanol can be easily oxidized by several ROS, generating $^\bullet CH_2OH$ radicals. Hence trapped radicals can be artefacts of solvent extraction. The anaesthetic **Ketamine** can be metabolized to radicals *in vivo* and should be avoided for anaesthesia of animals for spin trapping experiments.[54a]

5.2.3 *Metabolism of spin traps*

Another problem is that some nitroxide radical adducts of several spin traps (including DMPO, Fig. 5.3) can be reduced by cellular reducing agents, such as ascorbate and constituents of microsomal and mitochondrial electron-transport chains, to give 'ESR-silent' species (i.e. products that no longer give an ESR signal). Redox interactions of DMPO with certain iron chelates can also occur. These problems limit the application of such spin traps to biological systems.

Caution should also be employed in using DMPO to trap OH$^\bullet$/$O_2^{\bullet-}$ in photochemical systems that also generate singlet O_2, since DMPO reacts with this species.[4] It has been suggested that $^1O_2$ adds to the C=NO bond to give a biradical C(OO$^\bullet$)–N$^\bullet$(O) which can decompose in complex ways to give products that include DMPO–OH and DMPO/$HO_2^\bullet$ radical adducts. DMPO can also be oxidized by HOCl and upon addition of ONOO$^-$. Much work is being carried out to develop improved spin traps, particularly ones that might better reflect the site-specific nature of biological free-radical damage, e.g. by targeting the transition-metal ions catalysing OH$^\bullet$ production.

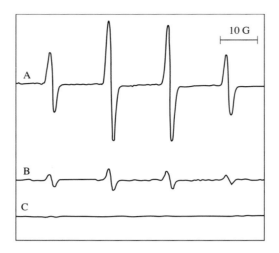

Fig. 5.3. Reduction of the DMPO–OH spin adduct by ascorbic acid. Hydroxyl radical was generated by reaction of Fe(III)–EDTA with H_2O_2. Reaction system (A) without ascorbate (gain 1.6×10^5), (B) with 0.66 mM ascorbate present, (C) with a crystal of solid ascorbate added after the spectra had formed. Unpublished data of the author (J.M.C.G.) with Lindsay Maidt, Gemma Brueggemann and Paul McCay. Ascorbate reduces DMPO–OH to an ESR-silent species. Caution should be employed when attempting to use spin trapping to study the reactions of reducing agents with radicals; it is essential to show that the reductant does not simply destroy the trap-radical adduct. For example, the thiol-containing drug captopril reduces DMPO–OH to an ESR-silent species, which may have led to overestimates of captopril's ability to scavenge ROS (*Free Rad. Res.* **24**, 391 (1996)).

PBN (Table 5.1) has been widely used as a trap *in vitro* (e.g. it has been used to trap radicals in beer in relation to the staling process—OH• plays an important role in beer oxidation).[66] PBN is not acutely toxic and thus has often been used in animal studies, frequently being reported to protect against oxidative damage. PBN is rapidly metabolized *in vivo*[8] and also decomposes when exposed to light; it has been suggested that NO• is a product.[6] NO• could conceivably contribute to some of the reported physiological effects of PBN, e.g. vasodilation. Both PBN and DMPO also bind to cytochrome P450, and PBN can inhibit various P450-dependent oxidations.[43]

5.2.4   Trapping of thiyl radicals[26]

Spin-trapping has also proved useful in detecting many free-radical species other than oxygen radicals. For example, Fig. 5.4 shows how DMPO can detect the cysteine thiyl radical. DMPO has been used to demonstrate the presence of this radical when cysteine oxidizes in the presence of transition metal ions, in cell culture media[56] or upon incubation with horseradish peroxidase or with peroxynitrite. In the presence of transition-metal ions, oxidizing cysteine also generates $O_2^{•-}$ and H_2O_2 which can be simultaneously trapped with DMPO. The DMPO adducts of different RS• radicals are

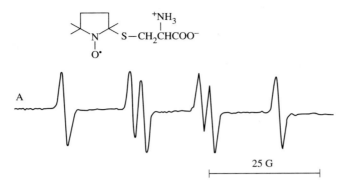

Fig. 5.4. ESR spectrum of the DMPO–cysteine thiyl radical adduct. The radical was formed during oxidation of cysteine by horseradish peroxidase. From *J. Biol. Chem.* **259**, 5606 (1984) by courtesy of Dr Ron Mason and the publishers.

distinctive and the rate constant for reaction is high, $\sim 10^8\,M^{-1}\,s^{-1}$. The DMPO adduct of the glutathione thiyl radical, GS$^\bullet$, superficially resembles DMPO–OH, giving a four-line spectrum. However, the intensity pattern is usually $1:1:1:1$ rather than $1:2:2:1$.

By contrast, PBN traps thiyl radicals but forms less-characteristic spectra than does DMPO.

5.3   Other trapping methods, as exemplified by hydroxyl-radical trapping

5.3.1   *Aromatic hydroxylation*[21,50]

The oxidation of aromatic compounds by metal ion–H_2O_2 mixtures has been known for over 100 years, and, since the pioneering work of Merz and Waters in 1949, has generated an enormous chemical literature. The reactions are complex. In the case of the simplest aromatic compound, benzene, there is a fast addition of OH$^\bullet$ to the aromatic ring to give a **hydroxycyclohexadienyl** radical (when a radical adds to a non-radical, the adduct must be a radical):

Two such radicals can join together to give a dimer, that can lose water to form **biphenyl**:

Hydroxycyclohexadienyl radical can also be oxidized to phenol, e.g.

H—OH (radical) →(loss of 1e⁻) H—OH (cation) →(loss of H⁺) OH (phenol)

and two such radicals can interact, forming a mixture of phenol and benzene (one hydroxycyclohexadienyl radical is oxidized to phenol, the other reduced to benzene in a disproportionation reaction). The presence of O_2, Fe(III) or Cu^{2+} tends to increase the yield of hydroxylation products. Hence the end-products obtained from attack of a given amount of $OH^\bullet$ upon benzene vary depending on reaction conditions.

If substituted benzenes are attacked by $OH^\bullet$, reactions become even more complex. For example, using aromatic carboxylic acids, decarboxylation reactions (loss of the carboxyl group) are favoured at low pH values in the absence of metal ions, whereas hydroxylated product formation is favoured if such metal ions are present. Hence under physiologically relevant conditions (pH 7.4, metal ions and O_2 present), hydroxylation will be a significant reaction pathway.

Both decarboxylation and hydroxylation reactions of aromatic compounds have been used to detect $OH^\bullet$. For example, decarboxylation of benzoic acid, labelled with ^{14}C in the carboxyl group, has been used to measure generation of $OH^\bullet$ in biochemical systems. The assay is sensitive, since small amounts of [14]CO_2 can be trapped in alkaline solutions and measured accurately by scintillation counting. An alternative approach is to use benzoic acid labelled with ^{13}C in the carboxyl group and measure production of [13]CO_2 with a mass spectrometer.[36] However, as explained above, decarboxylation can often be a minor reaction pathway under physiological conditions, i.e. only a small fraction of the $OH^\bullet$ is measured. Also, some $RO_2^\bullet$ radicals appear able to decarboxylate benzoate.[36]

Hydroxylated aromatic products are more often measured. They can be detected colorimetrically, but the colour reactions often do not detect all the isomeric hydroxylated forms and thus are semiquantitative. Ideally, the various products should be separated by gas chromatography (GC) or high performance liquid chromatography (HPLC) and measured by electrochemical detection, fluorescence spectra, mass spectrometry or simple 'colour' reactions. For example, HPLC separation combined with sensitive electrochemical detection has been used to measure the products formed by attack of $OH^\bullet$ on aromatic compounds, as a sensitive assay to measure $OH^\bullet$ generated by cells, organelles and perfused organs. Suitable aromatic 'detectors' must be chosen. Salicylate is popular; attack of $OH^\bullet$ upon salicylate (2-hydroxybenzoate) produces two dihydroxylated products (2,3- and 2,5-dihydroxybenzoates), together with a small amount of the decarboxylation product catechol (Fig. 5.5). Attack of $OH^\bullet$ upon phenylalanine produces three dihydroxylated products: 2-hydroxyphenylalanine (*o*-tyrosine), 3-hydroxyphenylalanine (*m*-tyrosine) and 4-hydroxyphenylalanine (*p*-tyrosine). Both the D- and L-isomers of

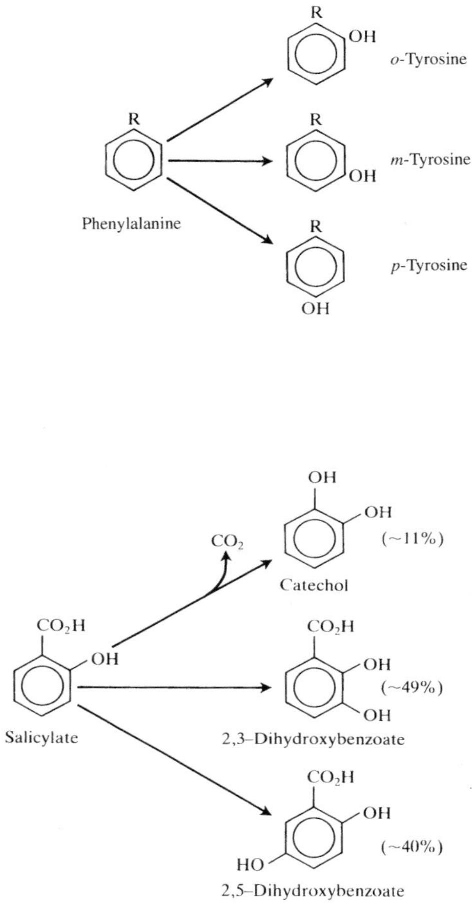

Fig. 5.5. Products of the reaction of hydroxyl radical with aromatic compounds. A typical percentage distribution of salicylate hydroxylation products is shown but ratios of end-products vary widely depending on reaction conditions. It should always be appreciated that initial products of attack of $OH^\bullet$ on aromatic rings are free radicals, that can have variable fates depending on what else is present in the reaction mixture.

phenylalanine react with $OH^\bullet$ with the same rate constant. Figure 5.6 shows an example of the use of the phenylalanine hydroxylation technique, measuring $OH^\bullet$ production resulting from ischaemia–reperfusion of the heart in an intact animal.

Precautionary measures

If an aromatic compound reacts with $OH^\bullet$ to form a specific set of hydroxylated products that can be accurately measured in biological material, and one or more of these products is not identical to enzyme-produced hydroxylated products, then formation of the 'unnatural' products could

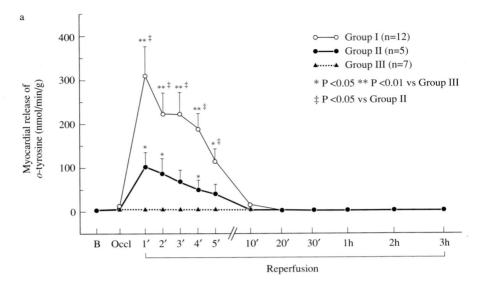

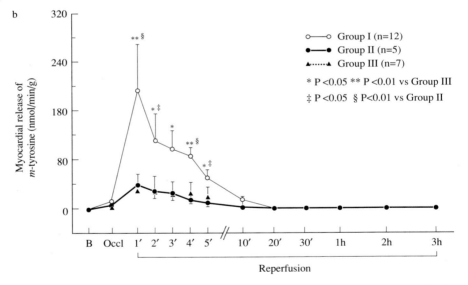

Fig. 5.6. Use of aromatic hydroxylation of phenylalanine to measure production of hydroxyl radicals after myocardial ischaemia–reperfusion *in vivo*. If part of the myocardium of an open-chest anaesthetized dog is made ischaemic for a brief period (too short to kill the tissue), subsequent reperfusion does not cause immediate return of contractile function. This myocardial stunning effect can be attenuated by several antioxidants and previous studies had suggested that OH$^\bullet$ was involved. Attempts were made to confirm this using phenylalanine (Fig. 5.5). Animals were pre-loaded to a high level of L-phenylalanine. No *o*- or *m*-tyrosines were detected in the circulation since these are not normal metabolites of L-phenylalanine (unlike *p*-tyrosine, produced by the enzyme phenylalanine hydroxylase). However, *o*- and *m*-tyrosine *were* produced by the heart after ischaemia–reperfusion (group I animals, ○). Data are shown for

conceivably be used to assess OH• radical formation *in vivo*.[21] Success in such studies also requires that the aromatic detector molecule be present at the sites of OH• radical generation at concentrations sufficient to compete with any other molecules that might scavenge OH•, and that any unnatural hydroxylated product is not immediately metabolized. For example, dihydroxybenzoates are rapidly metabolized in perfused liver, so that salicylate is not a good detector of OH• in this organ. On the other hand, it seems to work well in the isolated rat heart.[52]

Both salicylate and phenylalanine have been used as 'probes' for the generation of OH• *in vivo*. Thus, 2,3-dihydroxybenzoate (2,3-DHB; Fig. 5.5) has not been reported as an enzyme-produced metabolite of salicylate in humans, whereas 2,5-DHB can be generated from salicylate by cytochromes P450. For example, hydroxylation of salicylate to 2,3-DHB was used to gain evidence for *in vivo* generation of OH• in the lungs of animals breathing silicate dusts.[57] Metabolism of L-phenylalanine by phenylalanine hydroxylase generates *p*-tyrosine, but attack of OH• also produces *o*- and *m*-tyrosines (Fig. 5.5). D-Phenylalanine is not a substrate for the enzyme, but is still attacked by OH• to give three isomeric tyrosines.

It must be stressed that ROS/RNS other than OH• can hydroxylate aromatic compounds. Singlet O_2 attacks salicylate,[27] but produces 2,5-DHB rather than 2,3-DHB (another reason why measurement of 2,3-DHB as well as 2,5-DHB is preferable as an index of OH• generation). Observation of only some of the expected products in aromatic hydroxylation studies (e.g. 2,5-DHB but not 2,3-DHB from salicylate) should thus be regarded as suspect. Addition of ONOO⁻ to salicylate or phenylalanine at pH 7.4 can generate hydroxylation products identical to those in Fig. 5.5.[21] This may be due to generation of OH• from decomposition of ONOOH, but could also involve direct attack of excited-state *trans*-ONOOH (Chapter 2) on aromatic rings. Thus results of aromatic hydroxylation experiments must be interpreted with caution if there is a chance that ONOO⁻ is being generated. However, ONOO⁻ addition also generates nitration products from salicylate and phenylalanine, whereas OH• cannot, so a search for nitration products can help distinguish hydroxylation due to ONOO⁻ from that due to 'real' OH•.

Use of hydroxyl-radical scavengers

Addition of OH• scavengers has often been used to gain further evidence that OH• is responsible for a particular reaction (e.g. Table 5.2, see p. 357). In simple systems this principle often works well, but in biological systems it is fraught with danger. In particular, scavengers are often ineffective when OH•

---

ortho-tyrosine (a) and *meta*-tyrosine (b). The amounts formed were decreased by antioxidants, including the thiol mercaptopropionylglycine (group II, ●) and a mixture of SOD, catalase and the iron ion chelator desferrioxamine (group III, {▲}) but not by inhibitors of nitric oxide synthase, i.e. NO• (and, by implication, ONOO⁻) were not involved in the OH• generation. Data adapted from *Circul. Res.* **73**, 534 (1993) by courtesy of Dr Roberto Bolli and the publishers.

is not formed in 'free solution' but instead is formed upon a biological target, e.g. when catalytic metal ions bind to a protein or to DNA. Reaction with the target is favoured over reaction with a scavenger in free solution. This principle is illustrated by the deoxyribose assay (see below). In addition, many $OH^\bullet$ scavengers are non-specific (e.g. thiourea reacts with $O_2^{\bullet-}$, H_2O_2, HOCl and $ONOO^-$ in addition to $OH^\bullet$).

5.3.2   *The deoxyribose assay for hydroxyl radical*[20]

Attack of $OH^\bullet$ on the sugar 2-deoxyribose produces a huge variety of different products, some of which are mutagenic in bacterial test systems (Chapter 2). Some of the fragmentation products, when heated at low pH, decompose to form malondialdehyde (MDA), which can be detected by adding thiobarbituric acid (TBA) to the reaction mixture, resulting in formation of a pink $(TBA)_2$-MDA chromogen. This can be used to detect $OH^\bullet$ production, although it is unclear whether or not some other ROS can also degrade deoxyribose.

Measurement of rate constants

Another use of the deoxyribose assay is to measure rate constants for reactions of $OH^\bullet$. In this version of the assay, $OH^\bullet$ is generated by a mixture of ascorbic acid, H_2O_2 and ferric EDTA (Fig. 5.7). Any $OH^\bullet$ not scavenged by the iron–EDTA itself attacks the deoxyribose, degrading it into TBA-reactive material (TBARS). Any added $OH^\bullet$ scavenger competes with deoxyribose for the $OH^\bullet$ and blocks generation of TBARS. A simple competition plot can be used to calculate the rate constant for $OH^\bullet$ scavenging (Fig. 5.7). Although the definitive technique for measuring radical rate constants is pulse radiolysis (Chapter 2), the deoxyribose assay produces comparable values provided that certain controls are employed:

1. Check that the substance under test does not react with H_2O_2, which would interfere with $OH^\bullet$ formation.
2. Check that the substance is not a powerful iron ion chelator, able to take iron ions from EDTA.
3. Check that attack of $OH^\bullet$ on the substance does not produce TBARS. A control should be performed omitting deoxyribose from the reaction mixture.
4. Check that the substance does not interfere with measurement of products. It should not inhibit chromogen formation when added to the reaction mixture at the end of the incubation, with the TBA and acid. **Note**: this is a general control that should be applied to all methods for detecting $OH^\bullet$ (or any other ROS/RNS), including spin trapping (Fig. 5.3).

Site-specific reactions

In vivo $OH^\bullet$ is often generated by reaction of H_2O_2 with iron or other transition-metal ions that are bound to a biological molecule. The $OH^\bullet$ generated then preferentially attacks the molecule to which the metal is bound, since the 'local concentration' of that molecule is overwhelmingly greater than

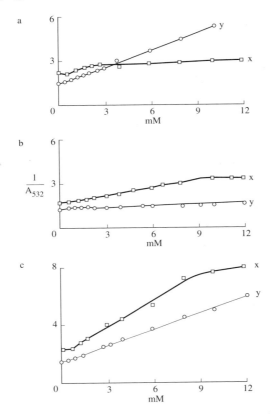

Fig 5.7. Examples of hydroxyl-radical scavenging in the deoxyribose assay. (A) Double-reciprocal plot showing the ability of Hepes, a powerful $OH^\bullet$ scavenger, to inhibit deoxyribose degradation by $OH^\bullet$ generation by a mixture of Fe(III)–EDTA, H_2O_2 and ascorbate at pH 7.4 (line y). Rate constant for $OH^\bullet$ scavenging calculated from the slope of the line as $> 10^9 \, M^{-1} s^{-1}$. Line x shows that Hepes is a much poorer inhibitor when EDTA is omitted. Our explanation is that Fe(III) binds to the deoxyribose, $OH^\bullet$ generation becomes site-specific and added $OH^\bullet$ scavengers cannot protect. (B) The identical experiment using citrate. Line y shows that citrate is a poor $OH^\bullet$ scavenger (rate constant $< 10^8 \, M^{-1} s^{-1}$). Line x shows it to be a good inhibitor of site-specific $OH^\bullet$ generation; citrate can chelate iron ions from deoxyribose and iron–citrate chelates are poorly redox-active. (c) The identical experiment, using ATP. ATP is a good $OH^\bullet$ scavenger (line y; rate constant $> 10^9 \, M^{-1} s^{-1}$). Line x also shows its ability to chelate iron ions from deoxyribose, as would be expected from the presence of sugar and phosphate groups. The iron–ATP complex is redox-active, but the ATP absorbs a large number of the $OH^\bullet$ produced, so that they do not escape into free solution. Data from the laboratory of one of the authors (B.H.). Calculation of rate constants from the slope of lines y is by the equation

$$\frac{1}{A} = \frac{1}{A^0}\left(1 + \frac{k_S [S]}{k_{DR}[DR]}\right)$$

where A is the absorbance in the presence of a scavenger S at concentration [S] and A^0 is the absorbance in the absence of a scavenger. Hence, a plot of $1/A$ against [S] should give a straight line of slope $k_S/k_{DR}[DR]A^0$ with an intercept on the y-axis of $1/A^0$, and the rate constant k_S for reaction of S with $OH^\bullet$ can be calculated (*Anal. Biochem.* **165**, 219 (1987)). The value of k_{DR} is $3.1 \times 10^9 \, M^{-1} s^{-1}$.

Table 5.3. Some methods for detection of hydroxyl radicals

Method	Principle of method	Comments
Hydroxylation of terephthalic acid	Non-fluorescent; hydroxylated by $OH^\bullet$ to fluorescent product.	Has not been widely used in biological systems (*Life Sci.* **56**, PL89 (1995)).
Bleaching of *p*-nitrosodimethylaniline (PNDA)	PNDA reacts rapidly with $OH^\bullet$ but not with $O_2^{\bullet-}$ or singlet O_2. Reaction is accompanied by bleaching of the yellow colour (*Eur. J. Biochem.* **95**, 621 (1975)).	PNDA is bleached non-specifically by many biological systems; confirmatory evidence for role of $OH^\bullet$ required.
Conversion of methional ($CH_3SCH_2CH_2CHO$) and related compounds (methionine or 2-keto-4-methylthiobutanoic acid, $CH_3SCH_2CH_2COCOOH$) into ethylene gas ($H_2C{=}CH_2$)	Measurement of ethylene by GC	Not specific for $OH^\bullet$; oxidized by $RO_2^\bullet$, decomposing $ONOO^-$ and some peroxidase enzymes; confirmatory evidence for role of $OH^\bullet$ required.
Tryptophan method	Reaction of $OH^\bullet$ with tryptophan produces a characteristic set of products (*Bull. Eur. Physiopath. Resp.* **17**, 31 (1981)).	Tryptophan also reacts with singlet O_2 but the products are different (Chapter 2).
Coumarin fluorescence	Coumarin-3-carboxylic acid (CCA) is hydroxylated at position 7 to a fluorescent product.	CCA has been covalently linked to various biomolecules and fluorescence changes used to measure $OH^\bullet$ generation in their vicinity (*Int. J. Rad. Biol.* **63**, 445 (1993)). For example, $OH^\bullet$ generation by copper ions bond to DNA could not be decreased by adding DMSO, methanol or ethanol but could be by histidine (which chelates copper). This is typical of a 'site-specific' reaction (Fig. 5.7) (*Free Rad. Biol. Med.* **18**, 669 (1995)).

Method	Description	Comments
Conversion of caffeine to 8-oxocaffeine Caffeine 8-Oxocaffeine (1,3,7-trimethyluric acid)	Caffeine is hydroxylated at position 8; product analysed by HPLC with electrochemical detection.	Analogous to analysis of 8-OH-deoxyguanosine in DNA. Formation of 8-oxocaffeine from endogenous caffeine has been used to measure free-radical generation during roasting and brewing of coffee (*J. Agr. Food. Chem.* **43**, 1332 (1995)).
Dimethylsulphoxide (DMSO) method	$OH^\bullet$ radicals react with DMSO, generating, among other products, methane gas, measured by gas-liquid chromatography (*Biochemistry* **20**, 6006 (1981)), or formaldehyde, measured colorimetrically. $$(CH_3)_2SO + OH^\bullet \rightarrow CH_3SO_2H + {}^\bullet CH_3$$ $${}^\bullet CH_3 + O_2 \rightarrow CH_3OO^\bullet$$ $$2CH_3OO^\bullet \rightarrow HCHO + CH_3OH + O_2$$ $${}^\bullet CH_3 + R{-}H \rightarrow CH_4 + R^\bullet$$	Not specific for $OH^\bullet$, e.g. oxidized by decomposing $ONOO^-$. Confirmatory evidence for role of $OH^\bullet$ required. Babbs *et al.* (*Free Rad. Biol. Med.* **6**, 493 (1989)) have suggested that oxidation of DMSO to CH_3SO_2H, methane-sulphinic acid (measured colorimetrically or by HPLC) is a means of detecting $OH^\bullet$ *in vivo*. Another approach is to trap the $CH_3^\bullet$ radicals (*Anal. Chem.*, **69**, 4295 (1997)).
Benzoate fluorescence	Reaction of benzoic acid with $OH^\bullet$ gives 3- and 4-hydroxybenzoates, which are fluorescent at 407 nm when excited at 305 nm.	Sensitive method (*Biochem. J.* **243**, 709 (1987)); confirmatory evidence for role of $OH^\bullet$ required
Spin trapping/HPLC	A combination of the principles of spin trapping and aromatic hydroxylation; HPLC is used to separate radical adducts of a spin trap, such as DMPO.	Electrochemical detection can allow high sensitivity (e.g. *Anal. Biochem.* **196**, 111 (1991)).

that of added scavengers. **Hence OH• scavengers do not behave in the way predicted from simple chemistry** (Table 5.3) **and may fail to inhibit**. This concept is also illustrated by the deoxyribose assay (Fig. 5.7). When iron ions are added to the reaction mixture as FeCl₃ (not chelated to EDTA), some of them bind to deoxyribose. The bound iron ions appear to participate in Fenton chemistry, but any OH• radicals formed immediately attack the deoxyribose and are not released into free solution. Hydroxyl-radical scavengers (such as Hepes, methanol or ethanol), at moderate concentrations, do not inhibit this deoxyribose degradation, presumably because they cannot compete with the deoxyribose for OH• generated by bound iron ions. The only substances that do inhibit in this assay are those that bind iron ions strongly enough to remove them from the deoxyribose. A good example is the organic acid citrate—a fairly poor OH• scavenger but a good inhibitor in this system (Fig. 5.7).

Confusion has arisen in the literature because some 'established' OH• scavengers can chelate metals. Examples are thiourea and (to a lesser extent) mannitol. Others, such as Good buffers, ethanol, methanol and DMSO, appear not to. Hence the protective efficiency of 'OH• scavengers' against oxidative damage in biological systems may appear unrelated to their rate constants for reaction with OH•, even if OH• generation is responsible for the damage being studied.

5.3.3   *Other trapping methods for hydroxyl radical*

Multiple other detector molecules for OH• have been described, the properties of some of which are described in Table 5.3.

5.4   **Detection of superoxide**

The question of detection of superoxide overlaps considerably with the methods available for assaying superoxide dismutase (SOD) activity, which were discussed in Chapter 3. DMPO spin-trapping is one available $O_2^{•-}$ detection method (Section 5.2.1 above). Other spin traps have also been used. However, 3,5-dibromo-4-nitrosobenzenesulphonic acid (DBNBS) (Table 5.1), although it reacts quickly with $O_2^{•-}$, does not appear to produce a stable adduct.[28] Fluorescent probes such as lucigenin are widely used to detect $O_2^{•-}$ production by phagocytes, but are subject to artefacts (Section 5.10.1 below). Often, $O_2^{•-}$ is detected by its ability to reduce cytochrome c or nitroblue tetrazolium (NBT). The former reaction is chemically simpler than NBT reduction (Chapter 2).

$$\text{cytochrome } c\text{–Fe(III)} + O_2^{•-} \rightarrow O_2 + \text{cytochrome } c\text{–Fe}^{2+}$$

However, many other substances can reduce cytochrome c (especially ascorbate and thiols) and interfere with $O_2^{•-}$ determination by removing the detector molecule. An '$O_2^{•-}$ electrode' has been described which measures reduction of bound cytochrome c.[44] Usually, addition of SOD is employed to show that reduction is $O_2^{•-}$-mediated.

NBT reduction is more complex, and $O_2^{\bullet-}$ can be generated during it (Chapter 2). Thus addition of NBT to a biological system not making $O_2^{\bullet-}$ but able to reduce NBT might create an artefactual $O_2^{\bullet-}$ generation. Superoxide can also oxidize molecules such as adrenalin (to an adrenochrome chromophore), and this has sometimes been used to detect $O_2^{\bullet-}$ (Chapter 2). Again, there are problems: once adrenalin has begun oxidizing (whether or not this oxidation was started by $O_2^{\bullet-}$), $O_2^{\bullet-}$ is generated and oxidizes more adrenalin. Thus it is difficult to relate adrenochrome formation to the actual rate of $O_2^{\bullet-}$ generation by the system under test. Another detector molecule sometimes used is **Tiron** (1,2-dihydroxybenzene-3,5-disulphonate). Like many diphenols, it reacts with $O_2^{\bullet-}$ to give a semiquinone, which can be measured by EPR. For example, this technique has been used to measure $O_2^{\bullet-}$ production by chloroplasts. One problem is that diphenols can be oxidized by many systems, including peroxidases and $RO_2^{\bullet}$ radicals, so confirmatory evidence is needed that $O_2^{\bullet-}$ is being detected.[25] Even inhibitions by SOD must be interpreted with caution in systems containing semiquinones (Chapter 2).

Rate constants for reaction of molecules with $O_2^{\bullet-}$ are ideally obtained by pulse radiolysis methods (Chapter 2), but simple test-tube systems are sometimes adequate. Indeed, pulse radiolysis is unsuitable for measuring most reactions of $O_2^{\bullet-}$ in aqueous solution, since the reaction rates are usually lower than the overall rate of non-enzymic dismutation of $O_2^{\bullet-}$.

For example, a mixture of hypoxanthine (or xanthine) and xanthine oxidase at pH 7.4 generates $O_2^{\bullet-}$ which reacts with cytochrome c and NBT with defined rate constants, 2.6×10^5 and $6 \times 10^4 \, M^{-1} s^{-1}$, respectively. Any added molecule capable of reacting with $O_2^{\bullet-}$ decreases the rates of cytochrome c or NBT reduction, and analysis of the inhibition produced allows calculation of an approximate rate constant. Essential controls are as follows:

1. Check that the substance under test does not inhibit $O_2^{\bullet-}$ generation (e.g. by inhibiting xanthine oxidase). One artefact that has caused some confusion is that many compounds being tested for $O_2^{\bullet-}$-scavenging ability absorb strongly at 290 nm, making spectrophotometric assessment of urate production by xanthine oxidase inaccurate. HPLC determination of urate, or measurement of O_2 uptake by the xanthine oxidase system, are alternatives.
2. Check that the substance under test does not directly reduce cytochrome c or NBT. This is often a problem with cytochrome c, but less so with NBT, provided that pH values of $\leqslant 7.4$ are used.
3. Consider the possibility that a radical formed by attack of $O_2^{\bullet-}$ on a substance could itself reduce cytochrome c or NBT. This will be revealed as deviations from linear competition kinetics at high scavenger concentrations.

5.4.1 *Histochemical detection*

Some histochemical methods to detect $O_2^{\bullet-}$ in tissues have been described. For example, perfusion with tetrazolium salts can lead to microscopically observable formazan precipitation at sites of $O_2^{\bullet-}$ generation. Figure 5.8 (see plate section)

Table 5.4. Some methods for detection of nitric oxide

Methods	Principle of method	Comments
A. Direct methods		
Light emission	Reaction of NO• with O_3 produces light, via excited-state nitrogen dioxide $NO + O_3 \rightarrow NO_2^* \rightarrow NO_2 + h\nu$ (excited)	Highly sensitive (nM range). Measures only gas-phase NO•. Potential interference by other light-emitting systems. NO• must be displaced from the biological material into the gas-phase for analysis, e.g. by flushing with N_2, sometimes the transfer is not quantitative.
NO• electrodes	Several types. In **porphyrinic sensors** NO• binds to a Ni^{2+}–porphyrin adsorbed on to an anode and is oxidized electrochemically: $Ni(P) + NO• \rightarrow Ni(P)NO \rightarrow Ni(P)NO^+ + e^-$ Electron flow is proportional to [NO•]. Another method is adaptation of the Clark O_2 electrode, in which O_2 diffuses through a membrane and is reduced. NO• can be detected by changing the relative potentials of the silver and platinum electrodes.	Used to study NO• exhalation in humans, and demonstrate elevated levels in asthma (*Thorax* **51**, 233 (1996)), ~ 10 nM sensitivity. Easy to use, but several reports of chemical interference with the electrodes. Clark probe is slower to respond to changes in NO• than the porphyrinic probes and less sensitive. Insertion of a porphyrinic probe into a hand vein has been used to detect NO• *in vivo* (*Lancet* **346**, 153 (1995)).
Haemoglobin trapping	NO• reacts with oxyhaemoglobin, eventually converting it to methaemoglobin, and the ΔA is measured.	Sensitivity: ~ μM. Simple to perform. Interference can occur with other redox agents. $ONOO^-$ leads to the same oxidation; NO_2^- (much more slowly) oxidizes oxyhaemoglobin to methaemoglobin. Myoglobin can also be used.

ESR spin-trapping

Various traps are available, including

- haemoglobin (ESR changes an alternative to ΔA measurement)
- other haem proteins (Fe^{2+} state)
- Fe^{2+}–dithiocarbamates
- Fe^{2+}–thiosulphate

$NO^\bullet$ is a good ligand to Fe^{2+} ions and the nitrosyl complexes produced have characteristic ESR spectra (e.g. HbNO, NoFe[DTC]$_2$, Fe[S$_2$O$_3$]$_2$[NO]$_2$).

- ribonucleotide reductase (removal of the tyrosyl radical at the active site by $NO^\bullet$ abolishes the ESR signal). Depends on availability of enzyme, thus not widely used.

Attempts to use DBNBS and DMPO to trap $NO^\bullet$ have given equivocal results. More success has been obtained with *nitronylnitroxides*, a group of organic compounds containing both nitrone and nitroxide groups. Nitronylnitroxide reacts with $NO^\bullet$ to give an imino nitroxide, detectable by ESR (Fig. 5.9) since it has a very different ESR spectrum. NNO has also been used to antagonize the action of $NO^\bullet$ in biological systems. Chelotropic agents such as 7,7,8,8-tetra-methyl-o-quinodimethane, have also been used to detect $NO^\bullet$. They bind $NO^\bullet$ to generate a nitroxide radical, detectable by ESR (*J. Am. Chem. Soc.* **116**, 2767 (1994)).

Formation of ESR–detectable nitrosylhaemoglobin in blood and tissues is often used as an index of $NO^\bullet$ production, e.g. in ischaemia–reperfusion of organs and in animals with septic shock. Injected Fe^{2+} dithiocarbamates have also been used to detect $NO^\bullet$ in animals with septic shock (*FEBS Lett.* **345**, 120 (1994)). Nitrothiolate ligand signals, probably involving non–haem iron–sulphur proteins have been detected by ESR in $NO^\bullet$-exposed cells and tissues.

Reduction of nitroxides to ESR–silent species by ascorbate and GSH (Fig. 5.3) can be a problem in biological systems. Superoxide might also reduce some of these species.

Table 5.4. (*Continued*)

Methods	Principle of method	Comments
B. Indirect methods		
NO_2^- measurement	Griess reaction: NO_2^- reacts with sulphanilamide in an acidic solution of N-(1-naphthyl)ethylenediamine to give a coloured azo product (ΔA at 548 nm) Fluorimetry: reaction of NO_2^- with 2,3-diaminonaphthalene forms a fluorescent product, 1H-naphthotriazole.	Sensitivity: μM. Easy assays. NO_2^- *in vivo* is rapidly oxidized to NO_3^-, which appears to be stable in body fluids. NO_3^- can be re-reduced for the assay using nitrate reductase enzymes or chemical reducing agents. NO_3^-/NO_2^- can also come from diet.[71] Thus in one human study basal plasma NO_3^- was $29 \pm 1\,\mu M$ and rose to $205\,\mu M$ 2 h after intake of NO_3^--rich food. NO_3^- in diet can be reduced to NO_2^- in the gut. In saliva, NO_3^- is converted to NO_2^- by oral bacteria. The ranges of NO_2^- and NO_3^- in plasma samples are quoted as 1.3–13 μM and 4.0–45.3 μM in healthy people (*Clin. Chem.* **41**, 892 (1995)). Inhaled $NO^{\bullet}$ is largely converted to NO_3^- in humans.
Measurement of other oxidation products	Based on formation of more reactive species on the pathway from $NO^{\bullet}$ to NO_2^- in presence of O_2. Thus oxidizing $NO^{\bullet}$ oxidizes ferrocyanide to ferricyanide (ΔA at 420 nm) and the colourless dye ABTS to the coloured ABTS$^+$ (ΔA at 660 nm). Nitrosation of sulphanilamide generates a coloured azo dye.	All simple spectrophotometric methods. Potential interference from other RNS, e.g. $ONOO^-$ or inhaled $NO_2^{\bullet}$.

| Use of NOS inhibitors | Can be 'general' inhibitors of all NOS isoforms or selective for a particular type. The former are generally analogues of the NOS substrate L-arginine, such as **N-monomethyl-L-arginine** (L-NMMA) or **N-nitro-L-arginine methyl ester** (L-NAME). Essential controls include showing that D-isomers do not inhibit and that inhibition is reversed by adding excess L-arginine. More selective inhibitors include *N*-iminoethyl-L-ornithine (eNOS), aminoguanidine (iNOS) and 7-nitro–indazole (nNOS). | Subject to suitable controls, can provide evidence that NO$^{\bullet}$ (*or species derived from it*) are involved. However, NO$^{\bullet}$ can arise by reactions not involving NOS, e.g. decomposition of NO_2^- by acid in the stomach. |

Adapted from Table 1 on p. 278 of *Analysis of Free Radical Reactions in Biological Systems* (Favier *et al.*, eds), Birkhauser-Verlag, Basle, 1995 by courtesy of Dr M. Fontecave and the publishers. Also see *Methods Enzymol.* **268** (1996).

shows an example. Another method employs conversion of **diamino-benzidine** (DBD) to an insoluble product. Tissues or organs are perfused with DBD and manganese (Mn^{2+}) ions. The superoxide oxidizes Mn^{2+} to $Mn(III)$, which in turn oxidizes the DBD. DBD can also be used to detect H_2O_2 under different reaction conditions (Section 5.8 below; Fig. 5.8, bottom).

5.5   Detection of nitric oxide[49]

Nitric oxide is involved in many physiological and pathological processes and so many methods have been developed to measure it. Some assays are direct trapping methods, involving reactions of $NO^{\bullet}$ with spin traps, haem compounds, other metal complexes or ozone (Table 5.4). An alternative approach is to implicate $NO^{\bullet}$ in a process if that process is inhibited by agents that block nitric oxide synthase (NOS) enzymes, just as inhibition by SOD is used to implicate $O_2^{\bullet-}$. NOS inhibitors are not necessarily specific; often they can scavenge $OH^{\bullet}$, HOCl or ONOOH and so controls are needed (explained in Table 5.4). Yet another technique is to measure end-products of $NO^{\bullet}$, the most common being the measurement of NO_2^- by the colourimetric **Griess test** (Table 5.4). Oxidation of $NO^{\bullet}$ in aqueous solution produces mostly NO_2^- (Chapter 2).

$$4NO^{\bullet} + O_2 + 2H_2O \rightarrow 4NO_2^- + 4H^+$$

5.5.1   *Interference by peroxynitrite*

$NO^{\bullet}$ generated *in vivo* at sites of inflammation/tissue injury can react with $O_2^{\bullet-}$ to give peroxynitrite, which can interfere with several methods used for

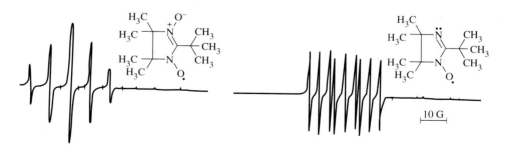

Nitronyl nitroxide (NNO)   Imino nitroxide (INO)

Fig. 5.9. ESR detection of nitric oxide. A nitronylnitroxide trap was used. Note the dramatic change in ESR spectra. Diagram kindly donated by Dr B. Kalyanaraman.

detection of NO$^\bullet$ (Table 5.4). For example, whereas NO$^\bullet$ under aerobic conditions eventually forms NO$_2^-$, breakdown of ONOO$^-$ generates mostly nitrate, NO$_3^-$. Thus measurement of only NO$_2^-$ in biological systems can seriously underestimate the total amount of NO$^\bullet$ that was generated. Both NO$_2^-$ and NO$_3^-$ should be measured (Table 5.4).[71]

5.5.2   *Calibration*

Regardless of the method used to measure NO$^\bullet$, it will need calibration. The direct methods (Table 5.4) can be calibrated using NO$^\bullet$ itself. This may be purchased as a gas, generated by acidification of nitrites or produced by the decomposition of 'NO$^\bullet$ donors'. NO$^\bullet$ donors are widely used by pharmacologists to investigate physiological and pathological effects of NO$^\bullet$ but care must be taken in interpreting results obtained with them (Section 2.4.6). For example, the decomposition of nitrosothiols is catalysed by metal ions, and thus the rate of NO$^\bullet$ generation will depend on what metal ions (especially copper) are added to the biological system, or contaminate the reagents. Other potentially reactive species can be produced from NO$^\bullet$ donors, e.g. RS$^\bullet$ radicals from nitrosothiols and ONOO$^-$ from SIN-1 (Chapter 2).

5.6   **Detection of peroxynitrite**[24,49]

The rapid reaction of O$_2^{\bullet-}$ and NO$^\bullet$ radicals (or of NO$^-$ and O$_2$) generates ONOO$^-$ (Chapter 2), whose formation has been implicated in many diseases (Table 5.5).

Generation of ONOO$^-$ by chemical systems and isolated cells (e.g. macrophages) and organelles has been assessed by several mechanisms. For example, ONOO$^-$ oxidizes the dye **dihydrorhodamine 123** (DHR 123; Fig. 5.10) into **rhodamine 123**, which fluoresces at 536 nm when illuminated at 500 nm. This oxidation does not require metal ions and is not inhibited by OH$^\bullet$ scavengers, but can be inhibited 60% by 100 mM HCO$_3^-$ (which reacts with ONOO$^-$; Chapter 2). Oxidation of injected DHR 123 was observed in rats suffering from endotoxic or haemorrhagic shock; oxidation was decreased by administration of NOS inhibitors.

Superoxide, NO$^\bullet$ and H$_2$O$_2$ appear not to oxidize DHR-123. Hydroxyl radicals can, but this is prevented by OH$^\bullet$ scavengers. However, H$_2$O$_2$/peroxidase mixtures and HOCl can also oxidize DHR, and so its oxidation is not a unique marker for ONOO$^-$. Hence careful controls are needed. ONOO$^-$ can also oxidize *o*-phenylenediamine.

5.6.1   *Nitration assays*[24,49]

It was reported in 1922 that an acidified mixture of H$_2$O$_2$ and NO$_2^-$ can nitrate aromatic compounds. Later work showed that addition of ONOO$^-$ leads to nitration of many aromatic compounds, including the amino acids tyrosine, tryptophan and phenylalanine. Most attention has been paid to tyrosine: both

Table 5.5. Some of the conditions in which the formation of nitrotyrosine or chlorotyrosine has been demonstrated

Condition	Representative reference
A. **Nitrotyrosine**	
Atherosclerosis	*J. Biol. Chem.* **272**, 1433 (1997)
Sporadic inclusion-body myositis	*Neuroreport* **8**, 153 (1996)
Rheumatoid arthritis	*FEBS Lett.* **350**, 9 (1994)
Inflammatory bowel disease	*Gastroenterology* **109**, 1475 (1995)
Neurodegenerative disease	*Neuroscientist* **3**, 327 (1997)
Acute inflammation	*Eur. J. Pharmacol.* **303**, 217 (1996)
Carbon monoxide toxicity	*J. Clin. Invest.* **97**, 2260 (1996)
Adult respiratory distress syndrome	*Am. J. Resp. Crit. Care Med.* **151**, 1250 (1995)
Gastritis (*Helicobacter pylori* infection)	*Cancer Res.* **50**, 3238 (1996)
Cystic fibrosis	*Adv. Pharmacol.* **38**, 491 (1996)
Endotoxic shock	*FEBS Lett.* **363**, 235 (1995)
Ageing of skeletal muscle	*FEBS Lett.* **379**, 286 (1996)
Viral infection	*Proc. Natl Acad. Sci. USA* **93**, 2448 (1996)
B. **Chlorotyrosine**	
Inflamed human kidney	*Am. J. Pathol.* **150**, 603 (1997)
Human atherosclerotic lesions	*J. Clin. Invest.* **97**, 1535 (1996)

the free amino acid and tyrosine residues within proteins can be nitrated upon addition of $ONOO^-$ (Fig. 5.10). For example, $ONOO^-$ nitrates tyrosine residue 108 on bovine CuZnSOD (Chapter 3). Usually, $ONOO^-$-dependent nitrations are accelerated by adding Fe(III) or Cu^{2+} ions and certain metalloproteins, including bovine CuZnSOD, MnSOD and horseradish peroxidase. Human CuZnSOD has no tyrosines and hence cannot be nitrated to give 3-nitrotyrosine.

Nitration of tyrosine residues can be measured spectrophotometrically (3-nitrotyrosine is yellow at alkaline pH), or by amino-acid analysis after hydrolysis of modified proteins. Gas chromatography/mass spectrometry or HPLC with absorbance or electrochemical detection have also been used. Electrochemical detection (which has high sensitivity) is easier if the nitrotyrosine in biological samples is first reduced to **aminotyrosine**, which is oxidized at much lower voltages, by addition of the strong reducing agent dithionite. Aminotyrosine, unlike nitrotyrosine, has a characteristic fluorescence spectrum.

An alternative approach is the use of polyclonal or monoclonal antibodies directed against nitrated proteins. Although less quantitative than the above approaches, these methods have the advantage that they permit the localization of nitrated proteins in tissues. Figure 5.8 (see plates section) shows an example. Essential controls include showing that the antibody binding is blocked by authentic 3-nitrotyrosine and/or by treatment of the sample with dithionite.

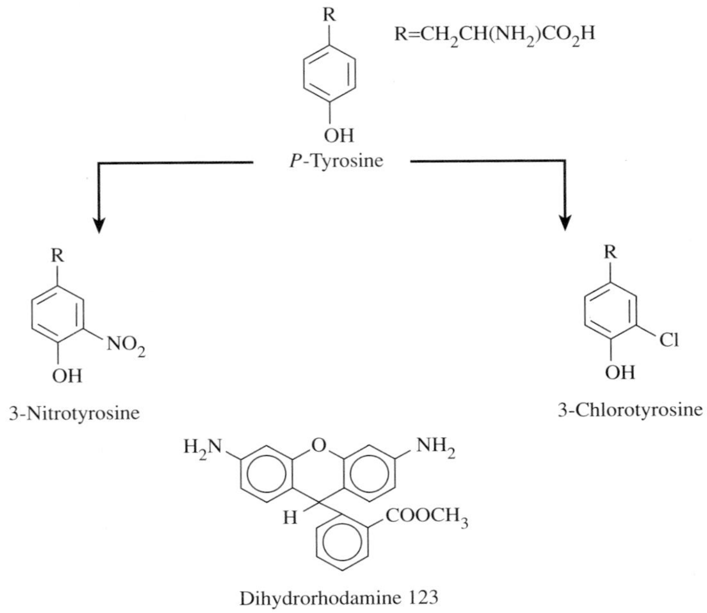

R=CH₂CH(NH₂)CO₂H

$R = CH_2CH(NH_2)CO_2H$

P-Tyrosine

OH

R

NO₂
OH

3-Nitrotyrosine

R

Cl
OH

3-Chlorotyrosine

H₂N   O   NH₂

H   COOCH₃

Dihydrorhodamine 123

Fig. 5.10. Some of the molecules used to detect peroxynitrite.

One key question is whether nitrotyrosine is specific as a marker for ONOO⁻. NO• appears not to nitrate aromatic compounds, but it can react rapidly with tyrosine radicals (TyrO•), presumably to form nitroso adducts that could conceivably rearrange to nitro products (Chapter 2). Nitrogen dioxide (NO₂•) can nitrate tyrosine both *in vitro* and *in vivo* in animals exposed to high levels of this toxic gas, but it appears generally less efficient as a nitrating agent than ONOO⁻. Mixtures of NO₂⁻ and HOCl can also nitrate tyrosine, apparently by the intermediate formation of **nitryl chloride** (nitronium chloride), NO₂Cl. Both HOCl and NO₂Cl can also lead to chlorination of tyrosine residues, and 3-chlorotyrosine (Fig. 5.8) has been used as a marker. Finally, myeloperoxidase (and other 'non-specific' peroxidases such as horse-radish peroxidase) can oxidize NO₂⁻ in the presence of H₂O₂ to a nitrating species, presumably one-electron oxidation to NO₂•, but conceivably two-electron oxidation to NO₂⁺. Hence observation of nitration *in vivo* should not necessarily be attributed to ONOO⁻ formation; it is more a general assay of 'reactive nitrogen species'.[20a]

5.7   Detection of chlorinating species

Hypochlorous acid is a powerful oxidizing and chlorinating agent used in bleaches and generated by the enzyme myeloperoxidase. Several assays have been described to measure it. The **taurine assay** is based on the conversion of taurine to taurine chloramine

$$\text{taurine–NH}_2 + \text{HOCl} \rightarrow \text{taurine–NHCl} + \text{H}_2\text{O}$$

which (like HOCl itself) oxidizes yellow thionitrobenzoic acid to a colourless oxidized product, 5,5′-dithiobis(2-nitrobenzoic acid) (DTNB, **Ellman's reagent**; see Fig. 5.20 below). Chlorination of **monochlorodimedon** to dichlorodimedon, with loss of absorbance at 290 nm, has been used to assay HOCl production by myeloperoxidase (Chapter 2). Since HOCl reacts with many biomolecules, these assays are not generally useful to measure its production *in vivo*.

Exposure of free or protein-associated tyrosine to HOCl leads to production of some 3-chlorotyrosine (Fig. 5.10), which has therefore been suggested as a biomarker of HOCl generation.[30] However, NO_2Cl can chlorinate tyrosine, so simultaneous measurement of nitrated and chlorinated products (e.g. by HPLC after hydrolysis of proteins) can be useful in distinguishing effects of $ONOO^-$ (nitration), HOCl (chlorination) and NO_2Cl (both reactions). It should be noted that in no case is the reaction a quantitative measure of the amounts of $HOCl/ONOO^-/NO_2Cl$ generated: all of these species react with multiple other cell targets (e.g. ascorbate, –SH compounds) whose levels will determine how much tyrosine modification occurs *in vivo*. Bityrosine is also generated when $ONOO^-$ is added to tyrosine at pH 7.4.

Antibodies against chlorinated proteins have been raised and used to demonstrate their existence *in vivo* (Table 5.5).

5.8   Detection of hydrogen peroxide

Several methods for measuring H_2O_2 exist, as summarized in Table 5.6. They need to be sensitive, since rates of H_2O_2 production by cells and organelles are often only in the nmol/min range (except for activated phagocytes). Perhaps the most commonly used assay employs the H_2O_2-dependent oxidation of scopoletin (Fig. 5.11) to a non-fluorescent product by horseradish peroxidase. Loss of the scopoletin fluorescence at 460 nm is measured. Other peroxidase substrates (e.g. conversion of guaiacol to a brown chromogen) are sometimes used. If the assay is used to study 'H_2O_2 scavengers', it is essential to check that the substance under test is not itself a substrate for peroxidase, that could compete with scopoletin to cause an artifactual inhibition. For example, ascorbic acid and thiol compounds can be oxidized by horseradish peroxidase, and they often interfere with peroxidase-based assay systems. Superoxide can decrease peroxidase activity (forming compound III) and may compromise measurement of H_2O_2 in systems generating $O_2^{\bullet-}$. This can be avoided by adding SOD.[31]

If an H_2O_2-scavenger does interfere with peroxidase-based systems, other assays for H_2O_2 can be used. Thus, H_2O_2 can be estimated by titration with acidified potassium permanganate ($KMnO_4$) or by measuring the O_2 release (1 mol of O_2 per 2 mol of H_2O_2) when a sample of the reaction mixture is injected into an oxygen electrode containing buffer and a large excess of catalase (Table 5.6).

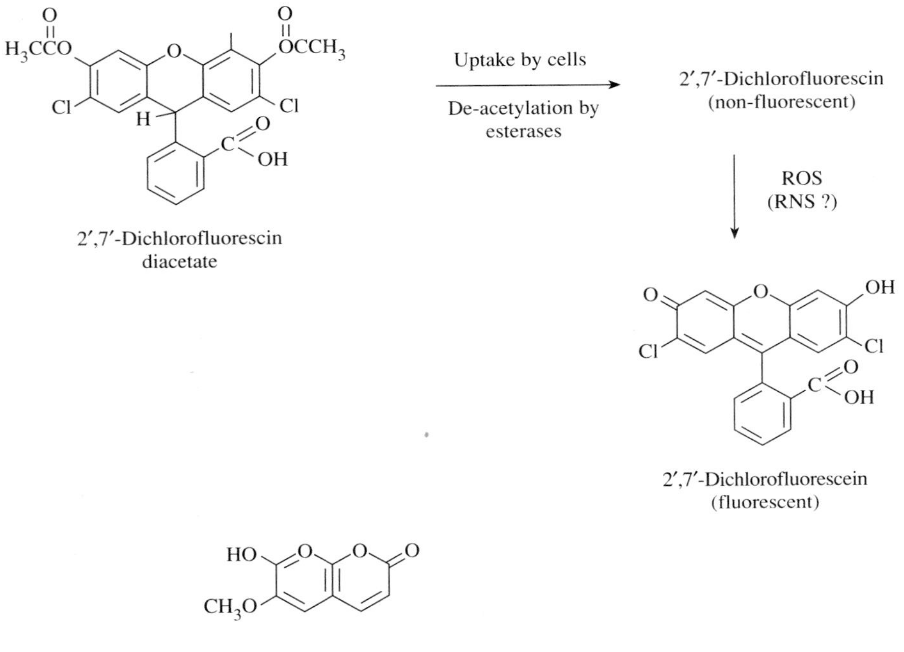

2',7'-Dichlorofluorescin
diacetate

Uptake by cells
———————————→
De-acetylation by
esterases

2',7'-Dichlorofluorescin
(non-fluorescent)

ROS
(RNS ?)

2',7'-Dichlorofluorescein
(fluorescent)

Scopoletin
(7-hydroxy-6-methoxy-coumarin)

Fig. 5.11. Structures of scopoletin and 2',7'–dichlorofluorescin diacetate.

Catalase is frequently used in assays of H_2O_2 production. In the presence of H_2O_2, catalase oxidizes methanol and formate, and this has been made the basis of various assay systems, e.g. decarboxylation of ^{14}C-labelled formate to $^{14}CO_2$. Observations of the spectral intermediates of catalase have been used to assess H_2O_2 production in whole organs (Fig. 5.12). Catalase is inhibited by **amino-triazole**, which reacts only with compound I (Chapter 3). Hence the extent of inactivation of catalase by aminotriazole has been used to calculate H_2O_2 production in isolated cells and organs. Indeed, the fact that aminotriazole inhibits catalase in animal and plant tissues shows that H_2O_2 is being generated *in vivo*.

Various cytochemical stains for H_2O_2 have also been developed, often relying on the ability of peroxidase to oxidize substrates such as **diamino-benzidine** when H_2O_2 is present (Fig. 5.8, plates section bottom photograph).

5.8.1 *Dichlorofluorescin diacetate*[37]

Conversion of the non-fluorescent compound **dichlorofluorescin diacetate** (DCFH-DA) to the fluorescent **2',7'dichlorofluorescein** (by horseradish peroxidase) was first described in 1965 as a fluorimetric assay for H_2O_2.

Table 5.6. Some methods for detecting H_2O_2 production in biological systems

Method	Principle of the method	Examples of systems to which it has been applied
Observation of intracellular catalase compound I	See Fig. 5.12	Bacteria, organs (perfused and *in situ*), organ slices, homogenates.
Oxidation of [^{14}C]methanol or [^{14}C]formate by catalase compound I		Difficult to use in whole animals as methanol can be oxidized by alcohol dehydrogenase and formate by formate dehydrogenase enzymes.
Cytochrome c peroxidase	Enzyme oxidizes reduced cytochrome c and is specific for H_2O_2. Forms a complex with H_2O_2 that absorbs at 419 nm whereas free enzyme absorbs at 407 nm (Chapter 3).	Animal and plant mitochondria, protozoa, peroxisomes, microsomes (*Biochem. J.* **128**, 617 (1972)). Assay often difficult to apply to biological systems because other cytochrome c reductases/oxidases present.
Horseradish peroxidase + scopoletin (or other substrates)	See text. Scopoletin is a popular substrate: others include homovanillic acid (3-methoxy-4-hydroxyphenylacetic acid) and Amplex Red (N-acetyl-3,7-dihydrophenoxazine), which are oxidized to fluorescent products.	Animal and plant mitochondria, sub-mitochondrial particles, phagocytes, protozoa, microsomes (*J. Biochem. Biophys. Methods* **18**, 297 (1989); *Anal. Biochem.* **253**, 162 (1997)).
O_2-electrode method	Add large excess of catalase and measure release of oxygen: $2H_2O_2 \rightarrow 2H_2O + O_2$	Chloroplasts, mitochondria. Used to study rate constants for reactions of H_2O_2, by measuring residual H_2O_2. Not very sensitive but useful in turbid systems.
Inhibition by catalase	If a reaction requires H_2O_2, then it should be inhibited by catalase	Catalase slow at destroying low concentrations of H_2O_2, so a large amount must be added. Often used to investigate the role of H_2O_2 in oxidative damage. Commercial catalases often contaminated with SOD, thymol (a phenolic antioxidant) and endotoxin. Catalase poorly active at low or high pH.

GSSG release	Chapter 3	Isolated organs. Measures GSH oxidation, not necessarily exclusively due to H_2O_2.
Aminotriazole inhibition of catalase	See text and Chapter 3	Fungi, various bacteria, *Mycoplasma*, erythrocytes, other animal tissues.
Dichlorofluorescin diacetate (DCFH-DA)	See text	Not always clear which species is detected; several ROS and RNS (including $ONOO^-$) can oxidize DCFH. Eye lens, human body fluids, peroxisomes (*Free Rad. Res. Commun.* **5**, 359 (1989)).
Decarboxylation of keto acids	Glyoxylate, pyruvate and α-ketoglutarate react non-enzymically with H_2O_2. Use of ^{14}C-labelled substrates permits sensitive measurement of $^{14}CO_2$ release.	
Histochemical staining method	Tissue is treated with Fe^{2+}-diethylenetriamine penta-acetic acid. This is oxidized by H_2O_2 to the ferric chelate which in turn oxidizes diaminobenzidine to an insoluble chromogen.	Perfused organs, cells, tissue slices. Would presumably detect other oxidizing agents (*Methods Enzymol.* **233**, 619 (1994)), but if due to H_2O_2 catalase should inhibit.
Oxidation of dimethylthiourea (DMTU) to a dioxide product	DMTU is oxidized by $OH^\bullet$, HOCl, $ONOO^-$ and H_2O_2, but only H_2O_2 is reported to give the dioxide product.	Enzymes, neutrophils, lung. See *Proc. Natl Acad. Sci. USA*, **85**, 3422 (1988). Not widely used.

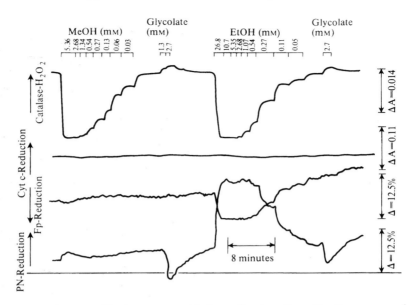

Fig. 5.12. Production of hydrogen peroxide in perfused rat liver. Rat liver was perfused with a bicarbonate–saline solution containing 2 mM L-lactate and 0.3 mM pyruvate at 30 °C. Light was shone through a liver lobe and the concentration of catalase compound I was measured by dual-wavelength spectrophotometry. There is a steady concentration of compound I from which the endogenous rate of H_2O_2 production can be calculated as being about 82 nmol H_2O_2/min/g of liver, in livers from normally fed animals. Inclusion of methanol (MeOH) in the perfusion medium decreases compound I concentration because it is a substrate for the peroxidatic action of catalase. Ethanol (EtOH) has a similar effect, but it also causes an increased reduction of pyridine nucleotides (PN) since it is a substrate for the alcohol dehydrogenases which convert NAD^+ to NADH. There is also some reduction of flavoproteins (Fp). Infusion of glycolate raises the steady-state concentration of compound I because it is oxidized in the peroxisomes to form H_2O_2 by glycolate oxidase:

$$glycolate + O_2 \rightarrow glyoxylate + H_2O_2$$

Cyt-c, cytochrome *c*. Data from *Arch. Biochem. Biophys.* **154**, 117 (1973), by courtesy of Professor Britton Chance and Academic Press.

DCFH-DA is taken up by cells and tissues, usually undergoing deacetylation by esterase enzymes (Fig. 5.11). Oxidation of DCFH within cells leads to fluorescent dichlorofluorescein, which can easily be visualized (strong emission at 525 nm with excitation at 488 nm). This technique is becoming popular as a means of visualizing 'oxidative stress' in living cells (Fig. 5.13). In addition to peroxidase/H_2O_2, several species cause DCFH oxidation, probably including $RO_2^\bullet$, $RO^\bullet$, $OH^\bullet$, HOCl and $ONOO^-$, but not $O_2^{\bullet-}$ or H_2O_2. Hence this 'fluorescent imaging' is an assay of 'generalized oxidative stress' rather than of production of any particular oxidizing species, and it is not a direct measure of H_2O_2 or $O_2^{\bullet-}$ levels. Other probes that have been used include **hydroethidine** (reduced ethidium bromide), which fluoresces upon oxidation to ethidium within the cell.[1]

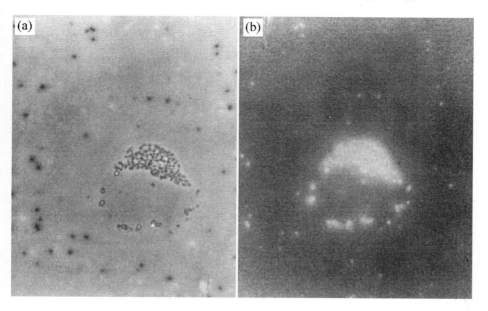

Fig. 5.13. Fluorescence from neutrophils loaded with dichlorofluorescin diacetate after stimulation to produce ROS by adding fMet–Leu–Phe, (a) light field and (b) fluorescence field. DCFH-DA was deacetylated by the cells to DCFH, which accumulated in granules, cytoplasm and nucleus. Excitation was at 500 nm; emission was measured at 530 nm. From *Biochem. J.* **248**, 173 (1987) by courtesy of Professor A.K. Campbell and the Biochemical Society.

5.9   Detection of singlet oxygen[16]

Singlet O_2 is often formed in biological systems (Chapter 2) but it is difficult to detect unambiguously. For example, photosensitization reactions can damage biomolecules by production of other ROS (e.g. OH•) and by direct interaction of excited states with the biomolecule (type I reactions). Hence it must never be assumed that $^1O_2$ is responsible for photochemical damage.

It is possible to distinguish between type I and type II mechanisms of photosensitized damage to a target by immobilizing the sensitizer on a solid support (e.g. a glass slide, suspended above a thin well of substrate solution), thus allowing the $^1O_2$ produced to diffuse a short distance to the target. Hence the target cannot come into contact with the excited-state photosensitizer. An alternative method is to pass a stream of O_2 over a bed of sensitizer and then into the test system.

5.9.1   *Direct detection*

Decay of $^1O_2$ yields two types of light emission: the weak so-called **monomol** emission at 1270 nm and the **dimol** emission resulting from collision of two $^1O_2$ molecules, which produces light at 630 and 701 nm. The 1270 nm (infrared) emission has been much used to study $^1O_2$ chemistry but special detectors are needed (e.g. germanium diode detectors) because most photo-multipliers are insensitive at this wavelength. The dimol emission wavelengths are prone to interference by other light-emitting reactions.

Hence the mere observation of light emission does not implicate $^1O_2$ unless the characteristic emission spectrum can be demonstrated. Many chemical reactions produce light, e.g. lipid peroxidation, Fenton reactions and reactions of certain haem proteins with H_2O_2. The decomposition of dioxetanes (Chapter 2) can produce excited-state carbonyl compounds, which emit light as they return to the ground state.

5.9.2 Use of scavengers and traps[16]

In investigating the role of $^1O_2$ in biological systems, scientists have often relied on the use of 'singlet-oxygen scavengers' and quenchers. Popular compounds have included DABCO, diphenylisobenzofuran, histidine and azide (Chapter 2). Addition of these should inhibit a reaction dependent on $^1O_2$. If, when added at high concentrations, none of them inhibits the reaction under study, one may conclude that $^1O_2$ is not required for that reaction to proceed. If they do inhibit, this does *not* prove a role for $^1O_2$. All these compounds react with $OH^\bullet$, often with a greater rate constant than for reaction with $^1O_2$. For example, azide reacts with $OH^\bullet$ to give a reactive **azide radical**, $N_3^\bullet$

$$N_3^- + OH^\bullet \rightarrow N_3^\bullet + OH^-$$

and both DABCO and histidine react with $OH^\bullet$ with rate constants of $> 10^9 \, M^{-1} s^{-1}$. DABCO and β-carotene also react with $RO_2^\bullet$ radicals.[48] Fortunately, the products of reaction of certain fatty acids, cholesterol and tryptophan with $^1O_2$ are different from those obtained on reaction with $OH^\bullet$ (Chapter 2), so isolation and characterization of them allows distinction. For example, observation of the 'ene' product, 5α-hydroperoxycholesterol, is considered good evidence for $^1O_2$. β-Carotene may give unique products when oxidized by $^1O_2$. These approaches need skill in analytical chemistry, however, and the use of $^1O_2$ scavengers, like that of spin traps, $OH^\bullet$ scavengers and fluorescent molecules, cannot be recommended to the chemically naive.

Another trap that has been employed is **9,10-diphenylanthracene** (or its water-soluble derivatives),[62] to which singlet O_2 adds to form an endoperoxide. Oxidation of diphenylanthracene bound to glass beads has been used to investigate $^1O_2$ production by phagocytes after engulfment of the beads into the phagocytic vacuole. Singlet O_2 also causes hydroxylation,[27] e.g. of phenol (to hydroquinone), aniline (to 4-hydroxyaniline) and salicylate (to 2,5-dihydroxy-benzoate). By contrast, $OH^\bullet$ causes formation of multiple hydroxylation products from aromatic substrates (Section 5.3 above). Attempts to develop spin traps specific for singlet O_2 are continuing.

5.9.3 Deuterium oxide

Another approach, particularly employed in studying photodynamic effects, is the use of deuterium oxide.[16] The lifetime of singlet oxygen is longer in D_2O

than in H_2O by a factor of ten or fifteen (Chapter 2). Thus if a reaction in aqueous solution is dependent on $^1O_2$, carrying it out in D_2O instead of H_2O should potentiate the reaction. Theoretically, a type I photodynamic reaction would be unaffected.

5.10 Studies of 'generalized' light emission (luminescence/fluorescence)

Following early research in Eastern Europe, many scientists have used photo-multipliers, or luminometers, to measure the light emitted by ROS-generating systems *in vitro*, such as peroxidizing lipids. These studies have been extended to animal tissues, injured plants, whole organs and whole organisms such as protozoa or parasitic worms.[67] Human breath shows low-level light emission (about 7000 counts per litre of breath per second),[72] as do blood and plasma.

Such **low-level chemiluminescence** can be detected from many cells and tissues. It is far too weak to be seen by the naked eye (hence it is sometimes called **ultraweak chemiluminescence** or **dark chemiluminescence**) and its source is hard to define. Light can arise from $^1O_2$, triplet excited states (e.g. carbonyls), reactions of $ONOO^-$, lipoxygenase action, haem protein/peroxide reactions and Fenton chemistry (see above). For example, the NOS inhibitor nitro-L-arginine has been reported to decrease basal light emission from perfused lung.

Nevertheless, the total photon count from tissues and organs (both isolated and *in situ*) often increases in response to oxidative stress, e.g. elevated O_2 concentration or introduction of ROS-generating drugs. Hence low-level chemiluminescence might be a useful 'general assay' of oxidative stress.

5.10.1 *Luminol and lucigenin*

The output of light can be increased by adding DCFH-DA (Section 5.8.1 above), **luminol** or **lucigenin** (Fig. 5.14): the latter two are frequently used in studies of ROS production by activated phagocytes (Chapter 6). With activated phagocytes, luminol seems to largely detect HOCl production, although it can also be oxidized by $OH^\bullet$ and ONOOH. Lucigenin appears

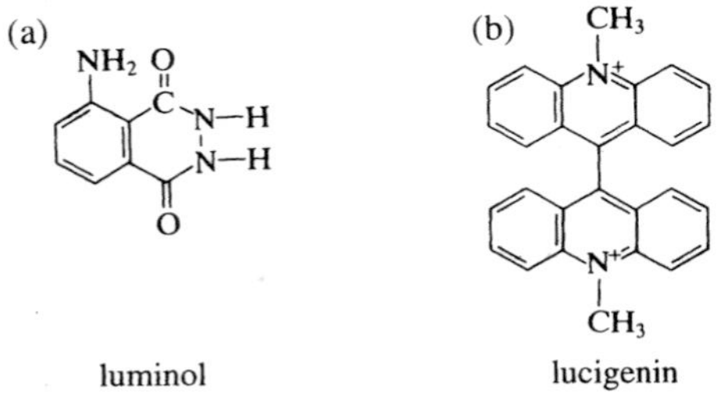

luminol

lucigenin

Fig. 5.14. Structures of luminol and lucigenin.

more specific for $O_2^{\bullet-}$, as does the molecule CLA, **2-methyl-6-phenyl-3,7-dihydroimidazo[1,2-a]pyrazin-3-one**. CLA reacts with $O_2^{\bullet-}$ to form a dioxetane, leading to light production. Another 'probe' for $O_2^{\bullet-}$ is **pholasin**, a luciferin from the bivalve mollusc *Pholas dactylus* (Chapter 6).

However, specificity can never be assumed with any of these light-emission-enhancing compounds.[38] For example, $O_2^{\bullet-}$ can be generated *during* the oxidation of luminol and lucigenin by other species. To detect $O_2^{\bullet-}$, lucigenin (LC^{2+}) must undergo reduction to the lucigenin cation radical, $LC^{\bullet+}$

$$LC^{\bullet+} + O_2^{\bullet-} \rightarrow \text{dioxetane} \rightarrow \text{light}$$

Superoxide is poor at reducing lucigenin to $LC^{\bullet+}$ and so other reducing systems (e.g. the phagocyte NADPH oxidase system) are needed for it to function. Even worse, $LC^{\bullet+}$ can react with O_2 to *generate* $O_2^{\bullet-}$. Thus use of lucigenin to detect $O_2^{\bullet-}$ in biological systems should be interpreted with caution. Luminol must undergo one-electron oxidation before it can react with $O_2^{\bullet-}$; the resulting luminol radical can both reduce O_2 to $O_2^{\bullet-}$ and react with $O_2^{\bullet-}$ to generate an unstable endoperoxide, whose decomposition generates light. Peroxynitrite addition can generate light from luminol but not apparently from lucigenin unless HCO_3^- is present.

Neither luminol nor lucigenin is directly oxidized by H_2O_2.

5.11   Fingerprinting methods: oxidative DNA damage

5.11.1   *Introduction*

ROS and RNS react in characteristic ways with DNA, proteins, lipids and certain low-molecular-mass antioxidants (e.g. ascorbate, urate). The products generated can thus be regarded as fingerprints (or 'footprints') of oxidative attack. For example, information from multiple chemical fingerprints comes together to show that oxidative damage occurs in normal aerobes (Table 4.1).

It is, of course, necessary to be sure that the products measured are unique to oxidative attack and are not generated by normal biological processes. For example, some groups have used ascorbyl radical, measured by ESR, as a bioindicator of oxidative stress (Fig. 5.1). Another approach (applicable only to primates) has been to measure allantoin and other products of ROS/RNS attack upon urate (Section 3.21.7).[29]

5.11.2   *Products of DNA damage*

The chemistry of DNA damage by several ROS/RNS has been characterized *in vitro*, although further studies are needed with $RO^{\bullet}$, $RO_2^{\bullet}$ and O_3 under physiological conditions. Oxidative damage to DNA produces strand breakage, damage to the deoxyribose sugar and modification of the purine and pyrimidine bases (Chapter 4). However, strand breakage can also result from nuclease activity and occurs during DNA repair, so it cannot be equated to oxidative DNA damage. Deoxyribose fragmentation generates multiple products that are hard to quantify on a routine basis. Hence most attempts to 'fingerprint' oxidative damage to DNA have measured modified bases. The most common technique is the assay of **8-hydroxydeoxyguanosine** (8-OHdG), 8-hydroxyguanine (8-OH-G) attached to deoxyribose.[15,27a] This can be formed by attack of OH$^\bullet$ upon DNA followed by one-electron oxidation of the resulting radical (Chapter 4)

$$\text{guanine} \atop \text{in DNA} \xrightarrow{\text{OH}^\bullet} [\text{8-OHdG}]^\bullet \xrightarrow[\text{(step B)}]{-1e^-} \text{8-OHdG}$$

However, the intermediate [8-OHdG]$^\bullet$ radical can have alternative fates, generating such products as FAPy-guanine (Chapter 4). The relative amounts of these two products (8-OHdG and FAPy-guanine) are affected by reaction conditions. Hydroxylated guanine can also result from attack of $^1O_2$, and possibly of $RO_2^\bullet/RO^\bullet$ radicals, upon DNA. Hence measuring only a single product gives limited information and the amount of hydroxylated guanine in DNA is not necessarily proportional to the amount of initial radical attack (i.e. changes in 8-OHdG levels could be due to alterations in the cell that favour or disfavour the second step in the above equation).

The distinctive feature of OH$^\bullet$ attack upon DNA is that it produces *multiple* products from all four DNA bases, whereas $^1O_2$ appears selective for attack upon guanine (producing, for example, 8-OHdG and FAPy-guanine). Several RNS can deaminate cytosine to uracil, guanine to xanthine and adenine to hypoxanthine, and exposure to ONOO$^-$ causes formation of 8-nitroguanine in DNA whereas HOCl generates chlorocytosine (Chapter 4). 8-Nitroguanine rapidly depurinates, forming an apurinic site in the DNA.

5.11.3   *DNA damage* in vivo

Oxidative DNA damage appears to occur continuously *in vivo*, in that low levels (presumably a 'steady state' balance between oxidative DNA damage and repair of that damage) of 8-OHdG and other oxidized bases have been detected in DNA isolated from the cells of every aerobe examined. The presence of multiple oxidized purine and pyrimidine bases suggests that OH$^\bullet$ formation occurs within the nucleus *in vivo*. However, if OH$^\bullet$ is attacking DNA, it must be made very close to the DNA since OH$^\bullet$ cannot diffuse from its site of formation. Background radiation may be one source. Other potential sources of OH$^\bullet$ include the reaction of $O_2^{\bullet-}$ with HOCl, but HOCl may preferentially chlorinate DNA bases. Fenton chemistry has also been considered. Although iron and copper can be detected in the nucleus, they are probably protein-bound and it remains to be established why and how 'catalytic' iron and copper ions reach the DNA. This has been discussed in detail in Sections 4.5 and 4.6.

Low levels of uracil, xanthine and hypoxanthine are also detected in DNA from aerobes: these may well arise by physiologically occurring deamination reactions.

5.11.4 *Measurement of oxidative DNA damage: basic principles*

There are two types of measurement of oxidative DNA damage. **Steady–state damage**, that found in DNA isolated from aerobic cells, presumably reflects the balance between damage and the activity of DNA repair enzymes. Hence a rise in steady-state oxidative DNA damage (as has been reported in some human cancerous tumours[46] and often occurs during oxidative stress) could be due to increased damage and/or decreased repair. Also, measurement of basal levels of modified DNA bases does not provide information as to whether this damage is in active genes or untranscribed DNA.

Attempts have also been made to estimate the total extent of oxidative DNA damage *in vivo*, by measuring products of DNA repair that are excreted in urine. Several base damage products are excreted in mammalian urine, including 8-OH-G, 8-OHdG, thymine glycol, 5-hydroxymethyluracil, 8-hydroxyadenine and 7-methyl-8-hydroxyguanine. However, the one most exploited is 8-OHdG, usually measured by a method involving HPLC with electrochemical detection after concentration of the urine on columns containing antibodies that recognize 8-OHdG. For example, in a study of 169 adults in Denmark,[39] the average 8-OHdG excretion was 200–300 pmol/kg per 24 h, which would correspond to the not inconsiderable average of 140–200 oxidative modifications to the guanine in the DNA of each cell of the human body every day. Furthermore, 32 smokers in this study excreted, on average, 50% more 8-OHdG than 53 non-smokers, suggestive of a 50% increased rate of oxidative DNA damage from smoking. 8-Hydroxyguanosine (hydroxyguanine attached to a ribose sugar) has also been detected in mammalian urine and may result from oxidative damage to RNA and/or RNA precursors.

The validity of these urinary measurements of oxidative DNA damage must be considered. The level of 8-OHdG in urine seems unaffected by diet, since the polar sugar molecule in nucleosides should deter absorption from the gut. The question of whether any 8-OHdG is metabolized to other products in humans has not been rigorously addressed, although 8-OHdG injected into pigs was completely excreted into urine. Additionally, it is possible that some or all of the 8-OHdG excreted in urine may arise not from DNA, but from sanitization of the dGTP pool (Section 4.8.1). Base excision repair pathways would not release 8-OHdG, but only 8-OH-G, from DNA (Chapter 4). However, an enzyme activity that does remove the nucleoside from DNA has been detected in human cells. Levels of 8-OH-G in urine are very much affected by diet: the oxidized base can be absorbed (e.g. from foods cooked at high temperature) and re-excreted. 8-OH-G can also arise from RNA oxidation.

5.11.5 *Measurement of guanine damage products in isolated DNA*

8-OH-G and 8-OHdG are the products most frequently measured as indicators of oxidative DNA damage. This is sensible, as these products arise when several different ROS attack DNA. It should be noted, however that addition of HOCl or ONOO$^-$ to DNA can destroy 8-OHdG, i.e. its levels are not a quantitative estimate of oxidative DNA damage.

Analysis of 8-OHdG using HPLC coupled to electrochemical detection (ECD) is a highly sensitive technique that is frequently used.[9,15] It is first necessary to release 8-OHdG from DNA, usually by enzymic hydrolysis. Gas chromatography–mass spectrometry (GC–MS) with selected ion monitoring (SIM) has also been used to characterize oxidative DNA base damage by the identification of a spectrum of products, including 8-OH-G (Table 5.7). For GC–MS, DNA is usually hydrolysed by heating in formic acid (which not only cleaves the DNA backbone but also hydrolyses the base–sugar bond, so producing 8-OH-G and not 8-OHdG). The DNA bases must then be converted to volatile derivatives by derivatization (often by trimethylsilylation). When GC–MS is used to measure modified DNA bases, another factor to be considered is that 8-OH-G could arise by hydrolysis of not only 8-OHdG in DNA, but of 8-hydroxyguanosine in oxidatively damaged RNA, so the DNA must be checked for RNA contamination.

One advantage of the GC–MS approach is that measurement of a wide range of base damage products allows accurate quantification of total DNA damage and can help to identify the ROS/RNS species that caused the damage (e.g. $^1O_2$ selectively attacks guanine whereas OH$^\bullet$ attacks all four DNA bases). It can also identify DNA base–protein cross-links, such as thymine–lysine. However, the levels of 8-OHdG measured in DNA by HPLC/ECD are often (but not always) less than the levels of 8-OH-G measured by GC–MS/SIM (Table 5.7). Why? Does HPLC underestimate damage, GC–MS overestimate it, or both? HPLC could underestimate the amount of 8-OHdG in DNA if the enzymic hydrolysis were incomplete; the action of the nuclease enzymes used to hydrolyse the DNA may be diminished by oxidative base damage, and

Table 5.7. Levels of hydroxylated guanine measured in DNA by different methods

Method	8-OHdG/10^6 DNA bases
HPLC analysis of 8-OHdG after enzymic hydrolysis of DNA	
Rat liver DNA (total)	3–20
Rat liver DNA (mitochondrial)	~33
Commercial calf thymus DNA	8–320
GC–MS after acid hydrolysis of DNA	
DNA or chromatin isolated from human cells	35–40
Commercial calf thymus DNA	159–318
Mouse liver mitochondrial DNA (HPLC–MS analysis after enzymic hydrolysis of DNA)	~3500

Data abstracted from *Free Rad. Res. Commun.* **16**, 75 (1992). Note that levels of 8-OHdG measured by HPLC tend to be lower than levels of 8-OHG measured by GC–MS (acid hydrolysis cleaves the base–sugar bond and so the base rather than the nucleoside is measured). Comparisons between results from different laboratories may be aided by the following conversion factors (by courtesy of Dr Miral Dizdaroglu). x 8-OHdG/10^5 guanosines is equal to $(x/4.65)$ 8-OHdG per 10^5 DNA bases (since about 21.5% of DNA bases are guanine in mammalian DNA) or $(x/33.6)$ nmol 8-OHdG/mg of DNA. Similarly, 1 nmol of 8-OHdG per mg of DNA is 156 8-OHdG/10^5 guanosines or 33.6/10^5 DNA bases.

the acid pH often used for nuclease digestions could cause hydrolysis of 8-OHdG to 8-OH-G, resulting in the loss of HPLC-detectable material. By contrast, GC–MS might overestimate 8-OH-G (and perhaps some other base damage products) as a result of their artifactual formation during the heating step involved in classical silylation-based derivatization procedures. The important factor is that the heating stages should be done anoxically: heating DNA bases in the presence of O_2 is bound to oxidize them! Hence some of the claimed artifacts are due to failure to remove O_2. However, it is difficult to remove O_2 *completely*. Pre-purification of the hydrolysate to remove undamaged bases before derivatization (so that they cannot be artifactually oxidized) is one solution.[12a] Low-temperature derivatization in the presence of antioxidants is another.[24a] Possible problems with derivatization can also be avoided using HPLC to measure oxidized base products in acid-hydrolysed DNA, or to use HPLC–MS and avoid derivatization altogether. Liquid chromatography–mass spectrometry techniques are under development in several laboratories.

The **comet assay**[9] (single-cell gel electrophoresis; Chapter 4) has been modified to allow an assessment of oxidative base damage. After lysis of the cells, the slides are exposed to DNA-repair enzymes (e.g. endonuclease III for oxidized pyrimidines and FAPy glycosylase for 8-OHdG and other guanine products). DNA is cleaved at sites of base modification, producing more strand breaks. Hence a comparison of the comets before and after enzymic treatment can estimate the oxidized bases present. This approach can also be combined with the other methods for measuring DNA strand breaks (Chapter 4). Yet other approaches include ^{32}P-post-labelling of nucleotides[12] in an enzymic DNA digest followed by radiometric detection after chromatographic separation, and the use of antibodies directed against 8-OHdG, or other lesions. Antibody staining for 8-OHdG can be applied to whole DNA, or antibodies can be attached to columns and used to purify 8-OHdG from digests of DNA or from urine (see above).

There is an urgent need to compare and standardize these various methods for assessing oxidative DNA damage. Thus data for 8-OHdG levels in rat tissues from ^{32}P-post-labelling are closer to values for 8-OH-G obtained by GC–MS. By contrast, results from the modified comet assay are often lower than those obtained with HPLC–ECD.

5.11.6   *DNA isolation problems*

A key problem to be considered for all these techniques is the possibility that DNA is oxidatively damaged during its isolation from cells and tissues. Classical phenol-based methods of DNA isolation are particularly suspect, since oxidizing phenols generate ROS (Chapter 8). It has been shown that phenol-based DNA purification procedures are capable of increasing 8-OHdG levels 20-fold in DNA samples exposed to air following removal of the phenol.[14] However, rigorous control of isolation procedures and avoidance of phenol in many laboratories (e.g. by studying isolated chromatin or by using different DNA isolation methods) does not abolish oxidative damage detected in isolated DNA, supporting the view that there is a low steady-state DNA damage *in vivo.* Indeed the presence of a DNA-repair enzyme system and the excretion of base damage products are observations that support the occurrence of oxidative DNA damage *in vivo.*

5.11.7 *DNA–aldehyde adducts*

Cytotoxic aldehydes are produced during lipid peroxidation and are also widespread in the environment. For example, cigarette smoke contains multiple aldehydes, including the toxic unsaturated aldehyde **acrolein**, and bananas contain hexenal. If they reach DNA, aldehydes can bind to DNA bases (Section 4.10). Several assays to measure such adducts have been described,[3] often employing GC–MS or HPLC–ECD after acid hydrolysis of DNA, or immunoassays after pre-separation of DNA hydrolysates by HPLC. For example, adducts of MDA with guanine have been detected at low levels in DNA from healthy animals, e.g. DNA from healthy human liver contained about 5400 adducts per cell ($\sim$9 adducts per 10^7 bases).[7] Liver DNA from rats treated with carbon tetrachloride to induce lipid peroxidation (Chapter 8) contained $\sim$4 adducts/10^7 bases compared with $\sim$2/10^7 bases in control animals. Deoxyguanosine–MDA adducts have also been identified in rat and human urine. Reaction of 4-hydroxynonenal (HNE) with DNA generates **N^2-ethenodeoxyguanosine**.[3]

5.12 Fingerprinting methods: lipid peroxidation

Lipid peroxidation is important *in vivo* for several reasons, especially its contribution to the development of atherosclerosis (Chapter 9). Hence a common test of the effectiveness of dietary antioxidants is to measure their effects on the 'peroxidizability' of low-density lipoproteins (LDLs) isolated from blood plasma and subjected to a pro-oxidant challenge *in vitro.* For example, Esterbauer *et al.* showed that dietary supplementation of humans with vitamin E increased the 'lag period' before peroxidation accelerated when LDL isolated from blood plasma was subsequently incubated with copper ions *in vitro* (Chapter 9).

5.12.1 *Measurement of lipid peroxidation; general principles*

Lipid peroxidation is a complex process and occurs in multiple stages. Hence many techniques are available for measuring the rate of peroxidation of membrane lipids, lipoproteins or fatty acids. Each technique measures something different, and no one method by itself can be said to be an accurate overall measure of lipid peroxidation. Table 5.8 summarizes various methods and we comment on some of them in more detail below.

5.12.2 *Loss of substrates*

Lipid peroxidation causes loss of unsaturated fatty-acid side-chains, so a simple way (in principle) to measure the overall rate of peroxidation is to measure the loss of each fatty acid (Table 5.9 shows an example). The system under study

Table 5.8. Some of the methods used to detect and measure lipid peroxidation in biological material[a]

Method	What is measured	Remarks
A. Loss of substrates		
Analysis of fatty acids by GLC or HPLC	Loss of unsaturated fatty acids.	Useful for assessing lipid peroxidation stimulated by different pro-oxidants that give different product distributions.
Oxygen electrode	Uptake of O_2 by carbon–centered radicals and during peroxide–decomposition reactions	Dissolved O_2 concentration is measured; useful when spectrophotometric interference occurs or toxic chemicals interfere with enzymic techniques; not very sensitive.
B. Peroxide assays: simple total peroxide measurements		
Iodine liberation	Lipid peroxides	One of the oldest methods, widely used in the food industry; lipid peroxides oxidize I^- to I_2. $ROOH + 2I^- + 2H^+ \rightarrow I_2 + ROH + H_2O$ In the presence of excess I^- the tri-iodide ion (I_3^-) can be measured at 358 nm. Useful for bulk lipids. H_2O_2 and protein peroxides also oxidize I^- to I_2. Method can be applied to extracts of biological samples if other oxidizing agents are absent. Levels in human plasma reported as 2.1–4.6 μM (*Anal. Biochem.* **176**, 360 (1989); *Anal. Biochem.* **176**, 353 (1989)).

| FOX (ferrous oxidation xylenol orange) assay | Absorbance change | Simple, easy to use, works well *in vitro*, e.g. for LDL peroxidation. Peroxides oxidize Fe^{2+} to $Fe(III)$, detected by xylenol orange (ΔA at 560 nm). Detects 3–4 μM 'lipid peroxide' in normal human plasma, much greater than HPLC-based assays perhaps due to 'amplification' in the assay since reaction of LOOH with Fe^{2+} will generate $LO^\bullet$ radicals that could propagate peroxidation. Antioxidants often added to try and stop this. Sensitivity low μM range (*Anal. Biochem.* **220**, 403 (1994)). Also detects H_2O_2 and protein peroxides. Addition of **triphenylphosphine** reduces lipid peroxides but not H_2O_2. |
| Glutathione peroxidase (GPX) | Fatty acid peroxides (GPX does not act on peroxidized fatty acids within membrane or LDL lipids) | GPX reacts with H_2O_2 and organic peroxides, oxidizing GSH to GSSG. Addition of glutathione reductase and NADPH to reduce GSSG back to GSH results in stoichiometric consumption of NADPH. Alternatively, GSSG can be determined directly, e.g. by HPLC (e.g. *Chem. Res. Toxicol.* **2**, 295 (1989); *Anal. Biochem.* **186**, 108 (1990)). Cannot measure peroxides within membranes unless phospholipases are first used. In principle, phospholipid hydroperoxide GP could also be employed. Peroxide levels in plasma quoted as approx. 1 μM (*Chem. Res. Toxicol.* **2**, 295 (1989)). |

Table 5.8. (*Continued*)

Method	What is measured	Remarks
Cyclooxygenase (COX)	Lipid peroxides (rate of COX oxidation of arachidonic acid stimulated by traces of lipid hydroperoxides)	Stimulation of COX activity (usually assayed as O_2 uptake) can be used to measure trace amounts of peroxide in biological fluids (*Anal. Biochem.* **145**, 192 (1985); **193**, 55 (1991)). Assay relates the presence of peroxides to one of their potential biological actions, i.e. stimulation of eicosanoid synthesis. Human plasma levels measured are ~0.5 μM. Cannot be used to identify individual peroxides and value for 'total peroxide' will depend to some extent upon what species are present (since different peroxides stimulate COX to different extents).
C. Peroxide assays: separation of products		
Haem degradation of peroxides (after HPLC separation)	Lipid peroxides (HPLC allows separation of phospholipid, cholesterol ester etc. peroxides)	Haem and haem proteins decompose lipid peroxides with formation of radicals that react with isoluminol to produce light. For example, **microperoxidase**, a haem-peptide produced by proteolytic degradation of cytochrome c, is often used. HPLC method measures ~40 nM levels of peroxides in human blood plasma (e.g. *Anal. Biochem.* **175**, 120 (1988); *Methods Enzymol.* **233**, 319 and 324 (1994)). Electrochemical or redox-dye detections of peroxides have also been described (e.g. *Free Rad. Biol. Med.* **20**, 365 (1996)). Chemical identity of peroxides after HPLC separation must be confirmed chemically and not based simply on retention time. **Diode array**

Method	Detects	Comments
GC–mass spectrometry	Lipid peroxides (also aldehydes/isoprostanes, cholesterol/cholesterol ester peroxides)	**detection**, a technique which records the absorbance spectrum of each peak, is a useful validation method. Another means of detecting peroxides after HPLC is to react them with derivatives of **diphenylphosphine**, oxidized to fluorescent **phosphine oxides** (*Anal. Chim. Acta* **307**, 97 (1995)). Peroxides are extracted, usually reduced (e.g. by borohydride) to alcohols, separated by GC, and identified by mass spectrometry. Several methodological variations exist (e.g. *Methods Enzymol.* **233**, 332 (1994); *Anal. Biochem.* **198**, 104 (1991)). Controls are needed to show that the alcohols were not present in the system *before* reduction, e.g. alcohols are generated by glutathione peroxidase activity.
D. Miscellaneous methods		
Spin trapping	Intermediate radicals	Spin traps (e.g. PBN, POBN, DMPO) intercept radicals intermediate in the chain reaction. Have been used in whole animals to detect carbon-centred as well as $RO^{\bullet}$ and $RO_2^{\bullet}$ radicals. Also important in mechanistic studies (e.g. *Proc. Natl Acad. Sci. USA* **78**, 7346 (1981); *J. Biol. Chem.* **267**, 5743 (1992)). Section 5.12.7
Light emission	Excited carbonyls, singlet oxygen, many other light-emitting systems	Section 5.12.4
Diene conjugation	—	

Table 5.8. (*Continued*)

Method	What is measured	Remarks
E. Measurement of end-products		
TBA test	TBA-reactive material (TBARS)	Section 5.12.10
GC–HPLC/antibody techniques	Cytotoxic aldehydes	Section 5.12.13 and Table 5.12
Fluorescence	Aldehydes	Section 5.12.8
Hydrocarbon gases	pentane and ethane	Section 5.12.6. Potentially a non-invasive measure of peroxidation *in vivo*. Results in practice have been variable; some authors have found that the technique works well and other have abandoned it. Many GC columns do not separate pentane from isoprene. Ethane can result from free-radical attack on certain amino acids (*Biochem. Pharmacol.* **39**, 1347 (1990)).
F$_2$-Isoprostanes	fatty acid peroxides (GC/negative ion chemical ionization mass spectrometry)	Peroxidation of PUFAs produces a complex mixture of prostaglandin isomers. Found at low levels (both free and esterified to lipids) in human and other animal tissues and body fluids (e.g. ~30–40 pg/ml in fresh human plasma; ~2 ng/mg creatinine in human urine) (Section 4.9.13). Also see Section 5.12.12.

[a] No single method is adequate by itself as an accurate index of the whole process of lipid peroxidation.

Table 5.9. Loss of fatty acids during peroxidation of the red blood cell membrane

No. of carbon atoms in fatty acid	No. of $>C=C<$ bonds in fatty acid	% of total fatty acids in membrane	
		Normal	After lipid peroxidation
16	0	21	21
18	0	14	14
18	1	12	11
18	2	10	8
20	4	15	5
22	0	3	3
22	4	2	1

Erythrocytes were stressed by incubation with high levels of H_2O_2 in the presence of azide to inhibit catalase.

must be disrupted (e.g. lipids extracted from cells or lipoproteins) and the lipids hydrolysed to release the fatty acids, which can then be measured by HPLC or converted into volatile products (e.g. by formation of esters with methanol) and separated by GLC. Great care must be taken to avoid peroxidation of fatty acids during the hydrolysis and extraction procedures: a common practice is to carry out the reactions under nitrogen gas. Additional information can be gained by separating the different classes of lipids before hydrolysis to release the fatty acids.

The other substrate for peroxidation is O_2. Hence measurement of the rate of O_2 uptake in an oxygen electrode is a useful overall index of the progress of peroxidation. Indeed, de Saussure used this method in his pioneering studies on oxidation of walnut oil (Chapter 4).

Most modern O_2 electrodes are Clark-type arrangements which measure dissolved O_2, unlike de Saussure's simple mercury manometer. Calibration of O_2 electrodes is achieved by exposing them to known concentrations of dissolved O_2, although it should be appreciated that, strictly speaking, the electrode measures the 'activity' of O_2 and not concentration. It is usually assumed that air-saturated pure water has $0.258\,\mu mol/ml$ of dissolved oxygen at $25\,°C$ and 1 atmosphere air pressure, although the presence of dissolved solutes decreases this solubility slightly (Chapter 1).

5.12.3 *Measurement of peroxides*

Several methods exist for direct peroxide measurement: they can be classified into those that measure 'total peroxides' (e.g. iodine liberation, the ferrous oxidation xylenol orange (FOX) assay, cyclooxygenase, glutathione peroxidase) and those that attempt separation of the different classes first (Table 5.8). One should also think about the significance of what is measured: the amount of peroxide present at a given time during lipid peroxidation will depend not only on the rate of initiation of peroxidation but also on how quickly peroxides are decomposing to give other products.

5.12.4   Diene conjugation

The oxidation of polyunsaturated fatty acids (PUFAs) is accompanied by the formation of conjugated diene structures (Chapter 4), with a double–single–double bond arrangement. They absorb ultraviolet (UV) light in the wavelength range 230–235 nm (Figs 5.16 and 5.17). Measurement of this UV absorbance is useful as an index of peroxidation in studies upon pure lipids and isolated lipoproteins (Chapter 9) and has the advantage that it measures an early stage in the peroxidation process. Diene conjugation measurements often cannot be carried out directly on tissues and body fluids because many other substances present, such as haem proteins, chlorophylls, purines and pyrimidines, absorb strongly in the UV. The breakdown of lipid peroxides also produces several UV-absorbing carbonyl compounds. Extraction of lipids into organic solvents (e.g. chloroform : methanol mixtures) before analysis is a common approach to this problem.

Greater sensitivity for the diene conjugation method can be achieved by using HPLC separation of different diene conjugates or by applying second-derivative spectroscopy (Figs 5.16 and 5.17). An absorbance spectrum plots absorbance (A) against wavelength (λ). A first-derivative spectrum plots rate of change of absorbance with wavelength ($dA/d\lambda$) against wavelength. The second derivative spectrum plots the rate of change of this rate of change ($d^2A/d^2\lambda$). Figure 5.16 shows the appearance of the first and second derivatives of an absorption band. Figure 5.17 applies this technique to linolenic acid; the changes in the second-derivative spectrum upon oxidation of the fatty acid are much clearer than those in the simple UV-absorption spectrum. Thus the 'hump' that appears in the UV-absorption spectrum as a result of peroxidation translates into a sharp minimum 'peak' at 233 nm in the second-derivative spectrum. The increased resolution of this technique can allow discrimination between different conjugated diene structures present.

5.12.5   Interpretation of conjugated diene assays

Although conjugated diene measurements are often used to study peroxidation in pure lipids, lipoproteins and membrane fractions, their application to animal body fluids has produced serious problems.[2,65] The use of HPLC to separate the UV-absorbing 'diene conjugate' material from human body fluids revealed that most or all of it consists of a non-oxygen-containing isomer of linoleic acid, **octadeca-9(*cis*),11(*trans*)-dienoic acid**. The origin of this product is uncertain. One suggestion is that it is produced by reaction of the carbon-centred radicals, obtained when H is abstracted from linoleic acid, with protein. However, reaction with O_2 is the preferred fate of carbon-centred radicals, except at very low O_2 concentrations. In addition, peroxidation of biological

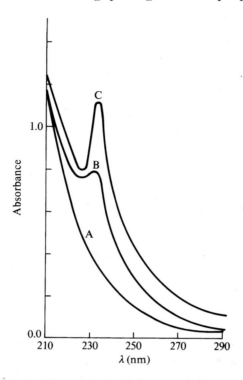

Fig. 5.15. Lipid peroxidation followed by the formation of conjugated dienes. Ethyl linoleate (an ester of linoleic acid with the alcohol ethanol) was purified, and its absorbance spectrum is shown (plot A). Plot B shows the sample after oxidation by air at 30 °C for 8 h; and plot C, a sample in which peroxidation has been speeded up by addition of nitrogen dioxide. In both cases the 'shoulder' of UV-absorbance due to conjugated diene formation is clearly visible. Data by courtesy of Professor Bill Pryor.

lipids should produce carbon-centred radicals from several fatty acids, not only linoleic acid, and would not be expected to give uniquely a 9(*cis*),11(*trans*) isomer. This UV-absorbing product is not found in the plasma of animals subjected to oxidative stress (e.g. rats given bromotrichloromethane, a potent inducer of lipid peroxidation; Chapter 8). Thus octadeca-9,11-dienoic acid may not arise by lipid peroxidation. It can be produced by bacterial metabolism. The material detected in human plasma may be largely or entirely absorbed from food or produced by the metabolism of bacteria, such as gut bacteria and those present in lung and cervical mucus. Other animals can also absorb this product.

It follows that application of simple diene conjugation methods to animal body fluids, or to extracts of them, is a questionable index of lipid peroxidation. Of course, diene conjugation methods are still very useful for isolated lipid/ lipoprotein fractions, such as LDLs. HPLC or GC–MS can be used to separate octadeca-9,11-dienoic acid from what appear to be 'real' conjugated diene products of lipid peroxidation.

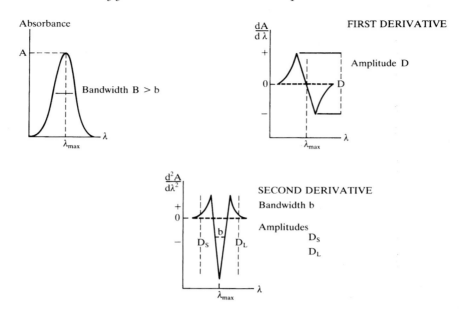

Fig. 5.16. Principal features of the first and second derivatives of an idealized absorption band. Diagram by courtesy of Philips Analytical.

5.12.6   *Measurement of hydrocarbon gases*[33]

Most assays of lipid peroxidation products in tissues measure the 'steady-state', i.e. the net balance between peroxidation and removal of peroxidation products. The hydrocarbon gas exhalation technique is an attempt to assess the overall rate of peroxidation *in vivo*.

It is based on the formation of hydrocarbon gases such as pentane and ethane during lipid peroxidation (Chapter 4). Ethane is derived from (*n*-3) PUFAs and pentane from (*n*-6) PUFAs. The latter PUFAs predominate in the human body, suggesting that pentane would be produced in greater amounts if all PUFAs peroxidize equally. Both gases are easily measured by gas chromatography. The expired breath is passed through an adsorbent at low temperature to adsorb and concentrate the hydrocarbons, which are then desorbed and assayed.

It must be emphasized that hydrocarbons are minor end-products of peroxidation, and their formation depends on the presence of transition-metal ions to decompose peroxides. Hence an increased rate of gas production might reflect increased availability of such metal ions rather than increased initiation of peroxidation. It has also been reported that, at least in isolated microsomal systems, the formation of hydrocarbons is affected by O_2 concentrations, being favoured at low O_2 concentrations. This may be because the carbon–centred radicals that lead to hydrocarbon production (e.g. the pentane radical, $CH_3(CH_2)_3\dot{C}H_2$) can also react with O_2 to form peroxyl radicals.

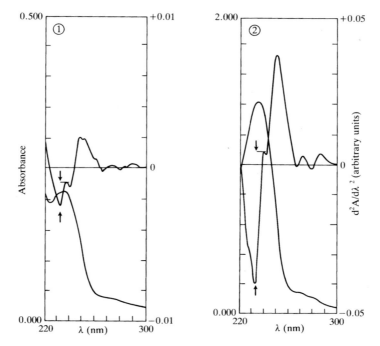

Fig. 5.17. Second-derivative spectroscopy of conjugated dienes. 1, UV and second-derivative spectra of linolenic acid (allegedly pure but in fact slightly oxidized). In the second-derivative spectra the arrows show the height of the minimum peak at $\lambda = 233$ nm that is expressed as $d^2 2A/d\lambda^2$, in arbitrary units. 2, UV and second-derivative spectra of linolenic acid after 24 h of oxidation. The arrows show the height of the peak $d^2A/d\lambda^2$ at $\lambda = 233$ nm. Data by courtesy of Professor F. Corongiu (also see ref[2]).

In addition, some hydrocarbons are metabolized in the liver, e.g. pentane is hydroxylated by certain cytochromes P450 to pentanol.[64] This can introduce artifacts, e.g. drugs that alter liver metabolism might affect pentane consumption by liver P450, and so might mistakenly be thought to be altering the rate of lipid peroxidation *in vivo*. A more serious problem is that many GC columns fail to separate pentane from **isoprene** (2-methyl-1,3-butadiene), a hydrocarbon present in large amounts in human breath that is apparently a side-product of cholesterol biosynthesis. Thus several reported 'breath pentane' levels are artifactually high and recent reports suggest that healthy humans exhale little, if any, pentane.[45]

Given the problem of pentane metabolism, perhaps ethane should receive more attention. The greatest care must be taken in experiments of this kind to control for hydrocarbon production from the bacteria always present on the skin (and fur) and in the gut (it is quite possible to ignite the flatulence from cows because of the rich variety of hydrocarbons present). 'Normal' air in large cities is contaminated with hydrocarbons from combustion processes, e.g. in motor vehicles and environmental tobacco smoke. Hydrocarbons probably partition into body fat stores and must first be flushed out by breathing hydrocarbon-free air before reliable measurements can be made.

5.12.7 *Light emission*[67]

Several groups have used 'low-level chemiluminescence' as an index of oxidative stress (Section 5.10 above). Stimulation of peroxidation in isolated membrane systems or in perfused animal organs (e.g. by infusing *tert*-butyl hydroperoxide) is usually accompanied by increased light emission. Hence one source of light *in vivo* is lipid peroxidation, although basal light emission is by no means only a measure of lipid peroxidation (Section 5.10).

Some of the light produced in peroxidizing lipid systems arises when two peroxyl radicals meet, by the Russell mechanism (Chapter 2)

$$\text{>CHO}_2^{\bullet} + \text{>CHO}_2^{\bullet} \longrightarrow \text{>C=O} + \text{>C-OH} + {}^1O_2$$

carbonyl

Alternatively, the carbonyl compound produced might be in the excited state and emit light as it decays to the ground state. Other possible sources of light are self-reaction of alkoxyl radicals to give excited-state ketones:

$$\text{>CHO}^{\bullet} + \text{>CHO}^{\bullet} \longrightarrow \text{>C=O}^* + \text{>C-OH}$$

and the formation of a dioxetane by reaction of singlet O_2 with unsaturated fatty-acid side-chains (Chapter 2): dioxetanes can decompose to give excited-state carbonyl compounds:

$$\text{>C=C<} + {}^1O_2 \longrightarrow \begin{array}{c} -\overset{|}{C}-\overset{|}{C}- \\ | \quad | \\ O-O \end{array} \longrightarrow \text{>C=O} + \text{>C=O}^*$$

Measurements of light emission in isolated cells or membrane fractions (such as microsomes) are reasonably well correlated with results from other techniques of measuring lipid peroxidation, except that the peak light emission often occurs slightly later than the 'peak' recorded by other techniques. This may be because maximum light production by the Russell-type mechanism will tend to occur during the later stages of lipid peroxidation, when sufficient peroxyl radicals are present in the membrane to increase the chance of collision between them.

5.12.8   Measurement of fluorescence[13,32]

Fluorescent pigments, partly derived from lipids, are known to accumulate in some tissues as a function of age, as the **age pigments** (Chapter 10).

The reaction of carbonyl compounds, such as malondialdehyde (MDA), with side-chain amino groups of proteins, amino acids, phosphatidylethanolamine, phosphatidylserine or nucleic-acid bases, can generate **Schiff bases**.

MDA, with two carbonyl groups, can cross-link two amino compounds to produce fluorescent molecules that have the general formula $R_1-N=CH-CH=CH-NH-R_2$, where R_1 and R_2 are the compounds to which the amino groups are attached. The products are called **aminoiminopropene Schiff bases** (Fig. 5.18) and they form readily at acidic pH values. However, their significance in the formation of fluorescent products in peroxidizing membranes is unclear, since several reports suggest that MDA is only a minor contributor to the formation of fluorescent products, and MDA is poorly reactive at pH 7.4 (Section 4.9.10). Other aldehydes, such as the unsaturated mono-aldehyde HNE, may be more important contributors to fluorescence development.

The formation of Schiff bases as a source of fluorescence accompanying lipid peroxidation at neutral pH values has been questioned; **dihydropyridines** may be the major fluorescent products formed. For example, in the case of MDA, a possible product is a 1,4-disubstituted 1,4-dihydropyridine-3,5-dicarbaldehyde.

4-methyl 1-substituted 1,4 dihydropyridine
3,5 dicarbaldehyde

If both MDA and an aldehyde with only one carbonyl group are present, the reaction may be written:

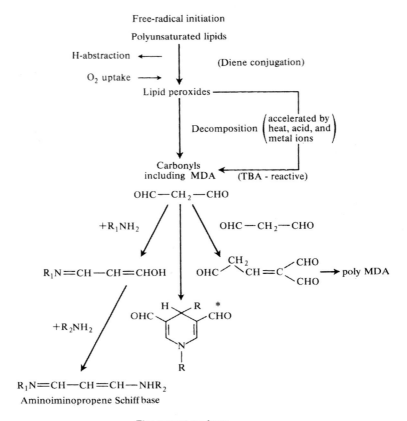

Fig. 5.18. Formation of fluorescent products during lipid peroxidation. The starred product (★) is a 1,4-disubstituted 1,4-dihydropyridine-3,5-dicarbaldehyde.

A common property of aldehydes is their polymerization—several molecules joining together to form a larger molecule. This can happen with MDA (Chapter 4), and the polymeric products are fluorescent (Fig. 5.18). Auto-oxidizing PUFAs develop fluorescence even though no compounds with $-NH_2$ groups are present, which may be due to polymers from MDA and other aldehydes.

Thus, as summarized in Fig. 5.18, mechanisms for the formation of fluorescent products are complex. Nevertheless, fluorescence is a sensitive method for measuring peroxidation and the results obtained correlate reasonably well with those of other methods, although fluorescence is, of course, a measurement of a late stage in the peroxidation process. The results of fluorescence studies are usually expressed in relative fluorescence intensity (RFI) units by comparison with a stable standard solution, such as quinine sulphate in dilute sulphuric acid, or tetraphenylbutadiene. Of course, problems can occur when applying the technique to biological systems; emitted light can be absorbed by other biomolecules and fluorescence can develop in proteins and DNA subjected to oxidative attack.

5.12.9   *Parinaric acid*

cis-**Parinaric acid** (*cis, trans, trans, cis*-9,11,13,15-octadecatetraenoic acid) is a fluorescent PUFA used by biochemists as a 'probe' to study lipid–protein interactions within membranes. When incorporated into lipids undergoing peroxidation, it is rapidly oxidized, accompanied by loss of its characteristic fluorescence (emission 413 nm when excited at 324 nm). Some groups have used parinaric acid for continuous monitoring of lipid peroxidation in liposomes, cell membranes, lipoproteins, microsomes and sub-mitochondrial particles, since parinaric acid incorporates readily into lipids.[35]

5.12.10   *The thiobarbituric acid test*

The thiobarbituric acid (TBA) test is one of the oldest and most frequently used tests for measuring the peroxidation of fatty acids, membranes and foods.[19,22] It is also one of the most criticized methods, although almost every scientist working in this field is an overt (or closet!) TBA user. One great advantage of the TBA test is its ease of use (the material under test is simply heated with TBA under acidic conditions, and the formation of a pink colour measured at or close to 532 nm), and it works on biological material. Unfortunately, the simplicity of performing the TBA test belies its chemical complexity.[19]

Small amounts of 'free' MDA are formed during the peroxidation of most membrane systems, especially microsomes (Chapter 4). MDA reacts in the TBA test to generate a coloured product, as shown below.

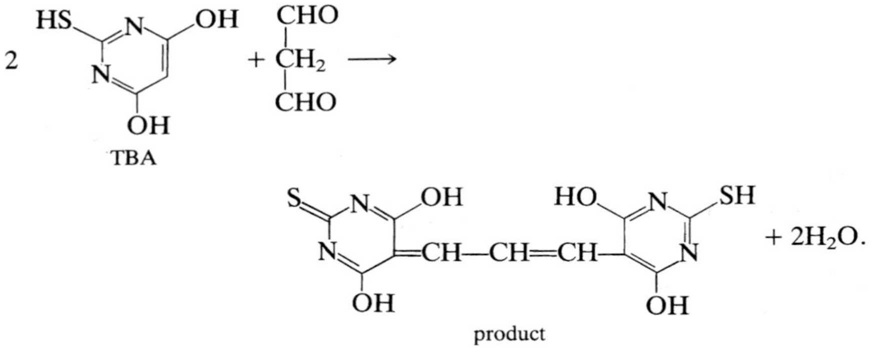

In acid solution the product absorbs light at 532 nm and fluoresces at 553 nm, and it is readily extractable into organic solvents such as butan-1-ol. Construction of a calibration curve for the assay is complicated by the fact that MDA is unstable and must therefore be prepared immediately before use by hydrolysing its derivatives,

1,1,3,3-tetramethoxypropane or **1,1,3,3-tetraethoxypropane**:

$$\underset{RO}{\overset{RO}{\diagup}} CH\!-\!CH_2\!-\!CH \underset{OR}{\overset{OR}{\diagdown}}$$

R = C_2H_5 (ethyl)
or CH_3 (methyl).

In studying the biological properties of MDA formed in this way, it is essential to ensure complete hydrolysis of these derivatives and to bear in mind that the solution will contain four molecules of ethanol (or methanol) per molecule of MDA. For example, some early reports that MDA is powerfully genotoxic and carcinogenic may have originated from the biological activities of partially hydrolysed derivatives of the above compounds. MDA does bind to DNA, however, and is probably weakly mutagenic (Chapter 4).

Problems in the TBA test: most TBARS detected is generated during the assay[19]

Because the TBA test is calibrated with MDA, the results are often expressed in terms of the amount of MDA produced in a given time. This can give the impression that the TBA test measures MDA in the peroxidizing lipid system, and the molar extinction coefficient of the MDA-TBA adduct $(1.54 \times 10^5 \, M^{-1} \, cm^{-1}$ at 532 nm) is often used to calculate the amount of 'MDA formed'.

However, the amount of free MDA produced in most peroxidizing lipid systems is low, insufficient to give a substantial colour yield. Indeed, it was shown as long ago as 1958, in studies with peroxidizing fish oil, that 98% per cent of the MDA that reacts in the TBA test was not present in the sample assayed but was formed by decomposition of lipid peroxides during the acid heating stage of the TBA assay. Peroxide decomposition generates $RO_2^\bullet$ radicals that can oxidize more lipid, i.e. the TBA test amplifies peroxidation and subsequent MDA formation. MDA production has been suggested to involve the formation of cyclic peroxides and endoperoxides that undergo fragmentation (Chapter 4). Peroxide breakdown is accelerated by the presence of iron salts in the reagents used in the TBA test, and removal of such iron salts decreases the TBA-reactivity detected. This can lead to artifacts in studies of the action of metal-chelating agents on lipid peroxidation: they will also affect colour development in the TBA test itself. Alternative approaches have been to add excess iron with the TBA reagents, or to add chain-breaking anti-oxidants (usually butylated hydroxytoluene, BHT) to suppress peroxidation during the test itself. Table 5.10 illustrates how different assay conditions can affect the results obtained. Thus the TBA-reactivity (or TBA reactive material, TBARS) content measured in a sample will differ according to the assay conditions used.

Problems in the TBA test: other chromogens

Several compounds other than MDA give products that absorb at, or close to, 532 nm on heating with TBA (Table 5.11). Measurement at 532 nm after a

Table 5.10. Variability of results in the HPLC-based TBA test depending on whether or not butylated hydroxytoluene (BHT) is added to the sample with the TBA reagents

Sample	TBA reactivity (μM MDA equivalents)	
	BHT not added	BHT added
Fresh human plasma		
healthy subjects	0.45 $\pm$ 0.35 (12)[a]	0.10 $\pm$ 0.10 (12)
hyperlipidaemic subjects[b]	0.88 $\pm$ 0.22 (17)	0.61 $\pm$ 0.25 (17)
Rat liver microsomes		
incubated with FeCl$_3$/ascorbate		
peroxidizing for 5 min	12.0 $\pm$ 0.8 (4)	5.8 $\pm$ 0.9 (4)
peroxidizing for 20 min[c]	14.7 $\pm$ 1.0 (4)	12.8 $\pm$ 1.0 (4)

[a] n values are given in parentheses.
[b] Elevation of TBARS persists even when expressed per unit cholesterol.
[c] By 20 min peroxidation was essentially complete: note how the absence of BHT appears to accelerate the rate of the peroxidation process as measured by TBARS. This is because of further peroxidation of the lipids during the assay, which is inhibited by BHT. When all the lipids have been oxidized the same TBARS values will result. Data from *Free Rad. Res. Commun.* **19**, 51 (1991).

TBA test could therefore include contributions from these substances, although fluorescence measurements can often distinguish the products they form from the 'real' TBA–MDA adduct. For example, TBA tests on plasma from jaundiced patients can be confounded by bile pigments.

An alternative approach is to separate the TBA–MDA adduct from the reaction mixture before measurement. This is often done by HPLC. Even so, it must be noted that exposure of several carbohydrates and amino acids to OH$^\bullet$, produced by ionizing radiation or metal-ion H$_2$O$_2$ systems, yields products that give a *genuine* TBA–MDA adduct on heating with TBA (Table 5.11). The exact amount of chromogen produced by these alternative compounds depends on the type and strength of acid used in the TBA test, and on the time of heating. Nevertheless, HPLC-based TBA tests are useful 'screening' methods for examining large numbers of biological samples for lipid peroxidation. However, it is difficult to use them to compare levels of peroxidation between tissues with a different fatty acid composition, since not all fatty acids generate MDA on oxidation (Chapter 4).

Application of the TBA assay to human body fluids will also measure MDA produced enzymically during eicosanoid synthesis (Chapter 6).[59] The lack of specificity of the simple TBA assay when applied to plasma is exemplified by the work of Lands *et al.*; by the cyclo-oxygenase method (Table 5.8) they measured a mean peroxide content in human plasma of around 0.5 μM, whereas expression of the results from a simple TBA test on the same samples in terms of 'peroxide equivalents' gave a mean value of 38 μM. Table 5.10 shows similar variabilities. MDA can also be determined directly by HPLC or GC–MS:[73] basal levels in human plasma were reported as 25–38 nM.

Table 5.11. TBA reactivity of various molecules before and after radical damage

Compound	Damaging system (if any)	TBA chromogen[a]	λ_{max}^{b} (nm)
Compounds forming chromogens on heating with TBA			
biliverdin		?	532, 560
acetaldehyde plus			532
sucrose	not required: direct	?	
β-formylpyruvic acid	reaction with TBA	?	510, 550
glyoxal		?	522, 550
gossypol		?	532
Compounds forming yellow chromogens (430–450 nm) on heating with TBA			
streptomycin		—	
aldehydes, e.g. hydroxymethyl-furfuraldehyde	not required: direct reaction with TBA	—	
some carbohydrates		—	
Lipids			
PUFAs	OH•, LO•, LO$_2^•$	MDA–TBA	532
Carbohydrates			
deoxyribose	OH•	MDA–TBA	532
deoxyglucose	OH•	MDA–TBA	532
deoxygalactose	OH•	MDA–TBA	532
ascorbate	OH•	?	550
Amino acids			
glutamic acid	OH•	MDA–TBA	532
methionine	OH•	MDA–TBA	532
homocysteine	OH•	MDA–TBA	532
aminobutyric acid	OH•	MDA–TBA	532
proline	OH•	MDA–TBA	532
arginine	OH•	MDA–TBA	532
Nucleic acids and nucleosides			
adenosine	OH•	MDA–TBA	532
inosine	OH•	MDA–TBA	532
DNA	Fe–bleomycin (oxo-iron species), Cu–phenanthroline (OH•), Cu–rifamycin (OH•), Cu–β-lactams (OH•)	MDA–TBA	532
Carboxylic acids			
benzoate	OH•	MDA–TBA	532
picolinate	OH•	MDA–TBA	532
sorbic acid	OH•	MDA–TBA	532
Antibiotics and other drugs			
polyene antifungals	LO•, LO$_2^•$	MDA–TBA	532
rifamycin	Cu, OH•	MDA–TBA	532

Table 5.11. (*Continued*)

Compound	Damaging system (if any)	TBA chromogen[a]	λ_{max}[b] (nm)
cephalosporin	Cu, OH$^\bullet$	MDA–TBA	532
cefazolin	OH$^\bullet$	MDA–TBA	532
lisinopril	OH$^\bullet$	?	532
captopril	OH$^\bullet$	?	532
enalaprilat	OH$^\bullet$	?	532

[a]MDA–TBA indicates that the absorption and fluorescence spectra were indistinguishable from authentic MDA–TBA adducts. ? indicates that the identity of the chromogen is uncertain.
[b]Wavelength of maximum absorption.

Finally, it must be pointed out that any free MDA that is formed *in vivo* will be rapidly metabolized (Chapter 4).

5.12.11   Urinary TBARS

Urinary TBARS has often been proposed as an index of whole body lipid peroxidation, e.g. in nutritional studies. For example, one metabolite of MDA purifed from rat urine was identified as arising by reaction of MDA with the amino acid lysine, eventually being excreted as N^α-acetyl-ε-(2-propenal)lysine (Chapter 4). Unfortunately, there are major problems in using urinary TBARS assays. First, there is an effect of diet: most of the lipid-derived TBARS appearing in urine seem to arise from lipid peroxides or aldehydes in ingested food, which are presumably largely generated during cooking.[5] For example, a diet rich in cooked meat promotes urinary TBARS excretion, to an extent depending on the temperature at which the meat was cooked. Hence urinary TBARS is not a suitable assay to assess whole-body lipid peroxidation in response to changes in dietary composition, although it could be used to look at effects of antioxidant supplementation of people on a 'fixed' diet.

In any case, HPLC must be used to separate the real (TBA)$_2$–MDA adduct; much TBARS in urine is not even lipid-derived[18] and much of the rest arises from aldehydes other than MDA. Adducts of MDA with DNA bases, such as guanine, are also found in urine (Chapter 4) but again an origin from the diet is possible.

5.12.12   Isoprostanes

Animal tissues and body fluids (including urine) contain low levels of F$_2$-isoprostanes, arising largely by peroxidation of phospholipids containing arachidonic acid (Section 4.9.13). Isoprostanes in plasma are largely esterified to phospholipids rather than 'free' and sensitive GC–MS assays to measure them have been described. Levels are increased under conditions of oxidative stress, e.g. in plasma and urine of cigarette smokers, in lung lining fluid of rats exposed to elevated O_2, in plasma of iron-overloaded rats, and in animals treated with CCl_4. It is likely that PUFAs other than arachidonate (including eicosapentaenoic and docosahexaenoic acids) give rise to different families of isoprostane-like compounds upon peroxidation. Measurement of these might be an approach to assessing the relative rates of peroxidation of different PUFAs *in vivo*. Families of E_2 and D_2 isoprostanes, 'isothromboxanes' and 'isoleukotrienes' have also been described. Of particular interest is the possibility that urinary isoprostanes might be a 'whole-body' marker of lipid peroxidation.

5.12.13   *Aldehydes other than MDA: 4-hydroxy-2-trans-nonenal*[13,68]

There are many 'end-products' of peroxidation in addition to isoprostanes, peroxides, hydrocarbon gases and MDA. For example, production of the saturated aldehyde **hexanal**[17] has been assayed by gas chromatography of the headspace gas in sealed vessels to follow LDL peroxidation. Aldehydes can be reacted with dinitrophenylhydrazine to give dinitrophenylhydrazones, which can be separated by HPLC. Alternatively, they can be converted into volatile products that can be separated by GC and identified by MS. A range of methods (Table 5.12) for the measurement of HNE, one of the most cytotoxic aldehydes, has been described.

However, it must be pointed out that these aldehydes arise by peroxide decomposition, which can often be controlled by the availability of metal chelates to decompose peroxides rather than by the rate of peroxide formation. Further, these reactive aldehydes, like MDA, are metabolized by cells, e.g. HNE is a substrate for some glutathione transferase enzymes (Chapter 4). When generated within cells, they also bind rapidly to proteins (largely by Michael addition of $-SH$ groups to the double bond); some of the derivatizing processes used for GC–MS do not displace such bound adducts. Hence antibody-based methods that detect such adducts within cells and tissues are of considerable value (Table 5.12). Immunohistochemical procedures can be used to localize HNE- or MDA-modified proteins in tissues subjected to oxidative stress. For example, HNE–protein adducts were shown to accumulate in the phagosomes of neutrophils activated by exposure to bacteria.[53]

Fluorescent reagents, such as **3-hydroxy-2-naphthoic acid hydrazide** have also been used to identify carbonyls *in vivo*. Figure 5.19 (see plate section) shows an example. Of course, carbonyls can arise by oxidation of proteins as well as lipids (Section 5.13 below).

5.12.14   *Summary*

What is the best method to measure lipid peroxidation? The simple answer is 'none of them'. Each assay measures something different. Diene conjugation tells one about the early stages of peroxidation, as does direct measurement of lipid peroxides. In the absence of metal ions to decompose lipid peroxides there will be little formation of hydrocarbon gases, carbonyl compounds or their fluorescent complexes; lack of observation of these does not necessarily mean that nothing is happening. Whereas most scientists studying peroxidation in isolated lipid systems add an excess (50–200 μM) of iron salt or iron chelates, the availability of metal ions to decompose lipid peroxides *in vivo* is very limited (Chapter 4).

The most *specific* assays of peroxidation (unfortunately also the most difficult to do) involve HPLC,[23] GC–MS or antibody-based determinations of such individual products as HNE, isoprostanes or peroxides. One issue that needs to be resolved is the disagreement over 'basal' levels of peroxides, e.g. in human blood plasma, using different assays that all allegedly measure 'lipid peroxides' (Table 5.9). Similarly, levels of measured 'HNE' vary between laboratories (Table 5.12).

Despite our criticisms, the TBA test still has value as a simple means of determining the rate of lipid peroxidation in isolated membrane systems[60] (e.g. liposomes, microsomes), although its value for studying LDL peroxidation is limited (Chapter 9). Even if peroxides do not decompose, the TBA test can still detect them because of decomposition of peroxides during the assay itself. Various HPLC-based methods for measuring TBARS directly give values in human body fluids not far from those obtained by more sophisticated techniques.

Whatever method is chosen, one should think clearly *what* is being measured and *how* it relates to the overall lipid peroxidation process. Whenever possible, two or more different assay methods should be used.

5.13 Fingerprinting methods: protein damage by ROS and RNS

Oxidative damage to proteins can affect the function of receptors, enzymes, antibodies, transport proteins and signal-transduction pathways. It can also cause secondary damage to other biomolecules, e.g. inactivation of DNA-repair enzymes and loss of fidelity of DNA polymerases (Chapter 4). Oxidative damage to proteins can generate new antigens that provoke immune responses. How can protein damage be measured?

5.13.1 *Reactive nitrogen species*

Attack of various RNS (e.g. $ONOO^-$, $NO_2^\bullet$, NO_2Cl) upon tyrosine (both free and in proteins) leads to production of 3-nitrotyrosine (Section 5.6 above) which can be measured immunologically (Fig. 5.8) or by HPLC or GC–MS techniques.[24] Hydrolysis of proteins is required to liberate nitrotyrosine for the latter assays. Nitrotyrosine metabolites are excreted in human urine,[47] although the possible confounding effect of dietary nitrotyrosine (if any) and of dietary nitrate/nitrite requires evaluation (e.g. NO_2^- reacts with acid in the stomach

Table 5.12. Detection and measurement of 4-hydroxy-2-nonenal (HNE) in biological systems: some examples

Method	Sample	Findings/values	Illustrative references
Nuclear magnetic resonance spectroscopy	Liposomes, oils, LDLs	Several aldehydes detected; not very sensitive; useful to study effects of heating on cooking oils	*Biochem. J.* **289**, 149 (1993), *Free Rad. Res.* **19**, 335 (1994)
Thin-layer chromatography, HPLC, GC-MS	Omega-6 PUFAs omega-3 PUFAs	HNE other aldehydes	*Biochem. Biophys. Res. Commun.* **169**, 75 (1990)
[U-^{14}C]arachidonic acid	NADPH-dependent microsomal lipid peroxidation	HNE comes mainly from arachidonic acid in polar phospholipids	*Biochim. Biophys. Acta* **876**, 154 (1986)
HPLC, GC-MS	Synovial fluid of arthritic patients	0.54 μmol/l (rheumatoid arthritis); 0.24 μmol/l (osteoarthritis)	*Ann. Rheum. Dis.* **51**, 481 (1992)
	normal plasma	106.3 ± 65.8 ng/ml	*J. Chromatog.* **488**, 329 (1989)
HPLC	Isolated rat hepatocytes after anoxia/reoxygenation	1400 nmol/l	*Free Rad. Biol. Med.* **15**, 125 (1993)
GC-MS	Cultivated cells/plasma frozen/thawed human plasma	45 μmol/1.5 × 10^6 cells; 82 nmol/l 462 nM hexanal, 703 nM nonanal, <1.3 nM HNE	*J. Chromatog.* **578**, 9 (1992) *Anal. Biochem.* **241**, 212 (1996)
HPLC	Rat liver	0.93 nmol/g	*Free Rad. Biol. Med.* **15**, 281 (1993)
	plasma from Watanabe rabbits (hyperlipidaemic) plasma from control rabbits	74 ± 10 nmol/l 47 ± 6 nmol/l	*Biochim. Biophys. Res. Commun.* **199**, 671 (1994)

Method	Material/Description	Result	References
Monoclonal or polyclonal antibodies	A wide range, recognizing different HNE-amino-acid residue adducts, e.g. the preferred antigen recognized by monoclonal antibody Ig4 is an HNE-histidine adduct		*FEBS Lett.* **359**, 189 (1995); *Free Rad. Res.* **25**, 149 (1996)
GC-MS	Alveolar macrophages exposed to NO_2	1.3 ng/10^6 cells	*Free Radic. Biol. Med.* **18**, 553 (1995)
	Plasma of ARDS[a] patients; compared with controls	HNE higher in ARDS (41 ± 2 nmol/ml) than in controls (21 ± 2 nmol/ml)	*Free Rad. Res.* **21**, 95 (1994)
	Plasma of coronary bypass patients	Greater % increase in HNE when the plasma transferrin is fully iron-saturated	*Biochem. Mol. Biol. Int.* **34**, 1277 (1994)
HPLC	Human plasma	Measurement of fragmented phospholipids, some of which have PAF-like activity (Section 4.9.14); 0.6 ± 0.2 µM short-chain oxidized phosphatidylcholine reported in human plasma	*J. Lipid Res.* **37**, 2608 (1996)

[a] Adult respiratory distress syndrome (Chapter 9).
HNE and similar products formed *in vivo* can rapidly bind to proteins: the ability of many of the GC-MS-based assays to detect all of these adducts is uncertain.

to form HNO_2). Formation of nitrotyrosine is often taken as evidence for attack upon proteins by $ONOO^-$, but other RNS can nitrate tyrosine. Environmental levels of $NO_2^•$ do not seem to cause significant tyrosine nitration in healthy animals (including humans) but the role of NO_2Cl needs further elucidation.

5.13.2   Reactive chlorine species

Addition of HOCl to proteins leads to multiple changes, including oxidation of thiol groups and methionine residues and chlorination of $-NH_2$ groups. In addition, the aromatic ring of tyrosine can be chlorinated. Chlorination can also be caused by NO_2Cl. It is possible that chlorotyrosines may be a 'marker' of attack upon proteins by 'reactive chlorine species' (Section 5.7 above).

5.13.3   Reactive oxygen species

ROS (especially $OH^•$ but also $RO_2^•$ and $RO^•$) produce a multiplicity of changes in proteins,[11] including oxidation of $-SH$ groups, hydroxylation of tyrosine and phenylalanine, conversion of methionine to its sulphoxide and generation of protein peroxides (Chapter 4). Several assays for damage to specific amino-acid residues in proteins have been developed (Table 5.13). The levels of any one (or, preferably, of more than one) of these products in proteins could in principle be used to assess steady-state levels of oxidative protein damage *in vivo*. This would presumably measure the balance between oxidative protein damage and the repair or (more likely) hydrolytic removal (Chapter 4) of damaged proteins.

For example, levels of *ortho*-tyrosine and dityrosine in human eye lens proteins have been reported in relation to age.[70] These products were also measured in hair from 'Alpine man', *Homo tirolensis*,[41] although whether they were formed during life or after death as a result of exposure of the body to sunlight or transition metals (a corroded copper axe was next to the body) is, of course, unknown.

5.13.4   The carbonyl assay[61]

The **carbonyl assay** is a 'general' assay of oxidative protein damage. It is based on the fact that several ROS attack amino-acid residues in proteins (particularly histidine, arginine, lysine and proline) to produce products with carbonyl groups, which can be measured after reaction with 2,4-dinitrophenylhydrazine. For example, $OH^•$ converts histidine to 2-oxohistidine and singlet O_2 converts tryptophan to *N*-formylkynurenine (among other products). Indeed, formation of protein carbonyls after irradiation of proteins in aqueous solution was shown some decades ago.[11] Emphasis has often been given to transition metal ion-catalysed 'site-specific' damage to proteins as a source of carbonyls, but

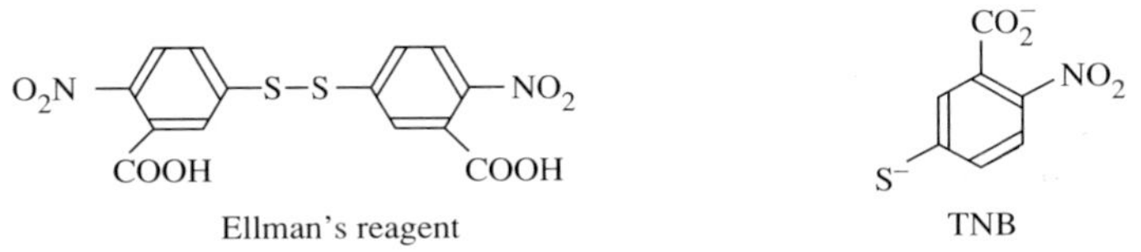

Ellman's reagent   TNB

Fig. 5.20. Above: Ellman's reagent, 5,5′-dithiobis(2-nitrobenzoic acid) (DTNB) is widely used to measure −SH groups in proteins or 'total −SH' in cells (the sum of GSH, cysteine and any other −SH compound). Below: thionitrobenzoate (TNB), absorbs strongly at 412 nm when ionized (pH ⩾ 8). Oxidation of TNB, with loss of absorption at 412 nm, has been used in assays for HOCl and ONOO⁻. See *Arch. Biochem. Biophys.* **82**, 70 (1959).

metal ions are not necessary for their formation, since irradiation and exposure to ozone generate carbonyls in proteins.

The carbonyl assay has become widely used and many laboratories have developed individual protocols for it (just as has happened with the TBA test). Sometimes the assay procedures used in a particular laboratory are not precisely specified in published papers. This point is important because there is a considerable variation in the 'baseline' levels of protein carbonyls in certain tissues, depending on how the assay is performed. For example, reported carbonyl levels for human brain cortex range from 1.5 to 6.4 nmol/mg protein.[42] By contrast, most groups seem to obtain broadly-comparable values for protein carbonyls in mammalian plasma, of 0.4–1.0 nmol/mg protein.

More work needs to be done to identify the molecular nature of the carbonyls, i.e. which amino-acid residues have been damaged and on what proteins. The binding of certain aldehydes to proteins (e.g. MDA, HNE) can generate carbonyls, as can glycoxidation (Section 9.4). Hence the presence of carbonyls is not necessarily indicative of oxidation of amino acid residues in proteins. Western-blotting assays based on the use of anti-dinitrophenylhydrazone antibodies have been developed in an attempt to identify oxidatively-damaged proteins in tissues and body fluids (Fig. 5.21). For example, exposure of human blood plasma to FeCl₃/ascorbate led to oxidation of several proteins. On a molar basis, fibrinogen suffered the most oxidative damage in this system.[58]

5.13.5 *Can 'total' oxidative protein damage be measured* in vivo?

Damaged (including oxidatively damaged) proteins are degraded *in vivo*. If some of the damaged amino acids were excreted unchanged, this might be a useful assay of 'whole body' oxidative protein damage. Little research has been carried out on the presence of oxidized amino acids and their metabolites in urine. Nitrotyrosine metabolites are present; bityrosine has also been detected and can be measured by HPLC with fluorescence detection. More work needs to be done in this area, and the possible confounding effects of oxidized proteins/amino acids in the diet (e.g. in cooked or irradiated foods) must be considered.

Table 5.13. Detection and measurement of oxidative damage to proteins: some methods

Method	Reaction/detection	Comments
A. General methods		
Amino acid analysis	Proteins hydrolysed and amino acids separated by ion-exchange chromatography	Gives an overall pattern of changes; measures loss of amino acids; can detect nitrotyrosine, chlorotyrosine and 2-oxohistidine
Protein carbonyls	2,4-dinitrophenylhydrazine: absorbance at 360–390 nm (after removal of excess DNPH by organic solvent extraction or HPLC), fluorescence, radiochemically after reduction by tritiated borohydride or 'blotting' (Fig. 5.21)	Carbonyls also arise from oxidized lipids, nucleic acids, protein glycation(Chapter 9) and binding of aldehydes such as MDA and HNE to proteins. Widely used 'general' assay; it is essential to remove nucleic acids in tissue studies
Protein peroxides	Several methods, including iodometry and FOX assay	See Table 5.8; peroxides on various biomolecules are measured
Increased susceptibility to proteolysis	Proteins more rapidly degraded by proteases	May be a common feature of oxidized proteins (Chapter 4)
(1) Alanine formation	Alanine dehydrogenase/NAD$^+$	Red blood cells cannot synthesize alanine, and its release indicates protein degradation, which is accelerated as a result of oxidative damage (Chapter 7)
(2) Increased/decreased exposure of amino groups (e.g. as a result of hydrolysis, or after modification of lysines by aldehydes)	Fluorescamine: excitation 390 nm, emission 475 nm; O-phthalaldehyde: excitation 340 nm, emission 450 nm	As a protein is degraded, one mole of -NH$_2$ is exposed for each mole of amino acid released; loss of -NH$_2$ can occur, e.g. when HNE reacts with protein

TBA-reactivity of oxidized amino acids	Thiobarbituric acid reactivity A_{532}, excitation 532 nm, emission 553 nm	Several amino acids yield TBARS upon oxidation (Table 5.11)
Amide bands	Infrared spectroscopy of the amide bands in the region 1900–1400 nm; absorbances vary depending on secondary structure	Sensitive method for measuring changes in secondary structure due to oxidative damage (or other mechanisms)

B. Specific changes to individual amino acids

Loss of -SH (thiol) groups	Ellman's reagent (Fig. 5.20)	Measures available thiol (-SH) groups
Tryptophan (1) Loss of fluorescence	Fluorescence changes: excitation 280 nm, emission 342 nm	Loss of the native fluorescence of tryptophan
(2) Formation of tryptophan oxidation products	Fluorescence: (acid) excitation 360 nm, emission 450 nm; (alkaline) excitation 327 nm, emission 342 nm; HPLC	Products such as N-formyl kynurenine are fluorescent oxidation products (*Amino Acids* **3**, 184 (1992))
Tyrosine	Bityrosine: fluorescence excitation 315 nm, emission 420 nm; HPLC separation with fluorescence, or ECD detection, GC-MS. Hydroxylation product L-DOPA can be measured by HPLC (DOPA present naturally in some proteins)	Bityrosine is highly fluorescent, but interference problems can occur in biological matrices (*Methods. Enzymol.* **233**, 363, 1994; Punchard, NA and Kelly, FJ eds (1996) *Free Radicals: A Practial Approach*, p. 171. IRL Press, Oxford)
Methionine	Oxidation to methionine sulphoxide and sulphone (HPLC or GC-MS)	Specific enzyme repairs methionine sulphoxide residues (Chapter 4)
Nitration of proteins	Antibodies, HPLC or GC-MS; usually measured on tyrosine, but nitration also occurs on phenylalanine and tryptophan	Provides evidence of damage by RNS

Table 5.13. (*Continued*)

Method	Reaction/detection	Comments
Chlorination of proteins	Chlorotyrosine; HPLC or GC-MS	Detects damage by reactive chlorine species such as HOCl or NO$_2$Cl
2-Oxohistidine	Attack of ROS upon histidine, especially OH$^\bullet$; HPLC (also detected in carbonyl assay)	Levels rise in SOD inactivated by H$_2$O$_2$, suggesting ROS production at active site (Chapter 3; see also *FEBS Lett.* **332**, 208 (1993))
Phenylalanine	Hydroxylated by OH$^\bullet$ to *ortho*-, *meta*- and *para*-tyrosines; nitrated by ONOO$^-$ and probably other RNS	HPLC or GC-MS measurement (*Free Rad. Res.* **26**, 71 (1997); *Anal. Biochem.* **244**, 270 (1997))
Valine/leucine	Peroxides and their reduction products (hydroxides) measured	*Biochem. J.* **324**, 41 (1997)
Arginine/proline	Can be oxidized to γ-glutamyl semialdehyde which on reduction and hydrolysis forms 5–hydroxy-2–aminovaleric acid	HPLC or GC-MS measurement (*Free Rad. Biol. Med.* **21**, 65 (1996))

Table 5.14. Some assays for total antioxidant activity

Method	Principle	Comments
TRAP assay	Peroxidation of endogenous or exogenous (added linoleic acid) lipids in body fluids on exposure to azo initiators.	Can be applied to whole fluids. Can measure O_2 uptake by O_2 electrode, or detect $RO_2^\bullet$ directly, e.g. by luminol-enhanced chemiluminescence (*Free Rad. Biol. Med.* **21**, 211 (1996)) or other chemiluminescence methods. Can be used to compare antioxidants as $RO_2^\bullet$ scavengers. Could also use proteins as targets, e.g. measure protein carbonyl formation or enzyme inactivation by $RO_2^\bullet$.
Phycoerythrin	Attacked by $RO_2^\bullet$ (or $OH^\bullet$) with loss of fluorescence. Assay usually uses emission 565 nm, excitation 540 nm and AAPH to generate $RO_2^\bullet$.	Phycoerythrin is a photosynthetic protein found in red algae. It carries 34 covalently-linked open-chain tetrapyrrole prosthetic groups, giving it high absorbance (λ_{max} 372, 497, 566 nm) and intense fluorescence (λ_{max} 578 nm) (*FASEB J.* **2**, 2487 (1988)).
ABTS method	2,2'-Azinobis(3-ethylbenzothiazoline 6-sulphonate) is oxidized to a radical cation, $ABTS^{\bullet+}$, which absorbs at 660, 734 and 820 nm. Oxidation of ABTS to $ABTS^{\bullet+}$ can be achieved by metmyoglobin/H_2O_2, horseradish peroxidase/H_2O_2 or simply by adding MnO_2. Antioxidants decolorize the radical.	Rapid method, easily automated (*Redox Report* **2**, 161 (1996)), measurement at high λ avoids interference from absorbance of most biomolecules. Can be applied to LDL.
FRAP (ferric reducing ability)	Measures ability of plasma antioxidants (vitamin E, urate, ascorbate, but not albumin) to reduce Fe(III) to Fe^{2+} at low pH.	Easily automated; simple. Assumes reducing ability equates to free-radical scavenging (*Anal. Biochem.* **239**, 70 (1996)).

In all these assays a radical is generated and scavenged by antioxidants. When the antioxidants are used up the radical reacts with a target molecule to produce a colour, fluorescence, chemiluminescence, loss or gain of an ESR signal or other observable change.

(h) (g) (f) (e) (d) (c) (b) (a)

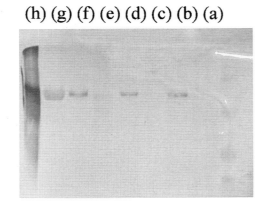

Fig. 5.21. Identification of protein carbonyls in albumin and human atherosclerotic lesions. Western blotting using anti-dinitrophenylhydrazone antibodies. Data by courtesy of Dr Pat Evans. Lanes a and c, control commercial albumin (0.4 μg loading); lanes b and d, albumin exposed to H_2O_2/Fe(III)−EDTA/ascorbate. Lane e, homogenate of normal human aorta; lane f, homogenate of atherosclerotic lesion; lane g, BSA, 40 μg (100 times greater than above) loading, to show traces of carbonyls in the commercial protein; lane h, oxidized BSA, 40 μg loading.

5.14   Fingerprinting methods: small molecules

5.14.1   *Ascorbate*

Ascorbate is a good scavenger of free radicals (Chapter 3) and the ascorbyl radical (Asc•) is usually formed. Ascorbyl is fairly unreactive and can be measured in biological systems by ESR (Fig. 5.1). Hence measurement of Asc• has been used as a biomonitor of oxidative stress, e.g. in body fluids and reperfused organs. Levels of Asc• are not linearly related to free-radical attack above a certain level, because as Asc• accumulates it can undergo disproportionation:

$$2 \text{ ascorbate}^{\bullet} \rightleftharpoons \text{ascorbate} + \text{dehydroascorbate}$$

5.14.2   *Uric acid*

Urate is an end-product of purine metabolism in primates and is thought to function as a free-radical scavenger (Section 3.21.7). Attack of various ROS upon urate generates allantoin, cyanuric acid, parabanic acid and other products. Levels of these products have been shown to rise in patients with rheumatoid arthritis, iron overload and copper overload diseases and their use as a 'biomonitor' of oxidative stress has been proposed.[29]

5.15   Assays of total antioxidant activity

Increased generation of ROS/RNS *in vivo* can lead to the depletion of one or more antioxidants. Loss of individual antioxidants (e.g. ascorbate or GSH) and/or generation of oxidation products from them (e.g. ascorbyl radical, GSSG or allantoin) can be measured as an index of oxidative stress. It should be noted that antioxidant depletion does not necessarily mean that oxidative damage has taken place: it might simply mean that the defence mechanisms have performed their normal function.

Several attempts have been made to assess the **total antioxidant activity** of body fluids rather than go to the trouble of specifically identifying what has happened to each component of the complex antioxidant defence system. The first assay of this type[69] to become popular was the **TRAP** (total (peroxyl) radical trapping antioxidant parameter) **assay**. The body fluid is incubated with AAPH, which decomposes to form peroxyl radicals (Chapter 2)

$$RN=NR \rightarrow N_2 + 2R^{\bullet}$$

$$R^{\bullet} + O_2 \rightarrow RO_2^{\bullet}$$

$RO_2^{\bullet}$ radicals react with antioxidants in the fluid: only when these have been depleted will the $RO_2^{\bullet}$ radicals attack lipids (lipids in the fluid, or lipids added to it) to cause peroxidation. By measuring the lag period before onset of peroxidation (e.g. assayed by O_2 uptake) and calibrating the assay with a known antioxidant (usually the water-soluble vitamin E analogue, **Trolox C**), a value for TRAP can be obtained as the number of micromoles of peroxyl radicals trapped per litre of fluid. Trolox traps two $RO_2^{\bullet}$ per molecule.

In human plasma TRAP values are around $10^3 \mu M$ (1 mM) $RO_2^{\bullet}$ trapped/litre: major contributors to TRAP are urate (35–65%), plasma proteins (10–50%), ascorbate (up to 24%) and vitamin E (5–10%). The known antioxidants do not account for the whole of the measured TRAP, i.e. there is an 'unidentified' component. It is essential to ensure that O_2 is not completely depleted during the TRAP assay because the carbon-centred radicals ($R^{\bullet}$) generated from AAPH can themselves react with certain antioxidants (Chapter 2).

Several variations on the TRAP assay exist. One is to use lipid-soluble $RO_2^{\bullet}$ generators (e.g. AMVN). Another is to use alternative detection methods, such as luminol-enhanced chemiluminescence (Table 5.14). Comparable assays have been developed using various sources of free radicals and different types of detector molecule (Tables 5.14 and 5.15). In both the phycoerythrin and ABTS-based assays, urate is again a major contributor to the total antioxidant activity observed in plasma. In human saliva urate is the major antioxidant present, albumin levels being low and ascorbate only $\sim 10 \mu M$. By contrast, human cerebrospinal fluid (CSF) has about 4-fold higher ascorbate concentrations and lower urate concentrations than plasma, and so ascorbate is a more significant contributor to TRAP.[40] Total CSF TRAP values are lower ($\sim 237 \mu M$) than those in plasma.

Total antioxidant assays can also be used to compare the antioxidant activities of different molecules. For example, Table 5.16 compares the antioxidant activity of various phenols found in red wine with the total activity of the wine. Many such assays have been employed for antioxidant characterization; one of

Table 5.15. Total antioxidant activity of human blood plasma/serum by various assays

Method	Typical results (μM)	Reference
TRAP	~1000	See text
Enhanced chemiluminescence	829±77	*Anal. Chim. Acta* **266**, 265 (1992)
Phycoerythrin assay	1162±265	*Free Rad. Biol. Med.* **18**, 29 (1995)
ABTS assay	1320–1580	*Clin. Sci.* **84**, 407 (1993); *Redox Report* **2**, 161 (1996)
FRAP (ferric reducing ability of plasma)	612–1634[a]	*Anal. Biochem.* **239**, 70 (1996)

Note that different assays give different values, not surprisingly, because the chemistry of each assay is different. It has been reported that the anticoagulant used to prepare plasma can affect determination of total antioxidant capacity (*Ann. Clin. Biochem.* **32**, 413 (1995)).
[a]Healthy Chinese adults.

Table 5.16. Contribution of identified constituents to the total antioxidant activity of red wine measured *in vitro* by the ABTS assay

	Antioxidant activity[a] of pure compound	Concentration in wine (mM)	Contribution to total antioxidant activity
Catechin	2.4±0.05	0.66	1.6
Epicatechin	2.5±0.02	0.28	0.7
Gallic acid	3.01±0.05	0.51	1.54
Cyanidin	4.42±0.12	0.01	0.04
Malvidin-3-glucoside	1.78±0.02	0.05	0.09
Rutin	2.42±0.12	0.01	0.02
Quercetin	4.72±0.10	0.02	0.09
Myricetin	3.72±0.28	0.03	0.09
Resveratrol	2.00±0.06	0.006	0.01
Caffeic acid	1.26±0.01	0.014	0.02
Total contribution to antioxidant activity			4.2
Mean TEAC for red wines			16.7

[a]Expressed by comparison with Trolox as **Trolox equivalent antioxidant capacity (TEAC)**. Data by courtesy of Prof. C. Rice-Evans. Note the very high *in vitro* antioxidant activity of red wines.

the first to be used employed **1,1-diphenyl-2-picrylhydrazyl** (DPPH; Fig. 5.22).[54] This is a stable free radical with a distinctive ESR signal. Its reactions with antioxidants can be followed by loss of the ESR signal or loss of absorbance at 540 nm.

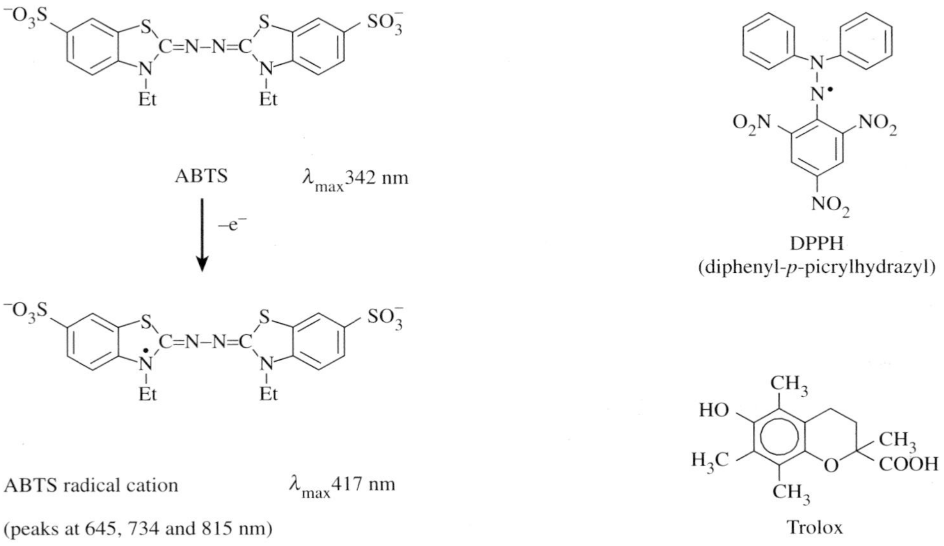

ABTS λ_{max}342 nm

$\downarrow$ -e$^-$

ABTS radical cation λ_{max}417 nm

(peaks at 645, 734 and 815 nm)

DPPH
(diphenyl-*p*-picrylhydrazyl)

Trolox

Fig. 5.22. Some of the compounds used in 'total antioxidant assays'. Note that ABTS$^{\bullet +}$ and DPPH are nitrogen-centred radicals. Trolox, used to calibrate many assays, is a water-soluble form of α-tocopherol: the hydrophobic side-chain is replaced by a $-$COOH group. One mole of Trolox can scavenge two RO$_2^{\bullet}$ radicals. DPPH is dissolved in ethanol, whereas ABTS is soluble in aqueous solutions.

5.15.1   *How useful are total antioxidant assays?*

Total antioxidant assays of body fluids are useful in getting a global picture of relative antioxidant activities in different fluids and how they change in clinical conditions. Results should be interpreted in the light of the chemistry of the assay (Table 5.15) and can sometimes be misleading. For example, rises in urate could obscure depletions of ascorbate and other antioxidants in certain diseases.

When ROS/RNS are generated *in vivo*, many antioxidants come into play. Their relative importance depends upon which ROS/RNS is generated, how it is generated, at what site, and which target of damage is measured (Chapters 3 and 4). Whatever assay is used, one will only see what one looks for. Antioxidant characterization methods are further considered in Chapter 10.

References

1. Al-Mehdi, AB *et al.* (1997) Intracellular generation of ROS during nonhypoxic lung ischemia. *Am. J. Physiol.* **272**, L294.
2. Banni, S *et al.* (1996) No direct evidence of increased lipid peroxidation in hemodialysis patients. *Nephron* **72**, 177.

3. Bartsch, H *et al.* (1994) Formation, detection and role in carcinogenesis of ethenobases in DNA. *Drug Metab. Rev.* **26**, 349.

4. Bilski, P *et al.* (1996) Oxidation of the spin trap DMPO by $^1O_2$ in aqueous solution. *J. Am. Chem. Soc.* **118**, 1330.

5. Brown, ED *et al.* (1995) Urinary MDA-equivalents during ingestion of meat cooked at high or low temperatures. *Lipids* **30**, 1053.

6. Chamulitrat, W *et al.* (1993) NO formation during light-induced decomposition of PBN. *J. Biol. Chem.* **268**, 11520.

7. Chaudhary, AK *et al.* (1994) Detection of endogenous MDA–deoxyguanosine adducts in human liver. *Science* **265**, 1580.

8. Chen, G *et al.* (1990) Excretion, metabolism and tissue distribution of a spin trapping agent, α-phenyl-*N-tert*-butyl-nitrone (PBN) in rats. *Free Rad. Res. Commun.* **9**, 317.

9. Collins, AR *et al.* (1996) Oxidative damage to DNA: do we have a reliable biomarker? *Environ. Health Perspect.* **104** (Suppl. 3), 465.

10. Connor, HD *et al.* (1994) New ROS causes formation of carbon-centered radical adducts in organic extracts of blood following liver transplantation. *Free Rad. Biol. Med.* **16**, 871.

11. Dean, RT *et al.* (1997) Biochemistry and pathology of radical-mediated protein oxidation. *Biochem. J.* **324**, 1.

12. Devanaboyina, U and Gupta, RC (1996) Sensitive detection of 8OHdG in DNA by ^{32}P-postlabelling assay and the basal levels in rat tissues. *Carcinogenesis* **17**, 917.

12a. Douki, T *et al.* (1996) Observation and prevention of an artifactual formation of oxidized DNA bases and nucleosides in the GC-EIMS method. *Carcinogenesis* **17**, 347.

13. Esterbauer, H *et al.* (1991) Chemistry and biochemistry of HNE, MDA and related aldehydes. *Free Rad. Biol. Med.* **11**, 81.

14. Finnegan, MTV *et al.* (1996) Evidence for sensitization of DNA to oxidative damage during isolation. *Free Rad. Biol. Med.* **20**, 93.

15. Floyd, RA *et al.* (1996) Hydroxyl free radical adduct of deoxyguanosine: sensitive detection and mechanisms of formation. *Free Rad. Res. Commun.* **1**, 163.

16. Foote, CS and Clennan, EL (1995) Properties and reactions of singlet O_2. In *Active Oxygen in Chemistry* (Foote, CS *et al.*, eds), p. 105. Blackie, London.

17. Frankel, EN *et al.* (1989) Rapid headspace gas chromatography of hexanal as a measure of lipid peroxidation in biological samples. *Lipids* **24**, 976.

18. Gutteridge, JMC and Tickner, TR (1978) The characterisation of TBA reactivity in human plasma and urine. *Anal. Biochem.* **91**, 250.

19. Gutteridge, JMC and Quinlan, GJ (1983) MDA formation from lipid peroxides in the TBA test: the role of lipid radicals, iron salts, and metal chelators. *J. Appl. Biochem.* **5**, 293.

20. Halliwell, B (1995) Antioxidant characterization. Methodology and mechanism. *Biochem. Pharmacol.* **49**, 1341.

20a. Halliwell, B (1997) What nitrates tyrosine? *FEBS Lett.* **411**, 157.

21. Halliwell, B and Kaur, H (1997) Hydroxylation of salicylate and phenylalanine as assays for OH$^\bullet$: a cautionary note visited for the third time. *Free Rad. Res.* **273**, 239.

22. Hartman, PE *et al.* (1983) Putative mutagens and carcinogens in foods. IV Malonaldehyde (malondialdehyde). *Environ. Mutagen.* **5**, 603.

23. Holley, AE and Slater, TF (1991) Measurement of lipid hydroperoxides in normal human blood plasma using HPLC–chemiluminescence linked to a diode array detector for measuring conjugated dienes. *Free Rad. Res. Commun.* **15**, 51.

24. Ischiropoulos, H *et al.* (1995) Detection of peroxynitrite. *Methods* **7**, 109.

24a. Jenner, A *et al.* (1998) Measurement of oxidative DNA damage by GC-MS: ethanethiol prevents artifactual generation of oxidized DNA bases. *Biochem. J.* **331**, 365.

25. Kahn, V (1989) Tiron as a substrate for horseradish peroxidase. *Phytochemistry* **28**, 41.

26. Kalyanaraman, B (1995) Thiyl radicals in biological systems: significant or trivial? *Biochem. Soc. Symp.* **61**, 55.

27. Kalyanaraman, B *et al.* (1993) Formation of 2,5-dihydroxybenzoic acid during the reaction between $^1O_2$ and salicylic acid. Analysis by ESR oximetry and HPLC with electrochemical detection. *J. Am. Chem. Soc.* **115**, 4007.

27a. Kasai, H (1997) Analysis of a form of oxidative DNA damage, 80HdG, as a marker of cellular oxidative stress during carcinogenesis. *Mutat. Res.* **387**, 147.

28. Kaur, H (1996) A water-soluble C-nitroso-aromatic spin-trap—3,5-dibromo-4-nitrosobenzenesulphonic acid. The 'Perkins spin trap'. *Free Rad. Res.* **24**, 409.

29. Kaur, H and Halliwell, B (1990) Action of biologically-relevant oxidizing species upon uric acid. *Chem.−Biol. Interac.* **73**, 235.

30. Kettle, AJ (1996) Neutrophils convert tyrosyl residues in albumin to chloro-tyrosine. *FEBS Lett.* **379**, 103.

31. Kettle, AJ *et al.* (1994) Assays using horseradish peroxidase and phenolic substrates require SOD for accurate determination of H_2O_2 production by neutrophils. *Free Rad. Biol. Med.* **17**, 161.

32. Kikugawa, K and Beppu, M (1987) Involvement of lipid oxidation products in the formation of fluorescent and cross-linked proteins. *Chem. Phys. Lipids* **44**, 277.

33. Kneepkens, CMF (1997) Assessment of oxidative stress and antioxidant status in humans: the hydrocarbon breath test. In *Antioxidant Methodology* (ed. Aruoma, OI), p. 23. American Oil Chemists Society, Indianapolis.

34. Knowles, PF *et al.* (1969) ESR evidence for enzymic reduction of oxygen to a free radical: the superoxide ion. *Biochem. J.* **111**, 53.

35. Kuypers, FA *et al.* (1987) Parinaric acid as a sensitive fluorescent probe for the determination of lipid peroxidation. *Biochim. Biophys. Acta* **921**, 266.

36. Lamrini, R *et al.* (1998) Oxidative decarboxylation of benzoic acid by peroxyl radicals. *Free Rad. Biol. Med.* **24**, 280.

37. Le Bel, CP *et al.* (1992) Evaluation of the probe 2′,7′-dichlorofluorescein as an indicator of ROS formation and oxidative stress. *Chem. Res. Toxicol.* **5**, 227.

38. Liochev, SI and Fridovich, I (1997) Lucigenin (bis-*N*-methylacridinium) as a mediator of $O_2^{\bullet-}$ production. *Arch. Biochem. Biophys.* **337**, 115.

39. Loft, S and Poulsen, HE (1996) Cancer risk and oxidative DNA damage in man. *J. Mol. Med.* **74**, 297.

40. Lönnrot, K *et al.* (1996) The effect of ascorbate and ubiquinone supplementation on plasma and CSF total antioxidant capacity. *Free Rad. Biol. Med.* **21**, 211.

41. Lubec, G *et al.* (1994) Racemization and oxidation studies of hair protein in *Homo tirolensis*. *FASEB J.* **8**, 1166.

42. Lyras, L *et al.* (1996) Oxidative damage and motor neurone disease. Difficulties in the measurement of protein carbonyls in human brain tissue. *Free Rad. Res.* **24**, 397.

43. Mason, RP (1996) *In vitro* and *in vivo* detection of free radical metabolites with ESR. In *Free Radicals: a Practical Approach* (Punchard, NA and Kelly, FJ, eds), p. 11. IRL Press, Oxford.
44. McNeil, CJ *et al.* (1992) Electrochemical sensors for direct reagentless measurement of $O_2^{\bullet -}$ production by human neutrophils. *Free Rad. Res. Commun.* **17**, 399.
45. Mendis, S *et al.* (1995) Expired hydrocarbons in patients with acute myocardial infarction. *Free Rad. Res.* **23**, 117.
46. Musarrat, J *et al.* (1996) Prognostic and aetiological relevance of 8-OHdG in human breast carcinogenesis. *Eur. J. Cancer* **12A**, 1209.
47. Ohshima, H *et al.* (1990) Nitrotyrosine as a new marker for endogenous nitrosation and nitration of proteins. *Food Chem. Toxicol.* **28**, 647.
48. Packer, JE *et al.* (1981) Free radicals and $^1O_2$ scavengers: reaction of a peroxy radical with β-carotene, diphenyl furan and DABCO. *Biochem. Biophys. Res. Commun.* **98**, 901.
49. Packer, L (ed.) (1996) Nitric oxide. Part A: Sources and detection of $NO^{\bullet}$, NO synthase; *Methods Enzymol.* **268**, whole volume; Part B: Physiological and pathological processes. *Methods Enzymol.* **269**, whole volume.
50. Packer, L (ed.) (1994) Oxygen radicals in biological systems. *Methods Enzymol.* **233** and **234**, whole volumes.
51. Pou, S *et al.* (1989) Problems associated with spin-trapping oxygen-centred free radicals in biological systems. *Anal. Biochem.* **177**, 1.
52. Powell, SR (1994) Salicylate trapping of $OH^{\bullet}$ as a tool for studying post-ischemic oxidative injury in the isolated rat heart. *Free Rad. Res.* **21**, 355.
53. Quinn, MT *et al.* (1995) Immunocytochemical detection of lipid peroxidation in phagosomes of human neutrophils: correlation with expression of flavocytochrome b. *J. Leuk. Biol.* **57**, 415.
54. Ratty, AK *et al.* (1988) Interaction of flavonoids with 1,1-diphenyl-2-picrylhydrazyl free radical, liposomal membranes and soybean lipoxygenase-1. *Biochem. Pharmacol.* **37**, 989.
54a. Reinke, LA *et al.* (1998) Free radical formation during ketamine anaesthesia in rats: a cautionary note *Free Rad. Biol. Med.* **24**, 1002.
55. Rice-Evans, CA *et al.* (1991) *Techniques in Free Radical Research.* Elsevier, Amsterdam.
56. Saez, G *et al.* (1982) The production of free radicals during the autoxidation of cysteine and their effect on isolated rat hepatocytes. *Biochim. Biophys. Acta* **719**, 24.
57. Schapira, RM *et al.* (1995) Hydroxyl radical production and lung injury in the rat following silica or TiO_2 instillation *in vivo*. *Am. J. Resp. Cell Mol. Biol.* **12**, 220.
58. Shacter, E *et al.* (1994) Differential susceptibility of plasma proteins to oxidative modification: examination by Western blot immunoassay. *Free Rad. Biol. Med.* **17**, 429.
59. Shimizu, T *et al.* (1981) Role of prostaglandin endoperoxides in the serum TBA reaction. *Arch. Biochem. Biophys.* **206**, 271.
60. Smith, MT *et al.* (1982) The measurement of lipid peroxidation in isolated hepatocytes. *Biochem. Pharmacol.* **31**, 19.
61. Stadtman, ER (1993) Oxidation of free amino acids and amino acid residues in proteins by radiolysis and by metal-catalyzed reactions. *Annu. Rev. Biochem.* **62**, 797.

62. Steinbeck, MJ *et al.* (1993) Extracellular production of $^1O_2$ by stimulated macrophages quantified using 9,10-diphenylanthracene and perylene in a polystyrene film. *J. Biol. Chem.* **268**, 15649.

63. Szabo, C *et al.* (1995) ONOO$^-$-mediated oxidation of dihydrorhodamine 123 occurs in early stages of endotoxic and hemorrhagic shock and ischemia–reperfusion injury. *FEBS Lett.* **372**, 229.

64. Terelius, Y and Ingelman-Sundberg, M (1986) Metabolism of *n*-pentane by ethanol-inducible cytochrome P-450 in liver microsomes and reconstituted membranes. *Eur. J. Biochem.* **161**, 303.

65. Thompson, S and Smith, MT (1985) Measurement of the diene conjugated form of linoleic acid in plasma by HPLC: a questionable non-invasive assay of free radical activity? *Chem.–Biol. Interac.* **55**, 357.

66. Uchida, M and Ono, M (1996) Improvement for oxidative flavor stability of beer—role of OH-radical in beer oxidation. *J. Am. Soc. Brew. Chem.* **54**, 198.

67. Vladimirov, YA (1996) Intrinsic (low-level) chemiluminescence. In *Free Radicals: A Practical Approach* (Punchard, NA and Kelly, FJ, eds), p. 65. IRL Press, Oxford.

68. Waeg, G *et al.* (1996) Monoclonal antibodies for detection of 4-HNE modified proteins. *Free Rad. Res.* **25**, 149.

69. Wayner, DDM *et al.* (1987) The relative contributions of vitamin E, urate, ascorbate and proteins to the total peroxyl radical-trapping antioxidant activity of human blood plasma. *Biochim. Biophys. Acta* **924**, 408.

70. Wells-Knecht, MC *et al.* (1993) Oxidized amino acids in lens proteins with age. *J. Biol. Chem.* **268**, 12348.

71. Wennmalm, A *et al.* (1994) NO formation in man as reflected by plasma levels of NO_3^-, with special focus on kinetics, confounding factors and response to immunological challenge. In *The Biology of NO* (Moncada, S *et al.*, eds), p. 474. Portland Press, London.

72. Williams, MD and Chance, B (1993) Spontaneous chemiluminescence of human breath. *J. Biol. Chem.* **258**, 3628.

73. Yeo, HC *et al.* (1994) Assay of MDA in biological fluids by GC–MS. *Anal. Biochem.* **220**, 391.

6

Reactive species as useful biomolecules

6.1 Introduction

We have already discussed many cases in which free radicals and other reactive oxygen and nitrogen species (ROS/RNS) are employed in living systems for useful purposes, such as the role of ferryl species at the active sites of cytochromes P450 (Chapter 1) and haem peroxidases (Chapter 3), the tryptophan radical present in cytochrome c peroxidase (Chapter 3) and the multiple physiological roles of NO$^\bullet$ (Chapter 2). ROS/RNS affect gene transcription and cell growth/proliferation, and some authors have suggested that ROS are intercellular signal molecules (Section 4.14). In this chapter, we examine in detail some other well-defined physiological roles for 'reactive species'.

6.2 Radical enzymes: ribonucleotide reductase

First, let us consider enzymes that use active-site free radicals to bring about their catalytic actions. In general, enzymes that employ radicals use them to remove hydrogen atoms from normally-unreactive positions in substrates.[51] For example, deoxyribonucleosides, the precursors of DNA, are made from ribonucleoside diphosphates by an enzyme called **ribonucleoside–diphosphate reductase** or **ribonucleotide reductase** (a ribonucleoside is a base attached to ribose). This enzyme replaces the OH group at position 2 on the ribose sugar by a hydrogen atom (generating 2′-deoxyribose). The overall reaction may be written as

$$\text{Ribonucleoside diphosphate} + R\begin{array}{c} \diagup SH \\ \diagdown SH \end{array} \longrightarrow$$

$$H_2O + \text{deoxyribonucleoside diphosphate} + R\begin{array}{c} \diagup S \\ \diagdown S \end{array}\Big|$$

where $R(SH)_2$ is a dithiol compound. In most cases $R(SH)_2$ is the dithiol protein **thioredoxin** (Section 3.13). Oxidized thioredoxin (RS_2) is re-reduced by thioredoxin reductase at the expense of NADPH (Fig. 6.1). However, in *Escherichia coli* and possibly some animal tissues, **glutaredoxin** (Chapter 3) can donate electrons to ribonucleoside–diphosphate reductase. Oxidized glutaredoxin is reduced at the expense of GSH, the GSSG then being recycled by glutathione reductase (Fig. 6.1).

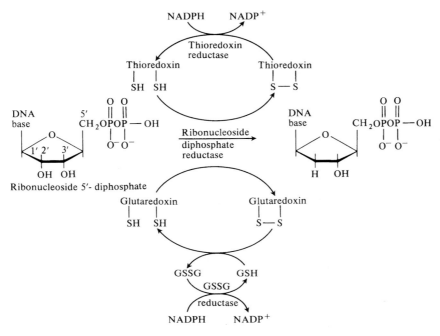

Fig. 6.1. Reduction of ribonucleoside diphosphates. The thioredoxin system operates in all cells containing ribonucleoside diphosphate reductase activity. Glutaredoxin can also supply reducing power, but is generally thought to be less important *in vivo*.

6.2.1 The enzyme mechanism[26,45,51]

In both aerobically-grown *E. coli* and eukaryotic cells the ribonucleoside diphosphate reductase enzyme contains two different subunits, **R1** and **R2**. The two catalytic sites of the enzyme are made up from parts of each subunit. R1 binds substrates and contains −SH groups which donate hydrogen to the substrate during catalysis. The oxidized groups are then re-reduced at the expense of thioredoxin or glutaredoxin (Fig. 6.1). Both R1 and R2 are dimeric. Each subunit of R2 contains two moles of Fe(III) ions bridged together by an oxygen ion. R2 gives an electron spin resonance (ESR) signal that arises from a **tyrosine phenoxyl radical**, derived by loss of hydrogen from a tyrosine residue in each amino-acid chain. The presence of TyrO$^\bullet$ is closely-linked to the presence of iron; the radical is lost on removal of iron from the enzyme. It re-forms on incubation of the R2 subunit with a Fe^{2+} salt in the presence of oxygen. Once bound to R2, the two Fe^{2+} ions are oxidized by O$_2$ to regenerate the diferric centre and cause TyrO$^\bullet$ production. The overall reaction is

$$\text{protein}-\boxed{\text{TyrOH}} + 2\text{Fe}^{2+} + \text{O}_2 + \text{e}^- + \text{H}^+$$

$$\rightarrow \text{protein}-\boxed{\text{Fe(III)}-\text{O}^{2-}-\text{Fe(III)}\text{TyrO}^\bullet} + \text{H}_2\text{O}$$

$$\text{diferric oxygen bridge}$$

Nucleotide reduction occurs on R1 where substrate is bound. The R2 tyrosyl radical acts as a chain initiator, but there is no direct communication between it and the substrate bound to R1. Instead, an electron-transfer process occurs, involving several amino-acid residues. In the active site of R1 are three cysteine residues; two provide the reducing equivalents needed for nucleoside-diphosphate reduction, being then oxidized to a disulphide. The third may be transiently converted into a **thiyl radical**, which abstracts H from the ribose sugar.

6.2.2   *Inhibitors of the enzyme*

$$\overset{O}{\overset{\parallel}{}}$$

Hydroxyurea, ($H_2NCNHOH$), a widely used inhibitor of ribonucleoside-diphosphate reductase (and hence of DNA synthesis in cells), inactivates ribonucleotide reductase by reacting with the tyrosine radical; high levels of $O_2^{\bullet-}$ and $NO^{\bullet}$ can inactivate the enzyme by the same mechanism (Chapter 2). Reaction with $O_2^{\bullet-}$ probably forms a tyrosine peroxide and that with $NO^{\bullet}$ probably forms a nitroso-tyrosine species. Hydroxyurea is sometimes used in cancer chemotherapy and in the treatment of sickle-cell anaemia. Depletion of cellular iron (e.g. by incubation with iron chelators such as desferrioxamine) can also lead to inhibition of ribonucleotide reductase and hence of DNA synthesis and cell proliferation.[13]

6.2.3   *An alternative radical*[51]

Under anaerobic conditions, *E. coli* uses a different ribonucleotide reductase, in which an iron–sulphur centre is used to form a free radical. **S-Adenosylmethionine** binds to the enzyme and undergoes one-electron reduction, producing a **5'-deoxyadenosyl radical**. This abstracts hydrogen from a glycine residue to form a **glycyl radical**. The glycyl radical then leads to modification of the substrate, probably again via an intermediate thiyl radical. Introduction of O_2 causes immediate loss of glycyl radical and inactivation of the enzyme.

6.3   Cobalamin radical enzymes[51]

Ribonucleotide reductases from some other bacteria use another source of 5'-deoxyadenosyl free radical, **5'-deoxyadenosylcobalamin**. **Cobalamin**, otherwise known as **vitamin B_{12}**, consists of a **corrin** ring structure with a central cobalt ion. Like the haem ring, the corrin ring is based on four pyrrole units (Fig. 6.2). The radical involved in the reductase reaction is formed by homolytic cleavage of the bond between the cobalt and the $-CH_2$ group in deoxyadenosylcobalamin (Fig. 6.2).

Cobalamin radicals are also involved in several other enzyme-catalysed reactions, including conversion of methylmalonyl-coenzyme A to succinyl-coenzyme A, a reaction important in fatty-acid metabolism. In essence,

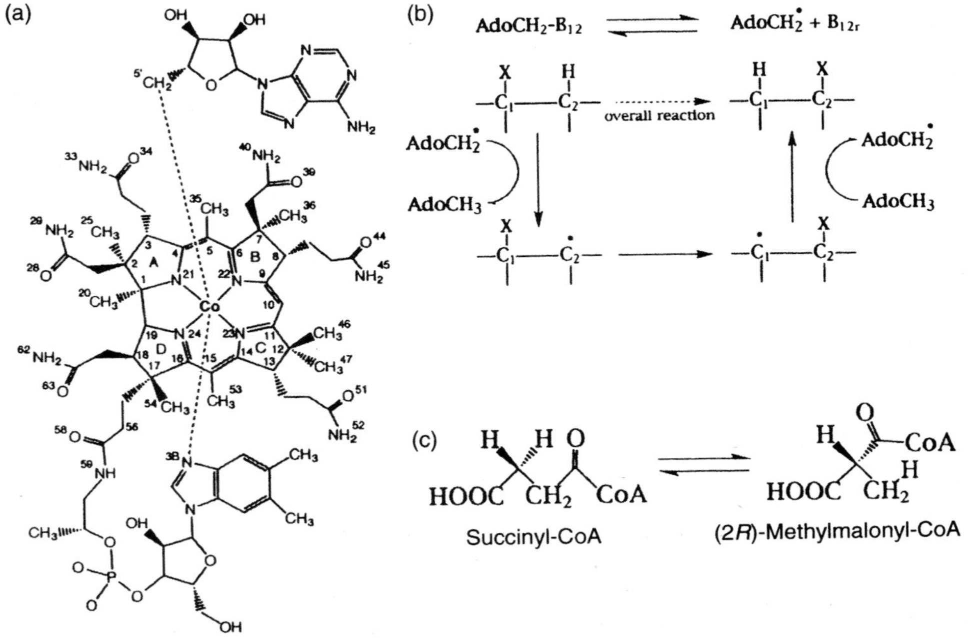

Fig. 6.2. Vitamin B_{12} and its derivatives. Co is cobalt, a transition metal (Appendix I). The corrin ring has four nitrogen atoms coordinated to cobalt. The fifth coordination position is used to attach a derivative of dimethylbenzimidazole which is also attached to a side-chain of the corrin ring. The sixth coordination position of the cobalt can be occupied by a number of groups, such as cyanide (CN^-) ion to give **cyanocobalamin** (cyanide is introduced during the isolation procedure and is not present *in vivo*). The cobalt atom in cobalamin can have a $+1$, $+2$ or $+3$ oxidation state. In **hydroxycobalamin**, OH^- occupies the sixth coordination position and the cobalt is in the Co(III) state. This form, called B_{12a}, is reduced by a flavoprotein enzyme to $B_{12r}(Co^{2+})$ which is reduced by a second flavoprotein to give $B_{12s}(Co^+)$. NADPH is the donor in both cases. B_{12s} is the substrate for a reaction with ATP that yields **5′-deoxyadenosylcobalamin**, whose structure is shown (a) and is almost unique in living organisms in having a carbon–metal bond (CH_2–Co). Impaired absorption of cobalamin from the human diet results in **pernicious anaemia**. (b) Mechanism of coenzyme B_{12} -dependent rearrangements. Ado-CH_2-B_{12} is 5′-deoxyadenosylcobalamin, Ado-$CH_2^{\bullet}$ and Ado-CH_3 are respectively the 5′-deoxyadenosyl radical and 5′-deoxyadenosine. X is the transferred group. (c) An example of (b): the rearrangement reaction catalysed by **methylmalonyl-CoA mutase**. The thioester group (O=C–CoA) is the group that migrates. From *Structure* **4**, 339 (1996) by courtesy of Dr Philip Evans and Current Biology Ltd.

cobalamin radicals react with substrates and abstract hydrogen, giving substrate radicals, which rearrange to the product and recreate a radical on the enzyme (Fig. 6.2).

Other enzymes acting by similar mechanisms are **ethanolamine ammonia-lyase**, **lysine-2,3-aminomutase** and **diol dehydratase**.

6.4 Pyruvate-metabolizing enzymes[51]

In animal cells, conversion of pyruvate produced by glycolysis into acetyl-coenzyme A (acetyl-S-CoA) is catalysed by a multienzyme complex, pyruvate dehydrogenase (Chapter 1). However, in many bacteria and in some parasitic protozoa of the trichomonad group (e.g. *Trichomonas vaginalis*) the reaction is brought about by other enzymes, such as **pyruvate–ferredoxin oxidoreductase**:

$$HS\text{-}CoA + pyruvate + 2\ ferredoxin_{ox} \rightarrow acetyl\text{-}S\text{-}CoA + CO_2 + 2\ ferredoxin_{red}$$

In trichomonad parasites, this enzyme is located in organelles known as **hydrogenosomes**. The reduced ferredoxin can then be used to reduce H^+ ions to hydrogen gas, a reaction accompanied by ATP synthesis.

Long-lived free radicals have been detected by ESR during the operation of bacterial and trichomonad enzymes. It appears that, after decarboxylation of pyruvate, one electron is transferred from it on to an iron–sulphur cluster on the enzyme, from which the electron eventually moves to ferredoxin. A possible mechanism is shown in Fig. 6.3, from which it may be seen that both carbon-centred (from pyruvate) and sulphur-centred (from coenzyme A) radicals could be involved. In the presence of O_2, the reduced iron–sulphur cluster might react to form $O_2^{\bullet-}$; reduced ferredoxin also reduces O_2 to $O_2^{\bullet-}$ (Chapter 7). Superoxide could participate in the inactivation of the enzyme that is observed under aerobic conditions.

Another pyruvate-metabolizing enzyme found in many bacteria (including anaerobically grown *E. coli*) is **pyruvate–formate lyase** (PFL), which converts

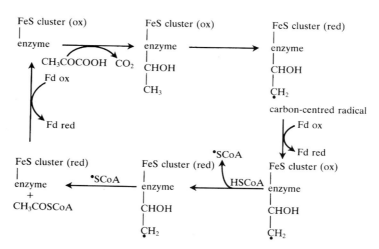

Fig. 6.3. A suggested mechanism for the action of pyruvate–ferredoxin oxidoreductase. Both carbon-centred and sulphur-centred radicals may be involved. HSCoA is coenzyme A. Adapted from *J. Biol. Chem.* **262**, 12417 (1987) by courtesy of Dr Ron Mason and the publishers.

pyruvate to acetyl-CoA in a different way:

$$\text{pyruvate} + \text{CoA} \rightarrow \text{acetyl-S-CoA} + \text{formate}$$

E. coli PFL has been shown to require a glycyl radical for its activity, similar to the anaerobic *E. coli* ribonucleotide reductase. The glycyl radical is generated by a specific activating enzyme, **PFL-activase**, which uses reduced flavodoxin (an iron–sulphur protein) and *S*-adenosylmethionine (SAM). The SAM is first cleaved to a 5′-deoxyadenosyl radical, which abstracts hydrogen from Gly734 on the PFL protein, giving a Gly$^{\bullet}$ residue.

6.5 Oxidation, carboxylation and hydroxylation reactions

As we saw in Chapter 1, the haem-containing enzyme **indoleamine dioxygenase**, found in most animal organs, catalyses the insertion of oxygen into the haem rings of tryptophan and its derivatives, such as 5-hydroxytryptophan, tryptamine and serotonin. The action of the enzyme *in vitro* under its usual assay conditions is inhibited by superoxide dismutase (SOD), but the exact role of $O_2^{\bullet-}$ in the catalytic mechanism is uncertain. Treating rabbit intestinal cells with the CuZnSOD inhibitor diethyldithiocarbamate, or the xanthine oxidase substrate hypoxanthine, increases tryptophan dioxygenation by these cells, suggestive of an interaction with $O_2^{\bullet-}$ *in vivo* (Chapter 1).

Similar inhibitory effects of SOD *in vitro* have been observed with the nitropropane dioxygenase from the yeast *Hansenula mrakii* and the galactose oxidase from the fungus *Dactylum dendroides*, an extracellularly secreted enzyme which catalyses a two-electron oxidation of the $-CH_2OH$ group on galactose, forming $-CHO$. The active site of galactose oxidase contains a copper ion coordinated by two histidine residues, a tyrosine and a covalently modified tyrosine, Tyr272. The latter is cross-linked to a cysteinyl residue (Cys228) to form a dimeric, cysteinyl-tyrosine residue. Initially, the reduced (Cu^{+}) protein undergoes two-electron oxidation by O_2 to give Cu^{2+} and a TyrO$^{\bullet}$ radical on the cysteinyl-tyrosine. Two electrons are then removed from the substrate to regenerate the resting form of the enzyme.[72]

The synthesis of prothrombin and factors VII, IX and X, essential constituents of the blood coagulation system, requires the attachment of carboxyl ($-COOH$) groups to glutamic acid residues to give γ-**carboxylglutamic acid** residues in the polypeptide chains. These have two $-COOH$ groups and are powerful Ca^{2+} chelators. These dicarboxylated proteins are also found in many tissues (e.g. **osteocalcin** in bone). Carboxylation is achieved by an enzyme located in the endoplasmic reticulum (especially in liver), which requires the fat-soluble **vitamin K** for its action.[64] As Fig. 6.4 shows, K-vitamins are quinones. They can be enzymically reduced by liver microsomal fractions to semiquinones at the expense of NADPH. This semiquinone, or possibly the fully reduced (hydroquinone) form, is required for the carboxylation reaction. The semiquinone can react reversibly with O_2 to form the quinone and $O_2^{\bullet-}$.

It has been observed that carboxylation of glutamate by isolated liver microsomes is inhibited by SOD and by copper chelates that can scavenge $O_2^{\bullet-}$.

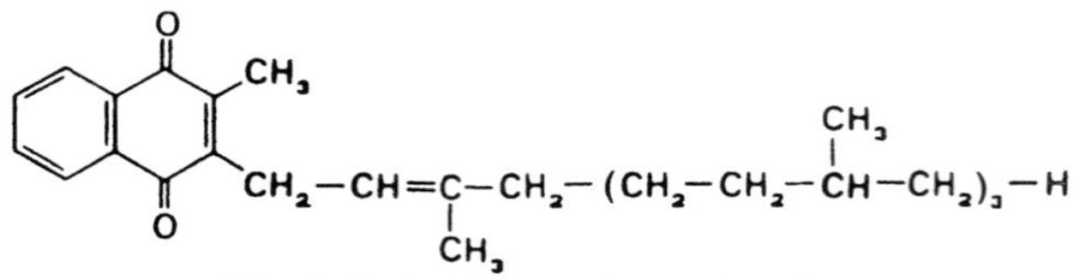

Vitamin K₁ in plants and algae (*n* = 3)

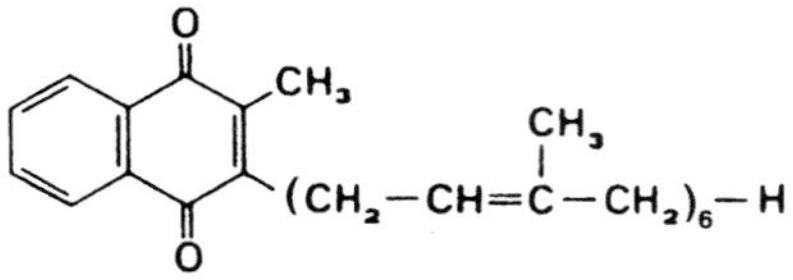

Vitamin K₂ in bacteria (*n* = 6)

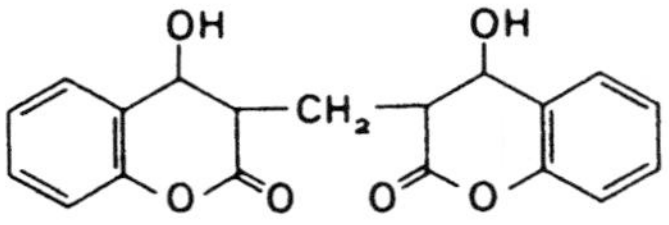

Dicoumarol
(Vitamin K antagonist)

Fig. 6.4. Vitamin K quinone. Vitamin K was discovered in Denmark in 1929 as a 'factor' that prevented abnormalities in blood clotting and resultant internal bleeding in chicks fed a fat-free diet (the 'K' was from the Danish word *koagulation*). There are different forms of vitamin K; the **phylloquinone** or K₁ form (*n*=3 in the side chain) derives from plants and the **menaquinone** series (*n*=6−11 in the side-chain) from bacteria (including gut bacteria). **Dicoumarol** antagonizes vitamin K action and interferes with blood coagulation. Dicoumarol is found in sweet clover and can cause haemorrhagic disease in animals grazing in fields where this clover grows.

A likely explanation is that SOD, by removing $O_2^{\bullet-}$, 'drains away' the required semiquinone by causing the equilibrium:

$$\text{semiquinone} + O_2 \rightleftharpoons \text{quinone} + O_2^{\bullet-}$$

to move towards the right (Section 3.3).

6.6   Yet more useful peroxidase enzymes

Peroxidases such as glutathione and cytochrome *c* peroxidases are important H_2O_2-removing antioxidant defence enzymes in animals and yeasts respectively (Chapter 3). 'Non-specific' peroxidases may play similar roles in plants, plus additional roles in such processes as auxin destruction (Chapter 3). In animals, non-specific peroxidases are less common, but they are present in some tissues and have important roles: examples are **myeloperoxidase** in neutrophils (Section 6.7 below) and salivary peroxidase. These and the other systems described below must be provided with H_2O_2 *in vivo* and so the enzymes that normally dispose of H_2O_2 cannot be completely effective in their vicinity.

Physiologically, a balance must be struck to allow them to operate yet prevent damage by excess H_2O_2 production.

6.6.1   *Thyroid-hormone synthesis*[17]

Formation of thyroid hormones involves iodination of tyrosine residues in the protein **thyroglobulin**, and coupling of the iodinated tyrosyls to form **iodothyronines** (Fig. 6.5). Both these processes are catalysed by a **thyroid peroxidase** enzyme, located in the endoplasmic reticulum of cells in the thyroid gland. This enzyme catalyses a **two-electron** oxidation[18] (unusual for peroxidases) of iodide ion (I^-), probably to form **iodonium ion** (I^+), which iodinates tyrosines. Thyroid peroxidase then catalyses one electron oxidation of di-iodo-tyrosine to tyrosyl radicals which combine to form dimers. Thyroid peroxidase can be inhibited by the anti-thyroid drug **mercaptomethylimidazole**. The H_2O_2 required by thyroid peroxidase seems to be generated by a flavoprotein enzyme that requires NADPH and Ca^{2+} ions.

   Thyroxine and tri-iodothyronine are phenols (Fig. 6.5). As such, they can exert chain-breaking antioxidant activity *in vitro*, including increasing the 'lag period' of low-density lipoprotein (LDL) oxidation. However, the levels needed to achieve such effects are greater than physiological levels. Hence the biological significance of their 'antioxidant activity' is uncertain, especially as in plasma both compounds are almost entirely bound to proteins.

6.6.2   *An 'anti-molestation' spray*

The ability of peroxidase to oxidize phenols into quinones in the presence of H_2O_2 is made use of by **bombardier beetles**,[2] which attack their enemies by spraying them with a hot, quinone-containing fluid (Fig. 6.6). A sac near the insect's 'ejection mechanism' contains a 25% aqueous solution of H_2O_2 plus 10% hydroquinone, and the spray is generated when the contents of the sac are pushed into a reaction chamber containing catalase and peroxidase. The hydroquinones are explosively oxidized to semiquinones and quinones, and part of the H_2O_2 is decomposed to O_2 by catalase. These events cause the temperature to rise up to $100\,°C$ and the pressure to build up, driving ejection of the hot spray to scare off whatever disturbed the beetle.

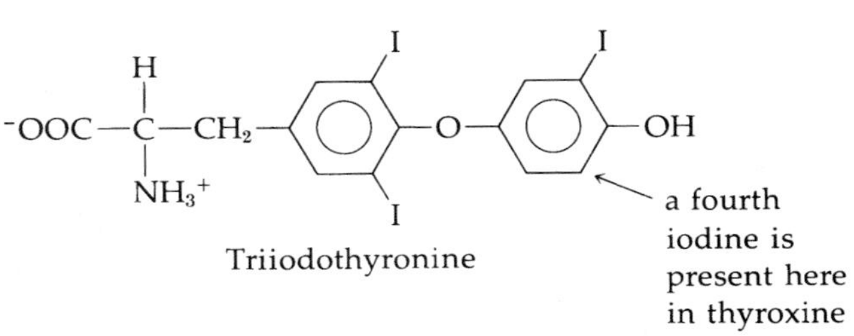

$$^-OOC-\underset{\underset{NH_3^+}{|}}{\overset{\overset{H}{|}}{C}}-CH_2-\text{(ring, I)}-O-\text{(ring, I)}-OH$$

Triiodothyronine

a fourth iodine is present here in thyroxine

Fig. 6.5. Structure of the thyroid hormones. These hormones act on many cell types by diffusing into the cell and binding to receptors, leading to increased gene transcription.

6.6.3   A fertilization membrane

When fertilized, eggs of the sea urchin *Stronglyotentrotus purpuratus* form a fertilization envelope, to block further spermatozoa from entering. This envelope contains cross-linked tyrosine residues, generated by H_2O_2-dependent reactions catalysed by **ovoperoxidase**, a peroxidase enzyme released from the egg. Fertilization is accompanied by a 'burst' of O_2 uptake, generating the H_2O_2 needed by the peroxidase by the action of an NADPH-dependent oxidase complex.[5] Any toxic effects of H_2O_2 produced at fertilization may be minimized by its scavenging by **ovothiol C** (Fig. 6.7), a thiol present at

Fig. 6.6. A bombardier beetle in action. Photograph by courtesy of Thomas Eisner and Daniel Aneshansley (also see Dean *et al.* (1990) *Science* **248**, 1219).

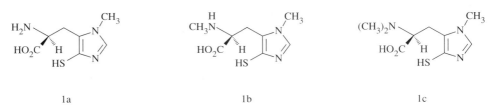

1a   1b   1c

Fig. 6.7. Structure of ovothiols. Ovothiols are mercapto-histidine derivatives. **Ovothiol C** (1c) and related compounds have been detected not only in sea urchins but also in other marine invertebrates. Related compounds include **ovothiol A** (1a) and **ovothiol B** (1b). For example, ovothiols A and B have been found in the eggs of trout and salmon. Ovothiol A has also been found in the parasites *Crithidia fasciculata* and *Leishmania donovani* (*FASEB J.* **9** 1138 (1995)). Diagram adapted from *Biochemistry* **29**, 1953 (1990) by courtesy of Dr Paul Hopkins and the publishers.

millimolar levels in sea-urchin eggs. The oxidized ovothiol may then be regenerated by reaction with GSH. Ovothiols have also been detected in other marine organisms and some parasites (Fig. 6.7).

6.6.4 *Lignification and ligninolysis*

The polymer **lignin**, a major constituent of wood, arises from the oxidation and polymerization of phenolic alcohols synthesized from phenylalanine and tyrosine (Fig. 6.8). Oxidation of these phenols to the phenoxyl radicals which polymerize is achieved by peroxidase enzymes bound to the plant cell wall. The source of the required H_2O_2 is not entirely clear, but a likely possibility is that the cell wall peroxidases simultaneously oxidize NADH. NADH oxidation does not require H_2O_2 to initiate it (Chapter 3); it generates $O_2^{\bullet -}$, and hence H_2O_2, which is used by the peroxidase to oxidize phenols. Reaction of superoxide with peroxidase generates compound III (Chapter 3). Despite the generally lower reactivity of compound III, it still participates in phenol oxidation.[31]

Lignin destruction can also involve peroxidases.[41] Some wood-destroying fungi release both H_2O_2 and peroxidase enzymes extracellularly to aid lignin degradation. For example, the white-rot fungus, *Phanerochaete chrysosporium*, can oxidize 60–70% of lignin to CO_2 and water, leaving the rest as small molecular fragments. This fungus produces at least two types of extracellular peroxidase. One type is haem-containing glycoproteins that resemble horseradish peroxidase, in that they have compound I, compound II and less-active compound III states. They catalyse H_2O_2-dependent oxidation of phenols in the lignin structure to free radicals, which can in turn attack other components of the lignin.

The second type of peroxidase, also haem enzymes, oxidize Mn^{2+} to $Mn(III)$ in an H_2O_2-dependent 'classical' peroxidase reaction sequence:

$$peroxidase + H_2O_2 \rightarrow compound\ I$$

$$compound\ I + Mn^{2+} \rightarrow compound\ II + Mn(III)$$

$$compound\ II + Mn^{2+} \rightarrow peroxidase + Mn(III)$$

The $Mn(III)$ binds to an organic acid, such as lactate, and the chelate is then believed to attack the lignin. In essence, both types of enzyme produce 'reactive species' which can diffuse away from the peroxidase and attack lignin.

The H_2O_2 needed by these peroxidases can be generated by the action of a **cellobiose oxidase**[54] enzyme (containing haem and flavin) also secreted by the fungus. This enzyme attacks cellulose in the plant cell wall, oxidizing it and simultaneously forming $O_2^{\bullet -}$, which gives H_2O_2 by dismutation. Another source of H_2O_2 is a secreted **glyoxal oxidase** enzyme, which oxidizes several aldehydes to acids. Two substrates for this enzyme, glyoxal ($OHC \cdot CHO$) and methylglyoxal (CH_3COCHO) are found in the extracellular environment of the fungus. Glyoxal oxidase contains copper and a cysteinyl-tyrosine dimeric residue at its active site: its reaction mechanism seems to resemble that of

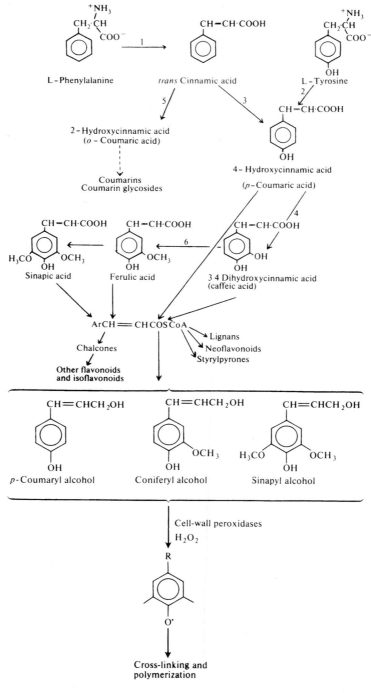

Fig. 6.8.

D. dendroides galactose oxidase (Section 6.5 above).[72] Intracellular glucose oxidase enzymes can also generate H_2O_2, which can diffuse out of the fungus to complement extracellular H_2O_2 generation.

A role for hydroxyl radical?

In the early days of research into lignin degradation by white-rot fungi, it was suggested that free $OH^\bullet$ radicals were involved, because certain $OH^\bullet$ traps (Table 5.3, paragraph 3) were oxidized to ethene gas. However, they can be oxidized directly by lignin peroxidases in the presence of phenols, an illustration of the potential artefacts in the use of methional and related compounds as $OH^\bullet$ detectors (Chapter 5). Nevertheless, the suggestion remains[37] that certain brown-rot fungi (e.g. *Coniophora puteana*) do employ $OH^\bullet$ to attack wood. They release oxalic acid, which chelates iron ions, which are then available to react with H_2O_2 to give $OH^\bullet$. These fungi degrade cellulose, leaving much of the lignin behind as an amorphous brown residue (hence their name).

6.6.5 Light production[8]

A peroxidase enzyme plays a role in light emission by the wood-boring mollusc *Pholas dactylus* (the common piddock). In most bioluminescent systems, light is produced when an enzyme (called a **luciferase**) acts on a low-molecular-mass substrate (a **luciferin**) to generate an excited state which then emits light as it decays. However, *P. dactylus* luciferin is itself a protein, and can be induced to emit light when exposed to several systems generating ROS, such as Fe^{2+} in the presence of O_2, or a mixture of xanthine oxidase and hypoxanthine.

The *P. dactylus* luciferase is a glycoprotein with peroxidase activity, but it contains copper ions and not a haem ring. Similarly, luciferase from the earthworm *Diplocardia longa* (a luminescent earthworm several inches long found in Georgia, USA) has also been discovered to be a copper-containing protein with peroxidase activity. Light emission from *P. dactylus* luciferin (often called **pholasin**) has been used as a sensitive method to detect ROS production by activated phagocytes. Bioluminescent marine scale-worms also contain a protein (**polynoidin**) that emits light when exposed to ROS.

Fig. 6.8. An outline of the biosynthetic pathways of phenylpropanoid compounds in plant tissues. Enzymes involved: 1, phenylalanine ammonia lyase (found in a wide range of plant tissues); 2, tyrosine ammonia lyase (found in some grasses); 3, cinnamate hydroxylase; 4, *p*-coumarate hydroxylase, 5, *o*-coumarate hydroxylase; 6, caffeic acid methyltransferase. Enzymes catalysing the later stages of the reactions are not shown in detail. Lignin is derived by the oxidation of various phenols into phenoxyl radicals which then polymerize in a non-enzymatic and random manner to give a three-dimensional polymer of high molecular mass with a variety of different linkages. Oxidation of phenols to phenoxyl radicals is catalysed by peroxidase enzymes bound to the plant cell walls. Some of the other phenols mentioned can act as antioxidants (Section 3.24). Adapted from Halliwell, B (1984) *Chloroplast Metabolism*, second edition, Oxford University Press.

Oxidation reactions are involved in light production by the jellyfish *Aequorea victoria*. When disturbed, intracellular Ca^{2+} levels change and cause blue light emission by the protein **aequorin** (widely used in the laboratory as a Ca^{2+}-detector). The blue light is absorbed by another protein, **green fluorescent protein (GFP)**, which emits a bright-green flash. Thus ships moving through waters in the north-west Pacific leave a trail of light in areas where this jellyfish congregates. GFP[6] generates its own chromogen: the peptide sequence serine–tyrosine–glycine (residues 65–67) in the protein undergoes autocatalytic oxidation to produce a chromophore.

Oxidative stress in molecular biology?

Luciferase genes have been widely used as 'reporter genes' to detect activation of gene transcription. Transcription and translation of the luciferase gene are readily detected by measuring the emitted light when a luciferin is added (e.g. the aldehyde *n*-decanal leads to production of blue-green light). For example, in one study the *E. coli soxS* promoter (Section 4.13.2) was linked to a gene encoding luciferase. Thus any agent capable of inducing *soxR* production should activate the *soxS* promoter, leading to luciferase synthesis. Experiments revealed, however, that luciferase itself appeared to generate ROS that could activate the *soxRS* system.[27] This must be borne in mind when using luciferase 'reporter genes' in bacteria. One potential use of this technology is to develop bacterial 'biosensors', which generate light when exposed to environmental toxins that generate ROS and/or deplete cellular antioxidant defences.

6.7   Phagocytosis

The phagocyte killing mechanism is probably the first process that comes to mind when one is asked to think of 'useful' roles for ROS. In the late 1800s, the Russian scientist Metchnikoff first reported the engulfment of bacteria by cells from the bloodstream of animals. This process is called **phagocytosis**— the cell 'flows around' the foreign particle and encloses it in a plasma membrane vesicle which is then internalized into the cytoplasm of the phagocytic cell.

Most of the phagocytic cells in the human bloodstream are **neutrophils** (Table 6.1), which have a multilobed nucleus (hence they are called **polymorphonuclear** cells) and multiple cytoplasmic granules, of several different types (Fig. 6.9). The **primary** or **azurophil** granules contain the enzymes myeloperoxidase and lysozyme, several proteases (e.g. the serine proteases cathepsin G and elastase), and a number of **granular cationic proteins**. The **specific** or **secondary granules** contain a protein that binds vitamin B_{12} (**cobalophilin**), the enzymes lysozyme and collagenase, and lactoferrin, an iron-binding protein (Chapter 3). A whole series of enzymes, including gelatinase, is housed within the so-called **tertiary granules**. Table 6.2 summarizes the types of proteinase enzyme found in animals.

When animal tissues are injured, an **acute inflammatory response** develops, characterized by swelling, warmth, pain, reddening and partial

Table 6.1. Cells in human blood

Type of cell	Cells/μl blood	Function
Erythrocytes	5×10^6	Transport O_2 and CO_2; regulate pH
White blood cells (leukocytes)		
Neutrophils	2500–7500	Phagocytosis
Eosinophils	40–400	Anti-parasite; allergic and hypersensitivity responses
Basophils	0–100	Hypersensitivity reactions; release histamine
Mononuclear cells		
monocytes	200–800	Precursors of macrophages (alveolar, splenic, peritoneal, microglial, skin, hepatic)
Lymphocytes		
B cells	$\sim$2000	Make antibodies[a]
T cells	$\sim$1000	Kill virus-infected cells, regulate activities of other leukocytes
Natural killer (NK) cells	$\sim$100	Kill virus-infected and some cancer cells
Platelets	$\sim 3 \times 10^5$	Blood coagulation

Leukocytes, monocytes, and lymphocytes differ in morphology and function; for example, eosinophils differ from neutrophils in having larger cytoplasmic granules and often a nucleus with only two lobes. An abnormally high content of white cells in the blood ($>11,000/\mu$l) is called **leukocytosis**, whilst a decrease below 4000/μl is termed **leukopaenia** or, since neutrophils constitute such a high proportion of white cells, **neutropaenia**. Monocytes are the precursors of macrophages at sites of inflammation. Most eosinophils reside in the tissues rather than circulating in the blood. All these cell types, and red blood cells, derive from the same **haematopoietic stem cells** in the bone marrow. Platelets arise from **megakaryocytes**.

[a]The principal function of B lymphocytes is to synthesize and secrete antibodies (**immunoglobulins**). When B cells are stimulated by encountering a foreign antigen, they undergo several cycles of cell division and then differentiate into specialized antibody-secreting **plasma cells**. The immune system can produce antibodies to almost any antigen encountered. When an animal comes into contact with an antigen, many B cells are stimulated. Each B cell carries on its surface immunoglobulin molecules with different antigen-binding sites. These act as the receptors for antigen and the secreted immunoglobulin of each cell has the same antigen binding site (specificity) as the membrane-bound immunoglobulin. Each B cell will divide to produce a **clone** of daughter cells, all producing immunoglobulin with the same specificity. Since many B cells bind the antigen the serum of the immunized animal will contain a mixture of the antibodies produced by many clones; **polyclonal antiserum**. (A **monoclonal antibody** is produced when a normal lymphocyte is fused with an immortalized cancer cell, producing a single clone which secretes monospecific antibody.)

T lymphocytes do not secrete a single major protein product such as immunoglobulin but carry out their functions either by direct contact with other cells or by producing cytokines (Chapter 3).

immobilization. The arterioles in and around the injured area relax, so that the capillary network becomes engorged with blood (hence the heat and redness). The permeability of the blood vessel walls increases so that more fluid leaks out, causing oedema. This fluid is rich in protein. As they enter the inflamed area, neutrophils often stop on the endothelial cells lining the blood vessels, a phenomenon known as **pavementing** or **margination**. Even in the normal

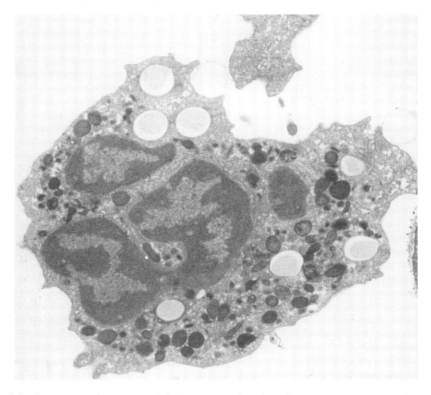

Fig. 6.9. Structure of a neutrophil as seen under the electron microscope. The cell is phagocytosing opsonized latex beads (the white circles). Some are being taken up by the cell flowing around them and others are already present within the cell in vacuoles ($\times 18\,000$). Photograph by courtesy of Professor Tony Segal.

circulation, a few neutrophils sometimes loosely and transiently adhere to the vessel wall and roll along it. However, neutrophils at sites of injury adhere firmly, push out cytoplasmic pseudopodia and squeeze through the gaps between endothelial cells, crossing the vessel wall and entering the inflamed tissue. The migration is induced by **chemotactic factors**, compounds formed in the inflamed area which attract neutrophils.[12]

At a later stage of inflammation, monocytes (Table 6.1) leave the circulation and enter the inflamed area. These cells are less actively mobile and phagocytic than neutrophils, but once in the inflamed area they undergo differentiation and change into **macrophages** (Fig. 6.10), which involves increases in their content of lysosomal enzymes, metabolic activity, motility, and phagocytic and microbicidal capacity. Apart from their formation during inflammation, macrophages can be found in the lymph nodes and spleen, and as scattered cells in connective tissue, brain, lungs and elsewhere. Indeed, the **alveolar macrophages** lie on the alveolar walls, and are a major defence of the lung against inhaled bacteria and other particles (Fig. 6.10). Pulmonary macrophages are

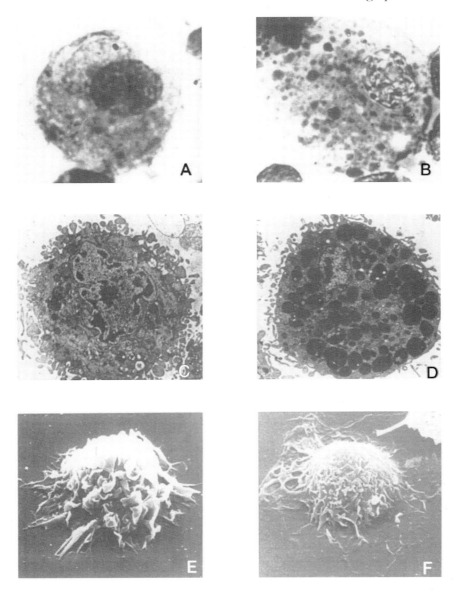

Fig. 6.10. Human pulmonary macrophages. The cells were obtained from the lungs of healthy volunteers. A, C and E are from non-smokers, and B, D and F from smokers. A and B are light micrographs. The cells vary from 15 to 50 μm in diameter. C and D are electron micrographs ($\times 6175$); and E and F are scanning electron micrographs which show the surface of the cells. Large numbers of vesicles are present. Cigarette smoke causes marked changes in the cells, which contain characteristic "smokers' inclusions", these include 'needle-like' or 'fibre-like' structures. These needle-like structures consist of kaolinite, an aluminium silicate present in cigarette smoke that the macrophages can engulf. Cigarette smoking increases the number of alveolar macrophages (Chapter 8). The plasma membrane of macrophages is highly ruffled. Photographs by courtesy of Professor W.G. Hocking.

also found in other parts of the lung. The **Kupffer cells**, which form part of the lining of the liver sinusoids, are macrophages.

Macrophages can phagocytose not only bacteria but also dead cells, e.g. neutrophils and erythrocytes. They can also ingest large amounts of insoluble material and retain it for months or years (Fig. 6.10). The term **mononuclear phagocyte system** is often used to encompass blood monocytes, their bone marrow precursors and tissue macrophages.

6.7.1   *Phagocyte recruitment and adhesion*[12]

Recruitment of phagocytes to sites of inflammation involves a complex network of proteins on both the neutrophil and vascular endothelial cell surfaces. As Metchnikoff said in 1893, 'next to leukocytes, the vessels and their endothelial lining play the most important role in inflammation'. Varying the levels of these **adhesion proteins** (e.g. in response to different cytokines) determines which phagocytes are recruited and how tightly they stick (Table 6.3).

A transient phagocyte contact, or loose adhesion, to the endothelium can occur without phagocyte activation and is usually mediated by a member of the **selectin** family. Selectins are cell-surface proteins that have a carbohydrate-binding (**lectin**) domain on the end of a protein 'stalk' that extends out from the cell surface. L-Selectin (CD62L) is commonly expressed by phagocytes and binds glycoproteins on the endothelial cell surface that are expressed at sites of inflammation. P-Selectin (CD62P) and E-selectin (CD62E) on the endothelial cell surface bind to carbohydrate on specific glycolipids or glycoproteins on the neutrophil. E-Selectin is expressed in response to cytokines (e.g. tumour necrosis factor alpha (TNFα) and interleukin-1 (IL-1)), whereas P-selectin is stored in cytoplasmic granules, from which it is released to the plasma membrane when cells are activated by histamine or thrombin.

This initial selectin-mediated 'loose' binding allows the leukocyte to monitor the local environment: if it senses a chemotactic factor it sticks more tightly to the endothelium and flattens out (Fig. 6.11). The chemotactic factors bind to specific receptors that promote activation of **integrin** proteins on the phago-cyte surface to change shape in preparation for binding to endothelial cell ligands. Integrins are dimeric glycoproteins, composed of α and β subunits. For example, lymphocyte function antigen 1 (LFA-1) on neutrophils binds to **intercellular adhesion molecules 1 and 2** (ICAM-1 and -2) on endothelial cells, whereas integrin $\alpha_4\beta_1$ binds to **vascular cell adhesion molecule 1** (VCAM-1).

Chemotactic factors include N-formyl-peptides (bacterial products), prod-ucts of complement activation, leukotriene B$_4$ (Section 6.10.8 below), platelet-activating factor (PAF; Section 4.9.14), and chemokines (Table 4.12). PAF is rapidly synthesized by endothelial cells after exposure to histamine, thrombin or leukotriene C$_4$. In general, the α-chemokines, including IL-8, act on neutrophils whereas the β-chemokines act on mononuclear phagocytes (Chapter 4).

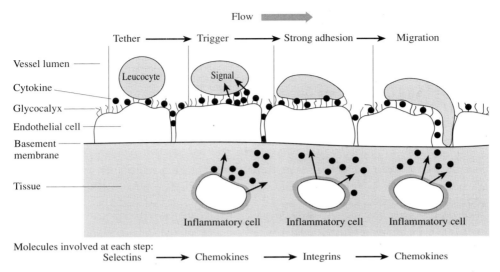

Fig. 6.11. Sequential steps in adhesion of leucocytes to endothelium. First, the flowing leucocyte is tethered and brought into contact with the endothelial wall by selectin-mediated interactions. Tethering allows cytokines—often members of the chemokine family—on the endothelial glycocalyx to bind receptors on the leucocyte and trigger integrins, which results in strong adhesion to the vessel wall. Subsequent migration into tissue is directed by chemokines, and possibly other cytokines, that are either secreted by inflammatory cells beneath the endothelium and transported to the endothelial surface or released by the endothelium itself. For simplicity, the figure shows one cytokine triggering adhesion and directing migration, although it is likely that several cytokines cooperate to regulate recruitment. From *Lancet* **343**, 832 (1994) by courtesy of Dr David Adams and the publishers.

Transmigration of phagocytes between the endothelial cells also involves adhesion molecules, including **platelet endothelial cell adhesion molecule 1 (PECAM-1)**, present on endothelial cells at sites of intercellular junctions and also on phagocytes. PECAM-1 molecules enable the phagocytes to interact with the endothelial cells as they pass through the junctions.

The fluid leaking into the inflamed area contains various antibodies which can bind to bacteria. Both neutrophils and macrophages have surface receptors that can recognize immunoglobulin G and the C3b component of reacted **complement**. The complement system plays a key role in allowing recognition of foreign material by phagocytes, and also in the process of inflammation itself. Approximately 15 liver-derived proteins make up the complement system. Once the system is activated (by exposure to antigen–antibody complexes or to pathogens) a cascade of reactions occurs, many involving proteolytic processes.

Coating of bacteria with such host-derived proteins, known collectively as **opsonins**, enables the neutrophils and macrophages to recognize them, although a few bacterial strains have coats that can be recognized directly. Once the bacteria have been engulfed, various cytoplasmic granules fuse with, and hence dump their contents into, the vacuole containing the engulfed particle (**phagocytic vacuole**). The engulfed particles are then killed and, if possible, digested within the phagocytic vacuole, the digestion products eventually being expelled from the cell.

After the bacteria causing the lesion have been destroyed (or, if injury was caused by another mechanism such as heat or chemicals after the insult ceases) there is usually reversal of the inflammatory changes. The vessel walls regain their normal permeability. Most of the emigrated neutrophils probably die (often by apoptosis) and the fragments are phagocytosed by macrophages. During phagocytic activity, neutrophils and macrophages release lysosomal enzymes into the surrounding fluid, where they contribute to the digestion of inflammatory debris.

If the bacteria are not completely eliminated, however, or the tissue injury continues, **chronic inflammation** can result. There is formation of vascular granulation tissue, which matures into fibrous tissue. Bacteria likely to cause chronic inflammation include those responsible for syphilis and tuberculosis; such inflammation can also be induced by silica dust (quartz) or silicate fibres (e.g. asbestos) inhaled into the lungs. The fibrous tissue produced during chronic inflammation causes loss of function. Chronic inflammatory lesions are especially rich in macrophages, several of which may fuse together to give **giant cells** with multiple nuclei. The various hydrolytic enzymes secreted by macrophages play an important role in tissue damage during chronic inflammation, as do ROS and RNS (Chapter 9). The function of giant cells is not clear, but they retain some phagocytic ability.

One important factor influencing susceptibility to infection (including HIV infection (Chapter 9)) and the risk of developing chronic inflammation is the relative amounts of pro-inflammatory (e.g. TNFα) and anti-inflammatory (e.g. IL-10) cytokines generated in response to tissue damage. This is to some extent genetically determined.[70]

6.7.2   *The killing mechanism of phagocytes*

The respiratory burst[5]

Most studies on the biochemistry of phagocytosis have been carried out upon neutrophils and macrophages, especially alveolar macrophages since they are the only macrophages that can readily be obtained from humans (by washing them out of the lungs in the technique of **bronchoalveolar lavage**). Resting neutrophils consume little oxygen, since they rely mainly on glycolysis for ATP production and are rich in stored glycogen. By contrast, macrophages possess more mitochondria, rely more on oxidative phosphorylation and hence consume more O_2. At the onset of phagocytosis, however, both cell types show a marked increase in O_2 uptake that is not prevented by cyanide or other inhibitors of mitochondrial electron transport.

This accelerated O_2 uptake is often called the **respiratory burst**. This is a bad name, since the extra O_2 uptake is not related to mitochondrial respiration.

The increase in O_2 uptake during the respiratory burst can be to ten or twenty times the 'resting' O_2 consumption of neutrophils. At the same time, there is an increased consumption of glucose by the cells, largely due to activation of the pentose phosphate pathway (Chapter 3). The respiratory burst can be triggered by opsonized bacteria, **opsonized zymosan** (a preparation of yeast cell walls), and several chemicals. These include phorbol esters such as **phorbol myristate acetate** (PMA), a co-carcinogen (Chapter 9), aluminium fluoride, peptide 'mimics' of parts of certain bacterial proteins (such as N-formylmethionylleucylphenylalanine; **fMet–Leu–Phe**), unsaturated fatty acids and **concanavalin A**, a lectin.[a] Hence phagocytosis of a particle is not necessary for the respiratory burst to occur.

There is considerable person-to-person variation in the extent and time-course of O_2 uptake by neutrophils isolated from human blood and activated *in vitro*.[21] Neutrophil responsiveness also varies when these cells are obtained from the same person at different times. Similarly, the size of the respiratory burst shown by macrophages depends on how, and from which tissue site, they are obtained. **Resident macrophages**, i.e. tissue macrophages that have not yet met foreign materials, are not very active. Macrophages with more vigorous respiratory bursts can be obtained by infecting animals with microorganisms such as the tuberculosis bacillus or by injecting irritating materials such as thioglycolate. The antibacterial behaviour of the macrophages obtained by these different procedures is very variable and papers concerning, for example, the relative importance of ROS or RNS in killing by macrophages should be read with this in mind.

Many cytokines affect macrophage function, and macrophages are also a rich source of cytokines, such as IL-1, IL-6 and TNFα (Section 4.16). Indeed, macrophages are complex multifunctional cells that are capable, under the right conditions, of secreting over 100 substances into their environment. These include not only ROS and cytokines but also eicosanoids, proteases (e.g. collagenase and macrophage elastase, an enzyme that has properties very different from neutrophil elastase and is not inhibited by α_1-antiproteinase), growth factors, enzyme inhibitors (e.g. the protease inhibitor α-macroglobulin; Table 6.2), some components of complement, angiogenesis factor (a compound that stimulates the migration and proliferation of fibroblasts and endothelial cells) and PAF. Macrophages also digest microbial antigens to produce peptides that bind to major histocompatibility complex (MHC) molecules; the complexes are 'presented' to T cells to activate an immune response.

Priming

Both macrophages and neutrophils can be 'primed' by prior exposure to certain agents; primed cells produce a more vigorous respiratory burst on subsequent stimulation. Priming agents include interferon γ, TNFα and bacterial endotoxin (lipopolysaccharide). The priming agents themselves (at physiological levels) usually induce little or no $O_2^{\bullet -}$ release.

Table 6.2. Animal proteinases and antiproteinases

Type of proteinase	Definition	Comments
Serine	Serine at active site	Inhibited by organophosphates: endogenous inhibitors include α_1-antiproteinase (formerly called α_1-antitrypsin) and α_1-antichymotrypsin. Examples are elastase, trypsin, chymotrypsin and cathepsin G.
Cysteine	Cysteine at active site	Inhibited by –SH modifying reagents and leupeptin. Endogenous inhibitors include cystatins and α_2-macroglobulins. Examples are cathepsins B, H, L and S. The caspases are involved in apoptosis (Chapter 4).
Aspartate	Aspartic acid at active site	Inhibited by pepstatin. Examples are cathepsins D and E.
Metalloproteinases	Metal ions at active site	Usually inhibited by EDTA. For example, the matrix metalloproteinases (collagenases, metalloelastase, stromelysins, matrilysin, gelatinases) are a family of zinc–containing enzymes response for turnover and remodelling of the extracellular matrix. They digest collagens and laminin, a protein of the basement membrane. Most are secreted as latent pro–enzymes, requiring proteolytic (or other) activation. Inhibited by batimastat; endogenous inhibitors include α_2-macroglobulin and tissue inhibitors of metalloproteinases (TIMPs).
Calpains	Ca^{2+}-dependent cysteine proteases	See Chapter 4.

Table 6.3. Phagocyte–endothelial cell adhesion molecules

Name	Present on	Examples of molecules it can bind to
Selectins		
L-(CD62L)[a]	PMN, lymphocytes	CD34, GlyCAM-1
P-(CD62P)	Activated endothelial cells, platelets	P-selectin glycoprotein ligand-1 (PSGL-1), sialyl Lewis[x]
E-(CD62E)	Cytokine-activated endothelial cells	Sialyl Lewis[x], PSGL-1, E-selectin ligand
Integrins		
LFA-1 (CD11a/CD18)	PMN, lymphocytes	ICAM-1, ICAM-2
Mac-1 (CD11b/CD18)	PMN, monocytes	ICAM-1 and others
VLA-4 ($\alpha_4\beta_1$ integrin) (CD49d/CD29)	Eosinophils, monocytes, B and T lymphocytes	VCAM-1, fibronectin
$\alpha_x\beta_2$ (CD11c/CD18)	Monocytes, PMN	Ic3b, fibrinogen
Immunoglobulin[b] 'superfamily'		
ICAM-1 (CD54)	Lymphocytes, endothelial cells (expression increases with activation)	LFA-1, Mac-1
ICAM-2 (CD102)	Lymphocytes, endothelial cells	LFA-1
VCAM-1 (CD106)	Endothelial cells (cytokine-exposed), macrophages	VLA-4
PECAM-1 (CD31)	Endothelial cells, PMN, lymphocytes, platelets	PECAM-1, (self-interaction), others

Sialyl Lewis[x] is a tetrasaccharide ligand bound by selectins: this and related structures are found on glycoproteins/glycolipids on most leukocytes and some endothelial cells.

Integrins are dimers of α and β chains: both are transmembrane proteins. Different types of cells assemble and express different $\alpha\beta$ complexes: the combination determines the ligand specificity.

Agents that up-regulate adhesion molecules on endothelial cells include TNFα, IL-1, interferon γ, endotoxin, histamine, thrombin, leukotrienes and H$_2$O$_2$. Agents up-regulating adhesion molecules on phagocytes include IL-8, TNFα, endotoxin, PAF, leukotrienes and fMet–Leu–Phe.

[a] CD stands for 'cluster of differentiation': a cell surface marker that is recognized by a particular group of monoclonal antibodies (defined in legend to Table 6.1) is given a CD number. For example, CD11a, CD11b and CD11c are three α chains that associate with the same β chain (CD18) to give three different integrins.

[b] Several adhesion molecules have amino-acid sequences similar to immunoglobulins and are considered to be part of the same protein family.

Oxygen-dependent killing[5]

If the respiratory burst of neutrophils is prevented by placing them under an atmosphere of nitrogen gas, the killing of some bacterial strains, such as *Bacillus fragilis* and *Clostridium perfringens*, is not impaired. Many of these organisms are anaerobes, and the neutrophil would not be expected to rely upon an O_2-dependent killing mechanism for species that it would be likely to encounter only under low-O_2 conditions. The killing is presumably achieved by the contents of the various phagocyte granules as they empty into the phagocytic vacuole. For example, the enzyme **lysozyme** digests cell-wall constituents of various bacteria, especially Gram-positive strains. Lysosomal enzymes attack some strains. Several of the granular cationic proteins have bactericidal activity, e.g. by acting as proteases or by opening 'channels' across bacterial cell membranes and dissipating ion gradients. Lactoferrin can bind iron essential for the growth of certain bacteria, and it has been reported to kill a few bacterial strains directly.

However, the killing of some other bacterial strains by neutrophils is greatly decreased under anaerobic conditions. Indeed, patients suffering from **chronic granulomatous disease** (CGD), an inherited condition in which phagocytosis is normal but the respiratory burst is absent, show persistent and multiple infections, especially in skin, lungs, liver and bones, by some of those bacterial strains whose killing by neutrophils requires oxygen (Table 6.4).

Biochemistry of superoxide production[35]

The O_2 uptake in both neutrophils and macrophages is due to the activation of an enzyme complex upon the plasma membrane. Activation involves migration of cytosolic components to the plasma membrane to assemble an active complex. The activated complex oxidizes NADPH provided by the pentose phosphate pathway in the cytosol into $NADP^+$, the two electrons thereby released being used to reduce O_2 to superoxide on the extracellular membrane surface. The overall equation can be written:

$$NADPH + 2O_2 \rightarrow NADP^+ + H^+ + 2O_2^{\bullet -}$$

Indeed, severe inborn deficiencies of glucose-6-phosphate dehydrogenase, the first enzyme in the pentose phosphate pathway (Chapter 3), also cause increased susceptibility to infections. Superoxide is easily detected outside stimulated neutrophils and other phagocytes by its ability to reduce cytochrome c or nitroblue tetrazolium (NBT), both reactions being inhibited by adding SOD. Formation of blue formazan from NBT can be observed under the microscope and its absence is one of the diagnostic tests for CGD. Alternatively, $O_2^{\bullet -}$ can be identified by spin trapping (Chapter 5).

The NADPH oxidase complex contains FAD and a **b-type cytochrome**. The latter has a sufficiently low reduction potential to be capable of reducing O_2 to $O_2^{\bullet -}$. The cytochrome b is a dimer, with a large β subunit (gp91^{phox}) and smaller α (p22^{phox}) subunit[b]. FAD is present in the β subunit. Usually, up to 30% of neutrophil cytochrome b is present in the plasma membrane and the rest in the cytoplasm, especially in the specific granules, which can supply more b-type

Table 6.4. Chronic granulomatous disease

Incidence	1 in 250,000 to 1 in 500,000
Male : female ratio	6 : 1
Inheritance	$\sim 60\%$ X-linked (gp91^{phox} in b-type cytochrome defective); rest mostly autosomal recessive; defects in gp22^{phox} of b-type cytochrome, or in p47^{phox} or p67^{phox} (the last two types have normal cytochrome b levels)
Carrier state	seen in X-linked CGD in mothers and sisters, usually asymptomatic
Age of onset	Congenital, first symptoms usually at <1 year of age (80%), rarely adult onset
Clinical manifestations	(a) recurrent bacterial and fungal infections; (b) sequelae of chronic inflammation (failure to thrive, anaemia, enlarged spleen, etc.); (c) obstruction due to granuloma formation
Diagnosis	Absence of $O_2^{\bullet -}$ and H_2O_2 production by activated phagocytes
Organ involvement[a]	(a) lung (80%)—pneumonia (*Staphylococcus aureus, Aspergillus*) abscess, chronic changes; (b) lymph nodes (75%)—suppurative lymphadenitis (*S. aureus*), adenopathy; (c) skin (65%)—infectious dermatitis, abscesses; (d) liver (40%)—abscess (*Staphylococcus aureus*); (e) bone (30%)—osteomyelitis (*Serratia*); (f) gastrointestinal tract (30%); (h) sepsis and meningitis (20%); (i) genito-urinary (10%)
Prognosis	Variable: death in infancy to middle age
Treatment	Prophylactic antibiotics

CGD is a defect in phagocyte $O_2^{\bullet -}$ production. Different defects exist, e.g. patients with X-linked CGD have a defective b-type cytochrome. Non-X linked CGD can affect the b-type cytochrome or other proteins involved in assembling the respiratory burst complex. Adapted from *Current Clinical Topics in Infectious Disease* **6**, 138 (1985) by courtesy of Professor Bernie Babior and the publishers.
[a] Transgenic mice lacking gp91^{phox} also show increased susceptibility to *S. aureus, Aspergillus* and *Listeria monocytogenes* infections (*J. Immunol.* **158**, 5581, 1997).

cytochrome to the plasma membrane upon activation of the respiratory burst. Activation of the NADPH oxidase requires several proteins normally present in the cytosol of 'resting' neutrophils, e.g. p47^{phox}, p67^{phox} and possibly p40^{phox}. They translocate to the membrane and help assemble the active complex. An electron transport then operates, the essence of which is indicated below:

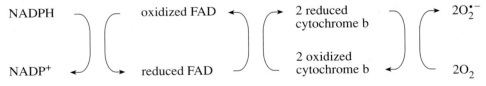

The NADPH oxidase complex is arranged vectorially in the membrane so that electrons pass across it from the NADPH-oxidizing site on the inside to the external O_2-producing site. Its operation can be inhibited by incubation of the cells with **diphenylene iodonium** and related compounds (Fig. 6.12).

Movement of electrons across the plasma membrane by the NADPH oxidase seems to be accompanied by outward movement of H^+ through proton-conducting channels in the membrane. The K_m for O_2 of the activated oxidase complex in rat (and probably human) neutrophils is within the range of O_2 concentrations in body fluids, and so it is possible that the amount of $O_2^{\bullet-}$ produced *in vivo* depends on O_2 concentration.[20]

The signal-transduction pathways leading to activation of the oxidase by physiological stimuli contain membrane receptors, GTP-binding proteins, a phosphatidylinositol-specific phospholipase C and a protein kinase C (PKC). Hydrolysis of phosphatidylinositol 4,5 bisphosphate produces inositol 1,2,5-triphosphate (IP$_3$) and diacylglycerol (DAG). IP$_3$ enters the cytoplasm and raises intracellular free Ca^{2+} (Chapter 4) whereas DAG participates in the activation of PKC, leading to phosphorylation of many proteins, including p47^{phox} (on serine). This phosphorylation promotes translocation to the membrane and contributes to activation of the oxidase, as does production of arachidonic acid. Receptor-dependent oxidase activation is short-lived; the oxidase rapidly 'switches off' unless the signal is continuous. PMA appears to stimulate the respiratory burst by activating PKC directly and so the resulting phagocyte response is Ca^{2+}-independent and much more prolonged than physiological responses. There is still considerable debate over the precise sequence of events in activation, which varies depending on the stimulus used. H_2O_2 from the respiratory burst can also increase tyrosine phosphorylation in neutrophils and surrounding cells (Chapter 4).

How does superoxide kill bacteria?

Since the bacteria or other engulfed particles are 'wrapped up' in a plasma membrane vesicle in the phagocyte cytoplasm, they are exposed to a high flux of $O_2^{\bullet-}$ (Fig. 6.13). In view of its low reactivity in aqueous solution (Chapter 2) and its inability to cross membranes, it is unlikely that $O_2^{\bullet-}$ directly kills many (if any) bacteria. It is often suggested that, at the acidic pH of the phagocytic vacuole, much $O_2^{\bullet-}$ will exist as the more reactive $HO_2^{\bullet}$ (Chapter 2). However, Segal *et al.*[62] used a pH indicator dye (fluorescein) conjugated to bacteria to measure intra-vacuolar pH. Despite several problems with these studies (e.g. oxidation of dye by myeloperoxidase-derived hypochlorous acid), the results

Fig. 6.12. Structure of diphenylene iodonium. This and related compounds inhibit $O_2^{\bullet-}$ production by the phagocyte NADPH oxidase (*Biochem. J.* **237**, 111 (1986)).

suggested that a few minutes after phagocytosis of bacteria the intra-vacuolar pH of human neutrophils *rose* from 7.4 to about 7.8 and then fell slowly, reaching 6.0–6.5 after 2 hours. The rise in pH might be accounted for by consumption of protons as $O_2^{\bullet-}$ dismutates to H_2O_2 in the phagocytic vacuole,

$$2O_2^{\bullet-} + 2H^+ \rightarrow H_2O_2 + O_2$$

If vacuoles are neutral to slightly alkaline initially, this would favour the operation of the NADPH oxidase, and some of the 'cationic antibacterial proteins', including the proteases. An acidic pH at later times would favour the operation of myeloperoxidase and hydrolytic enzymes with acidic pH optima. More of any $O_2^{\bullet-}$ that was still being generated at that time would also exist as $HO_2^{\bullet}$ (Chapter 2).

Several strains of bacteria are quickly killed by H_2O_2, which will be formed from $O_2^{\bullet-}$ by non-enzymic dismutation. Unlike $O_2^{\bullet-}$, H_2O_2 crosses membranes readily. It could be toxic to bacteria by causing damage directly, or by being converted into highly reactive $OH^{\bullet}$ within the bacterium. For example, increasing the intracellular iron content of bacteria sensitizes them to H_2O_2. Indeed, dimethylsulphoxide, a scavenger of $OH^{\bullet}$ that can penetrate into cells, can partially protect *Staphylococcus aureus* against killing by human neutrophils (Chapter 3). During these experiments, the formation of methane, a product of the reaction between dimethylsulphoxide and $OH^{\bullet}$ (Chapter 5) was observed. The variability of bacterial sensitivity to H_2O_2 is probably due both to their different levels of H_2O_2-metabolizing systems and to the different intracellular availabilities of transition-metal ions that can convert H_2O_2 into $OH^{\bullet}$.

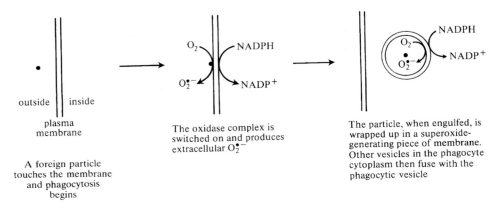

Fig. 6.13. A schematic representation of the respiratory burst. The engulfed particles are exposed to a flux of $O_2^{\bullet-}$ inside the phagocytic vacuole. Activation of the oxidase involves translocation of proteins from the cytosol to the developing phagocytic vacuole. It only occurs at the point of contact with the particle recognized as foreign.

The role of hydroxyl radical

Bacterial killing by H_2O_2 may involve intracellular $OH^\bullet$ formation. However, killing by converting H_2O_2 into $OH^\bullet$ in the phagosome outside the bacteria is intrinsically unlikely; extracellular $OH^\bullet$ probably attacks the bacterial cell wall and does little damage.

Can neutrophils make $OH^\bullet$ from the $O_2^{\bullet-}$ and H_2O_2 that they generate? Several scientists have detected $OH^\bullet$ in suspensions of activated neutrophils by using a range of techniques, including deoxyribose degradation, benzoate decarboxylation, oxidation of methional to ethene gas, and spin trapping with DMPO (Chapter 5). Other experiments with DMPO or aromatic detectors for $OH^\bullet$ (Chapter 5) have given negative results. When neutrophils are isolated, they are separated from their normal plasma environment and suspended in laboratory buffers, many of which are contaminated with iron ions. Thus neutrophil-derived $O_2^{\bullet-}$ and H_2O_2 can react with such iron to form $OH^\bullet$, accounting for some of the variable results.

Another question has been the role of lactoferrin. This iron-binding protein is secreted into the phagocytic vacuole of activated neutrophils, and also released from the cells. Lactoferrin binds two moles of Fe(III) per mole of protein, but its affinity for iron is high and it needs a very acidic pH (pH 2–3, compared with pH 5–6 for transferrin) in order for iron to be rendered labile. Iron bound to the two specific binding sites of lactoferrin is unable to promote $OH^\bullet$ formation (Chapter 3). The authors believe that the $O_2^{\bullet-}$ and H_2O_2 produced on activation of the respiratory burst will only lead to $OH^\bullet$ formation detectable outside the neutrophil if the environment of the neutrophils contains catalytic metal ions. Such ions are not present in normal human body fluids (Chapter 3) but can become available at sites of oxidative stress (Chapter 4). For example, the killing of cultured bovine pulmonary artery endothelial cells by activated human neutrophils was apparently mediated by $OH^\bullet$, but the iron required came from the endothelial cells and not from the neutrophils. Since neutrophil lactoferrin has only a low percentage iron saturation *in vivo*, its secretion by neutrophils may help to *minimize* iron-dependent radical reactions, such as $OH^\bullet$ generation, by binding any available iron ions at sites of inflammation. Lactoferrin iron could only promote $OH^\bullet$ formation if it were released from the protein by some mechanism (Chapter 3).

However, activated neutrophils make both $O_2^{\bullet-}$ and hypochlorous acid; reaction between these species (Chapter 2) could explain $OH^\bullet$ formation:

$$HOCl + O_2^{\bullet-} \rightarrow O_2 + OH^\bullet + Cl^-$$

Myeloperoxidase[40,58]

Once the phagocytic vacuole is formed, fusion of it with other granules in the neutrophil cytoplasm empties, among other things, the enzyme **myeloperoxidase** over the engulfed particle. Myeloperoxidase is not present in macrophages.

Myeloperoxidase was first purified in 1941 from large volumes of human pus and named verdoperoxidase, meaning a 'green peroxidase'. Indeed, it is the presence of myeloperoxidase that gives pus and other infected body fluids their green tinge. Myeloperoxidase is a haem–containing enzyme that shows 'non-specific' peroxidase activity (Chapter 3), being able to oxidize a wide range of substrates, including thiocyanate (CNS^-). In the presence of H_2O_2, and chloride or iodide ions, myeloperoxidase kills many bacteria and fungi *in vitro*. Since much more Cl^- than I^- ion is present in the phagocyte cytoplasm and in extracellular fluids, it is generally thought that myeloperoxidase kills by oxidizing Cl^- into **hypochlorous acid**, HOCl.[69] Hypochlorous acid is highly reactive; able to oxidize (directly or via chorine gas) many biological molecules (Chapter 2). It also decomposes to liberate toxic chlorine (Cl_2) gas:

$$HOCl + H^+ + Cl^- \rightarrow Cl_2 + H_2O$$

In addition, HOCl reacts with $O_2^{\bullet-}$ to give $OH^\bullet$ and with H_2O_2 (at alkaline pH values) to form singlet O_2 (Chapter 2). However, the latter reaction is much slower at pH 7.4 and its significance in phagocytes is uncertain.

$$HOCl + H_2O_2 \rightarrow {}^1O_2 + Cl^- + H_2O + H^+$$

A significant percentage of the H_2O_2 made (via $O_2^{\bullet-}$ dismutation) by activated neutrophils is used by myeloperoxidase to make HOCl.

Despite its antibacterial and antifungal actions *in vitro*, the evidence that HOCl plays an important antibacterial role *in vivo* is not strong.[58] Patients with inborn deficiencies of neutrophil myeloperoxidase show only minor decreases in resistance to infection, implying that this enzyme is of less importance in bacterial killing than is the respiratory burst. Several different mutations in the myeloperoxidase gene have been identified. Of course, in CGD $O_2^{\bullet-}$ is not made, so no H_2O_2 is provided for myeloperoxidase, e.g. there is a dual defect in ROS generation.

For example, one study examined the fate of *E. coli* cells engulfed by human neutrophils.[32] The bacterial envelope was quickly perforated and bacterial enzymes were then inactivated. Perforation was slower in neutrophils from patients with CGD. However, neutrophils from myeloperoxidase-deficient patients showed a normal rate of perforation, but a lower rate of enzyme inactivation. Thus HOCl produced by myeloperoxidase might help to destroy bacteria after they have been damaged, but does not, at least in this case, appear responsible for the initial damage to the bacteria. Nevertheless, neutrophils from myeloperoxidase-deficient patients often kill bacteria more slowly than normal *in vitro*, although the difference seems insignificant *in vivo*. Myeloperoxidase may be of somewhat greater importance in protection against fungi such as *Candida albicans*, and HOCl is a powerful anti-fungal agent *in vitro*.

Myeloperoxidase, like horseradish peroxidase (Chapter 3), reacts quickly with $O_2^{\bullet-}$ to form an 'oxyferrous' enzyme,[40] known as compound III (rate constant $2 \times 10^6 \, M^{-1} \, s^{-1}$):

$$\text{enzyme–Fe(III)} + O_2^{\bullet -} \rightarrow \text{enzyme–Fe}^{2+} - O_2$$

Thus secretion of myeloperoxidase into the phagocytic vacuole might decrease the amount of $O_2^{\bullet -}$ available. *In vitro*, compound III reacts with H_2O_2 to form compound II, and $O_2^{\bullet -}$ can regenerate myeloperoxidase in the Fe(III) form from compound II. The extent to which these various reactions occur in the phagocytic vacuole has yet to be established.

Nitric oxide and peroxynitrite

Rat, mouse and some other animal neutrophils and macrophages (especially if primed) can generate NO$^{\bullet}$. Nitric oxide at micromolar levels can kill certain bacteria and other cells, both directly (e.g. by inhibiting the function of respiratory chains) and indirectly,[4,36] by reacting with $O_2^{\bullet -}$ and generating peroxynitrite (ONOO$^-$) which attacks iron–sulphur proteins, essential –SH groups and many other targets. Many parasitic protozoa are directly killed by NO$^{\bullet}$, as well as by ONOO$^-$. However, it is difficult to persuade human neutrophils or macrophages to make NO$^{\bullet}$ *in vitro*,[1] so its significance as a bacterial killing mechanism in humans is uncertain. ONOO$^-$ can react with H_2O_2 to give $^1O_2$ (Chapter 2), an alternative source of the $^1O_2$ detected in activated phagocytes under certain circumstances.

However, human monocyte/macrophages do contain the gene for inducible nitric oxide synthase (iNOS) and transcribe it into mRNA if treated with certain stimuli, such as interferon γ and lipopolysaccharide. Some of the mRNA is translated into protein. RNS production (as assessed by NO_2^-/NO_3^- levels, or nitrotyrosine production), is markedly elevated in many human diseases, showing that there is increased NO$^{\bullet}$ generation *in vivo*. iNOS has been found in neutrophils isolated from the urine of patients with urinary tract infections.[71]

Extracellular myeloperoxidase, peroxynitrite and α_1-antiproteinase

During phagocytosis, some myeloperoxidase is released from neutrophils and can react with H_2O_2 and Cl$^-$ to form extracellular HOCl. HOCl *in vitro* at low concentrations rapidly inactivates α_1-antiproteinase, the main inhibitor of serine proteases (Table 6.2), such as elastase, in extracellular fluids.[69] Elastase is present in the neutrophil azurophilic granules and, after being released, can hydrolyse important connective-tissue components such as elastin and collagen: over-activity of elastase can contribute to the damage caused by excess phagocyte activation in adult respiratory distress syndrome (ARDS; Chapter 9) and in certain skin diseases, such as psoriasis. However, neutrophil elastase may play some role in bacterial killing, since transgenic mice lacking it are more susceptible to infection with gram-negative bacteria.[5a]

α_1-Antiproteinase is a member of the **serpin** (serine proteinase inhibitor) superfamily (Table 6.2). HOCl inactivates α_1-antiproteinase by oxidizing an essential methionine residue to methionine sulphoxide. Peroxynitrite inactivates α_1-antiproteinase in the same way. The extent to which inactivation could happen *in vivo* depends on the environment of the neutrophil. Thus

HOCl or ONOO$^-$ generated in plasma do not damage plasma α_1-antiprotease, because they both react preferentially with albumin, present at concentrations of 50–60 mg/ml (as compared with 1–3 mg/ml for α_1-antiprotease). In many other biological fluids (e.g. cerebrospinal fluid, synovial fluid or the fluid that lines the alveoli of the lung) albumin concentrations are much lower than in plasma, and any α_1-antiprotease present could conceivably be attacked by ONOO$^-$ or HOCl.

Ascorbic acid can also scavenge ONOOH and HOCl (Chapter 3), and the concentrations present in several extracellular fluids (40–100 μM in plasma and synovial fluid, more than this in alveolar lining fluid and cerebrospinal fluid) can protect α_1-antiprotease against these species, until the ascorbate has been depleted. In the inflamed rheumatoid joint, ascorbate concentrations are sub-normal and inactivated α_1-antiprotease can be detected (Chapter 9). GSH, present at high levels in alveolar lining fluid, can also scavenge HOCl and ONOO$^-$ to protect α_1-antiproteinase.

In vitro, HOCl can *activate* two enzymes present within neutrophils in a latent form—collagenase (which digests native type I collagen) and gelatinase (which attacks denatured collagen and native types IV and V collagen).[69] HOCl can *inactivate* several other neutrophil enzymes, including lysozyme and the myeloperoxidase molecule itself, as well as destroy chemotactic factors for neutrophils. Thus the overall effect of myeloperoxidase secretion and HOCl generation at sites of inflammation depends on what else is present at the site in question and is difficult to predict. A similar comment can be made about ONOO$^-$.

Taurine[69]

Neutrophils, other phagocytes and several other tissues (e.g. nerves) are rich in the compound **taurine** (2-aminoethanesulphonic acid), an end-product of the metabolism of sulphur-containing amino acids. Intracellular taurine concentrations in human neutrophils have been estimated as 10–50 mM. There have been several proposals that the biological role of taurine is to act as an antioxidant. However, its precursor, hypotaurine (intracellular concentration usually <1 mM), is a much better scavenger of ROS *in vitro* than is taurine,[3] and the products (Fig. 6.14) of reaction of taurine with HOCl (**taurine-N-chloramines**), although less oxidizing than HOCl itself, are still capable of inactivating α_1-antiprotease. Both chloramines and HOCl can inactivate α_2-macroglobulin (Table 6.2), an antiproteinase found in extracellular fluids that inhibits almost all types of proteolytic enzymes (including elastase, acting as back-up to α_1-antiproteinase). It may be that, in general, chloramines are still efficient antibacterial agents whilst being less damaging than HOCl to host tissues.

Light emission by phagocytes[8]

Activated phagocytes produce a weak 'background' light emission (Section 5.10), which can be greatly enhanced by addition of luminol or lucigenin. The background light emission is due to production of $O_2^{\bullet-}$ and H_2O_2 by the phagocytes, leading to the chemiluminescence that accompanies ROS production generally (Chapter 5).

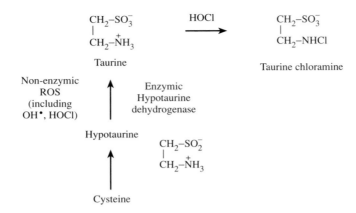

Fig. 6.14. Structure of taurine and its reaction product with hypochlorous acid. Taurine is one product of attack of certain ROS on hypotaurine.

By contrast, the enhanced light production in the presence of luminol appears largely[53] to involve the myeloperoxidase–H_2O_2–Cl^- system. For example, it is inhibited by the myeloperoxidase inhibitor **salicylhydroxamic acid** and is greatly diminished in cells from subjects with myeloperoxidase deficiency. Myeloperoxidase can also be inhibited by **4-aminobenzoic acid hydrazide**. Under some experimental conditions (e.g. see Fig. 6.15), luminol-enhanced light production is biphasic: the first phase seems to depend on interaction of luminol with extracellular myeloperoxidase–H_2O_2–Cl^- and the second to be due to entry of luminol into the cells and its measurement of intracellular ROS generation, presumably also involving myeloperoxidase. By contrast, the enhanced light emission produced by addition of lucigenin to activated neutrophils seems to involve $O_2^{\bullet-}$ rather than the myeloperoxidase system. The chemistry of light emission from luminol or lucigenin is complex (Chapter 5).

Despite this complexity, light emission in the presence of luminol or lucigenin has become widely used to measure the respiratory burst of isolated neutrophils, and has also been used to measure phagocyte activation in samples of body fluids (e.g. blood) from patients with various diseases. However, it should be noted that the time-course and extent of enhanced light emission by activated phagocytes depend on pH, temperature, the presence of compounds that might scavenge HOCl and thus depress the 'extracellular' part of luminol-dependent light emission (e.g. ascorbate and albumin in plasma), the stimulus used to activate the cells, and the cell density. For example, Fig. 6.15 shows how cell density and the nature of the activating stimulus (fMet–Leu–Phe or PMA) can alter the kinetics and the amount of light produced by human neutrophils in the presence of luminol.

Some scientists have suggested that activated neutrophils generate singlet O_2, but attempts to use various $^1O_2$ scavengers (Chapter 2) in experiments have

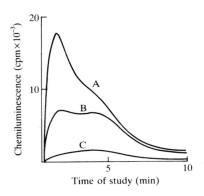 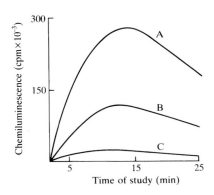

Fig. 6.15. Time course of luminol-dependent chemiluminescence from human neutrophils. Cells were exposed to fMet–Leu–Phe (left-hand graph) or to phorbol myristate acetate (right-hand graph) at 37 °C. (A) 10^5 cells; (B) 5×10^5 cells; (C) 10^6 cells. Diagram adapted from *Agents and Actions*, **21**, 104 (1987) by courtesy of Dr C. Dahlgren and the publishers. In the left-hand curve, the first peak is probably related to extracellular events (such as HOCl production by myeloperoxidase) and the second to intracellular events.

been frustrated by the fact that HOCl can oxidize a number of them directly, including DABCO, 2,5-diphenylfuran, histidine and β-carotene. The significance of $^1O_2$ in neutrophils is uncertain.

Other phagocytes

Formation of $O_2^{\bullet-}$ by a respiratory burst has been observed not only in monocytes and neutrophils, but also in the microglial cells of the brain, the osteoclasts of bone, the Kupffer cells of the liver (all macrophage lineage), basophils, mast cells and eosinophils. Indeed, isolated human monocytes show a more marked respiratory burst than do 'unprimed' macrophages, although it is not as large as in neutrophils. Eosinophils are highly active in $O_2^{\bullet-}$ production when stimulated.[66a] Monocytes contain a myeloperoxidase-like enzyme which disappears as they differentiate into macrophages.

Eosinophils and basophils are rare in the circulation (Table 6.1). Eosinophils are more usually found in tissues, especially the bone marrow and intestinal mucosa. Circulating numbers increase in allergic conditions, such as bronchial asthma and hay fever, and during infestations by some parasites, such as schistosomes. Eosinophils phagocytose and kill bacteria sluggishly, but they seem to play a role in defence against parasitic worms. Damage to the parasites can be achieved by ROS,[38] by release of an eosinophil peroxidase activity, and by an eosinophil basic protein that attacks cell membranes (worms are much bigger than eosinophils). The human eosinophil peroxidase preferentially uses bromide (Br^-) as a substrate to generate **hypobromous acid** (HOBr),[52] despite the fact that chloride levels in body fluids are much greater than Br^- levels (140 mM versus 20–100 µM). All these ROS are released when the eosinophils contact the parasite.

Basophils and mast cells, which are present in connective tissue, are rich in **histamine**, and both may be activated by allergens which combine with immunoglobulin E (IgE) antibody bound to their surface. Such activation produces not only $O_2^{\bullet-}$ but also histamine and 5-hydroxytryptamine. These products contribute to the symptoms produced when patients with allergies are exposed to the allergen that their bound IgE recognizes, such as grass pollen in the case of hay fever, or the house dust mite in many cases of bronchial asthma. Histamine has been reported to stimulate $O_2^{\bullet-}$ production by human eosinophils, and can promote phagocyte–endothelial interaction (Section 6.7.1).

6.7.3   Significance of extracellular ROS/RNS production by phagocytes

A useful process

The syndrome of CGD (Table 6.4) shows that production of $O_2^{\bullet-}$ by phagocytes is essential for the killing of some (but by no means all) strains of engulfed bacteria. Myeloperoxidase has a less important role.

Release of ROS/RNS can help phagocytes to attack extracellular targets, such as opsonized cancer cells (Chapter 9) or parasitic worms (which often try to protect themselves by producing extracellular SODs or glutathione peroxidases). The extracellular production of certain ROS/RNS might also serve intercellular signalling roles. H_2O_2 can affect the proliferation of surrounding cells (Chapter 4) and ROS may promote increased vascular permeability, although the extent of their contribution *in vivo* is uncertain. Many years ago, Del Maestro *et al.*[16] found that perfusion of a hamster cheek pouch with $O_2^{\bullet-}$-generating systems caused increased leakage of material from the vascular network and increased adherence of neutrophils to the vessel walls. Leakage was decreased by adding SOD or catalase to the perfusion medium.

Superoxide radical may play a role in chemotaxis: it has been claimed to react with a component of plasma to form a chemotactic factor, although the detailed chemistry of this process is not understood. By contrast, other leukotriene or cytokine chemotactic factors can be destroyed by $OH^{\bullet}$, ONOOH or HOCl (at least *in vitro*). H_2O_2 produced by activated phagocytes can facilitate phagocyte adherence to endothelium, by up-regulating expression of such adhesion molecules as E-selectin, ICAM-1 and VCAM-1 (Table 6.3).[46] Activation of their synthesis by ROS or by cytokines often involves the transcription factor NF-κB (Section 4.13.5). By contrast, NO$^{\bullet}$ can down-regulate expression of adhesion molecules and deter phagocyte adherence (under certain conditions), in part by decreasing NF-κB activation.[44]

H_2O_2 has also been reported to stimulate synthesis of the chemokine MIP-1α by macrophages.[65] Further, adherent neutrophils activate a respiratory burst much better in response to certain stimuli than non-adherent ones. Various ROS and RNS stimulate production of the chemokine IL-8 by fibroblasts and epithelial cells. PAF (synthesized by leukocytes, endothelial cells, platelets and several other cell types) induces phagocyte activation, platelet aggregation and degranulation, and PAF acetylhydrolase can be inactivated by ROS. In contrast, the nucleoside **adenosine** (adenine–ribose), released by a variety of normal, stimulated and injured cells (including endothelium, platelets and leukocytes) *suppresses* the respiratory burst in response to fMet–Leu–Phe, zymosan, and TNFα, and diminishes phagocyte adherence to endothelium.[14] Adenosine is a potent biological mediator that affects several cell types and exerts its effects by binding to at least four different types of receptor.

A mildly damaging process

ROS/RNS generated by activated phagocytes might also damage the surrounding environment, especially as extracellular fluids in the human body have only low activities of SOD or H_2O_2-removing enzymes (Chapter 3). For example, by inactivating α_1-antiproteinase, HOCl and $ONOO^-$ can allow elastase released from neutrophils to digest elastin. HOCl also activates collagenase and gelatinase from neutrophils. However, the interactions are complex: extracellular antioxidants such as ascorbate and albumin might scavenge $ONOO^-$ and HOCl. ROS and RNS can attack the phagocytes producing them, as well as released enzymes, e.g. HOCl can inactivate lysozyme *in vitro*. Indeed, oxidation of methionine residues in proteins to the sulphoxide, DNA strand breakage, and DNA base oxidation and deamination have been observed to occur within activated phagocytes. The DNA damage probably reflects diffusion of H_2O_2 (to form $OH^\bullet$ close to DNA) and RNS into the nucleus.

Such damage to neutrophils may be of limited physiological significance, since they are probably destined to die at sites of inflammation.[33] Nevertheless, phagocytes do have defences against ROS/RNS. They contain a CuZnSOD enzyme in the cytosol, a MnSOD in the mitochondria, catalase and glutathione peroxidase activities, together with GSH at millimolar concentrations. Neutrophils rapidly take up dehydroascorbate and re-convert it to ascorbate, accumulating millimolar levels (Chapter 3). Methionine sulphoxide reductase is also present.[22] Indeed, neutrophils deficient in glutathione reductase activity are more rapidly inactivated during phagocytosis than normal, presumably because H_2O_2 can easily diffuse out of the phagocytic vacuole into the cell cytoplasm and the absence of the reductase precludes GSH regeneration for glutathione peroxidase activity. Indeed, activated phagocytes show higher rates of GSH turnover and pentose phosphate pathway activity. Catalase activity seems to be less important in protection against H_2O_2 in phagocytes than the glutathione system, although activities vary considerably between species. For example, human and guinea-pig neutrophils contain more catalase than cells from rats and mice.

In view of the complex interactions between ROS, RNS and all the other species present at sites of inflammation, it seems likely that tissue 'damage' is limited during the 'normal' acute inflammatory response. The increased blood flow and vascular permeability that occurs at sites of inflammation will tend to replenish supplies of α_1-antiproteinase and may permit more albumin and other antioxidants to enter the site of inflammation, leading to more ROS/RNS scavenging. Concentrations of several acute-phase antioxidant proteins (e.g. haptoglobin, caeruloplasmin) also increase during inflammation (Chapter 4), giving more antioxidant protection.

A seriously damaging process

However, if activation of phagocytic cells at a site of inflammation is excessively prolonged, or an abnormally large number of cells is activated at a particular site, then significant tissue damage might result. For example, ROS/RNS can damage DNA not only in the phagocytes but also in surrounding cells, and chronic inflammation is well-known to pre-dispose to cancer (Chapter 9). Cigarette smoke is directly damaging to the lung, but its effects are worsened by the activation of macrophages and the recruitment of neutrophils (Chapter 8).

Damage by inhaled particles: silica and asbestos[10,23]

Silica particles and **asbestos fibres** can be phagocytosed by pulmonary macrophages. They then appear to cause rupture of the phagocytic vesicles, leading to release of proteolytic enzymes into the macrophage cytoplasm. If the macrophage is killed, the particles will be released to be taken up by other cells with the same effect. The proteases will thus be released into the surrounding lung tissue.

Asbestos fibres and many other minerals have been shown to contain iron, some of which is in a form that can stimulate $OH^\bullet$ formation, lipid peroxidation and oxidative DNA damage. Instillation of silica into the lungs of animals leads to $OH^\bullet$ generation, as measured by salicylate hydroxylation (Chapter 5). Iron may be an integral part of the silicate crystal structure (e.g. in **crocidolite asbestos**, which has the overall formula $Na_2O \cdot Fe_2O_3 \cdot 3FeO \cdot 8SiO_2$) or may be present as a contaminant on the surface of fibres, as with **chrysotile asbestos** ($3MgO \cdot 2SiO_2 \cdot 2H_2O$), where varying amounts of magnesium can be replaced by iron. It seems that penetration of asbestos fibres of a particular length and shape (e.g. longer than 8 μm and less than 0.36 μm diameter for rats) into cells leads to iron-dependent radical reactions that contribute to cancer development. Indeed, transfection of a gene encoding MnSOD into hamster tracheal epithelial cells[56] caused them to over-express this protein and become more resistant to the toxic effects of crocidolite. Inhalation of asbestos fibres for prolonged periods causes not only lung fibrosis but also two types of cancer, mesothelioma and bronchogenic carcinoma. In addition, several mineral dust fibres, including asbestos, are capable of activating the respiratory burst of phagocytes.

Dialysis leukopaenia and organ rejection

In 1968, it was discovered that within the first few minutes of blood dialysis, leukopaenia (Table 6.1) occurs, which was traced to the fact that contact of plasma with Cellophane in the dialyser caused activation of the complement system, leading to accumulation, aggregation and activation of neutrophils and monocytes in the lung. This can cause damage by interfering with blood flow, and perhaps by producing ROS. Indeed, accumulation and activation of neutrophils in the lung plays an important part in the pathology of adult respiratory distress syndrome (Chapter 9).

ROS/RNS may also be involved in rejection of transplanted organs, a process involving chronic inflammation.[49] For example, rejected kidney transplants were found to contain nitrotyrosine, indicative of damage by RNS. Purification of the nitrated proteins revealed that one of them was MnSOD, which had been inactivated by RNS attack. The potential for damage by excess ROS/RNS is dramatically illustrated by their ability to corrode plastics, a process which may cause damage to implantable biomedical devices such as heart valves if inflammation develops around them.

6.7.4   *Bacterial and fungal avoidance strategies*

It has been suggested that the fungus *Cryptococcus neoformans* produces high levels of intracellular mannitol as an $OH^\bullet$ scavenger to protect it against phagocyte killing.[11] Other strategies[30] used by certain microorganisms to avert phagocytic killing include production of toxins that inactivate the phagocyte (e.g. **leukocidin** by *S. aureus*), rapid induction of heat shock and other protective proteins, e.g. in *Salmonella* and *Mycobacterium tuberculosis* (Chapter 4), very high endogenous levels of antioxidant defence enzymes (e.g. SOD in *Mycobacterium leprae* and catalase in the gonococcus: Chapter 3), generation of thiol-containing capsules that hinder recognition and uptake of the organism by the phagocyte and can absorb ROS, and production of inhibitors of the respiratory burst, e.g. by *Legionella pneumophila* (the causative agent of legionnaire's disease) and *Francisella tularensis* (the causative agent of tularaemia).

6.8   NAD(P)H oxidases in other cell types

6.8.1   *Endothelial cells*

Many cells other than phagocytes have been found to release $O_2^{\bullet-}$ and H_2O_2, albeit at much lower levels.[15] Vascular endothelial cells obtained from several species, including humans, have been shown to release $O_2^{\bullet-}$ and H_2O_2 in culture, but it is not clear if they do this all the time, or only after exposure to cytokines or after an insult, such as ischaemia–reperfusion (or, in the case of studies on cultured cells, the trauma of the cell isolation process or the exposure to the 'hyperoxia' of ambient pO_2).

Suggested sources of the extracellular $O_2^{\bullet-}$ produced by vascular endothelial cells include the enzyme xanthine oxidase (although its significance in human cells has been questioned), NADH oxidase enzymes (e.g. described in bovine coronary artery endothelium) and NADPH oxidases. The physiological significance (if any) of this extracellular endothelial ROS production is unknown: one possibility is that $O_2^{\bullet-}$ and $NO^\bullet$ antagonize each other's actions as a vasoregulatory mechanism.[28] Their reaction would produce potentially-cytotoxic $ONOO^-$, however. Arterial smooth muscle cells from some species have also been claimed[43] to generate $O_2^{\bullet-}$ when treated with certain hormones, such as angiotensin II (a peptide hormone that increases blood pressure), an effect that could lead to $NO^{\bullet}$ inactivation and hypertension if it occurred *in vivo*. Another factor to be considered is the presence of EC-SOD in vessel walls (Chapter 3). Superoxide generation by vascular endothelium may be involved[57] in the development of 'tolerance' to organic nitrates, i.e. their ability to cause vasodilation decreases after prolonged use. Many mechanisms probably account for this, but studies in rabbits suggest that one of them is up-regulation of an $O_2^{\bullet-}$-producing NAD(P)H oxidase system in vascular endothelium.

6.8.2   *Lymphocytes and fibroblasts*[50]

Some cells (including B-lymphocytes and fibroblasts) have been shown to contain NADPH oxidase components resembling those in phagocytes, such as *b*-type cytochromes. B-lymphocytes have been most studied, and contain an $O_2^{\bullet-}$-generating system essentially identical to that of phagocytes but some 10- to 30-fold less active. The system can be activated by PMA or by the cross-linking of surface antigens, and is defective in CGD patients. Its physiological role is unknown. By contrast, the fibroblast $O_2^{\bullet-}$-generating system in not defective in CGD. Transforming growth factor beta (TGFβ) a cytokine that promotes fibrosis, has been reported to increase ROS production by human lung fibroblasts.[67]

6.8.3   *Sensing of hypoxia*[7]

Generation of ROS by an NADPH oxidase complex containing cytochrome *b* may be involved in the ability of the **carotid body** (Chapter 1) to sense O_2 concentration in the arterial blood; the carotid body initiates nervous signals controlling respiration and circulation in order to avoid hypoxia. The Ca^{2+}-dependent NADPH oxidase in thyroid plasma membrane that generates H_2O_2 for thyroid peroxidase (Section 6.6.1 above) is inhibited by diphenylene iodonium and may be similar, although it has been claimed to make H_2O_2 directly rather than via $O_2^{\bullet-}$.

The hormone **erythropoietin** is synthesized in liver and kidney under hypoxic conditions, and stimulates production of extra erythrocytes. The molecular mechanism of O_2 sensing has been studied in hepatocyte cell lines such as **HepG2**, and again may involve an NADPH oxidase system containing a *b*-type cytochrome. In this case, H_2O_2 production inhibits hormone release.

6.8.4   *Platelets*[50]

Platelets have been reported to generate $O_2^{\bullet-}$ at a low continuous level, which does not increase on activation. Platelets play a vital function in preventing blood loss from injured vessels; they rapidly adhere to collagen and basement membrane exposed when endothelial cells are damaged, followed by aggregation. The processes of adhesion and aggregation are triggered by changes in surface adhesion molecules, and by the release of pro-aggregatory mediators, including ADP (released from the 'dense granules' of the platelets), and thromboxane A_2 (Section 6.10.6 below). Exposure of platelets to ROS potentiates aggregation induced by thrombin, and it has been suggested that $O_2^{\bullet-}$ and H_2O_2 produced by platelets or cells around them may synergize with pro-aggregatory stimuli. The physiological importance of these effects is uncertain as yet.

6.8.5   *Other cells*

There are many literature descriptions of NAD(P)H oxidases in the plasma membranes of multiple cell types, including erythrocytes, adipocytes and renal brush border membranes. Often they have been identified by using artificial electron acceptors such as ferricyanide ($Fe(CN)_6^{3-}$), indophenol or cytochrome *c*. The adipocyte and thyroid membrane systems will also function with $Fe(CN)_6^{3-}$ as an electron acceptor, suggesting that these other reductase systems might also generate ROS. More work is required to elucidate their chemical nature, and even more work to establish their physiological function.

Iron uptake in the yeast *Saccharomyces cerevisiae* involves a plasma membrane ferric reductase, which appears to be a *b*-type cytochrome.[63]

6.9   Fruit ripening and the 'wound response' of plant tissues[25,34,68]

The ripening and senescence of fruits is a controlled oxidative process. Studies on pears have shown that, as ripening proceeds, the concentration of free −SH groups in the fruit decreases, and there is an accumulation of H_2O_2 and lipid peroxides. Ripening of pears, or the senescence of rice plant leaves, can be speeded up by treatments that stimulate formation of H_2O_2 within the tissue. As pears and bananas ripen, fluorescent compounds, which may arise by lipid peroxidation, accumulate within them. Membrane lipid fluidity decreases in cell and organelle membranes from many senescing plant tissues; this might occur as a result of lipid peroxidation and of the action of lipases that become activated during senescence. Senescing root nodules in leguminous plants show rises in 'catalytic' iron content. A role for leaf peroxidase activity has been suggested in the breakdown of cell walls.

6.9.1   *Lipoxygenases*

Many non-green plant tissues, such as tubers, fruits, and seeds, contain **lipoxygenase** enzymes.[48] Lipoxygenases are iron-containing dioxygenases that catalyse a direct reaction of polyunsaturated fatty acids (PUFAs) with oxygen to give 13- and 9-hydroperoxides (Fig. 6.16). Low activities of lipoxygenase are sometimes present in green leaves. The subcellular location of lipoxygenases is variable; they are usually found in several subcellular fractions. Indeed, contamination of plant mitochondrial fractions with lipoxygenase has led to confusion in studies of 'mitochondrial' O_2 uptake.

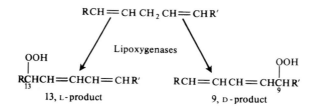

Fig. 6.16. Reactions catalysed by plant lipoxygenases. Purified lipoxygenases from plant tissue catalyse the peroxidation of fatty acids with a *cis,cis*-1,4-pentadiene structure. For example, linoleic acid may be converted into 13-L- or 9-D-hydroperoxy derivatives or both, depending on the enzyme. The reactions are stereospecific.

The first lipoxygenase to be purified was from soybean. It is now known as **soybean lipoxygenase 1**, since at least two other lipoxygenases are now known to exist in this plant (lipoxygenases 2 and 3). Lipoxygenase 1 has a pH optimum of 9 and its preferred substrate is linoleic acid, upon which it acts stereospecifically to produce almost entirely the 13-L-hydroperoxide (Fig. 6.16). Other lipoxygenases have different pH optima, substrate specificity and produce different ratios of products. For example, soybean lipoxygenase 2 has a pH optimum of 6.8 and prefers arachidonic acid as a substrate, but will convert linoleic acid into approximately equal amounts of the 13- and 9-hydroperoxides.

Lipoxygenases contain one (non-haem) iron ion per molecule, and begin their action by abstracting a hydrogen atom from a fatty-acid substrate in a stereospecific manner. Bond rearrangement and oxygen insertion follow. Often, hydroperoxide products interact further with the enzyme. The action of lipoxygenases on PUFAs can produce a 'co-oxidation' of other added materials, such as thiols, carotenoids and chlorophylls. Indeed, lipoxygenases are employed commercially to bleach wheat-flour carotenoids during the bread-making process: as carotenoids are oxidized they lose their yellow-orange colours. Chlorophyll is also bleached in the presence of lipoxygenase and a fatty-acid substrate.

It seems that the action of lipoxygenase generates peroxyl radicals. They are normally reduced to hydroperoxides but can also interact with an oxidizable co-substrate to cause damage. If LH is the fatty-acid substrate and XH the co-oxidizable molecule, the mechanism may be written:

$$L-H \xrightarrow{\text{enzyme}} L^\bullet \xrightarrow{O_2} LOO^\bullet \xrightarrow{\text{enzyme}} \underset{\text{product}}{LOOH}$$

$$LOO^\bullet + XH \rightarrow LOOH + X^\bullet$$

$$X^\bullet + O_2 \rightarrow \text{co-oxidation products}$$

Oxidation of linoleic acid by soybean lipoxygenase isoenzymes 2 and 3 has been reported to be accompanied by singlet O_2 production, which might contribute to co-oxidation reactions under certain circumstances. The singlet O_2 could arise by a Russell-type mechanism:

$$2 \text{ } \text{\textbackslash}CHOO^\bullet \longrightarrow \text{\textbackslash}CHOH + \text{\textbackslash}C{=}O + {}^1O_2$$

Hence lipoxygenase action may cause damage to surrounding tissues.

6.9.2   *The wound response*[25,34]

Many fruits, tubers, seeds and leaves respond to tissue damage by initiating a series of complex biochemical reactions (Fig. 6.17). First, hydrolytic enzymes attack membrane lipids and release fatty acids, many of which are polyunsaturated. For example, linoleic and linolenic acids represent about 75% of the total fatty-acid side-chains in potato tuber lipids. The released fatty acids are then acted upon by lipoxygenases, and the resulting hydroperoxides cleaved, both non-enzymically in the presence of metal ions (Chapter 4) and by the action of cleavage enzymes. This results in a wide variety of products, including volatile aldehydes and hydrocarbon gases, such as ethane and pentane.

   Some of the aldehydes so produced have characteristic smells which are responsible for the aroma of damaged plant tissues, such as new-mown grass and sliced cucumbers (see legend to Fig. 6.17). The odour of crushed green leaves is caused by processes similar to these in Fig. 6.17 and involves generation of such products as 2-hexenal (**leaf aldehyde**) and 3-hexenol (**leaf alcohol**). The reaction sequence shown in Fig. 6.17 can be extremely swift: over 30% of the lipids in potato tuber slices are hydrolysed in less than 15 min even at 3 °C. Hence slicing a potato greatly increases its uptake of O_2, both in the lipoxygenase reaction and in the subsequent metabolism of fatty acid

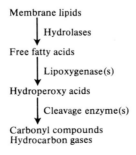

Fig. 6.17. Breakdown of membrane lipids induced on wounding plant tissues. The volatile products from the 9-hydroperoxides of 18 : 2 and 18 : 3 fatty acids are *cis*-**3-nonenal** and *cis*-**3-**, *cis*-**6-nonadienal** respectively. These are the two main components of the odour of sliced cucumbers. The 13-hydroperoxides give **hexanal** and *cis*-**3-hexenal**.

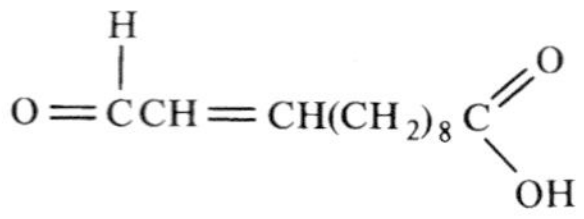

$$O = \overset{\overset{\text{H}}{|}}{C}CH = CH(CH_2)_8 C \overset{\displaystyle\nearrow O}{\underset{\displaystyle\searrow OH}{}}$$

Fig. 6.18. Structure of traumatic acid (12-oxo-*trans*-10-dodecenoic acid). Both the aldehyde form (shown) and the dicarboxylic acid form produced by oxidation of the −CHO groups are biologically active.

products. Indeed, such enzyme oxidation of disrupted plant material can give rise to problems of rancidity and 'off flavour' during processing and storage, that have to be controlled by the use of antioxidants to scavenge peroxyl radicals (Chapter 10). Plant lipoxygenase activity can be inhibited by several antioxidants, such as propyl gallate, certain flavonoids and nordihydroguiaretic acid, which may act by scavenging peroxyl radicals formed during the catalytic cycle and/or by reducing ferric iron at the active site of lipoxygenases to the catalytically inactive Fe^{2+} state.

Formation of lipid peroxides and aldehydes when plant tissues are damaged may play an important role in killing fungi and bacteria attempting to enter the wound, since several of these products have been shown to be toxic. Damage to certain plant tissues results in formation of **traumatic acid** or **wound hormone** (Fig. 6.18), a compound which induces proliferation of new cells. Traumatic acid is formed by oxidative degradation of PUFAs by a process similar to that shown in Fig. 6.17. It seems that the 'wound response' of plant tissues is an example of controlled lipid peroxidation put to a useful purpose.

6.9.3 *The hypersensitive response*[74]

Plants have developed a wide range of other mechanisms to protect themselves against infection by bacteria, fungi or viruses. Often these are associated with the death of a few plant cells at the site of infection, the **hypersensitive response**. The response is triggered by signal molecules (**elicitors**) originating from the invading organism or synthesized by the plant itself. An early effect produced by elicitors is the rapid generation of ROS at the cell surface by the so-called **plant oxidative burst**. The H_2O_2 produced may help to kill pathogens and also to promote later events, such as the synthesis of **phytoalexins** (small molecules with antimicrobial activity), the strengthening of cell walls to 'wall in' the pathogen (e.g. by cross-linking of proteins and more lignin synthesis) and the up-regulation of defences that render the plant resistant to attack by the same pathogen, often for weeks or months (**systemic acquired resistance**, SAR). Both salicylate and catalase seem to be involved in SAR; one (controversial) suggestion[19] is that salicylate decreases activities of catalase and ascorbate peroxidase, allowing H_2O_2 to accumulate, which leads to activation of genes encoding defence systems in surrounding tissues. The SAR signal may even be transmitted between plants[66] by production of

methyl salicylate (oil of wintergreen), a volatile liquid. Injury to tobacco plants by tobacco mosaic virus caused release of methyl salicylate, which could act on surrounding tobacco plants to up-regulate their salicylate synthesis.

The mechanism of ROS production in the plant oxidative burst is not yet fully elucidated, but involvement of NADPH oxidases and/or peroxidases has been suggested.

6.10 Animal lipoxygenases and cyclooxygenases: stereospecific lipid peroxidation

Further examples of useful lipoxygenases can be found in animals. For example, activation of a lipoxygenase enzyme in rabbit reticulocytes has been suggested to initiate the degradation of mitochondria that occurs as these cells mature to form erythrocytes.[60] However, the best-established examples come from studies of leukotriene production.

6.10.1 *Eicosanoids: prostaglandins and leukotrienes*

The prostaglandins and leukotrienes comprise a large and complex family of biologically active lipids derived from PUFAs by insertion of molecular oxygen, achieved by stereospecific free-radical mechanisms at the active sites of enzymes. Prostaglandins and leukotrienes have potent and varied actions in the body; they are involved in regulation of numerous physiological processes and play key roles in inflammation. The prostaglandins and leukotrienes, and other related substances such as the thromboxanes and various families of hydroxy-fatty acids, are often collectively referred to as **eicosanoids**, because they are synthesized from PUFAs with 20 carbon atoms (*eikosi* is Greek for 20). The most important precursor of eicosanoids is arachidonic acid (20:4; eicosatetraenoic acid). The eicosanoids act as local hormones. They are synthesized and released from cells in response to chemical or mechanical stimuli, interact with receptors on adjacent target cells, and are quickly inactivated.

6.10.2 *Prostaglandins and thromboxanes*

Prostaglandins and thromboxanes (sometimes collectively called **prostanoids**) comprise those products of the metabolism of arachidonate and similar fatty acids in which insertion of oxygen leads to ring formation (Fig. 6.19). The term 'prostaglandin' was coined in the 1930s to describe compounds present in semen that could affect blood pressure in animals and cause contraction of smooth muscle. In fact the term prostaglandin is a misnomer, because they arise not from the prostate gland but from the seminal vesicle. This is the richest known source of prostaglandin-synthesizing enzymes, and seminal fluid is the only biological material in which prostaglandins accumulate in any significant concentration. However, every body tissue (except erythrocytes) can make prostaglandins.

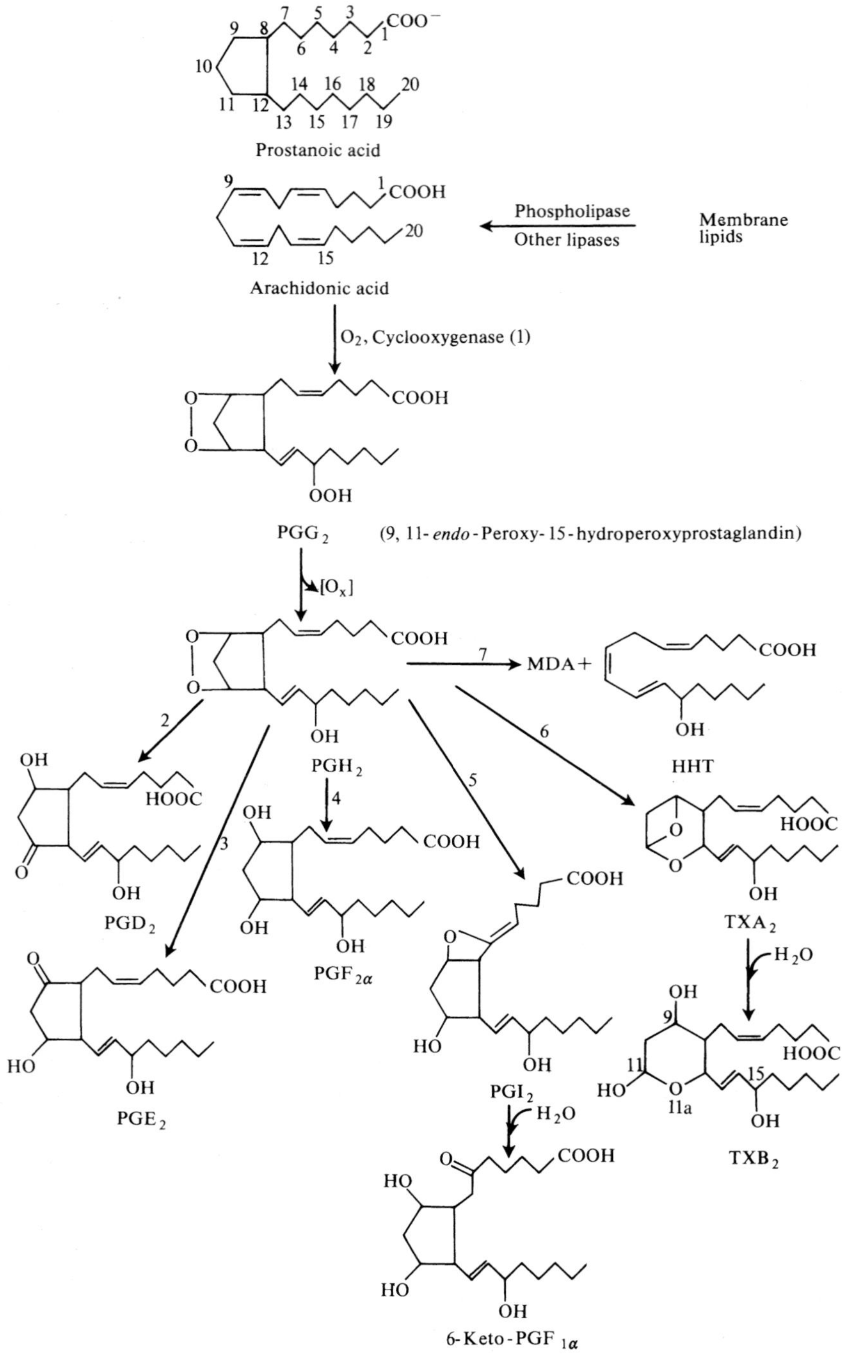
Prostanoic acid
COO⁻
Arachidonic acid
COOH
Phospholipase
Other lipases
Membrane lipids
O2, Cyclooxygenase (1)
OOH
PGG2
(9, 11-endo-Peroxy-15-hydroperoxyprostaglandin)
[Ox]
PGH2
MDA+
HHT
PGD2
PGE2
PGF2α
PGI2
H2O
6-Keto-PGF1α
TXA2
TXB2
HOOC
OH
HO

Fig. 6.19.

The first prostaglandins to be isolated were the chemically stable species PGE_2 and PGF_2 (Fig. 6.19), but better analytical techniques soon demonstrated a series of unstable prostanoids such as PGG_2, PGH_2, thromboxane A_2 (TXA_2), and PGI_2 (**prostacyclin**); all of these compounds possess an oxygen–containing ring structure (Fig. 6.19).

6.10.3   *Prostaglandin structure*

Chemically, prostaglandins may be regarded as derivatives of **prostanoic acid**, a hypothetical C_{20} acid containing a saturated five-membered (**cyclopentane**) ring. There are several types of prostaglandin, distinguished by the chemical nature and geometry of the groups attached to the ring (e.g. E, F, D) and by the number of double bonds in the side-chain (e.g. E_1, E_2, F_1, F_2); all these variations in chemical structure affect biological activity. Prostaglandins are synthesized *in vivo* from fatty acids that contain *cis* double bonds at positions 8, 11 and 14. The most important fatty acid of this type in humans is arachidonic acid (which has *cis* double bonds at positions 5, 8, 11 and 14), but others include 8,11,14-eicosatrienoic acid ($C_{20:3}$) and 5,8,11,14,17-eicosapentaenoic acid ($C_{20:5}$). This last fatty acid is unusually prevalent in the blood and membrane lipids of Eskimos, probably due to their fish-rich diet. It is thought that the relatively low incidence of coronary heart disease shown by Eskimos might, in part, be due to the formation of anti-thrombotic prostanoids from this $C_{20:5}$ fatty acid. However, we will confine further discussion to products derived from arachidonic acid.

6.10.4   *Prostaglandin synthesis*

The first stage in the formation of prostaglandins (and other eicosanoids) is to provide a fatty-acid substrate. Release of arachidonic acid from phospholipids occurs either by activation of phospholipase A_2, or from DAG (generated by activation of the phosphoinositide cascade) by the action of a lipase (Chapter 4). Hence agents increasing intracellular free Ca^{2+} concentrations are promoters of eicosanoid production.

Control over this vital early step can be exerted by a family of Ca^{2+}-binding anti-phospholipase proteins, including the **lipocortins**. Lipocortin was discovered in the early 1980s (and called macrocortin, or lipomodulin) and production is increased in cells exposed to certain steroid hormones. This helps to explain the powerful anti-inflammatory effect of pharmacological doses of these steroids; the increased synthesis of lipocortin stops many (but not all) cells

Fig. 6.19. Structure of prostanoic acid, and of thromboxanes and prostaglandins derived from arachidonic acid. Enzymes involved: 1, prostaglandin endoperoxide synthetase (cyclo-oxygenase); 2, glutathione S-transferase; 3, prostaglandin endoperoxide E isomerase; 4, prostaglandin endoperoxide reductase; 5, prostacyclin synthetase; 6 and 7, prostaglandin endoperoxide: thromboxane A isomerase (thromboxane synthetase). Reactions 2 and 3 require glutathione. Prostaglandins E_3 and E_1 come from other fatty acids. For simplicity, stereochemistry is not shown.

from generating prostaglandins and other pro-inflammatory eicosanoids. Cell injury by various mechanisms can raise intracellular free Ca^{2+} (Chapter 4) and cause formation of excess arachidonic acid; the resulting excess production of eicosanoids can sometimes contribute towards aggravating the injury.

Once arachidonic acid is available, the enzyme **cyclooxygenase I** (sometimes called prostaglandin G/H synthase I) acts upon it to form two endoperoxides, PGG_2 and PGH_2 (Fig. 6.19). Cyclooxygenase 1 (**COX-1**) is constitutively present in almost all mammalian tissues, uses molecular oxygen and is a haem protein. It can be inhibited by aspirin, which acetylates a serine residue at the active site to inactivate the enzyme. A second type of cyclooxygenase, **prostaglandin synthase-2** (**COX-2**)[73] has been detected in many cells (e.g. macrophages), but usually only if expression of the gene encoding it has been induced. Effective stimuli include PAF, IL-1, TNFα and lipopolysaccharide. However, some cells in the brain and kidney may produce COX-2 constitutively.

The gene encoding COX-1 in humans is on chromosome 9, whereas that for COX-2 is on chromosome 1. COX-2 induction may be responsible for a large part of the prostaglandin production at sites of inflammation, whereas COX-1 may synthesize the prostaglandins needed for normal cell functions. Hence selective COX-2 inhibitors might be efficacious in controlling inflammation without the side-effects (e.g. renal damage and gastric ulceration) commonly seen with drugs such as aspirin and indomethacin, which inhibit both enzymes.

6.10.5   *Regulation by 'peroxide tone'*[9]

If crude or purified preparations of COX-1 and its substrate are treated with GSH and glutathione peroxidase to remove traces of lipid peroxides, added arachidonic acid is not immediately oxidized by the enzyme: there is a lag period before rates of oxidation reach maximum. This lag period may be shortened or abolished by adding the peroxide product PGG_2. Other peroxides are also effective, including those formed by the action of lipoxygenase (Section 6.10.8 below) and during the non-enzymic peroxidation of arachidonic acid (Chapter 4). Traces of peroxide react with iron(III) haem at the active site of the enzyme, forming an oxo–haem species and an adjacent tyrosyl radical (on Tyr385 in the ovine enzyme). The latter is believed to abstract stereospecifically a hydrogen atom from arachidonic acid to start the process of PGG_2 formation (Fig. 6.20). Indeed, commercially available arachidonic acid often contains variable levels of peroxides and can therefore produce different initial rates of reaction when used in cyclooxygenase assays. Cumene hydroperoxide and $ONOO^-$ can activate cyclooxygenase, as can H_2O_2, although higher concentrations of the latter are required. For example, the spawning of abalones (marine snails) can be induced by addition of 5 mM H_2O_2 to their sea-water, apparently by stimulation of prostaglandin synthesis.[58]

Hence there is an intimate relationship between non-enzymic lipid peroxidation and prostaglandin metabolism. Efficient peroxide removal or prevention of peroxide formation by antioxidants will slow down prostaglandin synthesis, at least until sufficient PGG$_2$ is formed to activate COX-1 maximally. By contrast, an excess of lipid peroxides can inactivate cyclooxygenase. Many years ago, it was suggested that the **peroxide tone** of the cell (i.e. its content of lipid peroxides, as determined by the balance between peroxide generation by enzymic and non-enzymic mechanisms and the rate of peroxide removal) can control the activity of cyclooxygenase and hence the rate of prostaglandin synthesis. Although the supply of arachidonic acid may be the key regulatory mechanism in normal cells, peroxide tone could be important at sites of inflammation.

PGG$_2$ is then converted into PGH$_2$ (Fig. 6.19) by the peroxidase activity of cyclooxygenase. This peroxidase action is non-specific, in that it can reduce several different hydroperoxides whilst co-oxidizing a variety of other substances. The enzyme operates by a cycle of reactions similar to that of horseradish peroxidase (Chapter 3). It reacts with peroxide, undergoing two-electron oxidation to give a ferryl haem and TyrO$^\bullet$ radical (Fig. 6.20). This 'activated enzyme' can attack arachidonic acid, but also other substrates, accepting one electron to give a compound II. For example, during oxidation of PGG$_2$ *in vitro*, the peroxidase activity can co-oxidize methional into ethene, oxidize adrenalin or diphenylisobenzofuran (another example of the lack of specificity of this compound as a 'singlet O$_2$ scavenger'), activate several aromatic amines to mutagenic products, convert the carcinogen benzpyrene into a quinone (Chapter 9), oxidize 13-*cis*-retinoic acid, cause the emission of light from

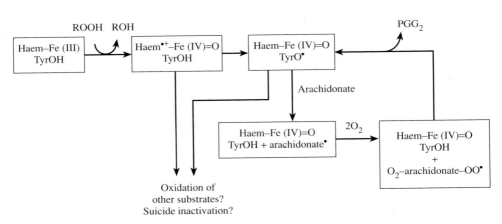

Fig. 6.20. A model of the cyclooxygenase and peroxidase active site chemistry. Note its resemblance to peroxidase chemistry. The 'seeding peroxide' oxidizes the resting, haem-Fe(III), enzyme to give a ferryl species plus haem cation radical. This radical then converts an adjacent tyrosine to TyrO$^\bullet$, which abstracts H$^\bullet$ from arachidonic acid, followed by O$_2$ binding to the resulting carbon-centred radical. In peroxidase catalysis compound I can accept one electron to give compound II (ferryl haem), which can accept a further electron to restore ferric haem enzyme. Adapted from *J. Biol. Chem.* **271**, 33157 (1996) by courtesy of Dr William Smith and the publishers.

luminol, and oxidize bisulphite ion (HSO_3^-) into sulphur trioxide anion (sulphite) radical ($SO_3^{\bullet-}$). The metabolic significance of these co-oxidations is unclear, however. Selenoprotein glutathione peroxidases can also convert PGG_2 to PGH_2.

If activated cyclooxygenase is not provided with an oxidizable substrate, then the haem–associated ferryl species and tyrosine radicals present in compound I can lead to inactivation of the enzyme. Thus appropriate concentrations of phenols, adrenalin and other substrates can cause enhanced prostaglandin synthesis in some tissue systems, by preventing this self-inactivation. Some anti-inflammatory drugs, such as sulindac or 5-aminosalicylate, might exert some of their actions *in vivo* by acting as substrates for the peroxidase (Chapter 9).

Thus antioxidants of different types can have variable actions on prosta-glandin synthesis, not only by regulating the 'peroxide tone' of the system but also by preventing self-inactivation of cyclooxygenase.[9] High concentrations of antioxidants can slow down cyclooxygenase by preventing formation of the traces of peroxides that shorten the lag period. For example, variations in dietary vitamin E intake have been shown to affect eicosanoid production by platelets; increased dietary vitamin E intake in rats also leads to decreased activity of phospholipase A_2 in platelets. There may also be differences in the effects of peroxide tone on COX-1 and COX-2.[9] Both require initiator peroxide but COX-2 seems to need lower levels and so could function at peroxide levels too low for COX-1 to operate at a significant rate.

Although most of the **isoprostanes** detected *in vivo* appear to arise by non-enzymic peroxidation (Chapter 4), the actions of COX-1 in platelets and monocytes appear to generate traces of 8-*epi*-$PGF_{2\alpha}$ in addition to their usual products.[59] 8-*epi*-$PGF_{2\alpha}$ is a powerful vasoconstrictor and also potentiates thromboxane-induced platelet aggregation. ROS released from monocytes (and other phagocytes) can also lead to non-enzymic formation of this and the other isoprostanes.

6.10.6   *Prostacyclins and thromboxanes*

Both PGG_2 and PGH_2 are unstable, with a half-life of only minutes under physiological conditions. They are rapidly transformed into other products such as PGE_2, PGD_2, $PGF_{2\alpha}$, TXA_2 and **prostacyclin** (Fig. 6.19). The major fates of PGG_2 and PGH_2 depend on the tissue. For example, platelets form predominantly TXA_2 and 12-hydroxy-5,8,10-heptadecatrienoic acid (HHT), endothelial cells generate predominantly prostacyclin, and mast cells make PGD_2. Although a detailed discussion of the tissue-specific complexities of prostaglandin biosynthesis is not appropriate here, it is worth saying a little about platelets and endothelial cells.

TXA_2 is formed upon platelet activation and promotes their aggregation. It is unstable (half-life about 20 s under physiological conditions) and quickly forms the stable, biologically inactive thromboxane B_2 (TXB_2; Fig. 6.19). TXA_2 is also a potent vasoconstrictor and aids the haemostatic process by assisting platelet clumping and decreasing blood flow. Its short half-life means that its action is localized. Low doses of aspirin prolong bleeding time after vessel injury by inhibiting cyclooxygenase and thus decreasing TXA_2 production. Platelets cannot synthesize cyclooxygenase; once the enzyme is irreversibly inhibited by aspirin, TXA_2 synthesis is blocked until the platelet is replaced. TXA_2 is also produced by activated neutrophils.

The enzyme **thromboxane synthase** (Fig. 6.19) cleaves PGH_2 to HHT, also producing malondialdehyde (MDA). MDA reacts with thiobarbituric acid (TBA) to give a pink chromogen, and prostaglandin endoperoxides decompose under the acid-heating conditions of the TBA test (Chapter 5) to form MDA. Indeed, the generation of TBA-reactive material (TBARS) by non-enzymically produced lipid peroxides also appears largely due to the formation and decomposition of cyclic peroxides (Chapter 4). Hence application of the TBA test to tissues synthesizing prostaglandins can overestimate lipid peroxidation (Chapter 5). For example, the amount of TBARS measured in rabbit serum decreased if the animals were pretreated with aspirin, suggesting that prostaglandins are a significant contributor to TBARS. This should be borne in mind in attempts to use TBARS as an index of lipid peroxidation *in vivo* in various disease states.

The vascular endothelium synthesizes prostacyclin (PGI_2) which dilates blood vessels and is a powerful inhibitor of platelet aggregation. Its actions therefore complement those of $NO^\bullet$ but oppose those of TXA_2, and the ratios between all the molecules are important in controlling platelet–vessel wall interactions and local blood flow. For example, nitric oxide has been shown[47] to lead to activation of both cyclooxygenase and peroxidase activities of COX-1, and to counteract enzyme self-inactivation. Prostacyclin is unstable (half-time of minutes), decomposing to give 6-*keto*-$PGF_{1\alpha}$ (Fig. 6.19). The capacity of vessel walls to synthesize prostacyclin *in vivo* seems less affected by aspirin than is the ability of platelets to synthesize TXA_2, so that low doses of aspirin may be useful in the prevention of unwanted thromboses. Prostacyclin has many important effects, such as to potentiate pain and swelling at sites of inflammation, and to regulate fluid transport in the kidney and gut and gastric acid secretion.

6.10.7   *Gene knockouts*[73]

Targeted gene disruption (Appendix 2) has been used to produce mice lacking COX-1 or COX-2. COX-1$^-$ mice seem healthy but show decreased platelet aggregation, decreased inflammatory response to arachidonic acid and less indomethacin-induced stomach ulceration, but the females produce few live offspring. This suggests a role for COX-1-derived prostaglandins in normal reproduction, e.g. regulation of uterine function. COX-2$^-$ mice show more obvious pathological changes: renal nephropathy, cardiac fibrosis, increased susceptibility to peritonitis and absence of corpora lutea in the ovary have been reported by various groups.

6.10.8   *Leukotrienes and other lipoxygenase products*[24]

In Section 6.9 we described the importance of lipoxygenase enzymes as initiators of 'controlled lipid peroxidation' in the wound response of plant tissues, an event known for many years. Not until 1974 was it discovered that a lipoxygenase enzyme is active in platelets, and similar enzymes have since been studied in many mammalian tissues. Like plant lipoxygenases, the animal enzymes contain non-haem iron.

Platelet lipoxygenase acts upon arachidonic acid to form 12-hydroperoxy-5,8,11,14-eicosatetraenoic acid (**12-HPETE**). This is unstable and can be reduced to the 12-hydroxy derivative (**12-HETE**) *in vivo*. Reduction appears to be achieved by platelet glutathione peroxidases in the presence of GSH, since 12-HETE formation is decreased in the platelets of selenium-deficient rats. Lipoxygenases present in other tissues introduce oxygen atoms at different places in the carbon chain (Fig. 6.21 gives some examples), e.g. formation of 12-HETE has been observed in skin, and of 5-HETE and 15-HETE in rabbit and human neutrophils, plus various 5,12-DHETEs (specific stereoisomers of 5,12-dihydroxy-6,8,10,14-eicosatetraenoic acids). At concentrations effective against cyclooxygenases, aspirin and many other cyclooxygenase inhibitors do not affect lipoxygenase. By contrast, anti-inflammatory steroids will inhibit

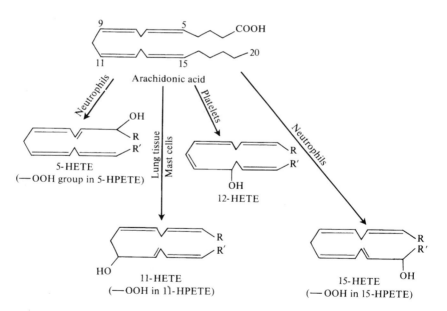

Fig. 6.21. Some products of arachidonic acid metabolism by mammalian lipoxygenases. In activated human neutrophils and alveolar macrophages both 5-lipoxygenase and a protein required for its activation (**FLAP**: 5-lipoxygenase activating protein) are present in the nuclear envelope.[75] LTA$_4$ can be released or converted to LTB$_4$, released as a chemotactic agent (Fig. 6.22).

both pathways by decreasing release of the fatty-acid substrate, as well as decreasing induction of COX-2. Lipoxygenases, like cyclooxygenases, require low levels of 'seeding peroxides' and are probably also affected by the 'peroxide tone' of the cell. The peroxides oxidize inactive Fe^{2+} enzyme to active Fe(III) enzyme.

Animal lipoxygenases, like those of plants, are inhibited by NDGA (Section 6.9 above) which appears to act by reducing Fe(III) at the active site of the enzyme to Fe^{2+}. Several phenolic antioxidants (e.g. some flavonoids; Chapter 3) are also inhibitors of lipoxygenase, but the relationship between their ability to inhibit lipoxygenase and their antioxidant and iron-binding properties has not yet been fully elucidated.[39] Lipoxygenases can also be inhibited by 5,8,11,14–eicosatetraynoic acid (**ETYA**), an analogue of arachidonic acid in which all the double bonds have been replaced by triple bonds, so preventing hydrogen atom abstraction at the enzyme active site. ETYA also inhibits cyclooxygenase.

The HPETE compounds are precursors of a range of chemicals with potent biological activity known as the **leukotrienes**. Leukotrienes differ structurally from prostanoids in that they have no cyclopentane ring. Instead they have a conjugated triene structure (three double bonds separated from each other by single bonds; Fig. 6.22); leukotrienes C_4, D_4, and E_4 also possess a GSH–derived substituent. Leukotriene C_4 (LTC$_4$) is formed by the conjugation of leukotriene A_4 (LTA$_4$) with GSH. Removal of a glutamate residue by glutamyltransferase activity yields leukotriene D_4 (LTD$_4$), which can be further degraded to leukotriene E_4, which is less biologically active.

Figure 6.22 shows the leukotrienes that arise from 5-HPETE but others can be generated from different HPETEs. LTA$_4$ is an unstable epoxide structure (Fig. 6.22) which, in some tissues, is enzymatically hydrolysed to leukotriene B_4 (LTB$_4$), a powerful chemotactic agent for neutrophils which promotes aggregation, degranulation and the 'respiratory burst'. Neutrophils themselves make LTB$_4$, attracting more cells to a site of inflammation. LTC$_4$ and LTD$_4$ increase vascular permeability. Macrophages produce LTB$_4$ and other leukotrienes, as do many other cell types, including mast cells. These mediators may contribute to the inflammation and bronchial constriction seen in allergic asthma (along with other mediators such as PGD$_2$, PAF, and histamine). Indeed, LTC$_4$ and LTD$_4$ are the principal components of 'slow–reacting substance A' (SRS-A), a mixture of compounds produced when lung tissue taken from humans allergic to a given substance is perfused with that substance. SRS-A increases vascular permeability and contracts bronchial muscles in a slow and sustained way (unlike the rapid and faster-acting action of histamine).

Of course, leukotrienes can also be destroyed at sites of inflammation, e.g. by OH$^\bullet$, HOCl and ONOOH. Targeted disruption[24] of the 5-lipoxygenase gene in mice produces apparently-normal animals, except that their response to inflammation induced by arachidonic acid or PAF is attenuated, there is somewhat less recruitment of neutrophils at sites of inflammation (in some experiments) and less eosinophil recruitment in the allergen-exposed lung.

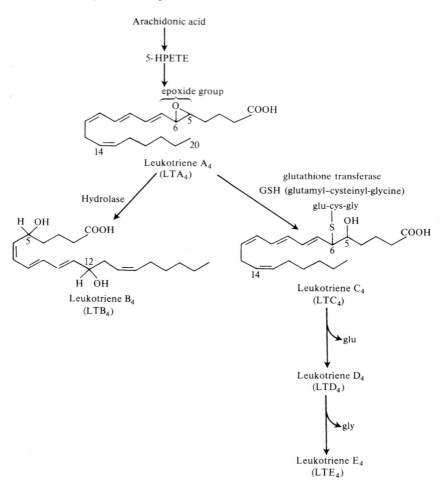

Fig. 6.22. The leukotrienes. Leukotriene A_4 (LTA$_4$) is an epoxide (full name: 5,6-epoxy-7,9,11,14-eicosatetraenoic acid). It can be hydrolysed enzymatically into leukotriene B$_4$ (LTB$_4$; 5,12-dihydroxy-6,*cis*-8,*trans*-10,*trans*-14,*cis*-eicosatetraenoic acid) or non-enzymically to other 5,12- and 5,6-dihydroxy acids. Only LTB$_4$ is shown for simplicity. LTA$_4$ is also converted by enzyme-catalyzed conjugation with GSH into LTC$_4$, which can be converted into LTD$_4$ and LTE$_4$ by successive removal of amino acids. Another group of products can be formed by initial oxygenation of arachidonic acid at C-15; for example, incubating neutrophils with 15-HPETE generates the **lipoxins**. Other hydroxylated compounds such as trihydroxy (20-OH-LTB$_4$) and dicarboxylate (20-COOH-LTB$_4$) derivatives have also been observed to be formed in neutrophils. The same enzyme (5-lipoxygenase) is thought both to generate 5-HPETE and to convert it into LTA$_4$.

6.10.9   *Commercial PUFAs: a warning*[29]

Before leaving this area, we wish to re-emphasize that PUFAs such as arachidonate are easily peroxidized by non-enzymic methods, giving multiple products, including isoprostanes (Chapter 4). Commercially available PUFAs are often extensively peroxidized, thus causing variable results in, for example, cycloxygenase and lipoxygenase assays. If such fatty acids are included in systems containing metal ions (either added deliberately, or present as contaminants in the reagents), then products can be formed not only by cycloxygenase and lipoxygenase activity but also by the attack of $RO_2^{\bullet}/RO^{\bullet}$ radicals formed by metal-dependent peroxide decomposition.

References

1. Albina, JE (1995) On the expression of NOS by human macrophages. Why no NO? *J. Leuk. Biol.* **58**, 643.
2. Aneshansley, DJ *et al.* (1993) Thermal concomitants and biochemistry of the explosive discharge mechanism of some little known bombardier beetles. *Experientia* **39**, 366.
3. Aruoma, OI *et al.* (1988) The antioxidant action of taurine, hypotaurine and their metabolic precursors. *Biochem. J.* **256**, 251.
4. Augusto, O *et al.* (1996) Possible roles of $NO^{\bullet}$ and $ONOO^{-}$ in murine leishmaniasis. *Brazil. J. Med. Biol. Res.* **29**, 853.
5. Babior, BM (1997) Superoxide: a two-edged sword. *Brazil. J. Med. Biol. Res.* **30**, 141.
5a. Belaaouaj, A *et al.* (1998) Mice lacking neutrophil elastase reveal impaired host defence against gram negative bacterial sepsis. *Nature Med.* **4**, 615.
6. Boxer, SG (1996) Another green revolution. *Nature* **383**, 485.
7. Bunn, HF and Poyton, RO (1996) O_2 sensing and molecular adaptation to hypoxia. *Physiol. Rev.* **76**, 839.
8. Campbell, AK (1989) Living light: biochemistry, function and biomedical applications. *Essays Biochem.* **24**, 41.
9. Capdevila, JH *et al.* (1995) The catalytic outcomes of the constitutive and mitogen inducible isoforms of PGH_2 synthase are markedly affected by GSH and glutathione peroxidase(s). *Biochemistry* **34**, 3325.
10. Chao, C *et al.* (1996) Participation of NO and iron in the oxidation of DNA in asbestos-treated human lung epithelial cells. *Arch. Biochem. Biophys.* **326**, 152.
11. Chaturvedi, V *et al.* (1996) Oxidative killing of *C. neoformans* by human neutrophils. *J. Immunol.* **156**, 3836.
12. Collins, T (1995) Adhesion molecules in leukocyte emigration. *Sci. Amer. Sci. Med.* November/December issue, p. 28.
13. Cooper, CE *et al.* (1996) The relationship of intracellular iron chelation to the inhibition and regeneration of ribonucleotide reductase. *J. Biol. Chem.* **271**, 20291.
14. Cronstein, BN (1994) Adenosine, an endogenous anti-inflammatory agent. *J. Appl. Physiol.* **76**, 5.
15. Darley-Usmar, V and Halliwell, B (1996) Blood radicals. *Pharmacol. Res.* **13**, 649.

16. Del Maestro, RF *et al.* (1981) Increase in macrovascular permeability induced by enzymatically generated free radicals. II. Role of $O_2^{\bullet-}$, H_2O_2 and $OH^{\bullet}$. *Microvasc. Res.* **5**, 423.

17. Deme, D *et al.* (1994) The Ca^{2+}/NADPH-dependent H_2O_2 generator in thyroid plasma membrane: inhibition by diphenyleneiodonium. *Biochem. J.* **301**, 75.

18. Dunford, HB (1995) One-electron oxidations by peroxidases. *Xenobiotica* **25**, 725.

19. Durner, J and Klessig, DF (1995) Inhibition of ascorbate peroxidase by salicylic acid and 2,6-dichloroisonicotinic acid, two inducers of plant defense responses. *Proc. Natl Acad. Sci. USA* **92**, 11312.

20. Edwards, SW and Lloyd, D (1988) The relationship between $O_2^{\bullet-}$ generation, cytochrome *b* and O_2 in activated neutrophils. *FEBS Lett.* **227**, 39.

21. Fletcher MP and Seligmann, BE (1986) PMN heterogeneity: long-term stability of fluorescent membrane potential responses to the chemoattractant N-formyl-methionyl-leucyl-phenylalanine in healthy adults and correlation with respiratory burst activity. *Blood* **68**, 611.

22. Fliss, H *et al.* (1983) Oxidation of methionine residues in proteins of activated human neutrophils. *Proc. Natl Acad. Sci. USA* **80**, 7160.

23. Fubini, B and Mollo, L (1995) Role of iron in the reactivity of mineral fibres. *Toxicol. Lett.* **82/83**, 951.

24. Funk, CD (1996) The molecular biology of mammalian lipoxygenases and the quest for eicosanoid functions using lipoxygenase-deficient mice. *Biochim. Biophys. Acta* **1304**, 65.

25. Galliard, T (1978) Lipolytic and lipoxygenase enzymes in plants and their action in wounded tissue. In *Biochemistry of Wounded Plant Tissues* (Kahl, G., ed.), p. 155. Walter de Gruyter & Co., Berlin.

26. Gaudu, P *et al.* (1996) The irreversible inactivation of ribonucleotide reductase from *E. coli* by $O_2^{\bullet-}$. *FEBS Lett.* **387**, 137.

27. Gonzalez-Flecha, B and Demple, B (1994) Intracellular generation of $O_2^{\bullet-}$ as a by-product of *Vibrio harveyi* luciferase expressed in *E. coli*. *J. Bacteriol.* **176**, 2293.

28. Gryglewski, RJ *et al.* (1986) Superoxide anion is involved in the breakdown of endothelium-derived vascular relaxing factor. *Nature* **320**, 454.

29. Gutteridge, JMC and Kerry, PJ (1982) Detection by fluorescence of peroxides and carbonyls in samples of arachidonic acid. *Br. J. Pharmacol.* **76**, 459.

30. Haas, A. and Goebel, W (1992) Microbial strategies to prevent O_2-dependent killing by phagocytes. *Free Rad. Res. Commun.* **16**, 137.

31. Halliwell, B (1978) Lignin synthesis: the generation of H_2O_2 and $O_2^{\bullet-}$ by horseradish peroxidase and its stimulation by manganese II and phenols. *Planta* **140**, 81.

32. Hamers, MN *et al.* (1984) Kinetics and mechanism of the bactericidal action of human neutrophils against *E. coli. Blood* **64**, 635.

33. Hannah, S *et al.* (1995) Hypoxia prolongs neutrophil survival *in vitro*. *FEBS Lett.* **372**, 233.

34. Hatanaka, A *et al.* (1993) The biogeneration of green odour by green leaves. *Phytochemistry* **34**, 1201.

35. Henderson, LM and Chappell, JB (1996) NADPH oxidase of neutrophils. *Biochim. Biophys. Acta* **1273**, 87.

36. Hurst, JK and Lymar, SV (1997) Toxicity of $ONOO^-$ and related RNS towards *E. coli. Chem. Res. Toxicol.* **10**, 802.

37. Hyde, SM and Wood, PM (1997) A mechanism for the production of OH• by the brown–rot fungus *Caniophora puteana*: Fe(III) reduction by cellobiose dehydrogenase and Fe(II) oxidation at a distance from the hyphae. *Microbiology* **143**, 259.

38. Kanofsky, JR *et al.* (1988) Singlet O_2 production by human eosinophils. *J. Biol. Chem.* **263**, 9692.

39. Kemal, C *et al.* (1987) Reductive inactivation of soybean lipoxygenase I by catechols: a possible mechanism for regulation of lipoxygenase activity. *Biochemistry* **26**, 7064.

40. Kettle, AJ and Winterbourn, CC (1997) Myeloperoxidase: a key regulator of neutrophil oxidant production. *Redox Rep.* **3**, 3.

41. Kirk, TK (1988) Lignin degradation by *Phanerochaete chrysosporium*. *ISI Atlas Sci. Biochem.* p 71.

42. Landino, LM *et al.* (1996) Peroxynitrite, the coupling product of NO• and $O_2^{•-}$, activates prostaglandin biosynthesis. *Proc. Natl Acad. Sci. USA* **93**, 15069.

43. Laursen, JB *et al.* (1997) Role of $O_2^{•-}$ in angiotensin II–induced but not catecholamine–induced hypertension. *Circulation* **95**, 588.

44. Lefer, AM (1997) NO•: Nature's naturally occurring leukocyte inhibitor. *Circulation* **95**, 553.

45. Lepoivre, M *et al.* (1994) Quenching of the tyrosyl free radical of ribonucleotide reductase by NO•. *J. Biol. Chem.* **269**, 21891.

46. Lo, SK *et al.* (1993) H_2O_2–induced increase in endothelial adhesiveness is dependent on ICAM-1 activation. *Am. J. Physiol.* **264**, L406.

47. Maccarrone, M *et al.* (1997) NO donors activate the COX and peroxidase activities of prostaglandin H synthase. *FEBS Lett.* **410**, 470.

48. Mack, A.J. *et al.* (1987) Lipoxygenase isozymes in higher plants: biochemical properties and physiological role. In *Isozymes: Current Topics in Biological and Medical Research*, Vol. 13, p. 127.

49. MacMillan-Crow, LA *et al.* (1996) Nitration and inactivation of MnSOD in chronic rejection of human renal allografts. *Proc. Natl Acad. Sci. USA* **93**, 11853.

50. Maly, FE and Schürer-Maly, CC (1995) How and why cells make $O_2^{•-}$: the 'phagocytic' NADH oxidase. *News Physiol. Sci.* **10**, 233.

51. Marsh, ENG (1995) A radical approach to enzyme catalysis. *BioEssays* **17**, 431.

52. Mayeno, AN *et al.* (1989) Eosinophils preferentially use Br^- to generate halogenating agents. *J. Biol. Chem.* **264**, 5660.

53. McNally, JA and Bell, AL (1996) Myeloperoxidase-based chemiluminescence of polymorphonuclear leukocytes and monocytes. *J. Biolum. Chemilum.* **11**, 99.

54. Morpeth, FF (1985) Some properties of cellobiose oxidase from the white-rot fungus *Sporotrichum pulverulentum*. *Biochem. J.* **228**, 557.

55. Morse, DE *et al.* (1977) H_2O_2 induces spawning in mollusks, with activation of prostaglandin endoperoxide synthetase. *Science* **196**, 298.

56. Mossman, BT *et al.* (1996) Transfection of MnSOD gene into hamster tracheal epithelial cells ameliorates asbestos-mediated cytotoxicity. *Free Rad. Biol. Med.* **21**, 125.

57. Münzel, T (1996) Hydralazine prevents nitroglycerin tolerance by inhibiting activation of a membrane-bound NADH oxidase. *J. Clin. Invest.* **98**, 1465.

58. Nauseef, WM (1988) Myeloperoxidase deficiency. *Hematol. Oncol. Clin. North Amer.* **2**, 135.

59. Praticó, D and FitzGerald, GA (1996) Generation of 8-epiPGF$_{2\alpha}$ by human monocytes. *J. Biol. Chem.* **271**, 8919.

60. Rapoport, SM and Schewe, T (1986) The maturational breakdown of mitochondria in reticulocytes. *Biochim. Biophys. Acta* **864**, 471.

61. Remick, DG and Villarete, L (1996) Regulation of cytokine gene expression by ROS and RN intermediates. *J. Leuk. Biol.* **59**, 471.

62. Segal, AW (1995) The NADPH oxidase of phagocytic cells is an electron pump that alkalinises the phagocytic vacuole. *Protoplasma* **184**, 86.

63. Shatwell, KP *et al.* (1996) The FRE1 ferric reductase of *S. cerevisiae* is a cytochrome *b* similar to that of NADPH oxidase. *J. Biol. Chem.* **271**, 14240.

64. Shearer, MJ (1995) Vitamin K. *Lancet* **345**, 229.

65. Shi, MM *et al.* (1996) Regulation of MIP-1α mRNA by oxidative stress. *J. Biol. Chem.* **271**, 5878.

66. Shulaev, V *et al.* (1997) Airborne signalling by methyl salicylate in plant pathogen resistance. *Nature* **385**, 718.

66a. Someya, A *et al.* (1997) Studies on the $O_2^{\bullet-}$-producing enzyme of eosinophils and neutrophils. *Arch. Biochem. Biophys.* **345**, 207.

67. Thannickal, VJ and Fanburg, BL (1995) Activation of an H_2O_2-generating NADH oxidase in human lung fibroblasts by TGFβ1. *J. Biol. Chem.* **270**, 30334.

68. Thompson, JE *et al.* (1987) The role of free radicals in senescence and wounding. *New Phytol.* **105**, 317.

69. Weiss, SJ (1989) Tissue destruction by neutrophils. *N. Engl. J. Med.* **320**, 365.

70. Westendorp, RGJ *et al.* (1997) Genetic influence on cytokine production and fatal meningococcal disease. *Lancet* **349**, 170.

71. Wheeler, MA *et al.* (1997) Bacterial infection induces NOS in human neutrophils. *J. Clin. Invest.* **99**, 110.

72. Whittaker, MM *et al.* (1996) Glyoxal oxidase from *P. chrysosporum* is a new radical–copper oxidase. *J. Biol. Chem.* **271**, 681.

73. Williams, CS and DuBois, RN (1996) Prostaglandin endoperoxide synthase: why two isoforms? *Am. J. Physiol.* **270**, G393.

74. Wojtaszek, P (1997) Oxidative burst: an early plant response to pathogen infection. *Biochem. J.* **322**, 681.

75. Woods, JW *et al.* (1995) 5-Lipoxygenase is located in the euchromatin of the nucleus in resting human alveolar macrophages and translocates to the nuclear envelope upon cell activation. *J. Clin. Invest.* **95**, 2035.

Notes

[a]Lectins are plant proteins that bind with high affinity to carbohydrate-containing substances, such as the glycoproteins found in cell-surface membranes.

[b]The nomenclature used is as follows: phox, phagocyte oxidase; p, protein; gp, glycoprotein; the numbers are the relative molecular mass (in thousands).

7

Oxidative stress and antioxidant protection: some special cases

7.1 Introduction

In previous chapters we have described the problems faced by aerobes in coping with O_2, and the ways in which they can counter or repair oxidative damage. For example, the lung is exposed to the highest concentration of O_2 of any body tissue and it uses a wide range of enzymic and non-enzymic defence systems (Chapter 3). The swim-bladders of certain deep-living fish (Section 1.6.4) have to tolerate high O_2 partial pressures, e.g. the effective O_2 concentration at a depth of 3000 m is 2500 times greater than ambient, yet the bladder remains undamaged. The fish as a whole cannot tolerate anywhere approaching this O_2 concentration, and so the bladder must be specially protected. The superoxide dismutase (SOD) activity of swim-bladders from several fish is higher than in other fish tissues examined, but there is nothing exceptional about their catalase and glutathione peroxidase activities.[63]

The purpose of this Chapter is to focus on some systems that have exceptionally difficult problems of antioxidant defence: erythrocytes, chloroplasts, the eye and premature babies. It is convenient here also to discuss the relation of reactive oxygen and nitrogen species (ROS/RNS) to problems of conception and embryonic development, and to evaluate the relation between oxidative stress and exercise.

7.2 Erythrocytes

Red blood cells, or erythrocytes, are biconcave discs averaging about 7.7 μm in diameter (Fig. 7.1). In mammals, they lack a nucleus. Their average lifespan in the human bloodstream is around 120 days. The plasma membrane of the erythrocyte contains some 15 major proteins, is selectively permeable and encloses the cytoplasm, of which about 33% is haemoglobin (up to 280 million molecules per erythrocyte). The phospholipid distribution across the erythrocyte membrane bilayer is asymmetrical, e.g. phosphatidylcholine is largely present in the outer leaflet, phosphatidylserine and phosphatidylethanolamine on the cytoplasmic half of the membrane bilayer. A **green haemoprotein**[123] has also been isolated from erythrocytes and reacts with H_2O_2 *in vitro* to give higher oxidation states of the iron. The physiological role (if any) of green haemoprotein is completely unknown. Green pigments can also be produced by oxidative denaturation of haemoglobin.

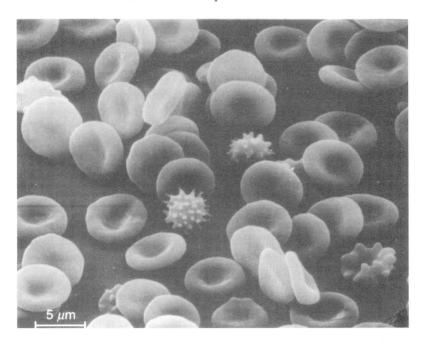

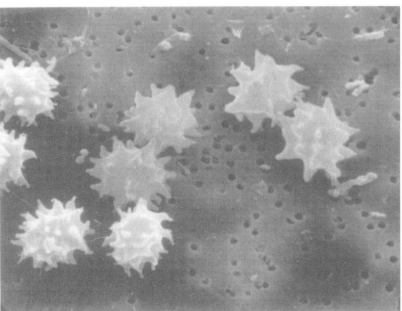

Fig. 7.1. Top: erythrocytes. Bottom: echinocytes produced by peroxide treatment. Electron micrographs by courtesy of Dr David Hockley and Prof. C Rice-Evans. Peroxide damage causes morphological change, increased membrane rigidity with consequent decreased deformability, lipid peroxidation and damage to membrane proteins, including spectrin and band 3. Peroxide-treated erythrocytes also show increased senescent antigen on their surfaces.

Adult males have 14–16.5 g of haemoglobin per 100 ml of blood, and females 12–15 g/100 ml. Haemoglobin has four haem groups, one on each of the four protein subunits (Chapter 1). As the erythrocyte passes through the lung capillaries, the majority of the haem rings coordinate with O_2. As we saw in Chapter 1, the iron in the haem ring of deoxyhaemoglobin is in the Fe^{2+} state. When O_2 attaches, an intermediate structure results, in which an electron is delocalized between the iron ion and the O_2

$$Fe^{2+}-O_2 \leftrightarrow Fe(III)-O_2^{\bullet-}$$

The bonding is intermediate in character between that in Fe^{2+} bonded to O_2, and that in $Fe(III)$ bonded to superoxide. Every so often a molecule of oxyhaemoglobin undergoes decomposition and releases $O_2^{\bullet-}$

$$haem-Fe^{2+}-O_2 \rightarrow O_2^{\bullet-} + haem-Fe(III)$$

The $Fe(III)$ product, **methaemoglobin** is unable to bind oxygen.

7.2.1   *What problems do erythrocytes face?*

Erythrocytes have to keep their membranes intact for long periods (Table 7.1) in the face of:

(1) A constant flux of $O_2^{\bullet-}$ from haemoglobin autoxidation. It has been estimated that about 3% of the haemoglobin undergoes oxidation every day. If this is so, complete oxidation should occur in 33–34 days and the cell would be unable to carry O_2.[120]

(2) Carrying high concentrations of a potentially pro-oxidant haem protein (haemoglobin) and O_2 inside a membrane rich in polyunsaturated fatty acid (PUFA) side-chains. Haem proteins can be degraded by excess peroxides to release haem and iron ions: at lower peroxide concentrations they produce ferryl species and amino-acid radicals that can oxidize many substrates (Section 3.18.3). Indeed, the phagocytosis of immunoglobulin G-coated erythrocytes by macrophages depresses macrophage function,[66] possibly by iron- or haem-dependent free-radical damage as the haemoglobin interacts with ROS produced by the macrophage. Pro-oxidant reactions of haemoglobin, as well as its ability to bind $NO^{\bullet}$, have caused problems[2] in the development of cross-linked cell-free haemoglobins as 'blood substitutes'. One attempt to overcome this has been to cross-link SOD and catalase with the haemoglobin.[26a]

(3) Repeated deformation and consequent physical stress upon the membranes as red blood cells squeeze through tiny capillaries at 37 °C. Deformation can aggravate the effects of lipid peroxidation in causing leakage of ions across the erythrocyte membrane.[78]

(4) Low metabolic activities, with no ability to synthesize new proteins or lipids. Unlike most cells, erythrocytes cannot replace oxidized lipids or proteins.

(5) It has been suggested that erythrocyte cytoplasm contains a low molecular mass 'catalytic' iron pool, but data are conflicting.

Table 7.1. Erythrocyte lifespan and antioxidant defence enzymes: a comparison between species

Species	Erythrocyte lifespan (days)	SOD (U/g Hb)	Glutathione peroxidase (U/g Hb)	GSH (μM/g Hb)	Catalase (IU $\times 10^{-4}$/g Hb)	Glutathione reductase (IU/g Hb)
Homo sapiens	120–150	2352	32	6.7	14	7.8
Macaca mulatta (Rhesus monkey)	100	2572	130	9.1	19	5.5
Canis familiaris (domestic dog)	115	2118	130	7.8	n.d.	3.2
Felis cattus (domestic cat)	77	2885	156	8.0	24	7.7
Oryctolagus cuniculus (rabbit)	67	3324	39	7.2	13	3.9
Ovis aries (sheep)	150	3132	124	8.9	2	1.2
Bos taurus (cattle)	175	3259	165	7.7	8	1.0
Rattus rattus (rat)	67	2967	141	8.0	11	1.8
Mus musculus (mouse)	51	3148	314	7.0	5	9.8
Misocricetus auratus (hamster)	79	2771	22	7.5	6	2.2

n.d., not determined. IU, international units.

Erythrocytes are long-lived cells in all animal species. SOD and GSH levels are broadly comparable between species, but activities of glutathione peroxidase and catalase vary widely. Data abstracted from Kurata, M *et al.* (1993) *Comp. Biochem. Physiol.* **106B**, 477 by courtesy of Dr M Suzuki and the publisher.

7.2.2 Solutions: antioxidant defence enzymes

Erythrocytes contain high concentrations of CuZnSOD; indeed, bovine erythrocyte CuZnSOD was the first SOD to be purified (Chapter 3). No MnSOD is present, correlating with the lack of mitochondria. SOD rapidly converts $O_2^{\bullet -}$ to H_2O_2, which is removed by catalase and selenoprotein glutathione peroxidase enzymes (Fig. 7.2). The relative importance of these two enzymes in catabolizing H_2O_2 may vary between species (Table 7.1): the conventional view is that glutathione peroxidase is more important at low physiological fluxes of H_2O_2 whereas catalase, with its higher K_m for H_2O_2, becomes more important at higher H_2O_2 levels (Chapter 3). Indeed, erythrocytes in blood can act as sinks for H_2O_2 and $O_2^{\bullet -}$ generated in the plasma or from activated phagocytes: H_2O_2 can cross their membranes easily and the erythrocyte has an anion channel through which $O_2^{\bullet -}$ can move;[66a] even some $ONOO^-$ can get through.[31a] **Thiol-specific antioxidant** has also been identified in erythrocytes.[23]

The NADPH needed for glutathione reductase is provided by the metabolism of glucose by the pentose phosphate pathway, which also supplies NADPH for a **methaemoglobin reductase** enzyme that converts methaemoglobin back to ferrous haemoglobin to permit continued O_2 transport. However, most methaemoglobin reductase activity in erythrocytes is NADH-dependent, using an enzyme system containing flavin adenine dinucleotide (FAD) and cytochrome b_5. NADH is provided by glycolysis.

Erythrocytes can also reduce certain extracellular electron acceptors, such as ferricyanide, and ferricytochrome *c*, apparently by exporting electrons across the membrane by the action of transmembrane NADH : acceptor oxidoreductases. The role of this in erythrocyte metabolism is unclear as yet.

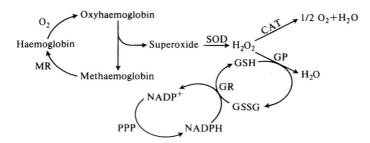

Fig. 7.2. Protection of erythrocytes against oxidative damage. MR, Methaemoglobin reductase (mostly NADH-dependent, some NADPH-dependent); SOD, Cu–Zn superoxide dismutase; CAT, catalase; GP, glutathione peroxidase; GR, glutathione reductase, PPP, pentose phosphate pathway (first enzyme: glucose 6-phosphate dehydrogenase). Haemoglobin is oxidized by H_2O_2 to damaging ferryl/amino-acid radicals and degraded by excess H_2O_2 with loss of haem and iron ions so it is important that erythrocytes dispose of H_2O_2 rapidly.

7.2.3   *Solutions: low-molecular mass antioxidants*

The erythrocyte membrane contains α-tocopherol as a chain-breaking anti-oxidant. Indeed, an early sign of inadequate vitamin E intake is the increased lysis of red blood cells (**haemolysis**) when treated with high (grossly unphysiological) levels of H_2O_2 in the presence of a catalase inhibitor (azide): the **peroxide stress haemolysis test**. Haemoglobin plays an important role in the peroxide stress test:[25] prior conversion of oxy- to methaemoglobin by chemical treatment of the cells diminishes the peroxidation observed on subsequent addition of H_2O_2. Older human erythrocytes in the circulation contain a higher α-tocopherol/arachidonic acid ester ratio than do younger ones, because the lipid content of the membrane decreases with age whereas vitamin E content remains approximately constant.

α-Tocopheryl radical is probably recycled in the membrane through the action of cytoplasmic (or perhaps even extracellular) ascorbate. The NADH–cytochrome b_5 reductase can also recycle α-tocopheryl radical to α-tocopherol, *in vitro* at least. Oxidized ascorbate in the erythrocyte cytoplasm can be regenerated by the action of GSH-dependent **dehydroascorbate reductase**,[70] although this activity is probably a property of such proteins as glutaredoxin (Chapter 3). Erythrocytes can also take up dehydroascorbate from the plasma and recycle it to ascorbate.

7.2.4   *Erythrocyte peroxidation in health and disease*

Despite the inability of the erythrocyte to synthesize new membrane lipids, it is difficult to persuade these cells to undergo lipid peroxidation when taken from healthy subjects. In the peroxide stress haemolysis test, some peroxidation can be measured (e.g. by the thiobarbituric acid (TBA) test), as can the generation of oxidatively modified and cross-linked membrane proteins. However, this test is a huge insult to the cell. Peroxidation of erythrocytes can also be induced by high levels of organic hydroperoxides, such as *tert*-butyl hydroperoxide or linoleic acid hydroperoxide. Figure 7.1 shows the **echinocytes** that can be produced under such conditions.

Susceptibility to lipid peroxidation is increased in erythrocytes from patients with several conditions, including the presence of certain mutant haemoglobins, as in thalassaemia and sickle-cell anaemia. It is also increased in glucose-6-phosphate dehydrogenase deficiency (Section 7.2.6 below) and autoimmune haemolytic disease. **Thalassaemias** are defects in which one of the haemoglobin chains is synthesized abnormally slowly, leading to an excess of the other (Chapter 3). **Sickle-cell anaemia** is an inborn defect in which a glutamic-acid residue at position 6 in the haemoglobin β-chains is replaced by valine. The protein binds O_2, but the deoxy form is unstable and tends to precipitate in the erythrocyte, interfering with its deformability and creating sickleshaped cells (hence the name of the disease). The sickled cells can block capillaries, haemolyse readily and are marked for destruction, leading to anaemia.

Unstable haemoglobins tend to precipitate on to the erythrocyte membrane, setting the scene for site-specific free-radical reactions that promote lipid peroxidation and damage to membrane proteins. Iron may also be released to bind to the membrane.[97] The accumulation of iron deposits on the cytoplasmic surface of sickle erythrocytes has been implicated in their destruction: the iron appears to be 'low molecular mass' since it can be chelated by iron ion chelators. Several mutant haemoglobins undergo oxidation and $O_2^{\bullet -}$ release much faster than normal, as do the isolated α or β chains that accumulate in thalassaemia.[18] Peroxidation of erythrocyte membranes is known to cause formation of cross-linked membrane proteins and to diminish the 'deformability' of the cells.

7.2.5   *Glucose-6-phosphate dehydrogenase deficiency*

Erythrocytes operate the pentose phosphate pathway to provide NADPH, mostly for GSSG reduction. However, almost 400 million people, principally in tropical and Mediterranean areas, have an inborn defect in the gene (on the X chromosome) that encodes glucose-6-phosphate dehydrogenase, so that its activity in erythrocytes is sub-normal.

This deficiency sometimes leads to damage to the erythrocyte membranes, but this is not usually severe enough to cause clinical symptoms unless the rate of H_2O_2 production in erythrocytes is increased, e.g. by certain drugs (Table 7.2). If the rate of H_2O_2 production exceeds the capacity of the residual enzyme to generate NADPH, then GSH/GSSG ratios fall and glutathione peroxidase activity is impaired. Lysis of the red blood cells may result, leading to anaemia and jaundice.[24]

7.2.6   *Solutions: destruction*

Erythrocytes can degrade oxidized and other abnormal proteins, apparently by the action of the proteasome system. For example, treatment with a large excess of H_2O_2 can lead to oxidative inactivation of CuZnSOD, which is then degraded[92] by erythrocytes (Chapter 4). Protein degradation can be followed in intact erythrocytes by release of the amino acid alanine, which is not metabolized in erythrocytes.

Senescent erythrocytes are removed from the circulation; the iron from haemoglobin degradation is recycled (Chapter 3). A **senescent cell antigen** appears on the surface of old cells. It binds immunoglobulin G and is then recognized by macrophages in the spleen. Recognition may involve receptors similar or identical to those that recognize oxidized low-density lipoproteins (LDLs; Chapter 9).[93] Senescent antigen arises by damage to **band 3**, a transmembrane protein involved both in anchoring the erythrocyte cytoskeleton to the membrane and also as an anion exchanger (HCO_3^-/Cl^-). Band 3 protein occurs in other tissues, including brain and lymphocytes, and damage to it there also generates senescent cell antigen. In vitamin E-deficient animals, senescent cell antigen appears earlier than usual, suggesting that its formation

Table 7.2. Inborn deficiencies of erythrocyte enzymes in relation to drug and toxin-induced haemolysis

Abnormality	Prevalence	Usual clinical feature	Drugs inducing haemolysis
Glucose-6-phosphate dehydrogenase deficiency (X-linked, recessive)	Common in patients in tropical or Mediterranean areas or their descendants	Some RBC damage is often detected in laboratory tests but severe haemolysis is rare *in vivo*	Fava beans, furazolidone, nitrofurantoin, nitrofurazone, pamaquine, primaquine, sulphonamides
Glutathione peroxidase deficiency	Very rare	Often none, sometimes severe haemolysis, infertility (Chapter 3)	Sulphonamides, nitrofurantoin
Glutathione reductase deficiency	Very rare, but lack of riboflavin in the diet can also decrease the activity (enzyme has FAD at the active site). The drug BCNU, used in cancer chemotherapy, is a powerful inhibitor of glutathione reductase and can cause erythrocyte damage	Often none, sometimes severe haemolysis (Chapter 3)	Sulphonamides
Abnormal haemoglobins (e.g. Hb Torino, Hb Shepherd's Bush, Hb Peterborough, Hb Zurich)	Rare	Often some haemolysis seen related to protein instability; drug administration can cause a severe haemolytic crisis	Sulphonamides

Data are mostly taken from the article by Gaetani, GF and Luzzatto, L. (1980) in *Pseudo-allergic Reactions. Involvement of Drugs and Chemicals*, Vol. 2 (Dukor, P *et al.*, eds), S Karger, Basle, Switzerland.

involves oxidative damage. The membrane protein **spectrin** can also undergo oxidative damage, e.g. by loss of −SH groups, and haemoglobin–spectrin cross-links are observed in peroxide-treated erythrocytes.

7.3   **Erythrocytes as targets for toxins**

7.3.1   *Nitrite*

The normally slow autoxidation of haemoglobin can be accelerated by several toxins. One is nitrite ion, NO_2^-. The presence of large amounts of nitrate (NO_3^-) in the water supply of some rural areas, due to excessive use of inorganic fertilizers, can cause problems in young bottle-fed babies: the NO_3^- in the water used to make up feeds is reduced by gut bacteria to NO_2^-, which is then absorbed and causes sufficient methaemoglobin formation to interfere with oxygenation of the body tissues (Chapter 2).

7.3.2   *Hydrazines*

Several toxic agents are known to cause haemolysis. One of the most studied is **phenylhydrazine** (and its derivative **acetylphenylhydrazine**).[120] Injection of these compounds into animals causes haemolysis and the bone marrow responds by putting immature erythrocytes into the circulation. Indeed reticulocytes, the precursors of erythrocytes, are often obtained for study by injecting animals with phenylhydrazine and removing reticulocyte-rich blood several days later.

Phenylhydrazine and its derivatives slowly oxidize in aqueous solution to form $O_2^{\bullet-}$ and H_2O_2, a reaction catalysed by traces of transition-metal ions. The first stages in the oxidation can probably be represented by the equations below, in which M^{n+} represents the metal ions and Ph symbolizes the benzene ring (Fig. 7.3).

$$Ph-NH-NH_2 + M^{n+} \rightarrow H^+ + Ph-NH-NH^\bullet + M^{(n-1)+}$$

$$Ph-NH-NH^\bullet + O_2 \rightarrow H^+ + O_2^{\bullet-} + PhN=NH$$

Addition of SOD slows down the oxidation process.

However, the damage done by phenylhydrazine to erythrocytes is not prevented by SOD.[120] Methaemoglobin, acting as a peroxidase, can oxidize phenylhydrazine in the presence of H_2O_2. Oxyhaemoglobin also oxidizes phenylhydrazine, but H_2O_2 is not required. These oxidase and peroxidase reactions of haemoglobin form a phenylhydrazine radical that can react with oxygen to give $O_2^{\bullet-}$, and can also lead to formation of phenyl radicals. The reactions may be written, in a simplified form, as:

$$Ph-NH-NH_2 \xrightarrow{\text{haemoglobin}} Ph-NH-NH^\bullet \xrightarrow{\text{haemoglobin}} PhN=NH$$

$$Ph-NH-NH^\bullet + O_2 \rightarrow PhN=NH + O_2^{\bullet-}$$

$$PhN{=}NH + O_2 \rightarrow \underset{\text{phenyl radical}}{Ph^{\bullet}} + O_2^{\bullet -} + N_2 + H^+$$

(via $PhN{=}N^{\bullet}$, phenyldiazine radical)

$$Ph^{\bullet} \xrightarrow{\text{H}^{\bullet} \text{ abstraction}} Ph{-}H \text{ (benzene)}$$

Hence the end-products of the oxidation are benzene and nitrogen gas. Of these various species, the most damaging seems to be the phenyldiazine radical, which can denature the haemoglobin molecule, and stimulate peroxidation of membrane lipids, causing eventual haemolysis. The haem group is converted into a green product; both cleavage of the ring and addition of phenyl groups to it occur. Oxidative denaturation of haemoglobin forms intracellular precipitates called **Heinz bodies**. Some damaged haemoglobin is degraded by the proteasome system.

Although $O_2^{\bullet -}$ and H_2O_2 are not required in the initial reaction of oxyhaemoglobin and phenylhydrazine, they are involved in the subsequent decomposition and precipitation reactions. If acetylphenylhydrazine is incubated with oxyhaemoglobin *in vitro*, addition of catalase to remove H_2O_2 decreases the rate of disappearance of oxyhaemoglobin, but SOD does not. Both ascorbate and GSH inhibit, possibly by directly scavenging intermediate radicals such as $Ph{-}N{=}N^{\bullet}$, $Ph^{\bullet}$, and $Ph{-}NH{-}NH^{\bullet}$. Hence, GSH, ascorbate, catalase and glutathione peroxidase, would seem to be the major defence

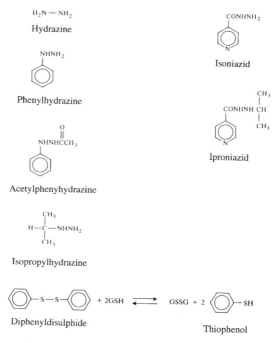

Fig. 7.3. Structures of some haemolytic drugs.

systems of the erythrocyte against phenylhydrazine. Consistent with this, the toxic effects of phenylhydrazine to mice are increased if they have been maintained on a selenium-deficient diet to lower glutathione peroxidase activity.

Hydrazine derivatives are widely used in industry, as rocket fuels, and in medicine. Many can be oxidized, not only by haemoglobin but also by cytochromes P450. For example, **hydralazine** is used to treat high blood pressure, but it has several side-effects, and prolonged use of it can produce symptoms of rheumatoid arthritis and of the autoimmune disease lupus erythematosus (Chapter 9). Like other hydrazines, hydralazine can be oxidized by haem proteins and 'free' transition metal ions to form $O_2^{\bullet-}$, H_2O_2, and nitrogen-centred radicals, which may participate in the side-effects of the drug. Paradoxically, it has been reported that hydralazine can inhibit vascular endothelial cell $O_2^{\bullet-}$ generation in rabbits and prevent the development of tolerance to the vascular relaxation produced by nitroglycerine, which has been postulated to involve up-regulation of $O_2^{\bullet-}$ generation and consequent antagonism of vasorelaxation by $NO^{\bullet}$ (Section 6.8).

The anti-tuberculosis drug **isoniazid** (isonicotinic acid hydrazide) and the anti-depressant **iproniazid** (Fig. 7.3) can similarly be oxidized to produce hydrazine derivatives and free radicals, which again might contribute to their side-effects. For example, iproniazid can give rise to the hepatotoxic compound **isopropylhydrazine** (Fig. 7.3). An antioxidant that has been used in the rubber industry, **N-isopropyl-N′-phenyl-*p*-phenylenediamine**, can cause oxidation and denaturation of haemoglobin (both free and in erythrocytes), at a rate about 40 times higher than that induced by phenylhydrazine.[120] Thus antioxidants designed for chemists are not always antioxidants in biological systems.

7.3.3 *Sulphur-containing haemolytic drugs*[75b]

Diphenyl disulphide administered orally to rats causes erythrocyte destruction. It can be reduced by GSH to form thiophenol (Fig. 7.3), which can be oxidized by oxyhaemoglobin with formation of methaemoglobin, thiyl ($RS^{\bullet}$) radicals, $O_2^{\bullet-}$, and H_2O_2; all these radicals probably contribute to its haemolytic action. Thiyl radicals combine with oxygen to yield damaging species such as $RSO_2^{\bullet}$ and $RSO^{\bullet}$ (Chapter 2).

7.4 Inborn defects in erythrocyte antioxidant defences: the link to malaria

7.4.1 *Favism*[24]

Many 'haemolytic agents' have only minor effects when small doses are given to healthy animals. Greater effects can be seen in patients with inborn defects in erythrocyte defence mechanisms, the commonest being a deficiency in glucose-6-phosphate dehydrogenase (indeed, worldwide this is the commonest human inborn error of metabolism). Most patients have incomplete loss of enzyme activity and show little haemolysis normally. However, haemolysis can be induced, often severely, by a few drugs (Table 7.2) and even by ingestion of certain foods such as the broad bean, *Vicia faba*. *Vicia*-induced haemolysis, a condition known as **favism**, is common in certain Mediterranean countries such as Sardinia and Greece, in the Middle East and in parts of south–east Asia. Its distribution follows that of glucose-6-phosphate dehydrogenase deficiency, but not all patients deficient in the enzyme are sensitive to the bean, for unknown reasons.

Soon after ingesting the bean, the erythrocytes of patients with favism show a rapid fall in $NADPH/NADP^+$ ratios and GSH concentrations, resulting in impaired functioning of glutathione peroxidase and glutathione reductase. The chemicals in *V. faba* responsible for this effect are the pyrimidine derivatives, **vicine** and **convicine**, which are present at about 0.5% of the total weight of the bean. They can be hydrolysed by β-glucosidase enzymes to give 'aglycone' products (Fig. 7.4) that react rapidly with O_2. They form H_2O_2, and a quinone that can be re-reduced by GSH, thus leading to more H_2O_2 generation. Drugs

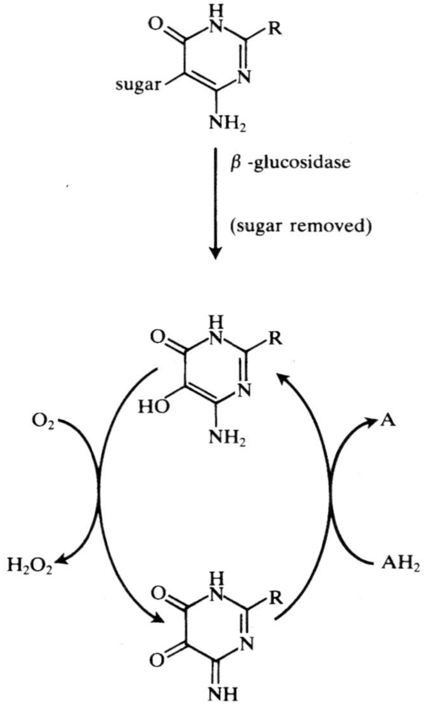

Fig. 7.4. The compounds causing favism. Redox cycling of the aglycones derived from hydrolysis of convicine (R = OH) and vicine (R = NH₂) from *Vicia faba* seeds. The aglycone of convicine is called **isouramil**, that of vicine is called **divicine**. Diagram by courtesy of Prof. G Rotilio. AH₂ signifies a reducing system (such as GSH).

that trigger haemolysis in glucose-6-phosphate dehydrogenase deficient patients include the antimalarials **primaquine** and **pamaquine** (Table 7.2). Mixtures of primaquine with NADPH and with oxyhaemoglobin have been shown to produce H_2O_2 *in vitro*, an observation perhaps relevant to its haemolytic effects. Indeed, the **blackwater fever** observed in some troops given prophylactic anti-malarials during World War II was probably related to glucose-6-phosphate dehydrogenase deficiency. The term 'blackwater' refers to the urinary discoloration produced by haemoglobin and its degradation products released from haemolysed erythrocytes.

Inborn defects in other erythrocyte enzymes can also cause haemolysis (Table 7.2), although such inborn errors are much rarer. The oxidation and precipitation of several abnormal haemoglobins, which produces $O_2^{\bullet-}$, is also accelerated by certain drugs. Infections frequently initiate haemolytic crises in carriers of unstable haemoglobins, which might be due to their faster denaturation during the elevated body temperatures that occur in fever.[121] Some of the denaturation products of haemoglobin, as well as released iron, might catalyse an interaction of $O_2^{\bullet-}$ and H_2O_2 to form $OH^{\bullet}$. Indeed, when mouse erythrocytes are incubated with phenylhydrazine, divicine or isouramil, there is a rapid release of iron in a form that can stimulate free-radical damage.[102]

7.4.2   *Malaria, oxidative stress and an ancient Chinese herb*

The prevalence of glucose-6-phosphate dehydrogenase deficiency, thalassaemia and sickle-cell anaemia in certain parts of the world has been suggested to be related to their ability to confer protection against malaria. Malaria is caused by infection with protozoan parasites of the *Plasmodium* genus. At least four species can infect humans, but *P. falciparum* is the most virulent. Worldwide, malaria affects more than 200 million people, of whom over 1 million still die each year, and the problem of resistance to current therapies is growing. Malaria is transmitted by the bite of an infected female mosquito, whereupon the parasites enter the blood, migrate to the liver and from there infect erythrocytes. They proliferate and release progeny to infect new red cells.

Evidence consistent with the 'malaria protection' hypothesis has been provided by the observation that transgenic mice expressing the gene for human β sickle-cell haemoglobin chain are protected against infection by murine *Plasmodium* species.[50] Perhaps sickling or haemolysis of infected cells helps to control the parasite. Studies in the south-western pacific island of Espiritu Santo have suggested increased incidence of a mild form of malaria, but decreased incidence of severe malaria in children with α-thalassaemia.[81]

Malaria-infected erythrocytes appear to be under an oxidative stress, perhaps as a consequence of parasite metabolism. Malaria parasites can be killed *in vitro* by systems generating ROS/RNS and by exposure to lipid peroxides (or cytotoxic aldehydes derived from peroxides).[84] Indeed, much of the pathology associated with malarial infections may be related to the host response, trying to eradicate the parasite but instead causing damage to host tissues (e.g. by excess production of cytokines such as TNFα).

Injection of alloxan or *tert*-butylhydroperoxide into malaria-infected mice kills a large number of the parasites. The effect of alloxan can be overcome by pre-treatment of the mice with the iron chelator desferrioxamine, an inhibitor of iron-dependent free-radical reactions. Phenylhydrazine and divicine also decrease parasitaemia in malaria-infected mice, and the anti-parasite action of divicine can be prevented by desferrioxamine. Another anti-malarial agent that may act by imposing oxidative stress is a constituent of a plant used in traditional Chinese medicine (Fig. 7.5). The active constituent is **qinghaosu (artemisinin)**, a sesquiterpene endoperoxide.[84]

Malarial parasites digest haemoglobin to obtain amino acids: they convert the haem to an inert haem polymer (**haemozoin**) in the food vacuole. Some antimalarial drugs (including chloroquine) block this and may damage the parasite by free-radical reactions involving haem liberated from degraded haemoglobin. Thus any defect in protection of erythrocytes against oxidative damage, or an increased rate of free-radical production *in vivo* might favour the eradication of malarial parasites within these cells.[65]

Although chelating agents can, under some experimental conditions, protect malaria parasites against oxidative damage, they can themselves suppress parasite growth, probably by depriving parasites of iron.[114] When grown in isolated human erythrocytes, the malarial parasite *P. falciparum* suffers complete growth inhibition at only 30 μM concentrations of desferrioxamine.

Several other protozoan parasites are sensitive to ROS/RNS, a concept being exploited in drug design. *Trypanosoma brucei*, for example, cannot synthesize haem and lacks catalase. It can be killed by increasing intracellular ROS levels, e.g. by adding menadione, and killing is accelerated by the addition of free haem. The trypanothione/trypanothione reductase system (Chapter 3) used by some parasites is another target for drug intervention. Several nitro-compounds have anti-parasite action (Chapter 8). Iron-chelators could conceivably be used as anti-parasite and even anti-bacterial agents.[99]

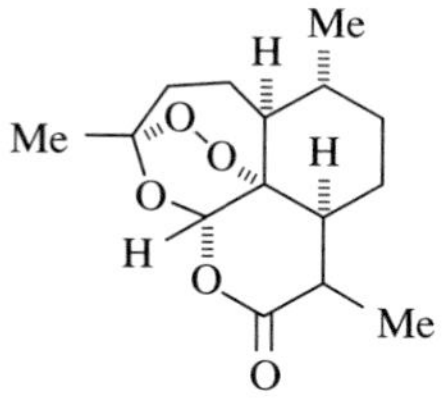

Fig. 7.5. Structure of artemisinin. This drug was first isolated from *Artemisia annua*, an ancient Chinese herb (*quinghao*, the 'blue-green' herb) used for treatment of malaria. Written descriptions of the use of quinghao date back to 168 BC. The action of artemisinin in animals is attenuated by desferrioxamine, suggesting that iron may decompose the peroxide to $RO_2^\bullet/RO^\bullet$ radicals that attack the parasite. Several chemically related synthetic compounds are being evaluated as anti-malarials (e.g. see Posner *et al.* (1994) *J. Med. Chem.* **37**, 1256). Artemether (a methyl ether) and artesunate (a sodium succinyl salt) are available in several countries.

7.5 Chloroplasts

7.5.1 *Structure and genetics*

The chloroplasts present in the leaves of higher plants can be seen under the electron microscope to be bounded by an outer **envelope** consisting of two membranes separated by an electron-translucent space of about $10\,nm$ (Fig. 7.6). The envelope encloses the **stroma** of the chloroplast, in which floats a complex internal membrane structure. The stroma is an aqueous solution containing various low-molecular-mass compounds plus a high concentration of proteins, most of which are the enzymes necessary to convert CO_2 into carbohydrate by a metabolic pathway known as the **Calvin cycle**. For its operation the Calvin cycle requires ATP and NADPH. The first enzyme in the Calvin cycle (**ribulose-bisphosphate carboxylase**) catalyses reaction of the five-carbon sugar ribulose 1,5-biphosphate with CO_2 to form two molecules of phosphoglycerate.

The internal membrane structure of the chloroplasts is complex. In the electron micrograph in Fig. 7.6, two distinct features may be recognized, i.e. regions of closely stacked membranes (**thylakoids**) interconnected by a three-dimensional network of membranes known as the **stromal thylakoids**. These membranes contain the photosynthetic pigments (chlorophylls *a* and *b*, green pigments that absorb light in the blue and red regions of the spectrum) together with **carotenes** and **xanthophylls**, yellow pigments absorbing blue light). The thylakoids produce the NADPH and, by the process of **photo-phosphorylation**, the ATP needed to drive CO_2 fixation by the Calvin cycle in the stroma.

Chloroplasts contain some DNA, but most chloroplast proteins are encoded in the nucleus and imported when required. Among the few proteins encoded in the chloroplast genome are the **D1** and **D2** proteins of the photosystem II reaction centre, the **A** and **B** proteins of the photosystem I reaction centre, cytochromes *f* and b_6 (Fig. 7.7, see plate section) and the large subunit of ribulose-bisphosphate carboxylase.

7.5.2 *Trapping of light energy*

Absorption of light energy by chlorophyll causes electrons to move into higher energy states (excited singlet states). Absorption of blue light by chlorophyll results in formation of a higher excited state than does absorption of red light, but this second excited state quickly loses its excess energy as heat, so that absorption of either red or blue light effectively produces the same first excited state of chlorophyll. The excitation energy can be lost by re-emission of light to give **fluorescence**; it can be lost as heat (**thermal energy dissipation**) or it can be transferred to another molecule, e.g. an adjacent chlorophyll, the first molecule returning to the ground state, whilst the second becomes excited. The rate of thermal energy dissipation is carefully regulated and the xanthophyll cycle (Section 7.5.10 below) appears to be involved.

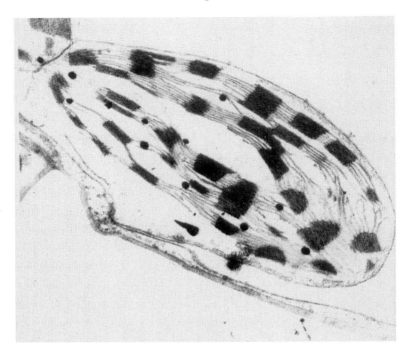

Fig. 7.6. Top: electron micrograph of chloroplasts in the leaves of higher plants. Bottom: structure of chlorophylls.

Special molecules of chlorophyll *a* in the thylakoid membrane (**reaction centre chlorophylls**) can, when excited, lose an electron to a neighbouring electron acceptor (A), producing a (chlorophyll$^+$ A$^-$) pair. This initial charge separation is the basic reaction of photosynthesis. Light energy absorbed by other chlorophyll *a* or chlorophyll *b* molecules in the thylakoid membrane is transferred to the reaction-centre chlorophylls. Indeed, each reaction-centre chlorophyll is associated with a **light-harvesting array** of other pigment molecules that channel energy to it.

Two different types of reaction centre, known as **P700** and **P680**, can be detected in the chloroplasts of higher plants. Each contains chlorophyll *a* and is served by its own light-harvesting pigment system; the complexes containing P700 and its light-harvesting 'antenna' pigments are referred to as **photosystems I** and those containing P680 as **photosystems II**. Stroma thylakoids are especially rich in photosystems I whereas the stacked membranes are enriched in photosystems II.

Electrons ejected from the P680 reaction centre of a photosystem II pass to a **phaeophytin *a*** molecule and then to acceptors Q_A (primary plastoquinone) and Q_B (secondary plastoquinone) (Fig. 7.7). Electrons then pass 'down' (i.e. to more positive reduction potentials) an electron-transport chain that contains plastoquinone, plastocyanin and a cytochrome *b–f* complex, eventually entering photosystems I to replace the electrons lost on excitation of the P700 reaction centres (Fig. 7.7). In photosystem I the electron produced by the light-induced charge separation is successively transferred to A_o (chlorophyll), A_1 (phylloquinone) and then to the A/B centre (an iron–sulphur cluster). Electrons then pass to other iron–sulphur proteins bound to the membranes, and then on to a soluble (stromal) iron–sulphur protein, known as **ferredoxin**. Soluble ferredoxin from higher plants contains two iron ions and two atoms of sulphur per molecule, and it acts as a one-electron acceptor. Reduced ferredoxin can then, in the presence of a reductase enzyme, donate electrons to NADP$^+$ to give NADPH (Fig. 7.7).

7.5.3   *The splitting of water*

The electrons ejected from the photosystem II reaction centres are replaced by the splitting of water molecules, accompanied by O_2 evolution. The detailed chemistry of the water-splitting reaction is not understood, but a 'charge-accumulating mechanism' is involved. It stores up to four units of positive charge, corresponding to loss of four electrons from photosystem II. When fully 'charged', it can then take four electrons from two water molecules to generate a molecule of oxygen. This charge accumulator is usually designated as 'S' so we can write:

$$S \xrightarrow{-1e^-} S^+ \xrightarrow{-1e^-} S^{2+} \xrightarrow{-1e^-} S^{3+} \xrightarrow{-1e^-} S^{4+}$$

$$S^{4+} + 2H_2O \rightarrow 4H^+ + S + O_2$$

The reaction centre of photosystem II has two different protein subunits, D1 and D2. The P680 primary electron donor is probably a dimer of chlorophyll *a*. A cluster of four manganese ions is intimately involved in charge accumulation. A tyrosine residue in the D1 protein, which forms a tyrosyl radical, is also involved.

7.5.4   *What problems do chloroplasts face?*

Chloroplasts are prone to oxidative damage for several reasons.[43,49]

1. Their internal O_2 concentration in the light will be greater than that in the surrounding atmosphere, because O_2 is produced by photosystems II.

2. The lipids present in the chloroplast envelope and thylakoids contain a high percentage of PUFA residues, and are thus susceptible to peroxidation.

3. Illuminated chlorophyll can sensitize the formation of singlet O_2, which is especially damaging to PUFAs (Chapter 2). When chlorophyll absorbs light energy, it enters an excited singlet state. Most of this energy is transferred to the reaction–centre chlorophylls, and some can be lost as heat or fluorescence, as explained previously. However, if energy is not completely dissipated by these mechanisms, **intersystem crossing** may occur to generate the triplet state of chlorophyll, which can transfer energy to O_2 and generate singlet O_2. Prolonged illumination of chloroplast thylakoids *in vitro* causes marked lipid peroxidation. Indeed, if PUFAs are simply mixed with chlorophyll and illuminated, they are rapidly peroxidized. The deterioration of isolated, illuminated chloroplasts is a complex process; in addition to lipid peroxidation and the formation of cytotoxic aldehydes by peroxide decomposition, there is hydrolysis of lipids to release fatty acids. Both esterified and released fatty acids undergo peroxidation, and the released fatty acids contribute to membrane damage. Lipid hydrolysis is due to the action of lipase enzymes which normally show little activity in chloroplasts, but are 'unmasked' during membrane deterioration.

4. If too much energy is absorbed in relation to the amount of NADPH needed, the reaction-centre chlorophylls may not be able to dispose of all their excitation energy, and so singlet O_2 formation should increase. For example, in photosystem II chlorophylls can be oxidized and proteins (especially D1) damaged. D1 can be replaced (Section 7.5.9 below) but if it is damaged too quickly for replacement to keep up, inhibition of photosynthesis (**photoinhibition**) can occur, e.g. in plants exposed to high light intensities.[5]

5. The electron-transport chain of chloroplasts, like that of mitochondria and the endoplasmic reticulum (Chapter 1), can 'leak' electrons to O_2. Isolated illuminated thylakoids slowly take up O_2 in the absence of added electron acceptors. This was first observed by A.H. Mehler, and is hence often referred to as the **Mehler reaction**. It appears to result from the reduction of O_2 to $O_2^{\bullet-}$ by the electron acceptors of photosystems I. Addition of the stromal protein ferredoxin increases O_2 uptake, since ferredoxin is rapidly

reduced by photosystems I and reduced ferredoxin can itself reduce O_2 (Fig. 7.8)

$$Fd_{red} + O_2 \rightarrow O_2^{\bullet -} + Fd_{ox}$$

Electron leakage will tend to be favoured by high local O_2 concentrations during photosynthesis. *In vivo*, however, reduced ferredoxin also passes electrons on to $NADP^+$ via ferredoxin–$NADP^+$ reductase (Fig. 7.8). Thus, electrons from photosystems I can pass through at least three routes, of which route C (Fig. 7.8) is preferred. If the supply of $NADP^+$ is limited, the rate of electron flow along pathway C would be expected to be decreased, and more $O_2^{\bullet -}$ should be made by route B and, to a lesser extent, by route A (Fig. 7.8). In some experiments, however, isolated intact chloroplasts have been shown to reduce O_2 even at low light intensities.

6. Oxygen has a direct effect on ribulose-biphosphate carboxylase. One of the intermediates formed during the catalytic action of the enzyme on ribulose bisphosphate can react not only with CO_2, but also with O_2 (Fig. 7.9). As a result of this **oxygenase** activity of the enzyme, ribulose bisphosphate is converted into phosphoglycerate and phosphoglycolate. The phosphoglycolate

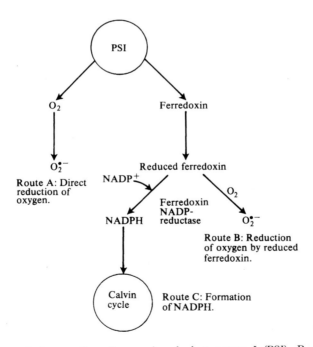

Fig. 7.8. Routes of electron flow from reduced photosystem I (PSI). Reduced ferredoxin reduces oxygen to $O_2^{\bullet -}$ as well as passing electrons on to $NADP^+$. The electron acceptors of photosystem I can themselves slowly reduce O_2 to $O_2^{\bullet -}$. Electrons might also escape from the ferredoxin–$NADP^+$ reductase enzyme.

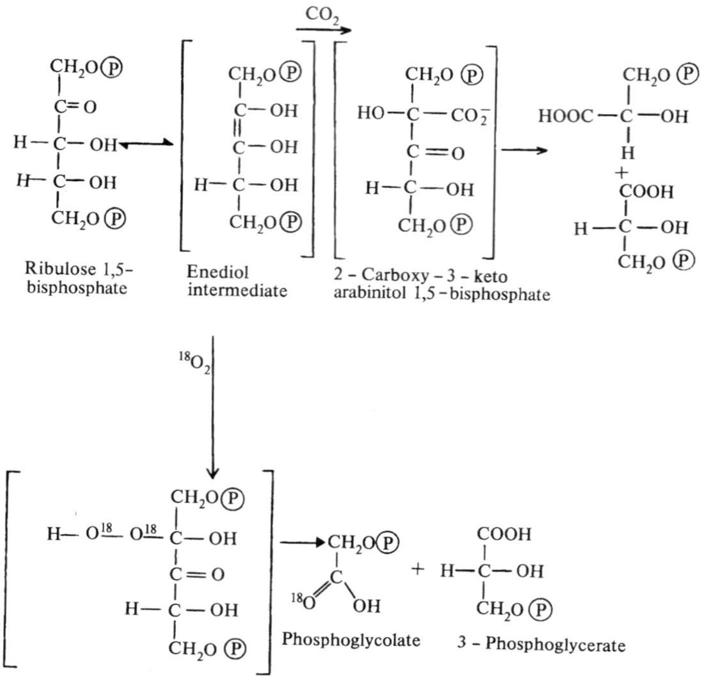

Fig. 7.9. Oxygenase and carboxylase activities of ribulose-bisphosphate carboxylase. This is the first enzyme of the Calvin cycle.

so produced is hydrolysed to glycolate and further metabolized by a series of reactions known as **photorespiration**. This metabolic pathway converts two molecules of glycolate into one molecule each of CO_2 and phosphoglycerate, which can re-enter the Calvin cycle. As a result of photorespiration, CO_2 previously fixed into the Calvin cycle intermediates is released and has to be refixed, with expenditure of more energy.

Oxygen and CO_2 compete for the enzyme intermediate (Fig. 7.9). The affinity of the carboxylase for CO_2 is much greater than that for O_2, so that carboxylation is favoured even at 21% O_2 and 0.03% CO_2, the levels in normal air. However, if O_2 is elevated and CO_2 depleted, CO_2 fixation into the Calvin cycle is decreased and CO_2 loss by photorespiration is increased. At some O_2/CO_2 ratio, a point will be reached at which the leaf is no longer achieving any net CO_2 fixation, and plant growth is inhibited. At even higher O_2/CO_2 ratios there can be net loss of carbon from the plant as photorespiration exceeds CO_2 fixation into the Calvin cycle: a situation which will result in death of the plant if continued for a long period. On the other hand, it can be argued that 'recycling' of CO_2 by photorespiration can help the plant use up excess light energy and decrease the risk of photoinhibition (point 4 above).

7. H_2O_2 can inactivate at least two enzymes of the Calvin cycle, apparently by a direct reaction with −SH groups essential for catalytic activity. Inactivation of **fructose bisphosphatase** and **sedoheptulose bisphosphatase** will block CO_2 fixation.

7.5.5 'Catalytic' metal ions in plants?

Little is known about the availability of metal ions 'catalytic' for $OH^\bullet$ formation and lipid peroxidation in chloroplasts. Chloroplasts often contain ferritin (**phytoferritin**)[15] from which $O_2^{\bullet-}$ could conceivably mobilize some iron ions. Ferritin levels in plants can be induced by several stresses, including ozone and iron overload.[54] Several iron-containing proteins (e.g. cytochromes; Fig. 7.7) could conceivably be damaged by excess H_2O_2, with release of 'catalytic' iron ions, although ferredoxin is resistant to H_2O_2.

Release of catalytic iron ions has also been implicated in the senescence of root nodules in leguminous plants.[9] For example, soybean root nodules contain **leghaemoglobin** to bind O_2 and maintain a low O_2 concentration, which facilitates N_2 fixation by the symbiotic bacteria (Chapter 1). Like other oxo-haem proteins, oxyleghaemoglobin can liberate $O_2^{\bullet-}$. Leghaemoglobin reacts with H_2O_2 to give ferryl species and protein radicals, and decomposes to release iron at high peroxide/leghaemoglobin ratios.[86] During root nodule senescence, accumulation of peroxides and degradation of leghaemoglobin to release 'catalytic' iron may be important events.

Exposing *Nicotiana plumbaginifolia* plants to excess iron in hydroponic culture led to rapid induction of H_2O_2-metabolizing enzymes in the leaves (catalase and ascorbate peroxidase: see below) but SOD did not rise.[54]

7.5.6 Solutions: antioxidant defence enzymes

Superoxide dismutases[3,14,95]

Superoxide produced in chloroplasts can be dealt with by SOD enzymes, whose importance to plants is illustrated by multiple experiments. (*In vitro*, $O_2^{\bullet-}$ can reduce plastocyanin and cytochromes, but the significance of this in the chloroplast is uncertain.)

For example, the SOD activity of developing tomatoes correlates with the resistance of the fruits to the 'sunscald' damage induced by exposure to heat and high light intensities (a nuisance for growers in hot countries who want to sell their tomatoes). In the cyanobacterium *Anabaena cylindrica*, nitrogen fixation is carried out in specialized cells (**heterocysts**) which lack the oxygen-evolving photosystems II. Heterocyst SOD activity is much lower than that of the photosynthetic cells. Root nodules (Section 7.5.5 above) contain SOD and H_2O_2-metabolizing enzymes (catalase and sometimes ascorbate peroxidase), presumably to minimize peroxide−leghaemoglobin interactions.[27]

In spinach chloroplasts, a CuZnSOD is present, some of which is bound to the thylakoids, and the rest free in the stroma. The overall CuZnSOD concentration may be as high as $10-30\,\mu M$. Plants also usually have one or more cytosolic (and sometimes peroxisomal) CuZnSODs.[20] No MnSOD has been detected in spinach chloroplasts. There have been reports of a MnSOD in the chloroplasts of some other plants, although the possibility of contamination of isolated chloroplast fractions by mitochondria must never be neglected, since plant mitochondria contain MnSOD. Plant FeSOD (Chapter 3), when present (e.g. in *Brassica campestris*, tomato, *Ginkgo biloba* and *Arabidopsis thaliana* leaves) is largely or entirely found in the chloroplast.

Removal of H_2O_2

SOD catalyses conversion of $O_2^{\bullet-}$ to H_2O_2; most or all of the H_2O_2 produced by illuminated chloroplasts seems to arise from $O_2^{\bullet-}$. Like other CuZnSODs (Chapter 3) the chloroplast enzyme is slowly inactivated on prolonged exposure to H_2O_2. Even worse, some Calvin cycle enzymes are rapidly inactivated by H_2O_2 (Section 7.5.4 above).

Thus it is essential that the chloroplast removes H_2O_2, yet it contains no catalase or selenoprotein glutathione peroxidase. Indeed, glutathione peroxidase enzymes are not commonly found in plants, although their presence has been suggested in some species, and genes similar to those encoding glutathione peroxidases in animals (including phospholipid hydroperoxide GPX) have been identified in several plant genomes.[34a] Glutathione transferases (present in many plants)[8] can reduce organic hydroperoxides (Chapter 3), but the physiological significance of this is uncertain.

As in animal tissues (Chapter 3), most leaf catalase is located in peroxisomes.[95] Indeed, the first step in metabolism of glycolate produced during photorespiration (Fig. 7.9) is oxidation in peroxisomes, generating H_2O_2 for disposal by catalase:

$$\begin{array}{c} CH_2OH \\ | \\ COOH \end{array} \; + \; O_2 \quad \xrightarrow[\text{oxidaze}]{\text{glycolate}} \quad \begin{array}{c} CHO \\ | \\ COOH \\ \text{glyoxylate} \end{array} \; + \; H_2O_2$$

This key role of catalase is supported by the properties of a barley mutant deficient in peroxisomal catalase; it would not grow well at low CO_2 levels (favouring photorespiration), but grew normally at high CO_2 levels.[104]

Some H_2O_2 may diffuse out of chloroplasts for metabolism in peroxisomes but the major chloroplast H_2O_2-metabolizing enzyme is a haem-containing **ascorbate-specific peroxidase**.[73] Ascorbate peroxidase in chloroplasts is present both as a stromal form and a thylakoid-bound form (an isoform of the enzyme is also present in leaf cytosol). Oxidized ascorbate can be regenerated by GSH (Fig. 7.10).[38] Considerable evidence for the role of antioxidant defence enzymes in helping plants resist stress is provided by transgenic studies (Section 7.6 below).

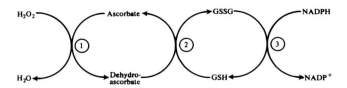

Fig. 7.10. The ascorbate–glutathione cycle in chloroplasts. The enzymes involved are as follows. 1, Ascorbate peroxidase. 2, Dehydroascorbate reductase; reaction 2 can also occur non-enzymically at high pH: the pH of the stroma during photosynthesis may rise to as high as 8, due to the formation of proton gradients for ATP synthesis. 3, Glutathione reductase. The first product of oxidation of ascorbate by ascorbate peroxidase is probably the semidehydroascorbate (SDA) radical. Two such radicals then react in a disproportionation reaction to form ascorbate and dehydroascorbate, (i.e. 2SDA→ascorbate + dehydroascorbate). NAD(P)H-dependent mechanisms for reducing semidehydroascorbate have also been described in chloroplasts. Thylakoid-bound ascorbate peroxidase can scavenge H_2O_2 as it is produced, and the resulting SDA can be reduced by electrons from photosystem I. This cycle has been called the Foyer–Halliwell–Asada cycle after the names of the two scientists who first proposed it and the third who has done much to establish evidence for its occurrence. However, this name ignores the contribution of Groden and Beck, who discovered ascorbate peroxidase in chloroplasts.

Ascorbate peroxidase purified from the algae *Euglena* can reduce not only H_2O_2 but also some artificial organic hydroperoxides. These enzymes might therefore be capable of acting on lipid peroxides *in vivo*, and further experiments are required on this point.

7.5.7 *Ascorbate and glutathione*

The chloroplast stroma contains GSH at concentrations in the range 1–5 mM and ascorbate at concentrations up to 20 mM. In both cases the ratios of reduced to oxidized form (GSH/GSSG and ascorbate/dehydroascorbate) are kept high under both light and dark conditions. This is essential, because incubation of ascorbate peroxidase with H_2O_2 in the absence of ascorbate inactivates this enzyme.[73] Ascorbate and GSH might also act as general ROS scavengers in chloroplasts, e.g. millimolar ascorbate could contribute to removal of $O_2^{\bullet-}$. Ascorbate is also involved in the xanthophyll cycle (Section 7.5.10 below).

Ascorbate is also present in all other parts of plants, including the space beneath the leaf cuticle and the walls of the cells below (**apoplastic space**). Here it may help to remove oxidizing toxins such as $NO_2^{\bullet}$ and O_3 (Section 7.6.3 below) before they reach their target cells. As in animals, ascorbate in plants is involved in several metabolic processes, such as hydroxylation of proline in proteins. Ascorbate synthesis in plants can be inhibited by the alkaloid **lycorine**.[6]

7.5.8 *Plant tocopherols*

The thylakoid membranes are rich in α-tocopherol, which inhibits the chain reaction of lipid peroxidation (Chapter 3) and scavenges singlet O_2. Indeed, α-tocopherol is synthesized in the chloroplast. The products of reaction of tocopherol with singlet O_2 and other ROS include tocopherylquinone (Chapter 3), which has been detected in chloroplasts and increases in amount during illumination.[43] Stromal ascorbate could presumably regenerate tocopherol from tocopheryl radicals.

As with animal tissues, however, it must not be assumed that α-tocopherol in plants serves *only* as a protection against membrane damage. Other roles are perhaps suggested by the fact that a **tocopherol oxidase**[40] enzyme, whose activity is under hormonal control, has been detected in many plant tissues. Whereas chloroplasts mainly contain α-tocopherol, other parts of the plant contain β-, γ- and δ-tocopherols in addition, whereas seeds and cereals are enriched in tocotrienols. γ-Tocopherols may play a role in helping plants resist air pollutants (Section 7.6.3 below). For example, stored grain in silos can generate high levels of $NO_2^\bullet$ (Chapter 8) which might be scavenged by γ-tocopherols.

In animal tissues, α-tocopherol has been suggested to be the major lipid-soluble, chain-breaking antioxidant present (Chapter 3). Whether this is also true of the thylakoid membrane, and other plant membranes, remains to be established. The other tocopherol isomers (β, γ and δ) may be important. In addition, phenolic compounds such as flavonoids are present. Many of these exert powerful antioxidant actions *in vitro* (Chapter 3). Examples include **silymarin** (a 3-oxyflavone isolated from the thistle *Silybum marianum*), quercetin, rutin and kaempferol (Chapter 3). However, many phenolics are located in the central vacuole of plant cells and not in the organelles, such as chloroplasts, that are subject to oxidative stress. The reduced form of plastoquinone (plastoquinol) can, like ubiquinol (Chapter 3), scavenge peroxyl radicals and inhibit lipid peroxidation *in vitro*.[51] Whether this is important in the plant *in vivo* is uncertain.

7.5.9   *Carotenoids*

Carotenoids are important light-harvesting pigments: they can transfer excitation energy to chlorophyll. In addition, they have an important antioxidant role; they quench singlet O_2 extremely rapidly and scavenge a range of ROS/RNS (Chapter 3).[112] The carotenoids present in chloroplasts are two main types: the carotenes (e.g. β-carotene) and xanthophylls (Chapter 3). For example, each photosystem II reaction centre contains two β-carotene molecules.

Carotenoids can also absorb energy from, and so diminish the concentration of, the triplet states of chlorophyll that lead to singlet O_2 formation. Hence they both decrease $^1O_2$ formation and help to remove any that is formed. Carotenoids may also react directly with $RO_2^\bullet$ and $RO^\bullet$ radicals, but the overall effect that this would have on lipid peroxidation at the high O_2 concentrations in the chloroplast is uncertain (Section 3.23.3).

There is considerable evidence for an antioxidant role of carotenoids in plants.[43] They are quickly oxidized on illumination of thylakoids *in vitro*.

Illumination of maize mutants that cannot synthesize carotenoids causes rapid bleaching of chlorophylls and destruction of chloroplast membranes. Illumination under anaerobic conditions causes much less damage, since singlet O_2 cannot be generated. Similar destructive effects of illumination in the presence of O_2 can be seen in normal plants if carotenoid biosynthesis is inhibited by certain herbicides, including **aminotriazole** (also a catalase inhibitor; Chapter 3), **pyriclor** and **norflurazon**.

The *Dunaliella* genus of green algae accumulate large amounts of β-carotene, both all-*trans* and 9-*cis*-β-carotene, apparently to protect against photodamage. Indeed, this alga has been used as a source of β-carotene for dietary 'nutritional supplements' (Chapter 10).

7.5.10   *The xanthophyll cycle*[31]

Higher plants and algae operate a series of reactions involving carotenoids, known as the **xanthophyll cycle** (Fig. 7.11). Epoxidation of zeaxanthin to violaxanthin, via antheraxanthin, is an O_2-dependent reaction that occurs in the thylakoids under both light and dark conditions. De-epoxidation is an ascorbate-dependent process that occurs only in the light, apparently because the de-epoxidase enzyme is located on the inner side of the thylakoid membrane and only functions at the low pH generated there when a proton gradient is established to drive ATP synthesis on illumination. The accessibility of violaxanthin to the de-epoxidase also appears to be increased by

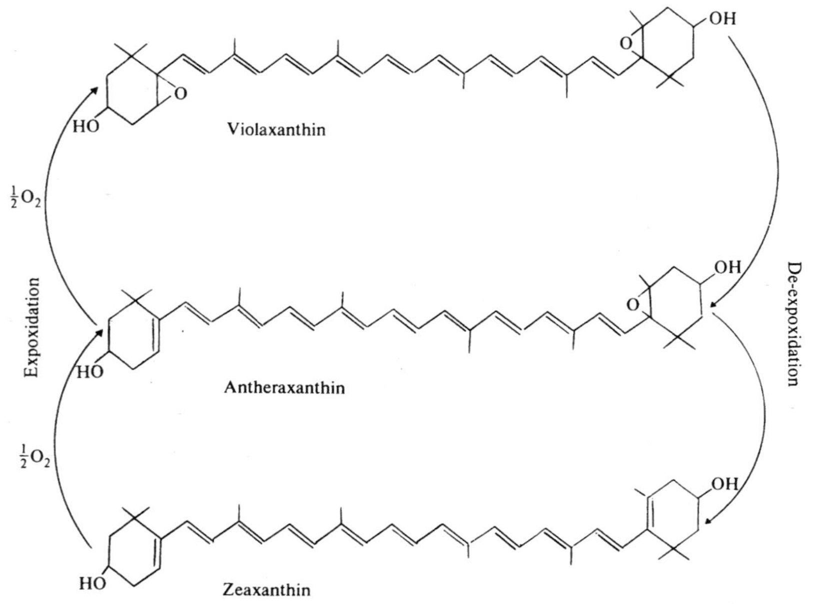

Fig. 7.11. The xanthophyll cycle.

light-dependent conformational changes in the membrane. Light-dependent formation of zeaxanthin appears to assist dissipation of excess energy, helping to prevent singlet O_2 formation in chloroplasts.

7.5.11   Solutions: repair and replacement

A large part of chloroplast protein-synthetic activity in the light is devoted to replacing the **D1 protein** in photosystem II.[5] It is destroyed rapidly (e.g. by singlet O_2) and the inactive protein is removed by proteases associated with photosystem II. Similarly, oxidation of $-SH$ groups on ribulose-bisphosphate carboxylase seems to 'mark' this enzyme for proteolysis, especially in senescing leaves.

Chloroplasts also possess multiple systems that can 'repair' oxidative damage. DNA-repair enzymes are present,[16] including **photolyases** that cleave UV-induced pyrimidine dimers. **Chaperones** exist to assist protein folding, including their re-folding after partial denaturation. The importance of **thioredoxin** was appreciated in plants long before its importance in animals became clear.[19] The disulphide form of thioredoxin in chloroplasts is reduced by a **ferredoxin-dependent thioredoxin reductase**. Reduced thioredoxin can re-activate Calvin cycle (and other) enzymes inactivated by ROS, such as the oxidized forms of fructose and sedoheptulose bisphosphatases. Thioredoxin is also involved in regulating the activity of Calvin cycle and other chloroplast enzymes by thiol–disulphide interchange. Chloroplasts have at least two thioredoxins, designated thioredoxins *f* and *m*.

A different thioredoxin is present in other plant organelles: mitochondria, endoplasmic reticulum and cytosol. In the latter cases, the thioredoxin reductase is NADPH-dependent (as in animal tissues) rather than ferredoxin-dependent. **Methionine sulphoxide reductase**, also thioredoxin-linked, is present in chloroplasts and helps repair oxidized methionine residues.[19]

7.6   Chloroplasts as targets for toxins

7.6.1   Inhibition of electron transport and carotenoid synthesis[43]

About 50% of all known herbicides act by inhibiting the electron-transport chain. They include **monuron** (CMU), **diuron** (DCMU) and **atrazine** (Fig. 7.12). They bind to the D1 protein of photosystem II and inhibit electron transport, stopping light-dependent generation of ATP and NADPH and preventing the plant from fixing CO_2. In green plants exposed to light, this inhibition is followed by carotenoid destruction, chlorophyll bleaching and membrane deterioration accompanied by increased lipid peroxidation and thought to be caused by singlet O_2.

Herbicide inhibitors of carotenoid biosynthesis lead to the same consequences. The **diphenylether** herbicides (e.g. **oxyfluorfen**; Fig. 7.12) appear to damage plants by generating singlet O_2 and other ROS, resulting in lipid peroxidation and other oxidative damage. One mechanism of their action is to

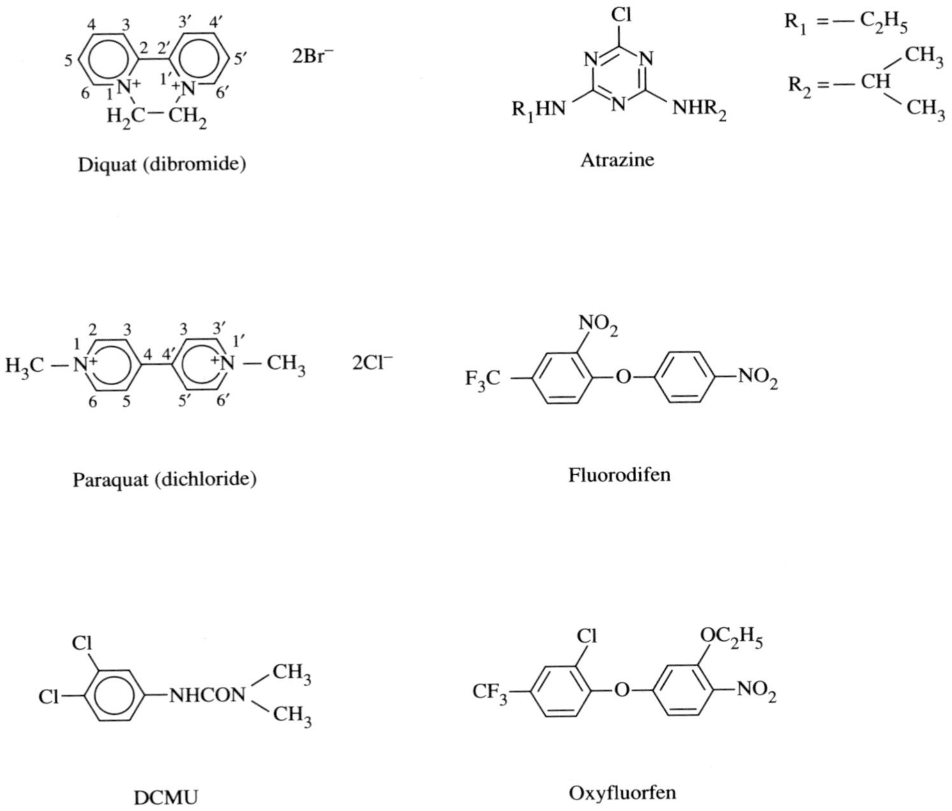

Diquat (dibromide)

Atrazine

$R_1 = -C_2H_5$

$R_2 = -CH\begin{smallmatrix}CH_3\\CH_3\end{smallmatrix}$

Paraquat (dichloride)

Fluorodifen

DCMU

Oxyfluorfen

Fig. 7.12. Structures of some herbicides.

inhibit the enzyme **protoporphyrinogen oxidase**[68] in mitochondria and chloroplasts, leading to accumulation of the photosensitizer protoporphyrin IX and inducing a plant version of porphyria (Chapter 2).

7.6.2 *Bipyridyl herbicides*

The bipyridyl herbicides **paraquat** and **diquat** act in a way different from the toxins considered in the previous section. Discovered by Imperial Chemical Industries in England, they are useful because of their broad toxicity to a wide range of plants, and because they do not accumulate in soil, being destroyed by various micro-organisms.[21]

'Bipyridyl' means that the structure contains two pyridine rings, aromatic rings in which one carbon atom is replaced by a nitrogen atom. The rings may be joined either by their number-2 carbon atoms, or by their number-4 carbon atoms (the nitrogen counting as atom number 1), to give 2,2'-bipyridyl or 4,4'-bipyridyl respectively. The prime is used to indicate an atom in the second ring. In paraquat, a methyl group is attached to each nitrogen; its full chemical name is **1,1′-dimethyl-4,4′-dipyridylium ion** (Fig. 7.12). Each nitrogen atom gains a positive charge because it has four bonds and is thus utilizing its lone pair of electrons (see Appendix I). In diquat, the two nitrogen atoms are joined by an ethylene group, to give **1,1′-ethylene-2,2′-dipyridylium ion** (Fig. 7.12). Paraquat is usually manufactured as a salt with chloride (Cl^-) ion, and diquat with bromide (Br^-) ion.

4,4′-Bipyridyl compounds form coloured products upon reduction. Indeed, several such compounds, under the general name **viologens**, have been used for this reason to study the chemistry of redox reactions. Hence paraquat is sometimes called **methyl viologen**. The colour is produced by one-electron reduction to form a stable (in the absence of O_2) radical that has a characteristic electron spin resonance (ESR) spectrum, and that absorbs visible light (λ_{max} in the visible range is at 603 nm for paraquat). The added electron is delocalized over both ring structures with partial neutralization of the positive charge on each nitrogen atom.

Ability to undergo redox cycling

Many years ago, it was observed that the killing of green plants by paraquat or diquat is accelerated by light and is slowed if O_2 is removed from the environment of the plant by flushing with nitrogen. Later work showed that paraquat and diquat cross the chloroplast envelope easily. Once inside, they can accept electrons from the non–haem–iron proteins associated with photosystems I and also from the flavin at the active site of ferredoxin–NADP$^+$ reductase (Fig. 7.7), in both cases being reduced to bipyridyl radicals. Indeed, in the absence of O_2, bipyridyl radicals can be identified in chloroplasts by their absorbance or ESR spectra.

On admission of O_2 (illuminated chloroplasts have a high internal O_2 concentration, of course) the radicals disappear because they react rapidly with O_2. If BP^{2+} is used to represent the bipyridyl herbicides, reaction with O_2 may be written as:

$$BP^{2+} \xrightarrow[\text{chain}]{\text{electron transport}} BP^{\bullet+}$$

$$BP^{\bullet+} + O_2 \rightarrow BP^{2+} + O_2^{\bullet-}$$

$$(k_2 = 7.7 \times 10^8 \text{ M}^{-1}\text{s}^{-1}, \text{ for paraquat})$$

Thus treatment of illuminated chloroplasts *in vitro* with paraquat or diquat leads to a rapid uptake of O_2 as the herbicides are continuously reduced and re-oxidized to generate $O_2^{\bullet-}$ in a **redox cycle**. Hence bipyridyl herbicides are examples of **redox-cycling** agents.

Much of the $O_2^{\bullet-}$ will be converted into H_2O_2 by chloroplast SOD. Since chloroplasts contain no catalase, H_2O_2 is dealt with by the ascorbate–glutathione cycle. However, the load of H_2O_2 is such that GSH and ascorbate are quickly oxidized and inactivation of Calvin cycle enzymes such as fructose

Table 7.3. Effects of paraquat on illuminated spinach chloroplasts

Time after paraquat addition (min)	[GSH] (mM)	[GSSG] (mM)	GSH/GSSG ratio	[Ascorbate] in reduced form (mM)	Fructose bisphosphatase activity (U/mg chlorophyll)
0	6.5	0.25	26	13	62
2	3	3.2	0.9	8	20
5	3.1	1.5	2.1	6	8
10	3.0	0.8	3.8	4	0

Spinach-leaf chloroplasts were isolated and treated with paraquat in the light. The stromal contents of GSH, GSSG and ascorbate were measured, as was the activity of fructose bisphosphatase, an enzyme essential to the Calvin cycle. Note that GSSG levels rise rapidly but it then disappears, possibly by mixed disulphide formation with proteins. Such rapid changes are not seen if paraquat is added to darkened chloroplasts, since the electron-transport chain is not then active to reduce paraquat. Data selected from *Biochem. J.* **210**, 899 (1983) by courtesy of the Biochemical Society.

bisphosphatase occurs, so that CO_2 fixation stops (Table 7.3). An additional reason for inhibition of CO_2 fixation is that diversion of electrons from photosystem I on to the herbicides will decrease the supply of NADPH both for the Calvin cycle, and for glutathione reductase.

In studies upon whole leaves and green algae, it has been observed that inhibition of CO_2 fixation is followed by leakage of ions from the leaves, and accumulation within them of TBA-reactive material (TBARS), indicative of lipid peroxidation. Analysis of lipids extracted from the leaves shows destruction of many fatty-acid side-chains. Electron microscopy reveals deterioration and breakdown of thylakoids and eventually of other membranes, including the membrane which surrounds the central vacuole of the plant cell. The central vacuole contains a number of hydrolytic enzymes, and often accumulates organic acids to a high concentration; so release of its contents into the rest of the cell will potentiate the damage (it will also release phenolics, some of which might exert antioxidant activity).[43]

Evidence that ROS are involved in paraquat toxicity

Evidence that ROS are responsible for the damaging effects of bipyridyl herbicides is extensive. Paraquat induces synthesis of MnSOD in the green alga *Chlorella*.[87] Some strains of ryegrass[44] are comparatively resistant to paraquat even though they still take up by the herbicide, and the leaves of such plants contain more SOD and catalase than the leaves of sensitive plants. Although leaf catalase is located in peroxisomes rather than in chloroplasts, H_2O_2 may diffuse across the chloroplast envelope to be dealt with by catalase in peroxisomes, which are often seen to be closely associated with chloroplasts in electron micrographs of leaf sections.

Paraquat-resistant strains of the fleabane, *Conyza bonariensis*, have been reported to have increased SOD, ascorbate peroxidase and glutathione reductase activities within their chloroplasts, although they may take up less paraquat into the chloroplasts than the sensitive strains, an additional reason for protection.[45] Addition of paraquat to mature cultures of the blue-green algae *Gloeocapsa*[32] inhibited its ability to fix nitrogen, an effect that could be mimicked by addition of H_2O_2.

Transgenic tobacco plants[3] engineered to express MnSOD or FeSOD activity in the chloroplast[a] were more resistant to paraquat. By contrast, resistance to paraquat was not enhanced if MnSOD was over-expressed in the leaf mitochondria. Expressing a fungal gene in tobacco chloroplasts that raised the intra-chloroplast mannitol concentration to ~ 100 mM also rendered the plants resistant to paraquat,[100] possibly by $OH^{\bullet}$ scavenging. The iron chelator desferrioxamine protects leaves of the pea, *Pisum sativum*, against damage by paraquat,[125] consistent with a role of iron in forming more-toxic species from $O_2^{\bullet -}$ and H_2O_2.

Paraquat and diquat are also toxic to non-green plant tissues.[43] For example, they cause increased membrane permeability and accumulation of TBA-reactive material in fungi such as *Aspergillus niger* and *Mucor hiemalis*. If this is due to redox cycling, what is the mechanism of formation of bipyridyl radicals? Several flavoprotein enzymes, including glutathione reductase,[107] can reduce paraquat and/or diquat; these herbicides take electrons from the flavin ring. Similar reactions can account for the more slowly developing toxicity of paraquat to leaves in the dark. Indeed, tobacco plants expressing chloroplast MnSOD were more resistant to paraquat in the dark as well as in the light. However, increased resistance was also observed if MnSOD was over-expressed in the mitochondria, suggesting that 'dark' paraquat reduction can involve enzymes (e.g. glutathione reductases) present in both organelles.

One disadvantage of paraquat is that it is toxic to animals: this toxicity, like that to plants, is mediated by oxidative stress (Chapter 8).

7.6.3   *Air pollutants*[49]

Antioxidants play a key role in the ability of plants to resist oxidizing environmental pollutants such as ozone (O_3) and nitrogen dioxide ($NO_2^{\bullet}$). Worldwide, **ozone** has been suggested to cause more damage to crops and forests than any other air pollutant. Exposure of plants to O_3 causes direct oxidation of lipids, aldehyde formation, lipid peroxidation and damage to proteins and DNA (e.g. levels of 8-hydroxyguanine in chloroplast DNA rise if beans or peas are exposed to high levels of O_3).[37] Low-level exposure of plants to O_3 can trigger the hypersensitivity response (Section 6.9.3). It may also lead to development of tolerance, in part by raising the levels of such antioxidants as ascorbate, GSH, phenolic compounds, ascorbate peroxidase and glutathione reductase. For example, transgenic tobacco plants over-expressing MnSOD in the chloroplast were more resistant to O_3.

Since the 'basal' levels of antioxidants appear to vary with leaf age and also seasonally in plants, pollution resistance may well depend upon time of year, among other factors.

Sulphur dioxide also damages plants. This toxic reducing gas dissolves in water to form sulphites and bisulphites, which can oxidize to generate $O_2^{\bullet -}$ and H_2O_2 (Chapter 8). Hence antioxidant defence systems contribute to protection against SO_2 toxicity. For example, growth of *Chlorella* in the presence of sulphite elevated SOD levels and rendered the cells more resistant to paraquat.[87] A paraquat-resistant strain of *Conyza boriensis* (with elevated SOD) is more tolerant to SO_2 than the wild-type. Spraying spinach leaves with diethyldithiocarbamate to inhibit SOD aggravated damage by SO_2.

Tocopherols may also help plants resist air pollutants. This is in part due to their chain-breaking antioxidant activity. It has been shown[26] that γ-tocopherol reacts with **nitrogen dioxide** ($NO_2^{\bullet}$) to form a nitro adduct (5-nitro-γ-tocopherol). By contrast, $NO_2^{\bullet}$ reacts with α-tocopherol to form a nitrosating product. The physiological significance of these effects is uncertain.

Nitrogen dioxide contributes to the **acid rain** that is devastating trees in some European countries. Acid rain causes damage to plants not only by its low pH and the presence of pollutants, but also by converting insoluble complexes of toxic metals in the soil (e.g. aluminium and possibly iron) into soluble metal ions that can then be taken up.

7.6.4   *Environmental stress*

Exposure of leaves, or isolated chloroplasts, to light intensities in excess of those required to saturate photosynthesis, produces photoinhibition.[5] In both leaves and isolated chloroplasts, photoinhibition is especially severe if illumination is performed in the absence of CO_2. Excessive loss of the D1 protein, due to damage by ROS faster than D1 can be resynthesized, is the key event. Catalase in peroxisomes may also be inactivated at high light intensities.[36]

Exposure of plants to extremes of temperature often causes damage, which can be more severe in the light than in the dark, e.g. 'sunscald' damage to tomatoes (Section 7.5.6 above) and the photo-oxidative death of blue-green algae (Section 3.2). At the other extreme, many temperate plants show depressed photosynthesis after exposure to low temperature for one or two days, especially if they are illuminated (**chilling injury**). Chilling injury probably initially involves changes in membrane structure as the lipids 'freeze' below their transition temperature and fluidity is decreased. Chilling injury seems to particularly affect photosystem II activity via damage to the D1 protein.

ROS are involved in chilling injury and chloroplast SOD activity may decrease during the chilling process.[122] For example, a strain of the green alga *Chlorella ellipsoidea* resistant to chilling injury showed greater SOD activity than did the chilling-sensitive strain. Increasing the SOD activity of this *Chlorella* strain by pre-treating it with paraquat conferred increased resistance to chilling injury. Chilling shoot cultures of rice (*Oryza sativa*) increased H_2O_2 levels and lipid peroxidation, but decreased the levels of GSH, catalase and glutathione reductase. Transgenic[3] tobacco plants expressing pea CuZnSOD in the chloroplast (so raising total chloroplast CuZnSOD) are more resistant to chilling injury, as are transgenic cotton plants expressing MnSOD in the chloroplast (Fig. 7.13, see plate section). Interestingly, in one experiment tobacco over-expressing CuZnSOD in the chloroplast was found to resist photo-oxidative damage whereas a strain expressing MnSOD in the chloroplast did not. Perhaps this is because the MnSOD is in the chloroplast stroma, whereas at least some of the extra CuZnSOD is thylakoid-bound.

Water deprivation induces oxidative stress in plants,[103] frequently lowering $-SH$/disulphide ratios. In some species, there is an adaptive response involving up-regulation of antioxidants such as SOD and ascorbate peroxidase. Transgenic alfalfa (*Medicago sativa*) plants over-expressing MnSOD were more-resistant to water-deficit stress than control plants.[72] The ability of many plant seeds to resist drying for long periods can involve accumulation of high levels of certain sugars, protective proteins and seed antioxidant defences.[48,64]

Paradoxically, not only drought but also waterlogging can cause oxidative stress in plants. Submerged parts of plants may become hypoxic. Upon re-exposure to the air an 'ischaemia/reperfusion damage' involving excess ROS production can occur (Chapter 9).

7.7   The eye

The eye (Fig. 7.14) combines a problem of the erythrocyte (long-lived proteins in the lens) with one of the chloroplast (the risk of damage by too much light). We have already described **retinopathy of prematurity** (ROP), a complication of the use of elevated O_2 concentrations in incubators for premature babies (Chapter 1). Although its incidence can be controlled by restricting the O_2 concentrations used, there is evidence that the frailest babies ($\leqslant 1\,kg$ birth-weight) need high O_2 in order to survive at all. Technological developments in our ability to maintain more-premature babies alive have resurrected the problem of ROP. Fortunately, there is evidence that administration of α-tocopherol to such babies lessens the risk of their developing severe ROP and (possibly) brain haemorrhage. However, the problem of ROP is not solved and interest in new therapies continues (Chapter 1).

Cataract, defined as a clinically significant opacity of the lens (Fig. 7.14) is one of the leading causes of blindness in the world.[111] The lens is surrounded by a capsule, and held in place behind the iris by ligaments. The outward-facing side of the lens is covered by a single layer of epithelial cells which are metabolically highly active and undergo cell division, elongation and development to form the **lens fibres**. Newly formed fibres push the older ones towards the centre of the lens to form the so-called **lens nucleus**. New fibre production is most rapid in young animals, but soon decreases to a low, constant level for the remainder of the lifespan. Thus the inner region of the lens is the oldest part of it. The nucleus, mitochondria and other organelles are lost during fibre maturation, so that fully differentiated lens fibres, which comprise a large proportion of the lens in adult animals, can no longer synthesize proteins to replace damaged ones. No blood vessels enter the lens, and it derives food and O_2 by diffusion (Fig. 7.14). Hence intra-ocular O_2

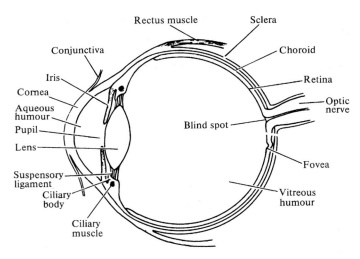

Rectus muscle   Sclera

Conjunctiva   Choroid

Iris   Retina

Cornea   Optic nerve

Aqueous humour

Pupil   Blind spot

Lens

Suspensory ligament   Fovea

Ciliary body   Vitreous humour

Ciliary muscle

Fig. 7.14. Structure of the mammalian eye. The **sclera** is a tough fibrous coat, inside which is the **choroid layer** containing a network of blood vessels that supply food and O_2 to the eye. The choroid is highly pigmented with melanin. The **aqueous humour** is a watery fluid whereas the **vitreous humour** is more jelly-like. Pressure of these fluids outward on the sclera maintains the shape of the eye. The cornea, lens and conjunctiva obtain food and O_2 by diffusion from the aqueous humour—they contain no blood vessels. The **lens** is suspended in position by ligaments and covered by a collagenous capsule. The **retina** contains neurones specialized to recognize light. These photoreceptors are of two types, rods and cones. Only the **cones** are sensitive to coloured light, but the **rods** are more responsive to dim light (hence the diminished perception of colour as light intensity decreases). The **macular region** is the area of greatest visual acuity, around the fovea. The light-absorbing material is an aldehyde, 11-*cis*-retinal attached to proteins called opsins, giving **rhodopsin**. 11-*cis*-Retinal is synthesized from all-*trans*-retinol, vitamin A. Any marked variation in the pressure of the humours can alter the blood flow in the eye. For example **ocular hypertension** (too much pressure) can restrict blood flow and cause ischaemic damage; **glaucoma** is the name of a group of diseases in which this occurs. **Cataract** is an opacity of the lens: it is divided clinically into different types depending on which part of the lens has become opaque. More than 17 million people worldwide are blind from cataract and in the USA cataract removal is the most frequently performed surgical procedure in old people. Age-related **macular degeneration** is the leading cause of permanent blindness in the old in the USA; there is atrophy of the retinal pigment epithelium (RPE), a layer of cells lying between the photoreceptors of the retina and the choroid capillary network. Atrophy is most pronounced in the central macular region. The RPE supplies the retina with O_2 and nutrients (which must pass through it from the capillaries) and removes waste products (such as worn-out shed photoreceptor membranes) from the rods and cones by phagocytosis. The RPE often accumulates lipofuscin with age and retinal is involved in its formation (*Exp. Eye. Res* **65**, 639, 1997). Degeneration of the RPE leads to the death of its associated rods and cones. The defective RPE can no longer absorb waste products, resulting in abnormal deposits (**drusen**) in the retina.

concentrations are low (unlike the chloroplast), which perhaps helps to minimize oxidative damage. Indeed, ROP illustrates the deleterious consequences of raising them. The lens epithelium actively transports sodium ions out of the lens, and potassium ions into it, using an ion-transporting ATPase enzyme.

7.7.1   *What problems does the eye face?*

1. UV light and singlet O_2 (which can be formed by photosensitizers endogenous to the lens; Chapter 2) can damage and cross-link lens proteins.[115] The tryptophan-derived products kynurenine, 3-hydroxykynurenine and 3-hydroxykynurenine glucoside may act as screens for UV light (below 400 nm) in the lens, preventing it reaching the retina. The downside is that they may also sensitize $^1O_2$ formation, as does the tryptophan degradation product N-formylkynurenine.

Lens proteins, especially the **crystallins**, have a very long turnover time and so damage to them is cumulative.[111] The crystallins account for about 90% of the soluble protein of the lens fibres and are divided into three main classes, α-, β- and γ-crystallins in mammalian eyes. The organization of the crystallins maintains the transparency of the lens: denaturation, oxidation and aggregation will lead to loss of transparency.[106] α-Crystallin may act as a 'chaperone' (Chapter 4), helping to protect the other crystallins from thermal insult and oxidative cross-linking.[117]

Ionizing radiation often induces cataract, perhaps by increased formation of $OH^\bullet$ and other ROS. Much evidence implicates prolonged exposure to UV light as a factor in cataract formation. Proteins isolated from cataractous lenses contain elevated amounts of methionine sulphoxide, *ortho*-tyrosine and leucine hydroxide residues, end products of oxidative damage (Chapter 5). UV-induced degradation of tryptophan has been reported in lens proteins. The hyperglycaemia of diabetes (Chapter 9) can lead to glycation of lens proteins (including CuZnSOD).[109]

2. The vitreous humour (Fig. 7.14) contains hyaluronic acid. Hyaluronic acid is depolymerized on exposure to $O_2^{\bullet-}$-generating systems, causing loss of viscosity (Chapter 9). The attacking species is $OH^\bullet$, formed from $O_2^{\bullet-}$ and H_2O_2 in the presence of transition metal ions. Indeed, introduction of iron ions or haem proteins into the eye leads to retinal and lens damage (sometimes sufficient to cause blindness);[116] this can happen as a result of penetration by iron objects or by haemorrhage. H_2O_2 levels in lens have been estimated as 20–30 μM normally and higher in cataract, although some authors have challenged the validity of the assays used, and the possibility of *post mortem* H_2O_2 generation in eyes must not be ignored.

3. Lens epithelial cells are readily damaged by H_2O_2, showing DNA strand breakage and abnormalities of ion transport, e.g. damage to the Na^+,K^+-ATPase.[106] Incubation of rat lens in culture with as little as 200 mM H_2O_2 caused loss of transparency within 24–48 h.

4. The lipids present in the membranes of retinal rod cells (Fig. 7.14) contain a high percentage of PUFA side-chains, especially of docosahexaenoic acid (DHA), and are thus susceptible to lipid peroxidation.

5. The 'visual pigment' rhodopsin can sensitize formation of singlet O_2 (Chapter 2). Exposure of isolated retinas to high light intensities induces lipid peroxidation. Lipid peroxides are increased in the rabbit retina after exposure of the animals to elevated O_2 concentrations or to ionizing radiation, and

injection of preformed lipid peroxides into the eyes of rabbits causes severe retinal damage. Dietary deficiencies of selenium and vitamin E in rats cause marked loss of PUFAs and accumulation of fluorescent products in the retinal pigment epithelium.[77]

7.7.2 Solutions

Antioxidant enzymes and their substrates

The eye has a lot of potential problems, and one would expect a corresponding degree of protection. One protective feature, already alluded to, is that intra-ocular O_2 tensions are low. Also, the concentration of GSH in the lens of several mammals is as high as in the liver, GSH being especially concentrated in the lens epithelium.[106] GSH may be especially important in protecting the thiol groups of crystallins, preventing them from aggregating to form opaque clusters. The GSH/GSSG ratio is normally kept high in the lens and other ocular tissues by the activity of a glutathione reductase enzyme, which obtains NADPH by operation of the pentose phosphate pathway. Glaucoma (defined in the legend to Fig. 7.14), by decreasing the supply of glucose to the lens, interferes with this pathway and so restricts NADPH supply. Glutathione peroxidase and catalase are present in all parts of the eye. Glutathione S-transferases have also been detected, but their role in lipid peroxide metabolism (if any) is unclear.

SOD is present in all eye tissues. The SOD activities of most eye tissues are largely inhibited by cyanide and thus most activity can be attributed to CuZnSOD. The retinal pigment epithelium additionally contains high levels of mitochondrial MnSOD.[77] CuZnSODs are susceptible to inactivation by H_2O_2. Indeed, feeding aminotriazole to rabbits raises H_2O_2 concentrations in the lens, induces cataract and causes a decrease in lens CuZnSOD activity, probably as a result of the accumulated H_2O_2. Hence, as usual, SOD activity has to be balanced with enzymes that remove H_2O_2.[106] The concentration of H_2O_2 in the aqueous humour of patients with cataract is elevated, and a common feature of all types of cataract is that lens GSH concentrations fall. Lens SOD can also be glycosylated in diabetes, causing loss of some enzymic activity. The redox-cycling bipyridyl herbicide diquat causes cataract in rats, apparently by production of $O_2^{\bullet-}$ and H_2O_2 after reduction to a bipyridyl radical by lens enzymes, including glutathione reductase.[107]

Low-molecular-mass antioxidants

GSH plays a key role in the lens as protector of −SH groups and a substrate for glutathione peroxidases. Injection of buthionine sulphoximine, an inhibitor of GSH synthesis, into newborn rats and mice causes them to develop cataract; high levels of ascorbate are protective (Chapter 3). It has been suggested that the ability to synthesize GSH decreases with age in the human lens, which would predispose to the development of cataract in lenses subjected to oxidative stress.

Rod outer segments and retinal pigment epithelium are rich in α-tocopherol;[77] retinal damage and lipofuscin accumulation in the retinal pigment epithelium are seen in vitamin E-deficient animals. Feeding extra vitamin E decreases the severity of the cataracts caused by aminotriazole administration to rabbits. In premature babies, vitamin E levels are sub-normal (a factor pre-disposing to retinal damage) but vitamin C levels greater than in adults (Table 7.4).

Ascorbic acid[111] is present at high concentrations in the lens, cornea, retinal pigment epithelium and aqueous humour of human, monkey and many other animals (e.g. 1–2 mM in human lens and aqueous humour). Its ability to recycle α-tocopheryl radical and scavenge $O_2^{\bullet -}$, singlet O_2, $OH^\bullet$ and other ROS may be of importance. For example, ascorbate protects the lens iontransporting ATPase against damage by a $O_2^{\bullet -}$-generating system *in vitro*. On the other hand, it has been suggested that light-induced degradation of ascorbate in the aqueous humour may be a source of H_2O_2. Once it has started to oxidize, ascorbate can cause 'glycation type' reactions on proteins, including crystallins.[80] Indeed, 'browning' of food and beverages related to ascorbate oxidation is an area well studied by food scientists. An ascorbate peroxidase enzyme has been identified in bovine RPE and choroid (Section 3.16.8).

Consistent with a protective role for ascorbate *in vivo* is the observation that nocturnal animals (e.g. cats) have lower ascorbate concentrations in the eye. Epidemiological evidence (Chapter 10) also supports the view that the overall effect of ascorbate in the eye is beneficial.[111]

Sequestration of metal ions

Ascorbate is a powerful antioxidant, but can become pro-oxidant if mixed with transition-metal ions (Chapter 3). Indeed, incubation of crystallins with ascorbate in the presence of iron ions *in vitro* produces oxidative changes

Table 7.4. Levels of vitamins E and C in retina from eyes taken from adult humans or premature babies

Source of retina	Vitamin E		Vitamin C	
	nmol/mg DNA	nmol/cm² retina	nmol/mg DNA	nmol/cm² retina
Premature babies (vascular region)[a,b]	1.4 ± 1.1	0.47 ± 0.32	244 ± 38	71.3 ± 15.2
Mature retinas[c] (central region)	10.9 ± 6.0	3.04 ± 1.34	162 ± 33	38.3 ± 8.8

[a]Levels of vitamin E but not vitamin C were lower in avascular regions.
[b]22–33 weeks gestation.
[c]13 samples, age range 1 month to 73 years.
Data abstracted from *Invest. Ophthalmol. Vis. Sci.* **29**, 22 (1988), by courtesy of Dr RE Anderson and the publishers.

resembling those seen with ageing in the human lens. Generation of ROS such as OH$^\bullet$ by iron/ascorbate mixtures (Chapter 3) may explain the damage produced by introduction of iron ions into the eye. GSH can also form GS$^\bullet$ radicals in the presence of iron or copper ions.

Under normal conditions, such reactions may be minimized by the presence in aqueous and vitreous humours of the iron-binding protein **transferrin**;[71] transferrin-bound iron cannot stimulate OH$^\bullet$ formation or lipid peroxidation (Chapter 3). **Lactoferrin** is present in tear fluid of most species (in rabbit it is replaced by transferrin).[13] Hence human tears are powerful inhibitors of iron-dependent free-radical reactions *in vitro* (so don't cry over free-radical experiments: it will make things worse).[60] Exposure of rats to high light intensities caused induction of **haem oxygenase I** in the retina,[61] presumably an antioxidant response (removal of haem, formation of biliverdin, but also creation of 'free' iron). Haemopexin can also be synthesized in retina.[52]

Melanin is present in the retinal pigment epithelium[77] and choroid (Fig. 7.14). It can bind iron, possibly as an antioxidant defence mechanism, scavenge $^1O_2$ and absorb light. It also binds zinc ions, which have been suggested to contribute (in a way as yet undefined) to antioxidant defence in the RPE. This tissue also contains metallothionein.

Repair of damage

Methionine sulphoxide reductase is present in the lens and presumably functions (in cooperation with thioredoxin) to repair oxidized methionine residues (Chapter 3). Partially oxidized crystallins can be degraded in lens epithelial cells by proteasome- and ubiquitin-dependent pathways, but it is possible that highly oxidized/cross-linked proteins can no longer be handled by this system and so accumulate.

7.7.3   *The question of carotenoids*

β-Carotene and some other carotenoids are important to vision as precursors of vitamin A and hence of the 'visual pigment' rhodopsin (Fig. 7.14). Lack of vitamin A is a major cause of blindness in the world. Carotenoids might also play a direct protective role in the eye of some species.[56] In the compound eye of the housefly, a carotenoid is present in large amounts (four to ten molecules per rhodopsin molecule). Carotenoid has also been reported in the lateral eye of the crab *Limulus*. The corneas of puffer fishes are clear in the dark but become yellow in the light, owing to the migration of carotenoid pigment in chromatophore cells. Some epidemiological studies show an inverse relation between plasma levels of certain carotenoids in humans and the incidence of cataract (Chapter 10). The retinal macula of primates is enriched in **lutein** and **zeaxanthin** (Chapter 3).[12a] Their role is uncertain; it has been suggested that they could act as blue-light filters and/or singlet O_2 and peroxyl radical scavengers.[111]

7.8   Reproduction and oxidative stress

7.8.1   *Pre-conception*

Spermatozoa: the problems

The first indication that oxidative stress could affect the viability and functioning of sperm came in 1943, when it was observed that human sperm rapidly lost their motility when incubated at elevated O_2 concentrations, and that addition of catalase offered some protection. Isolated washed mammalian spermatozoa readily undergo oxidative damage. Antioxidants in the seminal plasma normally help to protect them[1,58] but the washing of sperm (e.g. in preparation for *in vitro* fertilization) can remove this protective capacity, especially if the washing and resuspension fluids are contaminated with transition-metal ions.

Lipid peroxidation seems to be particularly important in causing sperm dysfunction.[1] The lipids of spermatozoa are rich in PUFA side-chains: almost 50% of the esterified fatty acid in a sperm cell is DHA. This high content of PUFAs gives the sperm plasma membrane considerable fluidity, needed for it to participate in the membrane fusion events associated with fertilization. In peroxidized sperm, fluidity decreases and the capacity to fertilize is diminished. Oxidative stress also causes rises in intracellular free Ca^{2+} (Chapter 4) and depletions of ATP, interfering with sperm motility.

Seminal fluid can generate ROS. The major source appears to be neutrophils, always present in the ejaculate.[1] An additional source is the spermatozoa themselves, which appear to generate $O_2^{\bullet-}$, although there is disagreement about how much is produced. Abnormal or non-functional sperm (e.g. from infertile men) seem to generate $O_2^{\bullet-}$ at greater rates than healthy ones. Excess generation of ROS in semen could thus be a consequence of infertility, but might then worsen the problem by damaging functional sperm.

Generation of ROS by human sperm may involve an NADPH oxidase system. However, it appears dissimilar to phagocyte NADPH oxidase, since monoclonal antibodies directed against phagocyte oxidase components do not recognize proteins in sperm extracts.[1] Low-level $O_2^{\bullet-}$ generation may play an important role in normal sperm function, but exactly how is not yet clear. It may contribute to **capacitation**, a collective term for the changes in sperm behaviour that occur as they pass through the female reproductive tract, such as the development of the hyperactivated movement needed to achieve penetration of the membrane that surrounds the oocyte (**zona pellucida**). It might also contribute to the **acrosome reaction**, a membrane fusion event involving the outer acrosomal and plasma membranes of the sperm. The acrosome reaction causes release of the acrosomal contents, including proteases that are thought to facilitate passage of the sperm through the zona pellucida.

As for the other sex, ROS have been suggested to be involved in regression of the corpus luteum[90] during the normal ovarian reproductive cycle. In sea-urchin eggs, ROS are involved in generation of a fertilization membrane (Chapter 6). Female transgenic mice lacking CuZnSOD (Section 3.4.2) show decreased fertility due to increased embryonic death.[49a]

Spermatozoa: the solutions

Sperm contain SOD, catalase, glutathione peroxidase/reductase and GSH, as well as ascorbate and α-tocopherol.[1] However, they have limited biosynthetic capacity, making it difficult to replace any molecules that do undergo damage. In addition, the enzyme antioxidants may be concentrated in the mid-piece of the sperm, leaving the large expanse of membrane overlying the head and tail less protected. Hence the seminal plasma is particularly important in protecting sperm against damage by ROS generated by the sperm themselves and also by the phagocytes in the ejaculate.

However, the antioxidant abilities of seminal plasma have not been fully characterized. Some SOD (mostly CuZnSOD plus some EC-SOD in humans)[81a] and glutathione peroxidase are present, the latter apparently secreted by the epididymis. Epididymal GPX lacks selenium, having cysteine at its active site instead of selenocysteine.[82] There are also higher levels of ascorbate than in plasma (usually 200–400 μM versus 40–100 μM). Hypotaurine (Section 6.7.2) is present and has been suggested to be a scavenger of several ROS, especially HOCl, which can be generated in ejaculate by the action of neutrophil myeloperoxidase. The levels of the oxidative DNA damage product 8-hydroxydeoxyguanosine (8-OHdG) in sperm DNA have been inversely correlated with seminal fluid ascorbate levels (Table 7.5). Iron-binding proteins may be present in seminal fluid, since it inhibits peroxidation of ox-brain homogenates, a reaction known to be dependent on iron ions liberated during tissue homogenization (Chapter 3).[58]

Spermatozoa as targets for toxins

Spermatozoa can be damaged by many mechanisms, including inflammation of the reproductive tract. Such damage can involve ROS (and possibly RNS) generated by activated phagocytes.

Gossypol is a yellow polyphenol found in cotton plants; it has been widely tested as a male oral contraceptive. Its toxic effects on sperm have been

Table 7.5. Levels of dietary ascorbate intake, semen ascorbate levels and 8-hydroxyguanine in DNA from sperm of human volunteers

	Ascorbate intake (mg/day)	Semen ascorbate (μM)	8-Hydroxydeoxyguanosine (fmol/mg DNA)
Baseline	250	399 ± 55	34.0 ± 2.4
Depletion	5	203 ± 72	66.9 ± 8.5
Marginal	10 or 20	115 ± 25	84.4 ± 22.3
Repletion	60 or 250	422 ± 100	53.8 ± 16.8

Human volunteers were maintained on controlled diets supplemented with various amounts of ascorbate. Dietary ascorbate was decreased from 250 mg/day to 5 mg/day, and then repleted after 28 days. Data abstracted from *Proc. Natl Acad. Sci. USA* **88**, 11003 (1991) by courtesy of Prof. Bruce Ames and the publishers.

suggested to be due to stimulation of ROS production.[10] If this is so, the damage caused by the oxidative stress is unlikely to be mediated by lipid peroxidation, since gossypol is a powerful inhibitor of this process (Chapter 3).

7.8.2 Post-conception

Problems of the early mammalian embryo[62]

The increasing popularity of assisted conception techniques in humans and other animals has focused attention on the problems of handling fertilized eggs and early embryos, especially as the 'take-home baby rate' after *in vitro* fertilization and embryo implantation is still low.

During the pre-implantation period (about 7 days in the human) prior to establishment of pregnancy, the fertilized egg spends abut 3 days in the oviduct and the rest in the uterus. Cell division gives a two-cell, four-cell and then up to 16-cell stage (Fig. 7.15). Compaction of the cells occurs at about the eight- to 16-cell stage; they flatten down upon one another, becoming wedge-shaped rather than spherical. Eventually a mature **blastocyst** is formed, consisting of an outer layer of polarized cells (**trophectoderm**) surrounding a fluid–filled cavity (the **blastocoel**) which contains a cluster of cells, the **inner cell mass**. The embryo proper, and thus the foetus, arises from some of the latter cells, whereas the remainder of the inner cell mass and the trophectoderm give rise to extra-embryonic structures such as the **placenta**.

Preimplantation development is initially controlled by mRNA from the egg: expression of the embryonic genome begins at the four- to eight-cell stage in humans, but at the two-cell stage in mice. Oocytes and embryos appear to contain the usual complement of antioxidant defences, including ascorbate, vitamin E, SOD, GSH, glutathione peroxidases, thioredoxin, catalase and metallothionein (levels of the last are induced by exposure to zinc or cadmium). Mice embryos cultured *in vitro* often cease division at the two-cell stage—the **two-cell block**. The block can be overcome by placing the embryos back into the oviduct, or by incubating in culture media containing EDTA and glutamine but lacking glucose. The two-cell block is representative of a series of 'blocks' that occur if one attempts to grow embryos in culture. Thus a four- to eight-cell block is often seen in human and cow embryos, and an eight- to 16-cell block for sheep and goat embryos. Pyruvate is the preferred metabolic substrate of the early stages of embryonic development, with glucose uptake increasing as the blastocyst is formed.

Several papers have suggested[76] that oxidative stress is involved in the 'cell blocks' observed during *in vitro* development, especially if embryos are cultured under 21% (hyperoxic for the embryo!) oxygen. The use of lower O_2 tensions or the addition of antioxidants[76] (including catalase, transferrin, SOD, thioredoxin or even whole erythrocytes,[69] which can absorb ROS) have been reported to facilitate embryonic development *in vitro*. Pyruvate could conceivably act as a H_2O_2-scavenger (Chapter 3) as well as a nutrient. Severe depletion of oocyte GSH blocks embryonic development. As with spermatozoa, however, there are suggestions that ROS play a useful role *in vivo*, e.g. by being

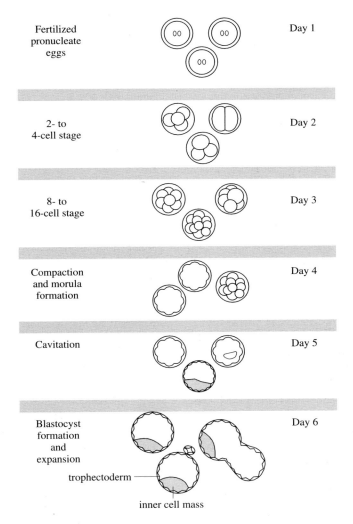

| | | Day 1 |
| Fertilized pronucleate eggs | | |

| | | Day 2 |
| 2- to 4-cell stage | | |

| | | Day 3 |
| 8- to 16-cell stage | | |

| | | Day 4 |
| Compaction and morula formation | | |

| | | Day 5 |
| Cavitation | | |

| | | Day 6 |
| Blastocyst formation and expansion | | |

trophectoderm

inner cell mass

Fig. 7.15. Embryonic development in humans. Diagram by courtesy of Drs MJ Conaghan, K Hardy and HJ Leese and the Biochemical Society (*The Biochemist*, April/May 1994, p. 9).

involved in the apoptosis of unwanted cells during embryonic development.[83] The role of ROS in apoptosis is a fiercely debated area, however (Chapter 4). Rabbit blastocysts have been reported to contain a cell–surface NAD(P)H oxidase that generates $O_2^{\bullet -}$.

Oviductal lining fluid probably has a low O_2 concentration and may contain some SOD and glutathione peroxidase, but its antioxidant defences have not yet been fully characterized, nor have those of uterine fluids. In any case, it has been estimated that over 50% of fertilized human eggs do not survive past the early embryonic stages even in the protective environment of the reproductive tract.

The embryo/foetus as a target for toxins[119]

Oxidative stress and glycation reactions have been suggested to contribute to the increased risk of foetal malformations that is observed in mothers with poorly-controlled diabetes. Indeed, induction of diabetes (by streptozotocin treatment) in pregnant female mice produced embryonic malformations, but the incidence was decreased if transgenic mice overexpressing CuZnSOD were used.[42]

Exposure of the pregnant female to several xenobiotics can cause embryonic death, lethal foetal malformations or birth defects (**teratogenesis**). Teratogenicity can include not only anatomical abnormalities but also biochemical ones, e.g. an increased risk of cancer development after exposure of the mother to certain compounds, an example being **diethylstilboestrol**.

Several teratogens have been suggested to act, in whole or in part, by imposing oxidative stress. They include the anti-convulsant **phenytoin** (diphenylhydantoin, Dilantin), **hydroxyurea** and possibly **cocaine** (Chapter 8). Fortunately the maternal drug-detoxification systems (cytochromes P450, glucuronidation, glutathione S-transferases) and the barrier action of the placenta do an excellent job in protecting against most such potential insults. Of course, cytochromes P450s can convert a few chemicals (including cocaine) to more reactive forms (Chapter 8), but levels of P450s in the placenta and foetus are low. However, placenta does contain a 'non-specific' peroxidase activity (Chapter 3), which could conceivably bioactivate toxins if a source of H_2O_2 was available. Human placenta contains the usual antioxidant defences and has also been reported to synthesize 'extracellular' glutathione peroxidase and secrete it into the maternal circulation.[7]

7.8.3   *Normal and premature birth*

Birth involves a hypoxia–reoxygenation process.[39] Newborn infants face a sudden transition from a warm, fluid-filled environment to a cooler, air-breathing environment, with exposure of the lung to O_2 concentrations from the inhaled air that are much greater than those experienced *in utero*. Often an even greater degree of hypoxia occurs during birth itself, making the reoxygenation even more of a shock. Preparations for birth occur in the later stages of development *in utero* and include rises in the levels of antioxidant defence systems (including SOD, catalase and glutathione peroxidase) in the lung. However, there are considerable differences in the time-courses and extent of these changes between different species.

Premature babies face greater problems: the continuing improvements in neonatal care have increased the survival rate of premature infants, but they often depend on respiratory support (with exposure to elevated O_2) for many weeks. Elevated O_2 contributes to the development of ROP and possibly haemorrhage in the brain (Chapter 1). **Bronchopulmonary dysplasia** (which resembles adult respiratory distress syndrome: Chapter 9) is also a significant cause of morbidity and mortality and there have been several reports of increased parameters of oxidative damage in affected babies, e.g. rises in allantoin/uric acid ratios[75] (Section 5.14.2). Some authors have grouped these three disorders together as different manifestations of **oxygen-radical diseases of prematurity**.[108] The immature lung is deficient not only in certain antioxidants but also in surfactant—the lipoprotein which decreases alveolar surface tension and permits normal lung expansion (Chapter 1). Surfactant can additionally be damaged by oxidative stress: such phagocyte-derived agents as $ONOO^-$ and $HOCl$ attack the surfactant proteins and diminish the surface tension-lowering ability. The lipid of surfactant is mainly **dipalmitoylphosphatidylcholine**, i.e. it contains saturated fatty acid side-chains and is fairly resistant to oxidative attack. Indeed, synthetic surfactants (often resistant to oxidative damage) have found a role in the treatment of bronchopulmonary dysplasia.

Antioxidants

Infants, especially premature ones, have low levels of vitamin E in plasma (Chapter 3) and tissues (Table 7.4). Plasma levels of vitamin C tend to be higher than in adults (Table 7.4) but fall rapidly after a few days. Levels of plasma caeruloplasmin, albumin and α_1-antiproteinase also tend to be lower than in adults.

Several papers have reported protective effects of various antioxidants or other agents against hyperoxia-induced lung injury in premature animals of various species, e.g. intraperitoneal GSH in rabbits,[17] intraperitoneal α_1-antiproteinase in rats, intraperitoneal N-acetylcysteine in guinea-pigs and intratracheal recombinant human CuZnSOD in piglets. Healthy, newborn animals of several species (including mice, rats and rabbits) are more tolerant to hyperoxia than adults, apparently because they can more rapidly induce antioxidant defence systems in the lung (Chapter 3). There is no evidence that this is the case in humans, however.[39]

Iron metabolism

The percentage loading of transferrin with iron is higher in newborn babies than in adults. In about 40% of low birthweight ($<1.5\,kg$) premature babies transferrin is completely saturated and non-transferrin-bound iron is present in the plasma. *In vitro* studies have shown that at least some of this iron is capable of catalysing $OH^\bullet$ generation and peroxidation of lipids, including surfactant lipids. Interestingly, assay of 52 samples of plasma from apparently-healthy full-term babies showed that 11 also contained non-transferrin-bound iron.[35,74]

During the last 3 months *in utero*, the foetus accumulates iron from the mother. About 80% of total foetal iron is in haemoglobin, but some is stored in ferritin. Birth is associated with a virtual cessation of red blood cell formation that persists for at least 6 weeks. Hence iron is not used up in erythropoiesis whilst red blood cell degradation continues (see below); both phenomena may contribute to increased transferrin iron saturation (and also increased plasma ferritin) seen in both full-term and pre-term babies after birth.

Foetal erythrocytes[53] differ from those of adults: they have a larger volume, are present in higher numbers ($54\pm10\,ml$ per $100\,ml$ of blood compared with

47 ± 5 or 42 ± 5 ml for adult males and females respectively) and contain predominantly **haemoglobin F**. Haemoglobin F, like adult haemoglobin, is a tetramer but the β chains are replaced by γ chains (i.e. $\alpha_2\gamma_2$ rather than $\alpha_2\beta_2$). Haemoglobin F has a higher affinity for O_2 under physiological conditions, optimizing transfer of O_2 from the maternal to the foetal circulation. *In vitro*, foetal erythrocytes are more susceptible to oxidative stress (e.g. in the peroxide stress haemolysis test: Section 7.2 above) than adult cells and they produce more $O_2^{\bullet-}/H_2O_2$. The latter observation is related both to a higher content of haemoglobin per cell and to higher autoxidation rates of haemoglobin F. In the membrane lipids, levels of linoleic and oleic acid esters are lower than in adult erythrocytes, but concentrations of arachidonic acid and DHA (22:6) are higher. Levels of vitamin E (both on a micromolar basis and per unit lipid), glutathione peroxidase, catalase, methaemoglobin reductase and SOD are all lower in foetal erythrocytes than in adult ones.

It is possible that the increased oxidative susceptibility of foetal erythrocytes contributes to their rapid destruction (Fig. 7.16). One result is increased plasma bilirubin levels. Although very high bilirubin levels are toxic, more moderate 'physiological jaundice' could be beneficial if proposals that bilirubin is a physiological antioxidant (Chapter 3) are valid.

Parenteral nutrition

It is not only the elevated O_2 exposure of premature infants that may cause problems. Parenteral nutrition has been used in the management of low-birth-weight babies for many years, yet the optimum protocols have not been

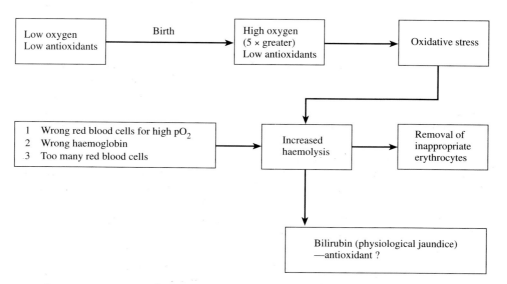

Fig. 7.16. Oxidative stress: is it a mechanism for destruction of foetal erythrocytes?

determined. Lipids used in infusions are often partially oxidized and contain (potentially toxic) lipid peroxides.[46] Exposure of some infused amino acid/vitamin/mineral solutions to light can generate H_2O_2. Although the net effect of parenteral nutrition is beneficial, preventing such oxidations would be expected to minimize pro-oxidant injury if peroxides are infused into babies with plasma non-transferrin-bound iron.

Antioxidants, PUFAs and iron

Maximal development of the brain and nervous system in children requires an adequate dietary intake of all major nutrients, including antioxidant vitamins, copper, iron and PUFAs. Severe anaemia in children can lead to impaired mental development, although the risk of this in advanced countries is small. Nevertheless, maternal iron deficiency should be avoided during pregnancy, although the widespread use of iron supplements may be unnecessary. Nutritional antioxidant deficiency in the mother may lead to similar problems in the foetus, perhaps rendering the baby more sensitive to the 'oxidative stress' of birth.[110]

An adequate supply of long-chain PUFAs, especially DHA, is needed for development of the central nervous system.[91] Infants appear to have some capacity to synthesize DHA from linolenic acid but may need to acquire more from the diet. Supplementation of infant formulas with DHA has to take into account the ease of peroxidation of this and other PUFAs, especially if simultaneous iron fortification is attempted. Breast milk is a good source of the required PUFAs and contains little, if any, 'available' iron because of the high iron-binding affinity of the lactoferrin present[28] (Chapter 3). Prolonged breast feeding alone might eventually lead to iron deficiency, however. More research is needed in this area to allow formulation of meaningful dietary requirements.

7.9 The skin

Mammalian skin consists of two major layers, the **dermis** and **epidermis**. The dermis contains multiple cell types, including fibroblasts and nerve endings, embedded in a fibrous network of connective tissue, containing elastin and collagen. The epidermis, unlike the dermis, has no nerves or blood vessels. Where it meets the dermis is a layer of actively dividing cells: cells from this layer (**keratinocytes**) migrate upwards in the epidermis and undergo differentiation, in the process of **keratinization**. The uppermost layer of the epidermis (**stratum corneum**) is the end result of this process and consists of a highly cross-linked keratin matrix of dead cells, forming an effective barrier. These cells are constantly shed and are a major contributor to household dust. They provide food for house mites, to which many people are allergic.

Most mammals have hair or fur to cover the skin, but apart from the top of the head and clothing, the skin of most humans is exposed to insults from the surroundings. These include ambient O_2 levels (as well as SO_2, O_3 and $NO_2^•$ in polluted air), applied chemicals (e.g. creams, lotions, cosmetics, greases, oils), water-borne toxins (e.g. chlorine, dissolved metal ions), UV light, other

radiation, deodorants (e.g. many antiperspirants are acidic and contain aluminium salts), products of skin bacteria (Fig. 7.17) and agents excreted in the sweat.
For example, sweat is rich in transition-metal ions (iron, copper) capable of
catalysing free-radical reactions; the presence of these metals also encourages
bacterial growth.[41] Skin can respond to many xenobiotics by increasing its
content of cytochromes P450, e.g. CYP1A1 levels increase in response to
carcinogenic hydrocarbons in coal-tar (Chapter 9). **Retinoic acid**, widely used
in the treatment of skin diseases, has the opposite effect: it down-regulates
CYP1A1, an effect which contributes to its chemopreventative effect in skin
cancer models.

7.9.1   *Insults to the skin*

Photosensitization

The skin is at risk of photo-oxidative damage due to the generation of singlet
O_2, $OH^\bullet$, H_2O_2 and other ROS from photosensitizers in applied cosmetics
(e.g. **musk ambrette**, **musk xylene**) or ingested photosensitizing drugs that
reach the skin, including some **phenothiazines** (used as tranquillizers), as well
as fluoroquinolone and tetracycline antibiotics (Chapters 2 and 8). Type I
reactions (direct damage by the excited state of the sensitizer) can also occur.
Porphyria patients can suffer photo-oxidative damage, as do animals consuming **sporidesmin** (Chapter 2). *Propionibacterium acnes*, a normal inhabitant of

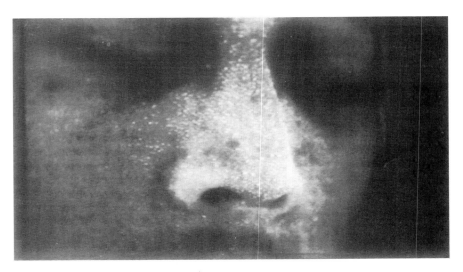

Fig. 7.17. Fluorescence of *P. acnes*-derived coproporphyrin III on human skin as seen under
UV light. This technique has been used to monitor the distribution of *P. acnes*, which is found
in increased levels in acne patients, on the skin. Photograph from *Biochem. Biophys. Res.
Commun.* **223**, 578 (1996) by courtesy of Dr Kumi Arakane and Academic Press.

the skin, produces trace amounts of coproporphyrin (Fig. 7.17); much larger amounts are made in patients with acne.

Many plants produce photosensitizers (Fig. 7.18), often to attack other plants or deter predators.[11] For example, *Cercospora* fungi produce **cercosporin**, which generates ROS in the presence of light to attack plant cell walls and allow the fungus entry. Marigolds produce **α-terthienyl**, apparently as a 'photodynamic insecticide'. The St John's wort (*Hypericum hirsutum*) produces **hypericin**, a photosensitizing compound that can cause phototoxicity in grazing livestock. This was recognized over a century ago by Arab herdsmen, who painted their white horses with henna or grazed only black sheep on pastures containing this plant.

Celery, parsnips and parsley contain **psoralens** (Chapter 2). There has been a report of this causing problems in a patient undergoing photodynamic therapy: it was suggested that the skin damage was caused by the additive effects of the psoralens in the vegetables she ate and those given during the therapy.[4]

Ultraviolet light[115]

The biggest problem the skin faces is probably UV light; the same may be true for the eye (Section 7.7 above). Chronic exposure to sunlight appears to be the major risk factor for skin cancers (basal cell carcinoma, squamous cell carcinoma or malignant melanoma) in humans and a contributor to the ageing of skin. Dermal photo-ageing is accompanied by wrinkling, loss of elasticity, increased skin fragility and slower wound healing.

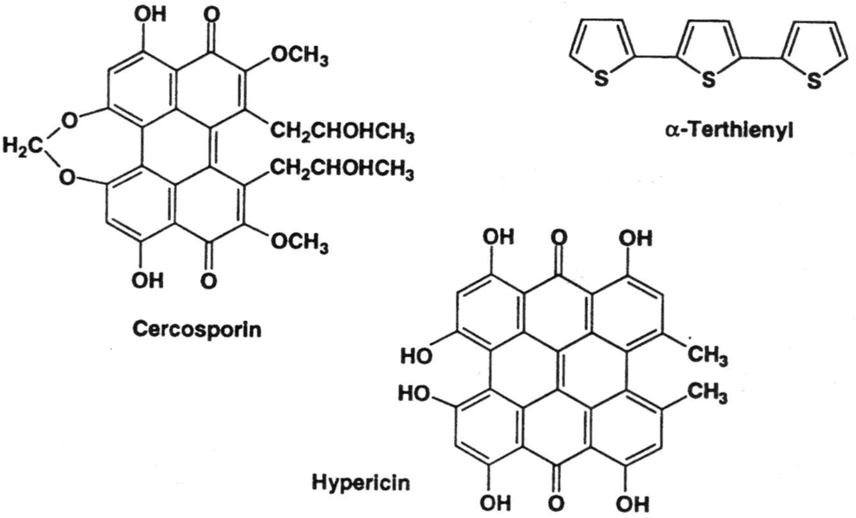

Fig. 7.18. Structures of some plant-derived photosensitizers. Diagram from *Active Oxygen in Chemistry*, Vol. 2 (1995) by courtesy of Dr Chris Foote and Blackie Academic and Professional Publishers.

In the UV region of the solar spectrum, it is only UV-A (320–400 nm)[b] and some UV-B (290–320 nm) that reach the surface of the earth. Fortunately, the more damaging UV-C (100–290 nm) is filtered out. The major barrier to UV-C (and largely to UV-B) is the ozone layer in the stratosphere. Depletion of this layer by photochemical reactions involving chlorofluorocarbons may result in more UV-B reaching the Earth, with a corresponding increase in photochemical damage to both aquatic (e.g. fish, amphibians) and terrestrial organisms, including plants and humans (Chapter 1). It has been estimated that each 5% depletion of stratospheric O_3 will raise UV-B flux at ground level by 10%.

Up to 10% of UV-B light falling on the skin can penetrate through the epidermis to the dermis. UV-A has greater penetration (e.g. about 20% at 365 nm). UV-B is much more damaging to skin than UV-A if equal exposures are carried out, but the deeper penetration of UV-A and its greater abundance in sunlight suggest that UV-A is a major contributor to photodamage.

In contrast to X-rays and γ-rays, UV light does not deposit sufficient energy in water molecules to ionize them. However, if H_2O_2 is available, UV-B causes homolytic fission:

$$H_2O_2 \xrightarrow{\text{UV}} OH^\bullet + OH^\bullet$$

to generate highly-reactive $OH^\bullet$ (Chapter 2). Absorption of UV light in the wavelength range 200–300 nm (UV-B range) by thymine or cytosine in DNA creates excited states that can react with water to form pyrimidine hydrates, or with an adjacent pyrimidine to produce cross-links (thymine–thymine, thymine–cytosine or cytosine–cytosine).[22] The dimers are chemically stable, but the hydrates are not and usually lose water again to reform the original pyrimidine (Fig. 7.19). Alternatively, the cytosine hydrate can lose its amino group in a deamination reaction. Defects in the ability to repair UV-induced lesions in DNA are seen in some human diseases such as *xeroderma pigmentosum* (Chapters 4 and 9).

DNA damage by these and other mechanisms probably contributes to photo-damage and skin cancer induced by UV.[115] UV-B exposure has been reported to raise levels of 8-hydroxyguanine in DNA from isolated keratinocytes and from the skin of hairless mice (possibly by attack of singlet O_2 or $OH^\bullet$). UV-induced pyrimidine dimers tend to cause $CC \rightarrow TT$ tandem double-base transition mutations in DNA (cytosine dimers are more mutagenic than thymine dimers). Mutations in the tumour suppressor gene **p53** (Chapter 9) appear to be early events in skin cancer development. Such double transition mutations are much more common in mutated p53 genes from skin cancers than in mutated p53 from tumours of internal organs. Although this evidence for a role of UV seems convincing, these tandem mutations are not absolutely diagnostic of UV-induced damage[113] since exposure of cells to ROS can also produce such mutations (Chapter 4), as can Ni^{2+} and high levels of sulphites (although the latter two are an unlikely explanation in most cases of skin cancer).

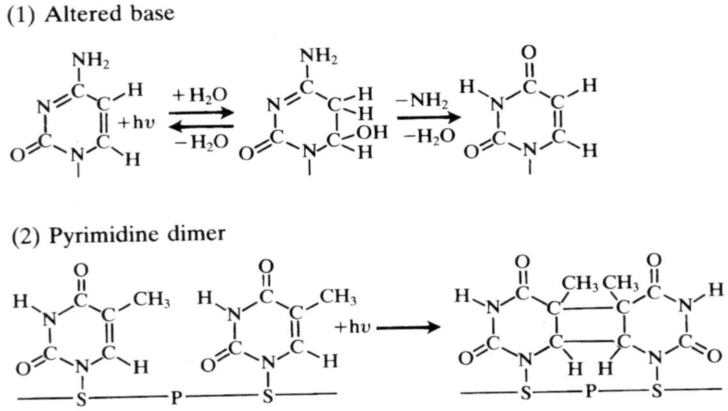

(1) Altered base

(2) Pyrimidine dimer

Fig. 7.19. Examples of damage to DNA by UV light. (1) Reversible addition of H_2O to cytosine is shown with the possibility of deamination to uracil. (2) Formation of a thymine–thymine cross-link. S, deoxyribose sugar, P, phosphate. Cross-linked thymine dimers are the main lesion in UV-exposed DNA, although they can be repaired efficiently and cytosine dimers are considered more mutagenic. Cytosine within photo-dimers undergoes accelerated deamination to uracil.

UV-A or -B can induce activation of a wide range of transcription factors in skin cells,[115] including AP-1 and NF-κB (Chapter 4). In skin, UV-B induces synthesis of metalloproteinase enzymes, which can degrade collagen and other connective tissue components. Induction of these enzymes may be mediated via NF-κB and AP-1, whose activation may be promoted by $^1O_2$ or other ROS.

7.9.2   *Inflammation*

ROS can also be generated in skin by inflammation, and the skin is affected in many autoimmune diseases (Chapter 9). Damage by UV light causes reddening and other aspects of a typical inflammatory response, including recruitment of neutrophils that can generate $O_2^{\bullet-}$, H_2O_2, HOCl and possibly NO$^\bullet$. Several skin diseases involve a prominent role for neutrophils, including **acne**, **psoriasis** and **Behcet's disease**. **Anthralin**, a drug used to treat psoriasis, has been suggested to exert some of its effects (and side-effects) by ROS generation.[75a]

Human keratinocytes have been shown to contain nitric oxide synthase,[30] and UV-B exposure was reported to increase NO$^\bullet$ production. NO$^\bullet$ might play a physiological role in stimulating melanocytes to produce more melanin, but an excess might be damaging.

Chronic exposure of hairless mice to low levels of UV-B increased the non-haem iron content of the skin;[12] a similar increase in iron has been observed in sun-exposed human skin. Topical application of certain iron ion chelators (e.g. 1,10-phenanthroline) to hairless mouse skin appeared to delay the onset of UV-B-induced damage to connective tissue. UV-B also induces synthesis of ICAM-1 and several cytokines by keratinocytes, including interleukin 1 (IL-1), IL-6, IL-8 and tumour necrosis factor alpha (TNFα), and may sometimes induce apoptosis. Skin also synthesizes leukotrienes (Chapter 6).

7.9.3 The solutions

Plant responses to UV-B exposure include increased synthesis of flavonoids and other phenolics in epidermal layers and induction of enzymes that repair DNA dimers by photo-reactivation (Chapter 4). How does skin protect itself? First, sunlight induces the synthesis of melanin (Chapter 3), which has a broad absorbance spectrum that ranges through the UV-B, UV-A and visible ranges. Eumelanin may be highly protective, but this is not necessarily true of pheomelanin (Chapter 3). Sunlight can also decrease levels of β-carotene and lycopene in skin (Chapter 2).[89]

Keratinocytes and fibroblasts contain millimolar levels of GSH and the usual complement of antioxidant defence enzymes, ascorbate and DNA repair enzymes. The epidermis also contains 8-oxo-dGTP hydrolase, which 'sanitizes' the DNA precursor pool (Chapter 4).[118] In mice, levels of catalase, glutathione peroxidase, glutathione reductase, α-tocopherol, ubiquinol, ascorbate and GSH (but not SOD) are higher in epidermis than in dermis.[101] Ascorbate, GSH, SOD, catalase and ubiquinol are depleted in UV-B-exposed skin, both dermis and epidermis. Levels of electron paramagnetic resonance (EPR)-detectable ascorbyl radical rise on UV exposure of skin. Thioredoxin and thioredoxin reductase in skin have also been suggested to contribute to antioxidant defence. In general, keratinocytes in culture are more resistant than most animal cells to killing by H_2O_2, organic peroxides or peroxynitrite.

In cultured human skin fibroblasts, UV-A induces 'heat-shock' proteins, including haem oxygenase-1 (Hsp32), co-upregulated with ferritin.[115] Keratinocytes are unusual in expressing Hsp72 constitutively, although levels rise under UV light. By contrast, keratinocytes have higher basal levels of haem oxygenase (HO-2), but little inducible (HO-1) activity. This may be explained by the fact that considerable UV radiation reaches the keratinocytes at the base of the epidermis whereas higher exposures are needed to reach the dermal fibroblasts.

An obvious way of avoiding UV-induced damage is to stay out of the sun. More usual practice is to coat the skin with a UV-absorbing oil or cream. However, one widely used sunscreen ingredient, **Padimate-O** (octyldimethyl *para*-aminobenzoic acid) was found to be converted to a mutagen after exposure to sunlight.[57] Constituents of sun creams could therefore conceivably be converted to toxic products by UV light and/or ROS in skin.

7.10   Exercise: an oxidative stress?

The amount of O_2 taken in by aerobes that leads to formation of ROS is estimated as 1–5% of the total (Chapter 1). During exercise, bodily O_2

consumption is greatly increased, up to 10- to 15-fold greater than resting levels. Oxygen uptake in the active skeletal muscle may increase up to 100-fold. If the same percentage for ROS formation holds true, exercise should lead to a large increase in total body ROS burden, and an even larger increase in the working muscles. This could conceivably cause muscle oxidative damage, especially interference with Ca^{2+} metabolism and impaired contractility. Yet levels of SOD, catalase, GSH and glutathione peroxidase in muscle are lower than in many other body tissues (Chapter 3). Skeletal muscle also generates $NO^\bullet$, which could conceivably lead to $ONOO^-$ formation if $O_2^{\bullet-}$ production were elevated.

7.10.1   *Does exercise cause oxidative damage?*

Is exercise accompanied by increased oxidative damage? Davies *et al.*[29] found that severe forced physical exercise in rats results in muscle damage, seen as a decrease in mitochondrial respiratory control, loss of structural integrity of sarcoplasmic reticulum and increased levels of some markers of lipid peroxidation. Later work showed increased protein carbonyls and 'catalytic' iron levels in the muscle. Vitamin E-deficient rats have markedly lower endurance capacity for exercise. Protection against oxidative damage might be offered by careful endurance training: such training was reported to increase the activities of glutathione peroxidase, glutathione reductase, catalase, and SOD in rat heart and skeletal muscle, although it seemed to decrease the content of vitamin E in muscle mitochondria.

Are these observations relevant to humans? Vigorous repeated excercise, especially in untrained individuals, produces muscle damage, as demonstrated by histological studies or by measuring release of muscle enzymes (such as **creatine kinase**) or of myoglobin into the circulation. Prolonged strenuous exercise (e.g. running a marathon) increases the number of circulating neutrophils[105] and may produce some features of an acute-phase response, e.g. falls in plasma zinc and iron, rises in IL-1 and C-reactive protein and mild fever. Elevated temperature in exercising muscle is also likely to trigger the heat-shock response (Chapter 4). Trained athletes have higher concentrations of caeruloplasmin in plasma than normal controls; caeruloplasmin is part of the extracellular antioxidant defences and is an acute-phase protein. Endurance exercise has also been reported to raise SOD and catalase activities in human muscle, as sampled by needle biopsy.[96]

Early work showed that some human subjects respond to exercise by increased exhalation of pentane (possibly arising from lipid peroxidation), an effect that could be diminished by pre-treating the subjects with excess oral vitamin E.[33] By contrast, other scientists have found no increase in serum TBA reactivity after maximal exercise by long-distance runners, although falls in muscle GSH/GSSG ratios are frequently reported.[96] Some of the GSSG may be exported to the plasma, since several authors report rises in plasma GSSG after exercise. It has also been suggested[94] that four hours of continuous running in well-trained runners caused subsequently-isolated plasma LDLs to be more susceptible to oxidation upon exposure to Cu^{2+}. Urate may be oxidized in human skeletal muscle during exercise, presumably as it intercepts ROS.[47] The falls in ATP levels during intense exercise lead to increased urate production, which is perhaps therefore serving a useful role.

What does all this mean? It is certain that excessive exercise can induce muscle damage. Damaged tissues often peroxidize more rapidly than normal (Chapter 9). Thus the extent of muscle damage may determine whether or not elevated lipid peroxidation can be detected in exercising humans. In any case, TBA reactivity and gas chromatography (GC) measurement of pentane are non-specific methods of assaying peroxidation (Chapter 5). Hence oxidative damage during exercise may be a consequence of muscle injury (and subsequent inflammation) rather than a cause of it. Whether antioxidant supplementation of athletes would be beneficial (either in minimizing oxidative damage secondary to trauma or in accelerating healing of damaged tissue) is an open question,[55] although it has been claimed that *N*-actylcysteine administration can decrease muscle fatigue in humans.[88]

7.10.2   *Exercise, health and free radicals*

Habitual physical activity in humans has several claimed overall positive effects, including improved cardiovascular function and glucose tolerance and lowered risks of obesity, hypertension and certain infections (by contrast, hard-training athletes may be more susceptible to certain infections). The extent to which the benefits of moderate exercise might be related to up-regulation of antioxidant defence systems in muscle, heart and elsewhere is unknown. There is a need to use modern methods for measuring ROS and oxidative damage (Chapter 5) to investigate the overall consequences of different types of exercise in humans. One factor that must always be considered is that prolonged vigorous exercise changes plasma volume, which must be corrected for in attempts to assess the significance of changes in antioxidant levels or markers of oxidative damage in plasma.

Strenuous exercise in both humans and dogs[79,85] has been reported to increase the urinary excretion of 8-OHdG; this could reflect more DNA damage (a bad thing) or activation of DNA repair processes (a good thing). In dogs, levels of 8-OHdG in lymphocytes and colon decreased after exercise, consistent with increased repair.[79]

Oxidative stress has also been implicated in the pathology of muscle **atrophy**.[59] Immobilization of rat hind limbs caused muscle wasting, accompanied by increases in protein carbonyls and non-protein-bound iron. The salicylate trapping technique revealed increased $OH^\bullet$ generation in the muscle.

7.10.3   *Muscle as a target for toxins*

Several toxins can damage muscle. For example, the ***para*-phenylenediamines** are widely used in the rubber, dyestuff and photographic industries. Some of them cause necrosis of skeletal and cardiac muscle in animals and it

Fig. 7.20. A suggested mechanism by which tetramethylphenylenediamine causes oxidative stress. Oxidation is catalysed by traces of metal ions. Damage to rat muscle by a series of *N*-methylated *para*-phenylenediamines was correlated with their rates of oxidation. From *Toxicology* **57**, 303 (1989) by courtesy of Dr Rex Munday and the publishers.

has been suggested that their oxidation to nitrogen-centred radicals is involved (Fig. 7.20).

Inhibition of acetylcholinesterase in muscle,[124] e.g. by organophosphates, delays the clearance of the neurotransmitter acetylcholine and causes abnormal contractions, muscle damage and sometimes necrosis. Rises in TBARS and F_2-isoprostanes in the damaged muscle have been reported: they were blocked by the lazaroid U78517F (Chapter 10), which also decreased the number of necrotic fibres. It seems that hyperactivity led to muscle damage and hence lipid peroxidation, which in this case was a significant contributor to further muscle injury.

References

1. Aitken, RJ (1995) Free radicals, lipid peroxidation and sperm function. *Reprod. Fertil. Dev.* **7**, 659.
2. Alayash, AI and Caston, RE (1995) Hemoglobin and free radicals: implications for the development of a safe blood substitute. *Mol. Med. Today* **1**, 122.
3. Allen, RD (1995) Dissection of oxidative stress tolerance using transgenic plants. *Plant Physiol.* **107**, 1049.
4. Anonymous (1996) Phototoxic celery. *Lancet* **348**, 742.
5. Aro, E-M *et al.* (1993) Photoinhibition of PSII. Inactivation, protein damage and turnover. *Biochim. Biophys. Acta* **1143**, 113.
6. Arrigoni, O *et al.* (1997) Lycorine: a powerful inhibitor of L-galactano-γ-lactone dehydrogenase activity. *J. Plant Physiol.* **150**, 362.
7. Avissar, N *et al.* (1994) Human placenta makes extracellular glutathione peroxidase and secretes it into maternal circulation. *Am. J. Physiol.* **267**, E68.
8. Bartling, D *et al.* (1993) A glutathione S-transferase with glutathione-peroxidase activity from *Arabidopsis thaliana*. *Eur. J. Biochem.* **216**, 579.
9. Becana, M and Klucas, RV (1992) Transition metals in legume root nodules: iron-dependent free radical production increases during nodule senescence. *Proc. Natl Acad. Sci. USA* **89**, 8958.
10. Bender, HS *et al.* (1988) Effects of gossypol on the antioxidant defense system of the rat testis. *Arch. Androl.* **21**, 59.
11. Berenbaum, M (1995) Phototoxicity of plant secondary metabolites: insect and mammalian perspectives. *Arch. Insect. Biochem. Physiol.* **29**, 119.

12. Bissett DL *et al.* (1991) Chronic UV radiation-induced increase in skin iron and the photoprotective effect of topically applied Fe chelators. *Photochem. Photobiol.* **54**, 215.

12a. Bone, RA *et al.* (1997) Distribution of lutein and zeaxanthin stereoisomers in the human retina. *Exp. Eye Res.* **64**, 211.

13. Boonstra, A and Kijlstra, A (1984) The identification of transferrin, an iron binding protein in rabbit tears. *Exp. Eye Res.* **38**, 561.

14. Bowler, C *et al.* (1994) SOD in plants. *Crit. Rev. Plant Sci.* **13**, 199.

15. Briat, JF and Lobréaux, S (1997) Iron transport and storage in plants. *Trends Plant Sci.* **2**, 187.

16. Britt, AB (1996) DNA damage and repair in plants. *Annu. Rev. Plant Physiol. Mol. Biol.* **47**, 75.

17. Brown, LAS *et al.* (1996) GSH supplements protect preterm rabbits from oxidative lung injury. *Am. J. Physiol.* **270**, L446.

18. Brunori, M *et al.* (1975) Formation of $O_2^{\bullet-}$ in the autoxidation of the isolated α and β chains of human haemoglobin and its involvement in hemichrome precipitation. *Eur. J. Biochem.* **53**, 99.

19. Buchanan, BB *et al.* (1994) Thioredoxin: a multifunctional regulatory protein with a bright future in technology and medicine. *Arch. Biochem. Biophys.* **314**, 257.

20. Bueno, P *et al.* (1995) Peroxisomal CuZnSOD. *Plant Physiol.* **108**, 1151.

21. Calderbank, A (1968) The bipyridylium herbicides. *Adv. Pest Control Res.* **8**, 127.

22. Carell, T (1995) Sunlight-induced DNA lesions. Lesion structure, mutation characteristics and repair. *Chimia* **49**, 365.

23. Cha, MK and Kim, IH (1995) Thioredoxin-linked peroxidase from human red blood cells. *Biochem. Biophys. Res. Commun.* **217**, 900.

24. Chevion, M *et al.* (1982) The chemistry of favism-inducing compounds. *Eur. J. Biochem.* **127**, 405.

25. Clemens, MR and Waller, HD (1987) Lipid peroxidation in erythrocytes. *Chem. Phys. Lipids* **45**, 251.

26. Cooney, RV *et al.* (1993) γ-Tocopherol detoxification of NO_2: superiority to α-tocopherol. *Proc. Natl Acad. Sci. USA* **90**, 1771.

26a. D'Agnillo, F and Chang, TMS (1998) Absence of hemoprotein associated free radical events following oxidant challenge of crosslinked hemoglobin-SOD-Catalase. *Free Rad. Biol. Med.* **24**, 906.

27. Dalton, DA *et al.* (1986) Enzymatic reactions of ascorbate and glutathione that prevent peroxide damage in soybean root nodules. *Proc. Natl Acad. Sci. USA* **83**, 3811.

28. Davidsson, L *et al.* (1993) Influence of lactoferrin on iron absorption from human milk in infants. *Pediat. Res.* **35**, 117.

29. Davies, KJA *et al.* (1982) Free radicals and tissue damage produced by exercise. *Biochem. Biophys. Res. Commun.* **107**, 1198.

30. Deliconstantinos G *et al.* (1995) Release by UV-B radiation of NO from human keratinocytes: a potential role for NO in erythema production. *Br. J. Pharmacol.* **114**, 1257.

31. Demmig-Adams, B and Adams, WW Jr (1996) The role of xanthophyll cycle carotenoids in the protection of photosynthesis. *Trends Plant Sci.* **1**, 21.

31a. Denicola, A *et al.* (1998) Diffusion of $ONOO^-$ across erythrocyte membranes *J. Biol. Chem.* **95**, 3566.

32. Dilek Tozum, SR and Gallon, JR (1979) The effects of methyl viologen on *Gloeocapsa* sp. LB795 and their relationship to the inhibition of acetylene reduction (N_2 fixation) by O_2. *J. Gen. Microbiol.* **111**, 313.

33. Dillard, CJ *et al.* (1978) Effects of exercise, vitamin E and O_3 on pulmonary function and lipid peroxidation. *J. Appl. Physiol.* **45**, 927.

34. Dorcy, CK *et al.* (1989) Superoxide production by porcine retinal pigment epithelium *in vitro*. *Invest. Ophthalmol. Vis. Sci.* **30**, 1047.

34a. Eshdat, Y *et al.* (1997) Plant glutathione peroxidases. *Physiol. Plant.* **100**, 234.

35. Evans, PJ *et al.* (1992) Bleomycin-detectable iron in the plasma of premature and full-term neonates. *FEBS Lett.* **303**, 210.

36. Feierabend, J *et al.* (1992) Photoinactivation of catalase occurs under both high- and low-temperature stress conditions and accompanies photoinhibition of PSII. *Plant Physiol.* **100**, 1554.

37. Floyd, RA *et al.* (1989) Increased 8-hydroxyguanine content of chloroplast DNA from O_3-treated plants. *Plant Physiol.* **91**, 644.

38. Foyer, CH *et al.* (1997) H_2O_2- and glutathione-associated mechanisms of acclimatory stress tolerance and signalling. *Physiol. Plant.* **100**, 241.

39. Frank, L (1985) Effects of O_2 on the newborn. *Fed. Proc.* **44**, 2328.

40. Gaunt, JK *et al.* (1980) Control *in vitro* of tocopherol oxidase by light and by auxins, kinetin gibberellic acid, abscisic acid and ethylene. *Biochem. Soc. Trans.* **8**, 186.

41. Gutteridge, JMC *et al.* (1985) Copper and iron complexes catalytic for oxygen radical reactions in sweat from human athletes. *Clin. Chim. Acta* **145**, 267.

42. Hagay, ZJ *et al.* (1995) Prevention of diabetes-associated embryopathy by overexpression of the free radical scavenger CuZnSOD in transgenic mouse embryos. *Am. J. Obst. Gyn.* **173**, 1036.

43. Halliwell, B (1987) Oxidative damage, lipid peroxidation and antioxidant protection in chloroplasts. *Chem. Phys. Lipids* **44**, 327.

44. Harper, DB and Harvey, BMR (1978) Mechanisms of paraquat tolerance in perennial ryegrass. Role of SOD, catalase and peroxidase. *Plant Cell Environ.* **1**, 211.

45. Hart, JJ and Di Tomaso, JM (1994) Sequestration and O_2 radical detoxification as mechanisms of paraquat resistance. *Weed Sci.* **42**, 277.

46. Helbock, HJ *et al.* (1993) Toxic hydroperoxides in intravenous lipid emulsions used in preterm infants. *Pediatrics* **91**, 83.

47. Hellsten, Y *et al.* (1997) Oxidation of urate in human skeletal muscle during exercise. *Free Rad. Biol. Med.* **22**, 169.

48. Hendry GAF (1993) Oxygen, free radical processes and seed longevity. *Seed Sci. Res.* **3**, 141.

49. Hippeli, S and Elstner, EF (1996) Mechanisms of O_2 activation during plant stress: biochemical effects of air pollutants. *J. Plant Physiol.* **148**, 249.

49a. Ho, YS *et al.* (1998) Reduced fertility in female mice lacking CuZnSOD. *J. Biol. Chem.* **273**, 7765.

50. Hood, AT *et al.* (1996) Protection from lethal malaria in transgenic mice expressing sickle hemoglobin. *Blood* **87**, 1600.

51. Hundal, T *et al.* (1995) Antioxidant activity of reduced plastoquinone in chloroplast thylakoid membranes. *Arch. Biochem. Biophys.* **324**, 117.

52. Hunt, RC *et al.* (1996) Hemopexin in the human retina: protection of the retina against heme-mediated toxicity. *J. Cell. Physiol.* **168**, 71.

53. Jain, SK (1989) The neonatal erythrocyte and its oxidative susceptibility. *Semin. Hematol.* **26**, 266.

54. Kampfenkel, K *et al.* (1995) Effect of iron excess on *Nicotiana plumbaginifolia* plants. *Plant Physiol.* **107**, 725.

55. Kanter, MM (1994) Free radicals, exercise and antioxidant supplementation. *Int. J. Sport. Nutr.* **4**, 205.

56. Kirschfeld, K (1982) Carotenoid pigments: their possible role in protecting against photo-oxidation in eyes and photoreceptor cells. *Proc. Roy. Soc. Lond.* **B216**, 71.

57. Knowland, J *et al.* (1993) Sunlight-induced mutagenicity of a common sunscreen ingredient. *FEBS Lett.* **324**, 309.

58. Kocak, T *et al.* (1990) The antioxidant activity of human semen. *Clin. Chim. Acta* **192**, 153.

59. Kondo, H *et al.* (1992) Role of iron in oxidative stress in skeletal muscle atrophied by immobilization. *Pflugers Arch.* **421**, 295.

60. Kuizenga, A *et al.* (1987) Inhibition of hydroxyl radical formation by human tears. *Invest. Ophthalmol. Vis. Sci.* **28**, 305.

61. Kutty, RK *et al.* (1995) Induction of heme oxygenase 1 in the retina by intense visible light: suppresion by the antioxidant dimethylthiourea. *Proc. Natl Acad. Sci. USA* **92**, 1177.

62. Leese, HJ (1994) Biochemistry of the early mammalian embryo. *Biochemist*, April/May 1994, p. 9.

63. Lemaire, P *et al.* (1993) Pro-oxidant and antioxidant processes in gas gland and other tissues of cod (*Gadus morhua*). *J. Comp. Physiol. B.* **163**, 477.

64. Leprince, O *et al.* (1993) The mechanisms of desiccation tolerance in developing seeds. *Seed Sci. Res.* **3**, 231.

65. Levander, OA and Ager, AL Jr (1993) Malarial parasites and antioxidant nutrients. *Parasitology* **107**, 595.

66. Loegering, DJ *et al.* (1996) Macrophage dysfunction following the phagocytosis of IgG-coated erythrocytes: production of lipid peroxidation products. *J. Leuk. Biol.* **59**, 357.

66a. Lynch, RE and Fridovich, I (1978) Permeation of the erythrocyte stroma by $O_2^{\bullet-}$. *J. Biol. Chem.* **253**, 4697.

67. Manes, C and Lai, NC (1995) Nonmitochondrial O_2 utilization by rabbit blastocysts and surface production of $O_2^{\bullet-}$. *J. Reprod. Fertil.* **104**, 69.

68. Matringe, M *et al.* (1989) Protoporphyrinogen oxidase as a molecular target for diphenyl ether herbicides. *Biochem. J.* **260**, 231.

69. Matsuoka, I *et al.* (1995) Impact of erythrocytes on mouse embryonal development *in vitro*. *FEBS Lett.* **371**, 297.

70. May, JM *et al.* (1996) Ascorbate recycling in human erythrocytes; role of GSH in reducing dehydroascorbate. *Free Rad. Biol. Med.* **20**, 543.

71. McGahan, MC and Fleischer, LN (1986) A micromethod for the determination of Fe^{2+} and total Fe-binding capacity in intraocular fluids and plasma using electrothermal atomic absorption spectroscopy. *Anal. Biochem.* **156**, 397.

72. McKersie, BD *et al.* (1997) Water-deficit tolerance and field performance of transgenic alfalfa overexpressing SOD. *Plant Physiol.* **111**, 1177.

73. Miyake, C and Asada, K (1996) Inactivation mechanisms of ascorbate peroxidase at low concentrations of ascorbate; H_2O_2 decomposes compound I of ascorbate peroxidase. *Plant Cell Physiol.* **37**, 423.

74. Moison, RMW *et al.* (1993) Induction of lipid peroxidation of pulmonary surfactant by plasma of preterm babies. *Lancet* **341**, 79.

75. Moison, RMW *et al.* (1997) Uric acid and ascorbic acid redox ratios in plasma and tracheal aspirate of preterm babies with acute and chronic lung disease. *Free Rad. Biol. Med.* **23**, 226.

75a. Müller, K (1997) Antipsoriatic and proinflammatory action of anthralin. *Biochem. Pharmacol.* **53**, 1215.

75b. Munday, R (1989) Toxicity of thiols and disulphides: involvement of free radical species *Free. Rad. Biol. Med.* **7**, 659.

76. Nasr-Esfahani, M *et al.* (1991) The origin of ROS in mouse embryos cultured *in vitro. Development* **113**, 551.

77. Newsome, DA *et al.* (1994) Antioxidants in the retinal pigment epithelium. *Prog. Ret. Eye Res.* **13**, 101.

78. Ney, PA *et al.* (1990) Synergistic effects of oxidation and deformation on erythrocyte monovalent cation leak. *Blood* **75**, 1192.

79. Okamura, K *et al.* (1997) Effect of endurance exercise on tissue 8OHdG in dogs. *Free Rad. Res.* **26**, 523.

80. Orthwerth, BJ *et al.* (1988) The precipitation and cross-linking of lens crystallins by ascorbic acid. *Exp. Eye Res.* **47**, 155.

81. Pasvol, G (1996) Malaria and resistance genes—they work in wondrous ways. *Lancet* **348**, 1532.

81a. Peeker, R *et al.* (1997) SOD isoenzymes in human seminal plasma and spermatozoa. *Mol. Hum. Reproduct* **3**, 1061.

82. Perry, ACF *et al.* (1993) Isolation and characterization of a rat cDNA clone encoding a secreted SOD reveals the epididymis to be a major site of its expression. *Biochem. J.* **293**, 21.

83. Pierce, GB *et al.* (1991) H_2O_2 as a mediator of programmed cell death in the blastocyst. *Differentiation* **46**, 181.

84. Postma, MS *et al.* (1996) Oxidative stress in malaria: implications for prevention and therapy. *Pharm. World Sci.* **18**, 121.

85. Poulsen, HE *et al.* (1996) Extreme exercise and oxidative DNA modification. *J. Sports Sci.* **14**, 343.

86. Puppo, A and Davies, MJ (1995) The reactivity of thiol compounds with different redox states of leghaemoglobin: evidence for competing reduction and addition pathways. *Biochim. Biophys. Acta* **1246**, 74.

87. Rabinowitch, HD and Fridovich, I (1995) Growth of *Chlorella sorokiniana* in the presence of sulfite elevates cell content of SOD and imparts resistance towards paraquat. *Planta* **164**, 524.

88. Reid, MB *et al.* (1994) N-Acetylcysteine inhibits muscle fatigue in humans. *J. Clin. Invest.* **94**, 2468.

89. Ribaya-Mercado, JD *et al.* (1995) Skin lycopene is destroyed preferentially over β-carotene during UV irradiation in humans. *J. Nutr.* **125**, 1854.

90. Riley, JCM and Behrman, HR (1991) *In vivo* generation of H_2O_2 in the rat corpus luteum during luteolysis. *Endocrinology* **128**, 1749.

91. Salem, N Jr *et al.* (1996) Arachidonic and docosahexaenoic acids are biosynthesized from their 18-C precursors in human infants. *Proc. Natl Acad. Sci. USA* **93**, 49.

92. Salo, DC *et al.* (1990) SOD undergoes proteolysis and fragmentation following oxidative modification and inactivtion. *J. Biol. Chem.* **265**, 11919.

93. Sambrano, GR *et al.* (1994) Recognition of oxidatively damaged erythrocytes by a macrophage receptor with specificity for oxidized LDL. *Proc. Natl Acad. Sci. USA* **91**, 3265.

94. Sanchez-Quesada, JL *et al.* (1995) Increase of LDL susceptibility to oxidation occurring after intense, long duration aerobic exercise. *Atherosclerosis* **118**, 297.

95. Scandalios, JG (1997) *Oxidative Stress and the Molecular Biology of Antioxidant Defenses.* Cold Spring Harbor Laboratory Press, Cold Spring Harbor, NY. Chapters beginning on pp. 343, 527, 587, 623, 715, 785, 815 and 861.

96. Sen, CK (1995) Oxidants and antioxidants in exercise. *J. Appl. Physiol.* **79**, 675.
97. Shalev, O and Hebbel, RP (1996) Extremely high avidity association of Fe(III) with the sickle red cell membrane. *Blood* **88**, 349.
98. Shang, F and Taylor A (1995) Oxidative stress and recovery from oxidative stress are associated with altered ubiquitin conjugating and proteolytic activities in bovine lens epithelial cells. *Biochem. J.* **307**, 297.
99. Shapiro, A *et al.* (1982) *In vivo* and *in vitro* activity by diverse chelators against *Trypanosoma brucei brucei*. *J. Protozool.* **29**, 85.
100. Shen, B *et al.* (1997) Increased resistance to oxidative stress in transgenic plants by targeting mannitol biosynthesis to chloroplasts. *Plant Physiol.* **113**, 1177.
101. Shindo, Y *et al.* (1993) Antioxidant defence mechanisms in murine epidermis and dermis and their response to UV light. *J. Invest. Dermatol.* **100**, 260.
102. Signorini, C *et al.* (1995) Iron release, membrane protein oxidation and erythrocyte ageing. *FEBS Lett.* **362**, 165.
103. Smirnoff, N (1993) The role of active O_2 in the response of plants to water deficit and desiccation. *New Phytol.* **125**, 27.
104. Smith, IK *et al.* (1994) Increased levels of glutathione in a catalase-deficient mutant of barley (*Hordeum vulgare* L). *Pl. Sci. Lett.* **37**, 29.
105. Smith, JA *et al.* (1990) Exercise, training and neutrophil microbicidal activity. *Int. J. Sports Med.* **11**, 179.
106. Spector, A (1995) Oxidative stress-induced cataract: mechanism of action. *FASEB J.* **9**, 1173.
107. Standcliffe, TC and Pirie, A (1971) The production of $O_2^{\bullet-}$ in reactions of the herbicide diquat. *FEBS Lett.* **17**, 297.
108. Sullivan, JL (1988) Iron, plasma antioxidants and the 'O_2 radical disease of prematurity'. *Am. J. Dis. Child.* **142**, 1341.
109. Takata, I *et al.* (1996) Glycated CuZnSOD in rat lenses: evidence for the presence of fragmentation *in vivo*. *Biochem. Biophys. Res. Commun.* **219**, 243.
110. Tan, S *et al.* (1996) Maternal infusion of antioxidants (Trolox and ascorbic acid) protects the fetal heart in rabbit fetal hypoxia. *Pediatr. Res.* **39**, 499.
111. Taylor, A (1993) Cataract: relationships between nutrition and oxidation. *J. Am. Coll. Nutr.* **12**, 138.
112. Telfer, A *et al.* (1994) β-Carotene quenches singlet O_2 formed by isolated photosystem II reaction centers. *Biochemistry* **33**, 14469.
113. Tkeshelashvili, LK *et al.* (1993) Nickel induces a signature mutation for O_2 free radical damage. *Cancer Res.* **53**, 4172.
114. Tsafach, A *et al.* (1996) Mode of action of Fe(III) chelators as antimalarials. *J. Lab. Clin. Med.* **127**, 574.
115. Tyrrell, RM (1996) Activation of mammalian gene expression by the UV component of sunlight—from models to reality. *BioEssays* **18**, 139.
116. Vergara, O *et al.* (1989) Posterior penetrating injury to the rabbit eye: effect of blood and Fe^{2+}. *Exp. Eye Res* **49**, 1115.
117. Wang, K and Spector, A (1995) α-Crystallin can act as a chaperone under conditions of oxidative stress. *Invest. Ophthalmol. Vis. Sci.* **36**, 311.
118. Wani, G and D'Ambrosio, SM (1995) Cell type-specific expression of human 8-oxodGTPase in normal breast and skin tissues *in vivo*. *Carcinogenesis* **16**, 277.
119. Wells, PG and Winn, LM (1996) Biochemical toxicology of chemical teratogenesis. *Crit. Rev. Biochem. Mol. Biol.* **31**, 1.
120. Winterbourn, CC (1985) Free-radical production and oxidative reactions of hemoglobin. *Env. Health Perspect.* **64**, 321.

121. Winterbourn, CC *et al.* (1981) Unstable haemoglobin haemolytic crises: contributions of pyrexia and neutrophil oxidants. *Br. J. Haematol.* **49**, 111.
122. Wise, RR (1995) Chilling-enhanced photooxidation: the production, action and study of ROS produced during chilling in the light. *Photosynth. Res.* **45**, 79.
123. Xu, F *et al.* (1993) H_2O_2-dependent formation and bleaching of the higher oxidation states of bovine erythrocyte green hemoprotein. *Arch. Biochem. Biophys.* **301**, 184.
124. Yang, ZP *et al.* (1996) Diisopropylphosphorofluoridate-induced muscle hyper-activity associated with enhanced lipid peroxidation *in vivo. Biochem. Pharmacol.* **52**, 357.
125. Zer, H *et al.* (1994) The protective effect of desferrioxamine on paraquat-treated pea (*Pisum sativum*). *Physiol. Planta.* **92**, 437.

Notes

[a]See Appendix II for an explanation of this technology.
[b]These wavelength ranges vary slightly in different publications.

8

Free radicals, 'reactive species' and toxicology

8.1 Introduction

8.1.1 *What is toxicology?*

Strictly speaking, toxicology is concerned with the mechanisms of injury to living organisms by *exogenous* chemicals (**xenobiotics**). In addition, endogenous molecules can often cause injury if present in excess: glucose, $O_2^{\bullet-}$ and $NO^\bullet$ are good examples. They are all useful at normal levels but toxic at higher ones.

However, we use the term **toxins** in the present chapter to mean molecules that are not synthesized by the organisms to which they are causing injury. Modern medicine and agriculture use a wide range of toxins to control unwanted organisms (bacteria, fungi, insects, rodents, weed plants, etc). We are exposed to some of these deliberately (e.g. antibiotics) and others because they are present in the environment. Plants and bacteria synthesize a wide range of compounds that animals do not, including aflatoxins, flavonoids, terpenoids and tocopherols. Some of these are useful to animals, but many are toxic.

8.1.2 *Principles of toxin metabolism*

Most xenobiotics can be metabolized by animals. Most are dealt with by a series of enzymes that have broad substrate specificity; in general, these enzymes detoxify xenobiotics and convert them into species that are more soluble in water and thus easier to excrete. There are exceptions to both these principles: we will meet some of them in this chapter and in Chapter 9.

Overall, drug metabolism can be considered in two phases. **Phase I** reactions introduce (or sometimes unmask) a polar functional group within the molecule, often by oxidations involving cytochromes P450. Other enzymes, such as esterases, monoamine oxidases and alcohol dehydrogenases, can sometimes do the same thing. Concentrations of 'total P450' vary widely between tissues, the liver having the highest. Substantial concentrations are also found in the Clara cells of the lung, in adrenal gland, in kidney and in small intestine.

Phase II reactions are conjugation reactions: an endogenous molecule is added to the phase I reaction product, or sometimes directly to the xenobiotic. In Chapter 3 we met two examples of phase II reactions, conjugations with GSH catalysed by **glutathione transferases** and conjugation with glucuronic acid, catalysed by **glucuronyl transferases**. The latter enzymes react uridine diphosphate glucose (**UDP–glucose**) with a wide range of substrates. We met glucuronyl transferases in the context of bilirubin metabolism, but they act on a wide range of xenobiotics also. Other examples of phase II reactions are sulphation (addition of a sulphate group, catalysed by **sulphotransferases**), **methylation** (addition of a methyl group, usually provided by *S*-adenosyl-methionine), **acetylation** (catalysed by *N*-acetyltransferases, an example being acetylation of the terminal amino group in **isoniazid** (Chapter 7)), and **glycine conjugation**. For example, benzoic acid is excreted in humans as a conjugate with glycine. The metabolism of **aspirin** (acetylsalicylic acid) in humans illustrates how a drug can often be metabolized by multiple pathways (Fig. 8.1).

Genetic variations affecting drug–metabolizing enzymes, especially cytochromes P450 and glutathione *S*-transferases, have been well described. For example, the activity of CYP2O6 affects the rate of metabolism of the

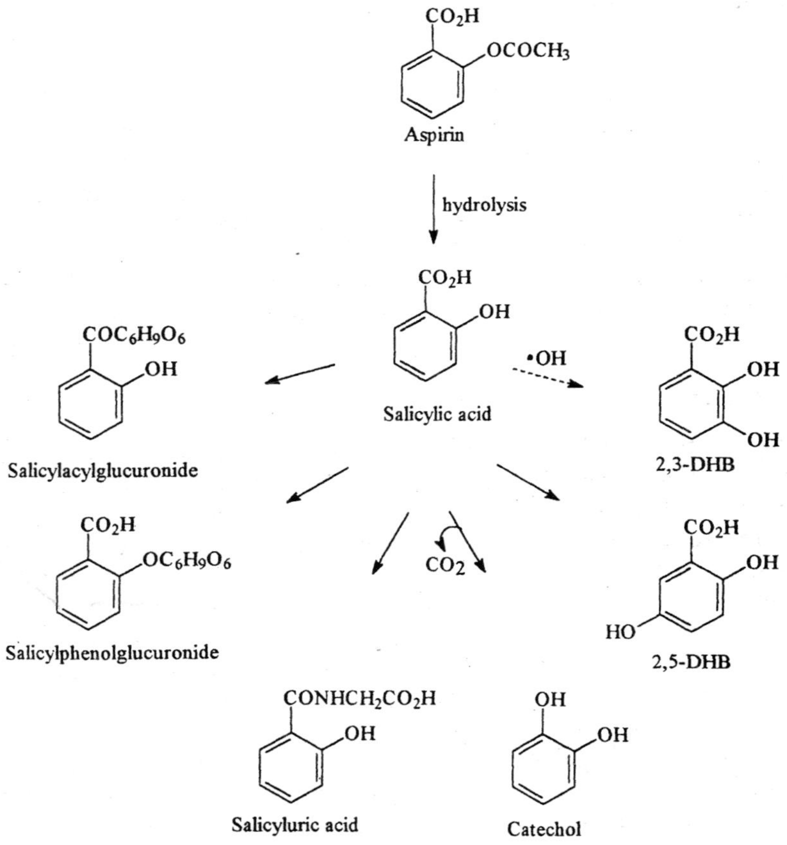

Fig. 8.1. Metabolism of aspirin. Aspirin can react with cyclooxygenase to acetylate the active site and inhibit the enzyme (Chapter 7). However, aspirin is readily deacetylated by esterase action *in vivo*. Metabolism by P450 produces 2,5-dihydroxybenzoate. 2,3-Dihydroxybenzoate does not appear to be metabolically produced and has been used as a 'biomarker' of the trapping of OH$^\bullet$ by salicylate (Chapter 5). Diagram by courtesy of Dr H. Kaur.

anti-hypertensive drug **debrisoquine**. Most people are 'extensive metabolizers' but about 10% of the UK population are 'poor metabolizers'. Exposure
to other agents affects drug metabolism, e.g. barbiturates and ethanol can raise
levels of certain cytochromes P450 (Chapter 1). Exposure to elevated O_2 can
decrease P450 synthesis in some organisms. Rat liver **arylsulphotransferase IV**
can be activated[86] by oxidation of a cysteine residue to form a disulphide
bridge with another cysteine in the same enzyme, or with added GSSG. Hence
there is growing evidence for 'redox regulation' of enzymes of xenobiotic
metabolism.

8.1.3   *How can ROS/RNS contribute to toxicology?*

Free radicals and other reactive oxygen and nitrogen species (ROS/RNS) have
often been suggested to be involved in the mechanism of action of toxins.
Several examples are discussed elsewhere in this book, such as haemolytic and
anti-malarial drugs, teratogenic agents and photosensitizers (Chapter 7). In
principle, the action of toxins involving ROS/RNS could be because:

(1) The toxin itself is a free radical (e.g. nitrogen dioxide, $NO_2^\bullet$).
(2) The toxin is metabolized to free radicals, e.g. carbon tetrachloride (CCl_4).
     The fact that a toxin can be demonstrated to form a free radical *in vivo* does
     not prove that the free radical causes the injury.
(3) The toxin undergoes **redox cycling**, i.e. it is reduced by a cellular system
     to give a molecule that is then oxidized by O_2, producing $O_2^{\bullet -}$ and
     regenerating the original compound. The cycle of reactions is then
     repeated. An excellent example is the toxicity of paraquat to plants
     (Chapter 7).
(4) The toxin increases oxidative damage by interfering with antioxidant
     defences. For example, many toxins are metabolized by conjugation with
     GSH (Chapter 3). A large dose may so deplete GSH that secondary
     oxidative damage occurs.
(5) The toxin stimulates endogenous ROS/RNS generation, e.g. by affecting
     mitochondrial electron transport, inducing iNOS or CYP2E1 (a 'leaky'
     form of P450; Section 8.8 below), raising intracellular free Ca^{2+} or
     activating phagocytes. One example is the neurotoxin MPTP (Chapter 9),
     which inhibits mitochondrial electron transport and causes secondary
     oxidative damage. Exposure of rats to **methyl chloride** (CH_3Cl), a widely
     used industrial gas, causes an acute inflammation of the epididymis plus
     sperm defects. It has been proposed that ROS/RNS resulting from the
     inflammation cause DNA damage in the sperm.[23] The toxin α-**naph-
     thylisocyanate** produces hepatitis in animals; it may provoke (by an
     unknown mechanism) the recruitment of neutrophils to the liver and their
     subsequent activation.[89]
(6) Any combination of the above mechanisms. For example, cigarette smoke
     contains free radicals ($NO^\bullet$, $NO_2^\bullet$), GSH-depleting agents (especially
     aldehydes) and redox-cycling agents, and it also irritates phagocytes in the

lung, causing increased phagocyte ROS/RNS production under certain circumstances.

Toxin-induced oxidative stress can damage cells by multiple mechanisms (Chapter 4), including direct oxidative damage to DNA, lipids and proteins, depletion of ATP and NAD^+, activation of poly (ADP–ribose) synthetase (PARP), falls in GSH/GSSG ratios and increases in 'free' intracellular Ca^{2+} and transition-metal ions. We must not forget, however, that oxidative damage often proceeds more rapidly in injured tissues, even if the damage was not initially caused by a free-radical mechanism. Rises in Ca^{2+} can stimulate proteases, nucleases and $NO^{\bullet}$ synthesis, by activating Ca^{2+}/calmodulin-dependent nitric oxide synthetases. Excess $NO^{\bullet}$ can damage cells directly, e.g. by inhibiting cytochrome oxidase and ribonucleotide reductase, and indirectly, e.g. by reacting with $O_2^{\bullet-}$ to form $ONOO^-$ (Chapter 4). Activation of calpains can lead to conversion of xanthine dehydrogenase to oxidase, and more ROS generation.

'Free-radical reactions' have been implicated in the action of many toxins, but published reports often do not take sufficient care to distinguish between compounds for which free-radical formation is directly responsible for the toxicity, and other compounds in which the free-radical formation is a later stage in the process of cell injury. For example, when inhibitors of ATP synthesis (e.g. the glycolysis inhibitor **iodoacetate** or the mitochondrial electron transport chain inhibitors **rotenone**, **MPTP** or **cyanide**) are added to nerve cells in culture, their primary mechanism of toxicity is by interfering with energy production.[91] Oxidative stress also occurs in these cells, as a secondary phenomenon. Often, addition of antioxidants does not prevent cell death. It may delay it, to an extent that depends on the system studied, indicating that the secondary oxidative damage can contribute to toxicity.

Even when increased ROS/RNS formation *is* a primary toxic mechanism, it is also important to ask what is the first target of oxidative damage—is it DNA, proteins or lipids? Answering this question is difficult because mechanisms of cell damage are inter-related (Chapter 4), and it is often hard to disentangle primary from secondary events. Low levels of oxidative stress may also inhibit communication between adjacent cells.[118]

Let us examine what is known about the mechanism of action of several xenobiotics, to see how far the claims that they act by 'free-radical mechanisms' can be justified.

8.2 Carbon tetrachloride

CCl_4 was the first toxin for which it was shown that the injury it produces is largely or entirely mediated by a free-radical mechanism.[16,88] Hence it is appropriate to discuss it at the beginning. CCl_4 is a colourless liquid, immiscible with water, that is used in industry as a 'degreaser' and organic solvent. Over 142 million kilograms were produced in 1991 in the USA. CCl_4 was formerly employed as a dry-cleaning agent, although the latter use was banned in the

peroxidation. Fatty-acid side-chains attached to phosphatidylserine (Chapter 4) are especially prone to attack, perhaps because this lipid is adjacent to P450 *in vivo*. Pre-treatment of animals with **phenobarbital**, an agent which induces certain P450s in liver (Chapter 1), renders the animals more susceptible to CCl_4, which is correlated with increased levels of peroxidation in liver microsomal fractions isolated subsequently.

Levels of F_2-isoprostanes, products of arachidonic acid peroxidation (Chapter 4), increase up to 50-fold in the plasma of rats given CCl_4; these increases are bigger in phenobarbital pre-treated rats.[96] Large rises in liver isoprostane levels precede those in plasma; smaller rises are seen in kidney, lung and heart and F_2-isoprostanes can be measured in bile from CCl_4-exposed rats. CCl_4-treated rats also exhale more pentane and ethane.

Administration of a wide range of antioxidants, (including vitamin E, promethazine, propyl gallate and GSH) or inhibitors of P450, decreases CCl_4 toxicity in parallel with decreased lipid peroxidation in animals. Vitamin E-deficient animals are more susceptible to CCl_4. Neonatal animals have low P450 levels (Chapter 7) and are more resistant to CCl_4 toxicity.

As a result of these observations, it is believed that CCl_4 is metabolized by the P450 system to give the **trichloromethyl radical ($CCl_3^\bullet$)**, a carbon-centred radical. Several P450s are involved, including CYP2E1, the 'ethanol-inducible' cytochrome P450.

$$CCl_4 \xrightarrow[\text{P450 system}]{\text{one-electron reduction}} {}^\bullet CCl_3 + Cl^-$$

Conversion of CCl_4 into $CCl_3^\bullet$ appears to be brought about by P-450 itself (after reduction by NADPH–cytochrome P450 reductase), although it is possible that the reductase might also interact with CCl_4. Destruction of P450 during the reaction makes CCl_4 toxicity, to some extent, a self-limiting event.

Spin-trapping experiments, initially upon isolated rat liver microsomes, have confirmed formation of $CCl_3^\bullet$. Various traps have been used, including phenyl-*tert*-butylnitrone, PBN (Chapter 5). PBN has also been used to trap $CCl_3^\bullet$ *in vivo* in rats given both CCl_4 and PBN through a stomach tube (Fig. 8.3). Experiments using PBN in perfused rat liver exposed to CCl_4 have trapped not only $CCl_3^\bullet$ but also the **carbon dioxide anion radical**, $CO_2^{\bullet-}$. This is derived from $CCl_3^\bullet$, possibly by reactions involving a GSH–$CCl_3^\bullet$ addition product.

8.2.3 *How does $CCl_3^\bullet$ cause damage?*[16]

The trichloromethyl radical might combine directly with biological molecules, causing covalent modification (Fig. 8.2) as well as abstracting hydrogen from membrane lipids, setting off the chain reaction of lipid peroxidation. Products of peroxidation are known to inhibit protein synthesis and the activity of certain enzymes. Indeed, liver microsomal fractions from CCl_4-treated rats contain more protein-bound 'cytotoxic aldehydes' than untreated rats.

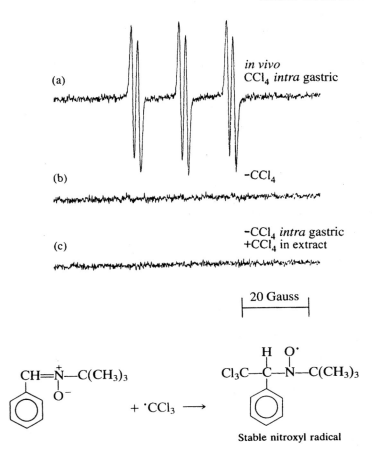

(a)   *in vivo* CCl$_4$ *intra* gastric

(b)   -CCl$_4$

(c)   -CCl$_4$ *intra* gastric +CCl$_4$ in extract

20 Gauss

Stable nitroxyl radical

Fig. 8.3. Trapping of trichloromethyl radical by the spin-trap PBN. From *Free Radicals: A Practical Approach* (Punchard, N.A. and Kelly, F.J., eds), p. 19, IRL Press, Oxford, by courtesy of Professor R.P. Mason and Oxford University Press. (a) ESR spectrum of a chloroform/ methanol extract of liver from a rat 1 h after administration of 0.8 ml/kg CCl$_4$ (intragastric) and 70 mg/kg PBN (intraperitoneal). (b) ESR spectrum of the extract of liver from a control rat given only the PBN. (c) As in (b), but liver was homogenized in a chloroform/methanol mixture to which 25 µl of CCl$_4$/g of liver tissue had been added; this is an essential control to check for radical formation during solvent extraction. Instrumental conditions: micro- wave power, 20 mW; modulation amplitude, 0.33 Gauss; time constant, 0.25 s; scan rate, 10 Gauss/min. The PBN radical adduct was also detected in bile of animals injected intraperitoneally with PBN and given CCl$_4$ into the stomach.

Promethazine, an inhibitor of lipid peroxidation, decreases peroxidation in liver cells and prevents loss of glucose 6-phosphatase activity, but not the loss of P450. Probably cytochrome P450 is directly attacked by CCl$_3^\bullet$ or other radicals derived from it, whereas the inactivation of glucose 6-phosphatase is brought about by products of lipid peroxidation. CCl$_4$-treated rats exhale chloroform vapour; $CHCl_3$ would be expected to be produced by combination of $CCl_3^•$ with a hydrogen atom abstracted from a membrane lipid. Hexachloroethane, another product of CCl_4 metabolism, could arise by the reaction

$$CCl_3^• + CCl_3^• \rightarrow Cl_3CCCl_3$$

Despite this evidence, questions have been raised about peroxidation of membrane lipids induced by $CCl_3^•$ as an explanation of CCl_4 toxicity. Pulse radiolysis studies of the reactivity of the trichloromethyl radical show that, like most carbon-centred radicals, its most rapid reaction is with molecular oxygen to form the **trichloromethylperoxyl radical**:

$$CCl_3 + O_2 \rightarrow CCl_3O_2^•   k_2 = 3.3 \times 10^9 \, M^{-1} s^{-1}$$

This is an especially reactive peroxyl radical: the presence of halogen in the side-chain increases the oxidizing capacity of $RO_2^•$ radicals. Indeed, $CCl_3O_2^•$ reacts much more rapidly with arachidonic acid ($k_2 = 7.3 \times 10^6 \, M^{-1} s^{-1}$), linolenic acid ($k_2 = 7.0 \times 10^6 \, M^{-1} s^{-1}$), ascorbate ($k_2 = 2 \times 10^8 \, M^{-1} s^{-1}$), thiol compounds, and the tyrosine and tryptophan residues of proteins than does the trichloromethyl radical, and so $CCl_3O_2^•$ would seem a more likely candidate for a damaging species. However, spin-trapping experiments have so far failed to reveal the presence of $CCl_3O_2^•$ in CCl_4-treated microsomes, although its high reactivity will make it difficult to detect. Formation of trichloromethylperoxyl radical could explain why small amounts of **phosgene** gas ($COCl_2$) are produced by CCl_4-treated microsomes. It might arise by the reactions:

$$CCl_3O_2^• + lipid-H \rightarrow CCl_3O_2H + lipid^•$$

$$CCl_3O_2H \rightarrow CCl_3OH \rightarrow COCl_2 + HCl$$

It seems likely that local O_2 concentration in different parts of the liver influences whether $CCl_3^•$ directly binds to biological molecules or combines with O_2 to form $CCl_3O_2^•$, which should be a much better initiator of lipid peroxidation (Fig. 8.2).

The free-radical stress imposed on the liver by CCl_4 can lead to rises in intracellular Ca^{2+}, GSH depletion and iron release. Pre-treatment of mice with the iron chelator desferrioxamine decreased the hepatotoxicity and depressed the exhalation of ethane caused by CCl_4. This suggests that the lipid peroxidation initiated by free-radical metabolites of CCl_4 is made worse by release of iron ions, a possible mechanism being illustrated in Fig. 8.2. Liver injury by almost any mechanism may provoke activation of resident macrophages (**Kupffer cells**) in this organ. Recruitment of neutrophils from the circulation can also occur; production of ROS by both types of phagocyte has also been suggested as a potential contributor to injury.

8.3   Other halogenated hydrocarbons

The role of free-radical reactions in the toxicity of CCl_4 is well established. This has naturally led to proposals that free radicals are involved in the toxicity of other halogenated hydrocarbons. For some, such as **trichloroethylene** and **vinyl chloride**, the evidence is not compelling. For others, the evidence is better: let us examine a few in detail.

8.3.1 *Chloroform and bromotrichloromethane*[16,107]

Chloroform is no longer used as an anaesthetic, but it is still widely employed in industry as a solvent, and small amounts are sometimes added to cough mixtures and mouth-washes. Traces (μg/litre) may contaminate chlorinated drinking water.[19]

Trichloromethane is much less damaging to the liver than CCl_4, consistent with the observation that it induces lipid peroxidation in isolated liver microsomes at a much lower rate. One factor contributing to this difference may be that the energy required to cause homolytic fission of trichloromethane to produce $CCl_3^{\bullet}$ is greater than for CCl_4. Consistent with this argument, compounds in which homolytic fission is easier, such as **bromotrichloromethane** ($BrCCl_3$) induce peroxidation more rapidly than does CCl_4 (Table 8.1). Nevertheless, liver damage is still a significant problem in humans exposed to excessive chloroform and several deaths due to liver failure occurred in the early days of chloroform anaesthesia. The toxicity again involves metabolism by the P450 system, and inducers of P450 potentiate $CHCl_3$ toxicity in animals. Again, there is covalent binding of metabolites and phosgene production.

Administration of $BrCCl_3$ (Table 8.1) to rats produced a different toxicity: damage to the **Clara cells** of the lung.[84] Lipid peroxidation appeared to be responsible. These cells are enriched in cytochrome P450 by comparison with other lung cells.

8.3.2 *Bromoethane and bromobenzene*[1,16]

1,2-Dibromoethane (**ethylene dibromide**, $CH_2Br \cdot CH_2Br$) is widely used as a 'lead scavenger' in petrol, as an industrial solvent, and in agriculture as a fungicide. Unfortunately it is mutagenic and carcinogenic, causing especial damage to the liver and kidneys. Free-radical reactions may contribute to this

Table 8.1. Lipid peroxidation induced by halogenated hydrocarbons

Reaction	Energy needed for homolytic bond fission (kcal/mol)	Relative rate of lipid peroxidation
$CCl_4^{\bullet} \rightarrow Cl^{\bullet} + {}^{\bullet}CCl_3$	68	100
$CHCl_3 \rightarrow H^{\bullet} + {}^{\bullet}CCl_3$	90	7
$BrCCl_3 \rightarrow Br^{\bullet} + {}^{\bullet}CCl_3$	49	3650

The rate of peroxidation induced in rat-liver microsomes in the presence of NADPH was measured by the thiobarbituric acid method. Results are expressed relative to the stimulatory effect of CCl_4. Data selected from Slater, T.F. and Sawyer, B.C. (1971) *Biochem. J.* **123**, 805.

damage: bromoethane can be metabolized by conjugation with GSH to form a mercapturic acid (Chapter 3) or by cytochrome P450-dependent oxidation to give bromoacetaldehyde via intermediate free radicals. Toxic amounts of dibromoethane cause rapid GSH depletion which is followed by lipid peroxidation. Although, unlike the case of CCl_4, lipid peroxidation does not seem to be a primary mechanism of tissue injury by bromoethane, its occurrence aggravates the damage to some extent.

The major toxicity of the aromatic hydrocarbon **bromobenzene** (C_6H_5Br) is associated with P450-dependent formation of an epoxide, which can react with GSH and protein thiols or be further metabolized to bromophenol. In experiments with mice, the vitamin E analogue Trolox C (Section 5.15) offered some protection against liver necrosis induced by toxic doses of bromobenzene, and desferrioxamine was also protective. It may be that the severe GSH depletion leads to 'secondary' lipid peroxidation. Rises in intracellular 'free' Ca^{2+} also occur in bromobenzene-treated hepatocytes (e.g. Fig. 8.2).

8.3.3   *Halothane*[49]

Halothane, a chlorofluorocarbon (Fig. 8.4) is frequently used as an inhaled anaesthetic gas. It is generally safe, although about 20% of patients exposed to it show very mild liver dysfunction, as detected by laboratory tests (**halothane hepatitis type I**). About one in 35,000 patients given halothane shows severe liver damage, leading to necrosis and liver failure (**halothane hepatitis type II**). Re-anaesthesia of patients previously exposed to halothane can produce severe liver damage in a greater number of cases (about one in 3700).

Incubation of halothane with rat liver microsomes in the presence of NADPH and the spin-trap PBN resulted in the formation of an ESR signal. The same signal was observed if the spin trap was fed in an oil emulsion to rats which were allowed to inhale halothane under hypoxic conditions (12% O_2), killed, and the liver lipids extracted and placed in the ESR spectrometer (see the legend of Fig. 8.3 for the principle of this). The radical was also detected in bile.

Incubation of halothane with liver microsomes has been observed to produce reactive metabolites that can bind to the membranes and might be able to stimulate lipid peroxidation. P450 is again involved: CYP2E1 is especially active in metabolizing halothane. One action of cytochrome P450 on halothane is to cause formation of the radical $F_3C\overset{\bullet}{C}CHCl$; this is favoured at low O_2 concentrations and was probably the species detected in the above ESR studies. Formation of $F_3C\overset{\bullet}{C}CHCl$ could account for the observation that the expired air of rabbits or humans exposed to halothane has been shown to contain chlorotrifluoroethane and chlorodifluoroethene (Fig. 8.4). The $F_3C\overset{\bullet}{C}CCl$ radical might also combine with oxygen to form a peroxyl radical. It therefore seems that radical reactions and lipid peroxidation could be involved in halothane toxicity, but there is no evidence that they are prominent in causing the hepatotoxicity. Exposure of phenobarbital-pretreated rats to 1% halothane caused small increases in liver and plasma F_2-isoprostane levels,

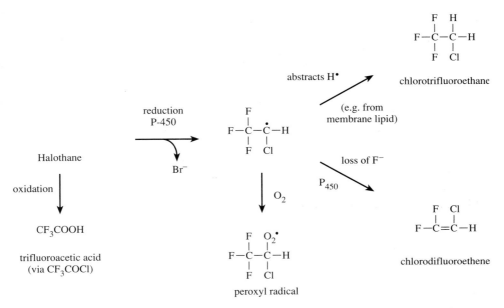

Fig. 8.4. The metabolism of halothane by P450 in a reductive pathway. This pathway is favoured at low O_2, $< 50\,\mu M$. Halothane (2-bromo-2-chloro-1,1,1-trifluoroethane), is also metabolized by an oxidative pathway, favoured at higher ($> 50\,\mu M$) O_2. The latter produces CF_3COCl, which is hydrolysed to trifluoroacetic acid (CF_3COOH) or can attack $-NH_2$ groups to generate adducts. The halothane-derived peroxyl radical can react with polyunsaturated fatty acids (PUFAs) but is approximately five times less reactive than $CCl_3O_2^\bullet$, consistent with data that halothane is a much weaker inducer of lipid peroxidation *in vivo* (see text). For example, $CCl_3O_2^\bullet$ reacts with arachidonic acid with a rate constant of $7.3 \times 10^6\,M^{-1}\,s^{-1}$; the value for the halothane-derived radical is $1.5 \times 10^6\,M^{-1}\,s^{-1}$ (*Chem.–Biol. Interac.* **45**, 171 (1983)).

which were exacerbated if the animals breathed 14% rather than 21% O_2. These changes were, however, much smaller than those produced by CCl_4.[6]

Probably both oxidative and reductive pathways of halothane metabolism (Fig. 8.3) occur in most, if not all, humans exposed to this gas; it has been estimated that about 20% of absorbed halothane is metabolized in the human body. Some halothane metabolites bind to proteins to form adducts, which can be detected in most people after halothane anaesthesia. The greatest attention has been paid to products from **CF₃COCl** (Fig. 8.4), a reactive acid chloride which binds to proteins and phospholipids to give CF_3CO- adducts.

How can the person-to-person variability of halothane toxicity, and the 'sensitization' from previous exposure, be explained? The favoured explanation is an exaggerated immune response to these antigenic CF_3CO- adducts. Proteins forming these adducts (**trifluoroacetylated proteins**) are mainly located in the endoplasmic reticulum and include such essential molecules as protein disulphide isomerase, chaperones and the Ca^{2+}-binding protein **calcireticulin**. Cytochromes P450, such as CYP2EI, can also be modified.[108]

8.3.4   Molecules similar to halothane

Alternative inhalation anaesthetics, such as **enflurane** (CHF_2OCF_2CHFCl), **isoflurane** ($CHF_2OCHClCF_3$) and **desflurane** ($CHF_2OCHFCF_3$) have been reported to cause liver damage in a few cases, perhaps by mechanisms similar to halothane, although they appear to generate fewer protein adducts *in vivo*. Enflurane and desflurane, unlike halothane, did not increase F_2–isoprostane levels in phenobarbital-pretreated rats.

It is interesting to note that some of the hydrochlorofluorocarbons that are being introduced as replacements for the ozone-depleting chlorofluorocarbons have similar structures to these anaesthetics and could conceivably be metabolized in the same way *in vivo* to cause liver damage.[59] Examples of industrial hydrochlorofluorocarbons include CF_3CHCl_2, CF_3CHFCl, CF_3CHF_2, $CF_3CF_2CHCl_2$ and CF_2ClCF_2CHFCl.

8.3.5   Pentachlorophenol and related environmental pollutants

Pentachlorophenol (PCP) is used as a wood preservative, herbicide and insecticide. It has been reported to cause liver cancer when administered at high levels to the B6C3F1 strain of mice. The major metabolite of PCP, **tetrachlorohydroquinone** (TCHQ), is mutagenic and may be the toxic species involved. Several studies have shown increased levels of the DNA base oxidation product 8-hydroxydeoxyguanosine (8-OHdG) in the livers of animals fed PCP at doses of 0.03% or more of the diet for up to 4 weeks. TCHQ can cause increases in 8-OHdG in isolated cells[30] and in the liver of B6C3F1 mice.[137] It has thus been hypothesized that oxidative stress, perhaps involving $O_2^{\bullet -}$ formation during oxidation of TCHQ (a diphenol) to a semiquinone and on to tetrachlorobenzoquinone, causes oxidative DNA damage that leads to cancer. Further work is needed to support this hypothesis, since even these very high levels of PCP or TCHQ produced only about 50% rises in 8-OHdG.

Other environmental contaminants which may cause oxidative stress include the chlorinated dioxin **TCDD (2,3,7,8-tetrachlorodibenzo-*p*-dioxin)**. It is a known carcinogen, being more effective as such in female rather than in male or ovariectomized rats.[133] TCDD toxicity is widely thought to involve binding to an intracellular receptor protein (the Ah or TCCD receptor) that subsequently affects gene transcription (including genes encoding several cytochromes P450). Giving high doses of TCCD to animals has been claimed to increase lipid peroxidation, although this could be a consequence of cell injury. A relationship also exists between iron and TCCD toxicity: iron-deficient animals show less injury. Increased levels of 8-OHdG were detected in the livers of female rats chronically treated with TCCD: elevations were less marked in ovariectomized rats. Again, the significance of these observations is unclear at present.

The fungicide **hexachlorobenzene** causes liver damage and interferes with haem biosynthesis, leading to the syndrome of liver damage and photosensitivity called **porphyria cutanea tarda**. Both the inherited version of this disease and the hexachlorobenzene-induced disease are aggravated by iron overload and are associated with inhibition of the hepatic uroporphyrinogen decarboxylase enzyme (Chapter 2). TCCD can have similar effects.

8.4 Redox-cycling toxins: bipyridyl herbicides

8.4.1 *Toxicity to bacteria*

We saw in Chapter 7 that the toxicity of paraquat and other bipyridyl herbicides to plants involves redox cycling: the herbicide is reduced in the chloroplast to a cation radical that reacts with O_2 to give $O_2^{\bullet -}$

$$PQ^{2+} + e^- \rightarrow PQ^{\bullet +}$$

$$PQ^{\bullet +} + O_2 \rightarrow PQ^{2+} + O_2^{\bullet -} \quad (k = 7.7 \times 10^8 \, M^{-1} s^{-1})$$

The same mechanism accounts for the toxicity of bipyridyl herbicides to bacteria and animals. Indeed, perhaps the most detailed studies of the role of $O_2^{\bullet -}$ in paraquat toxicity have been performed with the bacterium *Escherichia coli*.[71] Low concentrations ($0.1-1.0 \, \mu M$) of paraquat halt the growth of *E. coli* (a bacteriostatic effect), whereas much higher ($>100 \, \mu M$) concentrations are required to kill the cells.

Paraquat added to a culture of *E. coli* is taken up by the cells and rapidly reduced. Under anaerobic conditions, the bipyridyl radical can be detected by observing its absorption spectra. Extracts of *E. coli* can reduce paraquat if NADPH is added, a reaction that is catalysed by at least four flavoprotein enzymes. One has been identified as thioredoxin reductase and another as $NADP^+$: ferredoxin oxidoreductase.[82] In the presence of O_2, as would be expected, the paraquat radical disappears, and $O_2^{\bullet -}$ is generated. Addition of paraquat to aerobically grown *E. coli* induces rapid synthesis of MnSOD, the same enzyme induced upon exposure of the bacteria to elevated O_2 (Chapter 3). Catalase activity increases as well. *E. coli* cells whose SOD activity has been increased by paraquat pre-treatment are more resistant to elevated $O_2^{\bullet -}$. Vice versa, cells with raised SOD due to previous exposure to increased O_2 or to transfection with SOD genes, are more resistant to paraquat. If induction of MnSOD by *E. coli* is prevented, either by adding puromycin (an inhibitor of protein synthesis) or by a poor growth medium, then the toxicity of paraquat is increased.

8.4.2 *Protection by extracellular SOD*

It has been reported that addition of SOD to the growth medium offers some protection to *E. coli* against damage by paraquat. This at first sight is surprising, since $O_2^{\bullet -}$ cannot cross the bacterial cell wall and membrane and the SOD protein cannot enter the bacterial cells. However, just as paraquat can easily enter the cells, some of the paraquat radical can leak out and react with O_2 in the surrounding medium to give $O_2^{\bullet -}$, leading to extracellular damage.[56]

The amount of leakage is inversely related to intracellular O_2 levels. Paraquat, like O_2, has been shown to be mutagenic to those *Salmonella typhimurium* strains that are used to test for mutagenic ability in the Ames test. *S. typhimurium* cells containing elevated SOD activities are more resistant to the toxic and mutagenic effects of paraquat.

Addition of paraquat to *E. coli* under aerobic conditions leads to inhibition of the dihydroxyacid dehydratase enzyme involved in the biosynthesis of branched-chain amino acids. Providing amino acids to the growth medium partially relieves the inhibitory effect of paraquat on bacterial growth. Addition of nicotinamide, a precursor of NAD^+, gives further relief. These results are similar to those of studies on the effects of high pressure O_2 on *E. coli* (Chapter 1),[18] consistent with the view that the actions of hyperbaric O_2, and the bacteriostatic action of paraquat, are mediated by a common mechanism, namely increased generation of $O_2^{\bullet-}$ and products derived from it.

However, the lethal effects of high paraquat concentrations upon *E. coli* are not prevented by amino-acid supplementation, which suggests that other targets within the cell can be attacked by $O_2^{\bullet-}$ and/or H_2O_2. The mutagenicity of paraquat in the Ames test suggests that one of these targets is DNA, especially as *E. coli* strains deficient in DNA-repair mechanisms show an enhanced sensitivity to paraquat. Neither $O_2^{\bullet-}$ nor H_2O_2 reacts with DNA, however. It is possible that the DNA damage is mediated by $OH^\bullet$ formation, since iron or copper ions have been reported to aggravate the toxicity of paraquat to *E. coli*.[73]

8.4.3   *Toxicity to animals*

The main problem in the agricultural use of paraquat and diquat is that they are poisonous to several animal species, including fish, rat, mouse, cat, dog, sheep, cow and humans. Many cases have been reported of children drinking herbicides carelessly stored in soft-drink bottles, and paraquat is sometimes used in suicide attempts.

Bipyridyl herbicides can be absorbed slowly through the skin, but most poisoning involves swallowing them. Oral intake of paraquat first results in local effects—irritation of the mouth, throat and oesophagus, and sometimes vomiting and diarrhoea. Fortunately, gut absorption of bipyridyl herbicides is fairly slow, and life may often be saved by washing out the stomach and intestines repeatedly with saline solutions. Administration of suspensions of clays that absorb the herbicides (e.g. bentonite or fullers' earth), and dialysis of blood are often also carried out.

The major organ affected by paraquat in animals is the lung. The type I cells that line the alveoli (Chapter 1) begin to swell and are eventually destroyed, leading to oedema, capillary congestion and inflammation. Type II cells are also damaged and the synthesis of surfactant decreases. As a result of both oedema and loss of surfactant, gas exchange is hindered. In animals which survive, the damaged lung tissue is replaced by inelastic fibres that cause permanent interference with lung expansion.

8.4.4   *Why is paraquat toxic to the lung?*

The lung is a major target of damage because several lung cell types actively accumulate paraquat, i.e. it is taken up against a concentration gradient.[125] The diamino compound **putrescine** ($H_3\overset{+}{N}(CH_2)_4\overset{+}{N}H_3$) has been shown to block this uptake in isolated rat lung slices. The lung also has a high internal O_2 concentration. Other tissues, including liver and kidney, are damaged by paraquat, but more slowly. By contrast, although large doses of diquat affect the lung, it is not the major target tissue. In most diquat-exposed animals the intestines and liver are damaged.

The 'targeting' of paraquat damage to the lung by selective uptake means that poisoning can occur not only by oral intake but also by inhaling paraquat droplets from crop spraying. Paraquat has allegedly caused lung damage in the USA to 'pot'-smokers who obtained their marijuana from paraquat-treated Mexican plants. Microsomal fractions from lung (and most other animal tissues) will reduce paraquat in the presence of NADPH, and the paraquat radical then reacts with O_2 to form $O_2^{\bullet-}$.

It seems likely that paraquat is reduced by the NADPH–cytochrome P450 reductase enzyme, since antibody directed against this enzyme inhibits paraquat reduction by microsomes *in vitro*. Intravenous injection of paraquat into rats causes a rapid activation of the pentose phosphate pathway in lung, presumably as SOD converts $O_2^{\bullet-}$ to H_2O_2 and NADPH is then consumed by glutathione reductase. NADPH is also required for the biosynthesis of fatty acids (needed to replace damaged membrane lipids) and surfactant. Hence falls in NADPH may contribute to damage: fatty acid synthesis is decreased in the paraquat-treated lung, and there is an accumulation of 'mixed disulphides' formed between protein −SH groups and GSSG.[68]

The damaging effects of paraquat on lung *in vivo*, and on isolated lung cells, are potentiated at high O_2 concentrations, consistent with a key role for reaction of the paraquat radical with O_2. Selenium-deficient rats are more sensitive to paraquat poisoning than are animals fed on a normal diet, which suggests that the selenoprotein glutathione peroxidases play an important protective role in removing H_2O_2. It has also been suggested that selenoprotein P (Chapter 3) might be involved in protection.[20] There have been some reports that injection of SOD into animals slightly ameliorates the symptoms of paraquat poisoning. Other scientists have not found this effect, however. Injected SOD is unlikely to enter lung cells and is rapidly cleared from the body by the kidneys, so a large effect would not be expected. It may be that, as in *E. coli* (Section 8.4.1 above), some paraquat radical diffuses out of lung cells and generates $O_2^{\bullet-}$ in the surrounding tissue fluid, where it is available for SOD to act. Extracellular SOD in the lung (Chapter 3) could also contribute to protection.

There is considerable evidence that intracellular SOD can protect against paraquat. Introduction of more SOD into Chinese hamster ovary (CHO) cells or mouse fibroblasts protected them against damage caused by paraquat, and paraquat-resistant HeLa cells were found to have an increased activity of

CuZnSOD and MnSOD. Injection of diethyldithiocarbamate into mice to inhibit CuZnSOD enhances paraquat toxicity. Pre-treatment of adult rats with bacterial endotoxin increases catalase, SOD, glucose-6-phosphate dehydrogenase and glutathione peroxidase activities in the lung; these rats are more resistant not only to elevated O_2 (Chapter 3) but also to paraquat.

8.4.5 *Paraquat, lipid peroxidation and hydroxyl radical formation*

It was reported in 1974 that incubating mouse lung microsomal fractions with paraquat plus NADPH increased lipid peroxidation, suggesting that lipid peroxidation might be a primary mechanism of paraquat toxicity. However, later evidence does not support this view.[123] For example, pre-treatment of mice with the antioxidant N,N'-diphenyl-p-phenylenediamine did not decrease the toxic effects of paraquat, although it did decrease levels of peroxidation induced by paraquat in microsome fractions isolated from the animals. Other workers have observed little or no stimulation of lipid peroxidation by paraquat, e.g. the exhalation of ethane by rats is only slightly increased by paraquat or diquat.

It may be concluded that lipid peroxidation is not a major contributor to damage by paraquat. Perhaps tissue damage by paraquat, mediated by increased formation of $O_2^{\bullet-}$ and H_2O_2, might sometimes lead to increased peroxidation of damaged tissues. Exactly this occurs in plants (Chapter 7); paraquat inhibits CO_2 fixation rapidly, but increases in detectable lipid peroxidation in leaves occur later. Similarly, administration of diquat even to selenium-deficient rats produced only small changes in plasma F_2-isoprostane levels[5] when compared with the effects of CCl_4. Cellular injury by oxidative stress imposed by paraquat can raise intracellular free Ca^{2+} levels, which could increase $NO^\bullet$ production and the activity of Ca^{2+}-stimulated proteases, some of which can convert xanthine dehydrogenase to xanthine oxidase (Chapter 4). Indeed, there are reports that rises in xanthine oxidase activity contribute to lung damage by paraquat in animals.[142]

The ability of paraquat to damage DNA in *E. coli*, and the potentiation of its toxicity by iron salts, suggest that $OH^\bullet$ radical may be involved. Treatment of animal cells with paraquat has also been reported to produce chromosome damage. The paraquat radical ($PQ^{\bullet+}$) is a powerful reducing agent ($E^{\circ\prime} = -0.45$ V), much more so than $O_2^{\bullet-}$ (see Table 2.3). Hence $PQ^{\bullet+}$ can reduce Fe(III) and most Fe(III) chelates directly (route A), or it can react with O_2 to form $O_2^{\bullet-}$ (route B):[27]

A: $Fe(III)\text{-chelate} + PQ^{\bullet+} \rightarrow PQ^{2+} + Fe^{2+}\text{-chelate}$

B: $PQ^{\bullet+} + O_2 \rightarrow O_2^{\bullet-} + PQ^{2+}$

$Fe(III)\text{-chelate} + O_2^{\bullet-} \rightarrow Fe^{2+}\text{-chelate} + O_2$

The Fe^{2+} chelate can then react with H_2O_2 to form $OH^\bullet$ by the Fenton reaction

$Fe^{2+}\text{-chelate} + H_2O_2 \rightarrow OH^\bullet + OH^- + Fe(III)\text{-chelate}$

The relative contributions of route A and route B to generation of Fe^{2+} will depend upon the O_2 concentration; higher O_2 favours route B over route A. Paraquat may also contribute to *providing* the metal ions needed for $OH^{\bullet}$ formation, since not only $O_2^{\bullet-}$ but also $PQ^{\bullet+}$ can reductively mobilize iron ions from ferritin.[130]

The iron-chelating agent desferrioxamine has been used to study the role of iron in mediating paraquat toxicity to animals.[21] Some authors have reported protective effects, but others have claimed an aggravation of paraquat toxicity. Desferrioxamine prevents $O_2^{\bullet-}$-dependent $OH^{\bullet}$ formation because it binds Fe(III) tightly, and the Fe(III) chelate cannot be reduced by $O_2^{\bullet-}$. However, it is possible that the highly reducing $PQ^{\bullet+}$ radical *can* reduce the Fe(III)−desferrioxamine complex, leading to Fe^{2+}−desferrioxamine, a much less stable complex than the Fe(III) chelate. Ferrous desferrioxamine, or Fe^{2+} released by it, could lead to $OH^{\bullet}$ generation by reaction with H_2O_2. In any case, the low rate of entry of desferrioxamine into cells (Chapter 10) suggests that it would not be a suitable therapy for use in the treatment of paraquat poisoning in humans.

The toxicity of bipyridyl herbicides to tissues other than the lung is probably also mediated by redox cycling, since NADPH−cytochrome P450 reductase is widely distributed in animal tissues. Even in tissues without this enzyme, reduction may be achieved by other flavoproteins. Feeding diquat to rats causes cataract formation, possibly due to its reduction by lens glutathione reductase (Chapter 7). The haemolysis sometimes seen in paraquat-poisoned humans might involve reduction by erythrocyte glutathione reductase.

8.5 Diabetogenic drugs

8.5.1 *Alloxan*

Another redox-cycling toxin is the compound **alloxan**, first described as a diabetogenic agent in 1943. Injection of alloxan into animals causes degeneration of the β-cells in the islets of Langerhans of the pancreas. Since these cells synthesize the hormone insulin, alloxan is often used to induce diabetes in experimental animals. Two-electron reduction of alloxan gives **dialuric acid**[146] (Fig. 8.5), a cytotoxic pyrimidine whose structure resembles that of the haemolytic agents divicine and isouramil (Section 7.4).

An intermediate radical, formed by one-electron reduction of alloxan (or one-electron oxidation of dialuric acid), also exists. Dialuric acid is unstable in aqueous solution and undergoes oxidation, eventually to alloxan, accompanied by reduction of O_2 to $O_2^{\bullet-}$. Dialuric acid oxidation is accelerated by traces of transition metal ions, and in their presence it leads to generation of $O_2^{\bullet-}$, H_2O_2 and $OH^{\bullet}$, the latter probably by Fenton-type reactions. Indeed, solutions of dialuric acid have been observed to stimulate lipid peroxidation *in vitro*, and to inhibit the growth of several strains of bacteria.

Hence, any body tissue that can take up and reduce alloxan will be at risk of oxidative stress. When alloxan is injected into rats, it accumulates in the islets of Langerhans and in the liver. Whereas the liver contains high activities of

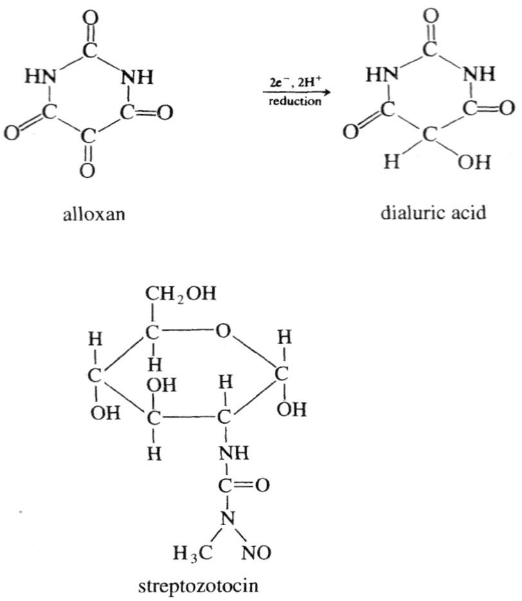

alloxan   dialuric acid

streptozotocin

Fig. 8.5. Structures of diabetogenic agents. The α form of streptozotocin (*N*-methylnitroso-carbamylglucosamine) is shown; in the β form the arrangement of the −H and −OH groups on carbon 1 (next to the oxygen in the ring) is reversed. The drug is normally a mixture of α and β forms.

SOD, catalase and glutathione peroxidase, the activities of these enzymes in the β-cells are moderate by comparison.[85] Further, isolated β-cells can reduce alloxan to dialuric acid at a high rate: reduction may involve GSH, thioredoxin and/or glutaredoxin[60,146] (these proteins are discussed in Chapter 3). It is probably this combination of fast alloxan reduction and mediocre antioxidant defences that makes β-cells sensitive to alloxan.[48] *In vitro*, β-cells are also sensitive to damage by excess NO$^\bullet$ and by ONOO$^-$.

Addition of alloxan to isolated β-cells causes membrane damage and cell death, effects which can be decreased by addition of SOD, catalase, OH$^\bullet$ scavengers (including mannitol and dimethylsulphoxide),[48] or iron chelators such as DETAPAC. Extensive DNA strand breakage has been observed in the β-cells of alloxan-treated rats. This DNA strand breakage, which is accompanied by an increased activity of PARP, may result from generation of OH$^\bullet$ by transition metals bound upon, or close to, the DNA (Fig. 8.6).

That these observations upon isolated β-cells are relevant to the action of alloxan *in vivo* is shown by several experiments. Injection of CuZnSOD, attached to a high-molecular-mass polymer to reduce its clearance by the kidneys, into mice protected them against the diabetes usually produced by a subsequent injection of alloxan, as did the iron-chelator DETAPAC. EDTA, which unlike DETAPAC allows rapid O$_2^{\bullet-}$-dependent reduction of iron bound to it, was not protective. The anti-cancer drug ICRF-159, a derivative

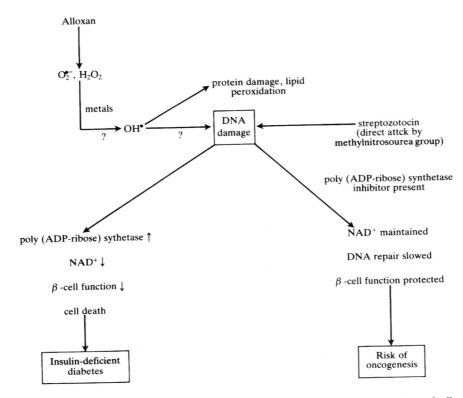

Fig. 8.6. Proposed mechanisms of the diabetogenic actions of streptozotocin and alloxan. Adapted from *BioEssays*, **2**, 19 (1985) by courtesy of Professor H. Okamoto and the publishers. Inhibitors of poly(ADP–ribose) synthetase used in such experiments include **benzamide**, **3–aminobenzamide** and **theophylline**. The effects of excess NO$^\bullet$ on islet cells may also involve DNA damage.

of EDTA, has also been observed to protect against alloxan, presumably by chelating iron. Transgenic mice over-expressing CuZnSOD in the islet cells were more resistant to alloxan than controls.[74]

As is the case with paraquat (Section 8.4 above), there are conflicting reports on the ability of desferrioxamine to protect animals against the diabetogenic action of alloxan. Some scientists have reported protective effects in short–term experiments, but others find that desferrioxamine exacerbates damage *in vivo*. The chelator 1,10-phenanthroline, which decreases DNA damage in mammalian cells treated with H$_2$O$_2$, has been reported to decrease the severity of alloxan-induced diabetes in rats.

8.5.2 *Streptozotocin*

Although it is not a redox-cycling agent, it is convenient to discuss here another drug that is often used to induce diabetes in experimental animals.

Streptozotocin is a nitrosourea compound produced by *Streptomyces ach-romogenes* (Fig. 8.5). It induces DNA strand breakage in β-cells. There are several suggestions that N-nitroso compounds from smoke-cured mutton may contribute to the increased prevalence of diabetes in Iceland, and that similar nitroso compounds in food (e.g. betel nuts and cycad (*Cycas circinalis*) plants) may relate to the high prevalence of diabetes on Pacific islands. Hence mechanisms of diabetogenicity of nitroso compounds are of considerable current interest.[32]

Injection of streptozotocin into rats has been observed to decrease the CuZnSOD activity of retina, erythrocytes and islet cells, but not that in other tissues. Vitamin E and the spin-trap PBN[129] have been reported to decrease the diabetogenicity of streptozotocin in animals. Injected SOD has been claimed by some scientists to decrease the diabetogenic effects of strep-tozotocin, but others have not confirmed this. Phenanthroline did not protect rats against streptozotocin-induced diabetes, unlike the case of alloxan. It is likely that streptozotocin damages DNA by mechanisms other than increased ROS formation. In any case, excessive DNA damage can lead to PARP activation and depletion of cellular NAD^+ to an extent that can kill the cell (Chapter 4). Inhibitors of PARP (Fig. 8.6) can keep streptozotocin-treated β-cells alive.

However, treatment of rats with both streptozotocin and PARP inhibitors led to the development of tumours of the β-cells (**insulinomas**) in a few animals. This is consistent with the concept of **lethal NAD^+ depletion** (Chapter 4): excessive PARP activation and severe NAD^+ depletion may help the organism by killing cells with excessive levels of DNA damage, so minimizing the occurrence of harmful mutations (Fig. 8.6).

It should be noted that isolated human islet cells seem less sensitive to oxidative damage (e.g. by alloxan) or to streptozotocin than rodent cells, perhaps because they have higher levels of SOD, catalase, heat-shock proteins and DNA-repair enzymes. They also appear less sensitive to $NO^\bullet$ and $ONOO^-$. Nevertheless, it seems likely that oxidative stress plays some role in both the origin of diabetes and in the pathology of established diabetes.[32] This is further discussed in Chapter 9.

8.6   Redox-cycling toxins: diphenols and quinones

8.6.1   *Interaction with O_2 and superoxide*

Superoxide can exist in an equilibrium with several diphenols and quinones, with the intermediacy of semiquinone radicals (Chapter 2).

Hence diphenol/quinone systems can act as scavengers of $O_2^{\bullet-}$ or generators of $O_2^{\bullet-}$, depending on the position of equilibrium[79,92] and also on the pH, since the degree of protonation affects the reduction potentials of these systems. Diphenols/semiquinones can also react with other oxygen radicals, including peroxyl ($RO_2^{\bullet}$), and so can often inhibit lipid peroxidation. Indeed, both ubiquinol (in mitochondria and low-density lipoproteins (LDLs) and plastoquinol (in chloroplasts) have been suggested to act as antioxidants protecting against lipid peroxidation (Chapters 3 and 7); their semiquinone forms do not appear to generate $O_2^{\bullet-}$ at a significant rate during the normal functioning of mitochondria and chloroplasts.[104]

8.6.2 *Formation of hydroxyl radical*

Semiquinones under physiological conditions do not appear to react with H_2O_2 to give $OH^{\bullet}$. However, many of them can reduce transition-metal ions and promote $OH^{\bullet}$ generation, e.g.

$$SQ^{\bullet-} + Fe(III) \rightarrow Fe^{2+} + Q$$

$$Fe^{2+} + H_2O_2 \rightarrow Fe(III) + OH^{\bullet} + OH^-$$

$$\text{Net:} \quad SQ^{\bullet-} + H_2O_2 \xrightarrow[\text{catalyst}]{Fe} OH^{\bullet} + OH^- + Q$$

This is a semiquinone-driven Fenton reaction, analogous to $O_2^{\bullet-}$-driven Fenton chemistry. If the $SQ^{\bullet-}$ reduces O_2 to $O_2^{\bullet-}$ an additional set of reactions is

$$SQ^{\bullet-} + O_2 \rightarrow Q + O_2^{\bullet-}$$

$$O_2^{\bullet-} + Fe(III) \rightarrow Fe^{2+} + O_2$$

Certain diphenols can reduce transition-metal ions, e.g. hydroquinone (Fig. 8.9 below) reduces Cu^{2+} to Cu^+.

Many quinones other than ubiquinone and plastoquinone are found in nature (Figs 8.7 and 8.8).[104] Some are used as anti-cancer drugs (Chapter 9). Others are present in cigarette smoke (Section 8.11 below) and yet others are used to repel predators—one example is the bombardier beetle (Chapter 6) and another is the walnut tree (Fig. 8.7).

Quinones can be toxic to aerobes by at least two mechanisms.[92] First, they or their semiquinones may react with GSH and protein −SH groups (ubiquinone and plastoquinone do not). Reaction with GSH may be non-enzymic and/or catalysed by GSH transferase enzymes. Second, quinones may redox cycle

$$Q \xrightarrow[\text{1e}^-\text{ reduction}]{\text{cellular reducing system}} SQ^{\bullet-}$$

$$SQ^{\bullet-} + O_2 \longrightarrow Q + O_2^{\bullet-}$$

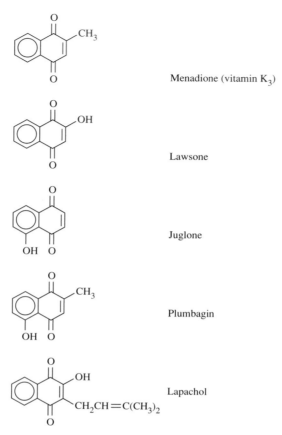

Menadione (vitamin K₃)

Lawsone

Juglone

Plumbagin

Lapachol

Fig. 8.7. Some quinones. Vitamins K are involved in blood coagulation (Chapter 6). **Menadione** is a synthetic compound lacking the isoprenoid side-chain, but still shows vitamin activity in animal tests; hence the name 'vitamin K₃'. **Lawsone** is a coloured pigment found in henna, a paste made from leaves of *Lawsonia inermis* that has been used to dye hair since ancient times. **Juglone** is exuded by the roots of walnut trees and has been suggested to prevent germination of other plant seeds in the vicinity of the tree. **Plumbagin**, juglone and menadione induce MnSOD activity in *E. coli* and the resulting cells are more resistant to paraquat. **Lapachol** is found in the wood of several trees and has been claimed to have anti-malarial and anti-cancer activities.

For example, **plumbagin** and **juglone**, redox-cycling quinones, are powerful inducers of MnSOD activity in *E. coli* (Fig. 8.7). In animal cells, enzymes in mitochondria, endoplasmic reticulum and cytosol can catalyse one-electron reduction of quinones.

8.6.3 *Menadione and quinone reductase (DT diaphorase)*

A popular object of laboratory studies has been the synthetic quinone **menadione** (Fig. 8.7).[147] In excess, menadione can cause haemolysis,

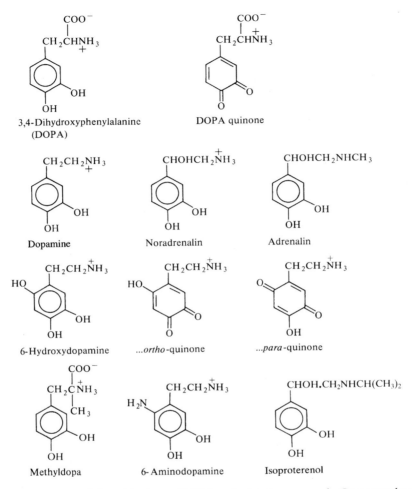

Fig. 8.8. Derivatives of phenylalanine: L-DOPA and related compounds. Compounds with the catechol ring structure plus an amino group are often called **catecholamines** and include dopamine and noradrenalin.

especially in erythrocytes deficient in glucose-6-phosphate dehydrogenase (Chapter 7). Reaction of menadione with oxyhaemoglobin causes oxidation and precipitation of the protein. It has been proposed[147] that menadione reacts with oxyhaemoglobin to give a semiquinone ($SQ^{\bullet -}$):

$$Hb(Fe^{2+})O_2 + Q \rightleftharpoons \underset{\text{methaemoglobin}}{Hb[Fe(III)]} + O_2 + SQ^{\bullet -}$$

The semiquinone can either re-reduce methaemoglobin or convert O_2 to $O_2^{\bullet -}$. Addition of SOD accelerates methaemoglobin formation, presumably by dismuting $O_2^{\bullet -}$ and thus altering the equilibrium to favour reaction of $SQ^{\bullet -}$

with O_2. In addition, menadione can react directly with the $\beta 93$ cysteine $-SH$ groups on haemoglobin. If these are written as protein $-SH$, then the reaction product is

Treatment of isolated erythrocytes with diethyldithiocarbamate to inhibit CuZnSOD has been shown to accelerate the rate of haemolysis induced by **naphthoquinone-2-sulphonate**, a derivative of menadione in which the $-CH_3$ group is replaced by a sulphonate group to make the compound more soluble in water and facilitate experimentation. Diethyldithiocarbamate also potentiates the toxicity of menadione to liver cells.

Several enzymes in liver, including NADPH–cytochrome P450 reductase, catalyse the one-electron reduction of quinones into semiquinones. Liver (and most other tissues) also contain high activities of a **quinone reductase (NAD(P)H: quinone oxidoreductase)** enzyme which, by contrast, catalyses a two-electron reduction of quinones into hydroquinones at the expense of NADH or NADPH.[10] It has been proposed that one function of quinone reductase is to decrease formation of $O_2^{\bullet-}$ *in vivo* by removing quinones, thus preventing their reduction to semiquinones by other enzyme systems. Hence quinone reductase is often regarded as a phase II detoxification system. Consistent with this hypothesis, transgenic "knockout" mice lacking this enzyme seem healthy, but are abnormally susceptible to menadione.[115a].

Quinone reductase contains FAD, is largely present in the cell cytosol, and has also been called **DT diaphorase**.[10] It reduces not only quinones (including menadione, coenzyme Q and 'natural' vitamins K) but also some artificial electron acceptors such as tetrazolium salts. It has been suggested to help maintain coenzyme Q in the reduced (ubiquinol) form, so 'recycling' it as an inhibitor of lipid peroxidation (Chapter 3).

Synthesis of DT diaphorase in liver and other tissues is accelerated by a wide variety of toxins, including carcinogenic aromatic hydrocarbons, dioxins, quinones and also by certain synthetic antioxidants such as butylated hydroxyanisole (BHA). Regulation of quinone reductase gene expression in animals involves two types of 'response element'. One is the **xenobiotic response element**, a DNA sequence also found in the promoters of the genes encoding CYPIAI and certain glutathione-S-transferase subunits. The second type[37] is the **antioxidant response element** (ARE; Section 4.13.6). A wide range of agents activate gene expression through the ARE, including H_2O_2, phorbol esters, β-naphthoflavone and *tert*-butylhydroquinone.

In hepatocytes, high concentrations of menadione deplete GSH, increase formation of $O_2^{\bullet-}$ and H_2O_2 by redox cycling and cause membrane blebbing, associated with increases in intracellular free Ca^{2+} (Chapter 4). Menadione also induces DNA strand breaks, but this seems largely due to the activation of Ca^{2+}-dependent nucleases rather than to direct oxidative damage to DNA, since rises in 8-OHdG levels have not been observed in menadione-treated hepatocytes despite extensive DNA strand breakage.[40] Menadione, plumbagin and several other quinones are mutagenic to bacteria.

8.6.4 *Substituted dihydroxyphenylalanines and 'manganese madness'*

Another class of biologically important diphenols is derived from the aromatic amino acid phenylalanine (Fig. 8.8). These products are important as hormones and/or as neurotransmitters, yet they can oxidize to generate $O_2^{\bullet-}$ and semiquinones. For example, oxidation of the hormone **adrenalin** and the hormone/neurotransmitter **noradrenalin** produces $O_2^{\bullet-}$ and H_2O_2, and the ability of SOD to inhibit adrenalin oxidation has been made the basis of an assay for this enzyme (Chapter 3).

The neurotransmitter **dopamine** and its precursor **L-DOPA** also oxidize to make $O_2^{\bullet-}$ and H_2O_2.[14] For all these compounds, the rate of oxidation is greatly accelerated by the presence of transition-metal ions such as iron and copper. In the presence of such metals, oxidation produces not only $O_2^{\bullet-}$, H_2O_2, semiquinones and quinones, but also $OH^{\bullet}$. Quinones and semiquinones can attack proteins (usually reacting with $-SH$ groups) and deplete GSH.

For example, treatment of animals or humans with large amounts of manganese promotes rapid dopamine oxidation in the brain. The disorder **manganese madness** or **locura manganica** has been observed in miners of manganese ores in parts of northern Chile and less often in Australia and Taiwan. Locura manganica in its later stages has a superficial clinical resemblance to Parkinson's disease and is similarly treated by L-DOPA administration, although the most extensive damage done to brain tissues is to the striatum and pallidum, and less to the substantia nigra, which is the main site of cell death in Parkinson's disease (Chapter 9). Manganese is added to many parenteral nutrition solutions and it has been hypothesized that excess uptake could cause brain damage in children.[38] However, intake of some manganese is essential for metabolism, e.g. for MnSOD. The symptoms of locura manganica may be due to loss of L-DOPA/ dopamine,[3] and perhaps also to damage done by ROS and by semiquinone/ quinone products of oxidation, e.g. GSH depletion.

Adrenalin is an important physiological regulator of cardiac contractility and metabolism, yet large doses of adrenalin, noradrenalin or the synthetic compound **isoproterenol** (Fig. 8.8) have been shown to produce an 'infarct-like' necrosis of heart muscle in animals. Oxidation products are thought to contribute to this myocardial damage, especially as the heart seems sensitive to oxidative stress (Chapter 3).[14]

An interesting example[148] of a useful role for dopamine auto-oxidation is provided by insects. During formation of the hard exoskeleton,

N-acetyldopamine and similar compounds undergo oxidation and reaction with proteins to toughen the structure.

8.6.5   *Neurotoxicity of 6-hydroxydopamine*

The rates of oxidation of adrenalin, L-DOPA, dopamine and noradrenalin would be expected to be slow *in vivo*, both because SOD is present to remove $O_2^{\bullet-}$ and break the 'chain reaction' of auto-oxidation and because concentrations of 'free' transition metals are very low (Chapter 3). By contrast, the related compounds **6-hydroxydopamine** and **6-aminodopamine** (Fig. 8.8) oxidize much more quickly, although minute traces of metal ions are still involved. During the oxidation, semiquinones and quinones are formed (Fig. 8.8 shows the two quinones that are produced by 6-hydroxydopamine), together with H_2O_2, $O_2^{\bullet-}$ and $OH^{\bullet}$. Superoxide participates in the oxidation of further molecules of 6-hydroxydopamine and hence addition of SOD decreases the observed rate of oxidation of 6-hydroxydopamine. Again, this has been made the basis of an assay for SOD (Chapter 3).

When injected into the brains of animals, 6-hydroxydopamine or 6-aminodopamine cause a rapid and specific damage to catecholamine nerve terminals. Hence, 6-hydroxydopamine is widely used as a research tool to investigate the physiological roles of such nerve terminals (Chapter 9). The selective action is probably due to the specific uptake of these compounds by catecholamine neurones, followed by increased intra-neuronal formation of $O_2^{\bullet-}$, H_2O_2, $OH^{\bullet}$ semiquinones and quinones. Consistent with this, the toxicity of various phenolic compounds to a line of neuroblastoma cells in culture was correlated with their rates of oxidation.

8.6.6   *Methyl-DOPA*

Figure 8.8 shows the structure of **α-methyl-DOPA**, a drug used to lower elevated blood pressure in humans. One of its side-effects is an impairment of liver metabolism. Incubation of liver microsomes with α-methyl-DOPA in the presence of NADPH causes covalent binding of the drug to the microsomal membranes that can be inhibited either by GSH or by SOD. It seems that $O_2^{\bullet-}$ produced by microsomes when NADPH is present can accelerate oxidation of methyl-DOPA, and the quinones and semiquinones formed attack −SH groups in the microsomal proteins, or react with added GSH.[31]

8.6.7   *Benzene and its derivatives*

The aromatic hydrocarbon benzene, a widely used industrial solvent and fuel additive, can damage most tissues if ingested in sufficient quantity, including the liver and brain. However, the bone marrow is particularly vulnerable, and excess exposure to benzene is a risk factor for the development of leukaemia. Levels of 8-OHdG have been reported to be elevated in the DNA from peripheral blood lymphocytes of benzene-exposed workers.[83]

The toxicity of benzene involves its conversion by cytochromes P450 into phenols (phenol, catechol, hydroquinone and 1,2,4–benzenetriol). These can undergo oxidation to quinones and semiquinones (Fig. 8.9), perhaps accelerated by the high levels of myeloperoxidase present in bone marrow. CuZnSOD accelerates the oxidation of hydroquinone to the more toxic benzoquinone (Fig. 8.9) by removing $O_2^{\bullet-}$ and driving the equilibrium between the intermediate semiquinone radical and O_2 to the right[79]

$$SQ^{\bullet-} + O_2 \rightleftharpoons benzoquinone + O_2^{\bullet-}$$

Halogenated benzenes seem less carcinogenic but may also lead to oxidative stress (Section 8.3 above).

8.6.8   *Toxic-oil syndrome*[121]

In 1981, a new syndrome was reported in Spain, the **toxic–oil syndrome** or **Spanish cooking–oil syndrome**. At the height of the epidemic more than 20,000 people in the central and north-western regions of Spain were affected, of whom 11,000 required hospitalization and over 800 eventually died. The symptoms of the disease in its acute stage were respiratory distress, fever, headache, itching, nausea and, sometimes, muscular pains and neurological disorders. Lung damage was responsible for most of the early deaths. When followed up later, about 10% of patients had developed muscular wasting and weakness, and another 49% had suffered some muscular impairment. All sufferers from the disease were found to have consumed 'olive oil' sold by door-to-door salesmen. Analysis showed that the oil in question had never seen

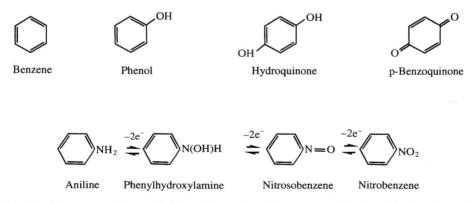

Benzene   Phenol   Hydroquinone   p-Benzoquinone

Aniline   Phenylhydroxylamine   Nitrosobenzene   Nitrobenzene

Fig. 8.9. Benzene and its metabolites. Conversion of benzene to phenol and hydroquinone is catalysed by cytochromes P450, largely in the liver. The bottom diagram (metabolism of aniline) is by courtesy of Professor Ron Mason.

an olive tree, and was basically oil obtained from seeds of the rape plant, *Brassica napus.*

Rapeseed oil imported into Spain is treated with **aniline** (Fig. 8.9) to deter its consumption. Aniline is widely used in the manufacture of explosives, dyestuffs and other chemicals. Excess exposure produces haemolytic anaemia and damage to the spleen, accompanied by splenic fibrosis with increased lipid peroxidation and oxidative protein damage.[70] It seems that aniline-damaged erythrocytes are rapidly sequestered by the spleen, which thus accumulates not only aniline and its metabolites but also excess iron. Damage to red blood cells by aniline appears to involve its metabolite **phenylhydroxylamine** (Fig. 8.9), which is oxidized by oxyhaemoglobin to give nitrosobenzene and methaemoglobin. Nitrosobenzene is then reduced by methaemoglobin reductase (or non-enzymically) back to phenylhydroxylamine, which oxidizes more oxyhaemoglobin in a cyclic manner.

Although aniline might have made some contribution to toxic-oil syndrome, the symptoms are different from those of simple aniline poisoning. It appears that attempts had been made to remove aniline from the rapeseed oil. During this 'purification', aniline-derived toxins were formed.

One good thing to come out of the toxic-oil affair was that it led to the setting up, largely through the efforts of Robin Willson and the late Trevor Slater, of the Society for Free Radical Research (SFRR) in the UK. From the seed they planted has grown a network of international societies with thousands of members: SFRR Europe, SFRR Asia, SFRR Australasia and SFRR North America (The Oxygen Society), plus two official society journals, *Free Radical Research* (SFRR Europe) and *Free Radical Biology and Medicine* (Oxygen Society). Ironically, however, there is only limited evidence that free radicals and oxidative damage had much to do with the toxicity of the adulterated oil.

8.7 Redox–cycling agents: toxins derived from *Pseudomonas aeruginosa*

Pseudomonas aeruginosa, a Gram–negative bacterium, can cause pneumonia and often chronically infects the respiratory tract of patients with cystic fibrosis (Chapter 9). Many *P. aeruginosa* strains secrete **pyocyanin** (5-methyl-1-hydroxyphenazine), a redox-cycling agent which induces MnSOD in *E. coli*. Production of $O_2^{\bullet-}$ by pyocyanin appears to account for its toxicity to *E. coli* and possibly also to the lung in patients with *P. aeruginosa* infection.[17] For example, the levels of pyocyanin present in sputum of cystic fibrosis patients are sufficient to induce ciliary dysfunction when added to human nasal epithelial cells.

P. aeruginosa also secretes a siderophore (**pyochelin**) which allows it to take up iron into the cell. Iron bound to pyochelin appears capable of catalyzing $O_2^{\bullet-}$-dependent $OH^{\bullet}$ formation. Superoxide reduces Fe(III)–pyochelin and the resulting Fe^{2+} chelate decomposes H_2O_2 to $OH^{\bullet}$. Hence the combination of pyochelin and pyocyanin is more cytotoxic than either agent alone.[17]

8.8 Alcohols

8.8.1 *Ethanol*

Most people drink solutions of ethanol (ethyl alcohol) as alcoholic drinks, and even in teetotallers it is formed in small amounts by gut bacteria. Ethanol is soluble both in water and in many organic solvents, and can cross cell membranes readily, incuding the blood–brain barrier. One beneficial effect that it has, apart from the social ones, is to partially protect experimental animals against alloxan-induced diabetes (Section 8.5 above). It may do this (at the very high concentrations used in the studies) by scavenging $OH^\bullet$ to form a much less reactive hydroxyethyl radical:

$$CH_3CH_2OH + OH^\bullet \rightarrow CH_3^\bullet CHOH + H_2O$$

Indeed, this reaction is made use of in spin-trapping experiments with DMPO (Chapter 2).

Ethanol metabolism and CYP2E1[80]

In animals, most ethanol is metabolized in the liver by **alcohol dehydrogenase** enzymes to form the aldehyde ethanal (**acetaldehyde**)

$$CH_3CH_2OH + NAD^+ \rightarrow NADH + H^+ + CH_3CHO$$

Ethanal is more toxic than ethanol and must be rapidly metabolized by an **aldehyde dehydrogenase** in mitochondria. A mutation in the gene encoding this enzyme results in a less active protein and subjects with this inborn error are less tolerant of alcohol, showing a **flush reaction** on alcohol consumption.

$$CH_3CHO + NAD^+ \rightarrow NADH + \underset{\substack{\text{acetic} \\ \text{(ethanoic) acid}}}{CH_3COOH} + H^+$$

Smaller quantities of ethanol can be oxidized by the peroxidatic action of catalase in peroxisomes (Chapter 3) and by certain cytochromes P450. The latter activity was originally identified as a **microsomal ethanol oxidizing system** (MEOS). It had been observed that liver microsomes can oxidize ethanol (and some other alcohols, including butanol) and that rates of oxidation were increased in microsomes from animals chronically pre-treated with ethanol. We now know that repeated exposure to ethanol induces synthesis of **CYP2E1**, the **ethanol-inducible cytochrome P450**. As well as oxidizing ethanol to acetaldehyde, CYP2E1 oxidizes many other substrates, including acetone, butanol, pentanol, aniline, paracetamol, benzene, pentane, halothane, isoflurane, chloroform, cocaine and CCl_4. Several xenobiotics that are metabolized by P450 to more-reactive products can thus exert toxic effects at lower levels than usual in chronic alcohol consumers: examples are paracetamol (Section 8.9 below) and cocaine.

The MEOS system has a high K_m for ethanol (8–10 mM) compared with 0.2–2 mM for alcohol dehydrogenase. However, CYP2E1 is 'leakier' than other P450 isoforms, i.e. during its operation more reaction intermediates release $O_2^{\bullet-}$ and H_2O_2 (Chapter 1). Indeed, induction of CYP2E1 has been proposed as one source of oxidative stress in alcoholics. Hydroxyethyl free radicals[62] appear to be generated during metabolism of ethanol by CYP2EI and could form adducts with proteins: antibodies against such adducts have been claimed to be present in blood sera from patients with alcoholic liver disease. Using POBN, hydroxyethyl radicals have been spin-trapped in both the liver and pancreas of animals given large doses of ethanol.[62]

Ethanol and oxidative stress[80,105]

Ethanol is remarkably non-toxic at millimolar levels, and small intakes may even be beneficial to the cardiovascular system (Chapter 10). Acceptable intakes have been quoted as 21 units per week in men and 14 in women (less in pregnant women). One unit (10 g) of alcohol is contained in about 1 oz of whisky, one glass of wine or half-a-pint of beer. Of course, it must be realized that a pregnant woman who drinks is exposing the foetus to equivalent levels of alcohol, since alcohol crosses the placenta readily. Prolonged intake of excessive amounts of ethanol by humans causes severe damage to many tissues, especially the liver, which may become cirrhotic. Hepatic iron overload is common in patients with alcoholic liver diseases. Alcohol abuse is also a major risk factor for pancreatitis, an inflammatory disease causing intense pain and extensive destruction of pancreatic tissue, and for brain damage. It may cause foetal damage in pregnant women.

In England, it is a criminal offence to drive a motor vehicle with a blood ethanol concentration greater than 80 mg per 100 ml, corresponding to a concentration of approximately 17.4 mM. Some heavy drinkers reach levels of 300 mg per 100 ml of blood, or 65.2 mM. These huge concentrations (by comparison with toxic levels of almost any other xenobiotic) have been shown in experimental animals to affect antioxidant protective systems. Injection of large doses of ethanol into rats has been reported to decrease the SOD activity measurable in brain homogenates by about 25%, and an even greater decrease is produced by repeated injection. Similarly, exposure of cultures of neuronal or glial cells from rats, mice, hamsters, and chicks to 100 mM ethanol *in vitro* decreased SOD activity. Whether these changes are sufficient to cause biological damage has yet to be determined.

By contrast, other scientists have reported increases in SOD activity in animals treated with ethanol for prolonged periods. CYP2E1 is present in various regions of the brain and can be induced by ethanol exposure; it may be an important route of alcohol metabolism since brain has only low levels of alcohol dehydrogenase activity.[93]

Both large doses of ethanol, and smaller doses given repeatedly, have been shown to increase lipid peroxidation in the livers of rats and baboons, as followed by accumulation of conjugated dienes or the production of ethane or F_2-isoprostanes (Chapter 5). Large doses of ethanol have also been observed to decrease the levels of GSH in liver, brain and kidney cells of animals. A fall in GSH might lead to increased lipid peroxidation, or be a consequence of it in view of the reactions catalysed by glutathione peroxidase (Chapter 3). Much ethanol toxicity (including mitochondrial damage) in liver may be due to acetaldehyde. This is mainly metabolized by aldehyde dehydrogenase in liver, but it is possible that some of it could be acted upon by the molybdenum-containing enzyme **aldehyde oxidase**, which is known to produce $O_2^{\bullet-}$. Xanthine oxidase can oxidize acetaldehyde, producing $O_2^{\bullet-}$ and H_2O_2, and the xanthine oxidase/dehydrogenase ratio has been claimed to increase in the liver of ethanol-treated rodents. However, the affinity of xanthine oxidase for ethanol is low, so that xanthine and hypoxanthine may out-compete acetaldehyde as a substrate *in vivo*. CYP2E1 may also oxidize some acetaldehyde.

Acetaldehyde undergoes reversible reactions with GSH (so decreasing GSH concentration) and also with protein $-SH$ and NH_2 groups, e.g. on plasma albumin and in mitochondrial proteins. Some of these adducts might provoke an immune response. It has been further suggested that acetaldehyde could react with dopamine and serotonin in the brain to form neurotoxic products, such as **tetrahydroisoquinolines** and **tetrahydro-β-carbolines**.[52] Other potentially toxic products generated from ethanol include **fatty acid ethyl esters**, which have been suggested to be neurotoxic. Withdrawal of ethanol from neurones 'adapted' to its presence in culture can also, paradoxically, cause damage (a situation perhaps analogous to ischaemia–reperfusion (Chapter 9)).[138]

The evidence that oxidative stress is responsible for ethanol toxicity is not compelling, although some contribution to tissue damage seems likely. For example, liver fibrosis in rats on a high fat diet infused with ethanol seemed to be associated with lipid peroxidation (perhaps because of stimulation of collagen synthesis by end-products of lipid peroxidation such as HNE and MDA) whereas the preceeding necrosis was not.[135] In this model, necrosis might lead to secondary lipid peroxidation, which then helped to induce fibrosis. By contrast, in minipigs chronically fed ethanol, MDA and HNE were detected before necrosis occurred.

Analysing the mechanisms of tissue damage by ethanol in chronic alcoholics is made more difficult by the fact that their diet is often inadequate and may lack enough vitamin E, selenium, or PUFAs.[143] Heavy drinkers are often also cigarette smokers, compounding the pathology of two different toxic agents. Iron overload in the cirrhotic liver may represent an additional mechanism of oxidative injury, depending on the ability of the extra iron to catalyse free-radical damage. The inflammation that can result from exposure of the liver and stomach to large intakes of ethanol may also involve damage by ROS/RNS generated by activated phagocytes. In the liver, this can involve resident Kupffer cells and/or neutrophils recruited from the circulation.

8.8.2 *Allyl alcohol and acrolein*

Alcohol dehydrogenase is not specific for ethanol as a substrate; other molecules it can oxidize include methanol (CH_3OH) to **formaldehyde** (HCHO) and **allyl alcohol** to **acrolein**. Administration of allyl alcohol to animals produces liver necrosis, preventable by supplying the alcohol dehydrogenase inhibitor **pyrazole**.[124] Acrolein is a toxic unsaturated aldehyde, $H_2C=CH-CHO$. Acrolein is also found in automobile exhaust gases, over-heated cooking oils, cigarette smoke (Section 8.11 below) and as a metabolite of **cyclophosphamide**, a drug used in cancer treatment. Acrolein reacts with GSH (largely by Michael addition of GSH to the double bond; Chapter 4), causing GSH depletion. It can also cause protein modification and lipid peroxidation: protection against damage can be achieved to some extent by thiol compounds or desferrioxamine. The lipid peroxidation may result from two phenomena caused by acrolein; GSH depletion and release of iron from cellular iron stores.[39]

8.9   Paracetamol (acetaminophen)

Paracetamol (called **acetaminophen** in the USA) is a mild painkiller that has found increasing use as a substitute for aspirin. Unlike aspirin, it does not irritate the stomach and appears safe in the recommended dosage. At high doses, however, paracetamol is toxic to both the liver and kidneys, and poisoning by paracetamol overdosage in suicide attempts is becoming increasingly common. Many patients requiring liver transplants have suffered paracetamol-induced fulminant hepatic failure. The paracetamol derivative, **phenacetin**, was originally introduced as a painkiller, but it is no longer marketed because of the risk of kidney damage. Phenacetin is converted into paracetamol *in vivo* by removal of the ethyl group.

Paracetamol can be metabolized by sulphation and glucuronidation. It is also a substrate for several cytochrome P450 isoforms, especially CYP2E1 (Section 8.8 above). Indeed, transgenic 'knockout' mice lacking CYP2E1 are much less sensitive to paracetamol.[77] The action of P450 on paracetamol produces a reactive **quinoneimine** (Fig. 8.10), which can attack proteins by combining with −SH groups. It also causes rapid depletion of GSH; this reaction occurs both spontaneously and is also catalysed by GSH transferase enzymes. Indeed, conjugation with GSH and subsequent mercapturic acid formation help detoxify paracetamol. Overdosage, resulting in severe GSH depletion, allows the quinoneimine to attack proteins, including the Ca^{2+}-ATPase of the endoplasmic reticulum, the Ca^{2+}-export system of the plasma membrane, and mitochondrial proteins. Rises in intracellular free Ca^{2+} result from this damage and contribute to hepatocyte death.[94]

Protection against the hepatotoxicity of paracetamol is afforded by administering sulphur-containing compounds such as **N-acetylcysteine** or methionine. They act by being metabolized to cysteine and maintaining intracellular GSH concentrations. Additional protection occurs because cysteine can scavenge the quinoneimine. The increased vulnerability of alcoholics to paracetamol does not occur when alcohol is in the body (since ethanol competes with paracetamol for metabolism by CYP2E1) but when alcohol is temporarily avoided yet the CYP2E1 is still present.

Some evidence for paracetamol-induced lipid peroxidation in the livers of animals has been reported, but there is no compelling evidence that lipid

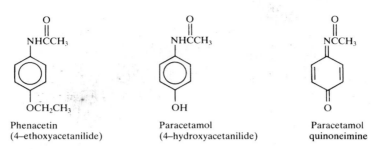

Phenacetin
(4–ethoxyacetanilide)

Paracetamol
(4–hydroxyacetanilide)

Paracetamol
quinoneimine

Fig. 8.10. Structures of paracetamol, phenacetin and paracetamol quinoneimine. The full name of the toxic metabolite is **N-acetyl-*p*-benzoquinoneimine**.

peroxidation is responsible for paracetamol hepatotoxicity.[5,96] For example, only small increases in F_2-isoprostanes were observed in the plasma of rats given hepatotoxic doses of paracetamol. Instead, the extensive GSH depletion and tissue injury caused by paracetamol may lead to secondary lipid peroxidation and other free-radical reactions. Rises in free Ca^{2+} can activate $NO^{\bullet}$ production, as well as the calpains that convert xanthine dehydrogenase to the oxidase.

The plasma of patients in paracetamol-induced liver failure contains iron and copper ions 'catalytic' for free-radical reactions, i.e. normal metal ion sequestration antioxidant defences (Chapter 3) break down,[36] which can lead to oxidative damage in other tissues. Hence oxidative damage by paracetamol is not a primary mechanism of toxicity, but a secondary event.

8.10   Air pollutants[8,29]

There is considerable interest in the possible deleterious effects of air pollution on health. To put this into perspective, it should be remembered that oxygen is a free-radical gas, and its appearance at high levels in the Earth's atmosphere as a result of the evolution of photosynthesis has been called 'the worst case of air pollution ever to occur on the planet' (Chapter 1). O_2 is, paradoxically, essential for aerobic life, but some of the other air pollutants have no such advantage. Let us consider three common ones.

8.10.1   Nitrogen dioxide

Damage by $NO_2^{\bullet}$

During combustion of organic material, the nitrogen atoms present react with O_2 to give both nitric oxide ($NO^{\bullet}$) and the toxic brown irritating gas nitrogen dioxide ($NO_2^{\bullet}$). Both are free radicals. $NO^{\bullet}$ forms more $NO_2^{\bullet}$, by reaction with O_2

$$2NO^{\bullet} + O_2 \rightarrow 2NO_2^{\bullet}$$

or with ozone. Unlike $NO^{\bullet}$, $NO_2^{\bullet}$ is highly reactive and its discharge into air from automobile exhausts, power plant emissions and general combustion processes has caused considerable concern.[8,29] In most large European and US cities, the main source of $NO_2^\bullet$ in outdoor air is vehicle exhaust emissions. The national ambient air quality standard (NAAQS) in the USA is an annual average of 0.053 p.p.m. $NO_2^\bullet$. This is regularly exceeded in several cities and $NO_2^\bullet$ levels in urban pollution episodes in the US have exceeded 1 p.p.m. In London in December 1991, air $NO_2^\bullet$ levels reached an all-time UK high of 0.42 p.p.m. $NO_2^\bullet$ is also found in indoor air (and can reach >1 p.p.m.). It can be produced by gas cookers, wood-burning stoves, coal fires, kerosene heaters and cigarette smoking (Section 8.11 below). Exposure to higher levels of $NO_2^\bullet$ can occur in acetylene welding, explosives manufacturing, military activities and in grain silos.

Healthy humans can tolerate short-term $NO_2^\bullet$ exposure at levels up to 4 p.p.m. without obvious acute lung injury, but asthmatic subjects might be affected by levels <0.3 p.p.m. (data are controversial). Levels above the NAAQS may pre-dispose to respiratory illnesses, especially in children, although data from different studies are conflicting. In animals, levels of $NO_2^\bullet$ in the 1 p.p.m. range can increase susceptibility to infections with such organisms as *Staphylococcus aureus*, probably by interfering with the normal defence functions of alveolar macrophages. For example, alveolar macrophages obtained from humans exposed to 2 p.p.m. $NO_2^\bullet$ for 4 h with intermittent exercise had impaired phagocytic activity and decreased $O_2^{\bullet-}$ production.

Higher levels of $NO_2^\bullet$ ($\geqslant 3$–4 p.p.m.) cause demonstrable lung damage, both by direct injury to cells (especially to type I alveolar cells) and possibly also by inactivating α_1-**antiproteinase**, sometimes called α_1-**antitrypsin**. This acute-phase glycoprotein (Section 4.16) is produced mainly in the liver and released into extracellular fluids, where it serves to inhibit several proteolytic enzymes. Among them is **elastase**, released by activated neutrophils. If elastase is not inhibited, it will hydrolyse several proteins, including elastin. Elastin is the major component of elastic fibres in the lung, which can stretch to several times their length and then rapidly return to their starting length when tension is released. This is important in allowing the correct expansion and contraction of the lung. An inborn deficiency of α_1-antiprotease in humans often causes a disease called **emphysema**, in which lung capacity is lost.

Exposure of humans to 3–4 p.p.m. $NO_2^\bullet$ has been reported to decrease α_1-antiproteinase activity in subsequently obtained lung lavage fluid, although 1.5 ppm $NO_2^\bullet$ did not. The methionine residue at position 358 on the amino-acid chain is crucial for the interaction of α_1-antiproteinase with elastase, yet this methionine is susceptible to oxidation by ROS/RNS. Similarly, $NO_2^\bullet$ may inactivate α_2-**macroglobulin**, an inhibitor of metalloproteinases released by alveolar macrophages. Loss of alveolar structures, microscopically similar to emphysema, has been observed in animals exposed to high levels of $NO_2^\bullet$ (e.g. 20 p.p.m. for 30 days), but whether air pollution levels of $NO_2^\bullet$ are a risk factor for emphysema development is uncertain.

$NO_2^\bullet$ can also attack **surfactant**, the proteolipid complex responsible for lowering surface tension in the lung to permit alveolar expansion and contraction. Nitration of surfactant proteins by $NO_2^\bullet$ (or by $ONOO^-$)[48a] impairs their function. Nitration of tyrosines in proteins involved in signal transduction could conceivably block this process by preventing tyrosine phosphorylation, a concept already discussed in connection with $ONOO^-$ (Chapter 2).

Nitrogen dioxide as a free radical[61]

$NO_2^{\bullet}$ dissolves in water to give nitric acid, which can irritate the respiratory tract by low pH.

$$2NO_2^{\bullet} + H_2O \rightarrow HNO_3 + HNO_2$$
nitric acid   nitrous acid

$$HNO_3 \rightarrow H^+ + NO_3{}^-$$

$$HNO_2 \rightleftharpoons H^+ + NO_2{}^-$$

Nitrous acid, although a weak acid, is a deaminating agent and can produce mutations, e.g. by converting cytosine to uracil, adenine to hypoxanthine and guanine to xanthine in DNA (Chapter 4). Indeed, $NO_2^{\bullet}$ is mutagenic in bacterial test systems. Reaction of $NO_2^{\bullet}$ with secondary or tertiary amines can generate carcinogenic nitrosamines. On the whole, however, there is little evidence that $NO_2^{\bullet}$ at low levels is carcinogenic in animals. $NO_2^{\bullet}$ can initiate lipid peroxidation by abstracting hydrogen atoms from PUFAs

$$LH + NO_2^{\bullet} \rightarrow L^{\bullet} + HNO_2$$

and by addition reactions

Both generate carbon-centred radicals that can react with O_2 to give peroxyl radicals that propagate peroxidation. However, the former mechanism ($H^{\bullet}$-abstraction) seems preferred[111] at low levels of $NO_2^{\bullet}$.

Increased lipid peroxidation has been observed (e.g. as increased conjugated dienes and ethane exhalation) in the lungs of animals exposed to p.p.m. levels of $NO_2^{\bullet}$ and in human body fluids exposed to $NO_2^{\bullet}$. Several antioxidants are protective. Ascorbate can react directly with $NO_2^{\bullet}$ with a rate constant $> 10^7 \, M^{-1}s^{-1}$; one-electron reduction by ascorbate would give $NO_2{}^-$. Vitamin E also reacts with $NO_2^{\bullet}$ (especially γ-tocopherol; Section 7.5.8). and can additionally inhibit lipid peroxidation induced by $NO_2^{\bullet}$. Indeed, supplementation of humans with ascorbate and vitamin E prior to $NO_2^{\bullet}$ exposure decreased the inactivation of α_1-antiproteinase *in vivo* by 4 p.p.m. inhaled $NO_2^{\bullet}$. Vitamin E deficiency in rats, or ascorbate deficiency in guinea pigs, aggravated lung damage by $NO_2^{\bullet}$. Uric acid, GSH and other thiols can also scavenge $NO_2^{\bullet}$; levels of urate and ascorbate in lung lining fluids fell rapidly in humans[69] breathing 2 p.p.m. $NO_2^{\bullet}$.

Lung antioxidants and air pollutants[29]

Levels of antioxidants in lung lining fluids differ from those in blood plasma and vary between species; Table 8.2 shows some representative values for humans. In general, GSH levels are higher than in plasma, urate and albumin levels lower, and ascorbate about the same. However, lung injury by toxic gases can allow leakage of plasma constituents into the lining fluid, replenishing ascorbate, urate and vitamin E and raising levels of albumin, perhaps thus augmenting antioxidant defences of the lung. Hence this transudation, often used as a marker of lung injury (e.g. as measurements of raised albumin levels in lung lavage fluid after pollutant exposure) may initially be a beneficial process.[51] The damage caused by inhaled noxious gases to lung cells may be directly mediated by the toxin, but it can also be mediated by products of reaction of the toxin with constituents of the lining fluids. Examples are lipid peroxides and aldehydes generated by oxidation of lipids in the fluids.[110]

$NO_2^{\bullet}$ reacts rapidly (rate constants $> 10^9 \, M^{-1} s^{-1}$) with other free radicals, e.g. in the gas phase (and probably in aqueous solution) with alkoxyl and peroxyl radicals[61]

$$RO^{\bullet} + NO_2^{\bullet} \rightarrow ROONO$$

$$RO_2^{\bullet} + NO_2^{\bullet} \rightarrow RO_2ONO$$

Table 8.2. Approximate mean values for the concentrations of non-enzymatic antioxidants in human plasma as compared with lung epithelial lining fluid

Antioxidant	Concentration (μM)	
	Plasma	Epithelial lining fluid
Ascorbic acid	40	40–100
Glutathione	1.5	100
Uric acid	300	90–250
α-Tocopherol	25	2.5
β-Carotene	0.4	0
Albumin-SH	500	70

Values are only approximate because lung lining fluid is sampled by washing out the lung (**bronchoalveolar lavage**). This causes considerable and variable dilutions, and complex calculations are needed to obtain values for the 'undiluted' fluid *in vivo*. Other antioxidants in lung lining fluids include low levels of transferrin, caeruloplasmin, EC-SOD, catalase and glutathione peroxidase (Chapter 3). Antioxidant levels differ in different parts of the respiratory tract, e.g. alveolar lining fluid has more GSH than fluid lining the upper respiratory tract. **Mucus** in the upper respiratory tract is a powerful scavenger of OH$^{\bullet}$, HOCl and probably ONOOH and may exert some antioxidant effects *in vivo*. However, excessive mucus secretion can lead to airway obstruction and increased risk of bacterial infection. Data abstracted from *Environ. Health Perspect.* **102** (Suppl. 10), 185 (1994) by courtesy of Professor Carroll Cross and the publishers.

Such reactions could modulate lipid peroxidation. As in the case of NO$^\bullet$ (Chapter 2), more work on the biological significance of nitrated lipid products is needed.

8.10.2 *Ozone*

Ozone is an irritating, colourless, powerfully oxidizing gas, poorly soluble in water. It has a characteristic pungent smell and its name is derived from the Greek *ozein,* which means 'to smell'. Unlike NO$_2^\bullet$, O$_3$ is not a free radical.

Ozone performs an essential 'antioxidant' function in the higher levels of the atmosphere by screening out UV light but it is a nuisance if formed at lower levels. It can arise by photochemical reactions between oxides of nitrogen and hydrocarbons: motor vehicle exhausts are a good source of the necessary gas mixture. Millions of people in the USA and Europe are regularly exposed to O$_3$ levels above the NAAQS of 0.12 p.p.m. (120 p.p.b.). Ozone levels as low as 0.5 p.p.m. can cause lung damage (especially loss of ciliated cells and type I alveolar cells) in a few hours. O$_3$ also induces inflammation, activating pulmonary macrophages and recruiting neutrophils to the lung; ROS/RNS produced by these cells are an additional source of oxidative stress after O$_3$ exposure.[29] Damage to macrophages can also decrease resistance to infections. Ozone irritates the eyes and can oxidize proteins (e.g. lysozyme) and lipids in tear fluids.[119]

Ozone is a powerful oxidizing agent.[110] It adds directly across double bonds in lipids to generate **ozonides**, which can decompose to cytotoxic aldehydes (Fig. 8.11). Ozone also oxidizes proteins,[9,29] attacking −SH, tyrosine, tryptophan, histidine and methionine residues among others. Both lipids and proteins (e.g. in surfactant) may be targets of direct attack by O$_3$ in the respiratory tract. In aqueous solution, O$_3$ decomposes to form some OH$^\bullet$, but this is a slow process at physiological pH (favoured at highly alkaline pH values). Nevertheless, aromatic hydroxylation studies[45,66] (Chapter 5) have detected OH$^\bullet$ in rats or humans breathing ozone. Reaction of O$_3$ with several biomolecules (NADH, NADPH, cysteine, albumin, methionine, uric acid or GSSG) has been shown[64] to produce singlet O$_2$ as a by-product, which could then cause further oxidations, as could the H$_2$O$_2$ produced when O$_3$ reacts with lipids (Fig. 8.11).

Much inhaled O$_3$ probably reacts with constituents of the respiratory tract lining fluids: ascorbate, GSH and urate (Table 8.2) seem especially effective as O$_3$ scavengers.[29] In humans the composition of nasal lining fluid is different from that of the fluid lining the rest of the respiratory tract, being relatively richer in urate (Table 8.3). Since urate reacts quickly with O$_3$ and NO$_2^\bullet$, it may act as a 'scrubber' of inhaled toxic gases in the nose. Unlike NO$_2^\bullet$, O$_3$ does not appear to directly induce lipid peroxidation, although free-radical products resulting from its reactions with PUFAs (Fig. 8.11) might be able to do so. Hence the contribution of vitamin E to protection against O$_3$ toxicity is uncertain. Vitamin E is only slightly depleted when human body fluids are exposed to O$_3$ *in vitro*.

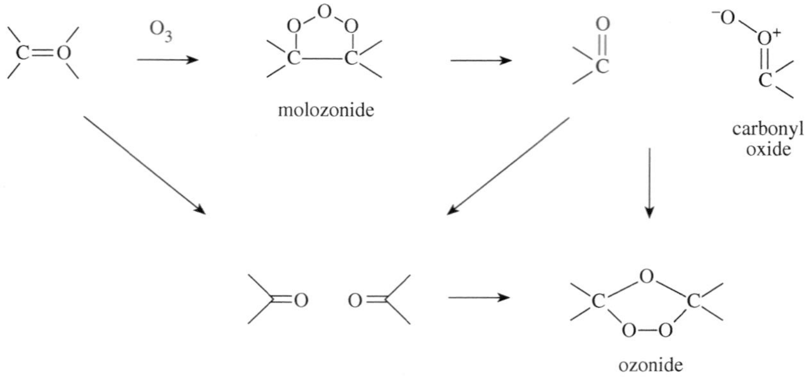

Fig. 8.11. Reaction of ozone with unsaturated compounds. Ozone reacts with PUFAs principally by addition across the double bond. The tri-oxygenated product can undergo heterolytic fission of the O–O bond (shown) or perhaps homolysis to give a diradical. The products decompose to liberate aldehydes and carbonyl oxides. In the absence of other molecules to react with, product recombination can occur to give **Criegee ozonides**, which can decompose to form radicals, carbonyl compounds and H_2O_2. For example, the C$_9$ aldehyde **nonanal** arises from ozonation of oleic acid, **heptanal** from palmitoleic acid and **hexanal** from arachidonic acid. Adapted from Foote, C.S. *et al.* (eds) (1995) *Active Oxygen in Chemistry*, Blackie, by courtesy of the publishers and Dr Chris Foote.

Table 8.3. Approximate mean concentrations of selected antioxidants in human nasal lining fluid compared with epithelial lining fluids (ELF)

Antioxidant	Concentration (µM)	
	Nasal lining fluid	ELF
Ascorbic acid	40	40–100
Uric acid	160	90
Glutathione	40	100
Albumin–SH	10	70

Data abstracted from *Environ. Health Perspect.* **102** (Suppl. 10), 185 (1994) by courtesy of Professor Carroll Cross and the publishers. The values are only approximate, for the reasons given in the footnote to Table 8.2.

Repeated daily exposure of animals to O_3 can lead to adaptation (**tolerance**), i.e. further O_3 exposures produce less damage. Increased levels of ascorbate in lung lining fluids have been suggested to contribute to adaptation in rats,[145] as have increases in lung cell SOD, catalase, glutathione peroxidase and glucose-6-phosphate dehydrogenase activities. Adaptation of rats to O_3 renders them more resistant to elevated O_2 concentrations.[63]

8.10.3 *Sulphur dioxide*

Another air pollutant found in urban areas is **sulphur dioxide** (SO_2), a colourless choking gas formed by the combustion of fuels containing sulphur (e.g. low-grade coals). Unlike O_3 and $NO_2^\bullet$, SO_2 is a reducing rather than an oxidizing agent. It contains no unpaired electrons, and thus is not a free radical. Sulphur dioxide dissolves in water to reversibly form the sulphite and bisulphite ions, also producing an acidic solution:

$$SO_2 + H_2O \rightleftharpoons \underset{\substack{\text{sulphurous}\\\text{acid}}}{H_2SO_3} \rightleftharpoons H^+ + \underset{\substack{\text{bisulphite}\\\text{ion}}}{HSO_3^-} \rightleftharpoons H^+ + \underset{\substack{\text{sulphite}\\\text{ion}}}{SO_3^{2-}}$$

In addition to inhalation of SO_2, humans are exposed to sulphite ion because it is used as a preservative in wine and in several foods, both as sulphites and metabisulphites (e.g. $K_2S_2O_5$, $Na_2S_2O_5$). Much ingested sulphite is oxidized to sulphate (SO_4^{2-}) by the mitochondrial molybdenum-containing enzyme **sulphite oxidase**, which uses cytochrome c as an electron acceptor. Sulphite is also generated during the metabolism of methionine and cysteine.

High levels of SO_3^{2-} can cleave disulphide bridges, e.g. in albumin, by the **sulphitolysis** reaction

$$R-S-S-R + SO_3^{2-} \rightarrow RS-SO_3^- + RS^-$$

Whether this occurs *in vivo* is uncertain. Sulphites can also oxidize to generate ROS and there is good evidence that $O_2^{\bullet-}$ is involved in the toxicity of SO_2 to plants (Section 7.6.3). Indeed, perhaps the earliest use of SOD to investigate a reaction mechanism was a study of the role of $O_2^{\bullet-}$ in sulphite oxidation.[87]

Horseradish peroxidase, among its many other reactions, will catalyse a one-electron oxidation of SO_3^{2-} to sulphite radical (**sulphur trioxide anion radical**), $SO_3^{\bullet-}$. Transition-metal ions can generate the same species. Sulphite has been shown to stimulate the peroxidation of linoleic and linolenic acid emulsions, microsomes and lipid extracts from rat lungs, possibly by the interaction of $SO_3^{\bullet-}$ and species derived from it with preformed lipid hydroperoxides and/or by abstraction of hydrogen atoms. Radicals such as $SO_3^{\bullet-}$, $SO_4^{\bullet-}$ (**sulphate radical**) and $SO_5^{\bullet-}$ (**peroxysulphate radical**) are formed during sulphite oxidation.[98] $SO_4^{\bullet-}$ and $SO_5^{\bullet-}$ are highly reactive; $SO_4^{\bullet-}$ can oxidize the 'OH$^\bullet$ scavengers' ethanol and formate and $SO_3^{\bullet-}$ will oxidize methionine, tryptophan and β-carotene *in vitro*. $SO_5^{\bullet-}$ is a peroxyl radical formed by reaction of $SO_3^{\bullet-}$ with O_2

$$SO_3^{\bullet-} + O_2 \rightarrow {}^-O_3SOO^\bullet \quad k > 10^9\,M^{-1}\,s^{-1}$$

It may then react with more SO_3^{2-} to continue sulphite oxidation

$$SO_3^{2-} + {}^-O_3SOO^\bullet \rightarrow SO_4^{\bullet-} + SO_4^{2-} \quad k > 10^7\,M^{-1}\,s^{-1}$$

$$SO_4^{\bullet-} + SO_3^{2-} \rightarrow SO_4^{2-} + SO_3^{\bullet-} \quad k = 2.6 \times 10^8\,M^{-1}\,s^{-1}$$

Inhalation of excess SO_2 might thus lead to lung damage by radical reactions as well as by low pH. Bisulphite can deaminate cytosine to uracil; when its mutagenicity was examined in a bacterial plasmid it was found to induce $C \rightarrow T$ and tandem $CC \rightarrow TT$ transition mutations.[24] However, bisulphite is an inefficient deaminating agent and there is no evidence that SO_2/sulphites at normal levels of exposure are carcinogenic in animals. The most prominent response to inhaled SO_2 in animals and humans is bronchoconstriction; asthmatics may be more susceptible than healthy subjects. Peroxynitrite can also oxidize SO_3^{2-} to $SO_3^{\bullet-}$ radicals.

8.10.4   *Mixtures*[76]

Polluted air contains a complex mixture of toxins, and interactions between them can aggravate toxicity. One of the best-known examples is the interaction of O_3 and $NO_2^{\bullet}$. There have been several reports that toxic effects of these two gases are synergistic rather than additive. For example, in one experiment, rats were exposed for 6 h daily to 0.2–0.8 p.p.m. O_3 or 3.6–14.4 p.p.m. $NO_2^{\bullet}$. Only minor effects were observed. Simultaneous exposure to both gases produced severe lung injury and several animals died. O_3 and $NO_2^{\bullet}$ are known to react:

$$O_3 + NO_2^{\bullet} \rightarrow O_2 + NO_3^{\bullet}$$
$$\text{nitrate radical}$$

$$NO_3^{\bullet} + NO_2^{\bullet} \rightleftharpoons N_2O_5$$

$$N_2O_5 + H_2O \rightarrow HNO_3$$

$$NO_3^{\bullet} + NO^{\bullet} \rightarrow 2NO_2^{\bullet}$$

Attention has focused on the high toxicity of **dinitrogen pentoxide**, N_2O_5, as a contributor to these synergistic effects. Interaction of oxides of nitrogen with SO_2 has also been reported, but there is no clear evidence for synergistic toxicity.

8.11   Toxicity of complex mixtures: cigarette smoke and other 'toxic smokes'

Cigarette smoking predisposes to emphysema, lung and several other cancers, atherosclerosis (increasing the risk of stroke and myocardial infarction) and many other diseases. On the other hand, smokers are at lower risk of developing Parkinson's disease and ulcerative colitis (Chapter 9).[42]

8.11.1   *Chemistry of cigarette smoke*[109]

Cigarette smoke is a hugely complex mixture of toxic agents, some of which are free radicals themselves, others redox cycling agents, others cytotoxic aldehydes and yet others carcinogens, examples being nitrosamines[102] and

Table 8.4. Some constituents of cigarette smoke

Phase	Examples of constituents
Gas	Ammonia (NH_3), carbon monoxide (CO), CO_2, $NO^\bullet$, $NO_2^\bullet$, hydrogen cyanide (HCN), volatile aldehydes (e.g. ethanal, formaldehyde, acrolein, crotonaldehyde), benzene vapour, acetone, vinyl chloride, unsaturated hydrocarbons (e.g. butadiene, isoprene)
Particulate	Tar, nicotine, metals (e.g. cadmium, lead, nickel, iron, chromium, arsenic), phenols/semiquinones/quinones, carcinogenic hydrocarbons (e.g. benzpyrene, benzanthracene, chrysene)

'Cigarette smoke consists of tarry particles suspended in a complex mixture of both organic and inorganic gases and contains >4,000 different known chemicals.' (Janoff, A *et al.* (1987) *Am. Rev. Resp. Dis.* **136**, 1058). The top of a burning cigarette reaches temperatures up to 900 °C, facilitating homolytic bond cleavage to generate radicals.

benzpyrene (Table 8.4). Smoke can be separated by the use of filters (e.g. a Cambridge glass fibre filter) into gas and particulate (sometimes called **tar**) phases, each with different chemistry. Overall, the gas phase is oxidizing and the particulate phase weakly reducing. Both contain free radicals.

The tar in smoke contains about 10^{17} ESR-detectable radicals per gram: most of them are highly stable, persisting for hours. Figure 8.12 shows typical ESR spectra. The tar contains over 3000 aromatic compounds and at least four different radical species. One prominent radical type consists of semiquinones embedded in a polymeric matrix and undergoing redox interconversions with quinones and hydroquinones (Section 8.6 above). Aqueous extracts of cigarette tar generate $O_2^{\bullet-}$ (presumably by reaction of semiquinones with O_2) and H_2O_2 and have been shown to damage isolated DNA. Inhibition of the DNA damage by the chelating agent DETAPAC suggests that $OH^\bullet$ generation is involved, presumably promoted by the metal ions present in cigarette smoke (Table 8.4) and derived from the tobacco. It has been estimated that $>1\,\mu g$ of iron is inhaled per pack of cigarettes. Lung macrophages and respiratory tract lining fluids in smokers have elevated iron contents.[131]

Gas-phase cigarette smoke radicals have much shorter lifetimes than those in the tar and have been studied using spin-trapping techniques (Chapter 4). Figure 8.12 shows a spectrum obtained with PBN. The gas phase contains over 10^{15} radicals per puff, alkoxyl, peroxyl and, to a lesser extent, carbon-centred radicals being present. The smoke can be drawn as much as 180 cm down a glass tube without a significant decrease in radical concentration, which has interesting implications for what happens in the lungs. Since the radicals have short lifetimes, they must be being constantly produced. Fresh cigarette smoke contains high concentrations of $NO^\bullet$ (~ 300–400 p.p.m.) and $NO_2^\bullet$, which can react with unsaturated hydrocarbons, such as isoprene in the smoke. Nitrates are added to cigarettes to improve burning properties and these are a major

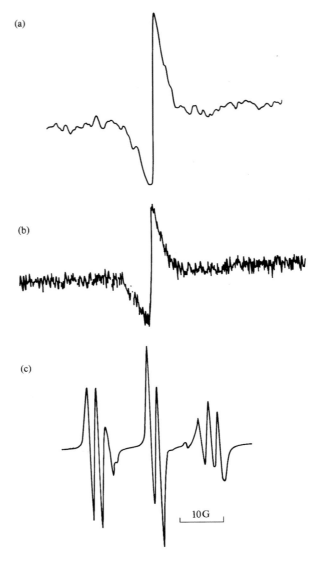

Fig. 8.12. Detection of radicals in cigarette smoke by electron spin resonance. (a) ESR signal from glass–wool filter after smoke from four cigarettes has been drawn through it. The g-value is 2.002. (b) ESR signal from a glass-fibre filter after smoke from four cigarettes has been drawn through it. The g-value of the centre of the spectrum is 2.002. (c) ESR spectrum of the spin adduct formed by the trapping of radicals in cigarette smoke with the spin-trap PBN in benzene. Spectra by courtesy of Prof. W.A. Pryor.

source of the oxides of nitrogen. Chemical reactions include

$$RH + NO_2^{\bullet} \rightarrow R^{\bullet} + HNO_2$$

$$R-CH=CH-R + NO_2^{\bullet} \quad \longrightarrow \quad R-\underset{\underset{NO_2}{|}}{CH}-\overset{\bullet}{CH}-R$$

$$R^{\bullet} + O_2 \rightarrow RO_2^{\bullet}$$

$$RO_2^{\bullet} + NO^{\bullet} \rightarrow ROONO$$

$$RO_2^{\bullet} + NO_2^{\bullet} \rightarrow ROONO_2$$

A steady-state level of radicals is quickly established. Reaction of oxides of nitrogen with $O_2^{\bullet-}$ generated from the tar phase might also occur

$$O_2^{\bullet-} + NO^{\bullet} \rightarrow \underset{\text{peroxynitrite}}{ONOO^-}$$

8.11.2   *Mechanisms of damage by cigarette smoke*[102,109]

Damage by cigarette smoke can occur by multiple mechanisms:

(1) $RO_2^{\bullet}$ and oxides of nitrogen can cause direct damage, stimulating lipid peroxidation and oxidizing DNA bases.[127] Plasma and urine from cigarette smokers show elevated levels of isoprostanes.[97] Cigarette smokers exhale more pentane[35] immediately after smoking (pentane was not detected in cigarette smoke itself in these studies). Levels of 8-OHdG are increased in the white blood cells, lung, sperm and urine of smokers.[4,43] Aqueous extracts of smoke stimulate peroxidation of LDLs *in vitro*, and exposure of human plasma to gas phase cigarette smoke also causes LDL peroxidation.[44] $NO_2^{\bullet}$ can nitrate tyrosine residues in proteins and deaminate DNA bases.[127]

(2) The aldehydes present, especially acrolein, other unsaturated aldehydes, acetaldehyde and formaldehyde, can cause GSH depletion and modify protein −SH and −NH$_2$ groups. The 'total aldehydes' generated by smoking one cigarette, if completely dissolved in the lung lining fluid, would be present at 2–3 mM.

(3) The hydroquinones/quinones in the tar phase may leach out into lung lining fluids, diffuse across cell membranes and undergo both extracellular and intracellular redox cycling to generate semiquinones, $O_2^{\bullet-}$ and H_2O_2.

(4) The carcinogens are partially absorbed and, usually after metabolism, can initiate and promote carcinogenesis.

(5) Cigarette smoke acts as an irritant to lung macrophages and may activate them to produce $O_2^{\bullet-}$, H_2O_2 and possibly $NO^{\bullet}$. It also promotes recruitment and retention of neutrophils in the lung; activation of neutrophils will generate HOCl in addition to the above species.[29] In normal lungs there are about 50–70 phagocytes per alveolus, largely macrophages;

in smokers the number of macrophages increases 2- to 4-fold and that of neutrophils 10-fold. The nicotine in smoke has been claimed to 'prime' neutrophils[47] so that the respiratory burst in response to stimuli is enhanced. However, cigarette smoke can also depress phagocyte function and increase risk of infection of the respiratory tract. In general, phagocyte recruitment/activation is favoured at lower smoking levels, but considerable inter-individual variation probably occurs.

(6) Both surfactant and α_1-antiproteinase can be inactivated by species within cigarette smoke or generated by activated phagocytes, e.g. $NO_2^\bullet$, $ROONO_2$, $ONOO^-$ and HOCl. Smoking predisposes to the development of emphysema. The most popular explanation is that smoking produces a localized inactivation of α_1-antiproteinase within the lung, allowing elastase released from neutrophils to remain active.[29] Lung lavage fluids from chronic smokers sometimes (but by no means always) show sub-normal antiprotease activity. Macrophages also contain elastase, not inhibited by α_1-antiproteinase. Hence the increased number of macrophages in the lungs of smokers, together with their activation by smoke to release enzymes, might represent an additional mechanism of elastin hydrolysis. Smoking also predisposes to bronchitis. The term **chronic obstructive pulmonary disease** (COPD) is often used to encompass both chronic bronchitis and emphysema, since the two conditions often co-exist to some extent in smokers. Cigarette smoking is the primary risk factor for COPD.

(7) Cigarette smoke not only contains iron but may also be capable of releasing iron from ferritin in lung lining fluids, possibly leading to $OH^\bullet$ formation from H_2O_2, stimulation of lipid peroxidation and faster auto-oxidation of phenols.[131] Cells damaged by smoke may also release iron (Chapter 3). The ferroxidase activity of caeruloplasmin in both plasma and lung lining fluids appears decreased in smokers, although total plasma caeruloplasmin levels are often increased. Hence the protein is present, but not functioning properly.

8.11.3   *Lung defences against cigarette smoke*[29]

The best defence against cigarette smoking is not to do it. The antioxidants in respiratory tract lining fluids offer some protection (Tables 8.2 and 8.3). For example, GSH can help remove acrolein, other aldehydes, $RO_2^\bullet$, $OH^\bullet$, HOCl, $ONOO^-$ and $NO_2^\bullet$. Mucus may scavenge some HOCl, $RO_2^\bullet$, $OH^\bullet$ and RNS, as well as binding particulates and removing them from the lung. As scavenging occurs, the mucus is degraded, but it is continuously replenished by the mucus-secreting cells of the trachea. Irritation of these cells by any mechanism causes them to increase mucus production, as is observed in smokers.

Ascorbate and urate react rapidly with $NO_2^\bullet$, and ascorbate also scavenges $RO_2^\bullet$ radicals and 'recycles' vitamin E. Indeed, peroxidation of LDLs is not observed in human blood plasma exposed to gas-phase cigarette smoke until ascorbate has been depleted.[44] Consumption of extra ascorbate by smokers decreased their F_2-isoprostane levels,[97] and administration of ascorbate to smoke-exposed hamsters prevented the adherence of neutrophils to their vascular endothelium.[78] Smokers seem to 'turn over' ascorbate faster than non-smokers and the recommended daily intake of ascorbate in both the UK and the USA is set at a higher level for smokers (e.g. 100 mg versus 60 mg in the USA). Vitamin E may also be important in protecting against smoke-induced lipid peroxidation. When equal amounts of *RRR*-α-tocopherol were fed to smokers and non-smokers, the former achieved lower blood levels, suggesting increased turnover of newly absorbed α-tocopherol in smokers.[99] An alternative explanation is decreased absorption through the gut.

Decreases in vitamins E and C and increased lipid peroxidation may help to explain increased atherosclerosis in smokers. Smokers also have decreased plasma levels of high-density lipoproteins (HDLs), which may be related to the ability of constituents of cigarette smoke (probably aldehydes) to inhibit the enzyme **lecithin–cholesterol acyltransferase**[13] (LCAT), which catalyses the formation of cholesteryl esters in HDLs (Chapter 4). Smoking also damages vascular endothelium, promotes phagocyte adhesion to endothelium, platelet aggregation, blood coagulation and acts as a vasoconstrictor.

8.11.4   *Adaptation*

There is evidence that parts of the lung antioxidant defence system can adapt to cigarette smoke. For example, activation of the AP-1 transcription factor (Chapter 4) leads to upregulation of γ-glutamylcysteine synthetase, the rate-limiting enzyme of GSH synthesis (Chapter 3).[116] Hence GSH levels in lung lining fluid and alveolar cells are often increased in smokers. Activities of repair enzymes that remove 8-OHdG also appear increased in white blood cells of smokers, although not sufficiently to prevent rises in steady-state levels of this base oxidation product.[4] Individual susceptibility to lung cancer in smokers may depend, among other factors, on the activity of the cytochromes P450 that activate carcinogens as well as upon the levels of detoxifying enzymes, such as quinone reductases and glutathione S-transferases. All of these enzymes are induced to varying extents by the 'xenobiotic exposure' of smoking.

However, it is unlikely that antioxidants can protect against all the damaging effects of smoking, especially not the actions of carcinogens such as benzypyrene. In the UK and USA, smoking is often associated with a poor diet and excess ethanol intake, compounding the effects. Exposure to both asbestos and cigarettes markedly increases the risks of developing lung cancer. The decreased risk of Parkinson's disease in smokers (seen even when corrections are made for the fact that smokers tend to die younger than non-smokers) is not fully explained. Suggested mechanisms include induction of GSH synthesis (although no data exist on whether smoking raises GSH levels in the brain) and decreases in the activity of **monoamine oxidase enzymes A and B**.[42] This would raise dopamine levels and might prevent bioactivation of neurotoxins by these enzymes (Chapter 9). However, to put this 'advantage' into perspective, one can quote the 1989 US Surgeon General's report: 'In 1985 [in the

USA] smoking accounted for 87% of lung cancer deaths, 82% of COPD deaths, 21% of coronary heart disease deaths and 18% of stroke deaths'.

8.11.5   Other tobacco usage

Cigarette smoking predisposes to cancer of the lung and upper digestive tract. Other uses of tobacco, which include snuff taking and tobacco chewing, are also associated with oral and upper digestive/respiratory tract cancer. The chewing of **betel quid**, sometimes with tobacco, is popular in certain parts of the world (e.g. Papua New Guinea, India and Taiwan) and has been associated with increased oral cancer.[134] Both carcinogenic nitrosamines and ROS can be detected in the oral cavities of subjects chewing tobacco and/or betel. Two major ingredients of betel quid, areca nut and catechu, oxidize at high pH (provided by the slaked lime, $Ca(OH)_2$, often added to the preparation chewed) to give $O_2^{\bullet-}$, H_2O_2 and $OH^{\bullet}$, especially as areca nut is rich in copper. Formation of $OH^{\bullet}$ has been demonstrated by adding phenylalanine to the betel quid and measuring formation of *ortho*- and *meta*-tyrosines (Chapter 4).

8.11.6   Fire smoke[109]

A major cause of death in persons trapped in burning buildings is smoke inhalation. It causes severe lung damage within minutes, often leading to death some days or weeks later, e.g. by adult respiratory distress syndrome (ARDS; Chapter 9). The smoke from burning buildings (which includes contributions from wood, cellulose, paint, polystyrene, polythene, other plastics and rubber) has some chemical similarities to cigarette smoke: CO, aldehydes, oxides of nitrogen, ESR-detectable free radicals, HCN (from nitrogen-containing polymers such as nylon) and particulates are all present. In addition, the combustion of halogenated plastics (such as polyvinyl chloride) can generate HCl, Cl_2 and phosgene.

One system often used to model smoke injury is the exposure of sheep to smoke from burning cotton. Reports have appeared that the resulting lung injury can be ameliorated with aerosols containing dimethylsulphoxide, or the iron chelator desferrioxamine attached to a starch polymer.[75] The use of antioxidants in the treatment of smoke-exposed patients is an area worthy of further investigation in view of the number of people worldwide who die from smoke inhalation.

8.11.7   Diesel exhaust

Diesel vehicles can emit 2–20 times more oxides of nitrogen and 30–100 times more particulates than petrol-engined cars. Diesel exhaust particles contain salts (mainly ammonium sulphate and nitrate) plus carbon particulates containing adsorbed carcinogenic hydrocarbons, nitro-aromatics and other toxins, including polyphenols/quinones that can generate ROS. Those particles with an aerodynamic diameter of 10 μm or less (called **PM10 particles**) seem to represent a particular threat because they penetrate deep into the lungs. They may generate ROS directly (including OH$^\bullet$) and also activate phagocytes, producing oxidative stress. These particles are mutagenic and carcinogenic in cell and animal models, show oxidizing ability *in vitro* and contain redox-active transition metals.[30a] Instillation of diesel exhaust particles into the lungs of animals raises levels of 8-OHdG.[101] Further research in this area is clearly warranted.

8.12 Toxicity of metals

8.12.1 *Cause or consequence?*

The possible role of oxidative damage in the pathology of iron and copper overload diseases has already been considered (Section 3.18.4), and metals within asbestos, cigarette smoke and other smokes may contribute to their toxicity. Increased rates of ROS generation have often been suggested[128] to contribute to the toxicity of high levels of several other metals, including lead, cobalt, mercury, nickel, cadmium, molybdenum, vanadium, chromium and aluminium, as well as such elements as **selenium** and **arsenic** (both in group V of the periodic table, with properties intermediate between those of metals and non-metals). For most or all of these elements, the evidence for a primary role of oxidative stress in their toxicity is not particularly convincing. For example, although increased lipid peroxidation has often been demonstrated in isolated cells exposed to metals, or in tissues from animals poisoned by metals, this peroxidation may be a consequence of tissue injury and GSH depletion caused by the metals rather than an early contributor to the metal toxicity.[128] This could then make cell injury worse, of course.

Let us briefly review, for selected metals, the data available about the role of xidative stress in the damage that they cause.

8.12.2 *Titanium*

Several transition metal ions other than iron and copper react with H_2O_2 to form OH$^\bullet$ (and possibly also oxidizing oxo-metal ion complexes). They include Cr(V), Cr^{2+}, titanium(III), certain Ni^{2+} chelates and possibly Co^{2+}, but not Mn^{2+}. Indeed, a mixture of titanium(III) and H_2O_2 has been used for decades as a laboratory source of OH$^\bullet$.

$$Ti(III) + H_2O_2 \rightarrow Ti(IV) + OH^\bullet + OH^-$$

Free radical reactions have been hypothesized to be involved in the ability of surgical titanium implants (e.g. hip replacements) to cause inflammation in surrounding tissues that may contribute to loosening of the implant.

8.12.3 *Aluminium*

Aluminium is the most abundant metal in the Earth's crust ($\sim 8\%$), and the third most abundant element (after oxygen and silicon). Animals and plants are constantly exposed to aluminium. It is slowly leached from aluminium cookware and cans, mobilized in soil by acid rain (Section 7.6), ingested from beverages such as tea and many processed foods, sprayed on skin as a constituent of deodorants, injected as a constituent of some vaccine preparations, and consumed in antacids containing the hydroxide $Al(OH)_3$.[41] It contaminates solutions used for parenteral nutrition as well as many laboratory reagents, including ATP and albumin.

Aluminium has a fixed valency of 3, so that aluminium compounds can release the hydrated Al^{3+} ion, $Al(H_2O)_6^{3+}$, in aqueous solution. Only small amounts of Al^{3+} exist in aqueous solution except at low pH; aluminium salts at pH 7.4 rapidly hydrolyse and form sparingly-soluble polyhydroxide species. Animal studies reveal that the gastrointestinal tract presents a formidable barrier to the entry of aluminium. Nevertheless, high doses of $Al(OH)_3$ administered orally to rats or to humans (as antacids) do raise tissue and urine aluminium levels. The possibility that inhaled aluminium compounds (e.g. in deodorants) might enter sensory neurones of the olfactory epithelium and pass into the brain has been raised. Some dietary organic acids, such as citrate, accelerate uptake of aluminium from the gastrointestinal tract. Absorbed Al^{3+} can bind to the iron transport protein transferrin and enter cells via the transferrin receptor. Aluminium ions can also be deposited within the iron-storage protein ferritin.

Until the early 1970s, the possible toxicity of aluminium was not considered. However, soon after the development of long-term haemodialysis therapy for patients with kidney malfunction, it was discovered that many patients maintained by dialysis develop a serious neurological syndrome, called **dialysis encephalopathy**.[41] The finding of high concentrations of aluminium in grey matter from the brains of patients dying with this disease led to the suggestion that dialysis encephalopathy is caused by aluminium intoxication. Dialysis encephalopathy is also associated with anaemia and demineralization of the bones, leading to increased risk of fracture. Subsequent work has confirmed the role of aluminium in causing these conditions and aluminium still represents a significant health hazard to dialysis patients in certain parts of the world.

Later work implicated aluminium in the pathogenesis of **Alzheimer's disease** (Chapter 9), following reports that the cores of the senile plaques characteristic of this disease are enriched in aluminium and silicon. There is no evidence that Alzheimer's disease is caused by aluminium and even the original observation of elevated aluminium has been challenged in several laboratories. The possibility remains that aluminium and other metals played some role in **Guam dementia**.[46] Until recently, the native Chamorro Indians of Guam and Rota, islands in the western Pacific, showed an elevated incidence of Parkinsonian and amyotrophic lateral sclerosis symptoms (these diseases are explained in detail in Chapter 9). This high incidence of neurodegenerative diseases has been declining since the 1950s, as the Chamorros have adopted an increasingly 'US-style' life. It has been proposed that a neurotoxin present in extracts from nuts of the false sago palm (*Cycas circinalis*), formerly a common part of the Chamorro diet but now consumed less often, could cause the disease. The neurotoxin appears to be an amino acid, **β-N-methylamino-L-alanine** (α-amino-β-methyl aminopropionic acid). It may kill neurones by an excitotoxic mechanism, binding to NMDA receptors (Chapter 9). Another proposal has been that the high levels of aluminium in the soil of Guam could contribute to neurodegeneration.

Aluminium toxicity can be a major problem for plants, especially if it is mobilized from soils or sediments by exposure to low pH. Similarly, acidification of water and subsequent Al^{3+} release can kill fish by damaging their gills.

Aluminium and free radicals

Several mechanisms have been proposed to explain the toxicity of aluminium, none supported by convincing data from *in vivo* experiments.[25,41] They include interference with Ca^{2+} uptake mechanisms, inactivation of glucose-6-phosphate dehydrogenase and inhibition of the enzyme **dihydropteridine reductase**. This enzyme catalyses the NADPH-dependent reduction of dihydrobiopterin to tetrahydrobiopterin, a cofactor required for biosynthesis of tyrosine (a precursor of dopamine, adrenalin, and noradrenalin) from phenylalanine.

Al^{3+} ions cannot stimulate lipid peroxidation or other free radical reactions, which is not surprising because of their fixed valency. However, if peroxidation in liposomes, erythrocytes, synaptosomes, myelin, or microsomes is stimulated by adding Fe^{2+} ions, the simultaneous addition of Al^{3+} increases the peroxidation rate (Table 8.5 shows illustrative data).[139] It may be that Al^{3+} ions bind to membranes and cause a subtle rearrangement of membrane lipids that aids the propagation of lipid peroxidation. This action of Al^{3+} might contribute to its neurotoxic properties, since the brain is sensitive to oxidative damage (Chapter 9). Injection of large doses of aluminium salts into animals has been claimed to increase brain TBARS levels,[106] and injection of aluminium-containing vaccines into mice caused a transient rise in brain aluminium levels.

Table 8.5. Action of Pb^{2+}, Cd^{2+} and Al^{3+} salts on peroxidation of human erythrocytes

Metal added to reaction mixture	Extent of peroxidation (A_{532})		
	Subject A	Subject B	Subject C
None (H_2O_2 only)	0.164	0.149	0.109
Al(III)	0.328	0.294	0.179
Pb^{2+}	0.219	0.367	0.159
Cd^{2+}	0.123	0.174	0.106

Erythrocytes were incubated with H_2O_2 in the presence of azide to inhibit catalase and, where indicated, 400 µM (final concentrations) of Al^{3+}, Cd^{2+} or Pb^{2+} were added. Peroxidation was measured by the TBA method. A, B and C are three different healthy adult male subjects. Metal ions in the absence of H_2O_2 did not cause any peroxidation. Data selected from Quinlan *et al.* (1988) *Biochim. Biophys. Acta* **962**, 196.

8.12.4   *Lead*[57]

Animals and plants can be exposed to lead from motor vehicle exhausts (although lead as a fuel additive has been phased out in many countries), exposure to lead-based paints (now banned in most countries), handling of lead-acid batteries and by contamination of food and drinking water. Lead has two common valencies, Pb^{2+} and $Pb(IV)$, and does not itself catalyse free-radical reactions. Nevertheless, like $Al(III)$, it can enhance Fe^{2+}-stimulated lipid peroxidation under certain circumstances (Table 8.5). Whether this contributes to lead toxicity is uncertain.

Lead is neurotoxic and also damages red blood cells, the immune system and the kidney. In children, blood lead concentrations in the micromolar range may be sufficient to impair development of the central nervous system. Mechanisms of lead toxicity include inhibition of δ-aminolaevulinic acid dehydratase (the first step in haem biosynthesis) leading to accumulation of δ-aminolaevulinic acid, an ability of Pb^{2+} to promote haemoglobin oxidation to methaemoglobin (with accelerated formation of $O_2^{\bullet-}$) and lead-dependent changes in Ca^{2+} homoeostasis, perhaps interfering with Ca^{2+}-dependent neurotransmitter release. Pb^{2+} can apparently enter cells by passing through Ca^{2+} channels. Lead combines quickly with $-SH$ groups on proteins and, at high concentrations, can cause GSH depletion. Low levels of Pb^{2+} have been shown to activate NF-κB in lymphocytes, perhaps contributing to its effects on the immune system.[112]

8.12.5   *Vanadium*[81]

The transition element vanadium exists in several oxidation states, e.g. $V(V)$, $V(IV)$, $V(III)$ and $V(II)$. Examples of each are (in decreasing order of oxidation state) sodium orthovanadate Na_3VO_4, the vanadyl ion VO^{2+}, and the vanadium chlorides, VCl_3 and VCl_2. Vanadium is used at the active site of some plant peroxidase enzymes (Chapter 3) and traces have been suggested to be essential in animal diets, apparently because it plays some (as yet undefined) role in glucose metabolism. Vanadate is a powerful inhibitor of ATPase enzymes, including the plasma membrane Na^+, K^+-ATPase. Indeed, commercial ATP samples can be contaminated with vanadium, as can preparations of albumin.[114] Vanadate inhibits phosphotyrosine phosphatases and thus potentiates phosphorylation of protein tyrosine residues in several cell types.

Vanadate is reduced to $V(IV)$ by ascorbate, GSH, other thiols and certain sugars. Vanadium (IV) can oxidize to make $O_2^{\bullet-}$

$$V(IV) + O_2 \rightleftharpoons O_2^{\bullet-} + V(V)$$

generate $OH^\bullet$ from H_2O_2

$$V(IV) + H_2O_2 \rightarrow V(V) + OH^- + OH^\bullet$$

and decompose lipid peroxides

$$LOOH + V(\text{IV}) \rightarrow V(\text{V}) + OH^- + LO^\bullet$$

In aqueous solution, vanadate catalyses the oxidation of NADH and NADPH by $O_2^{\bullet-}$ (in a way similar to that of Mn^{2+}; Chapter 2). Superoxide converts vanadate to a peroxovanadyl species

$$V(\text{V}) + O_2^{\bullet-} \rightleftharpoons V(\text{IV})-OO$$

$$V(\text{IV})-OO + NAD(P)H \rightleftharpoons V(\text{IV})-OOH + NAD(P)^\bullet$$

$$V(\text{IV})-OOH + H^+ \rightleftharpoons V(\text{V}) + H_2O_2$$

$$NAD(P)^\bullet + O_2 \rightarrow NAD(P)^+ + O_2^{\bullet-}$$

Hence vanadium salts are capable of promoting oxidative damage. Indeed, the use of vanadium compounds as inhibitors of tyrosine phosphatases in studies of signal transduction must not ignore their potential to cause oxidative stress.[33,67]

It is possible that oxidative stress contributes to the toxic effects of excess vanadium, which have been observed in miners of vanadium ores and workers in some other industries. Vanadium is used in the manufacture of sulphuric acid, steel alloys and several polymers.

8.12.6   *Molybdenum*

Molybdenum is essential to animals as a cofactor for several enzymes, including xanthine dehydrogenase, sulphite oxidase and aldehyde oxidase. Administration of **tungsten** salts to animals antagonizes molybdenum uptake and metabolism. Indeed, this is a popular technique to obtain animals lacking active xanthine dehydrogenase/oxidase for studies of the role of this enzyme in ischaemia–reperfusion (Chapter 9). Like vanadate, molybdate catalyses oxidation of $NAD(P)H$ by $O_2^{\bullet-}$ and similar mechanisms are involved.

8.12.7   *Chromium*[122]

Like vanadium, chromium has been suggested to be essential in the diet of animals in trace amounts for normal glucose metabolism. Chromium has multiple oxidation states; Cr(VI), Cr(V), Cr(III) and Cr(II). Cr(III) is the most stable. Examples of each are sodium chromate (Na_2CrO_4) and dichromate ($Na_2Cr_2O_7$), both Cr(VI) species, and the chlorides $CrCl_5$, $CrCl_3$ and $CrCl_2$. Cr(V) can apparently react with H_2O_2 to form $OH^\bullet$

$$Cr(\text{V}) + H_2O_2 \rightarrow Cr(\text{VI}) + OH^- + OH^\bullet$$

and Cr(VI) can be reduced to Cr(V) by ascorbate, GSH (forming $GS^\bullet$), $O_2^{\bullet-}$ and other cellular reducing systems. Chromates apparently enter cells via the sulphate (SO_4^{2-}) transport system. They are well-known to be toxic, carcinogenic and mutagenic in animals and in exposed human workers, e.g. in manufacture of chrome steel. It is possible that reduction to Cr(V) followed by

$OH^\bullet$ generation causes DNA damage which contributes to carcinogenicity. Another source of $OH^\bullet$ may be reaction of Cr^{2+} with H_2O_2, followed by re-reduction of Cr(III) to Cr^{2+} by cellular reductants

$$Cr^{2+} + H_2O_2 \rightarrow Cr(III) + OH^\bullet + OH^-$$

8.12.8   *Nickel*[65]

The transition element nickel (major oxidation states Ni^{2+} and Ni(III)) comprises about 0.01% of the Earth's crust. Nickel is used by several organisms as an enzyme cofactor, e.g. for urea-hydrolysing enzymes (**ureases**) in plants and bacteria and hydrogenases in many bacteria. In excess, nickel is a well-documented carcinogen in humans and other animals, especially when inhaled as insoluble particulates such as nickel subsulphide, Ni_3S_2. This can occur in miners of nickel ores and refinery workers.

Hydrated Ni^{2+} ions react only slowly, if at all, with H_2O_2 to form $OH^\bullet$, nor are they efficient at decomposing organic peroxides. However, chelation of the Ni^{2+} to certain biomolecules (including the 'antioxidants' anserine and carnosine; Section 3.21.8) alters the reduction potential to facilitate $OH^\bullet$ generation and organic peroxide decomposition

$$Ni^{2+}\text{-chelate} + H_2O_2 \rightarrow Ni(III) \text{ chelate} + OH^\bullet + OH^-$$
$$\text{(ROOH)} \qquad\qquad\qquad\qquad\qquad \text{(RO}^\bullet)$$

Nuclear protein–Ni^{2+} complexes can generate $OH^\bullet$ and incubation of chromatin with Ni^{2+} and H_2O_2 leads to extensive DNA base modifications characteristic of $OH^\bullet$ attack on DNA.[100] Such base modifications are also observed in DNA from the livers and kidneys of female rats injected with nickel. Hence it is widely thought that oxidative DNA damage is involved in nickel genotoxicity. Nickel may additionally interfere with DNA repair processes. Toxic doses of nickel also induce lipid peroxidation and protein carbonyl formation in animals.

8.12.9   *Cobalt*[65]

Cobalt (a transition metal, principal oxidation states Co^{2+} and Co(III)) is widely used in the hard-metal industry and cobalt salts are used as colouring agents for ceramics and glass. Cobalt is known to be toxic (especially to the heart) and suspected to be carcinogenic in animals when given in large quantities. However, trace amounts are needed in the diet because cobalt is a constituent of vitamin B_{12} (Chapter 7). Reaction of Co^{2+} with H_2O_2 produces a 'reactive species' that can degrade deoxyribose and hydroxylate phenol and salicylate (to 2,3- and 2,5-dihydroxybenzoates). The simplest explanation would be formation of $OH^\bullet$

$$Co^{2+} + H_2O_2 \rightarrow Co(III) + OH^\bullet + OH^-$$

but it has not proved possible to trap $OH^\bullet$ in mixtures of Co^{2+} and H_2O_2 using DMPO; instead, $O_2^{\bullet-}$ appears to be trapped.[53] However, incubation of chromatin with Co^{2+} and H_2O_2 or injection of Co^{2+} into rats led to a pattern of oxidative DNA base damage characteristic of $OH^\bullet$ attack.[100] As in the case of Ni^{2+}, it may be that Co^{2+} itself does not form $OH^\bullet$ from H_2O_2 but that certain Co^{2+} chelates can. In addition, cobalt may interfere with DNA repair processes.

8.12.10   *Mercury*[150]

Mercury is well-known to be toxic to humans and other animals. This was dramatically illustrated in 1956 by **Minamata disease**, an outbreak of poisoning by methyl mercury (CH_3Hg) ingested in fish and shellfish from the Shiranui Sea around Kyushu Island, Japan. The marine life had been contaminated by high levels of mercury discharged in waste water from a chemical plant. Fish and shellfish died, as did local cats and many people: pathological findings included damage to the brain cortex and cerebellum and peripheral nerve destruction. There was a significant increase in the birth of children with cerebral palsy. Organic mercurials such as CH_3Hg are lipophilic and cross the blood–brain barrier.

The mechanism of CH_3Hg neurotoxicity is unknown, but there are several suggestions that oxidative stress is involved. Hg^{2+} binds avidly to $-SH$ groups and can, under certain circumstances, accelerate Fe^{2+}-induced lipid peroxidation. Oxidative stress may occur as a result of mitochondrial damage induced in the brain by CH_3Hg. Low concentrations of Hg^{2+} ($\leqslant 10^{-9}$ M) have been reported[28] to stimulate $O_2^{\bullet-}$ production by human neutrophils, although higher concentrations diminish it by causing injury to the cells. $HgCl_2$ can displace Cu^+ ions from metallothionein *in vitro*, which might perhaps potentiate oxidative damage if it occurred *in vivo*.

8.12.11   *Cadmium*

Cadmium is a highly toxic metal. In Japan, **Itai–itai disease** (severe cadmium poisoning), was observed when cadmium was discharged from a mine into a river used to supply drinking water. As with mercury, there are several suggestions (but not a great deal of hard data) that oxidative stress plays a role in cadmium toxicity.[149] The divalent ion, Cd^{2+}, binds to $-SH$ groups. Administration of cadmium to animals rapidly induces metallothionein synthesis (Chapter 3). Yeast strains deficient in SOD were hypersensitive to Cd^{2+} toxicity, and cadmium raised levels of 8-OHdG in a lymphocyte cell line.[90]

8.12.12   *Arsenic*[34]

Arsenic is widespread in the environment, e.g. in soil, seawater and river water, fortunately usually at concentrations well below 0.1 p.p.m. However, higher arsenic levels occur in drinking water in some areas of Taiwan, India and Argentina. Its two oxidation states are As(III) and As(V).

Since its discovery in the Middle Ages, arsenic (usually as **arsenic trioxide**, As_2O_3) became one of the most popular poisons for disposing of rodents as well as unwanted rivals and spouses. The last murder conviction for poisoning by arsenic in the UK was in 1958. A dose of 200–250 mg is usually lethal, but tolerance can occur upon repeated exposure to lower doses, possibly involving metallothionein induction. This has caused problems to poisoners throughout history.

Arsenic in high doses is carcinogenic and p.p.m. levels may predispose to atherosclerosis. Arsenic compounds bind to −SH groups and can inhibit several enzymes, including glutathione reductase. Indeed the antidote developed to one of the first toxic gases used in chemical warfare (the arsenical **lewisite**) is a dithiol, **British anti-lewisite** (2,3-dimercapto-1-propanol). Arsenate (AsO_3^-) can be taken up by cells in mistake for phosphate and affects signal transduction systems. For example, in pig aortic vascular endothelial cells, micromolar levels of arsenite (AsO_2^-) caused increased intracellular ROS formation (as monitored by DCFH; Chapter 5) and activation of NF-κB.[7] Arsenite also interferes with DNA-repair processes.[54]

8.13   Antibiotics

Antibiotics are substances produced by living organisms that, at low concentrations, halt the growth of (or kill) fungi, bacteria or other microorganisms. Antibiotics have a wide range of chemical structures and many can participate in the formation of ROS, at least *in vitro*. For example, the β-**lactam antibiotics** (**penicillins** and **cephalosporins**) oxidatively damage DNA and deoxyribose in the presence of iron and copper salts,[113] apparently by producing OH•. The lipophilic polypeptide **polymyxin** antibiotics, and **bacitracin**, can stimulate lipid peroxidation *in vitro*.

8.13.1   *Peroxidation of antibiotics*[50]

The group of antifungal antibiotics known as **polyenes** (since they contain numerous double bonds) bind to sterols in the membranes of fungal cells. This group includes **amphotericin, candicidin, natamycin** and **nystatin**. Their polyunsaturated structures give them a propensity to oxidize with the formation of peroxides, alkoxyl/peroxyl radicals and cytotoxic aldehydes. It has been suggested that their toxic effects involve not only increases in membrane permeability due to binding to sterols, but also oxidative damage. Oxidation causes polyene antibiotics to deteriorate on storage.

8.13.2   *Tetracyclines as pro- and anti-oxidants*

The **tetracyclines** (Fig. 8.13) act by inhibiting ribosomal function in bacteria, but they are also sensitizers[55] of singlet O_2 formation (Chapter 2). The **quinolone** antibiotics such as **sparfloxacin, ciprofloxacin** and **ofloxacin** are also photosensitizers.[141] Tetracyclines chelate calcium, aluminium, iron and

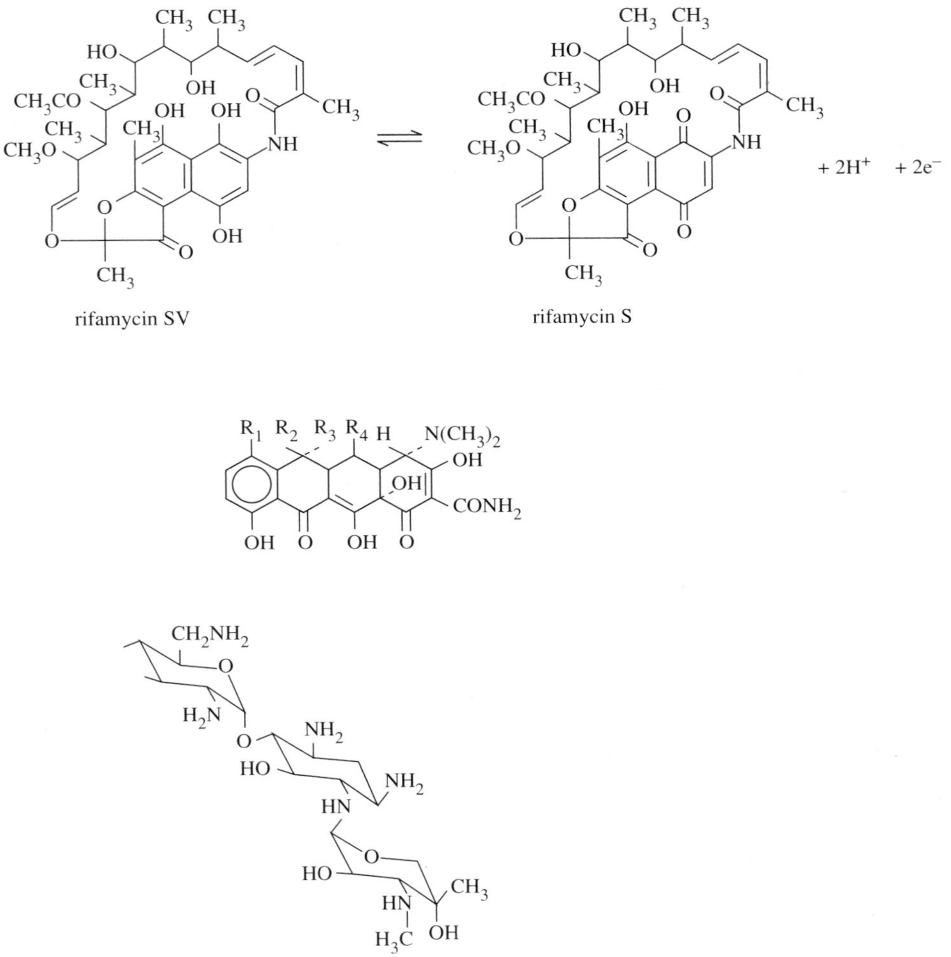

Fig. 8.13. Structures of some antibiotics. Top: the diphenol rifamycin SV and the quinone rifamycin S. Middle: the tetracyclines. Tetracycline: $R_1 = H$, $R_2 = OH$, $R_3 = CH_3$, $R_4 = H$; oxytetracycline: $R_1 = H$, $R_2 = OH$, $R_3 = CH_3$, $R_4 = OH$; chlortetracycline: $R_1 = Cl$, $R_2 = OH$, $R_3 = CH_3$, $R_4 = H$; doxycycline: $R_1 = H$, $R_2 = H$, $R_3 = CH_3$, $R_4 = OH$. Bottom: structure of gentamicin C1a.

copper ions, and the chelates with the last two metals can generate OH$^\bullet$ from H_2O_2. On the other hand, tetracyclines are powerful scavengers of hypochlorous acid and ONOO$^-$,[22,144] and there are several reports that they decrease ROS production by phagocytes and suppress induction of iNOS.[2] Hence their effects at sites of inflammation will be complex. Indeed, it must never be assumed that an effect exerted by any antibiotic *in vivo* is necessarily mediated by anti-bacterial action. Any oxidative stress induced by antibiotics could contribute to bacterial killing (sometimes called a **phago-mimetic** effect) but it might also damage host tissues.

8.13.3   *Quinone antibiotics*

Many anticancer antibiotics are quinones that undergo redox cycling to produce $O_2^{\bullet-}$, semiquinones and H_2O_2 *in vivo* (Chapter 9). An example is **streptonigrin**, which induces MnSOD in *E. coli*. **Rifamycin SV** (Fig. 8.13), an antibiotic that has been used to treat tuberculosis, oxidizes in the presence of transition metal ions to give a quinone (called **rifamycin S**), with intermediate formation of a semiquinone, and $O_2^{\bullet-}$

$$\underset{\text{rifamycin SV}}{QH_2} + O_2 \overset{\text{metal ions}}{\rightleftharpoons} \underset{\text{semiquinone}}{QH^{\bullet}} + O_2^{\bullet-} + H^+$$

$$QH^{\bullet} + O_2 \rightleftharpoons Q + O_2^{\bullet-} + H^+$$

$$QH^{\bullet} + \text{metal ion} \rightarrow Q + H^+ + \text{reduced metal ion}$$

The bactericidal action of rifamycin SV may involve increased ROS generation within the bacteria,[72] in addition to prevention of bacterial RNA synthesis.

8.13.4   *Aminoglycoside nephrotoxicity and ototoxicity*

Free-radical reactions have been suggested to be involved in the side-effects produced by several antibiotics. Thus high doses of aminoglycosides such as **gentamicin** (Fig. 8.13) can damage the kidneys (especially renal mitochondria), accompanied by increased lipid peroxidation. The cephalosporin antibiotic **cephaloridine** can have a similar renal toxic effect. **Tobramycin** and gentamicin react with HOCl *in vitro* to form chloramines that can still oxidize −SH groups. Again *in vitro*, it has been shown that gentamicin chelates iron ions and the chelate is a potent catalyst of lipid peroxidation.[136]

Whether lipid peroxidation is an early, causative, stage in aminoglycoside-induced renal injury or merely a consequence of tissue injury is as yet uncertain.[117] For example, vitamin E administration to rats depressed gentamicin-induced lipid peroxidation in the kidney, but did not prevent the kidney damage. However, desferrioxamine has been reported to diminish gentamicin-induced kidney damage in rats. It has been suggested that damage to renal mitochondria by gentamicin leads to iron ion release and formation of a redox-active iron–gentamicin complex.[136]

Iron chelators also decreased ear damage (**ototoxicity**) by high doses of gentamicin in guinea pigs. Ototoxicity of gentamicin appears more severe in malnourished animals, possibly due to lowered GSH levels.[126] Damage to the inner ear by excessive noise and ultrasonic sound have also been suggested to involve ROS.[115]

8.14   Nitro and azo compounds[11]

Several compounds containing nitro ($-NO_2$) groups are used therapeutically, e.g. in cancer treatment (Section 8.16.3 below) and as vasodilators (e.g. nitroglycerine). However, nitro-compounds can be toxic to animals and plants. Toxicity can occur from overdoses or side-effects of therapeutic nitro-compounds, or from over-exposure to nitro-compounds used in industry, such as 2-nitropropane (Chapter 3) and trinitrotoluene. The manufacture and military use of explosives have provided many cases of toxicity induced by nitro-compounds.

8.14.1   *Nitro radicals and redox cycling*

To take one example, the antibacterial drug **nitrofurantoin** (Fig. 8.14), sometimes used to treat infections of the urinary tract, can produce lung damage as a side-effect. It has been suggested that its toxicity to bacteria and to lung involves redox cycling. The drug is reduced (e.g. by NADPH–cytochrome P450 reductase in lung endoplasmic reticulum):

$$RNO_2 \xrightarrow[\text{electron}]{1e^-} RNO_2^{\bullet -}$$

followed by re-oxidation of the resulting nitro-radical to produce $O_2^{\bullet -}$

$$RNO_2^{\bullet -} + O_2 \rightarrow RNO_2 + O_2^{\bullet -}$$

Consistent with this mechanism, feeding chicks on selenium-deficient diets to decrease glutathione peroxidase activity potentiates the toxicity of nitrofurantoin. Nitrofurantoin is one of the agents that can induce haemolysis in patients with glucose-6-phosphate dehydrogenase deficiency (Chapter 7), again possibly by oxidative mechanisms. Reductions of nitro-compounds to $RNO_2^{\bullet -}$ radicals can also be achieved by mitochondria, by xanthine and aldehyde oxidases, and, in some cases, by glutathione reductase. Fish, crabs and mussels contain enzymes capable of catalysing one-electron reduction of nitro-compounds, producing $RNO_2^{\bullet -}$ and hence $O_2^{\bullet -}$. Hence oxidative stress has been implicated in the damage that can be caused to aquatic organisms by nitrocompounds in polluted waters.[58]

Similar redox-cycling reactions have been suggested to account for the toxicity of nitrofurazone (Fig. 8.14) and its derivatives to *Trypanosoma cruzi*. Attack of RNS on tyrosine produces 3-nitrotyrosine, which may also be converted to a nitro radical[73a] *in vivo*, again capable of reducing O_2 to $O_2^{\bullet -}$.

8.14.2   *Further reduction of nitro radicals*

In the absence of O_2, nitro-radicals can undergo further reduction (e.g. by GSH) or disproportionation to form nitroso (RNO) compounds:

$$2RNO_2^{\bullet -} + 2H^+ \rightarrow R-N=O + RNO_2 + H_2O$$

$$RNO_2^{\bullet -} + GSH \rightarrow R-N=O + GS^{\bullet} + OH^-$$

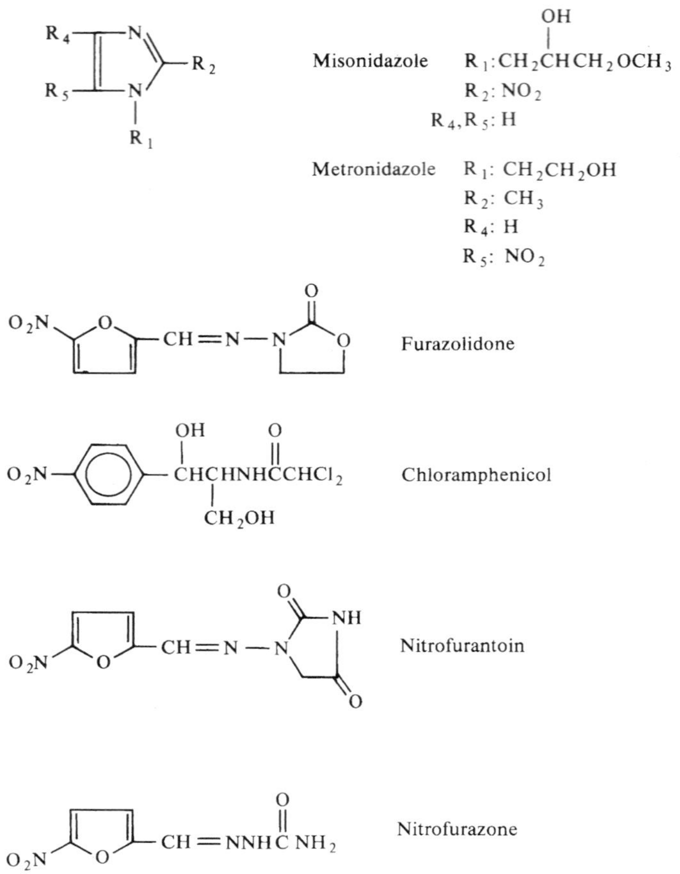

Fig. 8.14. Some nitro-compounds used in the treatment of disease. **Furazolidone** is used in veterinary medicine as an antibacterial agent: its toxicity to bacteria has been suggested to involve free-radical mechanisms. The antibiotic **chloramphenicol** is active against a wide range of bacteria, but its use is restricted since it can affect the bone marrow and sometimes produces severe and irreversible anaemia. Bone marrow damage may involve reduction of the nitro group to nitrosochloramphenicol as well as oxidation of the side-chain by cytochrome P450 to generate NHCOCOCl species that can form adducts with proteins.

RNO might then be further reduced to hydroxylamines (RNHOH) and perhaps eventually to amines, RNH_2. The extent of reduction, and which product is the primary toxin, depends upon the nitro-compound administered and the organism studied.

$$RNO_2 \xrightarrow{1e^-} RNO_2^{\bullet -} \xrightarrow[H^+]{1e^-} RNO \xrightarrow[H^+]{2e^-} R\text{-}NHOH \xrightarrow{2e^-} RNH_2$$

One-electron reduction of nitroso compounds will generate nitroxide radicals

$$RNO + e^- + H^+ \rightarrow RNHO^{\bullet}$$

which can also be obtained by one-electron oxidation of hydroxylamines

$$RNHOH \rightarrow e^- + RNHO^{\bullet} + H^+$$

Nitroxides can be reduced back to hydroxylamines by ascorbate, thiols and various enzyme systems. Nitroxide radicals can also be generated by addition of free radicals to nitrones and nitroso compounds. Indeed, this is the basis of the use of such compounds as spin traps (Chapter 5). An amino cation radical ($RNH_2^{\bullet+}$) might be generated by one-electron oxidation of RNH_2, or one-electron reduction of RNHOH.

For many nitro-compounds, the reduction products ($RNO_2^{\bullet-}$, RNO, etc.) are more cytotoxic than is $O_2^{\bullet-}$, so that the compounds cause more damage in the absence of O_2 than under aerobic conditions. For example, the nitro-compound (Fig. 8.14) **metronidazole** (Flagyl) is used to treat infections caused by anaerobic bacteria and protozoa. Two of the most important human intestinal parasites are protozoa; *Giardia lamblia* causes giardiasis and *Entamoeba histolytica* causes amoebiasis. Both are aerotolerant anaerobes.

8.14.3 *Cocaine teratogenicity*

Several teratogens have been suggested to act, in whole or in part, by imposing oxidative stress (Chapter 7). Fortunately the maternal drug detoxification systems (cytochromes P450, glucuronidation, glutathione S-transferases) and the barrier action of the placenta do an excellent job in protecting against most such potential insults. Of course, cytochromes P450s can convert a few chemicals into more reactive forms. One such substrate is cocaine.[14a]

Cocaine is metabolized in the adult liver mainly by hydrolysis by esterases (detoxification), but about 10% can be acted upon by cytochromes P450 in various tissues to give **norcocaine** (the **N-oxidation pathway**), and **N-hydroxy-norcocaine**, which can form a nitroxide radical. This can generate $O_2^{\bullet-}$ by redox cycling as well as being directly damaging to biomolecules, possibly by generating a RNO species. Cocaine and/or its metabolites also affect the foetus. For example, administration of the GSH precursor 2-oxothiazolidine-4-carboxylate or the spin trap PBN decreased the incidence of embryo toxicity in pregnant mice given cocaine. The foetotoxicity of cocaine may involve not only direct damage by the above reduction products, but also cocaine-induced vasoconstriction (itself possibly via ROS generation). Vasoconstriction can subject the embryo to ischaemia–reperfusion injury.[151]

8.14.4 *Azo compounds*

Reductive metabolism of certain **azo compounds** (which contain the –N=N– group) can generate free radicals.[95] Azo reduction can be achieved by NADPH–cytochrome P450 reductase and several other enzyme systems, including xanthine oxidase. Reduction can generate an azo anion free radical, capable of reducing O_2 to $O_2^{\bullet-}$

$$R-N=N-R' \xrightarrow{1e^-} [R-N-N-R']^{\bullet-}$$

$$[R-N-N-R']^{\bullet-} + O_2 \rightarrow R-N=N-R' + O_2^{\bullet-}$$

Further reduction of azo compounds yields hydrazines, R–NH–NH–R, whose toxicity can also involve oxidative damage (Chapter 7).

Azo compounds are widely used in the pharmaceutical, food and cosmetic industries and also as research tools. For example, **arsenazo III** is used to measure Ca^{2+} movements because it changes colour when Ca^{2+} binds to it. However, arsenazo III can be reduced by cytosolic enzymes, mitochondria and microsomes, leading to $O_2^{\bullet-}$ generation by the above reactions.[95] Oxidative stress is well known to alter Ca^{2+} homoeostasis, so that the Ca^{2+} 'probe' can perturb what it is trying to measure (Chapter 4).

8.15   3–Methylindole[15]

3-Methylindole (**skatole**) is a compound formed from the amino acid tryptophan in the rumen of cattle, by the action of anaerobic bacteria of the *Lactobacillus* group. If excess skatole is made, it can enter the circulation and cause lung damage in ruminants. This might involve metabolism of skatole by cytochromes P450 in the lung to a toxic free radical (Fig. 8.15) which can bind to macromolecules and react with GSH. Some skatole is made in the human colon, and skatole has been detected in cigarette smoke.

8.16   Radiation damage[140]

Whenever the energy of a photon exceeds the energy needed to remove an electron from a molecule, a collision with that molecule might then lead to ion formation (**ionization**). Visible and even UV light are insufficiently energetic to ionize most biomolecules, but ionization can be achieved by

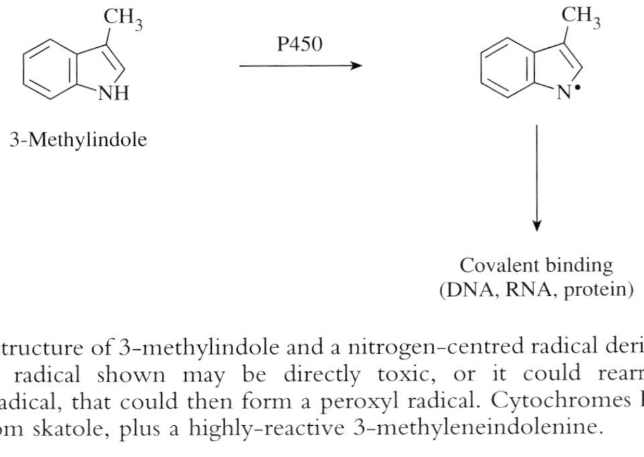

3-Methylindole

Covalent binding
(DNA, RNA, protein)

Fig. 8.15. The structure of 3-methylindole and a nitrogen-centred radical derived from it. The nitrogen-centred radical shown may be directly toxic, or it could rearrange to give a carbon-centred radical, that could then form a peroxyl radical. Cytochromes P450 also form a toxic epoxide from skatole, plus a highly-reactive 3-methyleneindolenine.

γ-rays, X-rays, high-energy electrons (β-particles), high-energy neutrons or fragments from nuclear fission, and α-particles (He^{2+} ions). These are often collectively called **ionizing radiation**.

When ionizing radiation passes through a biological system, energy is not deposited uniformly along its path but in small packages: these **spurs** have different sizes and may contain one (or several) excited molecules, ions and electrons. For example, in water exposed to γ-rays there is both ionization and excitation

$$H_2O^{\bullet+} + H_2O \rightarrow H_3O^+ + OH^{\bullet}$$

$$e^- + nH_2O \rightarrow e^-_{(aq)}$$

$$H_2O\star \rightarrow H^{\bullet} + OH^{\bullet}$$

The resulting products (Chapter 2) can recombine with each other within a spur or diffuse into the bulk water. A significant part of the initial damage done to cells by ionizing radiation is due to formation of $OH^{\bullet}$, which reacts with almost all cellular components to produce organic radicals (Chapter 2). DNA is a particularly important target, suffering double- and single-strand breaks, deoxyribose damage and base modification (Chapter 4). In patients undergoing radiotherapy, increased levels of 8-hydroxyguanine in lymphocyte DNA and in urine have been observed.[12]

Radiation dose is measured in **rads** or **Grays** (1 Gy = 100 rad). One rad is 100 ergs of energy absorbed by each 1 g of tissue at the site being examined. At whole-body doses of 100 rad, acute radiation sickness to humans results, manifested as nausea, vomiting, fatigue, fever, diarrhoea and other symptoms. The LD_{50} is about 400 rad in humans. Lower-level exposures pre-dispose to cataract and some forms of cancer.

8.16.1   *The oxygen effect*

It has been known for many years that the damaging effects of ionizing radiation on cells are aggravated by the presence of O_2, by a factor (**the oxygen enhancement ratio**) of 2- to 3.5-fold (Chapter 1). This **oxygen effect** can involve several mechanisms.

Attack of $OH^{\bullet}$ and other reactive radicals upon biomolecules often proceeds by $H^{\bullet}$ abstraction. The resulting radicals can sometimes be 'repaired' by reaction with GSH or possibly ascorbic acid. If $R^{\bullet}$ is used to denote such radicals, the 'repair' can be written as

$$R^{\bullet} + GSH \rightarrow RH + GS^{\bullet}$$

It was proposed many years ago by radiobiologists that O_2 'fixes' the damage by forming other radicals that cannot be repaired, e.g.

$$R^{\bullet} + O_2 \rightarrow RO_2^{\bullet} \text{ (peroxyl radical)}$$

$$RO_2^{\bullet} + GSH \rightarrow RO_2H + GS^{\bullet}$$
$$\text{(oxidized}$$
$$\text{biomolecule)}$$

Oxygen can also prevent recombination of oxidized and reduced radicals (which might sometimes recombine anoxically to 'cancel each other out').

GSH is known to play a role in protecting cells against ionizing radiation, e.g. treatment with buthionine sulphoximine to inhibit GSH synthesis (Chapter 3) increases radiosensitivity. Indeed, GSH, GSH precursors and other thiol compounds (such as **cysteamine**, $HSCH_2CH_2NH_2$ and mercaptopropionylglycine),

$$
\begin{array}{c}
SH \\
| \\
CH_3CHCH_2NHCHCOOH
\end{array}
$$

have been used as radioprotectors.

However, it seems unlikely that a single mechanism accounts for the radioprotective action of thiols, and the possible biological actions of sulphur-centred radicals produced by loss of $H^\bullet$ from thiols (Chapter 2) must not be ignored. GSH may be radioprotective not only because of 'repair', but also by providing a substrate for glutathione peroxidase and possibly by scavenging $OH^\bullet$ radicals directly. Thiols can react with $RO_2^\bullet$ radicals, preventing them from propagating chain reactions.

8.16.2   *The role of superoxide*

In the presence of O_2, the hydrated electrons formed by ionizing radiation can reduce it to $O_2^{\bullet-}$ (Chapter 2). In addition (although quantitatively less important), some peroxyl radicals formed by attack of $OH^\bullet$ on biomolecules can decompose to give $O_2^{\bullet-}$ e.g. α-hydroxyalkylperoxyl radicals:

$$
\begin{array}{c}
R^1 \\
| \\
R-C-O_2^\bullet \\
| \\
OH
\end{array}
\longrightarrow
\begin{array}{c}
O \\
|| \\
R-C-R'
\end{array}
+ \; H^+ \; + \; O_2^{\bullet-}
$$

Hydroxyalkylperoxyl radicals formed from some carbohydrates (e.g. glucose) and certain '$OH^\bullet$ scavengers' (such as ethanol, methanol and mannitol) can decompose in this way.

However produced, $O_2^{\bullet-}$ can dismute to H_2O_2, with the possibility of extra $OH^\bullet$ production, e.g. by Fenton chemistry. Consistent with this, iron loading increases the damaging effects of ionizing radiation upon macrophages, and tissue damage by radiation may lead to increased availability of transition metal ions *in vivo* (Chapter 4). Addition of SOD to the growth medium can partially protect *Acholeplasma laidlawii* or *E. coli* cells against damage by ionizing radiation, an effect also seen with several animal cell lines in culture. Treatment of animal cells with diethyldithiocarbamate to inhibit CuZnSOD increases their sensitivity to radiation. In such experiments, it is essential to remove the inhibitor completely before irradiating because diethyldithiocarbamate is a thiol compound and an efficient radioprotector. Several radioresistant bacterial strains such as *Arthrobacter radiotolerans* and *Micrococcus radiodurans* contain exceptionally high activities of SOD. Injection of CuZnSOD into mice has been reported to diminish mortality after X-irradiation, whereas injection of inactivated enzyme cannot. The effects observed depend on the time and dose of SOD administered, and on the intensity of radiation used, but SOD appears to be particularly protective towards the bone marrow. The mechanism of this radioprotective effect of SOD remains to be established. It may be an action in diminishing the inflammatory response to radiation–induced tissue damage rather than on the primary radiation damage.[34a]

Does $O_2^{\bullet-}$ contribute to oxygen enhancement of radiation damage? A mutant of *E. coli* lacking both FeSOD and MnSOD showed normal O_2 enhancement ratios, suggesting that $O_2^{\bullet-}$ does not contribute significantly in this organism.[91a]

8.16.3   *Hypoxic-cell sensitizers*[11]

The fact that hypoxia enhances resistance to radiation is relevant to the treatment of cancerous tumours by radiotherapy. As tumours grow, areas within them may no longer receive an adequate blood supply and thus becomes deprived of O_2. Whilst radiation treatment may destroy most of the tumour, the hypoxic cells are more resistant, and can serve as 'nuclei' for subsequent regrowth. There has therefore been some interest in therapies combining increased O_2 exposure with radiation (Chapter 1), and in various drugs that, like O_2, make hypoxic cells more sensitive to ionizing radiation. Such drugs are collectively known as **hypoxic–cell sensitizers**.

In 1973, it was proposed that the antibacterial agent metronidazole (Fig. 8.14) might be useful as a hypoxic-cell sensitizer. Exposure of *E. coli*, plant tissues, and animal cells in culture to metronidazole was observed to increase their susceptibility to ionizing radiation under anaerobic conditions, and experiments on tumour-bearing animals confirmed this effect *in vivo*. Several related compounds have been introduced for chemotherapy of bacterial and protozoal infections, although Flagyl is still very popular. By contrast, mis-onidazole (Fig. 8.14) and other sensitizers seemed more promising than metronidazole as adjuncts to radiotherapy. Unfortunately, a side-effect of misonidazole—damage to peripheral nerves—limits the amounts that can safely be given to patients. Research has continued to find less toxic sensitizers and/or drugs which minimize the neurotoxicity. More recently, however, the focus has shifted to find drugs that are directly toxic to hypoxic cells, including some of the anti-tumour antibiotics such as mitomycin C (Chapter 9).

As discussed above, one reason for the 'oxygen effect' is prevention of the 'repair' of organic radicals.

Metronidazole ($MN-NO_2$) and other hypoxic cell sensitizers might act in a similar way to O_2, forming a nitro radical adduct

$$R^{\bullet} + MN-NO_2 \longrightarrow MN-\overset{\bullet}{\underset{\underset{O}{\|}}{N}}-OR$$

or a radical anion

$$R^{\bullet} + MN\text{-}NO_2 \rightarrow MN\text{-}NO_2^{\bullet -} + R^+$$

These reactions could prevent 'repair' and the products might also injure hypoxic cells, either directly or after reduction to nitroso species or further reduction products. Nitroxide radicals have also been investigated as sensitizers

$$R^{\bullet} + \text{nitroxide}^{\bullet} \rightarrow R-\text{nitroxide}$$

8.16.4   *Food irradiation*[120]

The treatment of foodstuffs with ionizing radiation for the purposes of sterilization, killing insects or preventing germination and ripening, is slowly becoming an accepted method of food processing. Strict laws govern the types of food that may be irradiated, and the dose of radiation used. However, enforcement of these laws requires suitable methods to detect irradiated food and measure the radiation dose used.

Developing such methods is an interesting problem in free-radical biology. Irradiation involves the generation of free radicals after exposure of the food to high-energy electrons, γ-rays or X-rays and part of its action involves generation of $OH^{\bullet}$ from water in the food. Hence the techniques used to detect irradiated foods are similar to those used to detect oxidative damage in living organisms (Chapter 5). For example, most irradiated spices and some other foodstuffs give more chemiluminescence when added to alkaline luminol solutions than do unirradiated control samples. Bone and calcified cuticle (e.g. in chickens, crustacea and some fish) develop long-lived ESR signals upon irradiation because the radicals are trapped in a solid matrix. Food scientists may measure end-products of lipid peroxidation (e.g. volatile hydrocarbons, carbonyl compounds), DNA damage (thymine glycol, 8-hydroxyguanine) and oxidized amino-acid residues, (*ortho*-tyrosine, formylkynurenine, bityrosine, 2-oxohistidine). As yet, however, no single method to detect irradiation is applicable to all foods, just as there is no single 'best' method of detecting oxidative damage *in vivo*.

8.17   General conclusion

For some toxins (e.g. ionizing radiation and CCl_4), free-radical generation is the major, if not the only, mechanism by which they cause damage. For others, such as redox-cycling quinones and nitro-compounds, it may be a significant contributor. For others, it may play no role whatsoever. Always, it is important to distinguish primary from secondary events. Several toxins (e.g. metals, paracetamol) can deplete GSH and thus predispose the cell to secondary oxidative stress, which may (or may not) further contribute to the overall toxic effects. Again, we stress that the mere demonstration of oxidative stress induced by a toxin, or the trapping of a toxin-derived radical, are not evidence that ROS are important in the overall toxic effects. Such considerations apply equally to questions about the role of oxidative stress in human disease, the subject of the next chapter.

References

1. Albano, E et al. (1984) Toxicity of 1,2-dibromoethane in isolated hepatocytes: role of lipid peroxidation. *Chem.—Biol. Interac.* **50**, 255.
2. Amin, AR et al. (1997) Post-transcriptional regulation of iNOS mRNA in murine macrophages by doxycycline and chemically modified tetracyclines. *FEBS Lett.* **410**, 259.
3. Archibald, FS and Tyrce, C (1987) Manganese poisoning and the attack of Mn(III) upon catecholamines. *Arch. Biochem. Biophys.* **256**, 638.
4. Asami, S et al. (1996) Increase of a type of oxidative DNA damage, 8-hydroxyguanine and its repair activity in human leukocytes by cigarette smoking. *Cancer Res.* **56**, 2546.
5. Awad, JA and Morrow, JD (1995) Excretion of F_2-isoprostanes in bile; a novel index of hepatic lipid peroxidation. *Hepatology* **22**, 962.
6. Awad, JA et al. (1996) Demonstration of halothane-induced hepatic lipid peroxidation in rats by quantification of F_2-isoprostanes. *Anesthesiology* **84**, 910.
7. Barchowsky, A et al. (1996) Arsenic induces oxidant stress and NF-κB activation in cultured aortic endothelial cells. *Free Rad. Biol. Med.* **21**, 783.
8. Bascom, R et al. (1996) Health effects of outdoor air pollution. *Am. J. Resp. Crit. Care Med.* **153**, 477.
9. Berlett, BS et al. (1996) Comparison of the effects of O_3 on the modification of amino acid residues in glutamine synthetase and bovine serum albumin. *J. Biol. Chem.* **271**, 4177.
10. Beyer, RE et al. (1996) The role of DT-diaphorase in the maintenance of the reduced antioxidant form of coenzyme Q in membrane systems. *Proc. Natl Acad. Sci. USA*, **93**, 2528.
11. Biaglow, JE et al. (1986) Biochemistry of reduction of nitroheterocycles. *Biochem. Pharmacol.* **35**, 77.
12. Bialkowski, K et al. (1996) 8-Oxo-2'-deoxyguanosine level in lymphocyte DNA of cancer patients undergoing radiotherapy. *Cancer Lett.* **99**, 93.
13. Bielicki, JK et al. (1995) Copper and gas-phase cigarette smoke inhibit plasma LCAT activity by different mechanisms. *J. Lipid Res.* **36**, 322.
14. Bindoli, A et al. (1992) Biochemical and toxicological properties of the oxidation products of catecholamines. *Free Rad. Biol. Med.* **13**, 391.
14a. Boelsterli, UA and Goldin, C (1991) Biomechanisms of cocaine-induced hepatocyte injury mediated by the formation of reactive metabolites. *Arch. Toxicol.* **65**, 351.

15. Bray, TM and Emmerson, KS (1994) Putative mechanisms of toxicity of 3-methylindole. *Annu. Rev. Pharm. Tox.* **34**, 91.

16. Brent, JA and Rumack, BH (1993) Role of free radicals in toxic hepatic injury. II. Are free radicals the cause of toxin-induced liver injury? *Clin. Toxicol.* **31**, 173.

17. Britigan, BE *et al.* (1992) Interaction of the *P. aeruginosa* secretory products pyocyanin and pyochelin generates OH$^{\bullet}$ and causes synergistic damage to endothelial cells. *J. Clin. Invest.* **90**, 2187.

18. Brown, OR and Seither, RL (1989) Paraquat inhibits NAD biosynthesis at the quinolinic acid synthetase site. *Med. Sci. Res.* **17**, 819.

19. Bull, RJ *et al.* (1995) Water chlorination: essential process or cancer hazard? *Fund. Appl. Toxicol.*, **28**, 155.

20. Burk, RF (1995) Pathogenesis of diquat-induced liver necrosis in Se-deficient rats: assessment of the roles of lipid peroxidation and selenoprotein P. *Hepatology*, **21**, 561.

21. Burkitt, MJ *et al.* (1993) ESR spin-trapping investigation into the effects of paraquat and desferrioxamine on OH$^{\bullet}$ generation during acute iron poisoning. *Mol. Pharmacol.* **43**, 257.

22. Cantin, A and Woods, DE (1993) Protection by antibiotics against MPO-dependent cytotoxicity to lung epithelial cells *in vitro*. *J. Clin. Invest.* **91**, 38.

23. Chellman, GJ *et al.* (1986) Role of epididymal inflammation in the induction of dominant lethal mutations in Fischer 344 rat sperm by CH_3Cl. *Proc. Natl Acad. Sci. USA* **83**, 8087.

24. Chen, H and Shaw, BR (1994) Bisulfite induces tandem double CC $\rightarrow$ TT mutations in double-stranded DNA. 2. Kinetics of cytosine deamination. *Biochemistry* **33**, 4121.

25. Choo, SW and Joshi, JG (1989) Inactivation of glucose-6-phosphate dehydrogenase isozymes from human and pig brain by Al. *J. Neurochem.* **53**, 616.

26. Church, DF (1994) Spin trapping organic radicals. *Anal. Chem.* **66**, 419A.

27. Clejan, L and Cederbaum, AI (1989) Synergistic interactions between NADPH-cytochrome P-450 reductase, paraquat, and iron in the generation of active oxygen radicals. *Biochem. Pharamcol.* **38**, 1779.

28. Contrino, J *et al.* (1988) Effect of mercury on human polymorphonuclear leukocyte function *in vitro*. *Am. J. Pathol.* **132**, 110.

29. Cross, CE *et al.* (1997) General biological consequences of inhaled environmental toxicants. In *The Lung: Scientific Foundations* (Crystal, RG *et al.* eds), p. 2421. Raven Press, Philadelphia.

30. Dahlhaus, M *et al.* (1996) Oxidative DNA lesions in V79 cells mediated by pentachlorophenol metabolites. *Arch. Toxicol.* **70**, 457.

30a. Donaldson, K *et al.* (1997) Free radical activity of PM_{10}: iron-mediated generation of OH$^{\bullet}$. *Environ. Health Perspect.* **105** (suppl. 5), 1285.

31. Dybing, E *et al.* (1976) Oxidation of α-methyldopa and other catechols by cytochrome P450-generated $O_2^{\bullet-}$: possible mechanisms of methyldopa hepatitis. *Mol. Pharmacol.* **12**, 911.

32. Eizirik, DL *et al.* (1996) Potential role of environmental genotoxic agents in diabetes mellitus and neurodegenerative diseases. *Biochem. Pharmacol.* **51**, 1585.

33. Elfant, M and Keen, CL (1987) Sodium vanadate toxicity in adult and developing rats. Role of peroxidative damage. *Biol. Trace Element Res.* **14**, 193.

34. Emsley, J (1995) Whatever happened to arsenic? *New Sci.*, December issue, p. 10.

34a. Epperley, M *et al.* (1998) Prevention of late effects of irradiation lung damage by MnSOD gene therapy. *Gene Therapy* **5**, 196.

35. Euler, DE *et al.* (1996) Effect of cigarette smoking on pentane excretion in alveolar breath. *Clin. Chem.* **42**, 303.

36. Evans, PJ *et al.* (1994) Metal ions catalytic for free radical reactions in the plasma of patients with fulminant hepatic failure. *Free Rad. Res.* **20**, 139.

37. Favreau, LY and Pickett, CB (1995) The rat quinone reductase antioxidant response element. *J. Biol. Chem.* **270**, 24468.

38. Fell, JME *et al.* (1996) Manganese toxicity in children receiving long-term parenteral nutrition. *Lancet* **347**, 1218.

39. Ferrali, M. *et al.* (1990) Iron release and erythrocyte damage in allyl alcohol intoxication in mice. *Biochem. Pharmacol.* **40**, 1485.

40. Fischer-Nielsen, A *et al.* (1995) Menadione-induced DNA fragmentation without 8-oxo-2'-deoxyguanosine formation in isolated rat hepatocytes. *Biochem. Pharmacol.* **49**, 1469.

41. Fladen, TP *et al.* (1996) Status and future concerns of clinical and environmental Al toxicology. *J. Tox. Environ. Health* **48**, 527.

42. Fowler, JS *et al.* (1996) Brain monoamine oxidase A inhibition in cigarette smokers. *Proc. Natl Acad. Sci. USA* **93**, 14065.

43. Fraga, CG *et al.* (1996) Smoking and low antioxidant levels increase oxidative damage to sperm DNA. *Mutat. Res.* **351**, 199.

44. Frei, B *et al.* (1991) Gas phase oxidants of cigarette smoke induce lipid peroxidation and changes in lipoprotein properties in human blood plasma. *Biochem. J.* **277**, 133.

45. Frischer, Th. *et al.* (1997) Aromatic hydroxylation in nasal lavage fluid following ambient O_3 exposure. *Free Rad. Biol. Med.* **22**, 201.

46. Garruto, RM *et al.* (1986) Intraneuronal co-localization of Si with Ca and Al in ALS and Parkinsonism with dementia of Guam. *Lancet* **315**, 711.

47. Gillespie, MN *et al.* (1987) Enhanced chemotaxis and $O_2^{\bullet-}$ production by PMNS from nicotine-treated and smoke-exposed rats. *Toxicology* **45**, 45.

48. Grankvist, K. *et al.* (1981) CuZnSOD, MnSOD, catalase and glutathione peroxidase in pancreatic islets and other tissues in the mouse. *Biochem. J.* **199**, 393.

48a. Greis, KD *et al.* (1996) Identification of nitration sites on surfactant protein A by tandem electrospray MS. *Arch. Biochem. Biophys.* **335**, 396.

49. Gut, J *et al.* (1993) Mechanisms of halothane toxicity: novel insights. *Pharm. Ther.* **58**, 133.

50. Gutteridge, JMC *et al.* (1983) Free radical damage to polyene antifungal antibiotics: changes in biological activity and TBA reactivity. *J. Appl. Biochem.* **5**, 53.

51. Halliwell, B *et al.* (1992) Interaction of $NO_2^{\bullet}$ with human plasma. *FEBS Lett.* **313**, 62.

52. Han, QP and Dryhurst, G (1996) Influence of GSH on the oxidation of 1-methyl-6-hydroxy-1,2,3,4-tetrahydro-β-carboline: chemistry of potential relevance to the addictive and neurodegenerative consequences of ethanol abuse. *J. Med. Chem.* **39**, 1494.

53. Hanna, PM *et al.* (1992) Oxygen-derived free radical and active oxygen complex formation from cobalt(II) chelates *in vitro*. *Chem. Res. Toxicol.* **5**, 109.

54. Hartwig, A *et al.* (1997) Interaction of arsenic(III) with nucleotide excision repair in UV-irradiated human fibroblasts. *Carcinogenesis* **18**, 399.

55. Hasan, T and Khan, AU (1986) Phototoxicity of the tetracyclines: photosensitized emission of singlet delta O_2. *Proc. Natl Acad. Sci. USA* **83**, 4604.

56. Hassan, HM and Fridovich, I (1979) Paraquat and *E. coli*. Mechanism of production of extracellular $O_2^{\bullet-}$. *J. Biol. Chem.* **254**, 10846.

57. Hermes-Lima, M *et al.* (1991) Are free radicals involved in Pb poisoning? *Xenobiotica* **21**, 1085.

58. Hetherington, LH *et al.* (1996) Two- and one-electron dependent *in vitro* reductive metabolism of nitroaromatics by *Mytilus edulis*, *Carcinus maenas* and *Asterias rubens*. *Comp. Biochem. Physiol.* **113C**, 231.

59. Hoet, P *et al.* (1997) Epidemic of liver disease caused by hydrochlorofluorocarbons used as O_3-sparing substitutes of chlorofluorocarbons. *Lancet* **350**, 556.

60. Holmgren, A and Lyckeborg, C (1980) Enzymatic reduction of alloxan by thioredoxin and NADPH-thioredoxin reductase. *Proc. Natl Acad. Sci. USA* **77**, 5149.

61. Huie, RE (1994) The reaction kinetics of $NO_2^{\bullet}$. *Toxicology* **89**, 193.

62. Iimuro, Y *et al.* (1996) Detection of α-hydroxyethyl free radical adducts in the pancreas after chronic exposure to alcohol in the rat. *Mol. Pharmacol.* **50**, 656.

63. Jackson, RM and Frank, L (1984) O_3-induced tolerance to hyperoxia in rats. *Am. Rev. Resp. Dis.* **129**, 425.

64. Kanofsky, JR and Sima, PD (1993) $^1O_2$ generation at gas–liquid interfaces: a significant artifact in the measurement of $^1O_2$ yields from O_3–biomolecule interactions. *Photochem. Photobiol.* **58**, 335.

65. Kasprzak, KS (1991) The role of oxidative damage in metal carcinogenicity. *Chem. Res. Toxicol.* **4**, 604.

66. Kaur, H *et al.* (1994) Aromatic hydroxylation of phenylalanine as an assay for $OH^{\bullet}$. Measurement of $OH^{\bullet}$ from O_3 and in blood from premature babies using improved HPLC methodology. *Anal. Biochem.* **220**, 17.

67. Krejsa, CM *et al.* (1997) Role of oxidative stress in the action of vanadium phosphotyrosine phosphatase inhibitors. *J. Biol. Chem.* **272**, 11541.

68. Keeling, PL and Smith, LL (1982) Relevance of NADPH depletion and mixed disulphide formation in rat lung to the mechanism of cell damage following paraquat administration. *Biochem. Pharmacol.* **31**, 3243.

69. Kelly, FJ *et al.* (1996) Antioxidant kinetics in lung lavage fluid following exposure of humans to $NO_2^{\bullet}$. *Am. J. Resp. Crit. Care Med.* **154**, 1700.

70. Khan, MF *et al.* (1997) Oxidative stress in the splenotoxicity of aniline. *Fund. Appl. Toxicol.* **35**, 22.

71. Kitzler, J and Fridovich, I (1986) The effects of paraquat on *E. coli*: distinction between bacteriostasis and lethality. *J. Free Rad. Biol. Med.* **2**, 245.

72. Kono, Y (1982) O_2 enhancement of bactericidal activity of rifamycin SV on *E. coli* and aerobic oxidation of rifamycin SV to rifamycin S catalyzed by Mn^{2+}: the role of $O_2^{\bullet-}$. *J. Biochem.* **91**, 381.

73. Korbashi, P *et al.* (1986) Iron mediates paraquat toxicity in *E. coli*. *J. Biol. Chem.* **261**, 12472.

73a. Krainev, AG *et al.* (1998) Enzymatic reduction of 3-nitrotyrosine generates $O_2^{\bullet-}$. *Chem. Res. Toxicol.* **11**, 495.

74. Kubisch, HM *et al.* (1994) Transgenic CuZnSOD modulates susceptibility to type I diabetes. *Proc. Natl Acad. Sci. USA* **91**, 9956.

75. La Londe, C *et al.* (1994) Aerosolized deferoxamine prevents lung and systemic injury caused by smoke inhalation. *J. Appl. Physiol.* **77**, 2057.

76. Last, JA *et al.* (1994) O_3, NO and NO_2: oxidant air pollutants and more. *Environ. Health Perspect.* **102** (Suppl. 10), 179.

77. Lee, SST *et al.* (1996) Role of CYP2E1 in the hepatotoxicity of acetaminophen. *J. Biol. Chem.* **271**, 12063.

78. Lehr, HA *et al.* (1994) Vitamin C prevents cigarette smoke-induced leukocyte aggregation and adhesion to endothelium *in vivo. Proc. Natl Acad. Sci. USA* **91** 7688.

79. Li, Y *et al.* (1996) Role of CuZnSOD in xenobiotic activation. 1. Chemical reactions involved in the CuZnSOD-accelerated oxidation of the benzene metabolite, 1,4-hydroquinone. *J. Pharm. Exp. Ther.* **49**, 404.

80. Lieber, CS (1996) Role of oxidative stress and antioxidant therapy in alcoholic and non-alcoholic liver diseases. *Adv. Pharmacol.* **38**, 601.

81. Liochev, SI and Fridovich, I (1990) Vanadate-stimulated oxidation of NAD(P)H in the presence of biological membranes and other sources of $O_2^{\bullet-}$. *Arch. Biochem. Biophys.* **279**, 1.

82. Liochev, SI *et al.* (1994) NADPH: ferredoxin oxidoreductase acts as a paraquat diaphorase and is a member of the *sox*RS regulon. *Proc. Natl Acad. Sci. USA* **91**, 1328.

83. Liu, L *et al.* (1996) The study of DNA oxidative damage in benzene-exposed workers. *Mutat. Res.* **370**, 145.

84. Lungarella, G *et al.* (1987) $BrCCl_3$-induced damage to bronchiolar Clara cells. *Res. Comm. Chem. Pathol. Pharmacol.* **57**, 213.

85. Malaisse, WJ *et al.* (1982) Determinants of the selective toxicity of alloxan to the pancreatic B cell. *Proc. Natl Acad. Sci. USA* **79**, 927.

86. Marshall, AD *et al.* (1997) Control of activity through oxidative modification at the conserved residue Cys66 of aryl sulfotransferase IV. *J. Biol. Chem.* **272**, 9153.

87. McCord, JM and Fridovich, I (1969) The utility of SOD in studying free radical reactions. *J. Biol. Chem.* **244**, 6056.

88. McGregor, D and Lang, M (1996) CCl_4: genetic effects and other modes of action. *Mutat. Res.* **366**, 181.

89. Mehendale, HM *et al.* (1994) Novel mechanisms in chemically induced hepatotoxicity. *FASEB J.* **8**, 1285.

90. Mikhailova, MV *et al.* (1997) Cd-induced 8OHdG formation, DNA strand breaks and antioxidant enzyme activities in lymphoblastoid cells. *Cancer Lett.* **115**, 141.

91. Mills, EM *et al.* (1996) CN^--induced apoptosis and oxidative stress in differentiated PC12 cells. *J. Neurochem.* **67**, 1039.

91a. Misra, H and Fridovich, I (1976) SOD and the O_2 enhancement of radiation lethality. *Arch. Biochem. Biophys.* **176**, 577.

92. Monks, TJ *et al.* (1992) Quinone chemistry and toxicity. *Toxicol. Appl. Pharmacol.* **112**, 2.

93. Montoliu, C *et al.* (1995) Ethanol increases cytochrome P4502E1 and induces oxidative stress in astrocytes. *J. Neurochem.* **65**, 2561.

94. Moore, M *et al.* (1985) The toxicity of acetaminophen and *N*-acetyl-*p*-benzoquinone imine in isolated hepatocytes is associated with thiol depletion and increased cytosolic Ca^{2+}. *J. Biol. Chem.* **260**, 13035.

95. Moreno, SNJ *et al.* (1985) Reduction of the metallochromic indicators arsenoazo III and antipyrylazo III to their free radical metabolites by cytoplasmic enzymes. *FEBS Lett.* **180**, 229.

96. Morrow, JD *et al.* (1992) Formation of novel non-cyclooxygenase-derived prostanoids (F_2-isoprostanes) in CCl_4 hepatotoxicity. *J. Clin. Invest.* **90**, 2502.

97. Morrow, JD *et al.* (1995) Increase in circulating products of lipid peroxidation (F_2-isoprostanes) in smokers. *N. Engl. J. Med.* **332**, 1198.

98. Mottley, C and Mason, RP (1988) Sulfate free radical formation by the peroxidation of (bi)sulfite and its reaction with $OH^\bullet$ scavengers. *Arch. Biochem. Biophys.* **267**, 681.

99. Munro, LH *et al.* (1997) Plasma *RRR*-α-tocopherol concentrations are lower in smokers than in non-smokers after ingestion of a similar oral load of this antioxidant vitamin. *Clin. Sci.* **92**, 87.

100. Nackerdien, Z *et al.* (1991) Ni^{2+} and Co^{2+}-dependent damage by H_2O_2 to the DNA bases in isolated human chromatin. *Cancer Res.* **51**, 37.

101. Nagashima, M *et al.* (1995) Formation of an oxidative DNA damage, 8OHdG, in mouse lung DNA after intratracheal instillation of diesel exhaust particles and effects of high dietary fat and β-carotene on this process. *Carcinogenesis* **16**, 1441.

102. Nair, J *et al.* (1996) Endogenous formation of nitrosamines and oxidative DNA-damaging agents in tobacco users. *Crit. Rev. Toxicol.* **26**, 149.

103. Nakamura, M *et al.* (1997) Metal-induced $OH^\bullet$ generation by Cu^+-metal-lothioneins from LEC rat liver. *Biochem. Biophys. Res. Commun.* **231**, 549.

104. Nohl, H *et al.* (1986) Quinones in biology: functions in electron transfer and O_2 activation. *Adv. Free Rad. Biol. Med.* **2**, 211.

105. Nordmann, R (1994) Alcohol and antioxidant systems. *Alcohol and Alcoholism* **29**, 513.

106. Ohtawa, M *et al.* (1983) Effect of Al ingestion on lipid peroxidation in rats. *Chem. Pharm. Bull.* **31**, 1415.

107. Pohl, LR (1979) Biochemical toxicology of chloroform. *Rev. Biochem. Toxicol.* **1**, 79.

108. Pohl, LR (1993) An immunochemical approach of identifying and character-izing protein targets of toxic reactive metabolites. *Chem. Res. Toxicol.* **6**, 786.

109. Pryor, WA (1992) Biological effects of cigarette smoke, wood smoke and the smoke from plastics: the use of ESR. *Free Rad. Biol. Med.* **13**, 659.

110. Pryor, WA *et al.* (1996) Detection of aldehydes in bronchoalveolar lavage of rats exposed to O_3. *Fund. Appl. Toxicol.* **34**, 148.

111. Pryor, WA and Lightsey, JW (1981) Mechanisms of $NO_2^\bullet$ reactions: initiation of lipid peroxidation and the production of HNO_2. *Science* **214**, 435.

112. Pyatt, DW *et al.* (1996) Inorganic Pb activates NF-κB in primary human $CD4^+$ T lymphocytes. *Biochem. Biophys. Res. Commun.* **227**, 380.

113. Quinlan, GJ and Gutteridge, JMC (1988) Oxidative damage to DNA and deoxyribose by β-lactam antibiotics in the presence of Fe and Cu salts. *Free Rad. Res.* **5**, 149.

114. Quinlan GJ *et al.* (1992) Vanadium and copper in clinical albumin solutions and their potential to damage protein structure. *J. Pharm. Sci.* **81**, 611.

115. Quirk, WS *et al.* (1994) Lipid peroxidation inhibitor attenuates noise-induced temporary threshold shifts. *Hearing Res.* **74**, 217.

115a. Radjendirane, V *et al.* (1998) Disruption of the DT diaphorase (NQ01) gene in mice leads to increased menadione toxicity. *J. Biol. Chem.* **273**, 7382.

116. Rahman, I *et al.* (1996) Induction of γ-glutamylcysteine synthetase by cigarette smoke is associated with AP-1 in human alveolar epithelial cells. *FEBS Lett.* **396**, 21.

117. Ramsammy, LS *et al.* (1987) Failure of inhibition of lipid peroxidation by vitamin E to protect against gentamicin nephrotoxicity in the rat. *Biochem. Pharmacol.* **36**, 2125.

118. Saez, JC *et al.* (1987) CCl_4 at hepatotoxic levels blocks reversibly gap junctions between rat hepatocytes. *Science* **236**, 967.

119. Schmut, O *et al.* (1994) Destruction of human tear proteins by O_3. *Free Rad. Biol. Med.* **17**, 165.

120. Schreiber, GA *et al.* (1993) Detection of irradiated food—methods and routine applications. *Int. J. Rad. Biol.* **63**, 105.

121. Schurz, HH *et al.* (1996) Products of aniline and triglycerides in oil samples associated with the toxic oil syndrome. *Chem. Res. Toxicol.* **9**, 1001.

122. Shi, X and Dalal, NS (1990) On the $OH^\bullet$ formation in the reaction between H_2O_2 and biologically generated Cr(VI) species. *Arch. Biochem. Biophys.* **277**, 342.

123. Shu, H *et al.* (1979) Lipid peroxidation and paraquat toxicity. *Biochem. Pharmacol.* **28**, 127.

124. Silva, JM and O'Brien, PJ (1989) Allyl alcohol and acrolein-induced toxicity in isolated rat hepatocytes. *Arch. Biochem. Biophys.* **275**, 551.

125. Smith, LL and Wyatt, I (1981) The accumulation of putrescine into slices of rat lung and brain and its relationship to the accumulation of paraquat. *Biochem. Pharmacol.* **30**, 1053.

126. Song, BB and Schacht, J (1996) Variable efficacy of radical scavengers and Fe chelators to attenuate gentamicin ototoxicity in guinea pig *in vivo*. *Hearing Res.* **94**, 87.

127. Spencer, JPE *et al.* (1995) DNA damage in human respiratory tract epithelial cells: damage by gas phase cigarette smoke apparently involves attack by RNS in addition to O_2 radicals. *FEBS Lett.* **375**, 179.

128. Sugiyama, S (1994) Role of cellular antioxidants in metal-induced damage. *Cell Biol. Toxicol.* **10**, 1.

129. Tabatabaie, T *et al.* (1997) Spin trapping agent PBN protects against the onset of drug-induced insulin-dependent diabetes mellitus. *FEBS Lett.* **407**, 148.

130. Thomas, CE and Aust, SD (1986) Reductive release of iron from ferritin by cation free radicals of paraquat and other bipyridyls. *J. Biol. Chem.* **261**, 13064.

131. Thompson, AB *et al.* (1991) Lower respiratory tract iron burden is increased in association with cigarette smoking. *J. Lab. Clin. Med.* **117**, 493.

132. Tindberg, N and Ingelman-Sundberg, M (1996) Expression, catalytic activity, and inducibility of CYP2E1 in the rat central nervous system. *J. Neurochem.* **67**, 2066.

133. Tritschler, AM *et al.* (1996) Increased oxidative DNA damage in livers of 2,3,7,8-tetrachlorodibenzo-*p*-dioxin treated intact but not ovariectomized rats. *Cancer Lett.* **98**, 219.

134. Trivedi, C *et al.* (1997) Copper content in *Areca catechu* (betel nut) and oral submucous fibrosis. *Lancet* **349**, 1447.

135. Tsukamoto, H (1993) Oxidative stress, antioxidants, and alcoholic liver fibrogenesis. *Alcohol* **10**, 465.

136. Ueda, N *et al.* (1993) Gentamicin-induced mobilization of Fe from renal cortical mitochondria. *Am. J. Physiol.* **265**, F435.

137. Umemura, T *et al.* (1996) Oxidative DNA damage and cell proliferation in the livers of B6C3F1 mice exposed to pentachlorophenol in their diet. *Fund. Appl. Toxicol.* **30**, 285.

138. Vallett, M *et al.* (1997) Free radical production during ethanol intoxication, dependence, and withdrawal. *Alcohol Clin. Exp. Res.* **21**, 275.

139. Verstraeten, SV *et al.* (1997) Myelin is a preferential target for Al-mediated oxidative damage. *Arch. Biochem. Biophys.* **344**, 289.

140. von Sonntag, C (1987) *The Chemical Basis of Radiation Biology.* Taylor & Francis, London.

141. Wagai, N and Tawara, K (1992) Possible reasons for differences in phototoxic potential of 5 quinolone antibacterial agents: generation of toxic oxygen. *Free Rad. Res. Commun.* **17**, 387.

142. Waintrub, ML *et al.* (1990) Xanthine oxidase is increased and contributes to paraquat-induced acute lung injury. *J. Appl. Physiol.* **68**, 1755.

143. Ward, RJ and Peters, TJ (1992) The antioxidant status of patients with either alcohol-induced liver damage or myopathy. *Alcohol and Alcoholism* **27**, 359.

144. Whiteman, M *et al.* (1996) Protection against $ONOO^-$ dependent tyrosine nitration and α_1-AP inactivation by some anti-inflammatory drugs and by the antibiotic tetracycline. *Ann. Rheum. Dis.* **55**, 383.

145. Wiester, MJ *et al.* (1996) Adaptation to O_3 in rats and its association with ascorbic acid in the lung. *Fund. Appl. Toxicol.* **31**, 56.

146. Winterbourn, CC and Munday, R (1989) Glutathione-mediated redox cycling of alloxan. *Biochem. Pharmacol.* **38**, 271.

147. Winterbourn, CC *et al.* (1979) The reaction of menadione with haemoglobin. *Biochem. J.* **179**, 665.

148. Xu, R *et al.* (1996) Characterization of products from the reactions of N-acetyldopamine quinone with N-acetylhistidine. *Arch. Biochem. Biophys.* **329**, 56.

149. Yang, JL *et al.* (1996) ROS may participate in the mutagenicity and mutational spectrum of Cd in CHO-K1 cells. *Chem. Res. Toxicol.* **9**, 1360.

150. Yee, S and Choi, BH (1996) Oxidative stress in neurotoxic effects of methyl-mercury poisoning. *Neurotoxicology* **17**, 17.

151. Zimmerman, EF *et al.* (1994) Role of oxygen free radicals in cocaine-induced vascular disruption in mice. *Teratology* **49**, 192.

9

Free radicals, other reactive species and disease

9.1 Introduction

The biomedical literature is full of claims that 'free radicals' and other 'reactive species' are involved in different human diseases (Table 9.1).[105,278] They have been implicated in over 100 disorders, ranging from rheumatoid arthritis and haemorrhagic shock through cardiomyopathy and cystic fibrosis to gastrointestinal ischaemia, AIDS and even male pattern baldness. In 1984, we pointed out[107] that this wide range of disorders implies that reactive oxygen/nitrogen species (ROS/RNS) are not something esoteric, but that their increased formation *accompanies* tissue injury in most, if not all, human diseases. This is because tissue injury leads to oxidative stress.[105] For example, **electropulsation** is widely used to introduce modified DNA into cells in molecular biology experiments. The membrane disruption caused by electrical pulses can lead to oxidative stress.[249]

Reasons for rises in oxidative damage in injured tissues are summarized in Fig. 9.1 and include liberation of 'catalytic' transition-metal ions and haem proteins, plus damage to mitochondria causing increased leakage of electrons from the electron-transport chain.[197]

Sometimes ROS/RNS make a significant contribution to the disease pathology; at other times they may not (Fig. 9.2). A major task of researchers in this field is to tell the difference between the two. One requirement in doing so is to have specific assays of ROS/RNS, and of the damage that they can cause, which are applicable *in vivo*. The lack of such assays in the past has impeded progress in our understanding of the role played by ROS/RNS in normal physiology and in disease but things are improving quickly (Chapter 5). However, proving that ROS/RNS (or any other postulated mediator of tissue injury) are important in any disease involves much more than demonstrating their formation by trapping assays or by measuring increased products of oxidative damage (Fig. 9.3). The criteria for implicating ROS/RNS (or any other agent) as a significant contributor to tissue injury in human disease are as follows:

- the agent should always be present at the site of injury
- its time-course of formation should be consistent with the time-course of tissue injury
- direct application of the agent to the tissue at concentrations within the range found *in vivo* should reproduce most or all of the damage observed

Table 9.1. Some of the clinical conditions in which the involvement of ROS/RNS has been suggested

Category	Examples
Inflammatory/immune injury	Glomerulonephritis, vasculitis, autoimmune diseases, rheumatoid arthritis, hepatitis
Ischaemia–reflow states	Stroke, myocardial infarction/arrythmias/angina/stunning, organ transplantation, inflamed rheumatoid joint, frostbite, Dupuytren's contracture, cocaine-induced fetal damage
Drug and toxin-induced reactions	See Chapter 8
Iron overload (tissue and plasma)	Idiopathic haemochromatosis, dietary iron overload (Bantu), thalassaemia and other chronic anaemias treated with multiple blood transfusions, nutritional deficiencies (kwashiorkor), alcoholism, multi-organ failure, cardiopulmonary bypass, fulminant hepatic failure, prematurity, alcohol-related iron overload, cancer chemotherapy/radiotherapy
Radiation injury	Consequences of nuclear explosions, accidental exposure, radiotherapy or exposure to hypoxic cell sensitizers or radon gas; cataract
Ageing	Disorders of premature ageing, ageing itself, age-related diseases, e.g. cancer
Red blood cells	Phenylhydrazine, primaquine and related drugs, lead poisoning, protoporphyrin photoxidation, malaria, sickle cell anaemia, favism, Fanconi's anaemia, haemolytic anaemia of prematurity, chemotherapy
Respiratory tract	Effects of cigarette smoke, snuff inhalation, other smoke inhalation, emphysema (COPD), hyperoxia, bronchopulmonary dysplasia, exposure to air pollutants (O_3, NO_2, SO_2, diesel exhaust), ARDS, mineral dust pneumoconiosis, asbestos carcinogenicity, bleomycin toxicity, paraquat toxicity, skatole toxicity, asthma, cystic fibrosis
Heart and cardiovascular system	Alcohol cardiomyopathy, Keshan disease (selenium deficiency), atherosclerosis, anthracycline cardiotoxicity, cardiac iron overload
Kidney	Autoimmune nephrotic syndromes, aminoglycoside nephrotoxicity, heavy metal nephrotoxicity (Pb, Cd, Hg), myoglobin/haemoglobin damage, haemodialysis, transplant storage/rejection

Table 9.1. (*Continued*)

Category	Examples
Gastrointestinal tract	Betel nut-related oral cancer, liver injury caused by endotoxins or halogenated hydrocarbons (e.g. bromobenzene, CCl_4), exposure to diabetogenic agents, pancreatitis, NSAID-induced gastrointestinal tract lesions, oral iron poisoning
Brain/nervous system/ neuromuscular disorders	Hyperbaric oxygen, vitamin E deficiency, exposure to neurotoxins, Alzheimer's disease, Parkinson's disease, Huntington's chorea, stroke, neuronal ceroid lipofuscinoses, allergic encephalomyelitis, aluminium overload, sequelae of traumatic injury, muscular dystrophy, multiple sclerosis, amyotrophic lateral sclerosis, Guam dementia; may also occur during preservation of fetal dopamine-producing cells for transplantation
Eye	Cataract, ocular haemorrhage, degenerative retinal damage/macular degeneration, retinopathy of prematurity (retrolental fibroplasia), photic retinopathy, penetration of metal objects
Skin	UV radiation, thermal injury, porphyria, hypericin, exposure to other photosensitizers, contact dermatitis, baldness

Abbreviations: ARDS, adult respiratory syndrome; COPD, chronic obstructive pulmonary disease; NSAID, non-steroidal anti-inflammatory drug.
For a full compilation plus comments *see Free Rad. Res. Commun.* **19**, 141 (1993) and discussions in this chapter and Chapters 7, 8 and 10.

- removing the agent or inhibiting its formation should diminish the injury to an extent related to the degree of removal of the agent or inhibition of its formation.

9.1.1 *Origin of oxidative stress in disease*

In principle, disease-associated oxidative stress could result from either (or both) of the following:[105]

(1) Diminished antioxidants, e.g. mutations affecting antioxidant defence enzymes (such as CuZnSOD, MnSOD and glutathione peroxidase) or diseases that deplete such defences. Many xenobiotics are metabolized by conjugation with GSH; high doses can deplete GSH and cause oxidative stress even if the xenobiotic is not itself a generator of ROS or RNS

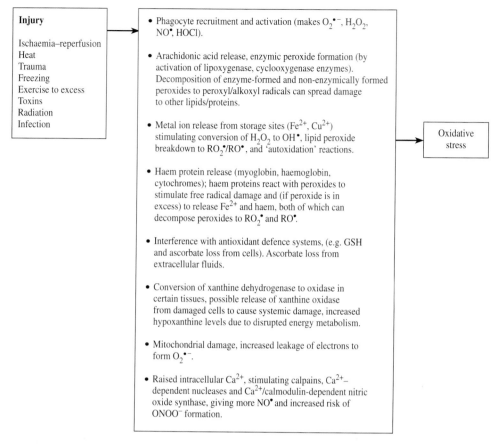

| Injury | | Oxidative stress |

Injury

Ischaemia–reperfusion
Heat
Trauma
Freezing
Exercise to excess
Toxins
Radiation
Infection

- Phagocyte recruitment and activation (makes $O_2^{\bullet-}$, H_2O_2, $NO^{\bullet}$, HOCl).

- Arachidonic acid release, enzymic peroxide formation (by activation of lipoxygenase, cyclooxygenase enzymes). Decomposition of enzyme-formed and non-enzymically formed peroxides to peroxyl/alkoxyl radicals can spread damage to other lipids/proteins.

- Metal ion release from storage sites (Fe^{2+}, Cu^{2+}) stimulating conversion of H_2O_2 to $OH^{\bullet}$, lipid peroxide breakdown to $RO_2^{\bullet}/RO^{\bullet}$, and 'autoxidation' reactions.

- Haem protein release (myoglobin, haemoglobin, cytochromes); haem proteins react with peroxides to stimulate free radical damage and (if peroxide is in excess) to release Fe^{2+} and haem, both of which can decompose peroxides to $RO_2^{\bullet}$ and $RO^{\bullet}$.

- Interference with antioxidant defence systems, (e.g. GSH and ascorbate loss from cells). Ascorbate loss from extracellular fluids.

- Conversion of xanthine dehydrogenase to oxidase in certain tissues, possible release of xanthine oxidase from damaged cells to cause systemic damage, increased hypoxanthine levels due to disrupted energy metabolism.

- Mitochondrial damage, increased leakage of electrons to form $O_2^{\bullet-}$.

- Raised intracellular Ca^{2+}, stimulating calpains, Ca^{2+}–dependent nucleases and Ca^{2+}/calmodulin-dependent nitric oxide synthase, giving more $NO^{\bullet}$ and increased risk of $ONOO^-$ formation.

Oxidative stress

Fig. 9.1. Some of the reasons why tissue injury can cause oxidative stress.

(Chapter 8). Depletions of dietary antioxidants and other essential dietary constituents can also lead to oxidative stress (Chapter 10).

(2) Increased production of ROS/RNS, e.g. by exposure to elevated O_2, the presence of toxins that are themselves reactive species (e.g. $NO_2^{\bullet}$) or are metabolized to generate ROS/RNS, or excessive activation of 'natural' ROS/RNS-producing systems (e.g. inappropriate activation of phagocytic cells in chronic inflammatory diseases).

Mechanism 2 is usually thought to be more relevant to disease and is frequently the target of attempted therapeutic intervention, but rarely is much attention paid to the antioxidant nutritional status of sick patients. For example, intensive care patients[260] often show low levels of ascorbate in body fluids—would it be beneficial to correct this?

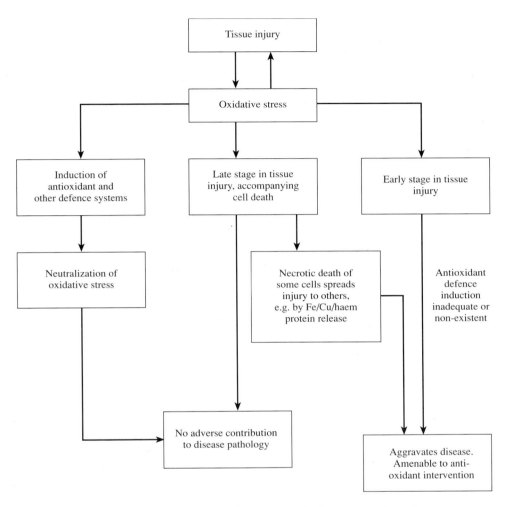

Fig. 9.2. What is the significance of oxidative stress in disease?

9.1.2 *Consequences of oxidative stress in disease*

Oxidative stress can result in:

1. **Adaptation** by up-regulation of defence systems, which may:

(a) completely protect against damage;

(b) protect against damage, but not completely;

(c) 'over-protect', e.g. the cell is then resistant to higher levels of oxidative stress imposed subsequently.

Antioxidant depletion does not prove oxidative damage, only that the defence system is working

(Total AOX potential, depletion
of specific AOX, measurement of
AOX-derived species e.g.
ascorbate radical, urate
oxidation products)

Induction of AOX enzymes Does not prove oxidative damage, only that the defence system is responding

Specific markers suggest that oxidative damage has occurred but rises in cell/tissue levels of markers could reflect either more damage and/or less repair

Lipid

peroxides   MDA, HNE and other
oxysterols   aldehydes
chlorinated/nitrated lipids   isoprostanes
                                   isoleukotrienes

DNA

oxidized bases in cells and urine
nitrated/deaminated bases in cells and urine
aldehyde/other base adducts in cells and urine

Protein

–SH oxidation
carbonyl formation
aldehyde adducts
oxidized tyr, trp, his, met, lys, leu, ileu, val
nitrated/chlorinated tyrosine, trp, phe
protein peroxides/hydroxides

ROS/RNS trapping PBN/other spin traps Aromatic probes (e.g. salicylate, phenylalanine) other detectors
ROS/RNS formation does not imply ROS/RNS importance. If they are important and the traps are efficient, the traps should be protective

Fig. 9.3. Some 'biomarkers' of oxidative stress that can be used in the study of disease.

As an example of (b), if adult rats are gradually acclimatized to elevated O_2, they can tolerate pure O_2 for longer than control rats, apparently due to increased synthesis of antioxidant defence enzymes and of GSH in the lung. However, the damage is slowed, not prevented (Section 3.4.4). As an example of (c), treatment of *Escherichia coli* with low levels of H_2O_2 increases transcription of genes regulated by the OxyR protein, and renders the bacteria resistant to higher H_2O_2 levels (Section 4.13.1). Examples of type (c) adaptation in animals are rarer, but a recent one is provided by work on **ischaemic preconditioning**.[286] A brief period of ischaemia in pig hearts led to depression of contractile function, and administration of antioxidants alleviated this. However, repeated periods of brief ischaemia led to quicker return of contractile function, but this adaptive response was blocked in the antioxidant-treated animals. Hence ROS produced by ischaemia–reperfusion were initially damaging, but also led to a response protective against subsequent insult.

2. **Tissue injury**. Oxidative stress can cause damage to all molecular targets: DNA, proteins, carbohydrates and lipids (Chapter 4). Prospects for therapeutic intervention may be confounded because it is often unclear which is the primary molecular target of oxidative stress. For example, DNA is an important early target of damage when H_2O_2 is added to many mammalian cells; increased DNA strand breakage occurs before *detectable* lipid peroxidation or oxidative protein damage. We emphasize the word 'detectable' because such conclusions are obviously dependent on the assays used to measure such damage. For example, measurement of protein carbonyls would not detect important early oxidative protein damage by oxidation of essential –SH groups on membrane ion transporters. However, if lipid peroxidation, for example, is a late stage in injury then therapies directed against it might have little beneficial effect. Their failure may lead to the erroneous conclusion that oxidative stress is not important.

3. **Cell death**. Cell death can occur by essentially two mechanisms, **necrosis** and **apoptosis**; both can involve oxidative stress (Chapter 4). In necrotic cell death, the cell swells and ruptures, releasing its contents into the surrounding area and affecting adjacent cells. Contents can include antioxidants such as catalase or GSH, and pro-oxidants such as copper and iron ions and haem proteins (Figs 9.1 and 9.2). Hence even if a cell dies by mechanisms other than oxidative stress, necrotic cell death can impose oxidative stress on the surrounding tissues (Fig. 9.2). In apoptosis, the cell's own intrinsic 'suicide mechanism' is activated; apoptosing cells do not release their contents and so apoptosis does not, in general, cause disruption to surrounding cells (Chapter 4). Apoptotic cell death may be accelerated in certain diseases, such as some of the neurodegenerative diseases, and oxidative stress has been implicated. However, evidence is growing that apoptosis and necrosis are two 'extremes' and elements of both can be found in some dying cells.

9.1.3 *Significance of oxidative stress in disease*

Some human diseases may be caused by oxidative stress. For example, ionizing radiation generates $OH^\bullet$ by splitting water molecules and many of the biological consequences of excess radiation exposure are probably due to oxidative damage to proteins, DNA and lipids (Section 8.16). The symptoms produced by chronic dietary deficiencies of selenium (e.g. Keshan disease) or of tocopherols (neurological disorders seen in patients with defects in intestinal fat absorption) might also be mediated by oxidative stress (Chapter 3). It is tempting to attribute the cardiomyopathy seen in Keshan disease to lack of active glutathione peroxidase causing a failure to remove H_2O_2, and perhaps lipid peroxides, at a sufficient rate *in vivo*. Although it is not proven that this is the mechanism of tissue injury, patients suffering from Keshan disease do show very low selenium-dependent glutathione peroxidase activity in blood and tissues.

However, in most human diseases, oxidative stress is a consequence and not a cause of the primary disease process. Tissue damage by infection, trauma, toxins, abnormally low or high temperatures, and other causes usually leads to formation of increased amounts of putative 'injury mediators', such as prosta-glandins, leukotrienes, interleukins and cytokines such as tumour necrosis factors (TNFs). All of these have, at various times, been suggested to play important roles in tissue injury.[105] ROS (and RNS, including NO•) can be placed in the same category, i.e. tissue damage will usually lead to increased ROS/RNS formation (Fig. 9.1).[105,119] If, in most human diseases, ROS/RNS are produced in increased amounts as a consequence of tissue injury, do they make a significant contribution to the disease pathology, or is their formation of little or no consequence? The answer probably differs in different diseases (Fig. 9.2). Indeed, the same question could be asked about any putative mediator of tissue injury.

In order to show that ROS/RNS (or any other mediators of injury) are important in a particular disease, it is necessary to fulfil the criteria listed on p. 617 and 619 using appropriate biomarkers (Fig. 9.3). For example, the increased lipid peroxidation demonstrated in the damaged muscles of patients with muscular dystrophy[57] may be a consequence of the tissue damage and make little or no further contribution to damage; there is no evidence that anti-oxidants are beneficial in this disease. To take a second example, human endothelial cells infected with *Rickettsia rickettsii* (the causative agent of Rocky Mountain spotted fever) show increased rates of lipid peroxidation,[269] but inhibition of this peroxidation does not prevent the cellular injury caused by the organism. Of course, these data show that lipid peroxidation is not important in these cases, which does not rule out oxidative stress by other mechanisms.

Although oxidative stress may often be a secondary event, it does play an important role in furthering tissue injury in several important human diseases. We have already considered the examples of porphyria (Chapter 2), vitamin E deficiency, emphysema, drug-induced haemolytic disease, cataract, macular degeneration and malaria (Chapter 7). Other possible examples are retrolental fibroplasia (Chapter 1), Wilson's disease and haemochromatosis (Chapter 3). The challenge for the future is to develop effective therapeutic antioxidants, demonstrate their benefit to patients and prove that they are working by an antioxidant mechanism. Indeed, some agents already in clinical use may exert part or all of their effects by antioxidant mechanisms (Chapter 10).

Let us now examine some diseases in detail.

9.2   Atherosclerosis

9.2.1   *Nature of atherosclerosis*[245]

Cardiovascular disease is the chief cause of death in the USA and Europe. Most heart attacks (**myocardial infarctions**) and many cases of localized cerebral ischaemia (**stroke**) are secondary to the condition of **atherosclerosis**, a disease of arteries that is characterized by a local thickening of the vessel wall that develops in the inner coat (**tunica intima**) (Fig. 9.4, see plate section). The clinical effects of atherosclerosis are mainly evident in mid-sized muscular arteries, e.g. coronary, carotid, femoral and iliac arteries, as well as in the aorta.

In general, three types of thickening are recognized. **Fatty streaks** are slightly raised, yellow, narrow, longitudinally lying areas. They are characterized by the presence of **foam cells** (Fig. 9.4), lipid-laden distorted cells that can arise both from endogenous smooth muscle cells and (more usually) from macrophages. Fatty streaks probably serve as precursors of **fibrous plaques**. These are approximately round, raised lesions, usually off-white in colour and often a centimetre or so in diameter, slightly obstructing the vascular lumen. A typical fibrous plaque consists of a fibrous cap (composed mostly of smooth muscle cells and dense connective tissue containing collagen, elastin, proteoglycans and basement membranes) covering an area rich in macrophages, smooth muscle cells and T lymphocytes. Often there is a deeper necrotic core, which contains debris from dead cells, extracellular lipid deposits and cholesterol crystals.

Complicated plaques are fibrous plaques that have been altered by necrosis, calcium deposition, bleeding and thrombosis. Plaques cause disease by limiting blood flow to a region of an organ such as the heart or brain. A stroke or myocardial infarction occurs when the lumen of an essential artery becomes completely occluded, usually by a thrombus forming at the site of a plaque. Thrombus formation is often triggered by plaque rupture, releasing noxious products into the bloodstream. Plaque disruption occurs most frequently where the fibrous cap is thinnest and most heavily infiltrated by foam cells; macrophage–derived protease enzymes[108] (e.g. collagenases, gelatinases, stromelysin, metalloelastase and matrilysin) may be involved.[a]

9.2.2   *The link to fat*[227,245]

Fatty streaks are frequently present in the arteries of those (even children) on Western diets; these diets are usually rich in fat. They are even found in the foetus, to an extent affected by maternal plasma cholesterol levels.[210a] Section 3.22 explains the important role of low-density lipoproteins (LDLs) in fat metabolism; they are largely produced from very-low-density lipoproteins (VLDLs) in the circulation by several processes, including removal of some triglyceride from VLDLs as they circulate through the tissues. LDLs are rich in cholesterol and its esters and they supply cholesterol to tissues requiring it. LDLs bind to receptors on the surface of cholesterol-requiring cells, and are internalized, releasing cholesterol within the cells.

In the disease **familial hypercholesterolaemia**, LDL receptors are defective, so that blood LDL (and hence cholesterol) levels become very high. Atherosclerosis is accelerated, and myocardial infarctions can occur as early as two years of age. This observation focused attention on the possible role of cholesterol in promoting atherosclerosis, especially as it has since been observed that people with high blood LDL cholesterol develop atherosclerosis at an accelerated rate and that lowering cholesterol levels can decrease the incidence of myocardial infarction.

LDLs enter the arterial wall both by transport in vesicles through endothelial cells, and by diffusion through the intercellular matrix.[214] The amount entering depends both upon the plasma LDL level and on the arterial wall permeability, so that endothelial injury by any mechanism (including ROS/RNS) facilitates LDL entry. Smoking, and genetic predisposition to 'permeable' vessel walls, may also increase LDL penetration. Another factor may be that, as body cells receive the cholesterol they need, they down-regulate expression of normal LDL receptors. Hence circulating LDL stays for a longer time in the circulation. In hypercholesterolaemia, LDL levels are higher and the LDL may be 'older', on average, perhaps facilitating its oxidation.[321]

Other errors of fat metabolism can also predispose to atherosclerosis, an example being **type III hyperlipoproteinaemia**. Plasma levels of cholesterol and triglyceride are markedly elevated. The disease is associated with defects in **apolipoprotein E** (apoE) or in receptors for it. ApoE is a major structural component of chylomicrons and VLDL and functions as a ligand for uptake of 'remnant particles', derived from VLDL and chylomicrons, by the liver.

In humans there are three common isoforms of apoE; apoE3, apoE2 and apoE4. They differ from each other by single amino-acid substitutions at residues 112 and 158 in the protein. The apoE2 form, with cysteine residues at both positions, is associated with impaired binding to the apoE receptors. Transgenic 'knockout' mice lacking apoE show accelerated atherosclerosis, facilitated by a high-cholesterol diet.

9.2.3   *What initiates atherosclerosis?*[212,245]

The origin of atherosclerosis is uncertain, but a popular current theory is that it begins with damage to the vascular endothelium, the single-cell-thick lining of the blood vessels. Damaging events could include mechanical damage, viral infection (herpes viruses and cytomegalovirus have been implicated), and exposure to blood-borne toxins, including both xenobiotics (e.g. from cigarette smoke) and elevated levels of normal metabolites, such as glucose or homocysteine. For example, endothelial damage by turbulent blood flow at bifurcations in the arteries could be worsened by high blood pressure and sites of such 'flow stress' are particularly prone to atherosclerosis. Damage may be aggravated by metabolic stresses such as hyperglycaemia and hypercholesterolaemia.

Endothelial injury is followed by attachment of monocytes from the circulation, which enter the vessel wall and develop into macrophages.

Activated monocytes and macrophages could injure neighbouring cells by secreting $O_2^{\bullet-}$, H_2O_2, hydrolytic enzymes and possibly $NO^{\bullet}$. Nitric oxide is always produced by vascular endothelium and it is also possible that endothelium generates low levels of $O_2^{\bullet-}$ (e.g. by xanthine, NADH or NADPH oxidases) (Chapter 7). Indeed, xanthine oxidase has been identified in human atherosclerotic lesions.[292] Endothelial $O_2^{\bullet-}$ production may be accelerated by injury, e.g. in hypercholesterolaemic rabbits it has been reported that $O_2^{\bullet-}$ formation, apparently involving xanthine oxidase, is elevated in endothelial cells. Whether $O_2^{\bullet-}$ arises from endothelial cells and/or from monocytes and macrophages,[53] it might then react with $NO^{\bullet}$ to give peroxynitrite, $ONOO^-$.

9.2.4 *What roles are played by ROS/RNS in atherosclerosis?*[53,227,284a]

It is now widely believed that ROS/RNS play an important role in the development, and perhaps sometimes in the initiation of, atherosclerosis.

1. Shear stress and turbulent blood flow can cause oxidative damage to vascular endothelial cells. They may respond by upregulating several enzymes, including CuZnSOD, MnSOD and endothelial NOS. The stress may lead to damage, or the enzyme up-regulation may render the cells more resistant to subsequent insult.[304] Perhaps this is one reason why regular exercise improves the functioning of the cardiovascular system.

2. Another risk factor for atherosclerosis (and hence for stroke and myocardial infarction) is high plasma levels of the amino acid **homocysteine**,[176] which arises from the dietary amino acid methionine. High homocysteine levels can result from inborn defects in enzymes involved in methionine metabolism, and/or insufficient dietary intakes of cofactors required by these enzymes, namely vitamin B_6 and folate (Chapter 10). The mechanism of homocysteine toxicity to vascular endothelium is unknown. *In vitro*, it can oxidize in the presence of transition-metal ions to generate $O_2^{\bullet-}$, H_2O_2, $OH^{\bullet}$ and sulphur radicals, but there is insufficient evidence to confirm the proposal that homocysteine damages endothelium by causing oxidative stress in humans.[66]

3. Activation of macrophages or their monocyte precursors to generate ROS/RNS in the vessel wall could injure neighbouring cells and lead to more endothelial damage and perhaps damage to smooth muscle cells. Damaged endothelial cells may allow more LDL into the vessel wall.

4. ROS/RNS might activate the latent forms of matrix metalloproteinases released by macrophages. This may contribute to weakening of atherosclerotic plaques and cap rupture, often the precipitating event in myocardial infarction.[108]

5. Macrophages possess some LDL receptors. However, LDL that has undergone lipid peroxidation is recognized by different receptors.[227] These were first identified as receptors that bound chemically-acetylated LDL (the **acetyl–LDL receptors**), but are now usually called the **scavenger receptors**. Two types of scavenger receptor (I and II) have been identified, one slightly smaller than the other but apparently functionally equivalent. Both are encoded by the same gene and are products of alternative splicing (see Appendix II for

an explanation if needed). Transgenic 'knockout' mice lacking these receptors seem less predisposed to atherosclerosis. They are also more susceptible to infection with *Listeria monocytogenes* or with herpes simplex virus, suggesting that scavenger receptors may play some role in host defence against pathogens.[289] Several other cell–surface receptors can also bind peroxidized LDL, including a macrophage receptor that appears to recognize oxidatively-damaged proteins on cell surfaces and may be involved in recognition of senescent erythrocytes (Chapter 7).[251]

LDL bound to any of these receptors is rapidly taken up by macrophages, so that intracellular cholesterol accumulates and may convert the macrophage into a foam cell (Fig. 9.4).

6.   Macrophages themselves, vascular endothelial cells, smooth muscle cells and lymphocytes can cause peroxidation of LDL if they are incubated with it *in vitro*.[227] The event during peroxidation that allows recognition by the scavenger receptors involves changes in the protein moiety of LDL (**apoprotein B**). Essentially, cytotoxic aldehydes, especially hydroxynonenal (HNE), bind to lysine residues on apoprotein B. Modification of lysine changes the surface charge on the protein (it becomes less positive). This can be detected by electrophoresis of LDL. HNE can also modify histidine residues on apoprotein B.

LDL modification by the above cell types *in vitro* requires traces of iron or copper ions, and is stimulated by certain thiol compounds.[66,337] Is this physiologically relevant? Injury to vascular endothelium *in vivo* might liberate 'catalytic' metal ions (Fig. 9.1), and certain plasma proteins entering the vessel wall, especially caeruloplasmin, may degrade to release transition metal ions. It seems unlikely that intact caeruloplasmin oxidizes LDL,[60] but it readily degrades at low pH and by the action of proteases.[102] Peroxides in LDL can release iron from haem proteins *in vitro*, and $ONOO^-$ releases copper from caeruloplasmin. 'Catalytic' metal ions have been detected in advanced human atherosclerotic lesions[77] and copper ions are pro-atherosclerotic in some animal model systems.[317] Haem oxygenase (HO–1) in lesions is a potential source of iron.[321a]

Incubation of LDL with aqueous extracts of cigarette smoke has also been shown to cause oxidative modification, leading to increased LDL uptake by isolated macrophages (Chapter 8).

7.   However, oxidative LDL damage that allows recognition by scavenger receptors need not necessarily involve transition–metal ions. Peroxynitrite can cause peroxidation of LDL, but excess $NO^\bullet$ can inhibit peroxidation, by scavenging $RO_2^\bullet$ radicals, although the biological significance of the resulting nitrosylated/nitrated lipids is unknown (Chapter 2).[161,247] HOCl can oxidize LDL. It directly oxidizes apoprotein B and converts lipids and cholesterol into chlorohydrins/chlorinated sterols (Chapter 2).[115]

As well as acting to form HOCl, the enzyme **myeloperoxidase** may make another contribution to LDL oxidation.[115] In the presence of H_2O_2, it can oxidize tyrosine into tyrosyl radicals, which appear to cause LDL oxidation. Consistent with the physiological relevance of all these processes, levels of bityrosine, 3–chlorotyrosine, 3–nitrotyrosine, protein carbonyls and several specific amino acid oxidation products have been reported as elevated in human atherosclerotic lesions. The extents of elevation are variable between laboratories and it seems likely to us that no two lesions will have exactly the same pattern of oxidative damage.

8. Peroxidized LDL might further injure endothelial cells. LDL that has begun peroxidation, but is not yet sufficiently oxidized to be taken up by macrophages (so-called **minimally modified LDL**) could be the major agent of damage.[212] Oxidized LDL may 'stick' within the vessel wall, passing through more slowly than normal LDL. Exposure of vascular endothelial cells to oxidized LDL in culture up-regulates expression of NF-κB, and hence of adhesion molecules such as ICAM-1 and VCAM-1 that facilitate binding of more phagocytes. Indeed, activated NF-κB can be detected in smooth muscle cells, macrophages and endothelial cells within atherosclerotic lesions,[28] but is rarely present in normal vessel walls. Oxidized LDL appears to impair net NO$^\bullet$ production by endothelial cells, e.g. NO$^\bullet$ can be consumed by reaction with $O_2^{\bullet-}$ and $RO_2^\bullet$ radicals. NO$^\bullet$ can down-regulate NF-κB (Section 4.13.5) and the lower levels of NO$^\bullet$ may contribute to increased NF-κB activation during atherosclerosis.

Vascular endothelial cells can take up and destroy limited amounts of oxidized LDL, through a receptor-mediated pathway, although the receptor(s) involved are not of the 'scavenger receptor' type.

9. Uptake of oxidized LDL by macrophages might be regarded as a 'defence mechanism' to protect the vascular wall. However, it imposes an oxidative stress on the macrophage, e.g. causing it to synthesize more 'stress proteins' and to raise intracellular GSH levels.[53] Oxidized LDL can also stimulate formation of macrophage-derived factors that recruit other cells and stimulate smooth muscle cell proliferation. Excess oxidized LDL can kill macrophages, either by initiating necrosis or by apoptosis.[109] Macrophage death can release proteolytic enzymes and transition–metal ions, and is a contributor to 'gruel' formation in advanced atherosclerotic lesions.

10. Peroxidized LDL may generate chemotactic factors for blood mono-cytes, encouraging their recruitment into an atherosclerotic lesion.[227] For example, HNE is chemotactic at low micromolar levels. Recruitment of monocytes is facilitated by induction of adhesion molecules in the endothelial cells (point 8 above), and of monocyte-activating proteins such as **monocyte chemoattractant protein–1** (MCP-1) and **macrophage colony-stimulating factor** (M-CSF) by endothelial and smooth muscle cells. M-CSF facilitates monocyte differentiation into macrophages in the vessel wall and up–regulates expression of scavenger receptors. Perhaps, therefore, the presence of oxidized LDL is 'signalling' the recruitment of macrophages to help remove it. Foam cells can also show increased expression of MCP-1 and of several other molecules, including matrix metalloproteinases (point 4 above), 15-lipoxygenase and the pro-inflammatory cytokine interleukin 8 (IL-8).

11. Peroxides can accelerate cyclooxygenase- and lipoxygenase-catalysed reactions in endothelium, phagocytes and in any platelets present, leading to enhanced formation of eicosanoids (Chapter 6).

12.   F$_2$-Isoprostanes generated by the peroxidation of arachidonic acid[236a] residues in LDL could have profound effects on vascular function because of their ability to mimic or antagonize the actions of some of the stereospecific products formed by cyclooxygenase and lipoxygenase enzymes, e.g. they can act as vasoconstrictors and potentiate platelet aggregation (Chapter 4).

13.   Oxidation products of cholesterol might be involved in atherogenesis.[40] Cholesterol is oxidized to multiple products in peroxidizing LDL. Some cholesterol oxidation products are toxic to arterial smooth muscle cells, to vascular endothelium and to monocyte/macrophages in culture, although the significance of this *in vivo* is uncertain (Section 4.9.15).

14.   Products of LDL oxidation may play a role in promoting calcification in advanced lesions[225] and may be pro-coagulant.[245a]

9.2.5   *Evidence relating to the 'oxidation theory' of atherosclerosis*[76,227,284a]

The oxidation theory states that oxidation of LDL in vessel walls is a significant contributor to the development of atherosclerosis. It should be noted that LDL oxidation occurs in the microenvironment of the blood vessel wall and not in the circulating blood. However, there have been several reports that LDL isolated from the plasma of people with extensive atherosclerosis shows a mild degree of oxidation. Of course, minimally modified LDL might escape from the vessel wall back into the circulation. Another factor is that the LDL might have lower antioxidant levels in such patients and tend to undergo more peroxidation during the isolation process. Oxidized LDL is, in any case, rapidly cleared from the circulation, mostly by uptake by the Kupffer cells of the liver.[75] These cells are essentially macrophages, and take up oxidized LDL using scavenger receptors.

There is much evidence consistent with the oxidation theory. It includes:

(1) The presence of oxidized cholesterol, oxidized lipids, F$_2$-isoprostanes, and decomposition products of lipid peroxides (such as 4-HNE and malondi-aldehyde (MDA)) in human and animal atherosclerotic lesions.

(2) LDL extracted from human and rabbit lesions resembles LDL that has been oxidized *in vitro*.

(3) Antibodies that recognize oxidized LDL are found in serum from patients with coronary artery disease. The antibody titre has been reported to correlate with progression of atherosclerosis of the carotid arteries (which can be measured by ultrasound techniques). It is uncertain whether or not production of these antibodies affects atherosclerosis *in vivo*, although the presence of T lymphocytes in atherosclerotic lesions suggests some immune response during atherosclerosis. In one study, immunization of rabbits lacking LDL receptors with MDA-treated LDL led to high plasma antibody levels and a decreased extent of atherosclerosis.[224]

(4) Epidemiological data are consistent with a protective action of dietary chain-breaking antioxidant inhibitors of lipid peroxidation (especially

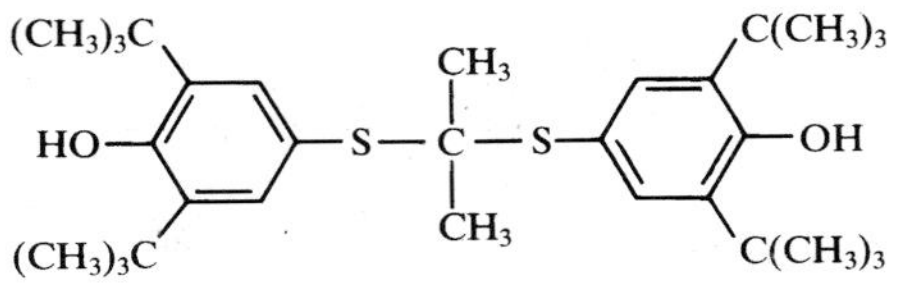

Fig. 9.5. Structure of probucol. The full chemical name is 4,4'-[(1-methylethylidene)-bis(thio)]bis-[2,6-bis(1,1-dimethylethyl)phenol]. The phenolic −OH groups confer chain-breaking antioxidant activity on the molecule.

α-tocopherol) against atherosclerosis (Chapter 10). Direct evidence for the importance of these antioxidants is provided by studies showing that dietary supplementation with α-tocopherol decreases the incidence of myocardial infarction (Chapter 10).

Dietary supplementation with α-tocopherol (or the synthetic, lipid-soluble antioxidant **probucol**) increases the resistance of LDL subsequently isolated from the blood to oxidation *in vitro*. Probucol was initially introduced as a drug to lower blood cholesterol levels. However, it is also a chain-breaking antioxidant, since its structure (Fig. 9.5) is similar to that of other phenolic antioxidants.[138] The anti-atherogenic effect of probucol in rabbits is greater than expected from its cholesterol-lowering ability, suggesting that its anti-oxidant activity might also be biologically relevant. Other antioxidants, such as lazaroids[55] and butylated hydroxytoluene (BHT), have also been shown to be anti-atherosclerotic in some animal studies, despite the fact that high-dose BHT raises plasma cholesterol in animals.

9.2.6 Chemistry of LDL oxidation[76]

One prediction of the oxidation theory is that the lower the resistance of LDL to oxidation, the greater the chance that it will become oxidized in the vessel walls and contribute to the development of atherosclerosis. Indeed, the sensitivity of LDL to oxidation *in vitro* has been measured and some studies have shown that it correlates to the risk of developing coronary heart disease.

Copper ion-stimulated oxidation

In these studies, LDL is first isolated from plasma by a lengthy process of centrifugation. It is then oxidized *in vitro*, usually by incubation with Cu^{2+} ions, e.g. as copper(II) sulphate, $CuSO_4$ (Fig. 9.6). What factors affect the oxidizability of LDL when measured by this approach? The experimental parameter defined by most investigators using Cu^{2+}-stimulated LDL oxidation is the **lag phase** or **lag period**, i.e. the time taken before peroxidation of a given sample of LDL subjected to oxidative challenge begins to accelerate. Factors affecting LDL resistance to oxidation in this model system include:

(1) Fatty acid composition; a greater proportion of saturated and mono-unsaturated fatty-acid side-chains increases oxidation resistance. For

example, LDL isolated from rabbits fed on a diet enriched in oleic acid were more resistant to oxidation *in vitro*. The reverse is true in animals fed on diets containing high levels of polyunsaturated fatty acids (PUFAs).

(2) The concentration of Cu^{2+}. The lag phase decreases with Cu^{2+} concentration up to a maximum. At least some of the added copper binds to apoprotein B, probably to histidine residues.

(3) The O_2 concentration.[113] Incubations are usually carried out under ambient O_2 levels (21%), whereas pO_2 in blood vessel walls may be as low as 2.5% O_2. At lower pO_2 values, the rate of LDL peroxidation is lower.

(4) The content of antioxidants. Dietary supplementation with antioxidants that intercept lipid peroxyl radicals, such as α-tocopherol or probucol, usually increases the length of the lag phase in LDL subsequently isolated from the plasma.

(5) The lipid peroxide content of the LDL particle. Copper ions appear largely or entirely to act by decomposing peroxides in the LDL to chain-propagating radicals

$$LOOH + Cu^{2+} \rightarrow LO_2^\bullet + Cu^+ + H^+$$

$$LOOH + Cu^+ \rightarrow LO^\bullet + Cu^{2+} + OH^-$$

Decomposition of lipid peroxides forms a wide range of carbonyl compounds, including MDA and HNE, which can modify lysine residues on apoprotein B. In addition $LO^\bullet$ and $LO_2^\bullet$ may react directly with the protein to cause fragmentation and modification of amino–acid residues.

(6) The assay used to measure LDL peroxidation. As expected from the complexity of lipid peroxidation, different assays give different results (Fig. 9.6) and it is customary to use at least two different assay methods when studying LDL oxidation.

The role of 'seeding peroxides'

The greater the content of peroxides in LDL, the greater—in general—is the rate of Cu^{2+}-stimulated oxidation. Levels of peroxides in LDL isolated from different individuals are very variable. It is not clear where these peroxides come from. There are several possibilities:

(1) Some may be introduced during the prolonged centrifugation procedures used to isolate LDL.

(2) Some may arise from peroxides in dietary fats. The question of absorption of peroxidized lipids from the diet has been investigated in several laboratories, with conflicting answers. The gut appears to be effective at detoxifying lipid peroxides (presumably involving glutathione peroxidases),[295] so that oxidized oils/fats are not acutely toxic to animals (Chapter 4). However, high levels do exert pathological effects and it is easy to envisage some dietary peroxidized lipids getting through as a component of the complex micellar structures absorbed from the gut during fat digestion, especially if a diet rich in cooked fatty foods is consumed.[283,312] The major PUFA in oils used for human nutrition (e.g. sunflower and

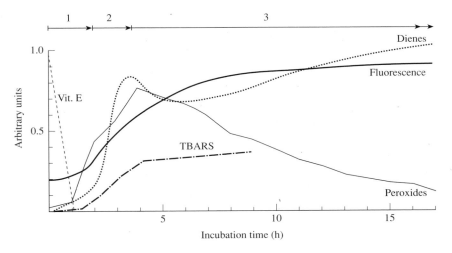

Fig. 9.6. Sequence of events during copper ion-stimulated oxidation of LDL. Usually, α-tocopherol is first consumed and the 'peroxide pool' is filled, whereupon peroxidation accelerates. Peroxidation can be measured as formation of conjugated dienes at 234 nm, fluorescence at 430 nm with excitation at 360 nm, formation of lipid hydroperoxide, or production of thiobarbituric acid-reactive material (TBARS), although TBARS alone is not a good index of LDL oxidation. Period 1 (top) is the lag phase, period 2 is the propagation phase. In phase 3, peroxides decompose to aldehydes and other products. Data provided by courtesy of the late Professor Dr Hermann Esterbauer.

soya) is linoleate, which can be oxidized to 9- and 13-hydroperoxides. Arachidonate (from meat) gives a more complex mixture of peroxides, and even more complex mixtures result from peroxidation of the eicosapentaenoic and docosahexaenoic acids present in fish and brain. Infusion of chemically oxidized chylomicrons into rats led to rapid uptake by hepatic Kupffer cells.[75]

(3) Some may be generated by the action of lipoxygenases on fatty acid side-chains in LDL; interest has focused on **15-lipoxygenase** in macrophages. The inhibitor **PD146176**, which blocks this enzyme, has been reported to decrease atherosclerosis in rabbits.[265]

Is copper relevant?

Cu^{2+} addition is only one model system for LDL oxidation *in vitro*, although it is the most popular.[243] Copper ions have been detected in advanced atherosclerotic lesions and in homogenates of normal vessel walls, but they are only one of the many pro-oxidants that could lead to LDL oxidation *in vivo*. Others that have been used to oxidize LDL *in vitro* include exposure to ionizing radiation, UV light, $ONOO^-$, $HOCl$, synthetic peroxyl radicals (e.g. generated from AAPH or AMVN), haem, and mixtures of haem proteins with H_2O_2. Nitric oxide can inhibit LDL peroxidation by scavenging chain-propagating $RO_2^{\bullet}$ radicals. Thus $NO^{\bullet}$ can be pro-atherogenic if the $NO^{\bullet}/O_2^{\bullet-}$

ratio is such as to favour ONOO⁻ formation, but it can inhibit LDL oxidation at high NO•/ROS ratios. Net NO• production appears lower in atherosclerotic lesions. Addition of extra L-arginine (the precursor of NO•) to the diet of hypercholesterolaemic rabbits decreases the development of atherosclerosis, an observation consistent with an overall antioxidant effect of high NO• levels *in vivo*.[45]

9.2.7   *Antioxidants and LDL oxidation*[76]

LDLs are approximately spherical particles with a diameter of 19–25 nm and a relative molecular mass between 1.8 and 2.8 million. Each particle in humans contains, on average, about 600 molecules of cholesterol, 1600 molecules of cholesterol ester (mostly cholesterol linoleate) and 170 molecules of triglyceride within a hydrophobic core (Table 9.2). In order to allow these particles to travel in an aqueous solution (the blood plasma), the core is wrapped in a monolayer of around 700 phospholipid molecules, mainly phosphatidylcholine, with their polar head groups oriented toward the aqueous phase. The

Table 9.2. Composition of native human low-density lipoprotein (LDL)[a]

Constituent	Mean amount (molecules/ LDL particle)	Constituent	Mean amount (molecules/ LDL particle)
Total lipids		Antioxidants	
Phospholipids	700	α-tocopherol	7–8 (3–15)[b]
Phosphatidylcholine	450	γ-tocopherol	0.50
Triglycerides	170	Ubiquinol-10	0.10–0.30[c]
Free cholesterol	600		
Cholesterol ester	1600	Putative antioxidants[d]	
		β-carotene	0.30
Fatty-acid esters		α-carotene	0.12
Palmitic	693	Lycopene	0.16
Palmitoleic	44	Lutein and zeaxanthin	0.04
Stearic	143	Phytofluene	0.05
Oleic	454		
Linoleic	1101		
Arachidonic	153		
Docosahexaenoic	29		
Total PUFAs	1282		

[a]Data abstracted from several publications of the late Professor Dr Hermann Esterbauer. The figures must not be taken too literally, since the LDL composition of a given individual will depend to a considerable extent on their diet. For example, consumption of large amounts of PUFAs will increase their levels in LDL.
[b]Range in parentheses for subjects not consuming vitamin E supplements.
[c]Ubiquinol is easily oxidized during isolation, and the true value has been suggested to be up to 0.5 molecules/LDL particle (Dr Roland Stocker, personal communication, 1993).
[d]Most of the lycopene, α-carotene and β-carotene in human plasma is found in LDL, but there is also some in HDL.

apoprotein B is embedded in the outer layer. Hence both the phospholipids and the cholesterol esters are potential targets of peroxidation.

When isolated LDLs are exposed to Cu^{2+} ions, peroxidation occurs, at first slowly (the lag period) and then more quickly (Fig. 9.6). During the lag period, the α-tocopherol is lost from LDLs, as are several other constituents (Fig. 9.7). The length of the lag period presumably reflects the natural lag period of autocatalytic lipid peroxidation (Chapter 4), the 'seeding peroxide' content of LDLs and the protective effects of chain-breaking antioxidants, especially α-tocopherol. Tocopherols react very much more quickly with peroxyl radicals than do PUFAs (Chapter 3) and thus contribute to the lag period:

$$TocH + LO_2^\bullet \rightarrow Toc^\bullet + LO_2H$$

The resulting tocopheryl radical ($Toc^\bullet$) does have a weak capacity to abstract hydrogen from PUFAs, but is far less reactive than are $LO_2^\bullet$ radicals. If isolated LDL is enriched with vitamin E, or LDL is isolated from subjects who have consumed vitamin E supplements, the length of the lag period is significantly increased. For subjects consuming RRR-α-tocopherol for 21 days, the oxidation resistance of LDL isolated from their plasma increased, on average, to 118%, 156%, 135% and 175% of control subjects for doses of 150, 225, 800 or 1200 IU/day, respectively (1 IU = 0.67 mg RRR-α-tocopherol). Six days after the supplements were stopped, resistance to Cu^{2+}-mediated oxidation, and α-tocopherol content, had returned to normal.

Tocopheryl radicals can be recycled to TocH if ascorbate is added to the LDL suspension:

$$Toc^\bullet + ascorbate \rightarrow TocH + semidehydroascorbate$$

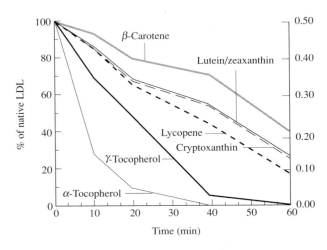

Fig. 9.7. Sequence of loss of antioxidants and putative antioxidants during the lag phase of LDL oxidation. Human LDL was isolated and exposed to copper ions. Data by courtesy of the late Professor Dr Hermann H. Esterbauer.

Added ascorbate delays the onset of LDL peroxidation even when it is started by adding Cu^{2+}, an exception to the general rule that Cu^{2+}/ascorbate mixtures stimulate oxidative damage (Fig. 9.8). Ascorbate can recycle probucol-derived phenoxyl radical in LDL treated with this synthetic anti-oxidant.[138]

LDL contains ubiquinol (Table 9.2), a lipid-soluble chain-breaking anti-oxidant (Chapter 3). When LDL is oxidized, ubiquinol disappears more quickly than do tocopherols, perhaps because it can participate in recycling of tocopherols.

$$CoQH_2 + Toc^\bullet \rightarrow CoQ^\bullet + TocH$$

However, the ubiquinol content of LDL is small (Table 9.2)—only ~ 10–50% of LDL particles contain it. Hence, although ubiquinol may exert antioxidant effects when it is present, it does not appear to be an essential antioxidant in LDLs, since up to 90% of LDL particles do not contain it. Oral consumption of coenzyme Q_{10} can be used to raise the concentrations of ubiquinol-10 in LDL to a limited extent.

LDLs also contain several plant-derived pigments, such as lycopene, β-carotene and lutein. These are also lost during the lag period (Fig. 9.7). The amounts present in LDL depend on diet, but in no case is there even approaching one molecule per LDL particle on average (Table 9.2). Again it may be suggested that they could exert antioxidant effects in the LDL particles containing them, but cannot be essential. Raising the β-carotene content of LDLs does not generally increase their resistance to oxidation *in vitro*,[113] but interest is now switching to the effects of lutein, lycopene, α-carotene and other carotenoids.

Plants contain many polyphenols with antioxidant activity that are capable of protecting LDLs against peroxidation *in vitro*; one example is the flavonoids (Chapter 3). It has been suggested that plant phenolics in red wine, cocoa and chocolate might exert anti-atherogenic effects. Further work is needed to assess

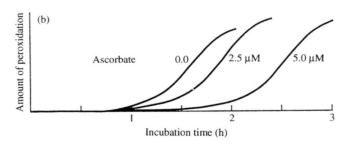

Fig. 9.8. Effect of ascorbate on the peroxidation of LDLs. LDLs isolated from human plasma were exposed to copper ions. Adding ascorbate delayed the onset of peroxidation, the lag period rising with increasing ascorbic acid concentrations. Diagram from *Free Rad. Res. Commun.* **6**, 67 (1989), by courtesy of the late Professor Dr Hermann Esterbauer and Harwood Academic Publishers.

the importance of plant phenolics in human nutrition and their possible contribution to antioxidant defence of LDL *in vivo*.

Pro-oxidant effects of vitamin E[27]

In some *in vitro* systems, α-tocopherol can be made to exert pro-oxidant activity. Thus, the tocopheryl radical is not completely unreactive with PUFAs. Tocopherols are reducing agents, able to convert Fe(III) to Fe^{2+} or Cu^{2+} to Cu^{+}. Reaction of α-tocopherol with Cu^{2+} (or Fe(III)) generates the tocopheryl radical, which can react with PUFAs at a low rate (Chapter 3). Thus tocopherols can accelerate metal-ion-dependent oxidative damage to substrates under certain circumstances since Cu^{+} and Fe^{2+} appear to decompose peroxides more quickly than Fe(III) and Cu^{2+}. However, the relevance of these effects to LDL oxidation *in vivo* is uncertain. *In vivo*, α-tocopherol is always present with co-reductants (such as ascorbate and ubiquinol) which remove the α-tocopheryl radical.

9.2.8   *The role of high-density lipoproteins*

High-density lipoproteins (HDLs) remove excess cholesterol from peripheral tissues, transport it to the liver (**reverse cholesterol transport**) and are considered to be anti-atherogenic. The beneficial cardiovascular effects of moderate intakes of ethanol and of PUFAs have been suggested to involve rises in blood HDL levels (Chapter 10). The enzyme **lecithin–cholesterol acyl-transferase** (LCAT) esterifies free cholesterol entering HDLs, facilitating storage in the core. Indeed, transgenic animals over-expressing LCAT show higher plasma HDL levels and are more resistant to diet-induced atherosclerosis.[123] Inactivation of LCAT, e.g. by constituents of cigarette smoke (Chapter 8), might thus be pro-atherogenic. By contrast, 'knockouts' of LCAT raise tissue cholesterol levels.

HDLs can undergo lipid peroxidation.[84a] However, incubation of mildly oxidized LDLs with HDLs can prevent some of the deleterious effects of the former *in vitro*. Apparently peroxidized lipids can transfer from LDLs to HDLs. They may then be degraded by two enzymes in HDLs, **platelet activating factor acetylhydrolase** (PAF acetylhydrolase; Section 4.9.14) and **paraoxonase**.[184] Paraoxonase is a Ca^{2+}-dependent esterase first discovered by its ability to catalyse the hydrolysis of organophosphate and aromatic carboxylic acid esters. The former are widely used as insecticides and have also been employed as nerve gases in chemical warfare. Examples are **sarin** (released by terrorists into the Tokyo subway system in 1995) and **soman**. The name 'paraoxonase' arises from the use of **paraoxon**, a toxic organophosphate, to assay the enzyme. Two major inherited forms of paraoxonase have been observed in European populations: a high activity (towards paraoxon) form, and a low activity form in which the amino acid arginine has been replaced by glutamine at position 192. Transgenic mice lacking paraoxonase develop more atherosclerosis on a high-fat diet.[267a]

PAF acetylhydrolase is found in both HDLs and LDLs and functions to hydrolyse PAF, a potent mediator of inflammation (Section 4.9.14). It can also hydrolyse peroxidized fatty acids arising from phospholipids during oxidation of LDLs. Peroxidation of phospholipid-bound fatty acid followed by fragmentation to release an aldehyde leaves behind a substrate for PAF, which removes the oxidized 'residue' to give a lysophospholipid. However, albumin can also remove toxic lipid oxidation products from LDLs. Hence in the complex mixture of proteins and lipoproteins found in plasma and tissue fluids, it is hard to be certain which reactions are most significant biologically.

Lipoprotein (a)[211] closely resembles LDL but contains an additional glycoprotein, **apo(a)**, linked to apoprotein B by a disulphide bridge. Increased plasma levels of lipoprotein (a) are an additional risk factor for atherosclerosis. Like LDL, lipoprotein (a) can be oxidized to a form recognized by macrophage scavenger receptors. The structure of apo(a) resembles that of **plasminogen**, the precursor of the enzyme plasmin that dissolves fibrin in blood clots. It has been suggested that apo(a) may antagonize plasminogen action and favour blood coagulation.

9.2.9   Conclusion

The evidence that lipid peroxidation occurs in mammalian atherosclerotic lesions and contributes to their development is strong and growing. Among the many questions that remain are:

(1) How exactly peroxidation is initiated in fatty streaks?
(2) Why do some lesions develop further and others do not?
(3) Is $NO^\bullet$ good or bad (or both)?
(4) What factors pre-dispose an advanced lesion to rupture and trigger thrombosis?
(5) How can peroxidation occur in vessel walls in the face of all the antioxidants that must diffuse into the wall from the plasma?
(6) Does a diet rich in PUFAs accelerate atherosclerosis by providing more substrates for free-radical attack?
(7) What contribution does caeruloplasmin make to LDL oxidation *in vivo*?
(8) How does peroxidation of other lipoproteins affect their biological function?

9.3   Hypertension[53]

High blood pressure is a well-established risk factor for cardiovascular disease, in part because it is pro-atherogenic. It presumably facilitates injury to the vascular endothelium by increasing stress caused by flowing blood.

The pathogenesis of hypertension can involve both genetic predisposition and exposure to toxins, including cigarette smoke. Normal blood pressure is maintained by a complex web of systems affecting cardiac output, sodium balance, vasodilation and renal function. One of the most important of these is the **renin/angiotensin** system. Inhibitors of **angiotensin-converting enzyme**, which converts angiotensin I to the hypertensive peptide angiotensin II, are frequently used to treat hypertension. One such inhibitor is **captopril**.

It has been suggested that one mechanism contributing to hypertension is increased generation of $O_2^{\bullet-}$ in the vessel wall, inactivating $NO^\bullet$ and causing $ONOO^-$ formation.[159,210] For example, injection of a form of SOD chemically modified to bind to blood vessel walls decreases blood pressure in spontaneously hypertensive rats, but not in normal rats. Angiotensin II may increase levels of $O_2^{\bullet-}$-generating oxidases in the vessel wall. Impaired endothelial response to vasodilators such as acetylcholine occurs in atherosclerosis and has been described in hypercholesterolaemic patients. Administration of ascorbate or vitamin E has improved vascular response in some studies,[225a] suggesting that increased ROS production is involved. 'Tolerance' to nitrovasodilation has also been suggested to involve increased endothelial $O_2^{\bullet-}$ generation (Section 6.8).

9.4   Diabetes

Diabetes mellitus (the Latin word *mellitus*, meaning 'honey-sweet', refers to the taste of diabetic urine) is a chronic disease marked by elevated blood glucose (**hyperglycaemia**) and urinary glucose excretion. It is caused by faulty production of, or tissue response to, insulin. Complications of poorly controlled diabetes include systemic vascular disease (accelerated atherosclerosis), microvascular disease of the eye causing bleeding and retinal degeneration (**diabetic retinopathy**), cataract, kidney damage leading to renal failure and damage to peripheral nerves (**peripheral neuropathy**).[128] Retina, kidney and nerves are all freely permeable to glucose and so suffer the most. By contrast, in the absence of insulin (or in failure to respond to it) muscle, adipose tissue and many other tissues show decreased glucose uptake, leading to elevated plasma glucose levels.

In the most common form of diabetes (**type II** or **non-insulin-dependent diabetes**) blood insulin levels are normal or elevated, yet tissue response to insulin is subnormal; there is said to be **peripheral resistance**. In the less common type I (**juvenile onset**) diabetes, insulin secretion is absent or impaired.

The extent of complications of diabetes appears to correlate with elevated blood glucose concentrations, and it is thus widely thought that excessive glucose is the major cause of tissue injury. Indeed, several mechanisms exist by which glucose and other sugars can damage tissues (see below). However, blood LDLs, VLDLs and lipoprotein (a) are also elevated in diabetes and this must be considered as a contributor to pathology, e.g. to accelerated atherosclerosis. In addition, HDL tends to be lowered.

9.4.1   *Oxidative stress and the origins of diabetes*[128]

Some diabetogenic agents (e.g. alloxan) appear to act by imposing severe oxidative stress on the β-cell (Section 8.5). It is uncertain as to whether oxidative stress contributes to the origin of diabetes in humans. There is a strong genetic predisposition to diabetes, which is clearest in the 1–2% of cases inherited in a simple autosomal dominant manner (**maturity onset diabetes of the young**, MODY). Genes defective in different MODY patients include those encoding the enzyme glucokinase (which converts glucose to glucose 6-phosphate) and the gene transcription factor HNF-1α (**hepatic nuclear factor-1α**). HNF-1α regulates not only gene transcription in the liver but also transcription of the insulin gene in β-cells (at least in rats).

In some animals that develop diabetes spontaneously, such as the **non-obese-diabetic (NOD)** mouse and the **biobreeding (BB)** rat there appears to be an autoimmune attack on the β-cells. It has been suggested that ROS (including $ONOO^-$)[285] generated by attacking phagocytes are involved in islet cell killing. Indeed, vitamin E, SOD and desferrioxamine have been reported to delay the onset of diabetes in NOD mice.[26] Type I diabetes in humans may also involve autoimmune events, one suggested autoantigen being the enzyme **glutamate decarboxylase**. There have been repeated proposals that some infectious agents might trigger diabetes by causing an exaggerated, tissue-damaging, host response.

9.4.2   *Oxidative stress in diabetic patients*[244a]

Diabetic patients are often stated to be under an oxidative stress. Indeed, the link between diabetes and oxidative stress has been extensively discussed for years, but rigorous experiments to elucidate its importance are still awaited.[128] Plasma lipid peroxides (even when corrected for the generally elevated plasma lipids) appear higher than normal in diabetics, although many of the most striking differences have been found using assays of questionable reliability, e.g. plasma TBARS. Nevertheless, levels of F_2-isoprostanes and lipid peroxides, which are more reliable biomarkers (Chapter 5), are also elevated. One interesting observation is that rats rendered diabetic by streptozotocin treatment absorbed peroxides from oxidized fats in the diet to a greater extent than controls, suggestive of a diminished ability of the gut to detoxify food lipid peroxides.[283]

There is disagreement as to whether plasma α-tocopherol levels in diabetic patients are sub-normal, but general agreement that vitamin C levels are lower than normal in plasma,[128] despite the fact that elevated blood glucose has been reported to inhibit the uptake of ascorbate and of dehydroascorbate into cells.[326] Erythrocyte GSH levels are also slightly sub-normal. Elevated levels of 8-hydroxydeoxyguanosine (8-OHdG)[51] and of activated NFκB[244a] in blood monocytes have also been reported in diabetics. The significance of oxidative stress in the disease pathology is uncertain, but it is frequently proposed to be related to the hyperglycaemia. Other possible sources include elevated plasma lipids leading to increased lipid oxidation (e.g. by peroxisomal β-oxidation, which generates H_2O_2) and decreased levels of the antioxidant defence systems.

Maternal diabetes increases the incidence of foetal abnormalities, and oxidative stress has been suggested to contribute. Consistent with this, transgenic mouse embryos over-expressing CuZnSOD showed fewer malformations than non-transgenic embryos when exposed to hyperglycaemia caused by administration of streptozotocin to their mothers (Section 7.8.2).

9.4.3 *Mechanisms of glucose toxicity: aldose reductase*[128,281]

It has been proposed that the enzyme **aldose reductase**, which is found in many mammalian tissues including lens and retina, contributes to the development of diabetic cataract. This enzyme converts glucose to the polyalcohol **sorbitol** by reducing the aldehyde group of glucose:

$$RCHO + NADPH + H^+ \rightarrow RCH_2OH + NADP^+$$

Inhibitors of aldose reductase seem to delay the onset of cataract in animals. Why excessive activity of aldose reductase should be toxic is unknown; one suggestion is an excessive osmotic stress in the lens, as sorbitol accumulates to high concentrations. Alternatively, high aldose reductase activity may drain away cellular NADPH (needed by glutathione reductase) and make the lens vulnerable to oxidative damage. However, cataracts from human diabetics do not appear to contain osmotically-active concentrations of sorbitol. Excess sorbitol formation in kidney and nerves may interfere with synthesis of *myo*-inositol and thus intracellular signalling involving inositol phosphates. Several inhibitors of aldose reductase, e.g. **sorbinil** and **epalrestat**, are undergoing clinical testing. *In vitro*, aldose reductase can also reduce HNE and HNE–GSH conjugates, i.e. it might contribute to detoxification of this cytotoxic aldehyde.

9.4.4 *Non-enzymatic glycation and glycoxidation*[302]

Glucose may also be toxic by virtue of its ability to behave chemically as an aldehyde. Aldehydes are reactive substances and can bind to proteins, the amino groups of phospholipids, and DNA. Although most glucose in solution exists as non-aldehydic ring structures, glucose in its straight-chain form is an aldehyde. Thus it reacts slowly with proteins, amino-lipids such as phosphatidylethanolamine, and DNA, modifying them in a process referred to as **non-enzymic glycation** (Fig. 9.9). Glycation of proteins is the first step in the 'browning process', often called **Maillard browning**, after Professor Louis Camille Maillard. In 1912 he described the brown colour formed when heating mixtures of carbohydrates and amines. Indeed, the golden-brown colours (and some elements of the taste) of cooked foods are due to Maillard products. Maillard chemistry can generate pleasant aromas, such as those associated with roasted coffee and cocoa beans. However, mutagenic heterocyclic amines can be formed by Maillard chemistry in meats and fish. Many phenolic anti-oxidants, including catechins and the synthetic antioxidants BHT and butylated hydroxyanisole, decrease formation of these mutagenic products.[83]

Glycation is slow and reversible at 37 °C, and only significant in the case of proteins with a very slow turnover, e.g. collagen in connective tissue, and especially crystallins, which may have been present in the lens of the human eye for decades. By contrast, glycation is facilitated at higher glucose levels and is detectable in many proteins from human diabetics. Indeed, glycated haemoglobin levels are used as an index of how well blood glucose has been

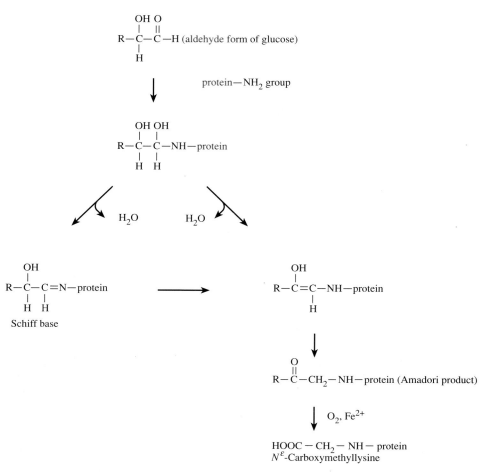

Fig. 9.9. Reactions of glucose (and other monosaccharides) with proteins. These are slow chemical reactions, producing glycated (glycosylated) products.

controlled over the previous month or so. Glycated haemoglobin (HbA$_{1C}$) contains a glucose Amadori product (Fig. 9.9) attached to the N-terminal valine of the β-chain. Glucose also modifies CuZnSOD in the erythrocyte, decreasing its activity; this may account for the lower SOD activity reported in the blood of some diabetics. Transferrin and caeruloplasmin can also be glycated, at least *in vitro*, resulting in decreased iron-binding capacity and loss of ferroxidase activity respectively. Both CuZnSOD and caeruloplasmin can fragment after glycation, to release copper ions.[133]

Glycation products can be oxidized (e.g. by ROS) to give **advanced glycation end-products** (AGEs; Figs 9.9–9.11). Indeed, ROS have been described as 'fixatives of glycation'. AGE accumulation on proteins is

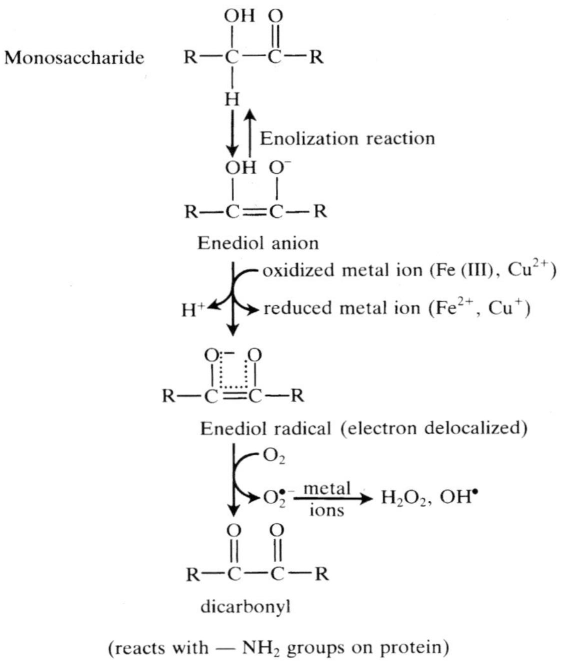

Fig. 9.10. Suggested involvement of radicals and metal ions in the oxidation of monosaccharides such as glucose. Dicarbonyls can react with the amino groups of proteins to form coloured products. Attack of OH$^\bullet$ produced in the above reactions upon further monosaccharide molecules can yield hydroxyalkyl radicals that decompose to produce more dicarbonyls.

accompanied by browning, increased fluorescence and cross-linking. AGE formation is irreversible, occurs over periods of months to years and can cause tissue damage. It is the animal equivalent of the browning of cooked foods. For example, accumulation of AGEs in collagen[71] may decrease the elasticity of connective tissue (impairing blood vessel function) and damage basement membranes of the kidney. In diabetic patients, AGEs are also present on circulating LDLs and in atherosclerotic lesions. AGE formation is thought to contribute to endothelial cell injury. Binding of glucose to the amino groups of LDL lipids can facilitate oxidation and eventual formation of cytotoxic aldehydes such as 4-HNE that modify apoB.[3] AGE products can also form directly on the apoB.

An additional way of making AGEs is to first oxidize the glucose and then allow the oxidation products to react with protein.[128] Monosaccharides can be oxidized, catalysed by traces of iron and copper ions, to produce $O_2^{\bullet-}$, H_2O_2, OH$^\bullet$ and toxic carbonyls which can damage proteins (Fig. 9.10); such reactions are involved in Maillard browning. Chemical structures found in AGEs include **carboxymethyllysine** and **pentosidine**, a fluorescent cross-link between lysine and arginine residues in AGE-modified proteins (Fig. 9.11). Levels of

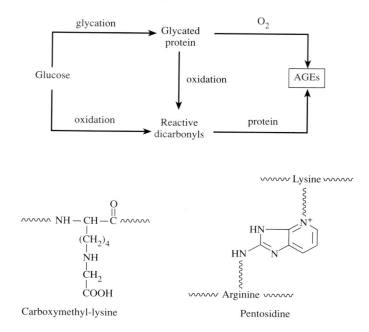

Fig. 9.11. Glyoxidation reactions. The combination of glycation and oxidation can result in the formation of advanced glycosylation end-products (AGEs) whose increased accumulation appears associated with tissue injury in diabetes mellitus. Glucose can be oxidized before binding to proteins, or glycated proteins can themselves oxidize releasing ROS. **Carboxymethyllysine** (N^ε) and **pentosidine** have been identified as glycoxidation-derived constituents of AGEs, but there are many others. Fig. 9.9 shows formation of N^ε-carboxymethyllysine from an Amadori product.

methylglyoxal,[301] formed from intermediates of glycolysis (Chapter 3), are increased in diabetes, and it can participate in AGE formation. It has been proposed that the glyoxylase system (Chapter 3) could be important in controlling such events *in vivo*, although its action could be compromised by falls in GSH levels in diabetes.

Both glycated and AGE-modified proteins can lead to oxidative stress. They may directly release $O_2^{\bullet-}$ and H_2O_2, and can also activate phagocytes. Several cells, including monocyte/macrophages, recognize AGEs using a cell-surface receptor, **RAGE** (receptor for AGE),[341] that is distinct from the scavenger receptors. RAGE is involved in monocyte migration and activation in response to AGEs. Normally, the RAGE may enable macrophages to recognize and engulf glycosylated cells, e.g. AGE-modified erythrocytes (which are cleared at a higher rate in diabetics). RAGE is also found on endothelial cells; exposure of them to AGEs activates NF-κB, up-regulates adhesion molecule production and may decrease GSH levels. AGE-modified proteins have been implicated in the excessive proliferation of blood vessels that occurs in diabetic retinopathy.

$$\overset{+}{\underset{||}{N}H_2}$$

The drug **aminoguanidine**, $H_2N-NH-C-NH_2$, a hydrazine compound which is reactive with dicarbonyls, is in clinical trials in diabetic patients: its ability to prevent AGE formation may help retard the development of kidney, eye, blood vessel and nerve damage. Lipoic acid (Section 3.21.5) also seems beneficial in decreasing some diabetic complications.[244a]

9.4.5 *How important is oxidative stress in diabetes?*[244a]

Possible sources of oxidative stress in diabetes include shifts in redox balance resulting from altered carbohydrate and lipid metabolism, increased generation of ROS (e.g. by glycation or lipid oxidation) and decreased levels of antioxidant defences such as GSH. There is a need to apply modern methods for assessing oxidative damage to diabetic patients. When improved methods are more extensively applied, it will be possible to answer some key questions:

(1) Are all patients with diabetes more oxidatively stressed, or only those patients with complications?
(2) Is the stress a cause or an effect of the disease or the complications? Does it respond to improvements in glycaemic control?
(3) Is it hyperglycaemia, hyperlipidaemia, other alterations in carbohydrate and/ or lipid metabolism, or the status of antioxidants which is most important in determining individual risk for development of complications?
(4) Can these parameters by decreased by antioxidant therapy?
(5) What is the effect of ascorbate—is it a beneficial antioxidant or does it become fragmented to products, such as L-threose and glyoxal, that can promote AGE formation?[208]

The 'browning' of food and beverages as a result of ascorbate oxidation/ glycation is well-known to food scientists. Dietary restriction of Lowry mice, a strain prone to age-related cataract development, delayed cataract onset and decreased levels of ascorbate in plasma, liver and kidney. Yet dietary ascorbate in humans appears to be protective against cataract (Chapter 10).

9.5 Ischaemia–reperfusion

Damage to the heart or brain by depriving a portion of the tissue of O_2 (**ischaemia**) is a major cause of death in Western society. Atherosclerosis, leading to rupture of a lesion, thrombosis and the blockage of an essential coronary or cerebral artery is usually the culprit (Section 9.4 above). Severe restriction of blood flow, leading to O_2 concentrations lower than normal (**hypoxia**), but not complete O_2 deprivation, can also result if the blocked artery is the major, but not the only, source of blood to the tissue in question. In fact, the terms hypoxia and ischaemia tend to be used interchangeably in the literature.

9.5.1 *Consequences of hypoxia*

Tissues made hypoxic or ischaemic survive for a variable time, depending on the tissue in question and the species it comes from. Thus skeletal muscle is fairly resistant to hypoxic injury, whereas the brain is more sensitive (Section 9.16 below). However, any cell made ischaemic for a sufficient period (except erythrocytes) will be irreversibly injured. Tissues respond to ischaemia in a number of ways. Early responses usually include increased rates of glycogen degradation and glycolysis, leading to lactate production and acidosis. ATP levels begin to fall, AMP is degraded to cause an accumulation of hypoxanthine (Fig. 9.12) and intracellular Ca^{2+} levels rise, activating Ca^{2+}-stimulated proteases and possibly nitric oxide synthase (NOS) (if present), although O_2 is a substrate for NOS and so complete O_2 deprivation will block $NO^{\bullet}$ formation. Membrane damage (e.g. blebbing) becomes visible under the microscope (Chapter 4).

9.5.2 *Reoxygenation injury*

If the period of ischaemia or hypoxia is insufficiently long to injure the tissue irreversibly, much of it can be salvaged by reperfusing the tissue with blood and re-introducing O_2 and nutrients. In this situation, reperfusion is a beneficial process overall. However, Parks, Granger, McCord and their colleagues in the USA showed in the early 1980s that re-introduction of O_2 to an ischaemic or hypoxic tissue could cause an additional insult to the tissue (**reoxygenation injury**) that is, in part, mediated by ROS.[94] The relative importance of reoxygenation (often called **reperfusion**) injury depends on the time of ischaemia/hypoxia. If this is sufficiently long, the tissue is irreversibly injured and will die, so reoxygenation injury is not important (one cannot further injure a dead tissue). Nevertheless, if a dead or dying tissue is reperfused *in vivo*, this can release potentially toxic agents, including xanthine oxidase and 'catalytic' transition-metal ions, into the systemic circulation, causing problems to other body tissues. For example, gut ischaemia can lead to depression of heart function and xanthine oxidase can bind to endothelial cells. It can produce $O_2^{\bullet-}$ to antagonize the action of $NO^{\bullet}$, as well as generating potentially cytotoxic species such as $ONOO^-$ and H_2O_2.

However, for a relatively brief period of ischaemia/hypoxia, the reoxygenation injury component may become more important and the amount of tissue remaining undamaged can be significantly increased by including scavengers of ROS in the reoxygenation fluid. The meaning of 'relatively brief' in this context depends on the tissue in question, whether one is dealing with ischaemia or hypoxia and, if the latter, what degree of hypoxia was achieved.

Although the enzyme xanthine oxidase is frequently used as a source of $O_2^{\bullet-}$ in experiments *in vitro* (Section 1.11), almost all the xanthine-oxidizing activity present in healthy animal tissues is a dehydrogenase enzyme that transfers electrons not to O_2, but to NAD^+, as it oxidizes xanthine or hypoxanthine

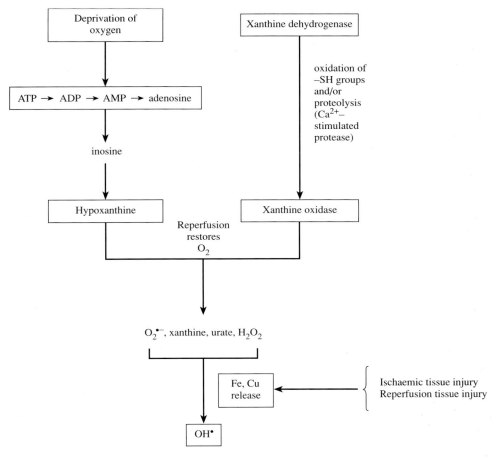

Fig. 9.12. A suggested mechanism for tissue injury upon reoxygenation of ischaemic or hypoxic tissues. Updated and modified from McCord, J.M. (1987) *Fed. Proc.* **46**, 2402. The enzyme that converts adenosine to inosine is **adenosine deaminase**. Adenosine modulates the activity of numerous cell types and may help to protect cells against some of the consequences of ischaemia–reperfusion.

into uric acid. When tissues are disrupted, some of the xanthine dehydrogenase can be converted into xanthine oxidase by oxidation of essential −SH groups or by limited proteolysis (e.g. involving Ca^{2+}-stimulated proteases). Xanthine oxidase produces $O_2^{\bullet -}$ and H_2O_2 when xanthine or hypoxanthine are oxidized (Chapter 1).[142] The depletion of ATP in hypoxic tissue causes hypoxanthine accumulation. This hypoxanthine can be oxidized by the xanthine oxidase when the tissue is reoxygenated, causing rapid generation of $O_2^{\bullet -}$ and H_2O_2, which might lead to severe tissue damage. Released transition-metal ions can then promote $OH^{\bullet}$ formation (Fig. 9.12).

9.5.3   *Adaptation to hypoxia: the role of transcription factors*[99]

Tissues can adapt to some degree of hypoxia. A well-known example is increased synthesis of **erythropoietin**, a hormone which stimulates erythrocyte production, in humans at high altitudes. As early as the seventeenth century, physicians noted that blood becomes more viscous at high altitudes. It has been suggested that the molecular mechanism for 'hypoxia sensing' involves ROS production (Section 6.8). Hypoxia leads to more erythropoietin production by causing increased synthesis of the transcription factor **HIF-1** (hypoxia-inducible factor 1) which binds to an enhancer element for the erythropoietin gene. HIF-1 also up-regulates expression of **vascular endothelial growth factor**, to help develop new blood vessels, and of some enzymes of glycolysis. Heat-shock proteins (including **haem oxygenase I**) are sometimes up-regulated in response to hypoxia and exert protective effects (Section 4.15).

Studies on the HeLa tumour cell line suggest that hypoxia activates the transcription factor AP-1 by a mechanism involving new protein synthesis. By contrast, reoxygenation of hypoxic HeLa cells induced NF-κB activation.

9.5.4   *Intestinal ischaemia–reoxygenation*[94]

The first evidence supporting the hypothesis outlined in Fig. 9.12 came from studies upon intestine. Partial occlusion of the artery supplying blood to a segment of cat small intestine (hypoxia), followed by reperfusion, causes gross, histologically observable damage to the tissue and increases intestinal vascular permeability. Intravenous administration of SOD, or oral administration of **allopurinol** (an inhibitor of xanthine oxidase) to the animals before removal of the arterial occlusion, decreased damage. Infusion of a mixture of hypoxanthine and xanthine oxidase into the arterial supply of a segment of normal cat intestine increased vascular permeability, an effect that was decreased by the presence of SOD or dimethylsulphoxide (DMSO) in the infusion. The effect of DMSO, a powerful scavenger of OH$^\bullet$ (Chapter 2), implicates this radical in the damage, although inhibition by a single scavenger is insufficient evidence to prove OH$^\bullet$ involvement. Desferrioxamine, which suppresses iron-dependent free-radical reactions, also protected against the injury.

Iron ions able to catalyse OH$^\bullet$ production could arise as a result of cell necrosis causing release of intracellular iron, proteolytic digestion of metalloproteins, release of iron from ferritin or iron–sulphur proteins by $O_2^{\bullet-}$ and/or breakdown of haemoglobin (e.g. by H_2O_2) liberated as a result of bleeding upon tissue reperfusion. Regional intestinal ischaemia in cats results in an accumulation of hypoxanthine in the tissue; the hypoxanthine disappears quickly on reperfusion and both lipid peroxidation and the formation of GSSG can be measured in the reperfused intestine. It has also been reported that SOD or desferrioxamine administration diminish the mortality consequent upon bowel ischaemia in rats. Intestine is rich in xanthine dehydrogenase activity, and some of this is converted to oxidase during ischaemia/hypoxia.

Hence, as far as intestinal ischaemia–reoxygenation is concerned, the essential features of the model in Fig. 9.12 are supported by experimental evidence. Although allopurinol is not specific as an inhibitor of xanthine oxidase, other evidence for the involvement of this enzyme has been obtained. **Pterinaldehyde** (2-amino-4-hydroxypteridine-6-carboxaldehyde), a powerful xanthine-oxidase inhibitor that often contaminates commercial preparations of the vitamin folic acid,[215] also offered protection. Feeding animals on a diet rich in tungsten decreases tissue concentrations of xanthine dehydrogenase (Section 8.12.6); intestinal segments from rats pre-treated in this way showed less re-oxygenation injury after hypoxia than segments from normal rats.

The hypoxanthine–xanthine oxidase system is probably not the only source of ROS to which reoxygenated intestine (and other tissues) are subjected *in vivo*. Mitochondria damaged by ischaemia may 'leak' more electrons than usual from their electron transport chain,[197] forming more $O_2^{\bullet -}$. Generation of ROS by activation of neutrophils entering (or already present within) reoxygenated intestine is another potential source.[94] Post-ischaemic tissues can generate increased amounts of leukotriene B_4, PAF and other chemoattractants for neutrophils, and up-regulate expression of adhesion molecules. Neutrophils then adhere to endothelium and may activate to release products ($O_2^{\bullet -}$, H_2O_2, HOCl, HOCl-derived chloramines, eicosanoids, proteolytic enzymes such as elastase) that can worsen injury. Indeed, neutrophil depletion of animals, or pre-treatment of them with antibodies that prevent neutrophil adherence to endothelium, has been reported to diminish reoxygenation injury to intestine and other tissues in whole animal studies.[146,314]

Intestinal ischaemia can lead to acceleration of heart rate and increased blood pressure. Apparently, ROS produced by reoxygenation are one of the factors that stimulate sensory nerve endings in the viscera to produce these cardiovascular reflexes.[282]

9.5.5 *Cardiac ischaemia–reoxygenation*

The phenomenon

The model in Fig. 9.12 naturally attracted the attention of cardiologists, since myocardial infarction (essentially ischaemic/hypoxic injury to heart muscle) is a major cause of death, and thrombolytic therapy is now widely used to dissolve blood clots and promote reperfusion. Heart muscle can survive up to 60–90 min of ischaemia, depending on the species.

After relatively short periods of ischaemia or hypoxia, reoxygenation of hearts has been shown to produce an additional 'insult' to the tissue that can be diminished by MnSOD or CuZnSOD, catalase, the spin-trap PBN, scavengers of $OH^\bullet$ radical such as mannitol, desferrioxamine, or allopurinol. The manifestations of the insult vary according to species and the experimental protocol. For example, reoxygenation of isolated rat hearts after brief ischaemia can produce irregular contractions (**arrhythmias**).[23] Reoxygenation after 15–20 min ischaemia can result in delayed return of contractile function (**myocardial stunning**) in pig and dog hearts, even within the animal.[286]

The **stunned myocardium** refuses to contract, but is not irreversibly injured and will eventually recover normal function (hours to days later).

Importance of the model used

Many studies showing protective effects of antioxidants have been done with hearts isolated from rats, rabbits, dogs or pigs and perfused with buffer. These studies are easy to control and standardize. Problems include possible contamination of buffers with transition-metal ions, and the absence of the range of antioxidants normally present in blood.

Some studies have been performed with organs *in situ*. For example, in 'open-chest' dogs (in which the chest is opened, the coronary-artery branch partially or completely occluded and the block removed after various periods) several studies showed that pre-treatment of animals with SOD plus catalase, or allopurinol could decrease the amount of tissue death (size of the **infarction**) produced by prolonged (often $\geqslant 60$ min) occlusion, followed by reoxygenation. This model has also been used to show the involvement of OH$^\bullet$ in myocardial stunning resulting from shorter (10–15 mins) ischaemia followed by reoxygenation in open-chest anaesthetized dogs (Fig. 5.6). However, the molecular mechanism of stunning is unknown: its rapid onset and the protective effects of thiol compounds have led to suggestions that OH$^\bullet$ oxidizes −SH groups on membrane ion-transport systems that are essential for maintaining membrane potential and for keeping intracellular Ca^{2+} at low levels. Normal heart function depends on periodic 'waves' of electrical activity to stimulate the heart muscle to contract in an ordered sequence, so it is easy to envisage how disruptions of membrane ion transport could provoke stunning or arrhythmias.

The extent of the protection by antioxidants has been very variable in different experiments, especially in studies of effects on infarct size, the most controversial area. At least three reasons account for this.

1.   The importance of reoxygenation injury as a fraction of total tissue damage declines as the period of ischaemia increases, so that in experiments with long periods of ischaemia/hypoxia so much tissue injury has been done by O_2 deprivation that antioxidants are unlikely to offer much protection; a tissue irreversibly injured by O_2-deprivation will die, however it is reperfused. In dog heart, for example, coronary occlusion for less than about 20 min often produces stunning on reperfusion, but essentially no infarct. Occlusion for more than 4 h followed by reperfusion gives an infarct about the same size as that resulting from permanent occlusion (i.e. damage is maximum and cannot be further increased). Whereas antioxidants are protective against stunning, they do not alter infarct size in the latter experiment.

2.   In 'whole-animal' studies (e.g. using open-chest pigs or dogs), the extent of O_2-deprivation during arterial occlusion can vary considerably because of collateral vessels. Thus 'ischaemia' is often hypoxia, to variable extents. If flow does not cease completely, not only does some O_2 enter the 'ischaemic' tissue, but also some metabolic products, such as H^+, are removed and acidosis is less severe.

3.   Some of the 'antioxidants' used in these studies show bell-shaped dose–response curves. For example, in studies upon production of arrhythmias in isolated rat hearts after a brief period of ischaemia, there is an optimal concentration of SOD for protection, and higher or lower concentrations give smaller protective effects. Similar effects are observed if release of enzymes from damaged cells is measured (Fig. 9.13). The reasons for these effects remain to be established. For example, a deleterious effect of impurities in some of the antioxidants studied might become significant at very high antioxidant concentrations.[129]

More recent studies have used the conscious dog (or pig) models in an attempt to eliminate possible confounding effects of anaesthesia, surgical trauma and mechanical ventilation.[263,286] In a first operation, probes that record myocardial function and a hydraulic 'balloon occluder' are implanted, the latter around the mid left anterior descending coronary artery. The animals are allowed to recover. Activation of the occluder in the awake animal for 15 min, followed by reperfusion, produces myocardial stunning which can be attenuated by thiol compounds and desferrioxamine. Spin adducts from pre-administered PBN can also be detected in coronary venous blood after hypoxia–reperfusion. Both the extent of the stunning and the amounts of PBN adducts detected are lower in 'conscious' than in 'open-chest' dogs, but the stunning phenomenon is still evident.

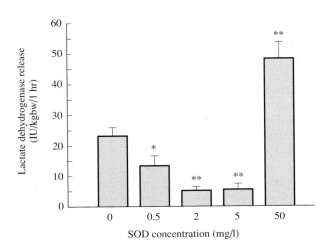

Fig. 9.13. Dose–response effect of human recombinant MnSOD on lactate dehydrogenase release in isolated rabbit hearts subjected to 1 h of global ischaemia followed by 1 h of reperfusion. MnSOD was not given (controls, $n = 5$) or given at 0.5 mg/L ($n = 3$), 2 mg/L ($n = 10$), 5 mg/L ($n = 5$) and 50 mg/L ($n = 5$). Means and standard errors are shown (*, $P < 0.05$; **, $P < 0.01$). Note that while MnSOD lowered enzyme release at lower doses, it *increased* the release at the higher dose. From *Free Rad. Biol. Med.* **9**, 465 (1990) by courtesy of Professor Joe McCord and Elsevier.

The relevance of xanthine oxidase[142]

The multiple reports that antioxidants and allopurinol protect against cardiac ischaemia–reoxygenation injury naturally led to proposals that the model shown in Fig. 9.12 explains reoxygenation damage in heart. Indeed, the adenosine deaminase inhibitor **erythro-9-(2-hydroxy-3-nonyl)adenine**, which should block formation of hypoxanthine (Fig. 9.12) also decreased free-radical formation in the isolated reperfused rat heart.[338] Exposure of myocytes or isolated hearts to ROS-generating systems produces severe injury, including inactivation of the ATP-dependent Ca^{2+}-sequestering system of cardiac sarcoplasmic reticulum. Damage not only to myocytes but also to vascular endothelial cells in the blood vessels of the heart can occur. Ca^{2+} activates the protease that converts xanthine dehydrogenase to the oxidase form (Fig. 9.12).

However, some problems have arisen. Rat and dog heart contain xanthine dehydrogenase activity, but the rate of proteolytic conversion of dehydrogenase to oxidase during ischaemia is slow (e.g. in rat heart one study reported that about 20% of enzyme is present as oxidase to start with and it takes about 4 h of ischaemia for this to increase to 30%). Of course, the basal level of xanthine oxidase could still lead to extra ROS generation because of increased hypoxanthine levels. The activity of xanthine dehydrogenase in rat heart has been reported to increase with the age of the animals, which may help to explain the observation that hearts from newborn rats are more resistant to ischaemia–reperfusion.[246]

Rabbit and pig hearts have been reported not to contain any xanthine oxidase activity.[67] If these reports are correct, how then can one explain the protective effects of allopurinol reported by several groups? It may be that allopurinol inhibits another enzyme involved in ROS generation. It might also protect by preventing the depletion of purine nucleotides, that can be used as substrates for resynthesis of ATP on reoxygenation.[158] Pre-treatment of animals with allopurinol causes formation of oxypurinol (Chapter 1). For example, administration of allopurinol to humans produces plasma concentrations of oxypurinol of 40–300 μM. Oxypurinol is a good scavenger of the myeloperoxidase-derived species HOCl.

Questions have also been raised about whether the established protective action of mannitol in isolated hearts is really due to $OH^\bullet$ scavenging.[23] Both mannitol and glucose react with $OH^\bullet$ with about the same rate-constant (Chapter 2), but mannitol is far more protective against ischaemia–reperfusion-induced arrhythmias in isolated rat heart than is equimolar glucose. Deleterious cell volume changes can occur in response to oxidative stress (Chapter 4) and mannitol may help prevent these by exerting osmotic effects.

Of course, *in vivo* sources of radicals other than xanthine oxidase, such as mitochondria[197] and infiltrating neutrophils,[146,314] may also be important. Neutrophils play little role in the arrhythmias and stunning seen after brief ischaemia followed by reperfusion, but they may be important after longer ischaemic periods. Vascular endothelial cell injury and neutrophil adherence may contribute to the **no-reflow phenomenon**, the inability to reperfuse areas of previously ischaemic tissue, often thought to be due to plugging of capillaries by swelling of endothelial cells and neutrophil accumulation. Thus depleting animals of neutrophils, or injecting them with an antibody that prevents neutrophil adherence to endothelium, before performing ischaemia–reperfusion studies *in situ* has been reported to decrease the size of the infarct produced in some studies.

The relevance of transition metals[36]

If $OH^\bullet$ is formed in the reoxygenated heart by Fenton chemistry (Fig. 9.12), then a source of transition-metal ions must be identified. Their importance is suggested by the generally protective effects of iron chelators (such as desferrioxamine) on ischaemia–reoxygenation-induced damage and the fact that hearts isolated from iron-overloaded animals are more susceptible to reoxygenation injury *in vitro*.

Iron and copper might simply be released as a result of cell injury (Fig. 9.1). Haem and 'free iron' could derive from bleeding into reperfused myocardium that has undergone vascular damage during ischaemia. Another possible source is myoglobin.[239] Incubation of myoglobin with excess H_2O_2 causes liberation of haem and iron ions. Myoglobin–H_2O_2 mixtures also exert direct pro-oxidant effects (Section 3.18.3). Of course, studies with isolated buffer-perfused hearts must not ignore the possibility of metal ion contamination of the perfusing solution.

Despite all the above questions, the protective effects of SOD, desferrioxamine and other antioxidants in many of the studies performed cannot be doubted. The biggest area of debate is whether antioxidants have any effect on infarct size. One additional area of uncertainty is whether or not $NO^\bullet$ and $ONOO^-$ play any significant role in ischaemia–oxygenation injury. They could cause damage, but $NO^\bullet$ might promote vasodilation and down-regulate NF-κB, deterring expression of adhesion molecules on endothelial cells and thus minimizing neutrophil recruitment.

Clinical relevance[150]

In 'classical' human myocardial infarction, it has been suggested that occlusion of a coronary artery is maintained for such a long period that ischaemic/hypoxic injury is the sole cause of tissue death, reoxygenation injury being insignificant. However, the early use of thrombolytic agents (such as **streptokinase** or **tissue plasminogen activator**) to produce clot dissolution is now commonplace in many countries. Combined use of an antioxidant (such as desferrioxamine or SOD) and a thrombolytic agent might be expected to give enhanced benefit, although results of controlled trials using human recombinant CuZnSOD have been disappointing.[80]

Severe arrhythmias rarely result from thrombolytic therapy. However, reoxygenation injury might play a significant role in the depressed 'stunning-like' myocardial function sometimes seen after open-heart surgery and heart transplantation. Increased TBARS, conjugated dienes and GSSG release have been reported during heart reperfusion after cardiac surgery, indicative (although not conclusive, given the problems with the former two assays; Chapter 5) of oxidative stress. During cardiopulmonary bypass surgery, blood passes through an oxygenator before being returned to the circulation. Damage to erythrocytes, complement activation and phagocyte stimulation (Section 6.7.3) can lead to the release of trace amounts of 'catalytic' iron and copper ions.[203]

There are several suggestions that patients with low tissue levels of antioxidant nutrients, such as vitamin E, suffer more side-effects from open-heart surgery and that administration of vitamin E or allopurinol prior to surgery might be beneficial. Since it is widely thought that xanthine oxidase is absent from human heart,[67] any protective effect of allopurinol is presumably due to other mechanisms. Another possibility is that the assays have missed the enzyme: if it were only in the vascular endothelial cells, the activity in a whole-tissue homogenate might be too low to measure.

Another area of interest is the re-opening of partially occluded coronary arteries by **angioplasty**. In balloon angioplasty a coronary artery is completely occluded when the balloon is inflated to 'squash' the atherosclerotic lesions. Abrupt reperfusion occurs upon balloon deflation, when blood flows through the 'cleared' artery. Usually the short duration of the occlusion (1–2 min) precludes significant reoxygenation injury, however. Nevertheless, coronary artery constriction, attenuated by SOD, has been observed in dogs treated by balloon angioplasty.[159] Presumably excess $O_2^{\bullet-}$ generation resulting from reoxygenation of the vascular endothelium transiently decreased net $NO^{\bullet}$ production.

A more important problem is the high incidence of **restenosis** ($>50\%$ of cases) after angioplasty. This appears related to endothelial damage caused by the balloon, leading to accelerated atherosclerosis marked by excessive proliferation of smooth muscle cells. There is evidence that high doses of probucol decrease the severity of restenosis somewhat. This effect might be due to antioxidant effects, or to direct effects in inhibiting cell proliferation.[297a]

An alternative treatment for coronary atherosclerosis is bypass grafting, and again one problem is accelerated atherosclerosis in the grafted vessels. Endothelial injury, perhaps caused by hypoxia (e.g. during the preparation and placement of the vessel), followed by reoxygenation, has been suggested to contribute.[42] Short-term animal studies have suggested that antioxidants (e.g. desferrioxamine–manganese and lazaroids; Chapter 10) might exert some protective effects.[55,89]

9.5.6   *Ischaemic preconditioning*[226,286]

Short periods of ischaemic stress on the myocardium can sometimes induce adaptations that result in decreased sensitivity to a subsequent ischaemic insult. There appear to be two distinct phases of protection: a transient one that occurs within minutes but disappears quickly, and a delayed phase that takes hours to become apparent but can then last for days or longer. For example, if dog hearts were preconditioned by 5 min episodes of coronary artery occlusion, a

40 min re-occlusion resulted in infarcts up to 75% smaller than in control dogs. The effects of preconditioning can also be observed against stunning and arrhythmias, and it may have clinical relevance in **angina pectoris**. In this disease, atherosclerosis decreases blood supply through the coronary arteries. Exercise can cause the heart's O_2 requirements to exceed its blood supply, causing acute chest pain. Repetitive brief ischaemia in angina patients may, paradoxically, help protect against myocardial infarction.

The mechanism of preconditioning is intensively debated. Evidence (largely from studies in rabbits) has been presented that **adenosine** released from ischaemic myocytes may be involved (legend to Fig. 9.12). There are also suggestions that $NO^{\bullet}$, prostacyclin, induction of heat-shock proteins and ROS are important. As discussed in Section 9.1.2 above, the ROS produced by pig heart during stunning seem to be involved in the development of preconditioning *in vivo*. Similar data have been reported from open-chest rabbits. One possibility is that the second 'window' of resistance is caused by induction of antioxidant defences such as MnSOD. Indeed, injection of a small dose of endotoxin into rats, or pre-treating them with interleukin 1, can increase resistance of the heart to ischaemia–reoxygenation and induction of anti-oxidant defences may again be involved. Preconditioning has been observed in other tissues, including skeletal muscle and skin.

9.5.7   *Shock-related ischaemia–reoxygenation*[294]

Severe bleeding can cause a rapid fall in blood pressure. This places many tissues at risk since it results in hypoperfusion with blood and hence hypoxia. When blood volume is restored, a 'shock' syndrome (**haemorrhagic shock**) can result. Stomach, gut, kidney and pancreas are among the organs that can be affected.

Pre-treatment of rats or cats with allopurinol or SOD before haemorrhagic shock has been reported to minimize the gastric lesions observed. Both allopurinol and desferrioxamine have been claimed to increase survival in dogs subjected to severe haemorrhagic shock.[252] Hence ROS make some contribution.

ROS, $NO^{\bullet}$ and eicosanoids may also play some role in shock induced by bacterial endotoxin, e.g. in sepsis caused by infection with Gram-negative bacteria. Here there is marked over-production of $NO^{\bullet}$, largely involving iNOS enzymes induced in various organs in response to such cytokines as TNFα. Several inhibitors of NOS are in clinical trial for the treatment of endotoxic shock. If $O_2^{\bullet-}$ is generated during shock, it might react with the 'excess' $NO^{\bullet}$ to form $ONOO^-$. Indeed, evidence for this has been presented (Fig. 9.14).

9.5.8   *Birth trauma*

Both full-term and premature babies can suffer hypoxia during delivery, and plasma hypoxanthine levels have been proposed as an important marker for

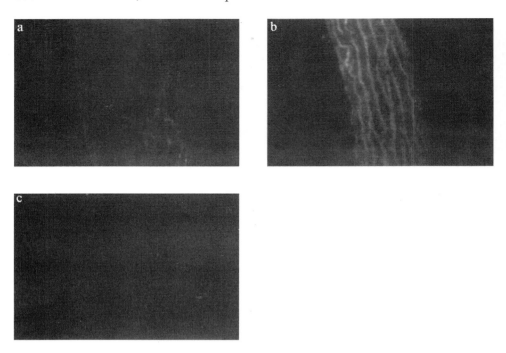

Fig. 9.14. Formation of nitrotyrosine in an endotoxic shock model. Nitrotyrosine, suggestive of ONOO⁻ generation, was demonstrated by immunostaining in the thoracic aorta of a rat after injection of *E. coli* lipopolysaccharide. Immunofluorescence photomicrograph of tissue from (a) control rats, (b) rats injected with endotoxin and (c) rats injected with endotoxin and a NOS inhibitor. Photograph by courtesy of Dr Csaba Szabó.

asphyxia during birth. The lungs of premature babies are ill-developed (Section 7.8.3) and this may create some degree of hypoxia. Xanthine oxidase activity can be detected in plasma from healthy full-term babies,[288] but levels of both this enzyme and its substrate hypoxanthine are increased in 'poor outcome' premature babies. Normal birth may be regarded as a mild hypoxia–reoxygenation that healthy babies can easily cope with, but difficult births or premature births can potentiate the problems.

9.5.9   *Kidney damage*[266]

Kidney is a metabolically active organ with high levels of ferritin and haem proteins. Its sensitivity to oxidative damage is well established. For example, kidney injury by the antibiotic **gentamicin**[73] may involve increased ROS production (Section 8.13) and the kidney can be damaged by circulating haemoglobin or myoglobin. Indeed, extensive haemolysis or severe muscle injury can lead to acute kidney failure involving released haemoglobin or myoglobin respectively. The isoprostane **8-*epi*-PGF$_{2\alpha}$**, produced during lipid peroxidation, is a powerful vasoconstrictor in the kidney.

Deposition of antigen–antibody complexes in the kidney glomeruli can lead to complement activation and infiltration of neutrophils. Resident phagocytic (**mesangial**) cells in the glomeruli can also respond to complement activation by secreting eicosanoids, proteases and ROS. For example, infusion into rats of an antibody directed against glomerular basement membranes causes severe glomerular inflammation involving both neutrophils entering the tissue and complement activation. Infusion of catalase was protective, but infusion of SOD was not. Both H_2O_2 and HOCl (from the action of myeloperoxidase) seem to be involved in the damage, as might the action of proteases such as elastase and collagenase. Other groups have reported that injection into rats of 'nephrotoxic serum' containing antibody to rat glomerular basement membranes produces renal injury that can be ameliorated by treating the animals with SOD. The acute phase of the glomerular injury produced by injecting this type of antibody into rabbits was reported to be suppressed by desferrioxamine. The variable results indicate the complexity of this model system of immune injury to kidney, but all point to some role of ROS in causing damage.

Reoxygenation of kidney after a brief period of ischaemia or hypoxia produces a reoxygenation injury that can contribute to tissue damage. By contrast, after prolonged periods of ischaemia, the ischaemic damage itself may be so extensive that reperfusion injury is insignificant. Several groups have reported that pre-treatment of whole animals or isolated kidneys with SOD, catalase, desferrioxamine or allopurinol diminishes reperfusion injury.

There are at least two clinical situations in which the kidney is subjected to temporary hypoxia: shock (when all body organs are affected to variable degrees) and renal transplantation. Some degree of hypoxia in organs stored for transplantation is difficult to avoid. One aspect is **cold hypoxia**, the tissue becoming hypoxic when stored at low temperature in an organ preservation fluid. Another is **warm hypoxia**, when the organ is placed inside the body cavity and warms up before reperfusion, while the blood vessel connections are made. Rat kidney contains xanthine dehydrogenase, although its irreversible (proteolytic) conversion to xanthine oxidase in ischaemic tissues occurs only slowly (with a half-life of about 6 h). There appears to be much less xanthine oxidase in dog or human kidney. Hence it is uncertain whether the model in Fig. 9.12 accounts for reperfusion injury in kidney.

Nevertheless, in transplanted human kidneys subjected to 25–28 h of cold hypoxia, SOD infusion into the renal artery at implantation appeared to produce significant improvement in short-term renal function: the difference was less marked with storage times below 25 h.[235] Similar data with allopurinol have been published. However, the overall clinical impact of this kind of therapy on renal transplant surgery may be minor: the priority is to obtain the freshest kidneys possible and improve the preservation fluids, which often already have added allopurinol (Section 9.5.11 below).

Transplanted kidneys are frequently infiltrated by host phagocytic cells, and it is possible that the ROS (and possibly RNS) they produce might contribute to transplant rejection. Enhanced immunostaining for nitrotyrosine has been measured in 'rejected' human kidney transplants. One of the nitrated targets was identified as MnSOD; exposure of this enzyme to ONOO⁻ causes inactivation (Section 6.7.3). Thus antioxidants may have a beneficial role not only in allowing organ preservation for longer periods, but also in diminishing graft rejection. These comments could equally be applied to other transplanted organs, such as liver or heart, and to transplants of pancreatic β-cells, dopamine neurones, corneas or skin flaps. For example, several groups have reported[59] that the 'survival time' of skin flaps is increased by SOD, desferrioxamine or allopurinol.

Patients awaiting new kidneys are maintained by dialysis. Dialysis may cause oxidative stress, e.g. by activating phagocytes and depleting plasma antioxidants such as ascorbate. The clinical significance of this has yet to be determined.

9.5.10 *Liver transplantation*

Several groups have been interested in the possibility of hypoxia–reoxygenation injury in transplanted livers. For example, combined transplantation of liver and intestine in pigs led to subsequent elevations in the urinary excretion of 8-OHdG, suggestive of increased oxidative DNA damage.[174] Human transplant patients have been reported to exhale more ethane (a putative marker of lipid peroxidation) at the time of organ reperfusion.[243a]

Rat liver contains xanthine dehydrogenase, but this is only slowly converted to oxidase during ischaemia, although both enzymes can be released into the systemic circulation from damaged rat liver and may contribute to extra-hepatic damage. However, dog and human liver have been reported to contain little of either form of the enzyme. Other sources of oxidative stress in reperfused livers include increased $O_2^{\bullet-}$ production by damaged mitochondria, adherence to the liver vascular endothelium of circulating neutrophils, and activation of the macrophages (**Kupffer cells**) present in the liver itself. The role of the latter can be investigated by selectively killing them using the agent **gadolinium chloride**, $GdCl_3$. Endothelial damage early in liver reoxygenation may involve activation of Kupffer cells, whereas neutrophils (and, in some animals, xanthine oxidase) may contribute to later damage. Transplant livers usually come from accident victims, who have sometimes been drinking heavily, and so their function might be somewhat compromised by previous ethanol exposure.

Over-expression of MnSOD in liver mitochondria decreased enzyme release from liver and the activation of AP1 and NFκB after hypoxia reoxygenation in mice, illustrating the importance of mitochondrial $O_2^{\bullet-}$ generation in causing injury in this model.[345a]

9.5.11 *Organ preservation fluids*[277]

As mentioned above, antioxidants are frequently added to organ preservation fluids. Examples are mannitol (although its major action may be as an osmotic agent that delays cell swelling), GSH, allopurinol and desferrioxamine. For example, the **University of Wisconsin preservation fluid** has millimolar levels of added GSH. However, by the time the fluid is used, this is largely or completely oxidized to GSSG. The fluid also contains high K^+ levels, 1 mM allopurinol, **lactobionic acid** (a weak Ca^{2+} and iron chelator), 5 mM adenosine (which may suppress phagocyte ROS production), the trisaccharide sugar raffinose and hydroxyethyl starch. Liver, pancreas and kidney can be preserved for many hours by flushing with this solution and storing at 0–5 °C.

9.5.12   *The eye*

Parts of the eye can suffer from ischaemia. The retina may become hypoxic as a result of atherosclerosis of ocular blood vessels (especially in diabetes), damage to the retinal pigment epithelium, or by vascular occlusion by sickled cells in patients with sickle-cell anaemia. Rises in intra-ocular pressure can also decrease blood flow (Section 7.7). In experimental animal systems, it has been shown that reperfusion of ischaemic retina leads to ROS production; for example, OH$^\bullet$ has been identified[222] by the salicylate trapping technique (Chapter 5).

9.5.13   *Limbs, digits and sex organs*

Limb ischaemia or hypoxia can occur as a result of several cardiovascular conditions, including severe atherosclerosis (especially in poorly controlled diabetics, who may suffer gangrene of the extremities), thrombosis, arterial injury resulting in rapid blood loss, and (to a controlled extent) in the use of tourniquets to provide a 'bloodless field' in orthopaedic surgery. Modern microsurgical techniques for re-attachment of severed digits, limbs and even penises often result in the re-oxygenation of tissue that has been hypoxic or ischaemic for a considerable time.

Skeletal muscle is, compared with most tissues, fairly resistant to ischaemic injury. However, after several hours of ischaemia, an ischaemia–reperfusion injury to muscle can be demonstrated.[325] Experiments with rat hindlimbs have shown some protective effects of SOD plus catalase against the injury, and other experiments with isolated dog gracilis muscle have shown some protection by SOD, allopurinol and DMSO. Levels of xanthine oxidase in pig or human muscle have been reported to be lower than in rat,[63] although allopurinol has been claimed to be beneficial in decreasing lower limb swelling after leg artery bypass surgery.[276]

It may not be the muscle cells themselves but the vascular endothelium that is the primary site of damage by ischaemia–reoxygenation of limbs and digits. This can lead to a 'no reflow' situation, e.g. as excess numbers of neutrophils adhere to damaged endothelium, and also as thrombosis occurs. Reperfusion of large areas of damaged muscle (e.g. crushed legs) can cause systemic injury to which both complement activation and ROS appear to contribute, e.g. by release into the bloodstream of xanthine oxidase, myoglobin and transition-metal ions.[37]

Dupuytren's contracture[206] has also been suggested to involve ischaemia–reoxygenation. Tendons in the hand shorten abnormally and pull the fingers into a closed position. The disease appears to be associated with elevated levels of hypoxanthine and xanthine oxidase in the palmar fascia.

9.5.14   Plants[50]

Plants can suffer hypoxia–reoxygenation, e.g. if the roots of a plant (or the whole plant) are covered by water during flooding, which restricts the access of O_2. The plant responds by shutting down the synthesis of most proteins, but up-regulates the synthesis of others, the **anaerobic polypeptides**. The most studied of these is the enzyme alcohol dehydrogenase.

Many plants appear, morphologically, to survive flooding only to die on subsequent emergence, a classical reperfusion injury. Indeed, lipid peroxidation after flooding was much higher in rhizomes of the flooding-sensitive plant *Iris germanica* than in rhizomes of the tolerant *Iris pseudocorus*. The latter species increases CuZnSOD activity during anoxia, but the former does not.[50]

9.5.15   Chemical ischaemia–reoxygenation: carbon monoxide poisoning[300]

Carbon monoxide, CO, is a colourless, odourless, tasteless gas produced by incomplete combustion of carbon-containing materials. It is found in wood-smoke, car exhausts, polluted air and cigarette smoke (Chapter 8). It can also be generated *in vivo* as a product of the enzyme haem oxygenase (Section 4.15) and as a minor end-product of lipid peroxidation.

Low levels of CO may be useful *in vivo*: a role as a neurotransmitter has been suggested, for example. However, high levels are clearly toxic. Toxicity results from the avid binding of CO to haem proteins, especially haemoglobin (where it out-competes O_2 for binding by a factor of about 220, and the resulting **carboxyhaemoglobin** dissociates only slowly). Cytochrome oxidase can also be inhibited. Coma can occur at carboxyhaemoglobin levels of $>40\%$ (levels in healthy non-smoking subjects are usually $\leqslant 0.7\%$) and there is concern that lower levels could adversely affect the activity of organs particularly sensitive to hypoxia, such as the heart and brain.

Treatment of CO poisoning is to remove the patient from the CO source and administer pure (or even hyperbaric) O_2. It is thus likely that ischaemia–reoxygenation occurs. Exposure to excess CO may also lead to abnormally high $NO^\bullet$ production, which might further contribute to the injury. Patients often appear to recover from CO poisoning only to suffer delayed neurological effects, to which ROS/RNS might contribute. Indeed, increased production of ROS (including $OH^\bullet$ and $ONOO^-$) has been reported in the brains of animals after CO exposure.[131]

9.5.16   Freezing injury

Blood circulation is impaired in severely chilled tissues and re-warming may create a reoxygenation injury. Thus it has been hypothesized that ROS could contribute to frostbite injury.

Several vertebrates naturally tolerate freezing, e.g. the red–sided garter snake *Thamnophis sirtalis parietalis* (found in North America) can be frozen for several hours at $-2.5\,°C$, with about 50% of the body water as ice.[118] This species can also endure ischaemia (under N_2 gas) for up to 2 days. One survival strategy may be a rapid up-regulation of antioxidant defences at the onset of freezing or anoxia. A similar strategy may be used in tissues of the ground squirrel *Citellus citellus*, in which an up-regulation of antioxidant defences is observed just before hibernation, presumably to minimize oxidative stress on re-awakening.[32]

9.6   Chronic inflammatory diseases: an introduction

We saw in Chapter 6 that the acute inflammatory response is beneficial to the organism in that it helps to deal with potentially dangerous microorganisms. However, inflammation does cause some degree of damage to surrounding tissues. ROS, prostanoids, leukotrienes, RNS and hydrolytic enzymes produced by neutrophils, macrophages and monocytes may all play a role in mediating inflammation.

9.6.1   *Anti-inflammatory effects of antioxidants*

For example, injection of SOD or other scavengers of ROS, has been observed to decrease inflammation in some animal model systems,[15] such as the **reversed passive Arthus reaction** in skin. The **Arthus reaction** is the name given to a local inflammation that results when an antigen is injected into the skin of an animal that has a high level of circulating antibody against that antigen. It is largely mediated by neutrophils. The **passive Arthus reaction** occurs if the antibody is injected into the animal rather than being formed by the animal itself. In the reversed passive reaction, the antigen is injected intravenously, and the antibody locally.

Similarly, injection of human serum into the bloodstream of rats, followed by injection of an antibody against it into the skin, causes swelling and heat, which is largely mediated by neutrophils. Intravenous injection of SOD into the animals has little anti–inflammatory effect, possibly because SOD is cleared from the circulation within minutes by the kidneys. However, if clearance is prevented by binding the SOD to a high-molecular-mass polymer (e.g. Ficoll), it has a marked anti–inflammatory effect. Ficoll-bound SOD also decreases inflammation induced by injecting **carrageenan** (an irritating substance derived from seaweed) into the feet of rats. Inhibitors of NOS are also anti-inflammatory in this model,[250] but Ficoll-bound catalase is not. Native SOD was observed to protect against kidney damage induced by the intravenous injection of preformed antigen–antibody complexes into mice (Section 9.5.9 above). The effectiveness of native SOD in this system may be due to its rapid accumulation in the kidneys prior to removal from the body.

These results may be due to the action of SOD in removing $O_2^{\bullet-}$, so preventing $O_2^{\bullet-}$-dependent formation of a factor chemotactic for neutrophils (Chapter 6).[15] There are many other reports of anti-inflammatory effects of SOD, and of other antioxidants, including caeruloplasmin. For example, both

SOD and catalase showed inhibitory effects against inflammation induced by the implantation of carrageenan-soaked sponges beneath the skin of rats. Injection of xanthine oxidase into the hind feet of rats produced a swelling that could be decreased by SOD, catalase or mannitol.

One criticism that can be levelled at many, but not all, of these experiments is the lack of suitable controls. Controls with inactivated SOD and catalase should always be performed to rule out 'non-specific' anti-inflammatory effects of proteins, or of contaminants within the enzyme preparations. These can include bacterial endotoxin[129] and **thymol**, an antioxidant phenol (Chapter 3) sometimes used as a preservative in commercial catalase solutions.

9.6.2   *Tissue damage by inflammation*[97,106]

Inflammation is normally a self-limiting event and its benefit outweighs the minor tissue damage it causes. However, anything causing abnormal activation of the immune system has the potential to provoke a devastating response. For example, the major biochemical feature of **gout** is an elevated concentration of urate in the blood. Inflammation is triggered by precipitation of sodium urate crystals within joints. These crystals provoke inflammation by a variety of mechanisms, stimulating the respiratory burst in neutrophils and causing production of leukotriene B_4, which will attract more neutrophils. Gout is treated with allopurinol, an inhibitor of xanthine dehydrogenase/oxidase.

Perhaps the most striking consequences of abnormal phagocyte action are seen in the **autoimmune diseases**. The body has mechanisms to prevent formation of antibodies against its own components. Any failure of these mechanisms allows formation of **autoantibodies** that can bind to normal biomolecules, and provoke attack by the immune system. In some autoimmune diseases, only a single tissue is attacked. For example, in **Hashimoto's thyroiditis**, infiltration of the thyroid gland by lymphocytes and phagocytes is accompanied by tissue damage and fibrosis, and the presence of circulating antibodies against certain thyroid constituents, such as thyroglobulin. In **myasthenia gravis**, a neuromuscular disorder characterized by weakness and fatigue of voluntary muscles, antibodies against the neurotransmitter acetylcholine are present. **Chronic autoimmune gastritis**, in which antibodies to gastric parietal cells are present, is a third example.

Autoantibodies are often demonstrable in small amounts in a few 'normal' members of the population, and their incidence increases with age. Several drugs, or products derived from them (in some cases by free-radical reactions; Chapter 8) can bind to proteins and create new antigens. Perhaps oxidative damage to tissues during the ageing process can create new antigens. Indeed, 'senescent antigen' is used by macrophages to recognize worn-out erythrocytes (Section 6.2).

In autoimmune diseases such as **systemic lupus erythematosus, scleroderma, dermatomyositis** and **autoimmune vasculitis**, lesions are widespread and autoantibodies are present against many tissues. Lupus mainly affects young women, and produces a wide variety of lesions involving the skin, kidneys, muscles, joints, heart and blood vessels. Among the wide range of autoantibodies produced are antibodies directed against DNA and RNA, against erythrocytes and even against subcellular organelles and plasma proteins. The kidney lesions are probably due to deposition of immune complexes on the basement membranes of glomeruli, followed by complement activation (Section 9.5.9 above). Scleroderma is a connective tissue disease characterized by fibrosis of the skin and often the internal organs.

Autoimmune diseases generally have active and quiescent phases, which makes the evaluation of medical treatment especially difficult. This should be borne in mind when assessing the effectiveness of any therapy, including the use of antioxidants such as SOD, in other than a fully randomized double-blind controlled clinical trial over a long period of time. How autoimmune diseases arise is not known, although there is an inherited predisposition to them, and infectious agents have often been suggested to be involved. Certain drugs can induce a condition resembling systemic lupus erythematosus, most significantly **hydralazine, isoniazid, chlorpromazine** and **procainamide** (Chapter 8). A few other drugs have been shown to induce conditions resembling lupus, but less frequently; these include penicillamine, α-methyl-DOPA and diphenylhydantoin. Metabolites of these drugs may bind to proteins to produce products that behave as 'foreign antigens'. In the cases of hydralazine, α-methyl-DOPA, penicillamine and isoniazid, this binding may involve free-radical reactions (Chapter 8). It is also possible that oxidative modification of proteins by ROS/RNS could create new antigens.

It has been reported[46] that exposure of DNA to an $O_2^{\bullet -}$-generating system (hypoxanthine plus xanthine oxidase) causes the DNA to become antigenic when injected into animals, possibly because $O_2^{\bullet -}$-dependent generation of DNA-damaging $OH^{\bullet}$ radicals leads to the formation of antigenic material. This observation could be relevant to the presence of anti-DNA antibodies in systemic lupus erythematosus.

9.6.3 *Are ROS/RNS important mediators of autoimmune diseases?*

It seems likely that ROS, RNS and released enzymes, such as proteases, play some role in tissue damage in the autoimmune diseases. Hence therapy directed against them might prove beneficial. Some evidence consistent with the former suggestion has been provided. For example, urinary excretion of F_2 isoprostanes is elevated in scleroderma patients; suggestive of increased lipid peroxidation.[284] Levels of nitrate plus nitrite tend to be higher in the plasma of patients with autoimmune diseases, indicative perhaps of more $NO^{\bullet}$ generation. By contrast, lupus patients were reported to excrete *less* 8-OHdG in the urine than controls, and appeared to show decreased repair of this oxidative DNA damage product in cells.[180]

9.6.4 *Clastogenic factors*[6]

Emerit *et al.* reported that the serum of patients with lupus contains **clastogenic factors**, agents that induce chromosome damage when added to cultures of human lymphocytes isolated from healthy subjects. Inclusion of SOD in the culture medium prevents this action of clastogenic factor. Lupus patients are light-sensitive and in about 40% of cases develop a characteristic 'butterfly' rash across the cheeks and the bridge of the nose when exposed to sunlight. Lymphocytes from lupus patients are damaged by exposure to light, in the wavelength range 360–400 nm, in the presence of clastogenic factor, an effect again prevented by SOD.

Clastogenic factors have also been reported in body fluids from patients with rheumatoid arthritis and from subjects exposed to ionizing radiation. One clastogenic factor appears to be lipid peroxidation end-products such as HNE. In addition, inosine triphosphate (ITP) and inosine diphosphate (IDP) have been detected in clastogenic factors. Indeed, addition of ITP or IDP can cause chromosome damage to cultured cells, again inhibitable by SOD. The significance of clastogenic factors to disease pathology is unknown.

A few case reports have appeared claiming benefit for antioxidants in the treatment of human autoimmune disease, but controlled trials have not been reported. For example, SOD and copper chelates that can react with $O_2^{\bullet -}$ have been reported as helpful in lupus and in dermatomyositis.

9.6.5   *Antiphospholipid antibodies*[126]

Antibodies apparently directed against phospholipids have been detected in patients with systemic lupus erythematosus and some other autoimmune diseases. At least some of these antibodies can cross-react with oxidized LDLs and it is possible that they are in fact directed against products of lipid peroxidation *in vivo*.

9.6.6   *Artefacts of sample storage: a cautionary note*

Scientists must always be aware that body fluids from patients can undergo lipid peroxidation on prolonged storage (even at $-70\,°C$) or mis-handling, in part because of release of copper ions from caeruloplasmin.[102] Several studies identifying 'cytotoxic factors' in the plasma of patients with various diseases have been led astray by this artefact. It is not always possible to allow for this using controls of normal body fluids stored for the same time. Plasma contains powerful preventative (metal-ion-binding) and chain-breaking antioxidants (Chapter 3) which limit lipid peroxidation; depletion of these protective mechanisms and rises in caeruloplasmin occur during several autoimmune diseases, so that the fluids are less stable on storage than those from healthy subjects. Hence studies of oxidative damage should, whenever possible, be carried out on freshly drawn samples.

9.7   Rheumatoid arthritis[68,106]

Rheumatoid arthritis (RA) is a disease characterized by chronic joint inflammation, especially in the hands and legs. It has many features of an autoimmune disease, although its cause is unknown, and T–lymphocytes seem to play a key role. This has led to the popular hypothesis that RA is due to a persistent, cell–mediated immune response to an unknown antigen. The blood plasma and joint fluid of RA patients often contain autoantibodies directed against **immunoglobulin G** (IgG), mostly against the Fc region of this protein. Something must cause IgG to become antigenic in RA; this may be because the carbohydrate side–chains are different and/or because IgG is damaged by ROS/RNS generated during the chronic inflammatory process.

9.7.1   *The normal joint*

In the normal synovial joint, articular cartilage covers the bone ends and both are enclosed by a thin synovial membrane. Articular cartilage is a dense tissue containing collagen fibres and cells (**chondrocytes**) embedded in a matrix of proteoglycan (largely **aggrecan**). Cartilage does not contain blood vessels and receives O_2 by diffusion from the bone and the synovial fluid.

The normal synovial lining of the joint exists as a thin fibrous tissue; it is made up of lining cells facing the joint cavity and overlying fatty and fibrous material. These lining cells (**synoviocytes**) are of at least two types: type A cells are essentially macrophages and type B cells are secretory, fibroblast–type cells. The synovial lining synthesizes the polymer **hyaluronic acid** (Fig. 9.15) and secretes it into the synovial fluid, where it is largely responsible for the viscosity of the fluid. The synovial lining also acts as a barrier to the free movement of proteins from plasma into the synovial fluid. It has been suggested that the macrophage–like cells may function to engulf debris produced as a result of 'wear and tear' in the joint. The total volume of synovial fluid in healthy synovial joints is small, e.g. only about 0.5 ml in human knee joints. Synovial fluid functions to lubricate the joint and to supply nutrients to (and remove waste from) the cartilage.

9.7.2   *The RA joint*

The onset of RA is usually slow. The synovial lining thickens and folds, there is infiltration by blood vessels and chronic inflammatory cells and its permeability

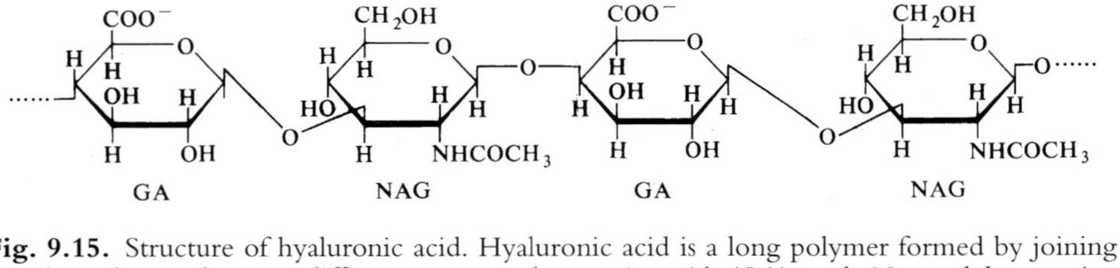

Fig. 9.15. Structure of hyaluronic acid. Hyaluronic acid is a long polymer formed by joining together alternately two different sugars: glucuronic acid (GA) and *N*-acetylglucosamine (NAG). The negative charge on the carboxyl groups of GA at physiological pH causes them to repel each other, so that the molecule extends out in solution. Hence solutions of hyaluronic acid are extremely viscous.

rises. Hence more proteins enter the joint from the plasma. Iron deposition occurs within the synovial cells; much of this iron is located in ferritin and haemosiderin. In many patients with RA, the disease leads to destruction of the articular cartilage, bone erosion and impairment of joint function. Bone is normally remodelled as required by a balance of opposing reactions: digestion by **osteoclasts** followed by deposition of new matrix by **osteoblasts** and then calcification of this matrix. This balance is altered in favour of net bone loss in RA.

Damage in the RA joint usually occurs from the periphery (where the synovial lining normally forms a junction with the articular cartilage) by the growth of the inflamed synovial lining over, and into, the cartilage. This 'overgrowing' tissue is called **pannus**; the word comes from the Latin for 'cloth', the pannus being said to resemble a reddish cloth spreading out over the cartilage structure. Pannus is often vascularized and contains iron deposits, and most cells in pannus are large and mononuclear, many having fibroblastic features. The majority of macrophages arise from blood monocytes attracted by chemotactic factors produced in the joint. Among the cytokines secreted by these macrophages is TNFα. TNFα induces activation of NF-κB in synoviocytes, leading to increased production of several products including IL-1, IL-8 and monocyte chemoattractant protein-1 (MCP-1). IL-1 and TNFα also promote bone resorption.

The synovial fluid of the inflamed rheumatoid joint swarms with neutrophils. Many are known to be activated, since the fluid contains increased quantities of products released by activated phagocytic cells, including the enzyme lysozyme, the iron-binding protein lactoferrin, and stable prostaglandins. Neutrophils are also present at the interface of cartilage and pannus. The volume of synovial fluid is greatly increased in the swollen inflamed rheumatoid joint, but its viscosity is lower than normal. This is because the hyaluronic acid present has a much lower average molecular mass than in normal synovial fluid. The cartilage wear particles, produced by increased friction in the joints, might activate neutrophils and make matters worse.

Joint inflammation often accompanies other autoimmune diseases, such as systemic lupus erythematosus, and is a complication of iron overload secondary to idiopathic haemochromatosis (Section 3.18.4).

9.7.3   Oxidative damage in RA

Interest in the role of free radicals in RA stems from the seminal work of McCord, who noted the decreased viscosity of synovial fluid in RA patients and showed that a similar decrease could be produced by exposing synovial fluid, or solutions of hyaluronic acid, to a system generating $O_2^{\bullet-}$ (xanthine/xanthine oxidase). Later work showed that the hyaluronate degradation is not caused by $O_2^{\bullet-}$, but by $OH^{\bullet}$ generated by $O_2^{\bullet-}$-driven Fenton chemistry. Nevertheless, his observations led to interest in the use of intra-articular injections of SOD as a treatment in RA. However, the clinical data presented did not convince many rheumatologists and over-enthusiastic interpretations of preliminary data may have led to unwarranted scepticism about the real role of ROS in RA. Indeed, the evidence that oxidative damage occurs in RA is very strong (Table 9.3). One advantage of studying RA is that it is easy to sample from the inflamed rheumatoid knee-joint, since aspiration of synovial fluid is a therapeutic procedure employed to reduce swelling and pain.

How could $OH^\bullet$ arise in the RA joint? Significant levels of 'catalytic' copper ions are not detected in fresh synovial fluid, but 'catalytic' iron can be measured by the bleomycin assay (Section 3.18) in about 40% of synovial fluids aspirated from inflamed RA knee joints. This iron has been shown to be capable of stimulating peroxidation of lipids *in vitro*. In addition, aspiration of synovial fluid from some RA patients into solutions of phenylalanine or salicylate produces patterns of hydroxylation products characteristic of $OH^\bullet$ attack upon the aromatic ring (Chapter 5), suggesting that constituents of RA synovial fluid can lead to $OH^\bullet$ formation.

The 'catalytic' iron could arise by liberation from necrotic cells, by H_2O_2-mediated degradation of haemoglobin (known to be released by microbleeding in the RA joint), and by the action of $O_2^{\bullet-}$ on synovial fluid ferritin. Indeed, release of iron upon exposure of synovial fluid to $O_2^{\bullet-}$, especially at acidic pH, has been demonstrated. The chemical pattern of damage to hyaluronate in RA synovial fluids (as demonstrated using nuclear magnetic resonance) is consistent with $OH^\bullet$ attack, though hyaluronate may additionally be secreted as abnormally short chains by the dysfunctional synovium in RA. Another possible source of $OH^\bullet$ is reaction of $O_2^{\bullet-}$ with HOCl, both produced by activated neutrophils:

$$O_2^{\bullet-} + HOCl \rightarrow O_2 + OH^\bullet + Cl^-$$

Reaction of $O_2^{\bullet-}$ with $NO^\bullet$ to form $ONOO^-$ is another potential source of injury in RA, since these patients show elevated levels of nitrate/nitrite in plasma and synovial fluid indicative of increased $NO^\bullet$ generation. Indeed, elevated nitrotyrosine levels have been detected in the rheumatoid synovium by immunostaining and in blood and synovial fluid from some (not all) RA patients by HPLC.

9.7.4   Sources of ROS/RNS in RA

Phagocytes, bone and cartilage[96,52,104]

One source of ROS/RNS is activated phagocytes. Activated neutrophils liberate $O_2^{\bullet-}$, H_2O_2, elastase, HOCl, eicosanoids and possibly $NO^\bullet$. ROS produced by activated phagocytes (e.g. $OH^\bullet$, HOCl, $ONOO^-$) could alter the antigenic behaviour of immunoglobulin G, producing fluorescent protein aggregates that can further activate phagocytic cells and are pro-inflammatory in animals.[96] Such radical-modified proteins could presumably provoke antibody formation. Hypochlorous acid, $ONOO^-$ and $O_2^{\bullet-}$ react with ascorbate, which may help to explain the low levels of ascorbate in RA body fluids (Table 9.3). Hypochlorous acid and $ONOO^-$ inactivate α_1-antiproteinase;

Table 9.3. Evidence consistent with oxidative stress in rheumatoid disease

Observation	Comment
Increased lipid peroxidation products in serum and synovial fluid	Decreased α-tocopherol (per unit lipid) in synovial fluid is consistent with lipid peroxidation, as are reports of 'foam cells' containing oxidized LDL in rheumatoid synovium and increased levels of 4-hydroxy-2-nonenal. LDL isolated from RA synovial fluid showed altered electrophoretic mobility consistent with oxidation (*Free Rad. Biol. Med.* **22**, 705 (1997)). High dose vitamin E reported to diminish pain but not other parameters of inflammation (*Ann. Rheum. Dis.* **56**, 649, 1997).
Depletion of ascorbate in serum and synovial fluid	Presumably results from oxidation of ascorbate during its antioxidant action. Activated neutrophils also take up oxidized ascorbate rapidly.
Decreased GSH in synovial T-lymphocytes	May result from oxidative stress or activation (*J. Immunol.* **158**, 1458 (1997)).
Increased exhalation of pentane	This is a putative end-product of lipid peroxidation, although its validity as an assay is debated (Chapter 5).
Increased concentrations of urate oxidation products	Products measured appear to be end-products of ROS/RNS attack upon urate.
Activation of NF-κB in rheumatoid synovia	May be mediated by oxidative stress (*Arth. Rheum.* **39**, 583 (1996); *ibid.* **40**, 226 (1997)).
Increase in formation of 2,3-DHB from salicylate	2,3-DHB appears to be a product of attack of OH• upon salicylate in patients taking high-dose aspirin.
Hyaluronic acid degradation by free-radical mechanisms	See text.
Formation of 'fluorescent' proteins	Fluorescence is probably caused by oxidative damage to amino-acid residues in proteins.
Increased steady-state levels (in cellular DNA) and increased urinary excretion of 8-OHdG	Indicative of oxidative DNA damage.
Increased levels of 'protein carbonyls' in synovial fluid	Indicative of oxidative protein damage, but could also result from binding of HNE and other aldehydes to proteins (Chapter 5).

Abbreviations: 2,3-DHB, 2,3-dihydroxybenzoate; HNE, hydroxynonenal; LDL, low-density lipoprotein; 8-OHdG, 8-hydroxydeoxyguanosine. For further references see *Ann. Rheum. Dis.* **54**, 505 (1995).

indeed, the amount of active α_1-antiproteinase is decreased in RA synovial fluid. HOCl also fragments collagen. Tissue inhibitors of metalloproteinases (TIMPs)[82] can be inactivated by $ONOO^-$ or HOCl.

The pannus contains many macrophage-like cells, presumably secreting $O_2^{\bullet-}$, H_2O_2 and possibly, $NO^{\bullet}$. It is as yet uncertain if neutrophils and macrophages in the human RA joint make $NO^{\bullet}$ although induction of iNOS in macrophages surrounding joint replacement implants has been reported.[204] Nevertheless, cells in the cartilage can produce $NO^{\bullet}$. Synovial fibroblasts, articular chondrocytes and osteoblasts cultured from tissues taken from patients undergoing hip replacement surgery, generate $NO^{\bullet}$ if treated with a mixture of cytokines (including IL-1 and TNFα), apparently by induction of iNOS.[93]

Oxygen concentration in the inflamed joint might regulate ROS production by activated phagocytes.[69] The K_m for O_2 of the respiratory burst oxidase of rat neutrophils is within the range of physiological O_2 concentrations in body fluids. If the same is true of human phagocytes, it follows that local O_2 concentrations at sites of inflammation could modulate ROS production. Human synovial fluid, like plasma, contains little, if any catalase, glutathione peroxidase or GSH and only traces of SOD activity (largely as EC-SOD (Chapter 3)). Thus $O_2^{\bullet-}$ and H_2O_2 generated by phagocytes in the inflamed rheumatoid joint would not be efficiently scavenged, and $OH^{\bullet}$ could form by reaction of $O_2^{\bullet-}$ with HOCl and of H_2O_2 with iron ions. However, one consequence of cell death may be increases in the levels of antioxidant enzymes due to their release from necrotic cells. Synovial membrane permeability is increased in RA, allowing more plasma antioxidants to enter.

Hypoxia–reoxygenation[68]

It has been proposed that the inflamed rheumatoid joint, upon movement and rest, undergoes a hypoxia–reperfusion cycle, which may result in ROS generation. Tensing an inflamed human rheumatoid knee joint can generate intra-articular pressures in excess of capillary perfusion pressure, resulting in sharp drops in O_2 concentration within the joint. Upon relaxing the joint, the O_2 concentration gradually returns to normal. It is interesting to note that one mechanism of ROS generation may be xanthine oxidase, relating back to the original work of McCord. Xanthine oxidase has been identified in synovial membranes, and hypoxanthine concentrations are elevated in synovial fluid from RA patients.

Drug treatment

Yet another source of oxidative stress in RA may be some of the drugs used in treatment of the disease. **Corticosteroids** lead to phospholipase A_2 inhibition, decreasing arachidonic acid release and hence diminishing the formation of eicosanoids. They also decrease production of IL-1, PAF, TNFα and iNOS by macrophages. Effective, naturally occurring corticosteroids include cortisol, and synthetic ones include **prednisolone** and **dexamethasone**. Steroids are powerful and effective drugs in RA, but side-effects limit prolonged use of high doses.

The drugs most commonly used in the day-to-day treatment of RA are cyclooxygenase inhibitors, blocking prostaglandin production and hence decreasing pain and swelling. The first drug of this type to be synthesized was **aspirin**.[134] The ancient Egyptians were aware of the analgesic effects of an extract of willow leaves. In 1763, the Reverend Edmund Stone of Oxford, England, read a report to the Royal Society on the anti-fever action of willow bark and in 1876 the Scottish physician MacLagan reported its effectiveness in arthritis. The active component was later extracted and shown to be salicylate and in 1899 the Bayer Company in Germany introduced acetylsalicylate (aspirin). Aspirin ingested by humans is quickly hydrolysed to salicylate by esterases in the digestive tract and in liver (Section 8.1).

Several other non-steroidal anti-inflammatory drugs (NSAIDs) have been developed as cyclooxygenase inhibitors: they include diclofenac sodium, indomethacin, ibuprofen, phenylbutazone, piroxicam and mefenamic acid. Major side-effects of cyclooxygenase inhibitors are gastrointestinal disturbances, irritation of the stomach mucosa and renal injury. In general, none of the these drugs decreases cartilage damage or bone erosion in RA, and some may even enhance it, perhaps by encouraging IL-1 formation by macrophages in the pannus or (in the case of salicylate) by inhibiting synthesis of cartilage proteoglycans.

Interest now focuses on the development of selective inhibitors of cyclooxygenase 2 (COX-2), the inducible cyclooxygenase (Chapter 6) which is up-regulated in several cell types in the RA joint as a result of exposure to cytokines. In principle, a selective inhibitor of COX-2 should not produce the typical side-effects of a COX-1 inhibitor.

Aspirin and many other NSAIDs have three major pharmacological properties:

(a) reduction of swelling, pain and redness associated with inflammation;
(b) ability to decrease elevated body temperature;
(c) an analgesic effect.

Some thiol compounds have been claimed to modify the underlying disease processes in RA; they include **penicillamine** (often given as its disulphide, which is reduced to penicillamine *in vivo*; Fig. 9.16), and gold–thiol complexes such as **aurothiomalate** and **aurothioglucose**. How much 'disease modification' they exert is a matter of debate. Gold complexes may act by depressing the activity of macrophages and lymphocytes, as well as by interfering with eicosanoid production.

Several drugs developed for other diseases have found some use in the treatment of RA. They include the anti-malarials **chloroquine** and **hydroxychloroquine**, the antibiotic **rifamycin**, and **sulphasalazine** (Fig. 9.16). Anti-malarials such as hydroxychloroquine have been reported to decrease interleukin-1-induced cartilage degradation *in vitro*. Sulphasalazine consists of **5-aminosalicylic acid** linked by an azo (−N=N−) bond to **sulphapyridine**. *In vitro*, millimolar levels inhibit both lipoxygenase activity and prostaglandin biosynthesis, although it is not clear if this accounts for the clinical actions of sulphasalazine. In RA, the effective agent seems to be sulphasalazine itself or sulphapyridine, rather than 5-aminosalicylate.

Several immunosuppressive agents have been used in the treatment of RA, since it is believed that lymphocytes contribute to joint damage. Those tried include **cyclophosphamide**, **chlorambucil**, **levamisole**, **azathioprine** and **methotrexate**.

Anti-inflammatory drugs as antioxidants[106,332]

The possibility that NSAIDs and other anti-inflammatory drugs have multiple mechanisms of action has raised the question as to whether they might exert antioxidant effects *in vivo*. Anti-inflammatory drugs could affect oxidative damage at sites of inflammation in several ways:

(1) Any drug that inhibits inflammation will lead to decreased ROS/RNS production, e.g. by reducing phagocyte numbers. In addition, steroids down-regulate production of TNFα and iNOS, among other inflammatory mediators.

(2) Drugs might directly scavenge such ROS as OH$^\bullet$, ONOOH and HOCl. Most, if not all, anti-inflammatory drugs react quickly with OH$^\bullet$ (rate constants in the range $5 \times 10^9 - 1 \times 10^{10} \, M^{-1} s^{-1}$). This would be expected from their structures (Fig. 9.16), since aromatic and thiol compounds are known to react very rapidly with OH$^\bullet$ (Chapter 2). Hence, any anti-inflammatory drug present at a site of inflammation would scavenge OH$^\bullet$ if the drug concentration were high enough. However, at least millimolar levels would be required since drugs would have to compete with the biomolecules present for any OH$^\bullet$ generated. Most drugs achieve nowhere near such concentrations. One exception is aspirin; in aspirin-treated RA patients salicylate concentrations in plasma and synovial fluid reach the millimolar range. Indeed, levels of 2,3-dihydroxybenzoate, a product of OH$^\bullet$ attack on salicylate (Chapter 5) are elevated in plasma from aspirin-treated RA patients. *In vitro*, many drugs can react with HOCl and ONOOH, but only in a few cases is the reaction fast enough for scavenging to be feasible at the drug concentrations found at sites of inflammation *in vivo*. Feasible HOCl and ONOOH scavengers include thiol compounds (penicillamine, gold sodium thiomalate) and phenylbutazone. This does not, of course, prove that these drugs do scavenge HOCl or ONOOH *in vivo*.

(3) Drugs may chelate transition metal ions needed for OH$^\bullet$ generation. Little information is available on whether this could happen at therapeutic drug levels, but it is feasible for salicylate.

(4) Thiols might depress activation of NF-κB in synovial cells, thus decreasing inflammation by depressing upregulation of COX-2, iNOS, and adhesion molecules involved in phagocyte recruitment.

(5) Drugs might decrease production of ROS by phagocytes. Again, there have been many claims, but few demonstrations that the drug levels present at sites of inflammation *in vivo* can do this.

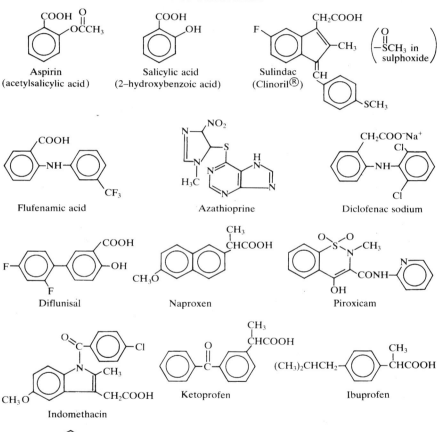

STEROIDS

Cortisol   (R^I=O; R^{II}–H; R^{III}–H;
                C 1–2 saturated)

Prednisolone   (R^I–OH; R^{II}–H; R^{III}–H;
                C 1–2 unsaturated)

Dexamethasone   (R^I–OH; R^{II}–F; R^{III}–CH$_3$;
                C 1–2 unsaturated)

NON-STEROIDALS

Aspirin
(acetylsalicylic acid)

Salicylic acid
(2–hydroxybenzoic acid)

Sulindac
(Clinoril®)

Flufenamic acid

Azathioprine

Diclofenac sodium

Diflunisal

Naproxen

Piroxicam

Indomethacin

Ketoprofen

Ibuprofen

Phenylbutazone

Mefenamic acid

Tenoxicam

Fig. 9.16.

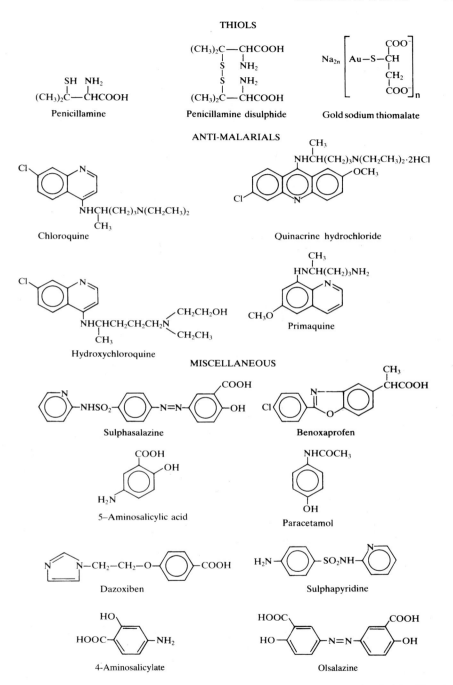

Fig. 9.16. Structures of some drugs used in the treatment of rheumatoid arthritis. Also shown are drugs used in the treatment of inflammatory bowel disease, including **4-amino-salicylate** and **olsalazine**, a pro-drug of 5-aminosalicylate.

Anti-inflammatory and other drugs as pro-oxidants[311]

In general, only a few anti-inflammatory drugs would appear capable of acting directly as antioxidants *in vivo*. Indeed, several drugs used in the treatment of RA might themselves be converted into free radicals *in vivo*, i.e. they might aggravate oxidative damage. For example, radicals derived from penicillamine, phenylbutazone, diclofenac, some fenamic acids and the aminosalicylate component of sulphasalazine can inactivate α_1-antiproteinase, oxidize ascorbic acid and accelerate lipid peroxidation *in vitro*. Such radicals might result from oxidation of drugs by myeloperoxidase/H_2O_2 (e.g. diclofenac, phenylbutazone), the peroxidase action of cyclooxygenase (e.g. phenylbutazone) or by reaction of the drug with ROS/RNS (e.g. aminosalicylate, penicillamine). Phenylbutazone is now rarely used to treat RA because of the high incidence of side-effects.

For example, a particular feature of penicillamine therapy in RA is the autoimmune reaction it often produces. Penicillamine thiyl ($RS^\bullet$) radicals and more reactive ($RSO_2^\bullet$ and $RSO^\bullet$) radicals formed from $RS^\bullet$ could combine with proteins and alter their antigenicity. Naproxen and benoxaprofen have a photosensitizing action and are able to stimulate singlet O_2 formation. This effect was bad enough in the case of benoxaprofen to cause its withdrawal from general clinical use.

Several drugs not used in the treatment of RA have also been suggested to generate free radicals *in vivo* that contribute to their side-effects. Halothane is one (Chapter 8). Another example may be **clozapine**, an anti-psychotic that can damage the bone marrow; its oxidation to a toxic product by myeloperoxidase may be involved.[171]

9.7.5 *Consequences of oxidative damage in RA*

There is little doubt that oxidative damage occurs in RA patients (Table 9.3). However, the key question is to the extent to which it contributes to the chronicity of the inflammation and the consequences of inflammation, such as the cartilage and bone destruction in RA. The answer is unclear. TNFα, an agent well-known to cause oxidative stress, seems to play an important role in the pathology of RA.[333] Much of the cartilage degradation taking place in RA involves increased release of proteolytic enzymes such as collagenase, gelatinase and stromelysin (Table 2.6) from the chondrocytes into the cartilage matrix. Secretion of IL-1 by macrophages stimulates both synoviocytes and chondrocytes to release proteases, and it can also promote bone resorption. Peroxynitrite and HOCl may facilitate this damage by inactivating TIMPs. TIMP-1 inhibits stromelysins, collagenases and gelatinases and this ability is lost after $ONOO^-$ or HOCl treatment. HOCl can also activate latent forms of neutrophil collagenases and gelatinase, although the extent to which this happens *in vivo* is uncertain.

Hydroxyl radicals degrade isolated proteoglycans and HOCl fragments collagen, but their effects on intact cartilage are probably limited. However,

H_2O_2 is very diffusible. It readily inhibits cartilage proteoglycan synthesis,[18] e.g. by interfering with ATP synthesis, in part by inhibiting the glycolytic enzyme glyceraldehyde-3-phosphate dehydrogenase in chondrocytes. Indeed, intra-articular injection of H_2O_2-generating systems causes severe joint damage in animals.[257] Hence inhibition by H_2O_2 of cartilage matrix resynthesis could aggravate the effects of proteolytic and free-radical-mediated cartilage degradation.

Low concentrations of H_2O_2, $O_2^{\bullet-}$ or both, accelerate bone resorption by osteoclasts, whereas $NO^{\bullet}$ inhibits it.[52] Osteoclasts (essentially macrophages) degrade bone by generating a microenvironment of low pH to dissolve the hydroxyapatite mineral, exposing the organic matrix to degradation by cysteine proteases such as the **cathepsins**. During bone resorption osteoclasts *produce* $O_2^{\bullet-}$ and H_2O_2, apparently using an NADPH oxidase system (Chapter 6) that can be inhibited by diphenylene iodonium.

In addition, ascorbate is essential for cartilage function[336] and the low concentrations found in RA synovial fluid (Table 9.3) might impair cartilage metabolism. For example, osteoblasts take up ascorbate in a Na^+-dependent process and concentrate it to millimolar intracellular levels.

In summary, we do not yet know the exact contribution made by ROS/RNS to joint damage in RA. The development of improved assays of oxidative damage that are applicable to humans should help to address this point and allow a rational selection of antioxidants for possible therapeutic application.

9.7.6   *Iron and rheumatoid arthritis*

It has been known for decades that RA is accompanied by changes in body iron metabolism. A rapid fall in the 'total iron' content of blood plasma at the onset of inflammation (Table 9.4) is followed by a marked drop in haemoglobin levels and increased deposition of iron proteins in the synovial membranes.

Table 9.4. Protein and iron concentrations (mean$\pm$SD) in plasma and synovial fluid from rheumatoid patients

	Plasma		Synovial fluid from rheumatoid knee-joint ($n=9$)
	Normal control ($n=8$)	Rheumatoid ($n=8$)	
Total non-haem iron (µM)	17.9 ± 6.7	7.5 ± 5.7	8.9 ± 4.1
Transferrin (g/l)	2.92 ± 0.38	2.76 ± 4.2	1.67 ± 0.31
% Transferrin saturation with iron	29.9 ± 10.2	16.1 ± 12.8	26.3 ± 11.7
Albumin (g/l)	50.3 ± 44	36.3 ± 7.9	17.2 ± 6.3
Caeruloplasmin (g/l)	0.259 ± 0.079	0.469 ± 0.085	0.256 ± 0.077

Data abstracted from *Biochim. Biophys. Acta* **869**, 119 (1986).

The drop in plasma iron correlates closely with the activity of the inflammatory process. It is an example of the **anaemia of chronic disease**,[199] also seen in chronic infections and in patients with cancerous tumours. Attempts to reverse this 'anaemia' by giving oral iron salts to RA patients are usually ineffective in the absence of improvement in their disease, because the 'anaemia' is due to an inhibition of erythrocyte formation by excess levels of cytokines, especially IL-1 and TNFα.

Indeed, oral iron treatment can occasionally worsen the symptoms of RA. Intravenous iron therapy, e.g. by injection of iron dextran, frequently causes problems. Although this has often been attributed to the dextran, it may be the iron that is at fault.[335] For example, exacerbations of joint inflammation in rheumatoid patients given iron dextran infusions occurred at exactly the time that plasma transferrin became saturated with iron and non-transferrin-bound iron was present in body fluids. Iron-overloaded patients suffering from idiopathic haemochromatosis frequently show joint inflammation, and bleeding into the joints of haemophiliac patients is also associated with inflammation.

Studies with desferrioxamine in animals[24a]

The action of this powerful chelator of Fe(III) has been examined in several animal models of inflammation. It was found that low doses of desferrioxamine aggravate acute inflammation in some rat models, but larger doses are anti-inflammatory. In a model of synovitis in guinea-pigs, desferrioxamine (100 mg/kg of body weight) aggravated the acute phase of the inflammation, but repeated administration depressed the chronic phase. A similar effect is seen in some other models, including the rat allergic air-pouch model of acute to chronic inflammation. Subcutaneous injection of sterile air in rodents results in formation of an 'air pouch' with a lining that superficially resembles synovium, generating a model useful for testing anti-inflammatory agents.

It is not clear why desferrioxamine should sometimes make acute inflammation worse. It must also not be assumed that the action of desferrioxamine in suppressing chronic inflammation is necessarily related to its effect on free-radical reactions. For example, it might block the proliferation of inflammatory cells by chelating essential iron, and iron may be involved in controlling the influx and efflux of lymphocytes in inflamed areas.

Studies in humans[24a]

The suppressive action of desferrioxamine on chronic inflammation in animals was sufficiently encouraging for preliminary trials with RA patients to be carried out (Chapter 10). In an Italian study, 1 g of desferrioxamine was injected intramuscularly and an abrupt rise in haemoglobin was observed, but no fall in the acute-phase response. The speed of the change in haemoglobin suggests that the effect might have been mediated by suppressing inflammation. No ill-effects were reported. Of seven rheumatoid patients given larger doses of desferrioxamine (up to 3 g/day for 5 days each week for 1–3 weeks), four developed ocular abnormalities that reversed on drug withdrawal. Two patients who received the sedative **prochlorperazine** to combat nausea during desferrioxamine therapy became unconscious for 48–72 h, possibly because this combination of drugs mediates iron transfer across the blood–brain barrier and achieves removal of iron essential to the functioning of the nervous system. Prochlorperazine is a weak metal-chelator and enters the brain, which desferrioxamine cannot. Hence doses of desferrioxamine suitable for treatment of iron overload are not necessarily safe in other diseases, and the combination of this chelator with phenothiazine drugs such as prochloperazine should be strictly avoided.

9.7.7   *Alkaptonuria*[191]

An illustration of the potential pro-inflammatory action of ROS may be provided by **alkaptonuria**. This disease is an inborn defect in an enzyme (**homogentisate oxidase**) involved in the metabolic degradation of the amino acid tyrosine. Alkaptonuric patients accumulate homogentisic acid, They show pigmentation of cartilage and connective tissue and may gradually develop severe inflammatory arthritis. The oxidation of homogentisic acid generates $O_2^{\bullet-}$ and H_2O_2, and also $OH^{\bullet}$ if iron ions are present, and it is possible that the arthritis of alkaptonuria is induced by increased production of these ROS. Auto-oxidation would be expected to produce melanin-like products, which could explain the pigmentation observed.

9.8   **Inflammatory bowel disease**[9,97]

ROS/RNS may be important contributors to tissue injury in the inflammatory bowel diseases (IBDs) such as **Crohn's disease** and **ulcerative colitis**. The former is a recurrent inflammation and ulceration of the whole digestive tract, although it is often most severe in the lower part of the ileum, and in the colon and rectum. In ulcerative colitis, the ulceration and inflammation affect the colon and rectum only. In both conditions autoantibodies to bowel components can be found in the blood plasma. Patients with ulcerative colitis (and probably also those with Crohn's disease) are at increased risk of developing colorectal cancer. Although ulcerative colitis and Crohn's disease can be clinically similar, it is not generally thought that they are just different manifestations of the same disease. For example, smoking appears to decrease the risk of developing ulcerative colitis, but may increase that for Crohn's disease.

As in the case of RA, the tissue injury and inflammation associated with these two diseases may be aggravated[167] by the extensive recruitment and activation of phagocytic cells, generating cytokines (e.g. TNFα), proteases, ROS and RNS, including HOCl and $ONOO^-$. Indeed, immunocytochemical evidence for formation of nitrotyrosine[272] has been obtained in several animal models of IBD and in biopsies from patients with ulcerative colitis and Crohn's disease, where it appears to colocalize with iNOS. Increased urinary nitrate excretion has been reported in IBD patients. It must not necessarily be assumed that increased $NO^{\bullet}$ is always bad;[94,162] although $ONOO^-$ derived from it can cause mucosal damage. $NO^\bullet$ itself may depress phagocyte recruitment, e.g. by down-regulating synthesis of adhesion molecules. H_2O_2, HOCl and chloramines could increase the permeability of the gut wall and promote electrolyte secretion, perhaps contributing to the diarrhoea. Bleeding is common in the inflamed gut and may provide a source of iron for $OH^\bullet$ generation. In addition, faeces often contain iron unabsorbed from the diet.

9.8.1   *The salazines*

One mainstay of treatment in IBD is sulphasalazine (Fig. 9.16); unlike the case of RA, the active constituent is 5-aminosalicylate (**mesalazine** or **mesalamine**) which is generated by hydrolysis of sulphasalazine by colonic bacteria. Indeed, both 5- and other aminosalicylates such as **4-aminosalicylate** are finding direct therapeutic use in IBD.

Aminosalicylates are powerful scavengers of $OH^\bullet$, $ONOO^-$ and HOCl *in vitro*. Since they accumulate in the bowel to millimolar levels during therapy, scavenging of ROS/RNS is a feasible mechanism of therapeutic action. Evidence consistent with such a mechanism is provided by the isolation from faeces of patients with IBD of what appear to be specific ROS-generated oxidation products of 5-aminosalicylate.[2] These products were not found in faeces from RA patients treated with sulphasalazine. However, it is also possible that oxidation products of aminosalicylates (e.g. iminoquinones and quinones) could cause damage,[172] e.g. by attacking proteins. Thus aminosalicylates may resemble penicillamine (Section 9.7.4 above) in scavenging ROS/RNS but forming noxious products.

The increased risk of cancer in ulcerative colitis may involve DNA damage by ROS/RNS in the chronically inflamed colon.[8] Indeed, levels of 8-OHdG in the inflamed colon have been reported as elevated. Patients with both Crohn's disease and ulcerative colitis have been reported to exhale more ethane than controls,[262] suggestive of increased lipid peroxidation *in vivo*, although changes in hydrocarbon gas production by gut flora could potentially confound such results.

9.9   Other chronic inflammations

It is likely that ROS/RNS play a role in other chronic inflammations. For example, **periodontal disease**[35] is the most common chronic inflammatory disease to affect adults. It would seem easy to take tissue samples for measurement of parameters of oxidative damage and to test the protective effects (if any) of topical antioxidants, but surprisingly few data exist. A similar comment may be made about chronic inflammatory skin diseases, such as **acne** and **psoriasis**.

9.9.1   *The pancreas*[258]

Inflammation of the pancreas can be provoked by several factors including excess alcohol intake, pancreatic-duct obstruction by gallstones, or by a period of ischaemia, e.g. after haemorrhagic shock. Several studies suggest that the injury induced by fatty-acid infusion, ischaemia or partial duct obstruction in isolated perfused, pancreas can be diminished by including SOD and catalase in the perfusing medium. Conflicting results have been obtained as to whether SOD and catalase are effective against pancreatitis in whole animals. The issue is important because current therapies for the treatment of pancreatitis in humans are inadequate. Double-blind trials in human pancreatitis sufferers suggest that supplementation with high levels of ascorbate can decrease the extreme pain associated with chronic pancreatitis.[310]

9.9.2   *Other parts of the gastro-intestinal tract*

ROS/RNS have been suggested to be involved in inflammation of the oesophagus[221] (e.g. after reflux of acid from the stomach) and in gastric damage secondary to haemorrhagic shock or inflammation.[232] An interesting recent suggestion is that ROS/RNS contribute to gastric injury in chronic infection by *Helicobacter pylori*. This organism infects the stomach of a large proportion of the population (from $\sim 40\%$ in Manchester to $\sim 80\%$ in Algeria) and, in a minority of cases, causes inflammation (with neutrophil and monocyte infiltration), predisposition to ulcer formation and elevated risk of gastric cancer. Comparable *Helicobacter* organisms are found in the stomach of other species. Studies in Korea, where gastric cancer is very common, showed elevated levels of 8-OHdG in the gastric mucosa of patients infected with *H. pylori*.[12] The presence of iNOS and nitrotyrosine in *H. pylori*-related gastritis has also been reported.[188]

   Gastric juice contains ascorbate, which appears to be secreted by gastric cells so as to achieve concentrations in the juice greater than those in plasma. Gastric inflammation, including chronic *H. pylori* infection, has been reported to decrease ascorbate levels. Gastric juice ascorbate may be an important scavenger of such species as $ONOO^-$ and $HOCl$, and it decreases the formation of carcinogenic *N*-nitroso compounds by reaction of RNS with certain dietary constituents. Saliva is rich in NO_2^-, which presumably forms HNO_2 when it enters the acid pH of the stomach. Phenolic compounds (e.g. catechins) in the diet may also scavenge RNS in the stomach.[224a]

9.10   Lung damage and the adult respiratory distress syndrome

9.10.1   *Oxygen and the lung*[147]

We began this book by discussing the toxic effects of elevated O_2 upon organisms and the involvement of free radicals and other ROS (Chapter 1). The impact of ROS on the lung may be especially important because this organ is exposed to greater O_2 tensions than other tissues, over a much larger surface area. The lung is a major target of air pollutants, several of which cause oxidative stress (Chapter 8). Over an average human lifespan, about 3×10^8 litres of air are inhaled.

Lung is histologically complex and different lung cell types show different susceptibilities to oxidative stress. For example, hyperoxia in experimental animals damages endothelial and type I alveolar epithelial cells, whereas the more resistant type II cells can proliferate to replace them (Chapters 1 and 3). As observed for other tissues, exposure of lung to elevated O_2 results in increased intracellular generation of $O_2^{\bullet-}$ (and hence of H_2O_2) by promoting autoxidation reactions and the leakage of electrons onto O_2 from the electron transport chains of mitochondria and endoplasmic reticulum. H_2O_2 can also be generated by oxidases in the lung, including urate (not in primates), mono-amine and xanthine oxidases.

9.10.2   *Phagocytes and adult respiratory distress syndrome*[178]

Another source of ROS within lung is the activation of phagocytes; both resident pulmonary macrophages and recruited monocytes and neutrophils. Few neutrophils are present normally in lung, but they accumulate during the later stages of O_2-induced lung damage, probably because lung macrophages release chemotactic factors.

Neutrophil involvement may be particularly important in some forms of the so-called **adult (or acute) respiratory distress syndrome**, ARDS. ARDS is severe hypoxia resistant to O_2 administration, caused by pulmonary oedema. It can arise as a complication of haemorrhagic or endotoxic shock, aspiration of stomach contents into the lung, cardiopulmonary bypass or severe tissue damage, e.g. caused by burns or accidents. It leads to the deaths of up to 50 000 patients per annum in the USA, and assumes considerable clinical and fiscal significance. For example, in the UK fewer than 2% of hospital beds are in intensive care units, yet they consume 15–20% of the resources. ARDS leads to death in about 50% of cases, little different from when it was first described in 1967. When patients die within the first 3 days of the onset of ARDS, their primary illness is usually thought to be responsible, but late deaths are more commonly attributable to infection and subsequent sepsis. Patients in whom the lung is the primary site of infection appear to have the worst prognosis, often responding poorly to antibiotic therapy.

Many reports document the dramatic appearance of neutrophils and neu-trophil-derived products in lung-lavage fluids of patients early in the course of ARDS. This neutrophil accumulation may well involve activation of the complement system, and it has been proposed that ROS produced by activated neutrophils may be of importance in producing the lung damage. For example, the acute lung injury[324] produced in rats after severe skin burns or injection of cobra–venom factor (both cause complement activation, and neutrophil infil-tration into lung) can be minimized by treating the animals with SOD, catalase, desferrioxamine or apolactoferrin, and worsened by infusion of iron(III) salts. These results imply that iron-dependent formation of $OH^{\bullet}$ radicals plays some part in the lung damage. Activated neutrophils additionally produce HOCl and secrete elastase. Lung elastic fibres can be degraded by elastase if α_1-anti-proteinase is inactivated by ROS such as HOCl or $ONOO^-$. Severe burns can lead to extensive fluid loss and shock syndrome, and administration of allopurinol or SOD to mice before thermal skin injury has been reported to increase survival times. Oxidative protein damage seems to be elevated in human burn blister fluid.[114a] Of course, neutrophils should not automatically be assumed to be the 'bête noir'; they also have beneficial effects: combating pulmonary infection and releasing lactoferrin, which can inhibit iron-dependent oxidative damage (Chapter 3).

Alveolar macrophages are another potential source of ROS and appear to contribute to lung damage in ARDS. Their role in the lung damage caused by silica and asbestos has already been mentioned (Section 6.7.3). For example, instillation of monoclonal immunoglobulin A into the lungs of rats, followed by intravenous injection of the antigen, produces a lung injury in which macrophages are important. Complement activation may accelerate ROS (and possibly RNS) production by alveolar macrophages. In this model, SOD, catalase or desferrioxamine again diminished the lung injury observed. Macrophages release an elastase that is not inhibitable by α_1-antiproteinase (Chapter 6).

However, it must be realized that ARDS is a blanket term for several closely related conditions provoked by different stimuli, and it is likely that the precise role played by oxidative damage differs in each. Thus ARDS has been observed to develop in neutropaenic human leukaemia patients, i.e. it can occur without significant neutrophil infiltration into the lung.

9.10.3 Oxidative stress and ARDS[178]

There is considerable evidence that oxidative stress occurs in ARDS (Table 9.5). It can result from the disease itself and from the elevated O_2 used in treatment. ROS/RNS could, for example, directly damage surfactant (e.g. $ONOO^-$ nitrates surfactant proteins) and α_1-antiproteinase and also impair their synthesis by alveolar type II cells. The antioxidant properties of the thiol **N-acetylcysteine** *in vitro*, together with the observed depletion of thiols in ARDS patients (Table 9.5), has provoked clinical trials of this substance. These have not so far demonstrated that thiol supplementation is of great benefit to ARDS patients. It may yet be shown, however, that some sub-groups of ARDS patients benefit from thiol supplementation whereas others do not. The complexity of ARDS and the large number of putative mediators of tissue injury involved (Table 9.6) make it unlikely that therapy directed against a single target will be highly beneficial in all cases. Combination therapies (Table 9.7) may be more useful.

Elevated levels of several cytokines have been implicated in the pathology of ARDS (Table 9.6) and particular attention has focused on TNFα. Elevated levels of TNFα are frequently reported in ARDS, and in septic patients at high risk of ARDS. TNFα appears to cause oxidative stress, e.g. its induction of apoptosis in certain cells can be prevented by antioxidants. TNFα-induced oxidative stress appears to involve interaction with mitochondria, inhibiting electron transport and facilitating $O_2^{\bullet-}$ generation; it can also prime neutrophils and other phagocytes to respond with an enhanced oxidative burst on subsequent stimulation. The transcription factor NF-κB is a key mediator of

Table 9.5. Some evidence consistent with oxidative stress in ARDS

Observation	Comments
Oxidized α_1-antiproteinase found in lung lavage fluids	Essential methionine residue oxidized, presumably inactivated by ROS, or such as ONOO$^-$
Elevated H_2O_2 levels in exhaled air	Little H_2O_2 present in exhaled air from healthy subjects. Elevated H_2O_2 in urine from ARDS patients has also been reported (*Chest* **105**, 232 (1994)).
Diminished levels of ascorbate in body fluids[a]	May be oxidized by ROS/RNS and taken up by activated neutrophils. Nutrition in very sick people is also compromised.
Reports of decreased α-tocopherol	Data conflicting: it is essential to standardize α-tocopherol levels against lipid before comparison as lipid levels in body fluids change in ARDS, e.g. cholesterol levels can fall sharply.
Oxidation of plasma proteins	Loss of $-SH$ groups, increased protein carbonyls, *chloro-*, *nitro-* and *ortho*-tyrosine levels (*Am. J. Resp. Crit. Care Med.* **142**, A28 (1994); *Free Rad. Res.* **20**, 289 (1994)).
Presence of 'catalytic' iron in plasma (BDI)	ARDS patients with impaired liver function and multiple organ failure show increased % transferrin saturation and 'catalytic' iron in plasma. Patients without such secondary damage show increased % transferrin saturation but not usually to an extent leading to BDI. Lung lining fluid transferrin levels rise in ARDS due to increased leakage from plasma, and can bind BDI (*Biochem. Biophys. Res. Commun.* **220**, 1024 (1996); *Thorax* **49**, 707 (1994)).
Diminished caeruloplasmin 'ferroxidase' activity	Plasma caeruloplasmin protein levels often elevated (acute phase response), also in lung lining fluid (increased leakage from plasma) but ferroxidase activity per unit protein is decreased. Could be due to increased release of proteases as caeruloplasmin is sensitive to proteolytic damage (*J. Lab. Clin. Med.* **124**, 263 (1994)).
Loss of plasma $-SH$ groups	Normal levels $\approx 500\,\mu M$; fall to $\sim 300\,\mu M$ in ARDS.
Decreased GSH levels in lung lining fluids	Levels in alveolar lung fluid are $\sim 20\,\mu M$ as compared with higher levels in normal subjects assayed under the same conditions. Levels of GSSG were also elevated (*Am. Rev. Resp. Dis.* **148**, 1174 (1993)).
Increased lipid peroxidation	Data are variable. Data consistent with increased peroxidation include decreases in the linoleic and arachidonic acid content of plasma lipids and increased plasma HNE levels (*Crit. Care Med.* **24**, 241 (1996)).

Table 9.5. (*Continued*)

Observation	Comments
Increased xanthine oxidase activity	Xanthine oxidase may be released from injured tissues to cause damage to other tissues by binding to cell surfaces (e.g. in lung); it may also be produced from xanthine dehydrogenase in the injured lung itself. Levels of plasma hypoxanthine increased, indicative of hypoxia. It is possible that an 'ischaemia–reperfusion' of lung can occur (*Am. J. Resp. Crit. Care Med.* **155**, 479 (1997)).
Nitrotyrosine formation	Immunostaining of lungs from ARDS patients revealed nitrotyrosine, scarcely detectable in normal lung. Suggestive of formation of, and damage by, $ONOO^-$ or other RNS (*J. Clin. Invest.* **94**, 2407 (1994)).

[a] Often $<5\,\mu M$ in plasma; normal values are $\sim 50\,\mu M$ in the same studies.
Abbreviation: BDI, bleomycin-detectable iron.
Remember that demonstration of oxidative damage is not a demonstration of its importance (Fig. 9.2). Oxidative stress in ARDS can result from the disease process and also from the elevated inspired O_2 used in its treatment. Lung previously injured by, for example, sepsis or aspiration of gastric contents, may be more susceptible to O_2 damage.

Table 9.6. Pathobiological considerations in ARDS

Humoral aspects	Complement activation, arachidonic acid metabolites (thromboxanes, leukotrienes, prostaglandins), circulating xanthine oxidase/raised hypoxanthine, platelet-activating factor (PAF), macrophage migration inhibitory factor (MIF), endotoxin, over-production of certain cytokines (TNFα, IL-1β, IL-6, IL-8)
Cellular aspects	Neutrophils; monocytes and macrophages; eosinophils; platelets; endothelial cells; epithelial cells; fibroblasts (e.g. stimulation of fibroblast proliferation by cytokines and ROS)
Other aspects	Sepsis syndrome (predisposing to ARDS or a consequence of ARDS); multiple organ failure; ROS; nitric oxide and other RNS

responses to TNFα, and its activation (at least in certain cell types) may be mediated by ROS (Chapter 4).

Paradoxically, TNFα can also lead to induction of antioxidant defence systems, in particular MnSOD, in a wide range of cell types including lung epithelial cells. Indeed, overexpression of MnSOD by gene transfection of cell lines increases their resistance to subsequent cytotoxic doses of TNFα. TNFα itself is not necessarily bad and is often useful overall: it seems more the *balance*

Table 9.7. Therapeutic strategies in ARDS

Approach	Strategies
Current approach	Prevention of ARDS risk factors, treatment of predisposing conditions, mechanical ventilation, positive end expiratory pressure, oxygen, fluids, Antibiotics, haemodynamic monitoring, measures to minimize risk of sepsis and/or multiple organ failure in ARDS
Experimental approaches	
Anti-inflammatory agents	Ibuprofen, prostaglandin E_1, pentoxifylline, lipoxygenase inhibitors, selective inhibitors of cyclooxygenase-2
Antibody therapy	Anti-C5a, anti-J5 antigen, anti-HA-1A, E5 Antibody, anti-TNFα, IL-1 receptor antagonist, blocking phagocyte adhesion (e.g. by antibodies against adhesion molecules)
Antioxidant therapy	Superoxide dismutase, *N*-acetylcysteine, ascorbic acid, iron chelators, catalase, α-tocopherol, Antiproteases, allopurinol, scavengers of $ONOO^-$ and other RNS
Exogenous surfactant therapy	—
New modes of mechanical ventilation	Pressure control-inverse ratio ventilation, extracorporeal CO_2 removal
Other	Lung transplantation

Adapted from *Adv. Pharmacol.* **38**, 457 (1996) by courtesy of Dr Samuel Louie and Academic Press.

of pro-inflammatory (including TNFα) and anti-inflammatory (e.g. IL-10) cytokines that may influence outcome in ARDS and sepsis generally (Chapter 6).

9.10.4   *Lung transplantation*[70,143]

Ischaemia–reoxygenation can occur during the preservation of lungs for transplantation, just as for other organs (Section 9.5 above). For example, reperfusion of isolated animal lungs after ischaemia leads to increased production of $O_2^{\bullet-}$ and H_2O_2. Thus addition of antioxidants (e.g. dimethylthiourea) to lung preservation fluids has been advocated.

Reoxygenation injury has also been suggested to occur upon re-expansion of collapsed lungs, and if pulmonary vessels are blocked by thrombi that then dissolve. Infiltration of neutrophils into transplanted or otherwise-injured lungs, with consequent ROS generation, is frequently suggested to be involved, but data are conflicting.

9.10.5   *Asthma*[112]

Asthma can to some extent be regarded as a chronic inflammatory disease of the airways, and thus a contribution to its pathology by ROS/RNS (e.g. produced by activated eosinophils) is feasible. Indeed, levels of H_2O_2 and $NO^\bullet$ in breath are reported to be increased in children sick with asthma. Induction of iNOS in bronchial epithelial cells has been reported. Low dietary vitamin C intakes are associated with increased incidence of pulmonary disorders, and asthmatic patients have sub-normal plasma ascorbate, but this does not, of course, imply a cause–consequence relationship. Supplementation studies suggest that extra ascorbate may have a modest benefit, although some studies showed none. It has been reported that the glutathione content of lung lining fluids is elevated in asthma patients (as in cigarette smokers; Section 8.11.4), suggestive of an adaptation to oxidative stress.[275]

In humans, levels of CuZnSOD, catalase and MnSOD in tracheobronchial epithelium seem to be low[74] and expression of their genes is not induced by hyperoxia, suggestive of a susceptibility of this tissue to damage by ROS. Some iNOS is present even in healthy subjects. Indeed, tracheobronchitis is an early symptom of O_2 toxicity in humans (Chapter 1). It is thus feasible for ROS/RNS to contribute to asthma pathology and, in view of the increasing incidence of this disorder, more work is urgently required.

9.11 Cystic fibrosis[316]

Cystic fibrosis (CF), one of the most common inherited diseases of humans (incidence about 1 in 3000 in Caucasians), is due to a single biochemical abnormality in a protein encoded by a gene located on chromosome 7. This protein (the **cystic fibrosis transmembrane conductance regulator**, CFTR) appears to couple ATP hydrolysis with chloride (Cl^-) transport across epithelial surfaces under the control of a cAMP-dependent protein kinase. CFTR probably has other physiological functions that remain to be clarified, such as transport of other anions (e.g. ATP) and regulation of other ion channels. Cystic fibrosis is thus an example of a disease in which the cause has nothing to do with oxidative stress. Why then should it be discussed in this book?

The defect in CF impairs the movement of water and electrolytes across various epithelial surfaces, most notably in the respiratory tract, but also in the liver, bile ducts, pancreas, gastrointestinal tract and skin (the sweat excretion system). In the respiratory tract there is decreased Cl^- secretion and increased Na^+ absorption, resulting in inadequate hydration of the respiratory tract mucus secretions. This makes them more tenacious and difficult to clear, and predisposes the CF patient to chronic lung infection. This infection and the ensuing inflammatory--immune responses lead to progressive lung destruction. Oxidative stress contributes to this injury, but, as with ARDS (Table 9.5), its precise significance is uncertain (Box 9.1). Some of the antibiotics regularly administered to CF patients may exert antioxidant as well as antibacterial actions (Section 8.13). The most common pathogen chronically infecting the lung is *Pseudomonas aeruginosa*, which releases siderophores and redox-cycling agents that aggravate oxidative damage (Section 8.7).

Box 9.1
**Evidence consistent with increased oxidative stress
in cystic fibrosis**

- elevated plasma TBARS[a]
- elevated levels of plasma lipid peroxides (measured by glutathione peroxidase assay or by HPLC)
- increased susceptibility of lipoproteins to peroxidation
- increased susceptibility of erythrocytes to peroxide-induced hemolysis
- elevated breath pentane levels[a]
- increases in myeloperoxidase in sputum
- increased phagocyte-derived 'long-lived oxidants' in sputum
- increased 'free' iron, catalytic for free-radical reactions, in sputum
- evidence of protein oxidation in sputum
- decreased GSH levels in bronchoalveolar lavage fluid
- increased amounts of pro-oxidant cytokines (e.g. TNFα)
- increased neutrophil and monocyte numbers in lung
- decreases in plasma selenium (and plasma and erythrocyte glutathione peroxidase levels)
- decreases in plasma ascorbic acid and α-tocopherol levels (in some but not all studies)
- decreased plasma β-carotene levels (in most studies)
- increased oxidative damage to DNA (assayed as 8-OHdG excretion)
- presence of nitrotyrosine in CF sputum
- elevated dityrosine levels in plasma and bronchoalveolar lavage fluids
- inactivation of α$_1$-antiproteinase
- decreased plasma TRAP values

[a]Specificity of assays questionable (Chapter 5).
For full references see *Adv. Pharmacol.* **38**, 491 (1996) and *Thorax* **49**, 738 (1994).

9.11.1   *Cystic fibrosis and carotenoids*

An interesting observation is the low plasma β-carotene levels in CF patients (often $\sim 0.04\,\mu M$ as opposed to normal levels of around $0.2\,\mu M$); levels of lutein, lycopene and α-carotene are also sub-normal. The decreases cannot easily be explained by decreased intake or by disordered fat absorption and (at least for β-carotene) are greater in more severe disease. It seems possible that this could relate to an increased turnover of carotenoids, possibly due to their reaction with ROS/RNS (Chapter 3). However, carotenoids have many biological actions. There is a need to measure products of carotenoid oxidation in CF and to carry out studies of carotenoid metabolism. Administration of β-carotene to children with CF was reported to decrease plasma TBARS (measured using HPLC) from $\sim 90\,nM$ to $\sim 70\,nM$ (close to the values found in normal children).[163]

9.12   Oxidative stress and cancer: a complex relationship

9.12.1   *The cell cycle*[4,72]

Most cells in an adult multicellular organism are not dividing and must receive growth signals (e.g. platelet-derived growth factor (PDGF), epidermal growth factor (EGF), insulin-like growth factor 1 (IGF-1), transforming growth factor β (TGF-β), fibroblast growth factor (FGF), nerve growth factors (NGFs), erythropoietin, IL-2 or IL-3) to start the process. Some growth factors act on a wide range of cells (e.g. PDGF, FGF, EGF, IGF-1, TGF-β) whereas others are more selective; for example, IL-2 acts on T-lymphocytes, NGFs on neurones, erythropoietin on bone marrow and erythrocyte precursor cells and IL-3 on bone marrow cells. Responsiveness depends on whether or not the target cell has receptors for the growth factor. Binding of growth factors to receptors activates intracellular signalling mechanisms, usually cascades of protein kinases, that produce changes in gene expression. **Early response genes** (e.g. c-*fos*) are expressed quickly, and can then be involved in activation of delayed response genes.

In adult humans, neurones and skeletal muscle cells do not normally divide at all, hepatocytes perhaps once every few years, whereas epithelial cells in the gut may divide twice daily as the gut lining is constantly renewed. In culture, cells supplied with serum as a source of growth factors will grow until a confluent monolayer is obtained—in general, when cells contact each other, further proliferation is inhibited. Both cell adhesion molecules and gap junctional proteins (e.g. **connexins**, Chapter 4) are involved in this 'contact inhibition'.[339]

To produce a pair of genetically identical daughter cells by cell division, the DNA must be accurately replicated and the chromosomes segregated into two different cells. Cell division begins with **mitosis (M phase)**, the process of nuclear division. The nuclear envelope breaks down, the chromatin condenses into visible chromosomes and the cellular microtubules form a **mitotic spindle** on which the chromosomes eventually separate. In **anaphase** the chromosomes move to the poles of the spindle and re-form intact nuclei and the cell then divides in two. In most cells M phase is short (often 60 min or less) by comparison with **interphase**, the gap between M phases.

Replication of nuclear DNA occurs in the **S phase** (S for 'synthesis') of interphase. Cells in S phase can be recognized experimentally because they will take up DNA precursors, e.g. tritium-labelled thymidine or bromodeoxyuridine, a thymidine analogue that can be recognized in the nucleus by antibody staining techniques. The interval between completion of mitosis and the start of a new round of DNA synthesis is the G_1 **phase**; the G_2 **phase** is the gap between finishing DNA synthesis and the onset of mitosis. G_1 and G_2 provide time for cell growth and synthesis of other components. Cells in G_1, if not yet committed to DNA replication, can halt their progress around the cell cycle and are then said to be in a resting, G_0, state. Withdrawal of growth factors from proliferating cells sends them into G_0. Cells in G_0 do not express early or late response genes.

The control of the cell cycle relies on multiple proteins: an important group is the **cyclin-dependent protein kinases**, whose ability to phosphorylate selected cellular targets is regulated by another important protein class, the **cyclins**. Ubiquitin and the proteasome (Chapter 4) play important roles in regulating cyclin levels. Also important are **checkpoints**: the cell must have fulfilled certain criteria (e.g. DNA replication completed, a certain minimum size), before moving on to the next stage. For animal cells, one major checkpoint is in G_1. A feedback control is essential to delay mitosis until DNA replication is complete in S phase. Damage to DNA (e.g. by γ-radiation) also generates a signal to delay mitosis, presumably until repair is complete. One protein essential for the G_1 checkpoint is **p53** (Section 9.12.5 below).

9.12.2   *Tumours*

A multicellular organism is essentially a clone of cells which collaborate, regulating their own proliferation in the interests of the whole organism. There are about 10^{16} cell divisions per lifetime in the human body, and about 10^{12} in the mouse. The 'spontaneous' mutation rate due to inevitable errors made by DNA polymerases has been estimated as 10^{-6} per gene per cell division. Hence each gene might have undergone 10^{10} mutations in the human (10^6 in the mouse) lifespan. As a result, mechanisms to repair damage and/or eliminate mutated cells are essential.[4]

A **tumour** may be defined as an abnormal lump or mass of tissue, the growth of which exceeds, and is uncoordinated with, that of the normal tissue, continuing after the stimuli that initiated it have ceased, i.e. normal intercellular collaboration has broken down. Most tumours form discrete masses, but in the **leukaemias** (tumours of myeloid or lymphoid cells) the tumour cells are spread through the bone marrow or lymphoid tissues, and circulate in the blood. Tumours vary widely in their growth rates. The most important classification of tumours is that of **benign** or **malignant**. The cells of benign tumours remain at the site of origin, forming a cell mass. When growing in a solid tissue, they usually become enclosed in a layer of fibrous material, the capsule, formed by the surrounding tissues. Benign tumours rarely kill, unless they press on a vital structure or secrete abnormal amounts of hormones.

Most fatal tumours are **malignant**,[4] or **cancerous**. Unlike benign tumours, the cells of malignant tumours invade locally, and often also pass through the bloodstream and lymphatic system to form secondary tumours (**metastases**) at other sites. The rates of growth and metastasis formation differ from tumour to tumour. For example, breast cancers often grow very slowly, as do many rodent ulcers. Rapidly growing malignant tumours usually lose their histological resemblance to their tissue of origin. Basal cell cancer of the skin epidermis (which arises from keratinocyte precursors) rarely metastasizes, whereas melanoma (which arises from pigment cells) frequently does. Cancers are classified according to the tissues from which they arose, e.g. **carcinomas** (such as adenocarcinoma) arise from epithelial cells and **sarcomas** from connective tissue or muscle.

9.12.3   *Carcinogenesis*

The process of conversion of a normal cell to the malignant state is called **carcinogenesis**, and agents that induce it are called **carcinogens**. Carcinogenesis is a complicated, multi-stage process; essentially, a small population of abnormal cells is generated and then increases in abnormality as a result of a series of mutations and changes in the patterns of gene expression.

Factors predisposing to malignancy include both inherited traits and environmental agents. For example, breast cancer risk is affected by genetic predisposition (e.g. two susceptibility genes, **BRCA1** and **BRCA2**, which appear to play some role in regulating DNA-repair processes, have been identified). It is also affected by diet: a high-calorie (especially fat-rich) diet is widely thought to be a risk factor for breast cancer. Some people are more sensitive to environmental carcinogens than are others, in part due to differences in the relative levels of their phase I and phase II xenobiotic-metabolizing enzymes (Section 8.1).[117a]

The development of cancer in animals takes many years. Progressive tissue and cellular changes are seen during the latent period. Often new cell populations appear that represent stages in evolution from normal cells through pre-neoplastic and pre-malignant to malignant cells (Table 9.8). This may be illustrated by **cervical cancer**.[4] Normal cell proliferation in the neck of the

Table 9.8. The stages of carcinogenesis

Stage	Description
Initiation	Cellular phenomenon characterized by genetic changes that are irreversible. Examples of initiators include agents that react directly with DNA (can be endogenous or xenobiotic) and ionizing radiation. Single initiated cells are not morphologically recognizable; may have subtle phenotypic changes after multiplication, depending on organ.
Promotion	Usually reversible process of gene activation; often the result of action of xenobiotics or endogenous substances; involves entire tissue (initiated and non-initiated cells); may produce a benign tumour from initiated cells. Examples of promoters include phorbol esters in the mouse skin, saccharin in the urinary bladder and phenobarbital in the liver. May have at least two stages (early reversible, and late irreversible).
Progression	Conversion of benign to malignant tumours, usually accompanied by more rapid growth, invasiveness, metastasis, increased genetic instability. May be associated with further irreversible genetic change (e.g. loss of a tumour-suppressor gene). Examples include progression of benign papillomas to squamous cell carcinoma in mouse skin, benign colon polyps to adenocarcinoma, and focal benign hepatic nodules to hepatoma.

Adapted from Table 1 in *Blood* **76**, 655 (1990) by courtesy of Prof. S.A. Weitzman and the publishers.

uterus occurs in the basal layer of epithelial cells. New cells move upwards, differentiate, become keratinized and eventually slough off the surface. Cervical screening frequently reveals **dysplasia**: dividing cells occur outside the basal layer and some disordered cell arrangement is visible. Dysplasia may persist harmlessly or even disappear, but sometimes it may progress to **carcinoma *in situ***: the whole epithelium appears disordered and contains proliferating cells, many of abnormal shape and size. Again, carcinoma *in situ* may persist without problems, but in 20–30% of cases it may give rise (over several more years) to a malignant **cervical carcinoma**, whose cells break out of the epithelium by crossing the basement membrane to invade the underlying connective tissue.

Studies upon the action of carcinogens in mouse skin (painting the skin with carcinogenic hydrocarbons such as **benzpyrene** or **dimethylbenzanthracene**) were the first to reveal that two distinct stages of carcinogenesis can be defined, and later work identified a third stage (Table 9.8). This three-stage model seems to apply (with reservations) to many cancers; it is outlined in the following sections.

Tumour initiation

Initiation, the first stage, is caused by irreversible DNA alteration, e.g. reaction of carcinogens with DNA. Thus a major protection against cancer initiation resides in efficiency of carcinogen detoxification by phase I/II enzymes, DNA repair, and elimination (e.g. by apoptosis) of cells with badly damaged DNA.

Successful initiation requires not only DNA modification but also some degree of DNA replication and cell proliferation to allow the mutation to be 'fixed' in the DNA before repair can remove it. The needed cell proliferation may be occurring normally in the tissue (e.g. epithelial cells), can be stimulated directly by certain carcinogens, or can occur secondarily as a result of the need to replace some cells killed by the carcinogen. Of course, if all the initiated cells are killed, initiation will not occur. Most human cancers are carcinomas, because epithelial cells proliferate rapidly in the healthy animal. In addition, these cells are often exposed to carcinogens, e.g. in inhaled air or from the diet.

Tumour promoters[144,308]

Initiation is followed by **promotion**. Tumour promoters cause expression of the latent phenotype of initiated cells (Table 9.8) by **selection** and **clonal expansion**. Many carcinogens, given in large doses, are both initiators and promoters. However, a low dose of a carcinogen, itself too small to induce a tumour, can be effective if supplied together with certain non-carcinogenic substances known as **tumour promoters**. For example, **croton oil**, a non-carcinogen, promotes the development of cancer by sub-carcinogenic doses of the hydrocarbon methylcholanthrene, but only if the oil is given simultaneously with, or after, the methylcholanthrene. Croton oil is obtained from seeds of the plant *Croton tiglium*. Fractionation of the oil has shown that the most powerful tumour promoter present is **phorbol myristate acetate** (PMA, Fig. 9.17), a compound frequently used to induce the respiratory burst in phagocytic cells (Section 6.7). Many tumour promoters, including PMA, are

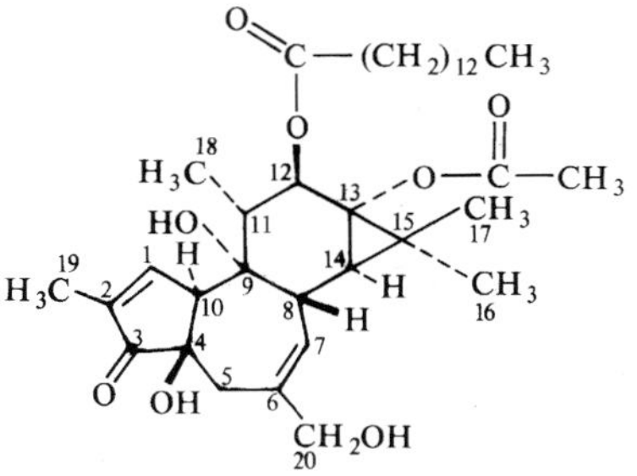

Fig. 9.17. Structure of phorbol myristate acetate (PMA). The full chemical name of this compound is 12-O-tetradecanoylphorbol-13-acetate.

potent inducers of inflammation, but there is as yet no single unifying principle that explains the action of all the different promoting agents. Another powerful promoter in animals is **2,3,7,8-tetrachlorodibenzo-*p*-dioxin**, TCDD (Section 8.3.5).

During tumour promotion, the altered genetic material of the initiated cell becomes expressed by changes in the expression of genes that regulate cell differentiation and growth. Tumour promoters can produce reversible changes in cell proliferation and phenotypic expression (Table 9.8). Removal of the promoter can often result in the tissue returning approximately to normal, although it will still be initiated.

Tumour progression

The final stage in carcinogenesis is the development of a pre-malignant lesion into a malignant one. This **tumour progression** involves additional changes in DNA. For example, colorectal cancers are common age-related cancers in the West; they arise from the epithelium lining the colon and rectum. Often they are preceded by a benign tumour, an **adenoma**, called a **polyp**.

9.12.4 Oncogenes[4]

The importance of changes in DNA to carcinogenesis is shown by several observations. Many viruses can induce tumours in infected animals after short latency periods; the first discovered was the **Rous sarcoma virus**, a retrovirus which causes sarcomas in chickens. From many of these viruses **oncogenes** have been isolated, specific genes that can transform cells in culture into a malignant state when the virus infects them (Table 9.9). Malignant (**transformed**) cells in culture lose contact inhibition and can proliferate without attachment to a substratum. Often there is down-regulation of connexin synthesis and hence of communication through gap junctions.

Rous sarcoma virus contains the oncogene *src*. Nucleic acid hybridization studies[b] showed that normal mammalian cells contain genes (**protooncogenes**) that are similar in base sequence to the viral oncogenes.

Table 9.9. Some oncogenes originally identified in transforming viruses

Oncogene	Function of gene product	Source of virus	Virus-induced tumour
src	Tyrosine kinase	Chicken	Sarcoma
abl	Tyrosine kinase	Mouse, cat	Leukaemia, sarcoma
erb-B	Tyrosine kinase: epidermal growth factor receptor	Chicken	Erythroleukaemia, fibrosarcoma
fos, jun	Fos and Jun proteins associate to form AP-1 (Chapter 4)	Mouse, chicken	Osteosarcoma, fibrosarcoma
raf	Protein kinase (serine/threonine) activated by Ras	Mouse, chicken	Sarcoma
myc	Transcription factor; normally acts in nucleus as signal for cell proliferation, collaborates with *max* gene	Chicken	Sarcoma, myelocytoma, carcinoma
H-*ras*, K-*ras*	GTP-binding proteins; cycle between active (GTP-bound) and inactive (GDP-bound) states; hydrolysis of GTP stops activity	Rat	Sarcoma, erythroleukaemia
v-*rel*	Gene-regulatory protein related to NF-κB (Chapter 4)	Chicken	Lymphoma in spleen and liver
sis	Platelet-derived growth factor B-chain	Monkey	Sarcoma

Proto-oncogenes can be activated to the oncogenic state by point mutation, gene amplification, deletion, chromosomal translocation to a transcription site or viral insertion. Most human cancers appear not to be related to viral infection, but some are, e.g. **Burkitt's lymphoma** (Epstein–Barr virus) and **adult T-cell leukaemia/lymphoma** (human T-cell leukaemia virus type I, HTLV-I). In general, viral infection alone does not produce cancer—many other events are involved. Thus <5% of adults infected with HTLV-I develop leukaemia. Some other viruses, e.g. hepatitis B virus, may indirectly lead to cancer by provoking chronic inflammation.

The normal cellular analogue of the viral *src* (**v-src**) gene is **c-src** (sometimes just written as **src**). Proto-oncogenes are usually related to cell growth; they encode growth factors, cell-surface receptors for such factors, proteins involved in transferring the growth 'signal' from surface receptors into the nucleus, or proteins involved in the control of cell death (Table 9.9). Many oncogenes have been identified by means other than the use of viruses, e.g. by direct identification in cancerous tumours or analysis of chromosomal abnormalities. Examples are **lck**, L-**myc**, **neu**, N-**ras** and **bcl**-2, whose proto-oncogene is involved in suppressing apoptosis (Chapter 4).

The exact role of proto-oncogenes in relation to initiation, promotion and progression of carcinogenesis is not entirely clear. Activation of proto-oncogenes could play a role in all three processes although, consistent with the multistage nature of cancer *in vivo*, it appears that activation of a single proto-oncogene cannot itself produce malignant transformation in whole animals (e.g. if introduced into transgenic animals), although it certainly can transform cells in culture and does predispose the animal to develop cancer. Activation of proto-oncogenes can involve mutation, e.g. the oncogene ***erbB*** (Table 9.9) is a mutated form of the EGF receptor that sends a signal all the time. It can also involve excessive expression of the normal gene, e.g. if chromosomal changes place it in a site where it is excessively transcribed.

Mutated proto-oncogenes are found in most human and other animal tumours. For example, most of the skin tumours evoked in mice by the carcinogen dimethylbenzanthrene have an $A \rightarrow T$ mutation in a ***ras*** oncogene. ***Ras*** oncogene protein products have diminished ability to hydrolyse GTP. The GTP-bound configuration is usually active in signal transduction and the GDP-conformation inactive (Table 9.9), so that loss of GTPase function promotes excessive signalling.

A few human tumours develop mainly, or wholly, as a result of an inherited genetic anomaly. **Familial adenomatous polyposis coli**, in which polyp development is enormously accelerated early in life, and colon cancer almost invariably develops is one example; it is associated with a defect in the ***APC gene*** on chromosome 5. *APC* seems related in some way to arachidonic acid metabolism, e.g. disruption of the *COX-2* gene in mice expressing an *APC* mutation decreases the number of tumours. Indeed, in humans the regular use of cyclooxygenase inhibitors such as aspirin seems associated with a 40–50% decrease in the risk of developing colorectal cancer.[328] *COX-2* is often expressed in adenocarcinomas. Aspirin and other NSAIDs may also be toxic to colon cancer cells by additional mechanisms.

Defects in DNA repair mechanisms also predispose to the development of cancer.[135,166] In **xeroderma pigmentosa**, an inborn recessive defect in the ability to repair UV-induced lesions in DNA, there is severe skin damage and a high risk of skin cancer on exposure to sunlight or to other sources of UV light. Patients with **Fanconi's syndrome**, in which DNA repair has also been suggested to be defective, show an elevated level of 8-OHdG in DNA and an increased incidence of malignancies. Elevated cancer risk is also seen in **Bloom's syndrome** (apparently caused by a genetic defect in a helicase enzyme involved in unwinding DNA to allow replication, transcription and repair) and in **ataxia telangiectasia**, caused by a mutation in the ***ATM gene***, whose normal product is involved in the G_1 cell-cycle checkpoint.[72] Cells from ataxia telangiectasia patients lack normal checkpoint control, e.g. after irradiation they can proceed into S phase without repairing DNA. Bloom's syndrome is an autosomal-recessive condition in which about 20% of patients develop cancer before the age of 20. Ataxia telangiectasia occurs in about one birth per 40 000, and about 10% of patients develop cancer at an early age. The disease is associated with progressive degeneration of the cerebellum and hypersensitivity of cells to ionizing radiation. In **Down's syndrome**, which is usually caused by the presence of three copies of chromosome 21 (Chapter 4), an increased risk of cancer has also been noted.

9.12.5   *Tumour-suppressor genes*[4]

Cell proliferation is regulated by the availability of 'growth factors' and by mechanisms which determine whether a cell is ready to pass through the various checkpoints of the cell cycle. In addition, cell numbers can be affected by regulating the pathways to the differentiated (non-dividing) state or to apoptosis. Thus mutations encouraging proliferation could be not only those that activate a stimulatory gene (e.g. converting a proto-oncogene to an oncogene) but also those that inactivate a gene that *halts* replication, a **tumour-suppressor gene**. Whereas mutations creating oncogenes are usually dominant, since they create active proliferation signals, mutations in suppressor genes are usually recessive, i.e. both copies of the gene must be inactivated in order for disease to result.

One important such gene is **p53** (so-called because the relative molecular mass of its protein product is 53 000). p53 is a transcription factor that acts to block cell division. The p53 gene is activated in response to DNA damage or abnormal changes in nucleotide precursor pools. Its product increases transcription of various other genes, including **p21** (a gene encoding a cyclin kinase inhibitor) and ***GADD45*** (growth arrest on DNA damage). Levels of p53 gene expression rise in cells subjected to stresses that result in DNA damage (e.g. radiation, excess NO$^\bullet$, UV light) and it halts DNA replication until DNA damage is repaired. Stimulation of p53 can also trigger apoptosis under some conditions. If both p53 genes are inactivated, cells can enter the cell cycle with damaged DNA. Mutations of p53 are common lesions in human cancers, found in over 50% of cases. For example, about 75% of colorectal cancers and >90% of squamous cell skin cancers have p53 mutations.

About 50% of human colon cancers have a point mutation in a *ras* proto-oncogene. Many also have *APC* mutations, which can be detected in small benign polyps: their effect seems to be to increase cell proliferation. Later mutations, e.g. in *ras*, alter cell differentiation and morphology (but may also induce p53 as a 'check' on proliferation). Finally, loss of p53 may be associated with full conversion of polyps to malignant tumours. Although this sequence of mutations (*APC, ras*, p53) is common in human colon cancer, it is not invariable: there are many routes to malignancy.

9.12.6   *ROS/RNS and carcinogenesis*[34,144,190,308]

In view of the importance of DNA damage in carcinogenesis, it is conceivable that any agent capable of chemically modifying DNA could be carcinogenic. ROS/RNS fall into this category.

Direct DNA damage

Exposure of organisms to ionizing radiation has long been known to favour development of cancer later in life. Radiation-induced carcinogenesis appears to involve both initiation and promotion; radiation can lead to activation of oncogenes and inactivation of tumour-suppressor genes such as p53. Some genetic damage by radiation occurs by direct absorption of energy by DNA, but some is mediated by ionization of water and formation of highly reactive $OH^\bullet$. Hydroxyl radical attack upon DNA generates a whole series of modified purine and pyrimidine bases (Chapter 4), many of which are known to be mutagenic. Attack of $OH^\bullet$ upon deoxyribose also yields a multiplicity of products.

The 'steady-state' levels of oxidative base lesions in DNA in normal cells vary according to the measurement method and also vary between laboratories (Chapter 5), but an average value is around 1 per 10^6 bases. By comparison with the levels of adducts of known carcinogens that are detected in carcinogen-exposed cells (e.g. levels of benzpyrene adducts in DNA from smokers range from 65 to 533 per 10^8 nucleotides), these levels of oxidative adducts are high and imply that endogenous damage to DNA by ROS is an important contributor to the age-related development of cancer.[173,305] It is well-known that increased production of ROS can cause DNA damage in cells.[330,345] For example, treatment of mammalian cells with H_2O_2 or exposure of them to activated phagocytes gives a pattern of DNA damage characteristic of attack by $OH^\bullet$. RNS can also attack DNA,[217] nitrating and deaminating bases to give mutagenic lesions (Chapters 4 and 5).

The incidence of most cancers rises with the fourth or fifth power of age (Fig. 9.18) in animals (including humans), e.g. about 35% of humans have cancer by age 85. At first sight this seems high, but the enormous number of mutations that occur over a lifetime caused by polymerase errors, spontaneous deamination and depurination of DNA, plus the actions of ROS,[173] RNS and exogenous carcinogens makes one admire the high efficiency of protective mechanisms in ensuring that the majority of animals do *not* develop cancer.

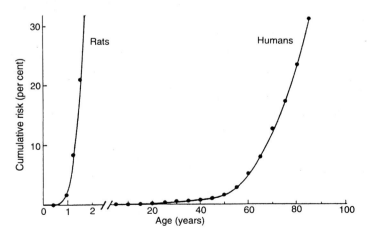

Fig. 9.18. The cumulative risk of death from cancer as a function of age in humans. Note the very sharp rise in cancer incidence in older people. Apart from a few rare cancers (e.g. testicular cancer, leukaemias), cancer is generally a disease of old age. The same phenomenon occurs, on a shorter time-scale, in other mammals such as rats. (From *Free Rad. Res. Commun.* **7**, 122 (1989), by courtesy of Professor Bruce Ames and Harwood Academic Press.)

As early as 1984 it was shown that exposure of mouse fibroblasts to ROS leads to malignant transformation.[330,345] Incubation of plasmids bearing proto-oncogenes (e.g. K-*ras*) with ROS followed by their transfection into cells has been reported to lead to cell transformation.[135] Similarly, incorporation of 8-OHdG into the first or second position of codon 12 in K-*ras* causes it to gain transforming ability. Several organic peroxides (e.g. **benzyl peroxide**) are tumour promoters in mouse skin; their conversion into free radicals is thought to be involved in their tumour-promoting effect.[144] Increased steady-state levels of multiple oxidative DNA base damage products (a pattern characteristic of $OH^•$ attack) have been found in DNA from cancerous breast tissue compared with DNA from surrounding areas. Similar elevations are found in lung, colon, stomach, ovary and brain (astrocytoma) cancers.[187] Whether these rises are due to increased oxidative DNA damage or decreased repair is as yet unclear. An exciting observation is that these DNA base oxidation products are also elevated in **benign prostatic hyperplasia**,[220] often thought to be related to later development of prostate cancer. Levels of 8-oxoGTPase, which sanitizes the nucleotide pool (Section 4.8.1) are elevated in some human tumours.[323a]

Do the mutations in cancer cells show the characteristics of ROS/RNS attack?[34]

One way of gaining evidence that ROS/RNS play a role in cancer would be to examine base changes in mutated oncogenes/tumour suppressor genes to see if they are of the type that could be generated by oxidative stress. Most attention has focused on 8-OHdG in relation to cancer, since this product is mutagenic and is produced in DNA in significant amounts by several ROS ($OH^•$, $RO_2^•$, singlet O_2 and, to a limited extent, $ONOO^-$). For example, G → T transversions are often seen in the p53 gene in primary liver cancers in Asia and Africa. These transversions could result from the mutagenic effect of elevated levels of 8-OHdG (Chapter 4). However, such lesions can also be generated by the carcinogen **aflatoxin** and by errors in the replication of DNA containing abasic sites. The same comment can be applied to G → T transversions observed in many other genes, e.g. in breast cancers, in K-*ras* in nickel-induced renal sarcomas in rats, in p53 in smoking-related lung cancers, and in colorectal cancer. Oxidative damage could also account for C → T and G → A transitions frequently seen in p53 genes in human cancer, but again these changes are not specific for attack by ROS/RNS. For example, C → T can result from deamination of 5-methylcytosine to thymine. By contrast, tandem mutations provide good evidence for UV-induced DNA damage to p53 in skin cancers (Section 7.9) and may also arise by attack of certain ROS on DNA.

Other mechanisms of ROS/RNS carcinogenicity[17,217]

Oxidative damage to lipids and to proteins (e.g. DNA repair enzymes) could also lead to mutagenic effects (Fig. 9.19). In addition, low levels of ROS can stimulate cell proliferation, whereas high levels decrease it (Chapter 4). ROS can increase net protein phosphorylation (often by decreasing the activity of

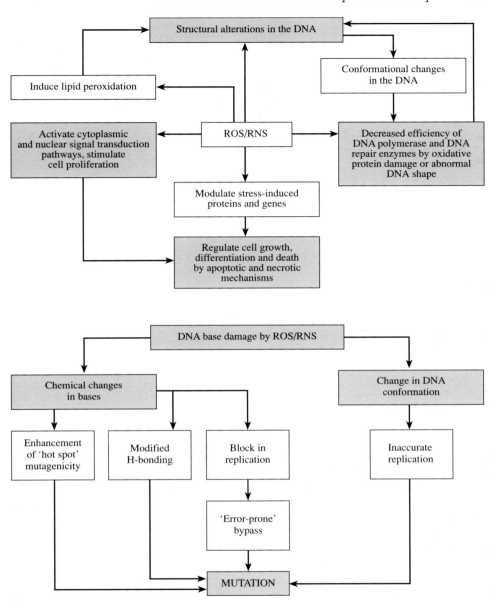

Fig. 9.19. Some of the ways in which reactive oxygen and nitrogen species could facilitate cancer development. Adapted from *Biochem. J.* **313**, 17 (1996) by courtesy of the *Biochemical Journal* (Portland Press). Top: aldehyde end–products of lipid peroxidation can form mutagenic adducts with DNA bases; peroxyl and alkoxyl radicals might also attack DNA. ROS/RNS could damage proteins, e.g. DNA repair enzymes and polymerases (increasing the error rate of replication). Bottom: how structural changes in the DNA can cause mutations.

protein phosphatases) and help promote proliferation and the expression of immediate early genes such as *c-fos* and *c-myc*. For example, NIH-3T3 fibroblasts[130] in culture transfected with a *ras* oncogene generated excess $O_2^{\bullet-}$, which made some contribution to abnormal growth.

Many transcription factors and some protein kinases are subject to redox regulation; NF-κB and AP-1 have already been discussed (Chapter 4). Redox regulation might also affect p21 and the p53 protein.[248] p53 has ten cysteine residues, some of which are involved in co-ordinating a zinc iron required for p53 activity. Oxidation of p53 inhibits its binding to DNA and activity as a transcription factor.

Another way in which ROS/RNS could affect the behaviour of tumour cells is to alter cell–cell communication.[339] Communication through gap junctions is generally decreased in tumour cells and this is thought to be involved in their excessive proliferation. One of the reported effects of several tumour promoters, and of oxidative stress, is to decrease gap-junctional communication (Chapter 4). Finally, ROS might be involved in metastasis.[153] Degradation of extracellular matrices by tumour cells involves proteolytic enzymes, some of which are secreted in latent forms (by both tumour cells and phagocytes) that can be activated by ROS.

Chronic inflammation as a carcinogen[8,9,217]

The multiple effects of ROS/RNS suggest that they could contribute to all stages of carcinogenesis. However, an excess of ROS/RNS can inhibit proliferation of cells (e.g. by up-regulating expression of p53 or cyclin genes)[48] and may kill them by inducing necrosis or apoptosis. Hence the net effect will depend upon how much ROS/RNS is generated, and the level of antioxidant defences.

There is little doubt that chronic inflammation is a risk factor for cancer. Examples of diseases where this inflammation–cancer link is observed include inflammatory bowel disease (Section 9.8 above), *Helicobacter pylori* infection (Section 9.9) and infection with the parasite *Schistosoma haematobium*, which produces chronic inflammation of the bladder and an increased risk of bladder cancer. In both Asia and Africa, hepatocellular carcinoma is a major cause of mortality. Primary hepatoma in these countries is often associated with chronic infection with hepatitis B or C viruses, liver flukes of the *Opisthorchis* genus or ingestion of the carcinogen aflatoxin. Chronic hepatitis is associated with the presence of inflammatory cells, presumably generating ROS and RNS. Indeed, patients with *Opisthorchis viverrini* infection show increased urinary excretion of nitrate, probably due to induction of iNOS. Increased levels of 8-OHdG have been detected in DNA from livers with chronic hepatitis[103] and in several neoplastic or pre-neoplastic tissues. Elevated levels of other base damage products are often also detected (see above).

One apparent exception to the general rule that 'chronic inflammation increases cancer risk' is rheumatoid arthritis (RA). In RA it is well established that there is increased oxidative damage to DNA, lipids and proteins (Table 9.3). However, there is no clear evidence that RA patients develop cancers at an increased rate, certainly not at the most intense site of oxidative stress, the inflamed joint. Perhaps this is linked to the nature of the cells of the synovium. Synovium seems to be a hostile environment for neoplastic cells, although it has been argued that the excessive synovial cell proliferation in RA patients resembles neoplastic transformation and might be related to the growth-promoting properties of ROS.[156] In addition, we have already commented that NSAIDs, the mainstay of treatment in RA, decrease the incidence of colon cancer.

9.12.7   *Changes in antioxidant defences in cancer*

Mitochondria from several malignant animal tumours, and from many tumour cells in culture, show low MnSOD activity.[287] Transfection of the MnSOD gene into cells has been reported to suppress the malignant phenotype in some (but not all)[148,165] cell lines and also to render the cells more resistant to radiation.[121] For example, transfection of cDNA encoding MnSOD into a human breast cancer cell line raised MnSOD levels and rendered the cells less malignant when injected into mice.

Low activities of CuZnSOD, catalase and glutathione peroxidase are frequently reported in transformed cell lines and sometimes in biopsies of animal tumours. Studies on human tissues have given variable results,[132,148,189,238] but in general there is no clear evidence for marked decreases in MnSOD or other antioxidant enzymes in freshly-analysed human cancerous tissue. Indeed, levels are sometimes reported as elevated. It should be borne in mind that areas within a large tumour mass often have a poor blood supply and so are often anoxic, which may down-regulate antioxidant enzymes as a consequence of hypoxia rather than malignancy. For example,[229] in one study of a rat breast cancer, SOD activity was found to be 54 ($\pm$10) μg/g tissue at the centre of the tumour, but 117 ($\pm$38) μg/g at the edge. Exposure of the tumour-bearing rats to elevated O_2 concentrations raised both values, to 162 ($\pm$73) and 286 ($\pm$103) μg/g, respectively. This variation, and the fact that rapidly growing malignant tumours frequently lose histological resemblance to their tissue of origin, make it difficult to obtain valid comparisons of tumour enzyme activities with the normal state. Transformed cell lines in culture often undergo rapid 'evolution' to allow fast growth *in vitro*, so one must be cautious in extrapolating their properties to those of cells in a malignant tumour *in vivo*.

Levels of lipid peroxidation are also altered in tumour cells.[90,274] For example, in a study on a baby hamster kidney cell line (BHK-21/C13) and its polyoma-virus-transformed malignant counterpart, the level of lipid peroxidation (as measured by HPLC-based determination of malondialdehyde) was higher in transformed cells than in non-transformed cells and the α-tocopherol content lower, suggesting that the level of lipid peroxidation is increased in the malignant state. By contrast, earlier work showed that susceptibility to lipid peroxidation is decreased in certain malignant hepatoma cell lines (e.g. Novikoff and Yoshida ascites hepatoma cells). The precise reasons for this decrease depend on the cells studied, but they include a lower content of PUFA side-chains in lipids, an increased accumulation of α-tocopherol, and falls in the activities of cytochromes P450 and NADPH–cytochrome P450 reductase.

Pre-malignant cells may also show decreased rates of lipid peroxidation. Dividing hepatocytes in partially hepatectomized rats appear to accumulate α-tocopherol during DNA replication,[274] suggesting that 'decreased per-oxidizability', at least as accounted for by this mechanism, may be a feature of normal cell division rather than a feature of the cancerous state. Damage to DNA during replication has more serious consequences than when the DNA is not replicating, so it is possible that cells take steps to increase their antioxidant defences at this time. Indeed, levels of 8-OHdG in regenerating rat liver were also decreased.[1]

The increased risk of cancer in Down's syndrome, Bloom's syndrome, ataxia telangiectasia and Fanconi's anaemia has prompted investigations of cellular antioxidant defences in these conditions, although in none of the latter three cases is it likely that the primary gene defect directly affects antioxidant defence. Lymphocytes from Fanconi patients, cultured *in vitro*, show fewer chromo-somal aberrations if the O_2 concentration in the culture medium is decreased, suggesting some abnormality in antioxidant defence.[136,166] Indeed, the SOD activity of erythrocytes in Fanconi patients has been reported to be decreased by 30–40%, whereas catalase and glutathione peroxidase are normal. Cultures of fibroblasts from patients with Bloom's syndrome have been reported to release a clastogenic factor that can induce chromosome breaks if it is added to cultures of normal lymphocytes. This clastogenic action was suppressed by including SOD in the culture medium.

On the other hand, in Down's syndrome the tissue content of CuZnSOD is increased by approximately 50%. There have been reports that mitochondrial MnSOD is slightly ($\sim 30\%$) decreased in platelets from Down's patients, but whether such a decrease could be biologically meaningful is unclear. Eryth-rocyte glutathione peroxidase activity has been reported to be increased by 30–50% in erythrocytes from Down's syndrome patients, possibly because the increased SOD activity leads to more H_2O_2 generation.

9.12.8 *Transition metals and cancer*[306]

Malignant disease, like chronic inflammation (Section 9.7 above), produces changes in body iron and copper distribution. In most cases, plasma caerulo-plasmin rises whereas iron levels fall. These 'acute-phase response' changes are induced by cytokines. One popular suggestion[329] is that these changes serve to withhold iron from the tumour and so slow its growth by restricting iron essential for cell division. In general, tumours do contain less 'total iron' and have a lower degree of iron saturation in ferritin than do normal cells. This is not always the case, however, since human breast tumours appear to accumu-late iron, and in Hodgkin's disease heavy deposits of iron and ferritin are seen surrounding the tumour nodules. In some cancers, including Hodgkin's disease, breast cancer and leukaemia, the normally low concentrations of ferritin protein present in the blood are greatly increased. In cancer patients, chemotherapy or radiotherapy produces sharp rises in 'total blood iron' and plasma-transferrin saturation, often approaching or reaching 100%. Indeed, non-transferrin-bound iron is sometimes present (Section 4.5.2).[33]

Iron and copper are well-known pro-oxidants. Incubation of rat liver nuclei with iron(II) salts causes extensive O_2-dependent DNA damage. Such oxidative damage may account for the increased risk of hepatoma in iron-overloaded patients with idiopathic haemochromatosis (Chapter 3). The **LEC rat**[290] (Long–Evans, cinnamon coat colour) is a mutant strain that develops hepatitis and hepatoma, apparently associated with increased copper accumulation in the liver. Several iron chelates can be carcinogenic when injected into animals. One example is **ferric nitrilotriacetate**, injection of which into rats or mice causes increased lipid peroxidation, oxidative DNA damage and, eventually, renal cancers.[306] Indeed, the first report of iron carcinogenicity was in 1959, when multiple injections of iron dextran into rats were shown to lead to tumours at the site of inoculation. It has been suggested that the iron content of asbestos relates to its ability to produce inflammation and cancer (Section 6.7.3) and iron has been implicated in lung damage by cigarette smoke (Section 8.11). Iron administration can aggravate the carcinogenicity of other compounds, e.g. hexachlorobenzene or polychlorinated biphenyls in mice (Chapter 8).

There are several suggestions that high iron stores in the human body predispose to the development of cancer and that pro-oxidant effects of iron are involved. However, iron is usually tightly sequestered in ferritin and transferrin, and so these 'increased stores' are not necessarily promoters of oxidative damage. A diet promoting high body-iron stores tends to be high in meat and low in vegetables, itself a risk factor for certain types of cancer (Chapter 10). However, when an iron-rich diet is eaten, most of the iron remains unabsorbed and enters the colon (Chapter 3). The normal colonic contents are usually thought to be anaerobic, perhaps fortunately since exposure of suspensions of faeces to air leads to generation of ROS, including $OH^{\bullet}$, at high rates, and iron appears to be involved.[8] Faeces can also contain **faecapentenes**,[234] unsaturated ethers produced by *Bacteroides* spp. Faecapentenes are highly mutagenic in *in vitro* testing systems. Mixtures of iron and faecapentenes generate 8-OHdG in isolated DNA and in the DNA of HeLa cells exposed to them.

Attempts have been made to design anti-cancer therapies based on 'iron withholding': they include the use of iron chelators (e.g. desferrioxamine) and antibodies directed against the transferrin receptor. The transferrin receptor has also been used as a target to deliver toxins into proliferating cells, e.g. a subunit of the toxin **ricin** has been covalently linked to transferrin, and toxins have also been linked to antibodies recognizing the transferrin receptor. Obviously, the presence of transferrin receptors on healthy cells, and their requirement for iron, can limit the success of this type of therapy.

9.13 Carcinogens: oxygen and others

The oxygen that surrounds us and is essential to aerobic life is probably, and paradoxically, carcinogenic. Its mutagenic ability is certainly well-established. Oxygen is not a direct carcinogen: it has first to be metabolized (to ROS). Exactly the same is true of many of the molecules that are more classically regarded as carcinogens. These often are, or are metabolized to, **electrophiles**, reagents that seek areas of high electron density such as the purine and pyrimidine bases of DNA.[17] In 1775, the English physician and surgeon Percival Pott reported that the occurrence of cancer of the scrotum in chimney-sweeps could be correlated with deposition of soot and tar in the scrotal creases. Later work showed that repeated application of coal tar to the skin of animals eventually produces malignant tumours at the site of application. Aromatic hydrocarbons such as **benz[α]pyrene** (Fig. 9.20), abbreviated in the rest of this chapter to benzpyrene, were isolated from coal tar and shown to be carcinogenic. Carcinogenic hydrocarbons are present in combusted organic matter, e.g. soot, vehicle-exhaust particulates and cigarette tar. In the late 1800s, the German physician Rehn noticed an association between exposure of dye workers to aromatic amines such as **benzidine** (Fig. 9.20) and the development of bladder cancer. Not until the 1930s was it shown that several aromatic amines could induce bladder cancer in animals.

A wide range of chemicals has since been shown to be carcinogenic; Table 9.10 lists some of them. About one-third of all cancer cases in Europe and North America are related to the presence of carcinogens in cigarettes and other tobacco products (Chapter 8).[116] Other carcinogens have been identified from epidemiological studies (e.g. vinyl chloride), from routine bioassays (e.g. nitrosamines), and from investigations into the cause of diseases in humans (e.g. asbestos, cycasin) or in other animals (e.g. the hepatotoxic aflatoxins).

9.13.1 *Carcinogen metabolism*

Repeated exposure to carcinogens is usually necessary to produce a malignant tumour, and its development often takes a considerable time because of the complex multi-step nature of carcinogenesis. Many carcinogens, e.g. benzpyrene, are **complete carcinogens**, i.e. both initiators and promoters of carcinogenesis, whereas others are initiators only ('**incomplete carcinogens**') and will produce malignant transformation only if promotion then occurs. Low doses of a complete carcinogen, themselves too small to produce carcinogenesis, can be caused to do so by another tumour promoter. How a carcinogen behaves depends on the dose, test species and mode of administration. For example, urethane has been suggested to act as a complete carcinogen in foetal animal lung but an incomplete carcinogen in mouse skin.

Most carcinogens act by attacking DNA. The resulting carcinogen–DNA adducts may cause mutations (e.g. activating proto-oncogenes). They can do this directly, by interference with DNA replication, or if they give rise to abasic sites. Some carcinogens, such as highly reactive *N*-methyl-*N'*-nitrosoguanidine (Fig. 9.20) can modify DNA directly, e.g. by methylating guanine. Most carcinogens, however, have to be converted into their active form, the '**ultimate carcinogen**'. Conversion of a **pro-carcinogen** to the ultimate carcinogen can proceed in one or more steps. If more than one step is required, the intermediates may be called **proximate carcinogens**. The necessary metabolism is sometimes carried out by gut bacteria, e.g. the hydrolysis of cycasin (Table 9.10) to the proximate carcinogen **methylazoxymethanol** (Fig. 9.20), which is then metabolically activated to a DNA-damaging product.

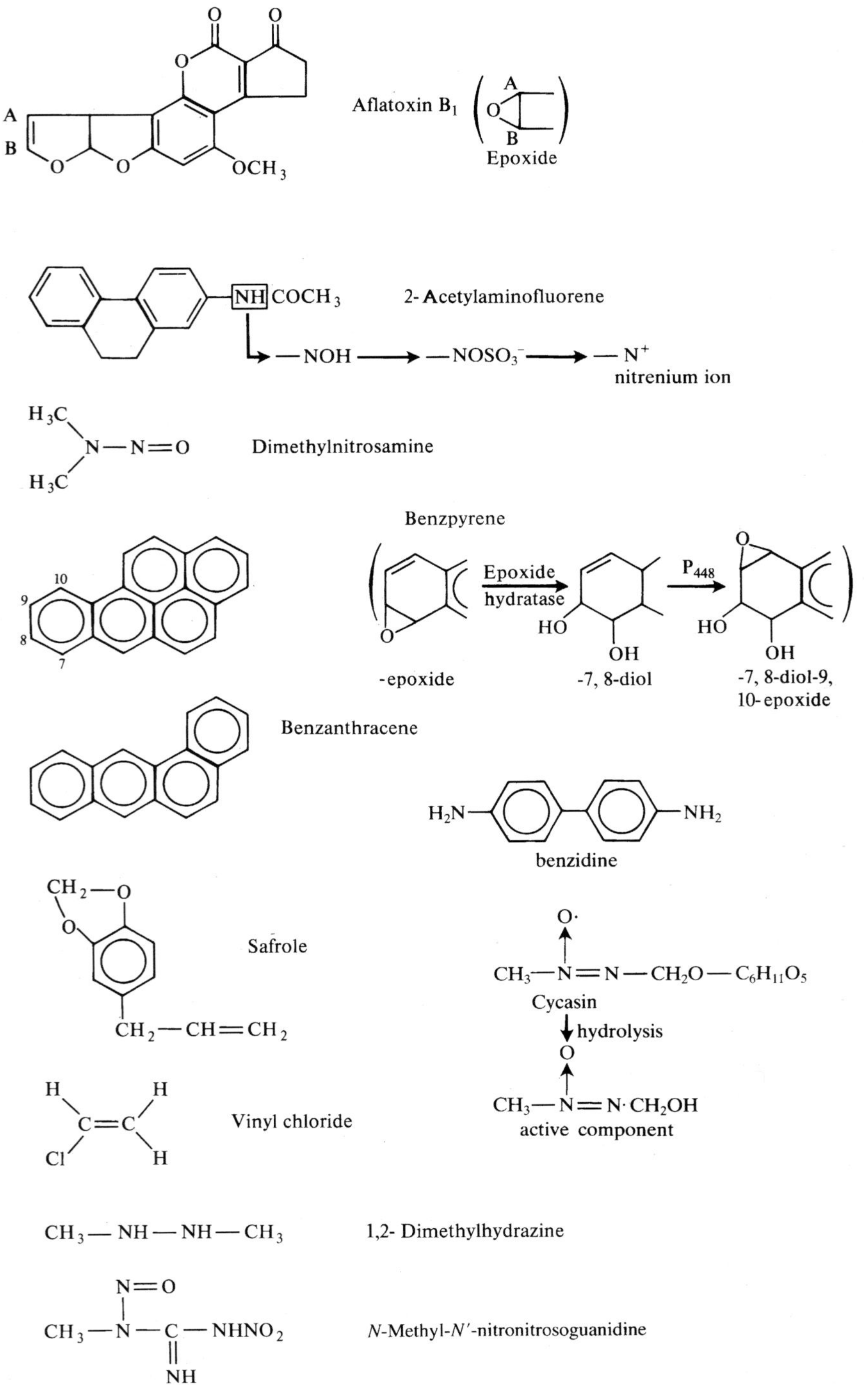

Aflatoxin B$_1$ (Epoxide)

2-Acetylaminofluorene

—NOH ⟶ —NOSO$_3^-$ ⟶ —N$^+$ nitrenium ion

Dimethylnitrosamine

Benzpyrene

-epoxide $\xrightarrow{\text{Epoxide hydratase}}$ -7, 8-diol $\xrightarrow{\text{P}_{448}}$ -7, 8-diol-9, 10-epoxide

Benzanthracene

benzidine

Safrole

Cycasin

CH$_3$—N=N—CH$_2$O—C$_6$H$_{11}$O$_5$

$\downarrow$ hydrolysis

CH$_3$—N=N·CH$_2$OH active component

Vinyl chloride

CH$_3$—NH—NH—CH$_3$ 1,2-Dimethylhydrazine

CH$_3$—N—C—NHNO$_2$ *N*-Methyl-*N'*-nitronitrosoguanidine

Fig. 9.20. The structure and metabolism of some carcinogens.

Table 9.10. Examples of chemical carcinogens

Chemical	Source	Site of cancer
A. Industrial chemicals		
Aromatic amines		Bladder
Asbestos[a]		Bronchus, pleura
Tars, oils		Skin, lungs
Diethylstilboestrol		Vagina
Arsenic[a]		Skin, bronchus
Benzene[a]		Bone marrow
Vinyl chloride		Liver
Aromatic hydrocarbons		Lung
B. Naturally occurring chemicals		
Cycasin	Cycads	Liver, kidney, intestine (rats)
Ptaquiloside	Bracken	Bladder, intestine (rats, cows)
Ochratoxin A	Fungi (several *Aspergillus* and *Penicillium* spp.)	Liver, kidney[b]
Nitrosamines and others	Betel quid[a]	Mouth
Safrole	Oil of sassafras	Liver (rats)
Aflatoxins[c]	The fungus *Aspergillus flavus*	Liver (several species)
Agaritine	The fungus *Agaricus bisporus*	Skin (in mouse models)

[a] The carcinogenicity of these molecules is considered further in Chapter 8. The purification and structure of ptaquiloside can be found in *Tetrahedron Lett.* **24**, 4117 and 5371 (1983). Agaritine in *A. bisporus* is metabolized to a phenylhydrazine, from which diazonium salts and various carbon-centred radicals can be generated (*Chem. Biol. Interac.* **94**, 21 (1995)). For details of ochratoxins see *J. Biol. Chem.* **271**, 27388 (1996).
[b] Presence in food has been implicated in the fatal renal disease **Balkan endemic nephropathy**, although its contribution is uncertain (*Q. J. Med.* **91**, 457 (1998)).
[c] Several are known; the most active is aflatoxin B1.

More often, enzymes of the body tissues are involved, and there is a considerable variation in the rates of activation between individuals, which is in part genetically determined. It depends on levels of the various cytochromes P450 and phase II enzymes.

The first example of metabolic activation to be discovered was the conversion of the aromatic amine derivative **2–acetylaminofluorene** (Fig. 9.20) to an *N*-hydroxylated product by cytochromes P450. This type of metabolism is common to several aromatic amines, including **2–naphthylamine**, which induces cancer of the bladder. The N—OH product can undergo several reactions, including sulphation to give an N—O—sulphate ester. This is probably the ultimate carcinogen; loss of sulphate produces a reactive **nitrenium ion** which combines with guanine residues in DNA. Peroxidase enzymes can also oxidize the N—OH product to an N—O· radical *in vitro* but whether this plays any role in carcinogenesis *in vivo* is unclear.

9.13.2   *Benzpyrene*

Cytochromes P450 are also involved in formation of the ultimate carcinogen, a **7,8-diol-9,10-epoxide** (Fig. 9.20), from benzpyrene. Cytochromes P-450 (especially CYP1A1 and 1A2) convert benzpyrene into a 7,8-epoxide. This is acted upon by the enzyme **epoxide hydratase** to form a diol, which is a substrate for further epoxidation (Fig. 9.20). The resulting 7,8-diol-9,10-epoxide combines with guanine in DNA. Epoxidation can occur at other positions on the ring, although the products are less carcinogenic. Phenobarbital, an inducer of cytochrome P450, accelerates carcinogenesis due to benzpyrene.

Benzpyrene in cigarette smoke is one of many contributors to the development of lung cancer.[116] Much of the total lung P450 is located in the non-ciliated bronchiolar epithelial cells, **Clara cells**, although some is also found in type II pneumocytes and other cell types. Levels of CYP1A1 are low in normal Clara cells, but are induced by exposure to cigarette smoke. **Aflatoxin B1** (Fig. 9.20) is also metabolized to an epoxide, which binds to guanine.

The peroxidase activities[137] of prostaglandin synthetase, myeloperoxidase and lactoperoxidase can oxidize several pro-carcinogens *in vitro*. For example, cyclooxygenase can convert benzpyrene into 6-hydroxybenzpyrene, the 7,8-diol into a 7,8-diol-9,10 epoxide, and several other pro-carcinogens into more active forms. 6-Hydroxybenzpyrene rapidly oxidizes into a mixture of quinones, with production of $O_2^{\bullet-}$ radical. Other substrates for the peroxidase action of cyclooxygenase include benzidine, 2-aminofluorene, 2-naphthylamine and 2,5-diaminoanisole. Lactoperoxidase can oxidize a wide range of products, including aromatic amines.

It is unclear how important these peroxidase-catalysed metabolic transformations are *in vivo*. Peroxidizing microsomes, or irradiated food lipids, have been reported[92] to convert benzpyrene into mutagenic products (including 6-hydroxybenzpyrene), and to form diol epoxides from benzpyrene-7,8-diol, probably by the action of peroxyl radicals. Oxidizing sulphite ions (which generate a large number of radical species; Chapter 8) can produce diol epoxides from the 7,8-diol,[241] perhaps by the oxidizing action of the peroxyl radical $^{\bullet}O_2SO_3^-$ formed by reaction of sulphite radical ($^{\bullet}SO_3^-$) with oxygen. This observation is interesting because SO_2 has been claimed to be a co-carcinogen for the development of lung cancer in animals treated with benzpyrene.

9.13.3   *Detoxification of carcinogens*

As well as forming diols, epoxides are substrates for glutathione S-transferase enzymes (Chapter 3), being converted into more soluble glutathione derivatives which can then be excreted. This, and other metabolic pathways, such as conjugation with glucuronic acid, usually represent detoxification mechanisms. Hence the overall effect of carcinogens *in vivo* depends upon the balance between activation and detoxification mechanisms, as well as upon the efficiency with which any DNA lesions produced by the ultimate carcinogen are repaired.

Several 'antioxidants', such as butylated hydroxyanisole and butylated hydroxytoluene, have been observed to decrease the carcinogenicity of benzpyrene and other pro-carcinogens to animals. This usually appears to involve increased synthesis of the enzymes involved in detoxification pathways, e.g. by acting on antioxidant response elements (Section 4.13.6). GSH is not only a co-substrate for transferase enzymes, but can also often combine directly with the ions or radicals that attack DNA. Other dietary agents, including vitamin A, β-carotene and selenium, have been suggested to have some anti-cancer effect (Chapter 10). Carotenoids, for example, may promote gap-junctional communication.[339]

Selenium can also protect against carcinogenesis induced by certain chemicals that require metabolic activation. Selenium-deficient animals are more sensitive to certain carcinogens, especially if the animals are fed a diet rich in PUFAs. It should be noted that the experimental results obtained vary depending on the species used, the doses of selenium and carcinogen administered, and the relative times of administration. It has been suggested that excessive consumption of PUFAs by humans might predispose to cancer,[92] but the evidence is not clear. Since many carcinogens are lipid-soluble, increased fat intake might lead to increased carcinogen intake. As we have seen, some organic hydroperoxides can act as tumour promoters.

9.13.4   *Carcinogens and oxidative DNA damage*

Despite the fact that specific chemical mechanisms of DNA damage have been determined for many carcinogens, a frequent observation in carcinogen-treated tissue is a rise in 8-OHdG in the neoplastic, and often the pre-neoplastic, lesions (Table 9.11). The implication is that oxidative stress is involved in carcinogenesis, whatever carcinogen started the process. Of course, a decrease in the rate of repair of 8-OHdG could also explain the rise. Some cytochromes P450 generate ROS, and their induction by carcinogens could make a contribution to oxidative stress (Chapter 1).

9.13.5   *Peroxisome proliferators*[38,44,255]

The ability of ROS such as H_2O_2 to induce DNA damage in cells, largely by site-specific $OH^\bullet$ formation, is of particular interest in relation to the mechanism of action of several rodent hepatocarcinogens that have been called **peroxisome proliferators**. A wide range of compounds, including trichloroethylene, drugs that lower blood lipid concentrations (e.g. **clofibrate** (Atromid-S)), plasticizers (e.g. di-(2-ethylhexyl)phthalate), and phenoxyacetic acid herbicides (e.g. 2,4-dichlorophenoxyacetic acid), produce enlargement of the liver, accompanied by marked increases in the number of hepatic peroxisomes, in several animal species. Peroxisomes[c] not only increase in number, but their overall balance of enzyme activities changes. Thus the activity of the peroxisomal system for β-oxidation of fatty acids often increases more than that of catalase.

Some peroxisome proliferators increase the incidence of malignant tumours in rat liver but neither they nor their metabolites react directly with DNA (the term **non-genotoxic carcinogen** is often used for agents of this type). It has been suggested that tumours can result because the excess H_2O_2 in peroxisomes (e.g. due to increased oxidation of fatty acids) cannot be fully metabolized by the catalase present and leaks out of these organelles. If some of the H_2O_2 survives to reach the nucleus, DNA damage would be expected. Indeed, perfusion of lauric acid, a substrate for peroxisomal β-oxidation, into livers of fasted rats produced only small increases in the efflux of GSSG (Chapter 3). However, if the rats had been pre-treated with the peroxisome proliferator nafenopin, livers responded to infusion of lauric acid by increased GSSG efflux. This observation is consistent with an escape of H_2O_2 from the peroxisomes in treated animals; its metabolism by glutathione peroxidase in the cytosol would account for the extra GSSG. Indeed, transfection of cells with a gene encoding the enzyme catalysing the first step in the peroxisomal β-oxidation pathway, or with a gene encoding urate oxidase, caused the cells to undergo transformation when exposed to fatty acids or to urate respectively. The above-mentioned hypolipidaemic agents also cause proliferation of hepatic endoplasmic reticulum and its associated cytochromes, which might also be sources of $O_2^{\bullet -}$ and H_2O_2 *in vivo*.

Levels of 8-OHdG have been reported by some (but not all) scientists to be increased in the livers of rats treated with peroxisome proliferators, but such increases are observed with several carcinogens (Table 9.11) and so their significance must not be over-interpreted.

9.13.6 *Reactive nitrogen species*[217]

Nitrous acid and $ONOO^-$ can deaminate DNA bases and are thus mutagenic and potentially carcinogenic (Chapter 4). Reaction of secondary or tertiary amines with nitrosating agents produces **N-nitrosamines** (in the case of a secondary amine, for example, N–H is replaced by N–N=O). Nitrosamines can be formed in the stomach by reaction of dietary amines with salivary (or food-derived) nitrites at the low gastric pH, and they can be generated at sites of inflammation when RNS from activated phagocytes react with amines. Nitrosamines are found in human faeces (levels increase on a high-meat diet)[24] and tobacco smoke is a rich source of nitrosamines, some of which are 'tobacco specific'. At least one of these, 4-(methylnitrosamine)-1-(3-pyridyl)-1-butanone (NNK; Table 9.11), is a powerful carcinogen, especially for the lung. Nitrosamines require metabolic activation (involving cytochromes P450) to agents that can methylate guanine. Ascorbic acid, and possibly various plant phenolics (such as the flavonoids), may be important scavengers of nitrosamines *in vivo*. One 'bioassay' for such reactions *in vivo* is to feed nitrate plus the amino acid proline; proline can be converted into the non-carcinogenic **nitrosoproline**,[318] which is excreted in the urine. For example, both ascorbate and, to a lesser extent, α-tocopherol decreased nitrosoproline excretion in humans given NO_3^- and proline.

Table 9.11. Some of the carcinogens shown to increase levels of 8-hydroxydeoxyguanosine in DNA *in vivo*

Agent	Species	Tissue/body fluid	Effect on [8-OHdG]	Comment
γ-Radiation	Several	Several	Levels rise ~24h after injection (*Japan J. Cancer Res.* **82**, 161, 1991)	Complete carcinogen; other base damage products also raised, pattern suggestive of OH• attack on DNA
Potassium bromate	Rat	Renal tumours	Levels rise ~75% 6h after injection (*Cancer Lett.* **91**, 139 (1995))	
2-Nitropropane	Rat	Liver		Hepatocarcinogenic; 8-aminoguanine also produced (*Chem. Res. Toxicol.* **6**, 269 (1993))
Ferric NTA	Rat	Kidney	Levels increase prior to renal cancer (see text)	
Benzene	Mouse	Bone marrow	5 × rise 1h after administration (*Cancer Res.* **53**, 1023 (1993))	Leukaemogenic agent
Benzpyrene	Rat	Liver/kidney	3.5 × rise in liver, 2 × rise in kidney following oral treatment (*Cancer Lett.* **113**, 205 (1997))	
NNK[a]	Mouse	Lung	~2 × rise (*Carcinogenesis* **13**, 1269 (1992))	Induces lung cancer
Arochlor[b] + iron	Mouse	Hepatoma	3–5 × rise (*Carcinogenesis* **13**, 247 (1992))	
4-Hydroxyamino-quinoline-N-oxide	Rat	Pancreas	Significant rises (*Cancer Lett.* **83**, 97 (1994))	Powerful carcinogen in rats
Cigarette smoke	—	Blood/urine	Elevated in blood cells and urine (Chapter 8)	
Nickel (Ni²⁺) ions	Rat	Kidney	8-OHdG and other base products rapidly elevated; pattern characteristic of OH• (*Chem. Res. Toxicol.* **5**, 811 (1992))	Induces kidney tumours

Compound	Animal	Tissue	Finding	Comments
Hydrogen peroxide	Trout	Liver		Dietary H_2O_2 enhances liver carcinogenesis by N-methyl-N'-nitro-N-nitrosoguanidine; H_2O_2-enhancing effect reported to be correlated with elevated levels of 8-OHdG (*Carcinogenesis* **13**, 1639 (1992)).
2-Fluorenylacetamide	Rat	Liver	Elevated	Elevations in 8-OHdG and several other base-oxidation products observed by GC–MS (*Chem.–Biol. Interac.* **94**, 135 (1995))
N-Nitroso-N-2-fluorenylacetamide	Rat	Liver	Elevated	
$NaNO_2$ + trimethylamine	Rat	Lung	Elevated	
Transgenic mice expressing HBV large envelope protein in liver[c]	Mouse	Hepatoma	Rises early in life and continues to rise as liver damage progresses (*Proc. Natl Acad. Sci. USA* **91**, 12808 (1994))	
Aflatoxin	Rat	Liver	Single ip injection raised liver 8-OHdG (*Carcinogenesis* **16**, 419 (1995))	Hepatocarcinogen
Dimethyl-benzanthracene	Mouse	Skin	Elevated up to 15 ×	Higher 8-OHdG levels correlate with shorter time to first appearance of papillomas (*Proc. Natl Acad. Sci. USA* **92**, 5900 (1995))

[a] 4-(Methylnitrosamine)-1-(3-pyridyl)-1-butanone, a 'tobacco-specific' nitrosamine.
[b] A polychlorinated biphenyl.
[c] Animal model for hepatitis B infection. HBV-hepatitis B virus.
There are also several negative reports in the literature, i.e. carcinogen treatments that failed to raise 8-OHdG levels, but the timing of measurements is often critical.

9.14   Cancer chemotherapy

The objective of cancer treatment is to kill cancer cells with as little damage as possible to normal cells. Both radiotherapy and chemotherapy often kill cells by apoptosis; this process is disfavoured in cells with defective p53, such as many malignant cells. In malignant tumours, an abnormally large proportion of cells is dividing and so most chemotherapeutic agents are designed to interfere with cell proliferation, often by blocking synthesis of DNA, RNA or protein. The first reported use of chemotherapy was in 1942, when the DNA-modifying agent **nitrogen mustard** was used in a patient with lymphoma. One major class of chemotherapeutic agents is thus provided by the alkylating agents; they or their metabolites bind to DNA, and chemically modify it, interfering with replication and transcription. The nitrosourea **carmustine** (BCNU) is also an inhibitor of glutathione reductase.

A second major class of drugs is designed to interfere with metabolic reactions. Thus **methotrexate** inhibits the enzyme dihydrofolate reductase and prevents transfer of methyl groups in several biosynthetic reactions, including synthesis of deoxythymidine in DNA. **5-Fluorouracil** is an analogue of thymine and prevents DNA synthesis by inhibiting thymidylate synthetase. **Cytosine arabinoside** contains an arabinose sugar instead of ribose, and it interferes with DNA polymerase activity. **Cisplatin** (*cis*-diamminedichloroplatinum(II), see Fig. 9.21) probably acts by cross-linking guanine residues in DNA. **Hydroxyurea** inhibits ribonucleoside diphosphate reductase (Section 6.2). **Vincristine**, **vinblastine** and **taxol** interfere with mitotic spindle formation. Anti-steroidal agents such as **tamoxifen** have been developed to treat cancers whose growth depends on the continued availability of steroid hormones.

Of course, any normal cells undergoing rapid division will also be damaged by most of these drugs; dividing cells are found in the intestinal epithelium, hair follicles, gonads and bone marrow. Damaging side-effects limit the doses of most chemotherapeutic agents that can be used. Most, if not all, of the agents that chemically modify DNA are themselves mutagenic, and thus can be regarded as potentially carcinogenic, meaning that they must be handled with care.

9.14.1   *Natural products in chemotherapy*

As well as such natural products as taxol, many antitumour antibiotics are produced by living organisms. These antibiotics usually act, at least in part, by binding to DNA and interfering with DNA replication and gene transcription, often causing strand breakage.[236] The antitumour antibiotics can be grouped into several distinct chemical types. These include the **anthracyclines** and other quinone-containing drugs (such as mitomycins, streptonigrin, daunomycin and doxorubicin (otherwise known as adriamycin)), the metal-chelators (such as tallysomycin and the bleomycins), the protein antitumour antibiotics (such as macromomycin and neocarzinostatin), and aureolic-acid-based antibiotics (such as mithramycin, chromomycins and olivomycins). Free-radical reactions have been suggested to be important both in the mechanism of anti-tumour action and in the side-effects of antibiotics. Let us examine a few cases in detail.

9.14.2   *Bleomycin*

Bleomycin is the collective name given to a family of glycopeptide antibiotics produced by the microorganism *Streptomyces verticillus*. The clinical preparation commonly used, **Blenoxane**, is a mixture of several bleomycins that differ slightly in structure, although bleomycins A_2 and, to a lesser extent, B_2 are the major components (Fig. 9.22). Bleomycins are active against several human cancers, including Hodgkin's disease, cancer of the head and neck and cancer of the testis.[100]

Bleomycins act by binding to DNA, especially adjacent to guanine residues. They cause single-strand, and some double-strand breaks; strand cleavage correlates with inhibition of cancer cell proliferation and bleomycin can induce apoptosis. Degradation of the deoxyribose sugar forms, among other products, **base propenals** (base–CH=CH–CHO), which further break down to release MDA. Since MDA reacts with TBA to give a pink chromogen, the TBA test can be used to follow DNA breakdown by bleomycin[100] and this has been made the basis of the 'bleomycin assay' for 'catalytic' iron (Chapter 4). Some of the base propenals are directly cytotoxic.

Bleomycins are powerful chelators of transition-metal ions, such as Cu^{2+}, Co^{2+}, Zn^{2+}, Fe^{2+}, Fe(III) or Ni^{2+}, by donation of electrons from nitrogen atoms and from a $>C=O$ group to the metal ion (Fig. 9.22). Indeed, degradation of DNA by bleomycin requires that a transition metal ion is present within the DNA–bleomycin complex, and that O_2 is also present. *In vitro*, complexes of bleomycin with Fe^{2+}, Cu^+, Co^{2+}, Mn^{2+} and vanadium(IV) have all been claimed to be capable of causing DNA degradation. However, only iron ions seem likely to mediate bleomycin-induced DNA damage *in vivo*. The ability of bleomycin to attack the DNA of tumour cells (and some normal cells) *in vivo* presumably means that it is able to chelate iron, e.g. from the intracellular low molecular mass iron pool.[100]

An Fe^{2+}–bleomycin complex capable of degrading DNA in the presence of O_2 can be generated by adding Fe^{2+} salts to bleomycin or by reducing a Fe(III)–bleomycin–complex with such biological reducing agents as ascorbate, GSH, $O_2^{\bullet-}$, the microsomal NADPH–cytochrome P450 reductase enzyme and a similar reducing system in the nuclear membrane. Fe(III)–bleomycin will also degrade DNA if H_2O_2 is added, and O_2 is not required. Whether DNA damage is produced by Fe^{2+}–bleomycin plus O_2 or by Fe(III)–bleomycin plus H_2O_2, the reactive intermediate appears to be a ferric peroxide that has a distinctive EPR spectrum (Fig. 9.23).

Incubation of bleomycin with Fe^{2+} salt in the absence of DNA causes chemical modification of the drug that destroys its activity.[100] It is protected by addition of DNA, which then becomes a target of attack. Superoxide dismutase, low-molecular-mass scavengers of $O_2^{\bullet-}$ or $OH^{\bullet}$, catalase, caerulo-plasmin, or vitamin E, offer little if any, protection to DNA against damage by the bleomycin–ferric peroxide complex (Fig. 9.23). Indeed, propyl gallate and some other phenolic antioxidants can make the damage worse *in vitro* by reducing Fe(III)–bleomycin back to the Fe^{2+} state.[101] DNA degradation can,

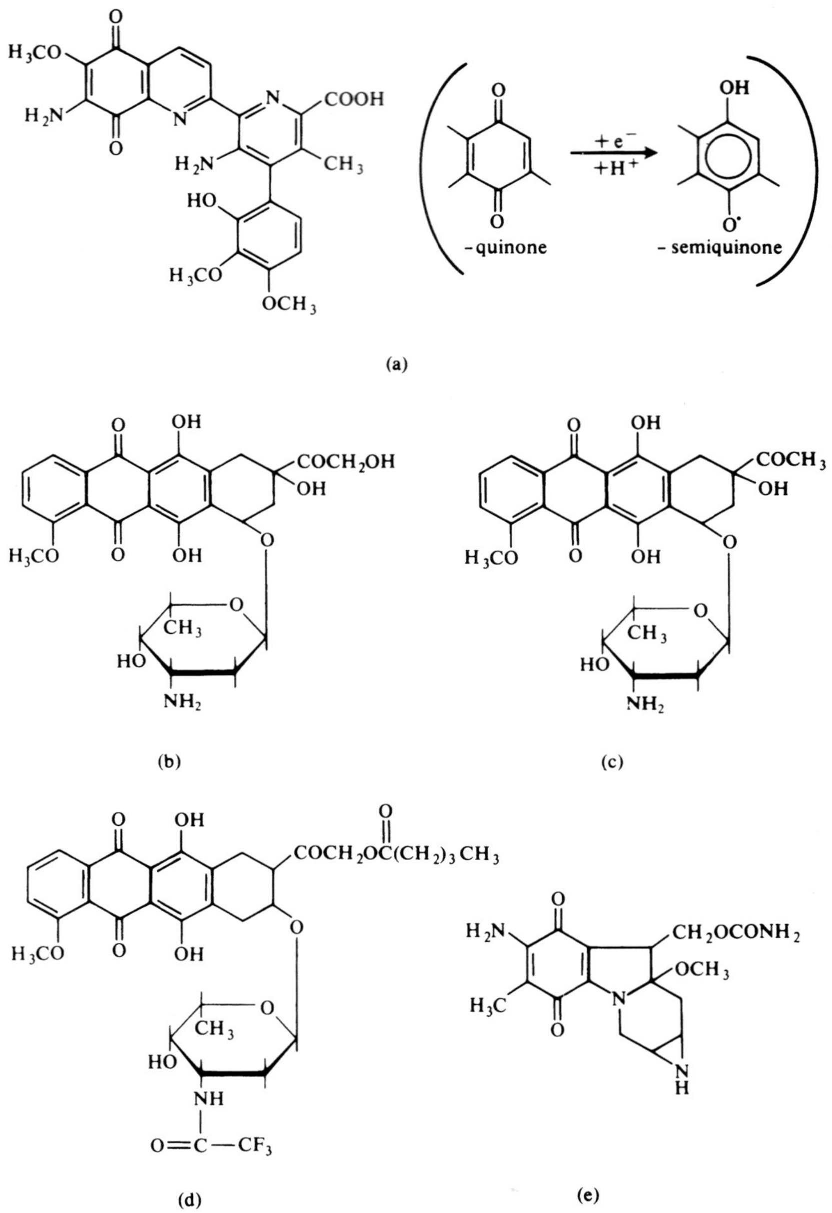

(a)

(b)

(c)

(d)

(e)

Fig. 9.21

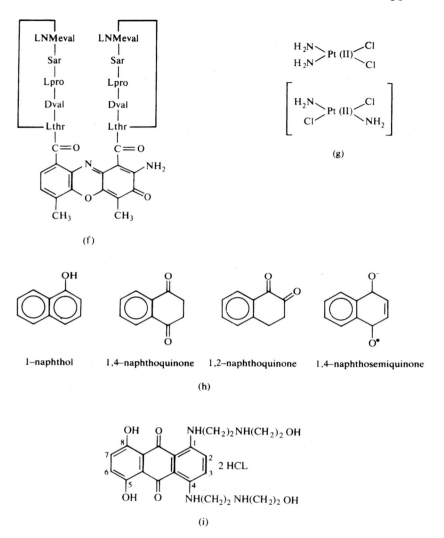

Fig. 9.21. Structures of some antitumour drugs. (a) Streptonigrin; (b) adriamycin; (c) daunomycin; (d) AD32; (e) mitomycin C; (f) actinomycin D (Lthr, L-threonine; Dval, D-valine; Lpro, L-proline; Sar, sarcosine; LNMeval, L-N-methylvaline); (g) *cis*-diamminedichloroplatinum(II); the *trans*-isomer (below, in brackets) is not an effective antitumour drug; (h) 1-naphthol and its metabolites; (i) mitoxantrone.

however, be inhibited by preventing iron binding to the bleomycin, using such chelating agents as EDTA, DETAPAC or desferrioxamine. However, if the iron has already bound to the bleomycin and DNA damage has begun, higher concentrations of chelators are required to stop it, since it takes time for the iron to transfer from bleomycin to the chelator.

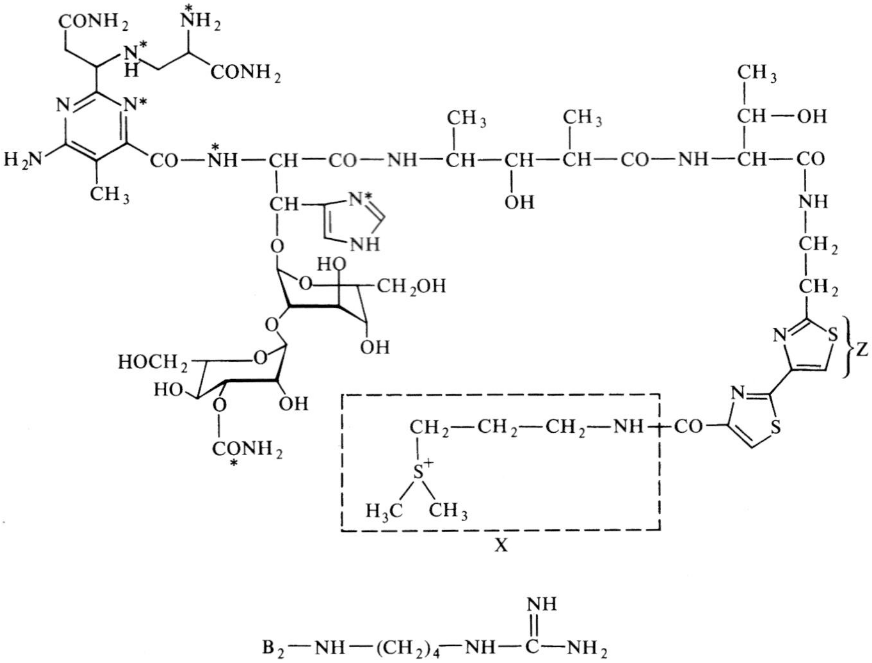

Fig. 9.22. Structure of bleomycin B$_2$. In bleomycin B$_2$ the substituent shown replaces the terminal group (X) of bleomycin A$_2$. The asterisks denote the atoms that interact with bound transition-metal ions. The **phleomycin** antibiotics differ from the bleomycins only in the absence of one of the double bonds in the ring marked Z. Bleomycin is usually supplied by the manufacturers as a sulphate salt.

The chemistry of DNA damage is summarized in Fig. 9.23. Hydroxyl radicals may be generated in small amounts, but analysis of the pattern of DNA base damage shows that they contribute little to the DNA damage.[84]

Side-effects of bleomycin

A major side-effect of therapy with bleomycins is lung damage, leading to pulmonary fibrosis. Over-production of **transforming growth factors** β by pulmonary macrophages and alveolar type II cells has been implicated.[145] The TGFβ1, 2 and 3 cytokines play an important physiological role in helping to repair tissue damage, e.g. in wounds, and are powerful chemoattractants for fibroblasts. The lung may be a target because it appears to have only low activities of **bleomycin hydrolase**,[79] a cysteine proteinase which breaks down bleomycin into an inactive form. The hydrolase is present in most body tissues and its level in malignant tumours is one factor that affects their sensitivity to bleomycin. Lung damage by bleomycin may also involve radical reactions, since microsomal fractions from rat lungs can reduce Fe(III)–bleomycin in the presence of NADPH, forming Fe^{2+}–bleomycin and hence ROS, which are known to stimulate fibroblast proliferation under certain circumstances (Chapter 4).

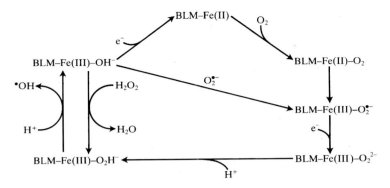

Fig. 9.23. Mechanism of bleomycin (BLM) action. The species that attacks DNA is thought to be a ferric peroxide, BLM–Fe(III)–O_2H^-. It can be formed by direct reaction of ferric bleomycin with H_2O_2, or from a BLM–Fe(III)–$O_2^{\bullet-}$ complex. Diagram adapted from Sugiura *et al.* (1982) *Biochem. Biophys. Res. Commun.* **105**, 1511, with permission. Under appropriate conditions, BLM–Fe(III)–$O_2^{\bullet-}$ might decompose to release $O_2^{\bullet-}$, and BLM–Fe(III)–O_2H^- to release OH$^{\bullet}$, explaining why these radicals have been detected (e.g. by spin-trapping) in some bleomycin-containing systems (e$^-$ represents a reducing system, e.g. thiols, ascorbic acid, $O_2^{\bullet-}$ or reductase enzymes). The mechanism of bleomycin action resembles the catalytic cycle of cytochrome P450 (Chapter 1). (*Chem. Rev.* **98**, 1153 (1998)).

Consistent with this, bleomycin-induced lung damage in animals is increased by exposure to elevated O_2 concentrations.

Another important factor relevant to bleomycin toxicity may be the availability of iron within the tissue. Thus intravenous injection of bleomycin into rats produces no lung injury, yet simultaneous exposure to hyperoxia or simultaneous tracheal installation of ferric ions leads to marked bleomycin-dependent lung damage.[114] However, attempts to use desferrioxamine to prevent bleomycin-induced lung fibrosis in experimental animals have given negative or marginally positive results. Of course, desferrioxamine does not penetrate easily into lung cells. The chelator ICRF-187 has also been tested and claimed to be protective against bleomycin-induced pulmonary toxicity in animals in some (but not other) studies. *N*-Acetylcysteine is not protective but intraperitoneal administration of nicotinamide decreased bleomycin-induced lung damage in hamsters, possibly by acting as a precursor of NAD$^+$ and preventing 'lethal NAD$^+$ depletion' due to poly(ADP–ribose) polymerase activation as a consequence of DNA damage (Chapter 4).[323]

The toxicity of bleomycin to some strains of bacteria might also involve ROS, since strains of *E. coli* K12 with increased SOD and catalase activities, due to previous exposure to paraquat, are more resistant to bleomycin.[195]

9.14.3   *Quinone antitumour agents*[236]

Several antitumour antibiotics are quinones (Fig. 9.21). It is thus possible that redox cycling and/or reaction of semiquinones with –SH groups (Chapter 8) could participate in their activity against malignant cells. Generation of $O_2^{\bullet-}$

and H_2O_2 by redox cycling could produce DNA damage if they led to generation of highly reactive OH$^\bullet$ close to the DNA molecule. Reduction of Fe(III) to Fe^{2+}, which facilitates OH$^\bullet$ generation, could be achieved not only by $O_2^{\bullet-}$, but also directly by semiquinones (SQ$^{\bullet-}$). Thus at least two pathways of Fe^{2+} generation are feasible.

A
$$SQ^{\bullet-} + O_2 \rightarrow Q + O_2^{\bullet-}$$

$$O_2^{\bullet-} + Fe(III)\text{-chelate} \rightarrow O_2 + Fe^{2+}\text{-chelate}$$

B
$$SQ^{\bullet-} + Fe(III)\text{-chelate} \rightarrow Fe^{2+}\text{-chelate} + Q$$

The balance between pathways A and B will depend on the rate constants for the appropriate reactions and on the concentrations of O_2 and biological Fe(III)-chelate. In general, route B is not favoured until O_2 concentrations become very low.

Let us now look at some examples of quinone antitumour agents.

Streptonigrin

Streptonigrin (Fig. 9.21), an antibiotic produced by *Streptomyces flocculus*, is one of the first quinone antibiotics whose action *in vivo* was shown to involve ROS. Streptonigrin is not generally used clinically because of serious side-effects. Its toxic action on *E. coli* requires O_2 and is decreased if the SOD activity of the cells is raised. The quinone part of streptonigrin can be reduced by bacterial enzymes to a semiquinone form which can then reduce oxygen to form $O_2^{\bullet}$.

Increased uptake of iron salts by *E. coli* enhances the bacteriocidal action of streptonigrin. This effect is inhibited by desferrioxamine, suggesting a role for iron in bacterial damage, presumably via OH$^\bullet$ formation.[111] Similarly, killing of the gonococcus by streptonigrin seems to depend on the availability of iron to promote Fenton-type reactions. Streptonigrin can bind several metal ions, including iron, copper, zinc, cobalt, cadmium and manganese.

Whether ROS are responsible for the toxic effects of streptonigrin to cancer cells (which include DNA damage) is uncertain, since the semiquinone itself is a reactive molecule. However, the toxicity of streptonigrin to mouse mammary tumour cells is decreased under hypoxic conditions, suggesting that ROS generation might be involved, at least in this cell line.[299] However, O_2 concentration might also affect the pathways of drug metabolism and so such studies must be interpreted with caution.

Actinomycin D

Actinomycin is a peptide antibiotic (Fig. 9.21), produced by *Streptomyces parvullus*, that has been used in the treatment of tumours in children. Its antitumour activity is generally attributed to its intercalation between DNA bases, so preventing replication and transcription. Intercalation occurs adjacent to guanine bases. Actinomycin D (dactinomycin) is only a weak redox-cycling agent. Its toxicity to mouse mammary tumour cells is decreased at low oxygen concentrations,[299] although this could be due to an effect on metabolism.

Mitomycin C

Mitomycin C, isolated from *Streptomyces caespitosus*, is used, in combination with other agents, in the treatment of many advanced human cancers. *In vitro*, it can be reduced by microsomal or nuclear electron-transport chains at the expense of NADPH. Reduction forms a semiquinone (Fig. 9.21) which can react with O_2 to generate $O_2^{\bullet-}$. NADPH–cytochrome P450 reductase also converts mitomycin C under hypoxic conditions into products that attack DNA directly, and cross-link the strands. Usually the toxicity of mitomycin C is greater under hypoxic conditions, suggesting that its reductive activation and DNA cross-linking is more important in damaging its target cells than is redox cycling. The action of mitomycin C against large 'solid' tumours, which often contain hypoxic areas (Section 8.16.3), is consistent with this conclusion.

Mitomycin C might also be a substrate for quinone reductase (**DT diaphorase**), an enzyme which catalyses two-electron reduction of many quinones (Chapter 8), apparently as a protective mechanism. In the case of mitomycin C, however, the reduction could generate the DNA-damaging product.

Naphthol[61]

The compound 1-naphthol (Fig. 9.21) has been reported to be much more toxic to isolated colon cancer cells than to normal colonic cells. It has been proposed that 1-naphthol may exert its toxic effects by metabolism to form 1,2- or 1,4-naphthoquinones. Reasons for the selective toxicity remain to be elucidated.

The anthracycline antibiotics

Anthracyclines are tetracyclic (i.e. have four joined ring structures) antibiotics produced by various strains of *Streptomyces*. The tetracyclic structure is linked to carbohydrate; removal of carbohydrate from anthracyclines usually leads to loss of antitumour activity. Anthracyclines are widely used in the treatment of leukaemia, breast cancer, Hodgkin's disease and sarcomas. The best known are **daunorubicin** (sometimes called daunomycin) and **doxorubicin** (often called adriamycin). Both are products of *Streptomyces peucetius*. In both these compounds, the attached sugar is **daunosamine** (Fig. 9.21).

Like all antitumour drugs, the anthracyclines produce a number of side-effects, the most serious being damage to the heart.[267] Acute cardiotoxicity occurs within minutes of drug administration and is revealed as arrhythmias and non-specific electrocardiographic changes. More serious than this is a chronic irreversible congestive heart failure, which becomes a significant problem once a certain total dose of anthracycline has been administered (e.g. 7% incidence at a dose of $550 \, \mathrm{mg/m^2}$ of body surface area). It is this cardiotoxicity that limits the doses of doxorubicin and daunorubicin that can safely be given to cancer patients.

The mechanism of action of anthracyclines is not entirely clear, and may involve multiple actions.[236] Doxorubicin and daunorubicin can intercalate between DNA bases, interfering with DNA replication and the transcription of RNA. They affect **topoisomerase II**, an ATP-dependent enzyme involved in unwinding DNA for replication; it passes an intact double helix through a transient double-stranded break in the DNA backbone.[47] Anthracyclines stimulate topoisomerase-dependent DNA strand breakage. They can also damage membranes and interfere with ion transport. Thus doxorubicin is toxic to at least one tumour cell line, even if it is prevented from entering the cells. Indeed, the doxorubicin derivative **AD32** (N-trifluoroacetyladriamycin-14-valerate; Fig. 9.21) will not bind to DNA, and yet it is still an antitumour agent.

Increased membrane permeability to Ca^{2+} ions may in part explain the cardiotoxic effects of adriamycin. Reports suggest that **doxorubicinol**, a major metabolite of doxorubicin, formed by NADPH-dependent reduction of the side-chain keto ($>C=O$) group to $-CHOH$ by a **carbonyl reductase** enzyme, is a much more powerful inhibitor of membrane-associated ion pumps than doxorubin and, in particular, interferes with cardiac Ca^{2+} metabolism.

Redox cycling of anthracyclines

Reduction of the quinone moiety can be achieved by NADPH–cytochrome P450 reductase, the electron transport chain located in the nuclear envelope, and by the mitochondrial NADH-dehydrogenase complex (hence interfering with the normal flow of electrons during mitochondrial electron transport). Indeed, administration of ubiquinone (Chapter 3) has been reported to protect rabbits against doxorubicin-induced heart damage to some extent.[313]

Damage could then be caused by semiquinones and/or by ROS. However, intercalation of anthracyclines into DNA prevents their reduction by enzymes (because of inaccessibility), and so it seems unlikely that ROS contribute to the DNA damage. Nevertheless, oxidative stress is widely thought to be involved in the cardiotoxicity of anthracyclines, since several newer anthracycline analogues with less cardiotoxicity (such as **mitoxantrone**, which also acts on topoisomerase II) appear to undergo less redox cycling *in vitro*.

It is not clear why heart should be a particular target, although the levels of its antioxidant defence enzymes are only moderate. Indeed, transgenic mice over-expressing catalase, or mitochondrial MnSOD, are resistant to the acute cardiotoxicity of adriamycin.[139,342] Treatments of various animals with ascorbate, the spin trap PBN, vitamin E, probucol[273] or the thiol compound N-acetylcysteine have often shown some protective effects against the acute toxicity of doxorubicin. In general, however, they seem less effective against chronic cardiotoxicity, a more serious clinical problem. Of course, some of these scavengers (e.g. thiols) might react directly with semiquinones. Mixtures of an α-tocopherol ester and DMSO have been claimed to offer some protection against ulceration induced by the direct application of doxorubicin to animal tissues,[291] but none of these scavengers has yet been shown to protect against the side-effects of doxorubicin in human patients. However, the chelating agent ICRF-187 (as **dexrazoxane**) appears to have limited cardioprotective effects in humans, although it is possible that it interferes with the antitumour action of anthracyclines. It is also cardioprotective in pigs, rabbits, mice and dogs. The flavanoid **7-monohydroxyethylrutoside** is also protective in doxorubicin-treated mice.[315]

Reductive metabolism of anthracyclines

Reductive metabolism of anthracyclines can lead to loss of sugar from the molecules, giving inactive **aglycones**. For example, if doxorubicin is incubated with NADPH and liver microsomes *in vitro*, NADPH is consumed as NADPH–cytochrome P450 reductase catalyses formation of doxorubicin semiquinone. The semiquinone reacts rapidly with O_2 to give $O_2^{\bullet-}$. As O_2 concentration falls to low levels, however, there is rapid formation of aglycone. As with mitomycin C, NADPH-dependent reduction of anthracyclines at low O_2 concentrations can generate reactive products that combine directly with DNA, proteins and membrane lipids. Quinone reductase may catalyse two-electron reduction of some anthracyclines.

Iron and anthracycline action

The effect of ICRF-187 suggests that iron may contribute to the actions of anthracyclines, e.g. by conversion of $O_2^{\bullet-}$ and H_2O_2 into $OH^{\bullet}$ (early claims that doxorubicin semiquinone reacts directly with H_2O_2 to form $OH^{\bullet}$ have not been substantiated.) Doxorubicin and daunorubicin can chelate several metal ions including Fe^{2+}, Fe(III) and Cu^{2+}. Chelates of doxorubicin with Fe^{2+} can oxidize in air to generate $O_2^{\bullet-}$ and $OH^{\bullet}$. Doxorubicin forms a tight complex with Fe(III), in which three doxorubicin molecules associate with each Fe(III). The Fe(III) is slowly reduced to Fe^{2+}, which can then re-oxidize, forming $O_2^{\bullet-}$ and $OH^{\bullet}$. *In vitro*, the doxorubicin–Fe(III) complex stimulates lipid peroxidation, possibly because it undergoes autoreduction and efficiently decomposes traces of lipid hydroperoxides; the observed peroxidation is not inhibited by scavengers of $OH^{\bullet}$. Increased rates of lipid peroxidation have been reported in the hearts of rats treated with doxorubicin.[181]

Thiol compounds such as cysteine or GSH can reduce the doxorubicin–Fe(III) complex, leading to increased formation of $O_2^{\bullet-}$ and H_2O_2. Doxorubicin semiquinone has been claimed to be sufficiently reducing to mobilize iron, as Fe^{2+}, from ferritin. Cancer chemotherapy can raise blood iron levels and sometimes create 'non-transferrin-bound' iron.[33] The origin of this iron is uncertain, but it may be released from damaged cancerous (and normal) cells. This could provide a ready source of iron for chelation by anthracyclines. Indeed, a combination of doxorubicin and iron, called **quelamycin**, was found to be highly toxic to heart and other tissues and clinical trials were abandoned.

Do ROS account for anthracycline action?

There is some evidence that the cardiotoxicity of anthracyclines may be mediated by ROS formation, but there are no convincing data that damage to cancer cells *in vivo* is achieved by the same mechanism. Low levels of anthracyclines can induce tumour cells to undergo apoptosis, possibly related to effects of topoisomerase producing excessive DNA damage.[47] The toxicity of anthracyclines to some tumour cells in culture is decreased under hypoxic conditions, *suggesting* a role for ROS. This is not always the case, however. For example, the toxicity of doxorubicin to mouse mammary tumour is greater under hypoxic conditions.

9.14.4 *Protein antitumour drugs*

Neocarzinostatin

Neocarzinostatin is a small protein, 109 amino acids long, secreted by a mutant strain of *Streptomyces carzinostaticus*. It has actions against a number of tumour cells *in vitro* and *in vivo*. A small fluorescent molecule bound to the protein is responsible for the antitumour effect. The protein does not attack DNA, and probably serves only as a vehicle for transporting the chromophore into cells.

The chromophore detaches from the protein and binds to the DNA of the target cells, mainly adjacent to adenine and thymine. It promotes damage to deoxyribose, leading to strand scission and formation of TBARS. Scission of duplex DNA by the neocarzinostatin chromophore *in vitro* requires the presence of both O_2 and a reducing agent, such as a thiol compound. The thiol converts the chromophore (Fig. 9.24) into a free-radical (detectable by ESR) that abstracts a hydrogen atom from carbon 5 of deoxyribose. The deoxyribose radical reacts with O_2 to give a peroxyl radical, whose decomposition leads to the strand breakage. Depleting the GSH content of tumour cells decreases the DNA damage observed on subsequent treatment with neocarzinostatin, suggesting that this mechanism is relevant *in vivo*. Incubation of the neocarzinostatin chromophore with thiols in the absence of DNA inactivates it, perhaps by self-attack of the radical species. DNA degradation by neocarzinostatin *in vitro* is not prevented by SOD, catalase, scavengers of OH$^\bullet$, or metal-ion-chelating agents.

Macromomycin and auromomycin[289a]

Macromomycin, a protein 112 amino acids long, with an amino-acid sequence similar to that of the neocarzinostatin protein, has been isolated from *Streptomyces macromomyceticus*. It has some antitumour effects, and induces DNA strand breakage. During purification of macromomycin, a small fluorescent molecule is released. Readdition of this greatly increases the antibacterial and antitumour

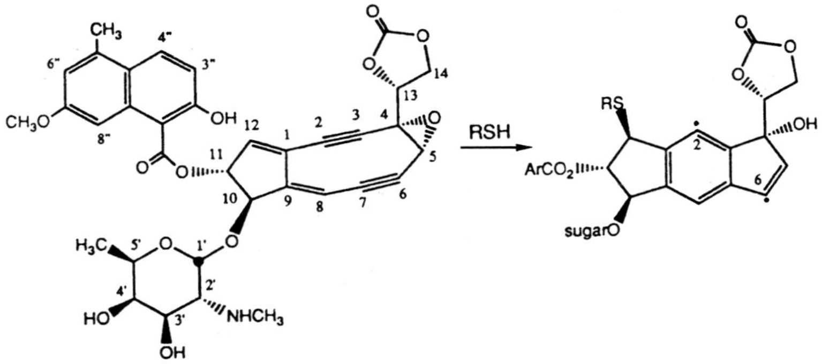

Fig. 9.24. Thiol activation of the neocarzinostatin chromophore to a free radical (diradical) species. Diagram adapted from *Science* **261**, 1319 (1993) by courtesy of Professor Irving H. Goldberg and the publishers.

activities of the protein, the complex of macromomycin and the fluorescent chromophore being known as **auromomycin**. The chromophore itself can induce DNA damage *in vitro*, again probably by some type of free-radical mechanism.

9.14.5   *Resistance to cancer chemotherapy*[25,160]

The cells in most malignant tumours are metabolically heterogeneous in many respects and can rapidly evolve resistance to many anti-cancer agents. Sometimes cells exposed to one drug become simultaneously resistant not only to that drug but also to others. This is called **acquired multidrug resistance** and can involve alterations to render the target of damage less sensitive (e.g. changes in topoisomerase II) and/or increased activities of drug-metabolizing enzymes such as glutathione S-transferases and quinone reductase.

Resistance often involves increased synthesis of proteins that pump drugs out of the cell. Examples are **P-glycoprotein** (a membrane glycoprotein, relative molecular mass 170 000) and **multi-drug resistance associated proteins** (MRPs). P-Glycoprotein action requires ATP and its substrates include doxorubicin, daunorubicin, mitoxantrone, vinblastine, vincristine, taxol, actinomycin D, etoposide (a topoisomerase II inhibitor), mithramycin and even the dyes rhodamine 123 and ethidium bromide. P-Glycoprotein is expressed in certain healthy tissues, e.g. bile canaliculi, proximal kidney tubules, pancreatic ducts, haematopoietic stem cells and brain capillaries. Hence cancers derived from these tissues may be particularly resistant to chemotherapy. Increased levels of P-glycoprotein after chemotherapy have been reported in breast and colon cancer and neuroblastoma. Several compounds that inhibit P-glycoprotein have been described, including calcium–channel blockers such as **verapamil** and **nifedipine**, as well as tamoxifen and cyclosporin A.

MRP-1 is a 190 000 molecular mass glycoprotein found in plasma membranes as well as in some organelle membranes. MRP in normal cells appears to be involved in the ATP-dependent transport of leukotrienes, especially the GSH conjugate leukotriene C_4 (Section 6.10.8). MRP-1 can also export GSSG, several xenobiotics conjugated with GSH (including cisplatin) and 17β-oestradiol glucuronides from cells. Like P-glycoprotein, MRP-1 can export multiple drugs, including daunorubicin and vincristine, a process in which GSH and ATP appear to be involved.

9.15   Oxidative stress and disorders of the nervous system: general principles[91,219]

9.15.1   *Introduction*

All living organisms can suffer oxidative damage, yet the animal brain is often said to be especially sensitive. One reason is its high O_2 consumption: in humans, the brain accounts for only a few percent of the body weight, but about 20% of basal O_2 consumption. Hence it processes a large amount of O_2 in a relatively small tissue mass.

The endothelium of the small blood vessels in the brain is in general much less permeable to molecules than other vascular endothelia, although essential molecules (such as glucose and ascorbate) and most lipid-soluble molecules (such as ethanol, other organic solvents and anaesthetics) can still penetrate. Many other molecules are excluded from the brain by this so-called **blood–brain barrier**. It also excludes circulating phagocytes from the healthy brain.

The adult brain contains about 10^{11}–10^{12} nerve cells (**neurones**), which are supported and protected by at least twice as many **neuroglial cells**. There are several types of glial cell, including **microglia**, **astrocytes** and **oligodendrocytes**. The latter synthesize and maintain the myelin sheath which insulates axons. Astrocytes provide structural support to assist the metabolism of, and help protect, neurones. For example, astrocytes can take up glutamate and convert it to glutamine, which is released for uptake by neurones (Section 9.15.4 below). A typical neurone in the cerebral cortex receives input from a few thousand fellow neurones and, in turn, passes on messages to a few thousand other neurones. Neurones do not normally divide and it is often said that the number of neurones in the brain is fixed before birth.

The neurones consist of a cell body, **dendrites** (short, branched extensions from the cell body), an elongated **axon** and a **nerve terminal**. The neuronal cytoskeleton (containing microtubules, neurofilaments and actin microfilaments) helps maintain the axon structure and facilitates communication between the cell body and the nerve terminal, which can be a considerable distance away. Neurones with long axons have to maintain large surface areas of membrane and must transport a wide range of materials from the cell body down the axon to its terminal (**anterograde transport**). All parts of the neurone contain mitochondria (presumably locally generating ATP) but the major site of protein synthesis is the cell body. Materials also move in the opposite direction (**retrograde transport**). Transport of some materials occurs quickly (50–200 mm/day) and that of others more slowly (0.1–3.0 mm/day).

Defects in transport can lead to axonal degeneration and death. If an axon is severed, the segment beyond the cut will degenerate. The neuronal cytoskeleton plays an important role in transport and also in maintaining neuronal structure. Hence damage to neurofilaments might lead to neuronal dysfunction. For example, oxidation or nitration[21] of neurofilament proteins could cause aggregation and interference with transport. Neurofilaments are assembled from three polypeptide subunits, NF-L, NF-M and NF-H. Abnormal accumulation and assembly of neurofilaments are seen in several types of neurodegenerative disease.

The nerve terminal forms part of a **synapse**, the communication mechanism between neurones. Synapses can be **excitatory** or **inhibitory**: the former releases neurotransmitters that decrease the membrane potential of the target cell and make it more likely to fire an action potential, whereas an inhibitory neurotransmitter raises the membrane potential of the target neurone.

9.15.2 *Energy metabolism*

A neurone uses much of the O_2 it takes up to make the ATP needed to maintain ion gradients (high intracellular K^+, low Na^+, very low 'free' Ca^{2+}). Na^+

influx is needed for propagation of the action potential; it must then be pumped out of the cell using Na^+,K^+-ATPases. The normal brain uses glucose for energy production, relying on glycolysis and the Krebs cycle, and needs about 4×10^{21} ATP molecules every minute. It is thus vulnerable to interruptions of the supply of glucose (e.g. injection of excess insulin produces **hypoglycaemic coma**) or of O_2: little glycogen is stored in the brain. Similarly, inhibitors of ATP synthesis can cause neuronal cell death. Examples include cyanide, **3-nitropropionic acid** (an inhibitor of mitochondrial succinate dehydrogenase) and **rotenone** (an inhibitor of complex I). Defects in mitochondrial function are associated with several neurodegenerative diseases (see below).

Neurones also use energy in secretion: arrival of an action potential at the nerve terminal opens ion channels to allow Ca^{2+} influx, leading to release of neurotransmitter into the synapse by exocytosis. Each sub-group of neurones is associated with a particular neurotransmitter or group of neurotransmitters. The released neurotransmitter binds to receptors on the target neurone, e.g. there are at least five types of receptor for the neurotransmitter **dopamine**, named D1 to D5). Other neurotransmitters include **acetylcholine**, **serotonin**, **noradrenalin**, **glycine** and **glutamate**. Neurotransmitter action on its target cells often involves Ca^{2+}; for example, activation of muscarinic cholinergic receptors by binding of acetylcholine activates a G protein, stimulates phospholipase C and releases inositol triphosphate to promote Ca^{2+} release from the endoplasmic reticulum. Glutamate is stored in synaptic vesicles in neurones. After release, it is taken up by both neurones and glia; the glia convert it into **glutamine** (the side-chain $-COOH$ becomes $-CONH_2$) by the ATP-dependent enzyme **glutamine synthetase**. Glutamine can be released and taken up by neurones for hydrolysis into glutamate.[49]

The general inability of neurones to divide renders the brain sensitive to some loss of function as neurones die during the normal ageing process, and also to neurotoxic agents, since if neurones are killed they cannot be replaced. Fortunately, many parts of the brain have considerable neuronal redundancy. Neurones in culture are also prone to injury (Fig. 9.25), often leading to necrosis or apoptosis, if treated with toxins that interfere with energy metabolism, exposed to ROS/RNS (e.g. $ONOO^-$), or if 'growth factors' (better called **neurotrophic factors**) are withdrawn from the culture medium. Sometimes apoptosis can be delayed or prevented by adding $-SH$ compounds and other antioxidants.

During development of the nervous system a wide range of growth factors regulates the proliferation of different neurones; they include nerve growth factor, brain-derived neurotrophic factor, ciliary neurotrophic factor, neurotrophins 3, 4 and 5 and even fibroblast growth factor. All of these factors have been shown *in vitro* to be capable of protecting cultured neurones against various types of injury, i.e. they are not only 'growth factors' but also 'support factors'.[213]

9.15.3   *Calcium and nitric oxide*[95a]

Most of the excitatory synapses in the brain use glutamate as a chemical messenger. To mediate fast excitatory transmission, glutamate binds to receptors;

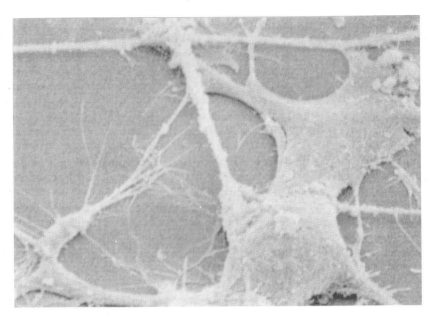

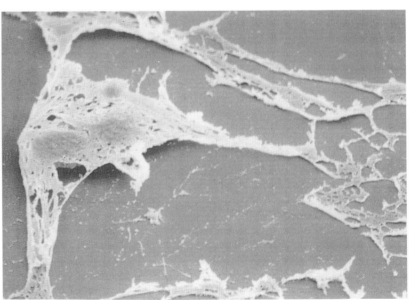

Fig. 9.25. Scanning electron micrographs of neurone-like cells to show the effect of H₂O₂. Top: PC12 cells, a phaeochromocytoma cell line, develop a neuronal phenotype after culture *in vitro* in the presence of NGF. Treatment with 100 μM H₂O₂ causes cytoskeletal damage (bottom) involving rises in intracellular free Ca²⁺. Note the fragmentation and flattening of the neurites. From *Brain Res.* **615**, 13 (1993) by courtesy of Elsevier and Drs Paul Hyslop and Dan Hinshaw.

NMDA, **AMPA** and **KA** receptors (named after molecules used in the laboratory to activate such receptors, i.e. **N-methyl-D-aspartate, α-amino-3-hydroxy-5-methyl-4-isoxazole-4-propionate** and **kainic acid**). Binding to these receptors opens cation channels and depolarizes the post–synaptic membrane (e.g. AMPA receptor binding allows in Na^+), increasing the likelihood that it will fire an action potential. The NMDA receptor requires both membrane depolarization and glutamate binding to allow Ca^{2+} and Na^+ into the cell (the fall in potential displaces Mg^{2+}, which blocks the 'resting' channel). Glutamate also binds to receptors coupled to GTP-binding proteins, the so-called **metabotropic** glutamate receptors. Binding of glutamate to its receptors (especially NMDA receptors) leads (directly or indirectly) to rises in intracellular free Ca^{2+}.[49] This Ca^{2+} is an important signalling molecule in neurones, which therefore have to expend considerable metabolic energy in maintaining low intracellular Ca^{2+} using pumps in the endoplasmic reticulum, plasma membrane and mitochondria (Chapter 4).

Nitric oxide[344] is used as an intercellular messenger by neurones in the peripheral nervous system (e.g. in regulation of penile erection and gut function) and by a few neurones in the brain, although its range of functions has not been completely elucidated. Both nitrite and nitrate can be detected in cerebrospinal fluid (CSF).[41] Nitric oxide plays a role in the development of **synaptic plasticity**; the strengthening or weakening of synaptic connections between neurones depending on how often they are used. Synaptic plasticity is particularly important in long-term memory. As in other tissues, $NO^\bullet$ usually acts by binding to the haem of guanylate cyclase to cause cyclic GMP formation. Thus some neurones (about 1–2% of the total in most brain regions, somewhat higher in cerebellum) contain **nNOS** (neuronal NOS), a constitutive form of NOS, which is Ca^{2+}/calmodulin-regulated.

Transgenic 'knockout' mice lacking nNOS show enlargement of the stomach, behavioural changes and seem more resistant to the toxicities of malonate (a succinate dehydrogenase inhibitor), the neurotoxin MPTP (Section 9.17 below) and to ischaemia–reperfusion damage to the brain.[259] Indeed, injection of 3–nitropropionate or of malonate into the brains of rats or mice leads to neuronal death in which $NO^\bullet$ and $ONOO^-$ appear to be involved. 3-Nitropropionic acid is produced by some fungi and plants that have caused illness in grazing animals. An apparently related illness has been reported from China in patients who ate sugar-cane infected with a fungus producing 3-nitropropionic acid.

The endothelial cells of brain blood vessels contain eNOS, which probably has an important vasodilatory role in regulating cerebral blood-flow. eNOS is also found in certain neurones. The compound **7-nitroindazole** is a nNOS inhibitor useful in neurological research, but it can also inhibit monoamine oxidase.[95a]

9.15.4 *Excitotoxicity*[49]

The term **excitotoxin** was coined in the 1970s when the ability of large excesses of glutamate or aspartate to kill neurones was observed. Similar neurotoxicity can be produced by NMDA, AMPA, kainic acid or **domoic acid** (a neurotoxin which binds to kainate receptors). Domoic acid is produced by the seaweed *Digenea domoi*; eating mussels contaminated with this toxin can induce acute gastrointestinal symptoms closely followed by neurological damage. Kainic acid was also first described as a product of the seaweed *Digenea simplex*. It has been suggested that damage to the inner ear by aminoglycoside antibiotics (Section 8.13) may involve excitotoxicity mediated via the NMDA receptor.[73]

Concentrations of glutamate in the brain extracellular fluids are normally $\sim 1\,\mu M$, while those inside neurones are millimolar. The death of cells or collapse of normal ionic gradients (e.g. due to severe energy depletion) in living neurones can cause massive glutamate release, leading to excessive and sustained increases in intracellular free Ca^{2+} and Na^+ in surrounding neurones. Cells treated with excess glutamate or other excitotoxins swell rapidly and appear to die by necrosis. Rises in Na^+ and Ca^{2+} play key roles in such death, the latter by mechanisms including interference with mitochondrial function, over-production of $NO^\bullet$ (in NOS-positive neurones) and increased ROS generation, in some cases involving conversion of xanthine dehydrogenase to the oxidase by Ca^{2+}-stimulated proteases. There is the usual debate about the significance of xanthine oxidase in human, as opposed to rodent, neuronal injury; Section 9.5 above). If produced in excess, $NO^\bullet$ and $O_2^{\bullet-}$ might form peroxynitrite. Rises in Ca^{2+} can also stimulate phospholipases, increasing concentrations of free fatty acids and eicosanoid production.

Hence studies on various neuronal or neurone-like cells in culture in different laboratories have shown variable protective effects against excitotoxic damage by adding NOS inhibitors, xanthine oxidase inhibitors and a variety of antioxidants, including transfection of the CuZnSOD gene. Interestingly, nNOS-containing neurones are fairly resistant to excitotoxicity, but can sometimes respond to excess glutamate or NMDA by releasing enough $NO^\bullet$ to injure adjacent neurones. The mechanisms of this resistance are uncertain, but one that has been suggested is a high level of MnSOD.

9.15.5   *Why should the brain be prone to oxidative stress?*

It is often said that brain and nervous tissue are prone to oxidative damage, for several reasons:

1.   The high Ca^{2+} traffic across neuronal membranes; interference with ion transport (e.g. by disruption of energy metabolism) can produce rapid rises in intracellular free Ca^{2+}, often leading to oxidative stress (Chapter 4).

2.   The presence of excitotoxic[49] amino acids (Section 9.15.4 above). It has been suggested that oxidative stress may promote the release of excitatory amino acids in neurones, generating a 'vicious circle' of events. Other possible relevant events are the ability of several ROS (including $ONOO^-$) to decrease glutamate uptake by glial cells and to inactivate the enzyme glutamine synthetase.

3. The high rate of O_2 consumption per unit mass of tissue. Like mitochondria in other tissues, those in brain can generate $O_2^{\bullet-}$ and levels of 8-OHdG, mutations and deletions increase with age in brain mitochondrial DNA.

4. Many neurotransmitters are autoxidizable molecules. Dopamine, its precursor L-DOPA, and noradrenalin react with O_2 to generate $O_2^{\bullet-}$, H_2O_2 and reactive quinones/semiquinones which can deplete GSH and bind to protein –SH groups (Section 8.6.4). Dopamine–GSH conjugates can be degraded by peptidase enzymes to produce dopamine–cysteine conjugates (e.g. **5S-cysteinyldopamine**), which can be detected in several brain regions. The striatum, a dopamine-rich region of brain, has been estimated to have an 'average' dopamine concentration of 65 µM and the level in striatal neurones has been approximated as 50 mM, although much of this will be stored in vesicles. The cytosolic concentration of noradrenalin in cell bodies of noradrenergic neurones has similarly been approximated as 0.6 mM.

Iron is found throughout the brain.[20] Important iron-containing proteins in brain include cytochromes, ferritin, aconitases, mitochondrial non–haem–iron proteins, cytochromes P450 and the tyrosine and tryptophan hydroxylase enzymes, which catalyse the first steps in the synthesis of dopamine and serotonin respectively. Several brain areas (e.g. substantia nigra, caudate nucleus, putamen, globus pallidus) have a high iron content. The molecular nature of most of the 'total' iron in healthy brain is however, uncertain; most is presumably in ferritin. What is clear is that damage to brain tissue readily releases iron (and copper) ions in forms capable of catalysing free radical reactions such as $OH^{\bullet}$ formation from H_2O_2, lipid peroxidation (Fig. 9.26) and autoxidation of neurotransmitters. Direct injection of iron salts into the brains of animals can produce epilepsy-like convulsions, dopamine depletion and death of neurones accompanied by lipid peroxidation and $OH^{\bullet}$ generation (as detected by aromatic hydroxylation of salicylate).

5. Unlike blood plasma, CSF has no significant capacity to bind any released iron. 'Total iron' values in CSF from normal humans have been reported to be in the range 0.2–1.1 µM. The transferrin content is ~ 0.24 µM. Since one mole of transferrin binds two moles of iron ions, these data suggest that CSF transferrin is often at, or close to, iron saturation. In adults, iron crosses the blood–brain barrier very inefficiently, e.g. in idiopathic haemochromatosis the peripheral iron overload (Chapter 3) is not reflected in the brain. Iron uptake into the brain is maximal during rapid brain growth in early life and if insufficient dietary iron is available at that time there may be permanent impairments of brain function. Transferrin is responsible for delivering iron across the blood–brain barrier, utilizing receptors located on the brain microvasculature.[20]

6. Neuronal membrane lipids contain a high content of highly polyunsaturated fatty-acid side-chains, especially those of eicosapentaenoic ($C_{20:5}$) and docosahexaenoic ($C_{22:6}$) acids. Simple homogenization of isolated brain tissue leads to high rates of lipid peroxidation (Fig. 9.26) which can be largely inhibited by iron-chelating agents such as desferrioxamine or by chain-breaking antioxidants.

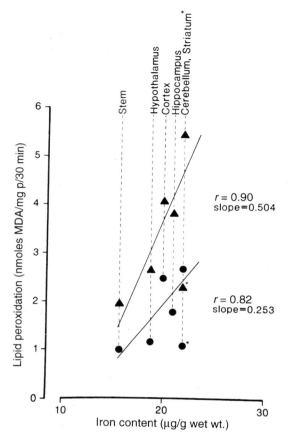

Fig. 9.26. Peroxidation of brain homogenates. Regions of rat brain were homogenized and the rates of TBARS formation measured (no iron or ascorbate was added). The iron content was also measured. Iron values for striatum and cerebellum were the same. Striatum values are designated by a superscript star. Lines were fitted by the least-square method. The closed circles represent values obtained from homogenates incubated under air and the triangles are the values obtained under 100% O_2. Note the correlation between iron content and rate of peroxidation. Data adapted from *Neurochem. Res.* **10**, 397 (1995) by courtesy of Professor Bob Floyd and the publishers.

7.   Brain metabolism generates H_2O_2. One example is the oxidation of dopamine by **monoamine oxidases**, flavoprotein enzymes located in the outer mitochondrial membrane.

$$\underset{(RCH_2NH_2)}{monoamine} + O_2 + H_2O \rightarrow \underset{(RCHO)}{aldehyde} + H_2O_2 + NH_3$$

For example, dopamine is oxidized to 3,4-dihydroxyphenylacetaldehyde, which an aldehyde dehydrogenase then converts to DOPAC (3,4-dihydroxy-phenylacetic acid). Monoamine oxidase exists in two forms (MAO-A and -B).

MAO-A preferentially oxidizes hydroxylated amines such as serotonin and noradrenalin and is found primarily in catecholamine neurones. MAO-B preferentially oxidizes non-hydroxylated amines and is localized in serotonergic neurones. Both oxidases are present in glial cells and both can oxidize dopamine. The ammonia generated can be disposed of by several mechanisms, including conversion of glutamate to glutamine by glutamine synthetase. **Clorgyline** is a drug that inhibits MAO-A, **selegiline** is selective for MAO-B and **pargyline** inhibits both.[91]

8. Antioxidant defences are modest. In particular, levels of catalase are low in most brain regions (Table 3.7). Catalase activity in rat or mouse brain is rapidly inhibited if aminotriazole is administered to the animals, showing that the brain generates H_2O_2 *in vivo* and that at least some of this reaches catalase.[271] The catalase appears to be located in small peroxisomes (**micro-peroxisomes**) and thus could probably not deal with H_2O_2 generated in other subcellular compartments (e.g. from MnSOD or monoamine oxidases, both located in mitochondria). In general, catalase is found at higher levels in hypothalamus and substantia nigra in mammals than in cortex or cerebellum.

9. Some of the glial cells are **microglia**, resident macrophages of the nervous system. Like other macrophages, they can produce $O_2^{\bullet -}$ and H_2O_2 (and probably $NO^{\bullet}$) upon activation[43] and are capable of secreting cytokines such as IL-1, IL-6 and TNFα.[64] TNFα and interferon γ may prime microglia to produce more ROS upon activation. Isolated human microglia seem to produce $NO^{\bullet}$ less readily than rodent cells but there is evidence that levels of nitrate and nitrite are elevated in CSF of patients with meningitis and some other infections, suggestive of increased $NO^{\bullet}$ production *in vivo*. This may involve cytokine-dependent induction of iNOS in microglia and perhaps astrocytes. The extra $NO^{\bullet}$ may perhaps be produced as an antibacterial mechanism, but too much could lead to neuronal damage.

10. Cytochromes P450 are present in certain brain regions.[307] Several isoforms have been detected. For example, in rats CYP2E1 is present in hippocampus, in substantia nigra and in the blood–brain barrier. Since CYP2E1 is 'leakier' and may produce more ROS than other cytochromes P450, this is another potential source of oxidative stress (although its magnitude may be small as P450 levels are low compared with those in, for example, the liver). CYP2E1 metabolizes ethanol, acetone, halothane, related anaesthetics and organic solvents such as CCl_4 and $CHCl_3$ (Section 8.8) and may be inducible by ethanol in the brain. Indeed, P450 and free radicals have been implicated in the neurotoxicity of several xenobiotics (Table 9.12).

9.15.6 *Consequences of oxidative stress*

As in other tissues (Chapter 4), oxidative stress can damage neurones and glia by several mechanisms:

(a) increased lipid peroxidation. For example, the aldehyde HNE is highly neurotoxic[196] and F_2-isoprostanes can constrict cerebral arterioles;[124]

Table 9.12. Some neurotoxic agents whose action may involve oxidative stress

Type of agent	Examples
Inhibitors of energy metabolism	Cyanide, carbon monoxide, iodoacetate, rotenone, picricidin, antimycin A, nitropropionic acid, malonic acid, MPP^+
Excitotoxic agents	Kainate, quinolinate, domoic acid, β-N-oxalylamino-L-alanine
Metals	Copper, lead, mercury, tin, aluminium, manganese, iron
Organic solvents	Ethanol, toluene, n-hexane, pyridine
Agents affecting antioxidant defences	Buthionine sulphoximine (GSH depletion), aminotriazole (catalase inhibitor), Chloropropionic acid (GSH depletion), mercaptosuccinate (glutathione peroxidase inhibition)
Free radical generators	6-hydroxydopamine, paraquat, CCl_4
Other agents	Cocaine, methamphetamine, methylenedioxymethylamphetamine[a], heroin, haloperidol, morphine, chlorpromazine, phenytoin

Many of these are considered in detail in Chapter 8 or in this section. In most cases, the contribution of oxidative stress to the overall neurotoxicity has not yet been elucidated.
[a] The neurotoxicity of this substituted amphetamine (**'Ecstasy'**) is minimized in transgenic mice over-expressing human CuZnSOD (*Synapse* **21**, 169 (1995)): its metabolite α-methyldopamine readily auto-oxidizes to generate ROS and products that form conjugates with GSH (Chapter 8).

(b) oxidative damage to DNA, causing base modification and strand break-age.[170] This leads to activation of poly(ADP–ribose) synthetase and possible 'lethal NAD^+ depletion' (Chapter 4);

(c) damage to proteins;

(d) induction of apoptosis and necrosis.

Thus, for example, failure to find lipid peroxidation in damaged nervous tissue cannot be equated with the absence of oxidative damage. Just as uncontrolled rises in intracellular Ca^{2+} can lead to oxidative stress, oxidative stress can produce rises in Ca^{2+} and derangements of energy metabolism (Chapter 4). One target of oxidative protein damage in the brains of some animals[81] has been identified as the enzyme glutamine synthetase. Another is neurofilaments, which, for example, can be nitrated by $ONOO^-$, leading to disruption of their organization. Oxidative stress can occur in vascular endothelial cells of the blood–brain barrier, which could increase its permeability.

Oxidative stress, if not too severe, might also cause adaptation, e.g. up-regulation of GSH synthesis. There are several suggestions that mild oxidative stress upon astrocytes causes them to up-regulate[213] synthesis of nerve growth factors (e.g. NGF and basic fibroblast growth factor), which may then exert protective effects upon adjacent neurones, e.g. by causing them to up-regulate their own antioxidant defences.

9.15.7   *Antioxidant defences in the brain*

One important antioxidant defence may be that of keeping intracellular O_2 levels as low as possible consistent with normal function (Chapter 1). Low O_2 decreases autoxidation reactions, leakage of electrons from mitochondrial electron-transport chains and the activity of monoamine oxidases. Thus intracellular brain pO_2 is low, but this of course renders the brain sensitive to interruptions of its blood supply.

Enzymes

All brain areas contain CuZnSOD, MnSOD, GSH (at millimolar levels) and glutathione peroxidases (Chapter 3). The brain also contains various types of glutathione S-transferase, but isoforms differ between neurones and glia.[231] Brain γ-glutamyl transpeptidase occurs in glial cells but apparently not in neurones.

    The importance of SOD is suggested by several experiments. Cultured neurones from transgenic mice over-expressing CuZnSOD (Chapter 3) live longer in culture, and transfection of CuZnSOD into normal neurones delays apoptosis triggered by removal of growth factors from the culture medium. By contrast, inhibiting SOD promotes neuronal apoptosis. A MnSOD 'knockout' mouse was reported to show extensive neurodegeneration (Chapter 3). Inhibition of glutathione peroxidase (e.g. by administering **mercaptosuccinate** to animals) produces neuronal death.

Glutathione and cysteine[152]

In rats (and possibly other animals), it seems that neuronal GSH levels are lower than in glial cells and that glial cells may assist neurones by supplying them with cysteine (as a GSH precursor) and efficiently detoxifying extracellular H_2O_2. In general, glial cells appear less susceptible to ROS (including $ONOO^-$) than neurones. The pentose phosphate pathway enzymes are also needed in all cell types to provide NADPH for glutathione reductase.

    Severe depletion of GSH is neurotoxic (Chapter 3). Ironically, excessive levels of cysteine can also be neurotoxic (e.g. newborn rats injected with cysteine develop brain atrophy). Cysteine seems to be weakly excitotoxic[237] and can act synergically with glutamate. Cysteine can also form ROS and thiyl radicals during its oxidation (Chapter 2).

Ascorbate[242]

There is a high concentration of ascorbate in the grey and white matter of the central nervous system in all species examined (Table 9.13). Ascorbate concentrations in the CSF are higher than in plasma ($>150\,\mu M$ in human CSF). Neurones and glia have an active transport system that concentrates intracellular ascorbate even more, apparently to millimolar levels. Like other cells, they can take up ascorbate and also dehydroascorbate, which can be reduced to ascorbate inside the cell. CSF ascorbate levels can be maintained high even in the face of a wide variation in plasma levels.

Table 9.13. Species comparison of ascorbate and GSH content in cortex

Species	Concentration (μmol/g tissue^a)		Ability of brain to tolerate anoxia
	Ascorbate	GSH	
Pond turtle (*Trachemys scripta*)	5.55 ± 0.29	2.34 ± 0.11	Good
Box turtle (*T. carolina triunguis*)	5.00 ± 0.27	1.98 ± 0.13	Good
Garter snake (*Thamnophis sirtalis*)	3.72 ± 0.18	1.83 ± 0.04	Moderate
Clawed toad (*Xenopus laevis*)	2.37 ± 0.16	2.53 ± 0.09	Poor
Rat	2.82 ± 0.05	1.97 ± 0.03	Poor
Guinea pig	1.71 ± 0.03	2.66 ± 0.06	Poor
Human	0.76 ± 0.11	1.97 ± 0.19	Poor

^{a}Wet weight.
Cortical ascorbate content was significantly ($P<0.001$) higher in each of the reptiles tested than in mammals or amphibia. The ascorbate content in turtle cortex was significantly higher than that in snake cortex ($P<0.001$ for pond turtle and $P<0.01$ for box turtle). In general, levels of ascorbate were higher in species able to tolerate anoxia. By contrast, GSH levels were not different between pond turtle and guinea pig cortex and were significantly higher in guinea pig than in box turtle ($P<0.01$) or snake ($P<0.001$) cortex. Data from *J. Neurochem.* **64**, 1790 (1995) by courtesy of Dr Margaret Rice and Raven Press.

Ascorbate in the absence of transition-metal ions has well-established anti-oxidant properties (Chapter 3). However, ascorbate–iron and ascorbate–copper mixtures generate free radicals. Thus, if 'catalytic' iron and copper were generated in the CNS as a result of injury, ascorbate might then stimulate oxidative damage within the brain and CSF. Indeed, injecting aqueous solutions of iron salts or haemoglobin into the cortex of rats has been shown to cause epileptiform discharges, lipid peroxidation and persistent behavioural and electrical abnormalities.[334] Haemoglobin is known to be toxic to the brain,[183] both by being directly pro-oxidant in the presence of peroxides (a reaction that is probably inhibited by ascorbate; Chapter 3) and also by releasing haem and iron ions on decomposition. Bleeding into the brain can thus cause oxidative damage. In addition, oxyhaemoglobin can bind NO$^\bullet$ and cause vasospasm.

Chain-breaking antioxidants[205]
Lipid-soluble chain-breaking antioxidants, especially vitamin E and presumably also ubiquinol, are very important in the brain. Severe and prolonged deprivation of α-tocopherol, as occurs in patients with fat malabsorption syndromes, produces neurological damage (Chapter 3). In general, the central nervous system is more severely affected than the peripheral one, and sensory axons more than motor ones.

However, it takes a considerable time (many weeks) to increase the α-tocopherol content of brain tissue in mammals supplemented with this vitamin. Similarly, vitamin E-deficient diets are very slow in lowering brain vitamin E levels. For example, it is necessary to feed rats a vitamin E-deficient diet for a year in order to decrease brain levels to <3% of control levels. Hence brain vitamin E levels seem tightly regulated.

Metal-binding and other protective proteins

Expression of the gene encoding caeruloplasmin[149] has been observed in mouse astrocytes, and it is interesting to note that the human disease acaeruloplasminaemia is associated with degeneration of the retina and basal ganglia (Chapter 3). Haptoglobin may help to bind any haemoglobin released into the CSF. Brain also contains metallothionein (MT)[5] and a 'brain-specific' MT-III isoform has been described in certain neurones, especially in hippocampus. The various heat-shock proteins (Chapter 4) can also be induced in the brain after stress, e.g. ischaemia.[193] **Haem oxygenase** is widespread in the brain: both constitutive (HO-2) and inducible (HO-1) forms have been detected.[296] Indeed, it has been suggested that one of the products of this enzyme, CO, may (at low concentrations!) have some function as a neurotransmitter. HO-1 levels can increase after ischaemia–reperfusion and in Alzheimer's disease.

'Thiol-specific antioxidant' proteins (Chapter 3) have been identified in brain and the gene encoding the human brain protein has been cloned. The brain contains ubiquitin and the proteasome system, so presumably has at least some ability to degrade oxidized proteins. DNA-repair enzymes are also present (Chapter 4). Carnosine and its derivatives in brain have also been postulated to exert antioxidant effects (Chapter 3).

Defence of the blood–brain barrier[298]

Another aspect to be considered in the maintenance of normal brain function is the resistance of the capillary endothelial cells constituting the blood–brain barrier to oxidative stress. Rat microvessels are enriched in GSH, glutathione peroxidase, glutathione reductase and catalase when compared with the rest of the brain.

9.16 Oxidative stress and ischaemic or traumatic brain injury

9.16.1 *Definition of terms*[270]

There are essentially two types of ischaemic injury to the brain. As is the case for other tissues (Section 9.5 above), the terms hypoxia and ischaemia are interchangeably used by neuroscientists. **Global ischaemia** occurs when O_2 supply to the whole brain fails or is severely limited. This occurs after asphyxia, myocardial infarction or poisoning by carbon monoxide. Cerebral blood flow in humans is about 50 ml/100 g of brain/minute; flows below 18 ml will lead to injury, and O_2 supply must be re-established within minutes to prevent neuronal damage.

Local ischaemia results after failure of the blood supply to a part of the brain, e.g. rupture of a blood vessel (**haemorrhagic stroke**) or formation of a thrombus in an atherosclerosed vessel (**thrombotic stroke**). The former has the additional problem that, as with any bleeding into the brain, haemoglobin is neurotoxic. For example, bleeding into the brain due to traumatic injury or rupture of a blood vessel is followed in a significant number of cases by **cerebral vasospasm** and thus a secondary ischaemia. In local ischaemia, cells in the centre of the ischaemic zone will be damaged most rapidly, but cells in the surrounding area (often called the **ischaemic penumbra**) will receive some O_2 from other blood vessels and thus will be less hypoxic.

Stroke is the third most frequent cause of medically related deaths and the second leading cause of neurological morbidity (after Alzheimer's disease) in the USA and Europe. Actions that help to prevent stroke include treatment of hypertension and a diet rich in fruits and vegetables, in which the antioxidants present may make a contribution by diminishing the incidence of atherosclerosis.

Global brain ischaemia for a short period leads to damage to selected groups of neurones, as can be seen in individuals resuscitated too long after a cardiac arrest. Most sensitive are the so-called **pyramidal neurones** in the CA1 region of the hippocampus. Other groups of neurones here and elsewhere (e.g. in layers 3, 5 and 6 of the cerebral cortex and in parts of the striatum, hypothalamus and cerebellum) have intermediate sensitivities. By contrast, the majority of neurones can tolerate 30 min or more of ischaemia (at least in animal studies). Death of susceptible neurones does not occur immediately but becomes visible 1–5 days after reoxygenation. By contrast, cerebral ischaemia lasting over about 60 min produces tissue infarction in which all cells within the most severely ischaemic region die. Ischaemia–reperfusion and traumatic injury can also damage the blood–brain barrier.

9.16.2 *Mediators of damage*

The simplest way of treating brain ischaemia is to restore blood supply as soon as possible. This is not always possible, of course, and even when it is, reoxygenation injury might occur and could cause damage in the ischaemic penumbra (in local ischaemia). Hence there has been much interest in attempts to develop therapies that could prevent neuronal death. Agents suggested to be important mediators of injury after ischaemia–reperfusion include ROS/RNS (especially $O_2^{\bullet -}$, H_2O_2, $OH^{\bullet}$, $NO^{\bullet}$ and $ONOO^-$), high glutamate levels producing excitotoxicity, acidosis, release of 'catalytic' iron, rises in intracellular free Ca^{2+} and an accumulation of free fatty acids due to phospholipase A_2 activation (via Ca^{2+}). For example, transgenic 'knockout' mice lacking nNOS show smaller infarcts in some global ischaemia models. The acidosis caused by lactate accumulation in anaerobic cells might accelerate lipid peroxidation and $OH^{\bullet}$ formation, by keeping iron in a soluble form. However, acidosis may also have beneficial effects, e.g. because it tends to decrease potentially-deleterious movements of ions such as Ca^{2+}.[303]

The role of ischaemia in causing a xanthine dehydrogenase to oxidase conversion in brain is uncertain. Brains from rat and gerbil,[168,228] species most often used to model ischaemia–reoxygenation, have been reported to contain low activities of xanthine dehydrogenase, but the activity of this enzyme in human brain does not appear to have been clearly established. Other sources of ROS include increased 'leakiness' of mitochondria damaged by ischaemia, and faster oxidation of neurotransmitters, e.g. by released transition-metal ions.

Extracellular glutamate levels in the brain increase rapidly during ischaemia and excitotoxicity thus contributes to injury.[49]

9.16.3   *Therapeutic interventions*

Multiple types of intervention in various animal models have been shown to decrease neuronal loss after ischaemia: lowering body temperature, Ca^{2+} blockers, NOS inhibitors, agents that prevent binding of glutamate to NMDA (or other) receptors and antioxidants such as SOD, Ebselen, desferrioxamine, lazaroids (Chapter 10) and the spin trap PBN. The success of these various interventions depends on the animal model used, the duration of ischaemia, the length of follow-up of the animals (some agents delay neuronal death rather than preventing it, an effect easily missed in a short-term study), the baseline levels of brain antioxidants (e.g. vitamin E-deficient animals suffer more damage after ischaemia–reperfusion) and other parameters of experimental design. Other observations showing that ROS/RNS play at least some role are that transgenic mice over-expressing CuZnSOD are more resistant to cerebral ischaemia–reperfusion and traumatic injury in at least some experimental protocols (Chapter 3), and that mice lacking nNOS are also more resistant to ischaemic injury. Of course, NOS requires O_2 and so cannot operate in a tissue completely deprived of O_2. Ischaemia–reperfusion in mouse cortex produces rises in several DNA base damage products,[170] a pattern suggestive of attack by $OH^\bullet$. Rises in protein carbonyls, lipid peroxides and HNE have been reported in other studies.

The fact that this wide variety of interventions can work (at least in some animal models) suggests that cell death is not due to a single event, but to multiple events acting in parallel and perhaps interacting synergically (Chapter 4). Thus hypothermia may protect by diminishing neuronal metabolic activity, which will tend to maintain ATP levels, decrease rises in intracellular Ca^{2+} and lower ROS (and possibly $NO^\bullet$) generation and glutamate release. Even a single type of injury mechanism can lead to different effects, e.g. ischaemia–reperfusion injury produces decreased glutamine synthetase activity and increased protein carbonyls in gerbil brain but not in rat brain,[81] suggesting that the molecular targets of oxidative damage may be different in the two cases.

Ischaemia–reperfusion also affects glial cells and the vascular endothelial cells of the blood–brain barrier, although the neurones are more sensitive. Suggestions as to why some neurones are more vulnerable than others include an increased propensity to undergo rises in intracellular free Ca^{2+} and mitochondrial damage.

9.16.4   *Traumatic injury*

Traumatic injury to the brain or spinal cord causes direct tissue damage (e.g. crushing, tearing, bruising and bleeding) but also leads to secondary damage involving many of the events described above, including release of transition metal ions and haemoglobin,[183] increased NO$^\bullet$ production and rises in glutamate. Ruptured blood vessels bleed, but in addition the parts of the brain they normally feed will become hypoxic. Oedema as a result of the injury can raise intracranial pressure and decrease blood flow.

Some hours after trauma, inflammatory and immune responses begin, involving cytokines, induction of transcription factors (e.g. NF-κB), activation of microglia and recruitment of circulating phagocytes (enabled to cross the blood–brain barrier by a series of changes in gene expression in the endothelial cells, e.g. up-regulating adhesion molecules). Traumatic injury can damage the blood–brain barrier, allowing molecules into the brain that do not normally enter it. The inflammatory/immune response is presumably an aid to repair of tissue injury,[64] but an excess response could lead to further damage, e.g. involving phagocyte-derived ROS and/or RNS. Interestingly, mice deficient in receptors for TNFα (a cytokine often picked as the 'bad guy' during inflammation) show *increased* neuronal damage after focal cerebral ischaemia or injection of the excitotoxin kainic acid.[29,64] Similarly, increased production of NO$^\bullet$ is not necessarily all bad; it can down-regulate adhesion molecules, improve blood flow (by vasodilation) and inhibit lipid peroxidation, for example.

Many of the agents tested for the treatment of stroke have been examined in models of traumatic injury to the CNS. For example, CuZnSOD conjugated to the polymer polyethylene glycol appears to be beneficial when administered to rats after percussive brain injury. It may be enabled to enter the brain only if the blood–brain barrier is damaged. The cerebral vasospasm that occurs in some patients several days after intracranial bleeding seems to involve oxidative damage and binding of NO$^\bullet$ by haemoglobin and its degradation products. During the first 5 days after an intracranial bleed, the erythrocytes in the CSF slowly haemolyse. Some of the oxyhaemoglobin is converted to bilirubin (presumably by haem oxygenase) but some undergoes non-enzymic oxidation and degradation.[319]

9.17   Oxidative stress in Parkinson's disease

9.17.1   *Pathology of the disease*[343]

There have been many proposals that ROS and/or RNS are involved in the pathology of neurodegenerative diseases. **Parkinson's disease** (PD) was first described by the English physician James Parkinson as the 'shaking palsy' in 1817. It usually appears in middle to old age (rarely before 50), often as a rhythmic tremor in a foot or hand, especially when the limb is at rest. As the disease develops, patients begin to have increasing problems in controlling movement. Movement is slow (**bradykinesia**), and initiation of movement is slow (**akinesia**) and there is muscle rigidity. The disease is slowly progressive, and can be detected in about 10% of people over age 65.

Parkinson's disease attacks a group of cells (the **substantia nigra**) in the upper part of the brainstem; the word 'nigra' derives from the fact that the cells are rich in a black pigment, **neuromelanin**. Neuromelanin appears to arise from dopamine by oxidation to quinones and semiquinones, followed by cross-linking. Like other melanins, it is redox-active and can bind metal ions (Chapter 3). It contains some residues derived from cysteinyl dopamine and appears to be[216] a 'mixed-type melanin', containing some sulphur (unlike eumelanins), but not as much as in phaeomelanins (Chapter 3).

The marked loss of neurones in PD is associated with microscopically observable changes, one of which is the appearance of **Lewy bodies** in the substantia nigra and elsewhere. These are electron-dense inclusion bodies, with a particularly dense core from which filaments radiate. Lewy bodies are found in about 10% of 'normal' persons above the age of 60 and in about 15% of Alzheimer patients, so they are not specific to PD.

9.17.2   *Treatment*[343]

The substantia nigra sends nerve fibres to a structure called the **striatum** (meaning 'striped') at the base of the brain; the terminals of these fibres secrete the neurotransmitter dopamine. Striatal cells relay the message to the cortex, helping to control movement. The progressive death of nigral cells means that less dopamine is available, which explains why parkinsonian patients benefit from therapy with L–DOPA, a precursor of dopamine. Unlike dopamine, L–DOPA crosses the blood–brain barrier. L–DOPA is administered together with an inhibitor of DOPA decarboxylase that does not cross the blood–brain barrier (e.g. **benserazide, carbidopa**) so that it is only decarboxylated in the central nervous system. The possibility of grafting foetal dopamine-producing cells into the brain is being tested. Cells also die in another part of the brain, the **locus coeruleus**, leading to lowered noradrenalin concentrations.

L–DOPA treatment produces several distressing side-effects and its efficacy diminishes as neuronal death continues. It is also possible that metabolism of excess dopamine by the monoamine oxidases could produce damaging levels of H_2O_2. Hence alternative or additional therapies are in use. One of these is selegiline (**deprenyl**), an inhibitor of MAO-B, which preserves dopamine levels and should also decrease H_2O_2 production. L–Deprenyl (Fig. 9.27) is rapidly metabolized to (−)methamphetamine. Prolonged administration of deprenyl to animals has been reported to increase levels of CuZnSOD in the substantia nigra and, to a lesser extent, in other brain areas. Hence it could be neuroprotective by mechanisms additional to MAO-B inhibition. *In vitro,*

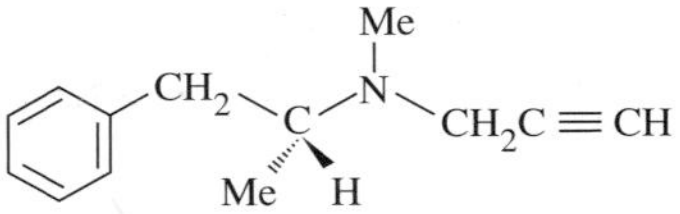

Fig. 9.27. The structure of deprenyl (selegiline). Deprenyl itself has no direct antioxidant ability.

L–deprenyl has been reported to protect neurones against damage by several toxins including pre-formed MPP$^+$ (i.e. after the MAO–B catalysed step) and to raise neuronal GSH levels. Its metabolite **desmethylselegiline**[207] is also neuroprotective *in vitro* even though it is only a weak inhibitor of MAO–B.

Yet another therapeutic approach is to use agents that directly stimulate dopamine receptors, such as **bromocriptine** and **pergolide**. The latter has also been reported to increase CuZnSOD levels in animal substantia nigra.[88]

9.17.3 *What is the cause of PD?*

We do not know what causes PD, but several clues have emerged. Genetic factors may be involved, but are hard to identify in the majority of patients. Nevertheless, a mutation on chromosome 4 has been identified in patients belonging to a large family of Italian descent who suffer inherited PD. The mutated gene encodes the protein **α-synuclein**,[280] a protein found in presynaptic nerve terminals and also reported to be present in Lewy bodies in PD.

Another clue came from studies of drug addicts.[157] In 1982, clinicians at hospitals in California were surprised by a sudden influx of young patients with severe parkinsonian symptoms. These patients were found to have used 'synthetic heroin' from a drug pusher, but this heroin substitute was contaminated with **1-methyl-4-phenyl-1,2,3,6-tetrahydropyridine** (**MPTP**). An earlier report, published in 1979, was then unearthed in which a 23-year old graduate student developed PD after trying to synthesize the drug Demerol, which became contaminated with MPTP as a by-product of the synthesis. Ironically, MPTP had been synthesized, tested (and discarded!) as an *anti*-Parkinsonian drug by a major pharmaceutical company in the late 1950s.

The structure of MPTP is shown in Fig. 9.28; it bears some resemblance to paraquat. MPTP crosses the human blood–brain barrier and is oxidized by MAO–B. In the substantia nigra, this enzyme is largely located in the mitochondria of glial cells. Oxidation produces MPDP$^+$, which then forms MPP$^+$ (Fig. 9.28). MPP$^+$ is recognized by the catecholamine presynaptosomal

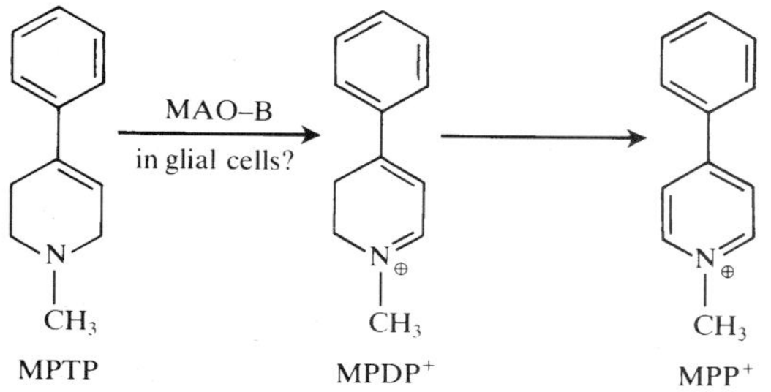

Fig. 9.28. MPTP and its metabolites. MPTP is 1-methyl-4-phenyl-1,2,3,6-tetrahydro-pyridine. It is oxidized to MPDP$^+$ (1-methyl-4-phenyl-2,3-dihydropyridine ion) which forms MPP$^+$, the 1-methyl-4-phenylpyridinium ion. Inhibitors of monamine oxidase, such as pargyline or deprenyl, diminish damage by MPTP. How MPDP$^+$ is converted into MPP$^+$ is not completely clear.

uptake system and so MPP$^+$ released by glial cells can be taken up by nigral neurones, a process that may be accelerated by the binding of MPP$^+$ to neuromelanin.

MPP$^+$ is neurotoxic, producing dopamine depletion and a rapid onset of parkinsonian symptoms. It is a powerful inhibitor of mitochondrial complex I (at or close to the rotenone-binding site); neuronal cell death is primarily due to interference with energy metabolism, since MPP$^+$ may accumulate in mito-chondria.[86,201] Secondary consequences probably include rises in Ca^{2+} and increased production of NO$^\bullet$ and of ROS. Indeed, some authors have reported that nNOS inhibitors (e.g. 7-nitroindazole) decrease MPTP neurotoxicity in animals and that transgenic mouse nNOS 'knockouts' are more resistant to MPTP than controls. 7-Nitroindazole may also inhibit monoamine oxidase as well as nNOS, however.[95a]

The inhibition of mitochondrial complex I by MPP$^+$ may involve ROS production; initial binding of MPP$^+$ to complex I may lead to ROS gener-ation and further damage to complex I. Desferrioxamine has been reported to protect against MPP$^+$ neurotoxicity when co-infused into the striatum of rats.

MPTP contains a pyridine ring (Fig. 9.28), as do many industrial chemicals, such as herbicides. This naturally raises the possibility that PD is caused by an environmental toxin, as is further suggested by studies of the Chamorro Indians on Guam and Rota (Section 8.12.3). Humans are born with a limited number of nigrostriatal neurones and this number declines with age (a common esti-mate is 8–10% per decade). Depletions of 70–75% seem to be the minimum for parkinsonian symptoms to appear. Hence an environmental insult to the nigra in early life, combined with the normal age-related death of nigral cells, might be sufficient to produce PD in late life. However, PD is found throughout the world, suggesting that if a single toxin were responsible, it must be present universally. It is also possible that many toxins can damage the substantia nigra and predispose to PD in later life. For example, consumption of the **yellow star thistle** has been identified as a cause of parkinsonism in horses in the USA.

Some investigators have tried to identify endogenous compounds that might mimic the action of MPTP.[192] One group of compounds identified is the **tetrahydroisoquinolines**, another is the **tetrahydro-β-carbolines**. Both can produce parkinsonian symptoms in animals after long-term treatment and both are found at low levels in certain foods (e.g. cheese, chocolate powder and wine). They may even be made endogenously in the brain, e.g. **salsolinol** is a tetrahydroisoquinoline that could be produced from dopamine.

Other groups have looked for defects in drug metabolizing-ability in PD subjects, e.g. altered P450 activities, glutathione S-transferases or other enzymes of sulphur metabolism. However, there is no clear evidence as yet that PD involves environmental toxins despite extensive epidemiological investigations. The attention of some groups is currently focused on organic solvents such as *n*-hexane,[230] a common constituent of adhesives, varnishes and fuels. Hexane is metabolized by a series of reactions to produce hexanol, hexanone, hexa-nediols and **2,5-hexanedione**, a neurotoxin.

9.17.4   *Oxidative stress and mitochondrial defects in PD*[91,202,219]

Comparison of the brains of patients who died with PD with those of neurologically normal brains shows several parameters consistent with increased oxidative stress (Table 9.14) and defective mitochondrial function. Decreases in complex I activity (NADH dehydrogenase) have been found in the substantia nigra (probably in both neurones and glia) but not in other brain regions, whereas activities of the other mitochondrial electron-transport complexes seem unaltered. Since the decreases are specific to complex I, they are unlikely to be related to cell death or loss of mitochondria.

Mutations in mitochondrial DNA are the basis of some diseases that cause defects in energy production and affect the basal ganglia (Chapter 1), including Leigh disease, Leber's disease with dystonia and sometimes MELAS. There is no evidence that PD involves mitochondrial DNA mutations, but these observations do show that mitochondrial damage can lead to neurodegeneration. Damaged mitochondria may generate more ROS than usual, and ROS/RNS (including $O_2^{\bullet-}$, $OH^{\bullet}$ and $ONOO^-$) can inactivate complex I. Hence it is possible that oxidative stress and mitochondrial defects form a 'vicious cycle' in PD.

9.17.5   *Distinguishing cause from consequence*

One major question is whether the changes in mitochondria or in oxidative stress parameters are primary to the disease process, a later consequence of it, or an effect of the treatment. Patients who die with PD will usually have been taking L-DOPA for many years. Another point to consider when pondering the data in Table 9.14 is that one is comparing a healthy substantia nigra with one in which >80% of the neurones have died and been replaced by other cell types. Among these are microglia, whose activation could be a source of ROS and NO$^{\bullet}$.

Both L-DOPA and dopamine can oxidize *in vitro* to generate semiquinones, quinones, $O_2^{\bullet-}$ and H_2O_2,[233] a process greatly facilitated by the presence of transition-metal ions such as iron and manganese (Section 8.6.5). In the case of iron, $OH^{\bullet}$ will also be generated whereas Mn^{2+} does not convert H_2O_2 to $OH^{\bullet}$. In addition, dopamine oxidation by monoamine oxidases produces H_2O_2. Indeed, 6-hydroxydopamine, which is oxidized more quickly than L-DOPA, causes degeneration of nigrostriatal neurones when injected into the brains of rats (Section 8.6.5). This damage is potentiated by iron injection or depletion of GSH (e.g. by intra-cerebral administration of buthionine sulphoximine) and decreased to some extent by pre-treating animals with vitamin E or desferrioxamine. 6-Hydroxydopamine-induced lesions in the rat striatum show increased iron levels. 6-Hydroxydopamine has also been reported to inhibit mitochondrial complex I.[87]

In PD, where ≥80% of the dopaminergic neurones have been lost, a rise in dopamine turnover in the remaining cells would be expected in the face of decreased GSH (Table 9.14). Treatment of neurones in culture with L-DOPA

Table 9.14. Evidence consistent with oxidative stress in Parkinson's disease (PD)

Observation	Comments	Selected representative references
A. Direct evidence (measurement of increased oxidative damage or other biomarkers of ROS/RNS production)		
Increased lipid peroxidation	Measured in SN as elevated TBARS and (more convincingly) as HNE–protein adducts, and increased peroxides (by HPLC/ chemiluminescence assays). No changes in SN vitamin E reported or any beneficial effect of administration of vitamin E in PD. HNE is neurotoxic: it can cause neuronal apoptosis.	*Movement Dis.* **9**, 92 (1994); *Proc. Natl Acad. Sci. USA* **93**, 2696 (1996); *New Engl. J. Med.* **328**, 176 (1993)
Increased oxidative DNA damage	Rises in 8-OHdG reported in mitochondria and 'total DNA'; little or no rise in other DNA base damage oxidation or deamination products, suggesting rise is not due to OH$^\bullet$ or ONOO$^-$ attack on DNA.	*Neuroscientist* **3**, 327 (1997); *Neurodegeneration* **3**, 197 (1994); *J. Neurochem.* **69**, 1196 (1997)
Increased protein damage	Rises in carbonyls observed in SN but also in several other brain regions, including those unaffected in PD.	*J. Neurochem.* **69**, 1326 (1997)
Increased nitrotyrosine	Core of Lewy bodies in parkinsonian SN stains with antibodies against nitrotyrosine	*Neuroscientist* **3**, 327 (1997)
B. Indirect evidence (evidence suggestive of a response to oxidative stress)		
Fall in GSH, no marked rise in GSSG	Decreases about 40% in SN, not in other brain regions. Lowering GSH by buthionine sulphoximine does not produce PD in animals but renders them more susceptible to neurotoxins.	*Pharmacol. Ther.* **63**, 37 (1994)
Increased iron content	SN zona compacta (but not other brain regions) have higher iron levels in PD, apparently with unaltered ferritin. 'Free' iron is neurotoxic but it is uncertain whether these extra iron deposits are 'catalytic' or not. No increase in Cu or Mn detected. Source of excess Fe unknown but increased expression of receptors for lactoferrin has been reported on neurones and microvessels in SN. Increased Zn also reported in substantia nigra.	*J. Neurochem.* **56**, 978 (1991); *Brain* **114**, 1953 (1991); *Proc. Natl Acad. Sci. USA* **92**, 9603 (1995); *J. Neurochem.* **52**, 1830 (1989)

Table 9.14. (*Continued*)

Observation	Comments	Selected representative references
Changes in SOD	SOD activity is elevated but it is unclear if CuZnSOD, MnSOD or both rise.	*Pharmacol. Ther.* **63**, 37 (1994)
Changes in enzymes of glutathione metabolism	Small decreases in catalase and GSHPx reported in SN and other brain regions by some (not all) scientists. γ-Glutamyl transpeptidase (involved in degradation and cellular translocation of GSH; Chapter 3) elevated. No marked changes in glutathione reductase or γ-glutamylcysteine synthase (usually the rate-limiting enzyme in GSH synthesis).	*Pharmacol. Ther.* **63**, 37 (1994); *Ann. Neurol.* **36**, 348 and 356 (1994)

C. Other evidence (consistent with the concept)

Observation	Comments	Selected representative references
Defects in mitochondrial function	See text. Coenzyme Q$_{10}$ levels reported as decreased in platelet mitochondria in PD.	*Ann. Neurol.* **42**, 261 (1997)
Increased glycoxidation	Formation of AGE products requires both glycation and oxidation.	*Brain Res.* **737**, 195 (1996)
Up-regulation of haem oxygenase I	—	*Brain Res.* **737**, 195 (1996)
Activation of NF-κB	Could be explained by oxidative stress.	*Proc. Natl Acad. Sci. USA* **94**, 7531 (1997)

SN, substantia nigra.

and dopamine has been reported to produce cell death (sometimes by apoptosis[320]) in some studies, but in others L-DOPA induced rises in GSH levels in embryonic neurones or glia (presumably a response to mild oxidative stress).[110] Parameters such as the O_2 tension, the monoamine oxidase level of the cells and the iron content of the medium could contribute to such conflicting results. A major question about the use of L-DOPA in the therapy of PD is whether L-DOPA/dopamine could be neurotoxic *in vivo*: does the therapy ameliorate the symptoms of the disease whilst aggravating the neuronal destruction? There is some *in vivo* evidence for L-DOPA-induced ROS formation, e.g. increased detection of OH$^\bullet$ (by aromatic hydroxylation of salicylate) in rats and increased 'lipid peroxidation' (as TBARS) after L-DOPA was administered to rats pre-treated with 6-hydroxydopamine. However, there is no evidence that feeding high doses of L-DOPA to monkeys causes oxidative damage parameters (e.g. protein carbonyls) to increase. Pre-existing oxidative stress (as might occur in PD) may be needed for L-DOPA toxicity; L-DOPA and dopamine can react with $O_2^{\bullet-}$, peroxyl radicals or 'catalytic' iron ions to initiate the chain reaction of their own oxidation. The resulting semiquinones/quinones could bind GSH (both non-enzymatically and catalysed by certain glutathione transferases) and cause GSH depletion.[279] *In vitro*, incubating L-DOPA or dopamine with H_2O_2 and iron or copper ions generates OH$^\bullet$ which can, for example, oxidize DNA bases. However, the pattern of base damage observed is that of OH$^\bullet$ (a wide range of products), unlike the selective rise in 8-OHdG seen in PD (Table 9.14).

One way of sorting out cause from consequence is to perform animal experiments. For example, injection of buthionine sulphoximine into the brains of rats decreased GSH by up to 70%, but did not alter SOD levels or mitochondrial complex I activity.[261] MPTP has provided a valuable tool for PD studies. Rats are much less sensitive than primates to MPTP, since rats have high levels of MAO-B in the blood–brain barrier. The enzyme may convert MPTP to MPP$^+$, but this charged species will not cross into the brain. However, MPP$^+$ or MPTP can be injected directly into the brain. Treatment of monkeys with MPTP has been shown to lead to iron deposition in the substantia nigra, but acute treatment did not cause falls in GSH or complex I activity.[86]

Another approach has been to use the observation that Lewy bodies are occasionally present in the substantia nigra of patients who died with no observable parkinsonian symptoms. Such 'incidental Lewy bodies' and nigral degeneration are found at post-mortem in 10–15% of patients over the age of 60. Many neurologists believe that these are patients in the pre-symptomatic stages of PD (and thus will not have been treated with L-DOPA). Some neurologists believe otherwise; it is impossible to resolve the question because incidental Lewy body disease is only diagnosed *post mortem*, so one cannot follow up cases to see if PD develops later.

However, if one accepts the former interpretation, one can ask which parameters of oxidative damage are still detected (Table 9.15). Yet another approach is to examine what effects L-DOPA produces in healthy animals or

Table 9.15. Oxidative stress in substantia nigra of 'incidental Lewy body' patients: a comparison with overt Parkinson's disease

Parameter	Comments
A. Features shared by ILBD and PD	
Decreased GSH, no rise in GSSG. Up-regulation of Bcl-2.	$\sim 40-50\%$ fall in GSH for both. Suggests fall is not due to L-DOPA treatment or gliosis (glial cells usually have high GSH in any case). Bcl-2 rise is perhaps a response to stress?
B. Changes observed in PD but not in ILBD	
Decreased mitochondrial complex I (but *trend* to lower values in ILBD) increased iron or zinc levels, changes in ferritin, raised protein carbonyls.	Could be that small rises are difficult to detect, but implication is that these events are later stages in the disease pathology and could be related to treatment with L-DOPA. Increased iron levels also found in several other neurodegenerative diseases, including MSA, PSN and Alzheimer's disease.

Abbreviations: ILBD, incidental Lewy body decrease; MSA, multiple system atrophy; PD, Parkinson's disease; PSN, progressive supranuclear palsy.
DNA base oxidation and lipid peroxidation in PD and ILBD still need to be investigated.
For further details see Olanow *et al.* (1996) in reference list; also *Ann. Neurol.* **35**, 38 (1994).

in humans given it for the treatment of other neurodegenerative diseases such as **progressive supra–nuclear palsy** (PSNP) and **multiple system atrophy** (MSA).[98] For example, L-DOPA-treated MSA patients are reported to show no decrease in mitochondrial complex I in substantia nigra. Iron deposition is increased in MSA and PSNP but falls in GSH have not been reported.

9.18   Oxidative stress in Alzheimer's disease

9.18.1   *Pathology of the disease*[127,264]

Dementia is defined by the World Health Organization as 'an acquired global impairment of higher cognitive functions, including memory, the ability to deal with daily life, the performance of sensorimotor and social functions, language communication and control of emotional reactions, without marked reduction of consciousness'. Up to 5% of people aged 65 and over suffer from dementia, and about half of these have **Alzheimer's disease** (AD). It is probable that the incidence of AD doubles every 4–5 years after the age of 60; the incidence seems higher in women.

Two of the pathological features of Alzheimer-type dementia, first reported by the German psychiatrist Alois Alzheimer in 1906, are the presence of

Fig. 9.29. Section of brain from a patient with Alzheimer's disease. Senile plaques (arrows) and neurofibrillary tangles (arrowheads) are shown. Holmes staining (silver impregnation) × 250. Photography by courtesy of Professor B.H. Anderton and Dr Jean-Pierre Brion. Senile plaques are round or oval, with diameters ranging from 1.5 to 20 nm: this is the 'classical plaque' of Alzheimer's disease. Also present are diffuse deposits of Aβ proteins, often called 'diffuse' or 'pre-amyloid' plaques, not accompanied by neurites or glia.

neurofibrillary tangles and **senile plaques** in the brain (Fig. 9.29). Several brain regions, especially the cortex, suffer massive neuronal loss. The tangles are fibrous masses inside affected neurones in several brain regions. They largely consist of pairs of filaments, each about 10 nm in diameter, twisted around each other with a cross-over roughly every 80 nm (**paired helical filaments**). They contain the microtubule-associated protein **tau** (τ) phosphorylated to an abnormally high extent and with bound ubiquitin. The normal function of tau seems to be to help maintain the microtubular structure.

Senile plaques are extracellular localized areas of degenerating and frequently swollen axons, neurites and glia surrounding a core of **amyloid** (Box 9.2). Many of the neurites contain the paired helical filaments described above. Plaques are most common in the amygdala, hippocampus, and neocortex, but they can also occur elsewhere. Amyloid protein is additionally deposited in blood vessel walls in the Alzheimer brain; this protein seems identical with the amyloid of plaque cores. Neurofibrillary tangles and plaques of the Alzheimer type are also found in young adults with Down's syndrome and in cases of Guam dementia.

Box 9.2

Some of the suggested mechanisms accounting for aggregation and precipitation of β-amyloid peptides

- oxidation by ROS/RNS (facilitated by iron and Al^{3+} ions)
- glycation/oxidation: AGE formation
- deamidation of asparagine; racemization of aspartate
- precipitation by metal ions (e.g. Zn^{2+}, Al^{3+})
- interaction with other proteins[a]: α_1-antichymotrypsin, apolipoprotein E, non β-AP protein, heparan sulphate, proteoglycan, serum amyloid P

'Amyloid' is a chemically non-specific term first used by the German pathologist Rudolf Virchow in 1853. It means 'starch-like', and is used to refer to tissue deposits of rigid, proteinaceous fibrils, 5–10 nm in diameter. Amyloids show whitish birefringence in unstained sections and a green birefringence under polarized light after staining with Congo Red. 'Amyloid' is a misnomer; the deposits are protein, vary widely in their composition and are the end-stage products of a variety of disease processes; the protein fibrils usually adopt a twisted β-pleated sheet conformation. Amyloid deposits are found in many conditions other than AD including tuberculosis, syphilis, rheumatoid arthritis and multiple myeloma. Many different polypeptides can form amyloid, including serum amyloid A, the enzyme lysozyme and certain immunoglobulins. In general, amyloid deposits occur when ordinarily-soluble proteins undergo structural changes that causes them to aggregate (e.g. the β-peptide in AD). Several rare inherited types of severe amyloid deposition have been described, such as **familial Mediterranean fever**, found principally in Mediterranean Jews and Armenians. This disease is eventually fatal, often because of damage to the kidneys by the amyloid deposits. This and other types of amyloid may be cytotoxic by mechanisms of the type identified for β-amyloid in AD (Section 9.18). For a review see *Blood* **75**, 531 (1990).

[a]It must not be assumed that all interactions are bad. For example, the glycoprotein **clusterin** (apoprotein J) found in AD plaques, can protect neurones against damage by β-peptide *in vitro*, as can apoE3 and apoE4 (*J. Neurotoxicol.* **67**, 1324 (1996)).

9.18.2 *The nature of amyloid in AD*[264]

Amyloid protein deposits in AD contain truncated products (**amyloid β-peptides**) of a family of much larger proteins (**amyloid precursor proteins**, or APPs). These precursor proteins are transmembrane proteins, expressed in all cells, with a short intracellular C-terminus and a longer extracellular N-terminus. They are encoded by a gene on chromosome 21 in humans and form a set of polypeptides from 563 to 770 residues in length. The most abundant of these, APP695, is predominantly expressed in neurones. APPs are made (and secreted) by many cell types and several functions have been ascribed to them; in neurones, they may facilitate survival and growth during brain development as well as neuronal responses to excitatory neurotransmitters. The normal secretion of APP involves proteolytic processing (by **α-secretase**) to release the N-terminal sequence and leave behind a small fragment in the membrane. By contrast, release of β-peptide is said to involve γ- and **β-secretase** enzymes (Fig. 9.30).

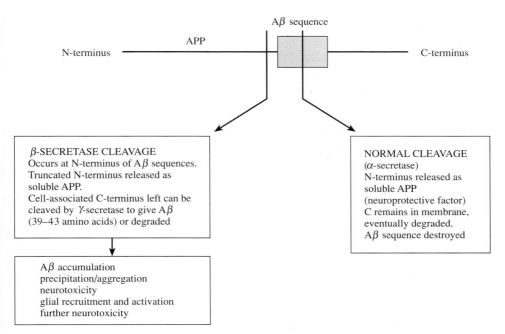

AβΑ sequence

APP

N-terminus

C-terminus

β-SECRETASE CLEAVAGE
Occurs at N-terminus of Aβ sequences.
Truncated N-terminus released as
soluble APP.
Cell-associated C-terminus left can be
cleaved by γ-secretase to give Aβ
(39–43 amino acids) or degraded

NORMAL CLEAVAGE
(α-secretase)
N-terminus released as
soluble APP
(neuroprotective factor)
C remains in membrane,
eventually degraded.
Aβ sequence destroyed

Aβ accumulation
precipitation/aggregation
neurotoxicity
glial recruitment and activation
further neurotoxicity

Fig. 9.30. Processing of the amyloid precursor proteins (APPs). The chemical identity of the secretases is as yet uncertain. APP is made in the endoplasmic reticulum and passes through the Golgi apparatus to reach the plasma membrane, undergoing glycosylation and other modifications during its passage. At the cell surface a minority of APP molecules are attacked by α-secretase to release soluble APP into the extracellular environment. Uncleared APP can be recycled by endocytosis and either completely degraded or recycled to the cell surface associated with β-secretase cleavage. Aβ; amyloid β-peptide.

In AD, correlations exist between the extent of dementia and the number of plaques present, and also with decreased synthesis of the neurotransmitter acetylcholine in the brain due to dysfunction and death of cholinergic neurones. It must be noted that neurofibrillary tangles and plaques also occur, to a much more limited degree, in non-demented old people. The amyloid core itself may be directly cytotoxic to neurones, as well as being secondarily cytotoxic by provoking an inflammatory response of microglial cells, with over-production of $O_2^{\bullet-}$, H_2O_2 and $NO^{\bullet}$. It is now widely believed that progressive cerebral accumulation of β-peptides is an early and necessary stage of AD, leading to neurofibrillary damage in neurones, neurotransmitter defects and the neuronal death that causes loss of cognitive function.

Thus the key element in senile plaques is amyloid β-peptide, which is cleaved from APP. The β-peptide can be 39–43 amino acids long, depending on the cleavage site. Synthetic β-peptides are initially non-toxic to cultured neurones but become toxic on incubation, a process related to their aggregation. Indeed, soluble β-peptide has been detected in CSF of normal subjects.

An obvious question is therefore the mechanism of aggregation (Box 9.2). The β-peptide containing 42 amino acids appears particularly neurotoxic (and may exert some direct effects on neurones as well as those related to aggregation), whereas the 40 amino acid β-peptide (the normal product of β-/γ-secretase activity: Fig. 9.30) is less damaging, probably largely because it aggregates less quickly.

Senile plaques contain a range of other proteins that may facilitate, decrease or have no effect on aggregation. Of special interest is **apolipoprotein E**, a protein found in chylomicrons, HDLs and VLDLs (Section 9.2 above). In the brain, apoE (like the antiproteinase α_1-antichymotrypsin; Box 9.2) is made by astrocytes and synthesis increases after traumatic injury and in several neuro-degenerative diseases. ApoE may play a key role in brain lipid metabolism and appears to be essential for neuronal and glial development.

Plaques also contain various metal ions, including aluminium, zinc and iron. Advanced glycation end products (AGE proteins) have been detected in senile plaques and paired helical filaments.

9.18.3   *Genetics of AD*[264]

There are several clear-cut genetic predispositions to AD. Other factors are also important, and may synergize with genetic pre-dispositions. For example, repeated brain trauma predisposes to AD-like pathology, e.g. **dementia pugilistica** ('punch-drunk syndrome') in boxers. **Familial Alzheimer's disease** cases with early onset and rapid progression have been linked to mutations in the APP gene on chromosome 21, in regions within or close to that encoding the β-peptide sequence. Transgenic mice over-expressing one of these mutant APP proteins (APP$_{695}$; Lys670 → Asn; Met^{671} → Leu, a dual mutation identified in a large Swedish family with early onset AD) seem normal at birth but by 9–10 months of age behavioural deficits appear and are accompanied by brain amyloid deposits in plaques. Presumably these various mutations favour production of the β-peptide. In Down's syndrome the presence of three copies of chromosome 21 may account for the early appearance of AD-type pathology; overexpression of normal human APP in transgenic mice is mildly neurotoxic.

The incidence of 'sporadic' AD is also affected by genetics. In particular, the gene encoding apoE is located on chromosome 19 and has three alleles. The most common isoform, apoE3, is a 299 amino acid protein with a single cysteine residue at position 112. ApoE2 and apoE4 differ in that the former has an extra cysteine (replacing an arginine) at 158 and the latter has lost the cysteine at 112, which is replaced by arginine. Since apoE4 lacks cysteine, it cannot form disulphide bonds with itself (to form homodimers) or with other proteins. The apoE4 allele appears to be a risk factor for development of AD, perhaps because the form of the protein encoded by this allele facilitates plaque formation (or microglial activation by plaques: Section 9.18.4 below) more than the other isoforms (Box 9.2). By contrast, the apoE2 subtype is negatively associated with AD. Mutations in the genes on chromosomes 1 and 14 encoding **presenilin** proteins are also linked to AD. The presenilins are transmembrane proteins perhaps involved in regulation of secretase action (Fig. 9.30).

9.18.4 *Mechanisms of plaque toxicity*[31,264]

Most studies have been carried out on neurones in culture. The neurotoxicity of β-peptides and of fragments derived from them can be manifested as cell injury or death (by apoptosis or necrosis). Toxicity has been variously attributed to increased intracellular free Ca^{2+} (perhaps involving the formation of 'cation channels' by insertion of the peptide into the plasma membrane), inhibition of mitochondrial function and ROS/RNS generation. Of course, all these events are linked (Chapter 4). Several scientists have shown that addition of β-peptide (or the 25–35 fragment of it) to neurones causes damage which is prevented by antioxidants. The Aβ (25–35) fragment aggregates and develops neurotoxicity exceptionally rapidly and is thus often used in experiments, although it is not found *in vivo*. Treatment of neuronal cultures with TNFα renders them more resistant to subsequent damage by β-peptides, apparently by inducing antioxidant defences.[16] In addition, ESR techniques have been used to show that the β-peptide itself generates free radicals, although the chemistry of this process is uncertain.

Evidence that increased oxidative damage occurs *in vivo* in AD comes from observations of elevations in activated NF-κB, 8-OHdG, other DNA base oxidation products, protein carbonyls, nitrotyrosine, 4-HNE and other products of lipid peroxidation in the brain,[31,182,196] although vitamin E levels seem normal. Increases in haem oxygenase-1 activity have also been reported. HNE is particularly toxic to neurones. Oxidative protein damage seems associated with senile plaques and paired helical filaments; the latter have also been reported to immunostain for nitrotyrosine, consistent with reports of the presence of iNOS in tangle-bearing neurones. The β-peptides may additionally interfere with the action of the proteasome complex,[95] perhaps contributing to the accumulation of ubiquitinated paired helical filaments in neurones. The iron in senile plaques (Box 9.2) may be capable of facilitating oxidative damage. An increase in lactoferrin activity in the AD brain has been reported,[140] although iron bound to lactoferrin is generally not redox-active (Chapter 3).

Plaques could also cause damage by provoking inflammation[198] with resultant ROS/RNS and cytokine production, consistent with the presence of microglial cells and various complement components in plaques. Activated microglia may secrete proteases that target amyloid deposits, whereas β-peptides may help to recruit microglia and provoke them to secrete cytokines such as IL-1. Cytokines might increase APP expression and APP may itself be processed to β-peptides by microglial proteases. The receptor for AGE products (RAGE)[341] is found on microglia and neurones, and might contribute to activation of the former and damage to the latter. AGE products have been found in both plaques and tangles. Glycoxidation of tau in paired helical filaments might also contribute to neuronal damage. Glia express macrophage scavenger receptors and might thus be able to take up oxidized lipids and proteins. Both iNOS and COX-2 could be induced in activated microglia.

Of course, the presence of oxidative damage does not prove it to be important. However, high doses of vitamin E (2000 units per day) were reported to produce a significant delay in the progression of AD.[253] Given that vitamin E enters the brain only slowly (Section 9.15.6 above) and does not appear to be depleted in AD, this gives hope that antioxidants designed to target the brain might have a significant therapeutic impact. In transgenic mice over-expressing amyloid-β precursor protein, deposition of β-amyloid was paralleled by rises in HNE, pentosidine, iron and HO-1.[275a]

Other changes reported in AD include decreases in α-ketoglutarate dehydrogenase activity,[194] a key enzyme of the Krebs cycle (Chapter 1). Such decreases are also seen in other neurodegenerative diseases (e.g. PD) and their significance is unknown, although interference with ATP synthesis is clearly possible. Decreases in mitochondrial cytochrome oxidase activity in AD have also been reported, perhaps related to mutations in the mitochondrial DNA genes that encode some of the cytochrome oxidase subunits.[56]

9.18.5 *Aluminium in Alzheimer's disease*[85,256]

Interest in the role of aluminium in AD arose from reports that the cores of senile plaques are enriched in aluminium and silicon (Section 8.12.3). Guam Indians with dementia–PD were also claimed to have increased deposition of calcium, silicon and aluminium in neurofibrillary tangle-bearing neurones.

Is excessive aluminium accumulation the *cause* of Alzheimer's disease? We do not think so. Although aluminium can induce the formation of intra-neuronal tangles when injected into cats, ferrets and rabbits, the fibrils produced are not identical to those of AD, although they may contain tau. No marked increase in the incidence of AD has been reported amongst workers in aluminium mines or smelting plants. Dialysis encephalopathy patients do not show neurofibrillary tangles, but only have the type of intra-neuronal tangles found in experimental animals after injection of aluminium around the brain. It may be that AD leads to an impaired blood–brain barrier that allows increased amounts of aluminium to reach the central nervous system. Since the neurotoxicity of aluminium is well-established from studies upon dialysis patients, it follows that aluminium accumulation, even if not causative of AD, could worsen the neurological damage being done, e.g. it could facilitate aggregation of the β-peptide (Box 9.2). The zinc and iron ions reported in plaques can also facilitate β-peptide aggregation *in vitro*, however, and one must be wary of focusing too much on aluminium.

The iron-binding agent desferrioxamine chelates Al^{3+}, and has been used successfully in the treatment of dialysis encephalopathy and also in a case of severe encephalopathy related to the presence of a piece of aluminium embedded in the spinal column of the patient.[122] However, its stability constant for Al^{3+} (about 10^{25}) is several orders of magnitude smaller than that for Fe(III) (about 10^{31}), and so prolonged treatment with desferrioxamine may lead to anaemia and other side-effects. Fortunately, chelating agents more specific for Al^{3+} are being developed. Aluminium ions can aggravate iron ion-induced lipid peroxidation (Chapter 8) and it is possible that this could contribute to elevated lipid peroxidation in AD, since iron ions have been detected in plaques.

9.19 Amyotrophic lateral sclerosis[309]

In PD, neurones of the substantia nigra are affected, whereas in AD there is marked neuronal cell death in cortex, hippocampus and associated areas. By contrast, in amyotrophic lateral sclerosis (ALS), motor neurones are the major targets affected, dying in the motor cortex, brainstem and spinal cord. The mean age of onset of ALS is 57 years; it begins as painless muscle weakness and impaired muscle tone, leading to muscle atrophy. Problems with speech and swallowing food and saliva follow. The disease is chronic and progressive, often leading to death within a few years of its first appearance, usually of respiratory failure secondary to muscle paralysis. Males are affected twice as commonly as females and the average incidence in the USA is about 1–2 per 100 000. Ninety per cent or more of ALS cases are sporadic, $\leqslant 10\%$ are inherited. The latter (**familial**) forms of ALS have an earlier mean age of onset and men and women seem equally at risk (unlike sporadic ALS). Motor neurone dysfunction is accompanied by neurofilament abnormalities and ubiquitin-positive inclusion bodies, and neuronal death follows.

Synonyms for ALS include **Lou Gehrig's disease** (after the New York Yankees baseball player who ended his distinguished career as a result of developing ALS), **Charcot's disease** (after the French neurologist who first reported it) and, by common usage, **motor neurone disease**.

One clue to the origin of ALS came from studies of the Chamorro Indians on Guam and Rota (Chapter 8). The parkinsonism–dementia associated with consumption of *Cycas circinalis* often has some clinical features of sporadic ALS, suggesting that a neurotoxin, perhaps acting as an excitotoxin, may be implicated. Indeed, the disease **lathyrism**,[22] which typically manifests itself as an irreversible spastic paralysis of the legs, has been observed in many third-world countries (especially India and Ethiopia) and is related to consumption of seeds of the legume *Lathyrus sativus*, the chickling pea. *L. sativus* seeds contain a neurotoxin, **β-N-oxalyl-α,β-diaminopropionic acid** (**β-N-oxalylamino-L-alanine**) (Fig. 9.31). Prolonged consumption of *L. sativus*, especially when the diet is poor in other nutrients, leads to death of motor neurones, again apparently by an excitotoxic mechanism.

9.19.1 *ALS and superoxide dismutase*[154,309]

There are several different mutations associated with familial ALS, but about 20% of cases of familial ALS (i.e. 2% of total ALS cases) have a defect on chromosome 21, in the gene encoding CuZnSOD. Multiple mutations in this gene have been identified in different families (Table 9.16). The Ala4 → Val mutation is the most common, in about 50% of cases, in the USA. The altered

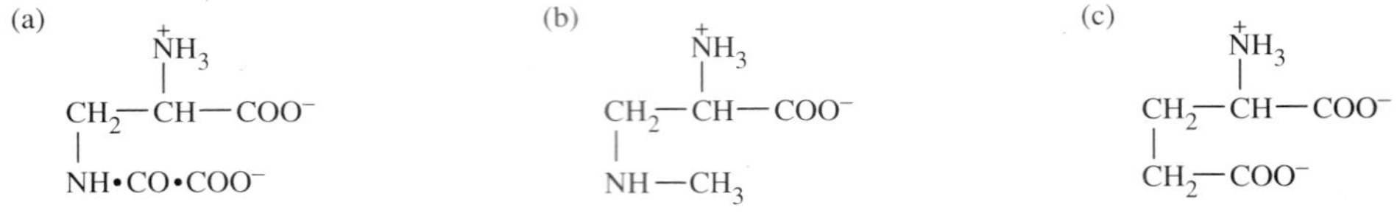

(a)

$$\overset{+}{N}H_3$$
$$|$$
$$CH_2-CH-COO^-$$
$$|$$
$$NH\cdot CO\cdot COO^-$$

(b)

$$\overset{+}{N}H_3$$
$$|$$
$$CH_2-CH-COO^-$$
$$|$$
$$NH-CH_3$$

(c)

$$\overset{+}{N}H_3$$
$$|$$
$$CH_2-CH-COO^-$$
$$|$$
$$CH_2-COO^-$$

Fig. 9.31. Structures of some plant neurotoxins. (A) β-*N*-Oxalylamino-L-alanine; (B) β-*N*-methylamino-L-alanine. Compound A is associated with lathyrism in several parts of the world; compound B with the increased incidence of ALS-PD-dementia syndromes in the Chamorro Indians (Chapter 8). (C) The structure of the excitatory amino acid glutamate, for comparison.

amino acids in the mutant CuZnSOD proteins are not located at the active site but at amino-acid residues that appear to play a role in enzyme stability, dimer interaction or affecting access to the active site. Most mutants have decreased enzyme activity (although at least one does not), but rarely does it fall below 40% of normal. For example, a Gly → Ala mutation at position 93 decreases SOD activity by about 64%.

Transgenic mice overexpressing certain mutant CuZnSODs (e.g. Gly93 → Ala or Gly85 → Arg) develop a degenerative disease of motor neurones, whose pathology and mode of progression resemble human ALS. By contrast, overexpression of normal human CuZnSOD in mice results in some subtle pathological changes (including limited neuromuscular impairment) but also increased resistance to several neurotoxins (Chapter 3).

Since transgenic mice expressing mutant CuZnSODs also express their own CuZnSOD, the ALS syndrome they develop cannot be due to a lack of $O_2^{\bullet-}$ dismutating activity. The greater the expression of the mutant transgene, the more quickly the mice develop disease. It thus seems that the mutant CuZnSODs have acquired a toxic property, a concept further supported by studies upon neurones in culture.[240] Raising the SOD level of neurones by transfecting the gene for normal CuZnSOD can delay the onset of apoptotic death induced by withdrawal of growth factors from the culture medium. However, transfection of the Gly → Ala93 mutant gene shortens the lifespan of neurones below that of controls upon growth factor withdrawal.

9.19.2 *Mechanisms of SOD toxicity*

The mechanism by which mutant SODs become cytotoxic is uncertain. More SOD might mean more H_2O_2 generation, but this would also be true in cells transfected with normal SOD. In general, the toxic mutant CuZnSODs appear less stable *in vitro* and might conceivably denature more quickly than normal *in vivo* to release zinc ions and potentially pro-oxidant copper ions.[254,327] They also appear more susceptible to degradation in the presence of H_2O_2, a process involving $OH^{\bullet}$ generation (Chapter 3). The metals at the active sites of the mutant CuZnSODs are more accessible than in normal CuZnSOD. This allows the mutant enzymes to catalyse H_2O_2-dependent peroxidase-type

Table 9.16. Some of the mutations in SOD associated with familial ALS

Exon[a]	Codon (site)	Base pair change	Substitution
1	4	GCC→GTC	Ala→Val
	4	GCC→ACC	Ala→Thr
	6	TGC→TTT	Cys→Phe
	14	GTG→ATG	Val→Met
	21	GAG→AAG	Glu→Lys
2	37	GGA→AGA	Gly→Arg
	38	CTG→GTG	Leu→Val
	41	GGC→AGC	Gly→Ser
	41	GGC→GAC	Gly→Asp
	43	CAT→CGT	His→Arg
	46	CAT→CGT	His→Arg
	48	CAT→CAG	His→Gln
4	84	TTG→GTG	Leu→Val
	85	GGC→CGC	Gly→Arg
	90	GAC→GCC	Asp→Ala
	93	GGT→GAT	Gly→Asp
	93	GGT→GCT	Gly→Ala
	93	GGT→TGT	Gly→Cys
	93	GGT→CGT	Gly→Arg
	93	GGT→GTT	Gly→Val
	100	GAA→GGA	Glu→Gly
	101	GAT→AAT	Asp→Asn
	101	GAT→GGT	Asp→Gly
	106	CTC→GTC	Leu→Val
	112	ATC→ACC	Ile→Thr
	113	ATT→ACT	Ile→Thr
	115	GCG→GGC	Arg→Gly
(4)[a]	(−10)[b]	T→G	+Phe−Leu−Gln
5	124	GAT→GTT	Asp→Val
	125	GAC→CAC	Asp→His
	126	TTG→−G	131 Stop
	133	GAA→—	Glu→—
	139	AAC→AAA	Asn→Lys
	144	TTG→TCG	Leu→Ser
	144	TTG→TTC	Leu→Phe
	145	GCT→ACT	Ala→Thr
	148	GTA→GGA	Val→Gly
	148	GTA→ATA	Val→Ile
	149	ATT→ACT	Ile→Thr

The CuZnSOD gene in humans has five exons.
[a]Numbers in parentheses designate introns.
[b]The base pair change occurs upstream of exon 5 and results in an alternatively spliced mRNA, generating a protein with three amino acids inserted between exons 4 and 5.
Adapted from *J. Neurol. Sci.* **139** (Suppl.), 10 (1996) by courtesy of Drs M.E. Cudkowicz and R.H. Brown Jr and Elsevier.

reactions and to form nitrating species from peroxynitrite at an increased rate.[21] SOD mutants may bind to neurofilaments, directing oxidative damage and nitration to neurofilament proteins. Another question is why motor neurones are affected. One obvious point is that these large neurones contain a lot of CuZnSOD, simply because of their size. It must also be remembered that large motor neurones may have axons more than 1 m in length. If transport of, for example, CuZnSOD from the cell body to the end of the axon is slow (Section 9.15 above), it is possible that denaturation, or association with neurofilaments, occurs during its travels. Slow transport through long motor neurones may take 1–2 years to reach the nerve terminus. Resultant oxidative and nitrative damage to the cytoskeleton might be less relevant to neurones with shorter axons. Neurofilament aggregation is a marked feature of ALS and presumably impairs delivery of essential molecules and organelles down the axon.

The various SOD mutations (Table 9.14) seem to be associated with different rates of disease progression in familial ALS. One set of mutations (e.g. Ala4 → Val) seems to be associated with rapidly progressive disease, another set (e.g. Gly93 → Ala) with average disease duration (2–5 years) and yet another set (e.g. His46 → Arg) with 'benign' progression. It will be interesting to compare the cytotoxicity of the proteins encoded by these various mutants. There appears to be no increased incidence of CuZnSOD mutations on Guam or Rota.

At least some of the altered genes in other familial ALS cases may relate to mutated neurofilament proteins.

9.19.3   *Oxidative damage in ALS*[309]

There is considerable evidence that oxidative damage occurs in sporadic ALS, although no agreement has been reached on its importance to the disease pathology. No changes in CuZnSOD, MnSOD or catalase activities have been reported in sporadic ALS, or alterations of mitochondrial function. Protein carbonyls are elevated in ALS spinal cord. Nitrotyrosine levels are elevated in lumbar spinal cord, as detected by immunostaining and HPLC analysis, and 8-OHdG levels are also raised. Motor neurones do not normally contain much NOS, but $NO^\bullet$ could originate from microglia recruited to sites of neuronal injury. Increases in the levels of iron and zinc in the spinal cord have been claimed in sporadic ALS.

The transgenic mouse model (overexpressing mutant CuZnSOD) has been used to examine potential therapies for ALS. Administration of coenzyme Q_{10} has been reported to extend lifespan slightly, and administration of vitamin E delayed onset of the disease to a small extent (but did not prolong lifespan). The drug **riluzole**, which inhibits glutamate release at presynaptic terminals, also extends lifespan slightly in these mice. These data suggest that both excitotoxicity and oxidative damage (which could, of course, be inter-related) make significant contributions to ALS pathology. Indeed, riluzole has been marketed for ALS treatment and extends survival slightly. However, as yet no convincing explanation of how excitotoxicity or oxidative damage could produce the selectivity of motor neurone cell death in both familial and sporadic ALS has been advanced. Over-expression of *bcl-2* prolongs lifespan in transgenic mice, suggesting a role for apoptosis.[152]

9.20   Other neurodegenerative diseases

9.20.1   *Down's syndrome*

Oxidative stress has been suggested to be involved in several other neuro-degenerative disorders, including **trisomy 21** or Down's syndrome (Chapter 3). Pathological changes similar to AD (senile plaques containing amyloid β-peptides, neurofibrillary tangles) are seen in Down's syndrome patients aged over 30 and it has been repeatedly postulated that an imbalance between CuZnSOD activity and H_2O_2-removing enzymes is involved in the pathology of this disorder. Of course, the gene encoding APP is also on chromosome 21 and is overexpressed in Down's syndrome.

It must be re-emphasized that the presence of oxidative damage does not necessarily mean that it is important in the disease pathology. For example, several papers in the 1980s reported elevated lipid peroxidation in degenerating muscles from patients with muscular dystrophy, or in animal models of this disease.[57] However, therapeutic administration of vitamin E or allopurinol had no clinical benefits. While it could be argued that these were the wrong antioxidants to test, it is also possible that the increased lipid peroxidation is secondary to muscle degeneration. Muscular dystrophy is caused by defects in the gene encoding the muscle protein **dystrophin**.

9.20.2   *Multiple sclerosis*

Multiple sclerosis (MS) is a disease characterized by impaired nerve conduction due to demyelination. Myelin electrically insulates segments of nerve axons, so permitting rapid transmission of action potentials. In MS, the myelin is destroyed, apparently by an inflammatory reaction involving attack by lymphocytes and macrophages. Attack by the latter involves both glial cells and macrophages recruited (initially as monocytes) from the circulation by the up-regulation of adhesion molecules on blood–brain barrier endothelial cells (which do not normally allow passage of circulating neutrophils or monocytes). MS is believed to be an autoimmune disease; suggested autoantigens include $\alpha\beta$-crystallin (a small heat-shock protein found in oligodendrocytes and astrocytes in MS lesions), the enzyme transaldolase (part of the pentose phosphate pathway) and certain myelin proteins, including **myelin basic protein**. Several myelin proteins can cause **allergic encephalomyelitis** (inflammation and demyelination of the central nervous system) when injected into animals.[125]

There have been repeated suggestions that oxidative damage contributes to MS pathology: its occurrence is likely because there is an over-production of TNFα and because activated macrophages will generate ROS and possibly

NO$^\bullet$ (induction of iNOS has been claimed in the MS brain).[11] Over-expression of TNFα by neurones or astrocytes in transgenic mice can produce an inflammatory disease with some resemblance to MS. Oligodendrocytes (the myelin-producing cells) may be more susceptible to damage by excess NO$^\bullet$ and ROS than other glia. Indeed, positive immunostaining of the MS brain for nitrotyrosine has been reported.[125] Sites of demyelination show increased iron accumulation and several studies have claimed that antioxidants alleviate some of the damage caused in animal allergic encephalomyelitis models, although results vary between laboratories. Further research is needed to evaluate the importance of oxidative stress as a tissue injury mechanism in human MS.

9.20.3   *Neuronal ceroid lipofuscinoses*[297]

The term 'neuronal ceroid lipofuscinosis' (NCL) refers to a series of recessively inherited disorders that occur worldwide in about 1 in 100 000 live births, but more frequently in Scandinavia. Some patients develop symptoms during infancy (**infantile NCL**), others in early childhood (**late infantile NCL**), others in late childhood (**juvenile NCL**) and yet others after adolescence (**adult NCL**). Juvenile NCL is sometimes known as **Batten's disease** and the adult form as **Kuf's disease**. The gene responsible for infantile disease is on chromosome 1, whereas that for the juvenile disease is on chromosome 16, suggesting genetically-different disorders. This is consistent with the accumulation of different proteins (see below).

The onset of NCL is marked by behavioural abnormalities which worsen to include disturbances of vision and speech, and muscular and mental deterioration, associated with seizures. In the terminal stages the brain is severely damaged and patients assume a contracted position. A similar disease has been observed in dogs, cattle and sheep. Attention was focused on NCL by those researching free radicals because the disease is accompanied by rapid accumulation of fluorescent pigments, resembling the 'age pigment' ceroid (Section 10.3), not only in the brain but also in other tissues. This autofluorescent pigment accumulates in 'storage bodies', which appear to be derived from lysosomes.

In the red setter dog, the genetic mode of transmission is autosomal recessive and provides a model for human juvenile NCL. Pigment formation in the dog's tissues begins at birth, if not *in utero*. Maximal accumulation in the brain is reached after 1 year of age, although at this time there is no clinical evidence of neuronal dysfunction. As pigment formation continues, accumulation is offset by an increasing rate of nerve-cell death (apparently, at least in part, by apoptosis).

NCL and lipid peroxidation

It was originally assumed that the 'age pigment' is largely peroxidized lipid. However, there is no clear evidence of increased oxidative damage or defects in antioxidant protection in NCL. Indeed, later analysis showed that in adult, juvenile and late infantile NCL a major fraction of the total lysosomal storage bodies is a specific protein, one of the protein subunits (subunit 9) of mitochondrial ATP synthetase. There are at least two different genes that code for subunit 9, but neither appears to be mutated in NCL, so the reason for the accumulation is unclear. Instead, the Batten's disease gene encodes a protein, predicted to be 483 amino acids long, that has been called **CLN3**.

Subunit 9 is a very hydrophobic protein, and its accumulation appears to result from an inability to degrade it at normal rates. In normal cells cytosolic proteins and organelles can be subjected to lysosomal enzyme degradation by first becoming enclosed in an **autophagic vacuole** that then fuses with lysosomes, but little is understood about the details of the process. Hydrolysates of subunit 9 reveal large amounts of ε-N-trimethyl-lysine (TML). The normal subunit 9 protein from mitochondria, however, does not appear to be methylated, suggesting that methylation plays a central role in the lysosomal storage of the protein. The pigment may result from defective degradation of this protein within lysosomes, although the molecular basis of this defect is uncertain.

In infantile NCL, different proteins (**saposins**) accumulate in the storage bodies. The defect has been identified as being in the gene encoding **palmitoyl protein thioesterase**, a lysosomal enzyme involved in the catabolism of lipid-modified proteins, e.g. it removes palmitate esterified to $-SH$ groups on proteins. Again, the link to saposins is unclear.

Thus any involvement of oxidative damage in NCL must, at best, be indirect. There have been several reports that patients with NCL respond to treatment with 'antioxidants', but double-blind controlled clinical trials have not been reported.

9.20.4   *Huntington's disease*[19]

Huntington's disease (HD; sometimes called **Huntington's chorea**) is an autosomal dominantly inherited disease characterized by psychiatric disorders, dementia and involuntary twitchings, writhings and other movements. It does not usually become apparent before the age of 30 and can last for 20 years or more, with progressively worsening symptoms. The prevalence in the USA and Europe has been estimated as 4–10 per 100 000. Death of neurones is especially marked in the striatum.

The disease is inherited as a defect in a gene on chromosome 4, which encodes a predicted protein product of molecular mass about 330 000 that is referred to as **huntingtin**. Gene sequencing revealed that, within the normal gene, the trinucleotide CAG occurs in sequences repeated 9–39 (mean 19) times. In HD the repeat frequency is much greater (range 36–121, mean 43). Repeat lengths of $\geqslant 40$ make it likely that a subject with even one copy of the abnormal gene will develop HD, whereas repeat lengths of $\leqslant 30$ make it unlikely. Repeat lengths vary from generation to generation, and overall tend to increase so that the disease tends to become more severe in each succeeding generation (**genetic anticipation**). Very high (>50) repeat lengths appear to lead to early disease onset (sometimes before the age of 30) in many, but not all, cases.

The gene encoding huntingtin is expressed in many mammalian tissues, especially neurones, testis, lung and ovary. The CAG repeat would be translated as a string of glutamine residues starting at residue 18: this and other features suggest that the protein may be a transcription factor. The glutamine repeat is followed by a string of proline residues. In another disease, **spinal-bulbar muscular atrophy**, an excessive number of CAG repeats occurs within the coding region of the androgen–receptor gene. Affected males have a syndrome resembling motor neurone disease as well as impaired response to androgen, yet point mutations in this gene do not give neurological symptoms. It seems that in both this disease and HD the mutant protein has gained a toxic function (somewhat analogous to mutated CuZnSOD in ALS) except that here the toxicity is related to glutamine repeats. Increased numbers of CAG repeats have also been identified in at least eight other neurodegenerative diseases, including **spinocerebellar ataxia type 1**.

What has all this to do with free radicals? In 1976, it was observed that injection of kainic acid into the brains of rats causes striatal damage reminiscent of that seen in HD. The hypothesis was therefore advanced that excitotoxicity contributes to HD, which would implicate ROS/RNS (Section 9.15.4 above).[219] Another excitotoxin, **quinolinic acid**, which binds to NMDA-type glutamate receptors, is normally present at nanomolar concentrations in the brain. It is a product of tryptophan metabolism, and the concentration of the enzyme synthesizing it (**3-hydroxyanthranilate oxygenase**) has been reported o be raised in HD. However, no increase in the concentration of quinolinic acid itself has been reported. Levels of 8-OHdG have been reported to be increased in the regions of the brain affected in HD, but little other information is available on parameters of oxidative damage. Thus the evidence for an important role of excitotoxicity in HD is incomplete, and that for a role of oxidative damage even more so.

Concentrations of **3-hydroxykynurenine**,[218] another tryptophan metabolite, have been reported as increased in the HD brain; *in vitro* this compound can undergo oxidation and is toxic to neurones, apparently by generating H_2O_2. Like most autoxidations, this probably depends on the availability of transition–metal ions.

Yet another clue came from the identification of HAP-1 (**huntingtin-associated protein 1**), first reported as a cellular protein that binds to the mutant form of huntingtin. The extent of binding increases as the length of the polyglutamine sequence is increased. HAP-1 seems to be specific to brain (especially in neurones expressing nNOS), unlike huntingtin itself. The glycolytic enzyme glyceraldehyde–3–phosphate dehydrogenase also binds to mutant huntingtin, suggesting that mutant proteins may interfere with neuronal energy metabolism.

9.20.5   *Friedreich's ataxia*[10]

Ataxia is impaired movement due to loss of motor coordination. **Friedreich's ataxia** (FA) is the most common hereditary ataxia, with an incidence of about

1 in 50 000. The disease primarily affects neurones with very long axons, which appear to die back from the periphery. Some patients also suffer cardiomyopathy.

Many years ago it was suggested that FA patients benefit from vitamin E supplementation, but this was not confirmed and no decreased levels of vitamin E have been reported in FA. One explanation could be the recent discovery that an inborn defect in the gene on chromosome 8 that encodes liver α-tocopherol transfer protein (Section 3.22) produces a clinically similar ataxia that *does* respond to vitamin E. The 'real' FA gene is on chromosome 9 and encodes a protein (**frataxin**) involved in mitochondrial iron metabolism. Mutations may prevent the formation of Fe/S clusters in mitochondrial aconitase and complex I. The gene defect consists of an expansion of a trinucleotide (GAA) repeat sequence. Unlike HD, the expanded repeat is outside the protein-coding domain of the gene and could conceivably affect gene expression. Also unlike HD, FA is recessively inherited.

9.20.6   *Tardive dyskinesia*

Tardive dyskinesia is a disorder characterized by abnormal involuntary movements of muscles in the face and limbs. Its cause is unknown, although long-term treatment of patients with neuroleptic drugs such as phenothiazines (e.g. for treatment of schizophrenia)[186] is a risk factor. It has been proposed that drug-related oxidative stress could be involved in its development and that vitamin E administration is beneficial to patients, but insufficient data have been published to reach clear-cut conclusions.

9.20.7   *Prion diseases*[13]

Scrapie, a neurodegenerative disease of sheep, is widespread in the UK. It was described some 250 years ago as a disease presenting with excitability, ataxia, itching and eventually paralysis and death. In the 1950s, Australian administrators exploring new territories in Papua New Guinea discovered a previously-unknown tribe (the **Foré**) in which death from the disease **kuru** was often observed, especially in women and children. Kuru and scrapie lesions in the brain look remarkably similar, and both were soon shown to involve an agent that could be transmitted to monkeys. The high kuru incidence was attributed to cannibalism; women and children eating the brains of infected subjects. The men were more likely to eat the flesh, thus developing kuru less often. In the 1920s Creutzfeldt and Jakob described the human neurodegenerative disease named after them as **Creutzfeldt–Jakob disease** (CJD), which usually occurs in late middle age.

Few readers will be unaware of the possible risk to health posed by the epidemic of **bovine spongiform encephalopathy** (BSE) in cattle in the UK, especially as eating tissue from cattle infected by this 'mad cow disease' has been suggested to be responsible for the occurrence of a small number of atypical cases of CJD, in young people. More than 75 cases of prion disease had previously occurred in young people given growth hormone prepared from human pituitary glands obtained at autopsy; this practice has now been abandoned since some of the glands must have come from cadavers with unrecognized prion disease. Some cases have occurred in patients who received grafts of meningeal tissue, derived from donors with unsuspected prion disease, during neurosurgery. The atypical, putatively 'BSE-related', CJD cases have clinical features closer to those of the growth hormone cases, and to kuru, and less similar to sporadic CJD.

Scrapie, BSE, CJD and kuru are all thought to be caused by **prions** (proteinaceous infectious particles). The prion protein gene is present in all mammals and encodes a glycoprotein present in neurones and glia. It may be involved in copper metabolism: PrP^c-knockout mice show less copper and decreased CuZnSOD in the brain[28a]. The problem arises when the normal protein (**PrP^c**) becomes converted into an abnormal form (**PrP^{Sc}**), which differs from the normal protein not in amino acid sequence but only in conformation. In most cases of CJD, this appears to be a rare random event. The conformational change involves creation of a β-sheet structure, typical of amyloidogenic proteins (Box 9.2). The abnormal protein cannot be degraded by proteolysis, and accumulates in tissues, apparently leading to dysfunction and death, and can sometimes precipitate to give PrP-amyloid plaques. In brain, spaces due to neuronal death create a sponge-like appearance, hence the name 'spongiform encephalopathies' often given to these diseases. At autopsy, the brains from the putative BSE-related CJD cases reveal spongiform encephalopathy and PrP plaque deposition.

When PrP^{Sc} gains access to a normal tissue, it alters the conformation of PrP^c to generate more PrP^{Sc}. Indeed, transgenic 'knockout' mice lacking PrP^c appear initially normal, but cannot be infected with scrapie, unlike normal mice. The transgenic mice show disordered sleep patterns, however, and may have abnormalities in synaptic transmission. They develop ataxia at about 70 weeks of age due to death of cells in the cerebellum. Several mutations in the PrP^c gene that greatly favour PrP^{Sc} formation, and cause autosomal dominant human prion disease, have been described. Fortunately, they are extremely rare, collectively accounting for no more than about 15% of cases of human prion disease.

A 21-residue fragment of PrP^c (residues 106–126) has been shown to be toxic to neurones in culture, although the relevance of this to the disease is as yet unclear. As with β-amyloid in AD, this toxicity may include formation of membrane ion channels, oxidative stress, activation of microglia, induction of apoptosis and rises in mRNA for haem oxygenase-1.[244]

9.20.8   *Does emotional stress lead to oxidative stress?*

Stress, especially psychological stress, is becoming increasingly common in the 'advanced' world and is known to predispose to several diseases, including cardiovascular disease, gastric ulcers[54] and infections, since stress can diminish the effectiveness of the immune system in combating foreign organisms. It has long been speculated that ROS are involved in stress, but only recently has this begun to be systematically evaluated. For example, the stress induced in rats by immobilizing them leads to falls in GSH and increases in protein carbonyls, lipid peroxidation and 8-OHdG in various regions of the brain and liver.[169] We have much to learn in this area and more research is needed.

9.21 Oxidative stress and viral infections

Viral infections start with the binding of virus to receptors on the plasma membrane of the target cell, followed by entry and uncoating of viral genetic material (DNA or RNA) within the cell. The virus uses the protein–synthetic machinery of the host cell to replicate its own proteins. Sometimes the host cell survives, in a chronically infected state, shedding virus periodically. Sometimes it is killed; cell lysis is an important mechanism of virus-induced damage. In either case, new virus particles are released from the cell to infect adjacent cells.

9.21.1 *Origin of oxidative stress*

Much attention has recently focused on the possibility that oxidative stress plays a role in the pathology consequent upon infection by the **human immuno-deficiency viruses** (HIV-1 or –2). However, any viral infections, like infection with bacteria, mycoplasmas, fungi or trypanosomes, can lead to oxidative stress. The phagocyte response of the host, aided by increased production of such cytokines as TNFα, can generate large amounts of ROS and RNS in an (often successful) attempt to eliminate the unwanted invader.[30] For example, NO• inhibits replication of many viruses in cells in culture. Mobilization of the host defence system involves activation of NF-κB, in lymphocytes and other cell types, but this can often be subverted by many viruses to their own end, since viral gene promoters often contain NF-κB-binding sequences that promote transcription.[331]

With some infections, as with chronic inflammations (Section 9.7 above), an over-exuberant host response can end up causing more damage to the host than to the infecting agent. For example, chronic hepatitis infection can lead to liver cancer in humans (Section 9.13 above). Infection of laboratory mice with a strain of influenza virus produces significant mortality. Oxidative stress contributes, since mortality can be decreased by administering SOD (linked to a polymer to prolong its lifetime in the circulation) or allopurinol to the animals.[185] Mortality is due to lung damage, and levels of xanthine oxidase, NO•, nitrotyrosine and activated macrophages are increased in the lungs of infected mice.

Several viral infections of the nervous system are associated with increased production of NO•, to an extent that might contribute to cell injury; examples include rabies and borna viruses in rats. Infection of endothelial cells with cytomegalovirus, hepatitis B virus and certain herpesviruses can lead to NF-κB activation and up-regulation of adhesion molecules that can promote phagocyte binding.[39,209] Indeed, viral infections of vascular endothelium may contribute to initiation of atherosclerosis (Section 9.2 above). Selenium deficiency in mice allows a normally fairly-benign coxsackievirus to cause significant heart damage (Chapter 3). **Canine distemper virus** causes demyelination in the nervous system of dogs, and it has been suggested that ROS are involved.

9.21.2   *HIV infection*[78]

The two known types of HIV virus (HIV-1 and -2) belong to a family of primate viruses whose other members infect monkeys. HIV-1 is more pathogenic than HIV-2. Cats can be infected by a **feline immunodeficiency virus** that eventually leads to a disease similar to human AIDS.

HIV virus particles contain a **capsid** enclosing the single-stranded RNA genome plus three viral enzymes: protease, reverse transcriptase and integrase. Reverse transcriptase copies viral RNA into DNA, and integrase incorporates the DNA into the host genome. The capsid is surrounded by a matrix protein which in turn is surrounded by an envelope consisting of a lipid bilayer membrane that contains proteins **gp120** and **gp41** (Fig. 9.32).

Initial infection with HIV-1 results from intimate exposure to the blood or other body fluids of an infected person. The first consequence is usually a mild,

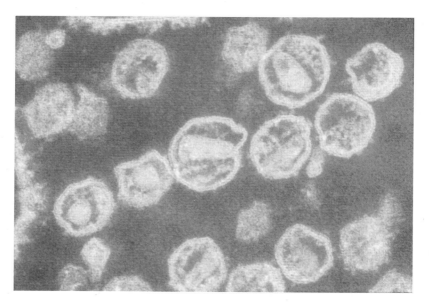

Fig. 9.32. Electron micrograph of sectioned human immunodeficiency virus (HIV). Each virus particle has a lipid bilayer envelope, derived from the plasma membrane of the host cell, and containing virus-specified glycoproteins. They consist of two non-covalently linked proteins, **gp120** (outer envelope protein) and **gp41** (transmembrane protein). Enclosed within the envelope is an elongated capsid which contains the viral RNA, nucleocapsid proteins and the viral enzymes protease, reverse transcriptase and integrase. Magnification: ×270 000. Photograph by courtesy of Dr David Hockley. The lipid bilayer envelope of the virus makes it susceptible to lipid-disrupting agents, including organic solvents, detergents, ozone and agents that stimulate lipid peroxidation.

short-term illness associated with a period of rapid viral replication, resulting in a high circulating viral load. HIV mostly infects CD4$^+$ lymphocytes, i.e. T-helper lymphocytes expressing the cell-surface antigen **CD4**. CD4$^+$ lymphocytes are essential to the functioning of the immune system. CD4$^+$ glycoprotein is recognized by the HIV-1 envelope protein complex (via gp120) to allow the virus to bind to host cells. Other receptors are also needed for virus to enter; normally these receptors bind **chemokines** such as MIP-1α, MIP-1β and RANTES (Table 4.12). Hence high levels of these chemokines can deter HIV infections at the cellular level.

After a brief illness, the viral load rapidly falls and the patient becomes asymptomatic. This low apparent viral load is the result of a dynamic equilibrium between production of new virus and destruction of infected CD4$^+$ T-lymphocytes. As many as 2×10^9 of these cells, about 5% of the total body pool of CD4$^+$ cells, are destroyed and replaced per day. An HIV-infected cell produces new virus for only about 2 days before being destroyed. Over many years, CD4$^+$ cell numbers tend to fall as the ability of the immune system to replace them fails. When they fall below the minimum level needed for proper immune system function, patients begin to develop a series of 'opportunistic' infections with organisms normally easily eliminated (e.g. *Pneumocystis carinii* pneumonia), and full blown **acquired immunodeficiency syndrome** (AIDS) appears. In general, the lower the steady-state viral load after the initial post-infection surge, the longer it takes a patient to develop AIDS. Progression to AIDS occurs on average about 10–11 years after infection in Western countries.

HIV is a **retrovirus**, i.e. it uses a reverse transcriptase^b enzyme to copy its RNA into DNA. The genome of HIV-1 has about 10 000 bases and, unlike DNA polymerase, reverse transcriptase has no error-correcting mechanism. It has been estimated that it makes one mistake per genome copied, i.e. mutations accumulate at an enormous rate. This is probably why the virus rapidly becomes resistant to therapeutic agents; there are an enormous number of variants to permit natural selection of resistant strains. It has been estimated that every possible single point mutation in the HIV-1 genome occurs 10^4–10^5 times per day in an infected individual. Most single antiviral agents do not decrease viral replication to zero and it is only a matter of time before resistance emerges.

Therapeutic agents include inhibitors of reverse transcriptase (e.g. **azidodeoxythymidine** (AZT), **dideoxyinosine** (ddI) and **dideoxycytidine** (ddC)) and inhibitors of the protease enzyme that is used by the virus to cleave newly-synthesized proteins into the proteins needed for viral assembly. Examples of protease inhibitors are **ritonavir, indinavir** and **saquinavir**. Hopefully, greater therapeutic success will be achieved by using triple combinations of agents acting in different ways.

The viral genome contains two genes that encode regulatory proteins, **Tat** and **Rev**. Tat activates gene transcription from the viral promoter and Rev facilitates export of viral mRNA from the cell nucleus. An additional approach to therapy is to develop agents that block the binding of Tat to the RNA, or antagonize the action of Rev. Yet another approach is to develop chemokine 'mimics' that block the co-receptor.

9.21.3   Oxidative stress in HIV disease?

As HIV directly infects cells of the immune system, it provokes a vigorous immune response accompanied by activation of NF-κB and elevated secretion of multiple cytokines, including IL-1β, IL-6 and TNFα, which might lead to oxidative stress.

HIV infection in the brain can lead to neuronal death and glial cell proliferation, resulting in cognitive and motor defects. The most severe form of this, **AIDS dementia complex**, occurs in 20–30% of AIDS patients. Neurological damage by HIV involves viral replication in brain macrophages and glia (but not neurones), leading to increased production of TNFα, IL-1β, other cytokines, NO$^\bullet$, ROS and eicosanoids. The viral coat protein gp120 is toxic to neurones[58] in culture; the toxicity requires external glutamate and Ca^{2+} and is blocked by agents that block glutamate receptors. Hence an excitotoxic mechanism seems important. Levels of brain 3-hydroxykynurenine (Section 9.20.4) have been reported to be elevated in HIV infection.

The first suggestions that oxidative stress might be an important pathological event in HIV infection came from measurements of GSH.[65] Blood plasma samples from HIV-infected patients show decreased levels of GSH, cysteine and cystine. Monocyte and lymphocyte levels of GSH are also subnormal. Levels of GSH in lung lining fluids also appear to fall. Although the size of the reported falls varies between laboratories (with a particular disagreement on whether or not there is a fall in lymphocyte, plasma or lung lining fluid GSH in asymptomatic HIV-positive patients),[14,223] it is possible that decreased thiol levels predispose lymphocytes to injury. In Rhesus macaque monkeys, plasma cysteine levels and intracellular GSH levels also fall 2 weeks after infection with simian immunodeficiency virus.

Different CD4$^+$ lymphocyte subtypes may show variable initial levels of, and depletions in, GSH. Hence analysis of GSH on total cell populations can be somewhat confused by changes in the various subtypes present in HIV-positive patients.[7]

The fall in lymphocyte GSH could be due to the decreased availability of precursors from the plasma. A rise in blood glutamate levels, reported in HIV-positive subjects, may further interfere with uptake since glutamate and cystine compete for one of the amino acid uptake carriers. The mechanism of the fall in plasma GSH is uncertain, but may reflect decreased output of GSH into the plasma from tissues: no evidence for accelerated removal of GSH was found when it was infused into healthy HIV-positive subjects.[117] Low plasma GSH levels are found in several other diseases, including advanced cancer. Low lymphocyte GSH levels seem to be associated with decreased survival in otherwise-indistinguishable HIV-positive subjects, and administration of N-acetylcysteine appeared to give limited clinical benefit in this sub-group.[120]

The falls in GSH and other thiols are most striking in patients with overt AIDS,[65] where malnutrition due both to poor diet and to impaired intestinal absorption of nutrients could contribute.[155] In AIDS patients, plasma concentrations of several vitamins and minerals appear subnormal, including those of zinc, magnesium, selenium, vitamin B_{12}, β-carotene and other carotenoids.

9.21.4   *Redox regulation of viral expression*[209,331]

Many studies on HIV-infected T-cell lines in culture have shown that imposing oxidative stress (e.g. by H_2O_2 or UV light) increases HIV gene expression, whereas antioxidants diminish it. Indeed, the HIV-1 promoter has binding sites for NF-κB and activation of NF-κB enhances viral genome expression. In some cells, expression of the HIV Tat protein decreases GSH and down-regulates MnSOD gene expression. Tat protein contains multiple cysteine residues and is thus susceptible to oxidation; oxidized Tat will not promote viral gene expression.

Thus a limited degree of ROS exposure might encourage HIV production (via NF-κB activation), whereas high levels of oxidative stress might deter it. Similarly, although ROS activate NF-κB, oxidized NF-κB will not bind to DNA. In addition, up-regulation of NF-κB is not all bad; it leads to increased synthesis of several cytokines, including RANTES, a chemokine that may block HIV entry to cells. High levels of oxidative stress might kill the HIV-harbouring cell. Hence the redox-balance is all important. Levels of thioredoxin, and the redox state of this protein, also seem important in regulating the behaviour of virally infected cells.

Effects of antioxidants in inhibiting viral replication in cultured cells have been observed for several other viruses, including herpes simplex type I, cytomegalovirus, murine Sindbis virus, hepatitis B virus and Epstein–Barr virus. Of course, beneficial effects of thiols such as GSH on lymphocyte function need not necessarily be mediated by antioxidant activity: GSH has multiple metabolic functions (Chapter 3) and thiols influence lymphocyte growth, proliferation and response to cytokines. HIV-induced apoptosis in lymphocytes may be *accompanied* by -SH depletion and increased ROS formation, like apoptosis in many other cell types (Chapter 4).

Several plant phenolics decrease HIV expression in cells in culture: this might involve their antioxidant activity but could also involve direct actions on enzymes such as reverse transcriptase. Iron chelators can also exert antiviral effects, at least *in vitro*.[39] For example, if ribonucleotide reductase is deprived of iron, DNA synthesis will be inhibited. Iron chelators might also decrease oxidative stress in infected cells.

9.21.5   *Drug toxicity*[164]

An additional way in which oxidative stress could be involved in HIV infection is by contributing to the side effects produced by drug therapy. For example, AZT treatment is associated with mitochondrial damage in about 20% of

Table 9.17. How important is oxidative stress in human disease?

Condition	Is oxidative stress a primary cause?	If oxidative stress is secondary, does it contribute significantly to disease pathology?	Is there evidence that antioxidants have, or will have therapeutic benefit?
Radiation-induced damage	Yes	[a]	Yes (Chapter 8)
Vitamin E deficiency[b]	Probably	—	Yes (Chapter 3)
Selenium deficiency	Very likely (animals); likely (humans)	—	Yes (Chapter 3)
Atherosclerosis	Sometimes[c]	Yes	Yes
Hypertension	Sometimes	Sometimes	Some
Diabetes	Probably not	Probably (phagocyte attack on islets)	Limited as yet
Rheumatoid arthritis	No[d]	Yes	Limited as yet
Autoimmune diseases	Probably no[d]	Uncertain	Very limited
Inflammatory bowel disease	Probably no[d]	Yes	Limited as yet
ARDS	Generally no[e]	Possibly	Limited as yet
Cystic fibrosis	No	Possibly some	Limited as yet
Cancer	In a few cases, yes[f]	Possibly	Very limited
Traumatic CNS Injury	No	Sometimes	Yes (Chapter 10)
Parkinson's disease	Probably not	Hard to decide as yet	Very limited (selegline)
Alzheimer's disease	No	Possibly	Limited as yet (vitamin E trial)
ALS	No	Hard to decide as yet	Limited (transgenic mice)
Neuronal ceroid lipofuscinoses	No	Possibly a little	Very limited as yet
Multiple sclerosis	No	Possibly	No
HIV infection	No	possibly a little	Limited
Drug side effects	In a few cases	In a few cases	Limited

[a] Secondary ROS/RNS production by inflammation may also contribute.
[b] Humans and other animals.
[c] ROS/RNS can damage vascular endothelium.
[d] Although oxidatively-modified proteins/DNA can be antigenic.
[e] But may contribute to phagocyte recruitment.
[f] Chronic inflammation-related.

patients, in part by inhibiting a mitochondrial DNA polymerase (pol-γ). Rises in mitochondrial 8-OHdG levels in AZT-treated animals, and increased urinary 8-OHdG excretion in patients, have been reported. Excretion of 8-OHdG in AZT-treated patients is decreased by oral supplements of vitamins E plus C (J. Viña, in press). Clinically, the most common side effect of AZT is damage to the bone marrow, producing anaemia and impaired white cell production. Skeletal myopathy, cardiotoxicity and hepatic toxicity can also occur. All these syndromes could be related to mitochondrial dysfunction, a phenomenon which often leads to secondary oxidative damage.

9.22   Conclusion

In reviewing the state of knowledge of the relationship between oxidative stress and human disease, it is important to keep in mind the criteria listed on pages 617 and 619. Table 9.17 is an attempt to assess how far these criteria are met by the various diseases we have considered.

References

1. Adachi, S *et al.* (1994) Increased susceptibility to oxidative DNA damage in regenerating liver. *Carcinogenesis* **15**, 539.
2. Ahnfelt-Ronne, I *et al.* (1990) Clinical evidence supporting the radical scavenger mechanism of 5-aminosalicylic acid. *Gastroenterology* **98**, 1162.
3. Al-Abed, Y *et al.* (1996) Hydroxyalkenal formation induced by advanced glycosylation of LDL. *J. Biol. Chem.* **271**, 2892.
4. Alberts, B *et al.* (1996) *Molecular Biology of the Cell*, 3rd edn. Garland Publishing, New York.
5. Aschner, M (1996) The functional significance of brain metallothionein. *FASEB J.* **10**, 1129.
6. Auclair, C *et al.* (1990) Clastogenic inosine nucleotide as components of the chromosome breakage factor in scleroderma patients. *Arch. Biochem. Biophys.* **278**, 238.
7. Aukrust, P *et al.* (1996) Markedly disturbed glutathione redox status in CD45RA$^+$CD4$^+$ lymphocytes in HIV type 1 infection is associated with selective depletion of this lymphocyte subset. *Blood* **88**, 2626.
8. Babbs, CF (1990) Free radicals and the etiology of colon cancer. *Free Rad. Biol. Med.* **8**, 191.
9. Babbs, CF (1992) Oxygen radicals in ulcerative colitis. *Free Rad. Biol. Med.* **13**, 169.
10. Babcock, M *et al.* (1997) Regulation of mitochondrial iron accumulation by Yfhlp, a putative homolog of frataxin. *Science* **276**, 1709.
11. Bagasra, O *et al.* (1995) Activation of iNOS in the brains of patients with multiple sclerosis. *Proc. Natl Acad. Sci. USA* **92**, 12041.
12. Baik, SC *et al.* (1996) Increased oxidative DNA damage in *Helicobacter pylori*-infected human gastric mucosa. *Cancer. Res.* **56**, 1279.
13. Baker, HF and Ridley, RM (1996) What went wrong in BSE? From prion disease to public disaster. *Brain Res. Bull.* **40**, 237.
14. Barditch-Crovo, P *et al.* (1995) Apparent deficiency of GSH in the PBMCs of people with AIDS depends on method of expression. *J. Acq. Immun. Def. Synd. Human Retrovir.* **8**, 313.

15. Baret, A *et al.* (1984) Pharmacokinetic and anti-inflammatory properties in the rat of superoxide dismutases (CuSODs and MnSOD) from various species. *Biochem. Pharmacol.* **33**, 2755.

16. Barger, SW *et al.* (1995) TNFα and β protect neurons against amyloid β-peptide toxicity: evidence for involvement of a κB-binding factor and attenuation of peroxide and Ca^{2+} accumulation. *Proc. Natl Acad. Sci. USA* **92**, 9328.

17. Bartsch, H (1996) DNA adducts in human carcinogenesis: etiological relevance and structure–activity relationship. *Mutat. Res.* **340**, 67.

18. Bates, EJ *et al.* (1985) Inhibition of proteoglycan synthesis by H_2O_2 in cultured bovine articular cartilage. *Biochim. Biophys. Acta* **838**, 221.

19. Bates, G (1996) Expanded glutamines and neurodegeneration—a gain of insight. *BioEssays* **18**, 175.

20. Beard, JL *et al.* (1993) Iron in the brain. *Nutr. Rev.* **51**, 157.

21. Beckman, JS (1996) Oxidative damage and tyrosine nitration from $ONOO^-$. *Chem. Res. Toxicol.* **9**, 836.

22. Bell, EA and Nunn, PB (1988) Neurological diseases in man—are plants to blame? *Biologist* **35**, 39.

23. Bernier, M and Hearse, DJ (1988) Reperfusion-induced arrhythmias: mechanisms of protection by glucose and mannitol. *Am. J. Physiol.* **254**, H862.

24. Bingham, SA *et al.* (1996) Does increased endogenous formation of *N*-nitroso compounds in the human colon explain the association between red meat and colon cancer? *Carcinogenesis* **17**, 515.

24a. Blake, DR *et al.* (1985) Cerebral and ocular toxicity induced by desferrioxamine. *Q. J. Med.* **219**, 345.

25. Bosch, I and Croop, J (1996) P-Glycoprotein multidrug resistance and cancer. *Biochim. Biophys. Acta* **1288**, F37.

26. Bowman, MA *et al.* (1994) Prevention of diabetes in the NOD mouse: implications for therapeutic intervention in human disease. *Immunol. Today* **15**, 115.

27. Bowry, VW *et al.* (1995) Prevention of tocopherol-mediated peroxidation in ubiquinol-10-free human LDL. *J. Biol. Chem.* **270**, 5756.

28. Brand, K *et al.* (1997) Role of NF-κB in atherogenesis. *Exp. Physiol.* **82**, 297.

28a. Brown, DR *et al.* (1997) The cellular prior protein binds Cu *in vivo*. *Nature* **390**, 684.

29. Bruce, AJ *et al.* (1996) Altered neuronal and microglial responses to excitotoxic and ischemic brain injury in mice lacking TNF receptors. *Nature Med.* **2**, 788.

30. Bürge, T *et al.* (1989) Antiviral antibodies stimulate production of ROS in cultured canine brain cells infected with canine distemper virus. *J. Virol.* **63**, 2790.

31. Butterfield, DA (1997) β-Amyloid-associated free radical oxidative stress and neurotoxicity; implications for AD. *Chem. Res. Toxicol.* **10**, 495.

32. Buzadzic B *et al.* (1990) Antioxidant defenses in the ground squirrel, *Citellus citellus*. 2. The effect of hibernation. *Free Rad. Biol. Med.* **9**, 407.

33. Carmine, T *et al.* (1995) Presence of iron catalytic for free radical reactions in patients undergoing chemotherapy: implications for therapeutic management. *Cancer Lett.* **94**, 219.

34. Cerutti, PA (1994) Oxy-radicals and cancer. *Lancet* **344**, 862.

35. Chapple, ILC (1996) Role of free radicals and antioxidants in the pathogenesis of the inflammatory periodontal diseases. *J. Clin. Pathol. Mol. Pathol.* **49**, M247.

36. Chevion, M *et al.* (1993) Copper and iron are mobilized following myocardial ischemia: possible predictive criteria for tissue injury. *Proc. Natl Acad. Sci. USA* **90**, 1102.

37. Chiao, JJC *et al.* (1994) Iron delocalization occurs during ischemia and persists on reoxygenation of skeletal muscle. *J. Lab. Clin. Med.* **124**, 432.

38. Chu, R *et al.* (1996) Transformation of epithelial cells stably transfected with H_2O_2-generating peroxisomal urate oxidase. *Cancer Res.* **56**, 4846.

39. Cinatl, J *et al.* (1995) Effects of desferrioxamine on human CMV replication and expression of HLA antigens and adhesion molecules in human vascular endothelial cells. *Transplant Immunol.* **3**, 313.

40. Clare, K *et al.* (1995) Toxicity of oxysterols to human monocyte–macrophages. *Atherosclerosis* **118**, 67.

41. Clelland, JD *et al.* (1996) Age dependent changes in the CSF concentration of nitrite and nitrate. *Ann. Clin. Biochem.* **33**, 71.

42. Coghlan, JG *et al.* (1994) Allopurinol pretreatment improves postoperative recovery and reduces lipid peroxidation in patients undergoing coronary artery bypass grafting. *J. Thorac. Cardiovasc. Surg.* **107**, 248.

43. Colton, C *et al.* (1996) Species differences in the generation of ROS by microglia. *Mol. Chem. Neuropathol.* **28**, 15.

44. Conway, JG *et al.* (1987) Role of fatty acylCoA oxidase in the efflux of GSSG from perfused livers of rats treated with the peroxisome proliferator nafenopin. *Cancer Res.* **47**, 4795.

45. Cooke, JP *et al.* (1992) Antiatherogenic effects of L-arginine in the hypercholesterolemic rabbit. *J. Clin. Invest.* **90**, 1168.

46. Cooke, MS *et al.* (1997) Immunogenicity of DNA damaged by ROS-implications for anti-DNA antibodies in lupus. *Free Rad. Biol. Med.* **22**, 151.

47. Corbett, AH and Osheroff, N (1993) When good enzymes go bad: conversion of topoisomerase II to a cellular toxin by antineoplastic drugs. *Chem. Res. Toxicol.* **6**, 585.

48. Corroyer, S *et al.* (1996) Altered regulation of G_1 cyclins in oxidant-induced growth arrest of lung alveolar epithelial cells. *J. Biol. Chem.* **271**, 25117.

49. Coyle, JT and Puttfarcken, P (1993) Oxidative stress, glutamate and neurodegenerative disorders. *Science* **262**, 689.

50. Crawford, RMM and Braendle, R (1996) O_2 deprivation stress in a changing environment. *J. Exp. Bot.* **47**, 145.

51. Dandona, P *et al.* (1996) Oxidative damage to DNA in diabetes mellitus. *Lancet* **347**, 444.

52. Darden, AG *et al.* (1996) Osteoclastic $O_2^{\bullet-}$ production and bone resorption: stimulation and inhibition by modulators of NADPH oxidase. *J. Bone Min. Res.* **11**, 671.

53. Darley-Usmar, V and Halliwell, B (1996) Blood radicals. *Pharm. Res.* **13**, 649.

54. Das, D *et al.* (1997) $OH^{\bullet}$ is the major causative factor in stress-induced gastric ulceration. *Free Rad. Biol. Med.* **23**, 8.

55. Davies, MG *et al.* (1996) Lazaroid therapy (methylaminochroman: U83836E) reduces vein graft intimal hyperplasia. *J. Surg. Res.* **63**, 128.

56. Davis, RE *et al.* (1997) Mutations in mitochondrial cytochrome *c* oxidase genes segregate with late-onset AD. *Proc. Natl Acad. Sci. USA* **94**, 4526.

57. Davison, A *et al.* (1988) Active oxygen in neuromuscular disorders. *Mol. Cell. Biochem.* **84**, 199.

58. Dawson, VL *et al.* (1993) HIV type 1 coat protein neurotoxicity mediated by NO$^\bullet$ in primary cortical cultures. *Proc. Natl Acad. Sci. USA* **90**, 3256.
59. Diaz, DD *et al.* (1992) Hematoma-induced flap necrosis and free radical scavengers. *Acta Otolaryngol. Head Neck Surg.* **118**, 516.
60. Di Silvestro, RA and Jones, AA (1996) High ceruloplasmin levels in rats without high lipoprotein oxidation rates. *Biochim. Biophys. Acta* **1317**, 81.
61. Doherty, MD *et al.* (1984) Mechanisms of toxic injury to isolated hepatocytes by 1-naphthol. *Biochem. Pharmacol.* **33**, 543.
62. Dohlman, AW *et al.* (1993) Expired breath H_2O_2 is a marker of acute airway inflammation in pediatric patients. *Am. Rev. Resp. Dis.* **148**, 955.
63. Dorion, D *et al.* (1993) Role of xanthine oxidase in reperfusion injury of ischemic skeletal muscles in the pig and human. *J. Appl. Physiol.* **75**, 246.
64. Douni, E *et al.* (1996) Transgenic and knockout analyses of the role of TNF in immune regulation and disease pathogenesis. *J. Inflamm.* **47**, 27.
65. Dröge, W *et al.* (1994) Functions of GSH and GSSG in immunology and immunopathology. *FASEB J.* **8**, 1131.
66. Dudman, NPB *et al.* (1993) Circulating lipid hydroperoxide levels in human hyperhomocysteinemia. *Arterioscler. Thromb.* **13**, 512.
67. Eddy, LJ *et al.* (1987) Free radical-producing enzyme, xanthine oxidase, is undetectable in human hearts. *Am. J. Physiol.* **253**, H709.
68. Edmonds, SE *et al.* (1993) An imaginative approach to synovitis—the role of hypoxic reperfusion damage in arthritis. *J. Rheumatol.* **20** (Suppl. 37), 26.
69. Edwards, SW *et al.* (1983) Decrease in apparent K_m for O_2 after stimulation of respiration of rat polymorphonuclear leukocytes. *FEBS Lett.* **161**, 60.
70. Egan, TM *et al.* (1993) Effect of a free radical scavenger on cadaver lung transplantation. *Ann. Thorac. Surg.* **55**, 1453.
71. Elgawish, A *et al.* (1996) Involvement of H_2O_2 in collagen cross-linking by high glucose *in vitro* and *in vivo*. *J. Biol. Chem.* **271**, 12964.
72. Elledge, SJ (1996) Cell cycle checkpoints: preventing an identity crisis. *Science* **274**, 1664.
73. Ernfors, P and Canlon, B (1996) Aminoglycoside excitement silences hearing. *Nature Med.* **2**, 1313.
74. Erzurum, SC *et al.* (1993) *In vivo* antioxidant gene expression in human airway epithelium of normal individuals exposed to 100% O_2. *J. Appl. Physiol.* **75**, 1256.
75. Esbach, S *et al.* (1993) Visualization of the uptake and processing of oxidized LDL in human and rat liver. *Hepatology* **18**, 537.
76. Esterbauer H *et al.* (1992) The role of lipid peroxidation and antioxidants in oxidative modification of LDL. *Free Rad. Biol. Med.* **13**, 341.
77. Evans, PJ *et al.* (1995) Metal ion release from mechanically-disrupted human arterial wall. Implications for the development of atherosclerosis. *Free Rad. Res.* **23**, 465.
78. Feinberg, MB (1996) Changing the natural history of HIV disease. *Lancet* **348**, 239.
79. Ferrando, AA *et al.* (1996) Cloning and expression analysis of human bleomycin hydrolase, a cysteine proteinase involved in chemotherapy resistance. *Cancer Res.* **56**, 1746.
80. Flaherty, JT *et al.* (1994) Recombinant human SOD (h-SOD) fails to improve recovery of ventricular function in patients undergoing coronary angioplasty for acute myocardial infarction. *Circulation* **89**, 1982.
81. Folbergrová, J *et al.* (1993) Does ischemia with reperfusion lead to oxidative damage to proteins in the brain? *J. Cereb. Blood Flow Metab.* **13**, 145.
82. Frears, ER *et al.* (1996) Inactivation of TIMP-1 by ONOO$^-$. *FEBS Lett.* **381**, 21.

83. Friedman, M (1996) Food browning and its prevention: an overview. *J. Agr. Food Chem.* **44**, 631.

84. Gajewski, E *et al.* (1991) Bleomycin-dependent damage to the bases in DNA: a minor side reaction. *Biochemistry* **30**, 2444.

84a. Garner, B *et al.* (1998) Oxidation of HDL. *J. Biol. Chem.* 273, 6080.

85. Garruto, RM and Brown, P (1994) Tau protein, Al, and Alzheimer's disease. *Lancet* **343**, 989.

86. Gerlach, M *et al.* (1996) Acute MPTP treatment produces no changes in mitochondrial complex activities and indices of oxidative damage in the common marmoset *ex vivo* one week after exposure to the toxin. *Neurochem. Int.* **28**, 41.

87. Glinka, Y *et al.* (1996) Nature of inhibition of mitochondrial respiratory complex I by 6-hydroxydopamine. *J. Neurochem.* **66**, 2004.

88. Glover, V *et al.* (1993) Effect of dopaminergic drugs on SOD: implications for senescence. *J. Neural Trans.* (Suppl.) **40**, 37.

89. Godfried, SL and Deckelbaum, LI (1995) Natural antioxidants and restenosis after percutaneous transluminal coronary angioplasty. *Am. Heart J.* **129**, 203.

90. Goldring, CEP *et al.* (1993) α-Tocopherol uptake and its influence on cell proliferation and lipid peroxidation in transformed and nontransformed baby hamster kidney cells. *Arch. Biochem. Biophys.* **303**, 429.

91. Götz, ME *et al.* (1994) Oxidative stress: free radical production in neural degeneration. *Pharmacol. Ther.* **63**, 37.

92. Gower, JD (1988) A role for dietary lipids and antioxidants in the activation of carcinogens. *Free Rad. Biol. Med.* **5**, 95.

93. Grabowski, PS *et al.* (1996) NO$^\bullet$ production in cells derived from the human joint. *Br. J. Rheumatol.* **35**, 207.

94. Granger, DN and Kubes, P (1994) The microcirculation and inflammation: modulation of leukocyte–endothelial cell adhesion. *J. Leuk. Biol.* **55**, 662.

95. Gregori, L *et al.* (1997) Binding of amyloid β-protein to the 20S proteasome. *J. Biol. Chem.* **272**, 58.

95a. Griffiths, C *et al.* (1998) NO in CNS physiology and pathology. *Neurotransmission* **14**, 3.

96. Griffiths, HR and Lunec, J (1996) The Clq binding activity of IgG is modified *in vitro* by ROS: implications for RA. *FEBS Lett.* **388**, 161.

97. Grisham, MB (1994) Oxidants and free radicals in IBD. *Lancet* **344**, 859.

98. Gu, M *et al.* (1997) Mitochondrial respiratory chain function in multiple system atrophy. *Movement Dis.* **12**, 418.

99. Guillemin, K and Krasnow, MA (1997) The hypoxic response: huffing and HIFing. *Cell* **89**, 9.

100. Gutteridge, JMC (1994) Bleomycin and metal interactions. In *Metal Compounds in Cancer Therapy* (Fricker, SP, ed.), p. 198. Chapman and Hall, London.

101. Gutteridge, JMC and Fu Xiao-Chang (1981) Enhancement of bleomycin-iron free radical damage to DNA by antioxidants and their inhibition of lipid peroxidation. *FEBS Lett.* **123**, 71.

102. Gutteridge, JMC *et al.* (1995) The behaviour of caeruloplasmin in stored human extracellular fluids in relation to ferroxidase II activity, lipid peroxidation and phenanthroline-detectable copper. *Biochem. J.* **230**, 517.

103. Hagen, TM *et al.* (1994) Extensive oxidative DNA damage in hepatocytes of transgenic mice with chronic active hepatitis destined to develop hepato-cellular carcinoma. *Proc. Natl Acad. Sci. USA* **91**, 12808.

104. Hall, TJ and Chambers, TJ (1996) Molecular aspects of osteoclast function. *Inflamm. Res.* **45**, 1.

105. Halliwell, B *et al.* (1992) Free radicals, antioxidants and human disease. Where are we now? *J. Lab. Clin. Med.* **119**, 598.

106. Halliwell, B (1995) Oxygen radicals, NO$^\bullet$ and human inflammatory joint disease. *Ann. Rheum. Dis.* **54**, 505.

107. Halliwell, B and Gutteridge, JMC (1984) Lipid peroxidation, oxygen radicals, cell damage and antioxidant therapy. *Lancet* **1**, 1396.

108. Halpert, I *et al.* (1996) Matrilysin is expressed by lipid-laden macrophages at sites of potential rupture in atherosclerotic lesions and localizes to areas of versican deposition, a proteoglycan substrate for the enzyme. *Proc. Natl Acad. Sci. USA* **93**, 9478.

109. Han, DKM *et al.* (1995) Evidence for apoptosis in human atherogenesis and in a rat vascular injury model. *Am. J. Pathol.* **147**, 267.

110. Han, SK *et al.* (1996) L-DOPA up-regulates GSH and protects mesencephalic cultures against oxidative stress. *J. Neurochem.* **66**, 501.

111. Hassett, DJ *et al.* (1987) Bacteria form intracellular free radicals in response to paraquat and streptonigrin: demonstration of the potency of OH$^\bullet$. *J. Biol. Chem.* **262**, 13404.

112. Hatch, GE (1995) Asthma, inhaled oxidants, and dietary antioxidants. *Am. J. Clin. Nutr.* **61** (Suppl.), 625S.

113. Hatta, A and Frei, B (1995) Oxidative modification and antioxidant protection of human LDL at high and low O_2 partial pressures. *J. Lipid Res.* **36**, 2383.

114. Hay, JG *et al.* (1987) The effects of iron and desferrioxamine on the lung injury produced by intravenous bleomycin and hyperoxia. *Free Rad. Res. Commun.* **4**, 109.

114a. Haycock, JW *et al.* (1998) Oxidative damage to protein and alterations to antioxidant levels in human cutaneous thermal injury. *Burns* **23**, 533.

115. Hazen, SL and Heinecke, JW (1997) 3-Chlorotyrosine, a specific marker of MPO-catalyzed oxidation, is markedly elevated in LDL isolated from human atherosclerotic intima. *J. Clin. Invest.* **99**, 2075.

116. Hecht, SS and Hoffmann, D (1988) Tobacco-specific nitrosamines, an important group of carcinogens in tobacco and tobacco smoke. *Carcinogenesis* **9**, 875.

117. Helbling, B *et al.* (1996) Decreased release of GSH into the systemic circulation of patients with HIV infection. *Eur. J. Clin. Invest.* **26**, 38.

117a. Henderson, CJ *et al.* (1998) Increased skin tumorigenesis in mice lacking pi class GST *Proc. Natl. Acad. Sci. USA* **95**, 5275.

118. Hermes-Lima, M and Storey, KB (1993) Antioxidant defenses in the tolerance of freezing and anoxia by garter snakes. *Am. J. Physiol.* **265**, R646.

119. Herold, M and Spiteller, G (1996) Enzymatic production of hydroperoxides of unsaturated fatty acids by injury of mammalian cells. *Chem. Phys. Lipids* **79**, 113.

120. Herzenberg, LA *et al.* (1997) Glutathione deficiency is associated with impaired survival in HIV disease. *Proc. Natl Acad. Sci. USA* **94**, 1967.

121. Hirose, K *et al.* (1993) Overexpression of mitochondrial MnSOD promotes the survival of tumor cells exposed to IL-1, TNF, selected anticancer drugs, and ionizing radiation. *FASEB J.* **7**, 361.

122. Hoang-Xuan, K *et al.* (1996) Myoclonic encephalopathy after exposure to Al. *Lancet* **347**, 910.

123. Hoeg, JM *et al.* (1996) Overexpression of LCAT in transgenic rabbits prevents diet-induced atherosclerosis. *Proc. Natl Acad. Sci. USA* **93**, 11448.

124. Hoffman, SW *et al.* (1997) Isoprostanes: free radical-generated prostaglandins with constrictor effects on cerebral arterioles. *Stroke* **28**, 844.

125. Hooper, DC *et al.* (1997) Prevention of experimental allergic encephalomyelitis by targeting NO$^\bullet$ and ONOO$^-$: implications for the treatment of multiple sclerosis. *Proc. Natl Acad. Sci. USA* **94**, 2528.

126. Hörkkö, S *et al.* (1996) Antiphospholipid antibodies are directed against epitopes of oxidized phospholipids. *J. Clin. Invest.* **98**, 815.

127. Hsiao, K *et al.* (1996) Correlative memory deficits, Aβ elevation, and amyloid plaques in transgenic mice. *Science* **274**, 99.

128. Hunt, JV (1995) Ascorbic acid and diabetes mellitus. In *Subcellular Biochemistry*, Vol. 25: *Ascorbic Acid: Biochemistry and Biomedical Cell Biology* (Harris, RJ, ed.), p. 369. Plenum Press, New York.

129. Iida, M and Saito, K (1996) Role of endotoxin-like contaminants in the apparent anti-inflammatory activity of bovine SOD. *Inflamm. Res.* **45**, 268.

130. Irani, K *et al.* (1997) Mitogen signalling mediated by oxidants in Ras-transformed fibroblasts. *Science* **275**, 1649.

131. Ischiropoulos, H *et al.* (1996) NO production and perivascular tyrosine nitration in brain after CO poisoning in the rat. *J. Clin. Invest.* **97**, 2260.

132. Ishikawa, M *et al.* (1990) Reactivity of a monoclonal antibody to MnSOD with human ovarian carcinoma. *Cancer Res.* **50**, 2538.

133. Islam, KN *et al.* (1995) Fragmentation of ceruloplasmin following non-enzymatic glycation reaction. *J. Biochem.* **118**, 1054.

134. Jack, DB (1997) 100 years of aspirin. *Lancet* **350**, 437.

135. Jackson, JH (1994) Potential molecular mechanisms of oxidant-induced carcinogenesis. *Env. Health Perspect.* **102** (Suppl. 10), 155.

136. Joenje, H *et al.* (1981) Oxygen-dependence of chromosomal abberrations in Fanconi's anaemia. *Nature* **290**, 142.

137. Josephy, PD (1996) The role of peroxidase-catalyzed activation of aromatic amines in breast cancer. *Mutagenesis* **11**, 3.

138. Kalyanaraman, B *et al.* (1992) Synergistic interaction between the probucol phenoxyl radical and ascorbic acid in inhibiting the oxidation of LDL. *J. Biol. Chem.* **267**, 6789.

139. Kang, YJ *et al.* (1996) Suppression of doxorubicin cardiotoxicity by overexpression of catalase in the heart of transgenic mice. *J. Biol. Chem.* **271**, 12610.

140. Kawamata, T *et al.* (1993) Lactotransferrin immunocytochemistry in Alzheimer and normal human brain. *Am. J. Pathol.* **142**, 1574.

141. Kazui, M *et al.* (1994) Visceral lipid peroxidation occurs at reperfusion after supraceliac aortic cross-clamping. *J. Vasc. Surg.* **19**, 473.

142. Kehrer, JP *et al.* (1987) Xanthine oxidase is not responsible for reoxygenation injury in isolated-perfused rat heart. *Free Rad. Res. Commun.* **3**, 69.

143. Kennedy, TP *et al.* (1989) Role of ROS in reperfusion injury of the rabbit lung. *J. Clin. Invest.* **83**, 1326.

144. Kensler, TW and Taffe, BG (1986) Free radicals in tumor promotion. *Adv. Free Rad. Biol. Med.* **2**, 347.

145. Khalil, N *et al.* (1994) Regulation of type II alveolar epithelial cell proliferation by TGF-β during bleomycin-induced lung injury in rats. *Am. J. Physiol.* **267**, L498.

146. Kilgore, KS and Lucchesi, BR (1993) Reperfusion injury after myocardial infarction: the role of free radicals and the inflammatory response. *Clin. Biochem.* **26**, 359.

147. Kinnula, VL *et al.* (1995) Generation and disposal of reactive oxygen metabolites in the lung. *Lab. Invest.* **73**, 3.

148. Kinnula, VL *et al.* (1996) MnSOD in human pleural mesothelioma cell lines. *Free Rad. Biol. Med.* **21**, 527.

149. Klomp, LWJ *et al.* (1996) Ceruloplasmin gene expression in the murine central nervous system. *J. Clin. Invest.* **98**, 207.

150. Kloner, RA (1993) Does reperfusion injury exist in humans? *J. Am. Coll. Cardiol.* **21**, 537.

151. Kostic, V *et al.* (1997) Bcl-2: prolonging life in a transgenic mouse model of familial ALS. *Science* **277**, 559.

152. Kranich, O *et al.* (1996) Different preferences in the utilization of amino acids for GSH synthesis in cultured neurons and astroglial cells derived from rat brain. *Neurosci. Lett.* **219**, 211.

153. Kundu, N *et al.* (1995) Sublethal oxidative stress inhibits tumor cell adhesion and enhances experimental metastasis of murine mammary carcinoma. *Clin. Exp. Metastasis* **13**, 16.

154. Kunst, CB *et al.* (1997) Mutations in SOD1 associated with ALS cause novel protein interactions. *Nature Genet.* **15**, 91.

155. Lacey, CJ *et al.* (1996) Antioxidant-micronutrients and HIV infection. *Int. J. STD and AIDS* **7**, 485.

156. Lafyatis, R *et al.* (1989) Anchorage-independent growth of synoviocytes from arthritic and normal joints. *J. Clin. Invest.* **83**, 1267.

157. Langston, JW (1985) MPTP and Parkinson's disease. *Trends Neurosci.*, Feb issue, p. 79.

158. Lasley, RD *et al.* (1988) Allopurinol enhanced adenine nucleotide repletion after myocardial ischemia in the isolated rat heart. *J. Clin. Invest.* **81**, 16.

159. Laurindo, FRM *et al.* (1991) Evidence for $O_2^{\bullet -}$-dependent coronary artery vasospasm after angioplasty in intact dogs. *Circulation* **83**, 1705.

160. Lautier, D *et al.* (1996) Multidrug resistance mediated by the multidrug resistance protein (MRP) gene. *Biochem. Pharmacol.* **52**, 967.

161. Leeuwenburgh, C *et al.* (1997) RNS promote LDL oxidation in human atherosclerotic intima. *J. Biol. Chem.* **272**, 1433.

162. Lefer, DJ *et al.* (1997) $ONOO^-$ inhibits leukocyte–endothelial cell interactions and protects against ischemia–reperfusion in rats. *J. Clin. Invest.* **99**, 684.

163. Lepage, G *et al.* (1996) Supplementation with carotenoids corrects increased lipid peroxidation in children with cystic fibrosis. *Am. J. Clin. Nutr.* **64**, 87.

164. Lewis, W and Dalakas, MC (1995) Mitochondrial toxicity of antiviral drugs. *Nature Med.* **1**, 417.

165. Li, J *et al.* (1995) Phenotypic changes induced in human breast cancer cells by overexpression of MnSOD. *Oncogene* **10**, 1989.

166. Liebetrau, W *et al.* (1997) Mutagenic activity of ambient O_2 and mitomycin C in Fanconi's anaemia cells. *Mutagenesis* **12**, 69.

167. Lih-Brody, L *et al.* (1996) Increased oxidative stress and decreased anti-oxidant defenses in mucosa of inflammatory bowel disease. *Dig. Dis. Sci.* **41**, 2078.

168. Lindsay, S *et al.* (1991) Role of xanthine dehydrogenase and oxidase in focal cerebral ischemic injury to rat. *Am. J. Physiol.* **261**, H2051.

169. Liu, J *et al.* (1996) Immobilization stress causes oxidative damage to lipid, protein, and DNA in the brain of rats. *FASEB J.* **10**, 1532.

170. Liu, PK *et al.* (1996) Damage, repair and mutagenesis in nuclear genes after mouse forebrain ischemia–reperfusion. *J. Neurosci.* **16**, 6795.

171. Liu, ZC and Uetrecht, JP (1995) Clozapine is oxidized by activated human neutrophils to a reactive nitronium ion that irreversibly binds to the cells. *J. Pharm. Exp. Ther.* **275**, 1476.

172. Liu, ZC *et al.* (1995) Oxidation of 5-aminosalicylic acid by HOCl to a reactive iminoquinone. *Drug Metab. Disp.* **23**, 246.

173. Loft, S and Poulsen, H (1996) Cancer risk and oxidative DNA damage in men. *J. Mol. Med.* **74**, 297.

174. Loft, S *et al.* (1995) Oxidative DNA damage after transplantation of the liver and small intestine in pigs. *Transplantation* **59**, 16.

175. Lomas, DA (1996) New insights into the structural basis of α_1-antitrypsin deficiency. *Q. J. Med.* **89**, 807.

176. Loscalzo, J (1996) The oxidant stress of hyperhomocyst(e)inemia. *J. Clin. Invest.* **98**, 5.

177. Loughpey, CM *et al.* (1994) Oxidative stress in haemodialysis. *Q. J. Med.* **87**, 679.

178. Louie, S *et al.* (1996) ARDS: a radical perspective. *Adv. Pharmacol.* **38**, 457.

179. Lundberg, JON *et al.* (1997) NO and inflammation: the answer is blowing in the wind. *Nature Med.* **3**, 30.

180. Lunec, J *et al.* (1994) 8OHdG. A marker of oxidative DNA damage in SLE. *FEBS Lett.* **348**, 131.

181. Luo, X. *et al.* (1997) Doxorubicin–induced acute changes in cytotoxic aldehydes, antioxidant status and cardiac function in the rat. *Biochim. Biophys. Acta* **1360**, 45.

182. Lyras, L *et al.* (1997) An assessment of oxidative damage to proteins, lipids and DNA in brain from patients with Alzheimer's disease. *J. Neurochem.* **68**, 2061.

183. Macdonald, RL and Weir, BK (1994) Cerebral vasospasm and free radicals. *Free Rad. Biol. Med.* **16**, 633.

184. Mackness, MI and Durrington, PN (1995) Paraoxonase: another factor in NIDDM cardiovascular disease. *Lancet* **346**, 856.

185. Maeda, H and Akaike, T (1991) Oxygen free radicals as pathogenic molecules in viral diseases. *Proc. Soc. Exp. Biol. Med.* **198**, 721.

186. Mahadik, SP and Mukherjee, S (1996) Free radical pathology and antioxidant defence in schizophrenia: a review. *Schizophrenia Res.* **19**, 1.

187. Malins, DC *et al.* (1996) Progression of human breast cancers to the metastatic state is linked to OH•-induced DNA damage. *Proc. Natl Acad. Sci. USA* **93**, 2557.

188. Mannick, EE *et al.* (1996) Inducible NOS, nitrotyrosine, and apoptosis in *Helicobacter pylori* gastritis: effect of antibiotics and antioxidants. *Cancer Res.* **56**, 3238.

189. Marklund, SL *et al.* (1982) CuZnSOD, MnSOD, catalase and glutathione peroxidase in normal and neoplastic human cell lines and normal human tissues. *Cancer Res.* **42**, 1955.

190. Marnett, LJ (1987) Peroxyl free radicals: potential mediators of tumor initiation and promotion. *Carcinogenesis* **8**, 1365.

191. Martin, JP Jr and Batkoff, B (1987) Homogentisic acid autoxidation and oxygen radical generation: implications for the etiology of alkaptonuric arthritis. *Free Rad. Biol. Med.* **3**, 241.

192. Maruyama, W *et al.* (1997) An endogenous dopaminergic neurotoxin, N-methyl-(R)-salsolinol, induces DNA damage in human dopaminergic neuroblastoma SH-SY5Y cells. *J. Neurochem.* **69**, 322.

193. Massa, SM *et al.* (1996) The stress gene response in brain. *Cerebrovasc. Brain Metab. Rev.* **8**, 95.

194. Mastrogiacomo, F *et al.* (1996) Brain protein and α-ketoglutarate dehydrogenase complex activity in Alzheimer's disease. *Ann. Neurol.* **39**, 592.

195. Matsuda, Y *et al.* (1982) Correlation between level of defense against active oxygen in *E. coli* K12 and resistance to bleomycin. *J. Antibiot.* **35**, 931.

196. Mattson, MP *et al.* (1997) Disruption of brain cell ion homeostasis in AD by oxyradicals, and signaling pathways that protect therefrom. *Chem. Res. Toxicol.* **10**, 507.

197. McCord, JM and Turrens, JF (1994) Mitochondrial injury by ischemia and reperfusion. *Curr. Top. Bioenerg.* **17**, 173.

198. McGeer, PL and McGeer, EG (1995) The inflammatory response system of brain: implications for therapy of Alzheimer and other neurodegenerative diseases. *Brain Res. Rev.* **21**, 195.

199. Means, RT and Krantz, SB (1992) Progress in understanding the pathogenesis of the anaemia of chronic disease. *Blood* **80**, 1639.

200. Metcalfe, T *et al.* (1989) Vitamin E concentrations in human brain of patients with Alzheimer's disease, fetuses with Down's syndrome, centenarians and controls. *Neurochem. Res.* **14**, 1209.

201. Miyako, K *et al.* (1997) The content of intracellular mtDNA is decreased by MPP^+. *J. Biol. Chem.* **272**, 9605.

202. Mizuno, Y *et al.* (1995) Role of mitochondria in the etiology and pathogenesis of Parkinson's disease. *Biochim. Biophys. Acta* **1271**, 265.

203. Moat, NE *et al.* (1993) Chelatable iron and copper can be released from extracorporeally circulated blood during cardiopulmonary bypass. *FEBS Lett.* **328**, 103.

204. Moilanen, E *et al.* (1997) NOS is expressed in human macrophages during foreign body inflammation. *Am. J. Pathol.* **150**, 881.

205. Muller, DPR and Goss-Sampson, MA (1990) Neurochemical, neurophysiological, and neuropathological studies in vitamin E deficiency. *Crit. Rev. Neurobiol.* **5**, 239.

206. Murrell, GAC *et al.* (1987) Free radicals and Dupuytren's contracture. *Brit. Med. J.* **295**, 1373.

207. Mytilineou, C *et al.* (1997) L-(−)-desmethylselegiline, a metabolite of selegiline [L-(−)-deprenyl] protects mesencephalic dopamine neurons from excitotoxicity *in vitro. J. Neurochem.* **68**, 434.

208. Nagaraj, RH and Monnier, VM (1995) Protein modification by the degradation products of ascorbate: formation of a novel pyrrole from the Maillard reaction of L-threose with proteins. *Biochim. Biophys. Acta* **1253**, 75.

209. Nakamura, H *et al.* (1997) Redox regulation of cellular activation. *Annu. Rev. Immunol.* **15**, 351.

210. Nakazono, K *et al.* (1991) Does $O_2^{\bullet-}$ underlie the pathogenesis of hypertension? *Proc. Natl Acad. Sci. USA* **88**, 10045.

210a. Napoli, C *et al.* (1997) Fatty streak formation occurs in human foetal aortas and is greatly enhanced by maternal hypercholesterolemia. *J. Clin. Invest.* **100**, 2680.

211. Naruszewicz, M *et al.* (1994) Oxidative modification of Lp(a) causes changes in the structure and biological properties of apo(a). *Chem. Phys. Lipids* **67/68**, 167.

212. Navab, M *et al.* (1996) The Yin and Yang of oxidation in the development of the fatty streak. *Arterioscler. Thromb. Vasc. Biol.* **16**, 831.

213. Naveilhan, P *et al.* (1994) ROS influence nerve growth factor synthesis in primary rat astrocytes. *J. Neurochem.* **62**, 2178.

214. Nielsen, LB (1996) Transfer of LDL into the arterial wall and risk of atherosclerosis. *Atherosclerosis* **123**, 1.

215. Nishino, T and Tsushima, K (1986) Interaction of milk xanthine oxidase with folic acid. Inhibition of milk xanthine oxidase by folic acid and separation of the enzyme into two fractions on Sepharose 4B/folate gel. *J. Biol. Chem.* **261**, 11242.

216. Odh, G *et al.* (1994) Neuromelanin of the human substantia nigra: a mixed-type melanin. *J. Neurochem.* **62**, 2030.

217. Ohshima, H and Bartsch, H (1994) Chronic infections and inflammatory processes as cancer risk factors: possible role of $NO^\bullet$ in carcinogenesis. *Mut. Res.* **305**, 253.

218. Okuda, S *et al.* (1996) H_2O_2-mediated neuronal cell death induced by an endogenous neurotoxin, 3-hydroxykynurenine. *Proc. Natl Acad. Sci. USA* **93**, 12553.

219. Olanow, CW *et al.* (eds) (1996) *Neurodegeneration and Neuroprotection in Parkinson's Disease.* Academic Press, London.

220. Olinski, R *et al.* (1995) DNA base modifications and antioxidant enzyme activities in human benign prostatic hyperplasia. *Free Rad. Biol. Med.* **18**, 807.

221. Olyaee, M *et al.* (1995) Mucosal ROS production in oesophagitis and Barrett's oesophagus. *Gut* **37**, 168.

222. Ophir, A *et al.* (1993) $OH^\bullet$ generation in the cat retina during reperfusion following ischemia. *Exp. Eye Res.* **57**, 351.

223. Pacht, ER *et al.* (1997) Alveolar fluid glutathione is not reduced in asymptomatic HIV^+ subjects. *Am. J. Resp. Crit. Care Med.* **155**, 374.

224. Palinski, W *et al.* (1995) Immunization of LDL receptor-deficient rabbits with homologous MDA-modified LDL reduces atherogenesis. *Proc. Natl Acad. Sci. USA* **92**, 821.

224a. Pannala, AS *et al.* (1997) Inhibition of $ONOO^-$-mediated tyrosine nitration by catechin polyphenols *Biochem. Biophys. Res. Commun.* **232**, 164.

225. Parhami, F *et al.* (1997) Lipid oxidation products have opposite effects on calcifying vascular cell and bone cell differentiation. *Arterio. Thromb. Vasc. Biol.* **17**, 680.

225a. Parker, JD and Parker, JO (1998) Nitrate therapy for stable angina pectoris *N. Engl. J. Med.* **338**, 520.

226. Parratt, JR (1994) Protection of the heart by ischaemic preconditioning: mechanisms and possibilities for pharmacological exploitation. *Trends Pharmacol. Sci.* **15**, 19.

227. Parthasarathy, S *et al.* (1992) The role of oxidized LDL in the pathogenesis of atherosclerosis. *Annu. Rev. Med.* **43**, 219.

228. Patt, A *et al.* (1988) Xanthine oxidase-derived H_2O_2 contributes to ischemia reperfusion-induced edema in gerbil brains. *J. Clin. Invest.* **81**, 1556.

229. Petkau, A *et al.* (1977) Modification of superoxide dismutase in rat mammary carcinoma. *Res. Comm. Chem. Pathol. Pharmacol.* **17**, 125.

230. Pezzoli, G *et al.* (1995) *n*-Hexane-induced parkinsonism: pathogenetic hypotheses. *Movement Dis.* **10**, 279.

231. Philbert, MA *et al.* (1995) Glutathione-S-transferases and γ-glutamyl transpeptidase in the rat nervous system: a basis for differential susceptibility to neurotoxicants. *Neurotoxicology* **16**, 349.

232. Phull, PS *et al.* (1995) A radical view of the stomach: the role of oxygen-derived free radicals and anti-oxidants in gastroduodenal disease. *Eur. J. Gast. Hepatol.* **7**, 265.

233. Pileblad, E *et al.* (1988) Studies on the autoxidation of dopamine: interaction with ascorbate. *Arch. Biochem. Biophys.* **263**, 447.

234. Plummer, SM and Faux, SP (1994) Induction of 8OHdG in isolated DNA and HeLa cells exposed to fecapentene12: evidence for the involvement of prostaglandin H synthase and iron. *Carcinogenesis* **15**, 449.

235. Pollak, R *et al.* (1993) A randomized double-blind trial of the use of human recombinant SOD in renal transplantation. *Transplantation* **55**, 57.

236. Powis, G (1989) Free radical formation by antitumor quinones. *Free Rad. Biol. Med.* **6**, 63.

236a. Praticò, D *et al.* (1997) Localization of distinct F_2-isoprostanes in human atherosclerotic lesions. *J. Clin. Invest.* **100**, 2028.

237. Puka-Sundvall, M *et al.* (1995) Neurotoxicity of cysteine: interaction with glutamate. *Brain Res.* **705**, 65.

238. Punnonen, K *et al.* (1994) Antioxidant enzyme activities and oxidative stress in human breast cancer. *J. Cancer Res. Clin. Oncol.* **120**, 374.

239. Puppo, A and Halliwell, B (1988) Formation of $OH^\bullet$ in biological systems. Does myoglobin stimulate $OH^\bullet$ formation from H_2O_2? *Free Rad. Res. Commun.* **4**, 415.

240. Rabizadeh, S *et al.* (1995) Mutations associated with ALS convert SOD from an antiapoptotic gene to a proapoptotic gene: studies in yeast and neural cells. *Proc. Natl Acad. Sci. USA* **92**, 3024.

241. Reed, GA *et al.* (1986) Epoxidation of $(\pm)$-7,8-dihydroxy-7,8-dihydro-benzo[α]pyrene during (bi)sulfite autoxidation: activation of a procarcinogen by a cocarcinogen. *Proc. Natl Acad. Sci. USA* **83**, 7499.

242. Reiber, H *et al.* (1993) Ascorbate concentration in human CSF and serum. Intrathecal accumulation and CSF flow rate. *Clin. Chim. Acta* **217**, 163.

243. Rice-Evans, C *et al.* (1996) Practical approaches to LDL oxidation: whys, wherefores and pitfalls. *Free Rad. Res.* **25**, 285.

243a. Risby, TH *et al.* (1994) Evidence for free radical-mediated lipid peroxidation at reperfusion of human orthotopic liver transplants. *Surgery* **115**, 94.

244. Rizzardini, M *et al.* (1997) Prion protein fragment 106–126 differentially induces heme oxygenase-1 mRNA in cultured neurons and astroglial cells. *J. Neurochem.* **68**, 715.

244a. Rösen, P *et al.* (eds) (1998) *Oxidative Stress and Antioxidants in Diabetes and its Complications*, Marcel Dekker, USA.

245. Ross, R (1993) The pathogenesis of atherosclerosis: a perspective for the 1990s. *Nature* **362**, 801.

245a. Rota, S *et al.* (1998) Atherogenic lipoproteins support assembly of the prothrombinase complex and thrombin generation: modulation by oxidation and vitamin E. *Blood* **91**, 508.

246. Rowland, RT *et al.* (1995) Mechanisms of immature myocardial tolerance to ischemia: phenotypic differences in antioxidants, stress proteins and oxidases. *Surgery* **118**, 446.

247. Rubbo, H *et al.* (1995) NO inhibition of lipoxygenase-dependent liposome and LDL oxidation: termination of radical chain propagation reactions and formation of nitrogen-containing oxidized lipid derivatives. *Arch. Biochem. Biophys.* **324**, 15.

248. Rupec, RA and Baeuerle, PA (1995) The genomic response of tumor cells to hypoxia and reoxygenation. *Eur. J. Biochem.* **234**, 632.

249. Sabri, N *et al.* (1996) Electropermeabilization of intact maize cells induces an oxidative stress. *Eur. J. Biochem.* **238**, 737.

250. Salvemini, D *et al.* (1996) Evidence of ONOO⁻ involvement in the carrageenan-induced rat paw edema. *Eur. J. Pharmacol.* **303**, 217.

251. Sambrano, GR and Steinberg, D (1995) Recognition of oxidatively damaged and apoptotic cells by an oxidized LDL receptor on mouse peritoneal macrophages: Role of membrane phosphatidylserine. *Proc. Natl Acad. Sci. USA* **92**, 1396.

252. Sanan, S and Sharma, G (1986) Effect of desferrioxamine mesylate (Desferal) in anesthetized dogs with clinical hemorrhagic shock. *Ind. J. Med. Res.* **83**, 655.

253. Sano, M *et al.* (1997) A controlled trial of selegiline, α-tocopherol or both as treatments for AD. *N. Engl. J. Med.* **336**, 1216.

254. Sato, K *et al.* (1992) Hydroxyl radical production by H_2O_2 plus CuZnSOD reflects the activity of free copper released from the oxidatively damaged enzyme. *J. Biol. Chem.* **267**, 25371.

255. Sausen, PJ *et al.* (1995) Elevated 8OHdG in hepatic DNA of rats following exposure to peroxisome proliferators: relationship to mitochondrial alterations. *Carcinogenesis* **16**, 1795.

256. Savory, J *et al.* (1996) Can the controversy of the role of Al in Alzheimer's disease be resolved? *J. Toxicol. Environ. Health* **48**, 615.

257. Schalkwijk, J *et al.* (1986) An experimental model for H_2O_2-induced tissue damage. *Arth. Rheum.* **29**, 532.

258. Schoenberg, MH *et al.* (1991) The involvement of O_2 radicals in acute pancreatitis. *Klin. Wochenschr.* **69**, 1025.

259. Schulz, JB *et al.* (1996) Striatal malonate lesions are attenuated in nNOS knockout mice. *J. Neurochem.* **67**, 430.

260. Scorah, CJ *et al.* (1996) Total vitamin C, ascorbic acid, and dehydroascorbic acid concentrations in plasma of critically ill patients. *Am. J. Clin. Nutr.* **63**, 760.

261. Seaton, TA *et al.* (1996) Mitochondrial respiratory enzyme function and SOD activity following brain GSH depletion in the rat. *Biochem. Pharmacol.* **52**, 1657.

262. Sedghi, S *et al.* (1994) Elevated breath ethane levels in active ulcerative colitis: evidence for excessive lipid peroxidation. *Am. J. Gastroenterol.* **89**, 2217.

263. Sekili, S *et al.* (1993) Direct evidence that OH• plays a pathogenetic role in myocardial 'stunning' in the conscious dog and demonstration that stunning can be markedly attenuated without subsequent adverse effects. *Circul. Res.* **73**, 705.

264. Selkoe, DJ (1994) Cell biology of the amyloid β-protein precursor and the mechanism of Alzheimer's disease. *Annu. Rev. Cell. Biol.* **10**, 373.

265. Sendobry, SM *et al.* (1997) Attenuation of diet-induced atherosclerosis in rabbits with a highly selective 15-lipoxygenase inhibitor lacking significant antioxidant properties. *Br. J. Pharmacol.* **120**, 1199.

266. Shah, SV (1995) The role of reactive oxygen metabolites in glomerular disease. *Annu. Rev. Physiol.* **57**, 245.

267. Shan, K *et al.* (1996) Anthracycline-induced cardiotoxicity. *Ann. Int. Med.* **125**, 47.

267a. Shih, DM *et al.* (1998) Mice lacking serum paraoxonase are susceptible to organophosphate toxicity and atherosclerosis. *Nature* **394**, 284.

268. Shires, TK (1982) Iron-induced DNA damage and synthesis in isolated rat liver muclei. *Biochem. J.* **205**, 321.

269. Silverman, DJ and Santucci, LA (1988) Potential for free radical-induced lipid peroxidation as a cause of endothelial cell injury in Rocky Mountain spotted fever. *Infect. Immun.* **56**, 3110.

270. Sims, NR and Zaidan, E (1995) Biochemical changes associated with selective neuronal death following short-term cerebral ischaemia. *Int. J. Biochem. Cell. Biol.* **27**, 531.

271. Sinet, PM et al. (1980) H$_2$O$_2$ production by rat brain *in vivo*. *J. Neurochem.* **34**, 1421.

272. Singer, II et al. (1996) Expression of iNOS and nitrotyrosine in colonic epithelium in inflammatory bowel disease. *Gastroenterology* **111**, 871.

273. Siveski-Iliskovic, N et al. (1995) Probucol protects against adriamycin cardiomyopathy without interfering with its antitumor effect. *Circulation* **91**, 10.

274. Slater, TF et al. (1990) Studies on the hyperplasia ('regeneration') of the rat liver following partial hepatectomy. *Biochem. J.* **265**, 51.

275. Smith, LJ et al. (1993) Increased levels of glutathione in BALF from patients with asthma. *Am. Rev. Resp. Dis.* **147**, 1461.

275a. Smith, MA et al. (1998) Amyloid-β deposition in Alzheimer transgenic mice is associated with oxidative stress. *J. Neurochem.* **70**, 2212.

276. Soong, CV et al. (1994) Reduction of free radical generation minimises lower limb swelling following femoropopliteal bypass surgery. *Eur. J. Vasc. Surg.* **8**, 435.

277. Southard, JH and Belzer, FO (1995) Organ preservation. *Ann. Rev. Med.* **46**, 235.

278. Southorn PA (1988) Free radicals in medicine. II: involvement in human disease. *Mayo Clin. Proc.* **63**, 390.

279. Spencer, J et al. (1995) Superoxide-dependent depletion of GSH by L-DOPA and dopamine. Relevance to Parkinson's disease. *Neuroreport* **6**, 1480.

280. Spillantini, MG et al. (1997) α-Synuclein in Lewy bodies. *Nature* **388**, 839.

281. Spycher, SE et al. (1997) Aldose reductase induction: a novel response to oxidative stress of smooth muscle cells. *FASEB J.* **11**, 181.

282. Stahl, GL et al. (1992) H$_2$O$_2$-induced cardiovascular reflexes. *Circul. Res.* **71**, 295.

283. Staprans, I et al. (1994) Oxidized lipids in the diet are a source of oxidized lipid in chylomicrons of human serum. *Arterioscler. Thromb.* **14**, 1900.

284. Stein, CM et al. (1996) Evidence of free radical-mediated injury (isoprostane overproduction) in scleroderma. *Arth. Rheum.* **39**, 1146.

284a. Steinberg, D (1997) LDL oxidation and its pathobiological significance. *J. Biol. Chem.* **272**, 20963.

285. Suarez-Pinzon, WL et al. (1997) Development of autoimmune diabetes in NOD mice is associated with the formation of ONOO$^-$ in pancreatic islet β-cells. *Diabetes* **46**, 907.

286. Sun, JZ et al. (1996) Evidence for an essential role of ROS in the genesis of late preconditioning against myocardial stunning in conscious pigs. *J. Clin. Invest.* **97**, 562.

287. Sun, Y et al. (1993) Lowered antioxidant enzymes in spontaneously transformed embryonic mouse liver cells in culture. *Carcinogenesis* **14**, 1457.

288. Supnet, MC et al. (1994) Plasma xanthine oxidase activity and lipid hydroperoxide levels in preterm infants. *Pediat. Res.* **36**, 283.

289. Suzuki, H et al. (1997) A role for macrophage scavenger receptors in atherosclerosis and susceptibility to infection. *Nature* **386**, 292.

289a. Suzuki, H et al. (1983) Evidence for involvement of a free radical in DNA-cleaving reaction by macromomycin and auromomycin. *J. Antibiot.* **36**, 583.

290. Suzuki, K et al. (1993) High Cu and iron levels and expression of MnSOD in mutant rats displaying hereditary hepatitis and hepatoma (LEC rats) *Carcinogenesis* **14**, 1881.

291. Svingen, BA *et al.* (1981) Protection against adriamycin-induced skin necrosis in the rat by dimethyl sulfoxide and α-tocopherol. *Cancer Res.* **41**, 3395.

292. Swain, J and Gutteridge, JMC (1995) Prooxidant iron and copper, with ferroxidase and xanthine oxidase activities in human atherosclerotic lesions. *FEBS Lett.* **368**, 513.

293. Swain, JA *et al.* (1994) Peroxynitrite releases Cu from caeruloplasmin; implications for atherosclerosis. *FEBS Lett.* **342**, 49.

294. Szabó, C (1996) The pathophysiological role of ONOO⁻ in shock, inflammation, and ischemia–reperfusion injury. *Shock* **6**, 79.

295. Tak Yee, AW *et al.* (1992) Absorption and lymphatic transport of peroxidized lipids by rat small intestine *in vivo*: role of mucosal GSH. *Am. J. Physiol.* **262**, G99.

296. Takahashi, K *et al.* (1996) Expression of heme oxygenase mRNAs in the human brain and induction of HO-1 by NO donors. *J. Neurochem.* **67**, 482.

297. Tanner, AJ and Dice, JF (1996) Batten disease and mitochondrial pathways of proteolysis. *Biochem. Mol. Med.* **57**, 1.

297a. Tardif, JC *et al.* (1997) Probucol and multivitamins in the prevention of restenosis after coronary angioplasty. *N. Engl. J. Med.* **337**, 365.

298. Tayarani, I. *et al.* (1987) Enzymatic protection against peroxidative damage in isolated brain capillaries. *J. Neurochem.* **48**, 1399.

299. Teicher, BA *et al.* (1981) Classification of antineoplastic agents by their selective toxicities towards oxygenated and hypoxic tumor cells. *Cancer Res.* **41**, 73.

300. Thom, SR *et al.* (1997) Release of glutathione from erythrocytes and other markers of oxidative stress in CO poisoning. *J. Appl. Physiol.* **82**, 1424.

301. Thornalley, PJ *et al.* (1996) Negative association between erythrocyte GSH concentration and diabetic complications. *Clin. Sci.* **91**, 575.

302. Thorpe, SR and Baynes, JW (1996) Role of the Maillard reaction in diabetes mellitus and diseases of aging. *Drugs Aging* **9**, 69.

303. Tombaugh, GC and Sapolsky, RM (1993) Evolving concepts about the role of acidosis in ischemic neuropathology. *J. Neurochem.* **61**, 793.

304. Topper, JN *et al.* (1996) Identification of vascular endothelial genes differentially responsive to fluid mechanical stimuli: COX-2, MnSOD and endothelial cell NOS are selectively up-regulated by steady laminar shear stress. *Proc. Natl Acad. Sci. USA* **93**, 10417.

305. Totter, TR (1980) Spontaneous cancer and its possible relationship to O₂ metabolism. *Proc. Natl Acad. Sci. USA* **77**, 1763.

306. Toyokuni, S (1996) Iron-induced carcinogenesis: the role of redox regulation. *Free Rad. Biol. Med.* **20**, 553.

307. Trindberg, N and Ingelman-Sundberg, M (1996) Expression, catalytic activity, and inducibility of CYP2E1 in the rat central nervous system. *J. Neurochem.* **67**, 2066.

308. Trush, MA and Kensler, TW (1991) Role of free radicals in carcinogen activation. In *Oxidative Stress: Oxidants and Antioxidants* (Sies, H, ed.), p. 277. Academic Press, London.

309. Tu, P *et al.* (1997) Oxidative stress, mutant SOD 1, and neurofilament pathology in transgenic mouse models of human motor neurone disease. *Lab. Invest.* **76**, 441.

310. Uden, S *et al.* (1990) Antioxidant therapy for recurrent pancreatitis: placebo-controlled trial. *Aliment. Pharm. Ther.* **4**, 357.

311. Uetrecht, JP (1989) Idiosyncratic drug reactions: possible role of reactive metabolites generated by leukocytes. *Pharm. Res.* **6**, 265.

312. Umeda, Y *et al.* (1995) Kinetics and uptake in vivo of oxidatively modified lymph chylomicrons. *Am. J. Physiol.* **268**, G709.

313. Usui, T *et al.* (1982) Possible prevention from the progression of cardiotoxicity in adriamycin-treated rabbits by coenzyme Q_{10}. *Toxicol. Lett.* **12**, 75.

314. Valen, G and Vaage, J (1993) Toxic O_2 metabolites and leukocytes in reperfusion injury. *Scand. J. Thor. Cardiovasc. Surg.* (Suppl.) **41**, 19.

315. van Acker, SABE *et al.* (1995) Monohydroxyethylrutoside as protector against chronic doxorubicin-induced cardiotoxicity. *Br. J. Pharmacol.* **115**, 1260.

316. van der Vliet, A *et al.* (1996) Oxidative stress in cystic fibrosis: does it occur and does it matter? *Adv. Pharmacol.* **38**, 491.

317. Völker, W *et al.* (1997) Copper-induced inflammatory reactions of rat carotid arteries mimic restenosis/arteriosclerosis like neointima formation. *Atherosclerosis* **130**, 29.

318. Wagner, DA *et al.* (1985) Effect of vitamins C and E on endogenous synthesis of *N*-nitrosoamino acids in humans: precursor-product studies with [^{15}N]nitrate. *Cancer Res.* **45**, 6519.

319. Wahlgren, NG and Lindquist, C (1987) Haem derivatives in the CSF after intracranial haemorrhage. *Eur. Neurol.* **26**, 216.

320. Walkinshaw, G. and Waters, CM (1995) Induction of apoptosis in catecholaminergic PC12 cells by L-DOPA. *J. Clin. Invest.* **95**, 2458.

321. Walzem, RL *et al.* (1995) Older plasma lipoproteins are more susceptible to oxidation: A linking mechanism for the lipid and oxidation theories of atherosclerotic cardiovascular disease. *Proc. Natl Acad. Sci. USA* **92**, 7460.

321a. Wang, LJ *et al.* (1998) Expression of HO-1 in atherosclerotic lesions. *Am. J. Pathol.* **152**, 711.

322. Wang, N *et al.* (1996) IL-8 is induced by cholesterol loading of macrophages and expressed by macrophage foam cells in human atheroma. *J. Biol. Chem.* **271**, 8837.

323. Wang, Q *et al.* (1991) Amelioration of bleomycin-induced pulmonary fibrosis in hamsters by combined treatment with taurine and niacin. *Biochem. Pharmacol.* **42**, 1115.

323a. Wani, G *et al.* (1998) Enhanced expression of the 8-oxodGTPase gene in human breast cancer cells. *Cancer Lett.* **125**, 123.

324. Ward, PA (1996) Role of complement in lung inflammatory injury. *Am. J. Pathol.* **149**, 1081.

325. Ward, PH *et al.* (1992) O_2-derived free radicals mediate liver damage in rats subjected to tourniquet shock. *Free Rad. Res. Commun.* **17**, 313.

326. Washko, P. and Levine, M (1992) Inhibition of ascorbic acid transport in human neutrophils by glucose. *J. Biol. Chem.* **267**, 23568.

327. Watanabe, Y *et al.* (1997) Instability of expressed Cu/Zn SOD with 2 bp deletion found in familial ALS. *FEBS Lett.* **400**, 108.

328. Watson, AJM and DuBois, RN (1997) Lipid metabolism and APC: implications for colorectal cancer prevention. *Lancet* **349**, 444.

329. Weinberg, ED (1996) The role of iron in cancer. *Eur. J. Cancer Prevent.* **5**, 19.

330. Weitzman, SA *et al.* (1985) Phagocytes as carcinogens: malignant transformation produced by human neutrophils. *Science* **227**, 1231.

331. Westendorp, MO *et al.* (1995) HIV-1 Tat potentiates TNF-induced NF-κB activation and cytotoxicity by altering the cellular redox state. *EMBO J.* **14**, 546.

332. Whiteman, M *et al.* (1996) Protection against ONOO⁻-dependent tyrosine nitration and α_1-AP inactivation by some anti-inflammatory drugs and by the antibiotic tetracycline. *Ann. Rheum. Dis.* **55**, 383.

333. Williams, RO *et al.* (1992) Anti-TNF ameliorates joint disease in murine collagen-induced arthritis. *Proc. Natl Acad. Sci. USA* **89**, 9784.

334. Willmore, LJ and Rubin, JJ (1981) Antiperoxidant pretreatment and iron-induced epileptiform discharges in the rat: EEG and histopathologic studies. *Neurology* **31**, 63.

335. Winyard, PG *et al.* (1987) Mechanism of exacerbation of rheumatoid synovitis by total-dose iron-dextran infusion: *in vivo* demonstration of iron-promoted oxidant stress. *Lancet i*, 69.

336. Wolf, G (1995) The mechanism of uptake of ascorbic acid into osteoblasts and leukocytes. *Nutr. Rev.* **54**, 150.

337. Wood, JL and Graham, A (1995) Structural requirements for oxidation of LDL by thiols. *FEBS Lett.* **366**, 75.

338. Xia, Y *et al.* (1996) Adenosine deaminase inhibition prevents free radical-mediated injury in the postischemic heart. *J. Biol. Chem.* **271**, 10096.

339. Yamasaki, H and Naus, CCG (1996) Role of connexin genes in growth control. *Carcinogenesis* **17**, 1199.

340. Yan, SD *et al.* (1994) Enhanced cellular oxidant stress by the interaction of advanced glycation end products with their receptors/binding proteins. *J. Biol. Chem.* **269**, 9889.

341. Yan, SD *et al.* (1996) RAGE and amyloid-β peptide neurotoxicity in Alzheimer's disease. *Nature* **382**, 685.

342. Yen, HC *et al.* (1996) The protective role of MnSOD against adriamycin-induced acute cardiac toxicity in transgenic mice. *J. Clin. Invest.* **98**, 1253.

342a. Yokota, T *et al.* (1997) Friedreich-like ataxia with retinitis pigmentosa caused by the his[101]gln mutation of the α-tocopherol transfer protein gene *Ann. Neurol.* **41**, 826.

343. Youdim, MBH and Riederer, P (1997) Understanding Parkinson's disease. *Sci. Amer.* **276**, 52.

344. Yun, HY *et al.* (1996) Neurobiology of NO•. *Crit. Rev. Neurobiol.* **10**, 291.

345. Zimmerman, R and Cerutti, P (1984) Active O_2 acts as a promoter of transformation in mouse embryo C3H/10T1/2C18 fibroblasts. *Proc. Natl Acad. Sci. USA* **81**, 2085.

345a. Zwacka, RM *et al.* (1998) Redox gene therapy for ischemia/reperfusion injury of the liver reduces AP1 and NFκB activation. *Nature Med.* **4**, 698.

Notes

[a]For an explanation of proteases see Table 6.2.
[b]Please see Appendix II if further explanation is required.
[c]For a basic description of these organelles see Chapter 3.

10

Ageing, nutrition, disease and therapy: a role for antioxidants?

10.1 Introduction

Chapter 9 showed that oxidative stress is involved in (but not necessarily important in) many, and possibly all, diseases. In some diseases, it makes a significant contribution to tissue injury. Such diseases should be amenable to therapeutic intervention with antioxidants. Additionally, diet-derived antioxidants might be important preventative agents against disease, as was briefly discussed in Chapter 3. Both these issues will be examined in more detail here.

The incidence of many of the diseases considered in Chapter 9, especially that of cardiovascular and neurodegenerative diseases and most of the common forms of cancer, increases with age.[1] In addition, it has been proposed for decades that oxidative stress could be involved in the ageing process itself.

10.2 Theories of ageing

The term 'ageing' is hard to define precisely. It has been described in general terms as a progressive decline in the efficiency of physiological processes (not necessarily all at the same rate) after the reproductive phase of life. It occurs in all multicellular organisms. In addition, the ability of organisms to recover from an insult, such as traumatic injury, decreases with age.[54] One of the many problems in studying ageing experimentally is the difficulty of separating ageing from age-related disease (indeed, it may be impossible). Most old animals (including humans) have some overt, or sub-clinical, disease (e.g. atherosclerosis, myopathy or β-amyloid plaques) that can influence parameters of oxidative damage.

10.2.1 General principles of ageing

As an organism ages, its chance of death increases, so that all individuals of a given species are dead by some age, characteristic of that species. It seems likely that this **maximum lifespan** is around 110–120 years for humans.

In many countries today (and, until very recently, in all countries), few individuals reach their maximum lifespan because of infectious diseases or lack of adequate nutrition. In Western societies, such deaths are now rare. Those few people who die under the age of 35 usually do so as a result of accidents, although deaths from AIDS and suicide are becoming more common. By contrast, older people more often die of cancer or cardiovascular diseases. The overall incidence of cancer varies strikingly with age; for most cancers the rate of incidence rises approximately proportionally to the fourth or fifth power of age (Chapter 9). This is seen both in short-lived species such as rats and mice (about 30% have cancer at the age of 2–3 years), and long-lived species such as humans (about 30% have cancer at the age of 85). Thus the average lifespan (or **mean lifespan**) of 'advanced' societies is greater than that of primitive ones, but the **maximum** lifespan is probably no different.

10.2.2   *What features of ageing must theories explain?*[87]

The nature of the ageing process has been the subject of many theories. Whatever theory is proposed, it must explain several phenomena. First, although maximum lifespan might be fixed for a species, the actual lifespan achieved by a member of that species can be altered by environmental conditions. Temperature is one: cold-blooded animals, such as insects and reptiles, live longer at lower temperatures. For example, the mean lifespan of the fruit fly *Drosophila* is 120 days at 10 °C, but only 14 days at 30 °C.

The most striking effects on lifespan in mammals are provided by studies of **caloric restriction** (eating fewer calories whilst maintaining intake of essential vitamins and minerals). Restriction of food intake during the early growth phase of life has been shown to produce significant increases in mean (and possibly maximum) lifespan of several species, including insects, mice, fish and rats. (It should be noted that the *ad libitum* food supply of laboratory animals is rarely representative of what happens in nature.) Indeed, caloric restriction is the only reproducible laboratory method for extending lifespan in small mammals. It can even delay the appearance of spontaneous cancers in transgenic mouse 'knockouts' lacking the p53 gene. In experimental studies, both the extent of caloric restriction and the age at which it is imposed are critical in obtaining the effect.

In 1987, similar studies began in primates, at the National Institute of Aging in the USA. Two hundred Rhesus and squirrel monkeys are under study, with half eating *ad libitum* and the rest receiving 30% less of the same diet. Data are so far consistent with rodent studies, in that calorie-restricted monkeys are smaller, mature later, and have lower blood glucose and insulin levels, lower body temperatures and increased daytime activity. Much of the evidence for the anti-ageing effects of caloric restriction points to metabolic readjustment in order to utilize energy more slowly and efficiently.

The most dramatic example of the effect of diet on lifespan can be seen with the queen honeybee (*Apis mellifera*). The queen can live for as long as 6 years whereas worker bees live for only 3–6 weeks, yet production of many hundreds of eggs each day requires a high metabolic rate in the queen. The difference in lifespan relates to whether or not the larvae were exclusively fed royal jelly and turned into queens.

Of the theories proposed to explain ageing, most fall into two groups, **genetic theories** and **damage–accumulation theories**. These are discussed further in Sections 10.2.3 and 10.2.6 respectively.

10.2.3   *Genetic theories of ageing*[55,87]

Genetic theories propose that ageing is a continuation of the process of development and differentiation, and is a sequence of events encoded into the genome. Genetic ageing has been suggested in the older literature to be a *purposeful* sequence of events but this seems unlikely. Humans have evolved for much longer with short mean lifespans (32 years in Roman times) than with long ones. Most animals rarely reach maximal lifespan in the wild. What then would be the selective pressure for purposeful genetic changes altering a maximum lifespan that is rarely achieved?

Ageing might also be the delayed effects of expression of genes selected because they enhance reproductive success. The power of natural selection to favour beneficial genes declines with the age at which they affect adult fitness, simply because young parents are the largest contributors to succeeding generations. Hence genes beneficial in early life but deleterious in later life (especially post-reproductively) will not be selected against. For example, the gene defect leading to idiopathic haemochromatosis may convey advantages in early life (Chapter 3). Experimental selection for longevity in *Drosophila* often produces flies with reduced early-life fitness components (e.g. reduced fecundity and ovary size, increased development time or decreased larval survival), consistent with this argument. One interesting observation is that flies selected for 'late-life fitness' often fail to live longer unless they are reared at high density.

Genetics clearly makes a considerable contribution to what is perceived as the ageing process. Work with *Drosophila* (see above), yeast cells, the fungus *Neurospora* and the nematode worm *Caenorhabditis elegans* has established the presence of genes that regulate lifespan. *C. elegans* mutations have been described that extend life expectancy by 40% to more than 100% (doubling lifespan). The first to be discovered were mutations of the ***age*-1** gene; mutations in this gene can increase the lifespan by about 100% but do not affect reproduction or movement. In yeast, increased replicative lifespan is achieved by deleting the gene ***LAG1***, which probably codes for a membrane-bound protein. Two other genes, ***RAS1*** and ***RAS2***, are also involved: deletions in *RAS1* prolong replicative life, but overexpression has no effect. Deletions of *RAS2* shorten lifespan, while overexpression of *RAS2* increases lifespan.

Environmental changes such as low temperature modulate expression of a wide range of genes. Cells cultured from caloric-restricted animals have exhibited reduced oncogene expression, mutation and transformation rates, as well as conserved replicative potential. Hence caloric restriction must also lead to a range of changes in gene expression. In recombinant inbred mice the chromosomal region with the strongest correlation with survival was found to be on chromosome 7, and included a locus encoding cytochrome P450.

10.2.4   *Human disorders of premature ageing*[55]

The human progeroid ('premature ageing') syndromes (**Hutchinson–Gilford syndrome** and **Werner's syndrome**) provide further evidence that some genes can have major effects on ageing. Hutchinson–Gilford syndrome (progeria) is a rare, dominantly inherited disease (one in every 4–8 million births). Affected individuals appear normal at birth but at about 12 months of age severe growth retardation is observed. Loss of hair and subcutaneous fat makes the skin appear aged (Fig. 10.1); pigmented 'age spots' are also present. The patients have an average weight of 25–30 lbs (~11–14 kg) and a height of around 40 inches (~100 cm) as teenagers. They have normal to above average intelligence with a median age of death at 12 years, often due to heart failure.

Werner's syndrome is a rare, autosomal-recessive mutation; it is somewhat more common in Japan. Affected individuals are usually normal during childhood but stop growth during their teens. Greying and whitening of the hair occur at an early age, and the skin appears old with a scaly appearance. The patients develop early cataracts, tumours, bone demineralization and diabetes and show peripheral muscular atrophy, poor wound healing, poor gonad development and accelerated atherosclerosis. They usually die in their 40s from atherosclerosis-related conditions. The defective gene maps to chromosome 8 and encodes a **helicase** enzyme that may be involved in

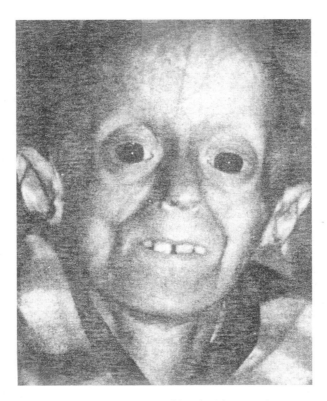

Fig. 10.1. A 12 year old girl with progeria.

unwinding of the DNA double helix to allow access to proteins required for transcription, replication and repair.

The progeroid syndromes individually, however, only show some features of ageing and cannot be cited as truly representative of 'accelerated human ageing'. Many features of normal ageing are not present in progeria; those that are absent include the increased frequency of malignant tumours, cataracts and bone demineralization. Similarly, Werner's syndrome has been called a 'caricature of ageing' rather than an acceleration of the real ageing process. **Bloom's syndrome** (Chapter 9) may also involve a helicase defect.

Transgenic mammalian models to test some of the genetic theories of ageing are being developed, and the results are awaited with interest. In addition, 'rapidly ageing' animal strains are becoming available. One is the **senescence accelerated mouse**, which has a shortened lifespan and shows early signs of ageing.[10]

10.2.5   *Telomeres and telomerase*[82]

Telomeres are structures found at the ends of the chromosomes in eukaryotic cells. Their presence allows complete replication of chromosomal DNA. DNA polymerases use the information in one DNA strand to synthesize another. However, they need an RNA primer to start the new strand off and there must be a place for the primer to attach at the 5' end of the DNA sequence being copied. Without telomeres, there would be a loss of genetic material at each round of replication since RNA is removed from the new strand and the information in the DNA to which it was bound will not be coded in the new DNA strand. Telomeres also protect the chromosome ends from damage or fusion. In most eukaryotes, telomeres are composed of variable numbers (several thousand in humans) of simple repeat sequences (TTAGGG in animals) that are generated by an RNA–protein enzyme called **telomerase**. The RNA provides a template for the addition of telomeric repeats onto DNA. Telomerase activity is regulated by **telomere repeat binding proteins** that decrease telomerase activity when the correct length has been obtained.

Mammalian cells undergo only a finite number of cell divisions when cultured in the laboratory. For example, fibroblasts taken from a human foetus will divide only about 50 times. As mammalian cells approach **senescence** (a terminal G_0 state of non-division), telomeres shorten: 25–200 base pairs are lost at each chromosomal division. Hence telomere shortening has been suggested to be a 'molecular clock' of the ageing process. Short telomeres have been proposed to trigger senescence by signalling a growth checkpoint. Telomerase is absent from most somatic cells (exceptions include bone marrow stem cells and the lymphocytes and other white blood cells derived from them). Telomere length is also maintained in vertebrate germline cells, which contain telomerase. By contrast, telomerase can be detected in many cancerous tumours, suggesting that telomere elongation is involved in inappropriate cell replication and tumour growth. If cultured cells are exposed to certain viruses (e.g. Epstein–Barr virus, human papilloma virus), many cells will die, but 'immortal' cells that continue to grow are soon selected; most such cells have telomerase and do not undergo telomere shortening. Thus inactivation of telomerase is one approach to cancer therapy. Genes affecting telomere length or the rate of telomere loss could obviously influence the ageing process at the cellular level. Patients with progeria, for example, have pronounced shortening of telomeres.

Is cultured cell senescence relevant *in vivo* or is it just an artifact of cell culture? The older an animal from which cells are taken, the fewer cell divisions occur in culture, suggesting some relationship. However, it is hard to answer the question because most cells in animals are not undergoing division and so cannot be identified as senescent or not. It has been suggested[22] that expression of a *β*-**galactosidase** enzyme is a marker of senescence in human cells: this enzyme has been detected in human skin to an extent increasing with age, suggesting that cell senescence may be relevant to the ageing process. Introduction of telomerase into human cell lines in culture can delay senescence and decrease *β*-galactosidase expression.[6a]

10.2.6   *Damage-accumulation theories of ageing*

Multiple 'damage-accumulation theories' have been proposed (Table 10.1) but they tend to follow similar themes. They all assume some progressive accumulation of damage because repair and maintenance are always less than

Table 10.1. Some of the damage-accumulation theories of ageing

Theory	Summary
Free radical	Random deleterious effects of free radicals are produced during normal aerobic metabolism.
Immediate survival	Ageing occurs because nature selects for genes that have immediate survival value, but with long-term damaging consequences.
Cross-linkage	Random cross-links of DNA and proteins disrupt function
Error catastrophe	Cumulative random errors in protein synthesis.
Glycation	Formation of glycated proteins and other molecules leads to AGE formation and serious disruption of cell function.
Longevity determinants	Ageing is caused by the products of cellular metabolism, and the rate of ageing is governed by protection against these damaging products.
Membrane hypothesis	Membrane damage leads to decreased elimination of waste products, decreased protein synthesis and loss of water from the cytoplasm leading to decreased enzymic activities.
Entropy	Mechanisms (e.g. caloric restriction) that reduce the rate of entropy production, liberating energy more slowly, delay molecular deterioration.

those required for 'indefinite' survival. Faulty macromolecules can accumulate through 'wear and tear', failure of systems that repair or degrade them and/or errors in synthesis. Some of these theories (e.g. immediate survival, and longevity determinants) overlap with the genetic theories of ageing. A popular theory in its time was the **error–catastrophe theory** of ageing, introduced in 1963. It proposed that errors in transcription of RNA and its translation into protein would, with ageing of cells, lead to accumulation of altered non-functional proteins, eventually reaching levels that ultimately lead to an 'error catastrophe' of complete failure to function. Some altered non-functional or less active enzymes have been isolated from aged cells (Section 10.3 below). For example, in humans, development of cataract is common after the sixth decade. The crystallins, which constitute more than 90% of the lens proteins, become cross-linked and aggregate.

However, there is no evidence for accumulation of altered proteins to very high ('catastrophic') concentrations in any aged tissue, and little evidence that the fidelity of transcription or translation decreases with age. A similar comment can be made about the 'membrane hypothesis' (Table 10.1).

10.3   Oxidative damage: a common link between all the ageing theories?

The **free-radical theory of ageing** was introduced in 1956 by Denham Harman,[38] who proposed that normal ageing results from random deleterious damage to tissues by free radicals. In subsequent papers, Harman focused on mitochondria as free-radical generators, as well as key targets of damage and Miquel *et al.* proposed a mitochondrial theory of ageing (progressive damage to mitochondrial DNA, e.g. by reactive oxygen species (ROS)). We now know, of course, that many ROS/RNS are not free radicals, examples being H_2O_2, HOCl and ONOO$^-$. Hence perhaps Harman's theory should be renamed the **oxidative damage theory of ageing**.

Harman's theory has some immediately attractive features in relation to explanations of ageing:

(1) ROS/RNS are produced during normal metabolism, sometimes accidentally and sometimes for useful purposes. Antioxidant defences do not protect completely against ROS/RNS-mediated damage and steady-state levels of damage to DNA, lipids and proteins can be detected in aerobic organisms (Table 4.1). Indeed, calculations suggest hundreds of 'hits' per day by ROS/RNS on the DNA of every cell in aerobes,[1,71] a risk factor for the age-related development of cancer in multicellular organisms.

(2) ROS/RNS production may be the consequence of genes selected because they confer benefit in early life (Table 10.1). Such events as phagocyte ROS/RNS production are beneficial in the short-term by preventing death from infections before or during the reproductive years. This benefit may lessen with age, particularly as failure of normal 'recognition of self' with age may provoke increasing 'self-attack' by phagocytes.

(3) Deterioration of tissues with age may lead to increased ROS/RNS production. For example, sub-mitochondrial particles isolated from tissues of aged rats produce more $O_2^{\bullet-}$ than from young rats, suggesting that mitochondrial electron transport gets 'leakier' with age.[87] Levels of H_2O_2-producing enzymes such as monoamine and xanthine oxidases may also increase with age in certain tissues.

(4) There is an inverse correlation between the basal metabolic rates of animals and their lifespan. In general, larger animals consume less O_2 per unit of body mass than do smaller ones, and they live longer. A simple explanation is a lower ROS burden per unit mass of tissue. We saw in Chapter 1 that cold-blooded animals are more resistant to O_2 toxicity at lower temperatures; presumably they make fewer ROS.

(5) Insects consume much more O_2 when flying than at rest. Prevention of houseflies from flying by removing their wings, or by confining them in small bottles, has been shown to produce a marked increase in lifespan (Fig. 10.2). Such effects must be taken into account when interpreting data from genetic manipulation experiments: any change that alters O_2 consumption (e.g. by increasing or decreasing activity) or food intake could produce indirect effects on lifespan.

(6) Lowered O_2 consumption could explain why queen bees (who do not fly for most of their lives) live 50 times longer than actively flying worker bees. However, they must do a lot of metabolic work in egg production.

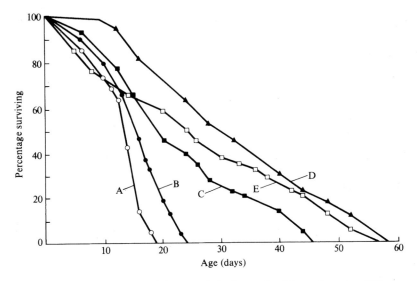

Fig. 10.2. Effect of activity on the lifespan of the housefly, *Musca domestica*. Fifty normal flies (A) or de-winged flies (B) were placed in a large cage. In group **C**, each fly was placed alone in a large cage; the insects are disturbed less often and so and fly less. In other groups, normal flies (**D**) or de-winged flies were placed in small bottles, one per bottle, to prevent flying. Data by courtesy of Prof. Raj Sohal.

(7) The great bulk of O_2 consumed by aerobic eukaryotes is used by mitochondria, so one might expect them to be a focus for oxidative damage.[53,102] Indeed, mitochondrial DNA rapidly accumulate mutations (especially deletions) with age, perhaps due to a combination of rapid oxidative damage and slow repair, and has been suggested to have higher steady-state levels of 8-hydroxydeoxyguanosine (8-OHdG) than nuclear DNA in rat liver, human brain and several other tissues.

(8) Calculations by Cutler *et al.* suggest that longer-lived species have better antioxidant protective mechanisms in relation to rates of O_2 uptake than shorter-lived species. For example, the activities of SOD in different mammals were found to correlate with the metabolic rate multiplied by the maximum lifespan of the species (Table 10.2). Thus the SOD activity of human liver, expressed per unit metabolic rate, is higher than that of other primate species and much higher than that of other mammals. This is not to say that absolute antioxidant levels are higher in human tissues: they may often be lower than in other species, depending on the antioxidant studied. The correlation is with antioxidant activity per unit metabolic rate. Since SOD appears to have a similar structure and function in all animals (Chapter 3), it was suggested that a change in gene regulation has allowed synthesis of higher *relative* cellular SOD concentrations in humans, contributing to longevity. In other words, at least some of the genes encoding antioxidant defences may be 'longevity-determining genes'. Cutler also reported some correlation between lifespan and the concentrations of uric acid, carotenoids and vitamin E in animals (Table 10.2). However, glutathione peroxidase and GSH were not positively correlated with lifespan—in fact, correlations tend to be negative. The negative correlation of glutathione S-transferase (a 'detoxifying' phase II system) with lifespan may relate to other observations showing that levels of cytochrome P450 per unit body mass tend to be lower in long-lived animals than in species such as the rat and mouse. Hence the metabolism of xenobiotics, which can sometimes result in formation of ROS and ultimate carcinogens (Chapter 8) will be lower (per unit body mass) in longer-lived animals. However, the negative correlation of lifespan with glutathione peroxidase, thought to be a key enzyme in the metabolism of peroxides, might be held to cast doubt on the validity of the whole analysis. Perhaps glutathione peroxidase is not as important as commonly supposed (Chapter 3).

(9) Certain species, e.g. rat and pigeon, have similar metabolic rates but different lifespans:[66] (the rat has a mean lifespan of 3 years and the pigeon 30 years). Mitochondria from pigeon tissues generate ROS *in vitro* more slowly than rat mitochondria, providing a possible explanation of the difference.

(10) Senescence in cultured cells can be induced[13] by H_2O_2 treatment (apparently associated with rises in 8-OHdG levels) and has also been observed in transfected cells over-expressing CuZnSOD, an effect preventable by raising the cell glutathione peroxidase activity. Of course, cell

Table 10.2. Relationship of antioxidants to longevity

Degree of positive correlation to maximum life span	Parameters with this degree of correlation
Excellent	SOD activity (expressed per unit metabolic rate) in liver, brain and heart of primate and rodent species
	Plasma uric acid (per unit metabolic rate) in primates
Weak to moderate	Ratio of MnSOD to total SOD activity in brain, but not liver, of primates
	Serum carotenoids in a range of mammals
	Plasma vitamin E (expressed per unit metabolic rate) in a range of mammals
	Plasma caeruloplasmin in primates
Poor or none	Serum vitamin A in a range of mammals
	Plasma ascorbic acid in a range of mammals[a]
	GSH concentrations in lens and whole blood in a range of mammals[a]
	Activity of glutathione peroxidase or glutathione S-transferase in brain, liver or blood of a range of mammals[a]
	Catalase[a]

[a]There is some suggestion that these parameters are negatively correlated with lifespan.
Results are abstracted from the article by Cutler, RG (1984) In *Free Radicals in Biology* (Pryor, WA, ed.), Vol. 6, p. 371. Academic Press, New York.
Great caution must be employed in comparing literature values of antioxidant levels in different species performed in different laboratories by different methods. It should also be noted that there is no relationship between actual concentrations of antioxidant levels and lifespan (for example, SOD activity in tissues of a range of animals does not correlate with lifespan), only levels corrected for metabolic rate.

culture is an abnormal stress (especially culture under ambient O_2) and cells that adapt to it are often not typical of cells *in vivo*. Cell culture media are frequently deficient in antioxidants (such as ascorbate) or in antioxidant precursors (such as selenium).

(11) Caloric restriction in rodents is associated with decreased levels of oxidative damage to DNA, lipids and proteins, increased DNA repair capacity and decreased production of $O_2^{\bullet-}$ by isolated sub-mitochondrial particles.[87] In general, antioxidant defence enzymes do not rise in such animals. Ascorbate levels often fall,[93] perhaps because ascorbate is made from glucose and this substrate may be less available.

10.3.1 *Experimental tests of the theory: altering antioxidant levels*[38]

Attempts have been made to test the role of ROS in the ageing process by supplying antioxidants to various organisms and examining their effect on longevity. Thiol compounds (such as glutathione and mercaptoethylamine) and chain-breaking lipid soluble antioxidants such as butylated hydroxytoluene, α-tocopherol, and **Santoquin** (ethoxyquin; see Table 10.11 for structure) have often been used. The most striking results have been obtained with lower organisms. Thus, in the laboratory, vitamin E prolongs the lifespan of several simple organisms such as *Drosophila*, nematode worms and the rotifer *Philodina* (Fig. 10.3).

However, the effects of administered antioxidants on the lifespan of mammals are small or zero. For example, high-dose vitamin E administration has no significant effect on lifespan in rodents, although it has sometimes been reported to diminish the decline in immune response with age. Early claims that antioxidants such as mercaptoethylamine and Santoquin raise the mean lifespan of mice by up to 18% have been challenged; apparently some of the control animals did not live as long as they should have done and so the apparent increase in lifespan might have been caused by a diminution of some environmental stress by the antioxidants rather than an effect on the ageing process itself. Thus the antioxidants might have acted to diminish tissue damage produced by free-radical reactions, e.g. involving toxins in the food or in the surrounding air, background exposure to UV light or other radiation, or from excessive amounts of polyunsaturated fatty acids (PUFAs) in the diet. We have

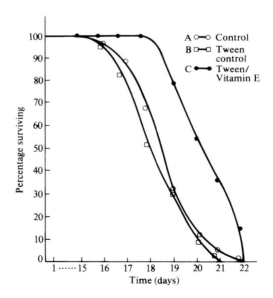

Fig. 10.3. Survival of the rotifer *Philodena* as affected by vitamin E. A rotifer was classified as dead if it did not move when prodded with a pipette. Three groups, each of 32 rotifers, were used: (A) control, (B) Tween (a detergent) added to the culture medium; (C) vitamin E dissolved in Tween was added to the culture medium. Data from *Exp. Gerontol.* 15, **335** (1980) by courtesy of Drs H. Enesco and C. Verdone-Smith and the publishers.

no real idea what is the *optimal* diet for small organisms such as *Philodina* (Fig. 10.3) and many laboratory animals may be overfed in terms of calories, as are most humans in the 'advanced' world. The antioxidant (e.g. vitamin E) content of laboratory animal diets is another important variable, and has tended to increase over the past three decades. In some early studies, effects of antioxidants might have been corrections of dietary deficiencies that shortened lifespan rather than real effects on lifespan in an optimally nourished animal.

Another problem is that excessive dosing of animals with an antioxidant might decrease the rate of synthesis or uptake of 'natural' antioxidants, so that the total 'tissue antioxidant potential' remained unaltered. Consistent with this, feeding female rats a diet rich in the synthetic antioxidant butylated hydroxytoluene (BHT) decreased liver α-tocopherol content.[84] Similarly, inhibiting catalase in houseflies (by feeding them aminotriazole) or inhibition of CuZnSOD (by diethyldithiocarbamate) led to increases in glutathione concentrations; neither aminotriazole nor diethyldithiocarbamate decreased the lifespan of the insects. One treatment that did decrease lifespan was giving the insects an excess of iron, which is particularly interesting in view of the stimulatory role of certain iron chelates in free-radical reactions (Chapter 3). Iron accumulates in most animal tissues with age. It may be safely sequestered in non-catalytic forms (e.g. ferritin),[74] but damage to tissues could release more iron from older tissues than younger ones (Chapter 4). One study reported that the rate of peroxidation of homogenates of brain and kidney from 24 different mammalian species (their **peroxidation potential**)[19] could be correlated with lifespan; peroxidation of tissue homogenates is highly dependent upon released iron.

Another point that must be made is that antioxidant inhibitors of lipid peroxidation, such as α-tocopherol and Santoquin, might not necessarily protect against free-radical damage to DNA and proteins, nor could any molecule acting only as a scavenger protect against 'site-specific' damage by radicals such as $OH^\bullet$.

Overall, therefore, it is difficult to interpret the data available on lifespan changes (or the lack of them) during antioxidant feeding trials on animals.

10.3.2   *Transgenic organisms*[55,87]

Perhaps the simplest way of evaluating effects on lifespan is to selectively manipulate antioxidant defences and examine the effects. 'Knockout' mice lacking MnSOD do not live long enough for ageing rates to be examined, but knockout mice lacking CuZnSOD and glutathione peroxidase do survive for longer periods and may become useful models (Chapter 3). Animals overexpressing genes, such as mice with human CuZnSOD (Chapter 3), may also be useful. It is, of course, essential before interpreting the data to check that genetic alterations do not lead to secondary changes in other antioxidant defences or behavioural changes (e.g. impaired appetite) that can themselves influence lifespan.

Despite the availability of transgenic mammals, few studies have yet employed them to study ageing questions. One intriguing piece of data is that treatment of rats with the monoamine oxidase inhibitor deprenyl (Section 9.17) raises CuZnSOD activity in the brain (but not glutathione peroxidase) and increases mean lifespan.[50]

Most antioxidant/lifespan transgenic work to date has been carried out on *Drosophila*. Overexpression of CuZnSOD by 32–42% above normal had, at best, only a minor effect on lifespan and the ability to withstand oxidative stress, although selective over expression in motor neurones did appear to increase the longevity of *Drosophila*.[65a] Overexpression of the catalase gene was also ineffectual. However, when genes for SOD and catalase were both overexpressed, the mean and maximum lifespans of *Drosophila* increased by up to one third. The normal age-related loss of function was decreased, and the flies showed lower levels of protein carbonyls and 8-OHdG, and increased resistance to DNA damage by ionizing radiation. The flies were more active than controls and their calculated 'lifetime O_2 consumption per unit body mass' had increased. By contrast, *Drosophila* mutants lacking SOD show a markedly decreased lifespan. Since oxidative damage may be particularly severe in mitochondria, it would be of interest to examine the effects of over-expressing MnSOD.

Several other observations are consistent with a key role for antioxidant enzymes in ageing. All 'extended lifespan mutants' of *C. elegans* are more resistant to ROS-generating agents (including UV light), and the ***age*-1** mutants show enhanced expression of SOD and catalase. The ***age*-1** gene does not encode either of these enzymes, but it could perhaps encode a factor that regulates genes encoding the enzymes. The requirement for high-density living conditions needed for phenotypic expression of 'extended longevity mutants' in *Drosophila* can be explained since these conditions favour enhanced transcription of antioxidant-encoding genes.

Scientists have examined progeria and Werner's syndrome patients for abnormalities of antioxidant defences, but none have been found.

10.3.3   *Does antioxidant protection fail with age?*

Generally, it seems that antioxidant protection does not fail with age,[48] but there are some specific exceptions. This does not rule out the free-radical theory of ageing since many other events are involved (Table 10.3).

In general, there are no marked falls in antioxidant defence with age in humans or other animals and some increases have been reported, e.g. vitamin E:PUFA ratios tend to rise with age in some human cells (e.g. erythrocytes) and antioxidant defence enzymes tend to rise in the brains of rats. By contrast, SOD purified from the livers of old rats is less stable to heating, and of lower activity, than the enzyme from young rats, apparently because of oxidative damage to histidine residues.[78] Since altered SOD enzymes can be toxic in ALS (Section 9.19), it is interesting to speculate that age-related changes might render the protein mildly neurotoxic and contribute to neuronal death in the

Table 10.3. Genetic changes that could affect oxidative damage

System/factor affected by genetic change	Examples
Antioxidant defence systems	SOD, catalase, glutathione peroxidases, thiol-specific antioxidants
Low-molecular-mass antioxidants	Enzymes synthesizing and catabolizing GSH, ascorbate (non-primates), urate (non-primates), histidine–dipeptides, etc.
Repair systems	Proteasome, other proteases, DNA repair, phospholipid hydroperoxide glutathione peroxidases, enzymes that metabolize cytotoxic aldehydes (e.g. glutathione-S-transferases)
Availability of transition-metal ions 'catalytic' for free-radical reactions	Transferrin, ferritin, caeruloplasmin, metallothionein, haemopexin, haptoglobin, Cu^{2+}-ATPases, lactoferrin
Targets of oxidative damage	Alterations in conformation of chromatin, membranes, lipoproteins, etc. that make it easier or harder for ROS/RNS to attack sensitive targets. Changes in DNA methylation (e.g. 8–OHdG in DNA alters methylation of adjacent cytosines; *Mutat. Res.* **386**, 141 (1995)).
Uptake or processing of dietary antioxidants	Vitamins C (primates) and E, rate and type of carotenoid cleavage in gut, flavonoid metabolism
Free-radical production	Types of cytochrome P450, 'leak rates' from mitochondrial electron transport (e.g. see *Nature* **394**, 694 (1998)).
Rate at which oxidatively damaged cells die	p53, *bcl-2*, *bax*, other genes affecting cell cycle and apoptosis/necrosis

In addition, changes in genes encoding transcription factors such as AP-1 or NF-κB could affect any or all of the above, e.g. they could affect the ability to up-regulate antioxidant defences in response to stress. The nuclear binding activity of NF-κB in mouse and rat tissues has been reported to rise with age, but to decrease during senescence of human fibroblasts in culture (e.g. *Biochem. J.* **318**, 603 (1996)).

ageing brain. Similarly, caeruloplasmin isolated from the plasma of old people shows distinct age-related changes to the copper centres that are associated with a loss of oxidase activity and higher levels of protein carbonyls.[63] It must not be forgotten that the *balance* of antioxidant activities is as important as the actual activities, e.g. high CuZnSOD:glutathione peroxidase ratios may be injurious to cells (Chapter 3).

Another factor to be considered is the ability to up-regulate antioxidant defence enzymes and other protective proteins[54] in response to oxidative and other stresses (Table 10.3). If this is insufficient, damage can occur even though basal levels of defences appear normal. There is evidence that the heat-shock response is attenuated in older animals. A fungal example may be provided by a rapidly ageing mutant of *Neurospora crassa*.[72] The mutant ages and accumulates fluorescent pigment (Section 10.3.6 below) more rapidly than does the wild-type strain. However, accumulation of pigment is decreased, and lifespan increased, by including in the culture medium antioxidants such as thiol compounds or α-tocopherol. This mutant has *increased* activities of SOD, catalase and glutathione peroxidase when compared with the wild-type fungus. Perhaps the mutant suffers an increased rate of ROS generation, which stimulates the synthesis of the above protective enzymes, but not to a sufficiently great extent.

10.3.4   *Does net oxidative damage increase with age?*

Measurements of 'markers' of damage in animal tissues suggest that, in general, net oxidative damage does not increase markedly with age, but there are some exceptions.[90] For example, the membranes of the oldest erythrocytes in human blood show decreased fluidity and increased cross-linking of membrane proteins. However, erythrocytes are unusual cells; studies on nucleated post-mitotic cells are more relevant. The ageing lens shows increased protein modification (Section 7.7).

Older rats have been claimed to exhale more ethane and pentane than younger ones.[76] Levels of protein carbonyls and 8-OHdG in the brains and other tissues of rats, Mongolian gerbils, mice and humans show a trend to rise with age and, in the old mouse, brain carbonyl levels correlate with loss of cognitive function.[28] The senescence–accelerated mouse (Section 10.2.4) accumulates brain protein carbonyls more quickly than wild-type.[10] Protein carbonyl levels increase with age in some cell lines and carbonyl levels in fibroblasts cultured from patients with progeria or Werner's syndrome are strikingly elevated.[90] AGE products slowly accumulate with age in such human proteins as collagen and crystallins, which have a slow turnover rate; formation of AGE products involves both oxidation and glycation steps. In rat liver, the enzyme **carbonic anhydrase III** seems to undergo exceptionally rapid oxidative modification with age.[11]

There is evidence that oxidative damage may be focused on mitochondria (Section 10.3 above). Thus large changes in mitochondrial oxidative damage parameters or antioxidant levels could be 'swamped out' by lack of changes in other cell compartments if the whole cell or tissue is assayed, since mitochondria make up only a small fraction of cell constituents. For example, mitochondrial DNA is less than 5% of total cell DNA. In houseflies, changes in metabolic rate are closely paralleled by changes in such parameters as 8-OHdG and protein carbonyls, e.g. decreases in physical activity decreased both parameters. Again, mitochondrial DNA accumulated 8-OHdG more quickly with age than nuclear DNA, and aconitase activity fell.

Steady-state levels of oxidative damage are a balance between rates of damage and rates of repair, or replacement, of damaged molecules (Chapter 4). Hence rises in damage levels could be due not only to more damage but also to failure of repair systems with age. A positive correlation between the efficiency of DNA-repair enzymes and species longevity has been claimed,[5] i.e. genes encoding repair systems may contribute to longevity (Table 10.3). The capacity of various cell lines in culture to degrade abnormal proteins and repair DNA seems to decline as they undergo senescence.

10.3.5   *Lipofuscin*[103]

One of the earliest pieces of evidence put forward to support the free-radical theory of ageing was the presence in old tissues of **age pigments**, thought to be end products of lipid peroxidation. The first description of the intracellular pigment known as **lipofuscin** was made in 1842 by Hannover, who reported its presence in neurones. The fluorescent properties of the pigment were described in 1911, and it has been found to accumulate, in amounts increasing with age, in many tissues, both in humans and in a wide variety of animals, including rats, nematodes, *Drosophila* and houseflies. It is also found in several fungi and in cells in culture.

In general, the most metabolically active tissues show most lipofuscin deposition. In houseflies, the lipofuscin content of muscles increases with flight activity; flies that are more active live for shorter times (Fig. 10.2) and accumulate this pigment more quickly. There is little or no lipofuscin in human heart muscle up to the second decade of life, but it then accumulates at a rate of about 0.3% of the total heart volume in each further decade of life. On the other hand, lipofuscin appears earlier in some tissues, such as the spinal cord. Large motor neurones of human centenarians may be more than 70% occupied by 'age pigments'.[a]

Lipofuscin varies in colour from red, through yellow, to dark-brown, and it occurs intracellularly as granules bounded by a single membrane, their diameters being in the range 1–5 µm. Both the number of granules, and their size, increase with age. Many different types of lipid are present in the pigments, including triglyceride, phospholipid and cholesterol, and an equally wide variety of proteins as judged from the amino-acid composition of hydrolysed pigments. For example, AGE-modified proteins can have fluorescent properties, and lipofuscin within the retinal pigment epithelium contains large amounts of retinal derivatives. Lipofuscin contains a high concentration of metal ions such as zinc, copper and, especially, iron.

Extraction of lipofuscin granules with a mixture of organic solvents (chloroform plus methanol) solubilizes part of the material, and the solution so obtained shows fluorescence characteristics quite similar to those of the conjugated Schiff bases, dihydropyridine dicarbaldehydes and aldehyde polymers formed during lipid peroxidation (Chapter 4). However, all of these extracted fluorophores display blue fluorescence, whereas microscopists observe green to yellow fluorescence from lipofuscin *in situ*. This suggests that the extracted material is not necessarily representative of the tissue fluorophores.

On the basis of the fluorophores extracted by organic solvents, it is widely thought that lipofuscin represents the end-product of the oxidative destruction of lipids, and their cross-linking with proteins and other compounds bearing amino groups. However, it may be misleading to emphasize a central role for lipid peroxides in the composition of lipofuscin. For example, the pigment accumulating in most neuronal ceroid lipofuscinosis patients is largely composed of the mitochondrial subunit protein 9 (Section 9.20). As we noted in Chapter 4, fluorescent material can be generated *in vitro* by oxidative damage to carbohydrates, DNA and proteins, as well as to unsaturated lipids.

10.3.6   *Ceroid*[103]

There is a stronger case for implicating increased lipid peroxidation as a key event in formation of the fluorescent pigment **ceroid**, a term first introduced in 1942. The amount of this pigment in animal tissues is greatly increased if the animals are fed on diets deficient in vitamin E or abnormally rich in PUFAs. Such feeding induces a so-called 'yellow fat disease' in, for example, pigs and minks that can be prevented by feeding excess vitamin E or other antioxidants such as N,N'-diphenyl-p-phenylenediamine (Chapter 3). Macrophages in culture accumulate ceroid when they are exposed to complexes of albumin with lipids rich in PUFAs (Fig. 10.4). Ceroid formation does not take place, however, when the lipids contain monounsaturated fatty acids, nor if lipid-soluble chain-breaking antioxidants are added to the culture medium with the lipids. Peroxidation of lipids by macrophages is involved in the pathogenesis of atherosclerosis (Chapter 9); both ceroid within macrophages and extracellular ceroid (presumably released from dead cells) are present in human atherosclerotic lesions.

Thus ceroid and lipofuscin must not be thought of as chemically identical, and the amount of lipofuscin accumulated does not necessarily reflect the rate of free-radical reactions in a tissue. Nor is there clear evidence that 'age pigments' are deleterious to cells. Another reason for caution comes from a consideration of the way in which fluorescent pigments arise. Histochemical studies show that they are derived from lysosomes. Lysosomes are continually digesting parts of the cell cytoplasm, a process known as **autophagy**. Proteins and lipids are taken into the lysosomes and degraded. It could be that lysosomes have a special affinity for peroxidized lipids and so gradually accumulate them,

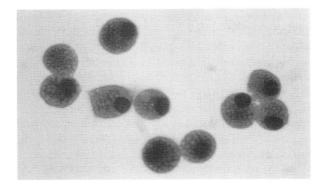

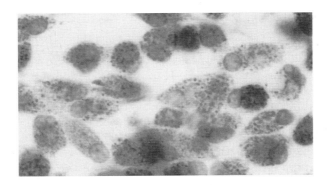

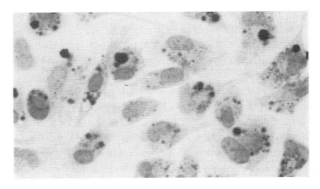

Fig. 10.4. Ceroid in mouse peritoneal macrophages. Top: cells maintained for 3 days in the presence of cholesterol linoleate ester. 'Rings' of ceroid are present with unstained centres; there are also some granules of ceroid. Magnification: × 1000. Middle: as top, except that cells were grown in the presence of cholesterol arachidonate. Cells contain multiple small granules of ceroid rather than rings. Bottom: as top, but with trilinolein (no albumin). Cells contain small granules and larger lumps of ceroid. Photographs by courtesy of Drs Keri Carpenter and M.J. Mitchinson. No ceroid was found within cells grown in the presence of cholesterol alone or a cholesterol oleate ester (not shown), i.e. ceroid accumulated only when the cells were incubated in the presence of a peroxidizable lipid material.

but it seems more likely that lysosomes accumulate 'normal' lipids that are peroxidized more rapidly once they are inside these organelles. Disruption of lipid organization by lysosomal hydrolytic enzymes should aid lipid peroxidation, as would the high internal concentrations of copper and iron salts within lysosomes. These metal ions are probably derived from ingested metalloproteins, including transferrin. The acidic pH of lysosomes should aid iron-dependent radical reactions, in part by helping to keep iron ions in solution.

Lipofuscin and ceroid deposition are promoted by several abnormalities of fat metabolism, including **abetalipoproteinaemia** (Chapter 3). Presumably more lipid than usual is degraded within lysosomes in such diseases. Hence accumulation of age pigments should not be taken as evidence that lipid peroxidation is important in the disease pathology.

10.3.7   *The free-radical theory of ageing: current status*[38,87]

The concept that ageing is purposefully encoded into the genome and most of the damage-accumulation theories can, we feel, be dismissed in the forms in which they were originally stated. Harman's free-radical theory, expanded in the light of modern knowledge, still persists and is gaining credence because it forms a link between most or all of the other suggested theories: errors in proteins, membrane damage and failure of repair can all be accommodated. Many longevity-determining genes could be, directly or indirectly, linked to oxidative damage (Table 10.3).

As the search for genetic determinants of ageing continues, some simple principles must not be forgotten. One is the close relation of ageing to age-related disease.[1] In one study, high immune responsiveness in mice was associated with longer lifespan. In a study of French centenarians, presence of the *apoE4* allele was associated with shorter life expectancy. But are these effects on ageing itself, or do mice with poor immune systems die earlier from infection, or humans with *apoE4* succumb to Alzheimer's disease (Section 9.18)?

Do ROS/RNS *cause* ageing? It is too early to say, but they seem to make a contribution. Remember, however, that as tissues deteriorate with age, more oxidative damage is a likely consequence of tissue disorganization and cell death (Chapter 9); rises in parameters of oxidative damage with age do not prove a cause–consequence relationship. A second important question is mechanism—are there specific molecular targets of oxidative damage or is there just a random attack on all biomolecules? More research is needed to answer this question.

10.3.8   *Differentiation*

An involvement of ROS in the process of cell differentiation has been suggested, based on studies with the slime mould *Physarum polycephalum*[88] and the fungus *Neurospora crassa*.[72,94] In both cases differentiation appears to be accompanied by increased oxidative stress (e.g. changes in cell redox state may be the trigger for differentiation). Indeed, addition of antioxidants can modulate the process.

10.4   Nutrition, health and oxidative stress

That food and health are intimately linked[100] is an old concept; in 400 BC the physician Hippocrates wrote, 'let food be your medicine and medicine be your food'. Indeed, it is widely thought that diet-derived antioxidants play a role in the prevention of human disease (Chapter 3). The evidence is strongest for a protective role of vitamin E against cardiovascular disease,[8,31,73] but there are multiple suggestions that carotenoids[9,31] and plant phenolics (e.g. in tea[59]) play important disease-preventing antioxidant roles *in vivo* (Chapter 3).

There is a need to assess these proposals experimentally, e.g. by testing not only the effect of antioxidants in preventing disease, but also the ability of different diets or 'antioxidant' supplements to modify parameters of oxidative damage in the human body.[36] Just because an antioxidant is found to prevent a disease, it must not be assumed that it does so by an antioxidant mechanism. For example, β-carotene shows limited antioxidant activity *in vitro* but is also a precursor of vitamin A and may exert effects on intercellular communication. Vitamin C intake may influence blood levels of fibrinogen; high plasma fibrinogen is a risk factor for cardiovascular disease.[49] Flavonoids in red wine (and in chocolate and cocoa) have been speculated to be cardioprotective,[39,77] but alcohol itself (in moderation) seems cardioprotective, in part by raising plasma levels of high-density lipoproteins (HDLs) and decreasing platelet aggregation.

By measuring net oxidative damage to DNA, proteins and lipids in the human body ('net' being the steady-state level, i.e. the balance between damage and repair) as well as indices of 'total' oxidative damage, the effects of dietary manipulations can be examined. Some of the methods available for such studies are reviewed in Chapter 5.

10.4.1   *Lessons from epidemiology*[92]

A common approach to gathering information about the putative disease-preventing role of antioxidants, and one that has provided much of the evidence for the importance of vitamin E, is **epidemiology**. Epidemiology is the science that deals with the distribution of, and the determinants of, health and illness in populations. The methods of this science were first applied to epidemic diseases (such as smallpox and cholera), from where the name arises. One early example is the discovery that cholera incidence is related to consumption of water from certain sources; a more recent one is the discovery of a strong association between the incidence of lung cancer and cigarette smoking.

Epidemiological studies may be essentially **descriptive** (e.g. relating the incidence of a disease to a particular diet) or **experimental** (e.g. taking a dietary supplement and examining what happens). The former type measures disease patterns and dietary habits in groups or populations, and correlates the data, searching for associations. The latter type tests directly the reality and importance of the correlations observed.

One example of the former type of epidemiological study is the **Rotterdam Study**.[45] The cognitive function of 5182 subjects was measured and correlated with diet as estimated by a food-frequency questionnaire. After adjustment of the data for age, education, sex and smoking, a lower dietary intake of β-carotene was associated with impaired cognitive function but no such association was observed for vitamins C or E. However, in a different study,[86] examination of a population of elderly nuns found no relation of 'dependency on care' to plasma β-carotene levels, although there was an inverse relationship to plasma lycopene levels.

Many countries conduct regular surveys of food consumption patterns, nutritional status indicators and incidence of various diseases. Epidemiologists can use these data in attempts to relate disease incidence to diet in different

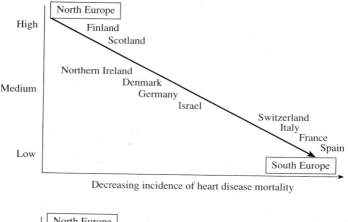

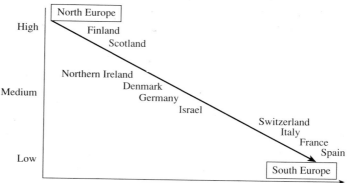

Fig. 10.5. A cross-cultural study: mortality due to heart disease (above) and plasma vitamin E levels (below) in different European countries. Diagram adapted from *Am. J. Clin. Nutr.* **53**, 3265 (1991) by courtesy of the publishers and Dr Fred Gey.

countries (**cross-cultural studies**), within regions of the same country, or over time (**longitudinal studies**)—looking for changes in disease incidence and attempting to relate them to changes in diet. For example, one can compare the incidence of myocardial infarction (Fig. 10.5)[33] or stomach cancer between different countries, and can also observe what happens to incidence rates when people move between countries.

10.4.2 *Problems of interpretation*

All these studies rely on accurate measurements of both diet and disease incidence, neither of which is easy. The accuracy of medical statistics can vary widely between countries. Dietary intake is commonly measured using 24 h dietary recalls, keeping diaries of food intake over several days, or food-frequency questionnaires. Food-composition tables can then be used to calculate nutrient intake. All these methods have their problems, as one can easily imagine. Too much random error in assessment of diet or disease incidence will obscure any relationship. Food-composition tables may be inaccurate and the composition of any given food may vary between countries and regions. Bias in recording can alter risk estimates (e.g. if obese people systematically under-record their calorie intake whereas non-obese people do not).

An additional approach is the measurement of 'biomarkers' of dietary intake, e.g. white cell ascorbate, plasma β-carotene, plasma α-tocopherol, or levels of α-tocopherol in adipose tissue or erythrocytes. More studies are needed on the relation between dietary intake and these various parameters. For example, concentrations of α-tocopherol in plasma change quickly after alterations in dietary intake whereas concentrations in adipose tissue do not. However, the plasma concentration of α-tocopherol correlates poorly with LDL α-tocopherol content.

Despite these problems, progress has been made. For example, there are clear differences in cardiovascular disease incidence in different European countries: in general the incidence is high in the north (e.g. Finland and Scotland) and low in the south (e.g. Southern Italy).[33] If populations are standardized on the basis of known risk factors (e.g. plasma cholesterol or smoking), there is an inverse correlation between coronary events and lipid-standardized plasma vitamin E levels (Fig. 10.5). Examination of data from this and several other studies has allowed determination of levels of plasma antioxidants that seem associated with low risk of cardiovascular disease (Table 10.4). Of course, these data alone do not *prove* that the higher concentrations of vitamins E and C and β-carotene are responsible for the lowered risk of cardiovascular disease. The **Zutphen elderly study**,[39] a longitudinal investigation of risk factors for chronic disease in old men, found a decreased risk of mortality from heart disease in subjects with elevated flavonoid intake, even after correction for intakes of vitamins C and E and β-carotene (Table 10.4). This is consistent with (but does not prove) a cardioprotective effect of flavonoids. If the effect is real, it would presumably be due to the ability of flavonoids to inhibit lipid peroxidation (Chapter 3), although this explanation must not be assumed.

Table 10.4. Relationship between risk of coronary heart disease (CHD) and plasma concentrations/intakes of certain antioxidants

Antioxidant	Concentration associated with	
	Moderate risk of CHD[a]	Small risk of CHD[a]
Carotene (β-carotene with about 15–25% α-carotene) (μM)	<0.3	>0.4–0.6
α-Tocopherol (μM)	<24	>27–28
α-Tocopherol:cholesterol ratio (μmol/mmol)	<4.1	>4.8–5.6
Ascorbate (μM)	<24	>35–60

	Daily flavonoid intake (mg) at baseline[b]		
	0–19	19.1–29.9	>29.9
Mortality from CHD	20.4	14.5	9.9
Relative risk after correction for intake of total energy, saturated fats, blood pressure, physical activity, weight, smoking, serum and HDL cholesterol	1.00	0.58	0.47

[a]Moderate increase in risk of CHD, >250 deaths/10^6; small risk, <130 deaths/10^6. Data are abstracted from Gey, KF (*Bibl. Nutr. Diet.* **51**, 84 (1994) and *J. Nutr. Biochem.* **6**, 206 (1995)) and are based on data from a wide range of epidemiological and intervention studies.
[b]From Hertog, MGL *et al.* (1997) *Lancet* **349**, 699.

10.4.3   *Types of study*[92]

In **cohort studies**, subjects are identified on the basis of their exposure to the factor of interest and followed to examine their future disease status. Cohort studies can be **prospective**, subjects being followed forwards in time (e.g. plasma antioxidant levels might be measured at the start of the study and then disease appearance in the subjects followed up). They may also be **retrospective**, where groups are identified on the basis of exposure in the past and then followed to the present to establish incidence of disease. An example of the latter would be relating mesothelioma incidence to prior exposure to asbestos; an example of the former would be measuring blood selenium levels in a group of subjects and then attempting to relate this to which subjects developed cancer.

In **case–control studies** subjects are recruited on the basis of the presence or absence of a particular disease. Controls are randomly selected from the same groups and matched for other criteria (e.g. age, sex and smoking status) as far as possible. An association between the disease and a specific factor is inferred if the frequency of the latter is greater in the cases; associations are usually expressed as an **odds ratio** (or **relative risk**). One problem to be considered is that the role of diet in affecting many chronic diseases may be most important before onset of the disease or in its early (often pre-diagnosis) stages, and current dietary patterns may be different from that time. Measurement of past diet using questionnaires suffers from obvious problems of inaccurate recall.

Nested case–control studies are case–control studies using patients from ('nested in') a cohort study, who develop a particular disease. Exposures measured as part of the cohort study are compared between patients and controls from within the same cohort. For example, the **Honolulu heart study** is examining a cohort of 8006 Hawaiian men whose vitamin E intake was assessed at enrolment. They are being followed to examine the development of coronary heart disease. A nested case–control study of 84 of these subjects who developed Parkinson's disease showed that, by comparison with 336 age-matched controls from the same cohort, there appeared to be no relationship of dietary E intake to the risk of developing PD. However, data from the Rotterdam study described above suggested a negative association between high vitamin E intake and risk of developing PD.[45,62a]

10.4.4   *Cause and consequence*

The main problem with any epidemiological study is that correlation does not imply causation (Table 10.5). Thus cross-cultural studies show that there is a relation between cardiovascular disease rates and the number of television sets or telephones per thousand members of the population. This is (presumably!) not a cause-and-effect relationship but due to the fact that TV sets and telephones are markers of increasing 'prosperity', itself usually associated with increased incidence of cardiovascular disease.

It is widely agreed by nutritionists that diets rich in vegetables and fruit are associated with lowered incidence of certain types of cancer (Table 10.6).[100] One common constituent of plants is β-carotene. In humans, the higher the blood plasma β-carotene level, the lower the risk (on a population basis) of developing some forms of cancer.[9,31] But does β-carotene itself protect against cancer? Or is it simply a marker for diet? Thus eating more vegetables might raise β-carotene levels, but it could be anything (or any combination of things) in the vegetables that is the real protective agent. Plants contain a huge range of potentially protective agents (Table 10.7).[46,70,97] Also, a vegetable-rich diet is often low in fat, and high fat intake has been postulated as a risk factor for some cancers.[100] Multiple **toxins** have also been identified in plants (Table 10.7); toxins may, ironically, be **protective** if they induce defence mechanisms.

Similarly, suppose that a group of subjects is known to consume large amounts of vitamins E and C and also shows lower incidence of heart disease than another group. Does this mean that vitamins E and C prevent heart disease? Not necessarily, since consumers of vitamin pills might be more interested in their health than average, so that they could also be less obese and smoke less. Poor people suffer more disease than richer people, and poor people may not be able to afford vitamins. Suppose that food-frequency

Table 10.5. Commonly used criteria for determining causation from epidemiological data: a critique

Proposed criterion	Comment
Biological plausibility, i.e. there are physiological reasons for expecting the correlation. For example, LDL oxidation is involved in atherosclerosis and is inhibited by vitamin E, so one would expect vitamin E to have an anti-atherosclerotic effect.	Runs the risks of ignoring phenomena which, at the time, have no obvious explanation yet for which the statistical evidence is equally strong. Also runs the risk of assuming what you are supposed to be trying to prove, as happened with β-carotene (Section 10.4.10).
Strength of the correlation: strong relation observed consistently in several population based studies conducted by different research groups in different countries.	The correlation is so clear-cut that it is difficult to dismiss it, e.g. that between cigarette smoking and lung cancer. In 1854 Dr John Snow was so convinced by the strength of the correlation between cholera and drinking water from a particular well that he removed the pump handle (an intervention trial which stopped the epidemic).
The effect is **dose-responsive**, e.g. the lower the intake of a given nutrient the higher the incidence of a particular disease. Often subjects are divided into tertiles (groups of 33.3%) or quartiles (25%) of intake, blood level, etc.	A negative result can occur if there is a correlation only over a given range of intakes. Suppose, for example, that 50 mg of vitamin C per day is sufficient to give its maximum protective effect against cardiovascular disease (Table 10.4). A study of vitamin C intake versus such disease in a country with a well-nourished population (average vitamin C intake well above 50 mg/day) will not show a relationship.
Trends in the population over time match changes in disease incidence.	An example is the alteration of lung cancer incidence with changes in smoking habits in several countries.

None of these criteria alone can prove causation: direct experimental intervention is required to do that.

Table 10.6. Summary of epidemiological studies of the relationship between fruit and vegetable intake and cancer prevention

| Site | No. of studies | No. reporting significant effect | | Odds ratio (mean) |
		Protective effect	Harmful effect	
All sites	170	132	6	
Lung	25	24	0	1.2–7.0
Larynx	4	4	0	2.1–2.8
Oesophagus	16	15	0	0.7–4.8 (2.0)[b]
Stomach	19	17	1	0.5–5.8 (2.5)[b]
Colorectal	27	20[a]	3	0.3–3.3 (1.9)[b]
Bladder	5	3	0	1.6–2.1
Pancreas	11	9	0	1.4–6.4
Cervix	8	7	0	1.2–4.7
Ovary	4	3	0	1.1–2.3
Breast	14	8	0	1.1–2.8
Prostate	14	4	2	0.6–3.5 (1.3)[b]

The data (adapted from Block, G (1992) *Nutr. Cancer* **18**, 1) review multiple studies of the relationship between intake of fruit and vegetables to cancer incidence. For many cancer sites (especially lung), persons in the lowest quartile of fruit/vegetable intake experience an approximate doubling in the odds ratio (relative risk) of cancer development.
[a] In two studies, results were significant in both the protective and harmful direction for different fruits and vegetables.
[b] Mean values in parentheses.

Table 10.7. Anti-carcinogenic factors in plant foods

Class	Examples
Antioxidants	Vitamins C and E; carotenoids? Flavonoids and other phenols (in wine, cocoa, chocolate)? Folic acid (may decrease homocysteine levels). Phytates (metal chelators). β-Carotene may decrease oxidative damage in porphyria patients (Chapter 2) and in cystic fibrosis (Chapter 9).
Agents scavenging potential carcinogens	Vitamin C reacting with nitrosamines in stomach. γ-Tocopherols may scavenge RNS (Chapter 6).
Agents interfering with metabolic activation or promoting metabolic deactivation, of carcinogens	Inhibitors of P450 (e.g. constituents of grapefruit juice and garlic).[b] Quinone-reductase inducers e.g. quercetin, coumarin, 1,2-dithiole-3-thiones (e.g. **Oltipraz**), sulphoraphane,[a] other isothiocyanates, rosemary extract. Glutathione-S-transferase inducers, e.g. quercetin, catechins, sulphoraphane. UDP–glucuronyltransferase and epoxide hydrolase inducers. Often compounds induce both these and other 'phase II' enzymes (e.g. epoxide hydrolases, sulphate-conjugating enzymes). Agents raising GSH levels.

Table 10.7. (*Continued*)

Class	Examples
Inhibitors of cell proliferation	Green tea phenolics (e.g. catechins), flavonoids. Several suggestions that green (and to a lesser extent, black) tea consumption has an anti-cancer effect.
Proteinase inhibitors	legumes (e.g. soybeans)
Antagonists of oestrogen's action in promoting growth of certain tumours	Isoflavonoids, indole-3-carbinol, lignans.
Inhibitors of metastasis	Flavonoids, catechins? Carotenoids may encourage cell–cell communication in transformed cells.
Inhibitors of angiogenesis	Genistein (an isoflavonoid).
Fibre	Increased speed of movement of faeces through colon; dilutes carcinogens and/or delays their formation.
Stimulation of immune response	Carotenoids, e.g. β-carotene. (*Am. J. Clin. Nutr.* **64**, 772 (1996); *J. Lab. Clin. Med.* **129**, 309 (1997))
Caloric restriction	Calorie-restricted diets increase lifespan and decrease cancer incidence in rodents. Diets rich in fruits and vegetables may have decreased fat and calorie content.
Selenium	Content in plants very variable; related to soil content. Has anti-cancer effect in several animal studies, apparently by inducing protective enzymes.
Phytoalexins	Resveratrol (grapes, wines). Antioxidant, anti-inflammatory, induce phase II enzymes.
Iron chelators?	Decrease iron uptake; has been speculated that high body iron stores are a risk factor for cardiovascular disease and cancer, e.g. phytates, other organic phosphates.
Down-regulators of transcription factors (e.g. NF-κB)	Capsaicin (found in *Capsicum* peppers; *J. Immunol.* **157**, 4412 (1996)). Caffeic acid phenethyl ester (*Proc. Natl Acad. Sci. USA* **93**, 9090 (1996)).

[a]Sulphoraphane, 1-isothiocyanato-4-(methylsulphinyl)butane, is present in broccoli and is a powerful inducer of quinone reductase and other phase II enzymes but not of cytochromes P450. Oltipraz is also found in cruciferous vegetables. Dithiolthiones have been reported to induce haem oxygenase and ferritin when administered to rats (*Carcinogenesis* **17**, 2291 (1996)). Structures are shown in Fig. 10.8. There is increasing evidence that quercetin and other plant phenolics can be absorbed from foods and beverages (e.g. *Anal. Biochem.* **248**, 41 (1997)).
[b]*Bergamottin* in grapefruit juice inhibits CYP3A4 in the gut and can increase the oral bioavailability of several drugs (*Chem. Res. Toxicol.* **11**, 252 (1998)).

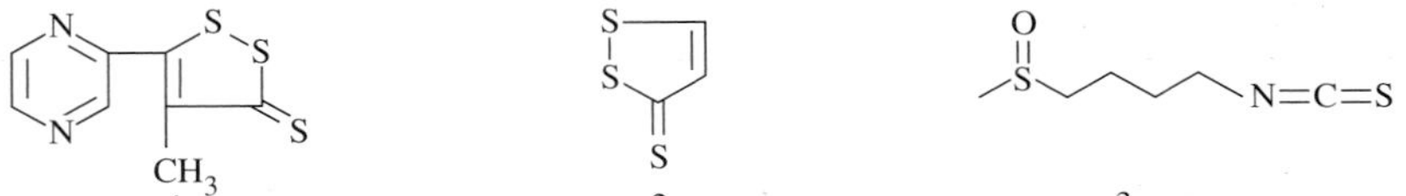

Fig 10.6. Structures of some of the compounds mentioned in Table 10.7: 1, oltipraz; 2, 1,2-dithiole-3-thione; 3, sulphoraphane.

questionnaires and food-composition tables are used to show in a particular study that intake of, say, folic acid is associated with less neurodegenerative disease. It is perfectly possible that foods rich in folic acid are also rich in other, perhaps-unidentified, protective constituents (e.g. vitamin C). The reverse argument can apply in attempts to investigate vitamin C as a protective factor: foods never deliver nutrients in isolation. Low folate intake in humans is associated with an increased frequency of chromosome breakage. Folic acid is required for the synthesis of DNA precursors and low intake may cause DNA damage by allowing mistaken incorporation of uracil instead of thymine (the CH_3- group in thymine arises from folate).[6]

All of these confounding factors, and others, must be taken into account in the interpretation of epidemiological studies. For example, the data in the Rotterdam study of cognitive function could be interpreted as a protective effect of β-carotene against loss of cognitive function, a protective effect of vegetables (the β-carotene being a marker) or even as a change in dietary habits that *resulted from* the cognitive impairment (e.g. people with poor cognitive function eat fewer vegetables, perhaps because they can no longer buy, prepare and cook them).

10.4.5 *Experimental epidemiology*[92]

One way around the 'cause–consequence' conundrum is to conduct **intervention trials**. In an intervention trial one could, for example, supplement a defined group of subjects with, say, selenium, or vitamin E and compare their fates with those of a 'control' group. One major task of the epidemiologist is the careful selection of control groups and the random allocation of subjects between control and test groups. The use of a **placebo**, ideally completely indistinguishable from the test compound, allows both subjects and investigators to be unaware of who is receiving what (**double-blinding**).

Population-based intervention trials are expensive to conduct and problems can occur if subjects do not comply with the regime (ideally, compliance should be monitored, e.g. by measuring blood levels of the administered substance), or if baseline parameters drift with time. For example, suppose that, over a five-year study testing the effects of supplementation with vitamin C on the incidence of cardiovascular disease, dietary intakes of C in *all* subjects increased, e.g. due to its addition to more foods or to a general drift in the population towards increased consumption of vitamin C-rich foods. One might then minimize any difference between those receiving vitamin C supplements versus placebo.

An extension of the above type of **randomized control trials** is to conduct **crossover studies**, in which all study subjects receive test compound or placebo for a given period. After a 'washout period', the subjects then receive the opposite treatment. Whether a subject receives placebo first and real treatment second or vice versa is randomized. An indicator of disease is measured at the beginning and end of each period and paired analyses are conducted where each subject acts as his/her own control. Crossover studies are popular in the pharmaceutical industry but, of course, can be confounded by a treatment that has lasting effects.

It is hoped that epidemiological research, in parallel with biochemical studies, will make it possible to make public health recommendations that will lead to decreased incidence of major diseases (Table 10.9). At present the recommended dietary allowances (RDAs) for various nutrients are set on a somewhat *ad hoc* basis and may need revision as more data come to light. For example, studies on vitamin C metabolism (Chapter 3) suggest that 200 mg/day might be a more appropriate RDA. Another aim may be the development of foods whose composition is altered (e.g. by supplementation or genetic engineering) to promote health ('functional foods', 'designer foods' and 'nutriceuticals' are terms used to describe this endeavour).

Table 10.8. Examples of toxins found in plants

Type	Main food sources	Potential toxic effect
Thiosulphinates	Onions, other *Allium* species	Haemolysis in domestic animals; reported in cows, sheep, cats and horses consuming onions (*J. Agr. Food Chem.* **42**, 959 (1994))
Protease inhibitors	Beans	Impaired food utilization
Cyanogens	Peas, beans, linseed, flax, cassava	Cyanide poisoning
Allergens	Potentially all food proteins, especially nuts and grains	Allergic response, anaphylactic shock
Carcinogens aflatoxin cycasin safrole sesamol	*Aspergillus flavus* *Cycas* sp. sassafras black pepper, sesame seeds	} Carcinogenic
some flavonoids and other phenolic compounds (at very high levels)		

Table 10.8. (*Continued*)

Type	Main food sources	Potential toxic effect
Carminic acid	A major component of cochineal[a]	Has close structural similarities to doxorubicin and can undergo redox cycling *in vitro*
Favism	—	Section 7.3
Alkaloids	Widely distributed	Hepatotoxic, cause lung damage and sometimes are carcinogenic
Neurotoxins	widely distributed	Chapter 9
Nitrates/nitrites	Spinach, other leafy vegetables[b]	Nitration and deamination reactions (e.g. by HNO_2 in the stomach)
α-Amanitin	*Amanita phalloides* toadstool	Salivation, vomiting, convulsions, death
Atractyloside	Thistle *Atractylis gummifera*	Inhibits mitochondrial ATP release to cytosol
Gossypol	Cottonseed	Section 7.8
Lathyrogens	—	Section 9.19
Psoralens	Celery, parsnips	Skin photosensitivity
Indolo(3,2-*b*) carbazole	Cruciferous vegetables	Induced CYP1A1 and raised formation of 8-OHdG in hepatoma cells in culture (*Proc. Natl Acad. Sci. USA* **93**, 2322 (1996))

Adapted from Pariza, MW (1996) In *Present Knowledge in Nutrition*, seventh edition (Ziegler, EE *et al.*, eds), Chapter 57 (IRL Press, Washington DC). Note that there is an overlap between 'protective agents' (Table 10.7) and toxins; low levels of toxins might in some cases be beneficial by inducing defence systems. Indeed, the famous physician Paracelsus (1493–1541) wrote, 'All substances are poisons: there is none which is not a poison. The right dose differentiates a poison and a remedy.'
[a]A red food colourant obtained from the insect *Napalea coccinellifera*.
[b]Kimchi, a food made from fermented cabbage, very rich in nitrites and salt, has been suggested as one cause of the high gastric cancer rates found in some Asian countries.

10.4.6 *The need for biomarkers*[36]

We believe that, whenever possible, epidemiological studies of the protective effects of antioxidant supplements against human disease should be accompanied by measurement of biomarkers of oxidative damage, to check that an antioxidant effect was actually achieved. Useful biomarkers include direct measurements of oxidative damage products (Chapter 5) and of effects on the 'peroxidizability' of plasma LDL (Fig. 10.7) when exposed to a pro-oxidant challenge after isolation (Chapter 9).

One must never *assume* that any beneficial effect of an antioxidant is necessarily due to its antioxidant activity. For example, it has been claimed that high-dose vitamin E supplementation might decrease the aggregation of activated platelets, the proliferation of arterial smooth muscle cells and/or the adhesion of monocytes to endothelium[21] and their secretion of IL-1. These

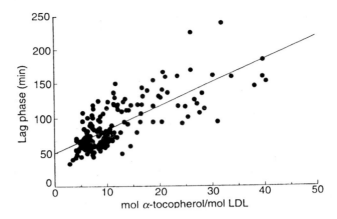

Fig. 10.7. Tocopherols and the peroxidizability of human LDLs: the relationship between the oxidation resistance of LDL and its α-tocopherol content. The peroxidation resistance of LDL isolated from human volunteers is not significantly correlated to the content of vitamin E (there are usually 3–15 molecules of α-tocopherol per LDL particle). However, if subjects consuming extra oral vitamin E (and so with higher vitamin E content of LDL) are included, a correlation is seen. For subjects consuming *RRR*-α-tocopherol for 21 days, the oxidation resistance of LDL isolated from their plasma increased, on average, to 118%, 156%, 135% and 175% of control subjects for doses of 150, 225, 800 or 1200 IU/day, respectively. Peroxidation resistance is measured as the length of the 'lag phase' before peroxidation accelerates after exposure to copper ions. Data are also included from isolated LDL that has been loaded with vitamin E by incubation in the test tube before analysis. (From *Ann. Med.* **23**, 580 (1991) by courtesy of the publishers and the late Prof. Dr H. Esterbauer.)

various effects could contribute to vitamin E's anti-atherosclerotic action, which is often assumed to be entirely due to its ability to inhibit LDL peroxidation (Fig. 10.6). Studies on Danish volunteers, for example, showed that urinary excretion of 8-OHdG, one putative biomarker of oxidative DNA damage (Chapter 5), was decreased about 28% by feeding male (but not female) subjects Brussels sprouts,[96] but not by supplementing them with β-carotene, vitamin C, vitamin E or coenzyme Q$_{10}$.[71] Of course, the basal levels of antioxidants in the subjects may have been sufficient to exert maximum inhibitory effects on oxidative DNA damage. For example, only very low ascorbate intakes have been associated with rises in tissue 8-OHdG levels (Chapters 3 and 7). Nevertheless, this study confirms the existence of food constituents able to decrease oxidative damage to DNA that are not the classical nutritional antioxidants.

10.4.7   Some examples of epidemiological studies[8,31,56,73]

The Nurses' Health Study and NHANES-I

The Nurses' Health Study began in 1976 with 121 700 female nurses. In 1980, 87 245 nurses aged 34–59, free of obvious cardiovascular disease, completed dietary questionnaires that aimed to assess their consumption of a wide range of antioxidants, including vitamin E. During follow-up of up to 8 years, 437 non-fatal heart attacks occurred as well as 115 deaths due to coronary disease. Those nurses in the top 20% of dietary vitamin E intake (as calculated from the questionnaire) showed 23–50% less heart disease after correction of the data for age and smoking. Women who took vitamin E supplements for short periods had little apparent benefit but those who took them for more than 2 years had a relative risk of major coronary heart disease of 0.38–0.91 after adjustment for age, smoking, risk factors and intake of other antioxidant nutrients. There was no relation to vitamin C intake, although this was above the US RDA of 60 mg/day even in the lowest quintile.

In the **First National Health and Nutrition Examination Survey** (NHANES-I), 11 349 American men and women were followed for a median period of 10 years. Cardiovascular mortality rates were found to be 34% lower than average among participants with the highest vitamin C intake, defined as 50 mg or more from the diet plus regular supplements. Hence in the Nurses' Study, vitamin C intakes may have been too large to see an effect. However, the NHANES-I data can be criticized because intake of other antioxidants (which could be correlated with vitamin C intake) was not assessed.

In another study on 34 486 post-menopausal women, the incidence of death from coronary heart disease decreased with calculated vitamin E intake from food, although no relation to vitamin C intake was observed.

Health professionals' study

In 1986, 39 910 US male health professionals (dentists, veterinary surgeons, pharmacists, opticians, osteopaths and podiatrists) 40–75 years old and free of obvious coronary disease, completed detailed dietary questionnaires designed to assess their intake of vitamin C, β-carotene and vitamin E. During 4 years of follow-up, 667 cases of coronary disease were observed. After controlling for age and other risk factors, a lower risk for coronary disease was observed among men with calculated high vitamin E intakes. For men consuming more than 60 International Units per day the relative risk was 0.49–0.83 compared with those eating less than 7.5 units per day. As compared with men who did not take vitamin E supplements, men who took at least 100 units per day for at least 2 years had a relative risk of 0.47–0.84 for coronary disease. By contrast, β-carotene intake was not associated with a lower risk among those who had never smoked, but among smokers, high β-carotene intake did appear to be associated with a lower risk. A high intake of vitamin C was not associated with a lower risk of coronary disease but again this was, on the whole, a well-nourished population. This study also showed that high dietary fibre intake appeared protective against myocardial infarction.

The authors of all the papers were careful to point out the limitations of their results, but their cautions were ignored in some lay publicity material. What are the problems? Here are some obvious ones:

1. The reliability of questionnaires about food/vitamin intake is a perennial problem in nutritional studies. It could be the concentration achieved in the body, rather than dietary intake, that is the main factor determining

antioxidant action. This may be especially the case with β-carotene, where the extent of absorption of the intact molecule from a given dietary intake varies widely between individuals. In the Health Professionals' study, a comparison of nutrient intake as specified on the questionnaire with weekly diet records in 127 subjects gave poor correlations for vitamin E in subjects not consuming supplements, so the errors in estimating vitamin E level by questionnaire in subjects not taking supplements may have been large. This means that real associations between vitamin E intake from the diet could be missed, and the effects of supplements thereby are over-emphasized.

2. People who consume vitamin E are more likely to be 'health-aware' and to have other aspects of their lifestyle (attention to diet, taking exercise, etc.) that could decrease heart disease. Perhaps healthier subjects select themselves for taking vitamin supplements. However, if self-selection were the explanation, it is difficult to explain why vitamin C did not appear to be protective.

3. Other diseases were not examined. We cannot be totally sure that perhaps vitamin E protects against heart disease but does not increase death from other causes.

4. In a European multicentre study (the EURAMIC study) published in *Lancet* (**342**, 1379 (1993)), vitamin E and β-carotene concentrations in human fat were measured. There was no difference in vitamin E levels between people who had had heart attacks or controls. Fat vitamin E levels were chosen as a measure of long-term vitamin E status, and are possibly a more accurate measure than dietary questionnaires. Low β-carotene levels were associated with a higher incidence of myocardial infarction, but only in patients who smoked.

5. By contrast, in an earlier study from Scotland published in *Lancet* (**337**, 1 (1991)) vitamin E concentrations in blood plasma were found to be lower in patients with angina pectoris than in controls (Chapter 9). β-Carotene and vitamin C concentrations were also lower in these patients, but this effect was largely due to the fact that they smoked more than controls, since smoking is associated with lower plasma levels of vitamin C and β-carotene (Chapter 3).

6. In the CHAOS (Cambridge Heart Antioxidant)[91] study, 2002 patients with established coronary atherosclerosis were randomized to receive supplementary α-tocopherol or placebo and followed. α-Tocopherol supplementation decreased the incidence of non-fatal myocardial infarctions by almost 50% but did not decrease that of fatal ones. Follow-up was short (median 510 days) and will be continued in the hope that mortality rates will be decreased over a longer time period.

10.4.8   *A status summary: antioxidants and cardiovascular disease*

It is easy to 'pick holes' in epidemiological studies. Questionnaires about diet are only an approximate way of measuring intake of antioxidants. But is measurement of the vitamin E content of a piece of body fat, or of blood plasma,[104] any better? It is difficult to be certain, e.g. different body fat depots may have different vitamin E contents.[40a] Overall, we feel that, on the weight of evidence and the consistency of results from various epidemiological studies, it is established that a high intake of vitamin E (perhaps involving the use of supplements) offers some protection against cardiovascular disease. Vitamin C seems not to affect cardiovascular disease much in the US populations studied (possibly because their vitamin C status was already good). Thus, avoiding smoking and maintaining a good intake of fruits and vegetables, rich in antioxidant nutrients, would seem to be the key to cardiovascular health, perhaps combined with moderate vitamin E supplementation.

10.4.9   *The Linxian study*

The rural county of Linxian, China, has a high death rate from cancer of the oesophagus and stomach. A clinical trial of vitamin supplementation has been published (*J. Natl. Cancer Inst.* **86**, 1483 (1993)). Subjects aged 40–68, 29 584 in total, were given various nutritional supplements or a placebo. It was found that a mixture of selenium, β-carotene and vitamin E produced significant falls in death from stomach cancer after 5 years, whereas supplementation with a mixture containing vitamin C did not.

The Linxian population was poorly nourished to start with (but then so, on average, are the poorest subgroups of the population of the USA and UK). Although the effect on cancer mortality is convincing, it is uncertain which component(s) of the supplement was responsible (it may have been the selenium, for example). A recent randomized placebo-controlled trial[15] examined the effects of 200 μg daily selenium on patients with a history of skin cancer, the hypothesis being that it might prevent further skin cancers from developing. This was not found to be the case, but there was a significant reduction in mortality from other cancers in the selenium group, including lung, colorectal and prostate cancers. Other studies with selenium (e.g. relating the selenium content of blood or nail clippings to cancer incidence) have given equivocal results. It is possible that there is an optimal selenium intake, levels above which give no further benefit (indeed, high selenium intakes are toxic; Chapter 3).

By contrast, intervention trials with vitamins E or C or carotenoids against cancer of the digestive tract or breast cancer in Western populations have given generally disappointing results. Thus high intakes of vitamin C or E did not decrease rectal polyp formation in patients with polyposis or protect against the development of breast cancer in women. On the other hand, trials of β-carotene and vitamin E suggest that both can cause regression of **oral leukoplakia**, a pre-malignant lesion of the oral cavity.[56]

10.4.10   *The Finnish study (α-tocopherol/β-carotene cancer prevention study) and CARET*

A Finnish study (*N. Engl. J. Med.* **330**, 1029 (1994)) attracted a great deal of attention as apparently showing a deleterious effect of β-carotene supplementation. The study examined 29 133 male smokers (age range 50–69 years, mean age 57 years) who had smoked (on average) 20 cigarettes a day for 36–37 years. Some received 20 mg of β-carotene per day, others 50 mg of racemic vitamin E (*dl*-α-tocopherol), some both, and some placebo. They were followed for 5–8 years and the incidence of lung cancer noted. Vitamin E had no significant effect on lung cancer, but (as in CHAOS) tended to decrease the incidence of non-fatal (but not fatal) myocardial infarction in subjects who had already suffered one myocardial infarction. However, after 3 years of the study, subjects receiving β-carotene showed a significantly greater incidence of lung cancer than controls (relative risk 1.16), an increased risk of myocardial infarction and an 8% increase in total mortality. Analysis of data from this study has also shown a decreased risk of coronary heart disease in subjects with high fibre intakes, and a decreased incidence of prostate cancer in subjects given vitamin E.

Previous studies have shown that low blood β-carotene levels are associated with higher levels of lung cancer (and some other forms of cancer). The obvious question is, is the β-carotene responsible or is high blood β-carotene simply a 'marker' of a diet rich in fruits and vegetables? The direct test of the effect of β-carotene in the Finnish study suggests that it is not protective and may be deleterious. However, patients who have smoked for 36–37 years are already well on the way to developing various pathologies, including lung cancer, and it is unlikely that β-carotene or vitamin E could reverse this. Hence, a logical result of this trial would be 'no effect', as was found with vitamin E and lung cancer. Indeed, the **Physicians' Health Study** in the USA examined the effect of 50 mg β-carotene or 325 mg aspirin versus placebo in 22 071 male physicians. After 12 years of follow-up, there was no significant effect (good or bad) of β-carotene on cardiovascular disease or cancer, although 11% of the participants were smokers.

It thus remains to be explained why β-carotene increased deaths from lung cancer in the Finnish study. Some suggested explanations are listed below:

(1) It could have been a random statistical 'fluke'. This seems unlikely because the data are supported by the **Carotene and Retinol Efficacy Trial (CARET)**. CARET was a double-blind trial of supplemental β-carotene plus retinol versus placebo in asbestos workers and smokers. It was terminated 2 years early because of a suggestion that lung cancer incidence was increased by up to 28% in the supplemented subjects. Of course, it could have been either retinol and/or β-carotene that was responsible.

(2) β-Carotene at high doses could be toxic or could interact with cigarette smoke to produce toxic products.

(3) The β-carotene may have been contaminated with, possibly-toxic, break-down products.

(4) The smokers receiving β-carotene had, on average, smoked a year longer than controls. In patients at high risk of developing lung cancer, this could be enough to skew the results. (In any case, the results suggest that the best way of protecting against smoke-induced pathology is to give up smoking rather than to keep smoking and take antioxidants.)

(5) The β-carotene could have interfered with uptake of other carotenoids (putatively more protective) from the diet. Much more attention needs to be given to the possible role of such molecules as lutein and lycopene as anti-cancer agents.

In general, and by contrast with findings on cardiovascular disease, we feel that the evidence for a protective role of dietary E or β-carotene against the development of cancer is not compelling, although a low intake of vitamin C may be a risk factor for the development of gastric cancer. It may be other factors in fruits and vegetables, or other combinations of these factors, that are responsible for the protective actions (Table 10.6). In the same way, data on antioxidants in relation to cataract and macular degeneration (Chapter 7) do not allow us to reach clear-cut conclusions as yet, although evidence for a link is growing.[44,93]

10.4.11   *Other dietary factors and cardiovascular disease*

The PUFA content of the diet has effects on cardiovascular disease. Increased consumption of PUFAs can render LDL subsequently isolated from humans more peroxidizable, at least *in vitro* (Chapter 9). However, (*n*-3) PUFAs such as linolenic and eicosapentaenoic acids may also decrease total LDL levels, as well as decreasing the tendency to thrombosis. In animals, dietary fish oil has been shown to decrease both the vulnerability of the heart to arrhythmias after ischaemia–reperfusion and the production of eicosanoids and cytokines by activated phagocytes. Overall therefore, the cardiovascular effects of PUFAs appear to be beneficial.[47] However, a high intake of PUFAs may increase dietary requirements for vitamin E (Chapter 3). The vitamin E content of various vegetable and fish oils varies widely and this is a point that must be considered in nutritional studies.

High plasma levels of the thiol **homocysteine**

$$\overset{\overset{+}{NH_3}}{\underset{|}{(HSCH_2CH_2CH - COO^-)}}$$

are a risk factor for cardiovascular disease (Chapter 9).[58] Plasma homocysteine arises by metabolism of the dietary sulphur-containing amino acid **methionine**. Homocysteine levels depend on the activity of three enzymes. The first is **cystathionine synthetase**, which requires vitamin B_6 for its activity. The second is **5,10 methylenetetrahydrofolate reductase**, which requires folic acid. The third is **methionine synthase**, whose activity requires both folate and vitamin B_{12}. Inadequate intake of any of these three vitamins can elevate plasma homocysteine. Mutations in the genes encoding these enzymes can also lead to increases in homocysteine. Since it has been suggested (although by no means proved)[97b] that homocysteine damages vascular endothelium and predisposes to atherosclerosis by generating ROS, the effects of inadequate intake of these B-vitamins illustrate how oxidative stress can result from deficiencies in nutrients that are not generally regarded as 'antioxidants'. Similarly, magnesium deficiency can lead to secondary oxidative stress (Table 4.2). Body iron status has also been suggested to affect cardiovascular disease (Section 10.3 above), but data are conflicting.[94a]

Exercise, in moderation, can also promote cardiovascular health; perhaps induction of antioxidant defences is involved (Section 7.10). Induction of phase II xenobiotic-metabolizing enzymes has also been reported in exercising rats.[24]

10.5   Antioxidants and the treatment of disease[35]

So far in this chapter we have considered the role of antioxidants in the prevention of disease. A good tissue antioxidant status may also help organisms resist traumatic injury. For example, open-heart surgery or organ transplantation have been reported in several studies to lead to drops in plasma and/or tissue antioxidant levels. Pre-treatment of kidney transplant recipients with a mixture of vitamins (including vitamins C and E) led to better early graft function, although this study does not prove that it was the antioxidants rather than the other vitamins present that were protective. Research in the livestock industry has shown that supplementation of pigs or cattle with vitamin E improves the 'keeping properties' of the meat. The extent of tissue damage after stroke may be decreased in patients who have previously eaten a diet rich in fruits and vegetables, and the incidence of stroke is also decreased in such subjects. Since tissue injury leads to increased oxidative stress (Chapters 4 and 9), it seems logical that good antioxidant nutritional status might diminish tissue damage in a wide range of diseases.

10.5.1   *Therapeutic antioxidants*

There is also considerable interest in the therapeutic use of antioxidants.[4] This may involve the use of naturally occurring antioxidants (with or without structural 'adaptations') or completely synthetic molecules (Table 10.10). In addition, there is evidence that some drugs already used clinically may exert part or all of their effects by antioxidant mechanisms (Table 10.10).[95] Examples already discussed include penicillamine in the treatment of rheumatoid arthritis and aminosalicylates for inflammatory bowel disease (Chapter 9). Of course, these compounds were not designed as antioxidants. Any antioxidant designed for therapeutic use must obviously be safe but must also meet other criteria. The following list of questions to ask when evaluating proposed therapeutic 'antioxidants' is adapted from reference 35:

(1) What biomolecule is the antioxidant supposed to protect? Does enough antioxidant reach that target *in vivo*?
(2) How does it protect—by scavenging ROS/RNS, preventing their formation, up-regulating endogenous defence systems or aiding repair of damage?
(3) If the antioxidant acts by scavenging, can the resulting antioxidant-derived radicals themselves cause damage? It may be advantageous if any antioxidant

derived radicals can be 'recycled' to the antioxidant, e.g. by ascorbate or GSH.

(4) Can the antioxidant cause damage in other biological systems?

(5) Some ROS/RNS have useful roles *in vivo* (e.g. phagocyte killing of foreign organisms). Could the proposed antioxidant interfere with these?

If an antioxidant acts as a scavenger of ROS/RNS, an important question is whether any resulting antioxidant-derived species can exert deleterious biological effects. For example, reactions of penicillamine, phenylbutazone or aminosalicylates with certain ROS/RNS can generate secondary products that might be cytotoxic. Sulphur and oxysulphur radicals produced by reaction of penicillamine with $OH^{\bullet}$ or $ONOO^{-}$ can, for example, inactivate α_1-antiproteinase and bind to proteins to create new antigens (Chapter 9).

A second point to be considered is the molecular mechanism of oxidative damage. In many early clinical tests of the importance of oxidative stress, vitamin E was administered to patients, with generally disappointing results (unless the patients were vitamin E-deficient to start with). This could mean that:

(1) Oxidative stress was not a significant contributor to the disease pathology (a common interpretation at the time of these studies).

(2) Not enough antioxidant reached the correct site of action (a particular problem with vitamin E; it takes time for concentrations in tissues, especially the nervous system, to increase).

(3) The oxidative injury may have involved damage not to lipids, but to protein and/or DNA, so that inhibiting lipid peroxidation will not protect tissues. We now know that oxidative damage to proteins and DNA can often be more important than lipid peroxidation as an early consequence of oxidative stress (Chapter 4).

(4) Vitamin E supplementation may down-regulate other antioxidants, so that there is no change in 'total' antioxidant status. An illustration of this concept is that feeding animals a diet with a high content of the synthetic antioxidant BHT (Table 10.11) decreased the α-tocopherol content of the liver.

In an analogous sense, one should also be wary of assuming that a clinically effective drug is acting by the mechanisms by which it was designed to act (Table 10.10, last row). Let us suppose that giving a therapeutic antioxidant to patients did improve their disease. This could mean that ROS/RNS make a significant contribution to disease pathology, or that the antioxidant has other beneficial pharmacological effects unrelated to its action on ROS/RNS. For example, many chain-breaking antioxidant inhibitors of lipid peroxidation also inhibit lipoxygenase enzymes, which are important in synthesis of leukotrienes (Table 10.11).

10.5.2   *Approaches to antioxidant characterization*

Perhaps a first step in evaluating putative therapeutic antioxidants is to establish *in vitro* what they are capable of doing. A battery of *in vitro* antioxidant

Table 10.9. Recommended dietary habits in adults
A. Recommended dietary allowances (RDAs) or reference nutrient intakes (RNIs) of various nutrients for adults[a]

	UK		USA	
	Males	Females	Males	Females
Vitamin A	700 µg	600 µg	1000 µg	800 µg
Vitamin E[b]	>4 mg	>3 mg	10 mg (15 IU)	8 mg
Vitamin C[c]	40 mg	40 mg	60 mg	60 mg
β-Carotene	not set	not set	not set	not set
Selenium	75 µg	60 µg	50–200 µg	50–200 µg
Iron	8.7 mg	14.8 mg	10 mg	10 mg
Zinc	9.5 mg	7.0 mg	15 mg	15 mg
Copper	1.2 mg	1.2 mg	2–3 mg	2–3 mg
Fibre[d]	12–14 g	12–14 g	—	—
Folic acid	200 µg	400 µg	—	—

B. Some of the US National Research Council's Dietary Recommendations (1989)
- decrease total fat intake to ⩽30% of calories, cholesterol to <300 mg/day and saturated fat to <10% of calories[e]
- eat at least five servings of fruits/vegetables (especially green and yellow vegetables and citrus fruits) per day
- increase starches and other complex carbohydrates by eating six or more daily servings of bread, cereal or legumes
- balance food intake with physical activity to maintain appropriate body weight
- maintain protein intake at moderate levels
- avoid taking dietary supplements in excess of the RDA in any one day

[a]Values are daily quantities needed to amply meet the known nutritional needs of healthy adults; RDA can also mean recommended daily amount.
[b]1.49 International units (IU) of vitamin E are equivalent to 1 mg of *d*-α-tocopherol. 1 mg of synthetic *dl*-α-tocopherol acetate (all-*rac*-tocopherol) equals 1.0 IU. The intake recommended is expressed as milligrams of *d*-α-tocopherol (*RRR*-α-tocopherol) equivalents. Multiplication factors are used to take into account the lower biological effectiveness of synthetic tocopherols (Chapter 3). Before taking these calculations too seriously, remember that the 'biological effectiveness' of vitamin E is based on a rat assay which is not necessarily relevant to humans.
[c]Smokers probably need a higher intake by up to 80 mg/day.
[d]The term 'dietary fibre' lacks precise definition. The UK RDA refers to total 'non-starch carbohydrate polymers'.
[e]For example, consumption of a meal containing 50 g of fat impaired vascular function for up to 4 h; pre-treatment with vitamins C (1 g) and E (800 Units) prevented this (*JAMA*, **278**, 1682 (1997)).

characterization methods has been developed (Table 10.12). Two important (but often forgotten) points are that:

(1) A compound should be tested at concentrations relevant to those achievable *in vivo*.
(2) In assaying putative antioxidants, one should use biologically relevant ROS/RNS and targets of damage. For example, in 'screening' peroxynitrite

scavengers[99] we use two types of assay: tyrosine nitration and inactivation of α_1-antiproteinase. Because the assays do not measure the same aspects of peroxynitrite reactivity, 'scavengers' do not always behave in the same way, e.g. uric acid inhibits $ONOO^-$-dependent tyrosine nitration, but not α_1-antiproteinase inactivation.

The results of *in vitro* tests may be used to evaluate the possibility (or impossibility) that a compound can exert *direct* antioxidant effects *in vivo*. Thus a compound needing millimolar levels to act as a free-radical scavenger *in vitro* is very unlikely to act as a scavenger if it is only present at micromolar levels *in vivo*. Many putative antioxidants have been tested *in vitro* and suggested for use *in vivo* as '$OH^\bullet$ scavengers'. However, it is extremely unlikely that they can work as $OH^\bullet$ scavengers *in vivo* since they will never be present at levels remotely approaching those of endogenous molecules that react at (or close to) diffusion-controlled rates with $OH^\bullet$ (Table 10.12).

Even an excellent *in vitro* antioxidant will not necessarily work as such *in vivo*, e.g. it could be rapidly excreted, or metabolized to inactivate products.

Table 10.10. Some antioxidants available for potential use in the treatment of human disease

Category of compound	Example
Naturally occurring	CuZnSOD, MnSOD, EC-SOD (recombinant or purified), α-tocopherol, lipoic acid, vitamin C, adenosine, transferrin, lactoferrin, cysteine, glutathione, carotenoids, other plant pigments, flavonoids, other plant phenolics, desferrioxamine, other 'natural' iron chelators, melatonin (?)
Synthetic	Thiols (e.g. mercaptopropionylglycine, N-acetylcysteine), synthetic metal-ion chelators (e.g. ICRF-187, hydroxypyridones), xanthine oxidase inhibitors, inhibitors of $O_2^{\bullet-}$ generation by phagocytes, lipid-soluble chain-breaking antioxidants, inhibitors of phagocyte adhesion/respiratory burst, SOD/catalase/glutathione peroxidase mimics, GSH donors
Those already in clinical use that might have antioxidant activity *in vivo* (but were not developed as antioxidants)	Penicillamine, aminosalicyates (alone or as components of sulphasalazine), apomorphine, selegiline, 4-hydroxytamoxifen, ketoconazole, tetracyclines, N-acetylcysteine, captopril, probucol, propofol, some β-blockers (e.g. carvedilol)[a], cimetidine, omeprazole, some Ca^{2+} channel blockers[a], phenylbutazone, nitecapone, entecapone, idebenone.

Adapted from *Drugs* **42**, 569 (1991).
[a]Several β-blockers/Ca^{2+} blockers inhibit peroxidation *in vitro*: it is uncertain if they do so *in vivo* at the therapeutic doses normally used. The antioxidant moiety of carvedilol may be its secondary amine group (Fig. 10.6).

Table 10.11. Some synthetic antioxidants

Compound	Structure	Comments
Butylated hydroxyanisole (BHA)		The major (90%) constituent (3-*tert*-butyl-4-hydroxyanisole) of commercial BHA is shown. It acts as an antioxidant by hydrogen donation, a mechanism common to all the phenolic (and amine) chain-breaking antioxidants. For example, addition of BHA to fat (e.g. butter) increases its storage life from a few months to a few years. Generally recognized as safe but huge doses ($\geqslant 1\%$ of diet) cause forestomach cancer in rats. High doses of BHA can also antagonize the action of other carcinogens in animals, by inducing phase II enzymes (*Crit. Rev. Toxicol.* **15** 109, 1985). Metabolized in animals by O-demethylation to the diphenol ***tert*-butylhydroquinone** (TBHQ) followed by glucuronidation and sulphation. TBHQ can undergo oxidation to a semiquinone radical *in vitro*, and then on to a quinone (*tert*-butylquinone).
Butylated hydroxytoluene (BHT)		Often added to foodstuffs; generally recognized as safe but very high doses can antagonize action of vitamin K and cause bleeding in animals. Does not cause forestomach cancer in rats.
Trolox		Water-soluble form of α-tocopherol; the hydrophobic side-chain is replaced by a hydrophilic –COOH group. Good scavenger of peroxyl and alkoxyl radicals, giving a Trolox radical that can be recycled by ascorbate.

Table 10.11. (*Continued*)

Compound	Structure	Comments
Propyl gallate	$COOCH_2CH_2CH_3$ (trihydroxybenzene ester structure with HO, OH, OH groups)	Fairly water-soluble; good inhibitor of lipid peroxidation; lipoxygenase inhibitor. It is sometimes added to foodstuffs. It binds iron ions and reduces Fe(III) to Fe^{2+}. Often used in foods in combination with chelating agents such as citrate to block this.
Nordihydro-guaiaretic acid (NDGA)	(structure with two catechol rings, HO/OH groups connected by a dimethyl-butane chain)	Occurs naturally in resinous exudate of *Larrea divaricata* (American creosote bush) and in some other plants. It is sometimes added to foodstuffs (but no longer widely used) and several polymers (e.g, rubber, lubricants). It binds iron ions and reduces Fe(III) to Fe^{2+}, and is a powerful lipoxygenase inhibitor.
N,N′-Diphenyl-*p*-phenylene diamine (DPPD)	HNC_6H_5 (benzene ring) HNC_6H_5	DPPD is popularly used as an antioxidant *in vitro* and is sometimes used for animal studies of *in vivo* lipid peroxidation. Aromatic amines are frequently used as antioxidants in the lubricant and polymer industries.
6 Hydroxy-1,4-dimethyl-carbazole (HDC)	CH_3 ... CH_3 (carbazole structure with HO and N–H)	Powerful inhibitor of lipid peroxidation *in vitro*.
Promethazine	$CH_2CHN(CH_3)_2$ (phenothiazine structure with N and S)	Inhibits lipid peroxidation *in vitro*; it is used as an antihistamine and sedative.

Chlorpromazine

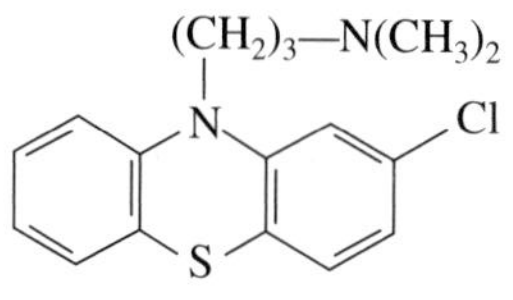

Used as a tranquillizer. Its antioxidant action in microsomes may partly depend on its enzymic conversion into hydroxylated products.

Ethoxyquin (Santoquin)

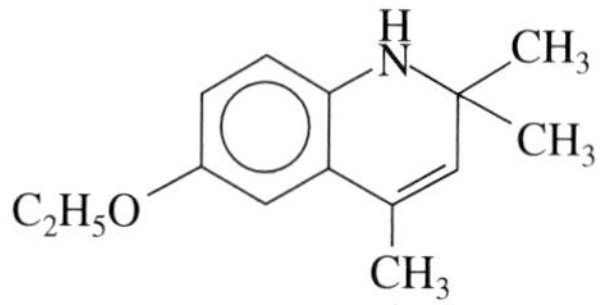

Has been used in fruit canning. Powerful enzyme inducer *in vivo*. Has been used in longevity experiments in animals (Section 10.3 above).

CS-045 (Troglitazone)

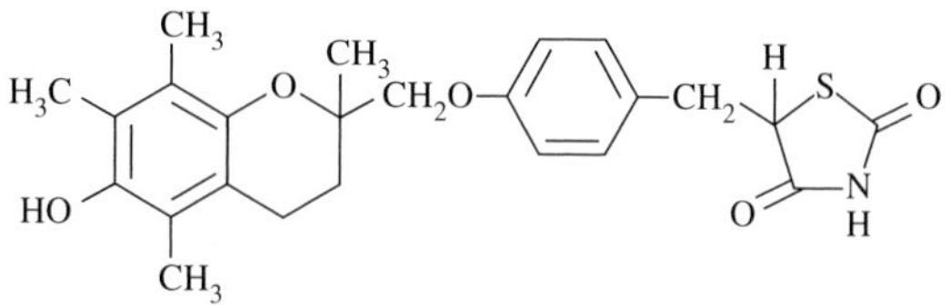

An analogue of α-tocopherol; it acts as a hypoglycaemic agent by mimicking or enhancing the effect of insulin. It may bind to a specific receptor on fat cells. Marketed in several countries for diabetic patients (*Biochem. Pharmacol.* **50**, 1109 (1995)).

MDL 74 405

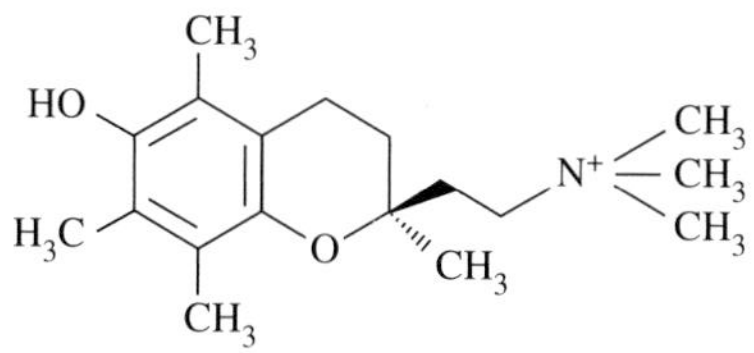

Quaternary ammonium compound. It is a powerful chain-breaking antioxidant; it apparently concentrates in the heart when administered to animals but also enters other tissues. It is protective in myocardial ischaemia–reperfusion models in several animals; also against models of inflammatory bowel disease in rodents. Related compounds containing a tertiary amine function are being evaluated in animals as neuroprotective agents (*J. Med. Chem.* **38**, 453 (1995)).

LY178002 (R = H), LY256548 (R = CH₃)

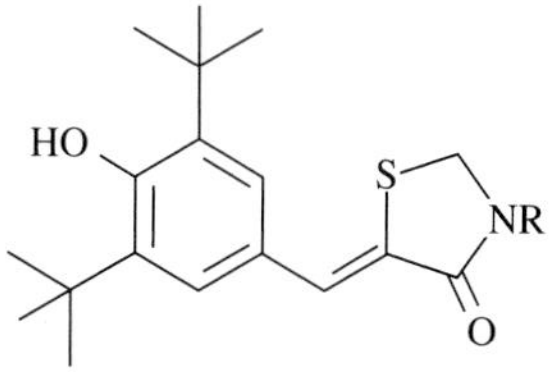

These powerful inhibitors of lipid peroxidation are orally active in animal models of rheumatoid arthritis and cerebral ischaemia–reperfusion. They and related compounds are under development for treatment of stroke and inflammatory bowel disease.

Table 10.11. (*Continued*)

Compound	Structure	Comments
ONO-3144		A chain-breaking antioxidant; it is anti-inflammatory in animal models.
MK-447		A chain-breaking antioxidant; it is anti-inflammatory in animal models.
Nitecapone		Designed to inhibit the enzyme catechol O-methyl transferase to preserve levels of L-DOPA/dopamine in patients with Parkinson's disease. Can act as an iron chelator and chain-breaking antioxidant *in vitro* (*Free Rad. Biol. Med.* **13**, 517 (1992)).
Carvedilol		Marketed as an antihypertensive drug. It is a moderate inhibitor of lipid peroxidation *in vitro*, but its hydroxylated metabolite (–OH at point X) SB211475 is much more potent (*Eur. J. Pharm.* **251**, 237 (1996)).

In addition, some compounds that have limited (or zero) direct antioxidant activity may exert antioxidant actions *in vivo* by up-regulating endogenous antioxidant defences (e.g. this has been suggested for selegiline (Chapter 9) and melatonin (Chapter 3)) and/or by inhibiting generation of ROS/RNS. It is important to accompany clinical testing of a putative 'antioxidant drug' with measurements of oxidative damage, to show that any benefit of a drug is correlated with its antioxidant activity. Such measurements should take account of different molecular targets: it is perfectly possible to have, for example, an antioxidant inhibitor of lipid peroxidation that has no effect on, or even aggravates oxidative damage to, DNA or proteins. For example, the carcinogen diethylstilboestrol is a powerful inhibitor of lipid peroxidation *in vitro*, but its carcinogenicity has been associated with increased oxidative DNA damage *in vivo*.[37]

A few synthetic antioxidants are already on the market (e.g. tirilazad and troglitazone) and more are in clinical trial, yet the majority are at present only at the stage of animal testing (Table 10.11). Let us look at some of the candidates.

10.5.3   *Superoxide dismutase*

The SODs available for potential therapeutic use include recombinant human CuZnSOD, MnSOD and EC-SOD. Indeed, transgenic mice that secrete human EC-SOD in milk have been obtained. A wide variety of CuZnSOD conjugates are available, including polyethylene glycol (PEG)–SOD, Ficoll–SOD, lecithinized SOD, polyamine-conjugated SOD, cationized SOD, genetically engineered SOD polymers, pyran–SOD and albumin–SOD complexes. All have longer circulating half-lives than the unconjugated SOD molecules. For example, intravenous administration of PEG–SOD and PEG–catalase has been shown to decrease ischaemia–reperfusion injury to the brain or spinal cord in several animal studies; injury to the blood–brain barrier may allow them to enter the damaged tissue.[20] SOD–heparin conjugates have also been developed which (like EC-SOD) can bind to endothelial cells.[42]

CuZnSOD has an anti-inflammatory effect in animal models of acute inflammation, in part because it can decrease the number of neutrophils entering sites of inflammation (Chapter 9). However, SOD enzymes of equal catalytic activity do not suppress acute inflammation to the same extent in rats, and it has been suggested that SOD must bind to specific sites to exert its anti-inflammatory activity. The surface charge distribution on the SOD protein appears to be an important factor.[79] Infusions of SOD in liposomes (usually with catalase) have been reported to protect animals against O_2 toxicity (Chapter 3). However, we have been unable to find any report in a good journal of a double-blind controlled clinical trial showing a positive effect of SOD in a chronic inflammatory or autoimmune disease in humans, although brief reports of uncontrolled studies continue to be published.

There has been considerable interest in the use of SOD in protection against reoxygenation injury. The reported effects of SOD in decreasing infarct size

Table 10.12. Some assays used for screening putative antioxidants *in vitro*

Species screened	Assays	Comments
Superoxide	Superoxide is generated (e.g. by xanthine/xanthine oxidase) or added (e.g. as KO_2) and allowed to react with a detector molecule. The inhibition of the reaction is measured, or pulse radiolysis is used to observe direct reactions with $O_2^{\bullet-}$.	For a list of generators/detectors see Chapter 3. For necessary controls see Chapters 3 and 5.
Hydrogen peroxide	The reaction of the compound with H_2O_2 can be measured.	Multiple methods are available (see Table 5.6).
Singlet O_2	Scavenging/quenching of $^1O_2$ can be measured.	See Chapter 5.
Nitric oxide	$NO^\bullet$ is generated. The reaction is followed directly, or as inhibition of the reaction of $NO^\bullet$ with a target molecule.	Table 5.4.
Hydroxyl radical	$OH^\bullet$ is generated and allowed to react with a target (e.g. deoxyribose or a spin trap). Inhibition is measured or pulse radiolysis is used to observe direct reactions with $OH^\bullet$.	Tables 5.1 and 5.3 list the targets available. Almost every biomolecule or synthetic antioxidant is a good $OH^\bullet$ scavenger. Hence, suggestions that 'antioxidants' act by scavenging $OH^\bullet$ *in vivo* are chemically unlikely. Their rate constants for $OH^\bullet$ scavenging may be high (often $>10^{10}\,M^{-1}\,s^{-1}$), but their molar concentrations *in vivo* are usually far lower than that of endogenous molecules capable of rapidly reacting with $OH^\bullet$, including albumin (rate constant $>10^{10}\,M^{-1}\,s^{-1}$) and glucose (rate constant $\approx 10^9\,M^{-1}\,s^{-1}$, but present at millimolar concentrations in body fluids).
Inhibition of Fenton chemistry[a]	The effects on iron/H_2O_2 damage to target molecules are measured.	An 'antioxidant' that affects $OH^\bullet$-dependent damage *in vivo* is more likely to act by blocking $OH^\bullet$ formation, e.g. by removing its precursors ($O_2^{\bullet-}$, H_2O_2, HOCl, etc.).

Peroxyl radicals	The reaction with a model peroxyl radical, e.g. AAPH-derived radicals or trichloromethylperoxyl ($CCl_3O_2^{\bullet}$) is measured, or the effects on lipid peroxidation are examined directly (Table 5.8).	Peroxyl radical scavenging ability does not necessarily parallel ability to inhibit lipid peroxidation in membranes or lipoproteins: other factors (e.g. lipophilicity and interaction with any endogenous antioxidants) are important. For example, dihydrolipoic acid does not inhibit iron/ascorbate–dependent peroxidation in liposomes, but it recycles vitamin E radical in microsomes to inhibit peroxidation.
Peroxynitrite	Nitration of tyrosine; inactivation of α_1-antiproteinase.	Good $ONOO^-$ scavengers inhibit both reactions. On reaction with $ONOO^-$, some molecules (e.g. urate and several thiols, including penicillamine and cysteine) generate oxidation products that also inactivate α_1-antiproteinase.
Hypochlorous acid	Oxidation of thionitrobenzoate; inactivation of α_1-antiproteinase.	On reaction with HOCl, some molecules (e.g. taurine) generate products that also inactivate α_1-antiproteinase.
Total antioxidant activity	For a list of available assays see Table 5.14.	Different assays may give different answers.

Whatever assay is used, an important control is to check for any effects of the putative antioxidant on the assay system itself. For example, does it interfere with generation of ROS/RNS (e.g. inhibiting xanthine oxidase) or react with the end-product detected (e.g. ascorbate reducing DMPO-OH$^{\bullet}$ to an ESR–silent species)? For more details of these assays see *Biochem. Pharmacol.* **49**, 1341 (1995).

[a]Inhibition of Fenton chemistry by iron chelators can involve either of the following. (i) Binding of iron ions to the 'antioxidant' alters their reduction potential and/or accessibility so as to stop them catalysing OH$^{\bullet}$ production; examples include the binding of iron ions to desferrioxamine, transferrin or lactoferrin. (ii) Their binding to the 'antioxidant' does not stop OH$^{\bullet}$ formation, but the OH$^{\bullet}$ is formed at the binding site, so that the 'antioxidant' absorbs it and 'spares' a more important target. For example, copper ions bound to albumin can still form OH$^{\bullet}$, and as a result the protein is damaged. To distinguish between these mechanisms, one can examine the fate of the antioxidant in the reaction mixture; it will be chemically modified if it is reacting with OH$^{\bullet}$. Similar principles apply to studies of effects of metal-chelating antioxidants on lipid peroxidation.

after prolonged myocardial ischaemia seem increasingly questionable (Chapter 9) and so research has concentrated on two areas:

(1) Does a combination of SOD and a thrombolytic agent produce increased functional activity and decreased mortality compared with a thrombolytic agent alone in cases of recent myocardial infarction?
(2) Would SOD be helpful in extending the preservation time of organs for transplantation or preventing graft rejection?

Data on the second question largely focus on studies of kidney transplantation and, in general, the beneficial effects of recombinant human CuZnSOD in double-blind trials seem to be small. Preservation solutions for kidneys already often contain antioxidants and agents such as allopurinol (Chapter 9).

Controlled clinical trials suggest that the answer to the first question is 'no', perhaps because reperfusion injury is not a significant problem in the clinical use of thrombolytic agents. One reason for this could be that reperfusion in most animal model systems is sudden, whereas thrombolytic reperfusion in humans is slow, and slow reoxygenation may minimize the injury. Nevertheless, markers of oxidative injury can be detected in blood from the hearts of patients reperfused after open-heart surgery (Chapter 9). Even if human myocardial reoxygenation injury is clinically significant, SOD need not be the ideal protective agent. For example, SOD failed to protect against myocardial stunning in open-chest dogs, although the combination of SOD and catalase was protective. Desferrioxamine and thiols were more effective.

Another problem is that both CuZnSOD and MnSOD show bell-shaped dose–response curves in isolated heart models of ischaemia–reperfusion injury, i.e. there is an optimum concentration for protection and concentrations higher than this show lower protective effects (Chapter 9). It is possible that negative effects in clinical trials could have resulted from the dose being too high or too low. Studies of the anti-inflammatory effects of MnSOD in rat models of arthritis have found a similar bell-shaped dose–response curve.[23]

Interest continues in the possible protective role of instilled recombinant human CuZnSOD or liposome-encapsulated SOD (with or without catalase) in protecting the lung against damage in infants with respiratory distress syndrome.[64] One approach is the conjugation of these enzymes to a monoclonal antibody that is directed against angiotensin-converting enzyme and accumulates selectively in the lung after systemic injection. SOD/catalase may also be emulsified with surfactant to promote their uptake by lung cells.[65] Some work continues in the use of PEG–SOD to treat traumatic injury to the nervous system and one double-blind clinical trial[20] has been positive. SOD linked to tetanus toxin fragments[30] may be selectively targeted to neuronal cells. Results of double-blind controlled clinical trails are awaited with interest.

10.5.4   *Mimics of SOD*[4]

Several low-molecular-mass compounds that react with $O_2^{\bullet-}$ have been described (Fig. 10.8; Table 10.13). Most contain transition-metal ions.

Examples include iron porphyrins, a complex of manganese ions with the chelating agent desferrioxamine and copper ions chelated to amino acids or to anti–inflammatory drugs, although many of the 'SOD-mimic' copper chelates described in the early literature probably readily dissociate to release copper ions (Table 10.13).

Superoxide scavengers that do not involve metal ions have also been described. **Nitroxides**[52] are stable free radicals (Fig. 10.8) that are often experimentally attached covalently to membranes or proteins so that the behaviour of macromolecules can be followed by changes in the ESR spectrum of the bound nitroxide. Several nitroxides react with $O_2^{\bullet -}$

$$R-\dot{N}-O + O_2^{\bullet -} + H^+ \rightleftharpoons O_2 + R-NOH$$

as can the resulting hydroxylamines

$$R-NOH + O_2^{\bullet -} + H^+ \rightleftharpoons R-\dot{N}-O + H_2O_2$$

If both reactions occur *in vivo*, then the nitroxide would achieve dismutation of $O_2^{\bullet -}$.

An alternative catalytic cycle is to move between nitroxide and oxo–ammonium cation:

$$R-\dot{N}-O + O_2^{\bullet -} + 2H^+ \rightleftharpoons R-\overset{+}{N}=O + H_2O_2$$

$$R-\overset{+}{N}=O + O_2^{\bullet -} \rightleftharpoons R-\dot{N}-O + O_2$$

In vivo, however, nitroxides undergo other redox reactions, e.g. they can oxidize ascorbate, thiols, NAD(P)H or Fe^{2+} ions. Nitroxides have radioprotective effects (Section 8.16) and have been shown to exert antioxidant effects in isolated cells and some animal model systems. It has been reported that some disposable plastic syringes can contaminate solutions with a nitroxide.[7]

Little information is available as yet on the therapeutic applicability or toxicity of any of these SOD mimics to humans.

10.5.5 *Spin traps*[26,27]

Spin traps are widely used to detect the formation of free radicals *in vivo* and *in vitro* (Chapter 5). If a spin trap is trapping significant amounts of a biologically damaging radical in an animal model of disease, then the trap ought itself to exert protective effects. Indeed, this concept applies to all types of radical-trapping agents (Chapter 5); they would be expected to protect against tissue injury if they were working correctly. If they do not protect against injury then

(a) the radicals trapped are not important contributors to damage and/or
(b) not enough of the radical is trapped for tissue protection to reach significant levels, and/or
(c) the trap, its metabolites and/or products of its reaction with free radicals cause tissue damage themselves.

Table 10.13. Some SOD mimics containing transition metals

Compound	Comments	References
Mn^{2+}-polyphosphate, –lactate, –succinate and –malate	Slowly catalyse $O_2^{\bullet-}$ dismutation, but in the absence of a complexing agent, Mn^{2+} is a poor catalyst. EDTA decreases catalysis by Mn^{2+}. Manganese complexes may be important catalysts of $O_2^{\bullet-}$ dismutation in some Lactobacillaceae (Chapter 3).	*Arch. Biochem. Biophys.* **214**, 452 (1982), *J. Phys. Chem.* **33**, 3111 and 6291 (1984)
Desferal-Mn(IV) and related Mn complexes	A green complex formed by reaction of desferrioxamine with manganese dioxide, MnO_2. 1 μM is equivalent to 1 unit of SOD in the cytochrome c assay. It has been shown in various animal models to protect against vascular injury. It has antioxidant effects in bacteria and protects rats against endotoxic shock. Other manganese-containing SOD mimetics such as EUK-8, EUK-134, SC-55858 and SC-55417 (Fig. 10.8) have been described as protective against myocardial ischaemia–reperfusion injury in animals and to inhibit neutrophil infiltration into injured tissues. The EUK compounds also have some H_2O_2-degrading ability and are neuroprotective in neuronal cultures, e.g. those exposed to β-amyloid. Some manganese chelates can undergo other redox reactions in vivo and may not simply be acting as SOD mimics	*Drug. Dev. Res.* **25**, 139 (1992), *Arch. Biochem. Biophys.* **310**, 341 (1994), *Eur. J. Pharmacol.* **304**, 81 (1996), *J. Biol. Chem.* **271**, 26149 (1996), *Cardiovasc. Drug Ther.* **10**, 331 (1996), *Transplantation* **62**, 1664 (1996), *Proc. Natl Acad. Sci. USA* **93**, 2312 (1996), *J. Am. Chem. Soc.* **118**, 4567 (1996), *Adv. Pharmacol.* **38**, 247 (1996); *Arch. Biochem. Biophys.* **288**, 215 (1991)

Copper complexes	$Cu^{2+}_{(aq)}$ and some chelates of it catalyse $O_2^{\bullet-}$ dismutation equally or more rapidly than equimolar amounts of CuZnSOD at low pH. Cu^{2+} is much less effective at physiological pH or when the copper is bound to proteins such as albumin. Some copper–amino-acid complexes (e.g. with lysine or with histidine) are still good catalysts at pH 7.4, as are complexes of copper with 3,5-diisopropyl salicylate or indomethacin. Many of these copper complexes are unstable, however, and problems can be caused due to the release of Cu^{2+} ions (e.g. as catalysts of $OH^{\bullet}$ formation). For example, a copper–penicillamine complex is a poor catalyst of $O_2^{\bullet-}$ dismutation (if a catalyst at all) under conditions where Cu^{2+} does not dissociate from it. EDTA decreases or abolishes reaction of $O_2^{\bullet-}$ with copper salts but has no effect on CuZnSOD. Several complexes, e.g. $Cu(II)_2(3,5\text{-diisopropyl salicylate})_4$ may be able to inhibit other enzyme systems, including NOS and the cytochrome P450 system.	*Biochem. Biophys. Res. Commun.* **81**, 576 (1978), *Arch. Biochem. Biophys.* **203**, 830 (1980), *Biochim. Biophys. Acta* **745**, 37 (1983), *J. Med. Chem.* **27**, 1747 (1984), *Arch. Biochem. Biophys.* **315**, 185 (1994), *J. Inorg. Biochem.* **60**, 133 (1995)
TMPyP[a] and other metalloporphyrins (Fig. 10.8)	Catalyses $O_2^{\bullet-}$ dismutation with about 2–3% of the rate of the SOD enzyme. The manganese and cobalt complexes are also effective, but the copper complex is not. Improved solubility can be achieved by replacing the four $-CH_3$ groups by carboxyl groups. They have antioxidant effects in *E. coli*, isolated mammalian cells, and mice lacking MnSOD but undergo complex intracellular redox reactions and may not simply be acting as SOD mimics. Some can interact with $ONOO^-$.	*J. Inorg. Biochem.* **15**, 261 (1981), *J. Biol. Chem.* **269**, 23471 (1994), *Arch. Biochem. Biophys.* **325**, 20 (1996) *Mol. Pharmacol.* **53**, 795 (1998) *Chem. Res. Tox.* **10**, 1338 (1997) *Nature Genet.* **18**, 159 (1998)

[a]Tetrakis (4-N-methyl)pyridyl porphine-Fe(III).

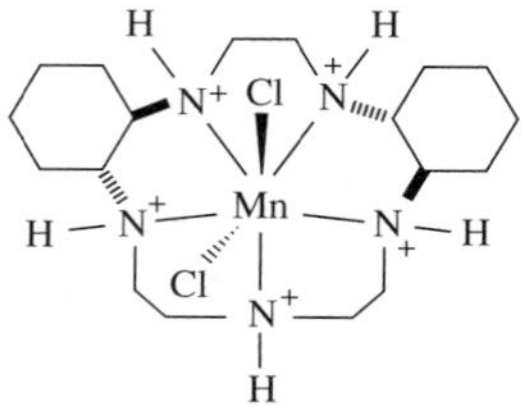

CH₃ ... OXANO

TEMPO

SC-55858

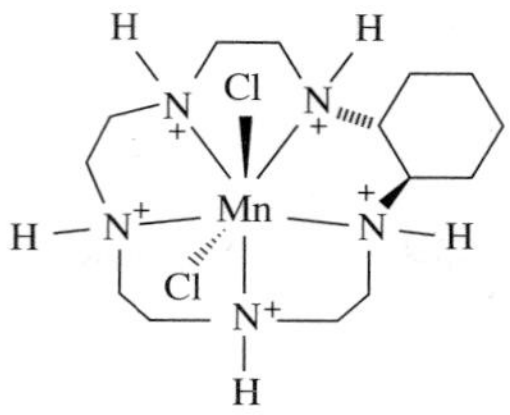

SC-54417

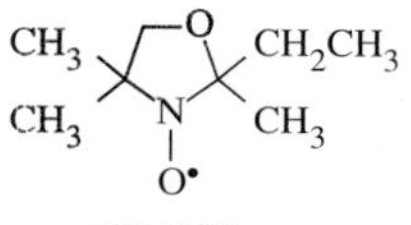

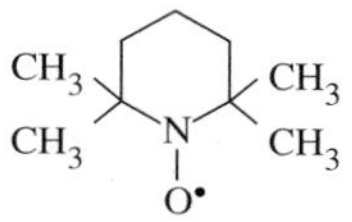

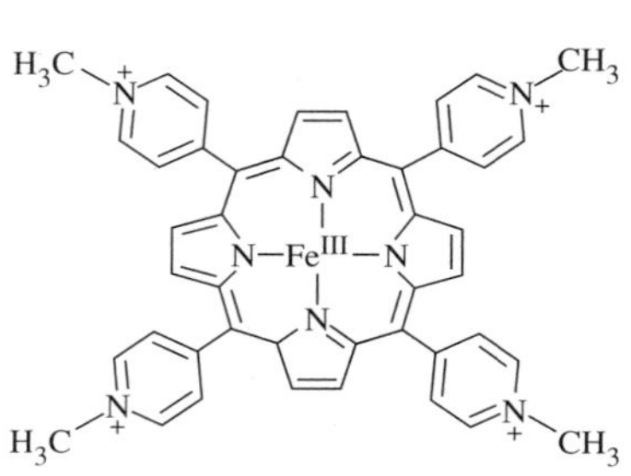

TMPyP

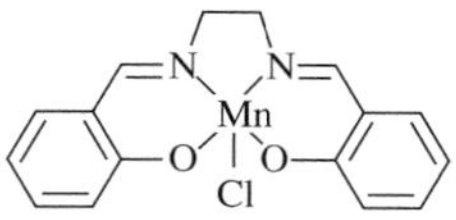

EUK-8

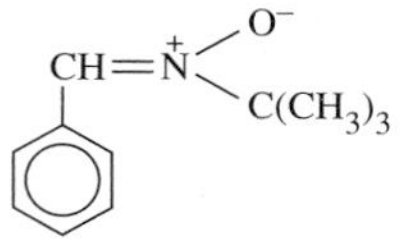

PBN

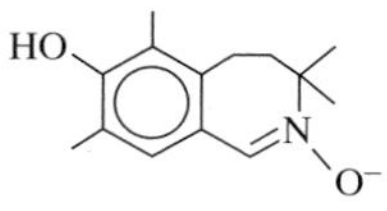

A cyclic nitrone
modelled on PBN

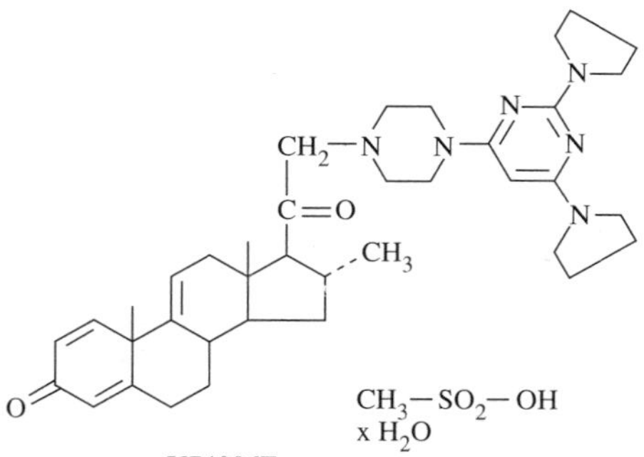

U74006F

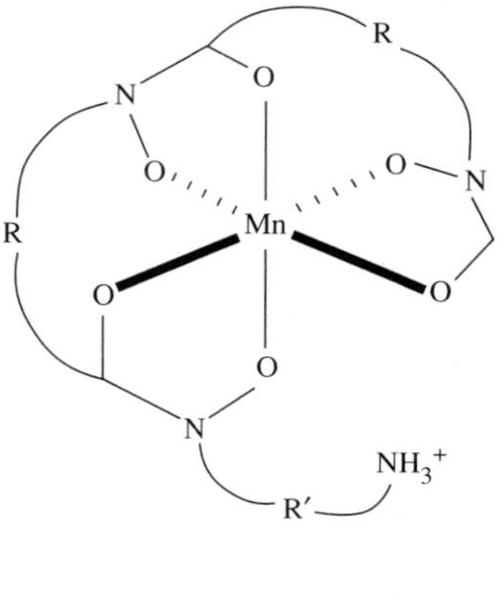

CH_3-SO_2-OH
$x\ H_2O$

U78517F

$R = -[(CH_2)_2CONH(CH_2)_5]-$
$R' = -(CH_2)_5-$

Mn–Desferrioxamine B

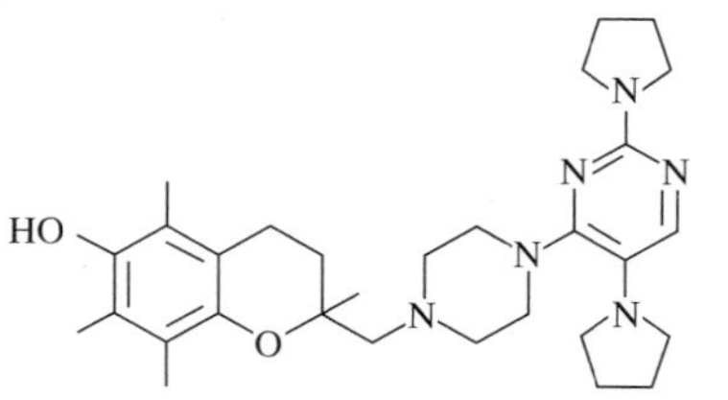

Fig. 10.8.

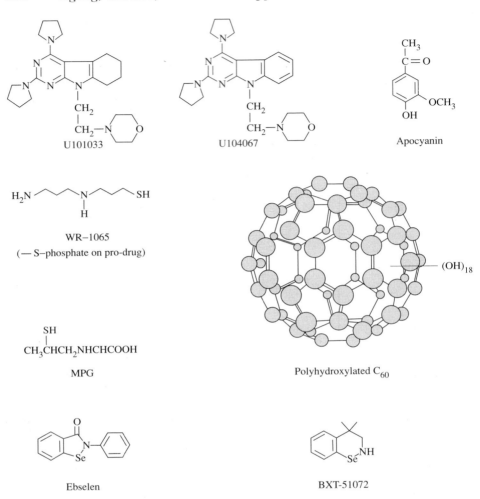

Fig. 10.8. Structures of some antioxidants mentioned in the text. Molecules such as **buckminsterfullerene** (made of 60 carbon atoms) show high reactivity towards organic radical addition ('radical sponges') and polyhydroxylated (18–20 –OH groups) or carboxylated fullerenes have been synthesized as antioxidants (*J. Pharm. Pharmacol.* **49**, 438 (1997); *Proc. Natl Acad. Sci. USA* **94**, 9434 (1997)).

The idea of using spin traps as therapeutic antioxidants arose from studies showing that phenyl-*tert*-butylnitrone (PBN) could protect rats from death due to shock induced by gut ischaemia–reperfusion or by endotoxin injection. Later work showed that PBN could protect against myocardial ischaemia–reoxygenation injury in rats and dogs. PBN given before cerebral ischaemia/reperfusion decreased mortality in gerbils, and it was noticed in the control experiments that PBN administration to old gerbils decreased levels of brain protein carbonyls and appeared to improve cognitive function. However, some similar studies on aged mice or rats have given less clear-cut (or in some cases negative) results.

In vitro characterization of PBN (Table 10.12) shows it to be a poor chain-breaking antioxidant or inhibitor of lipid peroxidation, so its mechanism of action is uncertain. After injection into animals, PBN is rapidly taken up by all tissues (including the brain); concentrations in gerbil brain are estimated to approach 0.5 mM after injection of 150 mg PBN/kg body weight. As well as scavenging ROS, PBN may also prevent induction of iNOS (an effect which could, of course, be secondary to ROS scavenging). As might be expected, several companies are developing derivatives of PBN, with improved radical trapping and chain-breaking antioxidant capacity, as potential therapeutic agents. An example of a cyclic nitrone with such improved antioxidant ability is shown in Fig. 10.8. More information is needed on the metabolism and toxicity of these compounds, but their ability to cross the blood–brain barrier makes them of potential interest for the treatment of neurodegenerative diseases.

Spin traps can have other metabolic effects, including metabolism to generate NO$^\bullet$ (Chapter 5). PBN, POBN and DMPO bind to cytochromes P450 and can inhibit some of their oxidase activities.[51]

10.5.6 *Vitamins C and E and their derivatives*[4,25,67]

Prolonged deficiency of α-tocopherol, e.g. due to fat malabsorption syndromes, produces severe neurological damage (Chapter 3). The antioxidant action of α-tocopherol in plasma lipoproteins helps to protect against atherosclerosis (Chapter 9). Hence, there are good reasons for maintaining an adequate dietary vitamin E intake and for giving supplementary vitamin E to patients with diseases affecting fat absorption. Therapeutic administration of vitamin E to premature babies has been reported to diminish the severity of retrolental fibroplasia (although there is some controversy) and the incidence of both haemolytic syndrome of prematurity and (possibly) intraventricular haemorrhage. Vitamin E supplementation has also been recommended as beneficial in patients suffering from haemolytic syndromes caused by an inborn lack of glutathione synthetase or glucose-6-phosphate dehydrogenase in erythrocytes. However, it has been reported that high doses of intravenous vitamin E can produce an increase in the infection rate in premature babies, probably by depressing phagocyte function. Depression of bactericidal function can also be demonstrated *in vitro* using phagocytes isolated from adults taking oral vitamin E. However, this effect does not seem to be clinically significant, in that the subjects showed no increased rates of infection.

However, attempts to use vitamin E to treat human diseases in which oxidative stress is thought to be important, such as anthracycline-induced cardiotoxicity, cancer or Parkinson's disease, have been largely disappointing. Its value in the treatment of diabetes is uncertain, although some studies are suggestive of benefit, as is a recent study of Alzheimer's disease (Chapter 9). One reason for this could be that it takes a considerable time to raise the vitamin E content of many tissues, especially that of brain, whereas oxidative damage can often be very fast (e.g. within minutes in some reoxygenation injury experiments). It is also difficult to administer α-tocopherol to the lung, where it has been proposed to be potentially useful in protecting against air pollution and inflammatory lung disease. The insolubility of α-tocopherol could potentially be overcome by administering it in liposomes. A second reason for the poor effectiveness of vitamin E is that oxidative damage frequently occurs by mechanisms other than lipid peroxidation.

High doses of vitamin E have been claimed to be beneficial in thrombotic vascular disease, and in prevention of restenosis after angioplasty. Even if true this might be due to effects other than antioxidant action. These could include partial inhibition of lipoxygenase, a weak suppression of platelet aggregation and decreased smooth muscle cell proliferation. It is also interesting to note that the reported protective effects of α-tocopheryl succinate against oxidative damage in isolated hepatocytes could not be explained simply by its hydrolysis to free tocopherol and may have involved 'cytoprotective effects' of the tocopheryl succinate itself (Chapter 3).

Several structural analogues of α-tocopherol, some with improved anti-oxidant activity (as assayed *in vitro* and/or tested in animals), have been described. A water-soluble analogue of α-tocopherol, Trolox C, is widely used *in vitro* (Table 10.11) and a 'cardioselective' vitamin E analogue, MDL 74 405 has been described. For most of these compounds, few (if any) human clinical trials have been reported. However, CS–045 (Table 10.11) is being used in the treatment of diabetes.

Ascorbic acid is essential in the human diet for a variety of reasons, one of which is probably its antioxidant effects *in vivo* (Chapter 3). Various ascorbate esters, e.g. **ascorbyl palmitate** and **2–octadecylascorbate**, have been syn-thesized as lipophilic versions of ascorbate. They have been used as food preservatives and tested as antioxidants in some animal studies, but do not appear to have attracted great interest for therapeutic use. *N*–Substituted uric acids and related compounds have also been evaluated as antioxidants *in vitro*.[29] High doses of vitamin C have been reported to improve vascular function in some cases (Table 10.9; Section 3.22).

10.5.7   *Other chain-breaking antioxidants: probucol*[32] *and ubiquinol*[61]

The drug **probucol** (Chapter 9) has antioxidant properties *in vitro* and has been reported to slow the progression of atherosclerosis in animal and in some human studies (Chapter 9). The probucol phenoxyl radical can apparently be recycled back to probucol in LDL by ascorbate. Probucol also protects rats against adriamycin cardiotoxicity (Chapter 9). It must not, of course, be assumed that the actions of probucol are necessarily due to its antioxidant activity. Probucol tends to decrease plasma HDL levels and so research is under way to develop antioxidants with comparable activity that do not affect HDL levels. Another approach is to combine antioxidant properties and ability to inhibit cholesterol biosynthesis in the same molecule.

Coenzyme Q (CoQ) is an essential component of the mitochondrial electron-transport chain (Chapter 1), and its reduced form, **ubiquinol**, may act as an antioxidant *in vivo* (Chapter 3). Consumption of CoQ supplements by humans has been advocated in some quarters because of their 'antioxidant' and 'mitochondria-protective' properties. Long-term supplementation of humans (3×100 mg CoQ/day) raised $CoQH_2$ levels in plasma and LDL by about 4-fold, and raised the peroxidation resistance of LDL subsequently isolated from the plasma.

Many papers have appeared reporting antioxidant action and/or lipoxygenase inhibition by a wide range of synthetic and natural phenolic compounds.[4,81] Most of these have not been fully evaluated as antioxidants, nor have they been tested on humans. The structures of some are illustrated in Table 10.11. Attempts are being made to add antioxidant (usually phenolic) moieties on to the basic structure of compounds already known to possess anti-allergic or other useful pharmacological properties, such as the oestrogens.[75a]

10.5.8   *BHA, BHT and plant phenolics*[81]

Chain-breaking antioxidants, especially BHA and BHT (Table 10.11) are widely used in the food industry to prevent rancidity (other strategies are to pack foods under nitrogen and to add metal-ion-chelating agents). The toxicity of BHA and BHT is extremely low; perhaps surprisingly, they have not been explored as therapeutic agents, although comparable phenolic structures can be seen in probucol and in many of the other agents that have been investigated (Table 10.11). Because of alleged toxicity problems of BHA and BHT (at absurdly high doses) there are attempts to replace them as preservatives in certain foodstuffs by 'natural' phenolic compounds such as flavonoids, hydroxytyrosol or rosemary antioxidants (Chapter 3). BHT is also present in the plastic film 'Parafilm' and can be released from it during laboratory experiments.[80]

The therapeutic potential of plant phenolics is also under investigation; most work has concentrated on mixtures such as **Ginkgo biloba** extract (Chapter 3) or **Daflon**, a mixture of diosmin (90%) and hesperidin (10%). More data on their clinical efficacy are needed. Flavonoids and derivatives of them are under investigation as anti-inflammatory agents and for the treatment of vascular problems, e.g. in diabetes, since some may inhibit protein glycation and aldose reductase activity as well as exerting antioxidant effects. Of course, the safety of high doses of any plant phenol should never be assumed just because it is a natural product (Table 10.8).

10.5.9   *The lazaroids*[20,34]

The lazaroids are a group of compounds developed by the Upjohn Company. The original aim was to add antioxidant activity to a steroid nucleus, since high doses of the glucocorticoid methylprednisolone had been claimed to be effective in diminishing brain lipid peroxidation after trauma and seemed to improve clinical outcome in patients. The most studied lazaroid is U–74006F, **tirilazad mesylate** (Fig. 10.8), which has undergone several clinical trials for the treatment of traumatic injury to the central nervous system. Others include U–74389 and U–74500. The lazaroids inhibit iron-dependent lipid peroxidation in brain homogenates *in vitro* and exert neuroprotective effects in various animal models of traumatic injury to the brain or spinal cord. Another compound that has been tested is U–78517F (Fig. 10.8), a tocopherol analogue; it seemed to offer no benefits over the lazaroids in animal testing.

To date, clinical trials of U–74006F (as **Freedox**) have shown significant benefit in cases of sub-arachnoid haemorrhage, a phenomenon associated with bleeding caused by traumatic brain injury or rupture of blood vessels (Chapter 9). In treated patients there was decreased vasospasm, decreased mortality and improved functional recovery. The benefits were shown in male patients but not in females, apparently because females metabolize the drug more quickly. Clinical trials in stroke patients have to date given less convincing evidence of benefit. It should perhaps be noted that there is no direct evidence that antioxidant action mediates any beneficial effects of the lazaroids *in vivo*.

Tirilazad appears to accumulate in the blood–brain barrier, protecting the microvascular endothelium, and its penetration into the brain is limited. An alternative is the **pyrrolopyrimidines** such as U–101033E and U–104067F (Fig. 10.8), antioxidants which appear to enter the brain more readily.

10.5.10   *Thiol compounds*

Glutathione[2]

Glutathione has multiple functions in human metabolism, including the detoxification of xenobiotics, radioprotection and antioxidant defence (Chapter 3). There have therefore been many suggestions that GSH will be therapeutically useful as an antioxidant and general cytoprotective agent. Areas of interest include the preservation of organs for transplantation (for example, GSH is added to 'University of Wisconsin solution' (Chapter 9)) and protection against tissue damage by cytotoxic drugs such as cyclophosphamide. Aerosolized GSH solutions have been suggested as a means of diminishing lung damage by ROS/RNS generated extracellularly by activated phagocytes in such diseases as ARDS and cystic fibrosis. GSH administered intra-tracheally to animals seems to be rapidly degraded and cleared from the lung, however. There is also evidence for abnormalities of GSH metabolism in HIV-infected subjects and the therapeutic use of GSH or its precursors has been suggested (Chapter 9).

GSH is not rapidly taken up by cells. However, methyl, isopropyl and ethyl monoesters (esterified on the carboxyl group of glycine) have been described, which can apparently cross cell membranes and be hydrolysed to GSH within the cell. Diethylesters (esterified on both the carboxyl group of glycine and the side-chain carboxyl group of the glutamate residue) have also been tested, and seem to be more efficient delivery systems than monoesters. However, there have been some reports of toxicity of GSH esters, possibly due to contaminating compounds arising during their preparation. The compound **L-2-oxathiazolidine-4-carboxylate** (OTC) has been reported to be hydrolysed to cysteine *in vivo*, which may lead to an increased synthesis of GSH. Hydrolysis requires the enzyme 5-oxoprolinase. OTC has been administered to patients with coronary artery disease and appears to improve vascular endothelial function.[97a]

N-Acetylcysteine[17]

N-Acetylcysteine has been used as an antioxidant in a wide variety of experiments and is effective in the treatment of paracetamol (acetaminophen) overdosage in humans (Chapter 8). It may protect by entering cells and being hydrolysed to cysteine, which stimulates GSH synthesis. Additionally, N-acetylcysteine may scavenge several ROS/RNS (including HOCl, ONOOH, $RO_2^{\bullet}$, $OH^{\bullet}$ and H_2O_2) directly. It has been widely used in humans for treatment of various respiratory disorders; it has limited toxicity but its therapeutic benefits in these conditions have been questioned.

N-Acetylcysteine has been reported to have protective effects in many animal models of lung injury. By contrast, it potentiated the lung damage produced by the anticancer drug bleomycin in hamsters (Chapter 9). *In vitro*, N-acetylcysteine did not potentiate iron ion–dependent damage to DNA by bleomycin, whereas cysteine did and so it is possible that the N–acetylcysteine acted as a precursor of intracellular cysteine, which could reduce the iron–bleomycin complex upon the nuclear DNA. N-Acetylcysteine has been reported to suppress activation of NF-κB in several cell types and has undergone some clinical testing in HIV-positive subjects and AIDS patients (Chapter 9).

Thiol toxicity

Mixtures of thiols with transition-metal ions can be cytotoxic because of reactions that produce both ROS ($O_2^{\bullet-}$, H_2O_2, and $OH^{\bullet}$) and sulphur-containing radicals such as $RS^{\bullet}$, $RSO^{\bullet}$ and $RSO_2^{\bullet}$ (Chapter 2). GSH is unstable in solution, particularly if iron or copper ions are present, and the oxidation products could conceivably cause damage. This must be borne in mind if solutions containing GSH or other thiols are to be stored for any length of time, e.g. in cell culture media or in fluids for organ preservation. Cysteine seems to be more toxic than GSH in such systems, perhaps because it is oxidized more quickly and because the cysteine–$S^{\bullet}$ radical is more reactive, e.g. it can abstract hydrogen atoms from fatty acids to initiate lipid peroxidation. Indeed, cysteine is much more toxic when administered to animals than are N-acetylcysteine or GSH. Similarly, high levels of plasma homocysteine are a risk factor for atherosclerosis (Section 10.4.11 above).

Other thiols

The disulphide **lipoic acid** (Chapter 3) can act as an antioxidant *in vitro* and is undergoing clinical trials in patients with type II diabetes. It has been suggested to decrease insulin resistance and protect against peripheral neuropathy. Interest has also been expressed in **ergothioneine** as an anti-oxidant (Chapter 3).

Many thiols have been tested for their ability to protect cells and animals against ionizing radiation (Chapter 8). They include GSH, cysteine, cysteamine, dimercaprol (British anti-Lewisite), penicillamine, mercaptopropionylglycine (MPG) and *S*-2-(3-aminopropylamino)ethyl phosphorothioic acid, a pro-drug which is hydrolysed to the free thiol (WR-1065; Fig. 10.8) *in vivo*. The 'WR' signifies that this radioprotector (along with many others) was developed at the Walter Reed Army Hospital in the USA.[75]

The thiol compound MPG has been shown to be protective against reperfusion injury after ischaemia in animal models of myocardial stunning and infarction (Chapter 9). It may act by its ability to scavenge ROS, but which species it acts upon *in vivo* is as yet uncertain. Several clinical studies have claimed that MPG is beneficial in the treatment of various diseases but well-controlled trials of its effects against reperfusion injury do not appear to have been published.

10.5.11   *Glutathione peroxidase 'mimics'*[83]

Selenocysteine is an essential part of the active sites of glutathione peroxidases (Chapter 3). Hence several attempts have been made to design low-molecular-mass selenium compounds with similar catalytic activity. The first to be developed was **Ebselen** (PZ51), 2-phenyl-1,2-benzisoselenazol-3(2*H*)-one (Fig. 10.8). In the presence of GSH, Ebselen decomposes peroxides in a catalytic manner. Ebselen also has direct antioxidant activities *in vitro*, scavenging HOCl, singlet O_2, ONOO$^-$ and RO$_2^{\bullet}$, for example. There is considerable debate about the mechanism of the catalysis of peroxide removal by Ebselen, and at least three different 'cycles' of reactions may be involved. However, it seems likely that GSH reacts with the Ebselen and the resulting product can remove H_2O_2 and organic peroxides (LOOH). Other thiols, including *N*-acetylcysteine and dihydrolipoate (DHLA), can replace GSH as cofactors, DHLA being especially effective. Ebselen can bind to −SH groups on proteins and much Ebselen in blood plasma is probably attached to albumin −SH groups.

Ebselen has shown protective effects in several cell culture systems and animal models of human disease, and its toxicity appears to be low. As well as showing antioxidant activity, however, it can also inhibit 5- and 15-lipoxygenases and nitric oxide synthases and may have direct inhibitory effects on phagocyte ROS production.

Other organoselenium compounds have been reported to possess peroxidase-like activity, including several diselenides (containing −Se−Se−groupings). However, diselenides may release selenium too easily. Some interest has focused on BXT-51072 (Fig. 10.8), one of a group of compounds which appear to exert greater protective effects than Ebselen in cell systems.

By analogy, several compounds containing the element **tellurium**[3] (which, like sulphur and selenium, is in group VI of the periodic table; Appendix I) can scavenge peroxides and show other direct antioxidant effects *in vitro*.

10.6   Iron chelators[40,41]

Iron is an essential element in the human body, but it is a powerful promoter of free-radical reactions (Chapter 3) and can be released in redox-active form at sites of tissue injury. A wide range of chelating agents has been used in attempts to inhibit oxidative damage *in vitro* (Table 10.14 gives some examples). The mechanisms by which chelators can act were discussed in Chapter 2 (Section 2.4.1) and are summarized in Table 10.12. The first chelating agent reported to decrease the rate of $O_2^{\bullet-}$-dependent $OH^\bullet$ generation *in vitro* was DETAPAC (Table 10.14). It slows $OH^\bullet$ generation because the Fe(III)–DETAPAC chelate is reduced only slowly by $O_2^{\bullet-}$. An Fe^{2+}–DETAPAC complex still reacts with H_2O_2 to form $OH^\bullet$, however, and more powerful reducing agents than $O_2^{\bullet-}$ (such as paraquat radical) *are* able to reduce Fe(III)–DETAPAC. Hence DETAPAC is not a *general* inhibitor of iron-dependent $OH^\bullet$ generation, only a partial inhibitor of the $O_2^{\bullet-}$-dependent $OH^\bullet$ generation. Complexes of iron salts with phytic acid, *o*–phenanthroline, bathophenanthroline sulphonate and desferrioxamine also show diminished reactivity in formation of $OH^\bullet$ from $O_2^{\bullet-}$ and H_2O_2.

Phytic acid, **inositol hexaphosphate**,[68] has been used as an antioxidant in foodstuffs and could conceivably be a protective agent in the human diet. It has been reported to be present in mammalian cells. Indeed, some research has taken place into the development of inositol triphosphates (e.g. inositol-1,2,3-triphosphate and inositol-1,2,6-triphosphate) and tetraphosphates as potential therapeutic agents for human disease. Several flavonoids and other plant phenolics can bind transition-metal ions (Chapter 3). For example, this may be the mechanism by which **7–monohydroxyethylrutoside** (Table 10.14) diminishes the cardiotoxicity of doxorubicin to mice. The chelator ***o*–phenanthroline**, which penetrates easily into cells, has been shown to diminish H_2O_2-dependent DNA damage in mammalian cells, perhaps by binding iron ions from the vicinity of the DNA and preventing DNA fragmentation by site-specific $OH^\bullet$ generation (Chapter 4).

10.6.1   *Desferrioxamine*

Desferrioxamine is a powerful chelator of Fe(III). In the presence of physiological buffer systems, it inhibits lipid peroxidation, and the generation of $OH^\bullet$ from $O_2^{\bullet-}$ and H_2O_2, in the presence of iron ions. Desferrioxamine is produced by *Streptomyces pilosus* and is widely and effectively used for the prevention and treatment of iron overload in patients who have ingested toxic oral doses of iron salts, or who require multiple blood transfusions, e.g. for the treatment of thalassaemia. Large doses of desferrioxamine can be injected into animals or humans: at least 50 mg/kg bodyweight per day appears fairly safe in thalassaemic patients provided that they do not become iron-deficient. Excess desferrioxamine can lead to auditory and visual problems, usually reversible when the excess is withdrawn, and an increased risk of serious infection with certain microorganisms, such as *Vibrio vulnificus* and *Yersinia enterocolitica*, and

Table 10.14. Some chelating agents

Chelating agent	Comments
Penicillamine HS—C(CH$_3$)$_2$ | H$_2$N—CHCOOH	Useful in promoting urinary excretion of copper salts in treatment of Wilson's disease. Also binds Fe^{2+}.
EDTA ion $^-$OOCCH$_2$ CH$_2$COO$^-$ N(CH$_2$)$_2$N $^-$OOCCH$_2$ CH$_2$COO$^-$	Chelates several metal ions. Manganese– and copper–EDTA chelates are usually less active than the unchelated metal ions in radical reactions, whereas chelates of EDTA with Fe^{2+} or Fe(III) still react with H$_2$O$_2$ or superoxide. Often EDTA promotes iron-dependent damage by keeping iron ions in solution in a redox-active form but it can inhibit 'site-specific' damage by removing metal ions from a specific target (e.g. DNA) and dissipating the free radicals over a broader range of targets. EDTA infusion has been used in some countries ('chelation therapy') to treat vascular problems but double-blind controlled clinical trials are needed to justify this use (e.g. *Circulation* **90**, 1194 (1994)). Fe^{2+}–EDTA chelates rapidly oxidize to Fe(III) chelates at pH 7.4, releasing $\overset{\bullet}{O}_2^-$.
DETAPAC ion $^-$OOCCH$_2$ CH$_2$COO$^-$ N(CH$_2$)$_2$N(CH$_2$)$_2$N $^-$OOCCH$_2$ CH$_2$COO$^-$ CH$_2$COO$^-$	Chelates several metal ions other than iron and copper. Little used clinically as it has several side effects, including zinc and magnesium depletion, but it has been used *in vivo* to remove lead and plutonium and to remove iron from thalassaemic patients who cannot tolerate desferrioxamine.
Rhodotorulic acid	Both this and desferrioxamine (below) are examples of **siderophores** (from the Greek words for 'iron' and 'carrier'). Siderophores are produced by microorganisms in order to chelate iron from the growth medium and bring it into the cell. Iron is usually bound as Fe(III). Siderophores are usually hydroxamates (such as desferrioxamine) or phenols.

2,3-Dihydroxy-benzoic acid ion

COO⁻ structure:

$$COO^-$$

A product of OH• attack on salicylate. Can be given orally.

Desferrioxamine B

A linear molecule that 'bends round' to complex Fe(III) with six ligands (i.e. it is a hexadentate chelator) forming a bright-red complex known as ferrioxamine. It is commercially available as **desferal** (desferrioxamine B methanesulphonate; mol. wt 657) and is often called **deferoxamine** in the USA.

PIH

It is an effective iron chelator in animals. Can be given orally.

o-Phenanthroline
(1, 10-phenanthroline)

Good chelator of Cu^{2+} ions (log stability constant[b] = 6.3). Also binds Fe^{2+} (log stability constant = 5.8) and zinc. Can prevent H_2O_2-mediated damage to DNA in some isolated mammalian cells. However, a Cu^{2+}-phenanthroline complex can stimulate DNA degradation and is the basis of the phenanthroline assay (Section 4.5).

1,2-Dimethyl-3-hydroxy-pyridin-4-one[a]

One of a series of chelators, the 1-alkyl-3-hydroxy-2-methylpyrid-4-ones, which can be given orally since they are readily absorbed from the gut. They also enter tissues much more readily than desferal and have shown promising preliminary results in the treatment of thalassaemia. They are bidentate; three molecules are needed to coordinate one iron completely.

DBED

Reaction of Fe^{2+} chelated to this compound with H_2O_2 hydroxylates site-specifically on aromatic ring (at★) to generate a better iron chelator (*Adv. Pharmacol.* **38**, 167 (1996)). X, axial ligands.

7-Monohydroxy-ethylrutoside

Its effectiveness in protecting against cardiotoxicity in mice is comparable to that of ICRF-187 (*Br. J. Pharmacol.* **115**, 1260 (1995)).

Abbreviations: DBED, *N,N*′-dibenzylethylenediamine *N,N*′-diacetic acid; DETAPAC, diethylenetriaminepentaacetic acid; PIH, pyridoxal isonicotinoyl hydrazone.

[a]Sometimes called CP20, deferiprone or L1. In CP94 both −CH₃ groups are replaced by ethyl groups.

Box 10.1

Properties of desferrioxamine[a] in relation to its action in radical-generating systems

- powerful chelator of Fe(III) (stability constant $\sim 10^{31}$)
- chelates several other metal ions with stability constants several orders of magnitude lower than for Fe(III) (e.g. Al(III), $\sim 10^{25}$; Cu^{2+}, $\sim 10^{14}$; Zn^{2+}, $\sim 10^{11}$); has been used to remove aluminium from dialysis patients (Chapter 8)
- reacts slowly with $O_2^{\bullet -}$ or $HO_2^{\bullet}$ ($k_2 \approx 10^3\,M^{-1}\,s^{-1}$) to form relatively stable nitroxide radicals that can inactivate yeast alcohol dehydrogenase and oxidize ascorbate, GSH and NAD(P)H
- reacts quickly with $OH^{\bullet}$ ($k_2 \approx 10^{10}\,M^{-1}\,s^{-1}$); nitroxide radicals are again produced, so desferrioxamine is an excellent $OH^{\bullet}$ scavenger; ferrioxamine (Fe(III) chelate) scavenges $OH^{\bullet}$ with the same rate constant
- inhibits iron-dependent lipid peroxidation and $OH^{\bullet}$ generation from H_2O_2 in most systems; ferrioxamine does not inhibit, so a control with this substance can distinguish protection by iron binding (and inhibition of $OH^{\bullet}$ formation) from protection by $OH^{\bullet}$ scavenging
- accelerates the oxidation of Fe^{2+} solutions, by binding the resulting Fe(III) more tightly than it does Fe^{2+}; this is the **ferroxidase** action of desferrioxamine, which will produce $O_2^{\bullet -}$:

$$Fe^{2+}-DFX + O_2 \rightarrow Fe(III)-DFX + O_2^{\bullet -}$$

- penetrates only slowly into most animal cells; it is poorly absorbed from gut but oral administration could conceivably be used to interfere with iron absorption
- moderately good scavenger of peroxyl ($RO_2^{\bullet}$) radicals in aqueous solution

[a]Most studies with desferrioxamine are carried out with the commercially available **desferal** (desferrioxamine B methanesulphonate). Adapted from *Free Rad. Biol. Med.* **7**, 645 (1989).

---

with fungi such as *Rhizopus* (sometimes seen in dialysis patients treated with desferrioxamine to remove aluminium). Desferrioxamine is highly (but not absolutely) specific for Fe(III). The Fe(III)–desferrioxamine complex (**ferrioxamine**) is difficult to reduce, not only by $O_2^{\bullet -}$, but also by more powerful reductants. Box 10.1 summarizes the properties of desferrioxamine that are relevant to its use as an antioxidant.

Desferrioxamine was originally developed for the treatment of iron-overload disease (Chapter 3) and later used to remove aluminium from dialysis patients (Chapter 8). Its limited toxicity allows it to be used *in vivo* to investigate the role of iron ions in animal models of some human diseases. Table 10.15 summarizes some of the results, which are consistent with a role for iron ions in promoting tissue damage *in vivo*, presumably via oxidative stress. Desferrioxamine can also react directly with several ROS, including $RO_2^{\bullet}$, $O_2^{\bullet -}$,

Table 10.15. Some of the animal models of oxidative stress in which desferrioxamine has been shown to be protective

Effect reported	Comments
Decreases neutrophil-mediated acute lung injury in rats after complement activation.	A model for some forms of ARDS in which $O_2^{\bullet -}$ and H_2O_2 produced by activated neutrophils in the lung cause damage (Chapter 9).
Anti-inflammatory in several acute and acute-to-chronic models of inflammation.	Section 9.7
Inhibits the toxic action of alloxan and paraquat to animals, which are thought to be mediated by increased generation of $O_2^{\bullet -}$ and H_2O_2 *in vivo*.	Results are variable (Chapter 8).
Decreases liver damage by CCl_4 in animals.	Iron-dependent decomposition of lipid peroxides may be inhibited (Chapter 8).
Protective against reoxygenation injury after ischaemia in lung, heart, kidney, gut and skin. Heart studies show protection against arrhythmias and stunning and perhaps decreased infarct size.	Reperfusion injury involves $O_2^{\bullet -}$ and H_2O_2 (Chapter 9).
Inhibits the progress of an autoimmune disease model (experimental allergic encephalomyelitis) in rats.	Mechanism of action unclear (see text). Results variable depending on experimental protocol (e.g. *J. Neuroimmunol.* **17**, 127 (1988)).
Decreases antigen–antibody-induced kidney damage in rabbits.	Damage may be due to $O_2^{\bullet -}$ and H_2O_2 produced by neutrophils, perhaps reacting with iron to form $OH^{\bullet}$.

Adapted from *Free Rad. Biol. Med.* **7**, 645 (1989).
Note: desferal has also been found protective in some plant systems.

$OH^{\bullet}$, HOCl, oxo-haem species produced by mixing haem proteins with H_2O_2, and $ONOO^-$.

Desferrioxamine is a hydroxamate (Table 10.14) and so its oxidation can generate nitroxide radicals which could oxidize other biomolecules.[101] However, the toxicity of nitroxides is fairly low—they have been used, for example, as SOD mimics (Section 10.5.4 above). Indeed, several hydroxamates (such as **N-methylhexanoylhydroxamate**),

$$CH_3(CH_2)_4 - \overset{\displaystyle O}{\overset{\displaystyle \|}{C}} - \overset{\displaystyle OH}{\overset{\displaystyle |}{N}} - CH_3$$

have been developed as antioxidants, e.g. that prevent damage mediated by myoglobin/H_2O_2 systems *in vitro*.[16]

Direct scavenging of ROS/RNS by desferrioxamine may be of limited significance *in vivo*, because the concentration achieved during normal therapeutic use (up to $20\,\mu M$ steady-state plasma levels during infusion) is too low for scavenging to be a feasible mechanism of action. Desferrioxamine has also been shown (again at high concentrations) to block the conversion of xanthine dehydrogenase to oxidase activity in cultured endothelial cells. Despite its high stability constant for iron, desferrioxamine is poor at removing iron ions bound to transferrin or lactoferrin, which may be advantageous since iron bound to these proteins does not stimulate free-radical reactions.

At therapeutically relevant concentrations ($20-100\,\mu M$) desferrioxamine can inhibit cell proliferation *in vitro*, and *in vivo*. It may do so by depleting the cells of iron, so inhibiting iron–dependent enzymes such as ribonucleoside-diphosphate reductase, but there is debate as to whether or not this is the only cytostatic mechanism. Hence, if desferrioxamine is administered by continuous infusion over several days, as has been done in some studies of its effect upon chronic inflammation or autoimmune disease in animals, it is possible (but remains to be proved) that any beneficial effects it exerts could relate to inhibition of the proliferation of inflammatory cells such as lymphocytes, and not necessarily be due to antioxidant action. The cytostatic action of desferrioxamine or other iron ion-chelating agents might conceivably be of therapeutic use in suppressing lymphocyte-mediated reactions and in the treatment of leukaemia and malaria or other parasitic infections (Section 7.4).[98]

Because iron is essential for normal cell function, prolonged administration of any powerful iron-chelating agent to humans (other than patients suffering from iron overload) will probably cause side effects (Section 9.7.6). Hence, proposals to use long-term desferrioxamine administration in the treatment of chronic disease in humans, such as Alzheimer's disease or rheumatoid arthritis, must be approached with caution, as must prolonged treatment of aluminium-overloaded patients with this drug. However, the side effects of chronic desferrioxamine administration should not preclude its acute use, e.g. in combination with thrombolytic agents in the treatment of myocardial reperfusion injury, in haemorrhagic shock or in ameliorating cell damage by toxins such as the anthracycline antitumour antibiotics. However, controlled clinical trials remain to be published in these areas. Some groups have added desferal (desferrioxamine B methanesulphonate) to organ preservation fluids, e.g. for heart transplantation. The University of Wisconsin preservation solution contains a modified carbohydrate, **lactobionic acid**,[43] which may exert antioxidant effects by chelating iron.

High-molecular-mass forms of desferrioxamine[60] have been described, in which it is attached to polymers such as dextran or hydroxyethyl starch. Conjugation to these polymers was reported not to diminish its Fe(III)-binding activity, but to decrease the toxicity of high doses and to increase circulating plasma half-life in animals. These conjugates have been reported as protective in several animal models of disease, including septic shock, and to delay the onset of diabetes in spontaneously diabetic rats.

10.6.2   *Other iron-chelating agents*

The gene encoding human **lactoferrin** has been cloned and expressed in female mice (which secrete the human protein in their milk),[69] but no clinical trials using this protein as an antioxidant have been reported. The chelator ICRF-187, which is hydrolysed *in vivo* to produce a metal ion chelator (ICRF-198) structurally similar to EDTA, has been reported to ameliorate the cardiotoxicity of anthracyclines in human breast cancer patients, while not affecting the therapeutic efficiency of these drugs (Chapter 9). Careful use of low concentrations of chelating agents may help to minimize anthracycline-dependent cardiotoxicity in treatment of other cancers, particularly the leukaemias, in which chemotherapy can cause transient iron overload.

The **hydroxypyridones** were introduced in an attempt to overcome a major drawback of desferrioxamine in the treatment of thalassaemia: it is not absorbed through the gut and has to be administered by intravenous or subcutaneous infusion, leading to poor patient compliance. Desferal is also expensive. There has therefore been a search for orally active chelating agents which, it is hoped, will be cheaper. Some work has been reported with pyridoxal isonicotinoyl hydrazone, desferrithiocin and 2,3-dihydroxybenzoate, but most attention has focused on the 3-hydroxypyrid-4-ones (Table 10.14). For example, 1,2-dimethyl-3-hydroxypyridin-4-one (**LI**) has undergone clinical trials in humans. It appears to be more toxic than desferrioxamine. Other hydroxypyridones (such as the 1,2-diethyl compound) are also undergoing evaluation.[40,41]

The apparent increased toxicity of hydroxypyridones, when compared with desferrioxamine, could be due to features of their chemistry. An iron ion can bind six ligands. Desferrioxamine is a **hexadentate** ligand, i.e. it occupies all six coordination sites. The desferrioxamine molecule 'wraps around' the iron ion, encasing it in an envelope of organic material. By contrast, hydroxypyridones can occupy only two coordination sites (i.e. they are **bidentate** ligands). Hence three molecules of hydroxypyridone are needed to complex one Fe(III) completely. Upon dilution, 3:1 hydroxypyridone:iron ion complexes might dissociate into 2:1 complexes, leaving available coordination sites on the iron that could allow catalysis of free-radical reactions. In addition, hydroxypyridones can mobilize iron ions from transferrin or lactoferrin, at least *in vitro*. Thus, one can envisage a scenario in which iron ions that were protein-bound (and so unable to stimulate free-radical reactions) could be mobilized and transferred to other sites in the body at which damage can be caused. However, there is no direct evidence that this effect occurs *in vivo* or contributes to the apparent increased risk of side-effects in patients treated with LI as opposed to desferrioxamine.

An interesting series of *in vitro* experiments has described 'oxidative stress-activatable' iron chelators. For example, in the presence of H_2O_2 an Fe^{2+}-chelate of DBED (Table 10.14) undergoes hydroxylation on the aromatic ring to generate a better iron chelator.

10.7   Inhibitors of ROS/RNS generation

Another approach to antioxidant protection is to inhibit the formation of ROS/RNS. For example, agents that prevent excessive glutamate release, or antagonize its action, can decrease excitotoxicity in the brain and subsequent oxidative stress (Chapter 9). Agents that down-regulate NF-κB (e.g. caffeic acid phenethyl ester (Table 10.7) and several thiols) may be protective by decreasing iNOS and adhesion molecule expression.

10.7.1   *Xanthine oxidase inhibitors*[4]

Allopurinol inhibits the production of $O_2^{\bullet-}$, H_2O_2 and uric acid by the enzyme xanthine oxidase. It inhibits by first acting as a substrate for the enzyme, so producing the true inhibitor, **oxypurinol**. Oxypurinol is also the major metabolite formed when allopurinol is administered to humans. When it was proposed that xanthine oxidase is the major source of the ROS generated when ischaemic tissues are reoxygenated (Chapter 9), interest naturally arose in the use of allopurinol or oxypurinol as protectors against tissue damage. Indeed, many animal experiments showed that allopurinol has protective effects against cardiac, cerebral or gastrointestinal reoxygenation injury as well as in some animal models of haemorrhagic shock. It might be thought more logical to use oxypurinol as a protective agent, since it is the true inhibitor of xanthine oxidase. However, allopurinol was reported to be more effective than oxypurinol in decreasing urinary uric acid excretion in humans. Similarly **B103U**, an analogue of oxypurinol in which the oxygen at position 6 is replaced by sulphur, did not inhibit the enzyme any better *in vivo* than did oxypurinol, although it was a more powerful inhibitor *in vitro*.

Since allopurinol is widely used to treat hyperuricaemia in humans, its toxicological profile is well known and it might therefore seem to be an agent of choice in testing for beneficial effects in diseases to which ischaemia–reperfusion is thought to contribute. Unfortunately, the evidence that xanthine oxidase makes a significant contribution to the generation of ROS in human heart and brain is weak. In addition, many studies in which protection has been shown have required pretreatment of animals with the drug, which is not usually possible in clinical cases of human ischaemia–reperfusion. However, xanthine oxidase activity has been detected in human gut, in the palmar fascia of patients with Dupuytren's contracture, in the circulation of patients after reperfusion of limbs following application of a tourniquet and in the synovia of patients with rheumatoid arthritis (Chapter 9). The therapeutic possibilities of treatment with allopurinol in these diseases have been discussed, but not yet evaluated clinically to date. Allopurinol and oxypurinol may also be beneficial in organ preservation for transplantation and have been added to some preservation solutions (Chapter 9).

As well as inhibiting xanthine oxidase, oxypurinol can scavenge HOCl *in vitro*, which might contribute to its protective effects under certain circumstances. *In vitro*, both allopurinol and oxypurinol react with OH$^{\bullet}$, but the concentrations achieved *in vivo* are probably too low for this to be a significant mechanism of protection. An additional mechanism by which allopurinol or oxypurinol could protect tissues is by preventing oxidation of hypoxanthine, so enhancing its salvage for reincorporation into adenine nucleotides when the tissue is reoxygenated.

Other inhibitors of xanthine oxidase that have been described include **lodoxamide, pterin-6-aldehyde** and **amflutizole**.

10.7.2 *Inhibitors of ROS generation by phagocytes*[14,57]

Generation of ROS/RNS by activated phagocytes contributes to inflammatory and reoxygenation injury. Thus, it is possible that inhibitors of phagocyte recruitment (e.g. agents that inhibit the production, or antagonize the action of, 'pro-inflammatory' cytokines such as platelet-activating factor and tumour necrosis factor alpha), blockers of phagocyte adherence to endothelium (e.g. agents that down-regulate expression of adhesion molecules, antibodies directed against such molecules or synthetic adhesion molecules that compete with phagocytes for binding to endothelium) or agents interfering with the respiratory burst could be therapeutically useful. Indeed, it has sometimes been argued that **adenosine**[18] (Chapter 9) is a 'natural' anti-inflammatory agent produced during transient injury that helps to prevent over-activation of the inflammatory response. Compounds acting in a way comparable to adenosine might be useful as anti-inflammatory agents and in protection against reoxygenation injury, to the extent that this phenomenon is clinically important. Thus the drug **acadesine** has been reported to raise adenosine levels in the ischaemic myocardium.

Several anti-inflammatory drugs have been suggested to suppress $O_2^{\bullet -}$ formation by phagocytes, but convincing evidence for this mechanism of action *in vivo* has been obtained only in a few cases. Some of the beneficial effects of Ebselen (Section 10.5.10 above) could be due to effects on phagocyte ROS production. **Diphenylene iodonium** compounds (Chapter 6) inhibit the respiratory burst in isolated neutrophils but do not appear suitable for therapeutic use as yet. **Apocynin**[85] (Fig. 10.8) a phenolic compound of plant origin, is oxidized to a toxic product by phagocytes containing peroxidase (such as neutrophils, which possess myeloperoxidase) and has an anti-inflammatory effect in rats. Only activated phagocytes (producing the H_2O_2 necessary for peroxidase action) are affected.

Recently, a family of thiazoles that appear to inhibit neutrophil $O_2^{\bullet -}$ production has been described, although their mechanism of action has not been elucidated.[14] **Melittin**,[89] a constituent of bee venom, has been reported to inhibit neutrophil $O_2^{\bullet -}$ production. **Tetrandrine**[12] is an alkaloid obtained from roots of the plant *Stephania tetrandria* and is used in China to treat silicosis in miners. It prevents NF-κB activation in macrophages and may act by decreasing the activity of iNOS and COX-2 in these cells, to which it is also cytotoxic at low levels.

10.7.3   *Inhibitors of nitric oxide synthase*[62]

Over-production of NO$^\bullet$, usually by iNOS enzymes, has been suggested to cause injury in many diseases, both by direct toxicity of NO$^\bullet$ and by production of ONOO$^-$. There is considerable therapeutic interest in developing selective iNOS inhibitors, which, as a secondary effect, should decrease ONOO$^-$ production. Table 5.4 gives more information. However, NO$^\bullet$ can have beneficial anti-inflammatory, antibacterial, antiviral and antioxidant effects and so it is possible to imagine scenarios in which NO$^\bullet$ donors could be beneficial. Diphenylene iodonium compounds have some NOS inhibitory effect and so it should be possible to obtain compounds that inhibit both iNOS and phagocyte ROS production. However, too much inhibition of NO$^\bullet$ generation might be deleterious, and it is also possible that excessive inhibition of phagocyte ROS production could predispose to infections.

References

1.   Ames, BN *et al.* (1993) Oxidants, antioxidants and the degenerative diseases of aging. *Proc. Natl. Acad. Sci. USA* **90**, 7915.
2.   Anderson, ME (1996) Glutathione and glutathione delivery compounds. *Adv. Pharmacol.* **38**, 65.
3.   Andersson, CM *et al.* (1994) Diaryl tellurides as inhibitors of lipid peroxidation in biological and chemical systems. *Free Rad. Res.* **20**, 401.
4.   Andersson, CM *et al.* (1996) Advances in the development of pharmaceutical antioxidants. *Adv. Drug. Res.* **28**, 65.
5.   Barnett, YA and King, CM (1995) An investigation of antioxidant status, DNA repair capacity and mutation as a function of age in humans. *Mutat. Res.* **338**, 115.
6.   Blount, BC *et al.* (1997) Folate deficiency causes uracil misincorporation into human DNA and chromosome breakage: implications for cancer and neuronal damage. *Proc. Natl. Acad. Sci. USA* **94**, 3290.
6a. Bodnar, AG *et al.* (1998) Extension of life-span by introduction of telomerase into normal human cells. *Science* **279**, 349.
7.   Buettner, GR and Sharma, MK (1993) The syringe nitroxide radical—part II. *Free Rad. Res. Commun.* **19**, S227.
8.   Buring, JE and Hennekens, CH (1997) Antioxidant vitamins and cardiovascular disease. *Nutr. Rev.* **55**, S53.
9.   Burri, BJ (1997) β-Carotene and human health: a review of current research. *Nutr. Res.* **17**, 547.
10. Butterfield, DA *et al.* (1997) Free radical oxidation of brain proteins in accelerated senescence and its modulation by PBN. *Proc. Natl. Acad. Sci. USA* **94**, 674.
11. Cabiscol, E and Levine, RL (1995) Carbonic anhydrase III. Oxidative modification *in vivo* and loss of phosphatase activity during aging. *J. Biol. Chem.* **270**, 14742.
12. Chen, F *et al.* (1997) Tetrandrine inhibits signal-induced NF-κB activation in rat alveolar macrophages. *Biochem. Biophys. Res. Commun.* **231**, 99.
13. Chen, Q and Ames, BN (1994) Senescence-like growth arrest induced by H_2O_2 in human diploid fibroblast F65 cells. *Proc. Natl. Acad. Sci. USA* **91**, 4130.

14. Chihiro, M *et al.* (1995) Novel thiazole derivatives as inhibitors of $O_2^{\bullet-}$ production by human neutrophils: synthesis and structure–activity relationships. *J. Med. Chem.* **38**, 353.

15. Clark, LC *et al.* (1996) Effects of Se supplementation for cancer prevention in patients with carcinoma of the skin. *J. Amer. Med. Assoc.* **276**, 1957.

16. Collis, CS *et al.* (1993) Comparison of N-methyl hexanoyl hydroxamic acid, a novel antioxidant, with desferrioxamine and N-acetylcysteine against reperfusion-induced dysfunctions in isolated rat heart. *J. Cardiovasc. Pharmacol.* **2**, 336.

17. Cotgreave, I (1996) N-Acetylcysteine: pharmacological considerations and experimental and clinical applications. *Adv. Pharmacol.* **38**, 205.

18. Cronstein, BN (1994) Adenosine, an endogenous anti-inflammatory agent. *J. Appl. Physiol.* **76**, 5.

19. Cutler, RG (1985) Peroxide-producing potential of tissues: inverse correlation with longevity of mammalian species. *Proc. Natl. Acad. Sci. USA* **82**, 4798.

20. Del Zoppo, GJ *et al.* (1997) Trends and future developments in the pharmacological treatment of acute ischaemic stroke. *Drugs* **54**, 9.

21. Devaraj, S *et al.* (1996) The effects of α-tocopherol supplementation on monocyte function. *J. Clin. Invest.* **98**, 756.

22. Dimri, GP *et al.* (1995) A biomarker that identifies senescent human cells in culture and in aging skin *in vivo. Proc. Natl. Acad. Sci. USA* **92**, 9363.

23. Dowling, EJ *et al.* (1993) Assessment of a human recombinant MnSOD in models of inflammation. *Free Rad. Res. Commun.* **18**, 291.

24. Duncan, K *et al.* (1997) Running exercise may reduce risk for lung and liver cancer by inducing activity of antioxidant and phase II enzymes. *Cancer Lett.* **116**, 151.

25. Fariss, MW (1997) Anionic tocopherol esters as antioxidants and cytoprotectants. In *Handbook of Synthetic Antioxidants* (Packer, L and Cadenas, E, eds), p. 139. Marcel Dekker, New York.

26. Fevig, TL *et al.* (1996) Design, synthesis and *in vitro* evaluation of cyclic nitrones as free radical traps for the treatment of stroke. *J. Med. Chem.* **39**, 4988.

27. Floyd, RA (1996) Protective action of nitrone-based free radical traps against oxidative damage to the CNS. *Adv. Pharmacol.* **38**, 361.

28. Forster, MJ *et al.* (1996) Age-related losses of cognitive function and motor skills in mice are associated with oxidative protein damage in the brain. *Proc. Natl. Acad. Sci. USA* **93**, 4765.

29. Fraisse, L. *et al.* (1993) Long-chain-substituted uric acid and 5,6-diaminouracil derivatives as novel agents against free radical processes: synthesis and *in vitro* activity. *J. Med. Chem.* **36**, 1465.

30. Francis, JW *et al.* (1995) CuZnSOD (SOD-1): tetanus toxin fragment C hybrid protein for targeted delivery of SOD-1 to neuronal cells. *J. Biol. Chem.* **270**, 15434.

31. Frei, B (ed.) (1994) *Natural Antioxidants in Human Health and Disease.* Academic Press, San Diego.

32. Fruelsis, J *et al.* (1994) A comparison of the antiatherogenic effects of probucol and of a structural analogue of probucol in LDL receptor-deficient rabbits. *J. Clin. Invest.* **94**, 392.

33. Gey, KF (1995) Ten-year retrospective on the antioxidant hypothesis of arteriosclerosis: threshold plasma level of antioxidant micronutrients related to minimum cardiovascular risk. *J. Nutr. Biochem.* **6**, 206.

34. Hall, ED *et al.* (1997) Pyrrolopyrimidines: novel brain-penetrating antioxidants with neuroprotective activity in brain injury and ischaemia models. *J. Pharmacol. Exp. Ther.* **281**, 895.

35. Halliwell, B (1991) Drug antioxidant effects. A basis for drug selection? *Drugs* **42**, 569.

36. Halliwell, B (1996) Oxidative stress, nutrition and health. Experimental strategies for optimization of nutritional antioxidant intake in humans. *Free Rad. Res.* **25**, 57.

37. Han, X and Liehr, JG (1995) Microsome-mediated 8-hydroxylation of guanine bases of DNA by steroid oestrogens: correlation of DNA damage by free radicals with metabolic activation to quinones. *Carcinogenesis* **16**, 2571.

38. Harman, D (1993) Free radical involvement in aging. *Drugs & Aging* **3**, 60.

39. Hertog, MGL *et al.* (1993) Dietary antioxidant flavonoids and risk of coronary heart disease; the Zutphen elderly study. *Lancet* **342**, 1007.

40. Hider, RC (1995) Potential protection from toxicity by oral iron chelators. *Toxicol. Lett.* **82/83**, 961.

40a. Hidiroglou, N *et al.* (1997) Vitamin E levels in superficial and intra-abdominal locations of white adipose tissue in the rat. *Nutr. Biochem.* **8**, 392.

41. Hoffbrand, AV (1994) Prospects for oral iron chelation therapy. *J. Lab. Clin. Med.* **123**, 492.

42. Inoue, M *et al.* (1991) Expression of a hybrid Cu/Zn-type SOD which has high affinity for heparin-like proteoglycans on vascular endothelial cells. *J. Biol. Chem.* **266**, 16409.

43. Isaacson, Y *et al.* (1989) Lactobionic acid as an iron chelator: a rationale for its effectiveness as an organ preservant. *Life Sci.* **45**, 2373.

44. Jacques, PF (1997) Nutritional antioxidants and prevention of age-related eye disease. In *Antioxidants and Disease Prevention* (Garewal, HS, ed.), p.149. CRC Press, Boca Raton, Florida, USA.

45. Jama, JW *et al.* (1996) Dietary antioxidants and cognitive function in a population-based sample of older persons. The Rotterdam study. *Am. J. Epidemiol.* **144**, 275 (also see *Arch. Neurol.* **54**, 762 (1997)).

46. Johnson, IT *et al.* (1994) Anticarcinogenic factors in plant foods: a new class of nutrients? *Nutr. Res. Rev.* **7**, 175.

47. Kang, JX and Leaf, A (1996) Antiarrhythmic effects of PUFAs. *Circulation* **94**, 1774.

48. Kellog, EW III and Fridovich, I (1976) Superoxide dismutase in the rat and mouse as a function of age and longevity. *J. Gerontol.* **31**, 405.

49. Khao, KT and Woodhouse, P (1995) Inter-relation of vitamin C, infection, haemostatic factors, and cardiovascular disease. *Brit. Med. J.* **310**, 1559.

50. Kitani, K. *et al.* (1994) (-)Deprenyl increases the lifespan as well as activities of SOD and catalase but not of glutathione peroxidase in selective brain regions in Fischer rats. *Ann. NY Acad. Sci.* **717**, 60.

51. Knecht, KT and Mason, RP (1993) *In vivo* spin trapping of xenobiotic free radical metabolites. *Arch. Biochem. Biophys.* **303**, 185.

52. Krishna, MC *et al.* (1996) Do nitroxide antioxidants act as scavengers of $O_2^{\bullet-}$ or as SOD mimics? *J. Biol. Chem.* **271**, 26026.

53. Lee, CM *et al.* (1997) Age-associated alterations of the mitochondrial genome. *Free Rad. Biol. Med.* **22**, 1259.

54. Liu, AYC *et al.* (1996) Attenuated heat shock transcriptional response in aging: molecular mechanism and implication in the biology of aging.

In *Stress-Inducible Cellular Responses* (Feige, U *et al.*, eds), p. 393. Birhauser-Verlag, Basle, Switzerland.

55. Martin, GM *et al.* (1996) Genetic analysis of ageing: role of oxidative damage and environmental stress. *Nature Genet.* **13**, 25.

56. Mayne, ST (1996) Antioxidant nutrients and cancer incidence and mortality: an epidemiologic perspective. *Adv. Pharmacol.* **38**, 657.

57. Miesel, R *et al.* (1996) Suppression of inflammatory arthritis by simultaneous inhibition of NOS and NADPH oxidase. *Free Rad. Biol. Med.* **20**, 75.

58. Miner, SES *et al.* (1997) Clinical chemistry and molecular biology of homocysteine metabolism: an update. *Clin. Biochem.* **30**, 189.

59. Mitscher, LA *et al.* (1997) Chemoprotection: a review of the potential therapeutic antioxidant properties of green tea (*Camellia sinensis*) and certain of its constituents. *Med. Res. Rev.* **17**, 327.

60. Moch, D *et al.* (1995) Protective effects of hydroxyethyl starch-deferoxamine in early sepsis. *Shock* **4**, 425.

61. Mohr, D *et al.* (1992) Dietary supplementation with CoQ$_{10}$ results in increased levels of ubiquinol-10 within circulating lipoproteins and increased resistance of human LDL to the initiation of lipid peroxidation. *Biochim. Biophys. Acta* **1126**, 247.

62. Moore, PK and Handy, RLC (1997) Selective inhibitors of nNOS—is no NOS really good NOS for the nervous system? *Trends Pharmacol. Sci.* **18**, 204.

62a. Morens, DM *et al.* (1996) Case-control study of idiopathic Parkinson's disease and dietary vitamin E intake. *Neurology* **46**, 1270.

63. Musci, G *et al.* (1993) Age-related changes in human ceruloplasmin. *J. Biol. Chem.* **268**, 13388.

64. Muzykantov, VR *et al.* (1996) Immunotargeting of antioxidant enzymes to the pulmonary endothelium. *Proc. Natl. Acad. Sci. USA* **93**, 5213.

65. Nieves-Cruz, B *et al.* (1996) Clinical surfactant preparations mediate SOD and catalase uptake by type II cells and lung tissue. *Am. J. Physiol.* **270**, L659.

65a. Parkes, TL *et al.* (1998) Extension of *Drosophila* lifespan by overexpression of human SOD1 in motor neurones. *Nature Genet.* **19**, 171.

66. Pérez-Campo, R *et al.* (1994) Longevity and antioxidant enzymes, non-enzymatic antioxidants and oxidative stress in the vertebrate lung: a comparative study. *J. Comp. Physiol. B.* **163**, 682.

67. Petty, MA *et al.* (1996) Design and biological evaluation of tissue directed α-tocopherol analogs. In *Handbook of Synthetic Antioxidants* (Cadenas, E and Packer, L, eds), p. 53. Marcel Dekker, New York.

68. Phillippy, BQ and Graf, E (1997) Antioxidant functions of inositol 1,2,3-trisphosphate and inositol 1,2,3,6-tetratrisphosphate. *Free Rad. Biol. Med.* **22**, 939.

69. Platenburg, GJ *et al.* (1994) Expression of human lactoferrin in milk of transgenic mice. *Transgenic Res.* **3**, 99.

70. Prestera, T and Talalay, P (1995) Electrophile and antioxidant regulation of enzymes that detoxify carcinogens. *Proc. Natl. Acad. Sci. USA* **92**, 8965.

71. Priemé, H *et al.* (1997) No effect of supplementation with vitamin E, ascorbic acid or CoQ$_{10}$ on oxidative DNA damage estimated by 8OHdG excretion in smokers. *Am. J. Clin. Nutr.* **65**, 503.

72. Rana, RS and Munkres, KD (1978) Ageing of *Neurospora crassa*. V. Lipid peroxidation and decay of respiratory enzymes in an inositol auxotroph. *Mech. Age Develop.* **7**, 241.

73. Rapola, JM *et al.* (1997) Randomised trial of α-tocopherol and β-carotene supplements on incidence of major coronary events in men with previous myocardial infarctions. *Lancet* **349**, 1715.

74. Rikans, LE *et al.* (1997) Age-associated increase in ferritin content of male rat liver: implication for diquat-mediated oxidative injury. *Arch. Biochem. Biophys.* **344**, 85.

75. Roberts, JC (1992) Amino acids and their derivatives as radioprotective agents. *Amino Acids* **3**, 25.

75a. Römer, W *et al.* (1997) Novel estrogens and their radical scavenging effects, iron-chelating, and total antioxidative activities: 17α-substituted analogs of $\Delta^{9(11)}$-dehydro-17β-estradiol. *Steroids* **62**, 688.

76. Sagai, M and Ichinose, T (1980) Age-related changes in lipid peroxidation as measured by ethane, ethylene, butane and pentane in respired gases of rats. *Life Sci.* **27**, 731.

77. Sanbongi, C *et al.* (1997) Polyphenols in chocolate, which have antioxidant activity, modulate immune functions in humans *in vitro*. *Cell Immunol.* **177**, 129.

78. Santa Maria, C. *et al.* (1995) Changes in the histidine residues of CuZnSOD during aging. *FEBS Lett.* **374**, 85.

79. Schalkwijk, J *et al.* (1985) Cationization of catalase, peroxidase and superoxide dismutase. Effect of improved intraarticular retention on experimental arthritis in mice. *J. Clin. Invest.* **76**, 198.

80. Selby, C *et al.* (1996) Inhibition of somatic embryo maturation in Sitka spruce [*Picea sitchensis* (Bong) Carr] by BHT, a volatile antioxidant released by parafilm. *Plant Cell Rep.* **16**, 192.

81. Shahidi, F and Wanasundara, PKJPD (1992) Phenolic antioxidants. *Crit. Rev. Food Sci. Nutr.* **32**, 67.

82. Shore, D (1997) Different means to common ends. *Nature* **385**, 676.

83. Sies, H and Masumoto, H (1996) Ebselen as a glutathione peroxidase mimic and as a scavenger of peroxynitrite. *Adv. Pharmacol.* **38**, 229.

84. Simán, CM and Eriksson, UJ (1996) Effect of BHT on α-tocopherol content in liver and adipose tissue of rats. *Toxicol. Lett.* **87**, 103.

85. Simons, JM *et al.* (1990) Metabolic activation of natural phenols into selective oxidative burst agonists by activated human neutrophils. *Free Rad. Biol. Med.* **8**, 251.

86. Snowdon, DA *et al.* (1996) Antioxidants and reduced functional capacity in the elderly: findings from the nun study. *J. Gerontol.* **51A**, M10.

87. Sohal, RS and Weindruch, R (1996) Oxidative stress, caloric restriction and aging. *Science* **273**, 59.

88. Sohal, RS *et al.* (1988) Oxidative stress and cellular differentiation. *Ann. N. Y. Acad. Sci.* **551**, 59.

89. Somerfield, SD *et al.* (1986) Bee venom melittin blocks neutrophil $O_2^{\bullet-}$ production. *Inflammation* **10**, 175.

90. Stadtman, ER and Berlett, BS (1997) Reactive oxygen-mediated protein oxidation in aging and disease. *Chem. Res. Toxicol.* **10**, 485.

91. Stephens, NG *et al.* (1996) Randomised controlled trial of vitamin E in patients with coronary disease: Cambridge Heart Antioxidant Study (CHAOS) *Lancet* **347**, 781.

92. Tarasuk, V (1996) Nutritional epidemiology. In *Present Knowledge in Nutrition*, 7th edition, (Ziegler, EE *et al.*, eds), p. 508. IRL Press, Washington DC.

93. Taylor, A *et al.* (1995) Dietary restriction delays cataract and reduces ascorbate levels in Emory mice. *Exp. Eye Res.* **61**, 55.

94. Toledo, I *et al.* (1994) Enzyme inactivation related to a hyperoxidant state during conidiation of *N. crassa*. *Microbiol.* **140**, 2391.

94a. Tuomáinen, TP *et al* (1998) Association between body iron stores and the risk of acute myocardial infarction in men. *Circulation* **97**, 1461.

95. Van Zyl, JM *et al.* (1993) Anti-oxidant properties of H_2-receptor antagonists. *Biochem. Pharmacol.* **45**, 2389.

96. Verhagen, H *et al.* (1997) Effect of Brussels sprouts on oxidative DNA-damage in man. *Cancer Lett.* **114**, 127.

97. Verhoeven, DTH *et al.* (1997) A review of mechanisms underlying anti-carcinogenicity by brassica vegetables. *Chem.−Biol. Interac.* **103**, 79.

97a. Vita, JA *et al.* (1998) L-2-Oxothiazolidine-4-carboxylic acid reverses endothelial dysfunction in patients with coronary artery disease. *J. Clin. Invest.* **101**, 1408.

97b. Wang, H *et al.* (1997) Inhibition of growth and $p21^{ras}$ methylation in vascular endothelial cells by homocysteine but not cysteine. *J. Biol. Chem.* **272**, 25380.

98. Weinberg, GA (1994) Iron chelators as therapeutic agents against *Pneumocystis carinii*. *Antimicrob. Agents Chemother.* **38**, 997.

99. Whiteman, M and Halliwell, B (1996) Protection against peroxynitrite-dependent tyrosine nitration and α_1-antiproteinase inactivation by ascorbic acid. *Free Rad. Res.* **55**, 383.

100. Willett, WC (1994) Diet and health: what should we eat? *Science* **264**, 532.

101. Willson, RL (1988) From nitric oxide to Desferal: nitrogen free radicals and iron in oxidative injury. In *Oxygen Radicals in Biology and Medicine* (Simic, MG *et al.*, eds), p. 87. Plenum Press, New York.

102. Yakes, FM and Van Houten, B (1997) MtDNA damage is more extensive and persists longer than nuclear DNA damage in human cells following oxidative stress. *Proc. Natl. Acad. Sci. USA* **94**, 514.

103. Yin, D (1996) Biochemical basis of lipofuscin, ceroid and age pigment-like fluorophores. *Free Rad. Biol. Med.* **21**, 871.

104. Ziouzenkova, O *et al.* (1996) Lack of correlation between the α-tocopherol content of plasma and LDL, but high correlations for γ-tocopherol and carotenoids. *J. Lipid Res.* **37**, 1936.

Notes

[a]The terms 'age pigment' and 'lipofuscin' are frequently used as synonyms for the fluorescent pigment that accumulates in older tissues. However, the term 'age pigment' should, strictly speaking, only be applied to the pigment that accumulates in neurones and in cardiac muscle fibres where, because of their inability to divide, pigment accumulation is roughly correlated with age. In other tissues, the normal cell turnover can result in lipofuscin removal.

[b]For an explanation of this term please see Appendix I.

Appendix I

Some basic chemistry for the life scientist

A1.1 Atomic structure

For the purposes of this book it will be sufficient to consider a simple model of atomic structure in which the atom consists of a positively charged nucleus that is surrounded by one or more negatively charged electrons. The nucleus contains two types of particle of approximately equal mass, the positively charged proton and the uncharged neutron. By comparison with these particles, the mass of the electron is negligible so that virtually all of the mass of the atom is contributed by its nucleus. The **atomic number** of an element is defined as the number of protons in its nucleus, the **mass number** as the number of protons plus neutrons. In the neutral atom, the atomic number also equals the number of electrons surrounding the nucleus. The simplest atom is that of the element hydrogen, containing one proton (atomic number equals one, mass number equals one) and one electron. All other elements contain neutrons in the nucleus.

Some elements exist as **isotopes**, in which the atoms contain the same number of protons and electrons, but different numbers of neutrons. These isotopes can be stable or unstable, the unstable ones undergoing radioactive decay at various rates. In this process, the nucleus of the radioactive isotope changes, and a new element is formed. For example, an isotope of the element uranium (atomic number 92) with a mass number of 238 undergoes nuclear disintegration to produce two fragments, one with two protons and two neutrons and the other with 90 protons and 144 neutrons, in fact an isotope of the element thorium. The elements with which we are largely concerned in this book, carbon, hydrogen, nitrogen and oxygen, exist almost exclusively as one isotopic form in nature (Table A1.1).

The electrons surrounding the atomic nucleus possess a negative charge. Since they do not spiral into the nucleus, they must possess energy to counteract the attractive electric force tending to pull them in. In 1900, Planck suggested that energy is quantized, i.e. that energy changes only occur in small, definite amounts known as 'quanta'. Application of Planck's quantum theory to the atom, by Bohr, produced a model in which the electrons exist in specific orbits, or 'electron shells', each associated with a particular energy level. The 'K'-shell electrons, lying closest to the nucleus, have the lowest energy, and the energy successively increases as one proceeds outwards to the so-called L-, M-, and N-shells. The K-shell can hold a maximum of two electrons, the L-shell, 8, M-shell, 18, and N-shell, 32. Table A1.2 shows the location of electrons in each of these shells for the elements up to atomic number 36.

Table A1.1. Isotopes of some common elements

Element	Isotope	Number of protons in nucleus	Number of neutrons in nucleus	Comments
Chlorine	$^{35}_{17}Cl$	17	18	Both isotopes are stable and occur naturally, ^{35}Cl being more abundant
	$^{37}_{17}Cl$	17	20	
Carbon	$^{12}_{6}C$	6	6	Over 90% of naturally occurring carbon is $^{12}_{6}C$. Small amounts of the radioactive isotope $^{14}_{6}C$ are formed by the bombardment of atmospheric CO_2 with cosmic rays (i.e. streams of neutrons arising from outer space). This isotope undergoes slow radioactive decay (50% decay after 5600 years)
	$^{13}_{6}C$	6	7	
	$^{14}_{6}C$	6	8	
Nitrogen	$^{14}_{7}N$	7	7	^{15}N is a stable isotope of nitrogen often used as a 'tracer' e.g. $^{15}NO_3^-$ can be fed to humans to trace its metabolism
	$^{15}_{7}N$	7	8	
Oxygen	$^{16}_{8}O$	8	8	Over 90% of naturally occurring oxygen is the isotope $^{16}_{8}O$
	$^{17}_{8}O$	8	9	
	$^{18}_{8}O$	8	10	
Hydrogen	$^{1}_{1}H$	1	0	Over 99% of hydrogen is $^{1}_{1}H$. Deuterium ($^{2}_{1}H$) is a stable isotope, whereas tritium ($^{3}_{1}H$) is radioactive. Deuterium oxide is known as 'heavy water', and is used in detecting the presence of singlet oxygen in biological systems.
	$^{2}_{1}H$	1	1	
	$^{3}_{1}H$	1	2	

The superscript number on the left of the symbol for the element represents the mass number, and the subscript the atomic number. All atoms of a given element have the same number of protons, but sometimes have different numbers of neutrons, giving rise to isotopes.

Subsequent developments of atomic theory have shown that an electron has some of the properties of a particle, and some of the properties of a wave motion. As a result, the position of an electron at a given time cannot be precisely located, but only the region of space where it is most likely to be. These regions are referred to as **orbitals**. Each electron in an atom has its energy defined by four so-called quantum numbers. The first, or **principal quantum number** (n) defines the main energy level the electron occupies. For the K-shell, $n = 1$; for L, $n = 2$; for M, $n = 3$; and for N, $n = 4$. The second, or **azimuthal quantum number** (l) governs the shape of the orbital and has values from zero up to $(n-1)$. When $l = 0$, the electrons are called 's' electrons;

Table A1.2. Location of electrons in shells for the elements with atomic numbers 1 to 36

Atomic number of element	Element	Symbol	Shell K	Shell L	Shell M	Shell N
1	Hydrogen	H	1			
2	Helium	He	2			
3	Lithium	Li	2	1		
4	Beryllium	Be	2	2		
5	Boron	B	2	3		
6	Carbon	C	2	4		
7	Nitrogen	N	2	5		
8	Oxygen	O	2	6		
9	Fluorine	F	2	7		
10	Neon	Ne	2	8		
11	Sodium	Na	2	8	1	
12	Magnesium	Mg	2	8	2	
13	Aluminium	Al	2	8	3	
14	Silicon	Si	2	8	4	
15	Phosphorus	P	2	8	5	
16	Sulphur	S	2	8	6	
17	Chlorine	Cl	2	8	7	
18	Argon	Ar	2	8	8	
19	Potassium	K	2	8	8	1
20	Calcium	Ca	2	8	8	2
21	Scandium	Sc	2	8	9	2
22	Titanium	Ti	2	8	10	2
23	Vanadium	V	2	8	11	2
24	Chromium	Cr	2	8	13	1
25	Manganese	Mn	2	8	13	2
26	Iron	Fe	2	8	14	2
27	Cobalt	Co	2	8	15	2
28	Nickel	Ni	2	8	16	2
29	Copper	Cu	2	8	18	1
30	Zinc	Zn	2	8	18	2
31	Gallium	Ga	2	8	18	3
32	Germanium	Ge	2	8	18	4
33	Arsenic	As	2	8	18	5
34	Selenium	Se	2	8	18	6
35	Bromine	Br	2	8	18	7
36	Krypton	Kr	2	8	18	8

when $l = 1$, they are 'p' electrons; $l = 2$, 'd' electrons; and $l = 3$ gives 'f' electrons. The third quantum number is the **magnetic quantum number** (m) and, for each value of l, m has values of l, $(l - 1)$, ..., 0, -1, ..., $-l$. Finally, the fourth quantum number, or **spin quantum number**, can have values of either $1/2$ or $-1/2$ only. Table A1.3 shows how various combinations of these four quantum numbers can fill the electron shells, and (hopefully!) makes the

Table A1.3. Orbitals available in the principal electron shells

Shell	Principal quantum number	Value of l (azimuthal quantum number)	Electron type	Value of m (magnetic quantum number)	Value of s (spin quantum number)	Maximum number of electrons in shell	
K	1	0	s	0	$\pm 1/2$	2 (1s-orbital)	
L	2	0	s	0	$\pm 1/2$	2 (2s-orbital)	$\left.\vphantom{\begin{matrix}a\\a\end{matrix}}\right\} 8$
		1	p	1, 0, −1	$\pm 1/2$	3×2 (three 2p-orbitals)	
M	3	0	s	0	$\pm 1/2$	2 (3s-orbital)	$\left.\vphantom{\begin{matrix}a\\a\\a\end{matrix}}\right\} 18$
		1	p	1, 0, −1	$\pm 1/2$	3×2 (three 2p-orbitals)	
		2	d	2, 1, 0, −1, −2	$\pm 1/2$	5×2 (five 3d-orbitals)	
N	4	0	s	0	$\pm 1/2$	2 (4s-orbital)	$\left.\vphantom{\begin{matrix}a\\a\\a\\a\end{matrix}}\right\} 32$
		1	p	1, 0, −1	$\pm 1/2$	3×2 (three 4p-orbitals)	
		2	d	2, 1, 0, −1, −2	$\pm 1/2$	5×2 (five 4d-orbitals)	
		3	f	3, 2, 1, 0, −1, −2, −3	$\pm 1/2$	7×2 (seven 4f-orbitals)	

above explanation a bit clearer. **Pauli's principle** states that 'no two electrons can have the same four quantum numbers'. Since the spin quantum number has only two possible values ($\pm\frac{1}{2}$), it follows that an orbital can hold only two electrons at most (Table A1.3).

In filling the available orbitals electrons will enter the orbitals with the lowest total energy content (**Aufbau principle**). The order of filling is:

1s　2s　2p　3s　3p　4s　3d　4p　5s　4d　5p　6s　4f　5d　6p　7s　5f

lowest energy　　　　　　　　　increasing energy　　　　　　　　　highest energy

Table A1.4 gives the electronic energy configurations of the elements with atomic numbers from 1 to 32. When the elements are arranged in the **periodic table** (Fig. A1.1), elements with similar electronic arrangements fall into similar 'groups' (vertical rows), e.g. the group II elements all have two electrons in their outermost electron shell, and the group IV elements have four. Since the 4s-orbital is of lower energy than the 3d-orbitals, these latter orbitals remain empty until the 4s-orbital is filled (e.g. see the elements potassium and calcium in Table A1.4). In subsequent elements the five 3d-orbitals receive electrons, creating the first row of the so-called **d-block** in the periodic table (Fig. A1.1). Some of these d-block elements are called **transition elements**, meaning elements in which an inner shell of electrons is incomplete (in this case these are electrons in the fourth shell, but all the d-orbitals of the third shell are not yet full). The term transition element, as defined above, applies to scandium and subsequent elements as far as nickel, although it is often extended to include the whole of the first row of the d-block.

If orbitals of equal energy are available, e.g. the three 2p-orbitals in the L-shell, or the five 3d-orbitals in the M-shell (Table A1.3), each is filled with one electron before any receives two (**Hund's rule**). Hence one can further break down the electronic configurations shown in Table A1.4. The element boron has two 1s, two 2s, and one 2p electrons. Three 2p-orbitals of equal energy are available (Table A1.3), and they are often written as $2p_x$, $2p_y$, and $2p_z$. If we represent each orbital as a square box and an electron as an arrow, boron can be represented as follows:

B　1s [↑↓]　2s [↑↓]　2p [↑ | |]

For the next element, carbon, the extra electron enters another 2p-orbital in obedience to Hund's rule:

C　1s [↑↓]　2s [↑↓]　2p [↑ | ↑ |]

Table A1.4. Electronic configuration of the elements

Element	Atomic number	Symbol	Configuration	Place in periodic table
Hydrogen	1	H	$1s^1$	uncertain
Helium	2	He	$1s^2$	group 0 (inert gases)
Lithium	3	Li	$1s^2 2s^1$	group I (alkali metals)
Beryllium	4	Be	$1s^2 2s^2$	group II (alkaline-earth metals)
Boron	5	B	$1s^2 2s^2 2p^1$	group III
Carbon	6	C	$1s^2 2s^2 2p^2$	group IV
Nitrogen	7	N	$1s^2 2s^2 2p^3$	group V
Oxygen	8	O	$1s^2 2s^2 2p^4$	group VI
Fluorine	9	F	$1s^2 2s^2 2p^5$	group VII (halogen elements)
Neon	10	Ne	$1s^2 2s^2 2p^6$	group 0
Sodium	11	Na	$1s^2 2s^2 2p^6 3s^1$	group I
Magnesium	12	Mg	$1s^2 2s^2 2p^6 3s^2$	group II
Aluminium	13	Al	$1s^2 2s^2 2p^6 3s^2 3p^1$	group III
Silicon	14	Si	$1s^2 2s^2 2p^6 3s^2 3p^2$	group IV
Phosphorus	15	P	$1s^2 2s^2 2p^6 3s^2 3p^3$	group V
Sulphur	16	S	$1s^2 2s^2 2p^6 3s^2 3p^4$	group VI
Chlorine	17	Cl	$1s^2 2s^2 2p^6 3s^2 3p^5$	group VII
Argon	18	Ar	$1s^2 2s^2 2p^6 3s^2 3p^6$	group 0
Potassium	19	K	$1s^2 2s^2 2p^6 3s^2 3p^6 4s^1$	group I
Calcium	20	Ca	$1s^2 2s^2 2p^6 3s^2 3p^6 4s^2$	group II
Scandium	21	Sc	$1s^2 2s^2 2p^6 3s^2 3p^6 4s^2 3d^1$	d-block
Titanium	22	Ti	$1s^2 2s^2 2p^6 3s^2 3p^6 4s^2 3d^2$	d-block
Vanadium	23	V	$1s^2 2s^2 2p^6 3s^2 3p^6 4s^2 3d^3$	d-block
Chromium	24	Cr	$1s^2 2s^2 2p^6 3s^2 3p^6 4s^1 3d^5$	d-block
Manganese	25	Mn	$1s^2 2s^2 2p^6 3s^2 3p^6 4s^2 3d^5$	d-block
Iron	26	Fe	$1s^2 2s^2 2p^6 3s^2 3p^6 4s^2 3d^6$	d-block
Cobalt	27	Co	$1s^2 2s^2 2p^6 3s^2 3p^6 4s^2 3d^7$	d-block
Nickel	28	Ni	$1s^2 2s^2 2p^6 3s^2 3p^6 4s^2 3d^8$	d-block
Copper	29	Cu	$1s^2 2s^2 2p^6 3s^2 3p^6 4s^1 3d^{10}$	d-block
Zinc	30	Zn	$1s^2 2s^2 2p^6 3s^2 3p^6 4s^2 3d^{10}$	d-block
Gallium	31	Ga	$1s^2 2s^2 2p^6 3s^2 3p^6 4s^2 3d^{10} 4p^1$	group III
Germanium	32	Ge	$1s^1 2s^2 2p^6 3s^2 3p^6 4s^2 3d^{10} 4p^2$	group IV

Groups

I II III IV V VI VII 0

s-block

I	II
1 H	
3 Li	4 Be
11 Na	12 Mg
19 K	20 Ca
37 Rb	38 Sr
55 Cs	56 Ba
87 Fr	88 Ra

d-block

21 Sc	22 Ti	23 V	24 Cr	25 Mn	26 Fe	27 Co	28 Ni	29 Cu	30 Zn
39 Y	40 Zr	41 Nb	42 Mo	43 Tc	44 Ru	45 Rh	46 Pd	47 Ag	48 Cd
57 La	72 Hf	73 Ta	74 W	75 Re	76 Os	77 Ir	78 Pt	79 Au	80 Hg
89 Ac									

p-block

1 H						2 He

III	IV	V	VI	VII	0
5 B	6 C	7 N	8 O	9 F	10 Ne
13 Al	14 Si	15 P	16 S	17 Cl	18 Ar
31 Ga	32 Ge	33 As	34 Se	35 Br	36 Kr
49 In	50 Sn	51 Sb	52 Te	53 I	54 Xe
81 Tl	82 Pb	83 Bi	84 Po	85 At	86 Rn

f-block

Lanthanides

58 Ce	59 Pr	60 Nd	61 Pm	62 Sm	63 Eu	64 Gd	65 Tb	66 Dy	67 Ho	68 Er	69 Tm	70 Yb	71 Lu

Actinides

90 Th	91 Pa	92 U	93 Np	94 Pu	95 Am	96 Cm	97 Bk	98 Cf	99 Es	100 Fm	101 Md	102 No	103 Lr

Fig. A1.1. The periodic table.

And for nitrogen we have:

	1s	2s	2p		
N	↑↓	↑↓	↑	↑	↑

Further electrons will now begin to 'pair up' to fill the 2p-orbitals, e.g. for the oxygen atom:

	1s	2s	2p		
O	↑↓	↑↓	↑↓	↑	↑

Hund's rule is particularly important in the d-block elements, e.g. Table A1.5 uses the same 'electrons-in-boxes' notation for the elements in the first row of this block. Each of the five 3d-orbitals receives one electron, before any receives two.

We shall now consider how atoms join together to form molecules in chemical reactions.

Table A1.5. Electronic configuration of the elements scandium to zinc in the first row of the d-block of the periodic table

		3d					4s
Scandium	Ar	↑					↑↓
Titanium	Ar	↑	↑				↑↓
Vanadium	Ar	↑	↑	↑			↑↓
Chromium	Ar	↑	↑	↑	↑	↑	↑
Manganese	Ar	↑	↑	↑	↑	↑	↑↓
Iron	Ar	↑↓	↑	↑	↑	↑	↑↓
Cobalt	Ar	↑↓	↑↓	↑	↑	↑	↑↓
Nickel	Ar	↑↓	↑↓	↑↓	↑	↑	↑↓
Copper	Ar	↑↓	↑↓	↑↓	↑↓	↑↓	↑
Zinc	Ar	↑↓	↑↓	↑↓	↑↓	↑↓	↑↓

'Ar' is used as an abbreviation for the argon configuration. $1s^2 2s^2 2p^6 3s^2 3p^6$, to simplify the table, i.e. each element begins with the argon configuration. The 'unusual' electronic configurations of chromium and copper seem to be due to the increased relative stability of atoms in which each 3d-orbital contains either one or two electrons.

A1.2 Bonding between atoms

A1.2.1 *Ionic bonding*

As with our consideration of atomic structure, the account of chemical bonding that follows is the simplest possible model consistent with the requirements of this book.

Essentially two types of chemical bond can be distinguished. The first is called **ionic bonding**. This tends to happen when so-called electropositive elements combine with electronegative ones. Electropositive elements, such as those in groups I and II of the periodic table (Fig. A1.1), tend to lose their outermost electrons easily, whereas electronegative elements (group VII, and oxygen and sulphur in group VI) tend to accept extra electrons. By doing so, they gain the electronic configuration of the nearest inert gases, which seems to be a particularly stable configuration in view of the relative lack of reactivity of these elements. Consider, for example, the combination of an atom of sodium with one of chlorine. Sodium, an electropositive group I element, has the electronic configuration $1s^2 2s^2 2p^6 3s^1$. If a sodium atom loses one electron, it then has the configuration $1s^2 2s^2 2p^6$, that of the inert gas, neon. It is still the element sodium because its nucleus is unchanged, but the loss of one electron leaves the atom with a positive charge, forming an ion or, more specifically, a **cation** (positively charged ion). For chlorine, configuration $1s^2 2s^2 2p^6 3s^2 3p^5$, acceptance of an electron gives the argon electron-configuration $1s^2 2s^2 2p^6 3s^2 3p^6$, and produces a negatively charged ion (**anion**) Cl^-.

In the case of a group II element such as magnesium, it must lose two electrons to gain an inert-gas electron-configuration Thus one atom of magnesium can provide electrons for acceptance by two chlorine atoms, giving magnesium chloride a formula $MgCl_2$, i.e.

$$\underset{1s^2 2s^2 2p^6 3s^2}{Mg} \rightarrow \underset{1s^2 2s^2 2p^6}{Mg^{2+}} + 2e^-$$
(neon configuration)

An atom of oxygen, however, can accept two electrons and combine with magnesium to form an oxide MgO:

$$\underset{1s^2 2s^2 2p^4}{O} + 2e^- \rightarrow \underset{1s^2 2s^2 2p^6}{O^{2-}}$$
(neon configuration)

Once formed, anions and cations are held together by the electric attraction of their opposite charges. Each ion will exert an effect on each other ion in its vicinity, and these effects cause the ions to pack together into an **ionic crystal lattice**, as shown in Fig. A1.2 for NaCl. Each Na^+ ion is surrounded by six Cl^- ions, and vice versa. Once the lattice has formed, it cannot be said that any one Na^+ ion 'belongs' to any one Cl^- ion, nor can 'molecules' of sodium chloride be said to exist in the solid. The formula of an ionic compound merely indicates the combining ratio of the elements involved. A considerable amount of energy is required to disrupt all the electrostatic forces between the many

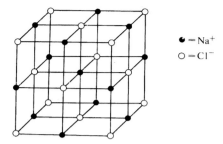

Fig. A1.2. Crystal structure of sodium chloride. The exact type of lattice formed by an ionic compound depends on the relative sizes of the ions. NaCl forms a cubic lattice, as shown.

millions of ions in a crystal of an ionic compound, so such compounds are usually solids with high melting-points. Ionic compounds are mostly soluble in water, and the solutions conduct electricity because of the presence of ions to carry the current. The properties of an ionic compound are those of its constituent ions.

A1.2.2 *Covalent bonding*

The covalent bond involves a sharing of a pair of electrons between the two bonded atoms. In 'normal' covalent bonding, each atom contributes one electron to the shared pair; but in **dative covalent bonding**, one atom contributes both of the shared electrons. The element hydrogen is usually found in nature as **diatomic** molecules, H_2. Two hydrogen atoms are sharing a pair of electrons. If we represent the electron of each hydrogen atom by a cross ($\times$) we can write:

$$H \times + \, ^\times H \rightarrow H \underset{\times}{\overset{\times}{}} H$$

where $\underset{\times}{\overset{\times}{}}$ is the shared pair of electrons. Many other gaseous elements, including oxygen and chlorine, exist as covalently bonded diatomic molecules.

When chlorine combines with hydrogen, the covalent compound, hydrogen chloride, is formed. If we represent the outermost electrons of the chlorine atom as circles, we can write:

$$\overset{\text{oo}}{\underset{\text{oo}}{\text{Cl}}}^{\text{oo}}_{}\!\!\!{}^{\circ} + H \times \longrightarrow H \, \overset{\text{oo}}{\underset{\text{oo}}{\text{Cl}}}^{\text{oo}}_{}\!\!\!{}_{\circ}$$

hydrogen chloride

where $\underset{\circ}{\times}$ is the shared pair of electrons. In reality, of course, one electron is the same as any other, in that once the bond is formed, the electron originating from the chlorine cannot be distinguished from that which came from the hydrogen. Similarly for the covalent compound ammonia, NH_3:

$$\overset{\circ\circ}{\underset{\circ}{^\circ_\circ N}} + 3H \times \longrightarrow \overset{\textstyle H}{\underset{\textstyle H}{^\circ_\circ \overset{\circ\times}{\underset{\circ\times}{N}}^\circ_\times}} H$$

In all the above cases, each atom has contributed one electron to the covalent bond. The compound ammonia also undergoes dative covalent bonding using the spare pair (**'lone-pair'**) of electrons on the nitrogen atom. For example, it forms a covalent bond with a proton (H^+), formed by the loss of one electron from a hydrogen atom and thus possessing no electrons of its own.

$$\overset{\textstyle H}{\underset{\textstyle H}{^\circ_\circ \overset{\circ\times}{\underset{\circ\times}{N}}^\circ_\times}} H + H^+ \longrightarrow \left[\overset{\textstyle H}{\underset{\textstyle H}{H^\circ_\circ \overset{\circ\times}{\underset{\circ\times}{N}}^\circ_\times} H} \right]^+$$

ammonia   ammonium ion, $NH_4{}^+$

Once formed, each of the four covalent bonds in NH_4^+ is indistinguishable from the others.

Covalent compounds do not conduct electricity and are usually gases, liquids, or low-melting-point solids at room temperature, because the forces of interaction between the molecules are weak (by contrast, covalent bonds themselves are usually very strong). Covalent bonds, unlike ionic bonds, have definite directions in space, and so their length, and the angles between them, can be measured and quoted.

The orbital theory applied to atomic structure (Section A1.1) can also be applied to covalent compounds, the bonding electrons being considered as occupying **molecular orbitals** formed by interaction of the atomic orbitals in which the electrons were originally located. The various possible interactions produce molecular orbitals of different energy levels, each of which can hold a maximum of two electrons with opposite values of the spin quantum number (i.e. Pauli's principle is obeyed). In the simplest case, the hydrogen molecule, there are two possible molecular orbitals formed by interaction of the 1s atomic orbitals. The lowest energy orbital is the **bonding molecular orbital** (often written as $\sigma 1s$) in which the electron is most likely to be found between the two nuclei. There is also an **antibonding molecular orbital** (written as $\sigma^\star 1s$) of higher energy in which there is little chance of finding an electron between the two nuclei. A bonding molecular orbital is more stable than the atomic orbitals that might give rise to it, whereas an antibonding molecular orbital is less stable. The two electrons in the hydrogen molecule have opposite spin, and both occupy the bonding molecular orbital. Hence H_2 is much more stable than are the isolated H atoms. By contrast, helium atoms have the electron configuration $1s^2$, and if they combined to give He_2, both the bonding and antibonding molecular orbitals would contain two electrons, and there would be no effective gain in stability. Hence He$_2$ does not form.

The combination of p-type atomic orbitals can produce two types of molecular orbital by overlapping in different ways. These are known as σ and π. Hence, for one of the 2p-orbitals (say 2p$_x$) combining with another such orbital, there will be two bonding molecular orbitals, σ2p$_x$ and π2p$_x$, and two antibonding molecular orbitals, σ*2p$_x$ and π*2p$_x$. Energy increases in the order:

$$\sigma2p_x < \pi2p_x < \pi\star2p_x < \sigma\star2p_x$$

With this in mind, we can consider bonding in three more-complicated cases: the gases, nitrogen, oxygen, and fluorine. The nitrogen atom has the configuration $1s^2 2s^2 2p^3$. If two atoms join together to form a diatomic molecule N$_2$, the four 1s-electrons (two from each atom) fully occupy both a σ1s bonding and a σ*1s antibonding orbital, and so there is no net bonding. The four 2s-electrons similarly occupy σ2s and σ*2s molecular orbitals, and again no net bond results. Six electrons are left, located in two 2p$_x$, two 2p$_y$, and two 2p$_z$ atomic orbitals. If the axis of the bond between the atoms is taken to be that of the 2p$_x$ orbitals, they can overlap along this axis to produce a bonding σ2p$_x$ molecular orbital that can hold both electrons. The 2p$_y$ and 2p$_z$ atomic orbitals cannot overlap along their axes, but they can overlap laterally to give bonding π2p$_y$ and π2p$_z$ molecular orbitals, each of which holds two electrons with different values of the spin quantum number. The 2p antibonding orbitals are not occupied; and the net result is a triple covalent bond N≡N, i.e. one σ covalent bond and two π covalent bonds. N$_2$ is thus far more stable than the individual N atoms.

The oxygen atom (configuration, $1s^2 2s^2 2p^4$) has one extra electron, and so when O$_2$ is formed there are two more electrons to consider. These must occupy the next highest molecular orbital in terms of energy. In fact, there are two such orbitals of equal energy, π*2p$_y$ and π*2p$_z$. By Hund's rule, each must receive one electron. Since the presence of these electrons in antibonding orbitals energetically cancels out one of the π2p bonding orbitals, the two oxygen atoms are effectively joined by a double bond, i.e. O=O.

The fluorine molecule contains two more electrons than does O$_2$, and so the π*2p$_y$ and π*2p$_z$ orbitals are both full. Since three bonding and two antibonding molecular orbitals are occupied, the fluorine molecule effectively contains a single bond, F–F.

A1.2.3   *Non-ideal character of bonds*

The discussion so far has implied an equal sharing of the bonding electrons between two atoms joined by a covalent bond. However, this is only the case when both atoms have a similar attraction for the electrons, i.e. are equally electronegative. This is often not the case. Consider, for example, the water molecule, which contains two oxygen–hydrogen covalent bonds:

$$2H\times + \overset{\circ\circ}{\underset{\circ\circ}{O}}\circ \longrightarrow \overset{H}{\underset{\circ\circ}{O\overset{\times}{\circ}H}}$$

Oxygen is more electronegative than hydrogen, and so takes a slightly greater 'share' of the bonding electrons than it should, giving it a slight negative charge (written as δ^-). The hydrogen similarly has a slight positive charge, i.e.

$$\underset{\delta^+}{H}\diagdown \underset{\delta^-}{O}\diagup \underset{\delta^+}{H}$$

where the dash between the atoms represents a covalent bond.

The existence of these charges gives water many of its properties. They attract water molecules to each other, so raising the boiling point to 100 °C at normal atmospheric pressure e.g.

weak electrostatic bond

These weak electrostatic bonds are called **hydrogen bonds.** The small charges also allow water to hydrate ions; water molecules cluster around ions and help to stabilize them, e.g. for A^+ and B^- ions;

The energy obtained when ions become hydrated is what provides the energy to disrupt the crystal lattice when ionic compounds dissolve in water. In those cases where the energy of hydration would be much smaller than the energy needed to disrupt the lattice, then the ionic compound will not dissolve in water.

A1.2.4 *Hydrocarbons and electron delocalization*

The element carbon has four electrons in its outermost shell (Table A1.4), and normally forms four covalent bonds. Carbon atoms can covalently bond to each other to form long chains. For example, the compound butane, used as a fuel in cigarette lighters, has the structure

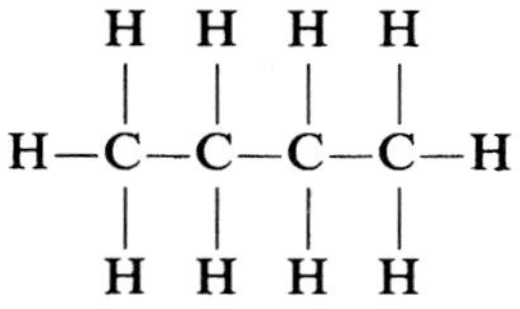

$$
\begin{array}{cccc}
H & H & H & H \\
| & | & | & | \\
H-C-C-C-C-H & & & (C_4H_{10}) \\
| & | & | & | \\
H & H & H & H
\end{array}
$$

each dash (—) representing a covalent bond. Butane is referred to as a **hydrocarbon**, since the molecule contains carbon and hydrogen only. Two other hydrocarbon gases, ethane and pentane, are released during the peroxidation of membrane lipids (Chapter 4). They have the structures:

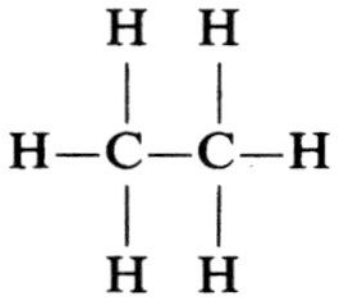

$$
\begin{array}{cc}
H & H \\
| & | \\
H-C-C-H & \text{(ethane)} \\
| & | \\
H & H
\end{array}
\qquad\qquad
\begin{array}{ccccc}
H & H & H & H & H \\
| & | & | & | & | \\
H-C-C-C-C-C-H & & & & \text{(pentane)} \\
| & | & | & | & | \\
H & H & H & H & H
\end{array}
$$

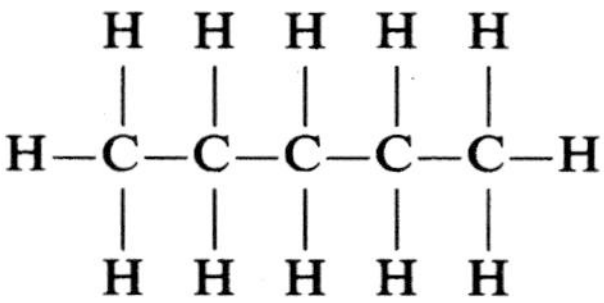

Carbon atoms can also form double covalent bonds (written as $>C{=}C<$) and triple covalent bonds ($-C{\equiv}C-$) with each other. A double bond consists of four shared electrons (two pairs), and a triple bond has six shared electrons (three pairs). The simplest hydrocarbon containing a double bond is the gas **ethene**, otherwise known as ethylene. It has the structure

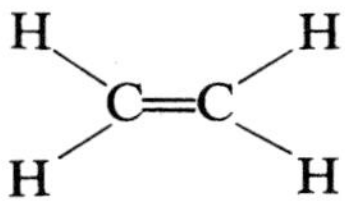

$$
\begin{array}{c}
H \qquad\qquad H \\
\diagdown \qquad \diagup \\
C{=}C \\
\diagup \qquad \diagdown \\
H \qquad\qquad H
\end{array}
$$

Ethene is produced in several assays for the detection of hydroxyl radicals (Chapter 2).

Ethyne, otherwise known as acetylene, contains a triple bond and has the structure **H—C≡C—H**.

Organic compounds containing carbon–carbon double or triple bonds are often said to be **unsaturated**. Many constituents of membrane lipids are of this type (see Chapter 4).

The organic liquid **benzene** has the overall formula C_6H_6. Given that carbon forms four covalent bonds, the structure of benzene might be drawn as containing three carbon–carbon single bonds, and three double bonds, i.e.

This structure cannot be correct, however, since benzene does not show the characteristic chemical reactions of compounds containing double bonds. A carbon–carbon single bond is normally 0.154 nm long (one nanometre, nm, is 10^{-9} metre), and a carbon–carbon double bond, 0.134 nm; yet all the bond lengths between the carbon atoms in benzene are equal at 0.139 nm, i.e. intermediate between the double and single bond lengths. The six electrons, which should have formed three double bonds, appear to be 'spread around' all six bonds. This is often drawn as:

or, in abbreviated form,

This abbreviated form is used in this book. Compounds containing the benzene ring are called **aromatic compounds**. This delocalization of electrons over several bonds greatly increases the stability of a molecule. Other examples can be seen in haem rings (Chapter 1) which show extensive delocalization of electrons, and in several ions such as nitrate (NO_3^-) and carbonate (CO_3^{2-}). In each case the negative charge is spread between each of the bonds, i.e.

(each O has, on average, one-third of the negative charge)

(each O has, on average, two-thirds of a negative charge).

Table A1.9. Miscellaneous useful data

Haemoglobin	concentration (in mg/100 ml) $= (A_{413} \times 13.5) - (A_{453} \times 8.4)$
Transferrin (apo)	1 mg will bind 1.45 µg ferric ion ($M_r = 77{,}000$); the $A_{280} : A_{470}$ ratio gives the % iron saturation (a ratio of <24 is indicative of $>95\%$ saturation)
Caeruloplasmin	oxidation of ferrous ions (ferroxidase): $2.7 \times 10^4 \ M^{-1} s^{-1}$. Purity is given by the $A_{610} : A_{280}$ ratio (100% pure protein gives a ratio of 0.046)

Appendix II
Some basic molecular biology for the chemist

A2.1 Introduction

Anyone reading this book will realize that modern cell and molecular biology are contributing significantly to our understanding of oxidative damage and antioxidants. For the sake of those chemists who may not be familiar with the concepts involved, we review here some of the basic principles.

DNA contains four 'bases', namely adenine (A), thymine (T), guanine (G) and cytosine (C). In double-stranded DNA, each base in one strand is hydrogen bonded to a base in the other strand. The pairing is specific: adenine pairs with thymine, and guanine with cytosine. Genetic information is stored in DNA in the sequence of bases along a strand; when a particular part of the information is to be used to synthesize proteins, a **messenger RNA** (mRNA) copy of part of the DNA sequence is made by **transcription** (see Section A2.2); the information in the mRNA is then decoded by **translation** into protein (Section A2.3)

The order of DNA bases in a **gene** specifies a single chain of amino acids, but usually in eukaryotic cells the parts of the DNA sequence that encode information for the polypeptide (**exons**) are separated by DNA base sequences that are not used to provide information (**introns**) (Table A2.1). Unusually, the genes encoding heat-shock protein hsp70 contain no introns. Many exons encode discrete structural or functional parts of proteins, e.g. the central exon of the myoglobin and haemoglobin genes encodes the region of the polypeptide chain that binds haem.

Table A2.1. Human genes encoding some of the proteins involved in antioxidant defence

Protein encoded	Exons	Introns	Comments
Plasma glutathione peroxidase	5	4	On chromosome 5 (*Gene* **145**, 293 (1994))
EC-SOD	3	2	*Genomics* **22**, 162 (1994)
CuZnSOD	5	4	On chromosome 21
MnSOD	5	4	On chromosome 6 (*DNA Cell Biol.* **13**, 1127 (1994))
Caeruloplasmin	19	18	*Biochem. Biophys. Res. Commun.* **208**, 128 (1995)

Prokaryotic organisms such as *Escherichia coli* rarely have introns in their genes. For example, the base sequences in the genes encoding FeSOD and MnSOD in *E. coli* match exactly the sequences of the protein products.

Almost all of the genes of a eukaryotic cell are present in the DNA within the nucleus. Mitochondria have some DNA (as do chloroplasts), but only a few genes are encoded there (Chapter 1). Mitochondrial proteins are not necessarily encoded by mitochondrial DNA; for example, MnSOD (a mitochondrial enzyme) is encoded in the nucleus, on chromosome 6.

A2.2 Transcription and editing[1]

The first stage in using the information in DNA is to activate **transcription**— the synthesis of new RNA from a DNA template. The location of a gene in eukaryotic DNA is important; if the promoter (see below) is buried in highly condensed chromatin (Chapter 4) the gene is unlikely to be transcribed. DNA **methylation** also deters transcription. Cytosine can be methylated to give 5-methylcytosine and about 70% of cytosine bases in C–G sequences in mammalian DNA are methylated.

If the gene is 'available' for transcription, RNA polymerase enzymes copy the whole of it (both exons and introns) into **pre-messenger RNA**. Eukaryotic nuclei contain three RNA polymerases: RNA polymerase II is the one that synthesizes pre-mRNA. The four major bases in RNA are adenine (A), guanine (G), cytosine (C) and uracil (U). During RNA transcription, uracil is inserted into the RNA opposite adenine in DNA, and guanine pairs with cytosine. The synthesis of RNA proceeds 5' to 3' as the polymerase copies the information in one of the strands of the DNA double helix. Once started, the polymerase only stops when it reaches a DNA base sequence signalling termination. Unlike DNA polymerases (Chapter 4), RNA polymerases have no error-correcting function and so their intrinsic error rate is higher.

The pre-messenger RNA is then edited in the nucleus. The intron sequences are removed by splicing the pre-mRNA; this occurs in the **spliceosome**, an assembly of pre-mRNA with small nuclear RNA molecules (**snRNA**) and proteins. For some proteins (e.g. human iNOS[3] and fibronectin), different types of splicing can occur, so that certain exons can sometimes be removed. This **alternative splicing** allows different protein products to arise from the same pre-mRNA. Some forms of thalassaemia (Chapter 3) are caused by **aberrant splicing**, in which a mutation creates an incorrect splice site and thus an abnormal polypeptide product.

A2.3 Translation

The edited mRNA (intron sequences missing) then leaves the nucleus and is translated into protein by the cell's **ribosomes**, some of which are free in the cytoplasm and others attached to the endoplasmic reticulum. The sequence of bases in the mRNA is a non–overlapping **triplet code**, each sequence of three bases (or **codon**) on mRNA specifying a single amino acid. The full genetic code is given in Table A2.2. For each possible codon there is a **transfer RNA** (tRNA) that can bind to that codon and also carries the corresponding amino acid. The ribosome passes along the mRNA, codon by codon, and the amino acids from each tRNA molecule are incorporated into a growing polypeptide chain. Three codons are **stop signals** and AUG can be a start signal for protein synthesis as well as encoding methionine. Some amino acids are encoded by only one base triplet but most are encoded by more than one codon. Hence, for example, if a mutation changes CAU to CAC the final protein product of the gene will not be affected, as both triplets code for histidine. In general, mutations can occur by the **substitution** of one base by another, the **deletion** of one or more bases or the **insertion** of one or more bases. The former type of mutation usually changes only one amino acid in the resulting protein (unless a stop signal is created, or a start signal deleted). For example, a point mutation in the mouse catalase gene replaced alanine (GCU) at position 117 by threonine (ACU) and produced a low-activity enzyme. If a purine is replaced by another purine, or a pyrimidine by a pyrimidine, a **transition** mutation occurs. If a purine is replaced by a pyrimidine, or vice versa, a **transversion** occurs. Some substitution mutations result in no change in the protein (see above). Deletions and insertions may shift the **reading frame**, altering many triplets, and so produce drastic changes in the protein product.

Table A2.2. The genetic code

Amino acid	Codon(s)	Amino acid	Codon(s)
Alanine	GCA, GCC, GCG, GCU	Lysine	AAA, AAG
Arginine	AGA, AGG, CGA, CGC, CGG, CGU	Methionine★	AUG[a]
Asparagine	AAC, AAU	Phenylalanine	UUU, UUC
Aspartate	GAC, GAU	Proline	CCA, CCC, CCG, CCU
Cysteine	UGC, UGU	Serine	AGC, AGU, UCA, UCC, UCG, UCU
Glutamate	GAA, GAG	Threonine	ACA, ACC, ACG, ACU
Glutamine	CAA, CAG	Tryptophan★	UGG
Glycine	GGA, GGC, GGG, GGU	Tyrosine	UAC, UAU
Histidine	CAC, CAU	Valine	GUA, GUC, GUG, GUU
Isoleucine	AUA, AUC, AUU	[Stop]	UAA, UAG, UGA[b]
Leucine	UUA, UUG, CUA, CUC, CUG, CUU		

★Amino acid encoded by only one base triplet.
[a]AUG is also used as the start codon.
[b]The stop codon UGA can also lead to incorporation of selenocysteine in glutathione peroxidases and other selenoproteins (Chapter 3).

For example, insertion of an extra adenine residue in exon 3 of the caeruloplasmin gene in a patient with hereditary caeruloplasmin deficiency changed all subsequent codons, produced a premature stop codon and no caeruloplasmin was synthesized.

The ribosome starts translation at an AUG codon on the mRNA, by incorporating methionine bound to a special transfer RNA. For example, mRNA encoding human CuZnSOD is initiated at AUG, but the methionine is then removed from the protein product and the adjacent alanine is N-acetylated to generate the mature protein.

The lifetime of mRNA in a cell varies widely. For example, mRNAs encoding cytokines and growth factors have short half-lives (often $\leqslant 30$ min) whereas mRNAs encoding widely-used or continuously expressed genes can be much longer, e.g. over 24 h for the mRNA encoding the β-globin chains of haemoglobin in erythroid cells. Sometimes regulation of gene expression is achieved at the mRNA level, e.g. by increasing or decreasing mRNA stability. An example is the binding of **iron regulatory proteins** to **iron responsive elements** on mRNA (Chapter 3). Another is stabilization of catalase mRNA in newborn rats exposed to hyperoxia (Section 4.13).

A protein destined to be imported into a particular subcellular organelle is usually synthesized with an N-terminal extension that functions as a targeting signal. For example, for proteins imported into the mitochondrial matrix, the signal peptide is 20–30 residues long. Once the protein has been imported, the targeting signal is usually removed.

A2.4   Regulation of transcription[1,11,12]

Although regulation at the level of mRNA stability can occur, most gene expression is regulated at the transcriptional level. Some genes are continuously expressed in most or all cells: these **housekeeping genes** encode essential structural proteins, receptors and enzymes in key metabolic pathways. Examples in animals are the genes encoding CuZnSOD, MnSOD, the structural protein actin and the glycolytic enzyme glyceraldehyde-3-phosphate dehydrogenase. Other genes are only transcribed in a few cell types, at a particular time during development, or in response to an extracellular signal.

The RNA polymerases which copy DNA into mRNA bind to a **promoter sequence** of bases on the DNA on the 5' side of the gene (Fig. A2.1). The detailed base sequences of promoters are unique to each gene. However, they do have some common features, regions of identical (or closely similar) bases. These are called **consensus sequences**. Promoters in mammalian cells usually contain a TATAAA consensus sequence (the **TATA box**). Many eukaryotic promoters also have a **CAAT box**, containing a CAATCT sequence and some have a **GC box** (consensus sequence GGGCGG). Housekeeping genes tend to have GC boxes in their promoters. All these DNA sequences facilitate RNA polymerase binding. For example, the promoter of the human CuZnSOD gene contains TATA and CAT boxes and GC-rich elements.

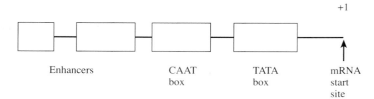

Fig. A2.1. Promoters and enhancers. The activities of many promoters in eukaryotes are greatly increased by binding of regulatory proteins to base sequences called **enhancers**. These enhancer sequences can often be distant (up to several thousand bases) from the start site, on either its 5' or 3' side. The translation start site in DNA is called nucleotide 1, the bases in the promoter are given negative numbers and are said to be **upstream** of the start site. The TATA box is usually about 25 base pairs upstream of the start site, the CAAT box about 75 base pairs upstream.

RNA polymerase II bound to the promoter is guided to the transcription start site by a set of **general protein transcription factors** which work on all (or most) genes. The first event is binding of the general protein transcription factor TFIID to the TATA box, followed by binding of the other factors. The promoter region can contain one or more **enhancers**, specific DNA base sequences which can further promote binding of RNA polymerase and synthesis of RNA. Sometimes enhancers are found within introns. However, many enhancers can be far away from the promoter in the genome. Specific gene transcription factor proteins bind to enhancer sequences and the DNA between the enhancer and promoter 'loops out' to allow the enhancer-bound proteins to interact with one of the general transcription factors or with RNA polymerase itself. Thus a **eukaryotic gene control region** consists of the promoter (where the general transcription factors and RNA polymerase gather) plus all of the enhancer and other regulatory elements (some of which can be inhibitory towards gene transcription). It is the balance between all the various transcription factors that determines whether a gene will be transcribed and, if so, to what extent.

For example, the promoter of the human MnSOD gene[12] contains a GC-rich sequence but no TATA box, plus binding sites for SP1 transcription factors. On the 3' side of the gene there is a binding sequence for the NF-κB transcription factor (Chapter 4), which helps to increase MnSOD levels in certain cells in response to stimuli that activate NF-κB, such as TNFα. The human glutathione S-transferase π gene promoter includes a TATA box and a sequence that recognizes the AP1 transcription factor (Chapter 4).[6] The mechanism by which xenobiotics activate glutathione transferases involves multiple transcription factors. For example, in the gene control region of the rat GST subunit 1 (Ya) gene **hepatocyte nuclear factor 1** (HNF1), a liver-specific transcription factor, binds to a DNA recognition sequence; a second enhancer element binds HNF4. Xenobiotic response elements (XREs), glucocorticoid response elements and antioxidant response elements (AREs) are also present. Activation of transcription of the Ya subunit gene by aromatic compounds involves transcription factor binding to the XRE, whereas H_2O_2 or *tert*-butylhydroquinone act via the ARE. XREs are also involved in the up-regulation of transcription of several cytochrome P450 genes in response to xenobiotics.

As another example, many of the actions of glucocorticoid hormones on cells are due to activation of the transcription of specific genes. The glucorticoid hormones diffuse across cell membranes and bind to cytoplasmic receptors. The activated complex moves to the nucleus and binds to enhancer sequences, often called **glucocorticoid-responsive elements**. These enhancer sequences affect the promoter regions of those genes whose expression is altered by glucocorticoids. A **heat-shock transcription factor** is expressed in *Drosophila* exposed to high temperatures. This DNA-binding protein attaches to a consensus sequence called the **heat-shock response element**, which is present upstream of the TATA box in the promoters of several genes. **Serum response factor** is a key regulator of many genes that respond to extracellular signals provoking cell growth and differentiation: it binds to DNA sequences called **serum response elements** in the regulatory regions of multiple genes involved in cell growth.

A2.5   Structure and regulation of transcription factors[1]

Four common structures are found in transcription factors that allow them to bind to DNA: zinc fingers, helix–turn–helix motifs, helix–loop–helix motifs (e.g. found in the protein product of the proto-oncogene c-*myc*) and leucine zippers. The activity of transcription factors is controlled in two ways: regulation of the amount present (often also by transcriptional regulation of the genes encoding the factors) and control of the activity of the pre-synthesized factor, e.g. by phosphorylation/dephosphorylation or oxidation–reduction of –SH groups (**redox regulation**).

A2.5.1   *Zinc fingers*

Zinc fingers are folded loops of protein stabilized by Zn^{2+} ions. They are among the most widely used types of DNA-binding regions within proteins that regulate gene transcription, and several types exist. A common type contains about 30 amino acids forming a 12 amino acid α-helix packed against a β-hairpin loop, stabilized by the Zn^{2+} ion liganded to two cysteines on the β-hairpin and two histidines on the α-helix. Residues on the α-helix hydrogen bond to DNA bases. The number of zinc fingers in proteins can range from one to >30, and each finger interacts with three base pairs on the DNA. Longer DNA sequences can be specifically recognized by bringing together a series of zinc fingers with different amino acid residues in the α-helix that recognize different base triplets. The enzyme poly(ADP-ribose) polymerase (PARP) contains zinc fingers that mediate its binding to DNA.

A second type of zinc finger is found in several proteins acting as hormone receptors which, once the hormone is bound, enter the nucleus and bind to

DNA; an example is the oestrogen receptor. They bind to **oestrogen response elements**, which resemble glucocorticoid response elements. The DNA-binding domain of the glucocorticoid receptor contains zinc, coordinated to cysteine, but with a different secondary structure.

If zinc is replaced by iron ions in zinc fingers, the receptor's zinc finger domain still binds to DNA but is then capable of causing free radical damage to DNA in the presence of H_2O_2. Hence zinc may have been a less 'risky' metal to select for transcription factor stabilization than redox-active metals such as iron or copper.

A2.5.2   *Leucine zippers*

Leucine zipper proteins are transcription factors whose distinctive feature is a residue of the hydrophobic amino acid leucine at every seventh position in a stretch of about 35 residues. These proteins form dimers held together by an α-helical coiled coil, stabilized by hydrophobic interactions of the leucines (hence the 'leucine zipper'). The N-terminal ends of the proteins bind to DNA, and the role of the leucine zipper is to bring together a pair of protein molecules to bind two adjacent DNA sequences. In higher eukaryotes, leucine zipper proteins mediate, among other effects, the actions of cyclic AMP on gene transcription. The regulatory sequences of genes affected by cAMP contain a **cAMP response element**. cAMP activates kinases leading to phosphorylation of **cAMP response element binding protein** (CREB). Phosphorylation promotes dimerization involving leucine zipper regions, facilitating binding to DNA and activation of transcription.

Leucine zippers can cause the dimerization of identical or non-identical protein chains. For example, active CREB is a homodimer, whereas AP-1 is a heterodimer containing Jun and Fos (Chapter 4) associated through their leucine zippers.

A2.6   Cell growth signals, kinases and immediate early genes[1]

Cell-surface receptors usually belong to one of three general classes. Some are linked to the opening of **ion channels**, e.g. the glutamate receptors in the brain (Chapter 9). Some are linked to **G-proteins** (Chapter 4). Enzyme-linked receptors function directly as enzymes, or are associated with enzymes. G-protein-linked and enzyme-linked receptors receive signals and relay them to the nucleus to alter the expression of specific genes. The relay system includes a range of proteins that become phosphorylated to activate them. Phosphorylation is achieved by **tyrosine kinases** (acting on tyrosine) or **serine/threonine kinases** (which phosphorylate serine or sometimes threonine residues).

The receptors for many cell growth factors are transmembrane tyrosine-specific protein kinases: the first to be recognized was the receptor for **epidermal growth factor** (EGF), a small protein that stimulates proliferation of epidermal and several other cell types. The exterior part of the EGF receptor recognizes the signal, and the interior part acts as a tyrosine kinase. Other such receptors are those for platelet–derived growth factor (PDGF), fibroblast growth factor, nerve growth factor and macrophage colony–stimulating factor. When the appropriate signal binds to a receptor, a cascade of **tyrosine kinases** is activated.

For example, binding of the appropriate growth factor causes the EGF (or PDGF) receptors to associate into dimers, phosphorylating each other on multiple tyrosine residues. Multiple cell proteins then bind, each to a different phosphorylated site on the activated receptor. Many of these bound proteins become phosphorylated (on tyrosine) themselves. The **Ras** proteins help relay the growth signals from receptor tyrosine kinases to the nucleus (and were first discovered as the hyperactive products of mutant *ras* genes, involved in cancer; see Chapter 9). Ras is active when GTP is bound, and inactive with GDP. Ras proteins perpetuate the signal by activating a serine/threonine phosphorylation cascade, in which the **mitogen activated protein** (MAP) kinases are especially important. MAP kinases are activated by a wide range of cell growth/ differentiation signals (including protein kinase C; Chapter 4) and their full activation requires phosphorylation of both threonine and tyrosine residues.

MAP kinases lead, via other protein kinases and gene transcription factors, to activation of the transcription of a set of **immediate early genes**. One such gene is *fos*, whose transcription is rapidly induced in response to every growth stimulus. MAP kinases may also phosphorylate the Jun protein, which combines with the newly made Fos to form the transcription factor AP-1 (Chapter 4). c-*Fos* mRNA is very short-lived, being rapidly degraded in the cytoplasm. The capacity of the transcription factor AP-1 to respond to a range of signals (Chapter 4) is thus largely mediated by the transcription factors that regulate the c-*fos* and c-*jun* promoters.

Many receptors for cytokines work through tyrosine kinases, although (unlike the above growth factor receptors), the kinase is not part of the receptor molecule but merely associates noncovalently with it. Often the Src family of tyrosine kinases is involved; members include **Src**, **Lck**, **Lyn** and **Blk**.

A2.7   Identifying DNA-binding proteins in the laboratory[1]

DNA is negatively charged and will therefore move towards the positive electrode (**anode**) on electrophoresis. If electrophoresis is carried out in polyacrylamide gels, smaller DNA molecules will move more quickly than bigger ones. Protein molecules bound to DNA will slow its movement. This provides the basis for the **gel mobility shift assay**. A DNA base sequence (**oligonucleotide**) is radioactively labelled, mixed with a cell extract and electrophoresed. Binding of proteins to the oligonucleotide will retard its progress. This type of assay can be used, for example, to study the activation of NF-κB in cells; activated NF-κB binds to DNA and slows its progress. Once the DNA sequence to which a transcription factor binds has been identified, the factor can be purified by passing a cell extract through a column to which that sequence has been attached (**DNA affinity chromatography**).

The DNA sequences to which proteins bind can be identified by **footprinting techniques** (Chapter 4), using either nuclease enzymes or chemical reagents to digest unbound DNA and isolate the 'protected' sequence to which a protein has bound.

A2.8   Reverse transcription[1]

Several viruses (including HIV, Chapter 9) contain RNA as the genetic material. This is copied into DNA by the viral enzyme **reverse transcriptase**. A primer, in the form of a host RNA, bind to the 3' end of the viral RNA and the reverse transcriptase elongates from this to give a DNA–RNA hybrid. The reverse transcriptase, acting as a nuclease, then destroys the RNA to leave a single-stranded DNA, which is converted into a DNA duplex by reverse transcriptase and incorporated into the host genome by an integrase protein (Chapter 9).

A2.9   Studying the genome[1]

The total information stored in the DNA of an organism is called its **genome**. To study the genome, DNA must first be isolated from cells by breaking them open whilst preventing nucleases from acting. Metal ion chelators (many nucleases are metal ion-dependent) and specific nuclease inhibitors are used in DNA extraction processes. In particular, when studying oxidative DNA damage, it is essential to be sure that further oxidative damage is not caused to DNA during isolation. Phenol is often used in DNA isolation to precipitate proteins, yet phenols can oxidize to generate ROS if they are impure, e.g. contaminated with metal ions (Chapter 8). Thus chelators may serve to inhibit not only nucleases but also to deter oxidation reactions.

Four factors have facilitated the development of modern genetic engineering: the availability of specific enzymes that can copy nucleic acids and others that can cleave nucleic acids at specific sites, the ability to determine DNA base sequences quickly, the ability to synthesize DNA sequences (**oligonucleotides**) chemically and the ability of nucleic acids to hybridize with each other by hydrogen bonding. Any two pieces of DNA and/or RNA will bind together (**anneal**) to form duplexes if they have sequences of approximately 20 complementary base pairs (the exact number of matching bases needed for annealing depends on the sequences).

In sequencing DNA, **restriction endonucleases**, enzymes that cut both strands of double-helical DNA at specific base sequences, are used to obtain manageable chunks. The name comes from their identification as bacterial factors that restricted infection by viruses, by chopping up the viral DNA. Restriction enzymes are named after the bacteria from which they are obtained, e.g. *Eco* enzymes come from *E. coli* and *Hin* enzymes from *Haemophilus influenzae*. After hydrolysis by restriction enzymes, DNA fragments can be separated on a size basis by gel electrophoresis. Their presence can be identified by autoradiography (if the DNA has been ^{32}P-labelled) or by using the dye **ethidium bromide**, which fluoresces under UV light when it is bound to DNA. DNA can be ^{32}P-labelled by copying it with a DNA polymerase in the presence of radioactive deoxynucleoside triphosphates, or by incubating with a kinase enzyme that attaches ^{32}P-labelled phosphate at the 5' end, from [^{32}P]ATP.

The presence of specific base sequences in a genome can be identified by the binding of complementary sequences of DNA (e.g. synthetic oligonucleotides), which will anneal to the sequence sought for. In **Southern blotting** (named after its inventor Ed Southern) DNA restriction fragments (i.e. fragments produced by digestion with restriction enzymes) are separated by gel electrophoresis, the gel soaked in alkali to separate the double helix and the fragments transferred ('blotted') to a nitrocellulose or nylon sheet to stop them re-annealing. The sheet is exposed to a ^{32}P-labelled single-stranded complementary sequence. A sheet of X-ray film is placed over the membrane to show the position of the binding sites of the oligonucleotide 'probes' from the radioactivity-induced dark spots on the film. Such techniques can be used to show, for example, that a given gene is (or is not) present in a particular genome. Point mutations in genes may be identified by the annealing behaviour of gene fragments with specific probes.

RNA molecules can be separated and identified from a total cell RNA extract by a similar technology (**Northern blotting**). This technique is very useful in obtaining specific mRNAs for production of cDNAs (see below) and

---

Box A2.1
Western blotting

1. Electrophorese protein mixture on sodium dodecylsulphate (SDS)–polyacrylamide gel. The SDS denatures proteins and gives them all a uniform negative charge so that proteins are separated on the basis of size only. This step may be preceded by a separation on a non-denaturing gel on the basis of charge followed by the above procedure at right angles (**two-dimensional gel electrophoresis**).
2. Blot proteins on to a polymer sheet.
3. Add antibody specific for the protein of interest.
4. Wash to remove unbound antibody.
5. Add a second antibody that recognizes the first (e.g. goat antibody that recognizes mouse antibody).
6. Wash to remove unbound antibody.
7. Develop the blot by one of the following methods:
   (a) by autoradiography if the second antibody is radioactively labelled;
   (b) by a colour reaction if an enzyme attached to the second antibody can generate a coloured product when a substrate is added, e.g. alkaline phosphatase.

to identify which tissues transcribe a given gene and how this is affected, e.g. by cytokines or hormones. **Western blotting** is a technique for detecting a particular protein by staining with a specific antibody (Box A2.1).

Some restriction enzymes produce staggered cuts, i.e. they do not cleave each strand of the duplex at exactly opposite positions. They thus leave short single-stranded tails at the two ends of each fragment. These cohesive ends (or **sticky ends**) can form complementary base pairs with the tail at any other end produced by the same enzyme. This allows **splicing** of any two DNA fragments that were generated using the same enzyme. DNA molecules produced by splicing DNA sequences in this way are called **recombinant DNA molecules.**

A2.10   Recombinant DNA technology[1]

The development of recombinant DNA technology has revolutionized the life sciences. New combinations of unrelated genes, or even completely synthetic genes, can be constructed in the laboratory. These novel combinations can be amplified many times (**cloned**) by inserting them into a suitable **vector** and introducing them into a suitable cell, where they are copied by the DNA-synthesizing machinery of the host and passed on to each daughter cell through many rounds of cell division. Hence the genetic endowment of the host cell can be permanently changed, and large quantities of the incorporated DNA can be obtained for study. The inserted genes are sometimes transcribed and the resulting mRNA translated within the host cell.

Plasmids, circular double-stranded pieces of DNA found in many bacteria, are often used as vectors. They can be opened with a restriction enzyme, e.g. *Eco*RI, which produces staggered cuts, leaving short sequences of single-stranded ends, then any DNA sequence can be inserted into the gap if it has compatible cohesive ends. After attachment of the new sequences, they can be covalently joined using a **DNA ligase** enzyme. A useful plasmid for cloning is pBR322, which contains genes encoding bacterial resistance to the antibiotics tetracycline and ampicillin. pBR322 can be cleaved at several sites. For example, insertion of a DNA sequence at the *Hin*dIII restriction site inactivates the gene for tetracycline resistance. Thus if insertion has occurred correctly and the modified plasmid has entered the host cell, the cell will be resistant to ampicillin but sensitive to tetracycline, enabling easy selection of the required strain.

Since plasmids are replicated independently from the main bacterial chromosome, they can be used to carry the genes of interest into the bacterial cell. Plasmids will pass across the bacterial cell wall and membrane when cells are placed in certain media containing calcium chloride, $CaCl_2$. This process is called **transformation**. Certain viruses that infect bacteria (**bacteriophages**) can also be used to introduce foreign DNA that has been inserted into the viral DNA. Alternatively, a plasmid can be packed inside a viral coat, giving a **cosmid**. When the phages infect bacteria, they carry the DNA with them. In all cases, the vector must carry not only the gene itself but also the promoter sequences needed for transcription, and sequences needed for DNA polymerase to attach and allow replication of the DNA.

Recombinant DNA technology also allows the generation of specific mutations in genes. Suppose that we want to replace base G by base T at a particular spot, to produce a change in a single amino acid when the gene is expressed as protein (**site-directed mutagenesis**). A primer is synthesized that binds to this region of the gene except that it contains a G where T should be. The two strands of the plasmid are separated and the primer allowed to bond to the complementary strand. The primer is elongated by DNA polymerase and the double-stranded DNA closed by DNA ligase. Subsequent replication of this duplex yields two kinds of progeny, half with the original sequence and half with the new one. For example, this type of technology was used to examine the role of Arg143 in CuZnSOD by replacing it with lysine or isoleucine residues.[2] Another approach is **cassette mutagenesis**. Restriction enzymes are used to remove a short segment of DNA, which is replaced by a synthetic double-stranded oligonucleotide (the cassette) containing the change required. Novel proteins can also be created by splicing gene segments, deleting large portions of a gene or inserting new bases.

A2.11 Libraries[1]

In early attempts to study the genome in eukaryotic cells, the DNA was mechanically sheared, or partly digested by restriction enzymes, into a random population of large fragments, separable by electrophoresis. The fragments were inserted into a vector, such as bacteriophage, which is used to infect a bacterium such as *E. coli*. After growth, the bacterial cells are lysed and the lysate contains fragments housed in a sufficiently large number of virus particles to ensure that the entire genome is represented. These phages are said to constitute a **genomic library**. Each phage can be replicated within bacteria to amplify the DNA. Genomic libraries can be used to study any parts of the genome, whether or not they encode proteins and whether or not they were being expressed in the material from which the DNA was extracted. Since genomic clones will still contain introns, they cannot be expressed in the prokaryotic host.

A different strategy is to begin the cloning process by extracting mRNA from the eukaryotic cell (i.e. selecting only the genes expressed, and *after* splicing of the pre-mRNA) and making a complementary DNA (**cDNA**) using reverse transcriptase. This generates cDNA clones and a cDNA library. Only the expressed genes will be cloned; genes not being translated into mRNA will not be cloned and hence a cDNA library will differ according to the tissue from which it is obtained. By contrast, a genomic library should be the same in all cells from an organism.

To identify which clone carries the DNA sequence of interest, a radioactive complementary DNA or RNA sequence is used as a probe. If the mRNA from the gene can be obtained (e.g. by purifying it from a cell in which the gene is highly active), then it can be copied into DNA using a reverse transcriptase enzyme and cloned to produce a highly specific probe.

Alternatively, if the partial or complete amino acid sequence of the protein product of the gene is known, DNA corresponding to a part of that sequence can be synthesized. Peptide sequences containing tryptophan and methionine are preferred, because these amino acids are specified by a single codon whereas other amino acids can arise from more than one codon (Table A2.2), so that alternative DNA base sequences are possible.

A2.12   Polymerase chain reaction[1]

The polymerase chain reaction (PCR) is a method that allows one to amplify specific DNA sequences rapidly without the need for a living cell. It is not necessary to know the sequence of the target gene, only the sequences of the DNA before and after it (the **flanking sequences**). Short pieces of DNA (**primers**) that will bind to these sequences in each strand of the double helix are synthesized.

PCR is carried out in a closed vessel containing the DNA and primers (usually 20–30 bases long) plus a DNA polymerase and its substrates. On heating to 95 °C the double helix denatures. On cooling the primers attach to the flanking regions on each single strand. On heating to 72 °C, the optimal temperature for **Taq DNA polymerase** (isolated from the thermophilic bacterium, *Thermus aquaticus*), elongation from both primers occurs in the direction of the target sequence. DNA polymerases need a primer in order to copy DNA: they will not copy a purely single-stranded DNA. The cycle of temperature changes is repeated and each new double helix provides a template, so that the amount of DNA of interest increases exponentially in subsequent cycles. PCR is thus highly sensitive: a single DNA molecule can be amplified and detected. PCR products can be cleaved by restriction endonucleases and separated by electrophoresis. RNA can be reverse-transcribed into DNA for PCR, a technique useful for the detection of RNAs present at only low levels in cells.

An important question, to which little attention has been given, is the effect of DNA base damage (e.g. by ROS/RNS) on the behaviour of *Taq* DNA polymerase, especially as PCR is being increasingly used to amplify DNA isolated from preserved or fossilized organisms (ancient DNA).[4] Some oxidative DNA damage products block *Taq* progression, so that only the undamaged DNA strand is amplified. However, others may lead to mutations (e.g. 8-OHdG). Ancient DNA is likely to be extensively oxidized and so results of its amplification should be interpreted with caution.

A2.13   Gene expression in mammalian cells[5,8]

The genes of bacteria such as *E. coli* do not contain exons and introns and the splicing machinery used in mammalian cells is absent. Hence bacteria cannot usually express eukaryotic genes as intron sequences cannot be removed. However, a cDNA can be attached to a vector and incorporated into a bacterium for expression. Indeed, pro-insulin is produced in *E. coli* from such

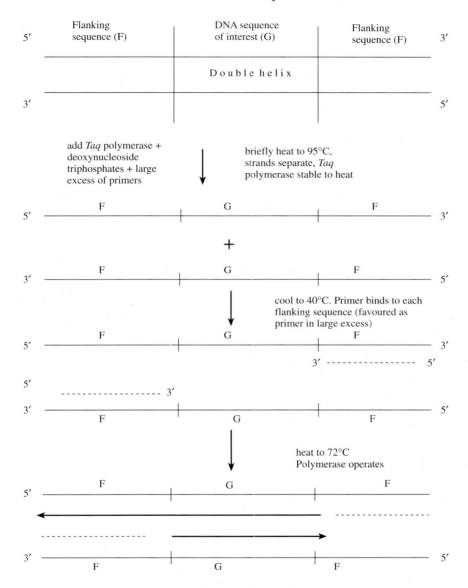

Taq polymerase extends from the primer to give two double
helices. Repeat the cycle using the two new double helices.

Fig. A2.2. The polymerase chain reaction.

a cDNA, which was obtained by reverse-transcribing proinsulin mRNA
isolated from mammalian pancreas.

However, many eukaryotic genes can only be correctly expressed in
eukaryotic cells, particularly if the protein produced is further cleaved or has

carbohydrate attached to it during final processing: bacteria cannot achieve these steps. DNA can be introduced directly into mammalian cells by several mechanisms. One is **microinjection**: a fine-tipped (0.1 μm diameter) glass micropipette containing DNA solution is inserted into the nucleus. Alternatively, DNA precipitated by calcium phosphate treatment can be taken up and incorporated (at low efficiency) into the chromosomal DNA. Viruses can also be used to carry DNA into mammalian cells. For example, a simian virus 40 expression vector was used to transfer the human EC-SOD gene into Chinese hamster ovary (CHO) cells.[10] Other viruses used include adenovirus, vaccinia and baculovirus. The latter infects insect cells, which are easily grown in culture. For example, adenoviral vectors[9] have been used to transfer a functional copy of the p47*phox* gene to cells cultured from patients with autosomal chronic granulomatous disease, resulting in restoration of NADPH oxidase activity in these isolated cells (Section 6.7). Herpes simplex viruses infect neurones and are being investigated for gene transfer to these cells.

It is possible to 'target' proteins to parts of the cell where they would not normally be expressed. Thus if a 'mitochondrial target sequence'[7] is attached to a gene encoding a non-mitochondrial protein, the translated protein product may be transported into mitochondria. For example, the gene encoding tobacco MnSOD was altered by removing its usual mitochondrial transit peptide sequence and replacing it with a sequence signalling uptake into the chloroplast. When the altered gene was expressed in tobacco plants, the result was plants expressing MnSOD in the chloroplasts; these plants were unusually resistant to paraquat, ozone or chilling injury (Chapter 6).

A2.14   Antisense technology[13]

In 1978, two scientists at Harvard University took a 13-strand DNA oligonucleotide that was complementary to part of the RNA of the Rous sarcoma virus (Chapter 9). They showed that the oligonucleotide would bind to the viral RNA and inhibit replication. This was the first use of **antisense oligonucleotides**, short singled-stranded sequences of DNA. They can be used to bind to DNA, generating a triple helix that interferes with gene transcription. More often, they are targeted to mRNA, binding the message and thus preventing protein synthesis. For example, oligonucleotides binding to the mRNA encoding Bcl-2 (Chapter 4) have been shown to promote apoptosis in isolated cells.[13]

A2.15   Transgenic organisms[5,8]

One striking aspect of the introduction of foreign DNA into mammalian cells is the production of **transgenic organisms**, organisms that have had their genetic make-up altered by receiving a gene from another species. With transgenic technology, it is possible to increase the level of (**overexpress**) foreign proteins in animals or plants and to examine the consequences.

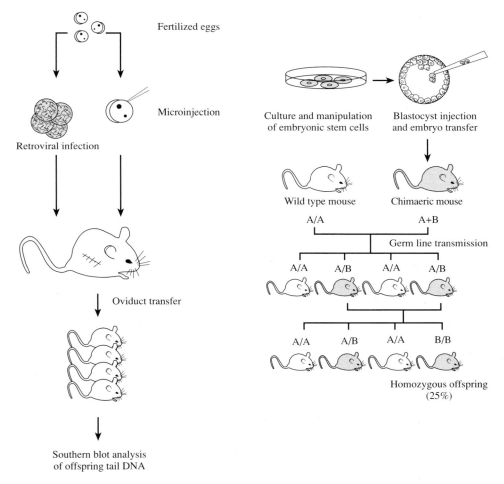

Fig. A2.3. Left: principles of generation of transgenic animals. Right: targeted mutation of genes using the 'knockout' technique. Embryonic stem cells are injected with DNA targeted to disrupt a particular gene and then returned to the embryo. Germline transmission of the disrupted allele produces heterozygous mice that may eventually be bred to homozygosity. From *Brain Pathol.* **4**, 3 (1994) by courtesy of Dr Adriano Aguzzi and the publishers.

Transgenic technology has most often been applied to *Drosophila* and to mice (which are cheap, very fertile, have a short gestation and mature rapidly) but in principle it can be applied to any species (examples of its application to plants are given in Chapter 6). The most commonly used technique employs direct injection of the DNA of interest into the pronucleus of fertilized mouse eggs. An alternative is to use a retroviral vector (Fig. A2.3). The coding DNA must be accompanied by a promoter and any other components needed for efficient expression. If a promoter from a ubiquitously expressed gene is used (e.g. β-actin) the transgene might then be expressed in most or all tissues.

An alternative is to use a promoter selectively expressed in certain tissues, e.g. myosin light-chain promoter for muscle and insulin promoter for pancreatic β-cells. For example, in one experiment the gene encoding diphtheria toxin was placed under the control of a promoter operative only in brown adipose tissue. The active gene destroyed this tissue specifically in the transgenic animals obtained, enabling the function of brown adipose tissue to be elucidated.

Linear DNA fragments injected into animal cells are rapidly ligated end-to-end by intracellular enzymes to give long tandem arrays, which can integrate into a chromosome at a random site. The embryos are cultured to the two-cell stage and then reimplanted in a 'false-pregnant' (induced by hormone treatment) foster mother and allowed to develop to term. The resulting pups are then screened to detect those animals which have incorporated the injected DNA into their genome. This is done by preparing DNA from the tails of transgenic mice and analysing it by Southern blotting. In experienced hands, the injected DNA is usually found in about 20% of the pups.

However, to be useful for experimental studies, the foreign gene must be expressed in the relevant organ. Even if it has been incorporated into the mouse DNA, it may have been integrated into a non-transcribed part of the mouse genome, so that no mRNA is made. Sometimes mRNA is made (e.g. revealed by Northern blot analysis) but not translated, so there will be no protein product. Sometimes too little protein is made to have an effect. If one is lucky, however, enough protein is made to change the phenotype of the cells expressing it. However, if the transgene inserts in (or close to) another gene, a 'knockout' of that gene with an unexpected phenotype can result (see below).

An early example of transgenic animal technology was the production of abnormally large ('giant') mice. The gene encoding growth hormone in rats was placed next to the mouse metallothionein gene promoter on a plasmid. Several hundred copies of the plasmid were injected into the pronucleus of a fertilized mouse egg. Several progeny animals expressed the gene for rat growth hormone: those that contained multiple copies of the gene (~ 30 per cell) grew much more rapidly than controls and reached twice normal weight at maturity. Examples of transgenic animals relevant to this book include the generation of animals over-expressing both normal (Chapter 3) and mutant (Chapter 9) CuZnSOD enzymes.

A variation on this technology is **gene knockout**. For example, *E. coli* mutants lacking MnSOD and/or FeSOD (Chapter 3) were obtained by insertions in the cloned structural genes, followed by exchange with the chromosomal normal genes. A yeast mutant lacking MnSOD was obtained by a similar procedure. The random integration of injected DNA into the mouse genome during normal transgenic manipulations can sometimes disrupt an endogenous gene. A more-targeted approach is the use of mouse **embryonic stem cells**, embryo-derived cells that grow in culture and are capable of producing cells of any tissue when replaced in the embryo. A DNA fragment containing the mutated gene is inserted into a vector and introduced into these cells. About one in a thousand times, the mutated gene will replace one of the two copies of the normal gene. The probability of this happening can be increased if both ends of the mutant gene are flanked by long DNA sequences homologous to those surrounding the target gene. The cells in which exchange has happened are selected (e.g. if the new DNA contains a gene encoding resistance to an antibiotic) and injected into a mouse blastocyst. The end-result should,it is hoped, be a normal-looking mouse, some of whose cells will have the mutant gene. If the germ cells contain it, breeding of these animals can generate **homozygous** animals, i.e. both copies of the gene are mutated. Studies of these homozygotes allow the function of the altered gene to be examined. For example, this method has been used to investigate the consequences of disrupting the MnSOD, CuZnSOD and glutathione peroxidase genes (Chapter 3).

References

1. Alberts, B *et al.* (1994) *Molecular Biology of the Cell.* Garland Publishing Inc, New York.
2. Beyer, WF Jr *et al.* (1987) Examination of the role of arg-143 in the human CuZnSOD by site-specific mutagenesis. *J. Biol. Chem.* **262**, 11182.
3. Eissa, NT *et al.* (1996) Alternative splicing of human iNOS mRNA. *J. Biol. Chem.* **271**, 27184.
4. Handt, O *et al.* (1994) Ancient DNA: methodological challenges. *Experientia* **50**, 524.
5. Majzoub, JA and Muglia, LJ (1996) Knockout mice. *New Engl. J. Med.* **334**, 904.
6. Rushmore, TH and Pickett, CB (1993) Glutathione S-transferases, structure, regulation and therapeutic implications. *J. Biol. Chem.* **268**, 11475.
7. Schatz, G (1996) The protein import system of mitochondria. *J. Biol. Chem.* **271**, 31763.
8. Shuldiner, AR (1996) Transgenic animals. *New Engl. J. Med.* **334**, 653.
9. Thrasher, AJ *et al.* (1995) Functional reconstitution of the NADPH oxidase by adeno-associated virus gene transfer. *Blood* **86**, 761.
10. Tibell, L *et al.* (1987) Expression of human EC-SOD in CHO cells and characterization of the product. *Proc. Natl Acad. Sci. USA* **84**, 6634.
11. Touati, D (1988) Molecular genetics of SODs. *Free Rad. Biol. Med.* **5**, 393.
12. Wan, XS *et al.* (1994) Molecular structure and organization of the human MnSOD gene. *DNA Cell Biol.* **13**, 1127.
13. Ziegler, A *et al.* (1997) Induction of apoptosis in small-cell lung cancer cells by an antisense oligodeoxynucleotide targeting the Bcl-2 coding sequence. *J. Natl. Cancer Inst.* **89**, 1027.

Index

The most extensive discussion of any indexed topic is indicated in bold type. The word '*structure*' in parentheses indicates the place in the text where the structure of the compound is shown. The abbreviation '*def.*' in parentheses indicates where a term is defined in the text.